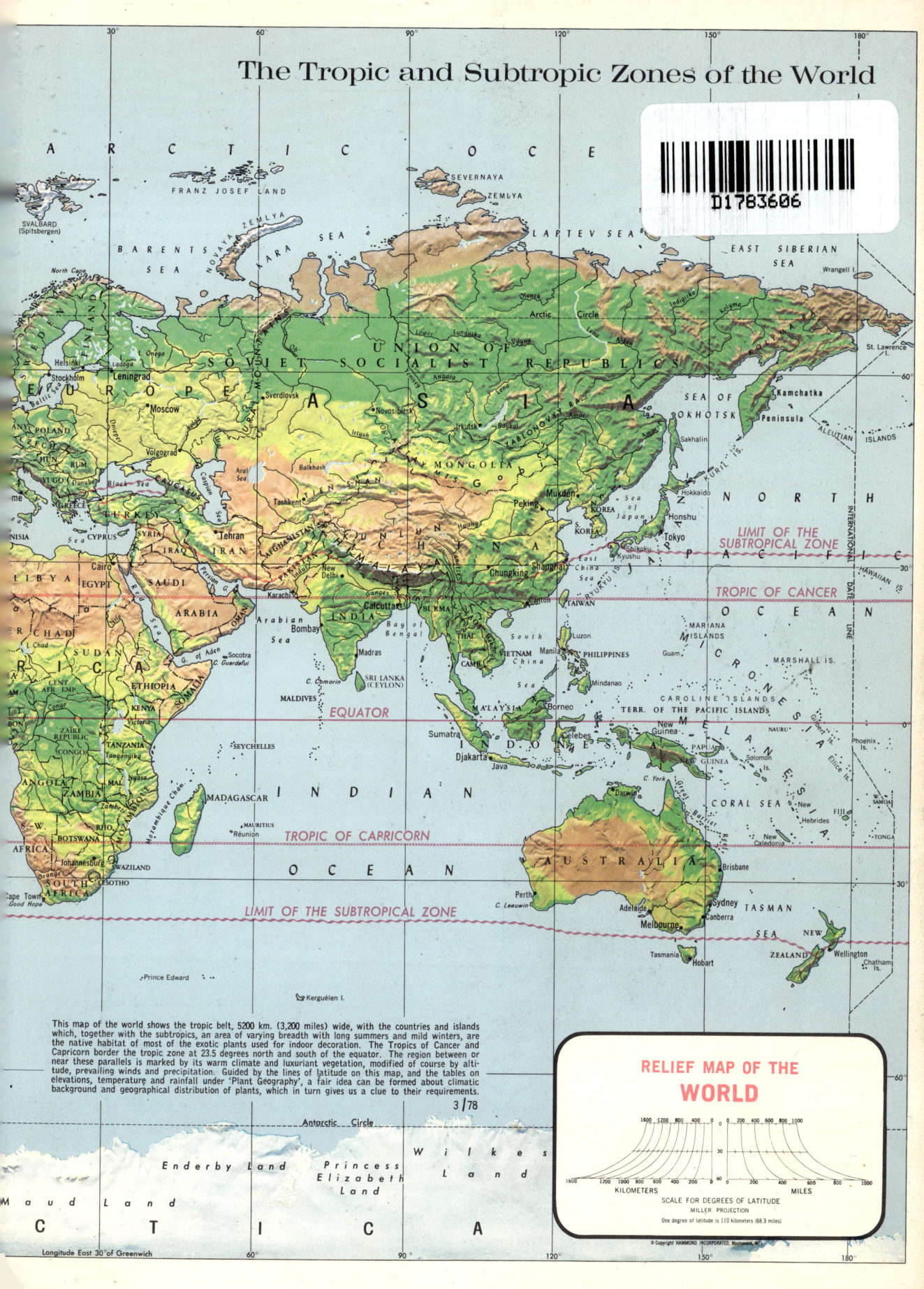

ABOUT THE AUTHOR:

Alfred Byrd Graf ranks with the great botanists and horticulturists of all time. His scholarly pursuits, carried out with great courage, determination and pioneering spirit, have opened new horizons for understanding our universe and for practical application of this knowledge in horticulture.

Dr. Graf is probably the most widely travelled plant explorer of the world's Tropics and Subtropics. Horticulturist, botanist and professional photographer, he has roamed the earth in the spirit of von Humboldt, Darwin, and David Fairchild earlier in the 20th and 19th centuries, in search of exotic botanicals to add to the enlarging horticulture of the world. From his headquarters at the Exotic Nurseries of the Roehrs Company, where he has presided over one of the world's most complete plant collections, Alfred Graf during the past forty years journeyed into the remote regions of the two hemispheres where tropical and subtropical plants abound.

Central and South America, the West Indies, Africa, India and Ceylon, China, Southeast Asia, Indonesia, New Guinea, and the South Pacific, have yielded to his painstaking search hundreds of new introductions to our ornamental horticulture. During this time he has created a vast and comprehensive collection of botanical photographs of plants in cultivation or in habitat, the best of which have been gathered in his horticultural books on ornamental plants, particularly the renowned and universally known Pictorial Cyclopedia EXOTICA, and now in his comprehensive Colorama TROPICA.

Among the honors received by the author are the award of the Large Gold Medal of the Massachusetts Horticultural Society, the Certificate of Merit of the Pennsylvania Horticultural Society, the Distinguished Service Award of the Horticultural Society of New York, a Citation Award of the American Horticultural Society, and the Tercentenary Medallion of the State of New Jersey, presented 1964 at the New York Coliseum. In 1967 he was invested with the Sarah Chapman Francis Medal of the Garden Club of America for outstanding literary achievement. During the National Convention in Pittsburgh 1972, Alfred Graf was elected to Horticulture's Hall of Fame, the highest distinction of the Society of American Florists, for his contributions to the advancement of floriculture in America. He is also a contributor to Encyclopedia Britannica. In cognizance of his research and published works on exotic plants, Fairleigh Dickinson University of New Jersey conferred on him the honorary degree of Doctor of Science.

TROPICA

COLOR CYCLOPEDIA
of EXOTIC PLANTS and TREES
from the TROPICS and SUBTROPICS

for Warm-Region Horticulture —
in Cool Climate the Sheltered Indoors

7,000 Photographs
in Living Color

ALFRED BYRD GRAF, D.Sc.

FIRST EDITION

ROEHRS COMPANY — Publishers
East Rutherford, N. J. 07073, U.S.A.

CONTENTS

Cover Girl: Tavaroa de Papeari, Tahiti

MAP of the Tropical World   *front*

The TROPICS and SUBTROPICS   4

Exotic ORIGINS and CLIMATE   6

WARM AREA HORTICULTURE   7

Climate, Light, Temperature   8
Shrubs in Gardening   9
Trees, Palms, Conifers, Fruit   10
Ground covers and Lawns   12
Vines, Basket plants, Bulbs   13
Annuals, Bromeliads, Succulents   14
Irrigation outdoors   15
Growing Materials and Soils   16

PLANTS INDOORS   17

Interior Decoration   20
Light, Temperature, Moisture, Soils   22
Orchids, Bromeliads   25
Cacti, Succulents; Ferns   26
Propagation Methods   27
Pest Control and Diseases   32

COLORAMA of PLANTS and TREES   33

Amaryllidaceae   54
Araceae   85
Begonias   159
Bromeliads   198
Cactaceae   238
Carnivorous Plants   292
Compositae   308
Conifers   333
Crassulaceae   361
Ericaceae   396
Euphorbiaceae   407
Ferns   431

Fruit   454
Gesneriads   491
Grasses and
    Bamboos   509
Leguminosae   537
Liliaceae   566
Moraceae   648
Musaceae   663
Myrtaceae   675
Orchids   708
Palms   764
Zingiberaceae   923

PLANT DESCRIPTIONS with Photo Index
  and Climate Guide   929

List of Corrections   1100

Pronunciation of Botanical Names   1100

BOTANICAL TERMS Chart   1101

GLOSSARY of Scientific Terms   1102

CHARACTERISTICS of PLANT FAMILIES   1104

BIBLIOGRAPHY and Literature References   1110

COMMON NAMES INDEX of Exotic Plants   1112

GENERIC BOTANICAL INDEX   1118

International
Standard Book Number
ISBN 0-911266-14-3

Library of Congress
Card No. 77-082461

© Copyright 1978
by Dr. A.B. GRAF

Art Director:
Philip Grushkin

Color Processing:
Eastman Kodak Co.
Rochester, New York

Color Separations:
Capper, Inc.
Knoxville, Tennessee

Lithography:
Rae Publishing Co.
Cedar Grove, N.J.

Map Makers:
C.S. Hammond & Co.
New York.

Manufactured by
A. Horowitz & Sons
Fairfield, N.J.

Printed in U.S.A.
First Edition, May 1978

PRELUDE to Tropical Beauty

TROPICA is my concept of plants beautiful and exotic, as I have found and photographed them under the steep sun of the world's warm-climate regions, or lived with them, in cultivation, all of my professional life.

Coddling humidity and an abundance of light have evolved grand and spectacular foliage in unusual patterns, with flowers a rainbow of colors, often of haunting fragrance, especially at night.

Creation must have been particularly fond of the Tropical and Subtropic zones, by favouring them with the greatest concentration and variety of recorded flora; more than 200,000 species, or perhaps three quarters of all higher land plants are denizens of equatorial environment. Most prominent of this vegetation are a multitide of flowering and fruited trees and shrubs; the palms, so symbolic of the Tropics, and treeferns of great size.

Forest giants are hosts to less imposing, but yet showy and often very curious smaller exotics—climbing their trunks, and clinging to lofty branches as epiphytes are a multitide of orchids and bromeliads, while the many begonias, gesneriads, bulbous plants and ferns live contentedly under their shadow on the forest floor. Out in the sunlight exists another, typically xerophytic world of cacti and other succulents, according to their preference, in forms developed by aeons in time and evolution.

Providence has given me the opportunity granted to few, to experience the magnitude of the fantastic plant kingdom, revealed in its manyfold shapes, throughout the wide world. During a lifetime of association with plants, and of travel for more than 30 years, I was able to accumulate some 100,000 photographs in color, and this bounty inspired me to select and compile the very choicest of them, into a comprehensive Colorama of Exotic Plants and Trees.

The object of this work is to have within one cyclopedic pictorial volume, in some 7,000 photographs, representing 1630 genera, all that is most beautiful or interesting in a cross section of warm-climate plants of all types, ready to draw us within their magic spell, making us want to live with them, in the garden or in our home.

Exploratory expeditions, and constant searching in botanical gardens, commercial nurseries and private plant collections have yielded to me not only the best of each area, but also a great many rare subjects to fill in, where I needed something specific, to round out this project.

Most species in this book, except such as Fruit trees, Conifers, Ferns, and Carnivorous plants, are classified within their natural systemic families. I have tried my best to give each plant shown its recognized botanical nomen, and have generally been guided by the latest HORTUS Third (1976). Many noted experts have assisted me in the difficult and often controversial task of plant identification and correct nomenclature, and this assistance is deeply appreciated. Some names of long horticultural usage have been retained, but synonyms and cross reference lines will help to track down each plant more easily, assisted additionally by an index of the most important common or vernacular names.

Text information, aside from scientific names and family affinity, includes native origin, their popular common names for those more familiar with them, and characteristic features of a plant or tree, with measurements given in international metrics. Primary how-to-do chapters cover both the tropical outdoors, as well as plants suitable for cultivation in home or greenhouse, or for interior decoration in Temperate regions.

All plants described in the text are classified according to their climatic backgrounds, ranging from Tropical to Temperate, and these are explained in the CLIMATE and TEMPERATURE Guide on pg. 929. Intended merely as a guide line as to their preferance, this does not necessarily limit their tolerance. In practical experience, most species can adapt to other surroundings—in fact it would be a challenge to try, and surprisingly, many plants succeed quite well under conditions very different from their native home.

During my travels from one end of the world to the other, I have found that plants used in gardens are apt to be very similar in all warm-climate regions, and do no longer reflect endemic flora. This is particularly evident on the islands of the South Pacific where showy trees of native origin are meager, yet the gardens there are lavish in their array because most subjects are introduced from elsewhere in the Tropics. This interchange of plants has created an integrated pattern of harmony and vividly demonstrates that most subjects within this cyclopedia may be adapted to widely varying conditions — outdoors or on the patio where climate permits — and again, where the cold of winter does not, we may enjoy their exotic presence indoors.

In thus scanning the far horizons of the world for the best it offers in showy or interesting plants and trees, this Colorama truly hopes to reflect their enchanting beauty for the pleasure of all plant fanciers in all climates.

First Edition, May 1978

19 Prospect Terrace (P.O. Box 125)
East Rutherford, New Jersey 07073

856 Mason Road (P.O. Box 397)
Vista (San Diego Co.), California 92083

ALFRED BYRD GRAF, D.Sc.

A quiet backwater or Igarape of the primeval forest of Amazonas, with giants of trees supported by buttressed trunks, lianes dangling from the lofty branches; slender Assai palms and flowering trees; on a slow-moving river the buoyant platter-like leaves of the Victoria regia, and their fragrant flowers floating on the warm waters. From the oil painting of a typical landscape by a local collector of Northwest Brazilian orchids in Manaus.

The TROPICS and SUBTROPICS

When we think about the Tropic zone, what first comes to mind are the geographical circles in latitude known as the northern Tropic of the Cancer and the southern Tropic of the Capricorn. These were drawn on the Greek maps for the first time by the ancient astronomer Eratosthenes of Alexandria in 230 B.C. He proved, long before Columbus, that the earth was round, and calculated its girth at 252,000 stadia or 40,000 km (about 25,000 miles). Next he marked the sphere with lines parallel to the equator. The most important were—aside from the equator itself—the northern Tropic circle where the sun is still directly overhead when the day is longest, and beyond this the "Winter" or "Arctic" circle from where the sun can be seen all night on June 21. Based on temperature, these three zones became known as the Torrid or Equatorial zone, the Temperate, and the Frigid or Arctic Zones. The Tropic of Cancer as marked by the ancients, crossed the Nile at Assuan in upper Egypt, an arid region of scorching heat, an example of what the climate of the Torrid Zone was thought to be. The word "tropic" is from the Greek tropikos, meaning the solstice—either of the two times a year, at the summer solstice and the winter solstice when the sun is at its greatest distance from the equator. It marks the time when the sun turns around in its annual north-south path—the "Place of Turning", the furthest north and the furthest south where the sun at noon is vertically, or directly overhead. This happens at about 23.5 deg. latitude or 2600 km (1600 miles) north or south of the equator.

Temperate Zone of the Greeks

The name Temperate Zone as known to the Greeks applied to a climate typical of the Mediterranean, an area which we would now call sub-tropic. As was discovered later, this broad area includes such far from "temperate" places between the Tropic and the Arctic (or Antarctic) circles as the American Midwest and high Western mountains, much of Siberia or, in the Southern Zone, Argentina. The resulting macroclimate is far from just temperate and even less so subtropic, with extremes in temperatures from far below minus 20°C (zero °F) to more than 40°C (100°F). However, these marked differences between winters and summers have created a flora adapted to seasonal growth and dormancy, a condition much less pronounced in the tropics.

The geographical lines of Cancer and Capricorn bordering the Tropic zone are very real in terms of hours of light each day. And typically, the hot and sticky climate we know as "tropical" would be equatorial lowland something like Jakarta (Java), Lagos (Nigeria), or Belém (Brazil). But in reality, the Tropic circles cannot confine tropical vegetation within straight borders. The region of tropical climate could theoretically be placed inside a boundary of a mean annual temperature or isotherm of 21°C (70°F). This would extend to as far north as St. Augustine, curving westward through Southern Texas and Arizona, and edging into Southern California, areas where plants strictly tropical will live if given some extra care.

It is natural for this reason that in areas along the Tropic zone where climatic conditions are similarly tropical, plants from the Equatorial belt spill over into such adjacent areas and acclimate themselves quite well, especially if they come from higher elevations where all kind of weather prevails. To deny these plants to be tropical would not be realistic.

Similarly, plants from outside the Tropical circles, or Subtropics, will adjust easily in higher elevations of the Equatorial zone. Thus the zonal limits set by geographers will become a question of semantics.

Subtropics

Because the borders between the Tropics and the actual Temperate zone are so difficult to define, ecological variations have created a zone of mild climate extending variously for 500 to 1200 km (300 to 700 miles) on both sides of the Tropic zone and now known as the "Subtropics". This line is determined by the presence of bodies of water and ocean currents, directions of mountains and their elevations and other meterological climatic conditions. In this region frosts become infrequent and the mean values of the warmest months exceed 20°C (68°F), and of the coldest months 5°C (43°F). Roughly, this line is winding, depending on local conditions, at about 30 to 40 deg. latitude and may also be known as Mediterranean climate, and which is also the limit for the cultivation of Oranges. Similarities prevail in parts of the American Southeast and Southwest, California, Chile, South Africa, Southern Australia and New Zealand. This includes the regions of Citrus and Fig trees and of the Olive tree. That the outline of the Subtropics does not run in a straight line is best shown by the location of Naples, in Southern Italy, a city of Palms and tropical flowers and a maritime, sheltered climate, considered well within the Semi-tropics, whereas New York City, at the same geographical latitude, though a place of tropical summers, experiences frigid and icy winters, a typical North temperate, Continental climate, because it is open to the landmasses of the cold north by the absence of sheltering East-West mountains.

The TROPICS and SUBTROPICS

WARM-CLIMATE ENVIRONMENTS

Imagine a place where the normal restrictions to plant growth do not exist; where there are no frosts to eliminate sensitive species, no drought to stunt development, a place where the only limiting factors are soil and the competition between plants. Such a place is warm and humid and green. The noon light intensity is uniformly high, averaging 80,000 to 115,000 lux (8,000 to 11,000 foot-candles) all year. Rain is frequent or mist is dense. Days are 12 hours long the year around. Trees crowd together, their crowns interlocking, pushed aside in places by the light-hungry vines coiling up out of the dimness below. The trees themselves are covered by other plants; mosses, ferns, bromeliads and orchids cling to their bark with probing roots. Not only is the vegetation dense, it is on a gigantic scale. Here, high rainfall, adequate soil and a steady warm temperature maintain a highly diverse plant community. In its ultimate form, this is what we know as rainforest. There is no other formation on earth with so many different life forms.

Tropical conditions prevail where the mean temperature of every month exceeds 18°C (65°F), and where the Coconut palm and the Breadfruit tree will mature and bear fruit. The lush growth of the true tropics is entirely due to a blend of comfortably warm climate with sufficient rainfall and atmospheric humidity which results in harmonious interdependence or symbiosis and well-being of "exotic" plants and people. This is enhanced by the fact that 4/10 hectar (1 acre) of tropical forests removes 100 tons of carbon from the atmosphere. Yet, rainfall or the lack of it, and elevation create big differences, and the term "tropical climate" has little meaning if we range from rainforest to desert, or from tropical sea level to mountains with alpine flora. Warm nights near the equator at sea level extend to 600 and 900 m elevation (2,000 or 3,000 ft.); between 1,200 to 2,000 m (4,000 to 6,500 ft.) climate is semi-tropical with cooler nights, and comfortable to hot daytime temperatures. Above 2,400 to 3,600 m (8,000 to 12,000 ft.) we find conditions approaching the climate of the temperate zones, chilly or with frost at night, warm in daytime, and with occasional snow. Higher, in the Andes of South America, at altitudes 3,800 to 5,000 m (12,000 to 16,000 ft.), extends the Paramo or Tierra Fria, cold at night, with occasional cacti, but otherwise bleak and comparable with the arctic tundra. Vegetation and the few stunted trees cease at 5,600 m (17,000 ft.), and at 6,000 m (18,500 ft.) there is perpetual snow and ice.

Tropical Rain Forest

A tropical rain forest is composed of hundreds of different species of trees, many of them belonging to families which in the temperate zone are represented as shrubs or herbaceous plants. Forests in temperate zones may have just as many trees in a given area, but these generally consist of only one to several kinds.

The immense equatorial rainforest, so typical of endless thousands of square-kilometers of northern South America, is at its best in lowland regions of steep sun and a high annual rainfall of 300 to 500 cm (120 to 200 inches), fairly evenly distributed throughout the twelve months. Such forest is quite open because the dense crowns of its majestic trees 40 to 60 metres high, cut out so much light that few low-growing plants can exist on the forest floor. These five metres of rain annually, come down in definite bursts with the force of hammer blows, after which the skies become clear and blue again. Evaporation causes a veritable steambath that favours luxuriant growth. Drizzly, all-day rains are more characteristic of higher elevation cloud forests, where tree growth is lower and light can penetrate, and which encourages an undergrowth dense with shrubs and vines creating what is best described as tropical jungle. Here, clothes will be forever damp, and like everything else, subject to mold, a paradise for fungi and parasites.

Forest Soils

The forest floor, contrary to popular belief, is not soft and humusy as might be common in higher elevation cloud jungles. The constant heat tends to recycle all humus most efficiently in a short time, and burnt out soils are hard as rock, and poor.

The primeval forest gives the appearance of unbounded soil fertility, but in fact, the Amazonian region is a desert covered with trees. Its ancient leached-out soils have released most of the nutrients stored in the rocks and have little more to contribute. Amazonia is less densely populated than the Sahara, a fact difficult to believe.

Romantic Aimeo, with Bali-Hai, Moorea, French Polynesia.

Epiphytes

One of the most striking characteristics of tropical vegetation is the multiplicity of plants that grow on other plants. All kinds of smaller species have found a luxuriant way of life as epiphytes or "air plants" in the canopy of giant trees. In their airy perches, a curious sky-top world nearer to optimum light and sun, is formed of an ecological community of Orchids, Bromeliads; Aroids large and small; Begonias and Ferns.

All tree dwellers have developed forms or mechanisms to hold on to what rain they need, any excess easily draining off, to keep their roots in health and from stagnation. The epiphytic habit poses some specific biological problems, as such plants grow in the absence of soil. They build up starch like all green plants, from carbon dioxide of the air but differ from others in that they cannot obtain water and minerals from the earth. Usually their root systems are strong, anchoring them to the bark on which they grow. These roots have a symbiotic relationship with fungi, which gather minerals and other nutrients from the trace elements that occur in rain, trickling down the bark. A supply of water itself from intermittent rains is carefully conserved and stored in epiphytes—the pseudobulbs of Orchids, or the "tanks" or "vases" of Bromeliads.

The TROPICS and SUBTROPICS

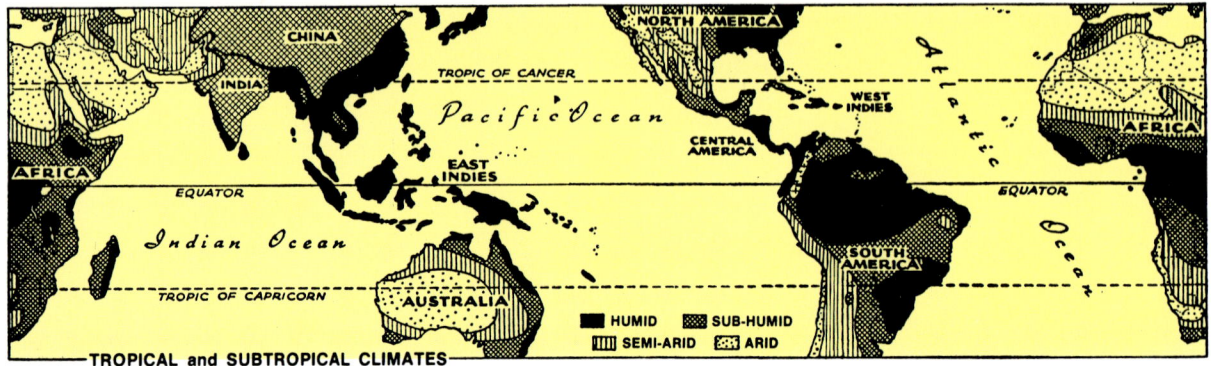

TROPICAL and SUBTROPICAL CLIMATES

EXOTIC ORIGINS and CLIMATIC BACKGROUNDS

The Tropics, because of the romance surrounding them, are frequently not truly understood. We imagine them as overflowing with color and exotic treasures, coddled by comfortable warmth, humidity, and sunshine. This is only true in part. The equatorial rain forest is not colorful. Typically it is a luxuriating scene of varied foliage in shades of green—an occasional bright spot the exception. Nor is the climate so ideal. The inner tropical belt 4,600 km wide, and the outer subtropical zones bordering it north and south for 480 to 1,100 km, are the home of most of our indoor plants. Depending on elevation, ocean currents and precipitation, various characteristic climates prevail, but they share a relative uniformity of mean monthly temperatures throughout the year, thus differing from the strong contrasts between summer and winter in temperate regions.

Climatic backgrounds also take a hand in shaping the relative wealth of species of tropicals in the various floristic regions of the world. Richest in tropical plants appears to be the western hemisphere with an estimated 112,000 species, in addition there is the rich near-tropical Cape flora with 16,000 species. The monsoon region of tropical Asia is immeasurably wealthy in beautiful, though often delicate plants but their number can only be estimated at upwards of 50,000 species, of which 35,000 are recorded for the Malaysian area including New Guinea.

To give a potential garden or indoor plant the best possible chance of success, we should know something about the climatic backgrounds that prevail in their habitats. Environment has shaped characteristics in plant types that make them tolerant or difficult to acclimate when taken into cultivation.

TEMPERATURE and RAINFALL at typical locations in the Tropic and Subtropic Zones

	LAT. deg.	ELEV. meters	TEMP. °C min. max.		RAIN cm
NORTH AMERICA					
California, San Diego	32.7 N	40	2	31	27
Florida, Miami	25.8 N	3	−3	35	140
Mexico, Mexico City	19.2 N	2310	−4	33	60
Mexico, Vera Cruz	19.1 N	16	9	35	172
WEST INDIES					
Cuba, Habana	23.8 N	49	10	35	120
Puerto Rico, San Juan	18.2 N	30	16	34	152
Jamaica, Kingston	18.1 N	7	14	37	82
CENTRAL AMERICA					
Guatemala, Guatemala City	14.3 N	1481	5	32	127
Costa Rica, San José	9.5 N	1147	8	34	177
Panama, Colón	9.2 N	8	19	35	312
SOUTH AMERICA					
Venezuela, Caracas	10.3 N	1043	7	33	80
Venezuela, Ciudad Bolivar	8.9 N	38	19	36	88
Guyana, Georgetown	6.5 N	21	20	33	225
Colombia, Buenaventura	3.5 N	12	23	33	975
Ecuador, Quito (Sierra)	0.1 S	2852	2	26	122
Ecuador, Mendez (Oriente)	2.4 S	698	16	32	255
Brazil, Manaos (Amazonas)	3.0 S	45	19	38	182
Brazil, Rio de Janeiro	22.5 S	64	11	39	108
Brazil, Sao Paulo	23.3 S	820	−2	38	140
Peru, Iquitos (Amazon)	3.7 S	90	18	31	258
Peru, Lima	12.3 S	156	4	32	5
Peru, Cuzco	13.3 S	3452	−2	26	80
Bolivia, La Paz	16.3 S	3660	−3	24	55
Chile, Santiago	33.2 S	520	−4	37	35
Argentina, Buenos Aires	34.3 S	25	−2	40	95
EUROPE					
France, Marseilles	43.1 N	75	−6	38	58
Italy, Palermo (Sicily)	38.1 N	70	3	36	75
Spain, Seville (Andalusia)	37.2 N	30	−5	46	47
AFRICA					
Egypt, Cairo	30.3 N	30	−1	45	3
Cameroon, Douala	4.0 N	10	19	32	395
Equat. Africa, Brazzaville	4.2 S	290	12	38	123
East Africa, Nairobi	1.1 S	1662	2	32	95
Tanzania, Amani (Usamb.)	4.5 S	945	7	30	125
Tanzania, Tanga	5.1 S	30	18	34	153
Madagascar, Tamatave	18.9 S	4	13	38	313
So. Africa, Johannesburg	26.1 S	1754	−5	32	80
South Africa, Cape Town	33.5 S	12	−1	40	63
ASIA					
Israel, Haifa	32.6 N	10	2	38	67
Japan, Nagasaki	32.4 N	133	−5	37	198
China, Yunnan-Fu	25.2 N	1943	−4	33	105
Sikkim, Manjitar, Rangit R.	27.1 N	249	10	35	437
India, Cherrapunji (Assam)	25.2 N	1289	9	32	1065
India, Madras	13.4 N	7	14	45	120
Taiwan, Keelung (Teipei)	20.1 N	10	3	33	337
Burma, Mandalay	21.6 N	76	9	41	82
Philippines, Baguio	16.5 N	1461	8	25	457
Philippines, Manila	14.3 N	14	14	38	200
Thailand, Bangkok	13.4 N	4	11	41	130
Vietnam, Saigon	10.4 N	11	15	40	175
Ceylon, Colombo	6.5 N	7	17	36	200
Borneo, Sandakan	5.5 N	3	20	36	300
Sumatra, Toba	2.5 N	1150	14	27	225
Malaya, Singapore	1.2 N	2	18	38	238
Java, Jakarta	6.1 S	8	19	36	180
Java, Bogor	6.6 S	280	18	32	430
New Guinea, Port Moresby	9.3 S	39	20	36	102
AUSTRALASIA					
Hawaii, Honolulu	21.2 N	4	11	32	90
Hawaii, Hilo	19.4 N	12	11	33	342
Fiji Is., Suva	18.8 S	13	14	36	280
Australia, Brisbane	27.3 S	42	2	42	112
New Zealand, Auckland	36.5 S	46	0	32	110

Freezing point zero degree Celsius (C) = 32 degrees Fahrenheit (F.)   *1 metre (m) = 40 inches (or 3.28 feet)*

CLIMATIC BACKGROUNDS

Some floristic regions of the world are wealthier in colorful plants than others, but generally in the wilds the separate kinds are diluted in an overwhelming mass of greens, with a scarcity of conspicuous blossoms.

A good example are the Pacific islands typified by Hawaii or Tahiti. We dream of, and experience them as tropical paradise overflowing with beautiful flowers. Yet, the fact is that Polynesian indigenous flora is quite poor in showy species, and very few can be regarded as interesting garden material, although some 1,800 species of plants have been recorded in the Hawaiian islands, and barely 500 different kinds in French Polynesia. Another, even more striking example is the flora of the Seychelles in the Indian Ocean, outwardly so colorful, where only 80 species endemic, or truly native have been discovered there.

Tropical islands, generally such as the West Indies of the Caribbean Sea, or South Sea isles of the Pacific, invite interchange of plants from other parts of the tropical world because they share a more equable climate than the interiors of large land masses or continents astride the equator. The world's oceans absorb the equatorial heat to hold it and give it off slowly so that island temperatures are rarely less than 21°C (70°F) or more than 28°C (85°F). Also, daylength varies very little with 12 hours more or less throughout the year. The intensity of the sun is an average of 100,000 metric candles or lux (10,000 f.c.), no more than New York on most any sunny day from spring to fall. Rainfall is more dependable on islands with high mountains as in Jamaica, Puerto Rico, Hawaii, Tahiti or New Caledonia, but I have found no lack of moisture even on coral isles without mountains such as fertile Tonga, with either rainclouds hovering over them or from subwater by osmosis through the coral floor.

For these reasons it is not difficult at all to create the ideal tropical garden full of color and bloom by taking advantage of introductions from distant warm-climate regions, inviting them to stay and share the company of those already familiar to us and at home.

WARM-REGION GARDENS

Tropical gardens are usually beautiful to behold with their bright flowers and gay foliage. This is not accidental and has not always been so. The endemic flora of any given region of the tropical world, in spite of its diversity can never present the composite colorama that we find assembled in any good warm-climate garden today. Plants from other tropical or mild climates have been brought together, through exchange or introduction, especially so during the last century. Fortunately, most are readily established and acclimated and now live happily together in symbiosis with their cultivators.

Horticulturists use the word "tropical" to designate characteristics and requirements of warmth-loving plants of lower altitudes in both the true geographical tropics and the adjoining semi-tropics (see map). Consequently, tropical flowers will be found growing outdoors in warm climates, or in planned tropical environments indoors whether in greenhouses of glass or of plastic, for office decoration, or nursed for their exotic beauty as a hobby in private homes.

The tropical gardens in Central and South America, the West Indies and the Pacific islands use basically the same plants as those of South-east Asia, but with numerous additions such as of Heliconias, Hedychium and other gingers, Philodendrons, and terrestrial Orchids. Bromeliads are grown more freely, both epiphytic and terrestrially on the ground. Ferns also are more frequent, with showy treeferns, epiphytic Polypodiums, Aspleniums, and Platycerium, the Staghorn-ferns.

It seems, however, that the greatest effort toward more varied plantings is made in regions with a dry-warm climate, such as Southern California, the Mediterranean Coast, Southern Africa, and Australia.

But climate alone does not create interesting gardens; there must be initiative and underlying industriousness, as well as popular interest in plants, response to beauty, and an urge to improve the surroundings of a home. If the easy-going, happy temperament so typical in nationals of Southern Italy or in Mexico prompts them to sing and play rather than to sweat in digging up a garden, this is understandable. It is far easier to put the varied plants offered in pots and tins at the mercado, into colorful terracotta jars for the patio, or hanging them from rafters where nature will take care of most of them. There they look picturesque and exotic, as on a picture postcard so typical of the countryside.

Heritage also plays a role, and much tropical gardening interest is shown by people transported from Northern, cooler climates, and to them a new home will be an attempted combination of the traditional garden so lovingly taken care of for generations, merged with the galaxy of tropical plants that will be found in neighboring gardens or in parks under the equatorial sun. A good example are the city-wide competitions for the best cared-for homegardens in Christchurch, New Zealand, that compare favorably with their prototypes in England, the mother country and yet have a blend of the tropics in their aspect.

It has been my observation that areas relatively poor in endemic ornamental flora, such as in three of our most colorful states, Florida, California and Hawaii, these have been very much enhanced by introductions from warm regions elsewhere. On the other hand, not much effort is expended in planning and planting a garden in climates favoring lush tropical growth such as Ceylon, Bali or Tahiti, and once the everyday shrubs or trees are planted, gardens are left pretty much to themselves and given a minimum of care. Often, a Croton, Hibiscus and Acalypha provides all the color the year round. Add to this a Palm tree and a Frangipani, with Allamanda, Bignonias and Bougainvilleas as typical shrubby climbers, and the tropical garden is complete.

Far from modern garden design, or the more naturalistic settings in the Tropics, the conservative style of landscape architecture in India and Pakistan was based entirely on Islamic rules brought to India by the Mughals from Central Asia. The gardens of the Taj Mahal are entirely formal and planned with the sentimental ideal of the paradise garden of the Koran: "a garden enclosed and of living waters, of sweet perfumes in the moonlight". They are equally divided in accordance with the concepts of the ancient Hindu four-fold "Paradise of Restfulness" by canals of water with fountains, so essential in an Indian garden for irrigation and cooling. These four parts are again divided into four, and in each section are grouped related plants or trees, fragrant roses, jasmine and frangipani; fruiting citrus, pomegranate, mangoes and palms, the bulbs of Kashmir; somber cypress pyramids, and bright annuals, to symbolize the marriage of the trees and flowers, and the paradise orchard.

The formal gardens of the Taj Mahal in Agra, India, symbolising the marriage of trees and flowers, planted with somber cypress trees and fruiting citrus, bright annuals, fragrant jasmine and roses, planned in true Hindu tradition, blended with ideals of the Paradise garden of the Koran -"a garden of living waters and sweet perfumes in the moonlight".

A palm-thatched cottage on stilts along the busy waterways or klongs of Bangkok, Thailand, festooned with hanging pots of orchids, lovingly tended and enjoyed.

WARM AREA HORTICULTURE

CLIMATE, LIGHT and TEMPERATURE

In colder climates, one has to contend with Spring, Summer, Autumn and Winter as definite seasons—in the tropics generally, the variations are far from showing any definite line of demarcation. One may describe them as the Wet Season and the Dry Season; the Wet Season occurring for about six months of the year, with a high rainfall; in the Caribbean and equatorial So. America between June and November, and the Dry Season for the other six months of the year, with a low rainfall, one merging gradually into the other. In any case the night to day temperature variation in the Dry Season—the cooler part of the year, is 19 to 31 deg. Celsius (67-88°F) with a comparatively low humidity; the temperature variation in the rainy season is from 20 to 33 deg. Celsius (C) (68-91°F) with a high relative humidity; these are typical of Trinidad, a location 10 deg. North of the Equator.

These of course are the averages for the low lying districts. There are however many factors which contribute to much wider variations in rainfall, temperature and humidity, in the same or different countries, governed by the presence of mountains, valleys or proximity to the sea.

Even at the Equator, the elevation above sea level makes a great difference to the type of plants one can grow satisfactorily. From 900 metres upwards it begins to become sub-tropical. A rough calculation would show that for every 100 metres in altitude, there is a fall of about 3/4°C (approx. 1°F) in temperature.

The position of the sun in the Temperate Zones shows a definite Northern or Southern variation, whereas in the tropics the sun varies little from overhead, except in East-West direction. Light intensity in the tropics is not much different from that in the Temperate Zone during summer, even at high altitudes except that it is richer in ultraviolet light, and ranging from 120,000 lux (12,000 fc.) in bright sunlight to 5,000 lux (500 fc) on a day with overcast sky and no shadows. It does change a little from one season to the next, as sunshine is less intense in the rainy season, due to water vapor in the sky. Ultraviolet rays of light have a tendency to cause sunburn to tender foliage, and to fade colors, but to my knowledge their exact effect on plant growth has not yet been determined.

In the cold climates there are marked differences in the length of daylight according to the seasons, whereas in the tropics, daylight variations are limited to an hour one way or the other throughout the year—sunrise to sunset.

Flowering Seasons Reversed

It is of interest to know that plants and trees brought from the Southern hemisphere—Australia, South Africa, South America, and vice-versa, will generally reverse their flowering season, after a period of adjustment. The Los Angeles Arboretum found that Eucalyptus trees planted in Arcadia, and blooming in Australia's summer, likewise flowered in summertime in California. Some plants such as imported orchids, may take several years to acclimate themselves entirely.

Daylength

Length of day is of course a major factor in influencing bud initiation, and is generally more pronounced further away from the equator. For instance, the epiphytic Rhipsalidopsis or "Easter cactus" will bloom in Southern Brazil in September, in New York during March.

CLIMATE and the GARDENER

Warm climates of the world properly fall into numerous specific groups defined by graphic climograms based on daily minimum and maximum temperatures, humidity and rainfall; duration or absence of sunlight, latitudes and altitudes. Papadakis (Argentina 1975) holds that the former application of mean temperatures are meaningless; the layman is more advanced and speaks of the minimum low and the maximum high in his daily usage. From this a graphic line can be drawn, which then allows the existence or cultivation of certain plants.

If the minimums are entirely frostless, this would be tropical climate, and with sufficient humidity, the usual heavy summer rainfall, and warm nights, is characterized by the presence of the Coconut, Oilpalm, Hevea rubber, Cocoa, Bananas and Breadfruit. At higher altitudes, where nights become cooler, climate is better adapted to fruit and many beautiful flowering trees, also most of the colorful annuals of temperate climate summer gardens. With diminishing rainfall the zone becomes more arid, and thornbush, cacti and other succulents become predominant. A maximum of light is less important to plant life than temperature and humidity. In photosynthesis Solar energy is converted into organic matter in plants. But while tropic sunlight may measure about 100,000 to 120,000 metre-candles, or lux (10,000 to 12,000 foot-candles), plants cannot use these high intensities, and often less than 1% of the radiation received by a leaf is transformed into chemical energy.

The most luxuriant vegetation and forests are encountered in cloudy climates with low light intensity. Light readings I have taken in deep shade for example on the rainforest floor in Veracruz, Southern Mexico, measured only 1,200 lux (120 fc), with intense sunlight above at 120,000 lux (12,000 fc), and yet such flowering plants as Impatiens, Begonias and Heliconias were in prolific bloom. Efficiency of energy transformation decreases as light intensity increases, especially where limited by the low content of carbon-dioxide (CO_2) in the atmosphere. In other geologic times when the atmosphere was richer in CO_2, vegetation was far more luxuriant.

PLANTS for Tropical Conditions

Ambition or nostalgia often prompts the uninitiated to try and grow the plants loved in the temperate zones. Some of them do willingly adapt themselves, and when one finds English forget-me-nots (Myosotis) growing and blooming with as much vigor on the shores of Trinidad in the West Indies as at the top of Alpine mountains in Switzerland, it gives one food for thought.

Generally, however, plants from "temperate" regions do not thrive under tropical conditions, because they are conditioned to seasons of cold and heat, or dryness, which spell a period of dormancy and unless they find conditions that allow them to rest, they will not grow well, if at all.

WARM AREA HORTICULTURE

Characteristic garden of massed colors designed and planted by the famous architect Roberto Burle-Marx at Terezopolis, in the Organ Mountains of Brazil.

PLANTS and TREES in Warm-Climate Gardening

Some gardeners want to try everything they may see, and difficult plants are a challenge they are unable to resist. But the best and most realistic rule in the choice of trees and other plants is to keep the species known to do well in the neighborhood, and offered by local nurseries.

The esteemed artist and landscape architect Roberto Burle-Marx of Rio de Janeiro has reduced the variety of species used in his garden designs, to sometimes less than a dozen. His concept is to employ a chosen few of characteristic plants, massed in groups or covering extended areas, each in a different shade of color. He considers color the basic element of a garden, and he uses whatever material is available, even stones. Roberto showed me some of his creations in Petropolis and Terezopolis, with vast expanses of colorful but often simple plants—red Iresines next to light-colored Chlorophytum and silvery Helichrysum, light green St. Augustine grass (Stenotaphrum), against stilted Pandanus, plumed tall pampas grass (Cortaderia), Phormium, Heliconias and a complex of tree Philodendrons for accent; Cannas and Caladiums, Dieffenbachias and islands of giant Bromeliads. In shapely pools, Nymphaeas, Cyperus and water hyacinths, with stately palms, Araucarias or somber pines as a backdrop. The total effect is that of a great painting, with lots of open breathing space, set in the beautiful landscape of the rugged Organ Mountains.

Exotic Evergreen Shrubs

In any tropical or subtropical garden the **Evergreen shrub** is the most important plant, especially if the area is small. We may find space for a flowering tree or two, and of course palms or treeferns are very desirable, because it is these which lend the tropical touch. Or, some fruit trees will be found as ornamental as any that are planted for their blossoms only. But generally, trees become important only if the landscaped space is large enough.

Shrubs come in many forms, with some more suited to humid-warm conditions, others for arid zones, conditioned to bright sunlight and neglect. There are shrubs that tolerate the saltspray along the seacoast, and others better suited for higher, cooler elevations. Some species lend themselves for use as hedges, or for background screening, others remain low and spreading.

Flowering and Berried shrubs in the garden are assets of permanent beauty, as they provide much of the color. They do not need the constant attention necessary to keep annual flower beds looking their best.

Normally, shrubs can be kept in shape by occasional pruning, preferably after blooming. Planting time is best with the onset of the rainy season. Older specimen benefit from feeding during the growing period, and a compost of humus mulch at the end of the rains prevents excessive evaporation and keeps the roots cool and damp.

Woody Evergreens most Popular:

Abutilon, Alpinia, Banksia, Beloperone, Bougainvillea, Brunfelsia, Buddleja, Caesalpinia, Calliandra, Callistemon, Cassia, Cestrum, Cissus, Clerodendrum, Clianthus, Clytostoma, Cuphea, Dais, Datura, Dombeya, Duranta, Ervatamia, Erythrina, Escallonia, Euphorbia, Feijoa, Galphimia, Gardenia, Hedychium, Hibiscus, Holmskioldia, Iochroma, Ixora, Jacobinia, Jatropha, Lagerstroemia, Lantana, Leonotis, Leptospermum, Malpighia, Malvaviscus, Medinilla, Megakepasma, Musa, Mussaenda, Nerium oleander, Nicolaia elatior, Odontonema, Pachystachys, Pentas, Plumbago, Protea, Ravenala, the Traveler tree; Roses, Ruellia, Russelia, Ruttya, Salvia, Sanchezia, Strelitzia, Streptosolen, Tamarix, Tecomaria, Thevetia, Tibouchina, Warscewiczia, Zingiber.

Shrubs with Colored Foliage

What makes tropical gardens especially exotic, is the availability of beautiful shrubs with variegated and multicolored leaves, and which have the advantage of showing year-round color, even while other shrubs are not in bloom. Most striking are the endless rainbow hues found in "Crotons" (Codiaeum), which thrive best in the hot tropics saturated with humidity, and which brings out to best advantage all their reds and yellows. Next are the Acalyphas in blazing red, and much used in hedges. Then Cordylines, known better as "Red dracaenas"; Graptophyllums and Pseuderanthemum, Pandanus, Polyscias, Sanchezia, Pedilanthus; variegated forms of Hibiscus, Bougainvilleas and Manihot.

For Cooler Locations:

Ardisia, Azalea, Camellia, Carissa, Ericas, Euonymus, Fatshedera, Fatsia, Fuchsia, Hebe, Hydrangea, Ilex, Ligustrum, Murraya, Pyracantha, Raphiolepis, Wisteria.

EVERGREENS grown for their Foliage:

Warm Climate

Acalypha, Brassaia, Breynia, Codiaeum (Croton), Coprosma, Cordyline, Dizygotheca, Dracaena, Graptophyllum, Kalanchoe beharensis, Leucadendron, Manihot, Philodendron, Polyscias, Pseuderanthemum, Ricinus, Sanchezia.

Semi-tropical

Aralia, Dodonaea, Euonymus, Fatsia, Ilex (Holly), Ligustrum, Mahonia, Osmanthus, Pittosporum, Schefflera, Tetrapanax, Trevesia, Tupidanthus.

Giant Grasses:

For background screening, as hedges, or individual specimen, the many Bamboos of the Grass-family (GRAMINEAE) offer wide possibilities. Good focal points because of their different structure are the plumy Pampas grass (Cortaderia), New Zealand flax (Phormium tenax), and the cluster-forming Yuccas or Palm-lilies.

Shrubs with Attractive Berries:

Planted in containers or the garden, such plants are colorful not so much for their flowers but for their showy berry-fruits, some of them edible. Some of the most ornamental are Acmena, Arbutus, Arctostaphylos, Ardisia, Berberis, Bucida, Carissa, Citrus mitis, Coffea, Cotoneaster, Fortunella, Ilex (Holly), Murraya, Nandina, Pernettya, Photinia, Pyracantha, Skimmia, Viburnum.

ORNAMENTAL TREES

No garden, save the very smallest, ought to be without trees. Shrubs may be most important in landscape planning—trees add a feeling of the forest, which draws every nature lover. No house feels like a home unless it is framed by trees which cast cooling shade in tropical settings. Properly placed, trees serve to hide unwanted views, telephone poles or neighboring buildings. Trees will also attract birds and become their home. And they are refreshing, giving off oxygen into the surrounding air. It has been figured that one tree produces enough oxygen to support one human life.

In planting a tree, it is of course important to know how large it is apt to grow. With limited space one should not plant trees that under tropical conditions are likely to become huge specimen, such as some Ficus, Bombax, Delonix, Amherstia, Albizia, Bischofia or Saman. Many flowering trees shed their leaves in winter, but eventually burst into glorious bloom; typical are Bombax, Cochlospermum, Erythrina and Tabebuia.

With most nursery stock now container-grown, trees can be planted in warm regions at any time of the year. However, under cloudy conditions or after a rain, while the soil is damp, they have the best chance to get started. In dry areas it is prudent to mulch the soil around the tree to prevent excessive evaporation and keep the root zone cool. To soak down a newly planted tree thoroughly with water goes without saying, and it is equally important to keep the root area moist until the tree shows signs of taking root and starts to grow.

Popular Tropical and Subtropical Trees

Acacia, Amherstia, Anacardium, Arbutus, Artocarpus, Bauhinia, Bombax, Brachychiton, Brassaia, Brownea, Butea, Calodendron capense, Cananga, Cochlospermum, Chorisia, Clusia, Coccoloba, Colvillea, Cordia, Cussonia, Delonix regia, Dombeya, Eucalyptus, Eugenia, Euphorbias, Ficus, Fraxinus uhdei, Grevillea, Jacaranda, Lagerstroemia, Leucadendron, Liquidambar, Magnolia, Melaleuca, Melia, Metrosideros, Montezuma, Nothofagus, Olea, Persea, Plumeria, Pyrus kawakami (Evergreen pear), Saraca, Schinus, Schotia, Spathodea, Stenocarpus, Stenolobium, Tabebuia, Tecoma argentea, Telopia, Terminalia, and Tree philodendrons.

A blend of the ornamental and the useful; delicious fruit - Phoenix dactylifera, the "Date palm", in the arid-hot climate of the Coachella Valley, California, an area near sealevel, with heavy irrigation from the Colorado River.

PALMS for Tropical Effect

Palms have been named the "Princes of plants", a noble family created by Linnaeus. Their majesty of stature and gracefulness whenever seen, sets them apart from all other, more lowly trees. Palms are evergreen and the most distinctive and regal foliage trees of the World's warm climates. Palms are easily divided into two main groups by their leaves—the "Feather palms" with pinnate fronds looking like enormous feathers; and the "Fan palms" with palmate, fan-like leafblades resembling giant hands.

Whenever a palm is featured in any garden, it immediately conveys the feel of the Tropics, and the romance of the jungle surrounding them. To me the real Tropics begins with the first Coconut palm. Very majestic in appearance are the Roystoneas or Royal palms, followed by Arecastrum, the "Queen palm" and popularly known as Cocos plumosa, in outdoor plantings of warm climates. The leading palms in the Subtropics are Washingtonias, Phoenix, Sabal, Trachycarpus.

Many palms are ideal decorators in containers. Best known is the long-suffering, reliable "Kentia palm", Howeia forsteriana, grown in Europe since Victorian times, after 1871. In later years, several species of Chamaedorea, the Parlor palm and the Bamboo palm, have become very popular, and all are known for their long-lasting qualities in interior decorative plantscaping, even with poor lighting, Such palms generally prefer relatively small containers and are not happy when overpotted, and attention is needed to keep them well watered for long and enduring life.

CYCADS or Fern Palms

Similar to Palms in appearance are the very tolerant and slow-growing Cycads, or "Fern palms": Cycas the "Sago palm", Dioon or "Virgin's palm", Encephalartos, Macrozamia and Zamia.

Tropical and Subtropical FRUIT TREES:

Warm-climate Fruit trees very often are also desirable as ornamentals. So why not combine beauty with usefulness especially in the backyard garden. The Mango, Avocado, Loquat, Rose apple, Macadamia nut, Sapote, Pomegranate and even Citrus are as attractive as any non-productive tree. Not only the foliage and flowers are handsome, but the colored fruit can be quite striking.

Those mentioned are the more tropical of fruits, but a number of warm temperate fruit may be adaptable to warmer climates, and I am thinking of Apples, Pears, Plums, Peaches and Apricots, but all these do better where winter chill will drop their foliage and induce best flowering and the setting of fruit. Such trees can or should be trimmed for size and shape in late autumn or winter.

Popular Fruit for Mild Regions:

Akee (Blighia), Avocado (Persea), Banana (Musa), Breadfruit (Artocarpus), Cashew nut (Anacardium), Cherimoya (Annona cherimola), Coconut (Cocos nucifera), Custard apple (Annona), Date palm (Phoenix dactylifera), Durian (Durio), Figtree (Ficus carica), Jackfruit (Artocarpus), Lemons (Citrus), Loquat (Eriobotrya), Lychee nut (Litchi), Macadamia nut (Macadamia), Mammee apple (Mammea americana), Mango (Mangifera), Mangosteen (Garcinia), Olive (Olea), Orange (Citrus), Papaya (Carica), Peach (Prunus persica), Pecan nut (Carya pecan), Persimmon (Diospyros), Pomegranate (Punica), Rambutan (Nephelium), Star apple (Syzygium), White sapote (Casimiroa).

CONIFEROUS Evergreens

The warm climates of the world are not usually associated with Conifers or "Needle trees". Their distribution in the Tropics is minimal, except at higher elevations of 2,000 m and up; in the Subtropics nearer sea level. Some representatives of this group are quite handsome and of noble if somber appearance, as are the cedar trees; Araucarias and Agathis; long-needle Pines; Podocarpus or Yellowwood; Junipers, Cunninghamia, Cupressus and Thujas. All these are occasionally planted as background or accent trees, also along streets.

A few of the subtropic conifers may be used with confidence for the cool indoors in containers; the best is Araucaria heterophylla or "Norfolk Island pine". Also Podocarpus are excellent decorators when grown in tubs. Dwarfed pines, cedars and junipers are grown as a hobby in Japan as "Bonsai" or in China as "Ming trees".

Typical of the Spanish interior courts open to the sky; plants in glazed or terracotta pots always decorate every nook and corner, or cascade down the stairway leading to the upper rooms, and enhanced by artful black iron railings, window grills, and hanging lanterns.

CONTAINER PLANTS Outdoors

Where not practical to make permanent outdoor plantings, trees or evergreens growing in containers offer an excellent alternative. Walk along a street in downtown San Francisco, and the sidewalks are shaded by handsome Ficus nitida trees growing in oversized white cement tubs. On New York's Fifth Avenue large fiberglass containers display the greenery of pyramid Taxus. Though in summer time only, graceful specimen of Ficus benjamina create a refreshing forest in the sunken plaza of Radio City's outdoor dining area, to the musical sound of splashing fountains.

On the open or sheltered patio of our homes, particularly in California and Florida; behind the garden walls in Hispanic-America or the Mediterranean; or on the veranda of tropical houses with English traditions, there exists another world of tropical gardening with great possibilities. Many a dwarfed Ornamental or Fruited tree or shrub can be grown in fancy pots of terracotta, fiberglass or stoneware, as well as in wooden tubs, on a floor of tile or brick. The walls covered by Bougainvillea or Passion vines; tiled benches or comfortable chairs help to form an outdoor extension of the living room, very homey and secluded from the outside world.

This concept goes back to Roman times, where the inner peristyle garden was part of every villa, where family and guests would gather on social occasions, the forerunner of today's patio. Such gardens would feature a water system of lead pipes, allowing the cultivation of evergreens, bulbs, annuals and creepers, grown in colored pots.

The Arabs, and especially the Moors in Spain during the Middle Ages, designed their houses and palaces on a similar theme, with fruiting trees and flowering, fragrant potted plants, cooled by splashing fountains and refreshing waters, in interpretation of their visions of the paradise of Islam.

Peristyle garden or open-air room of a Roman villa in Pompeii, until buried by volcanic ash in 79AD, was the sheltered refuge of the family, and forerunner of the present day patio.

The Generalife Gardens of the Alhambra Palace in Granada, Andalucia, built by the Moorish kings beginning in 1230, according to Islamic ideals, with pools of water lilies and flowering plants in terracotta pots.

Charming example of the Mexican's love of plants, decorating the walls of a home with potted succulents, fastened to the white stucco, in Fortin de las Flores, Veracruz.

WARM AREA HORTICULTURE

Practical demonstration of various grasses and ground covers suitable for subtropic conditions, at Los Angeles Arboretum, Arcadia, California.

Colorful garden walk bordered by well-groomed, heat tolerant Bermuda grass (Cynodon); the winding beds of lush Coleus kept to ground cover level by constant pinching, at Peradeniya Botanic Garden on Ceylon.

Always refreshingly cool is the greenery of perennial rye grass (Lolium) on a summer day in the Tuileries gardens at the Louvre, in Paris, France.

GROUND-COVERS and LAWNS (see pg. 510)

Grasses

Beautiful green grass, such as the lush Kentucky blue, is part of the refreshing garden color, in open spaces around the homes and in parks of moist-cool climates of the world. It is only natural that gardeners in warmer and tropical regions would also want a similarly pretty lawn. All depends on climate, and with care there are many possibilities to succeed—if not with grass, then with grass-like substitutes. Where neither will be practical, then one of many available ground-covers may have to be planted instead.

The more tropical the climate, the more difficult it will be to have an expanse of real grass, so easily ruined by bugs, disease and high acidity. In dry subtropic areas as in California several grasses are fairly satisfactory, such as the cool-season Perennial Rye (Lolium), and the fine-leaved Colonial or Creeping Bent (Agrostis); for sub-tropical, hot-weather lawn, the slow-growing, fine-textured Korean grass (Zoysia), the aggressive, silky-leaved Bermuda (Cynodon), and the coarser broad-bladed St. Augustine (Stenotaphrum) grasses. These latter three are also popular in Florida, although St. Augustine grass is at times annihilated by the small black chinch bugs. The many well-kept gardens in Tropical Brazil are also planted with the St. Augustine grass. In South Africa the favorites are Bermuda and St. Augustine, as well as Pennisetum, the "Kikuyu grass"; in Australia St. Augustine (locally the Buffalo grass), Bermuda grass and Pennisetum; in hot India I noticed primarily the tough and usually very persistent Bermuda grass (Cynodon) as lawns in public parks.

Lawn Substitutes

Several grass-like creepers and grass substitutes are seen in California and Florida nurseries. Best-known is Dichondra, the "Lawn-leaf", with little spoon-shaped leaves, perfect and springy to walk on. It is a ground-hugging plant that spreads by rooting surface runners, and may be mowed like grass. Arenaria verna caespitosa, the "Irish moss" with hair-like, lush green leaves and of compact, moss-like growth, and tiny white flowers; Sagina subulata, the "Scotch moss" with golden-yellow mossy foliage. Herniaria or "Green carpet" forms a thick, mat-like carpet of bright green, the runners with tiny oval, fern-like leaves. The "Blue fescue" (Festuca) is a somewhat taller silvery blue grass, growing in tufts.

Ground Covers

While grasses are an invitation to be walked on, other ground-covers form low matted areas, often more colorful than the greens of lawns. Broad-leaf creepers are more adaptable to areas where grass is difficult to grow, in spots severely arid, in the shade of trees or on steep, rough hillsides. A favorite in Florida is the yellow-flowered Wedelia which is easy to maintain even in full sun.

Most low plants with a tendency to spread may be used for ground cover planting, either all of one kind, or in massed color patches of contrasting varieties, in the manner of the noted architect Roberto Burle-Marx of Rio de Janeiro. I have seen him use brilliant red or yellow Iresines next to silvery Helichrysum petiolatum; Santolinas and pink Wax Begonias with blackish Ophiopogon, golden Lantanas, variegated Chlorophytum, purple Zebrina and Setcreasea purpurea or "Purple heart" against sky-blue Forget-me-nots (Myosotis).

Some of the best and most favored Creepers are: Ajuga (Carpet bugle); Alternantheras in several colors; Artemisia (Silver spreader); Catharanthus (Vinca rosea); Cymbalaria (Basket ivy), Euonymus radicans; Fragaria (Wild strawberry); Gazania (South African daisy); Hedera (Ivy in variety, for shade); Helxine (Baby tears, in shady spots); Iberis (Candy-tuft); Isotoma (Blue-star creeper); Creeping Juniperus (dwarf Conifer); Lantana sellowiana (Trailing lantana); Liriope (Lily turf); Lobularia (Sweet alyssum); Lotus berthelotii (Parrot's beak); Osteospermum (Trailing African daisy); Ophiopogon (Mondo grass); Pelargonium peltatum (Ivy geranium); Polygonum capitatum (Pink clover-blossom); Potentilla verna (Cinquefoil); Verbena peruviana (Crimson creeper); Vinca minor (Running myrtle).

For very sunny, arid areas various succulent "Ice plants" in many colors, such as Carpobrotus, Delosperma alba or "Disney creeper", Drosanthemum, Hymenocyclos, Lampranthus, and Portulaca, the annual "Rose moss"; also creeping Sedums.

VINES and BASKET PLANTS:

Vines and Trailers are actually weeping plants with stems too weak to support themselves. Climbers and Creepers are found in every plant family. In small sizes there is an endless selection which may be utilized indoors, and because of reduced light conditions, the various foliage types will do best. Tradescantias, Scindapsus, Pilea or Plectranthus dangling from a pot, on a stand or the office desk; Philodendron climbing on a pole or trellis, or Hoyas and Ivies around the window following a wire, bring into a room a touch of graceful greenery and "lived-in" feeling.

For real flowering effect, vines known for their colorful blooms develop their best if grown outdoors. They may be in large containers, on the patio or veranda during the warm season, or planted in the garden of tropical or subtropical climes. This may be on pergolas, against fences, or on walls with wire supports. Under these preconditions, scandent woody shrubs, or wiry climbers will become rampant and develop their flowering capacity best. The showiest and most popular of the tropicals are Bougainvilleas in shades of red, orange and purple, golden-yellow Allamandas and flame-colored Pyrostegias, and they all rejoice in sunshine. To keep in shape, vines should be pruned after blooming; in fact, close trimming will induce heaviest flowering in Bougainvilleas. I remember seeing Bougainvilleas draped around Roman columns on the isle of Capri, Italy, sheared and clipped severely to the old wood, and leaving only short spurs, but these were covered entirely with flowering bracts hiding even the foliage. In shady locations, flowering climbers should be avoided, and instead choose foliage types such as Philodendrons, Scindapsuses, Algerian ivy, Cissus, Parthenocissus, Pipers and Ficus pumila.

Popular Flowering Ramblers and Climbers:

Allamanda, Antigonon, Aristolochia, Beaumontia, Begonia limmingheiana, Bignonia, Bougainvillea, Browallia, Calliandra, Calonyction, Campanula, Campsis, Clematis, Clerodendrum, Clitorea, Cobaea, Columnea, Crataegus, Cryptostegia, Distictis, Epiphyllum, Gelsemium, Gloriosa, Hoya, Hibbertia, Hylocereus undatus (Queen of the Night), Ipomoea, Jasminum, Kennedia, Lapageria, Lathyrus, Lonicera, Mandevilla, Manettia, Mucuna, Pandorea, Passiflora, Pelargonium peltatum, Petrea, Podranea, Pyrostegia, Quisqualis, Russelia, Solanum wendlandii, Solandra, Stephanotis, Streptosolen, Stigmaphyllon, Strongylodon, Swainsona, Tecoma, Tecomaria, Thunbergia, Trachelospermum, Tropaeolum, Vinca, Wisteria.

Foliage Vines and Creepers:

Callisia, Ceropegia, Chlorophytum, Cissus, Coleus rehneltianus, Cymbalaria, Davallia, Dichorisandra, Dioscorea, Fittonia, Gibasis, Glechoma, Gynura, Hedera (ivy), Hemigraphis, Lamium, Lygodium, Mikania, Monstera, Muehlenbeckia, Parthenocissus, Pellionia, Peperomia, Philodendron, Pilea, Piper, Plectranthus, Rhipsalis, Rubus, Saxifraga sarmentosa, Scindapsus, Sedum, Selaginella, Senecio mikanioides, Setcreasea, Syngonium, Tradescantia, Vitis, Wedelia, Zebrina.

Hanging Baskets:

The hanging basket has become very popular for trailing plants, although many non-vining, spreading species are being utilized as well.

These suspended containers may be made of redwood slats or wire, lined with moss or polyethylene; terracotta or plastic pots on wires, or decorative jardinieres, to place a pot inside and hanging from chains, on overhead beams or wall brackets.

The watering of hanging baskets has always been a problem if not a deterring factor against installing them, especially indoors. In the tropics when outdoors, the frequent rains may be sufficient, but under shelter, their care is very much facilitated by the application of a trickle system of watering through small caliber tubing, strung along the ceiling rafters, and turned on as needed.

Plants frequently planted in hanging baskets are:

Asparagus densiflorus, Begonia tub. pendula, Begonia limmingheiana, Browallia speciosa, Campanula fragilis, Ceropegia, Chlorophytum, Cissus, Episcia, Fuchsia hybrida, Hedera (ivy), Lamium, Lantana montevidensis, Lotus berthelotii, Nephrolepis, Pelargonium peltatum, Petunia, Pilea depressa, Sedum sieboldii, Sedum morganianum, Senecio mikanioides, S. rowleyanus, Tolmiea, Tradescantia, Tripogandra multiflora, Tropaeolum, Vinca major, Zebrina.

BULBS and RHIZOMES

Bulbous plants are undoubtedly among the most rewarding kinds of plants, and their beauty and versatility is unlimited. However, with some exceptions, they do not fit easily into a tropical garden unless planted into a segregated border, or still better in pots or tubs. After what is often a short flowering season, one must leave the plants undisturbed, so the foliage can fully develop in order to mature the underground storage organs. Pots are more movable and besides, with proper timing, such container-grown plants may be brought into bloom throughout the year, depending on species, storage and temperature.

What we know as "bulbs" is a horticultural term and may properly include true bulbs, corms, tubers, rhizomes, or tuberous rootstocks. Examples of bulbous plants are Amaryllis, Haemanthus, Lilies, Tulips. Corms are bulb-like but solid—Freesias, Gladiolus, Tritonia. Tubers are thickened subterranean stems looking like an irregular sphere, found in tub. Begonias, Caladiums, Cyclamen, Gloriosa, Sinningia. A rhizome is an underground stem bearing buds; such are Cannas, Iris, Zantedeschia, Achimenes and Kohleria.

Most bulbous plants need a time of rest and dormancy, more so the "hardy" kinds such as Tulips, Convallaria and Narcissus, which are pre-chilled, being held in cold storage for a few weeks after they are dug from the nursery fields in Holland in autumn, especially if they are sent to opposite-season tropical areas of the Southern hemisphere. There they may be kept in a refrigerator for a month and when planted they will bloom within a few weeks.

Of the tender tropical and subtropic species most will grow well in pots or tubs and if potted in autumn they may be forced into winter-bloom, or brought along in stages depending on storage temperatures. Amongst the first to flower are Hippeastrum, the florists "Amaryllis". Others are Freesias, Lachenalias, Veltheimias and Lilies, also Clivia and Vallota.

Better planted into the garden are Agapanthus, the "Lily of the Nile"; Crinum, Crocosmia, Dietes, Gladiolus, Haemanthus, Hemerocallis, Kniphofia, the "Red-hot poker"; many Lilies, Iris, Tigridia, Tritonia, Watsonia, Zephyranthes, and Zantedeschia, the "Calla lilies".

Some bulbous or tuberous plants are good pot plants the year round—amongst the most popular are Clivia, some Hippeastrums, Lachenalia, Hymenocallis, Tuberous begonias, Veltheimia, Crinum, Eucharis, Tuberoses, Gloriosas, Haemanthus, Neomarica, Scilla, Oxalis, Ornithogalum; in gesneriads Achimenes and Sinningia. Much loved for their dainty, colorful foliage are the tuberous Caladiums; these however are quite tropical and when in growth, need constant warmth.

The trailing ivy geranium, Pelargonium peltatum, is equally colorful on the balconies of this Inn in the Bavarian Alps during summertime, or the year around in gardens from California to South Africa.

WARM AREA HORTICULTURE

ANNUALS and PERENNIALS

Most summer-blooming annuals we know in our cool-climate gardens or in window boxes, are actually of tropical or warm highland origin. For this reason they are well suited to planting in warm-region gardens, or in boxes on tropical veranda or the subtropic patio. Also, some cold-climate perennials such as Delphinium, Dianthus or Vinca are grown as annuals in frost-free regions during summertime.

Annuals in the tropics however should be planted at the beginning of the cool, dry season: in South Florida for instance, in October. In the more arid, subtropic California, the best planting time is after the rainy or foggy winter season, in late spring when the weather warms up. This is especially important for hot weather Zinnias. Planting when the rainy season begins invites fungus diseases and insect pests attacking young plantlets. During rainy periods "Bedding plants" will be poor as they all love sunshine which brings out their best in growth and color.

Where space is limited, there is nothing that offers so much variety and temptation than a summer planting of choice annuals, quickly grown from seed or cuttings, for the most lavish display of color, and offering to the gardener a wonderful opportunity for expressing his or her artistic skills.

Colorful beds of assorted annuals and biennials, including Petunias, Salvias, Begonias, Impatiens, and Alyssum, in King's Park, overlooking subtropic Perth, Western Australia.

Feathery treeferns, Cibotium chamissoi, transplanted from tropical Hawaii, into the typical humid-warm summer of New York, at Radio City, and enjoying it, bringing with them a touch of the rainforest.

Popular Garden Annuals:

Ageratum, Amaranthus, Begonia semperflorens, Calendula, Catharanthus (Vinca rosea), Celosia, Dianthus, Eschscholtzia (Calif. poppy), Euphorbia marginata (Snow-on-the-mountain), Gaillardia, Godetia, Gomphrena, Impatiens (Balsam), Ipomoea, Lathyrus (Sweet pea), Limonium (Statice), Nicotiana, Nemesia, Papaver (Poppy), Petunia, Portulaca, Salvia, Tagetes (Marigold), Tropaeolum (Nasturtium), Verbena and Zinnia.

PERENNIALS: Hardy vs. Tropical.

Perennials technically are enduring herbaceous plants that persist year after year, increasing in size of rootstock, which may then be divided for propagation. Hardy perennials are common in cold climate gardens, where the tops freeze down in winter, with new growth commencing in spring.

In the absence of frosts such perennials do not do well in the tropics, and partly because they have difficulty surviving the long warm, humid summers. For this reason, what the tropical gardener considers to be perennials, are different types of plants. They are almost all the same subjects that are grown, for their ornamental foliage or strange flowers, in pots or tubs under indoor conditions in the more frigid zones, and commonly known as Exotic plants.

EXOTIC PLANTS

Under this heading, we find some of the most beautiful plant subjects, and they are equally suitable outdoors in warm regions, and for home and greenhouse in temperate climate. The best known of the Tropicals are:

Aglaonema, Alocasia, Alpinia, Anthurium, Aphelandra, Begonias, Beloperone, Coleus, Colocasia, Costus, Dieffenbachia, Dracaena, Heliconia, Nicolaia, Pandanus, Pentas, Philodendron, Sanchezia, Spathiphyllum, Strobilanthes, Syngonium, Tacca.

From semi-tropic, somewhat cooler backgrounds we have:
Bamboos, Catharanthus (Vinca), Cordyline, Dimorphotheca, Fuchsia, Gerbera, Hedychium, Pelargoniums.

Easy warm-climate garden **Orchids** are Cymbidium, Epidendrum ibaguense (radicans), Spathoglottis, Vandas.

In **Bromeliads:** Aechmea, Billbergias, Guzmania, Tillandsia, often grown on trees.

Moist and shady places are ideal for **Ferns,** from treeferns such as Alsophila, Cibotium, Cyathea, and Dicksonia to smaller Asplenium with Birdsnest, Cyrtomium or Hollyfern, Davallia, the "Rabbit's-foot"; "Maidenhair" (Adiantum); Nephrolepis—the "Boston fern"; Polypodiums, or the epiphytic "Staghorn ferns"—Platycerium, usually grown on wall plaques or in basket planters, although they may be mounted directly onto the bark of trees.

SUCCULENTS, including CACTI.

Best adapted for drier, sunnier areas are almost all **Cactus;** in other **Succulents,** Aeonium, Agave, Aloe, Crassula, Euphorbias, Furcraea, Kalanchoe, Lampranthus, Sansevierias, and the Yuccas. The large CACTUS family is at home in the Western hemisphere, primarily in arid regions of North and South America. They are xerophytic, typified by the barrel cactus and the column types of Cereus, with exception of the more moist tropical epiphytic genera including the Epiphyllums or "Orchid cactus", and the clambering "Queen of the night."

Old world succulents are dominated by the Euphorbias, represented by giant Candelabra trees in Eastern and Southern Africa; a few species are at home in the Western hemisphere such as the "Poinsettia" (Euph. pulcherrima) of Mexico. Other large succulent families are the Aizoaceae, better known as Mesembryanthemums, and which include the "Living stones" and the "Ice plants" from Africa; and the Crassulaceae, known for the "Jade plant" and the Echeverias or "Hen-and-chickens". All other succulents belong to diverse families and are found in the Amaryllidaceae (Agaves), Asclepiadaceae, Bromeliaceae, Compositae, Geraniaceae, Liliaceae (Aloe, Sansevierias), Portulacaceae and Vitaceae.

In arid regions of the world many totally unrelated plants have been forced to adopt some method to exist through long periods of deficient rainfall. Plants that were once delicate and with abundant foliage have, in their struggle for survival in areas that have turned arid, evolved smaller leaves and developed stems or leaves capable of storing water, becoming succulent in the process and changing character and appearance.

Old-fashioned, fatiguing hand-watering of vegetables in the cool mountain area of Nuwara Eliya, on Ceylon.

IRRIGATION Outdoors

Watering today may be had almost automatic, from a trickling faucet to irrigation sprinklers, piped underground through plastic pipe, from plant to plant and tree to tree. In this manner, a gardener is able to grow many a sensitive tropical plant, even in arid climate. I had this experience in our own Southern California garden, 12km inland from the Pacific Ocean, where this particular climate, halfway between coast and desert, presents a challenge to experiment with moist-tropical shrubs and trees; ferns, vines, and palms, bromeliads and even orchids.

It is typical of most desert soils that they are quite rich, if not in humus but in minerals. In Arizona or Sonora, where desert land is irrigated by canals from the Colorado River, the ranchos are fruitful with agricultural and Citrus crops. When travelling along the Nile in Egypt, the green of irrigated fields contrasts sharply with the harsh sands of the waterless Libyan desert.

Water can perform miracles. The steep and rocky foothills of the Sierra Madre in San Diego County, are increasingly planted with oranges, lemons and avocados, supplied by drip irrigation through a piped trickle system. This has been tried with success in Israel on similar arid hills in Judea and Galilee, where water in rigid or flexible plastic pipe travels from tree to tree.

At each tree or shrub, "risers" with bubblers, or else small water spouts ("emitters"), attached to elastic small caliber plastic ("spaghetti") or rubber hose, will leak or trickle a slow flow of water into a basin, and directly to the roots where needed, for several hours at a time, to drench the soil thoroughly and deeply. Depending on the needs or size of trees or other plants, there may be more than one emitter attached at each location. This system also conserves precious water which is often wasted by overhead sprinklers. Such systems are simultaneously used for periodic feeding with liquid fertilizers and needed chemicals.

Where spot bubblers or tricklers are not practical, and more of an area such as flowerbeds or lawns must be covered, a great variety of fan sprinklers or overhead whirlers adapted to small or larger areas may be used with more efficiency.

It has been found in California that hill plantings of fruit trees, will allow the water to drain more freely, and avoid the fungus diseases which affected the roots of Avocados and Citrus orchards planted on more level ground with hard-pan underneath.

Conditions are different in Florida, which is based on maritime coral, generally with ample water in the subsoil. This will support a typical subtropical garden scene without much care and only occasional watering.

It is however possible to create a tropical rainforest scene, with lush climbing Philodendrons and Treeferns, when aided by tree-top whirling sprinklers. These, when turned on for short intervals, will provide the necessary conditions of humidity. Fairchild Tropical Gardens in Miami have installed such a system which is turned on from time to time for several minutes, and which encourages the growing of Staghorn ferns, Aspleniums, Davallias and Polypodium ferns, as well as an assembly of epiphytic Orchids and Bromeliads.

Controlled watering of flats of Zoysia grass, by small water spouts attached to elastic thin caliber plastic tubing, and designed to irrigate an entire area simultaneously.

Bubble-watering into concrete basin with Neodypsis palm, supplied by underground pipe and riser, in Southern California.

The most efficient manner to water fruit trees is by "trickle irrigation", from flexible black, 2cm plastic tubing running from tree to tree, where "bubblers" will leak a slow flow of water (or fertilizer) directly to the roots.

GROWING MATERIALS and SOILS

Soil is produced by a meeting of the mineral and vegetable kingdoms, a mass of inorganic particles mixed with living or dead organic matter. If the mineral particles are very fine, we have clay or adobe; larger particles may be limestone rock, decomposed granite, or sand. The vegetable matter may be in the form of ancient peatmoss, swamp or lake mulch or decaying leaves.

Arid climate zones are notorious for their alkaline soils high in calcium carbonate and other mineral salts. Acidity is most pronounced in areas of heavy rainfall, and may be found in volcanic ash and sand, carbonized sphagnum moss (peat), woods-soil or leafmold in tropic forests, or pine needles higher in the mountains.

Zones with prevailing alkaline soils (above pH 7) favor growth of cacti and other succulents; Hibiscus, Cassias and many other legumes; Eucalyptus and West Australian as well as South African trees and shrubs. Soils of acid reaction (pH 6 or less) is what Azaleas, Camellias, Rhododendron, Gardenias and Ixoras need.

In a general garden planting the elements of both should combine into friable, porous earth. Most organic matter is full of microscopic bacteria that are constantly at work, to break down fresh leaves, as well as inorganic soils; they make nutrient elements in fertilizers available, and introduce it to the roots as food.

A porous, well aerated soil composed of both coarse mineral particles or sand, to permit the access of oxygen, and enriched with life-giving humus able to retain water, will give plants the best chance for healthy root development and growth. If acidity must be increased, the addition of Ammonium sulphate or sulphur will lower the pH; while acid soils can be sweetened by light dressings of ground limestone.

Soil in its preparation, should have incorporated with it large quantities of unrotted organic matter—a most important factor. With this, heavy mulching should be practiced twice a year, before the rains and before the dry season. Experience has shown that mulching prevents the washing away of the top soil during rains, and the unrotted organic matter checks the cracking of the soil, and allows free drainage. In the dry season the mulch retains moisture, keeps the top soil cool, and slows up leaching. It is surprising to see how quickly organic material disappears in the tropics—as snow does under the sun.

The addition of finely ground limestone in areas with acid soils does help. Stable manure if obtainable, should be well matured, or if fresh, buried deep under the surface. Small amounts of Ammonium sulphate added to the soil will be found useful. Normal fertilizing practices will usually suffice, but should be applied in small amounts at more frequent intervals. It has been my experience that applications of diluted solutions of organic fertilizer appear to produce better results.

Plants in Pots

Soil for plants in pots or other containers in the Tropics, and also hanging baskets must be porous and have incorporated with it more organic material, in the form of rough scrapings from forest soil, tree-fern fiber, coconut bass (the left-overs after the extraction of the fiber from the coconut husk); chopped bagasse (the waste fiber from the sugar cane), or any other form of organic material which may be obtained locally from coffee hull, cocoa, rice or corn husks. These could also replace peat moss, a very expensive commodity in the tropics. A common practice in rainy, humid tropical Singapore is to grow potted Orchids, Araucarias, Dracaenas, Crotons, etc., in broken brick or pebbles of burnt clay; in Hawaii and elsewhere, planting is done into chips of tree bark or in rubble of volcanic rock, materials that will drain excess water quickly and prevent the root-system from becoming stagnant.

Pots may be devised from bamboo, cut into sections, or plaited straw or even well decorated containers from the vast number of tinned foods that are now widely consumed. Lately, plastic bags have been used with satisfactory results. Underpotting is far less harmful than overpotting, and it has been our experience that certain plants that are difficult to flower in pots, do so readily when pot-bound. Of course at that stage it will be necessary to fertilize regularly to keep them going, depending on the type of plant.

Vandaceous orchids potted in broken bricks and chunks of cow manure at nursery in Penang, where coddling humidity will quickly form roots.

Potted Crotons (Codiaeum) growing in pebbles of baked clay for best drainage, in rainy, humid-warm Malaya.

Porous clay flower pots with perforated side walls for rapid drainage during the frequent monsoon rains in Singapore.

A broad window sill in my wife's and chief assistant's ancestral home in Germany, really invites houseplants; a plate of polished marble extends the whole width of the living room, the heat radiator underneath, airy windows and short curtains for optimum light, combine to provide near greenhouse conditions for anything in pots.

PLANTS INDOORS

How to Care for Plants in Home, Office or Greenhouse Environment:

According to their respective climatic backgrounds, plants intended for indoors are often divided into (A) House- or Decorator plants and (B) Greenhouse plants, suitable for Warm, Intermediate, or Cool surroundings.

"House plants" are such exotics as are satisfactory for home and Interior decoration, tolerant of the reduced light and artificially dry atmosphere of the living room or office.

Those designated as a "Greenhouse plant" require the higher relative humidity prevailing in a glass or plastic greenhouse, but may also be grown within the home in a glass terrarium (converted aquarium) or a shelter of translucent plastic where extra atmospheric moisture is provided. In humid-warm regions a slatted shade house or a frame covered with open-meshed saran cloth will do just as well for many "Conservatory" plants.

"Warm" or "Tropical" greenhouse plants are known as "Stove plants" in England—and usually grown best there in humid-warm greenhouses, because living rooms are frequently kept much cooler than in the United States.

"Intermediate" or "Subtropic" plants generally enjoy fresh air and good light—a place near a window open in the summertime; a sun-porch glassed-in during winter; or a light, well ventilated home greenhouse; in warmer climate the sheltered patio, or under a tree outdoors.

"Cool" or "Temperate" plants of course also like the same basic environment as "Warm temperate" plants but could be left almost entirely outdoors; if kept indoors they require good light.

Flowering vs. Foliage Plants:

Blooming plants such as we receive at holidays require much more water, and stronger light, than the average foliage plant, for the continuous opening of their buds, and longer enjoyment of their blooms. Foliage plants, not having to support a demanding inflorescence, get along on less light, and are much easier to maintain. It is usually safer to keep them on the dry side rather than too wet, especially if they are not too well established in their pots or tubs.

Home Greenhouse

A small greenhouse of glass or plastic—flexible polyethylene or rigid fiber glass, artificially heated where necessary, is an ideal accessory to grow and maintain the more delicate, warmth loving tropical plants. In such a sheltered place, light, temperature and humidity can be regulated as desired, and the culture of exotics may become a real and relaxing hobby for the plant fancier.

In comparing the merits of a glasshouse against the modern plastics, it may be well to know that glass filters out ultraviolet light while fiberglass or polyethylene does not. One can get a mild sunburn from long hours under fiberglass.

Other than the old "white-wash" for shading of greenhouse glass, a good suggestion is to use water-based latex paint (1 part to 10 parts of water), applied with a long-handled roller; this can be washed off with soap and water. On smaller glasshouses, a shading of slatted window-shades, which may be rolled up or down as needed, is of course ideal, especially during periods of cloudiness.

Plants for Sunny and for Shady Windows

For gardening indoors the window is best. Not only is natural daylight better than artificial light, but the exposure from different directions offers an ideal place for almost any type of plant.

Seldom, however, are all four sides in a home equally available and suitable for plants. But the suggested types offer the opportunity to make a wise selection for any window. This should not, however, prevent us from trying any plant in whatever spot that may happen to be otherwise ideal. A north window can, if necessary, be artificially lighted, and a south window easily shaded.

For ease of care, attractive looks, and flourishing growth of plants try grouping them in boxes. For convenient access to the window, these could be mounted on legs with rollers. Fill a window box with moist peatmoss, and plunge your pots rim-deep. If the peatmoss is kept moist, plants absorb by capillary action, whatever moisture they may need, and practically take care of themselves, though some pots may have to be watered individually on occasion. The roots begin to ramble, and the big plants and the little ones, the climbers and the creepers, flowering plants and foliage, together form a mutually beneficial ecology with a micro-climate that favors luxuriant, natural growth. In close symbiosis, they enjoy each other's company, and will soon create the image of a natural tropic mini-landscape, cover the other's blemishes, and sickly plants take on new life. Miraculously, plants begin to flower, and even delicate exotics seem to forget their being supersensitive. The secret lies in the complementary give-and-take, and intimate climate, set up by such an harmonious community of partners, and a window filled with beautiful exotics, brings both restfulness and life into the home.

PLANTS INDOORS

NORTH WINDOW (Northern Hemisphere)
(10 metric lux = 1 footcandle)

(Reversed south of the Equator) (Sunny day in winter near glass, 1,500-5,000 lux; average 3,000 lux). Sun-less windows are not for most flowering plants. But foliage plants including tender tropicals, do well in a fully light north window. Weather-stripping or a protective sheet of plastic to prevent cold drafts, is advisable during the cold season. A recommended list of satisfactory subjects would be:

For WARM conditions—Aroids such as Aglaonema, Dieffenbachia, Philodendron, Scindapsus and Syngoniums; Cissus, Dichorisandra, Dizygotheca, Dracaenas; ferns including Platycerium; Ficus elastica and lyrata; Fittonias, Hoffmannias, Marantas, Peperomias and Pileas; even Sansevierias.

For INTERMEDIATE temperatures—Amomum, Araucaria, Aspidistra, Spider plants (Chlorophytum), Hollyferns (Cyrtomium), Helxine, Saxifraga, Tolmiea, Tradescantias; also most Palms including Kentias (Howeia).

At a COLD window—Ardisia, Fatshedera, Fatsia (Aralia), Laurus, Ophiopogon, Rohdea and Skimmia; also all kinds of Ivies (Hedera helix).

SOUTH WINDOW (Northern Hemisphere)
(Midday sun, near glass 50,000 lux winter, to 75,000 lux summer; cloudy day 10,000-20,000 lux). An exposure to the potent Southern sun offers a wonderful chance to grow the many plants that are sun-lovers. This includes practically all the Holiday blooming plants; flowering shrubs and vines; miniature subtropical fruit trees; many bulbs; and of course the exciting world of succulents.

Plants responding to both sun and WARMTH are: Acalyphas and Croton; Amaryllis and other tropical bulbs; Coleus and Iresine; Beloperone, Gardenia, Hibiscus and Strelitzia; Tropical clamberers such as Allamanda, Bougainvillea, Clerodendrum, Gloriosa, Jasmine, and Passiflora; flowering Succulents like Euphorbia splendens and pulcherrima (Poinsettia), and Kalanchoe, also sunny Dendrobium and Oncidium orchids.

INTERMEDIATE and COOLER plants in need of sun but with the window open when it is warm outside are: Abutilon, Campanula isophylla, and most bedding stock; Geraniums, Petunias, Wax begonias, Lantanas, Azaleas, Chrysanthemums, Heather, Hydrangeas, Pot roses; Lachenalia and Veltheimia, also Dutch bulbs; fruited Jerusalem cherry; Kumquats and other Citrus. Most Cacti and other Succulents enjoy the sun; want to be warm in daytime but cool at night.

EAST or WEST WINDOWS
(Winter sun, near glass, 20,000-40,000 lux, off-sun 2,000-5,000 lux; average 20,000 lux). The partial sunshine of several hours daily at an East or West window is ideal for many flowering and foliage plants that normally grow in partial shade and dislike the intense sun of windows facing South. East windows are generally cooler and receive the clear morning sun; Western exposures tend to be warmer, but often with hazy or moderated light.

WARM PLANTS doing well East or West with the benefit of a limited amount of sun are: Anthurium scherzerianum, Alpinia, Aphelandra, Brassaia, Brunfelsia, Caladium, Calathea, Clivia, Croton (Codiaeum), Costus, Eucharis; Birdsnest and Boston ferns; Ficus benjamina, Hoyas, Impatiens, Jacobinia, Oleander, Pandanus, Pseuderanthemum, Spathiphyllum, Torenia, Zebrinas. Great variety is also offered in the many Begonias, Bromeliads, and warm Orchids. More or less hairy Gesneriads, such as African violets (Saintpaulia), Achimenes, Aeschynanthus, Columneas, Sinningias including "Gloxinias", Smithianthas, and Streptocarpus enjoy the good light East or West but in sunny hours this light must be diffused to prevent burning.

INTERMEDIATE or COOLER PLANTS for partial sun are: Araucaria, Aucuba, Asparagus, Camellias, Cinerarias, Crinum, Cyclamen, Euonymus, Fuchsia, Grevillea, Kaempferia, Myrtus, Neomarica, Osmanthus, Oxalis, Plectranthus, Podocarpus, Primula, Vinca and variegated Ivies.

Inexpensive greenhouse in Fallbrook, California, for growing of Cactus seedlings; framework of bent aluminum pipe, covered with polyethylene film; fan for cooling at far end, and lights to lenghten days.

Shade frame simple and effective, with covering of hardwood slabs, for shade-loving plants, Botanic Garden of tropical Rio de Janeiro, Brazil.

A window to the East or West for typical exotic plants enjoying warmth and moderate light; this includes Episcia, Saintpaulia, Palms, Dieffenbachia, Philodendron, Alocasia, and Aphelandra.

PLANTS INDOORS

A lean-to glassed greenhouse snuggled against a tree-shaded home at Willowwood Arbor, Pottersville, New Jersey. Both attractive and useful, such a sheltered area knows no seasons. Cozily protected from winter's snow, out-of-season flowering bulbs and plants may be enjoyed even more because of the contrasts in climate indoors and outside.

What an invitation to open a sliding glass door from the living room into the small home greenhouse, attached directly to the house, and enter into an intimate area of sunshine and the fragrance of growing plants, and ideally suited for the hobbyists collections of fine exotics such as Orchids, but also seasonable blooming plants, kitchen herbs, or seedlings and cuttings for spring planting.

A warm, well-ventilated conservatory is conceded to provide the best atmospheric environment for the growing and maintaining of tender exotics, larger ornamental foliage plants, Palms and Ferns; the display of seasonable flowering plants and rambling climbers; it also offers peaceful relaxation in comfortable chairs, expecially during winter. Hanging baskets luxuriate in this "Winter garden" at Banff Springs Hotel, in the Rocky Mountains of Canada.

PLANTS INDOORS

An exhibition hall at the DeYoung Museum in San Francisco, where an attractive corner planting of large Dracaena warneckei, Philodendron selloum, Monstera deliciosa, and Rhapis palms frame an unobtrusive marble bench; there the visitor may rest and contemplate a French sculpture of the goddess "Pomona" Mellow sunshine diffused through a glass ceiling, and supplementary incandescent spotlights accent and benefit these handsome foliage plants with warm light.

Year-round tropical planting indoors, throughout this covered shopping mall at Cherry Hill, New Jersey, features tall "Umbrella trees" (Brassaia), Philodendrons, and "Areca" palms (Chrysalidocarpus), feeling right at home with the proper balance of temperature and sufficient light through a glass ceiling above. The designer understood to give prospective shoppers the exhilarating atmosphere of swaying palms and exotic greenery.

The well-lighted main office of a bank in Metropolitan New York, with low room divider boxes planted with Ferns, Spathiphyllum, Birdsnest fern (Asplenium), as well as flowering plants periodically changed, tactfully and graciously directing traffic; along the wall recessed boxes and panels, illuminated by concealed incandescent lamps, a tracery of green against the pastel tints.

PLANTS for INTERIOR DECORATION

Decorative plants in built-in stone boxes, fiberglass or ceramic bowls have become quite a standard feature, in urban offices and public buildings. Modern structures of glass and marble, forcefully invite the soft tracery of green plants. This need has established a definite trend in interior decor. However, unless plants are placed near windows, supplementary or artificial illumination must be provided for most. Present-day architectural planning includes provision for incandescent spot lights from the walls or ceiling, usually concealed, or in handsome reflectors which can be directed to throw a warm light on decorative plants, accentuating their living greenery, at the same time providing the radiant energy needed, to maintain foliage and long life.

Fortunately, through extended testing, many very satisfactory exotic evergreens have become available, and which over a long period may be maintained, in good condition, without much care, often replacing artificial plants of plastic.

Air-Conditioning.

Most container plants in office plantscaping adapt readily to airconditioning at the usual automatic daytime control of 21-23°C (70-75°F), but benefit immensely if the temperature for the night can be lowered to 16-18°C (60-65°F).

Floor Plants:

A list of dependable decorators as specimen plants are:
Brassaia, Bischofia, Clusia, Coccoloba, Dizygotheca, Dracaena marginata and massangeana; Ficus lyrata, benjamina, nitida, elastica; Monstera, Philodendron, Pittosporum, Polyscias, Podocarpus, Yucca; in Palms, Caryota, the "Fishtail palm", Chamaedorea, the "Bamboo palm"; Chrysalidocarpus (Areca), Howeia (Kentia), Phoenix or "Date palms", Rhapis or "Lady palm" and Veitchia, the "Christmas palm".

Table Plants:

Araucaria, Aglaonema, Aspidistra, Cissus, Dieffenbachia, Dracaena warneckei, Fatsia, Pandanus, Philodendron, Spathiphyllum, various Bromeliads; in Succulents—Cereus, Crassula, Sansevierias; in Ferns—Asplenium, Cyrtomium, Cibotium, Davallia, Nephrolepis, Polypodium, Pteris, Platycerium, Phyllitis, Rumohra, the "Leatherfern".

Flowering Plants.

Holiday blooming plants in pots require much more water and strong light to keep in good condition, than the average foliage plant. Also, a cooler temperature near 15°C (60°F) will markedly prolong their flowering season.

Plants for Cool Locations

Some of our decorative plants feel comfortable in surroundings that we would consider too cool for comfort. Many cool-type, broad-leaved evergreens come from warm-temperate areas with insular climate, such as New Zealand or Japan, notably Podocarpus, Pittosporum, Araucaria, Aucuba, Bamboos and many Araliads including Ivies, Fatsias, Ligustrum, Panax and others. These species have proven ideal for decoration in cooler parts of the house or office building. Various temperature requirements should be comparatively easy to determine, if we know the climatic background of a plant, and accordingly, provide it with similar surroundings, when growing as a house plant, to make it feel at home and happy.

Maintenance of Decorator Plants

Clean foliage improves the appearance of a decorative plant and assists it in maintaining its life processes. To wipe the foliage with soft cloth, will cleanse the breathing stomates, and a touch of milk adds luster to the leaves. Commercial leafshiners should be used with caution as they may clog the pores.

If a plant stands a long time, say a year, in the same pot, and the roots have filled the same to capacity, mild feedings once a month will sufficiently sustain it. It may be more desirable to confine an indoor plant intended for decoration, to as nearly as possible its original size by merely "maintaining" it, watering it sparingly and withholding growth-stimulating nutrition. Such a plant will subsist on a minimum of care, although in the "weaning" process from the humid greenhouse to a relatively dry living room, a few leaves may initially drop, before the subject becomes adjusted to its new surroundings. But after that, a container plant so conditioned will be less troublesome, not always crying for attention, and continue for a long time to grace the spot for which it was intended.

Covered Peristyle of a suburban residence in Westchester, New York, ideally suited for naturalistic planting with handsome exotic plants, and West Indian treeferns (Cyathea arborea) for tropical accent. The area features a pool and flagstone walks, as well as a patio area for dining and restful reading. Roofing of frosted glass admits diffused light, and the plants are thriving in each other's company under optimum conditions of humidity and temperature.

Colorful Easter basket with seasonal flowering plants, Azaleas, Polyantha roses, Lilies and Hydrangeas.

PLANTS INDOORS

LIGHT is VITAL

Light is the most important environmental factor influencing photosynthesis. Without light, life cannot exist. Light triggers the processes that manufacture sugar in the leaf, and without its presence all activity stops. In the absence of light, the plant utilizes reserve foods for energy. If conditions are unfavorable for photosynthesis and production of new sugars, but favorable for respiration, as during periods of low light intensity, combined with high temperature, the plants may use all of their stored food and perish.

Natural Light

Adequate daylight is the best type of light for house plants. They require light, but not all need the same amount. Most foliage plants do perfectly well in limited sunlight and react unfavorably to intense sun. Some highly colored foliage plants such as Coleus and Crotons, retain their brilliant colors best in bright sunlight. Sun-lovers, too, are most succulents and many flowering plants including Geraniums, Abutilons, Hibiscus and Gardenias. Sun rooms and glass-enclosed porches—and in a limited way the windows facing South, are naturally ideal for light-hungry plants. These include most of those timed to flower for our holidays like Roses, Hydrangeas, Azaleas, Lilies, Chrysanthemums, Poinsettias, and bulbous stock. Blooming plants may, of course, be placed into any other part of the home for temporary enjoyment, but to prolong their flowering period should be rotated to the lightest, and somewhat cooler places.

Light Intensity—the Metric-candle vs. Footcandle

To properly measure the energy of a light source significant to plant life, readings should be made in watts per 100 Angstrom bands. However, the old footcandle method of reading light, directly from the source, is relatively simple, although the current light meters are color-blind, and primarily read the yellow-green bands of the spectrum—the light made for reading. One foot-candle is the amount of light cast by a candle on a white surface 30 cm away in a completely dark room. In European practice, the light intensity is expressed in the metric term of Lux: 1 foot-candle = 10.76 metric candle or Lux. Bright moonlight may measure 6-7 Lux or 0.02 fc.

Light Requirements of Plants

Light, which gives energy to the plant, is like the force of electricity running a motor. Just like volts indicate the amount of electricity received by the motor, footcandles indicate the amount of light received by an object. Below a certain minimum of about 200 fc. plants fail to grow actively, although 10 footcandles in many species will sustain life. But too much light can be harmful too.

As we can burn up a motor, it is possible to burn up a plant by unneeded light. Depending on the type of plant, a single leaf can assimilate up to 2,000 fc. of light efficiently, and the total plant can use considerably more because many of the leaves will be shaded. But excessive light intensity and resulting high leaf temperature will cause the photosynthetic system to break down and food manufacture is impaired. Most foliage plants enjoy optimum levels of 1,000-2,000 footcandles; low energy flowering plants like Gloxinias 1,000-2,000 fc.; high energy flowering plants including Azaleas 3,000-5,000 fc.; Chrysanthemums and Roses to 10,000 fc. (100,000 lux).

Artificial Illumination

The single illuminant most capable of duplicating sunlight is an incandescent source. Combinations of fluorescent and incandescent lamps are the next best simulated daylight sources.

Most fluorescent tubes are high in blue (5,000-6,000 Angstroms), but being cooler may be placed nearer to the growing plants. A 3,500 degree Kelvin tube lamp is a good balance light. Where the close proximity of a light fixture near the plants is not objectionable, low energy flowering plants such as Saintpaulias (African violets) and gloxinias, also foliage plants, will grow well under it. The cost per footcandle of light received by the plant is less than that of incandescent light, and these tubes have their place in the growing of smaller plants, seed cultures, and vegetative propagation. But the decorative effect of fluorescent lamps is flat and cold, while incandescent lighting is orange in characteristic, and all the flower colors with a large amount of yellow and into red, appear rich and vibrant. (For details see Exotic Plant Manual pg. 22-26).

Lighting in the Home

It would be so easy to combine the comfort of a reading lamp in the home with the vast amount of good this will accomplish, in the maintenance of plants, if every day at dusk a 100 watt bulb from an overhead bridge lamp, could shine on a planter or the window garden. This, costing very little in a week, provides not only welcome added light, but will give accent to any plant or planting. For fluorescent illumination, a simple rule to follow, when planning a house plant corner, is to provide 15-20 watts of fluorescent light per 30 cm square of growing area.

Contrary to popular belief, 24 hours of continuous light will do no harm to growing plants not photosensitive provided that the temperature is changed and lowered for the night. However, such lengthening of the day by additional illumination, during hours of natural darkness, may play tricks on some flowering plants that are sensitive to daylength. Poinsettias or Chrysanthemums, which are long night (short day) plants, would continue in vegetative growth as long as illumination is over 12 hours, and even a fraction of light shining on a poinsettia, during its normal bud initiation time end of September (Northern hemisphere), would prevent it from coming into flower.

People find more than 100 footcandles indoors excessive for reading; a level of 15-20 footcandles, equivalent to fair reading light is sufficient to many foliage plants for survival, if combined with reduced moisture. However, in evaluating quantity of light, intensity alone is not all that matters, but footcandles should be multiplied by daily duration. Ten hours of light at 30 fc. is as effective as 15 hours at 20 fc. Incandescent light is rich in far red, needed particularly by energy-hungry blooming plants and fruit.

Plants for Poor Light

House plants recommended as most tolerant to poor light, down to 5 or 10 footcandles (50-100 lux), and yet long-lasting—include Aglaonema, Araucaria, Aspidistra, Bromeliads, Dieffenbachia, Dracaena marginata, sanderiana and warneckei; Ficus decora, Hoya, Philodendron, Chamaedorea and Kentia palms; Pittosporum, Podocarpus, and many Succulents including Sansevieria. Somewhat better light is needed by Begonias, Boston ferns and others, Brassaias, Cissus, Ivies, and Scindapsus. Most Ficus relish light and require at least 50 fc. and more. But more light will benefit all of these, and an incandescent spotlight flooding any interior plantings with warm and glowing light will enrich all colors, and dramatically accent their decorative effect.

Light cannot be measured for its intensity alone, but growth or flowering is usually regulated by length of day or night. The metabolism of a plant—its rate of growth—is in direct relation not only to temperature but also the quality and duration of light received. During fall and winter when the days are short, and temperature is low, growth virtually ceases. Most plants will not bloom, unless the intensity of light approximates that in their native habitat. Foliage plants are quite tolerant of low light intensity, but if the environment of a flowering plant is too dark, sufficient supplementary light must be provided.

PLANTS INDOORS

Wall planters with Peperomias and Philodendron illuminated by incandescent spotlights in handsome, adjustable bronze reflectors; from the ceiling fluorescent panels for balanced light; photographed at U.S.D.A. Experiment and Research Center, Beltsville, Maryland.

Fluorescent tube lamps are normally difficult to employ in decorative lighting, and are of more efficient use to propagators and plant fanciers in home cultivation of specialty plants. However, this ingenious planter box with small exotic plants is attractively disguised by hardwood shingles of African woods, and a center of attraction in the living room of a house in Pretoria, South Africa.

Fascinating setting of a spacious planter box holding a metal tray with a variety of handsome foliage plants in pots, from tall Ficus lyrata and Coccoloba, to Dieffenbachias, Dracaenas, Polyscias, Calatheas and smaller Begonias - all plunged into moist peatmoss for easy maintenance. The planting is set against a ceiling-high reflecting mirror, and is lighted from above with ornamental incandescent floods, in an art collector's study, Sydney, Australia.

PLANTS INDOORS

Self-watering in the Home: left - Aphelandra standing on a saucer and placed inside polyethylene bag which will hold moisture for weeks; right - Dieffenbachia 'Rud. Roehrs' drawing water by capillarity through a spun-glass wick from nearby pitcher.

The Influence of TEMPERATURE

Plant activities and growth generally increase with higher temperatures, and are retarded if the temperature is low. However, excessive heat may result in injury from desiccation, and a rate of respiration so high, that the consumption of food materials tends to exceed production by photosynthesis. If temperatures are steadily too high, the plant uses up the stored food manufactured during daytime and may exhaust itself, unless followed by an alternate period of lower temperature, to restore its energy. Temperature affects growth through its influence upon all metabolic activities: photosynthesis, respiration, digestion, transpiration, absorption of water, and root growth.

Variation in Room Temperatures

Room temperatures are not uniform throughout the house. Checking with a thermometer will prove that some rooms are cooler than others, and locations in the same room may vary tremendously. Positions close to radiators or hot air vents are often hot and very dry and distressing to plants. A shallow tray with pebbles, vermiculite, or peat onto which to set the plants, and with a trace of water at the bottom, would help to counteract excessive dryness of the air. On the other hand, exposure to cold, close to a window or an air-conditioning duct, can be just as harmful. There is always a difference of temperature near the glass and the interior of a room, and cool flowering plants such as Cyclamen, Azaleas, Begonias, Cinerarias, Primulas, and Dutch bulbs would love such a cool window. But in freezing weather, ice may form inside within 30 cm of the window glass, and differences 8 to 12 deg. C lower than room temperatures are quite common. Heat radiates from the foliage to the cold glass, causing a chilling and possible freezing of the leaves. Pull down the curtains, or place some layers of newspapers between the window and plants that could be damaged during times of temperature extremes.

SOIL MOISTURE and HUMIDITY

Soil moisture should be evenly maintained. Water a pot when the surface begins to get dry. Generally, plants with coarse roots and growing in heavier, loamy soil should be allowed to get on the dry side, then water well, by thoroughly soaking the pot or tub. But root-balls full of very fine fibrous roots, and growing in humusy soil mix, must be kept more evenly moist. To determine the moisture condition simply "feel" the soil; if it feels dry and hard to the touch, and it looks light-colored, then it is high time to water; if the soil feels damp or muddy, and looks dark, then better wait a day or longer. Generally, more watering is required in a heated room during winter time, possibly daily, while in summer, when the radiators are off, watering two or three times a week may be sufficient. No plant should stand for long in water unless a water plant, or is suitable for hydroponic culture.

Watering Systems

To avoid the tedious watering by hand, commercial growers have installed systems of piped water directly to plants, either by using thin plastic tubes (spaghetti), or bubblers close to the soil. Adaptations of these can be installed for use in house plant corners, on patio or peristyle, or the small greenhouse.

Soils and Feeding

The selection of a proper soil mix for any given plant, is not as important as many specific recommendations would make it seem. Proof of this is found in the fact that, based on University of California experiments, a universal mix could well be recommended for nearly all plants, consisting of equal parts of peatmoss with fine sand. Such a mix would of course have to be enriched by fertilizer, because it would be lacking in basic plant food. But in areas with good garden soils, fancy trimmings are not necessary, as friable loam mixed with some humus, will suit most any house plant satisfactorily. Eastern experiment stations recommend a basic mix of 1 loam, 1 peatmoss, and 1 pebbly perlite. Remember that the more delicate, fragile exotics with very fine fibrous roots, will want the addition of more humusy soil, leafmold, peatmoss, sphagnum, shredded fir-bark or humus-compost, with sand or perlite for drainage. Such soils have the advantage of not needing as a rule, the addition of much plant food, as the decomposing organic matter will provide for average needs. A bit of rotted barn manure or fish fertilizer for nitrogen, and possibly a little ground bone to add phosphate, will be of long-lasting benefit.

Glass fish-bowl with terrarium planting of small tropical plants thriving in the humid atmosphere generated through moisture transpired by the leaves, and reducing the need for watering.

Controlled watering system employed by commercial greenhouses, using thin flexible medical tubes (spaghetti) to conduct water into each flower pot, and fed by 2cm black plastic pipe, turned on as needed.

ORCHIDS—Queen of Exotics

Orchid growing is an interesting, rewarding, and relaxing hobby. A family of aristocrats world-wide—to know of their origin, conjures up visions of tropical islands or humid mountain forests, but also gives some indication of their requirements in cultivation. 35,000 species of orchids in 800 genera, plus countless hybrids have been recorded, of which 85% dwell within the tropics and the subtropical belt.

Success with Orchids in the Home

All orchids prefer the hospitable and controlled conditions of the greenhouse, but many can be grown and flowered with success and pleasure in the home if given the environment that meets their needs. A sunny, well ventilated room and a place, if possible, next to the window where the temperature is normally lowest, plus a relatively high humidity, preferably between 50 and 70%, is ideal. As the habitats of various orchids are found in different climatic zones, specialist's greenhouses aim to provide similitudes through several ranges of temperature for their cultivation under glass; warm 16-27°C (62-80°F), temperate or intermediate 10-18°C (50-65°F), and cool 5-15°C (40-60°F) and up.

Rest and Growth Periods

Orchids are plants of tropical and semi-tropical origin and tolerate both wet and dry seasons in their native habitats. During the wet season, the plants make their new leaves and pseudobulbs; in the dry season, they are resting. Flowering begins during the rest period just before new growth commences. To assure annual flowering, most orchids must have these periods of rest and growth.

Light

Most orchids, particularly the species with hard pseudobulbs, such as Cattleyas, want considerable sunlight, if possible 4-5 hours per day and need at least 15,000-30,000 meter-candles (lux) to insure good flowering. To place them outdoors during the summer time, hung under a tree or arbor, or set onto a slatted bench, will tend to rebuild jaded tissues, and favor the development of new growth, and bud initiation. The pseudobulbless, or softer-growthed kinds will prefer a location protected from direct rays of the sun and an east window would be best during the summer season, if grown entirely under artificial light. 6000 lux extended for 16 hours have proven sufficient for many normally light-hungry species.

Watering and Moisture

When they are in full growth and with active roots, orchids should be watered copiously, usually once or twice a week, but only when the pot is becoming dry, especially the varieties with pseudobulbs. Epiphytes with their spongy roots hold a great deal of water, and if in constant state of saturation these roots will rot. If potted in bark chips, a good drenching may be needed every day; Osmunda keeps moist longer. Most orchids require a period of rest with reduced moisture immediately after flowering.

Humidity is Important

A relative humidity of 50%, which can easily be checked with a hygrometer, is fairly satisfactory for most orchids, but difficult to maintain in winter, when humidity often drops to 30%. Frequent misting of the foliage with a hand-sprayer will materially increase it; so would a plastic enclosure, with the top left open; or a space humidifier. Gentle movement of fresh air is always good, either by opening a window or with the aid of a small fan.

Potting and Feeding

Diced roots of the Osmunda fern are an old, reliable growing medium for epiphytic orchids. But many growers pot them into coarsely shredded firbark or bark chips, frequently with addition of some peat-humus or perlite for better moisture retention. Deficient in nitrogen, porous bark materials need additional feeding with diluted liquid high nitrogen fertilizer or fish-emulsion monthly. Epiphytes generally grow well on tree fern slabs, provided they can be dunked into water almost daily, but terrestrials often prefer a humusy compost, even sphagnum moss. In the South, treefern fiber lasts better than bark.

Colorful planting of long-lasting Bromeliads in shallow bowl.

BROMELIADS—the "Pineapple Family"

Companions of orchids and aroids from the mountains and rain forests, or the rocky seasides of Central and South America, "Bromeliads", peculiar to the Western hemisphere, are a distinguished group of plants, dwelling as epiphytes on trees and rocks, or as terrestrials on the forest floor, and to which belong some of our most fascinating and decorative ornamentals. Their shape is usually in form of a rosette of leathery, concave leaves, sometimes plain, and others with a bizzare design, or strikingly variegated. Their flowers may be hidden deep in the center, surrounded by brilliant inner leaves, or carried high on showy spikes, feathery racemes, or brush-like heads between highly colored bracts, or on panicles of bright, long lasting berries, and even bring forth delicious fruit, such as our pineapple.

Notwithstanding their beautiful coloring and leaf designs, bromeliads are easy to maintain. They can be classed as succulents, since they also store an emergency supply of water, not inside fleshy leaves, but within a natural vase-like center formed by their durable foliage. Their root system, particularly in the epiphytic varieties, serves largely as a means of attaching themselves to trees, rocks, or other convenient hosts. In fact, it has been found that as long as bromeliads receive their needed moisture through their center funnel, they can get along for a considerable length of time without any roots at all. It is a most important characteristic of this family, that these plants absorb their food through the scales found at the base and the surfaces of the leaf. This is also why some oil sprays can easily cause damage. Bromeliads are therefore ideally suited for home decoration, even in unfavorable corners, in hanging pots, wall pieces, dishgardens, on driftwood, and for table adornment. In Europe, these interesting plants have been used in this manner for over a hundred years.

Bromeliads in the Home

Bromeliads tolerate a wide range of climatic conditions, from near freezing to high room temperature. Most of them prefer filtered sunlight, particularly the types with highly variegated leaves. As a growing medium, almost any light and porous material rich in humus, will be found satisfactory; best is peatmoss, also leafmold or shredded fir bark, with broken pots, sand, charcoal, perlite, and some organic fertilizer added. Orchid fibre (Osmunda) is also recommended but not essential. The roots should be kept moist, but they don't like continuously wet feet; however, water should stand in their funnels, preferably fresh. Occasional mild feeding, perhaps at monthly intervals, with some organic, or water soluble, complete synthetic plant food, will result in stronger growth. Chemical fertilizer should be used more diluted than recommended, and may be sprayed or poured freely over the foliage, as the leaves are capable of absorbing and utilizing it direct.

PLANTS INDOORS

Antique copper pan planted as a miniature desert scape with an interesting variety of cacti - Cereus, Opuntias and Cephalocereus, also exquisite succulents represented by Aloe, Pachyveria, Echeveria, Stapelia, Portulacaria and the pretty "Panda plant", Kalanchoe tomentosa, all excellent keepers under dry conditions and requiring little attention.

SUCCULENTS as House Plants

Because of their variety of form, beauty of coloring, and oddity of behavior, succulents are highly prized by hobbyists. They are durable and need a minimum of attention, and interesting collections in small pots can be kept close together in a very small space. As "hot weather plants" most of them prefer a place in the sun which brings out the best of their inborn coloring. Tones of glowing red or copper, glaucous blue or silver, will attract the eye wherever on display. When planted into glazed pottery as "Dishgardens", artistically arranged by character and color, or as individuals in novelty containers, such planters are perfect for use on furniture or the office desk.

MINIMUM of CARE - Succulent plants are from areas of wide temperature variation, native to regions that are arid or semi-arid for at least part of the year. It is in the study of their habitat therefore that gives the cue to the best cultivation and handling of succulents. Although from tropical and subtropical areas of intense daytime heat, temperatures in the tropics usually drop to a pleasant coolness or even chilliness at night. While rains may be absent, morning mist or fog is common, refreshing to shallow-rooted leaf succulents such as Aeoniums or Echeverias. It is fortunate that we can approximate climate conditions of their natural habitats right on our window sill, the sun-porch, or the open patio. At a sunny South window, it is warm and bright during day and cooler at night—ideal for Aloe, Crassulas, Echeverias, Kalanchoe, Sedums, Senecio; also the various Mesembryanthema, such as Lithops, although the latter need to be kept particularly dry. Haworthias and Gasterias prefer an East window, and perhaps a little warmer, as do Euphorbias, Stapelias, and Sansevierias.

Soil and Watering

Succulents need a porous, gritty soil that readily drains superfluous water, and yet is sufficiently compact and retentive, to hold the moisture the plant requires. A good mixture is two-thirds loam and one-third sand, with some leafmold or peatmoss added.

Succulents need less water than most other plants. Leafy kinds can take more than stem succulents. Water requirements are greater in the summer than in the dark days of winter unless much artificial heat should dry them out prematurely. A good rule is to water thoroughly twice a week during the growing season, but not to water on cloudy or rainy days. During winter months or when the plants are dormant, watering once every two weeks would be sufficient.

FERNS (FILICES)

For decorating your home indoors; for making it feel cool, woodsy and alive; for softening stark, harsh architectural lines or unifying a decorative scheme—few plants rival the beauty of graceful green ferns. Here are plants with such purity of form, one can't go wrong however they are used. In addition to decorative value, ferns give a great deal of growing satisfaction, they stay green and lush the year around. Ferns are always cool, calm and refreshing.

Ferns as House Plants and for Decoration

Treeferns of the tropics are the giants of the great fern group, and some species raise their slender trunks to the height of a 6 story building. Most graceful are the West Indian treeferns, Cyathea arborea, and for exotic decor they always excite the imagination of architects. Container-grown they are unfortunately difficult to keep indoors for more than several weeks, unless in a covered atrium with relatively high humidity. Their roots must never be allowed to dry. We have found it a wise precaution to place such tubs inside an outer container with peat, or wrapping the tub in plastic to protect the delicate roots from the danger of desiccation.

The best of the Hawaiian treeferns is the "Man fern", Cibotium chamissoi, in the trade as C. menziesii, and of great appeal to decorators. More stout of habit, they are usually sold as sawed-off tops, and can be grown in a dish of shallow water; lumps of charcoal may be used for support if necessary. However, we have been more successful growing them in soil.

Best Indoor Ferns

The best indoor fern always was and is today the "Boston fern". It enjoys good light and is fairly tolerant of dry air. From it originated many excellent mutations frilled and lacy; the "Lace fern", Nephrolepis 'Whitmanii' is still a favorite. These ferns keep best if root-bound which allows them to be watered freely without danger of stagnation. Any repotting necessary is best done in the spring when active growth begins, using porous soil.

Another old-fashioned friend is the "Holly fern", Cyrtomium falcatum with glossy-dark leaves as tough as leather. Also leathery but lacy is the "Leather fern", Rumohra adiantiformis, whose long-lasting fronds are cut and used in flower arrangements. The "Birdsnest fern", Asplenium nidus, is in a class by itself, a rosette of fan-like light green glossy leaves. If kept shady, warm and moist, it will prove a good and interesting houseplant.

There are many Polypodiums, and we know them as "Hare's foot fern" because of their paw-like woolly rhizomes creeping along the surface. Probably the best for indoor use is the "Blue fern", Polypodium aureum, in its crisped form 'Mandaianum'. The wavy, shimmering blue fronds develop best in warmth and shade, but otherwise they are not difficult at all.

The Davallias, or "Rabbit's-foot ferns" likewise have slender creeping, furry rhizomes which bear on wiry stalks their lacy but leathery fronds. They take much neglect and tolerate both light or shade and are often used in hanging baskets, and are small enough for a windowsill.

Amongst our best-loved ferns are the dainty "Maidenhairs", but they are not easy under house conditions and do best in a terrarium or behind a plastic shelter. Most satisfactory in this genus are Adiantum raddianum, better known as A. cuneatum, and the larger-leaved Adiantum tenerum 'Wrightii', but even these need shade, warmth and moisture.

The so-called "Table ferns" are a varied group of mainly Pteris and Pellaea. Some are frilly, others variegated, and in their small stage widely used in pottery bowls, or together with other plants as fillers. One of the most attractive is Pteris 'Victoriae', a dainty-looking but sturdy little fern of white and silver.

"Staghorn ferns" always have aroused the greatest curiosity because of their unusual shapes. Growing as epiphytes on trees, their sterile fronds cling snugly against branch or trunk, while their much-divided fertile leaves resemble the antlers of deer or elk. Platycerium are not difficult if grown on fibrous cushions fastened to a board or basket, and their root system is submerged in water, whenever it is beginning to get dry. Platycerium bifurcatum and its cultivar, P. 'Netherlands' are less tropical, and quite tolerant even if abused.

METHODS OF PROPAGATION

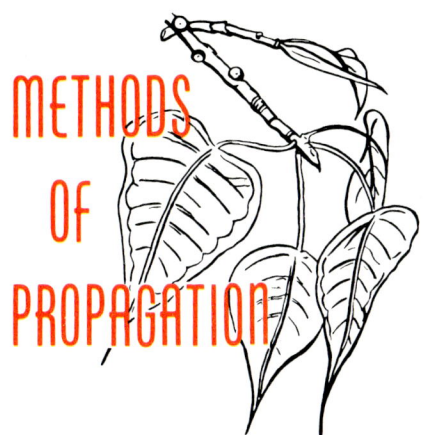

Propagation of Exotic Plants

Propagation is the miracle of rebirth—the creation of a new plant. This can be accomplished in numerous ways—some plants lend themselves better to one method, others to another technique.

There are two primary methods of reproducing plants—seminal and vegetative propagation. Seminal propagation is known as the sexual method of reproduction, because it involves sexual union to form the seeds of phanerogamic (flowering) plants, and the spores of cryptogamous plants such as ferns.

In asexual, or vegetative propagation, however, the new plant will reproduce the same character as the one from which it was taken. Both methods of reproduction can be used for most plants.

Unlike some hardy trees and shrubs from Temperate regions, seed of tropical origin must always be fresh and sown promptly when ripe, as its viability may be extremely fleeting. Old seed is usually the reason why exotics, for instance Aroids, Araucarias and Dizygotheca, and most other tropical trees and vines, fail to germinate despite the best of care. In ferns a cause of non-germination is, that they cannot survive if the spores were exposed to freezing, as in a sense they are living, delicate, and unprotected embryos.

Seeds are best sown in sterile clay pans with drainage such as broken pots on the bottom, and sandy soil to near the top. Friable soil, with sand and some peatmoss added, should be pasteurized to prevent damp-off, simply by baking it in a moderate hot oven in moist condition, at 60-80°C (140°-180°F) for 3/4 to 1 hr. Formaldehyde may also be used for soil sterilization, particularly to lessen nematode infestation. Use 1 part of 40% Formalin to 39 parts of water.

Seed boxes may be made comparatively free from infection, by pouring boiling water into the soil previous to sowing, and covering it with newspaper. When cool, and the seeds are sown, again a paper covering kept moist by watering the surface, encourages a very satisfactory germination. At the first sign of germination the cover is removed.

The Vegetative or "Asexual" Method

While almost all plants would grow from seeds, there are many that are with advantage, or must be, reproduced vegetatively. Trees that never become mature enough to reach the flowering stage in cooler clime, others valuable but sterile, and won't bear seeds at all, must be propagated asexually. For home needs many bulbs and rhizomes such as orchids are divided much more easily, than to wait for seedlings to mature. Depending on their makeup, plants may be reproduced asexually using various vegetative parts: softwood or hardwood stem cuttings, unrooted tips, joints, or leaves, root-cuttings, basal separations, runners, offsets or suckers; also through air-layering, cane sections, rhizomes, division of rootstocks and tubers, or by bulblets, or scales from bulbs, or even grafting.

Meristem Tissue Culture Propagation

Clonal multiplication of select plants by laboratory process has come into use by commercial growers of orchids, ferns and the better exotic plants in recent years. Embryonic cell tissue from growing tips is isolated and incubated in nutrient agar-culture flasks, under controlled and sterile conditions, for rapid mass production of virus-free plantlets of outstanding cultivars.

Most Orchids are quite easily propagated by cutting apart the surface rhizome between joints, or otherwise dividing them. For large scale production, the tiny seeds are germinated by laboratory process in test tubes or sterile culture flasks on nutrient-enriched agar jelly. When large enough to face the outside world, seedlings are transplanted into small pots with finely chopped osmunda fiber or other fibrous media. Then follow years of size increase before flowering strength is reached.

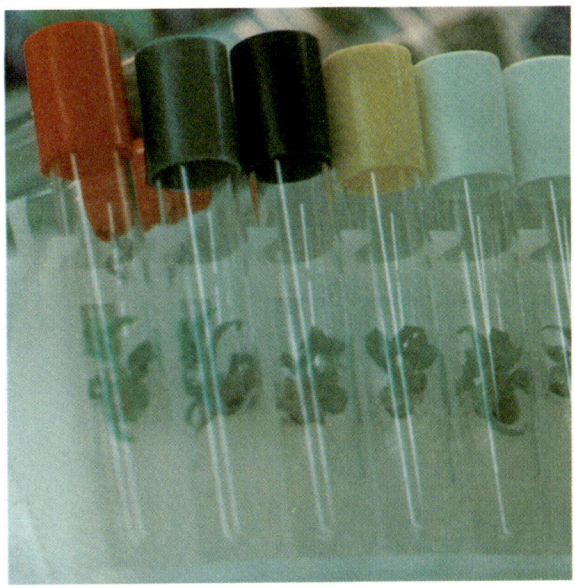

Sterilized Meristem cultures on jellied nutrient-agar in laboratory test tubes, in the first stage of developing plantlets, forming from dissected bud tissue, in California, a method of propagation in a minimum of space.

PROPAGATION

Propagating bench for the rooting of semi-woody cuttings, in a mixture of sand, sphagnum peatmoss and perlite, covered with wire frame supporting a sheet of transparent polyethylene film to provide humid atmosphere, in glass house of Arnold Arboretum, Boston, Massachusetts.

Top sections of vandaceous orchids with forming aerial roots, cut off their mother plants in the shade house, and potted in pebbles of broken burnt earth; this allows water to drain freely, preventing stagnation, and admitting oxygen to encourage a healthy root system; at a nursery in Penang, Malaysia.

Hardwood cuttings of Hibiscus rosa-sinensis stuck deeply into moist earth, in criss-cross fashion, to form a hedge, in tropical Fortin de las Flores, Veracruz, Mexico. Propagation by hardwood cuttings of various shrubs and vines is common practice in tropical and subtropical regions, especially where daily rains provide sufficient moisture.

PROPAGATION

Bromeliad seeds can be germinated remarkably well on layers of sterile Kleenex tissue soaked in water, in a shallow dish and covered by a pane of glass, at a warm 18 - 21°C (65 - 70°F).

Softwood cuttings in sharp builder's sand held fresh and turgid by an intermittent, automatic mist system, and when needed, a covering of open-mesh cotton screening, at Roehrs greenhouses.

Air-layering the top portion of an old, overgrown "Spider aralia" (Dizygotheca); the incision into the woody stem is covered with moist sphagnum and sealed tight with a sheet of plastic.

Juvenile cultures of Pteris ferns growing on sterilized humus, in primary shield-like prothallia stage barely germinated from spores (right), developing into the second stage sporophyte, or true fern plant (on left).

The miracle of young plantlets sprouting at the base of leaves of Pachyphytum compactum, a means of propagation characteristic of the Crassulaceae, also other succulents including Sansevieria.

Tropical plants possessing thick-succulent foliage, or with fleshy veins, are frequently reproduced by leaf cuttings or sections of leaves; this includes Peperomias (above), Saintpaulias, many rhizomatous Begonias and Begonia rex.

PROPAGATION

Many plants can easily be propagated by the time-honored home method of placing a cutting into a glass jar or drinking cup partially filled with water, where they will quickly develop new roots, even on a window sill. Proven subjects are Coleus, Saintpaulias, Geraniums, Begonias, Fuchsias, Impatiens, Tradescantia, Cissus, Cyperus, ivy, Philodendron, Aucuba, Dracaena, Gardenia, Hibiscus, oleander, and even succulents such as Crassula.

Softwood Cuttings

Cuttings should be taken from plants in active growth, most likely in the spring after awakening from winter rest. The green growth used for softwood cuttings should be half-ripened and crisp, so that it would snap between fingers, and cut to a length of about 8 to 12 cm. When propagating tropical plants from cuttings, keep in mind that these are not usually the hardwood type, so characteristic of hardy shrubs and trees but often softer, and with herbaceous leaves. Growth thus deprived of roots has extremely limited ability, and those with thin, membranous leaves may quickly wilt. It means winning half the battle toward root formation, if any wilting is prevented from the start, by high humidity around their foliage to keep it turgid. Although cuttings must be kept steadily moist, watering is not enough. Wilting can be prevented in several ways, easiest by covering such a cutting with a tent of transparent Polyethylene film, a glass jar, plastic cake cover, or in a glass-covered aquarium tank. Cuttings root much faster when given a close, humid atmosphere.

Hardwood Cuttings

Propagation by hardwood cuttings in fairly common practice in temperate climate nurseries, using well-ripened wood of pencil thickness 15-20 cm long of deciduous shrubs and trees, when dormant, taken after the leaves have fallen. Such cuttings are inserted two-thirds in damp sand, peatmoss or sandy soil, and will slowly form new roots during the cool winter season.

Similarly, certain subtropic and tropical shrubs, woody vines and trees can be propagated by the hardwood method. Bougainvilleas in Hawaii have been rooted in vermiculite by using 10 cm woody cuttings with their leaves removed, their stems dipped in hormone and kept moist, in partial shade. In tropical Fortin de las Flores, Veracruz, Mexico I have seen woody branches of Hibiscus 40-50 cm long, stuck criss-cross along a garden path, in time to form a hedge. In the South Pacific Marquesas islands, the New Hebrides and even in the mountain villages of New Guinea, it is common practice to push pieces of Cordyline cane into the ground and they will form beautiful living screens around their huts. On Timor I even saw a sizeable branch of the Royal Poinciana stuck into the ground, sprouting leaves, and blooming. Others that may be rooted by the hardwood method are Acalypha, Allamanda, Bauhinia, Cestrum, Citrus, Codiaeum (Croton), Erythrina, Euphorbia, Fig trees, Grapes, Jasmine, Hydrangea, Lagerstroemia, Malvaviscus, Mussaenda, Nerium, Poinsettias, Punica, Roses, Tamarix and Tecomaria.

Home Propagation

A simple home-propagating method is to place semi-hard or succulent cuttings (Geraniums, Ivy, Coleus, Crassula) into a glass of water where they will form roots. Softer cuttings may also be inserted into a pot with porous propagating mix (sand or perlite mixed with peatmoss), watered down, and placed inside a small, airtight plastic bag.

Cuttings with more leathery foliage, or waxy Begonias, Geraniums or Peperomias do not necessarily require any glass or glass-like shelter. However, such coverings help to keep the environment more warm and even, and a temperature of 18-21 °C (65-70 °F) is vital to the rooting of most exotic plants. Tropicals such as Croton and Ficus root best in 24 °C but should not go over 26 °C.

Root Inducement Hormones

Much scientific advice has recommended and proven the advantage of using root-inducing hormones, usually indolebutyric or naphthalene-acetic acid, or "Rooting powders". This is hardly needed with quick-rooting cuttings, but to dip their freshly cut base into powder or solution, can assist harder, slower-rooting cuttings in stimulating speedier rooting; in hardwood the butt ends may be wounded or slit open for more roots to form along the cambium. It has been our experience that hormone compounds are most effective if the bottom temperature is at least 16 °C (60 °F) or more. Light is also a factor in the rooting of cuttings. While in the beginning it is prudent to shield the propagating container from bright sun, with a layer of newspaper, cheesecloth, or Venetian blinds, light intensity for most cuttings, except pronounced shade lovers, should be as high as possible without causing the plants to wilt.

Since the advent of polyethylene plastic bags, new methods of propagation have become possible. Cuttings that need a close, humid atmosphere, may be dipped into hormone rooting powder, and placed into a mixture of peatmoss or sphagnum moss with sand or perlite, well wetted down, and sealed. Placed into a light window, plastic film allows for the exchange of oxygen and carbon-dioxide, without which roots do not develop properly.

PROPAGATION

Layering or Marcottage

Air-layering (bagging or marcotting) is one of the commonest and surest forms of propagation in the tropics. Cutting into a woody stem at an angle, or removing a ring of bark from a semi-mature part of the stem for about 3 cm, below a joint or leaf—the bared area is surrounded with wet sphagnum moss, or similar fibrous material. This is then enclosed within a translucent plastic sheet, or aluminum foil, tying it firmly above and below the unbarked strip or spliced cut. Watch for roots appearing through the moss (in about 4-6 weeks), when the rooted cutting may be cut off and transplanted into a pot or into open ground, shaded for about a week and kept damp. It is surprising how many plants can be propagated in this manner.

Air-layering is a time-honored and fairly fool-proof practice since ancient times and known around the world. The Romans first used the word mergus, for the merging of a notched branch with the ground (ground-layering), where it would strike roots, while still attached to the parent plant. The principle is the same in both methods. They are easy, and tops or branches from large plants or trees such as Mango, Lychee, or Ficus can be rooted in this manner without taking many risks. When a plant indoors has grown too tall, such as Ficus, Croton, Dieffenbachia or Dracaena, or even Chamaedorea palm, these may be reduced in size by mossing along their stems, to suit the space for which they were intended. Ground-layering may be helpful in plants otherwise difficult to propagate, such as woody Beaumontia vine. Part of a prostrate branch sometimes partially cut, is covered with soil and kept moist, and roots will form at the covered nodes. As soon as these are able to support growth, the layer can be clipped off from the mother plant.

Propagation by Division or Suckers

An easy method of plant increase and reproduction is by division. A great many tropical herbaceous, woody, suckering rhizomatous, bulbous and tuberous plants may be divided. These include Bamboo, Bananas, Heliconias, Gingers, Rhapis palms; Spathiphyllum, Aglaonemas, Calathea; Bromeliads; Cannas, Amaryllis, Agapanthus, Clivias; many Ferns.

Rooting Materials

The oldest rooting medium is good clean, sharp builders SAND, mined from the pits. Beware of salty sand from ocean beaches. A bit of sterile sphagnum peat if added to the sand, will hold moisture longer, and aid in rooting woody cuttings.

Milled SPHAGNUM moss is widely used, since it retains moisture well, and it is a natural fungus inhibitor, without discouraging the formation of roots. Some plant fanciers have successfully rooted their cuttings, simply by placing them in moist sphagnum inside a plastic bag. The exploded volcanic glass-like PERLITE is most promising. This is a white, granular material used as a plaster aggregate. It is light, clean and sterile, does not break down, and holds more than 10 times its weight or 60% by volume of water, without becoming water-logged, and plants root readily in it. This medium may be used alone but preferably in mixture with peatmoss. Perlite lets air into this mixture and prevents the peatmoss from becoming too wet. Horticultural grade VERMICULITE, a light-weight, spongelike heat-popped mica mineral, is used as medium for the rapid rooting of Chrysanthemums, Coleus and similar soft or succulent plants, but its harmonica-like kernels must not be compressed or over-watered. However, for plants slow to root, including Azaleas, peatmoss with sand or perlite is considered better, and conifers still do well in plain sand.

Containers for Cuttings and Seed Sowing

Various containers are suitable for cuttings or seeds. Clean flower pots, preferably shallow, with drainage in the bottom are good, especially as they can be set into a dish, holding some moisture in its base. Tin cans with holes punched in the bottom will do. Plant flats or little wood boxes are also used, all provided with a plastic hood or covering pane of glass to hold moisture. A glass tank, a plastic food or refrigerator box makes a miniature greenhouse which is translucent, sterile and moisture-holding. Glass-covered Pyrex casseroles also are ideal for seed or soft cuttings. Even a plastic polyethylene bag may be used for the rooting of cuttings. Gentle bottom heat will accelerate the formation of roots on cuttings, or the germination of seed. While electric cables are available, a simple alternative is to place an electric light bulb under an inverted clay pot, on top of which the container is set. (For detailed information on propagation see Exotic Plant Manual pg. 48 to 64.)

In Belgium, where horticulture is highly developed, older specimens of the decorative Cordyline australis 'Variegata' are rejuvenated by "mossing" their woody trunks, after removing a ring of bark with cambium, wrapping the area with moist sphagnum moss, then enclosing it with a flower pot split in half. Soon the container will fill with roots and can be sawed off.

Air-layering is an ancient art practised for centuries in India in the propagation of otherwise difficult-to-root trees such as Ficus and Mango. A slant incision made deep into the wood of a branch is filled and covered with wet moss or other fibrous material, wrapped with burlap and kept moist; photographed in the gardens of the Taj Mahal in Agra.

PEST CONTROL

CONTROL of INSECTS and DISEASES
Common Insect Pests of House Plants

1. *Aphids,* tiny green or black plant-lice 1-2 mm long, sucking juices around growing tips or underside of leaves.
2. *Mealybugs,* small pinkish crawlers 3-4 mm long dusted white, hiding under foliage or in leaf axils, often in a white cottony mass.
3. *Red Spider,* minute spider-like mites on underside of leaves, in time spinning gauzy webs, multiplying rapidly in dry-warm rooms. Causes discoloration of foliage.
4. *White Fly,* moth-like scales 2 mm long, with and without wings. Often on Bedding plants, Lantanas, Geraniums, where they suck juices from underside of foliage.
5. *Scale,* small, sucking turtle-shaped or shield-like bodies 3-5 mm long, mostly stationary, usually black or brown; white on ferns.

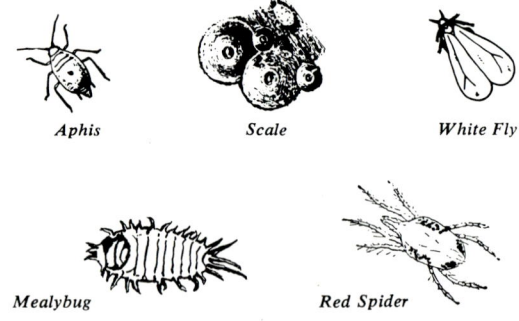

Simple Home Control of Pests

House plants should be diligently observed for the possible appearance of insect enemies, and to prevent any population build-up that is difficult to control. Where plants are few, it is good practice to wipe the foliage with a damp cloth, or to wash with water under a faucet, to remove possible pests, to keep the breathing pores open, and for cleaner and better appearance. When it becomes necessary, clean or spot-spray for specific problems as they arise. A light spray of rubbing alcohol, half diluted with water, in a hand atomizer, causes any mealybugs present to turn brown. A touch with a cotton-tipped toothpick or artist's brush dipped in alcohol (including whisky) or ether (nailpolish remover), is sure death for a mealybug, scale or aphid, but care must be taken or these concentrates may burn the foliage.

General infestations on house plants can be discouraged by the simple hygienic means of forcefully syringing each plant every week or two with clear water—especially the undersides of leaves. This is best done in sink, bathtub, the shower, or outdoors, and should dislodge any unwanted guests. Washing keeps foliage clean for breathing and a plant more healthy besides.

Some of the old-fashioned home remedies are still useful today. A bath in warm soap suds (2 tablespoons soap flakes or octagon soap in 4 liter (1 gal.) water is a mild remedy for foliage plants. If the plants are large and unwieldy, such a soap solution can be used for washing of the leaves with a sponge or a toothbrush.

Of course, washing with water alone is not always the complete answer, if pests should entrench themselves in hiding places of leaf axils, growing tips or in dense foliage. If badly infested, the entire foliage of a plant may be immersed for 30 seconds in a bath of a warm soapy solution, containing a teaspoon of nicotine sulphate to each 4 liter of water, and rinsed a few hours later; this should control green or black aphids and scale. Nicotine spray, if not available, can be homemade by soaking tobacco from cigars, or from pipe tobacco in water for several days; dilute to the color of weak tea for use.

If in doubt about chemical insecticides there are, for household use, reliable, relatively safe leguminous insect sprays containing Rotenone, a Derris extract, and Pyrethrum; usually in emulsion with vegetable or aromatic mineral oils. These normally will control aphids, mealybugs, scales, exposed thrips, white fly, red spider, and other mites; beetles, lace bugs, cutworm, and more. Small hand atomizers are available for wet-spray formulas. Refined miscible oils combined with Rotenone-Pyrethrum or with nicotine as a foliar spray, are especially effective in penetrating the waxy coat of mealybugs, also to loosen the hard shells of young scale, and smothering their young crawling stage.

CONTROLS in Warm Region Gardens

Insect pests outdoors are a natural hazard, and need even more constant attention in the tropics. The usual insecticides should be applied, diluted but more frequently. Scale insects are easily scrubbed off with a tooth brush and soap suds—sprays containing soapy water also help. There is a particular ant called the Parasol ant, which is a leaf cutting insect attacking plants chiefly at night. It is easier then to track them on their trails with a light and follow, while they carry pieces of leaf to their nests, as their path may be obliterated during the day when the ants are all within the colony. Ant bait can be used successfully by placing it in their tracks or around the base of the plant which is being attacked; and this is carried by them back to the nests, destroying the colony. Chlordane, when used in this manner, has proven most effective in the past, but may not be generally available, or be prohibited.

Insect and Disease controls vary in different warm-climate areas of the world. In the tropics some insect pests have 4 or 5 breeding cycles a year instead of only one.

DISEASES, except for mildew, are generally not a problem with house plants in temperate climate—but fungus diseases thrive in a moist-warm atmosphere.

Chemical insecticides and fungicides change frequently and may be superseded by newer products while other, old reliables are being prohibited for reasons of health. It has been my experience all over the world that it is best to check the shelves of garden supply stores for the current pest remedies available for local application.

EPITAPH

When, on occasion, we feel discouraged by the persistence of insect pests, or by the damage to our plants caused by some new commercial product, we might look back to similar experiences a century ago, before the advent of chemical insecticides. I quote Henry Williams in his book "Window Gardening" (1871): *"Years ago, when we had the charge of a small conservatory, we tried the effect of sulphur thrown upon hot coals to kill infested plants. Every insect succumbed before its direful fumes; so also did the plants; hardly a leaf remained on the stems the following day, and the poor leafless branches spake to me in terms of sad reproach through their mute lips."*

Plants partially defoliated by an infestation of mealy bugs and the tools that may be used for Insect and Disease Control: from left to right - a rubber bulb for sulphur and other dusting powders; sprayer-mister with suction pump action; rust-proof plastic hand sprayer for mist, with lever action; pushbutton-valve aerosol pressure sprayer for commercial insecticides.

ACANTHACEAE

Aphelandra fascinator
(*Colombia*)

Aphelandra squarrosa 'Louisae'
"Zebra plant"

Aphelandra nitens
(*Colombia*)

Aphelandra aurantiaca
"Fiery spike" (*Mexico*)

Acanthus pubescens
in Kenya

Aphelandra chamissoniana
"Yellow pagoda" (*Brazil*)

Aphelandra tetragona
(*West Indies*)

Aphelandra fusco-punctata
(*Colombia*)

Aphelandra sinclairiana
(*Panama*)

ACANTHACEAE
34

Fittonia verschaffeltii
"Mosaic plant" (*Perú*)

Chamaeranthemum igneum
(*Perú*)

Fittonia verschaffeltii
'Argyroneura'

Barleria cristata
"Philippine violet"

Barleria lupulina
"Hop-headed barleria"

Barleria albostellata
(*Rhodesia*)

Ruellia amoena
"Red Christmas-pride"

Crossandra pungens
(*Tanzania*)

Hypoestes sanguinolenta
"Freckleface"

Hypoestes aristata
"Ribbon bush" (*So. Africa*)

Pseuderanthemum alatum
"Chocolate plant" (*Mexico*)

Crossandra infundibuliformis
'Mona Wallhead'

ACANTHACEAE

Beloperone guttata
"Shrimp plant"

Adhatoda vasica
(*India*)

Crossandra infundibuliformis
"Firecracker flower"

Strobilanthes maculatus (*Himalayas*)
at Longwood Gardens, Pennsylvania

Strobilanthes dyerianus
"Persian shield" from Burma

Chamaeranthemum venosum
(*Brazil*)

Aphelandra squarrosa
'Fritz Prinsler'

Sanchezia nobilis glaucophylla
(*Ecuador*)

ACANTHACEAE

Graptophyllum pictum 'Tricolor'
"Caricature plant"

Graptophyllum pictum
(*New Guinea*)

Odontonema strictum
(*Justicia coccinea hort.*)

Hemigraphis repanda
(*W. Singer in Zuerich*)

Peristrophe hyssopifolia
'Aureo-variegata'

Hemigraphis colorata
"Red ivy"

Acanthus mollis
"Greek akanthos"

Acanthus ilicifolius
(*Tropical Asia*)

Acanthus montanus
"Mountain thistle"

ACANTHACEAE

Pseuderanthemum atropurpureum
'Tonga' (*South Pacific*)

Pseuderanthemum reticulatum
(*New Hebrides*)

Pseuderanthemum atropurpureum
'Variegatum' (*Brazil*)

Ruellia makoyana
"Monkey plant"

Pseuderanthemum sinuatum
(*New Caledonia*)

Ruellia amoena
(*Stephanophyllum longifolium*)

Ruellia affinis
(*Brazil*)

Ruellia macrantha
"Christmas pride"

Ruellia colorata
(*Brazil*)

ACANTHACEAE

38

Ruellia squarrosa
in South Florida

Thunbergia alata
"Black-eyed Susan"

Eranthemum nervosum
"Tropical blue sage"

Porphyrocoma pohliana
from Brazil (J. Bogner, Munich)

Hemigraphis colorata 'Exotica'
"Purple waffle plant" (Tonga)

Sanchezia nobilis glaucophylla
(Ecuador)

Chamaeranthemum venosum
(Brazil)

Stenandrium lindenii
(Perú)

Strobilanthes lactatus
(Brazil)

Ruttya fruticosa
(South Africa)

Mackaya bella (Asystasia)
(South Africa)

Ruttya speciosa scholesei
(Kenya)

ACANTHACEAE

Justicia rizzinii
(*Jacobinia pauciflora*)

Jacobinia carnea
"Flamingo plant"

Justicia betonica
"White shrimp plant"

Justicia pohliana obtusior
(*Sydney Botanic Garden*)

Jacobinia velutina
"Brazilian plume"

Pachystachys lutea
"Lollypops" or "Super Goldy"

Megakepasma erythrochlamys
"Brazilian red-cloak"

Pachystachys coccinea
"Cardinal's-guard"

Justicia aurea
in South Mexico

ACANTHACEAE

Thunbergia grandiflora alba
in Tahiti

Thunbergia mysorensis
(*South India*)

Thunbergia laurifolia
(*India*)

Thunbergia grandiflora
"Blue trumpet-vine"

Thunbergia fragrans
(*Sri Lanka*)

Thunbergia erecta
"King's-mantle"

Ruttya speciosa scholesei
(*Kenya*)

Ruttya fruticosa
in Tsavo Park, Kenya

Thunbergia alata
"Black-eyed Susan"

ACERACEAE, ACTINIDIACEAE

Acer platanoides
"Norway maple"

Acer palmatum
"Japanese maple" as Bonsai

Acer saccharum
"Sugar maple"

Acer negundo 'Variegatum'
"Variegated box-elder"

Acer palmatum
winged fruit

Acer palmatum 'Atropurpureum'
as Bonsai in Brooklyn Bot. Garden

Acer palmatum 'Sansokaku'
in Dallas, Texas

Actinidia kolomikta
"Kolomikta vine"

Actinidia chinensis
"Kiwi vine"

AIZOACEAE

Fenestraria aurantiaca
"Window-plant", in bloom

Dorotheanthus bellidiformis
colorful carpet in Simonstown, So. Africa

Aptenia cordifolia (Mesembryanthemum)
"Baby sun-rose", (South Africa)

Lithops villetii
in the desert, Bushmanland, So. Africa

Conophyllum tenuifolium
at Stellenbosch Botanic Garden, Cape Prov.

Mesembryanthemum crystallinum
"Ice plant" along California coast

AIZOACEAE

Lampranthus aureus (*left*), L. zeyheri (*right*)
at Mission San Luis Rey, California

Mesembryanthemum crystallinum
"Ice-plants" along Pacific Ocean, Baja California

Lithops: *top left:* lesliei; *right:* pseudotruncatella
bottom left: fulviceps, *bottom right:* bromfieldii

Lampranthus spectabilis
"Red ice-plants" near Escondido, California

AIZOACEAE

Drosanthemum hispidum
(*Namibia*)

Glottiphyllum fragrans
(*Karroo, Cape Prov.*)

Glottiphyllum depressum
"Tongue-leaf" (*So. Africa*)

Gibbaeum petrense (Argeta)
(*Karroo, Cape*)

Trichodiadema stellatum
25 yrs. old, in Hawaii

Cephalophyllum 'Red Spike'
Los Angeles Bot. Garden

Delosperma 'Alba'
"Disneyland iceplant"

Ruschia granitica
(*Namaqualand, Cape*)

Fenestraria rhopalophylla
"Baby toes" in bloom

AIZOACEAE

Hymenocyclus latipetalus
"Carmine iceplant"

Lampranthus spectabilis rosea
"Rosy-red iceplant"

Lampranthus aurantiacus glaucus
"Yellow bush ice-plant"

Lampranthus roseus
"Pink ice plant"

Lampranthus aureus
"Orange ice-plant"

Lampranthus zeyheri
(Cape Prov., So. Africa)

Carpobrotus edulis
"Hottentot's fig"

Lampranthus spectabilis
"Red iceplant"

Hymenocyclus luteola (Malephora)
"Yellow trailing ice-plant"

AIZOACEAE

Argyroderma roseum
(*Namaqualand, South Africa*)

Delosperma brunnthaleri
(*Natal, South Africa*)

Conicosia communis
on the Cape Flats, South Africa

Carpobrotus edulis
"Hottentot's fig" at Cape of Good Hope

Dorotheanthus bellidiformis
"Livingstone-daisy", in Germany

Drosanthemum speciosum
in Stellenbosch Botanic Garden, South Africa

AIZOACEAE

Lampranthus aureus, *"Orange ice-plant"*
at Mission San Luis Rey, California

Lampranthus spectabilis *from So. Africa,*
"Red ice plant" feels at home in California

Lampranthus aurantiacus, *"Golden ice plant"*
on the Cape Flats, South Africa

Lampranthus productus
"Purple ice plant" from Cape of Good Hope

Lampranthus roseus, *"Pink ice plant"*
in the Transvaal veld north of Johannesburg

Lampranthus tricolor, *in Transvaal,*
"Copper ice plant"; flowers open to the warming sun

AIZOACEAE

Conophytum auriflorum
(*Namaqualand, Cape*)

Conophytum elishae
(*Cape Province*)

Conophytum griseum
(*South Africa*)

Conophytum pearsonii
(*Cape Province*)

Pleiospilos nelii
"*Cleft stone*" (*Cape Prov.*)

Conophytum luisae
(*Namaqualand*)

Lithops bella
"*Pretty stoneface*"

Conophytum taylorianum
(*Namibia*)

Frithia pulchra
"*Purple baby-toes*"

Faucaria tuberculosa
"*Tiger jaws*"

Fenestraria aurantiaca
"*Baby toes*"

Faucaria tigrina
(*Cape Province*)

AIZOACEAE

Pleiospilos bolusii
"Living rock-cactus"

Hereroa granulata
(*Cape Province*)

Pleiospilos nelii
"Cleft stone"

Lithops turbiniformis monstrose
at Roehrs greenhouses

Center: Hereroa muirii
(*Karroo, Cape Prov.*)

Fenestraria aurantiaca
"Baby toes"

Aptenia cordifolia
at Cefalu, Sicily

Aptenia cordifolia variegata
Kew Gardens, London

Hymenocyclus latipetalus
(*Malephora*), *Cape Prov.*

Rhombophyllum nelii
"Elkhorns"

Trichodiadema densum
(*Karroo, Cape Prov.*)

Psammaphora modesta
(*Namibia*)

AMARANTHACEAE

Alternanthera amoena
"Parrot-leaf"

Alternanthera bettzickiana aurea
"Yellow calico plant"

Alternanthera bettzickiana
"Red calico plant"

Iresine herbstii
"Beefsteak plant"

Iresine herbstii aureo-reticulata
"Chicken gizzard"

Iresine herbstii acuminata
"Painted bloodleaf"

Alternanthera versicolor
"Copperleaf"

Iresine lindenii formosa
"Yellow bloodleaf"

Alternanthera dentata 'Ruby'
"False globe-amaranth"

AMARANTHACEAE

Amaranthus paniculatus cruentus
"Princes feather" at Hotel Mediterranée on Rhodes

Amaranthus paniculatus, *"Purple amaranth"*
in Napoleon's garden, Longwood Plantation, St. Héléna

Celosia argentea plumosa 'Flame'
"Burnt plume", in the Domain at Auckland, New Zealand

Celosia cristata, *"Coxcombs"* in brilliant color
thrive in the humid-warm climate of Sri Lanka

AMARANTHACEAE

52

Amaranthus caudatus
"Love-lies-bleeding"

Amaranthus tricolor
"Tampala"

Gomphrena globosa
"Globe amaranth"

Amaranthus tricolor hybrida
"Joseph's coat"

Celosia argentea 'Golden Spire'
"Feather celosia"

Iresine lindenii formosa
"Yellow bloodleaf"

Celosia argentea pyramidalis
"Burnt plume"

Celosia cristata
"Coxcomb"

Celosia argentea plumosa
"Plume celosia"

AMARANTHACEAE, AMARYLLIDACEAE

Alternantheras *and other bedding plants along the promenade on the Neva, Leningrad, Soviet Union*

Alternantheras *in carpet bedding at the old Orangerie, Missouri Botanic Garden, St. Louis*

Bomarea shuttleworthii, *one of the curious vines collected in the Finisterre Mountains, New Guinea*

Cyrtanthus mackenii, *the "Ifafa lily" from So. Africa, in experimental plastic containers, Kwangtung, China*

AMARANTHACEAE, AMARYLLIDACEAE

Carpet bed *with* Alternantheras *by Roehrs,
International Flower Show, New York Coliseum*

Floral clock *including* Alternantheras
in the art of carpet-bedding, in Geneva, Switzerland

Amaranthus caudatus, *known as "Love-lies-bleeding",
at Nymphenburg Botanic Gardens, Munich, Germany*

Celosia argentea 'Castle Gould', *strain from Long Island,
with large feathery plumes, grown by Roehrs, Rutherford*

Brunsvigia josephinae, *free-blooming "Josephine's lily"
at height of season in February, Adelaide, So. Australia*

Crinum augustum, *a fragrant "Spider lily"
in the tropical Seychelles Islands, Indian Ocean*

AMARYLLIDACEAE

Agave americana (*inflor.*)
Huntington Botanic Gardens, San Marino, California

Agave attenuata
flowering at Quail Gardens, California

Furcraea gigantea
at South Coast Botanic Gardens, Los Angeles

Agave sisalana
in Sisal plantation near Moshi, Tanzania

AMARYLLIDACEAE

Furcraea selloa marginata
in succulent house, Botanic Garden Munich

Manfreda maculosa
(Mexico: Chiapas, Morelos)

Furcraea gigantea 'Striata'
at Sibpur Botanic Garden, Calcutta, India

Agave victoria-reginae,
in habitat in Huasteca Canyon, Mexico

Agave parviflora
"Little princess agave" (Tucson)

Agave parryi
"Mesqual" in Southern Arizona

AMARYLLIDACEAE

Agave americana
overlooking the Acropolis, Athens, Greece

Agave americana *with* A. angustifolia marginata
near Government Palace, Habana, Cuba

Agave shawii
in habitat near Ensenada, Baja California

Agave atrovirens
"Pulque agave" in Central Mexico

Agave americana marginata
at South Coast Botanic Garden, Los Angeles

Agave sisalana
on plantation near Mt. Kilimanjaro, Tanzania

AMARYLLIDACEAE

Agave americana 'Medio-picta'

Agave filifera
"Thread agave"

Agave angustifolia 'Marginata'
"Variegated Caribbean agave"

Agave horrida
(*Morelos, Mexico*)

Agave striata nana
"Lilliput agave"

Agave victoriae-reginae
(*Northern Mexico*)

Agave geminiflora
from Mexico

Furcraea gigantea
in Southern Brazil

Agave albicans 'Medio-picta'
at Kew Gardens, London

AMARYLLIDACEAE

Anigozanthos manglesii, *the "Red and green kangaroo paw" at home along Shark Bay, Western Australia*

Beschorneria yuccoides *from Mexico with striking inflorescence of green flowers and red bracts*

Anigozanthos flavidus, *the "Tall kangaroo paw" in the southernmost districts of Western Australia*

Anigozanthos rufus, *the "Red kangaroo paw" in sandy areas of the Stirling range, Western Australia*

AMARYLLIDACEAE

Alstroemeria pelegrina
"Lily of Lima"

Alstroemeria caryophyllaea
"Peruvian lily"

Alstroemeria aurantiaca
"Lily of the Incas"

Bomarea carderi
(*Colombia*)

Brodiaea coccinea
"Firecracker flower"

Chlidanthus fragrans
"Perfumed fairy lily"

Bravoa geminiflora
"Twin flower"

Crinum asiaticum
"Asiatic poison bulb"

Calostemma purpureum
(*Southeastern Australia*)

AMARYLLIDACEAE

Doryanthes excelsa, *"Globe spear-lily", or "Gymea lily"*
a bold succulent from New South Wales and Queensland

Furcraea foetida 'Medio-picta', *the "Mauritius hemp",*
with colorful foliage, in Calcutta, Bengal

Lycoris radiata, *"Red spider lily" from China and Japan*
popularly known as Nerine sarniensis; in Palermo, Sicily

Alstroemeria pelegrina, *"Lily of the Incas"*
with beautiful azalea-like flowers, from Chile

Hymenocallis 'Sulphur Queen', *a showy "Basket flower";*
very fragrant summer blooming bulbous plant

Bomarea multiflora, *interesting twining plant*
from the warmer Andean valleys of Colombia

AMARYLLIDACEAE

Crinum moorei album
in South Africa

Crinum augustum
"Queen Emma lily"

Crinum moorei
"Longneck swamp lily"

Lycoris squamigera
the "Magic lily"

Crinum yemense
in Kew Gardens, London

Crinum x powellii
'Roseum'

Crinum x powellii
"Powell's swamp lily"

Crinum x powellii album
at Kew Gardens, England

Crinum procerum var. splendens
in Honolulu

AMARYLLIDACEAE

Haemanthus katherinae, *the "Blood lily" from So. Africa with white bulbs stained as if with blood, near Mombasa*

Crinum x powellii 'Roseum', *with evergreen foliage at National Flower Show, Hamburg, Germany*

Crinum amabile, *a "Spider lily" from Sumatra with very fragrant, showy flowers*

Crinum asiaticum, *the "Poison bulb" in Melanesia with giant bulb, used medicinally in Asia*

AMARYLLIDACEAE

Clivia nobilis
"Greentip Kafir lily"

Clivia miniata
the "Kafir lily"

Stenomesson variegatum
(*Ecuador to Chile*)

Eucharis korsakoffii
"Miniature Amazon lily"

Sprekelia formosissima
"Jacobean lily"

Eucharis grandiflora
the "Amazon lily"

Urceolina peruviana
"Urn flower" (*Perú*)

Polianthes tuberosa
the fragrant "Tuberose"

Vallota speciosa
"Scarborough lily"

AMARYLLIDACEAE

Hippeastrum procerum (Worsleya rayneri)
the "Blue amaryllis" from Brazil

Hippeastrum vittatum hybrids
Foster Botanic Garden, Honolulu

Hippeastrum reticulatum striatifolium
blooming in February in Singapore

Amaryllis belladonna (Brunsvigia rosea), *"Belladonna"
or "Naked-lady lily" flowering after leaves disappear*

AMARYLLIDACEAE

Hippeastrum 'Vittatum hybrid'
commercial "Amaryllis"

Nerine curvifolia fothergillii
"Curveleaf Guernsey lily"

Hippeastrum vittatum
'Giant White'

Hippeastrum puniceum
"American belladonna"

Eucharis grandiflora
"Eucharist lily"

Hippeastrum reginae
"Royal amaryllis"

Narcissus odorus
'Orange Queen'

Hymenocallis harrisiana
"Mexican Spider-lily"

Narcissus cyclamineus
'February Gold'

Haemanthus multiflorus
"Salmon blood lily"

Petronymphe decora
(*Mexico*)

Haemanthus albiflos
"White paintbrush"

AMARYLLIDACEAE

Urceolina peruviana
"Urn flower", in Bolivia

Hippeastrum leopoldii hybrid
a Dutch "Amaryllis"

Eucharis grandiflora
at Royal Palace, Tonga

Habranthus tubispathus
(Argentina)

Vallota speciosa
"Scarborough lily"

Habranthus bagnoldii
Kew Gardens, London

Manfreda maculosa (*inflor.*)
(Mexico)

Cyrtanthus herrei
a "Fire lily"

Hippeastrum vittatum
'Equestre Red'

AMARYLLIDACEAE

Polianthes tuberosa; *garlands of "Tuberoses" at Buddhist shrine, Bangkok, Thailand*

Polianthes tuberosa, *fragrant "Tuberoses" offered in prayer at Schwe Dagon Pagoda, Rangoon, Burma*

Narcissus 'Gustav Mahler' *in early spring at Longwood Gardens, Kennett Square, Pennsylvania*

Narcissus 'Gold Medal', *a daffodil grown in pots for Easter, at Roehrs greenhouses, New Jersey*

Crinum japonicum 'Variegatum', *with variegated foliage at Kosobe Experiment Station, Osaka University, Japan*

Narcissus poeticus 'Actaea', *from southern Europe, a "Poet's narcissus" or "Pheasant's eye" in spring bloom*

AMARYLLIDACEAE

Haemanthus multiflorus
"Blood lily"

Haemanthus natalensis
"Natal paintbrush"

Haemanthus puniceus
"Pink paintbrush"

Nerine krigei
at Kirstenbosch, South Africa

Nerine bowdenii
at Encinitas, California

Nerine sarniensis hybrids
"Guernsey lilies"

Nerine bowdenii
"Spider lily"

Amaryllis belladonna 'Alba'
Sydney Botanic Garden

Nerine masonorum
at Hamburg Show, Germany

ALISMATACEAE, AMARYLLIDACEAE

Hymenocallis speciosa
the "Winter spice" (West Indies)

Hymenocallis littoralis
in Nuku Hiva, Marquesas

Hymenocallis narcissiflora
"Peruvian daffodil"

Zephyranthes citrina
a "Rain lily"

Hymenocallis caribaea
"Caribbean spider-lily"

Hymenocallis longipetala
"Crown beauty" of Perú

Leucojum vernum
"Spring snowflake", in Germany

Galanthus nivalis
"European snowdrop"

Sagittaria latifolia
an aquatic "Arrowhead"

ANACARDIACEAE

Anacardium occidentale
the "Cashew-nut"

Mangifera indica
"Mango" in flower (Tahiti)

Mangifera indica
"Mango" in fruit (Marquesas)

Pistacia atlantica
"Mt. Atlas mastic tree"

Mangifera indica
"Mango" in Encinitas, California

Cotinus coggygria
"Smoke tree"

Spondias venulosa
"Spanish plum" (Rio de Janeiro)

Spondias dulcis
"Ambarella" (Oahu)

Pistacia lentiscus
"Mastic tree" on Capri

ANACARDIACEAE, ANNONACEAE

Cananga odorata
"Ylang-Ylang" (Tonga)

Schinus molle
on St. Héléna

Shinus terebinthifolius
"Brazilian pepper tree"

Monodora tenuifolia
"African nutmeg"

Polyalthia longifolia pendula
"Ashoka tree"

Monodora myristica
in Jardim Botanico, Rio

Annona cherimola
"Cherimoya"

Annona muricata
"Sour sop"

Annona squamosa
"Sugar apple"

ANACARDIACEAE, APOCYNACEAE

Adenium obesum multiflorum, *the succulent "Impala lily" or "Desert rose", in Krueger National Park, Transvaal*

Pistacia vera, *the "Pistachio-nut tree" from Syria; female tree bearing the edible nuts, in Florida*

Allamanda cathartica, *the "Golden trumpet-vine" as a hedge literally covered with flowers, in Honolulu*

Beaumontia grandiflora, *the "Easter-lily vine" in the rainforest belt on Mt. Kilimanjaro, Tanzania*

Mandevilla sanderi, *"Rose dipladenia" from Brazil tropical climber with exquisite, showy trumpet flowers*

Tabernaemontana (Ervatamia) corymbosa, *from India the "Flower of love", fragrant at night, in Singapore*

APOCYNACEAE

Cerbera odollan
Pacific Islands

Acokanthera spectabilis
"Wintersweet"

Alstonia scholaris
"Devil tree"

Cerbera manghas
(*Polynesia to India*)

Carissa grandiflora
"Natal plum"

Carissa humphreyi 'Variegata'
in Florida

Vinca major 'Variegata'
"Band plant"

Catharanthus roseus 'Bright Eyes'
(*Vinca rosea*)

Catharanthus roseus
"Madagascar periwinkle"

APOCYNACEAE

Adenium obesum multiflorum
(*Transvaal to Kenya*)

Adenium obesum
"Impala lily"

Adenium swazicum
"White Impala" (*Mozambique*)

Pachypodium lamerei
"Club foot"

Allamanda violacea
"Purple allamanda"

Vinca minor, *"Periwinkle"*
in Jardin Botanico, Mexico City

Allamanda cathartica
"Golden trumpet" (*Tahiti*)

Allamanda cath. 'Hendersonii'
"Golden trumpet" (*Curacao*)

Allamanda neriifolia
"Oleander allamanda"

APOCYNACEAE

Pagiantha dichotoma
"Forbidden fruit of India"

Tabernaemontana crassa
in Peradeniya, Ceylon

Tabernaemontana crassa
"Adam's apple"

Tabernaemontana divaricata
(*Ervatamia coronaria*)

Tabernaemontana coronaria plena
"Fleur d'amour"

Tabernaemontana coronaria
"Butterfly gardenia"

Kopsia flavida (Ochrosia)
in Peradeniya, Ceylon

Tabernaemontana corymbosa
"Pinwheel flower" in Borneo

Holarrhena antidysenterica
"Easter tree" (*Trop. Asia*)

APOCYNACEAE

Mandevilla sanderi
"Rose dipladenia"

Mandevilla x amabilis
'Alice Dupont'

Mandevilla boliviensis
"White dipladenia"

Mandevilla suaveolens
"Chilean jasmine"

Nerium oleander
'Variegatum plenum'

Nerium oleander 'Roseum'
in Cologne, Germany

Nerium oleander 'Variegatum'
in Auckland, New Zealand

Nerium oleander
"Common oleander"

Nerium oleander 'Album'
"Sister Agnes oleander"

APOCYNACEAE

Mandevilla splendens
(*S.E. Brazil*)

Mandevilla sanderi 'Rosea'
"Brazilian jasmine"

Monodora grandiflora
"African orchid nutmeg"

Nerium oleander
'Carneum florepleno'

Echites rubro-venosa
(*Rio Negro, Amazonas*)

Nerium oleander
'Variegatum'

Plumeria rubra
"Frangipani tree"

Trachelospermum jasminoides
'Variegatum'

Holarrhena pubescens
(*Trop. Africa*)

Ochrosia elliptica
in fruit (*Queensland*)

Thenardia floribunda
"Petatillo" (*Mexico*)

Thevetia peruviana
"Yellow oleander"

APOCYNACEAE

Nerium oleander 'Album, the "White oleander" in romantic Cypress Gardens, Winter Haven, Florida

Nerium oleander 'Mrs, Roeding', *double-flowered oleander, in stone jar, in summer-bloom outdoors, Maplewood, N.J.*

Pachypodium lamerei, *a spindle-shaped succulent in the dry Fort Dauphin region, Amboasary, Madagascar*

Plumeria rubra; *fragrant "Frangipani" blossoms strung into leis by Polynesian dancers, in Tahiti*

APOCYNACEAE, ASCLEPIADACEAE

Holarrhena antidysenterica
in Mahé, Seychelles

Kopsia fruticosa
"Shrub vinca"

Trachelospermum jasminoides
"Star jasmine"

Rauwolfia verticillata
(China)

Thevetia thevetioides
(Mexico)

Thevetia peruviana
"Be-still tree"

Cryptostegia grandiflora (ASCLEP.)
"Purple allamanda"

Beaumontia grandiflora
"Easter-lily vine"

Strophanthus speciosus
"Corkscrew flower"

APOCYNACEAE

Plumeria rubra acutifolia
Polynesians garlanded with frangipani, Moorea

Plumeria obtusa, *from the West Indies as "Temple tree" in Honolulu*

Plumeria rubra acutifolia
"Frangipani" in Belém, Brazil

Plumeria rubra, *"Flor de mayo" from Mexico to Panama*

Plumeria 'Singapore hybrid'
"Pink plumeria"

Plumeria 'Rubra hybrid'
Diamond Head, Hawaii

APOCYNACEAE, APONOGETONACEAE

Pachypodium lamerei var. lamerei
a "Club foot" (So. Africa)

Pachypodium baroni var. Windsori
from Madagascar

Stemmadenia galleottiana
"Lecheso" in Central America

Beaumontia grandiflora
in Madras, South India

Thevetia peruviana
on Viti Levu, Fiji

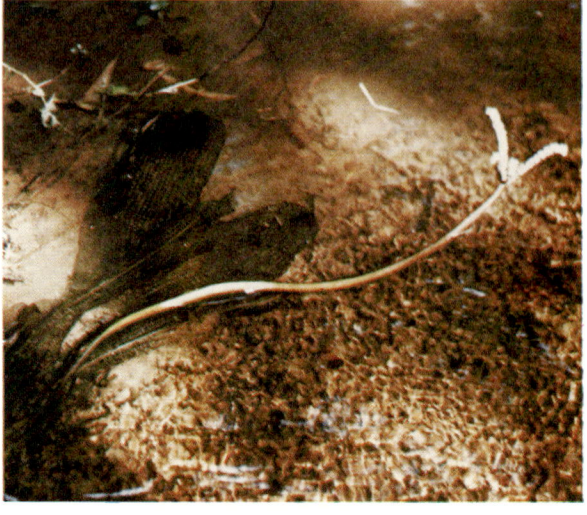

Aponogeton madagascariensis
"Laceleaf" in running stream (Madagascar)

APOCYNACEAE

Plumeria rubra acutifolia
with Rosaline, on Rarotonga

Plumeria rubra
"Frangipani" in Fiji

Plumeria pudica
in Curacao

Strophanthus gratus
"Climbing oleander"

Plumeria rubra acutifolia
"West Indian jasmine"

Plumeria rubra 'Tricolor'
in Caracas, Venezuela

Stemmadenia galleottiana
at Hilo Lagoon

Pachypodium namaquanum
"Ghostman" (Namibia)

Pachypodium lamerei
"Elephant's foot"

AQUIFOLIACEAE

Ilex rotunda
"Kurogane holly"

Ilex aquifolium 'Marginata'
"Christmas holly"

Ilex wilsonii
(China)

Ilex aquifolium
"English holly"

Ilex verticillata
"Winterberry"

Ilex cornuta rotundifolia
"Dwarf Chinese holly"

Ilex cornuta 'Dazzler'
in San Diego

Ilex crenata
"Japanese holly"

Ilex opaca
"American holly" (So. Carolina)

ARACEAE

Aglaonema 'Pseudo-bracteatum' at Roehrs greenhouses, Rutherford, New Jersey

Aglaonema 'Pseudo-bracteatum', *known as "White Rajah", handsome "Golden evergreen", and a good houseplant*

Alocasia sanderiana, *the exotic but delicate "Kris plant" enjoying the balmy climate of Sri Lanka*

Alocasia lowii grandis, *a tropical beauty from Borneo, growing from tubers, requiring warmth and high humidity*

Colocasia gigantea (indica), *an "Elephant's ear plant" from tropical Java and the Malay Peninsula*

Alocasia macrorhiza, *the "Giant alocasia" of Malaysia, at wood-carvers factory in Bien Hoa, Vietnam*

ARACEAE

Aglaonema modestum 'Variegatum'
at Pennock's, Puerto Rico

Aglaonema rotundum
"Red aglaonema"

Aglaonema costatum 'Foxii'
in Jogyakarta, Java

Aglaonema treubii
"Ribbon aglaonema"

Aglaonema 'Pseudobracteatum'
in bloom

Aglaonema pictum 'Tricolor'
(Malaya)

Aglaonema simplex
in Bogor, Java

Aglaonema commutatum
'Tricolor'

Aglaonema commutatum elegans
(marantifolium macul. hort.)

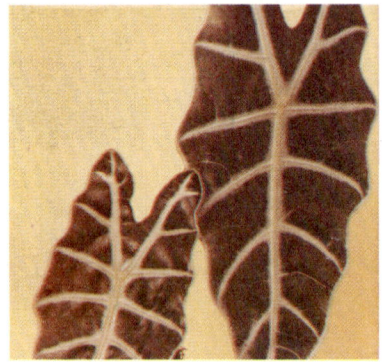
Alocasia x amazonica
(sanderiana x lowii grandis)

Aglaonema siamense
(Thailand)

Aglaonema costatum
var. immaculatum

ARACEAE

Aglaonema commutatum
(*Borneo*)

Aglaonema crispum
(*Schismatoglottis roebelinii*)

Aglaonema modestum
"Chinese evergreen"

Aglaonema 'Silver Queen'
(*Florida hybrid*)

Aglaonema marantifolium
in Borneo

Aglaonema modestum
'Medio-pictum'

Aglaonema; *top:* simplex
bottom: modestum

Aglaonema 'Pseudo-bracteatum'
"Golden evergreen"

Aglaonema 'Silver King'
(*curtisii hybrid*)

ARACEAE

Alocasia sanderiana
"Kris plant"

Alocasia cuprea
"Giant caladium"

Alocasia odora
in Taipo, Hongkong

Alocasia macrorhiza 'Variegata'
in Jamaica

Alocasia x mortefontanensis
at Kew Gardens

Alocasia watsoniana
(*Sumatra*)

Acorus gramineus 'Variegatus'
"Miniature sweet flag"

Acorus calamus variegatus
"Myrtle flag"

Alocasia cucullata
in Taipei, Taiwan

ARACEAE

Alocasia macrorhiza, *the "Giant elephant's ear",
near Dili, Timor in the Sunda Islands, Indonesia*

Alocasia odora *from the Philippines and Taiwan,
growing into tall tree, in Foster Gardens, Honolulu*

Anthurium andraeanum 'Rhodochlorum',
developing huge showy, long-lasting bracts; in Hawaii

Anthurium veitchii, *the "King anthurium",
with magnificent satiny, pendant leaves to 1 m long*

ARACEAE

Amorphophallus bulbifer
a "Devil's tongue"

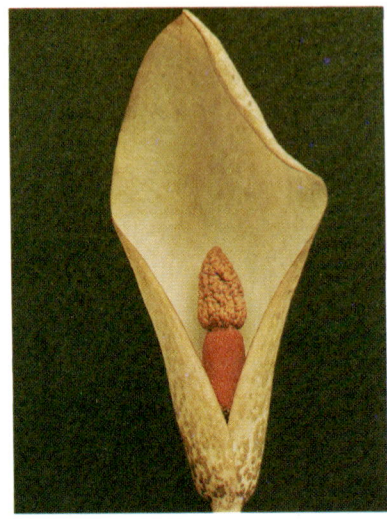

Amorphophallus corrugatus
"Snake palm"

Amorphophallus longituberosus
(Thailand)

Amorphophallus campanulatus
(East Indies)

Amorphophallus longituberosus
Munich Botanic Garden

Amorphophallus hildebrandtii
"Dragon lily"

Anchomanes difformis
(Tropical Africa)

Anchomanes welwitschii
(Trop. W. Africa)

Acorus calamus
"Sweet flag"

ARACEAE

Amorphophallus corrugatus, *known as "Snake palm" found in Thailand by J. Bogner, Munich Botanic Garden*

Amorphophallus campanulatus, *a "Devil's tongue" with gigantic spadix, collected near Lae, New Guinea*

Hydrosme rivieri (Amorphophallus) *the "Voodoo plant", its sinister inflorescence appearing before foliage*

Hydrosme rivieri, *also known as "Leopard palm" with leaf stalk spotted like a leopard; at Roehrs, N.J.*

ARACEAE

Anthurium andraeanum
'Rubrum'

Anthurium andraeanum
'Album'

Anthurium scherzerianum
"Flamingo flower"

Anthurium gladifolium
(Venezuela)

Anthurium scherzerianum
'Rothschildianum'

Anthurium bakeri
(Costa Rica)

Anthurium clarinervium
"Hoja de Corazon"

Anthurium pedatoradiatum
(Mexico)

Anthurium papilionensis
(Colombia)

Philodendron verrucosum
(inflorescence)

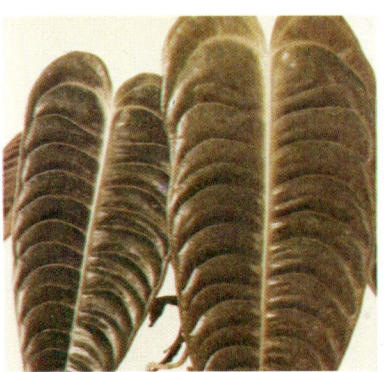
Anthurium veitchii
"King anthurium"

Anthurium warocqueanum
"Queen anthurium"

ARACEAE

Anthurium x ferrierense
"Oilcloth-flower"

Anthurium scherzerianum
(*Costa Rica*)

Anthurium andraeanum
"Tailflower"

Anthurium andraeanum
'Rhodochlorum'

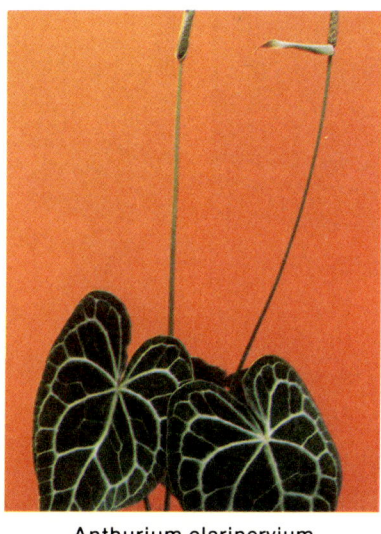
Anthurium clarinervium
(*Mexico*)

Anthurium andraeanum
'Guatemala'

Anthurium magnificum
(*Colombia*)

Anthurium hookeri
"Birdsnest anthurium"

Anthurium coriaceum
in Brazil

ARACEAE

Anthurium andraeanum 'Giganteum', *in Hawaiian cemetary lovingly decorated with exotic blossoms.*

Anthurium andraeanum, *at first imported from Colombia, greenhouse grown for a century at Roehrs, N.J.*

Anthurium scherzerianum, *the "Flamingo flower" from highland Costa Rica, an area of cool nights*

Anthurium andraeanum 'Rodochlorum', *Hawaii giants, with curiously beautiful spathes to 20 cm or more long*

Anthurium leuconeurum, *from tropical southern Mexico with handsome velvety, silver-veined foliage*

Anthurium holtonianum, *from Panama and Colombia, with fantastic, compound foliage, at Longwood Gardens*

ARACEAE

Anthurium panduratum
(*Northern Brazil*)

Anthurium warocqueanum
(*Colombia*)

Anthurium veitchii
(*Colombia*)

Anthurium hookeri
(*inflor.*) (*Guyana*)

Arum dioscoridis
(*Eastern Mediterranean*)

Anthurium corrugatum
(*Colombia*)

Arum maculatum
"Lords and ladies"

Arum italicum
"Italian arum"

Arum italicum
at Kew Gardens, London

ARACEAE

Arum palaestinum (sanctum), *"Solomon's lily" or "Purple lily of Bethlehem", in Israel*

Arisarum vulgare, *from southern Europe and Algeria subtropic tuberous plant, with curious hooded spathe*

Caladium humboldtii (argyrites) *from Pará, Brazil tropical "Miniature caladium" with diminutive, dainty leaves*

Caladium bicolor, *parent of our fancy-leaved caladiums, growing in alluvial clay of the Guyana rainforest*

Homalomena pygmaea var. purpurescens, (*Philippines*), *small clustering plant with purple spathe*

Homalomena wendlandii, *from Costa Rica cross-roads of North and South American tropical flora*

ARACEAE

Arisaema candidissimum
(*Japan, W. China*)

Arisaema fargesii
(*China: Szechwan*)

Arisaema sikokianum
(*So. Japan*)

Arisarum vulgare
(*No. Africa*)

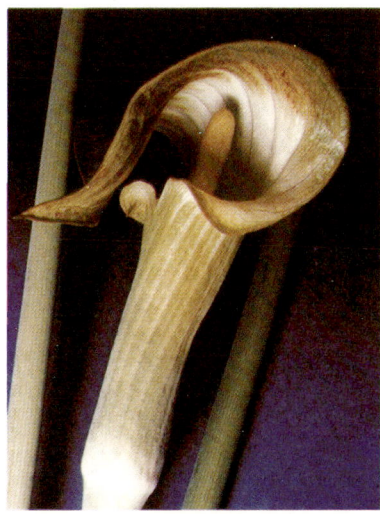
Arisaema triphyllum
"*Jack-in-the-pulpit*"

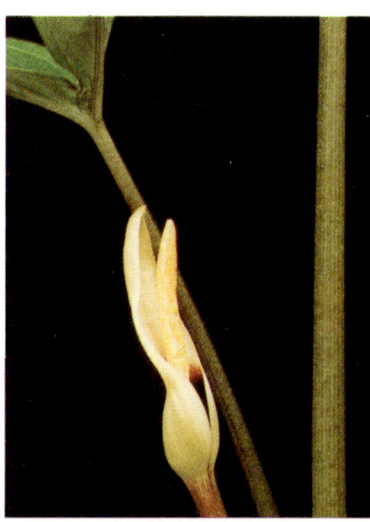
Arophyton tripartitum
(*Madagascar*)

Caladium hortulanum 'Rio'
"*Fancy-leaved caladium*"

Arophyton crassifolium
(*Madagascar*)

Culcasia rotundifolia
(*West Africa*)

ARACEAE

Caladium hort. 'Cleo'
"Texas Wonder"

Caladium hortulanum
'John Peed'

Caladium hort. 'Candidum'
"White caladium"

Caladium hort. 'Lord Derby'
"Transparent caladium"

Caladium hortulanum
'Poecile Anglais'

Caladium hortulanum
'Gen. W. B. Halderman'

Caladium hortulanum
'Keystone'

Caladium hortulanum
'Porto Novo'

Caladium hortulanum
'Bleeding Heart'

Caladium hortulanum
'Pink Festivia'

Caladium humboldtii (argyrites)
"Miniature caladium"

Caladium hortulanum
'White Princess'

ARACEAE

Caladium hortulanum
'Ace of Hearts'

Caladium hortulanum
'Candidum'

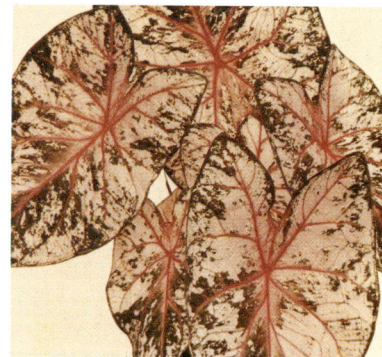
Caladium hortulanum
'Attala'

Caladium hortulanum
'A. B. Graf'

Caladium hort. 'Citation'
(*lanceleaf*)

Caladium hortulanum
'Debutante'

Caladium hortulanum
'Edith Meade'

Caladium hortulanum
'Elizabeth Dixon'

Caladium hortulanum
'E. O. Orpet'

Caladium hortulanum
'Hortulania'

Caladium hortulanum
'Jody'

Caladium hortulanum
'Itacapus'

ARACEAE

Caladium hortulanum 'Postman Joyner'

Caladium hortulanum 'White Queen'

Caladium hortulanum 'Pothos'

Caladium hortulanum 'Red Flare'

Caladium hortulanum 'Red Frill'

Caladium hortulanum 'Roehrs Dawn'

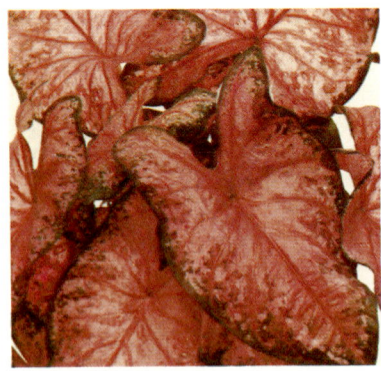
Caladium hortulanum 'Scarlet Beauty'

Caladium hortulanum 'Scarlet Pimpernel'

Caladium hortulanum 'Spangled Banner'

Caladium hortulanum 'The Thing'

Caladium hortulanum 'Thos. Tomlinson'

Caladium hortulanum 'Triomphe de l'Exposition'

ARACEAE

Caladium hortulanum
'Jessie Thayer'

Caladium hortulanum
'Frieda Hemple'

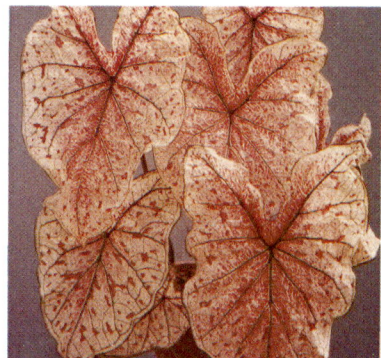
Caladium hortulanum
'Lady Chris'

Caladium hortulanum
'L. L. Holmes'

Caladium hortulanum
'Luella Whorton'

Caladium hortulanum
'Macahyba'

Caladium hortulanum
'Maid of Orleans'

Caladium hortulanum
'Marie Moir'

Caladium hortulanum
'Miss Muffet'

Caladium hortulanum
'Mrs. Arno Nehrling'

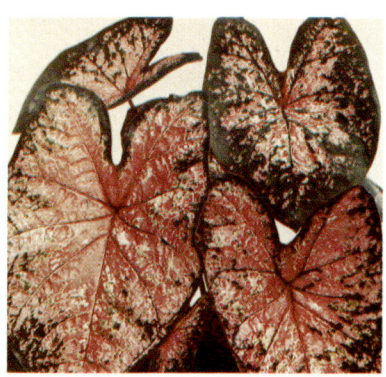
Caladium hortulanum
'Mrs. W. B. Halderman'

Caladium hortulanum
'Pink Blush'

ARACEAE

Cyrtosperma senegalense
in Sénégal

Colletogyne perrieri
in Madagascar

Asterostigma vermicidum
spathe partly removed

Carlephyton glaucophyllum
(*Madagascar*)

Gonatopus boivinii
(*Zanzibar*)

Dracunculus vulgaris
(*Mediterranean*)

Gonatanthus pumilus
(*Sikkim*)

Nephthytis afzelii
(*Liberia, Sierra Leone*)

Calla palustris
"Water arum"

ARACEAE

Colocasia esculenta
"Dasheen"

Dracontium asperum
(*Venezuela: Orinoco*)

Colocasia fallax
(*Assam, Sikkim*)

Orontium aquaticum
"Golden club"

Colocasia antiquorum
"Egyptian taro"

Colocasia antiquorum illustris
"Black caladium"

Cryptocoryne petchii
(*Sri Lanka: Ratnapura*)

Cryptocoryne affinis
(*Malay Peninsula*)

Cryptocoryne willisii
"Water trumpet"

ARACEAE

Cryptocoryne wendtii, *from Thailand tropical aquarium plant for underwater and above*

Cryptocoryne ciliata, *"Fragrant tape grass", aquatic herb usually growing under water, from India*

Epipremnopsis media (Epipremnum medium) *wiry tree-climber, from aequatorial Borneo*

Gonatopus rhizomatosus, *from Natal to Mozambique; with constricted spathe, by J. Bogner, Bot. Garden Munich*

Lagenandra ovata (Cryptocoryne) *from Western India tropical waterside plant, the spathe with black scales*

Lagenandra koenigii, *from Singaraja in Ceylon tropical aquatic plant of fresh-water rivers*

ARACEAE

Epipremnum pinnatum
the Eastern hemisphere "Philodendron", on Timor

Epipremnum elegans
branch with inflorescence, in New Guinea

Dieffenbachia reginae
at Burle-Marx facenda near Rio

Colocasia esculenta
"Taro" growing in shallow water in Tonga

ARACEAE

Dieffenbachia amoena 'Tropic Snow'

Dieffenbachia 'Exotica Perfection'

Dieffenbachia reginae *collected near Rio*

Dieffenbachia maculata (picta) 'Roehrs Superba'

Dieffenbachia maculata (picta) 'Rudolf Roehrs'

Dieffenbachia seguina *in Amazonas, Brazil*

Homalomena wallisii *"Silver shield"*

Homalomena sulcata (*Borneo*)

Lasia spinosa (*Malaysia*)

ARACEAE

Dieffenbachia x bausei (maculata x weirii)
warmth-loving hybrid with beautifully painted leaves

Dieffenbachia maculata (picta), *"Spotted dumbcane"*
growing along the broad Amazon, near Manaos, Brazil

Dieffenbachia 'Sao Antonio', *in Pará, Brazil*
handsome, showy plant with white leaf-stalks

Dieffenbachia maculata (picta) 'Superba' (*Roehrs 1950*)
compact mutant with foliage highly variegated

Dieffenbachia 'Wilson's Delight', *bold plant,*
leaves with white midrib; Fantastic Gardens, Miami

Dieffenbachia amoena 'Tropic Snow' *or* 'Tropical Topaz'
excellent Florida (Chaplin) mutation of compact habit

ARACEAE

Dieffenbachia x bausei
(*maculata x weirii*)

Dieffenbachia 'Exotica'
(*Costa Rica*)

Dieffenbachia amoena
(*Colombia, Costa Rica*)

Dieffenbachia fournieri
(*Colombia*)

Dieffenbachia maculata jenmanii
(*Guyana*)

Dieffenbachia x splendens
(*leopoldii x maculata*)

Dieffenbachia wallisii
(*No. S. America*)

Homalomena picturata
(*Colombia*)

Homalomena wallisii
(*Colombia*)

Rhektophyllum mirabile
(*Nigeria*)

Cyrtosperma johnstonii
(*Solomon Islands*)

Schismatoglottis novo-guineensis
(*New Guinea*)

ARACEAE

Epipremnum pinnatum, *"Taro vine", in forests of Java Eastern hemisphere parallel to Western philodendron*

Gonatanthus pumilus (sarmentosus) *near Darjeeling, in the Himalayas of North Bengal*

Philodendron melinonii, *like a "Red birdsnest" growing epiphytic in interior Surinam*

Montrichardia arborescens, *tree-like aroid in coastal swamps along the Demerara River in Guayana*

ARACEAE

Monstera deliciosa
'Albo-variegata'

Monstera deliciosa
"Mexican breadfruit"

Monstera deliciosa
'Marmorata'

Raphidophora decursiva (*inflor.*)
(*Southeast Asia*)

Raphidophora decursiva
at Bot. Garden Rio

Monstera pertusa
(*Panama, Guyana*)

Peltandra virginica
"Arrow arum"

Pycnospatha soerensenii
(*Thailand*)

Pseudohydrosme gabunensis
(*Equatorial Africa*)

ARACEAE

Monstera epipremnoides, *from Costa Rica, (Longwood Gardens) tall climber with leaves pinnately parted*

Monstera friedrichsthalii, *with perforated foliage in jungle habitat at Turrialba, Costa Rica, C. America*

Monstera deliciosa 'Borsigiana', *small "Splitleaf" as decorator plant, in hostelry of Salamanca, Spain*

Monstera deliciosa, *the "Ceriman" or "Mexican breadfruit" with maturity-stage character foliage, in Puerto Rico*

ARACEAE

112

Monstera obliqua
(*Alto Amazonas*)

Monstera deliciosa minima
made of plastic

Monstera deliciosa
with inflorescence

Raphidophora decursiva
(*Vietnam*)

Philodendron colombianum
inflorescence

Philodendron x corsinianum
"Bronze shield"

Philodendron cruentum
"Red leaf"

Philodendron domesticum
(*hastatum*), *inflorescence*

Philodendron distantilobum
collected in Amazonas

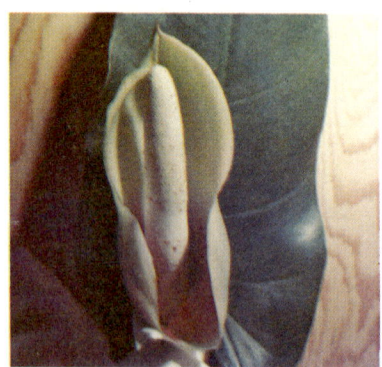
Philodendron
'giganteum x imbe' (*inflor.*)

Philodendron bipinnatifidum
(*Rio to Mato Grosso*)

Philodendron x evansii
in California

ARACEAE

Philodendron eichleri
"King of tree-philodendrons", in Rio

Philodendron giganteum
"Giant philodendron", in West Indies

Philodendron fragrans
(*Southern Brazil*)

Philodendron elegans
(*Trop. South America*)

Monstera obliqua expilata
(*leichtlinii*), "Window-leaf"

Philodendron selloum (*inflorescence*)
with Boa constrictor, Paraná, Brazil

ARACEAE

Philodendron 'Florida'
(*laciniatum hyb.*)

Philodendron domesticum
'Variegatum'

Philodendron
x 'Florida compacta'

Philodendron 'Goldiana'
(*imbe hybrid*)

Philodendron melanochrysum
"Black-gold"

Philodendron ilsemannii
(*Brazil*)

Philodendron karstenianum
"Mexican philodendron"

Philodendron linnaei
(*Surinam*)

Philodendron erubescens
"Blushing philodendron"

Philodendron x mandaianum
"Red-leaf philodendron"

Philodendron
'Emerald Queen'

Philodendron mamei
"Quilted silver-leaf"

ARACEAE

Philodendron andreanum (*juv.*)
"*Velour philodendron*"

Philo. scandens oxycardium
(*cordatum of florists*)

Philodendron andreanum (*mature*)
(*Colombia*)

Philodendron poeppigii
(*Alto Amazonas*)

Philodendron lacerum
(*Cuba, Haiti, Jamaica*)

Philodendron cannifolium
"*Flask-philodendron*"

Philodendron glaziovii
(*So. Brazil: Guanabara*)

Philodendron pinnatifidum
(*Venezuela, Amazonas*)

Philodendron x magnificum
(*selloum x eichleri*)

ARACEAE

Philo. 'Santa Leopoldina'
in Brazil

Philodendron erubescens
"Blushing philodendron"

Philodendron borsighiana hort.
Longwood Gardens, Penna.

Philodendron 'Majesty'
(*Bamboo Gardens*)

Philodendron
'Emerald Duke'

Philodendron
'Emerald King'

Philodendron
'Red Princess'

Philodendron
'Red Duchess'

Philodendron
'Royal Queen'

ARACEAE

Philodendron tweedianum
with Señora Camuyrano, Buenos Aires, Argentina

Philodendron 'Lynette'
"Quilted birdsnest" (Miami)

Philodendron longistilum
from Brazil (grown in California)

Philodendron 'Golden Erubescens'
by De Mello, São Paulo

Philodendron martianum
in São Paulo, Brazil

Philodendron warmingii
at Blossfeld's, São Paulo, Brazil

ARACEAE

Philodendron 'Mandaianum'
(*inflorescence*)

Philodendron panduriforme
"Fiddle-leaf"

Philodendron nobile
(*Venezuela, Guyana*)

Philodendron
'Red Emerald'

Philodendron pittieri
(*Costa Rica*)

Anthurium pedato-radiatum
from Mexico

Philodendron scandens
oxycardium 'Variegatum'

Philodendron sodiroi (*juv.*)
"Silver-leaf philodendron"

Philodendron variifolium
(*Perú*)

Philodendron
x 'Wend-imbe'

Philodendron wendlandii
"Birdsnest philodendron"

Philodendron mamei
"Quilted silver-leaf"

ARACEAE

Philodendron imbe 'Variegatum'
at Burle-Marx facenda, South of Rio de Janeiro

Philodendron imbe
in habitat, Pernambuco, Brazil

Philodendron imperialis
at Golden Gate Park, San Francisco, California

Philodendron lacerum
at Botanical Garden Durban, Natal

ARACEAE

Philodendron warscewiczii
in Honduras habitat, C. America

Philodentron undulatum,
arborescent species, in Paraguay

Philodendron selloum
"Lacy tree-philodendron", Paraná

Philodendron imbe
in Sao Paulo state, Brazil

Stenospermation sessile.
in the Cordilleras of Ecuador

Pothos scandens,
the true Pothos, in Sri Lanka

ARACEAE

Philodendron mello-barretoanum,
a tree philodendron in Acre, Western Brazil

Philodendron selloum
in habitat at Iguassu Falls, near Paraguay border

Philodendron pinnatifidum
(Venezuela, Amazonas)

Philodendron pseudoradiatum
in Cuernavaca, Mexico

Philodendron sagittifolium
near Cordoba, Mexico

Philodendron scandens oxycardium (cordatum)
mature form, in Costa Rica habitat

ARACEAE

122

Philodendron speciosum
"Imperial philodendron" (Brazil)

Philodendron sellowianum
in Cristobal, Panama Canal Zone

Philodendron speciosissimum
in Guanabara habitat, Brazil

Philodendron williamsii
in Rio Negro forest, Amazonas, Brazil

ARACEAE

Philodendron sodiroi
"Silver leaf philodendron"

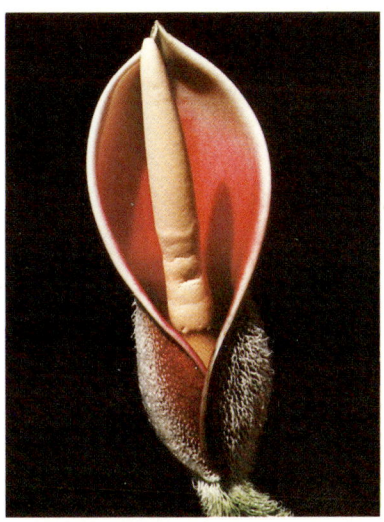
Philodendron verrucosum
(inflorescence)

Philodendron verrucosum
"Velvet-leaf"

Philodendron frits-wentii
(C. America)

Philodendron williamsii
(syn. 'Espiritu Santo')

Philodendron undulatum
(juv. stage), Minas Gerais

Philodendron lacerum
in Panama

Philodendron selloum
"Lacy tree-philodendron"

Philodendron bipinnatifidum
(C. Brazil)

ARACEAE

Scindapsus aureus (*mature*)
in the Marquesas, French Polynesia

Zantedeschia 'Striped hybrida'
at the Zapotec market, Oaxaca, So. Mexico

Xanthosoma sagittaefolium
giant "Yautia" in Costa Rica

Schizocasia portei
near Manila, Luzon, Philippines

ARACEAE

Scindapsus aureus (*juvenile*)
"Devil's ivy"

Scindapsus aureus
(*maturity stage*)

"Philodendron pertusum";
juvenile Monstera deliciosa, *in Holland*

Syngonium podophyllum
'Variegatum'

Syngonium podophyllum
'Imperial White', *in California*

Xanthosoma lindenii
'Magnificum'

Rhektophyllum mirabile
in habitat, Nigeria

Pothos scandens,
near Port Moresby, New Guinea

Pothos scandens,
showing winged petioles

ARACEAE

Spathiphyllum floribundum
"Spathe flower"

Spathiphyllum 'Clevelandii'
"White flag"

Spathiphyllum 'Marion Wagner'
in Hollywood, California

Spathiphyllum 'Mc Coy'
in Hawaii

Spathiphyllum wallisii
(Colombia, Venezuela)

Spathiphyllum blandum
(West Indies)

Spathiphyllum 'Mauna Loa'
in Los Angeles

Spathiphyllum cannaefolium
(Guyana)

Spathiphyllum commutatum
(Philippines)

ARACEAE

Typhonodorum lindleyanum
near Tamatave, Madagascar

Typhonodorum lindleyanum (giganteum),
Jardim Botanico, Rio de Janeiro

Xanthosoma jacquinii lineatum
"Malanga", in Southern Brazil

Syngonium macrophyllum
"Big-leaf syngonium" (Panama)

Pothos scandens
mature form in Thailand

Zamioculcas zamiifolia
with inflorescence, Munich Botanic Garden

ARACEAE

128

Callopsis volkensis
"Miniature calla"

Spathicarpa sagittifolia
"Fruit-sheath plant"

Spathiphyllum phryniifolium
"Peace plant" (*Costa Rica*)

Spathiphyllum 'Clevelandii'
"White anthurium"

Spathiphyllum ortgiesii
(*Mexico: Vera Cruz*)

Spathiphyllum 'Mc Coy'
(*Hawaii*)

Scindapsus aureus
'Marble Queen'

Scindapsus pictus argyraeus
"Satin pothos"

Scindapsus aureus 'Wilcoxii'
"Golden pothos"

Zantedeschia elliottiana
"Yellow calla"

Zantedeschia aethiopica
'Childsiana' (*dwarf*)

Zantedeschia rehmannii
"Pink calla"

ARACEAE

Zantedeschia aethiopica
"White calla" (*Transvaal*)

Sauromatum venosum
(*East Indies*)

Zantedeschia aethiopica
'Green Lily'

Typhonodorum lindleyanum
in Palmengarten, Frankfurt

Taccarum weddellianum
(*Paraguay*)

Schizocasia lauterbachiana
(*New Guinea*)

Synandriospadix vermitoxicus
(*Northern Argentina*)

Pistia stratiotes
"Water lettuce"

Typhonium trilobatum
(*Malaya*)

ARACEAE

Xanthosoma lindenii
"Yautia" (*Colombia*)

Xanthosoma lindenii
'Magnificum'

Syngonium xanthophilum
'Green Gold'

Syngonium podophyllum
'Albo-lineatum'

Syngonium erythrophyllum
"Copper syngonium"

Syngonium podophyllum
'Albo-virens'

Syngonium podophyllum
'Emerald Gem variegated'

Syngonium podophyllum
'Trileaf Wonder'

Syngonium wendlandii
(*Costa Rica*)

Zomicarpa riedeliana
(*Brazil*)

Nephthytis gravenreuthii
(*Cameroon*)

Hapaline brownii
(*Malaysia*)

ARACEAE

Spathiphyllum cannaefolium
a "Spathe flower", in Guyana

Zantedeschia aethiopica
growing in a tropical pond, Bogor, Java

Scindapsus sp. 'Exotica',
undetermined species which I found in Vietnam

Scindapsus aureus
the florists "Golden pothos"

Zantedeschia aethiopica;
wild "Pig lilies" near the Cape, South Africa

Zantedeschia elliottiana
Roehrs greenhouse of "Yellow callas" for Easter

ARALIACEAE

Boerlagiodendron
novo-guineense (*Papua*)

Brassaia actinophylla
flowering in Queensland

Fatsia japonica
in bloom, Los Angeles Arboretum

Aralia chinensis
in South China

Meryta sinclairi
on the Norwegian 'Sagafjord'

Cussonia paniculata
in South Africa

Tupidanthus calyptratus
facing the River Nile, Cairo, Egypt

Dizygotheca elegantissima
"False aralia" (*juvenile*)

Dizygotheca 'Castor',
broadleaved cultivar

ARALIACEAE

Cussonia paniculata,
in the 1000 hills of Zululand, Natal

Schefflera volkensii,
from the cloud forest, Mt. Kilimanjaro, Tanzania

Schefflera delavayii,
a tree aralia in Vietnam habitat

Boerlagiodendron eminens,
in the Finisterre Mountains of N.E. New Guinea

Fatsia japonica (Aralia),
in the suburbs of Tokyo, Japan

Neopanax arboreus,
in habitat at Wairakei, New Zealand

ARALIACEAE

134

Cussonia holstii, *from Kenya, East Africa ornamental tree with compound foliage, in Los Angeles*

Cussonia spicata, *the "Spiked cabbage tree" evergreen tree from Natal and the Transvaal mountains*

Cussonia paniculata, *the "Cabbage tree" from South Africa with grayish foliage, growing in tuft at top of trunk*

Dendropanax trifidus (Gilibertia japonica), *from Japan small decorative evergreen tree, with black berries*

ARALIACEAE

Hedera colchica, *"Persian ivy" from the Caucasus growing on the church of Santa Lucia, Santiago, Chile*

Hedera helix, *the "English ivy" old arborescent form at Neanderthal, Germany*

Meryta sinclairii
in habitat, Taranga Island, New Zealand

Hedera canariensis arborescens 'Variegata'
the "Ghost-tree" ivy, at Manhattan Beach, California

ARALIACEAE

136

Hedera canariensis
'Gloire de Marengo'

Hedera canariensis
arborescens 'Variegata'

Hedera canariensis 'Variegata'
"Variegated Algerian ivy"

Hedera helix
'Glacier'

Hedera helix
'238th Street'

Hedera canariensis
"Algerian ivy"

Hedera helix 'Harald'
in Germany

Hedera helix
"English ivy"

Hedera helix
'Patricia'

ARALIACEAE

Hedera helix 'Minima'
trained into a living topiary animal

Brassaia actinophylla
at Sunken Gardens, Radio City, New York

Hedera helix
"English ivy" on Fifth Avenue, New York

Hedera helix baltica
"Baltic ivy" on Bornholm, Denmark

Tetrapanax papyriferus 'Variegata'
"Variegated Rice-paper plant"

Fatsia japonica 'Moseri'
a French "Aralia"

ARALIACEAE

Hedera colchica 'Dentato-variegata'

Hedera helix 'Abundance'

Hedera hel. 'Curlilocks' *grafted on Fatshedera*

Hedera helix 'Star' *"Star ivy"*

Hedera helix 'Ripples'

Hedera helix 'Harald'

Hedera helix 'Maculata'

Hedera helix 'Jubilee'

Hedera helix 'Williamsiana'

x Fatshedera lizei 'Variegata'

Hedera helix 'Manda's crested'

Hedera helix 'Hibernica variegata'

ARALIACEAE

Schefflera arboricola, *"Miniature umbrella plant" or "Hawaiian Elf", a scandent miniature from Taiwan*

Brassaia actinophylla, *the "Octopus tree" in flower growing epiphytic in tropical Sogeri, Papua*

Schefflera farinosa, *from Malaya and Indonesia, tropical tree with compound leaflets in formal circle*

Schefflera digitata, *in the chilly climate of Fiordland Province, Milford Sound, New Zealand*

ARALIACEAE

Oreopanax salvinii
(*Mexico*)

Pseudopanax 'Adiantifolius'
(*New Zealand*)

Oreopanax capitatus
'Nymphaeifolius'

Nothopanax filicifolia
in Singapore

Pseudopanax ferox
(*New Zealand*)

Meryta denhamii
(*New Caledonia*)

Neopanax arboreus
"*Five fingers*"

Nothopanax laetus
(*New Zealand*)

Oreopanax xalapensis
(*Mexico*)

ARALIACEAE

141

Pseudopanax crassifolius
in habitat, Fiordland Prov., New Zealand

Polyscias fruticosa
"Ming aralia" as a decorator plant

Polyscias filicifolia
"Fernleaf aralia" in Tonga, South Pacific

Polyscias balfouriana
"Dinner plate aralia" in New Caledonia

ARALIACEAE

Polyscias guilfoylei 'Quinquefolia'

Polyscias guilfoylei 'Laciniata'

Polyscias paniculata 'Variegata' *"Variegated roseleaf"*

Polyscias balfouriana 'Variegata'

Polyscias balfouriana 'Marginata'

Polyscias guilfoylei 'Victoriae' *"Lace aralia"*

Polyscias balfouriana 'Blackie' *in Florida*

Polyscias balfouriana 'Pennockii'

Polyscias scutellaria (*Nothopanax*) *in Hawaii*

ARALIACEAE

Tetrapanax papyriferus
"Rice paper plant" (*South China*)

Tetraplasandra meiandra
"Hawaiian Ohe tree", on Oahu

Tupidanthus calyptratus
in Royal Botanic Gardens, Sydney

Trevesia palmata 'Micholitzii'
"Snowflake plant" (*Yunnan, China*)

ARALIACEAE

144

Schefflera venulosa
(*Heptapleurum venulosum, in Sydney*)

Schefflera venulosa
var. erythrostachys (*Kew Gardens*)

Schefflera octophylla
in Hongkong habitat

Tupidanthus calyptratus
in San Diego, California

Schefflera venulosa
juvenile, in Los Angeles

Schefflera volkensii
in Tanzania

Schefflera arboricola
"*Hawaiian Elf*"

Brassaia actinophylla
"*Queensland umbrella tree*"

Schefflera digitata
in habitat, New Zealand

ARISTOLOCHIACEAE, ASCLEPIADACEAE

Aristolochia veraguensis
(*Panama*)

Cryptostegia grandiflora
"*Indian rubber vine*"

Cryptostegia madagascariensis
(*Madagascar*)

Aristolochia leuconeura
(*Colombia*)

Aristolochia elegans
"*Calico flower*"

Aristolochia fimbriata
(*Argentina*)

Stephanotis floribunda
"*Madagascar jasmine*"

Hoya 'New Guinea Red'
(*Finisterre Mts.*)

Hoya purpureo-fusca
(*'Silver Pink'*)

ASCLEPIADACEAE

Calotropis gigantea
"Crown plant" in Timor

Calotropis procera
"Felt plant" (Morocco)

Dischidia benghalense
(India)

Cryptostegia grandiflora alba
in Western Australia

Cryptostegia grandiflora
"Indian rubber vine", on Java

Asclepias physocarpa
"Butterfly flower"

Asclepias currasavica
"Bloodflower"

Asclepias fruticosa
in South Africa

ASCLEPIADACEAE

Caralluma praegracilis
in the Transvaal

Echidnopsis archeri
(Kenya)

Edithcolea grandis
(Somaliland)

Huernia macrocarpa cerasina
a "Dragon flower"

Caralluma europaea
(Mediterranean reg.)

Huernia zebrina
"Owl eyes" or "Zebra flower"

Huernia reticulata
(Cape Prov., So. Africa)

Huernia namaquensis
(South Africa)

Hoodia rosea
"African hat plant"

ASCLEPIADACEAE

Ceropegia aristolochioides
(*Sénégal to Ethiopia*)

Ceropegia armandii
(*S.W. Madagascar*)

Ceropegia ballyana
(*Kenya*)

Ceropegia haygarthii
"Wine-glass vine"

Ceropegia krainzii
(*Canary Islands*)

Ceropegia fusca
"Carnival lantern"

Ceropegia sandersonii
"Parachute plant"

Ceropegia serpentina
(*Marnier-Lapostolle, France*)

Ceropegia woodii
"Rosary vine" in flower

ASCLEPIADACEAE

Dischidia platyphylla
in Foster Gardens, Honolulu

Dischidia rafflesiana
"Malayan urn vine", in New Guinea habitat

Ceropegia woodii
"String of hearts", from Natal

Fockea capensis
in Royal Botanic Gardens, Kew

ASCLEPIADACEAE

Hoya carnosa 'Rubra'
"Krimson Princess"

Hoya 'New Guinea White'
(*Wantoat, New Guinea*)

Hoya bella
"Miniature wax plant"

Hoya carnosa
compacta 'Regalis'

Hoya carnosa
'Silver Princess'

Hoya carnosa 'Tricolor'
"Krimson Queen"

Hoya coronaria
(*Java*)

Hoya bandaensis
(*Moluccas*)

Hoya darwinii
(*Philippines*)

ASCLEPIADACEAE

Hoya gigas, *with giant flower collected in Finisterres, New Guinea*

Aristolochia grandiflora (gigas) *"Dutchman's pipe" in Sri Lanka*

Asclepias physocarpa *"Silkweed", in Kruger Park, Transvaal*

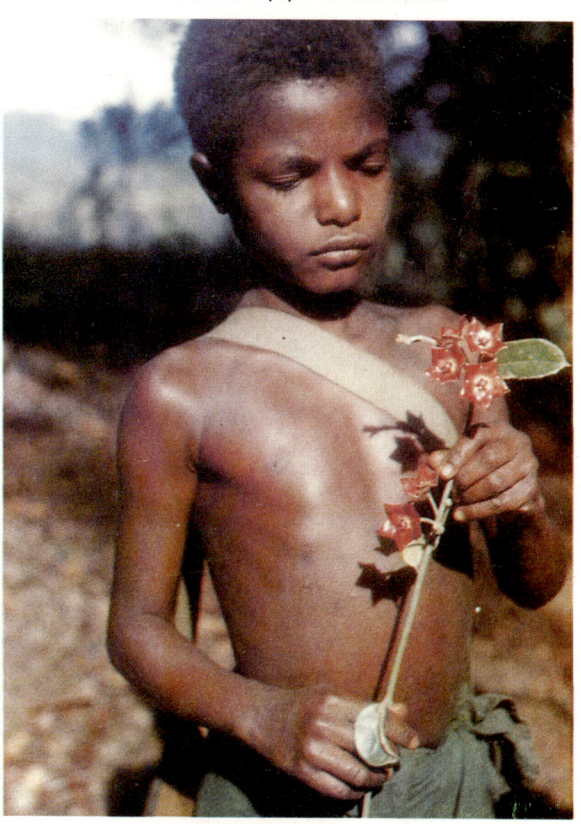

Hoya rubida *on New Guinea expedition, Finisterre Mountains*

ASCLEPIADACEAE

Hoya imperialis
"Honey plant", from Borneo

Hoya polyneura
(*Himalayan region*)

Hoya longifolia
(*Central Himalayas*)

Hoya carnosa 'Variegata'
"Variegated wax plant"

Hoya sikkimensis
from subtropic Himalayan valleys

Hoya globulosa
(*Sikkim Himalayas*)

ASCLEPIADACEAE

Hoya carnosa
"Wax plant"

Hoya bella
"Miniature wax plant"

Hoya australis
"Porcelain flower"

Hoya carnosa 'Marginata'
(*Queensland*)

Hoya carnosa
'Hummel's compacta'

Hoya carnosa 'Compacta'
"Hindu rope plant"

Hoya kerrii
"Sweetheart hoya"

Hoya carnosa 'Variegata'
"Variegated wax vine"

Hoya carnosa
'Verna Jeanette'

Hoya purpureo-fusca
"Silver-pink vine"

Hoya 'Silver Pink'
in Hawaii

Stephanotis floribunda
"Floradora"

ASCLEPIADACEAE

Stapelia gigantea, *red form*
"Zulu giants", in Vista, California

Hoodia gordonii
in Namib desert, Southwest Africa

Pseudolithos cubiforme (Lithocaulon)
from Somalia, East Africa

Hoodia lugardii
(*Botswana to Namibia*)

Caralluma mammillaris
in habitat, Great Karroo, So. Africa

Trichocaulon meloforme
in N.W. Cape Province desert

ASCLEPIADACEAE

155

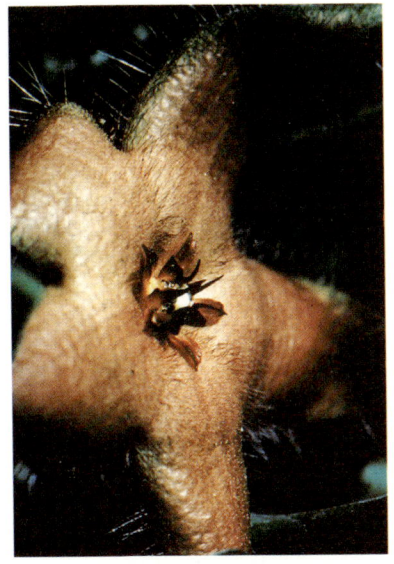
Stapelia bella
(*revoluta x deflexa?*)

Stapelia variegata
"Carrion flower"

Stapelia ambigua
in California nurseries

Stapelia gigantea pallida
"Giant toad-plant"

Stapelia hirsuta
"Hairy star-fish flower"

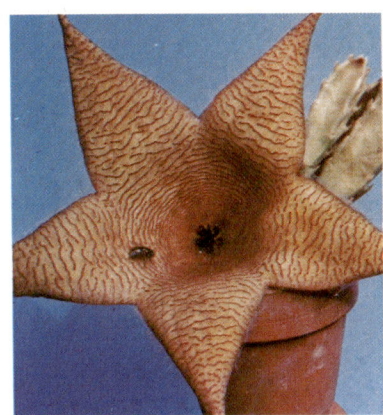
Stapelia nobilis
from Mozambique

Tavaresia angolensis
(*Luanda, Angola*)

Stapelia semota lutea
from Tanzania

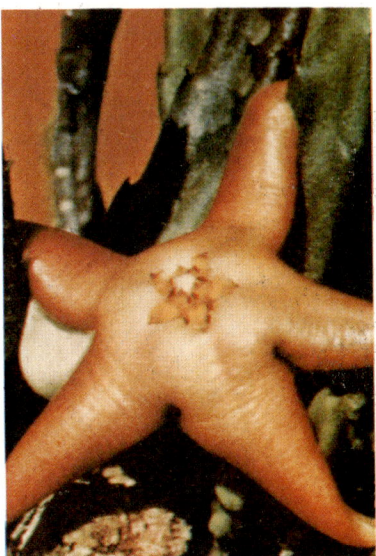
Stapelia divaricata
(*Cape Prov., So. Africa*)

BALSAMINACEAE

Impatiens balsamina
"Garden balsam"

Impatiens balsamina plena
"Rose balsam"

Impatiens glanduligera
a *"Touch-me-not"*

Impatiens balsamina
'Beijo de Frade'

Impatiens walleriana
on Marakaya rock, Kenya

Impatiens walleriana 'Holstii'
"Busy Lizzie"

Impatiens walleriana
'Variegata' (sultanii)

Impatiens repens
"Creeping impatiens"

Impatiens tuberosa
(*Madagascar*)

BALSAMINACEAE

Impatiens linearifolia
found on Chimbu Pass at 2500 m, New Guinea

Impatiens linearifolia hybrida
in hanging basket, Disney World, Florida

Impatiens Kilimanjari
in rainforest habitat, Mt. Kilimanjaro, Tanzania

Impatiens hawkeri
brought back to Australia from New Guinea 1960

BALSAMINACEAE

Impatiens walleriana
"Patient Lucy"

Impatiens hawkeri hybrida
"New Guinea hybrid"

Impatiens oliveri
"Giant touch-me-not"

Imp. platypetala aurantiaca
collected in Celebes

Impatiens hawkeri hybrida
variegated orange-red

Impatiens platypetala
hyb. 'Tangerine'

Impatiens walleriana
'Red Ripple'

Impatiens 'Uganda Red'
(*Mountains of the Moon*)

Impatiens hawkeri hybrida
in San Diego, California

BEGONIACEAE

Begonia semperflorens, *"Wax begonias"* in hanging baskets, Disney World, near Orlando, Florida

Begonia rex 'Helen Teupel' at Floristeria Camuyrano, Buenos Aires, Argentina

Begonia tuberhybrida, *"Tuberous begonias"* on display, Golden Gate Park, San Francisco

Begonia semperflorens, *bedding begonias* along walk in the Kremlin, Moscow

BEGONIACEAE

Begonia x argenteo-guttata
"Troutleaf begonia"

Begonia 'Bow-arriola'
(*boweri hybrid*)

Begonia bowerae
"Eyelash begonia" (*miniature*)

Begonia cathayana
(*China: Yunnan*)

Begonia 'Cleopatra'
"Mapleleaf begonia"

Begonia compta
cane-stem (*Brazil*)

Begonia cooperi
fibrous-rooted (*Costa Rica*)

Begonia 'Dancing Girl'
angelwing (*fibrous-rooted*)

Begonia acida
rhizomatous (*Brazil*)

Begonia x erythrophylla (feastii)
"Beefsteak begonia"

Begonia schulziana
rhizomatous (*Haiti*)

Begonia x heracleicotyle
"Mrs. Townsend" begonia

BEGONIACEAE

Begonia 'Cleopatra'
"Maple leaf begonia" (Maphil x Black Beauty)

Begonia 'Exotica'
collected 1960, from Baiyer River, New Guinea

Begonia repens Vellozo (*Brazil*)
by Rud. Ziesenhenne, Santa Barbara

Begonia serratipetala
"Pink spot angelwing", collected in New Guinea

Begonia micranthera (*Argentina*)
tuberous plant, at Logee's, Danielson, Conn.

Begonia listida
miniature angelwing, from Brazil

BEGONIACEAE

Begonia circumlobata
rhizomatous (China)

Begonia nelumbifolia
"Pond-lily begonia", in Honolulu

Begonia albo-picta
"Guinea-wing begonia"

Begonia epipsila
fibrous-rooted (Brazil)

Begonia 'Crestabruchii'
"Lettuce-leaf begonia"

Begonia manicata 'Aureo-maculata'
"Leopard begonia"

Begonia dichotoma (vitifolia)
"Kidney begonia" (Venezuela)

Begonia x ricinifolia
"Castor-bean begonia"

Begonia 'Alto-scharff'
(scharffiana x bradei)

BEGONIACEAE

Begonia bracteosa (roezlii)
in the Inca fortress of Machu Picchu, Perú

Begonia x hiemalis 'Rieger's Schwabenland'
outdoors during summertime, Duesseldorf, Germany

Begonia tuberhybrida pendula florepleno
offered at flower market, Bergen, Norway

Begonia 'Corallina de Lucerna'
an "Angelwing" begonia at Olympia, Greece

BEGONIACEAE

164

Begonia x hiemalis 'Schwabenland'
for Christmas bloom

Beg. x cheimantha 'White Christmas'
white "Lorraine-begonia"

Begonia x hiemalis 'The President'
an "Elatior" begonia from Holland

Begonia x cheimantha 'Lady Mac'
"Improved Lorraine" begonia

Begonia x cheimantha 'Marina'
"Scandinavian winter-begonia"

Begonia x hiemalis 'Emily Clibran'
English hybrid, winter-blooming

Begonia 'Tingley Mallet'
cane-stem begonia

Begonia froebelii
tuberous species from Perú

Begonia x speculata
"Grapeleaf begonia"

Begonia 'Sachsen'
free-blooming angelwing

Begonia 'Lulu Bower'
everblooming cane-stem

Beg. limmingheiana (glaucophylla)
"Shrimp begonia" (Brazil)

BEGONIACEAE

Begonia x hiemalis (elatior)
in hanging baskets, Del Mar, California

Begonia tuberhybrida fl. pl., *"Tuberous begonias"*
in window box, Interlaken, Switzerland

Begonia x cheimantha 'Lady Mac'
for Christmas bloom in Roehrs greenhouses

Begonia tuberhybrida multiflora plena maxima
in Haerens nursery, Ghent, Belgium

Begonia semperflorens, "Wax begonias"
planted in mossed wire fence, Hamburg, Germany

Begonia tuberhybrida pendula plena
at flower market, Oslo, Norway

BEGONIACEAE

Begonia x erythrophylla 'Bunchii'
"Curly Kidney-begonia"

Begonia x erythrophylla
"Beefsteak begonia"

Begonia 'Exotica'
from New Guinea

Begonia 'Orange rubra'
cane-stemmed "Angelwing"

Begonia x fuscomaculata
rhizomatous

Begonia 'Frutescans'
fibrous-rooted

Begonia haageana *in hort.*
(*scharffii*) *"Elephant-ears"*

Begonia heracleifolia nigricans
"Star-begonia"

Begonia hispida cucullifera
"Piggy-back begonia"

Begonia 'Piccolo'
(*imperialis x aridicaulis*)

Begonia 'Beatrice Haddrell'
(*boweri x sunderbruchii*)

Begonia 'Alzasco'
tall cane-stem, angelwing

BEGONIACEAE

Begonia deliciosa
rhizomatous, from Borneo

Begonia kenworthyae
(*Mexico: Chiapas*)

Begonia cathayana
(*China: Yunnan*)

Begonia x margaritae
(*echinosepala x metallica*)

Begonia margaritacea
(*Arthur Mallet x coccinea*)

Begonia roxburghii
(*Himalayas*)

Begonia grandis (evansiana)
tuberous "Hardy begonia"

Begonia kellermanii (peltata)
"Lily-pad begonia" (*Guatemala*)

Begonia incana
(*Mexico*)

BEGONIACEAE

Begonia 'Maybelle'
at New York Botanic Garden

Begonia sutherlandii
semi-tuberous, from Natal

Begonia floccifera
rhizomatous (India)

Begonia 'Mrs. Fred Scripps'
(*scharffiana x luxurians*)

Begonia fernandoi-costa
fibrous-rooted (Brazil)

Begonia 'Gwen Lowell'
(*olsoniae x obscura*)

Begonia 'Frutescaria'
fibrous-rooted

Begonia 'Skeezar'
rhizomatous

Begonia parilis
"*Zig-zag begonia*"

Begonia involucrata
rhizomatous (Costa Rica)

Begonia hispidivillosa
rhizomatous (Mexico)

Begonia echinosepala
cane-stem (Brazil)

BEGONIACEAE

Begonia macdougallii
rhizomatous (Chiapas)

Begonia leptotricha
"Wooly Bear" (Paraguay)

Begonia masoniana
"Iron cross begonia"

Begonia nelumbifolia
"Pond-lily begonia"

Begonia x sunderbruchii
"Finger-leaf begonia"

Begonia pustulata
"Blister begonia"

Begonia glabra (scandens)
trailer (West Indies)

Begonia manicata
'Aureomaculata crispa'

Begonia bradei (laetevirides)
"Alto da Serra" hort.

Begonia mazae viridis
"Stitchleaf begonia"

Begonia metallica
"Metallic-leaf begonia"

Begonia versicolor
"Fairy-carpet begonia"

BEGONIACEAE

Begonia imperialis
"Carpet begonia"

Begonia 'Manda's Eyelash'
miniature bowerae cultivar

Begonia imperialis smaragdina
"Green carpet-begonia"

Begonis 'Ricky Minter'
(*manicata crispa x mazae*)

Begonia hydrocotylifolia
"Pennywort begonia"

Begonia 'Fischer's ricinifolia'
rhizomatous

Begonia heracleifolia
"Starleaf begonia" (Mexico)

Begonia egregia
fibrous-rooted (Brazil)

Begonia luxurians
"Palm-leaf begonia"

Begonia 'Pinafore'
(*Elaine seedling*)

Begonia 'Di-Anna'
angelwing

Begonia 'Jinny May'
(*coccinea seedling*)

BEGONIACEAE

Begonia metallica
fibrous-rooted (Brazil)

Begonia 'Thurstonii'
(*metallica x sanguinea*)

Begonia masoniana
"*Iron-cross begonia*"

Begonia olsoniae
(*vellozoana in hort.*)

Begonia velloziana
Brazilian epiphyte, rhizomatous

Begonia conchifolia rubrimacula
"*Zip begonia*"

Begonia hispida cucullifera
"*Piggyback-begonia*"

Begonia limmingheiana
"*Shrimp begonia*"

Symbegonia sanguinea (strigosa)
from New Guinea (by J. Bogner)

BEGONIACEAE

Begonia rex 'Salamander'
(*diadema hybrid*)

Begonia rex 'Glory of St. Albans'
(*English hybrid 1838*)

Begonia rex 'President'
(*"President Carnot"*)

Begonia rex 'Can-can'
"Oakleaf" (*diadema hyb.*)

Begonia rex
'Silver Queen' (*1875*)

Begonia rex 'Thrush'
upright miniature rex

Begonia rex 'Mikado'
"Purple fan-leaf"

Begonia rex 'Fireflush'
(*'Bettina Rothschild'*)

Begonia rex 'Yuletide'
(*from Man-Holland 1964*)

Begonia 'Silver Star'
(*caroliniaefolia x liebmannii*)

Begonia 'Sir Percy'
(*Silverstar x speculata*)

Begonia decora
rhizomatous

BEGONIACEAE

Begonia rex 'Fairy'
(*diadema hybrid*)

Begonia rex 'Helen Teupel'
(*rex x diadema 1928*)

Begonia rex 'Merry Christmas'
(*'Ruhrtal'*)

Begonia rex 'Comtesse Erdoedy'
"*Corkscrew begonia*"

Begonia rex 'Curly Fireflush'
"*Spiral begonia*"

Begonia rex 'Baby Rainbow'
miniature rex

Begonia rex 'Filigree'
upright-stemmed

Begonia rex 'Frosty Dwarf'
(*Brooklyn Bot. Garden miniature*)

Begonia rex 'Happy New Year'
Roehrs selection 1964

Begonia rex 'His Majesty'
(*English hybrid 1903*)

Begonia rex 'Black Knight'
(*'Midnight'*) *1916*

Begonia rex 'Her Majesty'
fine English hybrid 1921

BEGONIACEAE

Begonia goegoensis
"Fire-king begonia"

Begonia paulensis
rhizomatous (Brazil)

Begonia rajah
dwarf rhizomatous (Malaya)

Begonia serratipetala
(New Guinea)

Begonia 'Braemar'
(scharffii x metallica)

Begonia 'Picta rosea'
angelwing cane-stem

Begonia 'Black Raspberry'
(acetosa x imperialis)

Begonia mannii
"Roseleaf begonia"

Begonia 'Maphil'
dwarf rhizomatous

Begonia 'Nelly Bly'
(metallica seedling)

Begonia bogneri
from Madagascar, by J. Bogner

Begonia macrocarpa (secreta)
from Cameroun

BEGONIACEAE

Begonia 'China Doll'
rhizomatous miniature

Begonia schmidtiana
fibrous-rooted (Brazil)

Begonia 'Silver Jewel'
(imperialis x pustulata)

Begonia 'Rogeri'
fibrous-rooted

Begonia x richmondensis
very floriferous

Begonia venosa
succulent (Brazil)

Begonia 'Corallina de Lucerna'
old angelwing 1892

Begonia sanguinea
fibrous-rooted (Brazil)

Begonia listida
miniature angelwing

BEGONIACEAE

176

Begonia semperfl. 'Cinderella'
"thimble" hybrid

Begonia semperfl. albo-foliis
"Calla-lily begonia"

Begonia semperflorens fl. pl.
'Curlilocks' (*"Thimble beg."*)

Begonia semperfl. 'Pink Pearl'
"Wax begonia"

Begonia semperfl. fl. pl. 'Firefly'
crested thimble

Begonia semp. fl. pl. 'Lady Frances'
"Rose begonia"

Begonia semperflorens 'Luminosa'
Wax bedding begonia

Begonia semperflorens
'Scandinavian White'

Begonia semperflorens
'Scandinavian Pink'

Begonia semperflorens
'Red Pearl'

Begonia semperflorens
'Flamingo' (*bicolor*)

Begonia semperflorens
'White Comet' (*F-1 hybrid*)

BEGONIACEAE

Begonia x tuberhybrida 'Rubro-marginata'

Begonia x tuberhybrida 'Fimbriata plena'

Begonia x tub. 'Crispa' (*Carnation-form*)

Begonia x tuberhybrida 'Picotee' (*bicolor*)

Begonia x tuberhybrida 'Camellia-type'

Begonia x tub. 'Triumph' "Rose bud" type

Begonia x tuberhybrida "Hybrid tuberous begonia"

Begonia x tuberhybrida multiflora 'Maxima'

Begonia x tub. 'Elegans' in Wellington, New Zealand

Begonia x tuberhybrida multiflora 'Tasso'

Begonia x tuberhybrida multiflora 'Helene Harms'

Begonia x tub. pendula 'Lloydii' "Basket begonia"

BEGONIACEAE

Begonia diadema
fibrous-rooted (Borneo)

Begonia 'Grey Feather'
cane stem angelwing

Begonia tomentosa
fibrous-rooted (Brazil)

Begonia 'Mrs. W. A. Wallow'
tall cane-stemmed

Begonia x weltonensis
"Maple-leaf begonia"

Begonia sparsipila
(Central America)

Begonia 'Gloire de Jouy'
(rex x incarnata)

Begonia 'Templinii'
"Crazy leaf begonia"

Begonia 'Interlaken'
(Lucerna seedling)

BEGONIACEAE

Begonia aconitifolia (sceptrum)
cane-stem (Brazil)

Begonia 'Tom Ment'
(Florida cane-stem 1965)

Begonia coccinea
"Angelwing begonia" (Brazil)

Begonia 'Gehrtii'
rhizomatous (Brazil)

Begonia bogneri
miniature from Madagascar 1969

Begonia 'Immense'
rhizom. ricinifolia seedling

Begonia coccinea 'Pink'
in Manhattan Beach, Calif.

Begonia scharffiana
fibrous-rooted (Brazil)

Begonia scharffii (haageana)
"Elephant-ear begonia"

BEGONIACEAE

Begonia odorata
fibrous-rooted (Jamaica)

Begonia rubra (Java)
(Los Angeles Arboretum)

Begonia suffruticosa
"Maple leaf begonia" (Natal)

Begonia fuchsioides
"Fuchsia begonia" (Mexico)

Begonia x drostii
(scharffii hybrid)

Begonia foliosa
"Fernleaf begonia"

Begonia ulmifolia
"Elmleaf begonia"

Begonia nitida (minor)
cane-stem (Jamaica)

Begonia chimborazo
rhizomatous (Guatemala)

BERBERIDACEAE, BIXACEAE

Mahonia aquifolium
"Oregon grape" in bloom

Mahonia aquifolium
"Mountain grape" fruiting

Mahonia (Berberis) repens
(*Brit. Columbia to California*)

Nandina 'Goshiki'
"Five color rainbow" in Hawaii

Nandina domestica longifolia 'Ito'
"String nandina" (*Oahu*)

Nandina domestica
"Heavenly bamboo" of China

Nandina domestica
"Sacred bamboo" in Taipo

Podophyllum emodi
"May apple" (*Himalayas*)

Bixa orellana
"Lipstick tree" (*West Indies*)

BETULACEAE

Betula pendula dalecarlica
European white birch, in Duesseldorf

Betula populifolia
"Gray birch" in Washington, D.C.

Betula papyrifera
"Paper birch" in the Kremlin, Moscow

Betula pendula
"White birch" in Southern California

BIGNONIACEAE

Dolichandrone lutea
(*Burma*)

Distictis buccinatoria
(*Bignonia cherere*)

Distictis x riversii
in Mahé, Seychelles

Millingtonia hortensis
"Indian cork tree", in Mombasa

Mansoa difficilis
in Rio de Janeiro

Doxantha unguis-cati
"Cats-claw" (Trop. America)

Pandorea jasminioides 'Alba'
(*Eastern Australia*)

Pandorea jasminioides (Tecoma)
"Bower plant" (Queensland)

Pandorea jasminioides 'Rosea'
"Manga del Niño"

BIGNONIACEAE

Anemopaegma chamberlaynii
in New Delhi, India

Anemopaegma subunculata
(*Brazil*)

Saritaea (Arrabidaea) magnifica
(*Colombia*)

Campsis grandiflora
"Chinese trumpet-creeper"

Campsis radicans
"Trumpet vine"

Catalpa bignonioides
"Indian bean"

Catophractes alexandrii
Namib Desert, S.W. Africa

Clytostoma callistegioides
"Argentine trumpet vine"

Eccremocarpus scaber
"Glory flower" (*Chile*)

BIGNONIACEAE

Crescentia cujete
"Calabash" tree in Panama

Kigelia pinnata
"Sausage tree" near Madras, South India

Jacaranda mimosaefolia
a "Green ebony" on St. Héléna Island

Parmentiera cereifera
"Candle tree" in Peradeniya Bot. Garden, Ceylon

BIGNONIACEAE

Crescentia alata, *the "Gourd tree"*
south of Taxco, Guerrero, Mexico

Parmentiera edulis
"Cuachilote tree" (Guatemala)

Podranea ricasoliana
"Port St. John's creeper" (So. Africa)

Tabebuia chrysantha
"Yellow Pui" along roadside in Puerto Rico

Jacaranda obtusifolia rhombifolia (filicifolia)
a "Green ebony" in Guyana

Rhigozum trichotomum
in Namibia, Southwest Africa

BIGNONIACEAE

Crescentia cujete
"Calabash" in Sri Lanka

Kigelia pinnata
"Sausage tree" in flower (Zambia)

Parmentiera cereifera
"Candle tree" (Panama)

Cydista aequinoctialis
"Garlic vine" in Oaxaca, Mexico

Saritaea magnifica (Arrabidaea)
on St. Thomas, Virgin Islands

Tecoma stans (Stenolobium)
"Yellow bells" in the Bahamas

Tecomaria capensis 'Aurea'
at Las Palmas, Canary Islands

Bignonia capreolata
"Crossvine" in Dallas, Texas

Spathodea campanulata
"African tulip tree"

BIGNONIACEAE

Pyrostegia venusta (Bignonia ignea)
"Flaming trumpet" in Las Quintas, Cuernavaca

Saritaea magnifica
on St. Thomas, Virgin Islands

Stereospermum chelonoides (*India*)
flowering tree in Jogjakarta, Java

Spathodea campanulata
"Flame of the forest" in Uganda

BIGNONIACEAE

Tabebuia donnell-smithii (Cybistax)
"Primavera", in Acapulco, Mexico

Tabebuia argentea (Tecoma)
in Lal Bagh Gardens, Bangalore, India

Tabebuia serratifolia
"Yellow Pui", blooming in Trinidad

Tabebuia rosea (pentaphylla)
"Pink Pui" in Puerto Rico

BIGNONIACEAE

Tabebuia argentea
"Silver trumpet tree" (Paraguay)

Tabebuia roseo-alba
(Brazil, Bolivia)

Tabebuia rosea (pentaphylla)
"Pink Pui" in Caracas, Venez.

Tabebuia donnell-smithii
"Primavera" on St. Thomas

Tabebuia chrysotricha
in Los Angeles Arboretum

Tabebuia pallida
"Cuban pink trumpet-tree"

Tabebuia rosea (pentaphylla)
"Salvador pink trumpet" in Honolulu

Tabebuia serratifolia
(Puerto Rico)

Tabebuia heterophylla
photographed in Singapore

BIGNONIACEAE

Tabebuia impetiginosa
(*Brazil*)

Tabebuia umbellata (fosteri)
(*Brazil*)

Tabebuia rosea (pentaphylla)
"Rosy trumpet-tree"

Tabebuia avellanedae (ipe)
(*Argentina*)

Tecomaria capensis
"Cape honeysuckle"

Campsis radicans (Tecoma)
"Trumpet creeper"

Tecomanthe dendrophylla
from New Guinea

Tecomanthe speciosa
(*Three Kings Isl., N.W. New Zealand*)

Tecomanthe venusta
(*New Guinea*)

BIGNONIACEAE

Tabebuia impetiginosa
showy "Trumpet tree" in Honolulu (Dr. Don Watson)

Tecomanthe dendrophylla
in Finisterre Mountains, N.E. New Guinea

Distictis buccinatoria (Tecoma cherere)
"Scarlet trumpet vine" at El Paseo, San Diego

Tabebuia umbellata (fosteri)
in Longwood Gardens, Kennett Square, Pennsylvania

BIGNONIACEAE, BOMBACACEAE

Bombax costatum
in Sénégal, West Africa

Tecoma stans (Stenolobium)
"Yellow elder, in Perú

Pseudocalymma alliaceum
"Garlic vine" (Guyana)

Bombax ellipticum 'Album'
in Foster Gardens, Honolulu

Chiranthodendron pentadactylon
"Mexican handflower"

Bombax ellipticum
"Shaving brush" (Mexico)

Chorisia speciosa
'Majestic Beauty' *(California)*

Chorisia insignis
"White floss silk tree"

Chorisia speciosa
"Floss-silk tree" (Argentina)

BOMBACEAE

Bombax ceiba (malabaricum)
in the Khmer temple of Angkor, Cambodia

Bombax ceiba (malabaricum)
"Red Kapok tree", in flower

Durio zibethinus
old "Durian" tree, Peradeniya, Ceylon

Durio testudinarum
"Durian tanah" in Malaysia (Dr. E. Menninger)

BOMBACACEAE

Adansonia digitata
ancient "Baobab" in Northern Transvaal

Bombax ceiba (malabaricum)
"Red silk cotton" in ruins of Angkor Wat, Cambodia

Ceiba pentandra
"Kapok" tree in Foster Gardens, Honolulu

Bombax ellipticum
"Shaving brush tree" (Mexico: Vera Cruz)

BORAGINACEAE

Cordia sebestena
"Geiger tree", in Aruba, W.I.

Cordia dodecandra
(Mexico, Guatemala)

Cordia lutea
on Nuku Hiva, French Polynesia

Echium fastuosum
"Pride of Madeira", in So. France

Cordia superba (elliptica)
in Rio de Janeiro

Cordia alba (dentata)
"Jackwood", on St. Thomas, V.I.

Bourreria ovata
"Strongback", in the Bahamas

Heliotropium arborescens
"Heliotrope" (Perú)

Pulmonaria saccharata
"Bethlehem sage" (So. Europe)

BOMBACACEAE, BORAGINACEAE

Chorisia speciosa (*Bomb.*)
"Floss-silk tree" in Los Angeles Botanic Garden

Chorisia insignis
"White floss-silk" or "Drunken tree"

Messerschmidtia argentea (*Borag.*)
"Tree heliotrope" in Tonga, South Pacific

Pachira aquatica (*Bomb.*)
"Guinea chestnut" or "Oje" in Recife, Brazil

BROMELIACEAE

198

Aechmea chantinii 'Vera'
in Arcadia, California

Aechmea chantinii
"Amazonian zebra plant"

Aechmea fulgens discolor
"Coral-berry"

Aechmea fasciata
"Silver vase" (*So. Brazil*)

Aechmea 'Bert'
(*orlandiana x fosteriana*)

Aechmea 'Black Prince'
(*Hummel hybrid, California*)

Aechmea fulgens discolor
'Magnificent', *in New York*

Aechmea coelestis 'Albomarginata'
(*David Barry, Los Angeles*)

Aechmea hoppei
in Amazonian Perú

BROMELIACEAE

Aechmea fosteriana
as epiphyte in Espirito Santo, Brazil

Aechmea dichlamydea var. trinitensis
at home in Trinidad

Aechmea chantinii
"Amazonian zebra plant" (Perú)

Aechmea mulfordii (Gravisia fosteriana)
(Atlantic coast of Southeastern Brazil)

BROMELIACEAE

Aechmea fulgens discolor
"Coral-berry" (*Pernambuco*)

Aechmea fasciata
"Silver vase" (*inflor.*)

Aechmea araneosa
(*Espirito Santo, Brazil*)

Aechmea caudata 'Variegata'
(*Brazil*)

Aechma caesia
epiphyte of So. Brazil

Aechmea caudata
from temperate So. Brazil

Aechmea 'Electrica'
(*Hummel hybrid, California*)

Aechmea bromeliifolia
"Wax torch" (*Honduras*)

Acanthostachys strobilacea
"Pinecone bromeliad"

Aechmea chantinii
"Amazonian zebra plant"

Aechmea filicaulis
"Weeping living vase"

Aechmea chantinii
'Silver Monarch'

BROMELIACEAE

Aechmea fasciata 'Variegata'
"Variegated silver vase"

Aechmea caudata 'Variegata'
at Roehrs greenhouses, New Jersey

Aechmea dichlamydea var. trinitensis
epiphyte in northern Trinidad

Aechmea fasciata
"Silver vase" in Rio de Janeiro

Aechmea fulgens
"Coral berry" from Pernambuco, Brazil

Aechmea 'Foster's Favorite Favorite'
(*victoriana discolor x racinae*)

BROMELIACEAE

Aechmea lueddemanniana
(*Guatemala to Honduras*)

Aechmea x maginali
(*miniata x fulgens*)

Aechmea mariae-reginae
"*Queen aechmea*" (*Costa Rica*)

Aechmea racinae
"*Christmas jewels*" (*Espirito Santo*)

Aechmea gracilis
(*Espirito Santo, Brazil*)

Aechmea miniata discolor
(*Bahia, Brazil*)

Aechmea ramosa
epiphyte in So. Brazil

Aechmea nudicaulis aureo-rosea
(*Southern Brazil*)

Aechmea nudicaulis
(*C. America to Brazil*)

Aechmea tillandsioides
(*Venezuela, Guyana*)

Aechmea phanerophlebia
(*Espirito Santo, Brazil*)

Aechmea mertensii
"*China-berry*" (*So. America*)

BROMELIACEAE

203

Aechmea chlorophylla
(*Organ Mountains, Brazil*)

Aechmea ornata nationalis
"*Porcupine plant*"

Aechmea magdalenae 'Quadricolor'
at Seaborn's, Escondido, California

Aechmea 'Redwing'
(*penduliflora x mutica*)

Aechmea ramosa x fulgens
(*David Barry, Los Angeles*)

Aechmea hystrix
(*Argentina*)

Aechmea ramosa
epiphyte of Southern Brazil

Aechmea orlandiana variegata 'Ensign'
at Seaborn's, Escondido, Calif.

Aechmea tillandsioides
(*Venezuela, Guyana*)

BROMELIACEAE

Bromeliads *in Tropical Planting
by Roehrs, Radio City, New York*

Aechmea mexicana
on the road to Turrialba, Costa Rica

Aechmea angustifolia
(Costa Rica to Colombia and Bolivia)

Aechmea mooreana
collected by Lee Moore, Amazonian Perú

Aechmea distichantha
in habitat, northern Argentina

Aechmea 'Black Wine'
at Wiley's, Palos Verdes, California

BROMELIACEAE

Aechmea weilbachii
from the forests of Southern Brazil

Ananas bracteatus striatus
"Variegated red pineapple"

Aechmea nidularioides
in humid forest of Amazonian Colombia

Aechmea mulfordii (Gravisia fosteriana)
from moist regions of S.E. Brazil

Ananas bracteatus striatus
terrestrial on the Serra do Mar, Brazil

Aechmea orlandiana x chantinii
in Arcadia, California

BROMELIACEAE

Aechmea tillandsioides 'Marginata'
in Escondido, California

Aechmea serrata
epiphyte in Martinique

Aechmea nudicaulis
epiphytic in Guanabara

Aechmea dichlamydea var. trinitensis
found as epiphyte on Barbados

Aechmea triangularis
(Espirito Santo, Brazil)

Aechmea victoriana
terrestrial in Espirito Santo

Aechmea tillandsioides var. Amazonas
in Amazonian Perú

Bromelia fastuosa (antiacantha)
at Jardin des Plantes, Paris

Aechmea tessmannii
(Perú, Colombia)

BROMELIACEAE

Ananas bracteatus
"Red pineapple" (Brazil)

Ananas comosus 'Porteanus'
"Golden rocket"

Ananas bracteatus
sprouting suckers from fruit

Ananas nanus
"Dwarf pineapple" 30 cm high

Bromelia balansae
"Piñuela" in Argentina

Ananas comosus 'Variegatus'
"Variegated pineapple"

Bromelia antiacantha
in Uruguay

Bromelia serra
(Eastern Paraguay)

Bromelia balansae
"Heart of fire"

BROMELIACEAE

Billbergia amoena viridis
(*Eastern Brazil*)

Billbergia nutans
"Queen's tears"

Billbergia 'Bob Manda'
"Manda's urn plant"

Billbergia euphemiae
(*Southern Brazil*)

Billbergia 'Albertii'
"Friendship plant"

Billbergia amoena
(*Eastern Brazil*)

Billbergia zebrina
"Zebra urn"

Billbergia 'Fantasia'
"Marbled rainbow plant"

Billbergia pyramidalis concolor
"Summer torch" (*Brazil*)

Billbergia pyramidalis
'Striata'

Billbergia seidelii
(*Estado do Rio*)

Billbergia saundersii
"Rainbow plant" (*Bahia*)

BROMELIACEAE

Ananas comosus
Maya women offer pineapple fruit in Guatemala

Aechmea tillandsioides
growing epiphytic in interior Surinam

Ananas comosus 'Smooth Cayenne'
in pineapple plantation on Maui, Hawaii

Ananas bracteatus 'Striatus'
"Variegated wild pinapple" in German garden-center

BROMELIACEAE

Billbergia tessmanniana
collected by Lee Moore, Amazonian Perú

Canistrum lindenii
in thorn thicket on the Banderantes Coast, Brazil

Billbergia 'Fascination'
at National Flower Show, Cologne, Germany

Bromelia serra
in Eastern Paraguay, by Dr. W. Rauh, Heidelberg B.G.

Bromelia balansae
"Piñuela" or "Heart of fire", in Argentina

Bromelia plumieri (*Mexico to Ecuador*)
in Kew Gardens, London

BROMELIACEAE

Billbergia buchholtzii
at *David Barry's, Los Angeles*

Billbergia 'Fantasia'
"Marbled rainbow plant"

Billbergia pyramidalis
"Foolproof plant"

Billbergia pyramidalis concolor
"Summer torch" (*Brazil*)

Billbergia nutans
"Queen's tears" or *"Indoor oats"*

Billbergia tessmanniana
"Showy billbergia (*Perú*)

Canistrum fosterianum
(*Brazil:Bahia*)

Canistrum lindenii
at *C. Wiley's, Palos Verdes, Calif.*

Aechmea aquilega
(*West Indies to Brazil*)

BROMELIACEAE

Cryptanthus bromelioides 'Tricolor'
"Rainbow-star" (*Brazil*)

Cryptanthus bivittatus minor
"Rose-stripe star"

Cryptanthus beuckeri
"Marbled spoon" (*Brazil*)

Cryptanthus diversifolius
"Vary-leaf star"

Cryptanthus fosterianus
"Stiff pheasant-leaf"

Cryptanthus 'It'
"Color band"

Cryptanthus zonatus 'Zebrinus'
"Zebra plant"

Bromelia pinguin
"Pinguin" (*West Indies*)

Cryptanthus x mirabilis
(*beuckeri x osyanus*)

Dyckia fosteriana
"Silver and gold dyckia"

Bromelia serra 'Variegata'
"Heart of flame" (*Argentina*)

Dyckia brevifolia
"Miniature agave" (*Brazil*)

BROMELIACEAE

Cryptanthus bivittatus 'Pink Starlite'
of Winter Garden, Florida

Cryptanthus zonatus 'Zebrinus'
"Zebra plant" or "Pheasant leaf"

Cryptanthus bromelioides 'Racinae'
in Frankfurt Bot. Garden, Germany

Cryptanthus bromelioides 'Tricolor'
"Rainbow star"

Cryptanthus bivittatus 'Tricolor'
at Longwood Gardens, Penna.

Cryptanthus 'It'
"Color-band", at Fantastic Gardens, Miami

Cryptanthus bromelioides 'Tricolor'
"Rainbow star" highly colored in the sun

BROMELIACEAE

Puya raimondii *in habitat*
Cordillera Quebrada Pacharoca, Perú (Dr. W. Rauh)

Neoregelia cruenta
in Guanabara, Southern Brazil

Billbergia pyramidalis concolor
epiphytic on trees in Bahia, Brazil

Deuterocohnia longipetala (*Perú*)
with Mr. Giridlian, Oakhurst Gardens, California

BROMELIACEAE

Guzmania craterifolia (sanguinea)
(*Costa Rica to Ecuador*)

Guzmania 'Fantasia'
at Palmengarten, Frankfurt, Germany

Guzmania insignis (*Colombia*)
at Les Cèdres, St. Jean-Cap Ferrat, France

Greigia sphacelata (*Chile*)
in Botanic Garden Essen, Germany

Guzmania sanguinea (*Ecuador*)
at Munich Botanic Garden, Germany

Hechtia stenopetala, *growing in lava rock
of the Pedregal, Jardin Botanico, Mexico City*

BROMELIACEAE

Guzmania lingulata 'Major'
at Broadview, New Rochelle, N.Y.

Guzmania lingulata
(Central America to Bolivia)

Guzmania berteroniana
"Flaming torch" (Puerto Rico)

Guzmania x 'Naranja'
at Carlsbad, California

Guzmania x magnifica
(lingulata x minor)

Guzmania lingulata minor
"Orange star"

Neoregelia carolinae 'Tricolor'
"Striped blushing bromeliad"

Neoregelia carolinae meyendorffii
'Variegata', at Seaborn's, Escondido

Neoregelia carolinae
from Teresopolis, Brazil

Neoregelia 'Mar-con'
"Marbled fingernail"

Neoregelia farinosa
"Crimson cup" (Espirito Santo)

Neoregelia concentrica 'Plutonis'
at Frankfurt Palmengarten

BROMELIACEAE

Guzmania zahnii
(*Colombia, Panama*)

Guzmania lindenii
(*Northeastern Perú*)

Guzmania lingulata 'Major'
"Scarlet star"

Guzmania berteroniana
"Flaming torch" (*Puerto Rico*)

Guzmania melinonis quitense
at Wiley's, Palos Verdes, Calif.

Guzmania monostachya
"Striped torch"

Guz. sanguinea brevipedunculata
in Brooklyn Botanic Garden

Guzmania 'Omar Morobe'
Longwood Gardens, Penna.

Guzmania 'Symphonie'
(*zahnii x peacockii*)

BROMELIACEAE

218

Neoregelia cruenta
on the Restinga coast, Guanabara, Brazil

Neoregelia spectabilis
"Fingernail plant" from Rio de Janeiro

Nidularium innocentii var. lineatum
grown by Dr. George Milstein, New York

Nidularium innocentii 'Maureanum'
cultivated in Belgium

Nidularium regelioides
terrestrial in Southeastern Brazil

Ochagavia lindleyana
a succulent plant from Chile

BROMELIACEAE

Neoregelia carolinae
terrestrial from Teresopolis, Brazil

Hohenbergia stellata
in cloudforest of Martinique

Neoregelia carolinae 'Marechalii'
"Blushing bromeliad"

Neoregelia carolinae 'Meyendorffii'
(Karatas) at flowering time

Neoregelia carolinae 'Volkaert's Favorit'
at German Flower Show, Cologne

Neoregelia carolinae 'Tricolor'
in Roehrs Exotic Nurseries, New Jersey

BROMELIACEAE

Neoregelia spectabilis
"Fingernail plant"

Neoregelia sarmentosa chlorosticta (*Rio de Janeiro*)

Neoregelia 'tristis x marmorata'

Nidularium billbergioides 'Citrinum'

Nidularium billbergioides 'Flavum' (*Brazil*)

Nidularium fulgens *"Blushing cup"* (*Brazil*)

Nidularium innocentii lineatum *"Striped birdsnest"*

Nidularium innocentii *"Black birdsnest"* (*Brazil*)

Nidularium innocentii innocentii (*Nidularium amazonicum*)

Nidularium innocentii nana *"Miniature birdsnest"*

Nidularium innocentii striatum *"Striped birdsnest"* (*Brazil*)

Nidularium innocentii viridis *"Green birdsnest"*

BROMELIACEAE

Nidularium
'Souvenir de Mme. Morobe'

x Orthotanthus 'What'
at Seaborn's, Escondido, Calif.

Nidularium billbergioides
(So. Brazil)

Portea petropolitana
at Kew Gardens, London

Pitcairnia heterophylla
by Doris Graf, California

Portea petropolitana extensa
Cologne Bot. Garden, Germany

Quesnelia arvensis
in Sao Paulo, Brazil

Puya raimondii
from Cordilleras of Perú

Streptocalyx fuerstenbergii
at Frankfurt Palmengarten

BROMELIACEAE

Puya alpestris (*Chile*)
at Huntington Botanic Garden, California

Pitcairnia tabuliformis
terrestrial from Chiapas, Mexico

Streptocalyx fuerstenbergii
Bot. Garden Marnier-Lapostolle, France

Orthophytum vagans
(*Espirito Santo, Brazil*)

Quesnelia quesneliana
in French Guiana

Streptocalyx holmesii
near Iquitos, Amazonas, Perú

BROMELIACEAE

Streptocalyx poeppigii
(*Amazonas to Bolivia*)

Streptocalyx poitaei
(*Amazonian Perú*)

Thecophyllum insigne *in hort.*
(*Guzmania 'Insignis'*)

Tillandsia ponderosa
from cloud forests of Guatemala

Tillandsia albertiana
from Salta, Argentina

Tillandsia andreana
(*Cordilleras of Colombia*)

Tillandsia plumosa
(*Mexico, Guatemala*)

Tillandsia oerstedii
in Costa Rica

Tillandsia duratii
(*Brazil, Bolivia, Argentina*)

BROMELIACEAE

Orthophytum maracasense
(*Bahia, Brazil*)

Pitcairnia corallina
"Palm bromeliad" (Colombia)

Pitcairnia andreana
(*Colombia:Choco*)

Tillandsia crispa
(*Panama to Perú*)

Tillandsia caput-medusae
(*Mexico and South*)

Thecophyllum (Vriesea) sintenisii
in Puerto Rico habitat

Tillandsia ionantha
"Sky plant" (C. America)

Tillandsia flexuosa
"Spiralled airplant"

Tillandsia geminiflora
in Petropolis, Brazil

Tillandsia cyanea
"Pink quill" (Ecuador)

Tillandsia lindenii (*Perú*)
"Blue-flowered torch"

Pitcairnia flammea
(*Organ Mountains, Brazil*)

BROMELIACEAE

Tillandsia recurvata
"Ball moss" on telephone wires in Jamaica

Tillandsia xerographica, *from Oaxaca, Mexico*
with A. Lau, Fortin de las Flores, Veracruz

Tillandsia fasciculata
"Wild pineapple" in Everglades National Park, Florida

Tillandsia fendleri
in the Inca ruins of Machu Picchu, Perú

BROMELIACEAE

Tillandsia flabellata (*Guatemala*)
in Nymphenburg Botanic Garden, Munich

Tillandsia bergeri (*Uruguay*)
with Frances Wiley, Palos Verdes, California

Tillandsia caput-medusae
(*Mexico and Central America*)

Tillandsia crispa
epiphyte from Panama to Perú

Tillandsia fasciculata
at San Diego County Fair, Del Mar, California

Tillandsia mooreana
in Amazonian Perú

BROMELIACEAE

Tillandsia imperialis
"Christmas candle" of Mexico

Tillandsia fasciculata
(Florida and Caribbean)

Tillandsia ionantha (erubescens)
at flowering time

Tillandsia wagneriana
"Flying bird" (Perú)

Tillandsia utriculata
"Big wild pine"

Tillandsia streptophylla
"Twist plant" (Jamaica)

Tillandsia stricta
"Hanging torch" (So. America)

Tillandsia jalisco-monticola
in Veracruz, Mexico

Tillandsia prodigiosa
in Fortin de las Flores, Mexico

BROMELIACEAE

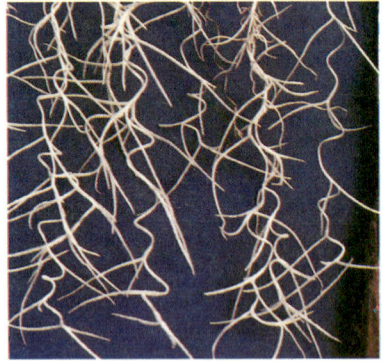
Tillandsia usneoides
"Spanish moss"

Tillandsia punctulata
(*Mexico to Panama*)

Vriesea ampullacea
(*syn. 'Espirito Santo'*)

Vriesea erythrodactylon
Serra do Mar, Brazil

Vriesea 'Favorite'
(*ensiformis hybrid*)

Vriesea carinata
"Lobster claws"

Vriesea fenestralis 'Variegata'
(*Frankfurt Palmengarten*)

Vriesea fosteriana 'Seideliana'
in Germany

Vriesea fosteriana
'Red Chestnut'

Vriesea imperialis
"Giant vriesea"

Vriesea fenestralis
"Netted vriesea" (*Brazil*)

Vriesea incurvata
"Sidewinder vriesea"

BROMELIACEAE

Tillandsia tricolor
epiphytic on palm tree, Sao Paulo

Tillandsia utriculata
Everglades National Park, Florida

Vriesea sintenisii (Thecophyllum)
Caribbean Nat'l. Forest, Sierra Luquillo, Puerto Rico

Vrieseas and Tillandsias
cover the rocks of the Cordilleras, Machu Picchu, Perú

BROMELIACEAE

Tillandsia usneoides
"Spanish moss" at Cypress gardens, Florida

Vriesea hieroglyphica
growing in low jungle, Serra do Mar, near Santos

Tillandsia multicaulis
in Botanic Garden Cologne, Germany

Tillandsia x 'Victoriae'
a miniature (ionantha x brachycaulon)

Tillandsia tenuifolia (pulchella)
in northern Argentina

Tillandsia cyanea
in the Moir collection, Honolulu

BROMELIACEAE

Vriesea hieroglyphica
"King of bromeliads" in Sao Paulo

Vriesea fenestralis
"Netted vriesea" as epiphyte in Southern Brazil

Vriesea bituminosa
on the Serra do Mar, near Santos

Vriesea gigantea
in Palmengarten, Frankfurt, Germany

Vriesea hieroglyphica
along the highway to Santos, Brazil

Vriesea fosteriana
as epiphyte in Espirito Santo, Brazil

BROMELIACEAE

Vriesea platyneura
(*West Indies to Brazil*)

Vriesea 'Polonia'
(*kitteliana x vigeri*)

Vriesea x poelmanii
(*gloriosa x vangeertii*)

Vriesea 'Polonia'
in Brooklyn Botanic Garden

Vriesea phillipo-coburgii
high on trees in So. Brazil

Vriesea ringens
Luquillo forest, Puerto Rico

Vriesea splendens
"Flaming sword"

Vriesea splendens 'Variegata'
(*Cologne Botanic Garden*)

Tillandsia multicaulis
from Costa Rica

BROMELIACEAE

Vriesea sintenisii (Thecophyllum)
epiphytic in Luquillo cloud forest, Puerto Rico

Vriesea 'Perfecta'
in floral shop. Hibiya Park, Tokyo

Vriesea imperialis
Sr. R. Burle-Marx collecting seed in Organ Mts., Brazil

Vriesea splendens 'Major'
from moist forests of Guayana and Surinam

BROMELIACEAE

234

Vriesea gigantea
(*tesselata*) *in Belgium*

Vriesea imperialis
"Giant vriesea"

Vriesea hieroglyphica
in Belgium

Vriesea imperialis
near Teresopolis, Brazil

Vriesea heliconoides
(*Guatemala to Bolivia*)

Vriesea grande *in hort.*
at German Flower Show, Cologne

Vriesea 'Intermedia'
(*hieroglyphica x fenestralis*)

Vriesea macrostachia
Toro Negro, Puerto Rico (1947)

Vriesea patula (Tillandsia)
(*Perú: Tarma to 2400 m*)

BROMELIACEAE

Vriesea 'Poelmanii hybrid'
at Rath Garden-Center, Duesseldorf, Germany

Vriesea 'Perfecta'
Sayonara to Expo 1970, Osaka, Japan

Vriesea sintenisii (Thecophyllum)
in Caribbean National Forest, Puerto Rico

Vriesea carinata x poelmanii
at Yamamoto Nursery, Kansai, Japan

Vriesea 'Rubin'
at bromeliad exhibition, Frankfurt, Germany 1973

Vriesea splendens
at Hummel's Exotic Gardens, Carlsbad, California

BROMELIACEAE

Vriesea x poelmanii
(*gloriosa x vangeertii*)

Vriesea retroflexa
epiphyte from Sao Paulo

Vriesea 'Polonia'
(*kitteliana x vigeri*)

Vriesea rodigasiana
"Wax-shells" (*So. Brazil*)

Vriesea scalaris
(*Santa Catarina, Brazil*)

Vriesea rubyae
from Brazil, by Geo. Kalmbacher

Vriesea imperialis
in Estado do Rio

Tillandsia recurvata
on telephone wires, Florida

Quesnelia testudo
on the Serra do Mar, Sao Paulo

Vriesea splendens
"Flaming sword"

Vriesea viridiflora
(*Costa Rica*)

Vriesea guttata
"Dusted feather"

BURSERACEAE, BUTOMACEAE, BUXACEAE

Buxus microphylla japonica
"California boxwood"

Sarcococca saligna
"Willow leaf" (Himalayas)

Canarium vulgare
"Java almond" in Peradeniya

Buxus sempervirens 'Bullata'
in Germany

Pachysandra terminalis
'Variegata'

Pachysandra terminalis
"Japanese spurge" (Japan)

Buxus microphylla japonica
topiary in Los Angeles

Buxus sempervirens
"Common boxwood" (in rear)

Hydrocleis nymphoides
"Water poppy"

CACTACEAE: tribe 1 Pereskieae, tribe 2 Opuntieae

Pereskia grandifolia
"Rose cactus" (*Brazil*)

Pereskia corrugata
(*Rhodocactus corrugatus*)

Pereskia bleo
"Wax rose" (*Panama*)

Opuntia vulgaris (monacantha)
"Irish mittens"

Pereskia corrugata
at Fairchild's, Miami, Florida

Opuntia pachypus
(*Tephrocactus*) (*Perú*)

Opuntia schickendantzii
"Lion's tongue" (*Argentina*)

Opuntia robusta
(*Central Mexico*)

Opuntia braziliensis
"Tropical tree-opuntia"

CACTACEAE: tribe 2 Opuntieae

Opuntia stricta, *from Cuba
in the arid region near Madras, South India*

Cactus *in naturalistic setting
at Nymphenburg Botanic Garden, Munich*

Opuntia spinosior
in habitat, Mohave desert, California

Opuntia polyacantha 'Rubra'
in the valley of Zion National Park, Utah

Opuntia fulgida
"Jumping cholla" in tree stage, Southern Arizona

Opuntia polyacantha
"Prickly-pears" amongst chips of petrified wood, Arizona

CACTACEAE: tribe 2 Opuntieae

Opuntia exaltata
"Eve's-pin cactus" in Andean valley of Bolivia

Opuntia ficus-indica
"Indian fig" grown for its fruit, Central Mexico

Opuntia basilaris, *"Beaver-tail"*
between lava rocks, Mohave desert, California

Opuntia compressa (humifusa)
winter-hardy along the Atlantic, New Jersey

Nopalea cochenillifera
"Cochineal plant" with cochineal scale, on Tenerife

Nopalea cochenillifera
along road on the South coast of Tenerife, Canary Is.

CACTACEAE: tribe 1 Pereskieae, tribe 2 Opuntieae

Opuntia erinacea ursina
"Grizzly bear" on the Mohave desert, California

Pereskia grandifolia
"Rose cactus" in Cape Town Botanic Garden

Opuntia ficus-indica
naturalized on the cliffs of Capri, Italy

Opuntia ficus-indica 'Burbank's Spineless'
in Luther Burbank's garden, Santa Rosa, California

CACTACEAE: tribe 2 Opuntieae

Opuntia linguiformis
(*South Texas*)

Opuntia littoralis
"Prickly pear"

Opuntia erinacea ursina
"Grizzly bear" in bloom

Opuntia clavarioides *in bloom*
"Sea-coral" from Chile

Opuntia dillenii
"Tuna", common in West Indies

Opuntia bigelovii
"Teddy bear cactus" in Southern Nevada

Opuntia basilaris
"Rose tuna" in Sonora, Mexico

Opuntia elata
"Orange tuna" from Paraguay

CACTACEAE: tribe 1 Pereskieae, tribe 2 Opuntieae

Pereskia grandifolia
"Rose cactus" (*Brazil*)

Opuntia vulgaris 'Variegata'
(*monacantha*) *in flower*

Opuntia vulgaris 'Variegata'
"Joseph's coat"

Opuntia microdasys 'Albispina'
"Polka dots"

Opuntia violacea santa-rita
"Dollar cactus"

Opuntia microdasys 'Albata'
"Angelwings"

Opuntia linguiformis 'Maverick'
monstrose "Maverick cactus"

Opuntia fulgida mamillata
'Monstrosa', *"Boxing glove"*

Opuntia fragilis
"Pigmy tuna" from W. Canada

Opuntia clavarioides
"Fairy castles" (*Chile*)

Opuntia elata elongata
"Green wax cactus"

Opuntia erinacea ursina
"Grizzly bear"

CACTACEAE: tribe 2 Opuntieae

Opuntia clavarioides
"Black fingers" (*Chile*)

Opuntia verschaffeltii
(*Northern Bolivia*)

Opuntia violacea santa-rita
"Blue blade" in Tucson

Opuntia arbuscula
(*Arizona - Sonora*)

Opuntia fragilis
in Wyoming (Tetons)

Opuntia ramosissima
a *"Pencil cactus"*

Opuntia decumbens
(*Mexico, Guatemala*)

Opuntia leptocaulis
"Tesajo" or "Christmas cactus"

Opuntia basilaris
"Beaver tail" in Arizona

CACTACEAE: tribe 3 Cereeae; subtribe 1 Cereanae

Trichocereus weberbaueri
in the desert near Arequipa, Southern Perú

Pachycereus pringlei
"Mexican giant" in Sonora, at sunset

Pachycereus pecten-aboriginum
giant "Hairbrush cactus" 10 m tall, in Sonora

Carnegiea gigantea
"Saguaro" near Gila Bend, Arizona

CACTACEAE: tribe 3 Cereeae: Cereanae

Cleistocactus baumannii
"Firecrackers", in Argentina

Cephalocereus fluminensis, *in company with Neoregelias on the thorny Restinga, S.E. Brazil*

Lemaireocereus marginatus
"Organ pipe" near Teotihuacán, Mexico

Trichocereus (Helianthocereus) pasacana
in habitat, Northern Argentina

Lemaireocereus thurberi
"Arizona organ pipe", near Tucson

Escontria chiotilla
"Little star big-spine" near Oaxaca, Mexico

CACTACEAE: tribe 3 Cereeae: Cereanae

Cephalocereus polylophus
(*Neobuxbaumia*) (*Mexico*)

Carnegiea gigantea
"Saguaro" in bloom, Arizona

Carnegiea gigantea 'Monstrosa'
in Organ-pipe National Monument

Cephalocereus polylophus
(*Neobuxbaumia*) (*Mexico*)

Cephalocereus palmeri
"Woolly torch cactus" (*E. Mexico*)

Cephalocereus senilis
"Old man cactus" (*Mexico*)

Cereus hexagonus
"South American blue column"

Cereus hildmannianus
night-blooming, in Kew Gardens

Cereus hildmannianus
tree-like, in Trinidad

CACTACEAE: tribe 3 Cereeae: Cereanae

Cephalocereus royenii
in bloom (West Indies)

Cephalocereus polylophus
(*Neobuxbaumia*) (*Mexico*)

Cephalocereus senilis
young seedling

Cereus peruvianus
in flower (Central Brazil)

Cereus peruvianus
fruit showing seed

Cereus hildmannianus
nocturnal flower

Boliviocereus samaipatanus
(*Santa Cruz, Bolivia*)

Arthrocereus campos-porto
Minas Geraes, Brazil

Calymanthium substerile
(*Northern Perú*)

Cleistocactus strausii
"Silver torch" (*Bolivia*)

Espostoa lanata
"*Peruvian old man*"

Pseudoespostoa melanostele
(*Chosica, Perú*)

CACTACEAE: tribe 3 Cereeae: Cereanae

Lemaireocereus weberi, *"Candelabra cactus"* in the Sierra Madre, Oaxaca, Mexico

Cereus hexagonus *"Blue columns", near Cali, Dep't. Valle, Colombia*

Cephalocereus senilis *"Old man" cactus, in Distrito Federal, Mexico*

Echinocactus grandis, *a giant "Visnaga" near Tehuacán, Puebla, So. Mexico*

CACTACEAE: tribe 3 Cereeae: Cereanae

Cereus peruvianus 'Monstrosus'
"Curiosity plant"

Cereus jamacaru
with fruit, in New Zealand

Cereus peruvianus hort.
"Column cactus" in Vista, Calif.

Cleistocactus smaragdiflorus
"Fire-cracker cactus"

Cleistocactus candelilla
"Small candle of the Andes"

Bolivicereus samaipatanus
(*Borzicactus*) *Bolivia*

Escontria chiotilla
(*Southern Mexico*)

Cleistocactus wendlandiorum
(*Andes of Bolivia*)

Cleistocactus strausii jujuyensis
"Silver torch" (*Argentina*)

CACTACEAE: tribe 3 Cereeae: Cereanae

Eulychnia floresii
"White fluff-post" (Chile)

Lophocereus schottii 'Monstrosus'
"Monstrose totem"

Lemaireocereus treleasii
(Mexico)

Harrisia jusbertii
"Moon cactus" (Paraguay)

Lophocereus schottii
"Totem-pole" in Sonora

Harrisia bonplandii
night-bloomer (Brazil)

Oreocereus celsianus
"Old man of the Andes"

Nyctocereus serpentinus
"Queen of the night"

Machaerocereus eruca
"Creeping devil" (Baja California)

Trichocereus pachanoi
"Night-blooming San Pedro"

Oreocereus celsianus
"Old man of the mountains"

Trichocereus johnsonii
(So. America)

CACTACEAE: tribe 3 Cereeae: Cereanae

Lemaireocereus marginatus, *"Organ-pipe cactus"* as fence at temples of Mitla, So. Mexico

Machaerocereus eruca *"Creeping devil"* in Baja California, Mexico

Trichocereus pachanoi *"Night-blooming San Pedro"* (Ecuador)

Trichocereus spachianus *"Torch cactus" in Western Argentina*

Cleistocactus strausii *"Silver torch"*, seedlings in Roehrs cactus house

Wilcoxia poselgeri *from Texas and Coahuila (by Chas. Glass)*

CACTACEAE: tribe 3 Cereeae: Cereanae

Lemaireocereus marginatus
"Organ pipe" (Mexico)

Carnegiea gigantea
"Saguaro" 12 m tall, Arizona

Cereus hexagonus
as windbreak, Cali, Colombia

Lemaireocereus thurberi
"Arizona organ pipe", Tucson

Lemaireocereus stellatus
Hope Botanic Garden, Jamaica

Nyctocereus serpentinus
"Snake cactus" (Mexico)

Lophocereus schottii
"Totem-pole" in Edinburg, Texas

Trichocereus pachanoi
night-blooming, in Vista, Calif.

Myrtillocactus geometrizans
at Jardin Botanico, Mexico City

CACTACEAE: tribe 3 Cereeae: Cereanae

Oreocereus hendrickseniana
(*Northern Chile*)

Myrtillocactus geometrizans
"Blue myrtle" (So. Mexico)

Neobinghamia mirabilis
in flower (Perú)

Trichocereus racaquiensis
at Palmengarten, Frankfurt, Germany

Cephalocereus leucostele
(*Stephanocereus*), (*Bahia*)

Trichocereus peruvianus
"Peruvian torch-cactus"

Pachycereus pringlei
"Mexican giant" (No. Mexico)

Maritimocereus nanus
(*Loxanthocereus*) *So. Perú*

Machaerocereus eruca
"Creeping devil" (Mexico)

CACTACEAE: tribe 3 Cereeae: Hylocereanae

Aporocactus flagelliformis
"Rat-tail cactus" in hanging basket (*Hidalgo, Mexico*)

Cryptocereus anthonyanus (*Chiapas, Mexico*)
"St. Anthony's rick-rack", a fragrant night-bloomer

Selenicereus grandiflorus
"Queen of the night", opening in the evening (*Antilles*)

Aporophyllum 'Starfire'
(*Aporocactus × Epiphyllum*) by Harry Johnson, Calif.

CACTACEAE: tribe 3 Cereeae: Hylocereanae

Cryptocereus anthonyanus
curious epiphyte from Chiapas

Selenicereus urbanianus
"Nightblooming cereus" (Cuba)

Selenicereus grandiflorus
"Queen of the night"

Deamia testudo
"Tortoise cactus"

Selenicereus pteranthus
"Princess of the night" (Mexico)

Selenicereus vagans
night-blooming climber (Mexico)

Hylocereus undatus
"Honolulu Queen"

Marniera chrysocardium
epiphytic night-bloomer

Hylocereus lemairei
night-blooming climber (Tobago)

CACTACEAE: tribe 3 Cereeae: Hylocereanae, Echinocereanae

Selenicereus hamatus
"Moon cactus" (Antilles)

Chamaecereus sylvestri
"Peanut cactus" (Argentina)

Rebutia (Aylostera) fiebrigii
"Crown cactus" (Bolivia)

Chamaecereus sylvestri
"Peanuts" in bloom

Chamaecereus sylvestri 'Aureus'
at Cap Ferrat, France

Chamaecereus sylvestri
'Monstrosus'

Echinocereus ehrenbergii
(Mexico)

Echinocereus dasyacanthus
"Rainbow cactus"

Echinocereus dasyacanthus
in bloom (W. Texas)

Echinocereus viridiflorus
"Green-flowered hedgehog"

Echinocereus reichenbachii
"Lace cactus"

Echinocereus rigidissimus
(pectinatus) from Sonora

CACTACEAE: tribe 3 Cereeae: Echinocereanae

x Chamaelopsis 'Firechief'
beautiful bigeneric (Chamaecereus x Lobivia)

Echinocereus engelmannii
"Hedgehog cactus" on the Mohave desert, California

Echinocereus longispinus
a "Pitaya" from Mexico

Echinocereus enneacanthus
a "Strawberry cactus" of Sonora, Mexico

Echinocereus triglochidiatus neo-mexicanus
floriferous "Hedgehog" from New Mexico

Echinocereus viridiflorus
"Green flower hedgehog" (South Dakota to Texas)

Echinopsis multiplex hybrida
"Lily cactus" hybrids in Fallbrook, California

Echinocereus triglochidiatus (polyacanthus)
"Hedgehog cactus" in Arizona

CACTACEAE: tribe 3 Cereeae: Echinocereanae

Echinocereus albispinus
"Lace cactus" in Oklahoma

Echinocereus chloranthus
"Brown-flowered pitaya" (W. Texas)

Echinocereus blanckii
"Hedgehog" in South Texas

Echinocereus papillosus
"Hedgehog cactus" in W. Texas

Echinocereus engelmannii
made of glass (Harvard University)

Echinocereus dasyacanthus
"Rainbow cactus" (W. Texas)

Echinocereus pentalophus
(South Texas)

Echinocereus sciurus
from Baja California, Mexico

Echinocereus pectinatus
"Pitaya" from Texas and Mexico

CACTACEAE: tribe 3 Cereeae: Echinocereanae

Echinocereus perbellus
a "Lace cactus" from Texas

Echinopsis hamatacantha
"Lily cactus" (Argentina)

Lobivia binghamiana
"Cob-cactus" from S.E. Perú

Lobivia bruchii
(Northern Argentina: Tucumán)

Lobivia marsoneri
(Northern Argentina) by Chas. Glass

Lobivia larabei
(Urubamba, Perú) found by H. Johnson, California

Rebutia grandiflora
"Scarlet crown cactus" (Argentina)

Rebutia violacaeflora
"Rosy crown cactus" (No. Argentina)

CACTACEAE: tribe 3 Cereeae: Echinocereanae

Rebutia deminuta
(*No. Argentina*)

Rebutia kupperiana
miniature in English tea cup

Lobivia cylindrica
(*Argentina: Cordoba*)

Echinopsis multiplex
"Easter lily cactus"

Lobivia huascha rubriflora
(*Trichocereus*) *W. Argentina*)

Rebutia hybrida
"Red crown cactus"

Lobivia (Rebutia) steinmannii
grown in New Zealand

Rebutia calliantha
(*Northern Argentina*)

Soehrensia (Lobivia) formosa
(*Argentina: Mendoza*)

Rebutia digitiformis
(*So. America*)

Lobivia aurea
"Golden lily cactus"

Lobivia huascha
(*W. Argentina*)

CACTACEAE: tribe 3 Cereeae: Echinocereanae

x Lobiviopsis 'Aurora'
(*Lobivia* x *Echinopsis*) *by C. Glass*

Lobivia pseudocachensis
(*Northern Argentina*)

Rebutia crispata
(*South America*)

Rebutia pseudodeminuta
"Wallflower crown" (*Argentina*)

Rebutia senilis
"Fire crown"

Rebutia miniscula
(*N.W. Argentina*)

Soehrensia ingens (Lobivia)
in collection Marnier-Lapostolle, France

Soehrensia oreopepon (Lobivia)
(*No. Argentina: Mendoza*)

CACTACEAE: tribe 3 Cereeae: Echinocereanae, Echinocactanae 263

Echinopsis multiplex
"Easter lily cactus"

Lobivia famatimensis
(*Northern Argentina*)

Ariocarpus trigonus
"Living rock" (*New Mexico*)

Epithelantha micromeris
"Button-cactus" (*Texas*)

Echinocactus grusonii
"Golden barrel" (*C. Mexico*)

Astrophytum capricorne
"Goat's horn" (*Mexico*)

Ferocactus orcuttii
near Tijuana, Baja California

Ferocactus horridus (*peninsulae*)
"Fishhook cactus" (*Baja California*)

Ferocactus fordii
(*Baja California*)

CACTACEAE: tribe 3 Cereeae: Echinocactanae

Acanthocalycium klimpelianum
"Argentina barrel cactus"

Ariocarpus trigonus
"Living rock cactus" (Tamaulipas, Mexico)

Ariocarpus retusus
"Seven stars" (Mexico: San Luis Potosi)

Ariocarpus trigonus elongatus
"Living star" (Mexico)

Astrophytum ornatum
"Monk's-hood" (Mexico)

Astrophytum myriostigma
"Bishop's cap" (C. Mexico)

Echinocactus conothelos (Thelocactus)
(Tamaulipas, Mexico)

Ferocactus acanthodes
"Firebarrel" (Nevada to Baja California)

CACTACEAE: tribe 3 Cereeae: Echinocactanae

Astrophytum myriostigma
"Bishop's cap" in bloom

Astrophytum asterias
"Sand dollar"

Astrophytum asterias
"Peyote" in flower (Nuevo Leon)

Astrophytum myriostigma
quadricostatum, "Bishop's hood"

Astrophytum myriostigma
"Peyote cimmaron" (Durango)

Astrophytum ornatum
"Star cactus" (Hidalgo)

Ariocarpus kotschoubeyanus
(San Luis Potosi)

Acanthocalycium violaceum
"Violet sea-urchin" (Argentina)

Coryphanta elephantidens
in Xochicalco, Mexico

Ferocactus covillei
"Bird cage" (Sonora)

Epithelantha micromeris
"Golf balls" (Texas, N. Mexico)

Ferocactus acanthodes (seedling)
"Fire barrel" (Nevada to Mexico)

CACTACEAE: tribe 3 Cereeae: Echinocactanae

Echinocactus grusonii *in flower*
"Golden barrel" (*Hidalgo, San Luis Potosi*)

Echinomastus macdowellii
(*Coahuila, Mexico*)

Thelocactus saussieri (Gymnocactus)
(*San Luis Potosi, Mexico*)

Echinocactus texensis
(*Texas, No. Mexico*)

Hamatocactus setispinus
"Strawberry cactus" (*No. Mexico*)

Gymnocalycium damsii
a "Chin cactus" from Paraguay

Gymnocalycium mihanovichii 'Multicolor'
by Roberto Seidel, Corupá, Brazil

Gymnocalycium bruchii
"Chin-cactus" from Argentina

CACTACEAE: tribe 3 Cereeae: Echinocactanae

Echinocactus grusonii
"Golden barrel" in bloom, Distrito Federal, Mexico

Ferocactus acanthodes
"Barrel cactus" in Devil's garden, So. California

Parodia aureispina
"Tom Thumb" miniatures in bloom

Gymnocalycium mihanovichii 'Rubra'
"Ruby balls" grafted on cereus, in Tokyo, Japan

Lophophora williamsii, "Peyote"
or "Sacred mushroom", flowering in Osaka, Japan

Uebelmannia pectinifera (grafted)
from So. Brazil, at Munich Botanic Garden, Germany

CACTACEAE: tribe 3 Cereeae: Echinocactanae

Ferocactus latispinus
"Fish-hook barrel"

Ferocactus acanthodes
"Fire barrel"

Gymnocalycium mihan. friedrichii
"Rose plaid cactus"

Gymnocalycium venturianum
(*Uruguay: Montevideo*)

Gymnocalycium mihanovichii
"Plain chin-cactus"

Gymnocalycium leeanum
"Yellow chin-cactus"

Gymnocalycium saglione
(*Argentina: Tucumán*)

Gymnocalycium fleischerianum
(*Paraguay*)

Gymnocalycium denudatum
"Spider cactus"

Hamatocactus setispinus
"Strawberry-cactus"

Gymnocalycium multiflorum
(*So. Brazil to Argentina*)

Gymnocalycium mihanovichii 'Rubra'
"Red cap" grafted on cereus

CACTACEAE: tribe 3 Cereeae: Echinocactanae

Gymnocalycium baldianum
(*Argentina*)

Gymnocalycium mihanovichii 'Rubra'
"Oriental moon", on Cereus

Gymnocalycium mihanovichii 'Rubra'
"Hibotan", on Cereus undatus

Leuchtenbergia principis
"Prism cactus" (*Mexico*)

Leuchtenbergia principis
"Agave cactus"

Hamatocactus setispinus
"Strawberry-cactus"

Neoporteria nidus senilis
(gerocephala), *in Vista, Calif.*

Lophophora williamsii (*juv.*)
"Mescal" or "Peyote"

Leuchtenbergia principis variegata
(*Tegelberg, Lucerne Valley, Calif.*)

CACTACEAE: tribe 3 Cereeae: Echinocactanae

Neoporteria nidus
"Birdsnest cactus" from No. Chile

Matucana haynei (Borzicactus)
from Matucana, Perú

Neoporteria nigrihorrida
from Coquimbo, Chile

Notocactus apricus
"Ball-cactus" from Uruguay

Notocactus concinnus
(Southern Brazil to Uruguay)

Neoporteria nidus
red-flowered "Birdsnest" (No. Chile)

Notocactus mammulosus
"Lemon ball" (Brazil to Argentina)

Notocactus ottonis
beautiful "Ball cactus" (Brazil to Argentina)

CACTACEAE: tribe 3 Cereeae: Echinocactanae

Notocactus mammulosus
"Lemon ball"

Notocactus magnificus hort.
(*Seaborn Del Dios, Escondido, Calif.*)

Notocactus scopa
"Silver ball" (*Paraguay*)

Wigginsia arechavaletai
(*Palmengarten, Frankfurt, Germany*)

Neolloydia conoidea
at *Fortin, Veracruz, Mex.*

Parodia sanguiniflora
"Red Tom Thumb"

Parodia camargensis
(*camblayana*) (*Bolivia*)

Obregonia denegrii
"Artichoke cactus"

Uebelmannia pectinifera
grafted plant

CACTACEAE: tribe 3 Cereeae: Echinocactanae

Notocactus rutilans
"Pink ball cactus" (Argentina)

Stenocactus coptonogonus
"Permanent-wave cactus" (Mexico)

Utahia sileri (Pediocactus)
from Pipe Springs, Southern Utah

Wigginsia tephracantha (Malacocarpus)
(So. Brazil, Uruguay, Argentina)

Turbinicarpus polaskii (Strombocactus)
near San Luis Potosi, Mexico

Pediocactus knowltonii
a miniature of the Colorado plateau

Strombocactus disciformis
from dry arroyos of Querétaro, Mexico

Uebelmannia pectinifera
mature plant, by J. Bogner, Munich

CACTACEAE: tribe 3 Cereeae: Echinocactanae

Notocactus haselbergii
"White-web ball"

Notocactus leninghausii
in flower (So. Brazil)

Notocactus leninghausii
"Golden ball" (older plant)

Notocactus scopa
"Silver ball" (Paraguay)

Notocactus rutilans
"Pink ball cactus"

Notocactus mammulosus
"Lemon ball", from the Pampas

Notocactus scopa 'Cristata'
"Spiralled silver ball"

Parodia mutabilis
(Argentina: Salta)

Parodia chrysacanthion
(No. Argentina: Jujuy)

Parodia aureispina
"Tom Thumb cactus"

Neoporteria nidus senilis
(Northern Chile)

Notocactus haselbergii
"White-web ball"

CACTACEAE: tribe 3 Cereeae: Echinocactanae, Cactanae, Coryphanthanae

Melocactus neryi
a "Red-cap" from Amazonas

Melocactus matanzanus
"Melon-cactus" from Cuba

Melocactus intortus
"Turk's cap" (West Indies)

Weingartia neocumingii
(Gymnocalycium) from Bolivia

Cochemiea poselgeri
"Long hooks" (Baja Calif.)

Mammillaria theresae
grafted plant in California

Coryphantha vivipara
"Spiny stars" in Alberta

Solisia pectinata
"Lace bugs" (Tehuacán)

Mammillaria zeilmanniana
"Strawberry cactus" (Guanajuato)

CACTACEAE: tribe 3 Cereeae: Echinocactanae, Cactanae, Coryphanthanae

Melocactus intortus, *mature "Turk's cap"*
at Tegelberg's, Lucerne Valley, California

Mammillaria parkinsonii
"Awl's eyes" in San Marino, California

Mammillaria geminispina (bicolor)
clustering at Huntington Botanic Garden, Calif.

Echinocactus grusonii
"Golden barrels" from Hidalgo, C. Mexico

CACTACEAE: tribe 3 Cereeae: Coryphanthanae

Mammillaria bocasana
"Snowball cactus" in seed

Mammillaria bocasana
"Powder puff" (*Mexico*)

Coryphantha vivipara
(*Manitoba, Canada to Texas*)

Mammillaria polythele (affinis)
"Pin-cushion" from Hidalgo

Mammillaria albilanata
(*Mexico: Guerrero*)

Dolichothele longimamma
"Finger-mound" (*Mexico*)

Mammillaria celsiana
"Showy pincushion"

Coryphantha palmeri
(*Coahuila to Durango*)

Mammillaria blossfeldiana
(*Baja California*)

Coryphantha vivipara arizonica
from Grand Canyon, Arizona

Mammillaria candida
"Snowball cactus"

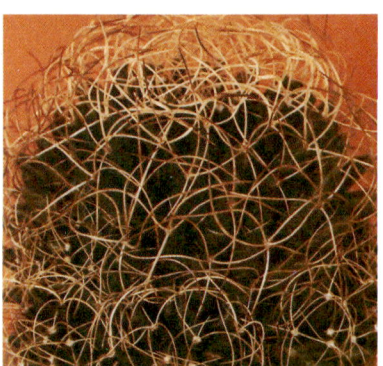
Mammillaria camptotricha
"Birdsnest" (*Mexico*)

CACTACEAE: tribe 3 Cereeae: Coryphanthanae

Mammillaria nejapensis, *crested*
"Smiling Jack", in Encinitas, California

Mammillaria parkinsonii, *crested*
"Sadsack", in Crestview Nursery, Encinitas

Mammillaria macracantha
with seed pods (San Luis Potosi, Mexico)

Mammillaria chionocephala
ringed with seed (Mexico: Coahuila)

Mammillaria bocasana 'Ed Hummel'
"Hummel's powder puff" in California

Escobaria tuberculosa (chisoensis)
from the Big Bend of the Rio Grande, Texas

CACTACEAE: tribe 3 Cereeae: Coryphanthanae

Mammillaria hahniana
"Old lady cactus" from Querétaro, Mexico

Escobaria chaffeyi (Coryphantha)
with brown flowers (Zacatecas, Mexico)

Mammillaria longicoma
(*Mexico: San Luis Potosi*)

Mammillaria magnimamma
"Mexican pincushion" (*Hidalgo, Mexico*)

Mammillaria pennispinosa
miniature from Coahuila, Mexico

Mammillaria lasiacantha
in Chihuahua, Mexico

Mammillaria sempervivi
a "Strawberry cactus" from Hidalgo, Mexico

Mammillaria wildii
"Fishhook pincushion" (*Querétaro and Hidalgo*)

CACTACEAE: tribe 3 Cereeae: Coryphanthanae

Mammillaria compressa
"Mother of hundreds"

Mammillaria hidalgensis
(Mexico: Hidalgo)

Mammillaria hemisphaerica
with flowers and seeds

Mammillaria geminispina
"Whitey" (C. Mexico)

Mammillaria elongata
"Golden stars" (Mexico)

Mammillaria geminispina nivea
glistening white

Mammillaria dolichocentra
"Ruby dumpling" (Querétaro)

Mammillaria pottsii
(W. Texas, Chihuahua)

Mammillaria seitziana
(Mexico: Ixmilquilpan)

Mammillaria parkinsonii
"Awl's eyes" (C. Mexico)

Mammillaria tegelbergiana
(Mexico: Chiapas)

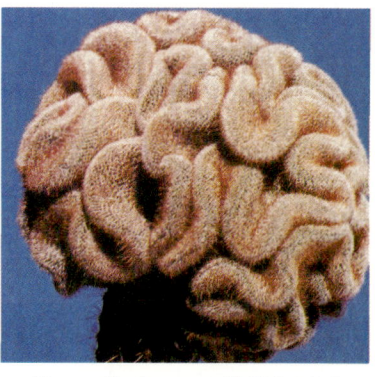
Mammillaria vaupelii 'Cristata'
"Silver brain"

CACTACEAE: tribe 3 Cereeae: Coryphanthanae

Thelocactus heterochromus
(Mexico: Chihuahua to Coahuila)

Neobesseya missouriensis
(Dakota and Montana to Oklahoma)

Pelecyphora asselliformis
"Hatchet cactus" (San Luis Potosi)

Pelecyphora pseudopectinata
from Palmillas, North Mexico

Thelocactus bicolor
(So. Texas, Mexico)

Thelocactus pottsii (heterochromus)
from Coahuila to Chihuahua

Mammillaria zeilmanniana
a "Strawberry-cactus" (Mexico: Guanahuato)

Mammillaria hahniana
"Old lady cactus" from Querétaro

CACTACEAE: tribe 3 Cereeae: Epiphyllanae

Epiphyllanthus obovatus
(*Zygocactus opuntioides*) (*C. Brazil*)

Epiphyllum oxypetalum
fragrant "Queen of the Night"

Epiphyllum hyb. 'Matador'
an "Orchid cactus" in Germany

Nopalxochia phyllanthoides
"Deutsche Kaiserin" (*Mexico*)

Zygocactus 'Llewellyn'
in bloom for Christmas

Zygocactus truncatus
"Thanksgiving cactus"

Heliocereus speciosus
the "Sun cactus" (*Mexico*)

Schlumbergera russeliana
"Shrimp cactus" (*Los Angeles Arb.*)

Epiphyllum hyb. 'Hermosus'
"Orchid cactus" in Pasadena

CACTACEAE: tribe 3 Cereeae: Epiphyllanae 282

Epiphyllum hyb. 'Argus'
rose-colored orchid cactus

Epiphyllum hyb. 'Flamingo'
bronze-red orchid cactus

Epiphyllum hyb. 'Brilliant'
a bright red Phyllocactus

Epiphyllum hyb. 'Elegantissimum'
"Dwarf orchid cactus"

Epiphyllum chrysocardium
"Golden heart" (Chiapas)

Nopalxochia ackermannii
"Orchid cactus" from Oaxaca

Zygocactus truncatus
"Crab cactus" (Rio de Janeiro)

Schlumbergera bridgesii
"Christmas cactus"

Zygocactus truncatus delicatus
flowers white when in shade

Nopalxochia phyllanthoides
floriferous basket plant

Zygocactus truncatus delicatus
flowers become pink in sun

Zygocactus truncatus
"Thanksgiving cactus"

CACTACEAE: tribe 3 Cereeae: Epiphyllanae

Disocactus eichlamii
showy epiphyte from Guatemala

Epiphyllum hybridum
"Orchid cactus" at window, in Duesseldorf, Germany

Nopalxochia phyllanthoides
known as "Empress", in Balboa Park, San Diego

Epiphyllum hyb. 'Hermosissimus'
Orchid cactus in Longwood Gardens, Kennett Square, PA.

CACTACEAE: tribe 3 Cereeae: Epiphyllanae, Rhipsalidanae

Nopalxochia ackermannii
epiphytic "Orchid cactus" from Southern Mexico

Rhipsalis cassutha
"Mistletoe cactus" as epiphyte in Brazil

Epiphyllum hyb. 'Grace-Ann'
in Santa Barbara, California

Epiphyllum hyb. 'Chas. Preston'
Orchid cactus with glistening, waxy flowers

Epiphyllum hyb. 'Peacockii'
iridescent Heliocereus hybrid

Zygocactus truncatus
"Thanksgiving cactus" from the Organ Mts., Brazil

CACTACEAE: tribe 3 Cereeae: Rhipsalidanae

Rhipsalidopsis gaertneri
"Easter cactus" (*So. Brazil*)

Rhipsalidopsis x graeseri 'Rosea'
in Sao Paulo, Brazil

Rhipsalidopsis x graeseri
(*gaertneri x rosea*)

Rhipsalidopsis gaertneri
(*Southern Brazil*)

Rhipsalidopsis x graeseri
close-up of flowers

Rhipsalidopsis rosea
"Dwarf Easter cactus" (*Paraná*)

Rhipsalis clavata
(*Rio de Janeiro*)

Rhipsalis capilliformis
"Old man's head"

Hatiora salicornioides
"Drunkard's dream" (*S.E. Brazil*)

Lepismium cruciforme
"Tree-and-rock cactus"

Rhipsalis quellebambensis
"Red mistletoe" (*Perú*)

Rhipsalis mesembryanthemoides
"Clumpy mistletoe cactus" (*Rio*)

CACTACEAE: tribe 3 Cereeae: Rhipsalidanae

Rhipsalis cassutha
"Mistletoe cactus" in Guanabara

Rhipsalis cassutha
close-up with berries

Rhipsalis zanzibarica
in Nairobi, Kenya, with Miss Powers

Rhipsalidopsis gaertneri
"Easter cactus" blooming in California

Rhipsalidopsis gaertneri, *on dinner table for Easter, out on the patio*

Rhipsalis neves-armandii
(*Brazil*)

Rhipsalis burchellii
in Brooklyn Botanic Garden

Rhipsalis cassutha
in Usambara Mts., Tanzania

CALYCANTHACEAE, CAMPANULACEAE

Campanula isophylla 'Alba'
"Star of Bethlehem" in hanging basket

Campanula isophylla mayii
"Italian bellflower"

Campanula medium 'Dean hybrid'
at Longwood Gardens, Penna.

Campanula isophylla 'Alba'
"Falling stars"

Campanula isophylla
in Hammerfest, Norway

Chimonanthus praecox
"Winter sweet" (China)

Campanula medium
"Canterbury bells" (So. Europe)

Canarina campanula
"Canary bellflower" (Tenerife)

CAMPANULACEAE, CANNACEAE

Canna generalis 'King Humbert'
"Bronze garden canna"

Canna generalis 'Rigoletto'
"Grand opera" canna

Canna generalis 'Brilliant'
blooming February in Cape Town

Siphocampylus manettiaeflorus (CAMP.)
from Cuba (by J. Bogner, Munich)

Canna generalis 'Red King Humbert'
at Botanical Gardens in Bogor, Java

Canna limbata
species from Brazil

Canna indica 'Moorea'
"Indian shot", in French Polynesia

Canna iridiflora
tall species from Perú

CANNACEAE

Canna gen. 'Richard Wallace'
growing on Tahiti

Canna generalis 'Striatus'
in Durban, Natal

Canna generalis 'President'
an old "Garden canna"

Canna generalis 'Lucifer'
on the road to Poona, India

Canna generalis 'City of Portland'
near Kanheri, Maharashtra, India

Canna generalis 'Confetti'
in Hilo, Hawaii

Canna generalis 'Cleopatra'
vari-colored beauty in Hawaii

Canna generalis 'Felix Ragout'
Botanic Garden Tuebingen, Germany

CAPPARACEAE, CAPRIFOLIACEAE

Cleome spinosa
"Spider flower" (W. Indies)

Abelia 'Eduard Goucher'
(grandiflora x schumanii) in Calif.

Viburnum tinus
"Laurustinus" (S.E. Europe)

Abelia x grandiflora
"Glossy abelia" from China

Abelia floribunda
"Mexican abelia", in New Zealand

Lonicera sempervirens
"Trumpet honeysuckle"

Lonicera heckrottii
"Gold-flame honeysuckle"

Lonicera hildebrandtiana
"Burmese honeysuckle"

CAPRIFOLIACEAE, CARIACACEAE

Viburnum tinus
"Laurustinus" in Lisbon, Portugal

Weigela florida 'Bristol Ruby'
in German garden center

Viburnum plicatum
"Japanese snowball" in Delaware

Viburnum lantana rugosum
"Wayfaring tree" in Denver

Viburnum odoratissimum 'Irvinii'
"Sweet viburnum" in San Diego

Viburnum hillieri
(erubescens x henryi)

Carica papaya
"Papaya" or "Melon tree"

Sambucus nigra
"European elderberry"

Viburnum opulus
"Cranberry bush" in fruit

Carnivorous Plants: BYBLIDACEAE, DROSERACEAE, LENTIBULARIACEAE

Dionaea muscipula
"Venus fly-trap"

Byblis gigantea
(*Western Australia*)

Dionaea muscipula 'Coccinea'
"Red fly-trap"

Drosera binata
"Twin-leaved sundew"

Drosera capensis
from South Africa

Drosera dichotoma
subtropical "Sundew" (*New Zealand*)

Pinguicula caudata
"Tailed butterwort" (*Mexico*)

Pinguicula vulgaris
"Butterwort" in Alberta, Canada

Pinguicula lutea
"Butterwort" from Louisiana

Carnivorous Plants: NEPENTHACEAE, SARRACENIACEAE

Nepenthes x mixta
a "Pitcher plant" in Munich Botanic Garden

Nepenthes maxima (Borneo)
at Kosobe Botanic Garden, Osaka, Japan

Sarracenia leucophylla
from Georgia, Florida and Mississippi

Darlingtonia californica
"Cobra plants" in Oregon habitat

Carnivorous Plants: NEPENTHACEAE, SARRACENIACEAE

tropical Nepenthes house
Nymphenburg Botanic Garden, Munich, Germany

Nepenthes mirabilis *in flower*
in habitat near Sogeri, Papua

Sarracenia purpurea
"Sweet pitcher plant" (Labrador to Maryland)

Sarracenia psittacina
"Parrot pitcher plant" (Georgia to Louisiana)

Nepenthes ampullaria
terrestrial pitcher plant from Malaysia

Nepenthes mirabilis
"Pitcher plants" growing terrestrially in New Guinea

Carnivorous Plants: NEPENTHACEAE

Nepenthes x mastersiana
(*sanguinea x distillatoria*)

Nepenthes rafflesiana
collected in Johore, Malaya

Nepenthes stenophylla (alata)
from tropical Borneo

Nepenthes x coccinea
(*distillatoria x mirabilis*)

Nepenthes villosa
at 3000 m, on Kina Balu, Borneo

Nepenthes 'Superba'
by M. Lecoufle, France

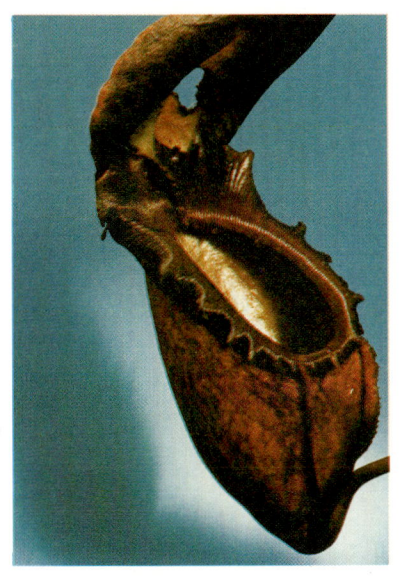

Nepenthes rajah
"Giant pitcher plant" *from Borneo*

Nepenthes bicalcarata (dyak)
from humid-warm Borneo

Nepenthes lowii
curious Malaysian pitcher plant

Carniv.: DROSERACEAE, LENTIBUL., NEPENTHACEAE, SARRACENIACEAE

Sarracenia 'Moorei'
(*drummondii x flava*)

Sarracenia purpurea
"Northern pitcher plant"

Sarracenia flava
"Yellow pitcher plant"

Sarracenia rubra
"Red pitcher plant" (N. Carolina to Florida)

Pinguicula lutea
"Yellow butterwort" growing in South Florida

Sarracenia psittacina
"Parrot pitcher plant"

Drosera schizandra
a "Sundew" in Queensland

Nepenthes sanguinea 'Superba'
in Munich Botanic Garden

Carnivorous Plants: NEPENTHACEAE, SARRACENIACEAE

Darlingtonia californica
"California pitcher plant"

Sarracenia 'Melanorhoda'
(*purpurea x stevensii*)

Sarracenia minor
"Hooded pitcher plant" in Carolina

Nepenthes 'Courtii'
(*psittacina x purpurea*)

Nepenthes x intermedia
(*rafflesiana x gracilis*)

Sarracenia purpurea
"Side-saddle flower"

Sarracenia x chelsonii
(*purpurea x rubra*)

Heliamphora nutans
"Sun pitchers" from Guyana

Sarracenia rubra
"Red pitcher plant" in flower

CARYOPHYLLACEAE

Dianthus caryophyllus 'Apollo'
cut-flower carnation

Dianthus caryophyllus 'White Sim', *fragrant*

Dianthus caryophyllus 'Red Sim'
excellent florists carnation

Dianthus caryophyllus 'Portrait'
for greenhouse culture

Dianthus caryophyllus minima
"Miniature carnation"

Dianthus caryophyllus 'Pink Littlefield'

Silene colorata
"Catchfly" on Delos, Greece

Dianthus 'Allwoodii'
(caryophyllus x plumarius)

Dianthus chin. 'Queen of Hearts'
"Chinese pinks" in California

Dianthus barbatus
"Sweet William"

Gypsophila paniculata
"Baby's breath"

Agrostemma gracilis
"Corn-cockle" from Greece

CARYOPHYLLACEAE

Dianthus caryophyllus
carnations grown by Amaro Indians in Cochabamba, Bolivia

Dianthus caryophyllus 'Arthur Sim'
"American Remontant" carnations at Cologne Flower Show

Dianthus caryophyllus chorus
"Clove pinks" in Norway

Dianthus caryoph. 'Pink Sim'
cut-flower carnation

Dianthus caryoph. 'Mini Queen'
exhibited in Malmö, Sweden

Dianthus chinensis 'Snowfire'
at Ball's, W. Chicago, Illinois

Dianthus gratianopolitanus
"Cheddar pinks", in Germany

Dianthus chinensis 'Heddewigii'
"Fringed rainbow pinks"

CASUARINACEAE

Casuarina equisetifolia
"Australian pine", on the shores of Tahiti

Casuarina decussata
Karri oak in Western Australia

Casuarina equisetifolia
"Horsetail trees" in the coral sand of Florida shores

Casuarina sumatrana
sheared "She-oaks" in Tiger Balm garden, Singapore

CELASTRACEAE

Euonymus japonica 'Aureo-marginata'

Euonymus japonica 'Medio-picta'

Euonymus japonica aureo-variegata 'Gold Spot'

Euonymus fortunei radicans *"Creeping euonymus"*

Euonymus japonica 'Argenteo-variegata'

Euonymus japonica 'Albo-marginata', *"Silver Queen"*

Euonymus japonica 'Albo marginata' *in California*

Euonymus fortunei vegeta *"Big-leaf winter creeper"*

Euonymus fortunei gracilis *"Silver-edge creeper"*

CELASTRACEAE, CHENOPODIACEAE, CISTACEAE

Cassine orientalis (juv.)
"False olive" (Madagascar)

Kochia scoparia
"Summer cypress" in Japan

Euonymus japonica
aureo-variegata 'Yellow Queen'

Combretum fruticosum
"Burning bush" in Jardim Botanico, Rio de Janeiro

Euonymus fortunei colorata
"Purple-leaf winter creeper" in Arnold Arboretum, Boston

Cistus salviifolius
"Sageleaf rock-rose"

Cistus incanus (villosus)
"Rock rose" (Corsica to Crimea)

Cistus ladanifer
"Laudanum" in Portugal

CLETHRACEAE, COCHLOSPERMACEAE, COMBRETACEAE

Quisqualis indica
"Rangoon creeper" (*Burma*)

Cochlospermum vitifolium plenum
(*Maximilianea*), "Brazilian-rose"

Cochlospermum vitifolium
"Buttercup tree"

Conocarpus erectus
"Button mangrove"

Licania tomentosa
the "Oiti", in Bahia, Brazil

Laguncularia racemosa
"White mangrove"

Combretum platypterium
a parasite of Tanga, Tanzania

Clethra arborea
"Lily-of-the-valley tree" (*Madeira*)

Combretum fruticosum
in Los Angeles Arboretum

BURSERACEAE, COCHLOSPERMACEAE, COMBRETACEAE

Canarium vulgare (commune)
"Java almond" along Kanari avenue, Bogor, Java

Cochlospermum vitifolium (Maximilianea)
"Buttercup tree" in Peradeniya Botanic Garden, Ceylon

Terminalia catappa
"Tropical almond" showing fruit, in Tahiti

Quisqualis indica
"Rangoon creeper" on Aruba, Netherlands Antilles

COMMELINACEAE

Geogenanthus undatus (R)
and Pilea cadierei (L) *at Roehrs greenhouses, New Jersey*

Dichorisandra thyrsiflora
"Blue ginger", in Brazil

Zebrina pendula
the "Silvery wandering Jew" of Mexico

Tradescantia fluminensis 'Variegata'
"Variegated wandering Jew" (Brazil)

Tradescantia sillamontana
"White velvet" creeper (Mexico)

Hadrodemas (Tripogandra) warscewiczianum
flowering rosettes in Guatemala

COMMELINACEAE

Geogenanthus undatus
"Seersucker plant"

Siderasis fuscata
"Brown spiderwort"

Palisota barteri
(Fernando Po, W. Africa)

Tradescantia sillamontana
"White gossamer"

Zebrina pendula 'Quadricolor'
"Happy wandering Jew"

Tradescantia flumin. 'Variegata'
"Speedy Henry"

Trad. albiflora 'Albo-vittata'
"Giant white inch plant"

Aplaleia multiflora
(Tropical America)

Tradescantia fluminensis
the *"Rio inch plant"*

Setcreasea purpurea
"Purple heart" vine

Campelia zanonia 'Mexican flag'
(Dichorisandra albo-marginata)

Dichorisandra reginae
"Queen's spiderwort"

COMMELINACEAE

307

Rhoeo spathacea (discolor)
"Moses-in-the-cradle"

Dichorisandra thyrsiflora
"Blue-flowered bamboo"

Rhoeo spathacea 'Vittata'
"Variegated boat-lily"

Tradescantia hirta
(*Setcreasea hirsuta*) (*Namibia*)

Setcreasea purpurea
"Purple heart", on St. Thomas

Cochliostema odoratissimum
(*jacobinianum*) (*Brazil*)

Gibasis geniculata
(*Tradescantia multiflora*)

Dichorisandra siebertii
in Sao Paulo, Brazil

Callisia repens
miniature "Turtle vine"

COMPOSITAE

Achillea filipendulina
"Fern-leaf Yarrow"

Arctotheca calendula
"Cape-weed" (South Africa)

Andryala pinnatifida
(Canary Islands)

Celmisia coriacea
"Mountain daisy" in New Zealand

Celmisia spectabilis
from sub-alpine New Zealand

Centaurea cyanus
"Cornflower" in Kwangchow, China

Calendula officinalis
"Pot-marigold" in W. Chicago

Chrysanthemum maximum
'Wirral Supreme', *"Shasta daisy"*

Bellis perennis fl. pl.
"English daisy" as bedding plant

COMPOSITAE

Senecio johnstonii, *"Giant groundsel"*
on Mt. Kilimanjaro at 3000 m elev., Tanzania

Argyroxiphium sandwicense
"Silver sword" in habitat, on Maui at 2600 m, Hawaii

Celmisia coriacea
"Mountain daisy" ascending Mt. Cook (3763 m), N.Z.

Chrysanthemum x morifolium
blooming in March, People's Park, Kwangchow, China

COMPOSITAE

Centaurea candidissima
"Dusty miller" (Sicily)

Senecio cineraria
"Cineraria Diamond"

Centaurea gymnocarpa
"Velvet centaurea" (Capri)

Artemisia arborescens
aromatic "Wormwood" of Southern Europe

Artemisia dracunculus
"Tarragon" or "Estragon", aromatic herb of Eurasia

Senecio cineraria 'Silver Dust'
dwarf "Dusty miller" in W. Chicago, Illinois

Senecio cineraria
(Cineraria maritima) in Kew Gardens, England

Helichrysum petiolatum
(Gnaphalium lanatum) an "Immortelle"

Calocephalus brownii
"Cushion bush" from Tasmania

COMPOSITAE

Chrysanthemum morifolium 'Turner'
large Exhibition variety, pot grown in England

Chrysanthemum morifolium 'Sungold'
florist's "Poodle" on Madison avenue, New York

Chrysanthemum morifolium 'Indianapolis Yellow'
treated with growth-retardant to shorten stem, by Roehrs

Chrysanthemum morifolium 'Decorative incurved'
florist's "Mum" gift-wrapped, Vaduz, Liechtenstein

COMPOSITAE

312

Chrysanthemum mor. 'Lorraine'
"Spider mum"

Chrysanthemum mor. 'Cathay'
"Spider-spoon"

Chrysanthemum mor. 'Golden Lace'
"Fuji-mum"

Chrys. mor. 'Peggy Ann Hoover'
"Spider-type"

Chrys. mor. 'Yellow Spoon'
"Spoon-mum"

Chrysanthemum mor. 'Venoya'
"Large anemone"

Chrysanthemum mor. 'Sweepstake'
"Pompon-mum"

Chrysanthemum mor. 'Real Mackay'
"Cascade-mum"

Chrys. mor. 'Golden Cascade'
"Daisy-cascade"

Chrysanthemum mor. 'Red King'
"Anemone-type"

Chrysanthemum mor. 'Precose'
"White anemone"

Chrysanthemum mor. 'Illini Bonbon'
"Cushion anemone"

COMPOSITAE

Chrysanthemum mor. 'Warhawk'
"Decorative type"

Chrysanthemum mor. 'Hawaii'
"Incurved Exhibition"

Chrysanthemum mor. 'Delaware'
"Incurved type"

Chrysanthemum mor. 'Imperial'
"Decorative form"

Chrysanthemum mor. 'Red Star'
"Decorative type"

Chrysanthemum mor. 'Princess Ann'
excellent pot-mum (decorative)

Chrys. mor. 'Beautiful Lady'
"Anemone mum"

Chrys. mor. 'Wilson's White'
"Decorative type"

Chrys. mor. 'Bonaffon de Luxe'
"Incurved" pot-mum

Chrys. mor. 'Yellow Princess Ann'
favorite "Decorative"

Chrysanthemum mor. 'Oregon'
"Semi-incurved"

Chrys. mor. 'Yellow Delaware'
"Semi-incurved"

COMPOSITAE

314

Ageratum houstonianum
"Floss-flower (C. America)

Callistephus chinensis
"China aster" in Bombay

Callistephus chinensis
'Dwarf double' (Pot-aster)

Aster novi-belgii
"New York aster"

Chrysanthemum frut. 'Mary Wootten'
in Napier, New Zealand

Aster amellus
"Italian aster"

Chrysanthemum frut. 'Mrs. Sanders'
"Tree marguerite"

Chrysanthemum frutescens
'Wellpark Beauty', in New Zealand

Chrysanthemum frut. 'Snowflake'
"New Zealand tree-marguerite"

Chrysanthemum frut. chrysaster
"Boston yellow daisy"

Chrysanthemum frutescens
"White marguerite" (Canary Isl.)

Chrysanthemum frut. 'Roseum'
"Pink marguerite"

COMPOSITAE

Emilia javanica (sagittata)
"Flora's paintbruch" (Pantropic)

Erigeron speciosus
"Fleabane" (Pacific coast)

Erigeron hybridus
in Cologne, Germany

Euryops pectinatus
"Resin bush" (So. Africa)

Eupatorium sordidum
"Boneset" from Mexico

Cosmos bipinnatus
"Garden cosmos"

Felicia fruticosa
"Asterbush" (So. Africa)

Felicia amelloides
'Astrid Thomas'

Felicia amelloides 'Variegata'
variegated "Blue daisy"

COMPOSITAE

Chrysanthemum x morifolium (*multistemmed*)
at People's Botanic Garden, Kwangchow, China

Dahlia pinnata 'Schweizerland'
"Garden dahlia", Essen Bot. Garden, Germany

Crassocephalum (Gynura) mannii
tree to 8 m tall, near Entebbe, Uganda

Dahlia pinnata, *growing in pots*
Repulse Bay Hotel, Hongkong

COMPOSITAE

Dahlia pinnata 'Siegerland'
(*cactus-flowered*)

Dahlia pinnata 'Apex'
(*straight cactus-flowered*)

Dahlia pinnata 'Jugendliebe'
(*formal decorative*)

Dahlia pinnata 'Pompon'
(*pompon or ball-type*)

Dahlia imperialis
"Tree dahlia" from Mexico

Dahlia pinnata 'Hit Parade'
(*semi-cactus*)

Dahlia pinnata
'Dwarf Collarette' (*bicolor*)

Dahlia pinnata
'Coltness Monarch' (*single*)

Dahlia pinnata
'Unwin's Dwarf' (*"Mignon"*)

COMPOSITAE

Gerbera jamesonii
"Transvaal daisy"

Gerbera jamesonii
"African daisies" in variety

Gerbera jamesonii
'Rosea'

Gerbera jamesonii plena
"Double Barberton daisies"

Gerbera viridifolia
from South Africa

Gazania rigens 'Fire Emerald'
superb strain planted in California

Gerbera jamesonii
in habitat, near Barberton, Transvaal

Oldenburgia arbuscula
in Kirstenbosch Botanic Garden, South Africa

Santolina chamaecyparissus
(Mediterranean region)

COMPOSITAE

Gazania rigens
'Gold Nuggets'

Gazania rigens 'Aztec'
a tricolor "Treasure flower"

Gazania rigens 'Copper King'
spectacular hybrid

Gazania uniflora
"Trailing gazania", in Durban, Natal

Gazania pavonia var. hirtella
in the Karroo, Cape Prov.

Gaillardia aristata
"Blanket flower" (Pacific Northwest)

Gazania nivea
in South Africa

Gaillardia x grandiflora
'Mandarin'

Gazania rigens hybrida
"Treasure flowers" in California

COMPOSITAE

Helenium bigelovii
"Sneezeweed" (California)

Helianthus annuus nanus
"Dwarf sunflower" in Trinidad

Helianthus annuus
"Common sunflower"

Helianthus laetiflorus
"Showy sunflower", in Colorado

Helianthus grosseserratus
in Western Texas

Heliopsis helianthoides
"Oxeye" (Eastern U.S.)

Helichrysum bracteatum
"Strawflower" (Australia)

Helichrysum belloides
"Immortelle" of New Zealand

Helipterum roseum (Acroclinium)
"Everlastings" of W. Australia

COMPOSITAE

Helichrysum petiolatum
"Immortelles" planted at base of tree, Rheinpark, Cologne

Tagetes patula, *"French marigold"*
in beds at Nymphenburg Palace, Munich, Bavaria

Gaillardia x grandiflora
planted in the casino garden in Estoril, Portugal

Gaillardia aristata
"Blanket flower" in native habitat, Montana

Onopordum nervosum (arabicum) (*Spain*)
ornamental thistle as biennial, Short Hills, New Jersey

Dahlia pinnata 'Hit Parade'
planted for exhibition, Gruga Park, Essen, Germany

COMPOSITAE

Phoenocoma prolifera
in Kirstenbosch, South Africa

Montanoa grandiflora (*Honduras*)
at Fairchild Gardens, Miami, Florida

Montanoa mexicana
"Daisy tree" in California

Carlina acaulis
ornamental thistle in Sweden

Senecio petasitis
"California geranium"

Jurinea sp. Wats
xerophytic shrub from Asia Minor

Liatris elegans
"Blazing star"

Oldenburgia arbuscula
up to Table Mountain, Cape Town

Baccharis pilularis
"Coyote bush", "Chapparal broom"

COMPOSITAE

Ursinia versicolor
in Joubert Park, Johannesburg, Transvaal

Vernonia colorata
an "Ironweed" in the Kigezi hills, Uganda

Senecio mikanioides
"Climbing senecio" near Mountains of the Moon, Uganda

Senecio macroglossus 'Variegatum'
"Variegated wax-vine" north of Nairobi, Kenya

Argyroxyphium sandwicense
"Silver sword" as grown in Longwood Gardens, Penna.

Leontopodium alpinus
"Edelweiss" growing on Schynige Platte, Swiss Alps

COMPOSITAE

Senecio macroglossus 'Variegatum'
"Variegated wax-vine"

Gynura x sarmentosa
"Purple passion vine"

Gynura aurantiaca
"Velvet plant"

Mikania ternata
"Plush vine"

Senecio picticaulis
(Tanzania and Sudan)

Senecio petraeus
(Tanzania)

Senecio serpens (Kleinia)
"Blue chalk sticks"

Senecio fulgens (Kleinia)
"Scarlet kleinia"

Senecio haworthii (K. tomentosa)
"Coccoon plant"

Senecio pendulus (Kleinia)
"Inchworm" from Arabia

Senecio rowleyanus
"String of pearls" (So. Namibia)

COMPOSITAE

Senecio herreianus
(*Kleinia gomphophylla*) (*Namibia*)

Senecio radicans
"*Creeping berries*" (*Karroo*)

Senecio citriformis
in Bot. Garden Edinburgh, Scotland

Silybum marianum
"*Holy thistle*", *on Delos, Greece*

Senecio rowleyanus
"*String of beads*"

Senecio confusus
"*Mexican flame-vine*"

Wedelia trilobata
"*Creeping daisy*" (*W. Indies*)

Senecio grandifolius
from southern Mexico

Senecio macroglossus 'Variegatum'
"*Variegated wax-vine*"

COMPOSITAE

Senecio hyb. 'Multiflora nana'
"Florists cineraria"

Senecio hyb. 'Grandiflora'
blue with white eye

Senecio hyb. 'Grandiflora'
giant red with eye

Senecio hybridus 'Maxima'
"Exhibition cineraria"

Cosmos bipinnatus
"Mexican aster"

Senecio hybridus
'Multiflora grandiflora'

Wedelia trilobata
"West Indian creeper"

Coreopsis tinctoria
"Calliopsis" in hort.

Sanvitalia procumbens
"Hussars-heads"

Ligularia tussilaginea
'Argentea' (*kaempferi*)

Ligularia tuss. 'Aureo-maculata'
"Leopard plant"

Senecio mikanioides
"Parlor" or "German ivy"

COMPOSITAE

Gynura x sarmentosa
"Purple passion vine"

Ligularia tussilaginea
'Aureo-maculata' (*Farfugium*)

Gynura aurantiaca
"Velvet plant"

L: Gynura sarmentosa (*Bot. Mag.*) C: Gynura bicolor
(*Bot. Mag.*) R: Gynura aurantiaca (*Ill. Hort.*) *"Velvet plant"*

Oyedaea verbesinioides
at University of California Bot. Garden, Los Angeles

Senecio johnstonii
"Giant groundsel" in Tanzania

Mikania hemispherica
in Sao Paulo, Brazil

Oldenburgia arbuscula
in Kirstenbosch Bot. Garden, Cape

COMPOSITAE

Senecio x hybridus (cruentus)
"Florist's cinerarias" at Roehrs greenhouses, N.J.

Rudbeckia fulgida
"Orange cone flowers" in containers, Stockholm, Sweden

Osteospermum fruticosum
"Burgundy mound" in Natal, South Africa

Osteospermum fruticosum 'Album'
"Trailing African daisy"

Rudbeckia hirta
"Blackeyed Susan" in Grindelwald, Switzerland

Senecio glastifoliius
"Maple aster" in Cape Prov., South Africa

COMPOSITAE

Mutisia clematis
(*Colombia, Ecuador*)

Astericus sericeus
(*Odontospermum*) (*Canary Is.*)

Vernonia *natalensis*
"Ironweed" in South Africa

Rudbeckia hirta
'Double Gloriosa'

Rudbeckia fulgida
"Coneflowers" in Cologne, Germany

Heliopsis helianthoides scabra
"Ox-eyes" (*Eastern U.S.*)

Solidago nemoralis
"Golden rod" (*Canada to Arizona*)

Othonna herrei
(*So. Africa: Namaqualand*)

Ligularia przewalskii (Senecio)
in Tivoli, Copenhagen

COMPOSITAE
330

Tagetes patula 'Spry'
"Bedding marigold"

Tagetes patula
'Diamond Jubilee'

Tagetes patula
'Petite Yellow' (*W. China*)

Tagetes patula 'Rusty Red'
"French marigold" developed from Mexican species

Tagetes patula 'Naughty Marietta'
"Single French marigold" at Ball's, W. Chicago, Illinois

Tagetes patula
'Gay Ladies' (*Ball, W. Chicago*)

Tagetes patula
'Gold Rush' (*double crested*)

Tagetes patula 'Orange Lady'
Kew Gardens, London

COMPOSITAE

Leontopodium alpinum
the "Edelweiss", in the Bernese Alps, Switzerland

Zinnia elegans
"Youth-and-old-age" in tropical Bali, Indonesia

Gerbera jamesonii, *"African daisies"*
cultivated strains at Flower Show in Malmö, Sweden

Cynara scolymus
"Artichoke", edible bud and open flower, in Vista, Calif.

COMPITAE

332

Venidium fastuosum
"Cape daisy"

x Venidio-arctotis 'Champagne'
(*Arctotis x Venidium*)

Acrtotis 'Hybrida'
"African daisy"

Zinnia elegans
'Red Sun'

Zinnia elegans
'Thumbelina' (*miniature*)

Tithonia rotundifolia
"Mexican sunflower"

Zinnia elegans
'Dwarf Salmon Rose'

Zinnia elegans
'Yellow Ruffles'

Zinnia elegans
'Cherry Ruffles'

Coniferae: ARAUCARIACEAE

Araucaria heterophylla (excelsa)
"Norfolk Island pine", in Petropolis, Brazil

Araucaria cunninghamii
"Hoop pine", in Wellington, New Zealand

Araucaria bidwillii
"Bunya-Bunya" pine, Queensland, Australia

Araucaria araucana (imbricata)
"Monkey-puzzle" tree in Chile

Coniferae: ARAUCARIACEAE

Agathis robusta
"Queensland kauri", near Maryborough, Queensland

Agathis australis
"Kauri pine", in Auckland, New Zealand

Agathis robusta
as a container plant, Willandra nursery, Sydney N.S.W.

Agathis australis
ancient "Kauri" giant, on North Island, N.Z.

Coniferae: ARAUCARIACEAE

Araucaria heterophylla (excelsa)
"Norfolk Island pines", in Radio City, New York

Araucaria araucana
old "Monkey-puzzle tree", in Vancouver, British Columbia

Araucaria columnaris
"Cook pine" in Peradeniya, Ceylon

Araucaria heterophylla (excelsa)
"Star-pine" on Oahu, Hawaii

Araucaria columnaris
near Vila, New Hebrides

Araucaria bidwillii, *"Bunya-Bunya"*
in the Southern Alps, Westland, New Zealand

Araucaria columnaris
"Cook pines" in habitat, near Noumea, New Caledonia

Coniferae: ARAUCARIACEAE

Araucaria heterophylla (excelsa)
"Norfolk Island pine" from cuttings, Bruges, Belgium

Araucaria columnaris (cookii)
at Foster Botanical Garden, Honolulu, Hawaii

Araucaria araucana (imbricata), *"Monkey-puzzle"*,
spiny leaves keep monkeys from climbing

Araucaria cunninghamii, *"Hoop pine"*,
young tree, at Royal Botanic Garden, Sydney

Coniferae: CUPRESSACEAE

Chamaecyparis nootkatensis 'Pendula'
weeping "Alaska cedar" on Long Island, New York

Juniperus chinensis 'Torulosa'
"Hollywood juniper", at Disney World, near Orlando, Fla.

Cupressus funebris
"Mourning cypress" in Peradeniya, Sri Lanka

Cupressus sempervirens
"Mediterranean cypress" in various forms, on Rhodes

Coniferae: CUPRESSACEAE

Cupressus sempervirens 'Stricta' at Yalta, Crimea, Soviet Union

Cupressus macrocarpa 'Aurea' "Golden Monterey cypress", in N.Z.

Juniperus chinensis 'Torulosa' "Hollywood juniper", Vista, Calif.

Cupressus sempervirens "Italian cypress" with cones

Juniperus communis "Common juniper", berry-like cones

Juniperus chinensis 'Torulosa' "Twisted Chinese juniper"

Cupressocyparis 'Leylandii' (Chamaecyparis x Cupressus)

Juniperus excelsa 'Stricta' "Pyramid Greek juniper"

Cupressus lusitanica "Portuguese cypress"

Coniferae: CUPRESSACEAE

Juniperus chinensis 'Sargentii'
trained as bonsai, at EXPO 70, Osaka, Japan

Libocedrus bidwillii (Calocedrus)
"Incense cedar" in Marlborough, South Island, N.Z.

Libocedrus plumosa
"Kawaka", in Christchurch, New Zealand

Thuja orientalis (Biota, Platycladus)
"Oriental arborvitae" in topiary, Disney World, Florida

Coniferae: CUPRESSACEAE

Widdringtonia whytei
"Cypress pine" in Tanzania

Thuja occidentalis
"American arbor-vitae"

Thuja orientalis
"Chinese arborvitae"

Juniperus scopulorum
'Tolleson's Weeping'

Juniperus chinensis
'Pfitzeriana'

Juniperus scopulorum
'Blue Haven', *in California*

Thuja occidentalis
"American arborvitae"

Thuja orientalis
'Aurea nana'

Thuja plicata
"Western arborvitae" (*Brit. Columbia*)

Coniferae: CUPRESSACEAE, PINACEAE

Cupressus macrocarpa
wind-blown "Monterey cypress" at Point Lobos, Calif.

Cedrus libanii
"Cedar of Lebanon" on Mt. Lebanon (Lebanon)

Juniperus sabina 'Tamariscifolia'
creeping "Tamarix juniper" in California

Juniperus virginiana 'Tripartita'
spreading "Red cedar" in Cologne, Germany

Thuja occidentalis 'Rheingold'
in Arnold Arboretum, Boston, Mass.

Picea pungens
"Colorado spruce" in Western Utah

Coniferae: PINACEAE

Abies concolor 'Compacta'
"Dwarf Colorado fir"

Abies concolor
"White fir" at Grand Canyon, Ariz.

Abies koreana
"Korean fir" in Germany

Callitris arenosa
"Cypress pine" (Tasmania)

Abies nordmanniana
"Caucasus fir"

Abies bracteata
"Bristle-cone" (Strybing Arboretum)

Cedrus atlantica 'Glauca'
"Blue Atlas cedar"

Cedrus atlantica
"Atlas cedar" in England

Cedrus atlantica 'Glauca'
"Atlas cedar" in Kew Gardens, London

Coniferae: PINACEAE

Cedrus atlantica 'Glauca pendula'
"Weeping blue cedar"

Cedrus atlantica 'Glauca'
in riverside nursery market, Amsterdam, Holland

Picea pungens
"Colorado spruce" within the Kremlin, Moscow

Cedrus deodara
"Himalayan cedar" in Yalta on the Crimea, Russia

Coniferae: PINACEAE

Cedrus deodara
"Deodar cedar" in Bonsall, San Diego County, Calif.

Cedrus libanii, *ancient tree*
on Mt. Lebanon, the cedar-wood for Solomon's Temple

Picea pungens
"Colorado spruce" in the Rocky Mountains of Colorado

Picea abies, *"Norway spruce"*
around ninth century stave church, Bygdoy, Norway

Coniferae: PINACEAE

Pinus thunbergiana
"Japanese black pine" as bonsai, Osaka, Japan

Pinus pinea
"Italian stone pine", in Vatican Gardens, Rome

Pinus halepensis
"Aleppo pine" near Olympia, Greece

Picea glauca densata
"Black Hills spruce" from South Dakota

Tsuga canadensis
"Hemlock spruce", in Lund University garden, Sweden

Pinus pinea, *the "Umbrella pine"*
along romantic Amalfi coast, Southern Italy

Coniferae: PINACEAE

Cedrus deodara
"Deodar cedar", young tree

Cedrus deodara 'Compacta'
"California Christmas tree"

Picea abies (excelsa)
"Norway spruce" in Oslo

Picea glauca albertiana
"Alberta white spruce"

Picea glauca 'Conica'
"Dwarf Alberta spruce"

Picea orientalis 'Aurea'
"Oriental spruce"

Picea pungens
"Blue spruce", Bryce Canyon, Utah

Picea pungens 'Kosteriana'
"Koster's blue spruce"

Picea pungens
"Colorado spruce" in New York

Coniferae: PINACEAE

Pinus cembroides
old weathered "Pinyon pine" in Bryce Canyon, Utah

Pinus halepensis stankewiczii
near Yalta on the Crimea Peninsula, Russia

Pseudotsuga menziesii
mighty "Douglas fir" in Bryce Canyon, Utah

Pinus montezumae, "Montezuma pine"
near Monte Alban, Oaxaca, So. Mexico

Coniferae: PINACEAE, PODOCARPACEAE

Pinus thunbergiana
as bonsai, Osaka, Japan

Pinus parviflora
with twisted trunk, in Japan

Pinus roxburghii
at Rancho Santa Fe, Calif.

Pseudolarix kaempferi (amabilis)
"Golden larch" in Germany

Microcachrys tetragona
(Mountains of Tasmania)

Dacrydium cupressinum
"Rimu" (juv.) in New Zealand

Tsuga canadensis
"Canada hemlock" in New Jersey

Podocarpus macrophyllus 'Maki'
as "Ming tree" in Hongkong

Pseudotsuga menziesii
"Douglas fir", North Rim, Arizona

Coniferae: PINACEAE

Pinus thunbergiana
"Japanese black pine"

Pinus canariensis
"Canary Island pine"

Pinus parviflora
trained as bonsai, in Japan

Pinus ponderosa
"Western yellow pine" with cone

Pinus palustris
"Southern pine" yields turpentine

Pinus halepensis
"Jerusalem pine" in Israel

Pinus cembroides
"Mexican stone pine"

Pinus radiata
"Monterey pine" in California

Pinus strobus
"Weymouth pine" in New Jersey

Coniferae: PODOCARPACEAE

Podocarpus nagi
"Broadleaf podocarpus"

Podocarpus macrophyllus
"Buddhist pine"

Podocarpus macrophyllus 'Maki'
"Southern yew", from China

Podocarpus falcatus
"Oteniqua yellowwood"

Podocarpus elongatus
"Fern podocarpus"

Podocarpus neriifolius
at University of Hawaii

Podocarpus gracilior
"African fern-pine"

Podocarpus henkelii
"Long-leaved yellow-wood"

Podocarpus milanjianus
in Uganda

Coniferae: PODOCARPACEAE

Podocarpus nerifolius
as decorator plant, De Young Museum, San Francisco

Podocarpus totara, *the "Totara"*
giant tree reputed 1000 yrs. old, Marlborough, N.Z.

Podocarpus macrophyllus, *a "Buddhist pine"*
at Kannon temple, Kamakura, Japan

Podocarpus macrophyllus 'Maki'
in topiary form, Disney World, Orlando, Florida

Coniferae: CEPHALOTAXACEAE, TAXACEAE, TAXODIACEAE

Taxus cuspidata
"Japanese yew"

Taxus cuspidata 'Capitata'
"Upright Japanese yew"

Taxus baccata
"English yew"

Cryptomeria japonica
"Japanese cedar"

Metasequoia glyptostroboides
"Dawn redwood"

Sciadopitys verticillata
"Umbrella pine", Mt. Rokku, Japan

Cephalotaxus harringtonia
"Harrington plum yew"

Sequoiadendron giganteum
"Giant redwood" (close-up)

Cunninghamia lanceolata
"Mao Chia" in Taiwan

Coniferae: PODOCARPACEAE, TAXACEAE, TAXODIACEAE

Taxodium mucronatum
ancient "Montezuma cypress" of El Tule, Oaxaca, Mexico

Cunninghamia lanceolata
in Central Mountains, near Taichung, Taiwan

Taxus baccata, *"English yew"*
shaped pyramids in the garden of Versailles, France

Taxus cuspidata 'Capitata'
artfully shaped, Longwood Gardens, Kennett Sq., Penna.

Cunninghamia lanceolata
close-up of "China fir", Morris Arboretum, Pennsylvania

Podocarpus macrophyllus 'Maki'
as pyramids, Keeline-Wilcox Nurseries, Irvine, California

Coniferae: TAXODIACEAE

Glyptostrobus pensilis
"Chinese water pine" at Golden Pavilion, Kyoto, Japan

Taiwania cryptomerioides
in temple garden of Heian Shrine, Kyoto, Japan

Cryptomeria japonica, *sacred "Japanese cedar"
in the temple compound, Kamakura, Japan*

Cryptomeria japonica 'Lobbii'
columnar form of "Japanese cedar" on Long Island, N.Y.

Coniferae: TAXODIACEAE

Sequoia sempervirens
"Coast redwoods" along Redwood highway, No. Calif.

Sequoiadendron giganteum, *"Giant sequoia"*
"General Sherman tree", Sequoia Nat'l. Park, California

Metasequoia glyptostroboides
a *"Fossil-age conifer"* 20 years old, in E. Rutherford, N.J.

Taxodium distichum, *"Bald cypress"*
hung with Spanish moss, in Winter Garden, Florida

CONVOLVULACEAE, CORNACEAE

Cornus florida
"Dogwood", in New Jersey

Cornus nuttallii
"Pacific dogwood", Yosemite, Calif.

Cornus kousa
"Kousa dogwood" in Japan

Cornus alba 'Argenteo-marginata'
"Tatarian dogwood", in the Rhineland

Cornus florida 'Welchii'
"Variegated dogwood" in Boskoop, Holland

Cornus florida 'Rubra'
"Pink dogwood"

Cornus florida 'White Cloud'
"Dwarf dogwood" in Dallas, Texas

Porana paniculata
"Snow vine" in Bombay, India

CONVOLVULACEAE, CORNACEAE

Aucuba japonica 'Dentata'
a "Japanese laurel"

Aucuba japonica 'Variegata'
"Gold-dust tree"

Aucuba japonica 'Picturata'
"Golden laurel"

Dichondra micrantha (repens)
"Lawn-leaf", in California

Aucuba japonica 'Crotonifolia'
in flower

Aucuba japonica 'Goldiana'
"Gold-leaf bush"

Griselinia lucida 'Variegata'
in Christchurch, New Zealand

Argyreia nervosa
"Silver morning-glory" (Borneo)

Cornus capitata
from the Himalayas

CONVOLVULAVEAE

Ipomoea batatas
"Sweet potato" or "Yam" cultivated in Kigezi, Uganda

Ipomoea batatas
"Sweet potatoes" in Igorot country, Philippines

Ipomoea cairica
in Kew Gardens, London

Ipomoea fistulosa
in Rio de Janeiro

Ipomoea crassicaulis
in Papeete, Tahiti

Ipomoea nil 'Imperialis'
"Imperial Japanese morning glory"

Ipomoea arborescens
"Casaguate" on Monte Alban, Mex.

Ipomoea murucoides
"Morning-glory tree" in Oaxaca

CONVOLVULACEAE

Stictocardia tiliifolia
in Hanging Gardens, Bombay

Quamoclit lobata (Mina)
"Star-glory" in Mexico

Ipomoea horsfalliae
"Princess vine"

Ipomoea tricolor
'Heavenly Blue'

Ipomoea mauritiana
in Kew Gardens, London

Ipomoea digitata
Royal Botanic Garden Kew

Ipomoea horsfalliae
(West Indies)

Maripa passifloroides
in Rio de Janeiro

Stictocardia beraviensis
from Tropical Africa

CONVOLVULACEAE

Merremia holubii
in Bot. Garden Heidelberg, Germany

Ipomoea tuberosa
seed pod known as "Woodrose"

Turbina holubii
at Botanical Institute Heidelberg

Calonyction aculeatum
"Moon flower" (Dr. J. Morton)

Ipomoea learii (acuminata)
in Xochicalco, So. Mexico

Ipomoea purpurea
on St. Héléna, Napoleon's exile

Ipomoea pes-caprae
"Beach morning-glory" (Aruba)

Ipomoea tuberosa
in flower on Oahu

Ipomoea palmata
in Nigeria, W. Africa

CRASSULACEAE

Adromischus festivus
"Plover eggs"

Adromischus maculatus
"Calico hearts"

Adromischus rupicolus
(*South Africa*)

Aeonium domesticum 'Variegatum'
"Youth-and-old-age"

Aeonium decorum (cooperi)
"Copper pinwheel"

Aeonium arboreum
'Atropurpureum cristatum'

Aeonium 'Pseudotabulaeforme'
"Green platters"

Aeonium tabulaeforme
"Saucer plant" (*Tenerife*)

Cotyledon orbiculata
(*Namibia*)

Echeveria chihuahuaensis
(*Mexico*)

Cotyledon orbiculata oophylla
(*Namibia*)

Cotyledon undulata
"Silver crown"

CRASSULACEAE

Aeonium canariense
"Velvet rose" (*Canary Is.*)

Aeonium arbor. 'Atropurpureum'
"Schwarzkopf"

Aeonium arboreum
(*Morocco*)

Aeonium holochrysum
(*Canary Islands*)

Aeonium glandulosum
(*Madeira*)

Aeonium haworthii
"Pinwheel" (*Tenerife*)

Dudleya traskiae
(*Santa Barbara Is., California*)

Crassula teres
"Rattlesnake tail"

Aeonium urbicum
(*Canary Isl.: Tenerife*)

CRASSULACEAE

Aeonium tabulaeforme
"Saucer plant" growing on rocks, Tenerife

Cotyledon orbiculata
in habitat in the Great Karroo, South Africa

Aichryson bollei
(Canary Islands: Palma)

Cotyledon heterophylla (*So. Africa*)
at Royal Botanic Gardens Kew, England

Cotyledon ladismithiensis (tomentosa)
"Cub's paws" in Cape Province

Aeonium nobile
grown at La Vera, Tenerife

CRASSULACEAE

Cotyledon macrantha
South Coast Bot. Garden, Los Angeles

Cotyledon orbiculata
at Kirstenbosch, South Africa

Crassula rubicunda
at Cap Ferrat, France

Crassula tetragona
"Miniature pine tree"

Crassula multicava 'Variegata'
at Seaborn Del Dios, Calif.

Crassula 'Flame'
at Quail Gardens, Encinitas, Calif.

Crassula obliqua
in col. Marnier-Lapostolle, France

Crassula argentea
(arborescens in hort.)

Crassula portulacea 'Convoluta'
(Paul Hutchison, Escondido, Calif.)

CRASSULACEAE

Dudleya farinosa, *in habitat*
Pacific Ocean coast of Monterey Peninsula, California

Kalanchoe synsepala, *"Cup kalanchoe"*,
at Kebun Raya Botanic Garden, Bogor, Java

Cotyledon orbiculata
in the arid Great Karroo, Cape Prov., South Africa

Cotyledon wallichii
in habitat in the desert near the border of Botswana

CRASSULACEAE

Crassula x portulacea
on the slopes of Table Mountain, Newlands, Cape

Cotyledon paniculata
"Botterboom" of the Boers, Worcester Veld, So. Africa

Crassula perfoliata
in Kirstenbosch Botanic Garden, Cape Town

Crassula rupestris
the "Rosary vine" in the Hottentot-Holland Mts., Cape

Dudleya farinosa
a silvery succulent at home on Point Lobos, California

Dudleya brittonii
in collection Marnier-Lapostolle, Cap Ferrat, France

CRASSULACEAE

Crassula argentea 'Sunset'
in Carlsbad, California

Crassula 'Tricolor Jade'
(argentea x lactea)

Crassula argentea
"Jade plant" (So. Africa)

Crassula falcata
"Scarlet paintbrush"

Crassula hemispherica
"Arab's turban"

Crassula 'Morgan's Pink'
(mesembryanthemopsis x falcata)

Crassula sladenii
(Cape Prov.)

Crassula hottentotta
(Namibia)

Crassula mesembryanthemopsis
(Namibia) by J. Bogner

Crassula pyramidalis
in Hout Bay, Cape

Rochea coccinea
in Botanic Garden Kirstenbosch

Sempervivum arachnoideum
"Cobweb hen-and-chicks"

CRASSULACEAE

Echeveria peacockii
Botanic Garden Essen, Germany

Echeveria elegans
"Mexican snowball"

Echeveria secunda
"Blue hen-and-chickens"

Echeveria 'Desmetiana' *hort.*
in Villach, Austria

Echeveria pulvinata
"Chenille plant" (inflor.)

Echeveria hyb.
'Perle von Nuernberg' (Germany)

Echev. gibbiflora 'Carniculata'
(col. Marnier-Lapostolle, France)

Echeveria ciliata
(Oaxaca, Mexico)

Echeveria x carnitricha
(carnicolor x leucotricha)

CRASSULACEAE

Echeveria 'Desmetiana' hort.
planted as border in carpet bed, Locarno, Switzerland

Echeveria elegans
with sedums painted blue, Golden Gate, San Francisco

Graptopetalum filiferum
from Chihuahua, Mexico

Echeveria shaviana (*Mexico*)
at Merry Gardens, Maine

Echeveria crenulata
"Scallop echeveria" (*Morelos, Mexico*)

Echeveria gibbiflora var. metallica
photographed in New Zealand with native Gecko-lizard

CRASSULACEAE

Echeveria crenulata
"Scallop echeveria"

Echeveria crenulata
'Roseo-grandis'

Echeveria 'Doris Taylor'
"Woolly rose"

Echeveria elegans
in Kew Gardens, London

Echeveria 'Hoveyi'
(*waltheri 'Zahnii' cv.*)

Echeveria derenbergii
"Baby echeveria" (*Oaxaca*)

Echeveria elegans 'Kesselringii'
in Botanic Garden Essen, Germany

Echeveria leucotricha
"White-plush plant"

Echeveria potosina
(*Mexico: San Luis Potosi*)

Echeveria lindseyana
from Mexico

Echeveria pallida
"Argentine echeveria"

Echeveria hyalina (*Mexico*)
Tegelberg, Lucerne Valley, Calif.

CRASSULACEAE

Echeveria agavoides (Urbinia)
"Moulded wax" (Mexico)

Echeveria albicans
in Kew Gardens, London

Echeveria agavoides corderoyi
"Urbinia red" in hort.

Echeveria 'Chantilly Lace"
at Del Mar Exposition, San Diego

Echeveria 'Lipstick'
in Vista, California

Echeveria derenbergii
"Painted lady" in bloom

Echeveria gibbiflora 'Flammea'
in San Diego, California

Echeveria x gilva
"Wax rosette"

Echeveria 'Hoveyi variegata'
at Kew Gardens, London

CRASSULACEAE

372

Echeveria 'Pulv-oliver'
"Plush plant"

Echeveria runyonii
from Mexico

Echeveria simulans
"True Mexican rose"

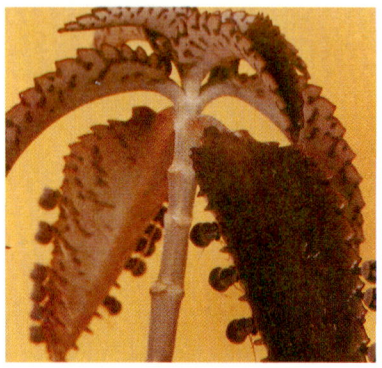
Kalanchoe daigremontiana
"Devil's backbone"

Kalanchoe fedtschenkoi
'Rosy Dawn'

Kalanchoe fedtsch. 'Marginata'
"Aurora Borealis plant"

Kalanchoe gastonis-bonnieri
"Life-plant" with plantlets

Kalanchoe flammea (*inflor.*)
from Somaliland

Kalanchoe marmorata
"Pen-wiper" (Ethiopia)

Kalanchoe tomentosa
"Panda plant"

Kalanchoe tubiflora
"Chandelier plant"

Kalanchoe paniculata
young plant at leaf-tip

CRASSULACEAE

Kalanchoe 'Roseleaf'
(*beharensis x pilosa*)

Kalanchoe farinacea
from Socotra, Arabian Sea

Kalanchoe blossfeldiana
"Flaming Katy" in Germany

Kalanchoe rhombopilosa
(*Malagasy Republic*)

Kalanchoe pumila
"Dwarf purple kalanchoe"

Kalanchoe tomentosa
"Pussy-ears" (Madagascar)

Kalanchoe pinnata (Bryophyllum)
"Miracle-leaf" or "Airplant"

Kalanchoe behar. 'Maltese Cross'
"Lace velvet-leaf"

Kalanchoe beharensis
"Elephant-ear" or "Felt-bush"

CRASSULACEAE

374

Oliveranthus elegans
"Red echeveria"

Crassula acinaciformis
(*Transvaal, Rhodesia*)

Orostachys erubescens japonicus
from W. Japan

Crassula rosularis
(*Cape Prov., Natal*)

Kalanchoe waldheimii
(*Madagascar*)

Graptopetalum paraguayense
(*Sedum weinbergii*)

Orostachys iwarenge fa. 'Fuji'
variegated cultivar from Japan

Crassula corymbulosa
(*S.E. Cape Prov.*)

Greenovia aurea
(*Canary Islands*)

CRASSULACEAE

Sedum sieboldii 'Medio-variegatum'

Sedum spathulifolium 'Capa Blanca', (*in Germany*)

Sedum sieboldii 'Variegatum' *an "October plant"*

Pachyphytum longifolium (*Mexico: Hidalgo*)

Sedum populifolium "*Hollyleaf sedum*" (*Siberia*)

Sedum dendroideum praealtum (*Mexico*)

Crassula acinaciformis *at Hummel's, Carlsbad, California*

Sempervivum arachnoideum "*Cobweb houseleek*"

Sedum spectabile "*Live-forever*" (*China*)

CRASSULACEAE

Kalanchoe beharensis
"Felt-bush" in bloom, near Nairobi, Kenya

Sedum cauticolum
winter-hardy "Stonecrop", flowering in Japan

Sedum spurium, "Dragon's blood"
in summer bloom near Interlaken, Swiss Alps

Rosularia pallida (Armenia)
Botanic Garden Marnier-Lapostolle, Cap Ferrat, France

Pachyphytum oviferum
"Pearly moonstones" in San Luis Potosi, Mexico

Sedum dendroideum praealtum
in Golden Gate Park, San Francisco

CRASSULACEAE

Sedum sieboldii (*Japan*)
"October plant" in hanging basket, Los Angeles

Sedum morganianum (*Mexico*)
"Burro-tails" on a tree stump, Kew Gardens, London

Sedum telephium
"European ice plant", summer-blooming in Sweden

Rochea coccinea 'Rosea'
(*Crassula rubicunda* of hort.) in Germany

CRASSULACEAE

Kalanchoe blossfeldiana 'Tom Thumb'
dwarf "Christmas kalanchoe"

Kalanchoe thyrsiflora
in hort. as "Vertical leaf"

Kalanchoe 'Yellow Darling'
"White kalanchoe"

x Pachyveria haagei
"Jewel plant"

Sedum pachyphyllum
"Jelly beans"

Pachyphytum bracteosum
"Moon-stones"

Sempervivum arachnoideum
"Cobweb chicks"

Sempervivum tectorum calcareum
"Hen-and-chicks"

Kalanchoe tubiflora
(*Bryophyllum*)

Sedum adolphii
"Golden sedum"

x Pachyveria nigra
"Black magic"

Sedum morganianum
"Burro tail" (Mexico)

CRUCIFERAE, CUCURBITACEAE

Iberis umbellata
"Globe candytuft"

Iberis sempervirens
"Edging candytuft"

Lobularia maritima
"Sweet alyssum"

Aethionema coridifolium
"Lebanon cress"

Lunaria annua (biennis)
"Honesty" or "Silver dollar"

Luffa cylindrica (*CUC.*)
"Vegetable sponge"

Arabis alpina 'Variegata'
"Mountain rock-cress"

Brassica oleracea aceph. crispa
"Flowering kale"

Brassica oleracea acephala
"Flowering cabbage"

CRUCIFERAE, CUCURBITACEAE

Matthiola incana
biennial "Stocks" on Capri

Matthiola incana 'Annua'
"Annual stocks" in Natal

Matthiola incana 'Rosea'
"Imperial stocks" in Germany

Matthiola sinuata
"Greek gilliflower" on Delos

Cheiranthus x allionii
"Siberian wallflower"

Cheiranthus cheiri
"English wallflower"

Seyrigia humbertii
from Malagasy Republic

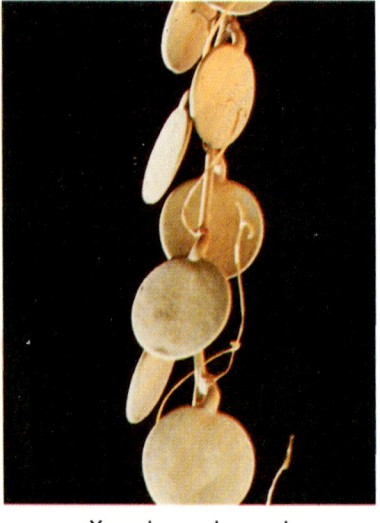
Xerosicyos danguyi
(Southwest Madagascar)

Dendrosicyos soccotrana
"Cucumber-tree" (Yemen)

CRUCIFERAE, CUCURBITACEAE

Brassica oleracea acephala
"Flowering cabbage" at Japanese inn, in Kobe

Matthiola incana
planting of "Stocks" in Golden Gate Park, San Francisco

Gerrardanthus macrorhizus (*East Africa*)
a desert monster in Botanic Garden Zuerich, Switzerland

Brassica oleracea crispa, "Flowering kale"
pot-grown in People's Botanic Garden, Kwangchow, China

CUCURBITACEAE

Gurania malacophylla (*flowers*)
from the upper Amazon

Gurania malacophylla (fol.)
in Botanic Garden Cologne

Cucumis anguria
"West Indian gherkin"

Lagenaria siceraria 'Rotunda'
in Palmengarten, Frankfurt

Seyrigia gracilis
Tulear, Madagascar

Ibervillea sonorae
climber from Mexico

Cucurbita pepo
'Lady Godiva'; *a pumpkin*

Citrullus lanatus 'Dixie Queen'
"Watermelon" (*Trop. Africa*)

Cucumis melo 'Reticulatus'
"Cantaloupe" (*W. Africa*)

CUCURBITACEAE

Benincasa hispida, *"Waxgourd"* or *"Chinese watermelon"*

Trichosanthes cucumeroides *"Snake gourd"*

Lagenaria siceraria 'Clavata' *"White-flowered Calabash gourd"*

Cucurbita pepo *"Gourds" at Hakone, Japan*

Cucurbita pepo *various gourds (Dr. D. Watson)*

Cucurbita pepo melopepo *"Banana squash" in California*

Cucurbita pepo 'Little Dumpling' *in England*

Cucumis sativus *a "Ridge cucumber"*

Cucurbita maxima *"Winter butternut squash"*

CYCADACEAE

Bowenia spectabilis
from North Queensland

Ceratozamia 'Hilda'
"Bamboo zamia" (Mexico)

Cycas media, with Karen Hoffbauer
at Roehrs greenhouses, New Jersey

Cycas neo-caledonica
"Bread palm", Leningrad Bot. Garden

Dioon edule
"Virgin's palm" (Mexico)

Cycas revoluta
"Japanese fern palm"

Encephalartos hildebrandtii
(male), Kew Gardens

Ceratozamia mexicana
"Mexican horncone"

Encephalartos altensteinii
"Prickly cycad" (So. Africa)

CYCADACEAE

Cycas media
"Australian nut palm" in North Queensland

Cycas revoluta
in glazed stoneware jar, Kwangchow, China

Cycas revoluta (*female tree*)
"Japanese sago palm" in Southern California

Cycas circinalis, *the "Queen sago" with Doris Graf in Longwood Gardens, Kennett Square, Pennsylvania*

CYCADACEAE

Encephalartos arenarius
(*South Africa*)

Encephalartos horridus
"Ferocious blue cycad"

Encephalartos latifrons
"Spiny Kaffir bread"

Encephalartos lehmannii
"Blue-leaved cycad"

Macrozamia moorei (*female*)
from Queensland

Encephalartos ferox
(*South Africa*)

Encephalartos kosiensis
in Zululand

Zamia furfuracea (pumila)
"Jamaica sago"

Stangeria eriopus
(*South Africa*)

CYCADACEAE

Encephalartos transvenosus
"Mujaji palm" in Mogadji village, No. Transvaal

Encephalartos laurentianum
on Lake Victoria, near Entebbe, Uganda

Cycas circinalis, the "Fern palm"
in Dili, Timor, in the Spice Islands

Macrozamia riedlei
in King's Park, Perth, Western Australia

Zamia fischeri (Mexico)
at Seaborn Del Dios, Escondido, California

Cycas revoluta, "Sago palm"
decorated with hibiscus, Nuku Alofa, Tonga

CYCADACEAE

Encephalartos transvenosus
on the High Veldt at 2000 m, North Transvaal

Cycas rumphii, *"Bread palm"*
(Malay Archipelago), at Foster Gardens, Honolulu

Macrozamia riedlei
"Zamia palm" in the Swan River area, Western Australia

Macrozamia denisonii (Lepidozamia)
in habitat on Mt. Tamborin, New South Wales, Australia

CYPERACEAE, DIDIEREACEAE

Didierea trollii
a curious succulent in arid Southwest Madagascar

Alluaudia ascendens, *in habitat*
near Amboasary, S.W. Malagasy Republic (by J. Bogner)

Alluaudia comosa, *a xerophyte*
near Lavanono, in the dry Southwest of Madagascar

Cyperus papyrus
the "Egyptian paper plant", on Lake Victoria, Uganda

CYPERACEAE

Cyperus diffusus
"Broadleaf umbrella palm"

Cyperus alternifolius
"Umbrella plant"

Carex foliosiss. 'Albo-mediana'
(*elegantissima*)

Cyperus alternifol. 'Variegatus'
"Variegated umbrella plant"

Carex morrowii 'Variegata'
"Japanese sedge grass"

Fuirema simplex (*aquatic*)
in Botanic Garden Munich

Scirpus cernuus
(*Isolepis gracilis*)

Cyperus diffusus 'Variegatus'
"Striped umbrella palm"

Cyperus papyrus, *landscaped by*
R. Burle-Marx, Teresopolis, Brazil

CYCLANTHACEAE, DIDIEREACEAE, DILLENIACEAE

Alluaudia comosa
flowering branch, in Madagascar

Alluaudia montagnacii
(S.W. Madagascar)

Alluaudia adscendens
in Cap-Ferrat, France

Hibbertia scandens
"Guinea goldvine"

Dillenia indica
as street tree in Singapore

Dillenia suffruticosa (Wormia)
in Mahé, Seychelles

Cunonia capensis
"African red alder"

Dillenia indica
"Elephant apple" (Java)

Carludovica palmata
"Panama hat plant"

DILLENIACEAE, DIOSCOREACEAE, EBENACEAE

Diospyros kaki *in fruit*
"Japanese persimmon" or "Kaki", in Vista, California

Dillenia indica
"Simpoh", in Summit Gardens, Panama

Dioscorea discolor
"Ornamental yam" in Amazonas habitat, Brazil

Dioscorea macrostachys (Testudinaria)
the weird "Tortoise plant", a xerophyte from Mexico

DIOSCOREACEAE, EBENACEAE, ELAEAGNACEAE

Dioscorea sansibarensis
in Kew Gardens, London

Dioscorea macrostachys
at Seaborn Del Dios, Escondido

Dioscorea discolor
"Ornamental yam" (Surinam)

Dioscorea elephantipes (Testudinaria)
"Elephant's-foot" or "Hottentot-bread" (South Africa)

Dioscorea bulbifera
"Air potato" or "True yam" from the Philippines

Elaeagnus pungens
"Silver-berry" in Istanbul, Turkey

Hippophae rhamnoides
"Sea-buckthorn" (Europe to Iran)

Diospyros malabarica
"Malabar ebony" on Ceylon

EPACRIDACEAE, ERICACEAE

Arbutus unedo
"Strawberry-tree" in California

Epacris impressa
in Taronga, New South Wales

Dracophyllum traversii
on South Island, New Zealand

Pernettya mucronata
"Chilean myrtle"

Pernettya mucronata 'Alba'
female plant with white berries

Arctostaphylos uva-ursi
"Common bearberry"

Rhododendron obtusum
in Nara, Japan

Pieris japonica (Andromeda)
"Lily-of-the-valley bush"

Vaccinium corymbosum
"Highbush blueberries" in New Jersey

EPACRIDACEAE, ERICACEAE

Rhododendron arboreum, *"Tree rhododendron"* at Lamaist monastery, in the Himalayas of Sikkim

Dracophyllum traversii *"New Zealand grasstree"* near Nelson, South Island

Arctostaphylos columbiana *"Manzanita"* in habitat, Bryce Canyon, Southern Utah

Arbutus unedo *"Strawberry tree"* or *"Cane apples"* in Vista, California

ERICACEAE

Erica baccans
"Berry heath" in South Africa

Erica melanthera
"Christmas heather" at Roehrs

Erica gracilis
"Rose heath" in Germany

Erica taxifolia
"Double-pink heath" in Kirstenbosch, South Africa

Erica baccans
"Berry heath" in Transvaal habitat

Kalmia latifolia
"Mountain-laurel", "Calico bush"

Kalmia angustifolia 'Rubra'
"Sheep laurel"

Erica vagans 'Rubra'
"Cornish heath"

ERICACEAE

Erica hyemalis
"French heather"

Erica 'Felix Faure'
"French hybrid" in San Diego

Erica 'Wilmorei'
"Prince of Wales heath"

Erica bauera
"Bridal heath" (So. Africa)

Erica speciosa
in Kirstenbosch, Cape Town

Erica regia
the "Royal heath"

Cavendishia acuminata
from the Andes of Ecuador

Erica mammosa
from South Africa

Pentapterigium serpens
(Eastern Himalayas)

EPACRIDACEAE, ERICACEAE

Rhododendron thomsonii
in the Sikkim Himalayas, on the border of Tibet

Dimorphanthera sp. 'Finisterre'
collected in the Finisterre Mountains of New Guinea

Rhododendron indicum 'Shinnyo-no-tsuki' (*Azalea*)
"Satsuki" (Fifth moon) azalea as bonsai, Hakone, Japan

Rhododendron auriganum
growing as epiphyte near Kikiepa village in New Guinea

Epacris longiflora
"Australian fuchsia", from New South Wales

Erica regia, *"Royal heath"*
or "Elim heath" from southwestern Cape Province

ERICACEAE

Rhododendron obtusum (Azalea)
"Kirishima azalea" at Buddhist shrine, Kamakura, Japan

Rhododendron indicum (lateritium)
growing in pots, Tiger Balm gardens, Victoria, Hong Kong

Rhododendron simsii 'Vittatum'
Spring-blooming at the Memorial temple, Taipei, Taiwan

Rhododendron macgregori
endemic species of the Goroka highlands, in New Guinea

ERICACEAE

400

Azalea 'Concinna'
(*Rhod. phoeniceum cv.*)

Azalea 'Polar Bear' (Rhod.)
(*evergreen hyb.*)

Azalea 'Pride of Detroit' (Rhod.)
(*Pericat hybrid*)

Azalea 'Double Coral Bells' (Rhod.)
(*Kurume hyb.*)

Azalea 'Sweetheart Supreme' (Rhod.)
(*Pericat hybrid*)

Azalea 'Redwing'
(*Belgian indica x Kurume*)

Azalea simsii 'Anytime' (Rhod.)
"Belgian indica" hyb.

Azalea rutherfordiana
'Constance' (Rhododendron)

Azalea simsii 'Brilliant'
"Southern sun azalea"

Azalea 'Kingfisher' (Rhod.)
(*Whitewater hyb.*)

Azalea rutherfordiana
'Alaska' (*evergreen*)

Azalea simsii 'Euratom'
(*Belgian Indica*)

ERICA

Azalea 'Hinodegiri' (Rhod.)
"Mist-of-the-rising-sun"

Azalea 'Snow' (Rhod.)
(*Kurume hybrid*)

Azalea 'Coral Bells' (Kirin)
(*Kurume hybrid*)

Azalea 'Rose Pericat' (Rhod.)
(*Pericat hyb.*)

Azalea 'Salmon Beauty'
(*Kurume hybrid*)

Azalea 'Purple Heart'
(*hardy evergreen*)

Azalea 'Peggy Ann' (Rhod.)
(*Roehrs-Baumann hyb.*)

Azalea 'Delaware Valley White'
(*mucronatum hyb.*)

Azalea 'Hinocrimson' (Rhod.)
(*Kurume hybrid*)

Azalea 'Marjorie Ann' (Rhod.)
(*Pericat hybrid*)

Azalea 'Madonna' (Rhod.)
(*Brooks California hybrid*)

Azalea 'Louise Gable'
(*Gable evergreen*)

ERICACEAE

402

Azalea simsii (Rhod.)
'Ernest Thiers'

Azalea simsii (Rhod.)
'Leopold-Astrid'

Azalea simsii 'Hexe'
(*Dwarf indica*)

Azalea simsii
'Mad. John Haerens'

Azalea simsii
'Haerens alba'

Azalea simsii
'Mad. Petrick'

Azalea simsii
'Haerens Beauty'

Azalea simsii
'Rose Queen' (*New Zealand*)

Azalea simsii
'Lentengroet'

Azalea simsii
'Southern Charm'

Azalea simsii
'Petrick alba'

Azalea simsii 'Brilliant'
a "Southern sun azalea"

ERICACEAE

Azalea simsii (Rhod.) 'Reinhold Ambrosius' (*grafted*)

Azalea simsii 'Niobe' (*Rhododendron simsii cultivar*)

Azalea simsii (Rhod.) 'Ambrosiana'

Azalea simsii 'Paul Schame'

Azalea simsii 'Chimes' (*Belgian indica*)

Azalea simsii 'Mad. Van der Cruyssen'

Azalea simsii 'Adventgloeckchen'

Azalea simsii 'Albert-Elizabeth'

Azalea simsii 'Eclaireur'

Azalea simsii 'Vervaeneana'

Azalea simsii 'Triumph' (*Belgian indica*)

Azalea simsii 'Simon Mardner'

ERICACEAE

Rhododendron javanicum
growing epiphytic near tropical Tjibodas, Java

Rhododendron 'Dexter hybrid'
(*fortunei cultivar*), *at Dupont's, Winterthur, Delaware*

Rhododendron 'Roseum elegans'
winter-hardy Catawba hybrid, at Longwood Gardens, PA.

Rhodendron maximum, *reproduction
made of glass, at Botanical Museum, Cambridge, Mass.*

Rhododendron mucronulatum
deciduous winter-hardy azalea, in New Jersey

Rhododendron luteum (*Azalea pontica*)
fragrant "Pontic azalea", winter-hardy in New Jersey

ERICACEAE

Rhododendron 'Bow Bells'
"Miniature rhododendron"

Rhododendron (Azalea)
'Ghent hybrid'

Rhododendron luteum
"Pontic azalea"

Rhododendron
'Christmas Cheer'

Rhododendron 'Eureka Maid'
(*Countess of Derby*)

Rhododendron
'Jean Marie de Montague'

Rhododendron 'Scintillation'
(*Dexter hybrid*)

Rhododendron 'Trilby'
(*Q. Wilhelmina x Derby*)

Rhododendron
'Unknown Warrior'

Rhododendron lochae
species from Queensland

Rhodo. ponticum 'Superbum'
on Mainau Isle, Germany

Rhododendron westlandii
on Kwan Yam Shan, China

ERICACEAE

Rhodo. pulchrum phoeniceum
in Bot. Garden Kwangchow, China

Rhododendron indicum
"Satsuki azalea" in Japan

Rhododendron canadense
"Rhodora" in the Poconos, Penna.

Rhododendron mucronatum
(*ledifolia alba*)

Rhododendron catawbiense
"Mountain rose-bay"

Rhododendron occidentalis
"Western azalea" (Oregon)

Rhododendron ferrugineum
"Alpine rose"

Rhododendron 'Kaempferi hybrid'
"Torch azalea"

Rhododendron luteum
(*Azalea pontica*)

ERYTHROXYLACEAE, EUPHORBIACEAE

Erythroxylum coca
"Cocaine plant" (*Bolivia*)

Aleurites fordii
"Tung oil tree"

Aleurites moluccana
"Candle-nut"

Cnidoscolus phyllacanthus
"Tread softly"

Breynia nivosa 'Roseo-picta'
"Leaf-flower"

Breynia nivosa (disticha)
"Snow-bush" (*Polynesia*)

Dalechampia spathulata (*Mexico*)
in Kew Gardens, London

Hippomane mancinella
'Manzanillo' (*Curacao*)

Bischofia javanica (*female*)
"Toog tree" (*Java*)

EUPHORBIACEAE

Acalypha hispida 'Alba'
"Philippine Medusa"

Acalypha hispida
"Chenille plant" (India)

Acalypha wilkesiana obovata
"Heart copperleaf"

Synadenium grantii 'Rubra'
"Red milk bush"

Euphorbia cotinifolia
"Hierba mala" (Mexico)

Acalypha wilkesiana 'Moorea'
in French Polynesia

Euphorbia marginata
"Snow-on-the-mountain"

Euphorbia neohumbertii
from Madagascar

Euphorbia leucocephala
"Flor de Niño" (Costa Rica)

EUPHORBIACEAE

Acalypha wilkesiana 'Macrophylla'
"Giant redleaf"

Acalypha wilkesiana 'Java white'
in Borobudur, Java

Acalypha wilkesiana 'Tricolor'
in Peradeniya

Acalypha wilkesiana 'Marginata'
"Copper leaf"

Acalypha wilkesiana 'Tahiti'
"Match-me-if-you-can"

Acalypha wilkesiana 'Ceylon'
"Fire dragon"

Acalypha godseffiana
'Heterophylla' (*Brooklyn B.G.*)

Acalypha wilkesiana
'Macafeana' *in Borneo*

Acalypha wilkesiana 'Hoffmanii'
on Nuku Hiva, Marquesas

EUPHORBIACEAE

Codiaeum variegatum 'Gloriosum superbum'

Codiaeum variegatum 'Appleleaf' ("Croton")

Codiaeum variegatum 'Clipper' ("Croton")

Codiaeum varieg. 'Elaine' "Lance-leaf croton"

Codiaeum varieg. 'Mona Lisa' "White croton"

Codiaeum variegatum 'Craigii'

Codiaeum variegatum 'Exotica'

Codiaeum variegatum 'General Paget'

Codiaeum variegatum 'Delaruye'

Codiaum variegatum 'Warrenii'

Codiaeum variegatum 'Gloriosum'

Codiaeum variegatum 'Aucubifolium'

EUPHORBIACEAE

Croton poilanii, *a true croton*
at the Tronc Bom Arboretum, Viet Nam

Codiaeum variegatum taeniosum
in Kebun Raya Bot. Garden, Bogor, West Java

Codiaeum variegatum 'Majesticum'
greenhouse-grown near Brussels, Belgium

Croton gossypifolius
economic croton tree near Port of Spain, Trinidad

EUPHORBIACEAE

Codiaeum variegatum 'Lord Belhaven' (*"Croton"*)

Codiaeum variegatum 'Delaware' (*"Croton"*)

Codiaeum variegatum 'Van Houtte' (*"Croton"*)

Codiaeum variegatum 'Jan Bier' (*"Croton"*)

Codiaeum variegatum 'Tapestry' (*"Croton"*)

Codiaeum variegatum 'Mortimer' (*"Croton"*)

Codiaeum variegatum 'Mad. Blanc' (*"Croton"*)

Codiaeum variegatum 'Mac Arthuri' (*"Croton"*)

Codiaeum variegatum 'Imperialis' (*"Croton"*)

EUPHORBIACEAE

Codiaeum variegatum 'Gloriosa'
highly colored in Miami Beach, Florida

Codiaeum variegatum 'Crispum'
in tropical Kebun Raya, Bogor, West Java

Codiaeum variegatum 'Baronne de Rothschild'
grown in broken brick for drainage, Singapore

Codiaeum variegatum ambiguum 'Stoplight'
in the Orchid gardens of Suva, Viti Levu, Fiji

Codiaeum variegatum 'Nobile'
at the flower market in Saigon, Viet Nam

Codiaeum variegatum 'America'
in Foster Botanical Garden, Honolulu, Hawaii

EUPHORBIACEAE

Codiaeum variegatum
'Mrs. Duncan Macaw'

Codiaeum variegatum
'Volutum', *"Ramshorn"*

Codiaeum variegatum
'Majesticum'

Codiaeum variegatum
'Illustris'

Codiaeum variegatum
'Dormannianum'

Codiaeum variegatum
'Punctatum aureum'

Codiaeum variegatum
'Aucubifolium'

Codiaeum variegatum
'Cornutum'

Codiaeum variegatum
'Picturatum'

EUPHORBIACEAE

Codiaeum variegatum 'Gloriosa'

Codiaeum variegatum 'Fred Sander'

Codiaeum variegatum 'Imperialis'

Codiaeum variegatum 'Punctatum aureum'

Codiaeum variegatum 'Lineatum', *in flower*

Codiaeum variegatum 'Interruptum', *on Ceylon*

Codiaeum variegatum 'Warrenii', *in Singapore*

Codiaeum variegatum 'Rubrum', *in Florida*

Codiaeum variegatum 'Norwood Beauty'

EUPHORBIACEAE

Codiaeum variegatum 'Mons. Florin'

Codiaeum variegatum 'Vulcan', *in England*

Codiaeum variegatum 'Ovalifolium'

Codiaeum variegatum 'Martha Brooks'

Codiaeum variegatum 'Cronstadtii'

Codiaeum variegatum 'Eburneum'

Codiaeum variegatum 'Excurrens'

Codiaeum variegatum 'Recurvifolium'

Codiaeum variegatum 'Stewartii'

EUPHORBIACEAE

Euphorbia ingens, *"Candelabra-tree"*
overlooking Explosion crater, Western Uganda

Euphorbia candelabrum
giant tree under the equator, in Toro Prov., Uganda

Euphorbia lactea, *"Dragon-bone tree"*
at Chinese Kwan Yin temple in Old Singapore

Euphorbia cooperi
fountain-like at Kirstenbosch, Newlands, South Africa

EUPHORBIACEAE

Pedilanthus tithym. 'Variegatus'
"Zigzag plant"

Pedilanthus tithymaloides
retusus (*Amazonas*)

Monadenium coccineum
in Masai country, Tanzania

Jatropha integerrima
"Peregrina" in Cuba

Euphorbia vigueri
in Cap Ferrat, France

Jatropha pandurifolia
(*West Indies*)

Jatropha podagrica
Kew Gardens, London

Jatropha multifida
"Coral plant" in Nigeria

Euphorbia punicea
(*Jamaica to Bahamas*)

EUPHORBIACEAE

Euphorbia leucocephala
"Pascuita" in Charlotte Amalie, St. Thomas, Virgin Is.

Jatropha multifida, *"Coral plant"*
as a roadside tree, Lagos, Nigeria, W. Africa

Euphorbia fulgens, *"Scarlet plume"*
flowering for Christmas, at Roehrs, New Jersey

Breynia nivosa 'Roseo-picta' (disticha)
"Foliage-flower" at Papeari Botanic Garden, Tahiti

Manihot peltata, *handsome foliage plant
in Brazil (by J. Bogner, Munich)*

Pedilanthus bracteatus, *from Mexico,
curious succulent with hood-shaped bracts*

EUPHORBIACEAE

420

Euphorbia
'Giant Christthorn' in California

Euphorbia lophogona
in Kew Gardens, London

Euphorbia milii (splendens)
"Crown of thorns"

Euphorbia milii hislopi (splendens)
in Kirstenbosch Bot. Garden, Cape Town, South Africa

Euphorbia milii (splendens) var. breonii
a "Christ-thorn" from Madagascar

Euphorbia x keysii
"Flamingo plant" in Florida

Euphorbia milii imperatae
"Mini-Christ-thorn"

Euphorbia atropurpurea
(Canary Islands: Tenerife)

EUPHORBIACEAE

Euphorbia grandicornis
"Cow-horn euphorbia"

Euphorbia clandestina
(South Africa)

Euph. mammilaris 'Variegata'
"Indian corncob"

Euphorbia neriifolia
'Variegata cristata'

Euphorbia obesa
"Baseball plant"

Euphorbia 'Zigzag'
(grandicornis x pseudocactus)

Euphorbia lactea
"Candelabra plant"

Euphorbia lactea 'Cristata'
"Elkhorn" or "Frilled fan"

Euphorbia 'Keysii'
"Dark Florida Christ-thorn"

Phyllanthus angustifolius
"Foliage flower"

Jatropha integerrima
"Spicy jatropha"

Excoecaria bicolor
"Picara" from Viet Nam

EUPHORBIACEAE

Euphorbia tirucallii
"Pencil cactus" in the Transvaal

Euphorbia stenoclada
"Silver thicket" in Nairobi

Euphorbia oncoclada
in Madagascar

Euphorbia cooperi
in Shingwidzi, Transvaal

Euphorbia quadrialata
in Nairobi, Kenya

Euphorbia candelabrum
"Candelabra tree" in Kenya

Euphorbia caerulescens
at Stellenbosch University

Euphorbia canariensis
"Hercules club"

Euphorbia lactea
"Dragon-bones"

EUPHORBIACEAE

Euphorbia leucocephala
"Rosa pomone" or "Pascuita" in Guatemala

Poinsettias (Euphorbia pulcherrima) *for Christmas as office decoration, on Avenue of Americas, New York City*

Acalypha hispida
"Chenille plant" in Frankfurt Palmengarten, Germany

Acalypha wilkesiana 'Tricolor'
"Match-me-if-you-can", in Peradeniya, Ceylon

EUPHORBIACEAE

Euphorbia pulvinata, *"Cushion euphorbia"* in Kirstenbosch Botanic Garden, Cape Town

Euphorbia caput-medusae "Medusa's head" in South Africa

Euphorbia myrsinites in Wellington Bot. Garden, New Zealand

Euphorbia polyacantha "Fish-bone cactus", in Cap Ferrat, France

Euphorbia mammillaris 'Variegata' "Indian corn-cob" in Santa Barbara, California

Euphorbia tuberosa in Hout Bay, Cape of Good Hope

EUPHORBIACEAE

Euphorbia canariensis, *"African cereus"* *in the lava rocks at Guimar, Tenerife*

Euphorbia ingens *"Candelabra tree" in the Veld, northern Transvaal*

Euphorbia tirucalii *"Milk bush" in Achole country, Uganda*

Euphorbia robecchii *lone tree in the Tsavo desert, Kenya*

Euphorbia cooperi *on High Veld at 2000 m, northern Transvaal*

Euphorbia stenoclada *"Silver thicket" along the Indian Ocean, Madagascar*

EUPHORBIACEAE

Euphorbia pulcherrima
'Annette Hegg Supreme'

Euphorbia pulcherr. 'Annette Hegg'
"Christmas poinsettia"

Euphorbia pulcherrima
'Oakleaf' (*Roehrs 1946*)

Euph. pulcherrima 'Ecke's White'
"White poinsettia"

Euphorbia pulcherrima 'Rosea'
"Pink poinsettia"

Euph. pulcherrima 'Variegata'
"Variegated-leaf poinsettia"

Euph. pulcherrima plenissima
"Double poinsettia", on Barbados

Euphorbia heterophylla
"Mexican fire plant"

Euph. pulcherrima plenissima
"Flaming sphere"

EUPHORBIACEAE

Euphorbia pulcherrima
"Flor de Noche Buena" or "Christmas star"

Euphorbia pulcherrima 'Praecox'
in the fertile Orotava Valley of Tenerife

Euphorbia pulcherrima 'Annette Hegg'
at Roehrs Exotic Nurseries for Christmas 1971

Euphorbia pulcherrima 'Mrs. Paul Ecke'
favorite winter-bloomer at Roehrs, in 1948

Euphorbia pulcherrima 'Henrietta Ecke'
"Double poinsettia" at Roehrs-Rutherford, winter 1946

Euphorbia pulcherrima 'Eckespoint C-1'
originated by Paul Ecke, Encinitas, California 1968

EUPHORBIACEAE

428

Manihot esculenta
"Cassava" in Ceylon

Manihot esculenta 'Variegata'
"Tapioca plant"

Manihot esculenta (utilissima)
"Manioc" in flower

Ricinus communis
"Castor oil plant"

Ricinus communis 'Coccineus'
"Red-leaf Palma Christi"

Jatropha gossypifolia
in Dili, Timor

Phyllanthus arbuscula (speciosus)
"Jamaica foliage flower"

Synadenium grantii 'Rubra'
"Red milk bush" (Tanzania)

Omphalea triandra
"Pignut" (West Indies)

EUPHORBIACEAE

Macaranga grandifolia
the "Coral tree" at the Aloha Tower, Honolulu, Hawaii

Phyllanthus acidus, "Otaheite"
or "Gooseberry tree", in Foster Gardens, Honolulu

Hevea brasiliensis
"Para rubber" on plantation in Johore, Malaya

Hevea brasiliensis
a rubber tree on Ceylon, tapped daily for its latex

FAGACEAE, FOUQUIERIACEAE

Nothofagus menziesii
"Silver beech" in New Zealand

Quercus agrifolia
"California live oak"

Quercus suber
"Cork oak"

Quercus dentata
"Daimyo oak", bonsai 100 yrs. old

Quercus robur
"English oak" in Germany

Cercis siliquastrum alba
"Judas tree" or "Love tree"

Idria columnaris
"Boojum tree" from Sonora

Idria columnaris
"Boojum tree" near Tucson, Arizona

Fouqieria splendens, *"Ocotillo"*
on the Mohave desert, California

Filices: CYATHEACEAE

Cyathea arborea
"West Indian treefern", Sierra Luquillo, in Puerto Rico

Alsophila cooperi (Cyathea australis)
"Australian tree-fern" in Royal Botanic Garden, Sydney

Cyathea spinulosa
a treefern in Kwangtung Prov., South China

Cyathea arborea
young treefern in Caribbean National forest, Puerto Rico

Filices: CYATHEACEAE, DICKSONIACEAE

Dicksonia squarrosa
treeferns near Franz-Joseph Glacier, S. New Zealand

Alsophila glauca
the "Blue treefern" in the volcanic mountains of Java

Dicksonia squarrosa
"Westland treefern" in Westland Province, New Zealand

Cyathea usambarensis
in the moist Usambara Mountains, near Amani, Tanzania

Dicksonia fibrosa
"Golden treefern", container-grown in New Zealand

Cyathea medullaris
"Black treefern" or "Mamaku", South Island, New Zealand

Filices: CYATHEACEAE, MARATTIACEAE

Cyathea lunulata
near Korolevu, on Viti Levu, Fiji

Alsophila glauca
fern forest in the mountains of Tjibodas, West Java

Angiopteris evecta
"Mules-foot fern" in Foster Gardens, Honolulu, Hawaii

Cyathea arborea
"West Indian treefern" in the Sierra Luquillo, Puerto Rico

Filices: CYATHEACEAE, DICKSONIACEAE

Cyathea medullaris
the "Mamaku" or "Sago fern" in Wellington, New Zealand

Dicksonia squarrosa
tree-ferns luxuriate in rain-rich Westland Province, N.Z.

Dicksonia antarctica
"Tasmanian treefern" in Royal Botanic Garden, Sydney

Cyathea arborea
enjoying the tropical summer, at Radio City, New York

Filices: DICKSONIACEAE, POLYPODIACEAE

Cibotium chamissoi (menziesii *in hort.*)
the "Man fern" of Hawaii, container-grown in California

Cibotium glaucum (chamissoi *in hort.*)
the "Blonde treefern" in Hawaii National Park

Polypodium subauriculatum 'Knightiae'
"Lacy pine fern" of Australia, in hanging basket

Cibotium chamissoi (menziesii)
Hawaiian treeferns in containers, Roehrs greenhouses, N.J.

Filices: CYATHEACEAE, DICKSONIACEAE, POLYPODIACEAE

Cyathea dregei
at Table Mountain, Cape Town

Alsophila latebrosa
in the foothills of the Himalayas

Cyathea dickenii
"Amani treefern" in Tanzania

Polypodium aureum
"Rabbit's-foot fern"

Cibotium schiedei
"Mexican treefern" in 20 cm tub

Cibotium barometz
"Lamb of Tartary" in Hongkong

Cyathea medullaris
"Mamaku" in New Zealand

Dicksonia fibrosa
"Golden treefern"

Hemitelia horrida (Alsophila)
in Puerto Rico

Filices: GLEICHEN., HYMENOPH., MARATT., OPHIOGLOSS., OSMUND.

Marattia salicina (fraxinea)
"King fern", giant frond to 10 m long, in New Zealand

Ophioglossum bulgatum
"Ribbon fern", an epiphyte in North Queensland

Gleichenia linearis (dichotoma)
"Savannah fern" in Northwest Borneo

Trichomanes reniforme
"Kidney fern", a filmy fern, Maitai valley, New Zealand

Osmunda cinnamomea
"Cinnamon-fern", the fertile fronds looking like flowers

Leptopteris superba (Todaea)
beautiful "Prince of Wales plume" in New Zealand

Filices: CERATOPT., MARATT., MARSIL., OPHIOGLOSS., OSMUND.

Angiopteris evecta
"Turnip-fern"

Ophioglossum pendulum
"Ribbon fern" in Queensland

Todea barbara
from New Zealand

Marsilea drummondii (vestita)
"Waterclover" (aquatic)

Ceratopteris thalictroides
"Water fern" (aquatic)

Asplenium lucidum
"Leatherfern" in New Zealand

Acrostichum danaeifolium
"Swamp-fern" in Florida

Acrostichum aureum
tropical "Leather-fern"

Sadleria cyathioides
"Pigmy cyathea"

Filices: POLYPODIACEAE

Adiantum macrophyllum
in Kew Gardens, London

Adiantum trapeziforme
"Giant maidenhair" (Longwood G.)

Adiantum trapeziforme
in Martinique

Adiantum raddianum (cuneatum)
"Delta maidenhair"

Adiantum tenerum 'Farleyense'
"Barbados maidenhair"

Adiantum tenerum 'Wrightii'
"Fan maidenhair"

Asplenium bulbiferum
in Kew Gardens, London

Asplenium bulbiferum
"Hen-and-chicken fern"

Asplen. viviparum (daucifolium)
"Mother fern"

Filices: POLYPODIACEAE

Nephrolepis exaltata 'Bostoniensis'
"Boston fern"

Nephrolepis exaltata 'Whitmanii'
a "Lacefern"

Drynaria quercifolia
"Oak-leaved fern" epiphytic in New Guinea

Asplenium mayii
in Brooklyn Botanic Garden, New York

Davallia griffithiana
epiphytic "Rabbit's foot" from India

Davallia mariesii (bullata)
"Squirrel's-foot fern" in Hakone, Northern Honshu

Filices: POLYPODIACEAE

Drynaria rigidula
as epiphyte in Fiji

Drynaria fortunei
in Taipo, New Territories

Drynaria quercifolia
"Oak-leaved fern"

Drynaria rigidula var. whitei
in Kew Gardens, London

Polypodium musifolium
(*Microsorium at Kew Gardens*)

Polypod. punctatum (Microsorium)
"Climbing bird's-nest fern"

Blechnum serrulatum, *"Saw fern",
Everglades Nat'l. Park, Florida*

Blechnum gibbum (Lomaria)
from New Caledonia

Blechnum brasiliense
Jardim Botanico, Rio

Filices: POLYPODIACEAE

442

Pellaea rotundifolia
"Button fern" (New Zealand)

Phyllitis scolopendrium
'Undulatum' (*Europe*)

Humata tyermannii
"Bear's-foot fern"

Nephrolepis exaltata 'Verona'
"Dwarf lace-fern"

Nephrolepis exaltata
'Bostoniensis compacta'

Nephrolepis exalt. 'Whitmanii'
a sturdy "Lace-fern"

Drymoglossum microphyllum
"Beanleaf fern"

Nephrolepis exalt. 'Norwoodii'
compact "Lace-fern"

Polypodium aureum
'Glaucophyllum'

Pteris ensiformis victoriae
"Victoria fern"

Pteris cretica
'Albo-lineata'

Pteris cretica 'Wimsettii'
popular "Table-fern"

Filices: POLYPODIACEAE

Asplenium nidus phyllitidis
growing epiphytic in subtropical Kwangtung, China

Asplenium nidus
"Birdsnest fern" on Fagraea tree in Malayan jungle

Drynaria rigidula, *growing epiphytic in hardwood forest of Trang Bom, Vietnam*

Ceterach officinarum
"Scale fern" in lava rock, on Easter Island, Chile

Filices: POLYPODIACEAE

Davallia fejeensis 'Plumosa'
"Dainty rabbit's-foot"

Davallia trichomanoides
"Squirrel-foot fern"

Davallia canariensis
on Tenerife, Canary Islands

Davallia fejeensis
"Rabbit's-foot" in Fiji

Pyrrosia nummularifolia
"Felt fern" from Celebes

Drymoglossum niphoboloides
from Madagascar

Pellaea viridis macrophylla
(*Pteris adiantoides* in hort.)

Pityrogramma chrysophylla
"Goldfern" (*West Indies*)

Microlepia strigosa
(*Trop. Asia to Polynesia*)

Filices: POLYPODIACEAE

Asplenium nidus plicatum
Kew Gardens, London

Asplenium nidus
"Birdsnest fern" as pot plant

Cyrtomium falcatum
"Holly-fern"

Pteris longipinna
in Taiwan

Phyllitis scolopendrium
"Heart's-tongue fern"

Phyllitis scolopend. 'Crispum'
"Crisped heart's tongue"

Pteris cretica 'Childsii'
in San Diego, California

Nephrolepis duffii
"Pigmy sword fern"

Nephrolepis biserrata furcans
"Fishtail fern"

Filices: POLYPODIACEAE

Phyllitis scolopendrium
growing at Neuschwanstein castle, Bavarian Alps

Polypodium punctatum 'Grandiceps' (polycarpon)
"Fish-tail fern" at Yamamoto nursery, Tokyo

Polypodium aureum 'Mandaianum'
attractive "Crisped blue fern"

Onoclea sensibilis
the "Sensitive fern", in New Hampshire

Salvinia auriculata (*Java*)
"Floating fern" at Munich Botanic Garden, Germany

Elaphoglossum pachycraspedon
epiphytic fern from tropical eastern Perú

Filices: POLYPODIACEAE

Polypodium bifrons
Frankfurt Bot. Garden, Germany

Polypodium punctatum (polycarpon)
"Climbing birdsnest"

Polypodium punctatum
'Grandiceps', *"Fishtail fern"*

Asplenium bulbiferum
"Motherfern"

Polystichum aculeatum
Kew Gardens, London

Polystichum tsus-simense
(*Aspidium*) *Japan*

Polystichum strigosum
"Wood-fern" in Calif.

Polystichum setiferum 'Proliferum'
viviparous "Filigree fern"

Polystichum setiferum, *"Hedgefern"*
Golden Gate Park, San Francisco

Filices: POLYPODIACEAE, SALVINIACEAE, SCHIZAEACEAE

Hemionitis palmata
"Strawberry-fern"

Lygodium circinatum
"Malay climbing fern"

Lygodium japonicum
"Climbing fern" (Japan)

Polypodium vulgare
"Common polypody" or *"Adder's fern"*, in Germany

Polypodium crassifolium
(West Indies to Perú and Brazil)

Pyrrosia lingua
"Tongue fern"

Polypodium persicarifolium
Kew Gardens, London

Salvinia auriculata
"Floating fern" (Trop. America)

Filices: POLYPODIACEAE

Platycerium andinum, *"American staghorn"*,
collected plant near Iquitos, Perú, with Lee Moore

Platycerium angolense
"Elephant's ear fern" as epiphyte, in Western Uganda

Platycerium stemmaria
in the rainforest near Ibadan, Nigeria

Platycerium hillii
growing on a wooden slab at Roehrs greenhouses in N.J.

Platycerium stemmaria
"Triangle staghorn", collected by Bwana Graf in Kenya

Platycerium madagascariense
in Madagascar habitat, by J. Bogner, Munich

Filices: POLYPODIACEAE

Platycerium bifurcatum
the "Common staghorn" from Australia

Platycerium bifurcatum cv. 'Netherlands'
introduced from Holland as "Regina Wilhelmina staghorn"

Platycerium "alcicorne" hort., *"Elkhorn fern"
grown by Roehrs Exotic Nurseries since 1930*

Platycerium "diversifolium" hort.
"Erect elkhorn"

Platycerium bifurcatum
"Common staghorn" (juv.)

Platycerium
wilhelminae-reginae

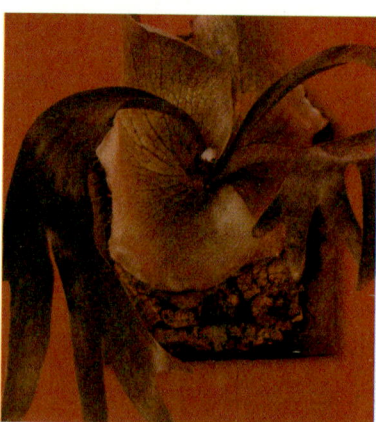

Platycerium bifurcatum 'Majus'
grown by Roehrs

Filices: POLYPODIACEAE

Platycerium bifurcatum
decorating the entrance to a village temple, in Bali

Platycerium grande
growing on its host, a stinging tree, in New South Wales

Platycerium coronarium
an old specimen, in a rubber-plantation in Malaya

Platycerium alcicorne
coll. by J. Bogner in Madagascar; in Bot. Garden Munich

Filices: POLYPODIACEAE

Platycerium andinum
"American staghorn"

Platycerium ridleyi
from Sumatra

Platyc. wilhelminae-reginae
Brown River, Papua

Platycerium stemmaria
growing in Nigeria

Platycerium superbum
Botanic Garden Munich

Platycerium quadridichotomum
(Madagascar)

Platycerium willinckii
"Silver staghorn"

Platycerium andinum
"American staghorn"

Platycerium alcicorne
from Madagascar

Filices: POLYPODIACEAE

Platycerium wilhelminae-reginae
in habitat near Port Moresby, Papua-New Guinea

Platycerium grande (superbum)
"Regal elkhorn", in northern New South Wales, Australia

Platycerium wallichii
epiphytic on forest tree, near Saigon, Vietnam

Platycerium madagascariense
with Cymbidiella orchid on tree branch, in Madagascar

FRUIT

454

Annona muricata, "Soursop" (*ANNON.*)
or "Prickly custard apple", on Nuku Hiva, Marquesas

Mangifera indica (*ANACARD.*)
the "Mango tree", in French Polynesia

Kigelia pinnata (*BIGNON.*)
"Sausage tree" on the Upper Nile river, Northern Uganda

Adansonia digitata, giant "Baobab" (*BOMBAC.*)
or "Monkey-bread tree" near Dakar, Sénégal, W. Africa

FRUIT

Spondias venulosa (*ANACARD.*)
"Mombin" or "Spanish plum"

Mangifera indica 'Joe Welch'
in Waimanalo, Oahu (ANAC.)

Mangifera indica 'Haden'
"Mango" in Hawaii (ANAC.)

Aegle marmelos (*RUTAC.*)
"Bengal apple" or "Bael fruit"

Spondias dulcis (*ANACARD.*)
"Otaheite apple" (Tahiti)

Annona muricata (*ANNON.*)
"Sour sop" or "Guanabara"

Annona squammosa (*ANNON.*)
"Sugar apple" or "Sweet sop"

Blighia sapida (*SAPIND.*)
"Aki tree" in Montego, Jamaica

Annona cherimola (*ANNON.*)
"Cherimoya" in Vista, California

FRUIT

Eriobotrya japonica (*ROSAC.*)
"Chinese loquat" or "Japan plum" (*China*)

Parmentiera edulis (*BIGNON.*)
"Guajilote" tree in Guatemala

Dovyalis hebecarpa (*FLACOURT.*)
"Ceylon gooseberry" (*Sri Lanka, India*)

Crescentia cujete (*BIGNON.*)
"Calabash tree", common in Tropical America

Cornus capitata (*CORNAC.*)
Himalayan tree, with strawberry-like fruit in N.Z.

Ficus carica, "Common fig" (*MORAC.*)
in the garden of Napoleon's exile, on St. Héléna

FRUIT

Mammea americana (GUTT.)
"Mammee apple" (W. Indies)

Garcinia mangostana (GUTT.)
"Mangosteen" (Malaya)

Monstera deliciosa (ARAC.)
"Mexican bread-fruit"

Mammea americana (GUTT.)
"South American apricot"

Crescentia cujete (BIGNON.)
tropical "Calabash"

Pachira aquatica (BOMBAC.)
"Malabar chestnut"

Nephelium lappaceum (SAPIND.)
"Rambutan" in Saigon

Litchi chinensis (SAPIND.)
"Lychee nut" (China)

Manilkara zapota (Achras) (SAP.)
"Chicle tree" or "Sapodilla"

FRUIT

458

Dovyalis x hybrida (*FLACOURT.*)
"Tropical apricot"

Dovyalis hebecarpa (*FLAC.*)
"Ceylon gooseberry"

Dovyalis caffra (*FLAC.*)
"Kei-apple" (So. Africa)

Olea europaea (*OLEAC.*)
"Olive" with ripe black fruit

Flacourtia indica (*FLACOURT.*)
"Governor's plum"

Malpighia glabra (*MALPIGH.*)
"Barbados cherry"

Arbutus unedo (*ERIC.*)
"Strawberry tree"

Punica granatum (*PUNIC.*)
"Pomegranate" in California

Punica granat. 'Nana' (*PUNIC.*)
"Dwarf pomegranate"

FRUIT

Carica papaya (*CARIC.*)
"Papaya" or "Melon tree"

Carica papaya 'Oblonga'
"Papaya" in Singapore

Ananas comosus (*BROM.*)
"Pineapple" in Bahia

Ficus carica 'Mission' (*MORAC.*)
"Black Mission fig"

Cereus peruvianus (*CACT.*)
"Peruvian apple"

Opuntia ficus-indica (*CACT.*)
"Indian fig" (Mexico)

Cordia sebestena 'Aurea' (*BORAG.*)
"Geiger tree"

Manilkara zapota (Achras)
"Sapodilla" or "Chicle tree"

Kigelia pinnata (*BIGNON.*)
"Sausage tree" (Trop. Africa)

FRUIT

460

Eugenia uniflora (*MYRT.*)
"Surinam cherry"

Acmena smithii (*MYRT.*)
"Lilly-Pilly" (E. Australia)

Syzygium jambos (*MYRT.*)
"Malay rose apple"

Syzygium samarangense (*MYRT.*)
(*Eugenia javanica*)

Syzygium malaccense (*MYRT.*)
"Rose apple" (Malaya)

Psidium cattleianum (*MYRT.*)
"Strawberry guava"

Psidium guajava (*MYRT.*)
"Common guava" in California

Psidium littorale (lucidum)
"Yellow strawberry guava"

Psidium guajava (*MYRT.*)
"Apple guava" (ripe fruit)

FRUIT

Durio zibethinus, *"Durian tree"*
in Thailand, bearing cauliflorous fruit from main branches

Durio zibethinus (*BOMBAC.*)
famous "Durian" fruit, in the Holy Forest of Sangeh. Bali

Diospyros kaki 'Fuju' (*EBEN.*)
non-astringent "Oriental persimmon" in Vista, Calif.

Casimiroa edulis 'Suebelle' (*RUT.*)
"White sapote" at Quail Bot. Garden, Encinitas

FRUIT

Arbutus unedo (*ERIC.*)
"*Strawberry tree*"

Marlierea edulis (*MYRT.*)
"*Cambuca*" (*Brazil*)

Myrica rubra (*MYRIC.*)
"*Chinese strawberry tree*"

Litchi chinensis (*SAPIND.*)
"*Lychee nut*" (*China*)

Elaeagnus philippinensis
"*Lingaros*" (*ELAEAG.*)

Catesbaea spinosa (*RUBIAC.*)
"*Lily-thorn*" (*Bahamas*)

Eriobotrya japonica (*ROSAC.*)
"*Loquat*" (*China*)

Carissa grandiflora (*APOC.*)
"*Natal plum*" (*Natal*)

Ziziphus mauritiana
"*Indian jujube*" (*RHAMN.*)

FRUIT

Averrhoa carambola (*OXAL.*)
"*Carambola*" (*India*)

Persea americana 'Fuerte'
"*Pear-avocado*" (*LAUR.*)

Diospyros 'Kaki' (*EBENAC.*)
"*Japanese persimmon*" or "*Kaki*"

Chrysophyllum cainito (*SAPOT.*)
"*Star-apple*" (*W.F. Whitman, Florida*)

Dillenia indica (*DILLEN.*)
"*Elephant-apple*" (*Java*)

Casimiroa edulis var. 'Vernon'
"*White sapote*" (*RUTAC.*)

Diospyros digyna (ebenaster)
"*Black sapote*" in Mexico

Licania tomentosa (*CHRYSOBAL.*)
the "*Oiti*" in Bahia, Brazil

Feijoa sellowiana 'Nazemetz'
"*Pineapple guava*" in California

FRUIT

Cereus peruvianus (*CACT.*)
night-blooming "Peruvian apple" with brilliant fruits

Punica granatum (*PUNIC.*)
"Pomegranate tree" with ripe fruit, in Greece

Ananas comosus 'Smooth Cayenne' (*BROM.*)
from French Guyana: since 1885 planted in Hawaii

Opuntia ficus-indica (*CACT.*)
"Opuntia fruit" sold in the markets of Rome, Italy

FRUIT

Vitis vinifera 'Riesling Sylvaner' (*VITAC.*)
"European wine-grapes" in the Moselle valley

Vitis vinifera 'Black Alicante'
"Dessert grape" grown under glass in Nelson, N. Zealand

Cocos nucifera (*PALM.*)
"Coconuts" taken to market on the Klongs of Bangkok

Musa acuminata 'Nana' (cavendishii) (*MUSAC.*)
grown in the fertile Orotava valley of Tenerife

Malus pumila 'Golden Delicious' (*ROSAC.*)
in Canton Valais along the Rhône in Switzerland

Ipomoea batatas (*CONVOLV.*)
"Sweet potatoes", carried by Igorot woman, Philippines

FRUIT

Myristica fragrans (*MYRIST.*)
"Nutmeg" in Trinidad

Piper nigrum (*PIPER.*)
"Black pepper" in Ceylon

Ochrosia marianensis (*APOC.*)
in Micronesia

Coffea laurifolia (*RUBIAC.*)
at Cap Ferrat, France

Coffea liberica (*RUBIAC.*)
"Liberian coffee"

Coffea arabica (*RUBIAC.*)
"Arabian coffee" in Sao Paulo

Cola nitida (*STERCUL.*)
"Cola nut" in Seychelles

Arctostaphylos uva-ursi (*ERIC.*)
"Bear-berry" (Pacific coast)

Arctostaphylos columbiana
"Manzanita" in Utah

FRUIT 467

Anacardium occidentale (*ANAC.*)
"Cashew nut" (W. Indies)

Pistacea vera (*ANACARD.*)
"Pistachio nut" (Syria)

Macadamia tetraphylla (*PROT.*)
"Queensland nut"

Carya pecan (*JUGLAND.*)
"Pecan nut" in California

Terminalia catappa (*COMBRET.*)
"Tropical almond" in Tahiti

Juglans regia (*JUGLAND.*)
"English walnut"

Anacardium occidentale (*ANACARD.*)
"Maranon" or "Cashew"

Parinari mobola (*ROSAC.*)
"Nef-fruit" in Sénégal

Coccoloba uvifera (*POLYGON.*)
"Seagrapes" in the Bahamas

FRUIT

Pyrus communis 'Lincoln'
in Bonsall, California

Pyrus communis (ROS.)
"Winter pear" in California

Pyrus comm. 'Doyenne du Comice'
in San Diego County, Calif.

Malus pumila 'Anna' (ROS.)
subtropic, from Israel

Malus pumila 'Winter Banana'
semi tropic apple in So. Calif.

Pachira aquatica (BOMBAC.)
"Guinea chestnut"

Prunus pers. 'Golden Jubilee'
"Freestone peach" in N. Jersey

Prunus persica 'Ventura' (ROS.)
"Freestone peach" in California

Prunus persica 'Sim's Cling'
"Cling peach" in San Diego

FRUIT

Prunus cerasus 'Montmorency'
"Sweet-tart cherry"

Prunus salicifolia 'Capulin'
subtropical "Capulin cherry"

Prunus cerasus (*ROS.*)
"Sour cherry"

Prunus armeniaca 'Royal'
"California apricot"

Prunus persica 'Santa Rosa'
"Freestone peach" in Vista

Prunus armeniaca 'Moorpark'
"Moorpark apricot"

Olea europaea (*OLEAC.*)
"Black olive", ripe fruit

Prunus salicina 'Great Yellow'
"Great yellow plum"

Prunus salicina 'Satsuma'
"Japanese plum"

FRUIT   470

Musa x paradisiaca (*MUS.*)
"French plantain" in Tahiti

Musa Fehi (troglodytarum)
"Cooking plantain"

Musa acuminata (cavendishii)
"Dwarf banana"

Chaenomeles japonica (*ROS.*)
"Japanese quince"

Musa x paradisiaca 'Champa'
"Lady fingers" (Kew Gardens)

Musa x paradisiaca (*MUSAC.*)
banana with flower and fruit

Prunus amygdalus (dulcis) (*ROS.*)
"Almond" in California

Cydonia oblonga 'Pineapple'
"Pineapple quince"

Cydonia sinensis (*ROSAC.*)
"Chinese quince"

FRUIT

Phoenix dactylifera (*PALM.*)
"Date palm" grove on Nile river, near Cairo, Egypt

Phoenix dactylifera 'Deglet Noor' (*PALM.*)
irrigated date palms in Coachella valley, California

Garcinia mangostana (*GUTT.*)
tropical "Mangosteen" fruiting in Thailand

Bactris gasipaes (Guilielma) (*PALM.*)
"Peach palm" or "Pejibaye" from Amazonian Perú

Coffea arabica (*RUBIAC.*)
"Arabian coffee" on high plateau of Sao Paulo, Brazil

Coffea robusta (canephora) (*RUBIAC.*)
"Robusta coffee" on Goroka highlands, New Guinea

FRUIT

472

Rubus laciniatus 'Prof. Rudloff'
"Thornless blackberry"

Rubus ursinus (ROS.)
"Pacific dewberry"

Arachis hypogaea (LEGUM.)
"Peanuts" in New Guinea

Vaccinium corymbosum (ERIC.)
"Highbush blueberry"

Rubus idaeus (ROS.)
"Red raspberry"

Ribes sativum (SAXIF.)
"Red currants"

Citrullus lanatus 'Family Fun'
"Water melon" (CUCURB.)

Capsicum annuum 'Longum'
"Cayenne pepper" (SOLAN.)

Lycopersicon lycopers. cerasiforme
"Cherry tomato" (SOLAN.)

FRUIT

Passiflora edulis fa. flavicarpa
form with fruit maturing yellow

Passiflora edulis (*PASSIFL.*)
"Purple granadilla"

Actinidia chinensis (*ACTIN.*)
"Kiwi vine" (China)

Vanilla fragrans (*ORCH.*)
"Vanilla" beans (Mexico)

Morus nigra (*MORAC.*)
"Black mulberry"

Cyphomandra betacea (*SOLAN.*)
"Tree tomato" or "Tamarillo"

Vitis vinifera 'Riesling'
"Moselle wine grape"

Vitis vinifera 'Black Alicante'
"Dessert grape" in New Zealand

Vitis vinif. 'Thompson Seedless'
seedless "Table grape"

FRUIT

474

Cocos nucifera 'Golden Malay'
"Dwarf coconut" (*PALM.*)

Areca catechu (*PALM.*)
"Betelnut palm"

Borassus flabellifer, (*PALM.*)
"Palmira palm" in Sénégal

Theobroma cacao (*STERCUL.*)
"Cocoa tree" in Trinidad

Phoenix dactylifera (*PALM.*)
"Date palm" fruit

Pandanus leram (*PANDAN.*)
on Awak highlands, New Guinea

Pometia pinnata (*SAPIND.*)
"Langsir", in the Marquesas

Ceratonia siliqua (*LEGUM.*)
"Carob" or "St. John's bread"

Canarium vulgare (*BURSERAC.*)
"Java almond" in Ceylon

FRUIT

Mammea americana (*GUTT.*)
"Mammee apple" of the West Indies

Carica papaya (*CARIC.*)
"Papaya" or "Melon tree", in Hawaii plantation

Artocarpus altilis (*MORAC.*)
the "Breadfruit", brought to Barbados from Tahiti

Artocarpus heterophyllus (*MORAC.*)
"Jackfruit", cauliflorous direct from trunk, in Hawaii

Lodoicea maldivica (*PALM.*)
"Double coconut", female tree on Mahé isle, Seychelles

Cocos nucifera 'Dwarf Samoan'
in the South Pacific island kingdom of Tonga

Cocos nucifera 'Dwarf Green'
bearing coconuts in Bahia, Brazil

Cocos nucifera (*PALM.*)
"Coconut palm" on the Caribbean isle of Barbados

FRUIT

Nephelium lappaceum (*SAPIND.*)
"Rambutan" on the market in Saigon, Vietnam

Pandanus odoratissimus (*PANDAN.*)
"Hala screw pine" near Noumea, New Caledonia

Theobroma cacao (*STERCUL.*) Byttneraceae
"Cocoa tree" with fruit, near Ibadan, Nigeria, W. Africa

Litchii chinensis (*SAPIND.*)
"Litchi nut" harvested in Kwangtung Prov., South China

FRUIT

478

Musa Fehi (troglodytarum) (*MUSAC.*)
"Cooking plantain" in Tahiti, French Polynesia

Musa acuminata (cavendishii) (*MUSAC.*)
"Dwarf banana" in Orotava, Tenerife, Canary Islands

Artocarpus heterophyllus (*MORAC.*)
"Jackfruit", cauliflorous on tree trunk, in Madagascar

Musa x paradisiaca (*MUSAC.*) *"Common banana"*
of the Cuna Indians, San Blas Islands, Panama

FRUIT

Citrus aurantifolia (RUTAC.)
"Mexican lime" in California

Citrus maxima (grandis)
"Shaddock" in Tahiti

Citrus aurantium
"Seville orange" in Spain

Citrus limon 'Ponderosa'
"American wonder lemon"

Citrus aurantium
"Sour Seville orange" in Curacao

Citrus aurantium myrtifolia (RUT.)
"Myrtle orange" in Argentina

Citrus limon 'Meyeri'
"Meyer lemon"

Citrus limon 'Eureka'
"Acid lemon" in Vista

Citrus maxima (RUT.)
"Pamplemousse" in Tahiti

FRUIT

Citrus paradisi 'Ruby Blush'
"Pink grapefruit" (RUT.)

Citrus sinensis 'Valencia'
"Valencia orange" in California

Citrus sinensis 'Washington Navel'
"Winter orange" in San Diego

Citrus x paradisi 'Marsh Seedless'
"Seedless grapefruit"

Citrus 'Dweet' (Tangor)
(Orange x Tangerine)

Citrus 'Minneola' (Tangelo)
(Grapefruit x Tangerine)

Citrus reticulata 'Dancy'
"Dancy tangerine"

Citrus reticulata 'Clementine'
"Tangerine"

Citrus reticulata 'Satsuma'
"Mandarin orange" (Japan)

FRUIT 481

Citrus reticulata (RUT.)
"Mandarin orange", in Kwangchow Bot. Garden, China

Citrus sinensis 'Valencia'
"Valencia orange", summer-fruiting in Florida sunshine

Citrus x paradisi (RUT.)
"Grapefruit" in great clusters in Vista, California

Citrus mitis (RUT.)
"Calamondin", dwarf mandarin pot-grown for Christmas

FRUIT

Fortunella japonica (*RUT.*)
"*Marumi kumquat*"

Fortunella margarita
"*Nagami kumquat*" (*China*)

Citrus microcarpa (*RUT.*)
"*Four-season tangerine*" (*Taipo*)

Citrus aurantium myrtifolia
"*Myrtleleaf orange*"

Citrus mitis
"*Calamondin*" (*Philippines*)

Fortunella margarita
"*Oval kumquat*"

Ximenia americana (*OLEAC.*)
"*Sour plum*" (*San Salvador*)

Fortunella hindsii (*RUT.*)
"*Dwarf kumquat*" (*Kwangtung*)

Cydonia oblonga (*ROSAC.*)
"*Fruiting quince*" (*Iran*)

GENTIANACEAE, GERANIACEAE

Gentiana lutea (GENT.)
"Yellow gentian" in the Bernese Alps, near Interlaken

Pelargonium x hortorum, "Zonal geraniums"
in terracotta jars, Copenhagen Bot. Gardens

Pelargonium peltatum 'L'Elegante'
"Sunset ivy", as hanging plant in San Diego

Nymphoides indica (GENT.)
"Water snowflake", aquatic in Frankfurt, Germany

FLACOURTIACEAE, FUMARIACEAE, GENTIANACEAE

Oncoba kraussiana
in Durban, South Africa

Hydnocarpus anthelmintica
"Chaulmoogra tree"

Gentiana acaulis
"Blue gentian" in Zermatt

Dicentra spectabilis (*FUMAR.*)
"Bleeding heart" (Japan)

Flacourtia indica
"Governor's plum"

Exacum affine
"Mexican violet"

Exacum affine (*GENT.*)
"Persian violet" in Germany

Nymphoides indica
"Floating heart" (aquatic)

Orphium frutescens
"Sticky flower" (So. Africa)

GERANIACEAE

Pelargonium x hortorum 'Olympic Red'
grown for Spring bloom, at Roehrs greenhouses

Pelargonium peltatum 'Ville de Paris'
"Strassbourg geranium" on the Moselle in Germany

Pelargonium x domesticum
"Regal geraniums", at Longwood Gardens, Pennsylvania

Pelargonium x hortorum 'Red Rambler'
a double-flowered bicolor geranium in Switzerland

Pelargonium inquinans
endemic "Geranium" species in the Transvaal, So. Africa

Pelargonium violarum
"Trailing pansy" or "Viola geranium" (So. Africa)

GERANIACEAE

Pelargonium x domesticum
'Marie Vogel'

Pelargonium x domesticum
'Lavender Queen'

Pelargonium x domesticum
'Grand Slam'

Pelargonium x domesticum
'Circus Day'

Pelargonium x domesticum
'Fire Dancer'

Pelargonium x domesticum
'Easter Greeting'

Pelargonium x domesticum
'Gay Nineties'

Pelargonium x domesticum
'Grossmama Fischer'

Pelargonium x domesticum
'Springtime'

Pelargonium x domesticum
'Jessie Jarrett'

Pelargonium x domesticum
'Orchid Edith North'

Pelargonium x domesticum
'Swabian Maid'

GERANIACEAE

Pelargonium x hortorum
'Mrs. Henry Cox' (*tricolor*)

Pelargonium x hort. 'Velma'
"Tricolor geranium"

Pelargonium x hortorum
'Miss Burdett Coutts' (*tricolor*)

Pelargonium x hortorum
'Mrs. Pollack' (*tricolor*)

Pelargonium x hortorum
'Mrs. Parker'

Pelargonium x hortorum
'Mr. Wren'

Pelargonium echinatum
"Sweetheart geranium"

Pelargonium x hortorum
"Skies of Italy" (*maple leaf*)

Pelargonium x hortorum
'Mrs. Strang' (*tricolor*)

Sarcocaulon rigidum
"Bushman's candles"

Pelargonium x domesticum
'Pansy' (*Pansy geranium*)

Pelarg. x domest. 'Mac Kay'
a *"Martha Washington geranium"*

GERANIACEAE

Pelargonium x hortorum
'Genie Irene'

Pelargonium x hortorum
'Enchantress Fiat'

Pelargonium x hortorum
'Penny Irene'

Pelargonium x hortorum
'Better Times'

Pelargonium x hortorum
'California Appleblossom'

Pelargonium x hortorum
'Olympic Red'

Pelargonium x hortorum
'Salmon Supreme'

Pelargonium x hortorum
'Radio Red'

Pelargonium x hortorum
'Daybreak Salmon'

Pelargonium x hortorum
'Red Fiat'

Pelargonium x hortorum
'Antares' (*miniature geranium*)

Pelargonium x hortorum
'Improved Ricard'

GERANIACEAE

Pelarg. pelt. 'Ville de Paris'
"Strassbourg geranium"

Pelarg. peltatum 'Mexico'
bicolor "Ivy geranium"

Pelarg. pelt. 'Lumière du Matin'
"Strassbourg geranium"

Pelargonium peltatum
"Ivy geranium" in California

Pelarg. peltatum 'Galilee'
double pink "Ivy geranium"

Pelarg. peltat. 'Variegatum'
"Variegated ivy geranium"

Pelargonium x hort. 'Irene'
excellent commercial "Geranium"

Pelarg. x hort. 'Mad. Salleron'
"Carpet-bed geranium"

Pelarg. x hort. 'Carefree' strain
grown from seed

GERANIACEAE

490

Erodium chamaedryoides
"Alpine geranium"

Pelargonium crassicaule
"Succulent geranium" (Kew G.)

Pelargonium zonale
"Zonal geranium" in Kirstenbosch

Pelarg. x limoneum 'Variegatum'
"English finger-bowl geranium"

Pelarg. 'Prince Rupert variegated'
"French lace"

Pelargonium tomentosum
"Peppermint geranium"

Pelargonium x domesticum
'Mrs. Mary Bard'

Pelargonium graveolens
"Rose geranium"

Geranium grandiflorum
in Sikkim Himalayas

GESNERIACEAE

Chirita lavandulacea
"Hindustan gentian"

Chirita sinensis
"Silver chirita"

Aeschynanthus parvifolius
(*Himalayas*)

Chrysothemis pulchella
(*Trinidad*)

Codonanthe gracilis
(*Brazil*)

Columnea hirta
(*Costa Rica*)

Sarmienta repens
pretty creeper from Chile

Columnea gloriosa
(*Costa Rica*)

Columnea gloriosa 'Variegata'
in Germany

GESNERIACEAE

Achimenes longiflora 'Alba'
(*Guatemala*)

Aeschynanthus lobbianus
(radicans) *from Java*

Achimenes longiflora 'Andersonii'
a "Monkey-faced pansy"

Aeschynanthus speciosus
(*Java, Malaya*)

Achimenes candida
"Mother's tears" (Guatemala)

Aeschynanthus javanicus
"Lipstick plant" (Java)

Aeschynanthus micranthus
(*Himalayan Region*)

Aeschynanthus marmoratus
"Zebra basket vine" (Burma)

Aeschynanthus pulcher
"Royal red bugler" (Java)

Alloplectus schlimii
(*Andes of Colombia*)

Alloplectus capitatus
"Velvet alloplectus"

Drymonia turrialvae
(*Costa Rica to Panama*)

GESNERIACEAE

Achimenes x hybrida
"Magic flower" or "Cupid's bower", in hanging basket

Aeschynanthus speciosus
(*Trichosporum splendens* in hort.), from Java

Chrysothemis pulchella
succulent gesneriad, from West Indies to Amazonas

Boea hygroscopica
"Oriental streptocarpus" from Queensland

Columnea teuscheri
(*Trichantha minor*) from Ecuador

Columnea campanulata
from tropical Colombia (J. Bogner)

GESNERIACEAE

494

Columnea linearis
"Pink goldfish plant"

Columnea crassifolia
(*Mexico, Guatemala*)

Columnea linearis
(*Costa Rica*)

Columnea nicaraguensis
(*Central America*)

Codonanthe crassifolia
(*Costa Rica to Venezuela*)

Columnea hirta
a "Goldfish plant" (*Costa Rica*)

Columnea panamensis
(*Panama*)

Columnea schiedeana
(*Eastern Mexico*)

Columnea tulae flava
(*Puerto Rico, Haiti*)

Columnea verecunda
(*Costa Rica*)

Columnea 'Vega'
"Goldfish bush" from Norway

Columnea x vedrairiensis
(*magnifica x schiedeana*)

GESNERIACEAE

Columnea microphylla
"Goldfish vine" in Longwood Gardens, Pennsylvania

Cyrtandra schraderi
on a jungle tree near Oomsis, Papua-New Guinea

Saintpaulia rupicola, an "African violet"
in habitat on rocks at Marakaya, Northern Kenya

Sinningia leucotricha (Rechsteineria)
the "Brazilian Edelweiss", in Western Paraná, Brazil

GESNERIACEAE

Episcia cupreata
a "Carpet plant" from Colombia

Episcia lilacina ('Fannie Haage')
"Blue-flowered teddy-bear"

Episcia cupreata 'Tropical Topaz'
"Canal zone yellow", from Panama

Episcia cupreata 'Shimmer'
exquisite iridescent foliage

Episcia lilacina 'Cuprea'
from Rio Reventazon, Costa Rica

Episcia lilacina 'Viridis'
in Costa Rica, Central America

Episcia reptans (fulgida)
a "Flame violet" from northern South America

Episcia x wilsonii 'Pinkiscia'
(cupreata x lilacina)

GESNERIACEAE

Episcia cupreata 'Acajou'
(*cupreata x variegata*)

Episcia cupreata 'Frosty'
silver leaf "Flame violet"

Episcia reptans
'Lady Lou'

Episcia cupreata
'Tropical Topaz' (*Panama*)

Episcia dianthiflora
"*Lace-flower vine*" (*Mexico*)

Episcia cupreata metallica
"*Kitty episcia*"

Episcia cupreata 'Musaica'
(*Panama*)

Episcia lilacina
('Mrs. Fannie Haage')

Episcia punctata
(*Guatemala, Mexico*)

Episcia cupreata 'Cleopatra'
at New York Botanical Garden

Chrysothemis friedrichsthalii
(*Guatemala and West Indies*)

Diastema vexans
from Colombia

GESNERIACEAE

498

Gesneria cuneifolia
"Fire cracker" (*Puerto Rico*)

Kohleria 'Bogotensis hybrid'
a "Tree gloxinia"

Hypocyrta nummularia
(*Nemathanthus or Alloplectus*)

Kohleria bogotensis
(*Isoloma pictum*)

Kohleria amabilis
from Colombia

Nematanthus fissus
(*Hypocyrta selloana*)

Kohleria 'Eriantha hybrid'
(*Isoloma hirsutum multiflorum*)

Kohleria 'Sciadotydaea'
(*Sciadocalyx x Tydaea*)

Kohleria eriantha
(*Isoloma hirsutum*)

Columnea teuscheri
(*Trichantha minor*)

Chirita sinensis
(*So. China*)

Nautilocalyx lynchii
"Black alloplectus"

GESNERIACEAE

Saintpaulia rupicola
species from Kenya

Saintpaulia ionantha
"African violet" (Tanzania)

Saintpaulia 'Snow Prince'
single white

Saintpaulia 'Double Delight'
floriferous double blue

Saintpaulia 'Corinne'
double white

Saintpaulia 'Flash'
double red

Saintpaulia "Kenya violet-blue"
(rupicola x ionantha)

Saintpaulia 'Fuchsia Red'
semi double red-purple

Saintpaulia 'Blue Peak'
double blue, edged in white

Saintpaulia 'Ruffled Queen'
frilled amethyst

Saintpaulia 'Pink Miracle'
single pink edged rose

Saintpaulia 'Blue Caprice'
double light blue

GESNERIACEAE

Saintpaulia 'Pocono'
"African violet" cultivar, with giant 5 cm flowers

Saintpaulia grotei
species, from the Eastern Usambara Mountains, Tanzania

Saintpaulia 'Show Queen'
variegated foliage

Saintpaulia 'Blushing Bride'
"Double pink African violet"

Saintpaulia
'Blue-boy-in-the-snow'

Saintpaulia 'Elfriede' in winter
(Rhapsodie strain)

Saintpaulia ionantha alba
in hanging basket

Saintpaulia 'Diana'
(Harmonie strain)

GESNERIACEAE

Sinningia (Rechst.) cardinalis
"Cardinal flower"

Sinningia (Rechst.) macropoda
"Vermilion helmet flower"

Sinningia leucotricha
(Rechsteineria)

Seemannia latifolia
from Bolivia

Sinningia macropoda
(Rechsteineria) from Brazil

Sinningia (Rechst.) 'Longiflora'
(cardinalis x eumorpha)

Streptocarpus caulescens (*inflor.*)
"Violet nodding bells"

Gloxinia perennis
"Canterbury bells" (Colombia)

Ramonda nathaliae
from Yugoslavia

Saintp. 'Savannah Sweetheart'
hollyleaf double

Saintpaulia 'Icefloe'
large double white

Saintpaulia 'Danube Waves'
double light blue

GESNERIACEAE

Kohleria 'Eriantha hybrid'
(*Isoloma hirsutum multifl.*)

Gesneria cuneifolia
"Fire cracker" (*Puerto Rico*)

Columnea teuscheri
(*Ecuador*)

Mitraria coccinea
vine from Chile

Kohleria eriantha
a "Tree gloxinia" (*Colombia*)

Nematanthus fissus
(*Hypocyrta selloana*)

Sinningia cardinalis
'Innocence' (Rechsteineria)

Sinningia cardinalis
(Rechsteineria) (*Brazil*)

Sinningia 'Rex'
(x Gloxinera)

GESNERIACEAE

Sinningia speciosa fyfiana 'Emperor Frederick' (*gloxinia*)

Sinningia speciosa 'Tom Thumb' *"Miniature gloxinia"*

Sinningia speciosa fyfiana 'Emperor William'

Sinningia speciosa fyfiana 'Defiance', *florist's gloxinia*

Sinningia speciosa fyfiana 'Tigrina'; *gloxinia of florists*

Sinningia regina *"Cinderella slippers"* (*Brazil*)

Sinningia pusilla *"Miniature slipper plant"*

x Streptogloxinia hybrida (*Streptocarpus x Sinningia*)

Sinningia 'Doll Baby' (*pusilla x eumorpha*)

Sinningia concinna *miniature from Brazil*

Sinningia speciosa fyfiana 'Double Chicago'

Sinningia pusilla *miniature in teacup*

GESNERIACEAE

Sinningia speciosa fyfiana 'Switzerland'
"Ruffled grandiflora gloxinia"

Sinningia speciosa fyfiana
the Gloxinia of florists, in Roehrs greenhouses

Sinningia pusilla 'White Sprite'
in Nymphenburg Botanic Garden, Munich

Epithema involucratum
in habitat near Bantimurong, Celebes, Indonesia

Saintpaulia 'Elfriede' (*German Rhapsody strain*)
on warm window-sill, sheltered from the snow outside

Rhabdothamnus solandri
unusual shrub, and sole gesneriad in New Zealand

GESNERIACEAE

Smithiantha 'Zebrina discolor' (*hybrida*)
"Temple bells" in full bloom, in New Zealand

Streptocarpus saxorum
clinging to rocks, in the Usambara Mountains, Tanzania

Streptocarpus saundersii
a single-leaf species, blooming in Rotorua, New Zealand

Streptocarpus phyllanthus
mono-leaved "Cape primrose" in flower, in Natal, So. Africa

GESNERIACEAE

Smithiantha hybrida 'Compacta'
"Temple bells"

Smithiantha 'Exoniensis'
formerly known as Naegelia

Smithiantha cinnabarina
(*Mexico, Guatemala*)

Smithiantha zebrina (Naegelia)
(*E. Mexico: Vera Cruz*)

Smithiantha (Naegelia)
'Golden King'

Smithiantha 'Orange King'
"Orange temple bells"

Streptocarpus caulescens
(*Trop. East Africa*)

Streptocarpus x hybridus
"Hybrid Cape primrose"

Streptocarpus rexii
"Cape primrose"

Streptocarpus x hybridus 'Veitch'
a compact English strain

Titanotrichum oldhamii
(*South China, Taiwan*)

Streptocarpus x hybridus
'Farbenwunder' (*German strain*)

GESNERIACEAE

Streptocarpus x hybridus (*blue*)
"Cape primrose"

Streptocarpus x hybridus (*red*)
in New Zealand

x Streptogloxinia hybrida
(*Bot. Garden, Frankfurt, Germany*)

Streptocarpus x kewensis
(*dunnii x rexii*)

Columnea campanulata
(*Tropical Colombia*)

Sinningia hirsuta
(*Brazil*)

Streptocarpus grandis (*Zululand*)
unifoliate species in Botanic Garden, Sydney

Streptocarpus synandrus
from Inyanga Range at 2400 m, Rhodesia

GINKGOACEAE, GOODENIACEAE, GUNNERACEAE, HYDROPHYLLACEAE

Scaevola frutescens
"Half-flower" in Tahiti

Ginkgo biloba
pyramidal form, in Delaware

Ginkgo biloba (*foliage*)
"Maidenhair tree" from China

Leschenaultia laricina
in Western Australia

Wigandia caracasana (*Venezuela*)
Bot. Garden Lisbon, Portugal

Gunnera manicata (*inflor.*)
(*So. Brazil*)

Gunnera manicata, *from Colombia*
in Queen Elizabeth Arboretum, Vancouver, Canada

Gunnera chilensis (*Ecuador, Chile*)
in Strybing Arboretum, San Francisco, California

GNETACEAE, GRAMINEAE

Welwitschia mirabilis (bainesii)
"Tree tumbo" with deep roots, in Munich Bot. Garden

Welwitschia mirabilis (*GNET.*)
female plant with seed cones, Stellenbosch, So. Africa

Arundinaria falconeri (*GRAM.*)
from the Nepal Himalayas, at Kew Gardens, London

Bambusa vulgaris
cut and rodded culms sprouting in water, Honolulu

GRASSES and LAWN SUBSTITUTES

Lolium perenne
"Perennial Rye-grass"

Stenotaphrum secundatum
"St. Augustine grass"

Agrostis tenuis
"Bent grass" in California

Paspalum notatum
"Bahia grass" in Florida

Zoysia tenuifolia
"Korean grass"

Oplismenus hirtellus varieg.
"Basket grass" or "Panicum"

Herniaria glabra
"Green carpet"

Arenaria verna caespitosa
"Irish moss"

Sagina subulata
"Scotch moss"

Festuca ovina glauca
"Blue fescue"

Cynodon dactylon
"Bermuda grass" in Bombay

Dichondra micrantha
"Lawn-leaf" in California

GRAMINEAE

Arundo donax, *the "Giant reed"* on the island of Madeira, Portugal

Cortaderia selloana (*Argentina*), the "Pampas grass", planted in Southern California

Saccharum officinarum "Sugar cane" ready for harvest, on Oahu, Hawaii

Yushania aztecorum "Evergreen Mexican bamboo", in Los Angeles Bot. Garden

GRAMINEAE

Pennisetum setaceum 'Cupreum'
"Red fountain grass"

Miscanthus sinensis 'Variegatus'
"Eulalia" or "Zebra grass"

Arundo donax versicolor
"Variegated reed" in Fiji

Arrhenatherum elatius 'Variegatum'
"Variegated oatgrass"

Oryza sativa
"Rice" heavy with grain

Saccharum officinarum
"Sugar cane" in Cuba

Zea mays gracillima
"Decorative maize" in San Diego

Zea mays rugosa
"Sweet corn"

Zea mays
"Maize" or "Indian corn"

GRAMINEAE

513

Oryza sativa, *tropical "Rice"*
yielding 3 crops a year, on irrigated terraces in Bali

Schizostachyum sp. 'Finisterre', *hollow bamboo
filled with water, carried by villagers of New Guinea*

Oryza sativa *in Indonesia*
preparing flooded rice paddy for planting of seedlings

Oryza sativa, *"Rice"*
Tamil women planting seedlings, south of Madras, India

Oryza clandestina
"Wild rice" as an aquatic plant at Munich Botanic Garden

Zea mays, *"Indian corn" or "Maize"*,
at harvest time in autumn, near Lincoln, Nebraska

GRAMINEAE

Pseudosasa japonica, *"Arrow bamboo"* or *"Metake"* in the gardens of Benrath Castle, Rhineland

Phyllostachys bambusoides, *a "running" species, "Japanese timber bamboo", in West Los Angeles*

Bambusa tuldoides *"Punting pole bamboo", in Fairchild Gardens, Miami*

Bambusa ventricosa, *"Buddha's belly bamboo", at entrance to ancient Ming Palace, Kwangchow, China*

GRAMINEAE

Bambusa (Sinocalamus) oldhamii
forming clumps, at Longwood Gardens, Pennsylvania

Phyllostachys sulphurea
"Yellow running bamboo", in Cap Ferrat, France

Phyllostachys bambusoides, *the "Madake"
in Pacific Tropical Botanic Garden, on Kauai, Hawaii*

Schizostachyum brachycladum (*Moluccas*)
tropical bamboo in the Botanical Gardens of Singapore

GRAMINEAE

Bambusa vulgaris, *"Feathery bamboo"*
on Lake Victoria, Entebbe Bot. Garden, Uganda

Olyra latifolia, *in Ituri forest,
Mountains of the Moon, Central Africa*

Sasa (Bambusa) tessellata
dwarf "running" bamboo in Peradeniya, Ceylon

Arundinaria (Sasa) pumila (*Japan*)
"Dwarf bamboo", sold in German Garden center

Sasa fortunei (*Japan*)
a miniature, at Finca Burle-Marx, near Rio de Janeiro

Phyllorhachis sagittata
an aquatic bamboo, in Botanic Garden Munich

GRAMINEAE

Arundinaria alpina, *"Mountain bamboo"* bamboo forest in Gorilla country, Kigezi, Uganda

Bambusa ventricosa, *"Buddha's belly"* bamboo with Fu-lion at Ming Tower, in Kwangchow, China

Dendrocalamus giganteus the *"Giant bamboo"*, in Peradeniya, Sri Lanka

Phyllostachys aurea, *"Otake"* or *"Fishpole bamboo"*, at Kyoto inn, a hooded monk soliciting food

GRAMINEAE

Bambusa vulgaris
in Takarazuka, Japan

Bambusa vulgaris
"Feathery bamboo" of Java

Phyllostachys nigra
"Black bamboo" (South China)

Phyllostachys aurea
"Golden bamboo" in Vista, Calif.

Sinoarundinaria nitida (*Korea*)
in Copenhagen, Denmark

Guadua angustifolia
a giant clump bamboo in Jamaica

Bambusa ventricosa
"Buddha's belly" in Puerto Rico

Yushania aztecorum
"Mexican bamboo" in California

Bambusa multiplex
"Fern-leaf bamboo" (Vietnam)

GRAMINEAE

Bambusa textilis, *"Wong Chuk"* in People's Botanic Garden, Kwangchow, China

Guadua angustifolia a giant clump-forming bamboo, Montego Bay, Jamaica

Bambusa vulgaris (*Java*) "Feathery bamboo", in Jardim Botanico, Rio de Janeiro

Dendrocalamus giganteus with culms 25 cm thick and 35 m tall, in Ceylon

GUTTIFERAE, HIPPOCASTANACEAE, JUGLANDACEAE

Kielmeyera coriacea
in Brazil

Clusia rosea (*fruit*)
"Balsam apple"

Mammea americana
"Mammee apple"

Garcinia xanthochymus
"Camboge tree" (India)

Garcinia mangostana
"Mangosteen" in fruit

Calophyllum inophyllum
"Indian laurel"

Aesculus x carnea
"Red horse-chestnut"

Juglans regia
"English walnut"

Mammea americana
at Fairchild Garden, Miami

GUTTIFERAE

Clusia rosea (*West Indies*)
"Pat pork tree" or "Autograph tree"

Garcinia xanthochymas (tinctoria)
"Camboge tree" in Peradeniya, Ceylon

Calophyllum inophyllum
"Kamani" in Laie, Oahu, Hawaii

Kielmeyera variabilis
in Jardim Botanico, Rio de Janeiro

Mammea americana (*GUTT.*)
"Mammee apple" in Waimanalo, Hawaii

Rhodeleia championi (*HAMAMELID.*)
in Kota Kinabalu, Sabah, Borneo

Aesculus pavia (*HIPPOCAST.*)
"Red buckeye" in Salt Lake City, Utah

Hypoxis rooperi (*HYPOXID.*)
yellow "Star flower" in South Africa

Hydrostachys pinnatifolia (*HYDROS.*)
aquatic plant in running streams, in Madagascar

Limnobium stoloniferum (Hydromystria)
floating water plant, from West Indies to Brazil

Hernandia ovigera (peltata) in Cairns, Queensland

Hernandia ovigera *in bloom* "Sea cups" on Tonga, Polynesia

Hernandia sonora with strange fruit (E. Indies)

Scaevola sericea (*GOODEN.*) "Half-flower" in Tahiti

Clusia rosea 'Marginata' "Monkey-apple"

Clusia rosea 'Aureo-variegata' "Variegated balsam apple"

Phacelia campanularia "California blue bell"

Xiphidium caeruleum in Jamaica

Liquidambar styraciflua "American sweet gum"

HIPPOCASTANACEAE, HYPERICACEAE

Hypericum calycinum
"Rose-of-Sharon"

Hypericum x moserianum
"Goldflower" in San Francisco

Hypericum patulum henryi
"St. John's-wort"

Hypericum x rowallane
in U. of Calif. Bot. Garden

Hypericum formosum scouleri
in Glacier Park, Montana

Hypericum leschenaultii
from Sumatra

Aesculus parviflora
"Dwarf horse-chestnut"

Aesculus x carnea
"Red horse-chestnut"

Aesculus hippocastaneum
"Horse-chestnut"

IRIDACEAE

Iris 'Wedgwood', *"Forcing iris"*
(*tingitana x xyphium*)

Neomarica caerulea, *"Apostle-plant"*
collected on the Serra do Mar, Sao Paulo, Brazil

Lapeirousia laxa (Anomatheca cruenta)
"Woodland painted petals" in the Transvaal

Geissorhiza rochensis
"Wine cups" in Cape Province, South Africa

Tigridia pavonia
"Tiger flower" from Guatemala

Crocosmia masonorum
"Golden swan tritonia" of the Transvaal

IRIDACEAE

Babiana stricta
"Baboon flower"

Colchicum autumnale (*LIL.*)
"Autumn crocus" near Zermatt

Crocus speciosus
from the Caucasus

Freesia x hybrida
of South African origin

Iris japonica 'Variegata'
from Japan

Dietes grandiflora
"Large wild iris" (So. Africa)

Neomarica bicolor
"Walking iris" from Brazil

Neomarica caerulea
"Twelve apostles"

Neomarica gracilis
"Apostle plant"

Tritonia crocata
"Kalkoentje" in So. Africa

Watsonia meriana
(South Africa)

Watsonia longifolia
(Cape Province)

IRIDACEAE

Crocosmia masonorum
"Golden swan tritonia" (Transvaal)

Tritonia crocosmaeflora
"Montbretia" of horticulture

Tritonia crocata
(S.W. Cape Prov.)

Watsonia pyramidata
"Pink watsonia"

Aristea ecklonii
"Blue stars" (Rhodesia)

Watsonia beatricis
in Kirstenbosch, So. Africa

Dierama pulcherrima
"East London harebell"

Cypella herbertii
(Uruguay, Argentina)

Babiana purpurea
"Baboon flower" in Cape Prov.

IRIDACEAE

Iris x germanica
"Bearded German iris"

Iris 'Ideal' (*forcing iris*)
U.S.D.A. Beltsville, Maryland

Iris pseudacorus
"Yellow flag" (W. Europe)

Ixia speciosa
"Corn lily" (So. Africa)

Iris pallida 'Variegata'
"Orris" (South Tyrol)

Gladiolus x hortulanum
"Painted lady" or "Garden gladiolus"

Neomarica bicolor (Marica)
"Apostle plant"

Tritonia crocosmaeflora
"Montbretia", Ola Pua, Kauai

Sparaxis tricolor 'Firebrand'
"Scarlet wand-flower"

IRIDACEAE

Watsonia pyramidata "Pink watsonia", Hottentot-Holland Mountains, So. Africa

Neomarica caerulea (Marica)
"Twelve apostles", Serra do Mar, Sao Paulo, Brazil

Dietes vegeta (Moraea iridioides)
"African iris" in Vista, San Diego Co., California

LABIATAE

Coleus rehneltianus
'Trailing Queen'

Coleus blumei
'Brilliancy'

Coleus blumei
'Beckwith Gem'

Coleus rehneltianus
'Red Trailing Queen'

Coleus blumei 'Klondyke'
"Butterfly coleus"

Coleus blumei
'Pride of Autumn'

Lavandula officinalis
"English lavender"

Coleus blumei
'Red Verschaffeltii'

Lamium galeobdolon variegatum
"Yellow archangel"

Scutellaria mociniana
"Scarlet skullcap" (Mexico)

Plectranthus oertendahlii
"Prostrate coleus"

Plectr. nummularius (australis)
"Swedish ivy"

LABIATAE

Stachys grandiflora
(*Caucasus*)

Leonotis leonorus
"Lion's ear" on St. Héléna

Physostegia virginiana
"Obedience" (*So. Carolina*)

Salvia involucrata
"Rose-leaf-sage" (*Mexico*)

Agastache cana (Cedronella)
"Mosquito plant" (*Texas*)

Moluccella laevis
"Irish bells" (*Syria*)

Lamium galeobdolon 'Variegatum'
(Lamiastrum) *"Yellow archangel"*

Coleus blumei
'Brilliancy'

Coleus blumei 'Verschaffeltii'
"Painted nettle"

LABIATAE

Plectranthus
'Variegated Mintleaf'

Plectranthus madagascariensis
"Mintleaf" in South Africa

Plectranthus coleoides
'Marginatus' "Candle plant"

Plectranthus saccatus
in Natal, South Africa

Plectranthus fruticosus
"African spurflower"

Iboza riparia
"Misty plume bush"

Plectranthus nummul. 'Variegatus'
"Variegated Swedish ivy"

Plectr. nummularius (australis)
"Creeping Charlie" (in Calif.)

Glechoma hederacea variegata
"Variegated ground ivy"

LABIATAE

Salvia officinalis
"Sage", in Yugoslavia

Salvia splendens
"Scarlet sage" (Brazil)

Salvia involucrata
"Roseleaf sage" (Mexico)

Salvia leucantha
"Mexican bush sage"

Salvia mexicana
"Ramona" (Mexico)

Salvia farinacea
"Mealy-cup sage"

Ajuga reptans purpurea
"Carpet bugleweed"

Salvia icterina
in Kew Gardens, London

Rosmarinus officinalis
"Rosemary" (Portugal)

LAURACEAE

Cinnamomum zeylanicum
"Cinnamon tree" in Singapore

Cinnamomum zeylanicum (*flowering*)
in Foster Gardens, Honolulu

Cinnamomum iners
(*Tropical Asia*)

Umbellularia californica
"Calif. laurel" or *"Myrtlewood"*

Laurus nobilis
"Sweet bay" in flower

Cinnamomum camphora
"Camphor tree"

Laurus nobilis, *"Baytrees"*
standards, some with twisted stems, in Germany

Laurus nobilis, *"Roman laurel"*
trained into pyramid shape, in Bruges, Belgium

LAURACEAE, LECYTHIDACEAE

Laurus nobilis
standard "Baytrees" in old Stockholm, Sweden

Laurus nobilis
"True laurel", with spiral stem, Essen Bot. Gard., Germany

Couroupita guianensis
"Cannonball tree" with fruit, in Summit Gardens, Panama

Couroupita guianensis
with cauliflorous flowers from trunk, in Hawaii

LAURACEAE, LECYTHIDACEAE

Lecythis ollaria (*Guyana*)
the "Monkey-pot tree" in Singapore

Barringtonia asiatica
"Fish poison tree" on Nuku Hiva, French Polynesia

Persea americana 'Fuerte'
"Avocado pear" at Quail Gardens, Encinitas, California

Barringtonia asiatica
the "Sea putat"; flowers and fruit, in Madagascar

LEGUMINOSAE

Acacia drepanolobium, *with black galls the "Whistling thorn", in the Rift Valley, Kenya*

Acacia heteracantha
"Umbrella thorn", in Krüger National Park, Transvaal

Acacia retinodes (floribunda)
everblooming "Wattle", in Adelaide, South Australia

Acacia woodii
flat-topped "Paperbark-thorn", near Kaptagat, Uganda

Delonix regia, *"Flamboyant" or "Royal poinciana" in glorious bloom, French Polynesia*

Ulex europaeus
"Gorse" in bloom, near Wellington, New Zealand

LEGUMINOSAE

538

Acacia armata
"Kangaroo-thorn"

Acacia cultriformis
"Knife acacia"

Acacia longifolia mucronata
"Narrow Sydney wattle"

Acacia latifolia
"Bush acacia"

Acacia alata
(Western Australia)

Acacia podaliriaefolia
"Pearl acacia"

Cytisus x racemosus
"Genista" of florists

Acacia sphaerocephala
from Mexico

Acacia retinodes
"Everblooming acacia"

Brownea coccinea
"Scarlet flame-bean"

Brownea macrophylla
Tahiti Botanic Garden

Cassia artemisioides
"Feathery senna"

LEGUMINOSAE

539

Acacia argyrophylla
in Cap Ferrat, France

Acacia armata
Longwood Gardens, Penna.

Acacia baileyana
"Golden mimosa"

Acacia decurrens
"Green wattle"

Acacia drummondii
(Western Australia)

Acacia longifolia
"Sydney golden wattle"

Acacia mearnsii
"Black wattle" (Uganda)

Albizia lebbeck
"Women's tongue" in St. John, V.I.

Acacia retinodes (floribunda)
(South Australia)

LEGUMINOSAE

Colvillea racemosa, *with Livistona decipiens palm in Brisbane, Queensland, Australia*

Acacia nigrescens *fragrant "Knobthorn" in northern Transvaal*

Delonix regia, *"Flamboyant" or "Royal poinciana", on Rarotonga, South Pacific*

Albizia lebbeck *"Mother-in-law tree" in Curacao, Netherlands Antilles*

LEGUMINOSAE

541

Bauhinia blakeana
"Hongkong orchid tree"

Bauhinia galpinii
"Pride of the Cape"

Bauhinia monandra
"Pink orchid tree"

Bauhinia corymbosa
"Phanera" (Vietnam)

Bauh. divaricata (lamarckiana)
(South America)

Bauhinia purpurea
"Butterfly tree"

Bauhinia variegata
"Orchid tree" (India)

Bauhinia corniculata (forficata)
"White camels-foot"

Bauhinia variegata 'Candida'
"White orchid tree"

LEGUMINOSAE

Amherstia nobilis
"Queen of flowering trees", in Burma

Brownea grandiceps
"Rose of Venezuela", in South America

Butia frondosa
"Flame of the forest", in India

Bolusanthus speciosus
"Tree wisteria", in Krüger National Park, Transvaal

LEGUMINOSAE

Brownea grandiceps
"Rose of Venezuela"

Amherstia nobilis
in Peradeniya, Ceylon

Delonix regia
"Peacock flower"

Dahlstedtia pinnata
in Guanabara, Brazil

Buddleja x madagascariensis
in Heraklion, Crete

Baikiaea insignis
(Nigeria)

Brownea macrophylla
from Panama

Calotropis gigantea (ASCL.)
in Hyderabad, So. India

Butea monosperma
"Flame of the forest"

LECYTHIDACEAE, LEGUMINOSAE

Clianthus formosus (dampieri)
"Glory-pea" (Australia)

Clianthus puniceus
"Parrot's bill" (New Zealand)

Ceratonia siliqua
"Carob" or "St. John's bread"

Chorizema cordatum
"Australian flame-pea"

Clitorea ternatea
"Butterfly pea" in Fiji

Canavallia ensiformis
"Jack-bean"

Eperua jenmanii
in Trinidad

Napoleona heudelotii
"Napoleon's button"

Camoensia maxima
in Angola

LEGUMINOSAE

Colvillea racemosa
a gorgeous tropical flowering tree, in Madagascar

Erythrina x bidwillii
the "Florida coral-bean" (crista-galli x herbacea)

Clianthus formosus (dampieri)
the "Glory-pea", in northwestern Australia

Geoffroea inermis
flowering in Balboa, Panama Canal Zone

Erythrina coralloides
"Naked coral-tree" of northeastern Mexico

Brownea grandiceps
"Rose of Venezuela" in northern South America

LEGUMINOSAE, ROSACEAE

Cytisus x racemosus
"Genista" of florists

Cytisus scoparius
'Windlesham Ruby'

Coronilla emerus
"Scorpion senna"

Crotalaria laburnifolia
"Queensland bird-flower"

Crotalaria retusa
"Rattle-box" in Trinidad

Cowania mexicana (ROS.)
"Cliff-rose" in Utah

Mimosa pudica
"Sensitive plant" (Brazil)

Christia vespertilionis
(Lourea), from Brazil

Trifolium dubium
"Irish shamrock"

LEGUMINOSAE

Cassia peltophora
in Peradeniya Botanic Garden, Sri Lanka

Caesalpinia coriaria
the wind-blown "Divi-Divi" tree on Aruba, Antilles

Cassia fistula
the "Golden shower" in the Marquesas, Polynesia

Cassia 'Rainbow Shower'
(fistula x javanica), in Honolulu, Hawaii

Calliandra haematocephala
"Pink powder-puff" in Southern Brazil

Calliandra inequilatera
"Red powder-puff" of Ecuador

LEGUMINOSAE

Caesalpinia pulcherrima
"Peacock flower"

Caesalpinia echinata
"Brazilwood" (Trop. America)

Caesalpinia gilliesii
"Bird-of-paradise shrub"

Caesalpinia vesicaria
(Cuba to Brazil)

Calliandra 'Minima' hort.
"Miniature powder-puff"

Calicotome villosa
in Kamira, Rhodos

Calliandra tweedii
"Mexican flame-bush"

Calliandra portoricensis
(West Indies)

Calliandra surinamensis
(Surinam)

LEGUMINOSAE 549

Chordospartium stevensonii (*on left*)
endemic leafless tree of South Island, New Zealand

Cassia moschata
"Bronze shower tree" in Panama

Cassia nodosa
"Pink and white shower" in Bengal, India

Cassia didymobotrya (nairobensis)
"Popcorn bush" in Quail Gardens, Encinitas, California

LEGUMINOSAE

Cassia bicapsularis (pendula) *in Escondido, California*

Cassia artemisioides *"Popcorn bush" in E. Australia*

Cassia angolensis *in Western Uganda*

Cassia fistula *"Golden shower" in India*

Cassia excelsa *"Crown of gold tree"*

Cassia corymbosa *"Flowering senna" (Argentina)*

Cassia 'Rainbow Shower' *in Honolulu, Hawaii*

Cassia alata *"Candle bush"*

Cassia surattensis (glauca) *"Scrambled egg bush"*

LEGUMINOSAE

Cassia fistula
the "Golden rain tree" on Nuku Hiva, French Polynesia

Cassia roxburghii marginata
"Red cassia" flowering in South India

Cassia grandis
"Pink shower tree" in Foster Garden, Honolulu

Cassia javanica (nodosa)
pink and white "Appleblossom shower" in Indonesia

LEGUMINOSAE

Cassia spectabilis
"Golden wonder"

Cassia siamea
"Coffee senna" in Sénégal

Cassia multijuga
"November shower" (W. Indies)

Cassia didymobotrya
"Peanut butter cassia"

Cassia sylvestris
Botanic Garden Rio de Janeiro

Cassia speciosa
in Southeastern Brazil

Cercis canadensis
"Eastern redbud" in Carolina

Cercis siliquastrum
"Judas tree" or "Love tree"

Albizia distachya
"Plume albizzia" (Australia)

LEGUMINOSAE

Mucuna bennettii
the magnificent "New Guinea creeper", in Fiji

Clianthus puniceus
"Parrot's-beak" or "Red kowhai" of New Zealand

Laburnum x watereri (vossii)
"Golden chain tree", in Winter Haven, Florida

Gliricidia sepium
"Madre de cacao" in Mysore, South India

LEGUMINOSAE

554

Erythrina caffra
"Kaffirboom coral tree"

Erythrina crista-galli
"Cockspur coral tree"

Erythrina humeana
at Leucadia Nursery, California

Erythrina indica picta
in Jogjakarta, Java

Erythrina falcata
(Minas Gerais, Brazil)

Erythrina vespertilio
"Bat-wing coral" with seeds

Erythrina lysistemon
"Kaffirboom" in Transvaal

Erythrina pallida
"Pink coral tree"

Erythrina humeana
Kirstenbosch Bot. Garden, So. Africa

LEGUMINOSAE

Erythrina falcata
under blue sky, in the Organ Mountains of Brazil

Erythrina indica picta
with painted foliage, Brisbane Bot. Garden, Queensland

Erythrina crista-galli, "Coral tree"
or "Crybaby tree", Bot. Garden Frankfurt, Germany

Erythrina abyssinica
"Red-hot poker-tree", in Entebbe Bot. Garden, Uganda

LEGUMINOSAE

Lupinus polyphyllus 'Roseus'
in Longwood Gardens, Penna.

Lupinus 'Russell hybrid'
in Edinburgh Bot. G., Scotland

Lupinus polyphyllus 'Carmineus'
"Garden lupine"

Robinia hispida
"Rose acacia" in Arizona

Podalyria caliptrata
"Sweet-pea bush"

Tamarindus indica
"Tamarind" in India

Sutherlandia frutescens
"Cancer bush" in Transvaal

Spartium junceum
"Broom" in Cuzco, Perú

Leucaena glauca
"Wild tamarind" in Hawaii

LEGUMINOSAE

Lathyrus odoratus
"Sweet pea" (Sicily)

Lathyrus odoratus
in variety, in California

Lathyrus clymeneus
articulatus, *on Delos, Greece*

Dolichos lablab
"Hyacinth bean" in Oaxaca

Strongylodon macrobotrys
"Philippine jade vine"

Kennedia rubicunda
"Coral pea" (E. Australia)

Phaseolus coccineus
"Dwarf scarlet runner"

Lotus berthelotii
"Winged pea" (Cap Verde Is.)

Phaseolus caracalla
"Snail flower"

LEGUMINOSAE

Gliricidia sepium
"Madre de cacao" (C. America)

Haematoxylum campechianum
"Bloodwood tree" (Yucatán)

Gleditsia triacanthos
"Honey locust" in California

Laburnum anagryoides
"Golden chain"

Schotia brachypetala
"Tree fuchsia" in the Transvaal

Prosopis glandulosa
"Mesquite" in Arizona

Lonchocarpus violaceus
"Lance-pod" in Trinidad

Liparia sphaerica
"Mountain dahlia" (So. Africa)

Lonchocarpus sericeus
in Sénégal, West Africa

LEGUMINOSAE

Schizolobium parahybum (*Panama*)
treefern-like "Bacurubu", in Vista, California

Maniltoa gemmipara
with unfolding leaves, in Kebun Raya, Bogor, West Java

Lathyrus odoratus, *"Guisante de olor"* or *"Sweet peas"*
grown in floating gardens of Xochimilco, Mexico City

Strongylodon macrobotrys
cascading *"Jade vine"* flowers, in the Philippines

LEGUMINOSAE

Strongylodon macrobotrys
"Philippine jade vine"

Sophora secundiflora
"Mescal bean" in Mexico

Wisteria sinensis
"Chinese wisteria"

Neptunia oleracea
"Water mimosa" (aquatic)

Hardenbergia violacea
"Purple coral-pea" (Australia)

Derris indica
(Tropical Asia)

Samanea saman
"Raintree" in Hawaii

Swainsona galegifolia
"Swan flower" (Australia)

Phaseolus caracalla (Vigna)
"Corkscrew flower"

LEGUMINOSAE

Parkia biglobosa
"African locust" (Sénégal)

Parkia javanica
Botanic Garden Singapore

Ceratonia siliqua
"Carob" or "Locust-bean"

Schotia brachypetala
"Tree fuchsia" (So. Africa)

Parkinsonia aculeata
"Jerusalem thorn" (Cap Verde)

Sabinea carinalis
"Caribe wood"

Caesalpinia pulcherrima
"Dwarf poinciana"

Peltophorum pterocarpum
"Yellow flame tree"

Caesalpinia pulcherrima flava
"Pride of Barbados"

LEGUMINOSAE

562

Lotus berthelotii
"Coral gems", creeping over arid soil, on Cape Verde Isl.

Mucuna bennettii
magnificent "New Guinea creeper", on Pali Dr., Honolulu

Mimosa pudica, *"Sensitive plant"*
leaves fold when touched, or in rain, and at night

Cytisus x racemosus
"Genista" of florists for Easter bloom

Prosopis glandulosa
"Honey mesquite" on the desert, Southern Arizona

Cercidium floridum (torreyanum)
blooming "Palo verde" on the desert, Tucson, Arizona

LEGUMINOSAE

Saraca indica
"Asoka tree" in Ceylon

Saraca thaipingensis
"Yellow saraca", in Ceylon

Tipuana tipu
"Rose wood" in Bolivia

Saraca declinata (*Malaya*)
"Red saraca"

Indigofera incarnata (decora)
"Indigo" in China

Sesbania grandiflora
"Hummingbird tree" (Cape Verde)

Samanea saman
"Rain tree" in Hawaii

Inga edulis
in Laie, Hawaii

Peltophorum pterocarpum
"Yellow poinciana"

LEGUMINOSAE, BIGNONIACEAE

Samanea saman
"Rain tree" or "Monkey-pod" near Port of Spain, Trinidad

Millingtonia hortensis (*BIGN.*)
"Indian corktree" in Mombasa, Kenya

Wisteria floribunda 'Alba'
white flowering "Japanese wisteria", in New Jersey

Wisteria floribunda 'Macrobotrys'
at Union buildings garden, Pretoria, Transvaal

Robinia hispida
"Idaho locust" or "Rose acacia", Grand Canyon, Arizona

Erythrina crista-galli
in Enoshima Botanic Garden, on Honshu, Japan

LEGUMINOSAE

Tamarindus indica
shaped "Tamarind" at Royal Palace, Bangkok, Thailand

Tamarindus indica
"Tamarind" in natural form, Botanic Garden, Trinidad

Schotia brachypetala, the "Tree fuchsia"
in the mountains of Transvaal, with native kraals

Saraca indica, "Asoka" or "Sorrowless tree"
at 13 century Ala-ud-Din tower, Old Delhi, India

LILIACEAE

Aloe arborescens, *in bloom*
overlooking the Bay of Funchal, Madeira

Aloe suarezensis
near Diego Suarez, northern Madagascar

Aloe eru, *green form*
in habitat near Hout Bay, Cape Prov. South Africa

Aloe arborescens, *spineless form*
on the Cape Peninsula, South Africa

Aloe comosa, *a tree aloe*
in the Great Karroo desert, on road to Botswana

Aloe mitriformis *in rock crevices,*
near Karoosport, Hottentot-Holland Mountains, So. Africa

LILIACEAE

Aloe bainesii, *giant tree*
in City Garden, overlooking Table Mountain, Cape Town

Aloe arborescens
in the mountains of Madeira Island, Portugal

Aloe chabaudii *in habitat*
at the Victoria Falls of the Zambezi River, in Zambia

Aloe plicatilis
in the wild Hottentot-Holland Mountains, South Africa

LILIACEAE

Aloe bainesii
young "Tree aloe" in Vista, San Diego County, California

Aloe marlothii
on the High Veld at 2000 m, in Botswana

Aloe vera, *the "Medicine plant"*
growing wild on St. Héléna, Napoleon's exile

Aloe littoralis *in full bloom*
Southern Angola habitat, Southwest Africa

LILIACEAE

Aloe salm-dyckiana
at Jardin Botanique 'Les Cèdres', Cap Ferrat, France

Aloe marlothii, *with yellow spires in southern Kenya, East Africa*

Aloe plicatilis, *the "Fan aloe" blooming in southwestern Cape Province, South Africa*

Aloe dawei
giant specimen in flower, near Entebbe, Uganda

LILIACEAE

Aloe dichotoma
"Dragon tree aloe" in So. Africa

Aloe peglerae
near Pretoria, Transvaal

Aloe vanbalenii 'Variegata'
Kew Gardens, London

Aloe pillansii
(Cape Province)

Aloe saponaria
"Soap aloe" (Fairchild G., Miami)

Aloe succotrina
in Palmengarten, Frankfurt

Aloe comosa
near Pretoria, Transvaal

Aloe ferox
in Worcester, Cape Province

Aloe ferox 'Variegata'
Kirstenbosch Bot. Garden, So. Africa

LILIACEAE

Aloe arborescens
in tree-like form, Stellenbosch, South Africa

Aloe vera, *the "Bitter aloe"*
together with Cephalocereus royenii, in arid Aruba

Aloe vera *in bloom*
in the shadow of Diamond Head, Hawaii

Aloe africana, *"Spiny aloe"*
in magnificent bloom, Los Angeles Arboretum

LILIACEAE

Aloe haemanthifolia
in Botanic Gardens Stellenbosch, South Africa

Aloe suprafoliolata
the curious "Propeller aloe" from Swaziland

Aloe humilis, *the "Spider aloe"*
in habitat near Port Elizabeth, So. Africa

Aloe melanacantha
collected in Namaqualand, in Stellenbosch Bot. Garden

Aloe brevifolia depressa
"Crocodile jaws" from the Caledon range, Cape Province

Aloe brevifolia at *Stellenbosch,*
(Aloe "humilis" in the California nursery trade)

LILIACEAE

Aloe striata
"Coral aloe" (Namibia)

Aloe ciliaris
"Climbing aloe"

Aloe arborescens
"Octopus plant"

Aloe jucunda
from Somaliland

Aloe variegata
in scarlet bloom

Aloe concinna
from Zanzibar

Aloe variegata
"Partridge-breast"

Aloe polyphylla
"Spiral aloe" (Basutoland)

Aloe vera chinensis
"Indian medicine plant"

LILIACEAE

Asparagus densiflorus 'Myers'
"Plume asparagus"

Asparagus densiflorus 'Sprengeri'
"Sprengeri fern"

Asparagus retrofractus
"Zigzag shrub"

Asparagus myriocladus
"Zigzag-asparagus (Natal)"

Asparagus scandens deflexus
"Basket asparagus"

Asparagus setaceus (plumosus)
"Fern-asparagus"

Asparagus densiflorus 'Sprengeri'
in hanging basket

Aspar. asparagoides myrtifolius
"Medeola" or "Smilax"

Asparagus setaceus 'Pyramidalis'
"Cypress asparagus"

LILIACEAE, AMARYLLIDACEAE 575

Zephyranthes grandiflora
"Zephyr lily" (AMARYLL.)

Agapanthus africanus minor
"Peter Pan lily"

Agapanthus africanus
"Lily-of-the-Nile"

Agapanthus inapertus pendulus
"Drooping agapanthus" in So. Africa

Arthropodium cirrhatum
(New Zealand)

Agapanthus orientalis 'Albidus'
Mission Carmel, California

Colchicum autumnale
"Autumn crocus"

Allium schoenoprasum
"Chives" or "Schnittlauch"

Hemerocallis aurantiaca
"Golden summer daylily"

LILIACEAE

Aloe humilis 'Globosa'
"Spider aloe"

Aloe sladeniana
from Namibia, S.W. Africa

Aloe eru (camperi)
from Eritrea

Aloe striata
"Coral aloe" *in So. Africa*

Aloe eru 'Maculata'
"Nubian aloe"

Aloe haworthioides
(*C. Madagascar*)

Cordyline term. 'Mme. André'
"Flaming dragon tree"

Cordyline terminalis 'Tricolor'
"Tricolored dracaena"

Chlorophytum bichetii
"St. Bernard's lily"

Cordyline terminalis 'Firebrand'
"Red dracaena"

Cordyline terminalis
'Hawaiian Bonsai'

Cordyline terminalis
'Calypso Queen'

LILIACEAE

577

Dracaena hookeriana 'Rothiana' *in San Francisco*

Dracaena deremensis 'Bausei' *with white median band*

Dracaena goldieana *"Queen of dracaenas"*

Dracaena godseffiana 'Florida Beauty'

Dracaena sanderiana 'Margaret Berkery'

Dracaena deremensis 'Warneckei' *"Striped dracaena"*

Dracaena sanderiana *"Ribbon plant" (Cameroun)*

Dracaena cantleyi *from Singapore*

Pleomele reflexa 'Variegata' *"Song of India"*

Dracaena fragrans 'Victoriae' *"Painted dragon lily"*

Dracaena fragrans *in bloom (Nigeria)*

Aspidistra elatior 'Variegata' *"Variegated bar-room plant"*

LILIACEAE

Cordyline australis 'Variegata'
"Tikouka" in New Zealand

Cordyline austr. 'Atropurpurea'
Strybing Arboretum, San Francisco

Cordyline indivisa 'Purpurea'
So. Coast Bot. Garden, Los Angeles

Cordyline banksii
near Nelson, New Zealand

Cordyline fruticosa
Papeari Bot. Garden, Tahiti

Cordyline terminalis 'TI'
"Hawaiian Good-luck plant"

Cordyline rubra 'Bruantii'
in Cologne, Germany

Cordyline stricta (congesta)
from Queensland

Cordyline 'Volckaertii'
Belgian terminalis cultivar

LILIACEAE

Cordyline terminalis 'Baby Ti'
Hawaiian miniature

Cordyline terminalis
"Tree of kings" with flowers

Cordyline terminalis
red cultivar on Nuku Hiva

Cordyline terminalis
'Hawaiian Flag'

Cordyline terminalis *in habitat
Cape York Peninsula, Queensland*

Cordyline terminalis 'Amabilis'
"Pink dracaena" in Puerto Rico

Cordyline terminalis 'Negri'
"Black dracaena" in Miami

Cordyline terminalis 'Liliput'
miniature in Longwood Gard., Penna.

Cordyline term. 'Schubertii'
in Trinidad

LILIACEAE

580

Dracaena marginata
an architect's delight

Dracaena deremensis 'Janet Craig'
container-grown decorator plant

Dracaena deremensis 'Warneckei'
"Striped dracaena"

Dracaena deremensis 'Compacta'
"Dwarf bouquet"

Pleomele reflexa
in terracotta jar, Agra, India

Dracaena deremensis
'Warneckei marginata'

Dracaena fragrans 'Lindenii'
in Florida lathhouse

Dracaena fragrans 'Massangeana'
"Cornstalk plant" with canes

Dracaena fragrans
in Rio Piedras, Puerto Rico

LILIACEAE

Dracaena deremensis
'Warneckei Roehrs Gold'

Dracaena deremensis
'Souv. de A. Schryver' *in Germany*

Dracaena fragrans 'Massangeana'
"Variegated dragon-lily" or "Cornstalk plant"

Dracaena goldieana
beautiful "Queen of dracaenas" from Equatorial Africa

Pleomele angustifolia honoriae
collected in Torres Straits off Northern Australia

Pleomele reflexa 'Song of India'
brought back from the ancient land of the Tamils

LILIACEAE

Pleomele reflexa angustifolia
in Borobudur, Central Java

Pleomele angustifolia honoriae
luxuriant in Tahiti

Dracaena marginata
"Madagascar dragon tree"

Dracaena marginata 'Tricolor'
"Rainbow tree"

Beaucarnea recurvata
"Pony-tail" (Mexico)

Lomatophyllum purpureum
(Mauritius)

Pleomele reflexa
"Malaysian dracaena"

Cordyline australis
in Christchurch, New Zealand

Cordyline indivisa
in Fiordland habitat, New Zealand

LILIACEAE

Dracaena fragrans 'Victoriae'
"Painted dragon lily"

Dracaena marginata 'Tricolor'
"Rainbow plant"

Dracaena goldieana
from Guinea coast

Dracaena hookeriana
in Natal habitat

Dracaena sanderiana var. 'Celes'
Puerto Rico miniature

Dracaena deremensis 'Roehrsii'
Cologne Flower Show, Germany

Dracaena godseff. 'Friedmannii'
"Milky way dracaena"

Dracaena sanderiana
"Ribbon plant" (W. Africa)

Dracaena godseffiana
"Gold-dust dracaena"

LILIACEAE

Asparagus densiflorus cv. 'Myers'
"Plume asparagus" from South Africa

Bulbine natalensis
aloe-like succulent, near Durban, Natal

Astelia solandri, *the "Kokaha",*
growing epiphytic in North Island forest, New Zealand

Chlorophytum ignoratum
of Tropical Africa; in St. Louis Climatron, Missouri

Chlorophytum comosum 'Vittatum'
"Spider plant" with young plantlets, in hanging pot

Chlorophytum comosum 'Variegatum'
"Ribbon plant" or "Green-lily"

LILIACEAE

Phormium colensoi 'Tricolor'
"Tricolored mountain flax" in New Zealand

Dracaena fragrans 'Victoriae'
in the humid-warm climate of Peradeniya, Ceylon

Asphodelus microcarpus
"Asphodel" in the ancient ruins on Delos, Greece

Beaucarnea recurvata
Mexican "Bottle palm" on Sepulveda Blvd., Los Angeles

LILIACEAE

Dasylirion wheeleri
"Desert spoon" (Mexico)

Dasylirion serratifolia
at City hall in Oslo, Norway

Dasylirion longissimus
"Mexican grasstree"

Dracaena draco
"Dragon tree" on Gran Canaria

Dracaena arborea
"Tree dracaena" from Guinea

Dracaena aurea
endemic in Hawaii

Beaucarnea recurvata
pot grown, in San Diego

Samuela carnerosana
"Palma de San Pedro" (Mexico)

Astelia nervosa
on Maitai River in New Zealand

LILIACEAE

Cordyline indivisa, *for summer decoration in Essen Botanic Garden, Germany*

Cordyline australis, *the "Grass-palm" overlooking the harbor of Wellington, New Zealand*

Cordyline terminalis 'Tricolor' *"Tricolored dracaena" in Fiji Orchid garden, Suva*

Cordyline terminalis 'Ti'; *their broad leaves fashioned into hula skirts for Polynesian dancers*

LILIACEAE

Phormium tenax 'Tricolor'
Auckland Domain, New Zealand

Phormium tenax
"New Zealand flax"

Phormium tenax 'Variegatum'
in Wellington, New Zealand

Phormium colensoi 'Tricolor'
"Mountain flax" in New Zealand

Ruscus hypoglossum
at Estufa Fria, Lisbon, Portugal

Pleomele angustifolia honoriae
in Papeari, Tahiti

Dasylirion glaucophyllum
at Amani, Usambara Mts., Tanzania

Nolina bigelovii
"Bear grass" in Baja California

Xanthorrhoea undulata (*New Zealand*)
in Zuerich Bot. Garden, Switzerland

LILIACEAE

Dracaena marginata
in landscape planting, Diamond Head, Honolulu

Dracaena draco, *ancient "Dragon tree"
at Icod, Tenerife, Canary Islands*

Xanthorrhoea preissii
an age-old "Blackboy" near Perth, Western Australia

Dracaena arborea
"Tree dracaena" south of Luanda, Angola, W. Africa

LILIACEAE

590

Aspidistra elatior 'Maculata'
"Milky way iron plant"

Aspidistra elatior 'Variegata'
"Variegated cast iron plant"

Tupistra macrostigma
(*Himalayas of India*)

Hosta sieboldiana
"Seersucker plantain-lily"

Chlorophytum bracteatum
from Zaïre, C. Africa

Hosta crispula
(*albo-marginata*), *from Japan*

Chlorophytum comosum
'Milky Way', *highly variegated*

Chlorophytum comosum 'Vittatum'
"Spider plant" with runners

Chlorophytum laxum
Kew Gardens, London

LILIACEAE

Kniphofia uvaria maxima
"Big poker plant"

Lapageria rosea 'Albiflora'
"White Chile bells"

Kniphofia uvaria
"Red-hot poker"

Gloriosa superba
"Crisped glory-lily" in Ceylon

Gloriosa verschurii
in New Caledonia

Gloriosa superba 'Aurea'
"Golden glory lily"

Kniphofia uvaria praecox
Strybing Arboretum, San Francisco

Gasteria batesiana (*inflor.*)
from Zululand, S.E. Africa

Lachenalia aloides (tricolor)
"Cape cowslip"

LILIACEAE

Gloriosa rothschildiana
"Glory-lily" (Kenya)

Gloriosa superba
"Crisped glory-lily"

Gloriosa carsonii
from Central Africa

Scilla peruviana
"Cuban lily" (Algeria)

Gloriosa verschurii
(East Africa)

Scilla violacea
"Silver squill" (So. Africa)

Veltheimia viridifolia
"Forest lily"

Tulbaghia violacea 'Variegata'
"Society garlic"

Ornithogalum caudatum
"False sea-onion"

Convallaria majalis
"Lily-of-the-valley"

Veltheimia viridifolia
in flower (So. Africa)

Muscari armeniacum
"Grape-hyacinth"

LILIACEAE

Gloriosa rothschildiana
"Glory lilies" in warm greenhouse, Rhineland, Germany

Gloriosa simplex (virescens or plantii)
"Dwarf glory-lily" in Jardim Botanico, Rio de Janeiro

Lapageria rosea, *"Chile bells"*
vining on a fence in Santiago, Chile, South America

Gloriosa superba, *climbing lilies*
at the Royal palace, Tonga-tapu, in the South Pacific

LILIACEAE

Gloriosa carsonii
on low jungle trees in Pigmy country, Central Africa

Gloriosa greeneae
in the Ruwenzori Range, Western Uganda

Lilium longiflorum 'Ace'
"Easter lilies" in Roehrs greenhouses, New Jersey

Gloriosa rothschildiana
along the north road out of Mombasa, Kenya

Erythronium grandiflorum
"Avalanche lilies" in melting snow, northern Montana

Lilium 'Imperial Crimson'
giant "Empress lilies" in Gresham, Oregon

LILIACEAE

Muscari armeniacum
"Grape hyacinths" from Turkey

Convallaria majalis
"Lily-of-the-valley" in pots, forced for early bloom

Tulbaghia fragrans
"Wild garlic" as known in South Africa, in Pretoria

Kniphofia uvaria praecox
blooming in Golden Gate Park, San Francisco

Lourya campanulata
curious aspidistra-like plant from Vietnam

Ornithogalum thyrsoides, a bowl of "Chincherinchee"
blooming in September, South Africa's spring

LILIACEAE

596

Haworthia limifolia
"Fairy washboard" from Swaziland, South Africa

Gasteria verrucosa
"Warty aloe" from So. Africa; in Kew Gardens, London

Sansevieria trifasciata 'Hahnii Variegata'
a colorful mutant from Puerto Rico

Sansevieria trifasciata 'Hahnii'
"Birdsnest"; low rosette sported in New Orleans 1939

Yucca aloifolia 'Quadricolor'
in Royal Botanic Garden Edinburgh, Scotland

Yucca filamentosa 'Variegata'
variegated "Adam's needle" in Christchurch, New Zealand

LILIACEAE

Gasteria verrucosa
"Warty aloe" (So. Africa)

Gasteria liliputana
a miniature "Ox-tongue"

x Gastrolea beguinii
"Pearl aloe"

x Gastrolea 'Spotted Beauty'
(Gasteria x Aloe)

Haworthia fasciata
"Zebra haworthia"

Haworthia papillosa
"Pearly dots"

Haworthia chalwinii
"Aristocrat plant"

Haworthia truncata
"Clipped window plant"

Haworthia "margaritifera"
in horticultural trade

Haworthia fasciata
"Zebra haworthia"

Haworthia cooperi
"Window haworthia"

Haworthia retusa
"Star cactus"

LILIACEAE

Ornithogalum caudatum
"Sea onion"

Ornithogalum saundersiae
"Star of Bethlehem"

Ornithogalum thyrsoides
"Wonder flower"

Milla biflora
"Mexican star flower"

Ornithogalum umbellatum
"Summer snowflake"

Ornithogalum montanum
in Pergamon, Asiatic Turkey

Tulbaghia violacea
"Society garlic" (So. Africa)

Fritillaria pudica
"Yellow bell" in Wyoming

Muscari commutatum
in Athens, Greece

LILIACEAE

Hyacinthus orientalis 'Blue Giant'

Hyacinthus orientalis 'Anne Marie'

Hyacinthus orientalis 'Blue Jacket'

Hyacinthus orientalis 'Delft Blue'

Hyacinthus orientalis 'Carnegie'

Hyacinthus orientalis 'Lord Derby'

Hyacinthus orientalis 'Ostara'

Hyacinthus orientalis 'Pink Pearl'

Lachenalia aloides (tricolor) *"Cape cowslip"*

Lilium longiflorum 'Croft' *"Easter lily"*

Tulipa saxatilis *from Crete*

Ornithogalum thyrsoides *"Chincherinchee"*

LILIACEAE

Lilium tigrinum
"Tiger lily" (China)

Lilium speciosum 'Album'
"White Japanese lily"

Lilium x aurelianense
"Aurelian lily"

Lilium dauricum
"Candle stick lily"

Lilium amabile 'Enterprise'
"Korean lily"

Lilium 'Limelight'
Longwood Gardens, Penna.

Lilium auratum
"Goldband lily" (Japan)

Lilium tigrinum
"Tiger lily" (China)

Lilium speciosum
"Japanese lily"

LILIACEAE

Lilium longiflorum 'Croft'
"Easter lily"

Lilium speciosum
'Lillian Wallace'

Lilium pumilum
"Coral lily" (*China*)

Lilium 'Harmony'
(*Mid-century*)

Lilium davidii
(*Western China*)

Lilium 'Prosperity'
(*Mid-century*)

Lilium cernuum
(*Korea, No. China*)

Lilium langkongense
(*Western China*)

Lilium 'Cinnabar'
(*Mid-century*)

Lilium 'Susan'
pink wax-lily

Lilium grayii
"Orange bell-lily"

Lilium auratum
"Goldband lily"

LILIACEAE

Littonia modesta
"Climbing lily"

Fritillaria imperialis
"Crown Imperial"

Sandersonia aurantiaca
"Chinese lantern lily"

Blandfordia nobilis
"Christmas bells"

Nomocharis mairei
(*Southwest China*)

Fritillaria meleagris
"Checkered lily"

Chionodoxa sardensis
"Glory-of-the-snow"

Tricyrtis hirta
"Toad lily"

Camassia esculenta
(*Pennsylvania to Texas*)

LILIACEAE

Eucomis undulata
"Pineapple flower" (So. Africa)

Eucomis pole-evansii
"Giant pineapple lily"

Bowiea volubilis
"Climbing onion"

Asphodelus microcarpus
in Mycaene, Greece

Urginea maritima
"Sea onion" on Cyprus

Eremurus elwesii
"Desert candle" (Turkestan)

Bulbinella floribunda
from South Africa

Galtonia candicans
"Giant summer hyacinth"

Rhodocodon urginioides
from C. Madagascar

LILIACEAE

Tulipa gesnerana 'Robinea'
a "Triumph tulip" in pots for Easter, at Roehrs

Tulipa gesnerana hybrids
cultivated tulips in Holland for bulb production

Tulipa greigii 'Royal Orange'
at New York International Flower Show

Tulipa gesnerana 'Peony'
a "Double tulip", at Mormon Temple, Salt Lake City, Utah

Tulipa gesnerana 'Bartigon'
an excellent "Darwin tulip" for pot culture

Tulipa gesnerana 'Apeldoorn'
a "Darwin hybrid" at Longwood Gardens, Pennsylvania

LILIACEAE

Tulipa gesn. 'Karel Doorman'
"Parrot tulip"

Tulipa gesn. 'Blizzard'
"Triumph tulip"

Tulip gesn. 'Blenda'
"Triumph tulip"

Tulipa gesn. 'Makassar'
"Triumph tulip"

Tulipa gesn. 'Robinea'
"Triumph tulip"

Tulipa gesn. 'Kees Nelis'
"Triumph tulip"

Tulipa gesn. 'Red Giant'
"Triumph tulip"

Tulip gesn. 'Paris'
"Triumph tulip"

Tulipa gesn. 'Princess Irene'
"Single early"

Tulipa gesn. 'Rose Beauty'
"Triumph tulip"

Tulipa gesn. 'Ursa Minor'
"Single early"

Tulipa gesn. 'United Europe'
"Triumph tulip"

LILIACEAE

Rohdea japonica
"Sacred lily of China"

Ophiopogon planiscapus nigrescens
"Black dragon" (Japan)

Liriope platyphylla
"Lilyturf" (China)

Ophiopogon japonicus
'Kyoto Dwarf' (Japan)

Drimiopsis kirkii
(Zanzibar)

Drimiopsis saundersii
in the Transvaal

Scilla violacea
"Silver quill" (So. Africa)

Ruscus hypoglossum
"Mouse-thorn" (S.E. Europe)

Semele androgyna
"Climbing butcher's broom"

LILIACEAE

Rohdea japonica 'Marginata'
"Sacred Manchu lily"

Ruscus hypoglossum
"Mouse-thorn"

Ophiopogon jaburan 'Variegatus'
"Variegated mondo grass"

Yucca aloifolia 'Marginata'
"Spanish bayonet"

Phormium tenax 'Variegatum'
"Variegated flax"

Yucca aloifolia 'Marginata'
as a tree

Sansevieria
trifasciata 'Craigii'

Sansevieria trifasciata
'Golden Hahnii'

Sansevieria trifasciata
laurentii 'Compacta'

Haworthia armstrongii
"Wart plant" (Cape)

Sansevieria trifasciata
"Snake plant" in flower

Sansevieria trif. 'Laurentii'
"Variegated snake plant"

LILIACEAE

Sansevieria caespitosa
from southwestern Africa; at Bot. Garden Munich

Sansevieria guineensis 'Marginata'
"Variegated bowstring hemp", in Singapore

Sansevieria dooneri (*Rift Valley, Kenya*)
in collection Kew Gardens, London

Sansevieria thyrsiflora
"Bowstring hemp", in Stellenbosch, South Africa

Sansevieria scabrifolia
from southeastern Africa, at Kew

Sansevieria senegambica
collected in Sénégal, West Africa

Sansevieria splendens
from Equatorial Africa (Kew Collection)

Sansevieria guineensis
in Lagos, Nigeria, West Africa

LILIACEAE

Sansevieria trifasciata
in Durban, Natal

Sansevieria trifasciata
'Golden Hahnii'

Sansevieria trif. 'Laurentii'
"Variegated snake plant"

Sansevieria raffillii
from Tsavo desert, Kenya

Sansevieria singularis
in Heidelberg Bot. Garden

Sansevieria kirkii pulchra
from Zanzibar

Sansevieria trifasciata
'Bantel's Sensation'

Sansevieria grandis zuluensis
in Kew Gardens, London

Sansevieria trif. 'Craigii'
Frankfurt Palmengarten

LILIACEAE

Sansevieria ehrenbergii
"Blue sansevieria" (Kenya)

Sansevieria stuckyi
from Rhodesia (Kew Gardens)

Sansevieria cylindrica
in Munich Bot. Garden

Sansevieria desertii
"Rhino-grass" (Botswana)

Sansevieria dawei
Heidelberg Bot. Garden

Sansevieria aethiopica
Munich Bot. Garden

Sansevieria fasciata
Kew Gardens, London

Sansevieria gracilis
from Mazeras, Kenya

Sansevieria grandis
"Grand Somali hemp"

LILIACEAE

Pleomele reflexa 'Song of India'
in Gampaha, near Colombo, Sri Lanka

Yucca brevifolia
"Joshua tree" on the Mohave desert, California

Phormium tenax, *"New Zealand flax"*
in containers, Rhine Park, Cologne

Cordyline indivisa
planted with ivy, in tubs, Germany

Sansevieria trifasciata 'Laurentii'
on Pennock's plantation, Rio Piedras, Puerto Rico

Sansevieria trifasciata 'Laurentii'
"Variegated snake plant", from Zaïre, C. Africa

LILIACEAE

Yucca elephantipes
"Spineless yucca" in bloom, in Guatemala

Beaucarnea gracilis
"Bottle palm" near Zapotitlán, Puebla, Mexico

Sansevieria ehrenbergii
in habitat, Teita Hills, Tanzania

Sansevieria arborescens
in the Tsavo game reserve, near Voi, Kenya

LILIACEAE

Dracaena fragrans
in the equatorial rainforest near Ibadan, Nigeria

Yucca elata, *"Soap wood",*
in Zion National Park, southern Utah

Yucca brevifolia
the ghostly "Joshua tree", on the Mohave desert, Calif.

Dracaena draco, *"Dragon trees"*
on the Pacific coast at La Jolla, California

LILIACEAE

Yucca gloriosa
"Palm lily" (Calif. to Florida)

Yucca aloifolia 'Tricolor'
"Red daggers"

Yucca baccata
in Mesa Verde, Colorado

Yucca filamentosa
"Adam's needle" in Copenhagen

Yucca elephantipes
"Bulb-stem yucca"

Yucca filamentosa 'Variegata'
in Christchurch, New Zealand

Yucca recurvifolia (pendula)
(Georgia to Mississippi)

Yucca whipplei
"Our Lord's candle"

Yucca schidigera
Mohave desert, California

LILIACEAE

Yucca aloifolia, *"Mata de Rosa blanca" in Puerto Rico*
egg shells on yuccas to avoid ill fortune

Kniphofia uvaria, *the "Torch lily"*
at the old Santa Barbara Mission, in California

Xanthorrhoea preissii
a "Black-boy" centuries-old, in arid Western Australia

Xanthorrhoea arborea, *"Grass trees"*
in the mountains of northern New South Wales, Australia

LOBELIACEAE, LOGANIACEAE

Lobelia erinus
"Bedding lobelia"

Hippobroma longiflora
(*Isotoma*), *in Bali*

Lobelia cardinalis
"Cardinal flower"

Buddleja cordata
Jardim Botanico, Rio de Janeiro

Isotoma fluviatilis
"Blue star creeper" in Calif.

Centropogon x lucyanus
French basket plant

Spigelia splendens
"Mexican pinkroot"

Gelsemium sempervirens
"Carolina yellow jessamine"

Fagraea fragrans
"Tembusu" in Malaya

LOBELIACEAE, LORANTHACEAE, LYTHRACEAE, MAGNOLIACEAE

Lobelia gibberoa, *the "Giant lobelia"*
tree-like, to 10 m tall, in mountains of Kigezi, Uganda

Psittacanthus calyculatus, *the "Mexican mistletoe"*
at Toltec fortress of Xochicalco, Morelos

Magnolia x soulangeana 'Verbanica'
in spring bloom, East Rutherford, New Jersey

Lagerstroemia speciosa
"Queen's grape-myrtle" or "Pride of India", in Tahiti

LORANTHACEAE, LYCOPODIACEAE, LYTHRACEAE

Viscum album, *"Mistletoe"* parasitic in Black Forest, Germany

Loranthus sansibarica *"Matchstick vine"* in Kenya

Phoradendron flavescens *"American mistletoe"* in Texas

Rotala macranda (*East Indies*)

Lythrum salicaria *"Purple loosestrife"*

Lycopodium phlegmaria *"Queensland tassel-fern"*

Lagerstroemia speciosa *in Tahiti, South Pacific*

Lagerstroemia indica 'Rubra' *"Red grape-myrtle"*

Lagerstroemia indica *"Grape myrtle" from China*

MAGNOLIACEAE

Magnolia x soulangeana
'Verbanica' *in New Jersey*

Magnolia x soulangeana
"Saucer magnolia"

Magnolia grandiflora
"Southern magnolia"

Magnolia liliflora
in Taipo, Hongkong

Michelia champaca
"Orange champak" (*Himalayas*)

Michelia alba, *"Champaca"*
from Java (*Dr. R. Criley*)

Liriodendron tulipifera
the fragrant "Tulip tree"

Magnolia tripetala
in Dortmund Bot. G., Germany

Magnolia stellata
"Star magnolia"

LYTHRACEAE, MALPIGHIACEAE, MALVACEAE

Heteropterys beecheyana
(*Mexico to Bolivia*)

Goethea strictiflora
(*cauliflora*), *from Brazil*

Tristellateia australasiae
"Galphimia vine" in Tahiti

Bombycidendron vidalianum
"Tree hibiscus" (*Philippines*)

Lawsonia inermis
"Mignonette tree" or "Henna"

Cuphea platycentra (ignea)
"Cigar flower"

Malope trifida
(*Spain, No. Africa*)

Malpighia glabra
"Barbados cherry"

Gossypium herbaceum
"Levant cotton (*Arabia*)

LYTHRACEAE, MAGNOLIACEAE, MALPIGHIACEAE, MALVACEAE

Magnolia x soulangeana
"Chinese magnolia" in spring bloom, in New Jersey

Magnolia grandiflora
breathing the romance of the Southern States

Cuphea melvilla
the "False heather" in tropical Guayana

Tristellateia australasiae
"Bagnit vine" on the island of Moorea, in French Polynesia

Abutilon x hybridum 'Souvenir de Bonn'
"Variegated flowering maple"

Abutilon x hybridum, *"Parlor maple"*
or "Flowering maple", an old-fashioned houseplant

LYTHRACEAE, MALVACEAE

Abutilon megapotamicum
"Trailing abutilon"

Abutilon x hybridum
'Yellow Belle'

Abutilon x hybridum 'Apricot'
"Chinese lantern"

Abutilon x hybridum
'Souvenir de Bonn'

Abutilon megapotamicum 'Variegata'
"Weeping Chinese lantern"

Abutilon pictum 'Thompsonii'
"Spotted flowering maple"

Hoheria populnea 'Argentea'
"Variegated lace-bark" (N.Z.)

Hoheria pop. 'Aureo-variegata'
on Moorea, Polynesia

Lagerstroemia hirsuta (reginae)
(fruit capsules) in Burma

MALVACEAE

623

Abutilon x hybridum
'Satin Pink Belle'

Abutilon x hybrid. 'Old Rose'
old-fashioned "Parlor maple"

Abutilon x hybridum
'Orange Red'

Abutilon pictum 'Thomsponii'
"Spotted parlor maple"

Abutilon x hybridum 'Savitzii'
"White parlor maple"

Abutilon x hybridum
'Souvenir de Bonn'

Abutilon mega. 'Variegatum'
"Weeping Chinese lantern"

Malvaviscus arboreus mexicanus
"Turk's cap" (*Mexico*)

Sida fallax
"Ilima" in Hawaii

Magnolia grandiflora (*seedpod*)
"Southern magnolia"

Hibiscus syriacus (Althaea)
"Rose-of-Sharon"

Hibiscus schizopetalus
"Japanese lantern"

MALVACEAE

Pavonia intermedia
(Brazil)

Pavonia multiflora
U. of Calif. Bot. Garden

Pavonia intermedia
'Kermesina'

Malvaviscus arboreus
"Wax mallow" (Mexico)

Malvaviscus arboreus mexicanus
"Turk's cap"

Sida fallax, "Ilima"
flower of Honolulu

Lavatera trimestris
"Tree-mallow"

Lagunaria patersonii
"Pyramid-tree" in Malta

Malva sylvestris
on Delos, Greece

MALVACEAE

Malvaviscus arboreus mexicanus
tropical "Turk's cap" in Mexico

Hibiscus arnottianus
native hibiscus or "Kokio-Keokeo" in Hawaii

Alcea rosea (Althaea)
"Hollyhock" in Anderson's Odense, Denmark

Pavonia intermedia 'Kermesina'
tropical shrub from Brazil, with showy red bracts

MALVACEAE

Hibiscus rosa-sinensis
in Nuku-Hiva, Polynesia

Hibiscus rosa-sinensis albus
in Rarotonga, Friendship Isles

Hibiscus rosa-sinensis 'Scarlet'
in the Marquesas

Hibiscus rosa-sinensis 'Cooperi'
"Checkered hibiscus"

Hibiscus rosa-sinensis 'Lutea'
in California

Hibiscus rosa-sinensis 'Mist'
Denver Botanic Garden

Hibiscus rosa-sin. 'California Gold'
in San Diego, California

Hibiscus rosa-sinensis 'Cheerful'
in Denver, Colorado

Hibiscus rosa-sinensis 'Hula Girl'
in Honolulu, Hawaii

MALVACEAE

Hibiscus rosa-sinensis cv. 'Natal'
common in hedges in South Africa

Hibiscus rosa-sinensis
worn by young Melanesians near Lautoka, Fiji

Hibiscus arnottianus
native hibiscus on the garden island of Kauai

Hibiscus rosa-sinensis 'Aurantiacus'
in the California nursery trade

Montezuma speciosissima (Thespesia grandiflora)
"Maga colorada" or "Purple haiti-haiti" in Puerto Rico

Hibiscus rosa-sinensis 'Matensis'
"Snow flake hibiscus" as a hedge in Panama

MALVACEAE

628

Hibiscus elatus
"Cuban bast" in Jamaica

Hibiscus huegelii
"Blue hibiscus" (Australia)

Hibiscus tiliaceus
the pantropic "Mahoe"

Hibiscus mutabilis
"Cotton-rose" (China)

Hibiscus brackenridgei
"Hawaii yellow hibiscus"

Thespesia populnea
"Portia tree" on Oahu

Thespesia lampas
in the Marquesas, Polynesia

Thespesia populnea acutiloba
in India

Thespesia lampas *with fruit*
Botanic Garden Rio de Janeiro

MALVACEAE

629

Hibiscus rosa-sinensis
'Scarlet' ('Brilliant')

Hibiscus rosa-sinensis
'White Wings'

Hibiscus rosa-sinensis
State flower of Hawaii

Hibiscus rosa-sinensis
'Crown of Bohemia'

Hibiscus rosa-sinensis plenus
"Double Rose of China"

Hibiscus rosa-sinensis
'Toreador'

Hibiscus diversifolius
pantropic "Mallow"

Hibiscus rosa-sinensis 'Cooperi'
"Checkered hibiscus"

Sida fallax, *"Ilima"*
used in Hawaiian leis

Hibiscus mutabilis
"Confederate rose"

Hibiscus brevipedunculata
(Islands of the Caribbean)

Montezuma speciosissima
"Hibiscus tree" of Puerto Rico

MALVACEAE

Hibiscus schizopetalus 'Pagoda'
"Flora en flora" (Colombia)

Hibiscus rosa-sinensis 'Lateritia'
at Garden center in Germany

Hibiscus schizopetalus
"Japanese lantern" in Mozambique

Hibiscus rosa-sinensis 'Cooperi'
in Indonesia

Hibiscus rosa-sinensis plenus
in Cartagena, Colombia

Hibiscus moscheutos 'Southern Belle'
"Confederate rose"

Hibiscus syriacus
"Rose of Sharon"

Hibiscus trionum
"Flower-of-an-hour" in N. Zealand

Hibiscus syriacus
"Double-flowered Althaea"

MARANTACEAE

Calathea tigrina hort.
Longwood Gardens, Penna.

Calathea aurantiaca
from Brazil (in Australia)

Calathea roseo-picta
(*Brazil*)

Calathea sp. 'Burle Marx'
in Longwood Gardens

Calathea princeps
(*Amazonas, Brazil*)

Calathea carlina
in Sao Paulo, Brazil

Calathea crocata
Col. Marnier-Lapostolle, France

Calathea insignis
in Brazil habitat

Calathea stromata
from Brazil (in Munich)

MARANTACEAE

Calathea concinna
(*N.W. Brazil: Acre*)

Calathea insignis
"Rattlesnake plant"

Calathea kegeliana
(*Brazil*)

Calathea lindeniana
(*Brazil: Amazonas*)

Calathea lietzei
(*Brazil*)

Calathea clossonii
(*Brazil*)

Calathea metallica 'Undulata'
(*Brazil*)

Calathea aemula
(*So. Brazil*)

Calathea 'Carlina'
(*Brazil*)

Calathea makoyana
"Peacock plant"

Calathea micans
miniature from Perú

Calathea wiotii
(*Brazil*)

MARANTACEAE

Calathea rotundifolia 'Fasciata'
in Bahia state, N.E. Brazil

Calathea leopardina
in col. Longwood Gardens, Pennsylvania

Calathea louisae (albertii)
in Munich Botanic Garden, Nymphenburg, Germany

Calathea makoyana
"Peacock plant" from Minas Geraes, Brazil

Calathea leopardina
at Roehrs Exotic Nurseries, in New Jersey

Calathea princeps *in habitat,
in Amazonas rainforest, near Manaus, Brazil*

MARANTACEAE

Calathea grandiflora (*Brazil*)
Longwood Gardens, Kennett Square, Pennsylvania

Calathea insignis
"Rattlesnake plant" in natural habitat, Brazil

Calathea leonii
an iridescent beauty from tropical Ecuador

Calathea bella
from Brazil (in the Longwood collection)

MARANTACEAE

Calathea zebrina, *"Zebra plant"*
in artistic terracotta pot, in Sao Paulo, Brazil

Calathea loeseneri
in Longwood Gardens, Kennett Square, Pennsylvania

Calathea makoyana, *the "Peacock plant"*
displaying the reverse pattern of their translucent leaves

Calathea warscewiczii, *from Costa Rica*
with large velvety, intricately patterned foliage

MARANTACEAE

Calathea roseo-picta
(*Brazil*)

Calathea ornata
'Roseo-lineata'

Calathea ornata
'Sanderiana'

Calathea picturata 'Argentea'
(*Venezuela*)

Calathea princeps
(*Amazonas*)

Calathea musaica
(*Espirito Santo*)

Calathea rotundifolia 'Fasciata'
(*Brazil: Bahia*)

Calathea undulata
(*Amazonian Perú*)

Calathea 'Tuxtla'
(*Southern Mexico*)

Calathea veitchiana
(*Ecuador, Perú*)

Calathea picturata 'Vandenheckei'
(*Amazonas, Colombia*)

Calathea zebrina
(*Brazil: Sao Paulo*)

MARANTACEAE

Maranta leuconeura kerchoveana
"Prayer plant"

Maranta leuconeura
"Rabbit's-foot"

Maranta leuconeura
'Massangeana' (*Brazil*)

Stromanthe sanguinea
(*Brazil*)

Stromanthe porteana
(*Brazil*)

Maranta bicolor
(*Brazil, Guyana*)

Ctenanthe setosa
(*Brazil*)

Maranta arundinacea 'Variegata'
"Arrow-root"

Ctenanthe oppenheimiana 'Tricolor'
"Never-never plant"

Ctenanthe oppenheimiana
inflorescence (*Brazil*)

Ctenanthe lubbersiana
(*Brazil*)

Maranta leuconeura erythroneura
"Red-veined prayer plant"

MARANTACEAE

Stromanthe sanguinea, *from Brazil, in Longwood Gardens collection, Pennsylvania*

Ctenanthe oppenheimiana 'Tricolor' *enjoying the tropical climate in Puerto Rico*

Maranta leuconeura erythroneura *beautiful carpet of foliage, in the Organ Mountains, Brazil*

Maranta arundinacea 'Variegata' (Phrynium), *the "Obedience plant", in Borobudur, Central Java*

MARANTACEAE

Calathea zebrina
tropical "Zebra plant" in Sao Paulo, Brazil

Calathea makoyana
"Peacock plant" from Minas Geraes State

Left: Calathea ornata 'Sanderiana'
Right: Ctenanthe oppenheimiana 'Tricolor'

Maranta leuconeura erythroneura
a charming and distinctive indoor plant

Left: Calathea bachemiana (*Brazil*)
Right: Calathea musaica (*Espirito Santo*)

Calathea veitchiana
on the forest floor in tropical Ecuador

MARANTACEAE

Calathea argyraea
"Silver calathea" (*Brazil*)

Calathea zebrina
(*Sao Paulo*)

Calathea stromata
in Brooklyn Botanic Garden

Maranta leuconeura erythroneura
from Petropolis, Brazil

Maranta leuconeura
"Rabbit's-tracks"

Ctenanthe compressa
(*Pernambuco*)

Ctenanthe kummeriana
(*Brazil*)

Calathea rotundifolia 'Fasciata'
in Dee Why, Australia

Maranta arundinacea 'Variegata'
"Variegated arrow-root"

MELIACEAE

Cedrela odorata (*West Indies*)
"Cigar box cedar" in Foster Botanic Gardens, Honolulu

Melia azedarach, *"Pride of India"*
or *"Bead tree"* in the ruins of ancient Corinth, Greece

Swietenia mahogani
"West Indian mahogany", a timber tree in Trinidad

Swietenia mahogani, *the "Mahogany tree"*,
showing fruit, and prized for its superb cabinet wood

MARTYN., MELAST., MELIAC., MELIANTH., MENISP.

Swietenia macrophylla
"Honduras mahagony" in Venezuela

Proboscidea fragrans
"Unicorn plant", Tlaxcala, Mex.

Melia azedarach umbraculifera
"Texas umbrella tree"

Melianthus major
"Honeybush" in So. Africa

Turraea obtusifolia
"South African honeysuckle"

Cocculus laurifolia
"Platter-leaf" (Himalayas)

Melastoma sanguineum
in West Java

Tibouchina semidecandra
"Princess flower" (Brazil)

Martynia annua
photographed in Timor

MELASTOMATACEAE, MELIANTHACEAE, MONOTROPACEAE

Phyllagathis rotundifolia (*Sumatra*)
Royal Botanic Garden Edinburgh, Scotland

Medinilla magnifica (*MELAST.*)
"Rose grape", from the Philippines

Tibouchina granulosa (*MELAST.*)
"Purple glory tree" in Durban, Natal

Tibouchina granulosa
"Glory bush" in habitat, Eastern Bolivia

Sarcodes sanguinea, *the "Snow plant"*
emerging in the redwood forest, Sierra Nevadas, Calif.

Greyia sutherlandii (*MELIANTH.*)
"Mountain bottle brush" in Transvaal habitat, South Africa

MELASTOMATACEAE

x Bertonerila houtteana
(*Bertolonia* x *Gravisia* x *Sonerila*)

Bertolonia maculata 'Wentii'
at Bot. Garden Essen, Germany

Bertolonia maculata (Brazil)
in Kew Gardens, London

Calvoa orientalis
ornamental tropical foliage plant from West Africa

Sonerila margaritacea 'Argentea'
from Java; in Botanic Garden Munich, Germany

Bertolonia 'Mosaica'
Palmengarten Frankfurt

Bertolonia marmorata (*Ecuador*),
Longwood Gardens, Penna.

Bertolonia marmorata 'Sanderiana'
Tuebingen Bot. Garden, Germany

MARCGRAVIACEAE, MELASTOMATACEAE

Medinilla magnifica (*MELAST.*)
"Rose grape", rainforest epiphyte from the Philippines

Norantea guianensis (*MARCGRAV.*)
striking "Red popcorn vine", in the rainforest of Guyana

Miconia calvescens (magnifica) (*MELAST.*)
the showy "Velvet tree", in tropical Southern Mexico

right: Medinilla sp. 'Rubra' (*MELAST.*)
and my "grasshopper" bearers, N.E. New Guinea

Marchantia polymorpha
"Liverwort" with spore-umbrella

Miconia calvescens
"Velvet-tree" (Mexico)

Medinilla magnifica
(Philippines)

Greyia radlekoferi
(Transvaal)

Miconia langsdorfii
(Trop. America)

Norantea brasiliensis
in Brazil

Schizocentron elegans
"Spanish shawl" (Mexico)

Cocculus laurifolia
"Platter-leaf" (Himalayas)

Melastoma sanguineum
(Malaya)

Sonerila margaritacea
"Pearly sonerila" (Java)

Martynia annua (proboscidea)
"Elephant's trunk"

Martynia annua 'Purpurea'
(Trop. America)

MELASTOMATACEAE, MELIACEAE

Tibouchina granulosa
"Purple glory tree"

Tibouchina grandifolia (*Brazil*)
Botanic Garden Sydney

Melastoma decemfidum
(*Sunda Islands*)

Clidemia vittata (*Perú*)
Jardin des Plantes, Paris

Heterocentron roseum
"Pearl flower" (*Mexico*)

Medinilla magnifica
in the Philippines

Monolena primulafolia
Selby Bot. Garden, Sarasota, Fla.

Tibouchina heteromalla (*Brazil*)
in Botanic Garden, Sydney

Miconia calvescens (*Mexico*)
in Hilo, Hawaii

MORACEAE

Artocarpus nobilis
endemic in Ceylon

Bocconia frutescens
"Plume poppy" (Mexico)

Brosimum galactodendron
"Palo de Vaca"

Dammaropsis kingiana
in New Guinea habitat

Cecropia peltata
"Guarumo" in Puerto Rico

Cecropia palmata
in Bahia, Brazil

Ficus aspera (parcellii)
"Clown fig" (So. Pacific)

Ficus benghalensis krishnae
"Krishna's buttercup" (India)

Ficus aurea
"Strangler fig" in Florida

MORACEAE

Artocarpus heterophyllus (integrifolius)
"Jackfruit" borne cauliflorous on tree trunk, in Java

Artocarpus altilis (communis)
"Breadfruit" at home on Nuku Hiva, French Polynesia

Ficus rubiginosa (australis)
buttressed "Rustyleaf", in New South Wales, Australia

Ficus benjamina
an old "Weeping fig" in Peradeniya, Sri Lanka

MORACEAE

Dorstenia yambuyaensis
in Zaïre, C. Africa

Ficus aspera (parcellii)
"Mosaic fig" with fruit

Ficus benjamina 'Exotica'
"Bali fig tree"

Ficus wildemanniana
(*panduriformis*)

Ficus elastica
"India rubber plant"

Ficus dryepondtiana
"Congo fig" (Zaïre)

Ficus religiosa
"Bo-tree" (India)

Ficus elastica 'Doescheri'
"Variegated rubber plant"

Ficus rubiginosa 'Variegata'
variegated miniature

Humulus japonicus 'Variegatus'
(scandens), *"Japanese hops"*

Ficus pumila (repens)
"Creeping fig" (Japan)

Ficus radicans 'Variegata'
"Variegated rooting fig"

MORACEAE

Ficus carica, *the "Common fig"*
in the Swedish town of Bornholm, isle of Gotland

Ficus religiosa, *"Peepul trees"*
on stupas, Schwe Dagon Pagoda, Rangoon, Burma

Ficus benjamina 'Exotica'
old tree with aerial roots in Sanur, Bali

Ficus benjamina, *sheared tree*
at Royal summer palace north of Bangkok, Thailand

Dammaropsis kingiana, *fruit*
harvested by native Melanesian of Kikiepa, New Guinea

Ficus benghalensis, *giant "Banyan tree"*
famous for its widespread stilt roots, in Sibpur, Calcutta

MORACEAE

Ficus elastica
"Rubber tree" in fruit

Ficus elastica 'Decora'
"Wideleaf rubber plant"

Ficus elastica 'Variegata'
"Variegated rubber plant"

Ficus petiolaris
"Blue Mexican fig"

Ficus rubiginosa 'Variegata'
variegated miniature

Ficus elastica 'Schryveriana'
in Belgium

Ficus rubiginosa
"Rusty fig" with fruit

Ficus lyrata (pandurata)
"Fiddle leaf fig"

Ficus macrophylla
"Mooreton Bay fig"

MORACEAE

Ficus barteri (*Nigeria*)
Kew Gardens, London

Ficus variolosa
on Victoria Peak, Hongkong

Ficus nekbudu
"Zulu fig" (Trop. Africa)

Ficus capensis
"Bush fig" (Africa)

Ficus benjamina comosa
Fairchild Garden, Miami

Ficus heteropoda
in Bulolo, New Guinea

Ficus triangularis
with fruit (Trop. Africa)

Ficus auriculata (roxburghii)
cauliflorous in India

Ficus diversifolia
"Mistletoe fig"

MORACEAE

Ficus benjamina (Java)
"Weeping fig" in container

Ficus retusa
"Chinese banyan"

Ficus nitida
"Indian laurel"

Ficus wildemanniana
(panduriformis) (Zaïre)

Ficus lyrata (pandurata)
"Fiddle leaf fig"

Ficus benjamina, *with fruit
U. of Calif. Botanic Garden*

Ficus nitida
Tiger Balm Garden, *Hongkong*

Ficus rubiginosa 'Variegata'
as standard, in Belgium

Ficus rubiginosa fa. australis
in Wellington, New Zealand

MORACEAE

Ficus rubiginosa fa. australis
hollowed out for electric lines, Newcastle, Australia

Ficus nitida, *old tree*
Tang Wai, Pearl River delta, South China

Ficus pumila (repens)
"Creeping fig", clinging to cement wall in Japan

Ficus benjamina, *"Weeping fig"*
wide-spreading tree 60 years old in Nehoa, Hawaii

Ficus nitida, *the "Laurel fig"*
in containers, lining street in downtown San Francisco

Ficus benjamina 'Exotica'
at the Prometheus fountain, RCA building, New York City

MORACEAE

Ficus saussureana
(*Guyana, Brazil*)

Ficus gibbosa batlinii
at Botanic Garden Munich

Ficus gilletii obovata
from Zaïre, C. Africa

Ficus mysorensis
(*South India*)

Ficus triangularis
(*Tropical Africa*)

Ficus benjamina 'Variegata'
in Del Dios, California

Ficus benghalensis krishnae
in Peradeniya, Ceylon

Ficus brandegei
from Mexico

Ficus palmeri
(*Baja California, Mexico*)

MORACEAE

Ficus benjamina, *the "Weeping fig"* growing as an epiphyte from a building, in Acapulco

Ficus gibbosa, *' "Strangler fig"* on the ruins of 9th century Angkor, Cambodia

Ficus nekbudu, *"Zulu fig",* attempting to dislodge an epiphytic seedling in Kenya

Ficus cordifolia, *strangler fig* embracing its host tree, near Madras, South India

MORACEAE

Ficus nitida, *"Laurel fig"*
standard trees, at Roehrs Exotic Nurseries, New Jersey

Ficus retusa, *"Chinese banyan"*
ancient tree in Kwangtung Province, South China

Ficus pumila, *"Creeping fig"*,
trained to wire frame, at Longwood Gardens, Penna.

Ficus elastica 'Decora'
as decorator tree, Keeline-Wilcox Nurseries, California

MORACEAE

Ficus sycomorus, *"Sycamore fig"* with edible fruit, on the road to Zambia, So. Africa

Ficus religiosa, *"Sacred Bo-tree"* at temple of the Emerald Buddha, Bangkok, Thailand

Ficus religiosa, *ancient "Peepul tree"* sanctified by Hindu lingam-yoni carving, in Benares, India

Ficus religiosa; *Tibetan girl with turquoise amulet, a sacred peepul leaf inside, in the Himalayas*

MORACEAE

Ficus carica, *"Common fig"*
on St. Héléna, Napoleon's exile

Morus nigra
"Black mulberry" (W. Asia)

Morus alba
"Silkworm mulberry" (China)

Ficus auriculata (roxburghii)
in India

Ficus religiosa
leaves with tail-like drip-tip

Ficus carica 'Mission'
with black fruit when ripe

Ficus radicans 'Variegata'
variegated creeper

Ficus pumila (repens)
miniature creeper

Humulus japonicus 'Variegatus'
"Ornamental hops", in Germany

MORACEAE

Ficus pseudopalma (*Philippines*)
"Dracaena fig" in Fairchild Gardens, Miami

Ficus retusa, *the "Chinese banyan"*
with aerial roots, as street tree in Kwangchow, China

Ficus retusa, *with weeping branches*
planted at the Music Center, Los Angeles, California

Ficus racemosa [*glomerata*] with edible fruit,
from Mysore, India; at U. of Calif. Bot. G., Los Angeles

MORINGACEAE, MUSACEAE

Ravenala madagascariensis (*MUSAC.*)
fan-shaped "Traveler's tree" with inflorescence, Honolulu

Moringa oleifera (pterygosperma)
"Horseradish tree" of many economic uses, in India

Ensete ventricosum (Musa ensete)
the ornamental "Abyssinian banana" in habitat, N. Kenya

Ensete maurelii (Musa)
distinctive, richly colored "Black banana" from Ethiopia

MUSACEAE

Heliconia caribaea
"Wild plantain" (Martinique)

Heliconia caribaea 'Purpurea'
on Kauai, Hawaii

Heliconia bihai
"Firebird" in Mexico

Heliconia latispatha
in Summit Gardens, Panama

Heliconia distans
from Perú

Heliconia jacquinii
Montego Bay, Jamaica

Heliconia humilis
"Lobster-claw" in Fiji

Heliconia revoluta
in Venezuela rain forest

Heliconia wagneriana
(Costa Rica, Panama)

MUSACEAE

Heliconia acuminata
'Espiritu Santo' (*Venezuela*)

Strelitzia reginae
"*Bird of paradise*" (*So. Africa*)

Heliconia psittacorum
"*Parrot flowers*" in Tahiti

Heliconia illustris 'Rubra'
dark-leaved California cultivar

Musa x paradisiaca 'Vittata'
"*Variegated banana*"

Heliconia spectabilis
(*Melanesia*)

Musa x paradiciaca 'Koae'
Hawaiian cultivar

Strelitzia nicolai (*inflor.*)
(*South Africa*)

Musa zebrina
"*Blood banana*" (*Java*)

Heliconia latispatha
(*Central America*)

Heliconia caribaea
"*Lobster claw*" (*Puerto Rico*)

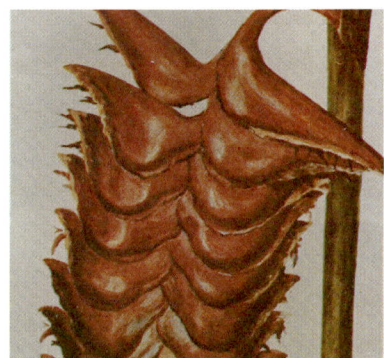
Heliconia mariae
"*Beefsteak heliconia*"

MUSACEAE

Heliconia brasiliensis
"False bird-of-paradise", in Brazil

Heliconia marginata
pendant inflorescense ready to unfold, in Venezuela

Heliconia aurantiaca
a small flowered species, in southern tropical Mexico

Heliconia caribaea 'Purpurea'
giant "Lobster-claws" in Venezuela

Strelitzia reginae, "Bird-of-paradise"
blooming mainly during cool season, in Vista, California

Strelitzia alba (augusta)
"Great white strelitzia" with inflorescense, in Natal

MUSACEAE

Heliconia mariae
Burle-Marx, Rio de Janeiro

Heliconia rostrata
from Perú

Heliconia collinsiana
(Guatemala)

Heliconia psittacorum
"Parrot flower" (Guyana)

Heliconia psittacorum 'Rubra'
in Barbados

Heliconia psittacorum rhizomatosa
"Parakeet flower" (C. Venezuela)

Musa coccinea
"Flowering banana" (Vietnam)

Ensete ventricosum
in Sao Paulo, Brazil

Heliconia wagneriana
Foster Bot. Garden, Honolulu

MUSACEAE

Heliconia humilis, *"Macaw flowers"*
at the foot of Iao-Needle Peak, on Maui, Hawaii

Heliconia rostrata, *pendant firebirds*
on Rarotonga, Friendship Isles, South Pacific

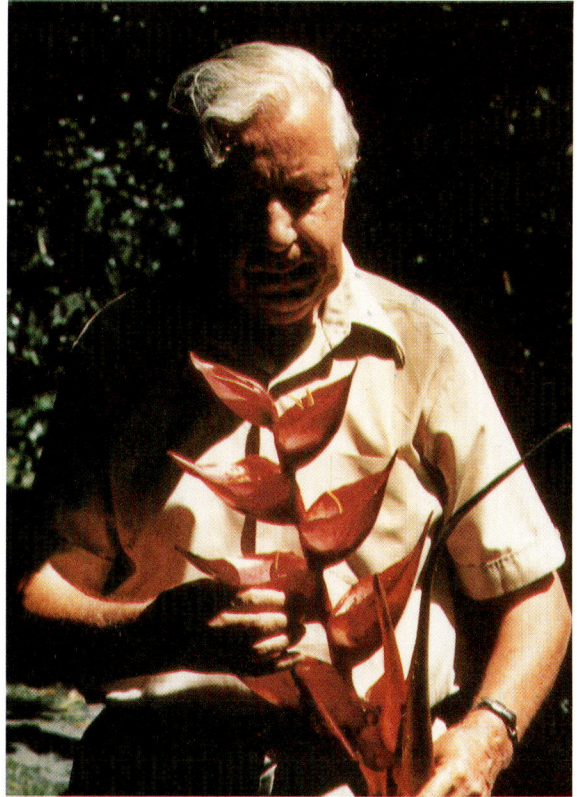

Heliconia bourgaeana *in the rainforest*
Mata Larga of Orizaba, Veracruz

Heliconia schiedeana, *at Posada Loma*
Fortin de las Flores, Veracruz, Mexico

MUSACEAE

Heliconia illustris
'Aureo-striata', *in Java*

Heliconia illustris 'Rubra'
Garfield Park, Chicago

Heliconia illustris rubricaulis
in Caracas, Venezuela

Heliconia spectabilis
'Edwardus Rex'

Ensete ventricosum 'Montbeliardii'
in Longwood Gardens

Heliconia indica
ornamental cultivar

Musa ornata (rosacea)
"Flowering banana" *(Bangladesh)*

Heliconia sharonii
from Ecuador

Musa mannii
"Dwarf banana"

MUSACEAE

Strelitzia alba (augusta)
"Great white bird-of-paradise" in Natal habitat

Ravenala madagascariensis
"Traveler's palm" at Raffles Hotel, Singapore

Musa sumatrana
"Blood banana" in Yogjakarta, Central Java

Strelitzia reginae, "Bird-of-paradise"
unfolding their wings in February, in Durban, So. Africa

MUSACEAE

Strelitzia reginae
"Crane lily"

Ravenala madagascariensis
inflorescence (Denver Bot. G.)

Strelitzia parvifolia juncea
"Rush strelitzia" in Cape Town

Musa basjoo
"Japanese fiber banana"

Strelitzia caudata
in Zwaziland, South Africa

Ravenala guayanensis
in Surinam habitat

Ravenala madagascariensis
near Diego Suarez, Madagascar

Musa zebrina
"Blood banana" (Java)

Musa sumatrana 'Rubra'
near Suva, Fiji

MUSACEAE

Musa x paradisiaca
heavily fruited banana on the garden island of Java

Musa uranoscopus
flowering banana in volcanic Central Java, Indonesia

Musa velutina
shapely banana with rosy-red ornamental fruit, in Assam

Musa acuminata (cavendishii or nana)
"Chinese dwarf banana" grown in vast areas of Queensland

MUSACEAE

Musa acuminata (nana or cavendishii)
"Chinese dwarf bananas" widely planted on Madeira

Ravenala madagascariensis
the "Traveler's tree" in habitat, Central Madagascar

Musa mannii
dwarf ornamental banana in flower in Assam, India

Musa coccinea
"Flowering banana" with scarlet inflorescence, in Vietnam

Orchidantha maxillarioides
"Orchid flower" mimicking a maxillaria orchid

Strelitzia parvifolia
small-leaved "Bird-of-paradise", Kirstenbosch, So. Africa

MUSACEAE, MYRICACEAE, MYRSINACEAE, NYSSACEAE

Musa velutina
ornamental banana in flower

Heliconia wagneriana
Munich Bot. Garden, Germany

Musa coccinea
scarlet inflorescence, in Vietnam

Myrica rubra
"Chinese strawberry tree"

Davidia involucrata
"Dove tree" (China)

Camptotheca acuminata
with winged fruit (China)

Ardisia solanacea (humilis)
berried tree from India

Ardisia humilis (*E. India*)
Brooklyn Botanic Garden

Ardisia crenata
"Coral berry" (Japan)

MYOPORACEAE, MYRISTICACEAE, MYRSINACEAE

Myoporum laetum
"Ngaio" or "Mouse-hole tree"

Myristica fragrans
branch with nutmeg fruit

Myristica fragrans
"Nutmeg tree" in Trinidad

Myrsine africana
"African boxwood"

Myrsine salicina
from Nelson, New Zealand

Myoporum parvifolium
ground cover from Tasmania

Jacquinia aculeata
from Cuba

Jacquinia pungens
"Cudjoewood" (Mexico)

Jacquinia armillaris arborea
in Barbados

MYRTACEAE

Eucalyptus ficifolia
"Scarlet flowering gum" in Southwestern Australia

Eucalyptus erythrocorys
spectacular "Red-cap gum" or "Illyarie" in W. Australia

Callistemon citrinus (lanceolatus)
"Crimson bottle brush" in New South Wales, Australia

Syzygium (Eugenia) malaccense
"Rose-apple" in flower, near Kikiepa village, New Guinea

MYRTACEAE

Callistemon citrinus
"Bottlebrush"

Beaufortia sparsa
"Swamp bottle-brush" (*W. Australia*)

Acca sellowiana 'Variegata'
(*Psidium*) *from Brazil*

Callistemon viminalis
"Weeping bottlebrush"

Calothamnus chrysantheros
"Crawflower" (*W. Australia*)

Calothamnus coccineus
"One-sided bottlebrush"

Callistemon montanus
in Western Australia habitat

Callistemon phoeniceus
"Fiery bottlebrush"

Callistemon rigidus
"Stiff bottlebrush"

MYRTACEAE

Syzygium jambos
"Malabar plum"

Eugenia uniflora
"Surinam cherry" (Guyana)

Syzygium samarangense
"Java apple"

Syzygium megacarpa
(India, Malaya)

Syzygium 'Jambeiro' (Eugenia)
Jardim Botanico, Rio de Janeiro

Acmena smithii (Eugenia)
"Lilly-pilly tree" (E. Australia)

Syzygium paniculatum
(Eugenia myrtifolia) (Australia)

Luma apiculata (Eugenia)
"Temu" in Chile

Syzygium malaccense
"Malay apple" (Eugenia)

MYRTACEAE

Eucalyptus ficifolia (S.W. Australia)
"Scarlet flowering gum", summer-blooming in California

Calothamnus chrysantheros
"Red crawflower" near Perth, in Western Australia

Eucalyptus caesia, the "Gungurru"
with its striking mealy-white pods, in Western Australia

Syzygium aromaticum (Eugenia);
"Clove" flower buds, used to make spice, in the Moluccas

Eucalyptus diversicolor, the "Karri gum", giant
felled timbertree in King's Park, Perth, W. Australia

Eucalyptus macrocarpa, "Mottlecah" or
"Rose of the West", in the Wongan Hills, W. Australia

MYRTACEAE

Eucalyptus ficifolia
"Scarlet flowering gum"

Eucalyptus calophylla
the "Marri" or "Red gum"

Eucalyptus erythrocorys
"Red-cap gum"

Eucalyptus lehmannii
"Bushy yate gum"

Eucalyptus globulus
"Tasmanian blue gum"

Eucalyptus caesia
"Gungurru" (W. Australia)

Eucalyptus globulus 'Compacta'
"Bushy blue gum" (California)

Eucalyptus cinerea
"Silver dollar", "Spiral eucalyptus"

Eucalyptus macrocarpa
"Bluebush" of W. Australia

MYRTACEAE

Eucalyptus camaldulensis (*rostrata*), "River red gum"

Eucalyptus rudis (*juv. stage*) "Desert gum"

Eucalyptus polyanthemos "Silver dollar tree"

Eucalyptus deglupta "New Guinea gum"

Eucalyptus citriodora graceful "Lemon-scented gum"

Eucalyptus nicholii "Willow peppermint"

Malaleuca quinquenervia "Cajeput" or "Paperbark tree"

Eucalyptus rhodantha "Rose mallee" (*W. Australia*)

Tristania conferta 'Aurea-variegata' "Brisbane box"

MYRTACEAE

Eucalyptus viminalis
"Manna gum" (S.E. Australia)

Eucalyptus maculata
"Spotted gum"

Eucalyptus rhodantha
"Rose mallee"

Eucalyptus woodwardii
"Yellow-flowered gum"

Eucalyptus citriodora
"Lemon gum" (Queensland)

Eucalyptus torquata
"Coral gum" (W. Australia)

Eucalyptus sideroxylon
"Red ironbark" (S.E. Australia)

Eucalyptus rudis (*mature stage*)
"Desert gum" of W. Australia

Eucalyptus sideroxylon rubra
"Ruby-red ironbark"

MYRTACEAE

Eucalyptus viminalis (S.E. Australia)
picturesque "Ribbon gum" in Vista, California

Eucalyptus camaldulensis (rostrata) of W. Australia
"River red gum" with colorful trunk in Cape Town, So. Africa

Pimenta racemosa (acris)
oil-yielding "Bay-rum" trees near Kingston, Jamaica

Eucalyptus salubris, "Gimlet-gum"
severely trimmed as a street tree in Perth, W. Australia

MYRTACEAE

Leptospermum scoparium 'Ruby Glow'
"New Zealand tea tree", container-grown in California

Leptospermum laevigatum
gnarled "Australian tea tree" in Balboa Park, San Diego

Leptospermum scoparium 'Rubrum plenum'
double-flowered "Manuka" in Wellington, New Zealand

Melaleuca leucadendron, "Paper bark" trees
along the Coral Sea, near Noumea, New Caledonia

MYRTACEAE

684

Pimenta dioica
"All-spice" (fruit) in Jamaica

Pimenta dioica
"Pimento" or "All-spice" in flower

Pimenta racemosa
"Bay-rum tree"

Myrtus pubescens (*Argentina*)
Palmengarten Frankfurt, Germany

Myrtus communis microphylla
"German myrtle" in Germany

Myrtus communis 'Compacta'
"Dwarf myrtle" in California

Tristania conferta 'Variegata'
in San Diego nursery

Rhodomyrtus tomentosa
"Hill guava" (Philippines)

Myrtus communis
"Greek myrtle"

MYRTACEAE

Metrosideros excelsus, *"Pohutukawa"* or *"New Zealand Christmas tree"*

Leptospermum scoparium 'Keatleyi' *in Auckland, New Zealand*

Leptospermum scoparium *"Manuka" in New Zealand*

Leptospermum petersenii (citratum) *"Lemon-scented tea tree"*

Leptospermum scoparium 'Album' *"White tea tree"*

Melaleuca quinquenervia *"Cajeput tree" in Florida*

Metrosideros collinus, *"Ohia-Lehua" on Kilauea Volcano, in Hawaii*

Melaleuca spathulata *"Honey-myrtle" (W. Australia)*

Melaleuca leucadendron *"River tea tree"*

MYRTACEAE

Feijoa sellowiana
with delicious fruit, in California

Feijoa sellowiana
"Pineapple guava" in flower

Metrosideros villosus 'Variegatus'
(*hermadecensis*)

Psidium guajava
"Common guava" of Mexico

Psidium cattleianum
"Strawberry guava"

Psidium cattleianum lucidum
"Yellow strawberry guava"

Psidium guineense
"Guyana guava"

Feijoa deflexa
"Sawtooth loquat" in California

Marlierea edulis
Jardim Botanico, Rio de Janeiro

MYRTACEAE

Melaleuca genistifolia, *"Fleece trees"*
in arid central Australia, near Alice Springs

Metrosideros excelsus (tomentosus)
"Pohutukawa" or *"New Zealand Christmas tree"*

Metrosideros tremuloides
a *"Rata"* along the Pali road in Honolulu

Leptospermum scoparium, *"Manuka"*
or *"Tea tree"* at home in New Zealand, north of Auckland

Regelia megacephala
strange xerophyte of the desert in Western Australia

Verticordia nitens
brilliant "Feather flowers" near Perth, Western Australia

MYRTACEAE

688

Myrtus communis 'Variegata'
"Variegated myrtle"

Eugenia myrtifolia 'Gracilis'
"Ornamental bush cherry"

Callistemon citrinus
"Crimson bottlebrush"

Eucalyptus torquata
"Coral gum"

Eucalyptus ficifolia
"Scarlet flowering gum"

Eucalyptus cinerea
"Silver dollar tree"

Eucalyptus erythrocorys
"Red helmet"

Eucalyptus rhodantha
"Rose mallee"

Eucalyptus macrocarpa
"Rose of the West"

Tristania conferta variegata
"Brisbane box"

Eugenia megacarpa
India, Malaya)

Syzygium samarangense
"Java apple"

NYCTAGINACEAE

Bougainvillea 'Alba'
Fairchild Bot. Garden, Miami

Bougainvillea x buttiana
'Barbara Karst', *in California*

Bougainvillea x buttiana
'Madonna'

Bougainvillea glabra
"Paper flower" (*Brazil*)

Bougainvillea 'Harrisii'
as beautiful foliage plant

Bougainvillea formosa
Singapore Botanic Garden

Bougainvillea spect. 'Rubra plena'
in Bangkok, Thailand

Bougainvillea poultonii
bracts glowing red, in Singapore

Ouratea groussordyi (OCHNAC.)
Bot. Garden Caracas, Venezuela

NYCTAGINACEAE

Bougainvillea x buttiana 'San Diego Red'
at Quail Botanic Gardens, Encinitas, California

Bougainvillea spectabilis 'Manila Magic Pink'
at Monrovia Nurseries, Azusa, California

Bougainvillea spectabilis 'Mary Palmer'
with both red and white bracts, Foster Gardens, Honolulu

Bougainvillea spectabilis 'Variegata'
planted along city streets in Nairobi, Kenya

Bougainvillea x buttiana 'Mrs. McLean'
in Longwood Gardens, Kennett Square, Pennsylvania

Bougainvillea spectabilis
in the Municipal Gardens of Nairobi, Kenya

NYCTAGINACEAE

Bougainvillea spectabilis 'Arborea'
Jardim Botanico, Rio de Janeiro, Brazil

Bougainvillea spectabilis
in the formal gardens of the Taj Mahal, Agra, India

Bougainvillea glabra 'Salmonea'
in old Arab garden of Dar-Es-Salaam, Tanzania

Bougainvillea spectabilis 'Milflores'
double-bracted flowers, at nursery in Manila, Philippines

Bougainvillea spectabilis 'Mary Palmer'
with bracts both white and red, in Hyderabad, South India

Bougainvillea glabra *in variety
sheared globes in Penang, Malaysia*

NYCTAGINACEAE, OCHNACEAE

Bougainvillea glabra 'Variegata'
in Botanic Garden of Port-of-Spain, Trinidad

Bougainvillea x buttiana 'Louis Wathen'
in Botanic Gardens Peradeniya, Sri Lanka

Ochna kirkii
"Mickey mouse plant" in Mozambique, East Africa

Heimerliodendron brunonianum 'Variegatum' (Pisonia)
variegated "Bird-catcher-tree" in Wellington, New Zealand

NYCTAGINACEAE

Bougainvillea spectabilis
'Tahitian Gold', *in California*

Bougainvillea spect. 'Mary Palmer'
in Bangalore, South India

Bougainvillea spect. 'Alba plena'
in Bangkok, Thailand

Bougainvillea spect. 'Carmencita'
in the Philippines

Bougainvillea spect. 'Rubra plena'
in Bangkok, Thailand

Bougainvillea x spectoglabra
in Kota Kinabalu, Sabah

Bougainvillea spectabilis 'Lateritia'
Jardim Botanico, Rio de Janeiro

Bougainvillea spectabilis
'Tahitian Maid'

Mirabilis longiflora
Trinidad Botanic Garden

NYCTAGINACEAE, OCHNACEAE, OLEACEAE

Bougainvillea 'Harrisii'
"Variegated paper-flower"

Bougainvillea glabra
"Paper flower"

Bougainvillea 'Braziliensis'
in California horticulture

Bougainvillea x buttiana
"Crimson Lake"

Bougainvillea x buttiana
'Praetoria'

Heimerliodendron brunonianum
'Variegatum' (Pisonia)

Jasminum rex
(*Southwest Thailand*)

Jasminum mesnyi
"Primrose jasmine" (*China*)

Jasminum nitidum
"Angelwing jasmine"

Ochna serrulata
"Birdseye bush" with fruit

Ochna serrulata
(*multiflora in hort.*)

Jasminum sambac
'Maid of Orleans', *"Arabian jasmine"*

OLEACEAE

Olea europaea, *the "Olive tree"*
sheared form, 115 years old, in Vista, California

Olea europaea, *ancient olive trees*
under the Golden gate and walls of Jerusalem, Israel

Jasminum grandiflorum, *"Poet's jasmine"*
near the old city gate, Damascus, Syria

Jasminum gracillimum
"Star-jasmine" in Peradeniya, Sri Lanka

Ligustrum japonicum, *as a tree*
at Union Buildings, overlooking Pretoria, Transvaal

Ligustrum japonicum, *"Japanese privet"*
sheared bush, in the white city of Casablanca, Morocco

OLEACEAE

Chionanthes virginicus
"American fringe tree"

Olea europaea, "Olive"
gnarled tree in Mallorca, Spain

Fraxinus uhdei
"Evergreen ash" in California

Osmanthus armatus
"Chinese osmanthus"

Osmanthus fragrans
"Sweet olive"

Olea europaea
"Olive" with ripe fruit

Osmanthus heterophyllus
'Variegatus' (*ilicifolius*)

Ligustrum ovalifolium
"California privet" (*Japan*)

Ligustrum japonicum
inflorescence

OLEACEAE

Jasminum nitidum (magnificum)
"Angelwing jasmine"

Jasminum rex
"King jasmine" (Thailand)

Jasminum multiflorum
"Star jasmine"

Jasminum undulatum
"Angel-hair jasmine"

Jasminum simplicifolium
"Little star-jasmine"

Jasminum sambac
'Grand Duke of Tuscany'

Jasminum mesnyi
"Primrose jasmine" in Chile

Jasminum floridum
(China, Japan)

Jasminum humile revolutum
"Italian jasmine"

NYCTAGINACEAE, OCHNACEAE, OLEACEAE

Ochna mossambicensis
"Birdseye bush" in Honolulu

Ochna serrulata (multiflora)
"Birdseye-bush"

Ochna madagascariensis
in Malagasy Republic

Heimerliodendron brunonianum
'Variegatum', "Para-Para"

Ochna kirkii
"Mickey mouse plant"

Noronhia emarginata
"Madagascar olive"

Ligustrum lucidum
"Glossy privet" of the South

Ligustrum lucidum 'Texanum'
"Wax-leaf privet"

Ligustrum lucidum 'Silver Star'
"Variegated wax-leaf"

NYCTAGINACEAE, OCHNACEAE, OLEACEAE

Syringa vulgaris, *the "Common lilac"*
along the Danube, near Vienna, Austria

Ligustrum lucidum, *in topiary shape*
at the Memorial temple near Taipei, Taiwan

Ligustrum lucidum, *the "Wax privet"*
trained in poodle shape, in Tallahassee, Florida

Jasminum sambac
the "Arabian jasmine" in Bahrein, Persian Gulf

NYCTAGINACEAE, OLEACEAE, ONAGRACEAE

Lopezia lineata
"Mosquito flower" (Mexico)

Jasminum rex
"King jasmine"

Osmanthus heterophyllus 'Variegatus'
"Variegated false holly"

Osmanthus fragrans
"Fragrant olive"

Ligustrum sinense 'Variegatum'
"Chinese silver privet"

Mirabilis jalapa
"Four-o'clock" (Perú)

Fuchsia x nybrida
'Winston Churchill'

Fuchsia x hybrida
'Lace Petticoats'

Fuchsia x hybrida
'Dollar Princess'

Fuchsia triphylla
'Gartenmeister Bohnstedt'

Fuchsia x hybrida
'Pink Cloud'

Fuchsia x hybrida
'Cascade'

ONAGRACEAE

Oenothera missouriensis
"Evening primrose" in Visby, on Gotland, Sweden

Fuchsia triphylla
from Hispaniola, parent of most hybrids

Fuchsia x hybrida 'Caledonia'
in the cool climate of Gold Beach, Oregon

Fuchsia excorticata (*inflorescence*)
the "Tree fuchsia" or "Kotukutuku" in New Zealand

Fuchsia lycioides
tall shrub 3 metres high, in Southern Chile

Fuchsia x hybrida 'Strawberry Queen'
in profuse bloom at an English cottage

ONAGRACEAE

Fuchsia splendens
from Mexico, at 3300 m alt.

Fuchsia x hybrida
'Checkerboard'

Fuchsia x hybrida
'Mrs. Victor Reiter', *as pyramid*

Fuchsia triphylla
'Gartenmeister Bohnstedt'

Oenothera fruticosa
day-blooming "Sundrops"

Fuchsia procumbens
"Trailing fuchsia" (New Zealand)

Clarkia elegans 'Rubra plena'
"Farewell-to-spring"

Epilobium angustifolium
"Fireweed" in Alaska

Clarkia elegans 'Rosea plena'
(unguiculata) from California

ONAGRACEAE

Fuchsia magellanica, *"Hardy fuchsia"*
in the fern forest of Kilauea volcano, Hawaii; also Chile

Fuchsia triphylla 'Gartenmeister Bohnstedt'
"Honeysuckle fuchsia" at the Domain, Auckland, N.Z.

Fuchsia x hybrida 'Golondrina' *as standard*
at Castle Egeskov on Fuenen, Denmark

Clarkia amoena whitneyi
(*Godetia grandiflora*) "Satin flower"

NYMPHAEACEAE

Nymphaea 'Mme. Julien Chifflot'
huge pink, hardy water-lily at Brooklyn Botanic Garden

Nymphaea 'Pink Sensation'
fragrant, free-blooming rich pink, hardy water-lily

Nymphaea caerulea, *"Blue lotus of Egypt",*
a tender species, in the warm waters of Kenya

Nymphaea gigantea, *"Giant water lily"*
tropical species with enormous flowers, in Australia

Nelumbo nucifera (Nelumbium nelumbo)
"East Indian lotus" in a temple pool, in Vietnam

Nymphaea rubra, *in Singapore*
beautiful night-blooming "India red water lily"

NYMPHAEACEAE

Euryale ferox, *"Gorgon"*
or *"Prickly water lily"* from India and Japan

Nymphaea lotus dentata
fragrant night-blooming "White lotus of Egypt"

Victoria cruziana, *in bloom*
"Santa Cruz water lily", from Paraná and Paraguay

Victoria cruziana, *in outdoor pond*
Jardin Botanique Cap Ferrat, on the French Riviera

Victoria regia, *"Royal water lily"*
in Kebun Raya Botanic Garden, Bogor, Java

Victoria regia, *"Amazon water lily"*
in a quiet igarapé on the upper Amazon, Brazil

NYMPHAEACEAE

Nymphaea micrantha
from West Arica

Nymphaea x daubeniana
viviparous plantlet from old leaf

Nymphaea colorata
Dar-es Salaam, Tanzania

Nymphaea capensis zanzibariensis
"Cape blue waterlily"

Nymphaea 'Tashkent'
Frankfurt Palmengarten, Germany

Nymphaea alba
"European white waterlily" (hardy)

Nymphaea capensis
in old Botanic garden, Cape Town

Nymphaea x marliacea 'Chromatella'
an old reliable, hardy water-lily

Nymphaea rubra 'Rosea'
in Caracas, Venezuela

NYMPHAEACEAE

Nymphaea x daubeniana, *viviparous*
"Pigmy water lily"; young plantlets sprout from old leaves

Nymphaea rubra
"India red water lily", in Palmengarten, Frankfurt

Nelumbo nucifera 'Alba plena', *"Shiroman"*
or "Sacred lotus" in Jardim Botanico, Rio de Janeiro

Nelumbo nucifera (Nelumbium nelumbo)
the "Sacred lotus" in year-round pond, Auckland, N.Z.

ORCHIDACEAE

Ada aurantiaca (Pleurothallis)
epiphyte from Colombia

Aerangis coriacea
epiphyte with fragrant blooms, in Kenya

Arachnis flos-aeris
the spectacular "Spider orchid", in Java

Angraecum filicornu (Mystacidium)
small epiphyte from tropical Madagascar

Brassavola cordata (subulifolia)
with fragrant flowers, from the West Indies

Brassavola acaulis
epiphytic with fragrant blooms; Guatemala to Panama

ORCHIDACEAE

Aerides falcata
(*India*)

Aerides fieldingii
"Fox-brush orchid" (*Assam*)

Aerides virens
(*odorata major*) *from Java*

Arpophyllum spicatum
"Hyacinth orchid" (*Mexico*)

Angraecum sesquipedale
"Star of Bethlehem" (*Madagascar*)

Angraecum eichlerianum
(*West Trop. Africa*)

Bifrenaria tyrianthina (Lycaste)
(*Brazil*)

Bifrenaria harrisoniae
(*Brazil*)

Arundina chinensis
terrestrial from China

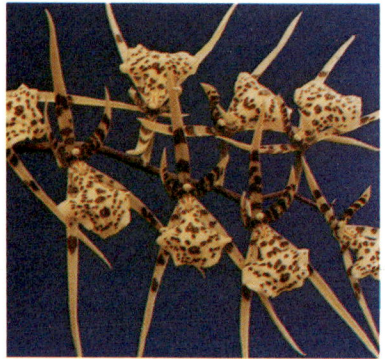
Brassia maculata
(*West Indies, Guatemala*)

Brassia allenii
(*Panama*)

Brassia verrucosa
(*Mexico to Venezuela*)

ORCHIDACEAE

Aerangis (Angraecum) kotschyana
(*Kenya to Nigeria*)

Angraecum eburneum
(*Madagascar*)

Angraecum eburneum longicalcar
long-spurred

Arachnis hookeriana
"Scorpion orchid" (*Malaya*)

Arachnis flos-aeris
"Spider orchid" (*Indonesia*)

Ansellia africana
(*Kenya to West Africa*)

Arachnis massoei
perforated pot in Malaya

x Aranda majala
(*Arachnis x Vanda*) *in Singapore*

Arpophyllum spicatum
"Hyacinth orchid" (*Mexico*)

ORCHIDACEAE

Ascoglossum calopterum
(*New Guinea*)

Ansellia nilotica
epiphytic in Kenya

Barlia longibracteata
on the Riviera, France

Brassavola acaulis
(*Central America*)

Bletilla striata
"Hyacinth orchid" (*Vietnam*)

Brassia antherotes
(*Ecuador*)

Cochlioda densiflora
(*Andes of Perú*)

Bulbophyllum dayanum
miniature from Burma

Bulbophyllum grandiflorum
hooded flowers, in New Guinea

ORCHIDACEAE

712

Calanthe 'Wm. Murray'
(*vestita x williamsii*)

Catasetum viridiflavum
(*C. America to Perú*)

Catasetum roseum
(*Mexico: Oaxaca*)

Catasetum trulla (socco)
epiphyte from Brazil

Brachycorythis kalbreyeri
(*West Africa to Kenya*)

Calanthe discolor sieboldii
terrestrial from Japan

Cattleya luteola
(*Brazil to Perú*)

Cattleya leopoldii
(*guttata var. from Brazil*)

Cattleya intermedia var. alba
(*Brazil*)

Cattleya intermedia superba
(*Brazil*)

Chysis aurea
(*Mexico to Venezuela*)

Cattleya aclandiae
(*Northern Brazil*)

ORCHIDACEAE

Brassavola cucculata
(*Trop. America*)

Brassavola digbyana
(*Honduras*)

Brassavola nodosa
"Lady of the night"

x Brassocattleya 'Carmen'
(*Brassia x Cattleya*)

Brassavola glauca
(*Mexico, Guatemala*)

x Brassocattleya 'Brittisa'
(*Brassavola x Cattleya*)

x Brassocattleya 'Heatonensis'
(*Brassavola x Cattleya*)

x Brassocattleya lehmannii
unique bearded flower

x Brassocattleya 'Illusion'
(*Rosa Bonheur x C. Admiration*)

x Brassocattleya veitchii
(*Brassavola digbyana x C. mossiae*)

Rodriguezia venusta
(*Burglingtonia fragrans*) (*Brazil*)

Bulbophyllum makoyanum
(*Malaya*)

ORCHIDACEAE

Calanthe 'Baron Schroeder'
(*regnieri x vestita*)

Calanthe furcata (veratrifolia)
(*India, Polynesia, Australia*)

Calanthe silvatica
(*Madagascar*)

Cypripedium acaule
"Pink ladyslipper" (*E. No. America*)

Brachycorythis kalbreyeri
(*Cameroun*)

Cypripedium calceolus pubescens
"Mocassin flower" (*No. America*)

Cymbidium 'Pixie'
(*Ceres x Landrail*)

Cymbidium 'Madonna'
at New York Flower Show

Cymbidium pumilum
in Taiwan

ORCHIDACEAE

Chysis bractescens
(*Mexico, Guatemala*)

Cochlioda rosea
(*Andes of Perú*)

Bulbophyllum makoyanum
(*Cirrhopetalum*) (*Malaya*)

x Laeliocattleya bella
(*Cattleya x Laelia*)

Coelogyne massangeana
(*Assam to Java*)

Coelogyne sparsa
(*Philippines*)

Cyrtopodium paranaensis
(*Brazil: Paraná*)

Coelogyne dayana
"Necklace orchid" (*Borneo*)

Coelogyne pandurata
"Black orchid" (*Sumatra*)

Coelogyne parishii
(*Southern Burma*)

Ceratostylis rubra
(*Philippines*)

Coelogyne flaccida
(*Nepal Himalayas*)

ORCHIDACEAE

Cattleya mossiae (*Venezuela*)
"Easter orchid" or "Flor de Mayo"

Cattleya labiata alba
pristine white "Autumn cattleya", from Brazil

Cattleya forbesii, *a "Cocktail orchid",
in habitat on the Restinga, south of Rio de Janeiro*

Cattleya gaskelliana
"Summer cattleya", from Venezuela and Brazil

Bulbophyllum lobbii
curious epiphyte, at home from Burma to Indonesia

Coelogyne pandurata, *the sinister "Black orchid"
from the remote jungles of Borneo and Sumatra*

ORCHIDACEAE

Cattleya citrina
"Tulip cattleya" of Mexico

Cattleya amethystoglossa
(*Brazil*)

Cattleya intermedia aquinii
(*Brazil*)

Cattleya intermedia
(*Brazil*)

Cattleya bicolor
(*Brazil*)

Cattleya granulosa
'Schofieldiana'

Cattleya granulosa
(*Venezuela, Brazil*)

Cattleya 'Enid'
(*gigas x mossiae*)

Cattleya mossiae
"Easter orchid" (*Venezuela*)

Cattleya skinneri
(*Guatemala*)

Cattleya 'Rex'
(*Amazonian Brazil and Perú*)

Cattleya 'Priscilla alba'
(*Enid x lueddemanniana*)

ORCHIDACEAE

Arachnis flos-aeris
popular "Spider orchid" of Monsoon Asia, Malaya to Java

Ansellia africana, *a "Leopard orchid" growing epiphytic on acacia trees, in Southern Uganda*

Cymbidiella rhodochila
nesting on Platycerium madagascariensis, in Malagasy

Cattleya hybrids *on display in private collection at Broadview, New Rochelle, N.Y.*

Masdevallia coccinea, *from Colombia to Perú, in collection of Greentree Estate, Manhasset, Long Island*

Dendrobium hybrids *in porous pots for best drainage, in the monsoon climate of Singapore*

ORCHIDACEAE

Arachnis flos-aeris, *"Spider orchids"*
luxuriate without effort in tropical Singapore

Cattleya mossiae, *"Flor de Mayo"*
as tree-dwelling epiphyte in the Andes of Venezuela

x Brassocattleya hybrida
in polychrome ivory, carved by craftsmen of Kyoto, Japan

Cattleya hybrids *of high excellence*
grown by Flandria Orchid nurseries in Bruges, Belgium

ORCHIDACEAE

Cattleya dowiana aurea
"Queen cattleya" (*Colombia*)

Disa uniflora (*So. Africa*)
"Pride-of-Table Mountain"

Cattleya dowiana
in Medellín, Colombia

Elleanthus capitatus
(*Mexico, W. Indies, Perú*)

Cyrtorchis praetermissa
(*Transvaal to Zaïre*)

Eulophia porphyroglossa
(*Kenya*)

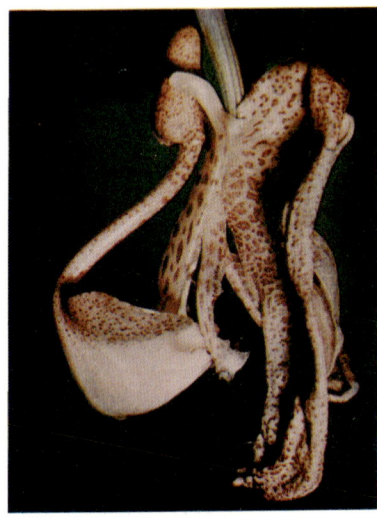

Coryanthes bicalcerata
"Bucket orchid" (*Perú*)

Cymbidiella humblotii
(*Madagascar*)

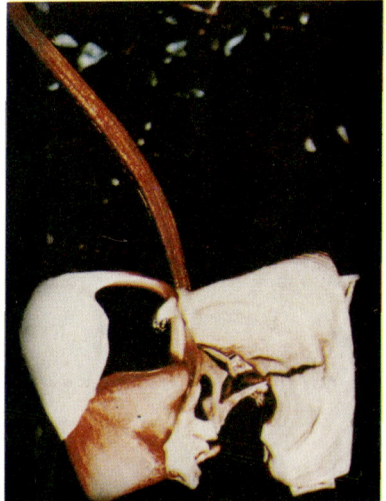

Coryanthes macrocorys
"Helmet orchid" (*Perú*)

ORCHIDACEAE

Dendrobium d'albertisii
"Rabbit's-ear orchid" as epiphyte, Port Moresby, Papua

Dendrobium macrophyllum
collected on expedition to the Finisterres, New Guinea

Disa uniflora, from South Africa
displayed at the German Horticultural show, in Hamburg

Coelogyne cristata from the Himalayan region,
blossoms in a cascade of flowers; used in home decor.

ORCHIDACEAE

Cymbidiella rhodochila (*Madagascar*)
a rare beauty, grown by M. Lecoufle, near Paris, France

Cryptophoranthus nigriflorus
the "Window orchid", a miniature of the West Indies

Coryanthes maculata punctata
remarkable "Helmet orchid" from Guyana and Venezuela

Bifrenaria tyrianthina alba
growing on trees and rock outcroppings in S.W. Brazil

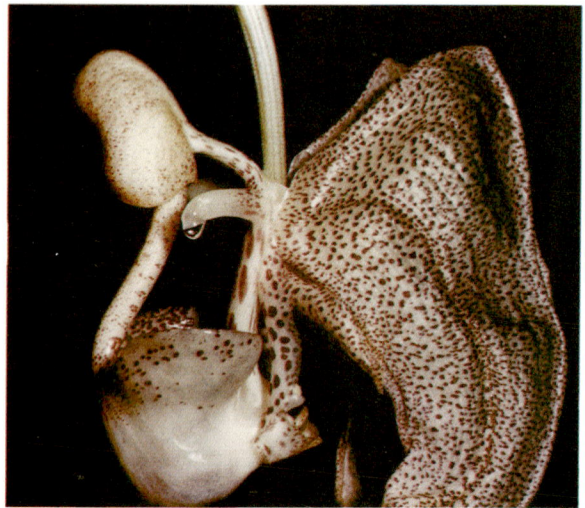

Coryanthes leucocorys
a weirdly exotic epiphyte from Amazonian Perú

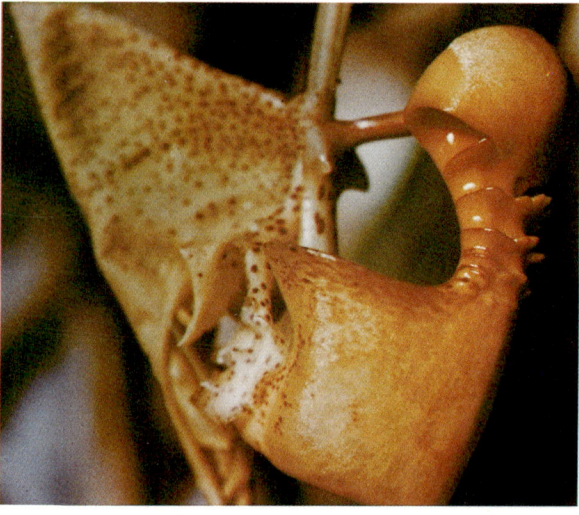

Coryanthes macrantha
another extraordinary "Helmet orchid" from Trinidad

ORCHIDACEAE

Cymbidium 'Swallow var. Broadview'
(*alexanderi x pauwelsii*)

Cymbidium 'Ceres'
(*l'Ansonii x insigne sanderi*)

Cymbidium 'Louis Sander'
(*alexanderi x Ceres*)

Cymbidium 'Alexanderi alba'
(*eburneo-lowianum x insigne*)

Eria merrillii, *from the Philippines*
tropical epiphyte with arching spray of fragrant flowers

Cymbidium 'Priscilla'
(*insigne x Yellow Hammer*)

ORCHIDACEAE

Cymbidium lancifolium
(*Himalayas to Malaysia*)

Cymbidium 'Doris'
(*insigne x traceyanum*)

Cymbidium canaligulatum
(*Australia*)

Cymbidium eburneum
(*North India, Burma*)

Cymbidium 'Flirtation'
charming semi-miniature

Epidendrum altissimum
(*Bahamas*)

Cymbidium virescens
var. angustifolium (*Japan*)

Cycnoches chlorochilum
(*C. America to Venezuela*)

Cycnoches pentadactylum
(*Brazil*)

Dendrochilum glumaceum
(*Platyclinis*) (*Philippines*)

Cycnoches ventricosum
"Swan orchid" (*Guatemala*)

Dendrochilum uncatum
(*Philippines*)

ORCHIDACEAE

Dendrobium senile
"Cowslip orchid" from Thailand and Laos

Dendrobium macrophyllum
strange beauty from Celebes and the Philippines

Dendrobium primulinum
an epiphyte of the subtropic Himalayas and Burma

Dendrobium x thwaitesiae
cane-stem hybrid, one of the showiest of the genus

Dendrobium pierardii (Burma)
epiphyte from steaming forest, with flowers in cascades

Dendrobium veratrifolium
collected near Wau, in the gold fields of New Guinea

ORCHIDACEAE

Dendrobium nobile
(*Assam to Yunnan*)

Dendrobium superbum
(*Celebes to Philippines*)

Dendrobium infundibulum
(*South Burma*)

Dendrobium phalaenopsis
(*Queensland to Timor*)

Dendrobium phalaenopsis 'Pompadour'
popular cut-flower in Singapore

Dendrobium parishii
(*Yunnan to Vietnam*)

Epidendrum ibaguense (radicans)
"Fiery reed orchid" (*Mexico*)

Epidendrum lanipes
(*Venezuela to Perú*)

Dendrobium secundum
(*Burma to So. Pacific*)

ORCHIDACEAE

Dendrobium densiflorum
(*Himalayas, Burma*)

Dendrobium johnsoniae
(*New Guinea*)

Dendrobium thyrsiflorum
(*Burma*)

Dendrobium secundum
(*Burma to Vietnam*)

Dendrobium chrysotoxum
(*Burma, Yunnan*)

Dendrobium infundibulum
(*So. Burma*)

Dendrobium gracicaule
(*Australia*)

Dendrobium formosum giganteum
(*Upper burma*)

Dendrobium fimbriatum
(*Himalayas, Burma*)

Dendrobium 'New Guinea'
(*macrophyllum x atroviolaceum*)

Dendrobium nobile
(*Assam to Yunnan*)

Dendrobium nobile
'Albiflorum'

ORCHIDACEAE

Epidendrum oncidioides
(*C. America to Brazil*)

Epidendrum falcatum
(*Mexico*)

Epidendrum paniculatum
(*Bolivia to Colombia*)

Epidendrum pseudoepidendrum
(*Costa Rica, Panama*)

Epidendrum ibaguense schomburgkia
(*Guyana to Perú*)

Epidendrum erubescens
(*Mexico: Oaxaca*)

Epidendrum ibaguense (radicans)
(*Perú to Mexico*)

Epidendrum ibaguense purpurea
"Purple reed orchid"

Epidendrum x o'brienianum
"Butterfly orchid"

ORCHIDACEAE

Epidendrum fragrans
(*W. Indies, C. America*)

Epidendrum ibaguense
(*Mexico to Perú*)

Epidendrum radiatum
(*Mexico*)

Epidendrum stamfordianum
(*Guatemala to Colombia*)

Epidendrum oncidioides
(*C. America to Brazil*)

Epidendrum atropurpureum
"*Spice orchid*" (*Mexico*)

Epidendrum vespa
(*variegatum*) (*Brazil*)

Epidendrum x o'brienianum
"*Scarlet baby orchid*"

Epidendrum ciliare
(*W. Indies to Brazil*)

Epidendrum pentotis
(*C. America, Brazil*)

Epidendrum mariae
(*So. Mexico*)

Epidendrum prismatocarpum
"**Rainbow orchid**" (*Costa Rica*)

ORCHIDACEAE

Grammatophyllum speciosum
"Queen of orchids" or "Giant orchid", from Malaya

Gastorchis luteus
a rare terrestrial from Madagascar

Dendrobium falcorostrum
free-flowering, showy epiphyte from Australia

Dendrobium infundibulum
from high elevations in Southern Burma and Thailand

Epidendrum pseudoepidendrum
one of the finest, from Costa Rica and Panama

Epidendrum lemorea
handsome epiphyte from the upper Amazon, Perú

ORCHIDACEAE

Grammatophyllum speciosum, *the "Queen of orchids"*, *giant epiphyte in the Botanical Gardens of Singapore*

Dendrobium farmeri var. albiflorum *with a shower of blooms, at the Palmengarten, Frankfurt*

Eulophia paivaeana *xerophytic terrestrial on the high steppe of Tanzania*

Cyrtopodium andersonii *in habitat on the Serra do Mar, above Santos, S.E. Brazil*

ORCHIDACEAE

732

Epigeneium lyonii
(*Phillipines*)

Grammatophyllum scriptum
(*Moluccas*)

Epidendrum dichromum
(*Brazil:Bahia*)

Eria spicata (*Burma*)
"Lily-of-the-valley orchid"

Graphorkis lurida
(*Trop. West Africa*)

x Epiphronitis veitchii
(*Epidendrum x Sophronitis*)

Gastorchis luteus
(*Madagascar*)

Erycina echinata
(*Mexico*)

Laelia anceps alba
(*Mexico*)

Haemaria discolor dawsoniana
"Jewel orchid" (*Malaya*)

Lockhartia lunifera
"Braided orchid" (*Brazil*)

Lockhartia acuta
(*Trinidad*)

ORCHIDACEAE

Laelia albida
(*Mexico, Guatemala*)

Laelia purpurata
(*Brazil*)

Laelia cinnabarina
(*Brazil*)

Laelia lundii (regnellii)
(*Brazil*)

Laelia rubescens
(*Mexico, C. America*)

Laelia anceps
(*Mexico, Honduras*)

Laelia x harpophylla 'Aurea'
(*Brazil*)

Laelia x harpophylla
natural hybrid of Brazil

Laelia grandis
(*Brazil: Bahia*)

Laelia pumila
(*Brazil*)

Laelia rubescens
"Flor de Jesú (*Mexico*)

Laelia gouldiana
(*Mexico*)

ORCHIDACEAE

Eriopsis biloba
(*Perú, Guyana, Brazil*)

Grammatophyllum scriptum
(*Moluccas*)

Gongora quinquenervis
(*Brazil*)

Leochilus labiatus
(*C. America*)

Govenia utriculata
(*Mexico, W. Indies, to Argentina*)

Houlletia lansbergii
(*Guatemala to Brazil*)

Huntleya meleagris (burtii)
(*Costa Rica to Brazil*)

Habenaria glazioviana
"Fringed orchid" (*Brazil*)

Laelia flava
(*Brazil*)

ORCHIDACEAE

Mediocalcar species, *from cloud forest of New Guinea epiphyte with strange balloon-shaped flowers*

Lepanthes pulchella, *daintily attractive dwarf epiphyte from the mountains of Jamaica*

Eulophiella x rolfei
lovely hybrid of roempleriana x elisabethae (Madagascar)

Isochilus linearis
grass-like rock-dweller in Mexico, Cuba and Argentina

Gongora quinquenervis (maculata)
Brazilian epi. with extraordinary, strongly scented flowers

Maxillaria rufescens
handsome, fragrant epiphyte from Cuba to Guatemala

ORCHIDACEAE

736

Masdevallia veitchiana
(*Perú*)

Miltonia x hyeana
"Pansy orchid"

Masdevallia rosea
(*Colombia, Ecuador*)

x Miltassia 'Cartagena'
(*Miltonia x Brassia*)

x Miltonidium 'Yellow Monarch'
(*Miltonia x Oncidium*)

Miltonia spectabilis 'Warneri'
(*Brazil*)

Oncidium ampliatum majus
(*Costa Rica, Guatemala*)

Gussonaea physophora
(*Madagascar*)

Oncidium sphacelatum
"Golden shower" (*Mexico*, **Honduras**)

ORCHIDACEAE

737

x Laeliocattleya hasselii
'Alba majestica'

x Laeliocattleya bella
(*Laelia x Cattleya*)

x Laeliocattleya canhamiana alba
(*L. purpurea x C. mossiae*)

Lycaste candida (brevispatha)
(*Costa Rica, Panama*)

Laelia purpurata
(*Brazil*)

Lycaste cruenta
(*Guatemala*)

Lycaste virginalis (skinneri)
(*Mexico, Guatemala*)

Lycaste aromatica
(*Mexico*)

Lycaste virginalis alba
"White nun orchid"

Masdevallia x measuresiana
(*tovarensis x amabilis*)

Masdevallia veitchiana
(*Perú*)

Masdevallia infracta
(*Brazil, Perú*)

ORCHIDACEAE

Miltonia 'Liberté'
a "Pansy orchid"

Miltonia vexillaria 'Volunteer'
(*Ecuador*)

Miltonia 'Storm'
(*Mokadem x Piccadilly*)

Masdevallia chimaera
(*Colombia*)

Maxillaria tenuifolia
(*Mexico*)

Meiracyllium trinasutum
(*Mexico*)

x Miltonidium 'Aristocrat'
(*Miltonia x Oncidium*)

x Miltonidium 'Manhasset'
Don Richardson, New York

Mormodes lineatum
(*Guatemala*)

Disa uniflora
(*South Africa*)

Menadenium labiosum (rostratum)
(*Venezuela, Guyana, Brazil*)

Maxillaria setigera (callichroma)
(*Colombia, Venezuela*)

ORCHIDACEAE

Mendoncella grandiflora
flamboyant epiphyte from Mexico to Panama

Masdevallia militaris
from the cloud forests of the Andes of Colombia

Maxillaria picta
with heavy-textured, fragrant flowers, from Brazil

Maxillaria punctata
small spring-blooming epiphyte from Brazil

Miltonia candida
epiphyte from Brazil with beautiful, waxy, fragrant flowers

Miltonia candida 'Grandiflora'
with showy, giant 12 cm rust-brown flowers

ORCHIDACEAE

Oncidium uniflorum
charming small epiphyte with waxy flowers, from Brazil

Oncidium flexuosum
"Dancing doll orchid" from Brazil and Paraguay

Neocogniauxia monophylla (Laelia)
brightly-hued, exquisite small epiphyte from Jamaica

x Odontioda charlesworthii
long-lasting hybrid of Odontoglossum x Cochlioda

Miltonia candida 'Shelter Rock'
in collection Greentree estate, Manhasset, New York

Oncidium globuliferum
a tree-climbing species in Costa Rica and to Perú

ORCHIDACEAE

Odontoglossum 'Elise'
(*triumphans x Ascaria*)

Odontoglossum crispum
(*Colombia*)

Neofinetia falcata (Angraecum)
(*Japan, Korea*)

Luisia teretifolia
"Bee orchid" (*Monsoon Asia*)

Haemaria discolor dawsoniana
"Jewel orchid" (*Malaya*)

x Odontioda 'Astargia'
(*Odontoglossum x Cochlioda*)

Neobenthamia gracilis
(*E. Africa: Zanzibar*)

x Odontocidium 'Surprise'
(*Odontoglossum x Oncidium*)

Odontoglossum 'Alispum'
(*Alorcus x crispum*)

Odontoglossum bictoniense
(*Guatemala*)

Odontoglossum cervantesii roseum
(*So. Mexico, Guatemala*)

Odontoglossum insleyi
(*Mexico*)

ORCHIDACEAE

742

Odontoglossum laeve reichenheimii (*Mexico*)

Odontoglossum rossii (*Mexico, Guatemala*)

Oncidium splendidum (*Guatemala, Honduras, Panama*)

Oncidium sarcodes (*Brazil*)

Oncidium marshallianum (*Brazil*)

Oncidium ornithorhynchum (*Mexico to Salvador*)

Ornithocephalus bicornis "Mealybug orchid" (*Panama*)

Oncidium pardinum (*Ecuador*)

Oncidium excavatum (*Perú, Ecuador*)

Oncidium flexuosum "Dancing doll orchid" (*Brazil*)

Oncidium varicosum (*Brazil*)

Oncidium sphacelatum "Golden shower" (*C. America*)

ORCHIDACEAE

Odontoglossum pulchellum
"Lily-of-the-valley orchid"

Oncidium papilio
"Butterfly orchid" (*Trinidad, Perú*)

Oncidium kramerianum
(*Ecuador, Colombia*)

Odontoglossum krameri
(*Costa Rica*)

Odontoglossum citrosmum
(*Mexico*)

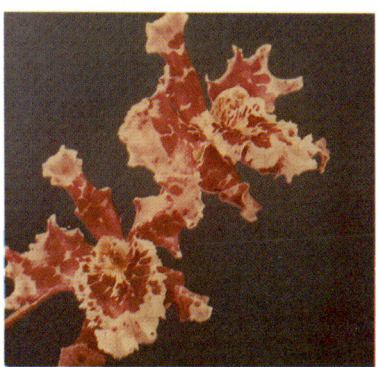
Odontoglossum 'Pedrito'
(*Aicard x Mrs. Sanders*)

Oncidium crispum
(*Brazil*)

Oncidium cheirophorum
(*Colombia*)

Oncidium lanceanum
"Leopard orchid" (*Guyana*)

Oncidium alatum (*in hort.*)
huge species, in Roehrs collection

Oncidium concolor
(*Brazil*)

Oncidium forbesii
(*Brazil*)

ORCHIDACEAE

744

Oncidium wentworthianum
(*Guatemala*)

Ophrys speculum
"Mirror of Venus" (*Greece*)

Oncidium reflexum
(*Mexico, Guatemala*)

Ophrys fusca
a *"Bee orchid"* (*So. Europe*)

Ophrys fuciflora
"Late spider orchid" (*Spain*)

Ophrys lutea minor
on *Delos, Greece*

Ophrys lutea (*Sicily*)
"Insect orchid"

Orchis papilionacea
"Pink butterfly orchid" (*Greece*)

Orchis latifolia
"Marsh orchid" (*Europe*)

ORCHIDACEAE

Megaclinium purpureorhachis
the sinister, fantastic "Cobra orchid" of Zaïre, W. Africa

Odontoglossum grande, *"Tiger orchid"*
with spectacular tigered flowers (Mexico and Guatemala)

Haemaria discolor dawsoniana
"Gold-lace orchid", one of the "Jewel orchids", in Burma

Oncidium grandiflorum, *a magnificent species,*
along the road from Quito to the Pacific, in Ecuador

ORCHIDACEAE

Paphiopedilum x maudiae
(*callosum x lawrenceanum*)

Paphiopedilum 'Olivia'
(*niveum x tonsum*)

Paphiopedilum x harrisianum
'C.S. Ball'

Paphiopedilum praestans
(*New Guinea*)

Phragmipedium caudatum
"Mandarin orchid" (*Ecuador*)

Polystachya stricta
(*Kenya, Uganda, Zaïre*)

Rodriguezia speciosa
(*Perú*)

Pterostylis baptistii
(*Queensland to Victoria*)

Pterostylis banksii
"Hooded orchid" (*New Zealand*)

ORCHIDACEAE

Paphinia cristata, *small epiphyte with spectacular blossoms (Colombia to the Guyanas)*

Paphiopedilum dayanum *known in hort. as a "Cypripedium", from N.E. Borneo*

Paphiopedilum sukhakulii *an unusual "Lady-slipper" flower from Thailand*

Pescatorea cerina *from Costa Rica and Panama epiphyte with exquisite waxy, heavily fragrant flowers*

Peristeria pendula, *from Panama and Guyana with cluster of hooded, fragrant flowers of unusual color*

Phragmipedium philippinense (Selenipedium) *a long-tailed "Mandarin orchid" from the Philippines*

ORCHIDACEAE

748

Paphiopedilum 'Albion'
(*Astarte x niveum*)

Paphiopedilum x maudiae
a charming, popular hybrid

Paphiopedilum exul
(*Thailand*)

Paphiopedilum villosum
(*So. Burma*)

Paphiopedilum niveum
(*Malaya*)

Paphiopedilum rothschildianum
(*Papua, Borneo*)

Phragmipedium grande
"*Spiralled lady-slipper*"

Phragmipedium longifolium
(*Costa Rica, Panama, Colombia*)

Phragmipedium vittatum
"*Selenipedium*" (*Brazil*)

Phaius tankervilliae (wallichii)
"*Nun's orchid*" (*Monsoon Asia*)

Pleione forrestii
(*China: W. Yunnan*)

Maxillaria densa (Ornithidium)
(*Mexico to Honduras*)

ORCHIDACEAE

Paphiopedilum lawrenceanum (*North Borneo*)

Paphiopedilum insigne "Lady slipper" (*Nepal*)

Paphiopedilum x maudiae (*callosum x lawrenceanum*)

Paphiopedilum insigne 'Sanderae' (*India: Assam*)

Paphiopedilum 'Camarin' (*Harefield Hall x spiceriana*)

Paphiopedilum x aureum 'Oedipe' (*spicerianum x nitens*)

Paphiopedilum x harrisonianum 'C.S. Ball'

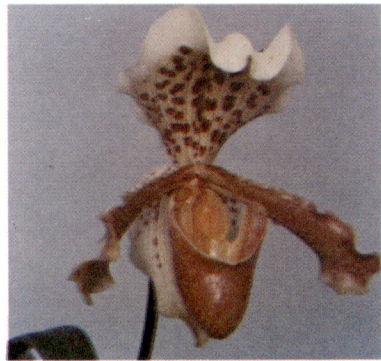

Paphiopedilum insigne 'Harefield Hall'

Paphiopedilum callosum (*Thailand, Vietnam*)

Paphiopedilum fairrieanum (*Himalayas: Bhutan*)

Paphiopedilum bellatulum (*Burma, Thailand*)

Paphiopedilum lowii (*N.E. Borneo*)

ORCHIDACEAE

750

Phalaenopsis lueddemaniana
(*Philippines*)

Phalaenopsis amboinensis
(*Moluccas*)

Phalaenopsis amabilis
"Moth orchid" (*S.E. Asia*)

Phalaenopsis 'Alice Bowen'
unusual rosy-red

Phalaenopsis x rothschildiana
(*schilleriana x amabilis*)

Phalaenopsis 'Hellé'
(*Marmouset x Adonis*)

Phalaenopsis schilleriana
"Rosy moth orchid" (*Philippines*)

Rodriguezia secunda
(*Colombia to Guyana*)

Rhynchostylis coelestis
"Fox-tail orchid" (*Thailand*)

Epigeneium lyonii (*Philippines*)
(*Sarcopodium acuminatum*)

Porphyrostachys pilifera
(*Ecuador, Perú*)

Pleurothallis sonderiana
(*Trop. America*)

ORCHIDACEAE

Phalaenopsis amabilis, *"Moth orchid"*
at Sun-Moon Lake, in the central Massif of Taiwan

Renanthera imshootiana, *vandaceous climber
with sprays of long-lasting scarlet flowers, in Burma*

Vanilla planifolia 'Marginata'
variegated tropical vanilla vine, on treefern fiber

Pleurothallis cardiothallis, *from Nicaragua,
epiphyte at Jardin Botanico, Mexico City*

ORCHIDACEAE

Phalaenopsis 'Doris'
(*Elizabethae* x *Katherine Siegwart*) *a large-flowered hybrid*

Phalaenopsis schilleriana
"Pink moth orchid" from the Philippines

left: Pterygodium catholicum (*South Africa*)
right: Satyrium odorum (*South Africa: Cape*)

Phymatidium tillandsioides
a grass-like miniature 6 cm high, from Brazil

Pleurothallis quadrifida (*W. Indies to Mexico*)
arching sprays of nodding, translucent little flowers

Rodriguezia secunda, *the "Coral orchid"*
floriferous epiphyte from Panama, Colombia and Guayana

ORCHIDACEAE

753

Taeniophyllum fasciola, *leafless epiphyte, resembles a spider web, on tree trunk in Fiji rainforest*

Vanilla planifolia (fragrans) (*Mexico*) *flowers are hand-pollinated to produce vanilla beans*

Spathoglottis plicata *free-blooming Malaysian terrestrial, near Saigon, Vietnam*

Sophronitis coccinea (grandiflora) (*Brazil*) *collected in rainforest of the Serra do Mar, Sao Paulo*

ORCHIDACEAE

Spathoglottis plicata pallida
in Mahé, Seychelles

Spathoglottis plicata
Foster Gardens, Honolulu

Spathoglottis vieillardii
in New Caledonia habitat

Stenoglottis longifolia
(So. Africa: Natal)

Thunia alba
(India, Burma, Thailand)

Spathoglottis papuana
(Papua - New Guinea)

Spiranthes nudicaulis
"Ladies-tresses"

Warrea costaricensis
(Costa Rica, Panama)

Vanda 'Sunset'
in Foster Gardens, Honolulu

ORCHIDACEAE

Vanda coerulea, *the "Blue orchid"*
long-time favorite from the Eastern Himalayan valleys

Stanhopea hasseloviana, *a "Horned orchid",*
pendulous flowers from the rainforest of Amazonian Perú

Vanda 'Miss Agnes Joaquim'
the popular "Corsage orchid", flourishing in Hawaii

x Vandaenopsis 'Frank Atherton'
bigeneric hybrid of Phalaenopsis x Vanda, in Hawaii

Vanda tricolor, *a grand epiphyte*
in the mountain forest of Tjibodas, West Java

Vanda tricolor suavis
collected on the slopes of the volcano Gedeh, E. Bali

ORCHIDACEAE

756

Vanda x rothschildiana
(*coerulea x sanderiana*)

Vanda tricolor
(*Indonesia*)

Vanda tricolor suavis
(*Java, Bali*)

Vanda parishii marriotiana
(*South Burma*)

Vanda x rothschildiana
(*coerulea x sanderiana*)

Vanda sanderiana
(*Philippines: Mindanao*)

Vanda 'Miss Agnes Joaquim'
Hawaiian "Corsage orchid"

Thelymitra ixioides (*Australia*)
"Women's cap orchid"

x Vandachnis 'Premier'
a "Spider orchid"

Zygopetalum mackayi
(*So. Brazil: Sao Paulo*)

Vanda coerulea
the "Blue orchid" (*Assam*)

Vanda 'Onomea'
(*rothschildiana x sanderiana*)

ORCHIDACEAE

Vanda 'Miss Agnes Joaquim'
"Corsage orchids", growing to perfection on Barbados

Vanda 'Josephine' *in Singapore*
(*teres aurorea x hookeriana alba*)

Vanda tricolor, *with fragrant blooms collected during June, in the mountains of West Java*

Vanda teres, *from upper Burma trained to treefern sections in Hawaii*

ORCHIDACEAE

Restrepia guttulata, *miniature epiphyte from the Andes of Venezuela to Ecuador*

Vandopsis lissochiloides (*Philippines, Bali, Moluccas*)

Vandopsis gigantea (*Burma, Thailand*) *with flowers lasting several months*

Warrea warreana (Aganisia tricolor) *large terrestrial from Colombia to the Guianas*

Cochleanthes discolor (Warscewiczella) *epiphyte from Cuba, Costa Rica and Panama*

Spathoglottis plicata var. pallida *in Mahé Botanic Garden, Seychelles*

Telipogon species, *a delicate miniature from mountain cloud forests in Colombia*

Rodriguezia strobelii *small epiphyte from the Cordillèras of Ecuador*

ORCHIDACEAE

759

Stanhopea tigrina
"El Toro" (*Mexico*)

Stanhopea graveolens straminea
"Horned orchid" (*Brazil*)

Schomburgkia undulata
(*Trinidad to Colombia*)

Sobralia macrantha
(*Mexico to Costa Rica*)

Sobralia leucoxantha
(*Costa Rica*)

Sophronitis coccinea
(*So. Brazil*)

Trichopilia suavis
(*Costa Rica*)

Trichocentrum albo-purpureum
(*No. Brazil*)

Trichocentrum tigrinum
(*Costa Rica, Ecuador*)

Vanilla pompona (*flowers*)
(*Mexico to Panama*)

Vanilla fragrans 'Marginata'
"Variegated vanilla"

Renanthera monachica
"Fire orchid" (*Philippines*)

ORCHIDACEAE

760

Restrepia antennifera
(*Andes of Colombia*)

Renanthera monachica
"*Fire orchid*" (*Burma, Philippines*)

Scuticaria mooreana
(*Amazonian Perú*)

Trichocentrum albo-purpureum
(*No. Brazil*)

Rodriguezia decora
(*Brazil*)

Sigmatostalix hymenantha
(*Costa Rica, Panama*)

Sobralia decora
"*Reed orchid*" (*Mexico*)

Sobralia xantholeuca
(*Central America*)

Scuticaria steelei
(*Brazil to Venezuela*)

ORCHIDACEAE

Vanda teres
(*N.E. India, Upper Burma*)

Vanda teres 'Alba'
white variety in Singapore

Vanda lamellata
(*Philippines*)

x Vandacostylis 'Dawn'
(*Rhynchostylis x Vanda*), *a bigeneric Hawaiian hybrid*

Holcoglossum quasipinifolium
rare miniature epiphyte from Taiwan

Vanilla ramosa
(*Guinea to Nigeria*)

Vanilla pompona *with fruit*
(*Mexico to Panama*)

Vanilla fragrans 'Variegata'
Marcel Lecoufle, France

NAPOLEONACEAE, OXALIDACEAE

Oxalis deppei
"Lucky clover" (*So. Mexico*)

Biophytum zenkeri
"Sensitive life plant" (*Zaïre*)

Napoleona heudelotii
"Napoleon's hat" (*W. Africa*)

Oxalis purpurea (variabilis)
"Grand Duchess oxalis"

Oxalis vulcanicola
(*siliquosa hort.*) (*Costa Rica*)

Oxalis braziliensis
(*Brazil*)

Oxalis martiana
'Aureo-reticulata', *"Gold-net"*

Oxalis hedysaroides rubra
(*Colombia to Ecuador*)

Oxalis hedysaroides rubra
"Fire fern" in flower

Oxalis hirta
(*So. Africa*)

Oxalis lasiandra
(*Mexico*)

Oxalis adenophylla
(*Chile*)

OXALIDACEAE

Averrhoa carambola
"Carambola tree" (India)

Oxalis hedysaroides rubra
"Fire fern" (Colombia)

Oxalis gigantea
(So. America)

Oxalis hirta
(South Africa)

Oxalis deppei
(So. Mexico)

Oxalis purpurea
"Grand Duchess" (So. Africa)

Oxalis deppei
"Good-luck plant"

Oxalis succulentum
(Chile)

Oxalis pes-caprae
"Bermuda buttercup"

PALMAE

Acrocomia intumescens (*Fairchild Garden*)
tall feather palm from Brazil

Acrocomia totai (*Fairchild Garden*)
"Gru-Gru" palm from Bolivia, Paraguay, No. Argentina

Archontophoenix alexandrae, *"King palm"*
in habitat on Mt. Warming, New South Wales

on left: Aiphanes caryotaefolia (*Colombia to Ecuador*)
"Spine palm" in Botanic Gardens Bogor, Indonesia

PALMAE

Archontophoenix cunninghamiana (*E. Australia*)
(*Seaforthia elegans in Calif. hort.*), *"Piccabeen palm"*

Arecastrum romanzoffianum (*W. Brazil*)
(*Cocos plumosa in hort.*), *"Queen palm"*

Areca catechu, *"Betelnut palms"*
in Kikiepa village, Finisterre Mts., N.E. New Guinea

Arecastrum romanzoffianum
in habitat, overlooking Iguassu Falls, Paraná, Brazil

PALMAE

Arecastrum romanzoffianum
(*Cocos plumosa*)

Areca langloisiana
(*Celebes, Indonesia*)

Archontophoenix cunninghamiana
(*Seaforthia elegans in Calif.*)

Bactris gasipaes
"Peach palm" (*Surinam*)

Bactris vulgaris (*Brazil*)
"Spiny-club" in Rio Bot. Garden

Borassus flabellifer
"Palmyra palm" (*India*)

Butia capitata
"Jelly palm" from Uruguay

Butia bonnetii
"Acorn butia" (*So. Brazil*)

Butia yatay
"Yatai palm" (*Argentina*)

PALMAE

Barbosa pseudococos
slender feather palms in Campinas, Southern Brazil

Arecastrum romanzoffianum, *the "Queen palm"*
planted along Wilshire Boulevard, Hollywood, California

Bismarckia nobilis (*Madagascar*)
majestic fan palm, at Fairchild Tropical Garden, Miami

Borassus aethiopum, *"African fan palm"*
with swollen trunk, in habitat at Kaptagat, Uganda

PALMAE

Arecastrum romanzoffianum
as a street tree in subtropical Kwangchow, China

Arenga pinnata (*Malaya*)
"Sugar palm", at Foster Botanic Garden, Honolulu

Orbygnia cohune (*Honduras*)
also known as *Attalea*, with erect fronds, in C. America

Attalea amygdalina, *"Cohune palm"*
from Brazil: short trunk and gigantic leaves (*Fairchild G.*)

PALMAE

Cocos nucifera, *"Coconut palm"*
with plentiful harvest of fruit, in Guayana

Cocos nucifera, *"Coconut palm"*
trunk used as a bridge, near Makassar, Celebes

Chamaerops humilis, *the "Dwarf fan palm"*
in its shrubby form, on the desert near Rabat. Morocco

Borassus flabellifer, *"Palmyra palms"*
along the Coromandel Coast, south of Madras, India

Astrocaryum jauari, *feather palms*
in habitat along the broad Amazon, near Manaos, Brazil

Corypha elata, *fan palms*
destined to flower only once; on Timor, Indonesia

PALMAE

Calamus ciliaris
"Dainty rattan" (Sumatra)

Calamus dealbatus
"Brazil rattan"

Caryota mitis
"Clustered fishtail" (Malaya)

Chamaedorea elegans
"Parlor palm" (Mexico)

Chamaedorea metallica (tenella)
"Miniature fishtail" (Mexico)

Chamaerops humilis
in Constanta, Romania

Chamaedorea stolonifera
"Climbing fishtail" (Guatemala)

Chamaedorea seifrizii
"Reed palm" (Mexico: Yucatán)

Chrysalidocarpus cabadae
"Cabada palm" (Madagascar)

PALMAE

Borassus aethiopum, *"African fan palm"* in the Hausa region near Ibadan, Nigeria

Butia capitata (*Brazil, Argentina*) *"Jelly palm"* in Buen Retiro Park, Madrid, Spain

Borassus flabellifer (*India to Malaya*) *"Toddy palm"* in the arid region near Cayar, Sénégal

Borassus flabellifer, *"Palmyra palms"* in South India are tapped for toddy; a Tamil woman carrying dung for fuel

Calamus dealbatus (*Brazil*)
vicious "Wait-awhile vine", Jardim Bot., Rio de Janeiro

Calamus scipionum (*Malaysia*)
thicket of "Malacca rattan cane", Bot. Gard. Singapore

Chamaerops humilis, *arborescent form*
"European fan-palm" in the Lateran area of Rome, Italy

Calamus rotang, *the "Rattan palm"*
source of rattan cane; rainforest climbers in New Guinea

PALMAE

Caryota mitis (*Burma to Indonesia*)
"Cluster fishtail" in fiberglass bowl, downtown Hongkong

Caryota urens, *"Fishtail palm"*
in the tropical countryside near Kandy, Sri Lanka

Chrysalidocarpus lutescens (Areca), *"Butterfly palm"*
in colorful jar, along the Paseo of Acapulco, Mexico

Chamaedorea erumpens (*Honduras*)
excellent indoor *"Bamboo palm"*, favorite of decorators

PALMAE

Howeia (Kentia) forsteriana
"Paradise palm" (Australia)

Chrysalidocarpus lutescens
"Areca palm" in hort.

Archontophoenix cunninghamiana
Seaforthia elegans in Calif.

Coccothrinax argentata
"Florida silver palm"

Euterpe edulis (*Brazil*)
"Assai palm" (young plant)

Coccothrinax fragrans
"Cuban silver palm"

Chambeyronia macrocarpa
with red fronds (New Caledonia)

Desmoncus horridus
"Wait-awhile palm" in Guayana

Ceroxylon andicola (*Colombia*)
"Wax palm", with Dr. Darian, Calif.

PALMAE

Cocos nucifera, *"Coconut tree"*,
one seed sprouted 7 trunks, on Rarotonga, South Pacific

Cocos nucifera, *"Coconut palms"*
swaying in the tradewinds on Nuku Hiva, French Polynesia

Cocos nucifera 'Golden Malay'
resistent to lethal yellowing, at U. of Florida, Davie

Cocos nucifera 'Dwarf Samoan'
a *"Midget coconut"*, in the South Pacific Kingdom of Tonga

PALMAE

Copernicia ekmanii (*Haiti*)
a handsome fan palm, at Fairchild Gardens, Miami

Copernicia macroglossa (torreana)
"Cuban petticoat palm" with underskirts of dead leaves

Copernicia baileyana
stately fan palm with swollen trunk, from Cuba

Copernicia rigida (*Cuba*)
forbidding palm with foliage like a wrapped package

PALMAE

Howeia (Kentia) belmoreana (*Lord Howe Island*)
"Sentry palm", on Italian oceanliner 'Raffaelo'

Chamaerops humilis, *the "European fan palm"
in the gardens of the Mormon Temple, Los Angeles*

Livistona chinensis, *"Chinese fan palm"
cristate, branched old curiosity at Icod, Tenerife, Spain*

Hyphaene thebaica, *"Dhoum palm",
a naturally forking palm, in the Rift valley of Kenya*

Licuala spinosa, *fan palm
with corrugated leaves, near Bien Hoa, Vietnam*

left: Chamaedorea elegans 'Bella' (Neanthe bella)
right: Chamaedorea elegans, *"Parlor palm" from Mexico*

PALMAE

Corypha elata (*Bengal to Indonesia*)
monocarpic palm, flowering once in life, then dying

Corypha umbraculifera, *"Talipot palm"*,
old enough to flower and die, in Trinidad Botanic Garden

Copernicia sueroana *from Cuba*
without visible trunk, in appearance like a fountain

Elaeis guineensis, *"African oil palm"*
important source of palmoil, near Lagos, Nigeria

PALMAE

Cyrtostachys renda, *"Sealing wax palm"*,
colorful crown shaft, Summit Gardens, Panama

Cyrtostachys lakka (*Malaysia*)
with scarlet leaf stalks, at Andromeda Gardens, Barbados

Dictyosperma aureum (*Réunion*)
"Hurricane palm" at Fairchild Tropical Garden, Miami

Dictyosperma album, *"Princess palm"*
on the tropical island of Mauritius, Indian Ocean

PALMAE

Brahea (Erythea) edulis
"Guadalupe palm" (*Mexico*)

Brahea (Erythea) armata
"Mexican blue palm" (*Baja Calif.*)

Brahea dulcis
"Rock palm" near Taxco

Latania loddigesii (*Mauritius*)
"Blue latan palm"

Kentiopsis olivaeformis
(*New Caledonia*)

Jubaea chilensis
"Chilean wine palm"

Howeia forsteriana
"Kentia palm", in Bruges, Belgium

Livistona chinensis
"Latania borbonica" in hort.

Latania verschaffeltii
"Yellow latan" (*Mascarene Isl.*)

Euterpe oleracea, *"Assai palm"* of Brazil
with slender single trunk and graceful fronds, Campinas

Euterpe edulis, *"Clustering assai palm"*
slender trunks symbolizing tropic beauty, Rio de Janeiro

Euterpe globosa, *"Palma de Sierra"*
in the Caribbean rainforest of Sierra Luquillo, Puerto Rico

Brahea brandegeei, (*Baja Calif., Mexico*)
"Daughter-of-the-West" in Los Angeles Arboretum

PALMAE

Hyphaene thebaica, *"Dhoum palm"*
luxuriantly branched in rainforest, Tanzania

Hyphaene ventricosa, *"Gingerbread palm"*
with bulging trunk, along the Zambezi River, in Rhodesia

Kentiopsis macrocarpa
lofty "Lucian palm", on New Caledonia, (South Pacific)

Jubaea chilensis (spectabilis), *"Syrup palm",*
huge trunk yields palm honey; Huntington Bot. G., Calif.

PALMAE

Howeia (Kentia) forsteriana, *"Paradise palm"*
seed developing in stages over 4 years, New Zealand

Mascarena lagenicaulis (*Hyophorbe*)
"Bottle palm" with grotesque fat trunk, from Mauritius

Hyphaene ventricosa
slender fan palm along the Zambezi River, Rhodesia

Hyphaene thebaica
bearing fruit having the taste of gingerbread, in Kenya

PALMAE

Licuala rumphii
(*Moluccas*)

Licuala peltata
(*India: Bengal*)

Licuala grandis
"Ruffled fan palm" on Tahiti

Linospadix monostachya
"Walking-stick palm" [N.So. Wales]

Livistona australis
"Australian fountain palm"

Livistona decipiens (*Australia*)
Los Angeles Arboretum

Pigafetta filaris
(*Celebes, Moluccas*)

Lodoicea maldivica, *female tree*
"Double coconut", in Mahé, Seychelles

Lodoicea maldivica
fruit of "Coco de mer"

PALMAE

Livistona australis
"Fountain palm" in Eucalyptus forest, North Queensland

Latania lontaroides (commersonii) (*Mauritius*)
"Red latan", in hort. as Latania borbonica or L. rubra

Licuala muelleri (*Australia, New Guinea*)
with handsome palmate fronds forming near perfect circle

Licuala grandis *in porcelain jars,*
at Sultan's Palace in Jogyakarta, Central Java

PALMAE

Livistona chinensis, *"Chinese fan palm"*
with younger stage fronds, in Zoo, Kwangchow, China

Livistona chinensis (*Central China*)
"Fountain palm" in maturity stage, Bot. Garden Sydney

Livistona saribus, a *"Sugar palm"*
in the ruins of ancient Khmer Temple of Angkor, Cambodia

Livistona mariae (*Central Australia*),
a blue-green fan palm, in Fairchild Gardens, Miami

PALMAE

Phoenix dactylifera, *"Date palms"*
fruiting trees along the river Nile, near Cairo, Egypt

Phoenix dactylifera 'Deglet Noor', *female trees bearing high quality fruit, Coachella valley, California*

Pigafetta filaris, *from Celebes at Dr. M. Darian's indoor Palmarium, Vista, California*

Phoenix canariensis, *"Canary Is. date palms"*
tubbed trees in the orangerie of Louis XIV, Versailles

Phoenix dactylifera, *"Date palms"*
in native habitat, near Guimar, Tenerife, Spain

Nypa fruticans, *"Nipa palms"*
in tidal waters, with water buffalo, near Bangkok, Thailand

PALMAE

Maximiliana maripa (regia) (*N.E. So. America*)
"Cucurite palm" with erect fronds, Bot. G. Rio de Janeiro

Metroxylon warburgii
a "Sago palm" in Papeari Botanic Garden, Tahiti

Metroxylon vitiense (*Fiji*)
feather palm with giant fronds, Foster Garden, Honolulu

Metroxylon sagus (*Malaya to Philippines*)
commercial "Sago palm", the pith of trunk yielding sago

PALMAE

Licuala muelleri, *with fan-shaped leaves,
in the rainforest of Oomsis, near Lae, New Guinea*

Mascarena revaughanii *in flower* (Mauritius)
"Bottle palm" with swollen base and twisted fronds

Lodoicea maldivica, *the "Double coconut"
female tree with fruit, on the Indian Ocean island of Praslin*

Lodoicea maldivica
also known as "Coco de Mer", in Mahé Bot. G., Seychelles

PALMAE

Neodypsis decaryi, *from southern Madagascar with fronds in three ranks, Fairchild Trop. Garden, Miami*

Microcoelum insigne (*Brazil*) *graceful small feather palm; Sydney Bot. G., Australia*

Nypa fruticans, *with flowers, in the brackish water of the Klongs, Bangkok, Thailand*

Aiphanes (Martinezia) caryotaefolia (*N.W. So. America*) *"Spine palm" in Botanic Garden Rio de Janeiro*

PALMAE

Orbygnia (Attalea) cohune (*Mexico to Costa Rica*)
"Cohune palm" fronds, 10 m long, in Bot. G. Singapore

Pelagodoxa henryana, *"Vahana palm"*
with undivided leaves, in the Marquesas, French Polynesia

Neodypsis lastelliana *from Madagascar*
with red crownshaft, at Dr. Darian Palmarium, Vista, Calif.

Pigafetta filaris (*Indonesia*)
beautiful, densely hairy palm, favorite of David Fairchild

Oncosperma tigillarium (*Monsoon Asia*)
the tropical "Nibong palm" in Bot. G. Peradeniya, Ceylon

Oncosperma filamentosum (bot. tigillarum)
graceful palm with pendant leaflets, in Singapore Bot. G.

Paurotis wrightii, (*So. Florida, W. Indies, C. America*)
"Everglades palm", at St. Petersburg, Florida

Oncosperma horridum (*Malaya to Philippines*)
the inland "Baya palm" with slender trunks, in Singapore

PALMAE

Phoenix canariensis, *the "Canary date palm"* in the crisp Mediterranean climate of Perth, W. Australia

Phoenix dactylifera 'Deglet Noor', *"Date palms each bearing 100 kg of fruit, in Coachella Valley, Calif.*

Phoenix reclinata, *the Sénégal date"*, spread across Central Africa, at Entebbe, Uganda

Phoenix roebelenii, *ornamental "Pigmy date palm"* at the Star ferry water front in Victoria, Hongkong

PALMAE

794

Phoenix canariensis
young tree in flower, Vista, Calif.

Phoenix loureirii (*India*)
cluster palm in Singapore

Phoenix rupicola
in Burma

Phoenix roebelenii
in stoneware jar, Kobe, Japan

Phoenix sylvestris
in Uttar Pradesh, India

Phoenix taiwaniana
in Taipei, Taiwan

Pritchardia hillebrandii
"Loulu-lelo" palm (*Molokai*)

Pritchardia arecina
(*Hawaii: Maui*)

Pritchardia beccariana
largest species (*Hawaii*)

PALMAE

Rhapis excelsa, *"Lady palm"*
on Sun-Moon lake, near Taichung, Taiwan

Nypa fruticans, *the "Nipa palm"*
showing inflorescence, along tidal waters, in Thailand

Serenoa repens, *"Scrub palmetto"*
in pine forest near Daytona Beach, Florida

Sabal palmetto, *the "Cabbage palm"*
at the Spanish Castillo de San Marcos, St. Augustine, Fla.

Trachycarpus fortunei, *"Japanese windmill palm"*,
along the mild-climate shore of Lago Maggiore, Switz.

Washingtonia filifera, *the "Pettycoat palm"*
in habitat, Palm Canyon near Palm Springs, California

PALMAE

Pinanga patula (*Sumatra, Borneo*)
dainty, very tropical feather palm in Rio de Janeiro

Pritchardia remota, *"Loulu palm"*
from Bird Is. N.E. of Kauai; in Foster Garden, Honolulu

Ptychosperma macarthurii (*New Guinea*)
"Hurricane palm" with jagged leaftips, Balboa, Panama

Ptychosperma elegans (*Queensland*)
"Solitaire palm" at shopping center in New Jersey

PALMAE

Phoenix reclinata (*Trop. Africa*)
"Sénégal date palm" at California Mission Santa Barbara

Phoenix dactylifera, *"Date palms"*
near the ancient city of Jericho, in the Jordan Valley

Phoenix canariensis *in decorative tubs*
at Bot. Garden orangerie, on Capitol Hill, Washington

Pritchardia pacifica (Eupritchardia),
the *"Fiji fan palm"*, in Recife, Pernambuco, Brazil

Phoenix sylvestris (*India*)
"E. Indian wine Palm", in Maharashtra state, near Bombay

Phoenix reclinata, *"Sénégal date"*
in mist forest, on the brink of Victoria Falls, Rhodesia

PALMAE

Rhapis excelsa (*China*)
"Lady palm" at Roehrs

Rhapis excelsa 'Variegata'
dwarf Japanese cultivar

Rhapis humilis (*China*)
"Slender lady palm"

Sabal minor (*Carolina to Texas*)
"Dwarf palmetto"

Pritchardia thurstonii
(*Fiji*)

Pritchardia macrocarpa
"Loulu palm" (*Polynesia*)

Syagrus coronata
"Licuri palm" (*Brazil*)

Syagrus weddelliana (Microcoelum)
"Baby cocos palm"

Syagrus vagans (*Brazil*)
Fairchild Trop. Garden, Miami

PALMAE

Raphia farinifera (ruffia)
in the Usambara Mountains, near Amani, Tanzania

Rhapis excelsa (flabelliformis)
at entrance to Japanese Inn, Gion district, Kyoto

Raphia pedunculata (*Madagascar*)
giant "Raffia palm" without trunk; Bot. Garden Heidelberg

Rhapis humilis, *"Reed palm"*
in People's Botanic Garden, Kwangchow, China

PALMAE

Roystonea venezuelana, *"Venezuelan Royal palm"*
at Capitol in Caracas, near Simon Bolivar's home

Roystonea regia (Oreodoxa)
"Royal palms" in October rain, Summit Gardens, Panama

Sabal causiarum (*Puerto Rico, Virgin Isl.*)
the massive *"Puerto Rican hat-palm"*

*La India, on Roystonea, honoring Victory of Carobobo
a monument of stone and bronze, in Caracas*

PALMAE

Roystonea regia (Oreodoxa)
"Cuban Royal palms", Peradeniya Bot. Garden, Ceylon

Rhopalostylis sapida, *the "Nikau palm"*
at latitude 42 south, on South Island, New Zealand

Roystonea elata, *the "Florida Royal palm"*,
endemic in Everglades National Park, South Florida

Roystonea oleracea, *majestic "Caribbee"*
or *"South American Royal palm"* in Jardim Botanico, Rio

PALMAE

802

Trachycarpus fortunei, *"Japanese windmill palms"* along the promenade of Lago Maggiore, Switzerland

Rhapis excelsa (flabelliformis), *"Lady palms"* in miniature forms, at World Exposition 1970, Osaka

Areca catechu, *"Betelnut palm"* at a village in the Himalayas of Sikkim

Elaeis guineensis, *"African oil-palm"* in the Niger delta, near Port Harcourt, Nigeria

Cocos nucifera, *"Coconut palm"* fruit sprouting leaves and roots, on Moorea, Polynesia

Cocos nucifera, *large frond* woven into wall siding by this Sinhalese woman in Ceylon

PALMAE

Scheelia osmantha (*Brazil*)
stately feather palms in Jardim Botanico, Rio de Janeiro

Scheelia liebmanii, *from Eastern Mexico with erect fronds 6 m long, at Fairchild Garden, Miami*

Syagrus oleracea (*Brazil*)
small feather palm in Montgomery Palmetum, Florida

Syagrus (Microcoelum) weddelliana, *mature tree; in hort. as Cocos weddelliana; in Petropolis, Brazil*

PALMAE

Trachycarpus fortunei 'Nana'
"Dwarf windmill palm" in Honolulu

Thrinax microcarpa (morrisii)
"Key palm", on Key Largo, Florida

Thrinax ekmanii (morrisii)
(West Indies: Navassa Isl.)

Verschaffeltia splendida
"Stiltroot palm" in Mahé, Seychelles

Diplothemium maritimum (*Brazil*)
on the Restinga south of Rio

Wallichia densiflora
(Eastern Himalayas: Assam)

Trachycarpus fortunei (*Japan*)
in Madrid Bot. Garden, Spain

Phoenix roebelenii
"Pigmy date" in decorator tub

Thrinax radiata (*Fla. to Honduras*)
"W. Indian thatch palm"

PALMAE

Trachycarpus fortunei, *"Japanese windmill palms"*
are part of the landscape, at mighty Osaka castle, Japan

Trachycarpus wagnerianus (takil in hort.)
from the Western Himalayas; in Vista, California

Verschaffeltia splendida, *"Stiltroot palm"*
from the Seychelles; in Kew Gardens collection, London

Wallichia disticha, *from India and Burma
with Cherie Darian, in Vista, San Diego County*

PALMAE

Veitchia arecina (Adonidia)
fruiting near Vila, New Hebrides, in the Coral Sea

Veitchia joannis (Adonidia)
graceful feather palm on Viti Levu, Fiji, in Melanesia

Veitchia (Adonidia) merrillii (*Philippines*)
"Christmas palm" with scarlet fruit, Ft. Lauderdale, Fla.

Veitchia winin (Adonidia)
tall, slender palm from New Hebrides, at Fairchild G.

PALMAE

Washingtonia robusta (*Baja California, Sonora*)
"Mexican fan palms" at Mormon Temple, Los Angeles

Washingtonia filifera, *the "Petticoat palm"*
in the desert, facing Mt. San Jacinto, Palm Springs, CA

Washingtonia robusta, *"Washingtonia palm"*
young tree, 10 years old, Sunset Ranch, Vista, Calif.

Washingtonia robusta, *"Mexican fan palm"*
with winter protection of rice straw, in Hibiya, Tokyo

PANDANACEAE

Freycinetia arborea (*South Pacific*)
"Climbing screw pine" in Korolevu rainforest, Fiji

Pandanus tectorius, *"Hala screw pine"*
in New Guinea, valued by Melanesians for their fruit

Pandanus sanderi (*Moluccas, Polynesia*)
ornamental screw-pine in Andromeda Bot. G., Barbados

Pandanus sanderi 'Roehrsianus'
highly colored cultivar, in Papeari Bot. Garden, Tahiti

PANDANACEAE

Pandanus odoratissimus, *the "Pandang" or "Breadfruit" of coral islands, in New Caledonia*

Pandanus rockii, *with stilt-roots at Lyons Arboretum, Manoa Valley, Oahu*

Pandanus brosimus, *the "Pandanus palm" source of food to natives of the mountains, in New Guinea*

Pandanus leram, *"Nicobar breadfruit" common in tropical forest on the way to Kandy, Sri Lanka*

PANDANACEAE

Pandanus baptistii (*Melanesia*)
"Blue screw pine"

Pandanus utilis (*Madagascar*)
"Common screwpine"

Pandanus veitchii (*Polynesia*)
"Variegated screw-pine"

Freycinetia funicularis (insignis), *male inflorescence woody climber with showy flowers, from Java, Indonesia*

Freycinetia multiflora (*Philippines*)
"Climbing pandanus"

Freycinetia rigidifolia
(*Thailand, Malaya, Borneo*)

Freycinetia stenophylla
in the Finisterres, New Guinea

Pandanus pristis
(*Madagascar, by J. Bogner*)

PLUMBAGINACEAE, POLYGALACEAE

Polygala x dalmaisiana
"Sweet pea shrub"

Limonium sinuatum (*Mediterranean*)
"Statice" or "Sea-lavender"

Plumbago auriculata (capensis)
"Cape leadwort" (*So. Africa*)

Ceratostigma willmottianum (*W. China, Tibet*)
"Chinese plumbago", at Bot. Garden Sydney

Plumbago indica (coccinea)
"Scarlet leadwort" (*E. Indies*)

Limonium suworowii
"Sea lavender" (*W. Asia*)

Polygala oppositifolia
"Purple broom" (*So. Africa*)

Polygala myrtifolia (*So. Africa*)
"Spring milkwort"

PAPAVERACEAE, PEDALIACEAE, PHYTOLACCACEAE

Eschscholtzia californica
"California poppy"

Papaver orientale
"Oriental poppy" (Iran)

Eschscholtzia californica 'Rubra'
"Red California poppy"

Rivina humilis
"Rouge plant" (W. Indies)

Argemone mexicana
"Prickly poppy" (Mexico)

Hunnemannia fumaraefolia
"Mexican tulip poppy"

Uncarina grandidieri
(Madagascar)

Papaver aurantiacum
"Alpine poppy" (Switzerland)

Papaver nudicaule
"Iceland poppy"

PAPAVERACEAE

Eschscholzia californica, *"California poppy"* covering the mountain sides in Ventura County, Calif.

Papaver orientale, *"Oriental poppy"* at Lake Louise in view of Victoria Glacier, Canada

Eschscholzia mexicana, *"Mexican poppy"* with lovely señorita of Sonora, northern Mexico

Papaver nudicaule, *the "Arctic" or "Iceland poppy"* under pine-tree at an ancient temple in Kyoto, Japan

PASSIFLORACEAE

Adenia subglobosa (*Tanzania*)
monstrous succulent with swollen base, in Nairobi, Kenya

Passiflora coriacea
curious "Bat-leaf" vine, from Mexico to Perú

Passiflora bomareifolia
a spidery passion flower in Venezuela

Passiflora capsularis
with strange bell-shaped flowers, from W. Trop. America

Passiflora vitifolia (*Nicaragua to Perú*)
"Crimson passion flower" at Longwood Gardens, Penna.

Passiflora holosericea
with fragrant flowers, from Central America

PASSIFLORACEAE

Passiflora edulis (*Brazil*)
"Purple granadilla"

Passiflora aurantia (mixta)
(*New Guinea to New Hebrides*)

Passiflora amethystina (*Brazil*)
on St. Héléna

Passiflora 'Imperatrice Eugenie'
(alata x caerulea)

Passiflora maliformis
"Conch apple" in Tonga

Passiflora racemosa
"Red passion flower" (*Brazil*)

Passiflora seemannii (*Trop. America*)
in Waimanalo, Oahu

Passiflora trifasciata (*Perú*)
"Three-banded passion vine"

Passiflora violacea (*Brazil*)
"Violet passion flower"

PASSIFLORACEAE

Passiflora caerulea (*Brazil*)
"Blue passion flower"

Passiflora cinnabarina
(*Australia*)

Passiflora coccinea
"Scarlet passion flower"

Passiflora racemosa (princeps)
"Red passion flower"

Passiflora maculifolia
(*Venezuela*)

Passiflora edulis (*Brazil*)
edible *"Purple granadilla"*

Passiflora coriacea
"Bat leaf" (*Mexico to Perú*)

Passiflora novo-guineensis
at Chimbu, New Guinea

Passiflora vitifolia
"Crimson passion flower"

PASSIFLORACEAE

Passiflora antioquiensis
"Banana passion fruit" from Colombia

Passiflora coccinea
"Red granadilla" in French Guiana

Passiflora caerulea (*Brazil*)
"Blue crown passion flower", a religious symbol

Passiflora x alato-caerulea
(pfordtii) (*alata x coerulea*)

Passiflora rubra (*West Indies*)
by J. Bogner, Munich Botanic Garden, Germany

Passiflora quadrangularis
"Giant granadilla" (*Trop. America*)

PIPERACEAE

Peperomia puteolata (Perú)
"Parallel peperomia"

Peperomia sandersii (Brazil)
"Watermelon peperomia"

Peperomia cuspidilimba (Perú)
Heidelberg Bot. G., Germany

Peperomia rauhii
(Trop. America)

Peperomia caperata
"Emerald ripple" (Brazil)

Peperomia campylotropa (Mexico)
Bot. Garden, Munich

Piper crocatum (Perú)
"Ornamental pepper"

Peperomia sarcophylla
(Colombia, Ecuador)

Peperomia caperata
'Red Ripple'

PIPERACEAE

819

Peperomia obtusifolia 'Variegata'
"Variegated peperomia"

Peperomia arifolia litoralis
(*Brazil: Sao Paulo*)

Peperomia caperata 'Variegata'
"Variegated ripple"

Peperomia sandersii (*Brazil*)
"Watermelon peperomia"

Piper sylvaticum
"Silver cissus" (*Burma*)

Peperomia glabella 'Variegata'
"Variegated wax-privet"

Peperomia galioides
(*Colombia*)

Peperomia verschaffeltii
"Sweetheart peperomia"

Peperomia bicolor (*Ecuador*)
"Silver velvet peperomia"

Peperomia cerea (*Perú*)
('Dr. Goodspeed' hort.)

Piper futokadsura
"Japanese pepper" (*Japan*)

Piper ornatum
"Celebes pepper" (*Sulawesi*)

PIPERACEAE

Peperomia obtusifolia
"Baby rubber plant" (Venezuela)

Peperomia scandens 'Variegata'
"Variegated philodendron-leaf"

Peperomia magnoliifolia
Desert privet" (W. Indies)

Peperomia clusiifolia (W. Indies)
"Red-edged peperomia"

Peperomia polybotrya (Colombia)
"Coin-leaf peperomia"

Peperomia incana
"Felted pepperface" (Brazil)

Peperomia maculosa
"Radiator plant" (Santo Domingo)

Peperomia obtusifolia 'Sensation'
Brooklyn Bot. Garden

Peperomia scandens (Perú)
"Philodendron peperomia"

PIPERACEAE

Peperomia verschaffeltii, *the "Sweetheart peperomia" found near Manaos on the Rio Negro, in Amazonas*

Piper nigrum, *the "Black pepper" of commerce derived from the dried unripe fruit; in Sri Lanka*

Piper betle, *a "Betel-leaf pepper" in India, where the leaves are chewed with Betel-nuts*

Piper crocatum (*Perú*) *at Limberlost Nursery in Cairns, North Queensland*

PIPERACEAE

Piper magnificum (*Perú*)
"Lacquered peppertree"

Macropiper excelsum
psittacorum (*New Zealand*)

Piper auritum (*Mexico*)
Longwood Gardens, Penna.

Piper crocatum
"Peruvian ornamental pepper"

Piper porphyrophyllum
"Velvet cissus" (*Indonesia*)

Peperomia cubensis
"Radiator plant" (*Cuba*)

Peperomia griseo-argentea
(*hederaefolia*) "Ivy peperomia"

Peperomia hoffmannii
(*Costa Rica to Colombia*)

Peperomia margaretifera
"Pepper face" (*Chile*)

PITTOSPORACEAE

Marianthus pictus
"Bell-climber" (*W. Australia*)

Marianthus ringens (*W. Australia*)
"Red bell-climber"

Pittosporum moluccanum
(*Indonesia*)

Pittosporum eugenioides 'Variegatum'
(*New Zealand*)

Pittosporum undulatum
"Sweet pittosporum" (*Australia*)

Pittosporum tobira
"Mock orange" (*Japan*)

Hymenosporum flavum
"Sweetshade" (*Queensland*)

Pittosporum wrightii
(*Seychelles*)

Pittosporum tobira 'Variegata'
"Variegated mock-orange"

PITTOSPORACEAE, POLYGONACEAE, PUNICACEAE

Coccoloba pubescens (grandifolia)
"Moralon" (*Puerto Rico*)

Pittosporum tobira
'Wheeler's Dwarf'

Coccoloba uvifera
"Sea-grape" (*Florida, W. Indies*)

Pittosporum tenuifolium 'Variegatum'
"Kohutu" (*New Zealand*)

Homalocladium platycladum
"Tapeworm plant" (*Solomon Is.*)

Coccoloba uvifera
with edible fruit

Punica granatum florepleno
"Double-flowering pomegranate

Punica granatum 'Nana'
"Dwarf pomegranate" (*Iran*)

Punica granatum 'Legrellei'
in northern Italy

POLEMONIACEAE, POLYGONACEAE, PONTEDERIACEAE

Cantua buxifolia, *"Sacred flower of the Incas*
with Aramayo Indian girl, near La Paz, Bolivia

Antigonon leptopus (*Mexico*)
"Coral vine" clambering over fence, in Timor, Indonesia

Rheum palmatum, *"Chinese rhubarb"*
as an ornamental in Visby, isle of Gotland, Sweden

Polygonum capitatum
"Knot-weed", pretty trailer of the Himalayas

Eichhornia azurea (*Brazil*)
the creeping "Peacock hyacinth", in Brazil

Eichhornia crassipes
the floating "Water hyacinth", in Madagascar

POLEMONIACEAE, POLYGONACEAE, PORTULACACEAE

Cantua buxifolia
"Magic flower" in Cuzoo, Perú

Portulaca grandiflora
"Rose-moss" (Brazil)

Cobaea scandens (Mexico)
"Monastery-bells"

Muehlenbeckia complexa
"Maidenhair vine" (New Zealand)

Phlox drummondii (Texas)
"Dwarf annual phlox"

Portulacaria afra 'Variegata'
"Rainbow-bush" (So. Africa)

Phlox paniculata
"Summer phlox (Eastern U.S.)

Phlox diffusa
in Utah

Phlox drummondii
'Sternenzauber'

POLYGONACEAE, PONTEDERIACEAE

Antigonon leptopus (*Mexico*)
"Love-vine" or "Corallita", in Rio Piedras, Puerto Rico

Coccoloba pubescens (grandifolia)
"Moralon", from the Sierra Luquillo in Puerto Rico

Eichhornia crassipes, *"Water hyacinths"*
for sale as floating aquatics, in Bombay, India

Coccoloba uvifera, *"Sea-grape" or "Platterleaf",*
a tempting fruit to youngsters, in Tahiti

PONTEDERIACEAE

Eichhornia azurea
"Creeping water-hyacinth"

Pontederia lanceolata
"Pickerel-weed" (Southeast U.S.)

Eichhornia crassipes
"Water hyacinth"

Heteranthera reniformis, *"Mud-plantain"*,
submersed aquatic from Pennsylvania to Argentina

Eichhornia crassipes (*Trop. and Subtropic America*)
"Floating water-hyacinth" in goldfish pond

Pontederia cordata
"Pickeral rush" (U.S. to Argentina)

Heteranthera dubia
"Water star-grass" (Mexico, Cuba)

Heteranthera osteniana
(*Argentina to Bolivia*)

PORTULACACEAE, PRIMULACEAE, PSILOTACEAE, PUNICACEAE

Psilotum nudum (triquetrum), *"Whisk-fern"*,
a club-moss, epiphytic on treefern, in Hawaii

Punica granatum (S.E. Europe to Himalayas)
"Pomegranate tree" with luscious fruit, in California

Cyclamen persicum, *"Alpine violet"*,
parent of florists cyclamen, in Karst Mts., Yugoslavia

Talinum guadalupense, *"Flame-flower"*
attractive succulent from arid Guadalupe Island, Mexico

RAFFLESIACEAE, RESEDACEAE, PORTULACACEAE, PRIMULACEAE

Reseda odorata
"Mignonette" (No. Africa, Egypt)

Dodecatheon meadia
"Shooting star" (Manitoba to Texas)

Lysimachia clethroides
"Loose-strife" (China)

Anacampseros lanceolata
"Love-plant", small succulent from South Africa

Rafflesia arnoldii, in Borneo
enormous fleshy parasite with flower 1 m across

Lysimachia nummularia (Europe)
"Moneywort" or "Creeping Jennie"

Portulacaria afra 'Macrophylla'
"Giant elephant bush"

Portulacaria afra
"Elephant bush" (So. Africa)

PORTULACACEAE, PRIMULACEAE

Lysimachia punctata
"Garden loose-strife", perennial from S.E. Europe

Primula x polyantha 'Pacific Giants'
"Polyanthus primrose" for outdoors, or indoor pots

Portulaca grandiflora florepleno 'Jewel'
"Double-flowered moss-rose" in Idaho

Portulaca grandiflora (*Brazil*)
"Sun plant" or "Eleven o'clock" opening to the sun

Primula malacoides, *"Baby primrose"*
or "Fairy primrose" at Roehrs greenhouses, New Jersey

Primula obconica 'Friesdorf Salmon', *"German primrose"*
with salmon flowers; Longwood Gardens, Pennsylvania

PRIMULACEAE, RHAMNACEAE

Paliurus spina-christii
"Jerusalem thorn", in England

Rhamnus alaternus 'Variegata'
young plant (So. Europe)

Colletia cruciata
"Anchor plant" (Brazil)

Emmenosperma alphitonioides
in Taronga, New South Wales

Ziziphus mauritiana
"Indian jujube"

Rhamnus alaternus 'Variegata'
"Italian buckthorn"

Cyclamen pers. 'Rose of Aalsmeer'
in Aalsmeer, Holland

Cyclamen persicum
in Kwangchow, China

Primula obconica 'Grandiflora'
"Florist's primrose"

PRIMULACEAE

Cyclamen persicum 'Bonfire'
elegant florist's "Alpine violet"

Cyclamen pers. 'Rose von Zehlendorf'
salmon pink with dark eye

Cyclamen persicum
'Perle von Zehlendorf'

Cyclamen persicum
'Fimbriatum'

Cyclamen persicum
giganteum 'Rococo'

Cyclamen persicum
'Vogt's Double'

Primula obconica
"German primrose"

Primula x polyantha
"English primrose"

Primula malacoides
'Glory of Riverside'

Cyclamen persicum
'Swiss Dwarf' (*on right*)

left: Cyclamen persicum (species)
right: Cyclamen persicum 'Perle von Zehlendorf'

PROTEACEAE

Grevillea robusta
"Silk oak" (E. Australia)

Banksia grandis (W. Australia)
sawtooth leaves of "Bull banksia"

Grevillea banksii
"Scarlet grevillea" (Queensland)

Grevillea asplenifolia
(New South Wales)

Hakea saligna (E. Australia)
"Willowleaf hakea"

Grevillea thelemanniana
"Hummingbird bush"

Banksia media
(Western Australia)

Grevillea robusta
as decorative foliage plant

Buckinghamia celsissima
"Ivory curl-flower" (Queensland)

PROTEACEAE

Banksia marginata, *the "Silver banksia"*, Cleland National Park, near Adelaide, South Australia

Banksia grandis, *"Bull banksia"* in King's Park Botanic Garden, Perth, Western Australia

Banksia ashbeyi, *with striking spikes*, in habitat near Shark Bay, Northwestern Australia

Banksia media, *"Southern plains banksia"* on the sand plains near Israelite Bay, Western Australia

PROTEACEAE

836

Leucospermum nutans, *"Nodding pincushion"*
with styles curled inward, in S.W. Cape Province

Banksia speciosa, *"Showy banksia"*
with saw-tooth leaves, in Western Australia

Leucospermum catherinae, *the "Catherine wheel"*
with styles curved sideways (So. Africa)

Leucadendron arboreum
on the slopes of Table Mountain, Kirstenbosch, So. Africa

Leucospermum reflexum, *"Rocket pincushion"*
flowering luxuriantly in Nelson, New Zealand

Hakea victoriae, *"Royal hakea"*
magnificent leaves of green and gold; Perth, Australia

PROTEACEAE

Mimetes cucullatus, *"Scarlet bottlebrush"*
rare beauty in Kirstenbosch Botanic Garden, So. Africa

Protea cynaroides, *the "King protea" (So. Africa)*
open flower, with oreole seeking nectar, in Vista, Calif.

Leucacendron argenteum, *the "Silver tree"*
with foliage glistening in the sun, Table Mt., So. Africa

Stenocarpus sinuatus, *"Fire wheel tree"*
a remarkable Australian tree, in Queensland

PROTEACEAE

Embothrium wickhamii
in Noumea, New Caledonia

Protea mundii
(Cape Prov., So. Africa)

Leucospermum conocarpum
"White sugar bush" (Cape Prov.)

Telopia speciosissima
"Crimson waratah" (E. Australia)

Serruria florida
"Blushing bride" (So. Africa)

Mimetes hirta
(Cape Mountains, So. Africa)

Protea eximia
"Ray protea" (So. Africa)

Leucospermum reflexum
"Rocket-pincushion" (So. Africa)

Protea repens (mellifera)
"Sugar bush" (Cape Prov.)

PROTEACEAE

Protea neriifolia (So. Africa)
"Oleander-leaved protea"

Leucadendron argenteum
"Silver tree" (Cape Prov.)

Leucospermum conocarpum
(carpodendron) *in Encinitas, Calif.*

Leucospermum prostratum
"Creeping pincushion" (So. Africa)

Protea neriifolia, (So. Africa)
"Oleander-leaved protea"

Protea barbigera, *"Queen protea"*
or *"Frilled panties" (So. Africa)*

Protea compacta
"Bot river protea" (Cape Prov.)

Protea cynaroides
"Giant honeypot"

Leucospermum nutans
"Pincushion", in Vista, California

PROTEACEAE

840

Protea repens (mellifera), *"Honey flower"*
or *"White sugar bush"*, *Cape of Good Hope, So. Africa*

Grevillea robusta, *the "Silky oak"*
striking masses of fiery brushes, in Queensland

Stenocarpus sinuatus, *"Fire wheel tree" (Australia)*
or "Rotary tree"; each spoke a scarlet and yellow flower

Protea susannae, *at home on limestone in S.W. Cape;*
Strybing Arboretum, San Francisco, California

Telopia speciosissima, *"Waratah"*
or "Australian protea", in New South Wales

Protea grandiceps, *"Peach protea"*
or "Princess protea", from Devil's Peak, S.W. Cape

RHIZOPHORACEAE

Bruguiera conjugata, *"Oriental mangrove"* near Port Moresby, on South coast of Papua

Bruguiera conjugata, *land-building* "Many-petaled mangrove" on Kaneohe Bay, Oahu, Hawaii

Rhizophora mangle, *"Red mangrove"* seedling in Leningrad Bot. Garden

Bruguiera conjugata mangrove, calyx flowers

Rhizophora mangle in Kew Gardens, London

Rhizophora mangle, *"American mangrove"* mangrove swamp on Key Largo, Florida

Rhizophora mangle, *"Red mangrove"*, forming thickets of trees on the coast of Brazil

RANUNCULACEAE

842

Anemone coronaria coccinea
on isle of Rhodos, Greece

Adonis aestivalis
"Pheasant's eye", in France

Anemone coronaria, *on Delos, Greece*
"Lily of the fields"

Clematis x lawsoniana 'Henryi'
"Virgin's bower"

Clematis x jackmannii
(lanuginosa x viticella)

Anemone hupehensis
"Japanese anemone" (China)

Ranunculus asiaticus 'Tecolote'
"Turban ranunculus" of florists

Clematis vitalba
"Traveler's joy" (Mediterranean)

Ranunculus asiaticus
"Persian buttercup"

RANUNCULACEAE

Ranunculus asiaticus (*Syria to Iran*)
"Persian buttercup" or "Turban", in Johannesburg

Anemone coronaria coccinea, "Poppy anemone" in habitat with Ornithogalum, in ruins of Pergamum, Asiatic Turkey

Helleborus orientalis 'Atropurpureus'
"Lenten-rose" (*Asia Minor*)

Helleborus niger
"Christmas-rose" (*Alps to Balkans*)

Aquilegia skinneri
"Columbine" (*Mexico*)

Paeonia lactiflora (albiflora sinensis)
"Chinese garden peony" in Pennsylvania spring

Clematis lanuginosa 'Nellie Moser'
summer-blooming with giant 15 cm flowers, in Delaware

RANUNCULACEAE

Paeonia suffruticosa 'Angelet'
"Yellow tree peony"

Paeonia lactiflora 'Festiva'
in Duesseldorf, Germany

Paeonia suffruticosa 'Vesuvian'
"Red tree-peony" in Brooklyn Bot. G.

Paeonia lactiflora 'Requiem'
"Garden peony" of Siberian origin

Paeonia officinalis 'Rubra-plena'
"Piney" of Southern Europe

Nigella damascena
"Love-in-a-Mist" (So. Europe)

Ranunculus lyallii
"Mount Cook lily" (New Zealand)

Delphinium elatum 'Pacific hybrid'
"Candle larkspur"

Delphinium ajacis (So. Europe)
"Larkspur" or "Annual delphinium"

RANUNCULACEAE

Clematis x lawsoniana 'Ramona'
a most beautiful flowering vine in Wilmington, Delaware

Clematis x jackmanii, *"Virgins bower"*
in the Fiskorgroend, Visby on Gotland, Sweden

Delphinium elatum, *"Candle larkspur"*
Nordic perennial in bloom, Visby, Isle of Gotland, Sweden

Delphinium elatum 'Pacific hybrid'
giant larkspur at Botanic Garden, Tuebingen, Germany

RANUNCULACEAE, ROSACEAE

Ranunculus asiaticus 'Tecolote', *florists ranunculus grown for production of tubers at Carlsbad, California*

Malus pumila paradisiaca, *"Paradise apple" as bonsai, in an alley at Nagoya, Japan*

Prunus persica, *"Persian peach" (China) as good luck cut flower, for the Chinese New Year, in Taipo*

Prunus serrulata, *"Japanese flowering cherries" in spring, in view of legendary Mt. Fuji, Kanagawa-ken, Japan*

Prunus persica flore alba-plena, *in September bloom* "Double white-flowering peach" *in Johannesburg*

Pyracantha coccinea 'Lalandei' *espaliers in bloom, on the South shore of Connecticut*

RANUNCULACEAE, ROSACEAE

Clematis cirrhosa, *"Virgin's bower"* clambering over ruins of ancient Pergamum, in Turkey

Aruncus sylvester, *"Goat's beard"*; on the steep slopes of majestic Geirangerfjord, Norway

Paeonia lactiflora 'Festiva', *"Double-fl. peony"* at the Orangerie, Missouri Botanic Garden, St. Louis

Malus 'Van Eseltine' *"Double flowering apple"* in Greenville, South Carolina

ROSACEAE

Aruncus sylvester (dioica)
"Goat's beard" in Germany

Chaenomeles japonica (*Japan*)
"Flowering quince" as bonsai

Chaenomeles speciosa (Lagenaria)
"Japanese quince" (*China*)

Cotoneaster dammeri radicans
"Bearberry" (*China*)

Cotoneaster glaucophyllus
"Bright-bead cotoneaster" (*China*)

Cotoneaster lacteus
(*parneyi*) (*W. China*)

Cotoneaster x watereri 'Cornubia'
(*frigida x henryana*)

Cotoneaster x watereri 'Pendulus'
in German nursery

Cotoneaster franchetii
in Palma de Mallorca

ROSACEAE

Cydonia oblonga
"Quince" (Iran)

Cydonia oblonga cv. 'Pineapple'
"Pineapple quince", in Calif.

Eriobotrya japonica
"Japan plum" or "Loquat"

Filipendula vulgaris
"Meadow sweet"

Eriobotrya deflexa
"Bronze loquat" (Taiwan)

Eriobotrya japonica
"Chinese loquat" in flower

Fragaria chiloensis (W. America)
"Beach strawberry"

Duchesnea indica
"Mock strawberry" (India)

Heteromeles arbutifolia (Calif., Mex.)
"Toyon" or "Christmas berry"

ROSACEAE

850

Prunus persica *in Taipo*
"Good-luck peach"

Prunus serrulata
"Japanese flowering cherry"

Prunus serrulata 'Kwansan'
"Double pink cherry"

Prunus cerasifera 'Atropurpurea'
"Purple-leaf plum"

Prunus armeniaca
"Apricot" (*China*)

Pyrus communis 'Kieffer'
"Kieffer pear tree"

Malus floribunda 'Scheideckeri'
"Flowering crab-apple"

Malus baccata
"Siberian crab-apple" in Sweden

Kerria japonica 'Pleniflora'
"Japanese rose" in France

ROSACEAE

851

Prunus mume
"Japanese apricot" in Yokohama

Prunus domestica (*Eurasia*)
"Dwarf Italian plum"

Prunus dulcis (amygdalus)
"Wild almond" in Mycaene, Greece

Prunus cerasus
"Sour cherry"

Prunus serrulata 'Kwansan'
"Japanese pink cherry"

Prunus triloba plena
"Double flowering almond" (*China*)

Malus pumila (*Europe*)
"Common apple" in flower

Prunus persica 'Elberta'
"Elberta peach" in bloom

Pyrus kawakamii
"Evergreen pear" (*Taiwan*)

ROSACEAE

852

Heteromeles arbutifolia (Photinia)
"California holly" or "Toyon"

Photinia x fraseri
"Redleaf photinia" in Australia

Parinari mobola
in Cayar, Sénégal

Pyracantha koidzumii 'Victory'
"Red firethorn" (Taiwan)

Pyracantha fortuneana 'Graberi'
"Chinese firethorn"

Rubus laciniatus 'Prof. Rudloff'
"Thornless blackberry"

Prunus laurocerasus 'Otto Luyken'
in Duesseldorf, Germany

Prunus laurocerasus
"English laurel" (S.E. Europe)

Mespilus germanica
"Medlar" (Europe to Iran)

ROSACEAE

Prunus dulcis (amygdalus), *"Wild almond"*
with glorious blossoms, in Mycaene of Agamemnon

Prunus persica 'Rubra plena', *"Double-flowering peach"*
September-blooming on street of Pretoria, Transvaal

Pyrus kawakamii, *"Evergreen pear"*
as specimen tree in Fashion Valley, San Diego, California

Pyrus kawakamii, *from Taiwan;*
willowy young branches fastened to trellis as espalier

ROSACEAE

854

Raphiolepis indica
"India hawthorne" (So. China)

Raphiolepis indica 'Enchantress'
in California

Sorbus intermedia (No. Europe)
"Swedish white beam"

Sorbus aucuparia
"Mountain ash", in Norway

Sorbaria aitchisonii
"False spiraea" (Kashmir)

Rosa canina (Europe, W. Asia)
"Dog rose" or "Brier"

Spiraea x vanhouttei
"Bridal wreath"

Rubus rosifolius 'Coronarius'
"Mauritius raspberry"

Potentilla verna
"Spring cinquefoil"

ROSACEAE

Raphiolepis indica, *"Hongkong hawthorn"* in habitat on Kwun Yum Shan mountain, Hongkong

Rosa x floribunda *in formal beds carefully tended, in Frogner Sculpture Park, Oslo, Norway*

Sorbus americana, *"Missey-moosey"* or *"American mountain ash"*, in Vancouver, B.C.

Rosa banksiae, *beautiful climber from China,* *"Lady Banks rose" with dbl. yellow blooms, in Dallas, TX*

ROSACEAE

Rosa polyantha 'Margo Koster'
"Baby rose" for Easter

Rosa 'Double Paul's Scarlet'
Roehrs climber

Rosa (floribunda) 'Spice'
(Goldilocks x Floradora)

Rosa (floribunda)
'Scarlet Marvel'

Rosa (grandiflora) 'Queen Elizabeth'
(hybrid tea x floribunda)

Rosa polyantha 'Mothers Day'
late-blooming pot rose

Rosa polyantha
'Triomphe Orléanais', *for Easter*

Rosa villosa
"Apple rose", in Germany

Rosa odorata 'Peer Gynt'
Kew Gardens, London

Raphiolepis indica 'Enchantress'
(Monrovia Nursery, Calif.)

Rubus reflexus (*Hongkong*)
"Trailing velvet plant"

Prunus persica 'Gold Jubilee'
a delicious freestone "Peach"

ROSACEAE, RHAMNACEAE

Rosa borboniana 'Magna Charta'
"Hybrid-perpetual" rose

Rosa 'Bonfire' (*Rambler*)
trained globe for Easter

Rosa odorata 'Mrs. W.C. Miller'
old "Hybrid-tea" (1909)

Rosa (floribunda)
'Crimson Rosette'

Rosa chinensis 'Minima' (roulettii)
"Fairy rose"

Rosa odorata 'Golden Rapture'
florist's cut flower

Rosa odorata 'Better Times'
greenhouse cut flower

Rosa (floribunda) 'Fashionette'
fragrant pot rose

Rosa odorata 'Happiness'
fancy greenhouse cut flower

Rosa odorata 'Mrs. Pierre S. Dupont'
fragrant hybrid tea

Colletia cruciata (*RHAMN.*)
"Anchor plant" (Uruguay)

Rosa banksiae (*China*)
"Lady Banks rose"

ROSACEAE

858

Rosa 'Eugene Jacquet' (*Rambler*)
in "Basket" shape, forced for Easter bloom at Roehrs

Rosa x floribunda *in formal plantings*
at Vigeland's controversial sculpture collection in Oslo

Rosa polyantha 'Triomphe Orléanais'
potted "Baby roses" forced for Easter bloom, at Roehrs

Rosa (floribunda) 'Korham'
large-flowered bicolor rose, Bot. G. Cologne, Germany

Rubus reflexus, *"Trailing velvet plant"*
rambling over the rocks on Victoria Peak, in Hongkong

Pyracantha coccinea 'Lalandei', *a "Firethorn"*
fruiting equally well in New Jersey as in California

ROSACEAE

859

Rosa 'Blaze' (*Rambler*)
everblooming "Scarlet climber"

Rosa (floribunda) 'Europeana'
a "Baby rose" from Holland

Rosa (floribunda) 'Rumba'
charming bicolor

Rosa rubiginosa 'Magnifica'
"Sweet brier" rose

Rosa odorata 'Joseph's Coat'
multicolor hybrid tea

Rosa polyantha 'Mother's Day'
late-blooming pot rose

Rosa noisettiana 'Maréchal Niel'
tea-rose on Santa Fe train to Calif.

Rosa odorata
'Red American Beauty'

Rosa 'Double Paul's Scarlet'
everblooming climber

RUBIACEAE

Bouvardia ternifolia 'White Joy'
albino cultivar

Bouvardia ternifolia
"Scarlet trompetilla" (*Mexico*)

Bouvardia ternifolia 'Giant Pink'
florist's cut flower

Bouvardia versicolor
(*South America*)

Bouvardia longiflora humboldtii
"Sweet bouvardia" (*Mexico*)

Coffea arabica
"Arabian coffee" (*Ethiopia*)

Gardenia jasminoides fortuniana
"Belmont gardenia"

Gardenia jasminoides 'Veitchii'
"Everblooming gardenia"

Mitchella repens (*No. America*)
"Partridge berry"

Gardenia jasminoides 'Veitchii'
potted plant for Easter

Gardenia radicans flore pleno
"Miniature gardenia"

Gardenia carinata
"Kedah gardenia" in Singapore

RUBIACEAE

Coffea arabica, *from Ethiopia and Angola,*
coffee on the high plateau of Sao Paulo State, Brazil

Coffea robusta *from Zaïre*
fruiting on the Goroka highlands of New Guinea

Alberta magna, *from Natal*
striking shrub in Kirstenbosch Bot. Garden, Cape Town

Gardenia jasminoides 'Veitchii'
"Everblooming gardenia", pot-grown to flower for Easter

Cinchona officinalis, *"Jesuit's bark"* or *"Quinine"*
on plantation at Chicacao, Pacific slope of Guatemala

Ixora macrothyrsa *from Sumatra*
"King ixora" or "Jungle flame" near Suva, Fiji

RUBIACEAE

862

Hoffmannia ghiesbreghtii 'Variegata'
"Variegated taffeta plant"

Hoffmannia ghiesbreghtii
"Tall taffeta plant"

Hoffmannia refulgens (*Mexico*)
"Quilted taffeta plant"

Ixora javanica (*Java*)
"Jungle geranium"

Ixora macrothyrsa
'Super King'

Randia maculata
(*W. Africa: Sierra Leone*)

Manettia inflata (*Paraguay*)
"Firecracker plant"

Morinda citrifolia
"Indian mulberry" or "Noni"

Rondeletia odorata
(*West Indies, Panama*)

Nertera granadensis (depressa)
"Coral-bead plant" (*Perú to N.Z.*)

Trichostachys aurea
(*Trop. Africa*)

Pentas lanceolata
"Egyptian star cluster"

RUBIACEAE

Ixora macrothyrsa, *"Jungle flame"*
standard tree in shopping center of Kahalui, on Maui

Luculia grandifolia (tsetensis) (*India*)
showy clusters of waxy flowers, in Auckland, New Zealand

Calycophyllum spruceanum, *"Pan-mulato"*;
trees with polished trunks, Bot. Garden Rio de Janeiro

Warszewiczia coccinea (*Tropical America*)
"Wild poinsettia", in Papeari Botanic Garden, Tahiti

RUBIACEAE

Coffea liberica
"Liberian coffee" (*W. Africa*)

Coffea laurifolia
(*col. Marnier-Lapostolle, France*)

Coffea robusta (canephora)
"Robusta coffee" (*Zaïre*)

Casasia clusiaefolia
"Seven-year apple" (*W. Indies*)

Catesbaea spinosa
"Lily-thorn" (*Cuba, Bahamas*)

Coprosma robusta
"Kakaramu" (*New Zealand*)

Coprosma repens 'Marginata'
(baueri) *"Mirror plant"*

Coprosma williamsii
(*New Zealand*)

Luculia grandifolia
(*Bhutan Himalayas*)

RUBIACEAE

Warszewiczia coccinea, *"Wild poinsettia"*, with scarlet bracts, in Papeari Botanic Garden, Tahiti

Pavetta opaca, *"Christmas bush"*, sweetly scented flowers in South Africa's summer

Nertera granadensis (depressa), *"Coral-bead plant"* on granite rocks at Milford Sound, Fiordland, N.Z.

Alberta magna (*South Africa*) brilliant scarlet flowers, at Table Mountain, Cape Town

RUBIACEAE

Gardenia taitensis
"Tiare Tahiti" (Polynesia)

Gardenia thunbergia
(So. Africa: Natal)

Gardenia jasminoides 'Veitchii'
"Everblooming gardenia"

Burchellia bubalina
"Wild pomegranate" (So. Africa)

Pavetta revoluta *(So. Africa)*
Kew Gardens, London

Pavetta lanceolata
"Christmas bush" at Kirstenbosch

Pavetta angustifolia
Cologne Bot. Garden, Germany

Pentas lanceolata
"Egyptian star-cluster"

Pentas lanceolata 'Pallida'
in Andromeda Garden, Barbados

RUBIACEAE

Hamelia patens (*West Indies to Bolivia*)
"Scarlet bush", in Jamaica, West Indies

Warszewiczia coccinea 'David Auyong'
inflorescence of fully double bracts, in Trinidad

Gardenia thunbergia, *large fragrant flowers,
in Kirstenbosch Botanic Garden, Cape Town*

Gardenia taitensis, *"Tiare" or "Symbol flower"
on unspoiled Nuku Hiva, Marquesas, French Polynesia*

Cephaelis tomentosa (*Mexico to Bolivia*)
remarkable shrub with bracted flowers, in Trinidad

Mussaenda erythrophylla, *"Ashanti blood"
yellow flowers with deep red bracts, in Durban, Natal*

RUBIACEAE

Rondeletia odorata
in Panama

Rondeletia strigosa
(Guatemala)

Ixora borbonica
with M. Lecoufle, France

Ixora chinensis 'Fraseri'
salmon color-form, in Sydney Botanic Garden, Australia

Palicourea rigida
curious tropical shrub with showy flowers, in Venezuela

Manettia inflata (bicolor)
"Firecracker vine" (Paraguay)

Morinda citrifolia 'Variegata'
Pacific Trop. Bot. Garden, Kauai

Warszewiczia coccinea
in Singapore Bot. Garden

RUBIACEAE

Ixora coccinea (*E. Indies*)
Andromeda Garden, Barbados

Ixora macrothyrsa (duffii)
in Curacao

Ixora chinensis
"Flame of the woods"

Ixora 'Herreras Pink'
Kew Gardens, London

Ixora odorata
(*Madagascar*)

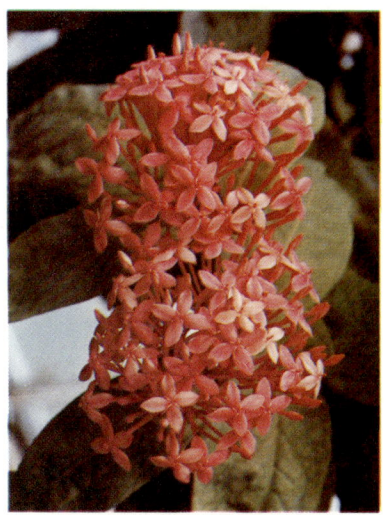
Ixora undulata (*Bengal*)
Kew Gardens, London

Ixora macrothyrsa
"King ixora" in Tahiti

Ixora borbonica
(*Réunion Is.*)

Ixora chinensis 'Alba'
Brooklyn Bot. Garden, New York

RUBIACEAE, MELASTOMATACEAE

Mussaenda erythrophylla
"Ashanti blood"

Mussaenda erythrophylla 'Rosea'
in Jogyakarta, Java

Mussaenda philippica 'Aurorae'
"Doña Aurora" in Manila

Portlandia grandiflora
"Glorias floridas de Cuba"

Mussaenda philippica
Manila, Philippines

Mussaenda philippica
close-up of flower and bracts

Morinda citrifolia
"Noni" in flower on Tonga

Blakea sanguinea (MELAST.)
Andes of Colombia

Calycophyllum spruceanum
"Pan-mulato" in Rio

RUBIACEAE

Mussaenda philippica 'Aurorae'
"Doña Aurora" in Luzon nursery, Manila, Philippines

Mussaenda frondosa (*East Indies*)
in Royal Botanic Garden, Sydney, Australia

Mussaenda erythrophylla, *from Central Africa*
in Mahé Botanic Garden, Seychelles Islands

Mussaenda philippica
collected on expedition into northeastern New Guinea

RUTACEAE

Casimiroa edulis 'Vernon'
in Encinitas, California

Choysia ternata
"Mexican orange-blossom"

Poncirus trifoliata
"Hardy orange" (China)

Murraya paniculata (India)
"Orange jessamine"

Boronia heterophylla
"Red boronia" (W. Australia)

Adenandra uniflora
"China flower" (So. Africa)

Correa alba (S.E. Australia)
"White correa"

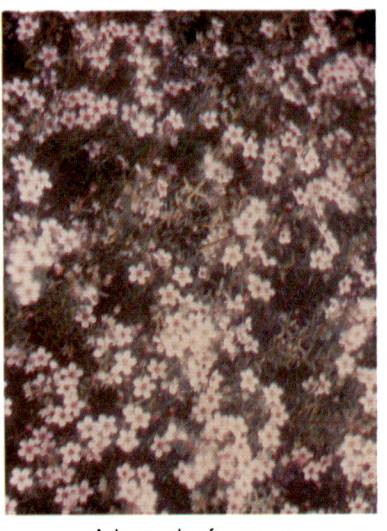
Adenandra fragrans
"Dwarf China flower"

Phellodendron amurense
"Amur cork" (Arnold Arbor. Boston)

RUTACEAE

Skimmia japonica 'Dwarf Female'
with handsome scarlet fruit

Calodendrum capense
"Cape chestnut" (So. Africa)

Boronia megastigma
"Scented Boronia" (W. Australia)

Severinia buxifolia
"Chinese box-orange"

Boronia elatior (*W. Australia*)
"Scented tall boronia"

Citrus limon 'Ponderosa'
"American wonder-lemon"

Fortunella margarita
in Kwangchow, China

Citrus taitensis
"Otaheite orange"

Citrus sinensis 'Washington Navel'
with winter fruit

RUTACEAE

Citrus sinensis 'Valencia'
"Valencia orange" (Spain)

Citrus limon 'Meyeri'
"Meyer lemon"

Citrus reticulata 'Dancy'
"Tangerine" in California

Citrus limon 'Eureka'
"Acid lemon" on Sunset Ranch, Vista, California

Citrus aurantium 'Myrtifolia', *"Myrtle-leaf orange"*
with Señora Camuyrano, Buenos Aires, Argentina

Citrus mitis
"Calamondin" (Philippines)

Fortunella japonica
"Marumi kumquat"

Citrus aurantium 'Marginata'
in Yokohama, Japan

SALICACEAE, SAPINDACEAE 875

Harpullia pendula
"Tulipwood" (E. Australia)

Harpullia arborea
"Puas tree" (Philippines)

Koelreuteria elegans
"Chinese raintree" (Fiji)

Harpullia zanguebarica
"Mgambo tree" in Kenya

Populus tremuloides, *"Quaking aspen",
in brilliant golden fall color, in Kaibab forest, Utah*

Litchii chinensis
"Lychee nut" (So. China)

Kolreuteria elegans (formosana)
"Shrimp tree" in Florida

Pometia pinnata
"Langsir" in the Marquesas

SALICACEAE, SANTALACEAE, SAPINDACEAE

Salix caprea 'Pendula'
"Kilmarnock willow", in Germany

Populus tremuloides
"Trembling aspen" in Utah

Salix matsudana 'Tortuosa'
"Corkscrew willow"

Cardiospermum halicacabum
"Balloon vine" on St. Héléna

Dodonaea viscosa 'Purpurea'
"Purple hop bush"

Dodonaea viscosa
"Sand olive" (Arizona)

Cupaniopsis anacardioides
"Carrot-wood" (Australia)

Blighia sapida
"Aki tree" (W. Africa: Guinea)

Santalum ellipticum
"Hawaiian sandalwood"

SAXIFRAGACEAE

Astilbe japonica, *"False spirea"* or *"Meadow sweet"* in Petrodvorets gardens, near Leningrad, Soviet Union

Astilbe japonica, *florist's "Spiraeas" in red and white, outdoors in Visby on Gotland, Sweden*

Saxifraga sarmentosa (stolonifera) 'Tricolor' "Magic carpet" or "Variegated strawberry geranium"

Tolmiea menziesii (*Pacific coast from Calif. to Alaska*) "Piggy-back plant", young plantlets form in base of leaves

Hydrangea macrophylla 'Otaksa', *old-time "Hortensia" the parent of all named florist cultivars; in Sweden*

Hydrangea macrophylla 'Kuhnert' *for Easter bloom, popularly known as "Snowballs", at Roehrs greenhouses*

SAXIFRAGACEAE

Hydrangea macrophylla
'Chaperon Rouge' (*France 1951*)

Hydrangea macrophylla
'Kuhnert' (*1926*)

Hydrangea macrophylla
'Stafford' ('Mad. Cayeux')

Hydrangea macrophylla 'Todi'
German "Hortensia"

Hydrangea macrophylla 'Regula'
white blooms for Easter

Hydrangea macrophylla 'Merveille'
French hortensia 1927

Hydrangea macrophylla
'Enziandom' (*Germany 1949*)

Hydrangea macrophylla
'Flamboyant'

Hydrangea macrophylla
'Soeur Thérèse' (*France 1945*)

Hydrangea macrophylla 'Variegata'
ornamental variegated foliage

Hydrangea macrophylla 'Rosabelle'
for Mother's Day

Hydrangea macrophylla 'Alpengluehn'
clear blue, aluminum treated

SAXIFRAGACEAE

Heuchera sanguinea, *"Coral bells" as window plant at Kandersteg, Bernese Alps, Switzerland*

Hydrangea macrophylla 'Merveille', *French "Hortensias" in bloom for Easter, at Roehrs greenhouses in New Jersey*

Hydrangea aspera (strigosa) *showy ornamental shrub from the Himalayas*

Astilbe japonica, *the florist's "Spiraea" in the flower market of Oslo, Norway*

SAPOTACEAE, SAXIFRAGACEAE

Ribes speciosum (*California*)
"Fuchsia-flowered gooseberry"

Mimusops commersonii
"Spanish cherry" (*Madagascar*)

Manilkara (Achras) zapota
"Chicle" or *"Sapodilla"*

Philadelphus x virginalis
"Mock-orange"

Escallonia laevis (organensis)
"Pink escallonia" (*Brazil*)

Deutzia gracilis
"Slender deutzia" (*Japan*)

Philadelphus lewisii (*No. America*)
"Wild mock-orange"

Hydrangea quercifolia
"Oakleaf hydrangea" (*Southeast U.S.*)

Chrysophyllum cainito
"Star-apple" in Florida

SAXIFRAGACEAE

Heuchera sanguinea
"Coralbells" (Arizona, Mexico)

Astilbe x arendsii
"False spiraea"

Saxifraga rosacea 'Arendsii'
"Kumomaso" in Tokyo, Japan

Carpenteria californica
"Tree anemone"

Bergenia purpurascens
in the Sikkim Himalayas

Bergenia crassifolia
"Winter begonia" in Mexico

Saxifraga longifolia
(Pyrenées)

Saxifraga sarmentosa 'Tricolor'
"Magic carpet"

Tolmiea menziesii
"Youth-on-age" or "Piggyback"

SCROPHULARIACEAE

Calceolaria herbeohybrida 'Grandiflora' (*red form*)

Calceolaria herbeohybrida 'Grandiflora' (*tigered form*)

Calceolaria herbeohybrida 'Multiflora nana'

Allophyton mexicanum "Mexican foxglove"

Hebe buxifolia 'Variegata' (*New Zealand*)

Mazus reptans "Wart flower" (*Himalayas*)

Antirrhinum majus "Floral carpet snapdragon"

Rehmannia angulata "Foxglove gloxinia" (*China*)

Antirrhinum majus 'Butterfly' "Pink Pixie" bedding type

Phygelius capensis "Cape fuchsia" (*So. Africa*)

Mimulus aurantiacus (*Diplacus glutinosus*)

Mimulus cupreus "Monkey flower" (*China*)

SCROPHULARIACEAE

Castilleja lanata, *"Indian paintbrush"*
and other wild flowers, blooming in April in Texas

Penstemon barbatus, *"Scarlet beard-tongue"*
up to Tioga Pass, in the Sierra Nevadas of California

Sutera grandifolia, *"Wild phlox"*
"Purple glory plant", in the mountains of Transvaal

Angelonia gardneri
showy spikes of purple flowers, in Brazil

Calceolaria herbeohybrida 'Multiflora nana'
compact type of *"Lady's pocketbook"* plant

Calceolaria herbeohybrida 'Multiflora nana'
"Slipper flowers" for spring bloom at Roehrs

SCROPHULARIACEAE

Antirrhinum majus
tall "Rocket snapdragon"

Antirrhinum majus 'Butterfly'
tall bronze butterfly type

Bowkeria gerrardiana
(South Africa)

Calceolaria pavonii
on Rio Urubamba, Perú

Paulownia tomentosa
"Empress tree" (China)

Nemesia strumosa
"Cape jewels" (So. Africa)

Calceolaria herbeohybrida
'Grandiflora'

Calceolaria integrifolia
"Chilean pouch flower"

Calceolaria herbeohybrida
'Multiflora nana'

SCROPHULARIACEAE

Verbascum thapsus (*Europe and Asia*)
"Common mullein" in summer, on Bornholm, Denmark

Nemesia strumosa, *"Cape jewels"*
in colorful planting on Emerald Lake, Canadian Rockies

Antirrhinum majus, *"Snapdragons" carefully tended
in People's Botanic Garden, Kwangchow, China*

Lamarouxia rhinanthifolia, *on Monte Alban,
at the Zapotec temples, Oaxaca, So. Mexico*

SCROPHULARIACEAE

Russelia sarmentosa
"Coral blow" (*Mexico*)

Torenia fournieri
"Wish-bone plant" (*Vietnam*)

Russelia equisetiformis
"Coral plant" (*Mexico*)

Phygelius aequalis
"River bells" (*So. Africa*)

Uroskinnera spectabilis
(*Mexico*)

Castilleja coccinea
"Scarlet paintbrush"

Mimulus x burnetii
"Monkeymusk"

Rhodochiton volubile
"Purple-bell-vine" (*Mexico*)

Asarina (Maurandya) erubescens
"Creeping gloxinia" (*Mexico*)

SCROPHULARIACEAE

Phygelius capensis
"Cape fuchsia" (So. Africa)

Hebe speciosa (New Zealand)
"Showy veronica"

Hebe x imperialis
(speciosa x salicifolia)

Alonsoa warscewiczii
"Mask flower" (Perú)

Lunaria annua (biennis) (CRUC.)
"Honesty" or "Moonwort"

Penstemon x gloxinioides
"Garden penstemon"

Digitalis purpurea 'Alba'
"White foxglove"

Digitalis purpurea
"Common foxglove" (W. Europe)

Penstemon hartwegii
"Beard-tongue" (Mexico)

SELAGINELLACEAE

Selaginella umbrosa
(*Yucatán to Colombia*)

Selaginella galeottei
(*Mexico, Costa Rica*)

Selaginella serpens
(*West Indies*)

Selaginella uncinata (*So. China*)
"Rainbow fern"

Selaginella plumosa
(*Ceylon, Burma*)

Selaginella martensii 'Variegata'
(*Mexico*)

Selaginella kraussiana 'Brownii'
"Cushion moss"

Selaginella willdenovii
"Peacock fern"

Selaginella apoda
"Basket selaginella"

SELAGINELLACEAE

on left: Selaginella tamariscina (*Honshu, Kyushu*)
during *Selaginella* exhibition in Tokyo, Japan

Selaginella caulescens japonica (*Japan*)
at Flower show in Hibiya Park, Tokyo

Selaginella kraussiana (denticulata) (*So. Africa*)
"Trailing Irish moss" or "Spreading clubmoss"

Selaginella kraussiana 'Aurea'
"Golden trailing club-moss"

Selaginella emmeliana (pallescens)
"Sweat-plant" from tropical Venezuela and Colombia

Selaginella lepidophylla (*Texas, Mexico to El Salvador*)
"Resurrection plant" or "Rose of Jericho"

SOLANACEAE

Browallia speciosa 'Blue Troll"
Longwood Gardens, Penna.

Brunfelsia americana (*W. Indies*)
"Lady-of-the-night"

Browallia speciosa
"Sapphire flower" (Colombia)

Brunfelsia pauciflora 'Macrantha'
"Yesterday-and-today"

Brunfelsia pauciflora 'Floribunda'
"Yesterday-today-and tomorrow"

Brunfelia pauciflora calycina
"Morning-noon-and-night"

Iochroma coccineum (*Perú*)
Quail Bot. Garden, California

Iochroma cyaneum (tubulosum)
"Violet bush" (Colombia)

Iochroma warscewiczii
(*Trop. America*)

SOLANACEAE

Cestrum elegans (purpureum) (*Mexico*)
in Horotane Valley, Christchurch, New Zealand

Datura arborea, *"Tree datura*
on the road to Esmeraldas, Ecuador

Datura suaveolens, *"Angel's trumpet"* (*Brazil*)
on the walls of the Alcazar, Sevilla, Spain

Solanum mammosum, *"Nipple-fruit"* (C. America)
with weird fruit, in Kebun Raya Botanic Garden, Java

Ailanthus altissima, *"Tree of Heaven"*
in autumn color along the Rhine, Bonn, Germany

Physalis alkekengi (franchetii), *"Chinese lanterns"*,
cut lanterns at flower market in Nuremberg, Franconia

SOLANACEAE

Datura mollis (Brugmansia) (*Ecuador*)
in Longwood Gardens, Kennett Square, Pennsylvania

Datura candida, *"Floripondio tree"* (*So. America*)
at Quail Botanic Garden, Encinitas, California

Datura aurea (Brugmansia) (*Colombia*)
University Botanic Garden, Dortmund, Germany

Datura chlorantha, *from the Andes of Perú*
in col. University Bot. Garden of Tuebingen, Germany

SOLANACEAE

Datura inoxia (Angels Trumpet, Thorn apple, Sacred datura)

Datura versicolor, *from Guayaquil, Ecuador in Nassau Botanic Garden, Bahamas*

Datura suaveolens (Brugmansia) (*So. Brazil*) "Angel's-trumpet" in Longwood Gardens, Pennsylvania

Datura arborea, "Tree datura" from Peruvian Andes; South Coast Botanic Garden, Los Angeles, Calif.

Datura sanguinea (*Andes of Perú*) Jardin Botanique 'Les Cèdres', Cap-Ferrat, France

SOLANACEAE

894

Solandra nitida
"Cup of gold" (Mexico)

Solandra longiflora
"Chalice vine" (Jamaica)

Datura rosei (Brugmansia)
from Ecuador

Cyphomandra betacea
"Tomato tree" (So. Brazil)

Solanum melongena esculentum
"Black bell eggplant"

Solanum capsicastrum 'Variegatum'
"Variegated Jerusalem-cherry"

Capsicum annuum conoides 'Fiesta'
ornamental "Christmas pepper"

Capsicum annuum grossum
"Italian bell pepper"

Lycopersicon lycopersicum
'Tiny Tim', *"Miniature tomato"*

SOLANACEAE

Cestrum diurnum
"Day jessamine" (*W. Indies*)

Cestrum nocturnum
"Night jessamine" (*W. Indies*)

Cestrum parqui spurium
"Willow jasmine" (*Chile*)

Quassia amara (*SIMAROUB.*)
"Bitterwood" (*Surinam*)

Cestrum aurantiacum
"Orange cestrum" (*Guatemala*)

Juanulloa aurantiaca
in Fortin, Veracruz, Mexico

Physalis alkekengi
"Japanese lantern" (*Japan*)

Nicotiana tabacum
"Common tobacco" (*W. Indies*)

Nicotiana alata
"Jasmine tobacco"

SOLANACEAE

Petunia hyb. grandiflora 'Crusader'
bicolored petunia

Petunia hyb. fl. pl. 'Caprice'
double pot petunia

Petunia hyb. grandiflora 'Bingo'
huge bicolor

Petunia hyb. 'Celestial Rose'
dwarf bedding petunia

Petunia hyb. grandiflora 'Popcorn'
giant ruffled

Petunia hyb. 'California Giant'
giant pot petunia

Petunia hyb. multiflora 'Glitters'
star petunia

Petunia hyb. 'Pink Magic'
single grandiflora

Petunia hyb. multiflora 'Satellite'
small star

Petunia x hybrida multiflora 'Summer Fun', *bright yellow*

Petunia hyb. fl. pl. 'Sonata'
double white frilled

Petunia hyb. grand. 'Elk's Pride'
"Blue balcony petunia"

SOLANACEAE

Nicotania tabacum, *"Common tobacco"*
on mysterious Easter Island, in the South Pacific

Physalis alkekengi (franchetii), *"Chinese lanterns"*
at the flower market, along the Gracht in Amsterdam

Solanum uporo, *an ornamental type of tomato*
at the People's Botanic Garden, Kwangchow, China

Solanum melongena 'Moorea', *"Ornamental eggplant"*
or *"Golden aubergine"* on Moorea, French Polynesia

SOLANACEAE

Schizanthus retusus
"Butterfly flower" (Chile)

Salpiglossis sinuata (Chile)
"Painted tongue" bronze

Salpiglossis sinuata
orchid-flowered, from Chile and Perú

Solanum martii (Brazil)
Jardim Botanico, Rio de Janeiro

Streptosolen jamesonii
"Marmalade bush" (Colombia)

Solanum giganteum
"African holly" (Ceylon)

Mandragora officinarum
"Devil's apples" on Delos

Hyoscyamus aureus
"Golden henbane" in Crete

Solanum hispidum
"Devil's fig" (Mexico)

SOLANACEAE

Solanum jasminoides
'Grandiflorum', *in Los Angeles*

Solanum seaforthianum
"Star potato vine"

Solanum jasminoides
"Potato-vine" (Brazil)

Solanum macranthum
"Potato tree" in Durban

Solanum rantonnetii
"Blue potato bush" (Argentina)

Solanum wendlandii
"Paradise flower" (Costa Rica)

Solanum macranthum
"Brazilian potato tree"

Solanum quitoense
"Naranjilla" (Andes of Ecuador)

Solanum aviculare
"Kangaroo-apple" (Australia)

SOLANACEAE

Cestrum elegans (purpureum)
"Red makahala" (*Mexico*)

Solanum rantonnetii
in Melbourne, Victoria

Solanum marginatum
(*No. Africa, Ethiopia*)

Brunfelsia pauciflora
'Floribunda' (*Brazil*)

Brunfelsia latifolia
"Kiss-me-quick"

Brunfelsia pauciflora calycina
"Morning-noon-and-night"

Brunfelsia americana
"Lady-of-the-night"

Solanum pseudo-capsicum
"Jerusalem cherry" (*Madeira*)

Capsicum annuum conoides
'Red Chile' (*ornamental*)

Solandra nitida
"Cup-of-gold" (*Mexico*)

Schizanthus x wisetonensis
"Poor-man's orchid"

Iochroma cyaneum (tubulosum)
"Violet bush"

STERCULIACEAE

Brachychiton rupestris, *"Queensland bottle tree"* in Royal Botanic Gardens, Sydney, Australia

Brachychiton acerifolius (Sterculia), *"Flame tree"* in brilliant bloom while without foliage, in Queensland

Dombeya spectabilis (rotundifolia), *the "Wild pear"* flowering on the High Veld, in the Transvaal

Heritiera fischeri, *the "Looking-glass tree"* with mirror-like foliage, Botanic Garden Leningrad, Russia

STERCULIACEAE, TACCACEAE

Tacca chantrieri
"Bat-flower" (Malaya)

Tacca plantaginea
"Cat's whiskers" (Thailand)

Firmiana simplex (platanifolia)
"Phoenix tree" (E. Asia)

Cola nitida (W. Africa)
"Cola nuts" in Seychelles

Theobroma cacao
"Cocoa tree" in Trinidad

Sterculia ceramica
"Fairchild's sterculia" (Celebes)

Pterospermum semisagittatum
(India: Assam)

Pterospermum lancifolium
(East Indies)

Pterospermum acerifolium
"Bayur tree" (Burma)

STERCULIACEAE, TAMARICACEAE

Fremontodendron mexicanum
"Flannel-bush" (Calif., Mexico)

Dombeya burgessiae
"Wedding flower" (S.E. Africa)

Dombeya wallichii
(E. Africa, Madagascar)

Dombeya elegans (Réunion)
at Fairchild Garden, Miami

Dombeya x cayeuxii
"Mexican rose" or "Pink ball"

Dombeya tiliacea (natalensis)
"Heartleaf dombeya" (S.E. Africa)

Sterculia foetida
"Indian almond" (E. Africa)

Tamarix parviflora
"Salt-cedar" (S.E. Europe)

Tamarix ramosissima (pentandra)
"Summer tamarisk" (E. Europe, Asia)

Camellia sinensis, *the commercial "Tea" plant; young leaves are picked weekly in Nuwara Eliya, Ceylon*

Eurya japonica 'Tricolor' on bed of Erica gracilis, Rhineland, Germany

Camellia sasanqua, *small-flowered "Sasanquas"; blooms are short-lived but numerous, in Osaka, Japan*

Camellia japonica (*Mountains of Japan and Korea*) *cut flowers last longest when floating on water*

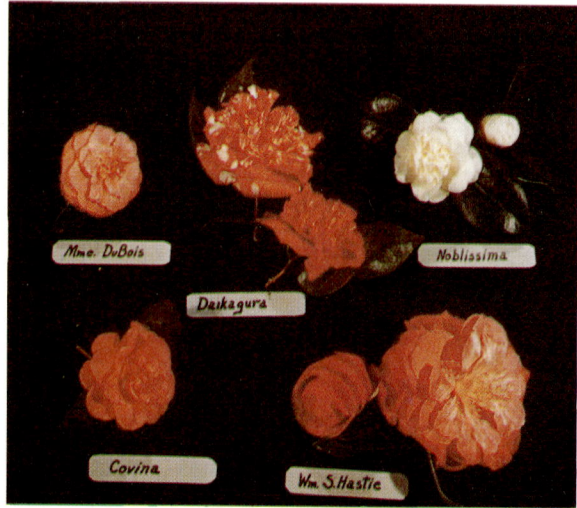

top: Camellia jap. 'Mme. Le Bois', 'Daikagura', 'Noblissima'
bottom: Camellia japonica 'Covina', 'William S. Hastie'

top: Camellia jap. 'Purity', 'Alba plena', 'Pr. Baciocchi'
bottom: Cam. jap. 'Debutante', 'Col. Firey', 'Delfosse'

THEACEAE

Camellia japonica
'Col. Firey'

Camellia japonica
'Daikagura'

Camellia japonica
'Debutante' ('Sara Hastie')

Camellia japonica
'Elegans'

Camellia japonica
'Jordan's Pride' ('Herma')

Camellia japonica
'Pearl Maxwell'

Camellia japonica
'Pink Perfection'

Camellia japonica
'Purity'

Camelia japonica
'William S. Hastie'

Gordonia lasianthus
"Black laurel" (*Southeast U.S.*)

Camellia sasanqua
(*Mountains of Japan*)

Camellia sinensis, *"Tea plant"*
glassflower in Harvard Museum

STACHYURACEAE, THEACEAE, THEOPHRASTACEAE

Camellia reticulata
"Temple flower" in China

Camellia sasanqua 'Yuletide'
brilliant single red

Stewartia sinensis (*China*)
Kew Gardens, London

Theophrasta americana (fusca)
(*Santo Domingo*)

Ternstroemia gymnanthera
(*India, Malaya, Japan*)

Eurya japonica
(*Himalayas*)

Jacquinia armillaris arborea
used for beads in Barbados

Jacquinia pungens
"Cudjoewood" (*Mexico*)

Stachyurus praecox
"Spiketail", in Yokohama

THEOPHRASTACEAE, TILIACEAE, TROPAEOLACEAE, URTICACEAE

Sparmannia africana, *"Indoor linden"*
popular house plant in Germany, outdoors during summer

Clavija tarapotana (*Perú, Bolivia*)
palm-like, fragrant subtropical tree from the Andes

Tropaeolum majus, *the "Garden nasturtium"*
framing an office entrance in Stockholm, Sweden

Helxine soleirolii, *"Baby tears" or "Irish moss"*
covering topiary frame, Mitchell Park, Milwaukee

THYMEL., TROCHOD., TROPAEOL., ULMAC., UMBELL.

Tropaeolum majus
"Nasturtium" (Perú)

Tropaeolum tricolor (Chile)
"Tricolored Indian cress"

Tropaeolum speciosum
(Chilean Andes)

Eryngium bromeliifolium
Strassbourg Bot. Garden, France

Angelica archangelica
"Angelica" or "Wild parsnip"

Heracleum mantegazzianum
"Giant hogweed" (Caucasus)

Peddiea africana
(Cape Prov. to Transvaal)

Trochodendron aralioides
"Wheel-tree" (Japan)

Ulmus parvifolia
"Evergreen elm" (China)

THYMELAEACEAE, TILIACEAE

Dais cotinifolia
"Pompon tree" (So. Africa)

Phaleria capitata (*Java*)
with cauliflorous fragrant flowers

Grewia caffra
"Lavender starbush" (So. Africa)

Daphne odora 'Marginata'
"Variegated daphne"

Daphne odora, *from China,*
intensely fragrant "Winter daphne"

Tilia x europaea
"Common lime tree"

Tilia x euchlora
"Crimean linden"

Sparmannia africana
"African hemp" in flower

TURNERACEAE, UMBELLIFERAE, VIOLACEAE

Actinotus helianthi
"Flannel flower" (Australia)

Aegopodium podagraria 'Variegatum'
"Bishop's weed" (Europe)

Viola tricolor
"Climbing greenhouse pansy"

Viola tricolor hortensis 'Maxima' (European origin)
the charming "Garden pansy", for early bloom outdoors

Viola hederacea
"Trailing violet" (Australia)

Muntingia calabura
"Panama berry" (Trop. America)

Turnera ulmifolia
"West Indian holly"

Apeiba tibourbon
(Guayana)

TAMARICACEAE, TROPAEOLACEAE, UMBELLIFERAE, URTICACEAE

Tamarix africana, *"Salt cedar"*
wind-blown tamarisk on arid south shore of Tenerife

Tropaeolum majus, *"Indian cress"*
in hanging basket along street in Bergen, Norway

Eryngium alpinum, *"Alpine thistle"*
in Royal Botanic Garden Edinburgh, Scotland

Helxine soleirolii, *from Sardinia and Corsica*
known as "Baby's tears" or "Mind-your-own-business"

Pilea grandis, *from Jamaica*
in tropical house of Munich Botanic Garden, Germany

Pilea mollis, *from Costa Rica and Colombia*
better known as 'Moon Valley', with exquisite carpet leaves

URTICACEAE

Pilea cadierei
"Aluminum plant" (*Vietnam*)

Pilea serpyllacea
"Creeping Charley" (*Mexico*)

Pilea spruceana 'Norfolk'
"Angelwings" (*Perú*)

Pilea 'Silver Tree'
"Silver and bronze" (*Caribbean*)

Pellionia pulchra
"Satin pellionia" (*Vietnam*)

Pellionia daveauana
"Trailing watermelon begonia"

Pilea depressa (*Puerto Rico*)
"Miniature peperomia"

Pilea superba hort.
in Florida nursery trade

Pilea microphylla
"Artillery plant" (*W. Indies*)

VERBENACEAE

Clerodendrum inerme, *the "Indian privet"* as topiary in temple garden, Bangkok, Thailand

Clerodendrum inerme, *the Hindu "Koyanal"* trained as elephant, in Mehta Hanging Gardens, Bombay

Clerodendrum fragrans pleniflorum (*China*) the fragrant "Glory tree", Bot. Garden Sydney

Petrea volubilis (*Mexico*) "Purple wreath", a most beautiful tender twiner, in Panama

Clerodendrum buchananii *of the West Indies,* along the north road to Bien Hoa, Vietnam

Clerodendrum paniculatum (*E. Trop. Asia*) "Kashmir bouquet", in Peradeniya Bot. Garden, Ceylon

VERBENACEAE

Oxera pulchella
(*New Caledonia*)

Holmskioldia sanguinea
"*Chinese hat plant*" (*Himalayas*)

Clerodendrum bungei
very fragrant, from China

Clerodendrum inerme
"*Indian privet*" (*India*)

Clerodendrum sahelangii
in Barbados

Clerodendrum wallichii
from India

Clerodendrum myricoides
in Kirstenbosch Bot. Garden

Clerodendrum splendens
(*Senegambia to Angola*)

Clerodendrum ugandense
in Tahiti Bot. Garden

VERBENACEAE

Petrea volubilis (*Mexico to Panama*)
"Purple wreath" vine, *flourishing in Kaptagat, W. Uganda*

Clerodendrum foxii hort.
enormous chandelier of scarlet flowers, in Florida

Clerodendrum speciosissimum (fallax)
the striking "Glory-bower", in the mountains of Java

Tectona grandis, *from monsoon forest in India, east to Java*
"Teak tree", *valued for its durable wood, in Durban, Natal*

TYPHACEAE, VALERIANACEAE, VERBENACEAE

Tectona grandis
"Teak tree" (Malaya)

Vitex lucens
"Pururi" in New Zealand

Duranta repens (W. Indies)
"Pigeon berry" in flower

Duranta lorentzii
(Argentina)

Duranta repens 'Variegata'
"Sky-flower" in Singapore

Duranta repens
"Golden dewdrop" in Martinique

Typha latifolia
"Cat-tail" of the marshland

Centranthus ruber
"Red valerian" (Europe)

Stachytarpheta speciosa
(Trop. America)

VERBENACEAE

Clerodendrum speciosissimum
Missouri Bot. Garden, St. Louis

Clerodendrum paniculatum
Peradeniya Bot. Garden, Ceylon

Clerodendrum x speciosum
"Glory-bower", Rio Bot. Garden

Clerodendrum speciosissimum
(fallax) *in Rio Bot. Garden*

Clerodendrum trichotomum (*Japan*)
in Bot. Garden Sydney

Clerodendrum thomsonae (*W. Africa*)
"Bleeding heart vine"

Holmskioldia sanguinea
"Chinaman's hat" (Himalayas)

Gmelina hystrix (*Philippines*)
Kew Gardens, London

Holmskioldia sanguinea 'Aurea'
"Mandarin hat"

VERBENACEAE

Congea tomentosa
"Shower orchid" in Ceylon

Petrea arborea (*Venezuela*)
"Queen's wreath tree"

Petrea volubilis
"Purple wreath" (*Mexico*)

Verbena tenera var. maonettii
(*Argentina: La Plata*)

Verbena peruviana 'Flame'
"Scarlet vervain" (*Perú*)

Verbena x hortensis
"Rainbow vervain"

Lantana camara
orange-red form (*W. Indies*)

Lantana montevidensis (*Uruguay*)
"Polecat geranium"

Lantana camara
"Yellow sage"

VITACEAE, VITIDACEAE

Cissus gongylodes (Vitis) *from Paraguay with long pendant aerial roots, in Munich Bot. Garden*

Cissus discolor, *known as "Rex-begonia vine", clambering over a sculptured Khmer Devata in Cambodia*

Cyphostemma laza, *curious xerophytic succulent in arid southwest Madagascar, by J. Bogner, Munich*

Cissus gongylodes (Vitis pterophora) *in the great conservatory at Kew Gardens, London*

VITACEAE

Tetrastigma (Cissus) voinierianum
"Chestnut vine" (*Indochina*)

Rhoicissus (Vitis) capensis
"Evergreen grapevine" (*So. Africa*)

Cissus juttae
(*Namibia, S.W. Africa*)

Leea rubra (*Burma*)
Bot. Garden Rio de Janeiro

Cissus amazonica (*Brazil*)
in Kew Gardens, London

Tetrastigma obovatum (*China*)
Kew Gardens, London

Parthenocissus quinquefolia
(Ampelopsis), *"Virginia creeper"*

Parthenocissus tricuspidata
"Boston ivy" (*China*)

Parthenocissus tricusp. 'Veitchii'
"Japanese ivy"

VITACEAE

Vitis vinifera 'Thompson Seedless', *leading table grape fruiting in sunny Vista, San Diego County, California*

Parthenocissus tricusp. 'Lowii', *"Miniature Japanese ivy" covering the impregnable walls of Osaka castle, Japan*

Cissus quadrangularis, *clambering 'Veld grape" on Euphorbia candelabrum, on the Achole steppe, Uganda*

Cissus juttae (Cyphostemma), *on the Namibian desert storing water for survival, in southwestern Africa*

VITACEAE

Cissus rhombifolia 'Mandaiana'
"Bold grape-ivy"

Cissus discolor
"Rex begonia vine" (Cambodia)

Cissus antarctica
"Kangaroo vine" (New So. Wales)

Cissus hypoglauca
(New South Wales, Victoria)

Cissus rotundifolia
"Arabian wax cissus" (Yemen)

Cissus (Vitis) rhombifolia
"Grape ivy" (W. Indies)

Cissus gongylodes
tropical climber of Brazil

Parthenocissus inserta
in Pink Zone, Mexico City

Cissus rhombifolia 'Ellen Danica'
introduced from Denmark

ZINGIBERACEAE

Alpinia purpurata, *the "Red ginger",
with brilliant red bracts, in Tahiti, French Polynesia*

Alpinia zerumbet (speciosa or nutans in hort.)
the "Shell ginger" or "Porcelain lily" from S.E. Asia

Curcuma roscoeana, *the "Hidden lily"
with vivid orange-scarlet spire, near Rangoon, Burma*

Tapeinochilus ananassae (*Malaysia to Queensland*)
"Giant spiral ginger" with majestic cones of scarlet

ZINGIBERACEAE

Kaempferia pulchra (*Burma*)
"Pretty resurrection lily"

Kaempferia galanga
(*India to Vietnam*)

Kaempferia grandiflora
(*East Africa: Kenya*)

Kaempferia gilbertii
"Variegated ginger-lily"

Alpinia calcarata
"Indian ginger" (*India*)

Alpinia calcarata
with split seed capsules

Costus igneus
"Fiery costus" (*Brazil*)

Brachychilum horsfieldii
with inflorescence (*Java*)

Costus sanguineus
"Velvet spiral flag" (*C. America*)

Zingiber capitatum
"Roundhead ginger" (*S.E. Asia*)

Alpinia sanderae
"Variegated ginger" (*New Guinea*)

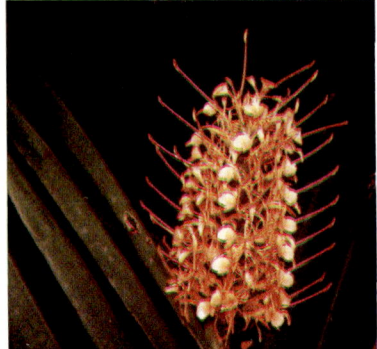

Hedychium coccineum
"Ginger-lily" (*Burma*)

ZINGIBERACEAE

Globba winitii, *an ornamental ginger with Mr. Somphong, collected near Ayuthia, Thailand*

Hedychium longicornutum, *"Scarlet ginger lily" a Malayan epiphyte, at W. Moir's garden in Honolulu*

Curcuma ionodora, *a tropical "Hidden lily" in rainforest habitat, Maharashtra state, east of Bombay*

Curcuma siamensis *collected by J. Bogner of Munich Bot. Garden, in Thailand*

Alpinia zerumbet 'Variegata' (speciosa in hort.) *the lovely "Variegated shell ginger" in Hawaii*

Kaempferia brachystemon, *a "Dwarf ginger lily", from the humid Usambara Mountains in Tanzania*

ZINGIBERACEAE

Costus spiralis
"Scarlet spiral flag" (Colombia)

Curcuma domestica
"Turmeric" from India

Costus spicatus (W. Indies)
"Indian head ginger"

Costus malortieanus
"Stepladder plant" (Costa Rica)

Kaempferia 'Pobeda'
(decora x kirkii elatior)

Costus speciosus, *in flower*
"Spiral ginger" (India)

Alpinia sanderae, *in flower*
Peradeniya Bot. Garden, Ceylon

Costus speciosus
with spiralling stem

Alpinia zerumbet 'Variegata'
(speciosa variegata in hort.)

ZINGIBERACEAE

Hedychium coronarium, *the "Butterfly lily"; lovely, fragrant "White ginger" in Bengal, India*

Hedychium gardnerianum, *the fragrant "Kahili ginger"; in the chilly Sikkim Himalaya at 2500 m altitude*

Zingiber zerumbet, *from India east to Polynesia the "Bitter ginger", cultivated for its aromatic roots*

Nicolaia elatior (*in hort. as* Phaeomeria magnifica) *the resplendent "Torch ginger" in Bogor, Java, Indonesia*

ZINGIBERACEAE, ZYGOPHYLLACEAE

Hedychium flavescens
"Cream ginger", in Melbourne

Hedychium gardnerianum
"Kahili ginger" (No. India)

Hedychium coronarium
"White ginger" in Florida

Zingiber zerumbet 'Variegata'
in Trinidad Botanic Garden

Alpinia taiwaniana
"Taipei ginger" in Taipei

Guaiacum sanctum
"Lignum vitae" (ZYG.)

Hedychium flavum
"Yellow ginger" (India)

Amomum cardamomum
"Cardamon ginger" (Java)

Zingiber officinale
"Canton ginger" (Malaya)

Descriptive Botanical Terms Illustrated
their Family, Origin, and Common Names

Combination Text-Index: Numerals at end of each listing indicate page numbers to plant illustrations, a short-cut to easier finding of photographs.

Plant Names other than species, in conformance with the International Code of Nomenclature, are generally distinguished as follows:

Names of Hybrids: Generic names of bigeneric hybrids, and Latinized specific names of hybrids derived from two species, are as a rule preceded by the x mark;

Cultivar names (horticultural sports or varieties, hybrids with fancy names and clonal selections) start with a capital initial letter, and are enclosed within single quotation marks (');

Names of uncertain standing, or where incorrectly used in horticulture, are shown in double quotes (''), and/or are followed by the abbreviation "hort".

Terms of measurement are generally given according to the International Metric System. Conversions to the Old English terms are:
1 centimeter (cm) = 0.4 inch;   2½ cm = 1 inch;   10 cm = 4 inches
1 metre (m) = 40 inches (or 3.28 feet). (1 cm = 10 mm; 1 m = 100 cm; 1 foot = 30 cm)
1 gram (g) = 0.035 oz; 1 kilogram (kg) = 2.2 lbs; 1 liter (l) = 1.06 quarts; 4 liters = 1.06 gal.

Temperature Conversion: Degrees Fahrenheit vs. Centigrade.
Freezing point zero deg. Centigrade = 32 deg. Fahrenheit (F.).
Boiling point 100 deg. Celsius (C) = 212 deg. Fahrenheit (F.).

CLIMATE and TEMPERATURE GUIDE, For General Orientation on Plant Requirements and Care.

Plants in the text following are designated **TROPICAL, SUB-TROPIC, WARM-TEMPERATE,** and **TEMPERATE** to indicate the environment of their native habitat, or that which they would prefer to best succeed in gardens or indoors. However, most living beings are very tolerant and flexible, and experience has shown that a suggestion of temperature preferences does not necessarily mean that plants would not adapt to different living conditions—cold tolerant subjects to warmer climate, or tropicals to more rigorous and cooler regions—once they are acclimated and reasonably sheltered.

TROPICAL: warm surroundings, with hot or humid-warm days 21-28°C (70-85°F), and balmy nights, where temperature should normally not go below a minimum of 15° Celsius (60°F), never with frost. Primary tropical climates are found round the world, except at higher elevations, in Central and No. South America, West Indies, West and East Africa, India, Malaysia, Philippines, Indonesia, South Pacific, Hawaii.

SUBTROPIC: the mild climate typical of Southern California, Florida, Mediterranean Region, South Africa, Southern So. America, Southern Australia; with warm to hot sunny days, the nights with temperatures down to 10°C or 8°C (50°F or 45°F), rarely with frost in winter. (U.S. Hardiness zones 9-10).

WARM-TEMPERATE: climate as prevailing in the Southern U.S., Oregon; England to S.E. Europe, Asia Minor, Japan, So. New Zealand; warm and often rainy in daytime; minimums at night generally down to 5° Celsius (40°F), with some frost and soil freezing occasionally in winter. (U.S. Hardiness zones 7-8).

TEMPERATE: cool climate, such as in the Northern United States, Southern Chile and Argentina, Northern Europe, Eastern China. Frigid winter temperatures dropping to under zero 0° Celsius (32°F or less), with heavy snow and hard freezes, especially nights. This chilling causes perennials and bulbous plants to go dormant, and to initiate flower buds and fruit wood in deciduous trees. (U.S. Hardiness zones 4-6).

HUMID-TROPICAL and **HUMID-SUBTROPIC** is for plants which need high humidity with their roots kept moist, and requiring more water and attention than those with **ARID-TROPICAL** or **ARID-SUBTROPIC** backgrounds. Such xerophytes from arid deserts and regions of sparse rains require bright light or tolerate burning sun.

TEMPERATE climates are usually blessed with adequate rainfall or snow to maintain plants and trees in moist condition sufficient for their needs.

PLANT DESCRIPTIONS with Photo Index

ABELIA *Caprifoliaceae*
   'Edouard Goucher' (grandiflora x schumannii); evergreen shrub, lower and lacier than grandiflora, 1-1½m high; small lilac-pink tubular flowers with orange throat. *Warm-temperate*   p. 290

   floribunda (Mexico), "Mexican abelia"; evergreen shrub 1-2m high, with arching wiry, downy red branches; ovate, glossy green 3cm leaves; from the ends of twigs the showy pendulous trumpet flowers 4cm long, reddish purple with white throat. *Subtropic*   p. 290

   x grandiflora (chinensis x uniflora) (China), "Glossy abelia"; attractive half-evergreen shrub with opposite ovate, glossy leaves; terminal clusters of small bellshaped fragrant flowers white flushed pink. *Warm temperate*   p. 290

ABIES *Coniferae: Pinaceae*
   bracteata (California), "Bristle-cone fir"; pyramidal tall tree 30-60 m high, with spreading lower branches and slender crown; flat, stiff needle-like leaves spine-pointed, 4-6 cm long, shining green; broad white lines beneath; female cones erect, roundish and bristly, 10 cm long. *Warm temperate.*   p. 342

   concolor (Colorado to Arizona and Mexico), "White fir"; noble timber tree 25 to 50 m tall, of symmetrical pyramidal habit, and with gray bark; branches horizontal in tiers; flat bluish-green, stiff needles 5 to 8 cm long; purplish glaucous, cylindric, erect cones to 14 cm long. *Temperate.*   p. 342

   concolor 'Compacta', "Dwarf Colorado fir"; rare form; dwarf, irregular, compact evergreen shrub, with stout glaucous blue needles, curving or sickle shaped, to 4 cm long, borne in bundles toward tips of branches. *Temperate.*   p. 342

   koreana (Korea), "Korean fir", sparry branches with blue stiff needles all around branch; long erect cones. *Warm-temperate*   p. 342

nordmanniana (Greece, Caucasus, Armenia), "Nordmann fir"; vigorous pyramidal fir 10 to 40 m tall, branches horizontal in dense tiers; thick-set with flat, grooved needles, rounded or notched at apex, shining bluish-green, 2-3 cm long, with silvery bands beneath; erect, ovoid green cones to 20 cm long. *Warm temperate* p. 342

ABUTILON Malvaceae

x hybridum, the "Flowering maple" or "Parlor maple"; an old-fashioned house plant, typical of the numerous hybrids of the tropical American species striatum, pictum and darwinii, resulting in this floriferous group of herbaceous shrubs with pubescent, soft green foliage varying from lobed to not lobed, often resembling maple-leaves, and with bell-like flowers in shades from white through yellow and salmon to red; in bloom the year round. *Tropical.* p. 621

x hybridum 'Apricot', "Chinese lantern"; one of the maple-leaved cultvars of the "Flowering maple", with pubescent green foliage and bell-like salmon flowers with darker red veining, flowering all year. *Tropical.* p. 622

x hybridum 'Old Rose'; a good strain of "Flowering maple" or "Parlor maple"; strong grower of sturdy habit, with soft-woody branches, and maple-like lobed, or unlobed leaves; the cup-shaped flowers a good rich rose, blooming 12 months of the year. Excellent both as pot plant and for the garden. *Tropical.* p. 623

x hybridum 'Orange Red'; soft-woody shrub with variable, lobed and toothed leaves; large and showy, solitary flowers deep salmon red with red veining. *Tropical.* p. 623

x hybridum 'Satin Pink Belle'; a good "Chinese lantern" of the Belle strain offered by Logee, Connecticut; selected hybrid with large cupped, luscious rose-pink flowers. *Tropical.* p. 623

x hybridum 'Savitzii', "White parlor maple"; colorful old cultivar of bushy habit, with its little maple-like lobed and toothed leaves grayish green and highly and irregularly variegated white from the margins in, some leaves are entirely white. *Tropical.* p. 623

x hybridum 'Souvenir de Bonn', "Variegated flowering maple"; herbaceous shrub with soft, long-stalked, maple-like leaves a grayish green bordered in creamy-white; bell-shaped flowers salmon veined with crimson. Grows to twice the size as 'Savitzii'. *Tropical.* p. 621, 622, 623

x hybridum 'Yellow Belle'; a strong-growing "Parlor maple" with fresh green leaves, and large bell-shaped vivid yellow flowers 3-5 cm across. *Tropical.* p. 622

megapotamicum (So. Brazil), "Trailing abutilon"; slender-branched shrub to 2 m high, with woody, wire-like arching twigs; leaves arrow-shaped, sometimes lobed, 4-6 cm long; pendulous lantern-like flowers, with tubular inflated red calyx, and yellow petals nearly closed. *Tropical.* p. 622

megapotamicum variegatum (Brazil), "Weeping Chinese lantern"; evergreen shrub of lax, graceful habit, the slender, drooping branches with small, arrow-shaped, crenate leaves fresh-green, with ivory to yellow variegation, and small pendulous flowers lemon-yellow with lantern-like red calyx. *Tropical* p. 623

'Thompsonii' (Guatemala), "Spotted flowering maple"; herbaceous shrub with slender branches; soft, maple-shaped leaves deeply 5-7 lobed, the middle lobe narrowed at base, dark green with chartreuse-yellow mottling; bell-like orange-salmon flowers; leaves not pubescent. *Tropical.* p. 622, 623

ACACIA Leguminosae

alata (Western Australia); flattened cladodes giving rise to small globular, yellow flower heads. *Subtropic.* p. 538

argyrophylla (Australia); tall shrub with gray-glaucous 3 cm phyllodes covered by silvery hair; flower heads yellow, in short clusters. *Arid-subtropic.* p. 539

armata (paradoxa) (Australia), "Kangaroo thorn"; dark green, densely branched erect shrub, to 3 m, with stems ribbed and bristly, dense phyllodia half-ovate, to 2½ cm long; flowers rich yellow in 1 cm globose heads, close to stem; willingly blooming for Easter, even as a small pot plant. *Subtropic* p. 538, 539

baileyana (New So. Wales), "Golden mimosa"; handsome spreading tree, to 10 m, with branches often pendulous, very leafy with fern-like bipinnate, glaucous, bluish-silvery leaves; the fluffy bright yellow, fragrant, globose flower heads massed in great sprays of clustering racemes, blooming into April. *Subtropic.* p. 539

cultriformis (New So. Wales, Queensland), "Knife acacia"; bushy shrub to 2 m, or more, with stiff, spirally set, triangular phyllodia silvery-gray and glaucous; small fluffy, bell-shaped yellow flowers in long, arching racemes forming a terminal panicle, during February-April. *Subtropic.* p. 538

decurrens (New South Wales to Tasmania), "Green wattle"; tree 18 m and upwards, densely leafy; leaves bipinnate, 8-12 cm long, green; flowers yellow, fragrant, in numerous globose heads 5 mm wide, forming large panicles. *Subtropic.* p. 539

drepanolobium (Kenya: Rift Valley; Uganda, Sudan). Known in arid regions of Africa as "Whistling thorn" or "Black-galled acacia". Low tree with horizontal twigs forming a flattened top, bark blackish, wth long straight white spines to 8 cm long, deciduous, bipinnate, glaucous leaves, and fragrant creamy-yellow flower heads; large hollow, black galls form at base of spines, said to be caused by ants who hollow them out, and when wind blows through the holes in the galls it sounds like a whistle. *Arid-subtropic.* p. 537

drummondii (Western Australia); free-blooming shrub to 3 m, with shoots furrowed, downy, and bipinnate leaves 5 cm long, main divisions 2 or 3, each with 2-6 pairs of oblong leaflets, smooth, pale bluish-green; flowers lemon-yellow, in dense cylindrical, drooping spikes to 3 cm long, in March-April. *Subtropic.* p. 539

heteracantha (Uganda, Tanzania, Botswana), "Umbrella thorn"; flat-topped tree of the bush veld to 10 m high, with stiff, needle-like thorns in pairs; compound 3 cm leaves of tiny leaflets; small white flower heads are sweetly scented. *Arid-Tropical.* p. 537

latifolia (No. Australia), "Bush acacia"; attractive flowering shrub with glaucous leaflike stems to 15 cm long and 5 cm wide; the cylindrical flower heads, in loose spikes, to 5 cm long, yellow. *Subtropic.* p. 538

longifolia (Australia, Tasmania), "Sydney golden wattle"; tree to 10 m of willowy, spreading habit, the arching branches with long linear, leathery dark green phyllodia to 15 cm, the bright yellow, small globose flowers in thin risps along stem during February-March, rather short-lived. *Subtropic.* p. 539

longifolia mucronata (Australia), "Narrow Sydney wattle"; small spreading tree with lightly drooping branches having narrow linear, stiffly thick phyllodia only 5-8 cm long, and the lemon-yellow flowers in long, 3 cm cylindrical, fluffy spikes along the stem, March blooming. *Subtropic.* p. 538

mearnsii (Australia), "Black wattle"; black-barked, bushy tree to 6 m high, naturalized in Africa where it is considered a "beastly" wattle; I have encountered these at 2,700 m in the mountains of Kigezi, in Central Africa, with finely cut lacy, somber green foliage and attractive ball-shaped lemon-yellow clusters of flowers. *Subtropic.* p. 539

nigrescens (Africa: Rhodesia, Mozambique, Transvaal), "Knobthorn"; a timber tree in Africa, to 12 m high, light green, with bipinnate, deciduous leaves, and sharp, recurved thorns, which will grow into thorny knobs; sweetly scented, elongate or cylindrical flower heads pale lemon-yellow. A striking flowering tree as observed in the landscape of the dry veld of the northern Transvaal. *Subtropic.* p. 540

podaliriaefolia (Queensland), "Pearl acacia"; tall, silvery-gray, pubescent shrub to 3 m, set with ovate, silvery glaucous phyllodia; flowers in axillary racemes, 5-10 cm long, carrying up to 20 globose 1 cm heads of golden yellow; December-January blooming. *Subtropic.* p. 538

retinodes, in California nurseries as longifolia floribunda, and commonly planted on the Riviera as A. "floribunda"; (South Australia), the "Everblooming acacia"; dense, upright tree to 8 m, having most of its foliage toward the ends of its branches, with narrow linear, dark green phyllodia to 12 cm long, and small ½ cm globular light yellow, fragrant flowers in loose, clustered racemes 15-25 cm long, blooming constantly from February nearly all year. *Subtropic* p. 537, 538, 539

sphaerocephala (Mexico), tree with bipinnate, feathery leaves, armed with hollow, horn-like, ant-tenanted thorns; globose flower-heads yellow. *Subtropic.* p. 538

woodii (Uganda to Natal), "Paperbark-thorn"; striking flat-topped tree to 8 m high, with corky, yellowish bark, which peels off in papery strips; leaves 10 cm long, of numerous small leaflets; covered with golden velvety hairs; small whitish flower heads. *Arid-Tropical* p. 537

ACALYPHA Euphorbiaceae

godseffiana 'Heterophylla' (New Guinea); form with drooping branches and ragged foliage sometimes reduced to shreds, these narrow, lacy leaves are green with pale yellow, wavy edges. *Tropical.* p. 409

hispida (sanderi) (India), "Chenille plant"; showy tropical shrub with broad ovate, bright green, hairy leaves with crenate

margins; bright red flowers in long pendant spikes resembling foxtails. *Tropical.* *p. 408, 423*
    hispida alba, "Philippine Medusa"; variety having drooping tassel-like, pistillate, creamy-white spikes, tinted pink. *Tropical* *p. 408*
    wilkesiana 'Ceylon', "Fire-dragon"; handsome tropical bush with woody stems covered toward the ends with curiously twisted roundish, coppery maroon leaves about 12-15 cm long, the margins prettily bordered with white to bright pink little lobes. *Tropical* *p. 409*
    wilkesiana 'Hoffmannii'; beautiful ornamental shrub with woody branches displaying a dense bouquet of twisted grass green leaves, fringed by short lobes of ivory white; photographed in Honolulu. *Tropical.* *p. 409*
    wilkesiana 'Java white'; broad-ovate leaves variegated or entirely creamy-white. *Humid-Tropical.* *p. 409*
    wilkesiana macafeana (New Hebrides), "Copper-leaf"; robust, branching shrub dense with red ovate leaves marbled crimson and bronze, margins serrate; the slender flower spikes are red. *Tropical.* *p. 409*
    wilkesiana 'Macrophylla' (South Sea Isl.), "Heart copperleaf"; form with large pointed leaves with cordate base, russet-brown alternating with metallic bronzy green and bright copper. *Tropical.* *p. 409*
    wilkesiana 'Marginata', "Copper leaf"; narrow lanceolate leaves coppery, with serrate margins pink or red. *Humid-Tropical* *p. 409*
    wilkesiana 'Moorea'; spectacular tropical bush photographed on Moorea in French Polynesia, with broad, blackish coppery, waxy leaves having crenate margins and twisted in coquette fashion; small greenish-red flowers in axillary spikes. *Tropical.* *p. 408*
    wilkesiana obovata (Polynesia), "Heart copperleaf"; large obovate leaves emarginate or notched at apex, green edged cream-white when young, later changing to copper with orange-rose margins. *Tropical.* *p. 408*
    wilkesiana 'Tahiti', "Match-me-if-you-can"; twisted broad leaves moss-green with yellow and cream, the scalloped margins cream or pink. *Humid-Tropical.* *p. 409*
    wilkesiana 'Tricolor'; spectacular color variegation in the foliage, shades of green with cream, pink, orange and red. *Humid-Tropical.* *p. 409, 423*

ACANTHOCALYCIUM   *Cactaceae*
    klimpelianum (Argentina), "Argentina barrel cactus"; small globe to 10 cm dia., dark green with numerous ribs and bundles of needle spines; flowers white. *Arid-Subtropic.* *p. 264*
    violaceum (Argentina), "Violet sea-urchin"; globular cactus resembling Echinopsis, to 20 cm high, deep olive green, with 15 acute ribs lightly notched, and straight yellowish needle spines; funnel form flowers pale violet. *Arid-Subtropic.* *p. 265*

ACANTHOSTACHYS   *Bromeliaceae*
    strobilacea (So. Brazil, Paraguay, Argentina), "Pinecone bromeliad"; epiphytic plant with long pendant, very narrow, succulent and channeled leaves deep green with gray scurf and spiny; inflorescence on reed-like stems bearing red cone-like fruit. *Subtropic.* *p. 200*

ACANTHUS   *Acanthaceae*
    ilicifolius (Trop. Asia); tender spiny shrub, aquatic in tropical brackish waters, to 1½ m high, forming stilt roots in bogs or shallow water; narrow oblong leaves shiny dark green with spiny lobes, 20-30 cm long. *Tropical.* *p. 36*
    mollis (So. Europe), "Greek acanthos"; perennial herb with large, glossy, lobed leaves cordate at base, not spiny. Flowers lilac. *Subtropic.* *p. 36*
    montanus (W. Trop. Africa), "Mountain thistle"; shrub with decorative, hard, black-green, pinnatifid, spiny leaves; flowers rose tinted, in terminal spike. *Tropical.* *p. 36*
    pubescens (Kenya); attractive semi-woody bush with leathery, deeply lobed, pointed leaves armed with spines; irregular brick-red flowers in showy spikes. *Tropical.* *p. 33*

ACCA   *Myrtaceae*
    sellowiana 'Variegata' (Psidium) (Brazil); scrambling shrub seen in Cologne Botanic Garden, Germany; with wiry stems, and small, leathery, obovate leaves 3 cm long, prettily bordered with cream; the inflorescence with bright red stamen bundles. *Subtropic.* *p. 676*

ACER   *Aceraceae*
    negundo 'Variegatum' (New England to California and Guatemala), "Variegated box elder" or "Ash-leaved maple"; deciduous shrub or small tree with handsome variegated foliage; distinctive pinnate leaves with 3 to 5 or more toothed leaflets, fresh green with broad cream-white margins; flowers appear before leaves. Photographed on Lago Maggiore, Switzerland. *Warm-temperate.* *p. 41*
    palmatum (Japan, Korea), "Japanese maple"; attractive deciduous small tree or shrub to 8 m or more, with membranous leaves deeply 5-9 lobed, in forms green to crimson; flowers purple. Frequently dwarfed and trained in containers as 'Bonsai' in Japan. *Temperate.* *p. 41*
    palmatum 'Atropurpureum', "Red Japanese maple"; favorite color-form in garden plantings; handsome saw-toothed leaves or rich crimson-purple or coppery bronze, deeply divided into lanceolate lobes; propeller-like winged fruit. *Temperate.* *p. 41*
    palmatum 'Sansokaku'; a rare Japanese cultivar of slender habit; finely and deeply cut light green, delicate leaves, carried in contrast on striking red branches. Photographed at North Haven Gardens, Dallas, Texas. *Warm temperate.* *p. 41*
    platanoides (Europe to Turkey and Iran), "Norway maple"; round-headed deciduous tree to 20 m or more high; smooth branches with 5-lobed, glossy 10-18 cm leaves, the lobes sharply pointed; small greenish yellow flowers before the foliage, followed by attractive fruit with spreading wings. *Temperate.* *p. 41*
    saccharum (No. America to Texas), "Sugar maple"; deciduous tree with silver-gray bark; leaves 10-15 cm dia, distinct with 3 to 5 lobes and deep rounded sinuses; flowers without petals, appearing before leaves; striking autumn foliage in colder regions, turning golden yellow to bright crimson. The sap yields sugar or syrup. *Temperate.* *p. 41*

ACHILLEA   *Compositae*
    filipendulina (Asia Minor, Caucasus), "Fern-leaf yarrow"; stiff, erect perennial herb 1-1½ m high, hairy leaves finely dissected; small flower heads in large, flat clusters, all yellow. *Warm temperate.* *p. 308*

ACHIMENES   *Gesneriaceae*
    candida (Dicyrta) (Guatemala), "Mother's tears"; low growing hairy herb with scaly rhizome, red-brown wiry stems and rugose, serrate, oblique leaves; small nodding, curved, funnel-shaped flowers white with yellow, buff outside, purple spotted throat. *Humid-tropical.* *p. 492*
    x hybrida: the "Magic flowers" are generally low, tropical American herbs with peculiar scaly rhizomes, slender stems that trail gracefully over the side of a pot or basket then bend and grow upright; showy axillary tubular to flat-faced flowers from 1 to 8 cm across cover the plant in summertime and fall, in great profusion. Many good, free-blooming hybrids have been produced since 1840. Achimenes make wonderful hanging plants but need warmth with humidity for growth until budded; after blooming they require a period of dormancy through fall and winter. *Humid-tropical.* *p. 493*
    longiflora 'Alba' (Guatemala); known as 'Jaureguia maxima', a graceful, free-blooming variety for hanging baskets, with large, snowy-white flowers with yellow throat marked purple, the large oblique limb 8 cm wide. *Humid-Tropical.* *p. 492*
    longiflora 'Andersonii', "Monkey-faced pansy"; with slender stems distantly bearing small hairy, opposite, grass-green, ovate leaves, and axillary, short stalked, dipping salverform flowers of purplish-blue. *Humid-Tropical.* *p. 492*

ACHRAS: see Manilkara
ACHYRANTHES: see Iresine

ACMENA   *Myrtaceae*
    smithii (Eugenia) (E. Australia), "Lilly-pilly"; evergreen tree to 18 m high; with thickish ovate leaves to 9 cm long, opening in rich bronze; terminal clusters of small snow-white flowers, followed by showy edible, pinkish or purple small berries. *Subtropic.* *p. 460, 766*

ACOKANTHERA   *Apocynaceae*
    spectabilis (So. Africa), "Wintersweet"; evergreen shrub with narrow-oval, leathery, glossy leaves and clusters of sweetly-scented, pure-white flowers. *Subtropic.* *p. 74*

ACORUS   *Araceae*
    calamus (No. and C. Europe, Temp. Asia, S. E. Canada, U.S.A.), "Sweet flag"; robust water-loving perennial herb of marshlands, to 2 m high; creeping rhizome containing aromatic cells, with flat iris-like leaves arranged in flat, shingled fans, the linear grass-green leaves parallel-veined, 2 cm wide and with thick midrib; the inflorescence on a cylindrical spadix 5-10 cm long thickly covered with minute greenish-yellow perfect

flowers. Winter-hardy. The candied rhizome was an old-time confection. *Warm temperate.* p. 90
  calamus variegatus (N. Hemisphere), "Sweet flag"; bog plant with flat, iris-like leathery leaves, green with broad, white, lengthwise variegation. *Warm temperate.* p. 88
  gramineus variegatus (Japan), "Miniature sweet flag"; water-loving perennial with creeping rhizomes and tufted, linear, flat, leathery leaves, light-green and white, spreading fan-like. *Tropical.* p. 88

ACROCOMIA *Palmae*
  intumescens (Brazil); bold feather palm with smooth trunk, swollen in lower part; dense spiny crown of feathery fronds gracefully arching, the leaflets glossy green and pendulous. *Tropical.* p. 764
  totai (No. Argentina, Paraguay, Bolivia), "Gru-Gru" palm; tall palm 15 to possibly 25 m, the trunk set with stout spines, smallish pinnate fronds smooth and green on both sides, the pinnae 2 cm wide. *Subtropic.* p. 764

ACROSTICHUM *Polypodiaceae (Filices)*
  aureum (Old and New World Tropics), coarse "Swampfern"; 1 to 3 m high, with stout rootstock and dark green, thick-leathery pinnate fronds, only the upper pinnae fertile, smaller than the barren pinnae, and covered with the deep reddish-brown coating of the spore-cases. A leather fern growing luxuriantly on the Florida Keys, in mangrove swamps and salt marshes, forming dense masses. *Humid-Tropical.* p. 438
  danaeifolium (excelsum) (Trop. America), tall "Swampfern"; with pinnate fronds 1-4 m high, with most or all of the pinnae fertile and bearing spores, on mature fronds. *Humid-Tropical.* p. 438

ACTINIDIA *Dilleniaceae*
  chinensis; (China, Taiwan); "Chinese gooseberry" or "Kiwi vine"; woody vine twining to 10 m, sparsely furnished with large 12 to 20 cm, rough, corrugated ovate leaves, dark green above, white velvety beneath; small 3-4 cm flowers cream turning to yellow in August, followed about November, in the Northern hemisphere, by pendant ovoid, brownish fruit 4-5 cm diameter; delicious to eat, with acid, gooseberry-like flavor; male and female flowers usually on separate plants. Hortus III refers this to Actinidiaceae. *Subtropic.* p. 41, 473
  kolomikta (Japan, Manchuria, China), "Kolomikta vine"; shrubby twining vine to 5 m, grown for their attractive foliage; the long-stalked heart-shaped leaves 8-12 cm long, variegated green with white or pink, some are all white, others with red, more so on male plants; for colorful arbor, trellis or wallcovering; cup-shaped white, axillary flowers; yellow or greenish fruit on female plants. Hortus III refers this to Actinidiaceae. *Warm temperate.* p. 41

ACTINOTUS *Umbelliferae*
  helianthi (E. Australia), "Flannel flower"; small subshrub 30-50 cm high, with compound white-woolly leaves; longstalked daisy-flowers with outer rays of cream "flannel", 5-10 cm across. *Subtropic.* p. 910

ADA *Orchidaceae*
  aurantiaca (Colombia); epiphyte with compressed pseudobulbs bearing 1-3 tapering leaves to 22 cm long; arching racemes of showy flowers red-orange 3 cm long, often spotted black; petals and sepals narrow, with shorter lip (WinterSpring). *Humid-Tropical.* p. 708

ADANSONIA *Bombacaceae*
  digitata (across Trop. Africa), the giant "Baobab" or "Sour gourd" or "Monkey-bread-tree"; one of the largest trees in the world, thought to become more than 2000 years old, and while only to 18 m tall, the swollen trunk attains a diameter of more than 10 m, is of pulpous wood without growth rings. Leaves deciduous, digitately compound with leaflets 12 cm long; large 18 cm solitary, scented, pendulous white flowers with purplish stamens; oblong, woody, hairy fruit 30 cm long, which also earns it the name "Dead rat-tree". *Tropical.* p. 195, 454

ADENANDRA *Rutaceae*
  fragrans (So. Africa), "Dwarf China flower"; small shrublet 30 cm high, green aromatic leaves 4 cm long; fragrant, bright pink flowers with 5 petals, 2 cm across. *Subtropic.* p. 872
  uniflora (Diosma) (So. Africa), "China flower"; small evergreen shrub with erect slender branches, 1 cm lanceolate leaves, and solitary flowers 3 cm wide, the 5 petals white with a deep rose streak, anthers brown-purple; late spring. *Subtropic.* p. 872

ADENIA *Passifloraceae*
  subglobosa (E. Africa); weird-looking succulent with massive swollen base to 1 m thick, looking like a gray-green stone, topped by pencil thick clambering gray twigs 3-4 m long, normally leafless and armed with long, stout thorns; during rains with narrow lanceolate leaves; grouped inflorescence with star-like, shining-red, odorous flowers. *Arid-Tropical.* p. 814

ADENIUM *Apocynaceae*
  obesum (coetanum) (E. Africa, Tanzania, Kenya, Uganda), the "Desert Rose"; spreading succulent bush 2 m high, with thick, fleshy, twisted base and short branches; deciduous obovate, fleshy leaves glossy dark green with pink midrib, 8 cm long, when young sometimes with minute hairs; numerous showy flowers with spreading petals pinkish edged with carmine or all carmine-rose. *Arid-Tropical.* p. 75
  obesum multiforum (E. Africa to So. Arabia), "Impala lily"; most attractive succulent shrub, fat, swollen trunk 1-3 m high, with poisonous milky sap; during the growing season the thick, stubby branches are covered with dark green, simple, alternate, fleshy, deciduous, spirally arranged leaves; the funnel-shaped flowers white, edged with crimson, looking like bright stars in the yellow, dusty vegetation of the steppe; much relished by elephants and monkeys. *Arid-tropical.* p. 73, 75
  swazicum (Transvaal, Swaziland, Mozambique), "White Impala lily"; succulent bush of arid areas, with short obovate, thick leaves; showy flowers in terminal clusters, pure white with blood-red eye. *Arid-Tropical.* p. 75

ADHATODA *Acanthaceae*
  vasica (India, Sri Lanka); tropical shrub to 2 m or more high, with long-ovate, corrugated leaves to 20 cm long and covered with fine pubescence; the inflorescence in terminal spikes with tubular, two-lipped flowers to 2 cm across, white striped with red. *Tropical.* p. 35

ADIANTUM *Polypodiaceae (Filices)*
  macrophyllum (Trop. America); showy fern with creeping rhizome, the pinnate fronds 30-40 cm long on shiny, blackbrown stalks, the lower pinnae of barren frond 5-10 cm long, ovate, toothed, papery, yellow-green; reddish when young; the fertile leaflets narrower. *Humid-Tropical.* p. 439
  raddianum (cuneatum) (Brazil), "Delta maidenhair"; an old greenhouse favorite because of its tolerance, sturdiness and simple elegance of the dark green fronds with many small, firm leaflets having a wedge-shaped base, with veins running into sinus between lobes. *Humid-Tropical.* p. 439
  tenerum 'Farleyense' (Barbados), "Barbados maidenhair"; magnificent feathery, but delicate fronds gracefully drooping, rose-tinted when young, pea-green later, the leaflets large and lacily cut and crisped, base articulated, veins run into teeth; infertile. *Humid-Tropical.* p. 439
  tenerum 'Wrightii' (W. Indies, Mexico, Venezuela), "Fan maidenhair"; good potplant with graceful medium-size fronds of good texture, pink when young, later fresh-green, the fanshaped leaflets occasionally growing large and lacily lobed. *Humid-Tropical.* p. 439
  trapeziforme (Mexico and W. Indies south to Brazil), delicate looking, yet bold-growing "Giant maidenhair" with slowly creeping rhizome and large 2-pinnate fronds on black stems, the stalked trapezoid leaflets to 5 cm long and brilliant green. *Humid-Tropical.* p. 439

ADONIDIA: see Veitchia

ADONIS *Ranunculaceae*
  aestivalis (France, Germany), "Pheasant's eye"; annual herb 30-50 cm high, with dark green leaves divided into fine segments; flowers crimson with black spot at base. *Warm temperate.* p. 842

ADROMISCHUS *Crasulaceae*
  festivus (Cape Prov.), "Plover eggs"; fascinating clustering succulent with cylindric or thick spatulate leaves flattened toward the crested apex, silvery green marbled maroon. *Arid-subtropic.* p. 361
  maculatus of hort. (form of rupicolus) (Cape Prov.), "Calico hearts"; succulent rosette with few flat leaves almost round, gray green heavily blotched red-brown except at edge; flowers tipped red-white. *Arid-subtropic.* p. 361
  rupicolus (So. Africa); attractive succulent with densely clustered, broad obovate fleshy leaves silvery with brownpurple blotches and along margins, the edge a pale line. *Arid-subtropic.* p. 361

AECHMEA *Bromeliaceae*
  angustifolia (Costa Rica to Colombia, Perú and Bolivia); sturdy rosette to 90 cm high, with strap-shaped leaves 6 cm

wide, gray with brown-scurfy scales, spiny-serrate at margins; the red spike with bipinnate inflorescence of red bracts and yellow flowers, followed by white, then blue berries, long lasting. *Tropical.* *p. 204*
aquilega (Gravisia), (West Indies, Costa Rica, Venezuela, Guayana, Brazil); spreading rosette with concave strap-shaped, recurving leaves almost 1 m long, gray-scaly and with marginal spines; dense inflorescence nearly globular, with deep yellow bracts and orange flowers, on erect stem with red leaf-bracts. *Tropical.* *p. 211*
araneosa (Brazil); attractive open rosette from Espirito Santo, with narrow bright green, shiny leaves to 50 cm long, and edged with brown spines; floral stem and bracts red, with branched, pyramidal cluster of numerous small prickly, greenish-yellow flowers. *Tropical.* *p. 200*
'Bert' (orlandiana x fosteriana); stocky rosette of short leathery leaves matte green marked with irregular purplish-brown crossbands, heavy dark spines; arching inflorescence with dense head of red bracts and pale flowers. *Tropical. p. 198*
'Black Prince'' showy rosette of broad, shiny blackish-green leaves; branched inflorescence with whitish berries; bracts scarlet, base and tips scarlet, on maroon stalk. *Tropical. p. 198*
'Black Wine'; broad, recurving leaves, glossy blackish purple; inflorescense heavy cluster of pinkish berries. *Tropical.* *p. 204*
bromeliifolia (B. Honduras, Guatemala to N.E. Argentina), "Wax torch"; large tubular rosette with variable leaves ½-1 m long, green with white-scaly coating, a few brown teeth toward apex, with tips curled under; erect, stout cylindric, long-lasting inflorescence, densely white-woolly, with leathery, broad floral bracts, the flower petals greenish-yellow soon turning black. *Tropical.* *p. 200*
caesia (So. Brazil); attractive epiphytic rosette to 15 leaves 35 cm long and 4 cm wide, green with dense black spines along margins, and rounded at tip; erect wiry floral stalk bears a dense inflorescence with rosy bracts and purplish-blue petals. *Subtropic.* *p. 200*
caudata (S.E. Brazil); robust rosette to 1 m high with leathery leaves 10-12 cm wide, glossy green, with weak marginal spines; inflorescence on wiry stem a pyramidal panicle with orange-yellow bracts and deep-yellow flowers. *Tropical. p. 200*
caudata variegata (Billbergia forgetii) (Brazil); big sparry rosette of rich green stiff leaves broadly banded creamy-yellow; bold inflorescence with white-mealy stem and panicle of yellow bracts and golden-yellow flowers. *Tropical.* *p. 200, 201*
chantinii (Venezuela, Amazonas, Amazonian Perú), "Amazonian zebra plant"; colorful open rosette of hard olive-green leaves with pronounced pinkish-gray cross-bands; inflorescence on branched spike with tight red bracts tipped yellow, supported by red bract leaves. *Tropical. p. 198, 199, 200*
chantinii 'Silver Monarch'; beautiful colorform with foliage cross-banded silver over bronze; the inflorescence with ribbonlike floral bracts scarlet-red, the flattened spikes yellowish; collected by Lee Moore in Amazonian Peru. *Tropical.* *p. 200*
chantinii 'Vera'; cultivar with rosette of leaves a lighter green than the species and with less pronounced cross-bands; seen in Arcadia, California. *Tropical.* *p. 198*
chlorophylla (So. Brazil); epiphytic or on rocks in the Organ Mountains; small species with glossy green strap-shaped leaves, covered with whitish scales; inflorescence with showy, glowing crimson bract leaves, the flowers with yellow petals, in slender cone. *Subtropic.* *p. 203*
coelestis 'Albomarginata' (S.E. Brazil); rosette of 12-20 narrow, concave, stiff gray green leaves to 50 cm long, differing from the species by having broad white margins; the reverse with gray-white cross-bands; few marginal spines; the inflorescence a branched reddish panicle, yellowish bracts, white calyx and blue corolla petals. *Tropical.* *p. 198*
dichlamydea var. trinitensis (Trinidad); handsome, majestic epiphyte with gray-green leaves 50-80 cm long; long, scandent inflorescence with pink, arching stalk and flattened lateral spikes bright coral, with densely shingled, blue bracts tipped with black-purple; flowers white with lilac, and deep blue berries. *Tropical.* *p. 199, 201, 206*
distichantha (So. Brazil, Bolivia to Argentina); upright dense rosette of stiff gray leaves; inflorescence a robust spike with faded rose bracts and purplish blue flowers. *Subtropic.* *p. 204*
'Electrica'; elegant Hummel-California hybrid involving Ae. dealbata x miniata x fasciata; large rosette with leaves dominated by wine and red; the bold inflorescence a compound head of orange bracts and lavender flowers. *Tropical* *p. 200*
fasciata (Billbergia rhodocyanea) (Rio de Janeiro), "Silver vase"; stocky rosette of leathery green leaves covered with gray scales and richly tigered silver-white; blackish spines; durable inflorescence in rose-colored globose heads with blue flowers. *Tropical.* *p. 198, 200, 201*
fasciata variegata, "Variegated silver vase"; variety with center of the channeled leaves attractively striped and banded ivory-white lengthwise through the regular green, and with silver cross-banding. *Tropical.* *p. 201*
filicaulis (Venezuela), "Weeping living vase"; open rosette with grass-green, thin-leathery, strap-shaped, oblanceolate leaves glossy on both sides and with dark mottling; tiny soft marginal spines; long pendulous flowering panicles on snaky, string-like axis with distant, red bract leaves and white flowers. *Tropical.* *p. 200*
fosteriana (Espirito Santo); striking tubular rosette of pale green to reddish green leaves with purplish brown irregular mottling and green spines; flower spike with panicle of crimson bracts and rich yellow petals. *Tropical.* *p. 199*
'Foster's Favorite Favorite'; a most beautiful plant, variegated sport of A. 'Foster's Favorite'; rosette with soft-leathery, glossy, strap-like leaves bordered by broad cream margins, the upper leaves tinted all over with glowing coppery rose to wine-red or maroon according to prevailing light conditions. *Tropical.* *p. 201*
fulgens (Pernambuco), "Coral berry"; loose rosette of stiff green leaves dusted gray; inflorescence in showy panicles with oblong red berries tipped with purple flowers. *Tropical.* *p. 201*
fulgens discolor (Pernambuco), "Coral berry"; free-growing rosette of soft-leathery dark olive-green leaves purple beneath, covered on both sides with glaucous gray crossbands; produces showy spikes with oval red berries tipped with violet flowers. *Tropical.* *p. 198, 200*
fulgens discolor 'Magnificent' (miniata x fulgens discolor cultivar); open rosette of broad olive green leaves lightly covered with gray scales; erect red floral stalk with pyramidal panicle of glistening oval red berries tipped by purple petals. *Tropical.* *p. 198*
gracilis (Brazil); small epiphytic rosette of dark green leaves 30 cm dia.; inflorescence a slender spike with tubular carmine red bracts tipped by blue purple flowers. *Tropical.* *p. 202*
hoppei (Perú); robust epiphyte of the Amazon region, with stiff erect, shiny green, strap-shaped leaves, the margins spiny; inflorescence with triangular boat-shaped, bright crimson bracts tipped with white; found by Lee Moore. *Tropical.* *p. 198*
hystrix (Argentina); long stiff erect channeled leaves; brush-like raceme of salmon-red flowers. *Subtropic.* *p. 203*
lueddemanniana (coerulescens) (So. America); stiff rosette with metallic green leaves mottled dark green and bronze base; flower spike with panicle of white berries turning a beautiful bright purple after flowering; petals lavender. *Tropical.* *p. 202*
magdalenae 'Quadricolor'; beautiful recurved spiny leaves with red to yellow in center. *Tropical.* *p. 203*
x maginali (miniata discolor x fulgens discolor); open rosette with broad soft-leathery olive-green glaucous leaves red-purple beneath; flower spike with oblong berry-like salmon-red bracts tipped by blue flowers. *Tropical.* *p. 202*
mariae-reginae (Costa Rica), "Queen aechmea"; robust rosette of broad, gray-green, leathery leaves, recurved and with toothed edge; stout spike with pendant, delicate pink bract-leaves, topped by cylindrical head of red-tipped berries and violet flowers. *Tropical.* *p. 202*
mertensii (Trinidad, Guayana, Venez., Colombia, Perú, No. Brazil), "China-berry"; epiphytic open rosette with few green leaves to 60 cm long, covered with white scales especially beneath, and having marginal spines; slender stalk with rose bracts, the inflorescence many-flowered, bipinnate with yellow or red petals, fruit blue. *Tropical.* *p. 202*
mexicana (Mexico); large rosette of broad leathery leaves pale green with darker green blotches becoming rose-tinted in the sun; bold flower stem with long panicle of white berries and red petals. *Tropical.* *p. 204*
miniata discolor (Brazil); open rosette of soft-leathery olive-green leaves, pale red reverse; inflorescence a panicled spike of orange-red rounded berries tipped by pale blue flowers. *Tropical* *p. 202*
mooreana (Amazonian Perú); very showy bromeliad with branched inflorescence similar to chantinii but with lower bracts carmine-rose, the upper, flattened bracts lime green tipped flame-orange. Foliage bronzy green. *Tropical.* *p. 204*
mulfordii (Gravisia fosteriana) (Atlantic coast of southeastern Brazil); large plant with stiff leaves; tall branched spikes of orange-red bracts, yellow flowers. *Tropical.* *p. 199, 205*

nidularioides (Colombia); epiphytic rosette of open habit, with strap-shaped, green leaves 60 cm long, at first white-scaly, and with broad marginal spines; globular inflorescence on short stalk, red-bracted and with white flowers. *Tropical.* p. 205

nudicaulis (Mexico and West Indies to Brazil); stiff tubular rosette of few, variable gray-green leaves to 40 cm long, armed with sharp teeth; arching slender red stalk carries a colorful inflorescence of rosy bract leaves and pinkish-yellow floral bracts and yellow flowers. *Tropical.* p. 202, 206

nudicaulis aureo-rosea (So. Brazil); close rosette of soft leathery, glossy deep green foliage having a natural fold near base of leaf; small flower spike with bright red bracts and flowers. *Subtropic.* p. 202

orlandiana x chantinii (Brazil); open rosette of broad, lacquered leaves, a burnt red and with blackish blotches. *Tropical.* p. 205

orlandiana variegata 'Ensign'; fancy open vase with spiny margined leaves, richly variegated with cream bands and margins, suffused with carmine areas. *Tropical.* p. 203

ornata nationalis (hystrix) (So. Brazil), "Porcupine aechmea"; beautiful variety, a formal agave-like rosette of stiff, vivid green leaves with cream-white margins and length-stripes, to 50 cm long and finely spiny; the inflorescence a dense elongate head to 20 cm high, carried on a stout stalk dressed with glossy rose-red bracts, floral head with green sepals tipped with brown spines, petals violet. *Tropical.* p. 203

phanerophlebia (So. Brazil); tightly closed rosette in habitat growing both epiphytic or on rocks; broad, rigid leaves to 50 cm long edged with stout black spines; erect cylindrical inflorescence with rosy-red bracts and blue flowers. Dr. L. B. Smith thinks the photo shown could be Ae. distichantha. *Subtropic.* p. 202

racinae (Espirito Santo), "Christmas jewels"; so called because of the striking orange-red berrylike inflorescence with yellow and black flowers, on pendant stem; shiny, friendly green, straplike leaves. *Tropical.* p. 202

ramosa (So. Brazil); large symmetrical rosette composed of many leathery medium green leaves coated with gray scurf; inflorescence a vermillion-red spike with loose panicle of greenish-yellow berries and yellow flowers. *Subtropic.* p. 202, 203

ramosa x fulgens; noble hybrid with broad, glossy green leaves and with inflorescence a large panicle of numerous yellow and red berries tipped by purple flowers. *Tropical.* p. 203

x 'Red Wing'; stunning Hummel hybrid involving Aechmea penduliflora x mutica; handsome rosette with large coppery leaves, dark purple beneath. The wine-red stalk bears an inflorescence of many berries first pink then purple, in heavy clusters; the flowers are straw-colored. *Tropical.* p. 203

serrata (Martinique); green rosette of stiff leaves; inflorescence an erect raceme with whitish bracts and lavender petals. *Tropical.* p. 206

tessmannii (Perú, Colombia); stiff ornamental rosette of grayish leaves with small spines; inflorescence on branched stem, the closed bracts pale orange-red each with subtended bract leaf, flowers yellow. *Tropical.* p. 206

tillandsioides (No. Brazil, Venezuela, Guayana); small epiphytic rosette with narrow, leathery, grayish leaves armed with marginal spines; inflorescence with serrated floral bracts green, yellowish or red; flower petals yellow, followed by berries first white then blue. *Tropical.* p. 202, 203, 209

tillandsioides var. 'Amazonas' (Perú); beautiful large epiphyte of the rainforest, with gray-green leaves; the stout inflorescence similar to chantinii, with spreading, flattened spikes arranged in candelabra fashion, the closely appressed bracts lavender pink and bright red, tipped by small yellow flowers; the long pendant bract-leaves at base of spikes rosy-red; collected by Lee Moore, 1962. *Tropical.* p. 206

triangularis (Brazil); epiphyte from the forests of Espirito Santo; squat, fluted rosette of stiff, broad leaves 8-10 cm wide and 40 cm long, dark metallic green and scaly, at margins with large black spines, the tips recurving; inflorescence cone-like, the red stalk furnished with showy red bracts; flower petals are purple, quickly turning black. *Tropical.* p. 206

victoriana (Espirito Santo); terrestrial; glossy flexible leaves; scandent inflorescence of red berries; purple flower. *Tropical.* p. 206

weilbachii (Brazil); rosette of coppery leaves, from center, several spikes with reddish bracts and purple flower. *Tropical.* p. 205

AEGLE Rutaceae

marmelos (India), "Bael fruit"; small spiny tree, with slender branches, the foliage divided into 3 slender leaflets with crenate edges; flowers with 4 or 5 narrow petals and showing numerous stamens; fruit is globular or pear-shaped, 5 to 10 cm dia., and covered with smooth, hard gray or yellow rind which contains orange, sweet aromatic pulp in 8-16 cells with seeds. The fruit pulp is used for drinks and conserves; the flowers are made into perfume in India. *Tropical.* p. 455

AEGOPODIUM Umbelliferae

podograria 'Variegatum', (Europe, natur. in No. America), "Bishop's weed" or "Goutweed"; coarse perennial herb 35 cm high, with creeping rootstock; compound herbaceous, wrinkled leaves milky-green with irregular cream-white margins; flowers white; attractive hardy border plant for the garden in shady places. *Warm temperate.* p. 910

AEONIUM Crassulaceae

arboreum (Morocco); erect succulent with thick stem to 1 m high, topped by a rosette of numerous thin, spatulate green leaves, ciliate white on margins; flowers golden yellow. *Subtropic.* p. 362

arboreum atropurpureum, "Black tree aeonium"; striking decorative variety with coppery to deep purple leaves. The species Ae. arboreum is Mediterranean from Morocco and Portugal east to Crete; an erect bold succulent to 1 m high, little branching, topped by a flaring 20 cm rosette of spatulate light green, fleshy leaves, fringed white at margins; flowers golden yellow. *Subtropic.* p. 362

arboreum 'Atropurpureum cristatum'; interesting fasciation of the gray-brown fleshy stem broadened obliquely fan-like, the curved apex supporting, comblike, a dense chain of tiny, deep lacquer-red rosettes, the little oblanceolate leaves with ciliate margin. *Subtropic.* p. 361

canariense (Canary Islands), "Velvet rose"; large rosette to 75 cm dia., with velvety leaves, light green, flushed red at tips. *Subtropic.* p. 362

decorum (cooperi) (Canary Isl.), "Copper pinwheel"; much branched shrublet topped by several open rosettes of small spatulate copper-colored succulent leaves and rose-tinted; flowers white with rose lines. *Subtropic.* p. 361

domesticum 'Variegatum' (Canary Isl.), "Youth and old age"; succulent intermediate between sedum and sempervivum, freely branching rosettes with rounded thin fleshy leaves light green with white margins; flowers yellow. *Subtropic.* p. 361

glandulosum (Madeira); attractive flat rosette of densely shingled spatulate and recurved leaves, green to deep coppery, sticky to the touch and with soft hairs; flowers golden yellow. *Subtropic.* p. 362

haworthii (Tenerife), "Pin-wheel"; bushy plant with short woody branches bearing rosettes of thick, obovate-acute gray-green leaves with ciliate, red margins; flowers pale yellow flushed rose. *Subtropic.* p. 362

holochrysum (Canary Islands); succulent rosette 20 cm across, or narrow spatulate, gray-green leaves tinted bronze, with red stripe and margins; stem forming; flowers golden yellow in long-stalked pyramids. *Subtropic.* p. 362

nobile (Canary Islands); magnificent large rosette of very fleshy, broadly channeled obovate, sticky and puckered leaves to 30 cm long, apple green; flowers coppery scarlet, in large showy clusters; a most beautiful aeonium. *Subtropic.* p. 363

'Pseudo-tabulaeforme' (tabulaeforme hybrid), "Green platters"; shrubby bush with distinct, thick, brown stem, with branches topped by flat open succulent rosettes, waxy green, leaves broad fan-like, ciliate at margins; golden flowers. *Subtropic.* p. 361

tabulaeforme (Tenerife), "Saucer plant"; circular, plate-like rosette of small spatulate leaves arranged flat like shingles, fresh green, margins ciliate; flowers yellow. *Subtropic* p. 361, 363

urbicum (Canary Isl.: Tenerife); unbranched, stem-forming large succulent rosette 25-30 cm dia., rather incurving, with spatulate leaves spoon-like, fresh light green and waxy, lightly keeled and with ciliate reddish margins; flowers greenish or pinkish; the plant dies after flowering. *Subtropic.* p. 362

AERANGIS Orchidaceae

coriacea (Kenya, Tanzania); vigorous epiphyte of vandaceous habit, with stem having alternate tongue-shaped, leathery leaves, widest at the apex and notched tips, glossy green and with marmoration above; floriferous with long axillary racemes of waxy star-shaped, fragrant flowers having long curving or spirally twisted spur to 18 cm long, creamy-white and tinted rose around the column, greenish toward tips, spur reddish-brown. *Tropical.* p. 708

kotschyana (Kenya, Uganda, S.E. Africa, Zaire, Nigeria); epiphyte resembling phalaenopsis in habit with oblongate leathery leaves 10-25 cm long, notched at apex, grayish-green often speckled; pendulous basal racemes of handsome, fragrant 5 cm flowers, white sometimes suffused with pink, lip rhomboid, the cord-like spur to 25 cm long, spirally twisted clockwise, and light brown. *Tropical.* p. 710

AERIDES   *Orchidaceae*
   falcatum (India); striking, free-growing epiphyte with 25 cm leathery bluish metallic leaves, closely ranged in 2 ranks, ½ to 1 m high; dense pendulous racemes of 2½ cm flowers creamy-white with a crimson spot at the apex of the sepals and petals, lip rose and ciliate (summer). *Tropical.* p. 709
   fieldingii (Assam), "Fox-brush orchid"; epiphyte to 1 m high with fleshy leaves; racemes often ½ m long, crowded with white flowers beautifully dotted and suffused with bright rose, the trowel-shaped lip rosy purple, (May-June). *Tropical.* p. 709
   virens (odoratum majus) (Java); handsome, free-growing epiphyte with 2-ranked, bright green, fleshy broad leaves, and waxy peach-pink flowers with yellow horns, very fragrant, on long, drooping racemes (April-July). *Tropical.* p. 709

AESCHYNANTHUS (TRICHOSPORUM)   *Gesneriaceae*
   javanicus (Java), "Lipstick plant"; trailing epiphyte with small ovate leaves slightly toothed; the flowers in terminal clusters, the tubular corolla downy, scarlet with yellow mouth, and the cup-like downy calyx purplish-red. *Tropical.* p. 492
   lobbianus (Trichosporum) (Java); epiphytic trailer with small elliptic, fleshy, dark green leaves; tubular, two-lipped flowers with hairy calyx cup soot-red glistening like silk, the downy corolla fiery red, creamy-yellow in throat, and only twice as long as calyx. *Tropical.* p. 492
   marmoratus (zebrinus) (Thailand), "Zebra basket vine"; epiphytic trailer with beautiful waxy leaves to 10 cm long, dark green with a reticulated network of contrasting yellow-green, and maroon underneath; tubular green flowers spotted brown. *Tropical.* p. 492
   micranthus (Himalayan reg.); epiphytic trailer with slender flexuous branches, small opposite elliptic leaves waxy green, and with 2-3 axillary, miniature slender tubular flowers deep purplish red edged blackish at petal tips. *Tropical.* p. 492
   parviflorus (ramosissimus) (Himalayas); trailer similar to grandiflorus, with deep glossy green lanceolate leaves 10 cm long; flowers scarlet, tipped yellow, the corolla 2½ cm long and scarcely contracted. *Tropical.* p. 491
   pulcher (Trichosporum) (Java), "Royal red bugler"; trailing epiphytic plant with small opposite ovate, waxy light green leaves and showy tubular flowers axillary or in terminal clusters; calyx green and smooth, the bilabiate corolla 3 times longer, vermillion red with yellow throat. *Tropical.* p. 492
   speciosus (Trichosporum splendens) (Java); strong straggler with stems to 60 cm long and large lanceolate, waxy-green leaves; showy tubular flowers in terminal clusters, corolla flame-orange, yellow at base and beneath; throat marked brown-red on yellow. *Tropical.* p. 492, 493

AESCULUS   *Hippocastanaceae*
   x carnea (hippocastaneum x pavia) (Midwestern U.S.), the "Red horse-chestnut"; ornamental flowering tree 10 to 15 m high; large compound leaves of 3 to 7 leathery green leaflets 8-15 cm long, with prominent ribs and serrate margins; in April-May a mature tree bears hundreds of 20 cm plumes of soft pink to red flowers, in center a yellow eye; brown prickly fruit 4 cm dia. *Temperate.* p. 520, 524
   hippocastaneum (Balkan Peninsula), "Horse chestnut"; large tree with coppery corrugated leaflets; large erect panicles of small white flowers blotched with red. *Temperate.* p. 524
   parviflora (Georgia and Alabama), "Dwarf horse-chestnut"; shrub to 5 m high; leaves with 5-7 leaflets; inflorescence tall pyramid of small white flowers. *Warm temperate.* p. 524
   pavia (Illinois to No. Carolina and Texas), "Red buckeye"; deciduous shrub or small tree to 3 m high; rather refined foliage with 5-7 leaflets, and dainty salmon-red flowers in large pyramids 10 cm tall. *Temperate.* p. 522

AETHIONEMA   *Cruciferae*
   coridifolium (Asia Minor); "Stonecress"; small subshrub 15-20 cm high with small fleshy, linear, glaucous leaves edged in pink, the flowers of 'Warley Rose' deep pink, in terminal racemes; rosy-purple in the species. *Subtropic.* p. 379

AGAPANTHUS   *Liliaceae*
   africanus (Cape of Good Hope), "Blue African lily"; summer blooming plant grown in tubs, with basal strap-like rich green leaves to 2 cm wide; funnel-shaped flowers pale porcelain-blue with darker center and margins, up to 30 in large umbels, on erect stalks. *Subtropic.* p. 575
   africanus minor, "Peter Pan lily"; a diminutive form of the "African lily"; fresh-green basal strap leaves; inflorescence on slender stalk 50 cm high, with trumpet flowers pale porcelain-blue. *Subtropic.* p. 575
   inapertus pendulus (So. Africa), "Drooping agapanthus"; robust plant; deciduous linear foliage; tall 1 m erect stalks with tubular 3 cm pendulous flowers violet-blue. *Subtropic.* p. 575
   orientalis 'Albidus' (So. Africa), handsome perennial herb with thick-fleshy roots; rather broad, arching succulent leaves in 2 ranks, to 5 cm wide; flower stalk 60 cm long with up to 110 flowers in large umbels, in this form white, otherwise blue; summer-blooming. *Subtropic.* p. 575

AGASTACHE   *Labiatae*
   cana (Cedronella) (New Mexico to Texas), "Mosquito plant"; perennial herb to 60 cm, ovate leaves 4 cm long; spikes with tubular pink flowers. *Warm temperate.* p. 531

AGATHIS   *Araucariaceae*
   australis (New Zealand: North Island), the famous "Kauri pine"; stately timber tree rising like a gray, straight column to 30 and 50 m high, with dia. of trunk recorded at 7 m, the thick bark peels off, tending to throw off epiphytes; leaves on young trees sparse, linear-oblong, bronze-green, to 6 cm long; the adult leaves oval 1-3 cm long; 8 cm erect ovoid cones. Produces kauri gum from resin. *Subtropic.* p. 334
   robusta (Queensland), "Queensland kauri"; massive, resinous evergreen tree to 50 m high with variable leaves; juvenile leaves elliptic, waxy dark green. *Subtropic.* p. 334

AGATI: see Sesbania

AGAVE   *Amaryllidaceae*
   albicans 'Medio-picta' (Mexico); beautiful rosette of broad succulent, recurving leaves to 10 cm wide, milky green or grayish with prominent central bands of creamy-white; softly fleshy, the margins with corky edge; inflorescence 1 m tall. *Subtropic.* p. 58
   americana (Mexico), "Century plant"; large, loose open, trunkless rosette of spreading, broad and thick-succulent, glaucous gray-green leaves sharply bend downward above the middle, with sharp brown hooks at margins, and ending in a spiny point; yellowish flowers on a tall spike to 12 m, produced when plant is 10 yrs. or more old. *Arid-Tropical.* p. 55, 57
   americana 'Marginata', "Variegated century plant"; large rosette with the broad, laxly recurved, glaucous gray leaves with showy, broad, yellow margins. Arid-Tropical. p. 57
   americana 'Medio-picta'; has broad yellow band down the center of leaves instead of along the margins. *Arid-Tropical.* p. 58
   angustifolia marginata (wilsonii, caribaea) (W. Indies), "Variegated Caribbean agave"; beautiful, freely suckering, densely formal rosette with stiff-erect, short, sword-shaped leaves, bluish-gray with broad white marginal bands and little brown spines. *Tropical.* p. 57, 58
   atrovirens (So. Mexico: Oaxaca), "Pulque agave"; large stemless rosette producing offsets; stout, fleshy smooth leaves spreading to 2m long and 25-30 cm wide, concave above, keeled convex beneath, dull dark green or blackish, margins wavy with small spines on horny base. Main source of pulque drink. *Arid-Tropical.* p. 57
   attenuata (Mexico), "Dragon-tree agave"; rosette, on 1 m stem when old; leaves wide in the middle, narrow at base, smooth, without teeth, gray-green; the beautiful inflorescence with greenish-yellow flowers in 3 m spikes gracefully arching, occasionally producing bulbils. *Arid-Tropical.* p. 55
   filifera (Mexico), "Thread agave"; many-leaved rosette of narrow, stiff, bright green leaves, white lines along edge which split into loose filaments. Leaves 20-25 cm long. *Arid-Tropical.* p. 58
   geminiflora (Mexico); dense rosette, branching when old, with 100-200 linear, stiff dark-green reed-like leaves 50 cm long, convex on both sides, the margins with white horny edge and loose curly threads; terminal spine 3-angled, inflorescence 4 m. *Arid-Tropical.* p. 58
   horrida (Mexico); full rosette with hard, stiff, fleshy leaves to 45 cm long and rather wide, glossy dark green, smooth, with brown horny edge and large marginal spines curved both directions, terminal spine twisted; plant dies after flowering. *Arid-Tropical.* p. 58
   parryi (Arizona, New Mexico, Chihuahua), "Mescal"; trunkless, leafy rosette with leaves 30 cm long x 10 cm wide, glaucous gray, smooth, with brown to gray terminal spine and small marginal spines. *Arid-Tropical.* p. 56
   parviflora (Sonora, Chihuahua, Arizona), "Little princess agave"; striking small rosette dense with stiff leaves 10 cm

long, with white lines above, the margins with white threads toward tip, lower part dentate; gray-brown terminal spine. *Arid-Tropical.* p. 56

shawii (Baja California); dense rosette forming clumps, short-stemmed when old; rigid leaves 25-50 cm long x 12 cm wide, dark-green glaucous; prominent lateral spines yellowish to brown; inflorescence to 3 m high. *Arid-Tropical.* p. 57

sisalana (Mexico: Yucatán), "Sisal-hemp"; large rosette forming 1 m trunks, developing offsets; with straight and stiff, sword-shaped leaves to 1½ m long and 10 cm wide, matte gray-green, or green , with horny edge and occasional deformed teeth, keeled beneath, and with small black-brown terminal spine; branched inflorescence to 6 m high, with greenish, odoriferous flowers. Young plants will form as suckers on the ground as well as on the flowering stalk, and plantlets or bulbils rise vegetatively from the floral bracts, according to observations I made in Tanzania. *Arid-Tropical.* p. 55, 57

striata 'Nana' (Mexico), "Lilliput agave"; small rosette of flat, pale gray-green, linear, short leaves, suckering as a young plant. *Arid-Tropical.* p. 58

victoriae-reginae (No. Mexico), "Queen agave"; shapely dense, many-leaved rosette with keeled, dark-green leaves, white margin and abrupt point, and short, blunt, terminal spine. *Tropical.* p. 56, 58

AGERATUM   *Compositae*
houstonianum, "Floss-flower"; the popular garden bedding plant, 25 cm high, a compact horticultural form of Mexican origin; bushy herbaceous perennial grown as an annual, with tassel-like heads of clear blue tubular florets, carried nicely above the fresh green, small foliage. From seed, or occasionally cuttings. *Tropical.* p. 314

AGLAONEMA   *Araceae*
commutatum (Philippines, Sri Lanka), "Silver evergreen"; durable plant with leathery, oblong-lanceolate leaves deep green with markings of silver-gray, waxy-white spathe; berries yellow to red. *Humid-Tropical.* p. 87

commutatum elegans (marantifolium hort.) (Moluccas), "Variegated evergreen"; robust-growing plant with long, lanceolate leaves deep green with greenish-gray feather design. *Humid-Tropical.* p. 86

commutatum 'Tricolor' (marantifolium tricolor); relatively broad deep-green leaves with silvery feathering, borne on pink petioles; waxy pinkish spathe and white spadix. *Humid-Tropical.* p. 86

costatum foxii (Malaya), a "Spotted evergreen"; short plant with leathery, shiny, cordate-pointed leaves with broad, white center band; slow-growing. *Humid-Tropical.* p. 86

costatum var. immaculatum (Thailand); handsome erect, spear-shaped leaves 20-30 cm long, glossy deep green and with prominent white midrib. *Tropical.* p. 86

crispum (Schismatoglottis roebelinii) (Malaya), "Painted droptongue"; large and showy, ovate-pointed, leathery leaves medium grayish-green, largely variegated with silver. *Humid-Tropical.* p. 87

marantifolium (Borneo); handsome species of compact habit; broad-lanceolate leaves with midrib and margins glossy dark green, the central area richly overlaid with silvery gray. Photographed in Kota Kinabalu, North Borneo. *Tropical.* p. 87

modestum (sinensis) (Kwangtung), "Chinese evergreen"; with durable, leathery, waxy-green leaves, ovate-acuminate and, to some extent, pendant, on a slender cane. Spathe green, spadix cream. *Humid-subtropic.* p. 87

modestum medio-pictum (pat. as 'Shingii'), "Mandalay plant"; sport of A. modestum (Mahaffey, Apopka, Florida about 1967); attractive and sharply distinct variegation of the leaf color, with yellowish chartreuse or apple-green and ivory along center, abruptly changing to dark green around outer edge of leaf, on variegated petioles; excellent ornamental plant with stocky cane and of compact habit, freely branching; ovate-pointed, leathery glossy foliage, with leaves 12-25 cm long; of good keeping qualities. *Tropical.* p. 87

modestum 'Variegatum'; cultivar from Pennock's Puerto Rico nursery, with the leathery, lanceolate deep green leaves dominated by creamy variegation. *Humid-Tropical.* p. 86

pictum tricolor (versicolor) (Sumatra); deep green, satiny leaves with patches of silver and mixed with yellow-green blotches; difficult. *Humid-Tropical.* p. 86

'Pseudo-bracteatum', "Golden evergreen"; colorful, free-growing mutation or hybrid with long, showy leaves deep green variegated with light green and yellow, center largely cream white; stem white and marbled green; cupped, waxy, greenish-white spathe, and cream spadix. *Humid-Tropical.* p. 85, 86, 87

rotundum (Malay Peninsula), "Red aglaonema"; compact, slowly growing beauty which I found in northern Thailand, with stout stem covered by sheathing short petioles, leaves broad-ovate or rhombic 10-15 cm or more long, thick-leathery, dark metallic glossy green to coppery with pronounced midrib and lateral veins, later paling; the reverse side glowing wine-red with rosy veins. *Humid-Tropical.* p. 86

siamense (Thailand); long, leathery, deep-green leaves with white center vein, on slender petiole. *Humid-Tropical.* p. 86

'Silver King' (nitidum 'Curtisii' x pictum 'Tricolor'); striking, excellent hybrid of lush habit, freely suckering; 20 to 35 cm leaves almost entirely silver-gray; petioles marked silver. *Humid-Tropical.* p. 87

'Silver Queen'; very ornamental and satisfactory decorative plant, from the same Florida hybrid complex as A. 'Silver King', with leaves narrower, about 22 cm long and 3 to 5 cm wide; green with gray marbling in feather pattern, petioles more green; freely branching and suckering, for bushy effect, and of excellent keeping qualities. *Tropical.* p. 87

simplex (Java); similar to A. modestum but leaf is more oblong and narrow, with a twist; texture more thin and papery; deep green with depressed veins. *Humid-Tropical.* p. 86, 87

treubii (Celebes), "Ribbon aglaonema"; slender plant with narrow, leathery, bluish-green leaves attractively marked with silver-gray; petioles marbled. *Humid-Tropical.* p. 86

AGROSTEMMA   *Caryophyllaceae*
gracilis (Greece, Turkey), "Corn-cockle"; silky annual or biennial with linear leaves; showy flowers 3 cm across, magenta-red, shading to white toward center and with red lines. *Subtropic* p. 298

AGROSTIS   *Gramineae*
tenuis (Europe), "Bent grass" or "Colonial Bent"; an especially fine-leafed, dense-clustered perennial grass spreading by stolons or surface runners, but no rhizomes; blades rich green and flat, 2-3 mm wide; needs much care and cutting, feeding, watering and disease control, and is used on golf course greens. *Temperate, but planted in California.* p. 510

AICHRYSON   *Crassulaceae*
bollei (Canary Islands: Palma); bushy succulent with white-hairy spatulate leaves to 5 cm long, margins wavy and toothed, marked chocolate; flowers pale yellow. *Subtropic.* p. 363

AILANTHUS   *Solanaceae*
altissima (China, naturalized in U.S.), "Tree of Heaven"; rapid-growing deciduous tree to 15 m high; leaves odd-pinnate to 75 cm long, divided into 13 to 25 leaflets 8-12 cm long; inconspicuous greenish flowers are followed by handsome clusters of red-brown winged fruit; foliage turning a showy red in autumn. A tree for adverse conditions, the "Tree that grows in Brooklyn". *Temperate.* p. 891

AIPHANES   *Palmae*
caryotaefolia (Martinezia) (Colombia, Venezuela, Ecuador) "Spine palm"; very unusual with slender trunk to 12 m ringed with long black spines, and a crown of light green pinnate leaves, with pinnae suddenly expanded and squared off and lobed at tip, clothed with long black spines beneath; yellow-red fruits. *Tropical.* p. 764, 790

AJUGA   *Labiatae*
reptans purpurea (Europe), "Carpet bugleweed"; ground cover spreading by runners; lustrous round-crenate leaves 5-8 cm wide, dark green in the species, milky green with bronze and purple in purpurea; flowers blue or white. *Temperate.* p. 533

ALBERTA   *Rubiaceae*
magna (Natal); evergreen shrub with willowy reddish branches and leathery, oblanceolate opposite leaves 12 cm long; erect terminal panicles of brilliant crimson flowers with long silky tubular corolla. *Subtropic.* p. 861, 865

ALBIZIA   *Leguminosae*
distachya (lophantha) (Australia), "Plume albizzia"; fast-growing shrub or tree to 6 m high, semi-evergreen; bipinnate foliage dark velvety green, fern-like; flowers greenish-yellow in fluffy, 5 cm spikes. *Subtropic.* p. 552

lebbeck (Tropics of the Old World); "Woman's-tongue"; tree to 25 m, bipinnate leaves 8-22 cm long; flowers yellowish-white, in globose heads 2 cm wide forming a terminal panicle; fruit bean-like to 30 cm long, rattling in the wind when ripe. *Tropical.* p. 539, 540

ALCEA *Malvaceae*
rosea (Althaea chinensis) (Asia Minor), "Hollyhock"; tall, straight, leafy-stemmed, hairy biennial to 3 m; rough leaves 5 to 7-angled; single or double flowers in long spike, red, rose, pink, yellow or white. *Warm temperate.* p. 625

ALEURITES *Euphorbiaceae*
fordii (C. Asia), "Tung oil tree" or "China wood-oil tree"; large tree to 12 m high, with broad cordate, leathery leaves 15-20 cm long, showing the yellow ribs; flowers white with red veins, to 3 cm long; the squarish fruit 5-8 cm dia. is source of tung-oil. *Subtropic.* p. 407

moluccana (Moluccas and South Pacific Is.), a "Candlenut" or "Varnish tree"; large tropical tree to 10 or 12 m high, with long-stalked ovate-pointed, or lobed pale green foliage 10-20 cm long, white downy above, rusty-pubescent beneath; small whitish flowers in large clusters; the green, rough fruit 5 cm dia., containing 1 or 2 jet-black seeds which are polished and made into jewelry; also oil is extracted for varnish or for lamps. *Tropical.* p. 407

ALLAMANDA *Apocynaceae*
cathartica (So. America: Brazil), "Golden trumpet"; robust shrubby, tropical clamberer with whorled, long, leathery, glossy leaves to 12 cm long; large wax-like, funnel-shaped golden flowers 6-8 cm across. With its showy flowers one of the most satisfactory summer bloomers in a sunny location. *Tropical.* p. 73, 75

cathartica 'Hendersonii' (So. America: Guayana); vigorous cultivar with longer leaves to 15 cm, and extra large golden-yellow flowers 10 cm across. *Tropical.* p. 75

neriifolia (Brazil), "Golden trumpet bush"; erect shrub with rough, oleander-like leaves in whorls; small, light-yellow flowers, streaked orange-red and swollen at base. *Tropical.* p. 75

violacea (Brazil), "Purple allamanda"; weak slender climber, branches hairy, with ovate leaves 8-12 cm in a whorl, downy above, hairy beneath; funnel-shaped flowers reddish purple, in axillary cymes. *Tropical.* p. 75

ALLIUM *Liliaceae*
schoenoprasum (Siberia, N. No. America), "Chives" or "Schnittlauch"; hardy tufted, clustered bulbs with onion-like hollow green leaves to 45 cm high, tipped by round heads of rose-purple flowers. Use: fresh leaves in soup, salads, omelet, hamburgers, for a mild onion flavor. *Temperate.* p. 575

ALLOPHYTON *Scrophulariaceae*
mexicanum (Tetranema) (Vera Cruz, Mexico), "Mexican foxglove"; darling little plant with short stem, long obovate, dark green, leathery, flexible leaves glaucous beneath, and angled purplish stalks with clusters of pretty little nodding, trumpet-shaped flowers, orchid-colored with large, lobed, whitish lip and purple-violet throat, blooming from summer on. *Tropical.* p. 882

ALLOPLECTUS *Gesneriaceae*
capitatus (Colombia, Venezuela), "Velvet alloplectus"; erect, 4-angled red tomentose stem to 60 cm with large velvety, olive-green, ovate leaves, reddish beneath; dense clusters of bright yellow flowers with red sepals. *Tropical.* p. 492

schlimii (Andes of Colombia); erect fleshy green to woody stem with large, broad-lanceolate leaves, bristly but shimmering light green, scalloped and fringed with red hairs; small golden-yellow flowers with crimson lines. *Tropical.* p. 492

ALLUAUDIA *Didiereaceae*
adscendens (S.W. Madagascar); curious xerophyte of the dry-tropical region near Fort Dauphin; succulent erect tree 5 to 12 m tall, with thick stem and few branches, of habit like cereus; the gray columns thorny, and with tiny obcordate, dark green leaves; small stalked flowers. *Arid-Tropical.* p. 389, 391

comosa (S.W. Madagascar); succulent shrub 2 to 6 m high, the branches set with pairs of long thin 4 cm thorns, small fleshy obovate leaves to 2½ cm long. *Arid-Tropical.* p. 389, 391

montagnacii (S.W. Madagascar); strange succulent tree 6-8 m tall, columnar, with annual constrictions; long silver-gray 2.5 cm thorns, tipped black, cover the stem densely; oval leaves 1.5 cm long. *Arid-Tropical.* p. 391

ALOCASIA *Araceae*
x amazonica (sanderiana x lowii grandis); free and bushy grower with leaves very dark green, veins contrasting white, and white scalloped margin. *Humid-Tropical.* p. 86

cucullata (Bengal, Burma), "Chinese taro"; shapely plant with smaller peltate, pointed leaves deep green with prominent veins, with base drawn together like a spoon. *Tropical.* p. 88

cuprea (Borneo), "Giant caladium"; Trop. Asian herb with large, peltate-cordate leaves with deeply depressed veins; dark, metallic, shining green; purple beneath. *Humid-Tropical.* p. 88

lowii grandis; broader than type, deep metallic brownish green, veins grayish-green; has a very large leaf, the veining more silvery than A. lowii veitchii; spathe whitish-green outside. Propagates from young tubers. *Humid-Tropical.* p. 85

macrorhiza (Malaya, Ceylon), giant "Elephant's ear"; to 4½ m high, with thick trunk, large 60 cm broadly arrow-shaped, fleshy leaves waxy green, with prominent ribs and wavy margin. *Tropical.* p. 85, 89

macrorhiza variegata (E. India), "Giant alocasia"; light green leaf blotched and mottled with white. *Tropical.* p. 88

x mortefontanensis (lowii x sanderiana); robust plant with handsome leaves to 50 cm long, oblong-sagittate, sinuately lobed along margins; glossy bluish green with contrasting silvery-white veins and border; purple beneath; stalks olive green with dark rings. *Tropical.* p. 88

odora (Philippines, Formosa); glossy, light green, stiff-fleshy, sagittate leaves to 1 m long, with elevated ribs and rounded lobes on vaginate stalks from thick stem; flowers fragrant. *Tropical.* p. 88, 89

sanderiana (Philippines), "Kris plant"; tuberous plant with graceful, sagittate leaf shining metallic, silver-green with grayish-white ribs; margins deeply lobed and white; reverse purple. *Humid-Tropical.* p. 85, 88

watsoniana (Sumatra); large, corrugated, leathery leaf blue-green with silver-white veining and margin; purple beneath. *Humid-Tropical.* p. 88

ALOCASIA: see also Colocasia, Xanthosoma

ALOE *Liliaceae*
africana (Transvaal), a "Spiny aloe"; a very pretty tree aloe, good compact rosette when young, though reaching 4 m or more in its habitat, the glaucous bluish leaves are hard and brown spined at the margins; flowers yellow, tipped green. *Arid-Subtropic.* p. 571

aborescens (So. Africa), "Candelabra aloe"; spreading rosette with sword-like, fleshy tapering leaves to 60 cm long, glaucous pale bluish-green, edged with yellow horny teeth; on stems reaching 4½ m; flowers red, in dense racemes. *Subtropic.* p. 566, 567, 571, 573

bainesii (So. Africa); tree to 18 m high, with trunk to 1½ m thick and head to 6 m across, one of the largest of the genus; sword-shaped green leaves 60-90 cm long, leathery and concave, with small, marginal prickles; flowers salmon-pink tipped with green. *Arid-Subtropic.* p. 567, 568

brevifolia (Cape Prov.); glaucous, robust 10-12 cm rosette with triangular-oblong leaves twice as broad as humilis, flat on top, rounded beneath to keeled, grayish-green and with white horny teeth at margins. Suckers much more freely than var. depressa, and when full grown rarely exceeds 20 cm in dia., with narrower leaves in proportion, it is also more spiny, especially on the keel below. Flowers red, in dense racemes. *Subtropic.* p. 572

brevifolia depressa (Cape: Caledon), (Aloe brevifolia of the trade), "Crocodile jaws"; open rosette with broad leaves having white cartilage edge, only very few spines on back of leaf toward tip only. 12-18 cm dia., but grows to about 30-40 cm, with leaves 8-10 cm wide at the base. *Subtropic.* p. 572

chabaudii (Rhodesia, Transvaal); clustering stemless rosettes with spreading fleshy concave leaves long tapering and curving, 40-50 cm long, glaucous blue to bronze, with small marginal teeth, branched inflorescence 1 m high with red flowers—a glorious sight with such plants clinging to the sheer cliffs alongside the foaming Victoria Falls, constantly bathed in the clouds of mist rising from the churning canyon of the Zambesi below. *Subtropic.* p. 567

ciliaris (So. Africa), "Climbing aloe"; weak, branching and scrambling stems with open spirals of not very fleshy, 15 cm leaves, dark green with white teeth along the edge; flowers scarlet with green tips. *Subtropic.* p. 573

comosa (Cape Prov.: Great Karroo); striking rosette becoming tree-like to 2 m high, with broad, fleshy, sword-shaped leaves glaucous blue, to purple in the strong sun of the Karroo, the margins with horny teeth; flowers greenish-white. *Arid-subtropic.* p. 566, 570

concinna (zanzibarica) (E. Africa); attractive scandent succulent rosette branching from base, with attractive fleshy recurving leaves 10 cm long, light green with pale spots; margins dentate; flowers scarlet tipped green. *Tropical.* p. 573

dawei (Uganda: Entebbe); stem-forming rosette branching from the base, to 1½ m high, with 45 cm sword-shaped, tapering fleshy leaves blue-green to purplish, sinuate-dentate; flowers red. *Tropical.* p. 569

dichotoma (Cape Prov. to Namibia), "Dragon-tree aloe"; bold succulent rosette with broad fleshy leaves 25 cm or more long, glaucous green edged in brownish-yellow and with triangular teeth; growing into monstrous trees to 9 m high, with branches forking, on a smooth trunk; flowers canary yellow. *Arid-Tropical.* p. 570

eru (Eritrea, Nubia, No. Kenya); branching rosette with stem 45 cm high, very fleshy, arching leaves to 75 cm long, concave above, glossy dark green striped with white, and red teeth; the 'Green form' with only faint markings; salmon-red flowers in branched racemes. *Arid-Tropical.* p. 576

eru 'Maculata', "Nubian aloe"; very attractive variety with leaves smaller and narrower than the type, bluish-glaucous and very much spotted with cream; prominent pale marginal spines. *Arid-Tropical.* p. 576

ferox (So. Africa), "Ferocious aloe"; bold rosette, developing a stem to 5 m high, with broad, fleshy leaves bronzy-green, hollow above, curved and warty beneath, margins with brown-red teeth; flowers orange-red. The Zulus use it as an unguent, and for stomach troubles. *Arid-subtropic.* p. 570

ferox variegata (S. E. Cape Prov.); a natural variety found in the wilds, and which I photographed at Kirstenbosch Botanical Gardens, Cape Town; striking, shapely rosette with broad, fleshy, upward-pointing leaves glaucous green variegated by many ivory-white longitudinal bands and stripes, marginal teeth brownish; branched raceme 60 cm long, with orange flowers. *Arid-subtropic.* p. 570

haemanthifolia (Cape Prov.); curious stemless rosette with tongue-shaped, broad fleshy leaves arranged in two series, olive-green with red margins, 10 to 20 cm long, the apex rounded. *Arid-subtropic.* p. 572

haworthioides (C. Madagascar); stemless, pretty rosette of dark green leaves to 6 cm long, densely set with soft, pale spines; flowers pale orange. *Tropical.* p. 576

humilis (E. Cape Prov.), "Spider aloe"; small, shapely, dense succulent rosette forming tufts by suckering, glaucous blue-green narrow, concave leaves with white marginal teeth, and tubercles on back, giving hairy-looking appearance; flowers red, tipped with green, in long racemes. *Arid-subtropic.* p. 572

humilis 'Globosa', a "Spider aloe"; a cultivar in the California nursery trade; small suckering rosette with glaucous bluish leaves more incurved and more numerous than humilis, the tips turning purplish in the sun, prominent horny teeth cream to pinkish. *Arid-subtropic.* p. 576

jucunda (Somaliland); small stemless, group-forming 8-10 cm rosette, with broadly ovate-tapering, recurving fleshy leaves, dark green with numerous pale green to white translucent spots; margins with horny red-brown teeth; flowers pale pink. *Arid-tropical.* p. 573

littoralis, (Angola, Namibia); stout stem to 2 m, rosette of sickle-shaped tapering, gray-green leaves to 90 cm long; flowers coral red, in many-branched panicle to 5 m tall. *Arid-Tropical.* p. 568

marlothii (Bechuanaland, Transvaal, Natal); attractive tree with single trunk, to 5 m high; leaves to 75 cm long and 18 cm wide, in dense rosette, pale glaucous blue or greenish, concave and very spiny with short thick purplish-brown thorns; flowers red-orange. I did, however, find a double-headed old tree in the northern Transvaal. *Arid-subtropic.* p. 568, 569

melanacantha (So. Africa); short stemmed, clustering rosette of narrow green leaves curved beneath, the margins with horny white spines, later black spines curved toward the tip; flowers scarlet. *Arid-subtropic.* p. 572

mitriformis (Cape Prov.); clustering rosette with broad concave, fleshy incurving leaves to 15 cm long and 8 cm wide, blue-green to green, with pale horny teeth; tubular flowers bright scarlet red. *Arid-subtropic.* p. 566

peglerae (Transvaal); stemless rosette of crowded, incurving leaves to 35 cm long, pale blue-green; flowers pink becoming lemon-yellow. *Arid-subtropic.* p. 570

pillansii (Cape Province); distinctive rosette with close-jointed thick fleshy, bluish glaucous leaves in symmetrical ranks, to 60 cm long and 20 cm wide at base, with yellow marginal teeth; flowers lemon-yellow; grows into a large branched tree 10 m high with trunk to 2 m thick. *Arid-subtropic.* p. 570

plicatilis (Cape Prov., So. Africa), "Fan aloe"; small forking tree to 5 m high, with closely packed, linear, fleshy 30 cm leaves of even width arranged in two ranks, tip rounded, pale glaucous blue with translucent yellow margin; large scarlet flowers in loose spike. *Arid-subtropic.* p. 567, 569

polyphylla (So. Africa: Drakensberg; Basutoland), "Spiral aloe"; striking low rosette 30-40 cm dia., dense with fleshy, broad leaves relatively short, but becoming 20-30 cm long, pale glaucous green and with a strong, red-brown tip; the foliage spirally arranged and overlapping; branched inflorescence bearing salmon-pink flowers. A succulent of unique beauty whether large or small, with its leaves regularly arranged as in shingles on a roof, in clockwise or counter-clockwise coils. Somewhat difficult in cultivation; in habitat at 2,400 m they are often under snow. *Arid-subtropic.* p. 573

x salm-dyckiana, (natural hybrid of arborescens x ferox); large rosette 2 m tall with heavy, thick-fleshy leaves grayish-green; bold inflorescence orange-red. *Subtropic.* p. 569

saponaria (Natal, Transvaal), "Soap aloe"; short stemmed, suckering rosette with fleshy, broad lanceolate leaves 15-20 cm long, bluish-green marked with yellow-green elongate blotches arranged in transverse bands; margins with large cream to brown thorns; flowers red-yellow. *Arid-subtropic.* p. 570

sladeniana (Southwest Africa); attractive stemless rosette forming clusters, with shapely pointed, short-channeled leaves to 10 cm long, keeled beneath, moss-green tinted pink, and richly covered with white oblong, translucent spots and with white edge, set with small white teeth; flowers rose. *Arid-tropical.* p. 576

striata (S. W. Africa, Cape Prov.), "Coral aloe"; large rosette, almost stemless, of broad fleshy leaves almost 60 cm long, pale glaucous gray faintly striped, the hard margins pink and without spines; pendulous coral-red flowers. *Arid-subtropical.* p. 573, 576

suarezensis (N. W. Madagascar); stem-forming rosette to 1 m high, with 10-20 spreading, fleshy, channeled leaves 40-60 cm long, and sharp, red marginal teeth; flowers vivid red. *Arid-tropical.* p. 566

succotrina (Cape Prov.); branching, stem-forming rosette to 1 m high, leaves to 50 cm long and 5 cm wide, sword-shaped, bluish-green to gray-green, lightly lined and spotted, with white prickles; flowers red with green tips. *Arid-subtropic.* p. 570

suprafoliata (So. Africa: Swaziland), "Propeller aloe"; very attractive succulent plant when young, when its thick, elegantly recurved, tapering leaves are arranged in two opposing series; glaucous blue but bronzy-red toward margins and dressed with reddish marginal spines; leaves become 30 cm long, set in rosettes with age; flowers coral-red with green tips. *Arid-subtropic.* p. 572

vanballenii 'Variegata' (Natal, Zululand); broad rosette with fleshy spreading leaves to 30 cm long, often curving sideways, milky green with darker nerves, cream margins and red spines; flowers red. *Arid-subtropic.* p. 570

variegata (Cape of Good Hope), "Partridge breast"; beautiful succulent to 30 cm high with triangular blue-green leaves arranged in 3 ranks and painted with oblong white spots in irregular crossbands, the margins horny and white-warty; tubular flowers salmon red, in loose racemes. *Arid-subtropic.* p. 573

vera (barbadensis) (Cape Verde, Canary Isl., Madeira), the "True aloe" or "Medicine plant"; the pulp of which is used to heal sore burns and cuts; widely cultivated; short-stemmed, suckering rosette with fleshy, dagger-shaped, channeled leaves to 60 cm long, gray-green and glaucous; spotted when young; edged with soft pale spines; nodding, cylindrical, yellow flowers. *Subtropic.* p. 568, 571

vera chinensis (India, Vietnam), "Indian medicine plant"; smaller Asiatic form, the fleshy, lanceolate leaves 30 cm long and recurved at tips, rounded beneath, blue-green with white markings and whitish teeth; flowers orange. *Tropical.* p. 573

ALONSOA Scrophulariaceae
warscewiczii (Peru) "Mask flower"; bushy herbaceous plant 60 to 90 cm high, grown for its attractive scarlet-red, or cinnabar winter-blooming flowers 1 cm or more in dia.; 2-lipped and carried on brownish stems with fresh green, toothed leaves. *Tropical.* p. 887

ALPINIA Zingiberaceae
calcarata (India), "Indian ginger"; rhizome bearing reed-stems, to 1 m, with spicy leaves and terminal panicle of waxy flowers greenish-white; lip yellow with maroon. *Tropical.* p. 924

purpurata (South Seas from Moluccas to New Caledonia and Yap), "Red ginger"; ornamental perennial herb with leafy stems ranging from 2-5 m each ending in a showy inflorescence, with a flower spike brush-like to 30 cm long, consisting of numerous large, boat-shaped bright red bracts each with a small white flower, normally erect but drooping if long; new plants germinate among the flower bracts. *Tropical.* p. 923

sanderae (New Guinea), "Variegated ginger"; very attractive ornamental ginger of dwarf habit, having creeping rhizome and clustered, leafy stems about 45 cm high, the lanceolate leaves arranged somewhat in 2 ranks, pale green, edged and obliquely banded from the center to the margin with pure white. *Humid-tropical.* p. 924, 926

taiwaniana (Taiwan), "Taipei ginger"; clump-forming perennial with strong canes 1-2 m high; well spaced glossy green, lanceolate leaves 30 cm long, along the stem; from near the apex a pendant or arching raceme of waxy, bell-shaped flowers creamy-white, orange yellow inside the 3-lobed corolla. *Subtropic.* p. 928

zerumbet (speciosa or nutans in hort.) (China, Japan), "Shell ginger"; majestic rhizomatous plant forming dense clumps of leafy, arching canes to 4 m high, with smooth, long-bladed, leathery leaves arranged in 2 rows; the striking fragrant, porcelain-textured flowers with bell-shaped, waxy-white calyx, the corolla white flushed with pink and tipped with red and yellow lip, in dense racemes to 30 cm long, becoming nodding. *Tropical.* p. 923

zerumbet 'Variegata', "Variegated shell ginger"; a variety seen in Hawaii, with the broadly lanceolate thin-leathery dark green leaves variegated in feather design with stripes and bands of creamy yellow. *Tropical.* p. 925, 926

ALPINIA: see also Hedychium

ALSOPHILA   *Dicksoniaceae (Filices)*
australis: see cooperi
cooperi (Cyathea australis) (Tasmania, Australia), "Australian treefern"; a noble treefern, with heavy trunk to 6 m high, with well proportioned, spreading crown even when small, but requiring lots of water, the arching fronds finely divided, metallic green, on rough stalks covered with small pale brown hair-like scales. *Humid-subtropic.* p. 431

glauca (Java, Malaysia, Assam), "Blue treefern"; slender treefern with trunks to 15 m, crowned by gigantic bipinnate fronds to 4 m long, bright glossy-green above, and bluish beneath, the stalks and heart clothed with white chaffy scales. *Humid-tropical.* p. 432, 433

latebrosa (North Bengal); noble treefern with slender trunk, covered with brown fiber; the crown of fronds rather flat, with bipinnate leaves herbaceous and the pinnae rather scattered. *Humid-Tropical.* p. 436

ALSTONIA   *Apocynaceae*
scholaris (Australia to India and Africa), "Devil tree", "Pali-Mara"; tall evergreen tree to 20 m; the branches with milky sap, at their ends the thick-leathery oblanceolate leaves to 20 cm long, arranged in whorls; numerous small greenish-white flowers in clusters at the tips of twigs; pod-like pendant, paired fruits. *Tropical.* p. 74

ALSTROEMERIA   *Amaryllidaceae*
aurantiaca (Chile), "Lily of the Incas"; herbacous perennial with fibrous roots from egg-shaped tubers which are attached to a common stem; showy, floriferous plant blooming in summer, with leafy floral stem nearly 1 m high; lanceolate leaves, and terminal clusters of beautiful azalea-like flowers about 4 cm across, orange with two upper segments streaked with orange-red, the outer tipped with green; long-lasting as cut flowers. *Subtropic.* p. 60

caryophyllaea (Brazil), "Peruvian lily"; flowering stem 20 to 30 cm high, the leaves spoon-shaped oblong; two-lipped flowers scarlet red or striped with red; very fragrant. *Subtropic.* p. 60

pelegrina (Chile), "Peruvian lily"; 40 cm high, with short, lanceolate leaves on stems, the numerous 5 cm lily-like flowers with spreading segments outside lilac, spotted and lined with red-purple inside. Horticultural forms in various colors. *Subtropic.* p. 60, 61

ALTERNANTHERA   *Amaranthaceae*
amoena (Telanthera) (Brazil), "Parrot-leaf"; small, bushy herb of robust growing habit with broad lanceolate leaves, brownish-red and carmine, to orange; small whitish flower. *Tropical.* p. 50

bettzickiana (Brazil), "Red calico plant"; dwarf, clustering herb with twisted leaves narrow and spatulate, blotched and colored cream-yellow to salmon-red; ideal for carpet-bedding. *Tropical.* p. 50

bettzickiana 'Aurea', "Yellow calico plant"; a pretty horticultural variety of slightly more open habit; bushy little plant with its small 2½-4 cm spoon-shaped leaves prettily variegated pale yellow and fresh green. The green counterpart in carpet-bedding to the red A. bettzickiana, and kept in shape by shearing. *Tropical.* p. 50

dentata 'Ruby' (Brazil), "False globe amaranth"; red-leaf form of dentata; tall species with metallic, wine-red leaves rather flat and uniform, purple beneath; inflorescence strawflower-like, greenish-white heads. *Tropical.* p. 50

versicolor (Brazil), "Copper leaf"; low compact herb with opposite broad leaves almost round and a short cusp at apex, crisped and corrugated, dark green with purplish-carmine veining edged pink and white. *Tropical.* p. 50

ALTHAEA: see Alcea, Hibiscus

AMARANTHUS   *Amaranthaceae*
caudatus (South Asia), "Love-lies-bleeding"; stout plant with long, ovate green leaves and showy, red, nodding, tail-like panicles. *Tropical.* p. 52, 54

paniculatus (Trop. America); widely planted summer-blooming annual to 1½ m high, usually hairy, with long-ovate leaves; the inflorescence erect, with branched spikes, dark crimson or red. *Tropical.* p. 51

paniculatus cruentus (Tropics), "Prince's feather"; magnificent large bush 1-2 m high, with dense foliage, and the plumy inflorescence with side branchlets spreading and nodding, rich blood-red; photographed on the Greek island of Rhodes. *Tropical.* p. 51

tricolor (gangeticus) (E. Indies), "Joseph's coat"; vigorous, erect plant with ovate, pointed leaves brilliantly colored in shades of red, green and yellow. *Tropical.* p. 52

AMARYLLIS   *Amaryllidaceae*
belladonna (Brunsvigia rosea) (So. Africa), the "Cape belladonna" or "Belladonna lily", the "true" Amaryllis; a bulbous plant also known as "Naked lady" because the lovely trumpet-shaped, lily-like clear rosy-pink fragrant flowers, 10 cm long, appear after the leaves die down, in clusters, on solid, reddish 60 cm stalks, bare and ahead of the new strap-like leaves, in August; forming clumps. *Subtropic.* p. 65

belladonna 'Alba' (Brunsvigia rosea cv.); a "Cape belladonna" with large cluster of white trumpet flowers having a yellow throat, carried on a bare solid floral stalk, and after summer dormancy, blooming before the foliage. *Semi-tropic.* p. 69

AMARYLLIS: see also Hippeastrum

AMHERSTIA   *Leguminosae*
nobilis (Burma, India), "Queen of flowering trees"; one of the showiest, most striking tropical trees when in flower; 10-20 m high, with dark green pinnate leaves, unfolding brownish-pink from drooping clusters. Graceful racemes 60-90 cm long pendulous from every branch, with 20 cm flowers, with bracts, stalks, sepals and smaller petals vermilion red, the larger petals red with white base and tipped with yellow; long protruding anthers red also. *Tropical.* p. 542, 543

AMOMUM   *Zingiberaceae*
cardamomum (Java), "Cardamon"; perennial herb with creeping rootstock, and clustering leafy stems which can attain 3 m, but usually grown as a durable foliage plant 60 cm high; linear-lanceolate, leathery leaves deep dull green, and finely hairy, giving off when rubbed, a spicy aroma; yellow flowers in cone-like spikes beneath the foliage; produces cardamon seeds. *Tropical.* p. 928

AMORPHOPHALLUS   *Araceae*
bulbifer (India, Afghanistan), "Devil's tongue"; tuberous plant with inflorescence on 30 cm stalk marked brown, spadix green and pink; spathe green with rose to red outside, yellow-green within; segmented leaves. *Tropical.* p. 90

campanulatus (E. Indies, New Guinea); a giant aroid with tuber 25 cm thick, dark green solitary lobed leaves pinnately cut; inflorescence with ovate spathe 20 x 25 cm, fleshy and funnel-shaped below, green spotted white, purple inside. *Tropical.* p. 90, 91

corrugatus (Thailand); showy spathe green with greenish-white spots and purplish margin, whitish inside, 8-10 cm wide, the short spadix with female portion purple, the corrugated appendix ochre-yellow; solitary dissected leaf greenish with brown blotches, segments to 15 cm long; stalk to 60 cm, from 5 cm tuber. *Tropical.* p. 90, 91

hildebrandtii (Madagascar, Africa, Trop. Asia); handsome dragon-lily with spathe unfurling into broad blade, white inside, greenish with purple spots outside, and long slender spiralling spadix; after bloom, a dissected leaf cut into narrow segments, arises from the tuber on a stalk handsomely marbled purple on light green. *Tropical.* p. 90

longituberosus (Thailand); spathe boat-shaped 8-10 cm long, pale green marked with darker spots, cradling the shorter spadix with fertile, warted lower half, and smooth cone-like appendix in upper part; from the tuber 5-12 cm long and 2-3 cm diameter, a long stalk carrying the divided, solitary leaf 30-50 cm across. *Tropical.* p. 90

AMORPHOPHALLUS: see also Hydrosme

AMPELOPSIS: see Parthenocissus

AMYGDALUS: see Prunus

ANACAMPSEROS *Portulacaceae*
lanceolata, (So. Africa), "Love plant"; small branching succulent with stems 8 cm long, numerous 3 cm narrow leaves, with long hairs in the axils; bell-shaped pink flowers. *Arid-subtropic.* p. 830

ANACARDIUM *Anacardiaceae*
occidentale (West Indies), the "Cashew-nut tree"; evergreen spreading tree to 12 m high with obovate leathery leaves 20 cm long, numerous small fragrant flowers yellowish-pink; the fruit a kidney-shaped nut borne on a pearshaped yellow or red fleshy receptacle. *Tropical.* p. 71, 467

ANANAS *Bromeliaceae*
bracteatus (So. Brazil), "Red pineapple"; large terrestrial rosette with flattened leathery leaves coppery green with spines larger and more widely spaced than comosus; produces large brown edible fruit; flowers lavender. *Tropical.* p. 207

bracteatus 'Striatus' (variegatus); open rosette of relatively broad colorful leaves predominantly cream-yellow with coppery-green center; marginal spines red. *Tropical.* p. 205, 209

comosus (sativus); (Bahia, Mato Grosso), "Pineapple"; large formal terrestrial rosette of stiff, tapering and spiny-edged leaves grayish to bronze-green, gracefully arching; violet flowers borne in dense heads with tufts of leaves, producing the edible fleshy fruit. *Tropical.* p. 209, 459

comosus 'Porteanus' (comosus hyb.), "Golden rocket"; large and stiff rosette with olive green leaves acquiring a reddish color, and having a broad creamy-yellow center band on upper surface. *Tropical.* p. 207

comosus 'Smooth Cayenne'; elegant rosette of broad recurving, gray-green leaves, the margins smooth and spineless; the large golden-yellow to brown-red fruit with luscious, sweet flavor; topped in time by another dense, smaller rosette of leaves. The 'Smooth Cayenne' was brought to Hawaii from Guyana in 1886 where it developed the big plantations. *Tropical.* p. 209, 464

comosus variegatus, "Variegated pineapple"; beautiful variety with the channeled leaves having broad ivory bands along the margin and red spines; center of rosette tinted rosy-red. *Tropical.* p. 207

nanus (Brazil), "Dwarf pineapple"; miniature species to 30 cm high, with recurved, serrate leaves; a slender stalk carries few-flowered small fruit less than 15 cm, topped by leafy crown. Grown for cultivation in pots indoors. *Tropical.* p. 207

ANCHOMANES *Araceae*
difformis (Trop. Africa); large aroid with tuberous rhizome and solitary 1 m pinnatisect leaf on spiny petiole, green with purple spots, 1-1½ m long; fleshy spathe 15-30 cm brownish, purple spotted black inside, and pale, cupped spadix on short stalk. *Tropical.* p. 90

welwitchii (Trop. West Africa); tuberous aroid from savannah country sending up a short-stalked fleshy, yellow-green spathe, shielding the stiff-erect club-shaped spadix, the clasping basis of the spathe covered with spines; later the tall foliage appears, on long fleshy stalk covered with tubercles, the leaf divided into 3 sections, each bipinnately cut into broad-angled segments tipped with sharp points. *Tropical.* p. 90

ANDRYALA *Compositae*
pinnatifida (Spain: Canary Islands); herbaceous perennial from Tenerife, with succulent stems fuzzy with white hair and with milky sap, the somewhat pubescent leaves irregularly lobed and divided; yellow daisy-like flower heads. *Subtropic.* p. 308

ANEMONE *Ranunculaceae*
coronaria (So. Europe to Cent. Asia), "Lily of the fields"; attractive perennial with tuberous roots, finely divided leaves both basal and sessile along the erect, hairy stalk which bears the large, solitary, poppy-like, 6 cm flowers, the showy sepals in combinations of red, blue, yellow, white; in spring. *Subtropic.* p. 842

coronaria coccinea (Greece, Asia Minor); variety with showy large flowers 4-6 cm across, glowing crimson-red, with white or yellow circle around base inside. *Subtropic.* p. 842, 843

hupehensis (China), "Japanese anemone"; long-lived fibrous-rooted perennial with graceful branching stems 1 m or more high; dark green, soft-hairy, 3-5 lobed leaves; semi-double flowers white, also pink or rose, 5-8 cm across. *Warm-temperate.* p. 842

ANEMOPAEGMA *Bignoniaceae*
chamberlaynii (Brazil); vigorous climbing tropical shrub with leaves of 2 leaflets to 16 cm long, and sometimes a terminal tendril; flowers funnel-form 6-8 cm long, in pendant clusters, the tube constricted near base, bright yellow in throat with purple or white. *Tropical.* p. 184

subunculata (Brazil); woody tropical climber with long-stalked leaves consisting of two firm, leathery, lanceolate leaflets about 15 cm long, the surface matte-green and quilted; tubular white flowers in clusters at leaf axils, 5-6 cm long; photo Jardim Botanico, Rio de Janeiro. *Tropical.* p. 184

ANGELICA *Umbelliferae*
archangelica (Europe, Asia), "Archangel" or "Wild parsnip"; stout swamp biennial herb to 2 m, leaves twice divided into threes; small white flowers in large clusters 25 cm across. Tender stems and petioles are often candied; the leaves cooked as vegetables. *Temperate.* p. 908

ANGELONIA *Scrophulariaceae*
gardneri (Brazil); subshrubby, hairy, perennial to 1 m high, with lanceolate, toothed leaves, and tall racemes with showy, 2-lipped tubular flowers purple with white center, dotted with red; upper lip 2-lobed, lower 3-lobed. *Tropical.* p. 883

ANGIOPTERIS *Marattiaceae (Filices)*
evecta (Japan to Australia and Madagascar), "Mules-foot fern"; fleshy, robust fern from swampy places, developing a stout stem, the bipinnate leaves on swollen stalk may grow to 5 m long, the linear-oblong pinnae are succulent, dark green, and finely toothed. *Humid-subtropic.* p. 433, 438

ANGRAECUM *Orchidaceae*
eburneum (Madagascar); epiphyte with very thick leaves on stems to 1 m high; flowers upside-down on stout spike, sepals and petals green, the broad lip ivory-white with green spur, (Dec.-March). *Tropical.* p. 710

eburneum longicalcar; variety with white spur to 30 cm long. *Tropical.* p. 710

eichlerianum (African West Coast); epiphyte with scandent stems developing aerial clinging roots, thick light green leaves; often waxy, 8 cm flowers with yellow-green sepals and petals, white lip green near base and greenish spur, (June-Sept.). *Tropical.* p. 709

filicornu (Mystacidium) (Madagascar); neat, small epiphyte with fleshy channeled leaves; flowers waxy white, 3 cm across, with segments in star-like effect, and with 10 cm spur; carried on long arching stems. *Tropical.* p. 708

sesquipedale (Madagascar), "Star of Bethlehem"; epiphyte with stems to 1 m, densely 2-ranked leaves; thick-fleshy flowers to 18 cm dia., ivory-white, with long spur; largest in the genus, (Nov.-March). *Tropical.* p. 709

ANGRAECUM: see also Mystacidium

ANIGOZANTHOS *Amaryllidaceae*
flavidus (S. W. Australia), "Yellow kangaroo paw"; odd perennial herb with thick rootstock, smooth lanceolate leaves, the inflorescence 1-1½ m high, with large woolly flowers having a long bent tube yellowish green, tinged red, 4 cm long, and with blue petals. *Arid-subtropic.* p. 59

manglesii (Western Australia), known in Australia as "Kangaroo paw"; rosette of onion-like leaves with erect, striking inflorescence 1 m tall with red-hairy stem topped by racemes of yellowish-green, woolly 8 cm tubular flowers, red at calyx, with dark green lines, the lip reflexed and tipped blue. *Arid-subtropic.* p. 59

rufus (Southwestern Australia), "Red kangaroo paw"; curious perennial with stiff woody stems to 1 m high, linear basal leaves; the branched inflorescence with red velvety tubes and starry petals pinkish to purple. *Arid subtropic.* p. 59

ANNONA *Annonaceae*
cherimola (Andes of Peru and Ecuador), the "Cherimoya", a custard apple; woody tree to 8 m high, briefly deciduous with sappy branches and large, luxuriant, leathery leaves 25 cm long,

dull green with pale veins, velvety on back; fragrant fleshy flowers 3 cm long, yellow or brown-tomentose outside, whitish with purple spot inside; followed by large green conical fruit 12 cm or more long, the skin looking like overlapping scales or knobby warts; the creamy white flesh contains large black seed, tasting like custard or bananas, and is eaten with a spoon; ripening winter into spring. For best success flowers are hand-pollinated. *Subtropic.* p. 72, 455

muricata (West Indies), the "Soursop"; evergreen tropical fruit tree 6 m high, widely planted and liked in Latin America; dark shiny green, leathery, obovate leaves 10-15 cm long; large yellowish flowers with 6 fleshy petals spreading 5 cm across; big ovoid or irregular oblong, pendulous fruit is the largest of the annonas, 15-20 cm long and weighing as much as 3 kilos, deep green and covered with short fleshy spines. The pulp is like white cotton from which a pleasing drink, custard, or chilled sherbet is made in tropical America; in season June through November or longer. *Tropical.* p. 72, 454, 455

squamosa (C. America, West Indies), "Sugar apple" or "Sweet-sop"; tropical fruit tree to 6 m high, partially deciduous with thin-leathery bluish-gray, soft oval leaves 15-20 cm long; small greenish fleshy 2 cm axillary flowers followed by ovoid, grayish-beige fruits 8 cm long, covered with large prominent knobs, divisions falling apart much broken up by its separable sections (carpels), ripening in August to winter; the flesh is sweet, custard-like but very perishable. Greatly liked in markets from the West Indies to Rio. *Tropical.* p. 72, 455

ANSELLIA   *Orchidaceae*

africana (Kenya, Uganda, Tanzania, Zaire to West Africa), strong growing, variable epiphyte with ribbed slightly swollen pseudobulbs to 60 cm high, bearing linear leaves and an erect and rigid terminal panicle of long-lasting fleshy flowers, the spreading, linear sepals and petals about 2½ cm long, yellow with cross bars of red-brown, lip bright yellow lined with brown (Winter). *Tropical.* p. 710, 718

nilotica (East Africa); in habit usually smaller than africana and with pendulous inflorescence, the brighter flowers yellow with chocolate flecks, golden lip marked red-brown in throat (Winter). *Tropical.* p. 711

ANTHERICUM: see Chlorophytum

ANTHURIUM   *Araceae*

andraeanum (S. W. Colombia), "Tailflower"; erect plant with long-lobed, heart-shaped, green leaves; the showy, cordate spathe waxy, coral red, puckered, with pendant spadix tipped yellow with white band marking the zone where stigmas are receptive. *Humid-tropical.* p. 93, 94

andraeanum album; round cordate, quilted, waxy-white spathe; spadix dipping, with base white, center area purplish-rose and tip yellow; glossy leaves. *Humid-Tropical.* p. 92

andraeanum giganteum; a highly developed variety with large corrugated, salmon-red spathe carried on wiry stems; spadix not so prominent, dipping, white to yellow. *Humid-tropical.* p. 94

andraeanum 'Guatmala'; large-flowered cultivar from Guatemala and grown in Germany under the name spelled 'Guatmala', very popular as a cut flower because of its elegant form, broad crimson-red, glossy and firm spathe of medium size, pendant small yellow spadix, and continuous production of blooms. *Humid-tropical.* p. 93

andraeanum rhodochlorum; a robust form with giant, deeply cordate, salmon-red spathe in which lobes or tip are green. *Humid-tropical.* p. 89, 93, 94

andraeanum rubrum (Colombia), "Wax flower"; a striking color form with large, waxy, quilted spathe of dark crimson; spadix white, yellow at tip. *Humid-tropical.* p. 92

bakeri (Costa Rica); on short stem, strap-like, leathery, elliptic, lanceolate leaves, deep green, with stout midrib; spathe and spadix green. *Tropical.* p. 92

clarinervium (Mexico), "Hoja de Corazon"; a dwarf, ornamental species found growing in the clay of the Chiapas mist forest; dark green, velvety, heartshaped leaves with clear, silvery-gray veins; spathe reddish-green. *Tropical.* p. 92, 93

coriaceum (So. Brazil, Argentina); creeping on rocks; upright standing, leathery, glossy gray-green, long oblanceolate leaves to 1 m on short stiff stalks; spathe green, reddish inside. *Subtropic.* p. 93

corrugatum (papilionense) (Colombia); ornamental leaf plant from Putumayo, the sagittate foliage light green, leathery, corrugated, with pointed basal lobes; green spathe, spadix purplish-black. *Tropical.* p. 95

x ferrierense (andraeanum x ornatum), "Oilcloth flower"; robust climber with lobed, heart-shaped leaves; ovate-cordate, rosy spathe waxy-smooth, carried upright, and erect spadix white to rose; willing bloomer. *Humid-tropical.* p. 93

gladifolium (Venezuela, Brazil); short-stemmed epiphyte with long thick-leathery, strap-like pendant leaves to about 1 m long, with prominent veins, on pale, stippled petioles; narrow spathe. *Humid-tropical.* p. 92

holtonianum (Panama, Colombia); striking plant with large, palmately compound, glossy green leaves, the long segments pinnately lobed and undulate. *Humid-tropical.* p. 94

hookeri (huegelii) (Guayana, W. Indies), "Birdsnest anthurium"; symmetrical rosette resembling birdsnest, with broad, obovate, grass-green leaves 60 cm long; spathe green. *Humid-tropical.* p. 93, 95

leuconeurum (Mexico); handsome tropical foliage plant with velvety deep green, broad heart-shaped leaves 12-18 cm long, with silvery-white veins, glaucous beneath, and of leathery texture; spathe light glaucous green, spadix gray-green. *Humid-tropical.* p. 94

magnificum (Colombia); large and showy heartshaped, velvety, olive-green leaves with prominent white veins; the petioles 4-angled; spathe and spadix green. *Tropical.* p. 93

panduratum (No. Brazil); showy plant from the forests of Amazonas, with large, palmately compound, glossy leaves, the segments pinnately lobed and with marginal vein; spathe purple. *Humid-tropical.* p. 95

papilionensis (Trop. America); handsome species with showy sagittate-cordate leaves deeply corrugated, the sunken veins forming a network pattern of fine lace, the raised areas glossy yellowish green; the inflorescence with small white spathe. *Humid tropical.* p. 92

pedato-radiatum (Mexico); large, graceful, glossy-green leaf palmately lobed, segments slender, oblanceolate; basal lobes broader; spathe reddish, spadix green. *Humid-tropical.* p. 92, 118

scherzerianum (Costa Rica), "Flamingo flower"; leathery plant with lanceolate, green leaves; spathe broadly ovate, subtended brilliant scarlet; spadix spirally twisted, golden-yellow. *Humid-subtropic.* p. 92, 93, 94

scherzerianum rothschildianum, "Variegated pigtail plant"; spathe more or less red with white spots. *Subtropic.* p. 92

veitchii (Colombia), "King anthurium"; unusual plant with pendant, showy leaves to 1 m long, cordate at base, rich metallic green; curved lateral veins sunken, giving a quilted look; pale midrib. Inflorescence with narrow green spathe. Beautiful but difficult. *Humid-tropical.* p. 89, 93, 95

warocqueanum (Colombia), "Queen anthurium"; climbing species with showy, long tapering, velvety leaves to 1 m long; deep green with ivory veins; small spathe green to yellowish. *Humid-tropical.* p. 92, 95

ANTIGONON   *Polygonaceae*

leptopus (Mexico), "Coral vine"; showy tendril climber, with tuberous roots, slender zigzag stems bearing arrowshaped, light green leaves and axillary racemes of bright rose-pink flowers with deeper center. *Tropical.* p. 825, 827

ANTIRRHINUM   *Scrophulariaceae*

majus (So. Europe, No. Africa), "Floral snapdragon"; normally summer-flowering, erect herb, with leafy terminal spikes of very showy, curious sac-shaped, bilabiate flowers of many colors and shades, velvety red to pink, yellow or white, the yellow mouth closed but forced open by the bees. Also known as "Dragon-jaws". *Subtropic.* p. 882, 884, 885

majus 'Butterfly'; the "Butterfly strain" has a unique flower form, the florets are open-tubular resembling Penstemon, not closing like the dragon jaws; available in cut-flower-types, 75 cm tall and in dwarf carpet-bedding cultivars, 15-20 cm high. *Subtropic.* p. 882, 884

APEIBA   *Tiliaceae*

tibourbon (Guayana); evergreen tropical shrub to 3 m high, covered with downy hair; ovate oblong leaves having cordate base, with yellow midrib, and toothed margins; golden yellow star-like flowers; fruit a bristly capsule. *Tropical.* p. 910

APHELANDRA   *Acanthaceae*

aurantiaca (Mexico), "Fiery spike"; erect shrub with ovate, smooth, green leaves gray in vein areas; showy, bracted spikes with brilliant, scarlet-red flowers, orange in throat and along tube. *Tropical.* p. 33

chamissoniana (Brazil), "Yellow pagoda"; erect plant with closely set, thin, slender, pointed leaves, silver-white area along midrib and veins; flowers clear yellow; bracts yellow with green tips. *Tropical.* p. 33

fascinator (Colombia); emerald-green, satiny leaves marked with silver, veins a silvery amethyst, purple beneath; large, scarlet flowers. *Tropical.* p. 33

fuscopunctata (Colombia); introduced as 'Epple's Findling'; evergreen, herbaceous shrub to 1 m high with green, round stems, small ovate leaves to 15 cm long, dark green above, pale beneath, hairy and ciliate; the dense terminal spikes with overlapping red bracts and sticky, the bilabiate flowers light brown spotted dark-brown. *Tropical.* p. 33

nitens (Colombia); upright shrub with ovate, leathery, shining, dark-olive leaves with silver center vein, purple beneath; flowers red. *Tropical.* p. 33

sinclairiana (Central America, Panama, Darien), "Coral aphelandra"; compact plant with limp, ovate leaves rich glossy-green; pale, depressed veins; inflorescence clustered, the cupped bracts orange-coral, the corolla rosy-pink. *Tropical.* p. 33

squarrosa 'Fritz Prinsler' (squarrosa leopoldii x sq. louisae), "Saffron spike"; valuable cultivar combining the remarkable white to straw-yellow leaf-veining on dark olive-green of louisae with the leaf form and willingness of leopoldii but remains smaller; floral bracts deep yellow, marked green, corolla canary yellow. *Tropical.* p. 35

squarrosa var. louisae (Brazil), "Zebra plant"; compact growing plant with shiny, emerald-green, elliptic leaves and prominent, white veins; bright yellow flowers, tipped green, on fleshy, terminal spikes of long-lasting, waxy, golden bracts; fall blooming. *Tropical.* p. 33

tetragona (W. Indies, So. America); straggling plant to 1 m, with broadly ovate, green leaves; inflorescence in clustered spikes terminal or axillary; slender tubular, bright scarlet flowers 5-8 cm long. *Tropical.* p. 33

APLALEIA   *Commelinaceae*
multiflora (Trop. America) rambling trailer with purplish, fleshy stem and long ovate, clasping leaves entirely green, waxy margins and depressed midrib; flowers scented of violets, with 3 stamens. Photographed in Botanic Garden of Zurich, Switzerland. *Tropical.* p. 306

APONOGETON   *Aponogetonaceae*
madagascariensis (fenestralis) (Madagascar), "Laceleaf"; curious aquarium plant, from running streams, rootstock like a small potato, with leaves ascending to a little below the surface, 15 to 40 cm long, only a skeleton network of veins, pale yellow to dark olive; flower on long stalk rises a little above water, is of pinkish color, dividing into two curved hairy tufts. *Humid-tropical.* p. 82

APOROCACTUS   *Cactaceae*
flagelliformis (Mexico), "Rat-tail-cactus"; slender creeping or long pendant stems close-ribbed, covered with small reddish spines; small crimson flowers. *Tropical.* p. 255

x APOROPHYLLUM   *(Cactaceae)*
'Star Fire' (Aporocactus x Epiphyllum); beautiful Johnson bigeneric hybrid between Orchid cacti and Rat-tail cacti, normally epiphytic and ideally suited for baskets. This and similar hybrids are free-flowering, earlier than orchid cactus and bloom at odd times. The slender stems are ribbed and pendant but may be trained on a trellis. The large rose-red flowers are flushed scarlet outside. *Subtropic.* p. 255

APTENIA   *Aizoaceae*
cordifolia (So. Africa), "Baby sun rose"; low succulent creeper of the eastern coastal deserts; freely branching with prostrate or pendant branches, small soft-fleshy cordate, fresh green leaves to 2½ cm long, covered with glands, later gray; small 1 cm rosy-purple flowers. *Arid-subtropic.* p. 42, 49

cordifolia variegata (Mes.) (So. Africa); soft, prostrate succulent with small, heartshaped leaves, shimmering gray-green, and cream margins; tiny, purple flowers. *Arid-subtropic.* p. 49

AQUILEGIA   *Ranunculaceae*
skinneri (Mexico, Guatemala), "Colombine"; handsome columbine from the mountains of Mexico and Guatemala, often to 1 m high; leaves 2 or 3 times compound; showy flowers, the sepals with long hollow spurs, red, the petals yellow. *Subtropic.* p. 843

ARABIS   *Cruciferae*
alpina 'Variegata', a variegated-leaf form of the "Mountain rockcress" native in the European Alps; low, densely clustering and slowly creeping perennial with green, pubescent obovate, crenate leaves 3-8 cm long, attractively edged in ivory-white; small white flowers. *Warm temperate.* p. 379

ARACHIS   *Leguminosae*
hypogaea (Brazil), the "Peanut vine" or "Ground-nut"; annual herb 30-45 cm high with compound leaves of two pairs of oval leaflets, without tendrils; yellow flowers of two kinds: one set pea-like, showy and sterile, the others fertile, on recurving stalks which touch the ground and force into it where the fertilized ovary ripens into the peanut, an oily, edible seed. *Subtropic.* p. 472

ARACHNANTHE: see Arachnis

ARACHNIS   *Orchidaceae*
flos-aeris (moschifera) (Malaya, Java, Borneo), "Spider orchid"; epiphytic; slender stems to 4 m, with scattered narrow leaves and erect spike of 10 cm blooms, greenish-yellow, heavily blotched red-brown, more so in basal flowers; musk-scented. (summer). *Humid-tropical.* p. 708, 710, 718, 719

hookeriana (Malaya), "Scorpion orchid"; vandaceous epiphyte with leaves in ranks; long arching sprays of pure white flowers. *Humid-tropical.* p. 710

massoei (Malaya); robust epiphyte with slender, leafy pseudobulbs, terminal racemes of yellow, waxy flowers. As customary such epiphytes are grown in Singapore nurseries in perforated clay pots in mixtures of broken brick, charcoal and bone meal, to allow good drainage in monsoon climate. *Tropical.* p. 710

ARALIA   *Araliaceae*
chinensis (South China), "Chinese angelica"; deciduous shrub or small tree to 10 m; with prickly stems and large leaves to 1 m long, divided into 5-15 toothed leaflets; inflorescence in huge clusters of small white flowers, followed by black berries. *Warm-temperate.* p. 132

ARALIA: see also Polyscias, Dizygotheca, Fatsia, Tetrapanax

x ARANDA   *Orchidaceae*
majala (Arachnis x Vanda); vandaceous bigeneric hybrid with fleshy leaves in 2 ranks; erect raceme of fleshy flowers, pinkish beige with brown crossbars and small purple lip. *Humid-tropical.* p. 710

ARAUCARIA   *Araucariaceae (Coniferae)*
araucana (imbricata) (Mts. of So. Chile, No. Patagonia), the so-called "Hardy monkey puzzle" tree; 15-30 m high, with resinous bark and whorled branches, sharp-pointed, ovate, thick-leathery, dark green leaves 3-5 cm long, densely and uniformly arranged around shoot. *Warm temperate.* p. 333, 335, 336

bidwillii (Queensland), "Bunya-Bunya", or "Monkey-puzzle"; tree to 45 m high, the long, glossy needles with depressed parallel veining, mostly arranged in two flat rows, and sharp pointed; supposed to keep monkeys puzzled because they can't climb up. *Tropical.* p. 333, 335

columnaris (cookii) (New Caledonia), the "Cook pine"; so named because Capt. Cook discovered it on the Isle of Pines, in silhouette appearing like tall pines, with somewhat leaning trunks, said to grow as high as 60 m; difficult to distinguish from A. heterophylla as young trees in their curved, needle-like leaves, very symmetric with about 5 branches radiating around the trunk in each tier, and each branch triangular in shape; later the Cook pine develops into a narrower column than heterophylla, and on its branchlets bears overlapping more numerous, scale-like sharp-pointed leaves 1 cm long; the cones are ovoid and 8-10 cm in diam. *Tropical.* p. 335, 336

cunninghamii (New So. Wales), "Hoop pine"; majestic evergreen tree to 60 m high; the juvenile leaves are short, needle-like and bluish-green. *Subtropic.* p. 333, 336

excelsa—see heterophylla

heterophylla (excelsa in hort.) (Norfolk Is.), "Norfolk Island pine"; evergreen tree to 70 m high, from the South Pacific; when juvenile very formal with branches parallel to the ground, in tiers of bright green, soft, awlshaped needles; best Araucaria for the home. *Subtropic.* p. 333, 335, 336

ARBUTUS   *Ericaceae*
unedo (So. Europe, etc.), the "Strawberry-tree"; small evergreen tree with rough, shreddy bark, sticky-hairy branches, oblong, shiny, toothed 10 cm leaves, and small white, or pinkish, bell-like flowers in drooping panicles; strawberry-like, orange-red fruit edible but without flavor. *Subtropic.* p. 394, 395, 458, 462

ARCHONTOPHOENIX   *(Palmae)*
alexandrae (Australia: Queensland), the tall-growing "King palm" or "Alexandra palm", with erect ringed trunk 15 cm thick and to 20 or even 30 m high, and bulging at base; bearing

a majestic crown of arching pinnate fronds, the narrow leaflets 4-5 cm wide, green above but prominently grayish-white beneath; flowers white or creamy; attractive red fruit. *Subtropic.* p. 764

cunninghamiana (Australia: Queensland, N.S.W.), "Seaforthia palm" or "Piccabeen palm"; tall feathery palm with slender trunk not enlarging below except at surface of ground; the gracefully arching fronds to 3 m long; broad leaflets dark green on both sides, 8-10 cm wide; pendulous inflorescence with lilac flowers; the coral fruit less than 2½ cm. In the California nursery trade erroneously as Seaforthia elegans; in Florida nurseries as Ptychosperma elegans. *Subtropic.* p. 765, 766, 774

ARCTOSTAPHYLOS *Ericaceae*
columbiana (Coastal Brit. Columbia to No. California), "Manzanita"; hard-wooded, often arborescent shrub to 3 m high, usually shorter, with twisted chocolate brown branches; leathery ovate leaves 6 cm long; small white or pink flowers; fruit like little apples green to bright red; eaten by the Indians. *Subtropic.* p. 395, 466

uva-ursi (California north to Alaska), "Bearberry" or "Trailing manzanita"; a popular ground-cover in Pacific Northwest; prostrate spreading branches and rooting as it creeps to 5 m; bright glossy green, leathery leaves 2-3 cm long, turning red in winter; flowers white tinged pink; bright red fruit. *Warm temperate.* p. 394, 466

ARCTOTHECA *Compositae*
calendula (South Africa), "Cape weed"; evergreen perennial, rapid-running ground-cover, less than 30 cm tall; gray green divided leaves; flowers yellow. *Subtropic.* p. 308

ARCTOTIS *Compositae*
'Hybrida' (acaulis and other species), "African daisy"; stemless perennial with stout rhizome; leaves in rosettes, pinnately lobed or lyrate, green and hairy above, lower surface white woolly; long-stalked flowers 8 to 10 cm across, in many colors, cream, yellow, orange, pink, red and purple, many with central zones of contrasting colors. *Subtropic.* p. 332

ARDISIA *Myrsinaceae*
crenata (Japan), "Coral berry"; smooth evergreen shrub to 2 m, waxy, deep green, leathery elliptic leaves with undulate crisped margins; small 1 cm white or pink flowers; clusters of attractive and long-lasting glossy red berries. *Subtropic.* p. 673

humilis (E. Indies); evergreen shrub with softly leathery, lanceolate leaves 5-12 cm long, and umbels of small rose-pink flowers; 1 cm fruit flattened-globular, first red then black. *Subtropic.* p. 673

solanacea (Sri Lanka); small tree to 6 m, with red-brown branches; leathery narrow-oblanceolate leaves 10 cm long, pink when new; rosy-pink to purple flowers, followed by many 1 cm berries ripening red to shining black. *Tropical.* p. 673

ARECA *Palmae*
catechu (Malaysia to Polynesia), the famed, graceful "Betelnut palm"; a very slender, erect palm with solitary trunk, 10 to 30 m high and 5-12 cm dia.; relatively small crown of pinnate fronds 1-2 m long, with many broad, rather soft pinnae irregularly notched at apex, the upper ones united; fragrant white flowers; fruit olive shaped to 5 cm long, orange red, with soft fibrous covering. Much cultivated for the nut which is chewed with lime and the leaf of Piper betle, as a mild stimulant. *Tropical.* p. 474, 765, 802

langloisiana (Celebes); a slow-growing, suckering palm with heavy feather fronds, the dark green, ribbed leaflets broad and leathery; showy with orange-colored crown shaft. The palm photographed at Dr. Darian's collection in Vista, California is 10 years old. *Tropical.* p. 766

ARECA: see also Chrysalidocarpus

ARECASTRUM *Palmae*
romanzoffianum (Cocos plumosa) (Bahia to Argentina and Bolivia), "Queen palm"; very handsome with straight smooth trunk to 12 m high and a graceful crown of long arching plumy fronds, the soft, dark shiny-green leaf segments pendant above the middle; edible orange fruit. *Humid-subtropic.* p. 765, 766, 767, 768

ARENARIA *Caryophyllaceae*
verna caespitosa (Spain to No. Russia), "Irish moss"; cushion-forming perennial, moss-like and used in California as lawn substitute; leaves linear to 2 cm long, grayish green; small pinkish-white, starry flowers. Needs more moisture than lawn; gets lumpy if crowded. *Warm-temperate.* p. 510

ARENGA *Palmae*
pinnata (saccharifera) (Malaya), the "Common sugar palm"; important economic palm to 12 m high, with robust solitary trunk covered with black fibers; erect, pinnate leaves to 6 m or more, the leaflets dark green above, whitened underneath, and with jagged apex. When pierced, the young inflorescence yields sugary sap. *Tropical.* p. 768

ARGEMONE *Papaveraceae*
mexicana (Mexico, W. Indies), "Prickly poppy"; prickly leaved annual with clasping foliage deeply lobed and spine-tipped; flowers yellow 4-6 cm across. Seeds yield Argemone oil. *Subtropic.* p. 812

ARGYREIA *Convolvulaceae*
nervosa (speciosa) (India), "Elephant creeper" or "Silver Morning-glory"; elegant, robust climber reaching up to 8 m; woody stems with large, overlapping, cordate, rich green leaves prominently nerved and silvery silky beneath, 15 to 30 cm wide; showy deep rose flowers with flaring limb, 5-8 cm long, violet in throat, white-hairy outside, the base enveloped by a large white-hairy calyx. *Tropical.* p. 357

ARGYRODERMA *Aizoaceae*
roseum (delaetii) (So. Africa: Cape Prov.); succulent leaves 3 cm thick, bluish-green, united about halfway; flowers violet-rose, 8 cm dia., petals lax. *Arid-subtropic.* p. 46

ARGYROXIPHIUM *Compositae*
sandwicense (Hawaii), the "Silver sword"; endemic on mountains to 3950 m altitude on volcanic cinders; narrow pointed leaves form a rosette 60 cm dia., clothed entirely with silvery-silky hair; nodding flower heads with purplish ray florets. *Subtropic.* p. 309, 323

ARIOCARPUS *Cactaceae*
kotschubeyanus (C. Mexico); plant sprouting from the beet-like taproot, forming groups; low fleshy, star-like rosette of angular dark green rugose leaves cleft at triangular apex; flowers from the woolly center purplish-carmine. *Arid-tropical.* p. 265

retusus (Mexico: San Luis Potosi), known as "Seven stars"; handsome star-like rosette 10 to 12 cm across, with silvery glaucous, triangular tubercles pointed outward, like the leaves of an haworthia; with yellow wool in axils, the areole near tips; flowers whitish to pink, 5 cm dia. *Arid-tropical.* p. 264

trigonus (Mexico), "Living rock"; hard fleshy rosette of numerous erect leaflike tubercles, grayish green, turnip-like root; pale yellow flowers. *Arid-tropical.* p. 263, 264

trigonus elongatus (Mexico), "Living star"; similar to trigonus, with longer, narrow numerous leaves; in center a cluster of yellow flowers and dark eye. *Arid-tropical.* p. 264

ARISAEMA *Araceae*
candidissimum (Japan, W. China); hardy tuberous plant with solitary leaf on petiole 75 cm long, the blade leathery, trifoliolate with leaflets widely ovate, each 7-20 cm wide and long; tubular spathe green at base, white tinged rose toward pointed tip and with white length-lines; slender green spadix. *Warm temperate.* p. 97

fargesii (China: Szechwan); rhizome-like tuber with solitary green, trisected leaf, the three broad smooth segments 15 to 20 cm long; spathe tube to 8 cm long, striped white or pale purple, the forward arching pointed blade greenish purple outside, deep violet inside with long slender point; spadix glossy purple, greenish and white; near hardy. *Warm-temperate.* p. 97

sikokianum (So. Japan); subtropic tuberous plant with attractive inflorescence with white-club-shaped spadix; spathe trumpet-shaped outside purple, with narrow greenish bands, the upper inside white, the long lanceolate extension greenish-red with white stripes; trifoliolate leaves painted with silver, along the pseudostem or stalk carrying the floral head; fairly hardy. *Subtropic.* p. 97

triphyllum (E. No. America), "Jack-in-the-pulpit"; tuberous woodland herb to 1 m, with usually a pair of trifoliate ovate leaflets to 18 cm long; the erect spadix is surrounded by the green spathe or "pulpit", striped with purple inside, its tip arched forward; berries red. *Warm temperate.* p. 97

ARISARUM *Araceae*
vulgare (So. Europe, Algeria); subtropic tuberous plant with 1 or 2 fleshy leaves triangular sagittate, deep green; the inflorescence on a spotted stalk, a light green cylindrical spathe vivid purple toward apex, and the pointed hood black-purple, curving forward. *Subtropic.* p. 96, 97

ARISTEA *Iridaceae*
ecklonii (So. Africa), "Blue stars"; perennial herb having a woody rootstock with long, strap-like, light green, arching leaves in basal rosette; blue, short-lived flowers with flattened lobes in loose clusters, during summer. *Subtropic.* p. 527

ARISTOLOCHIA *Aristolochiaceae*
elegans (Brazil), "Calico flower"; graceful climber with kidney-shaped leaves and flowers a yellowish, inflated tube and expanded cup rich purplish-brown inside with white markings. *Tropical.* p. 145

fimbriata (Argentina), a "Birthwort"; stems scarcely climbing with glabrous, kidney-shaped leaves and bent, odd-shaped irregular tubular calyx simulating a corolla; tube swollen at base; one-sided limb greenish-brown outside, purple-brown inside, veined yellow, fringed. *Subtropic.* p. 145

grandiflora (gigas) (Jamaica), "Pelican flower"; climber with heartshaped, downy leaves and large flowers like an expanded bent pipe, limb veined and spotted purple, with long tail. *Tropical.* p. 151

leuconeura (Colombia), "Ornamental birthwort"; ornamental-leaved climber with glossy, yellow-green, heart-shaped leaves with wide sinus, attractively veined cream-white. *Tropical.* p. 145

veraguensis (Panama); rampant climber with wiry stems; ornamental cordate leaves with rugose surface, deep coppery green with network of yellow veins; peculiar flowers with bent tube. *Tropical.* p. 145

AROPHYTON *Araceae*
crassifolium (Humbertina) (Madagascar); tuberous tropical aroid with smooth broad hastate-cordate bluish-green leaves 8-16 cm long; fleshy ovate reflexed spathe greenish with reddish lines, 9-13 cm long, subtending the slender erect spadix swollen in the middle. *Tropical.* p. 97

tripartitum (Madagascar); creeping rhizome 4-7 cm long; fleshy green leaves divided into 3 elliptic segments; the central one 7-14 cm long; young foliage is undivided and hastate; floral stalk 10-20 cm long, with small 4-5 cm spathe, in a concave, pointed shield behind the 3 cm spadix. *Tropical.* p. 97

ARPOPHYLLUM *Orchidaceae*
spicatum (Mexico, Nicaragua), "Hyacinth orchid"; epiphyte with long leathery leaves, forming dense raceme of cupped, carmine-rose flowers. *Tropical.* p. 709, 710

ARRABIDAEA: see Saritaea

ARRHENATHERUM *Gramineae*
elatius 'Variegatum' (Mediterranean), "Variegated oatgrass"; perennial grass with stems to 1½ m; flat, soft leaf blades 1 cm wide, fresh green with white margins; panicle pale green or purplish. *Subtropic.* p. 512

ARTEMISIA *Compositae*
arborescens (Mediterranean region), aromatic "Wormwood"; shrubby perennial to 1 m with finely divided silvery-hairy leaves; small bright yellow flower heads in leafy clusters. *Subtropic.* p. 310

dracunculus (Europe), "Tarragon" or "Estragon"; hardy herbaceous perennial with green linear leaves scented like anise; panicles of whitish flowers. Use: in salads, steaks, and other cookery; yields flavor to pickles. *Warm temperate.* p. 310

ARTHROCEREUS *Cactaceae*
campos-portoi (Minas Geraes, Brazil); slender column with short spines; long-necked white flowers with narrow petals, outside reddish. *Arid-tropical.* p. 248

ARTHROPODIUM *Liliaceae*
cirrhatum (New Zealand); tufted perennial herb with fleshy roots to 1 m high, numerous spreading flexible, lanceolate, light green, clasping leaves grayish beneath, with a narrow translucent edge and parallel veins; clusters of white flowers. *Subtropic.* p. 575

ARTOCARPUS *Moraceae*
altilis (communis, incisa) (Malaysia to Tahiti), "Breadfruit"; widely planted in tropical countries because of the large, yellow, prickly fruit, which resembles bread when baked; large, milky-juiced tree to 20 m high, with beautiful, huge, deeply lobed leathery leaves, luxuriantly green with yellowish veins. *Tropical.* p. 475, 649

heterophyllus (integrifolia) (India to Malaya), the "Jackfruit"; interesting tropical tree to 15 or 20 m high related to the breadfruit, with milky juice; glossy oblong leaves, lobed on younger branches; remarkable for its enormous fruit dangling directly from the trunk and biggest branches of the tree. This fruit is one of the largest in the world, ⅓ to 1 m long and weighing up to 18 kg.; a green knobby rind encloses a soft sweet or acid pulp of unpleasant odor, and is eaten raw or cooked, the seed roasted, in tropical Southeast Asia. *Tropical.* p. 475, 478, 649

nobilis (Sri Lanka); sparry tree with handsome leaves 30 cm or more long, thick leathery and glossy green, the ribs yellow and margins wavy. *Tropical.* p. 648

ARUM *Araceae*
dioscoridis (Eastern Mediterranean); handsome tuberous plant with hastate or sagittate 25-30 cm leaves on long petioles; the colorful spathe pale yellowish-green with large brown-purple areas and spots inside. *Subtropic.* p. 95

italicum (Europe, No. Africa), "Italian arum"; robust, tuberous plant with hastate, fresh-green leaves with whitish veining; spathe green, white inside with reflexed purple limb. *Subtropic.* p. 95

maculatum (England to S.E. Europe and No. Africa), "Lords and ladies" or "Aaron's staff"; fleshy tuberous herb 30 cm high, with arrow-shaped leaves, usually black-spotted, withering in summer; long-stalked spathe 15-25 cm long pinched at the middle, light green and purple-spotted or margined, longer than the spadix; end of July, on stiff-erect stalks, the ripening berries turn scarlet red. All parts poisonous. *Subtropic.* p. 95

palaestinum (sanctum) (Israel, Syria, Afghanistan), "Black calla"; tuberous plant with arrow-shaped, green leaves, followed by flower spike with dark spadix, and spathe green outside, black-purple within and the tapering limb. *Subtropic.* p. 96

ARUNCUS *Ranunculaceae*
sylvester (choicus, vulgaris) (Eurasia, No. America), "Goat's beard"; perennial 1-2 m high with 2-3 times pinnate leaves; tall stems with pyramidal panicle of small flowers. *Temperate.* p. 847, 848

ARUNDINA *Orchidaceae*
chinensis (China); terrestrial with slender reed-like stems, to 1 m high; narrow leaves; flowers to 5 cm across, pale lilac with crimson blotch. *Subtropic.* p. 709

ARUNDINARIA *Gramineae*
alpina (Mountains of Africa), "Mountain bamboo"; hollow-stemmed bamboo to 15 or more metres high, spreading densely but not in clumps from woody rhizomes, the culms 5-10 cm thick, turning yellow with age; the linear leaves to 20 cm long. I photographed this species in the bamboo forest at 2,430 m on the Kivu border of Zaire, typical gorilla country. *Subtropic.* p. 517

falconeri (Thamnocalamus) (Himalayas); handsome bamboo with dark green culms from 8 to 20 m high, and 4 to 5 cm thick, at nodes forming clusters of branches; the leaves to 10 cm long, pale green and finely toothed, glaucous beneath. Photographed in Kew Gardens, London. *Subtropic.* p. 509

pumila (Sasa) (Japan); dwarf bamboo or woody grass spreading by rhizomes, with slender stems to 60 cm; narrow leaves 15 cm long, bright green and slightly hairy on both sides. *Subtropic.* p. 516

ARUNDO *Gramineae*
donax (Mediterranean reg., naturalized in Southern U.S.), "Giant reed"; one of the largest grasses, and resembling bamboo, to 6 m or more high; perennial with woody stems; two-ranked flat green leaves to 60 cm long and 8 cm wide and with rough margins; inflorescence in slender, greenish or purplish plumes. *Warm-temperate.* p. 511

donax versicolor (So. Europe), the "Variegated giant reed"; majestic perennial grass, only 2 m, with knotty rootstock and stout stems almost woody; arching ribbon-like leaves gray-green striped creamy white; to 60 cm long, alternately arranged on canes, and topped by showy plume-like panicles at first reddish, then white. Very impressive with its bold, bamboo-like leafy canes; quite ornamental in containers. *Warm temperate.* p. 512

ASARINA *Scrophulariaceae*
erubescens (Maurandya) (Mexico), "Creeping gloxinia"; strongly vining, hairy plant with alternate, triangular, toothed, downy leaves and twining flowerstalks, bearing large, 8 cm trumpet-shaped blossoms having broad green sepals and carmine-rose corolla with pale throat spotted rose, blooming into Nov. *Tropical.* p. 886

ASCLEPIAS *Asclepiadaceae*
currassavica (Trop. America), "Blood-flower; showy perennial to 1 m with woody base, stems with milky sap, oblanceolate leaves 5-15 cm long, the flowers in umbels with reflexed, 5-parted corolla

brilliant red-purple, exposing the crown of 5 orange horned hoods. *Tropical.* p. 146

fruticosa (Gomphocarpus) (So. Africa); dense bush with willowy stems 1-2 m high, narrow linear leaves 12 cm long; small white axillary flowers; spiny ovoid, inflated seed pods 8 cm long. *Subtropic.* p. 146

physocarpa (South Africa), "Butterfly flower"; white-hairy shrubby perennial to 2 m; linear to lanceolate leaves 10 cm long; axillary inflorescence with white corolla and 4-angled hood; curious yellow-green, inflated bristly seedpods 6 cm dia. *Subtropic.* p. 146, 151

ASCOGLOSSUM *Orchidaceae*
calopterum (New Guinea); remarkably showy epiphyte; robust stem to 60 cm long; thickly leafy; leaves rigidly stiff, bi-lobed at apex; branched raceme 60 cm long, with many small, pretty 3 cm flowers rose pink to purplish. *Tropical.* p. 711

ASPARAGUS *Liliaceae*
asparagoides myrtifolius (Medeola) (Cape of Good Hope), the "Baby smilax" used for wedding decorations, etc., graceful twining vine with thread-like stems and dainty, fresh glossy-green, little ovate leaves, usually allowed to climb on strings for cutting. *Subtropic.* p. 574

densiflorus cv. 'Myers' (So. Africa), "Plume asparagus"; showy plant dense with stiffly erect, plume-like branches 60 cm or more high, the dense needle-like "foliage" rich green. *Subtropic.* p. 574, 584

densiflorus cv. 'Sprengeri' (Natal), "Sprengeri fern"; much branched from tuberous roots, scarcely climbing, the fluffy branchlets set with soft, fresh green needles, true leaves reduced to thorns; small fragrant flowers white, followed by bright red berries. *Subtropic.* p. 574

myriocladus (So. Africa: Natal), "Zigzag asparagus"; erect, much branched sinuous shrub to 1½ m high, from swollen roots; gray stems with zigzag branches, the cladodes (branches simulating leaves) thread-like 1-2 cm long, in dense clusters, bright green. *Subtropic.* p. 574

retrofractus (So. Africa), "Zigzag shrub"; shrubby plant with beige-gray woody stem to 2 m long, with stiff, thorny zigzag twigs at angles, bearing dense bundles of curved needle-like cladodes (branchlets) bright green; true leaves reduced to scales; small white flowers. *Subtropic.* p. 574

scandens deflexus (ramosissimus) (So. Africa), "Basket asparagus"; climber, with round green stems, much branched above, the flat leaf-like unarmed, light green cladodes linear and curved, to 2 cm long. In var. deflexus the branches are deflexed and very zigzag. *Subtropic.* p. 574

setaceus (plumosus of hort.) (So. Africa), "Fern-asparagus"; climber with lacy, fern-like, rich green fronds of needle-like branchlets, arranged on a horizontal plane, on thin wiry stems with sharp prickles. *Subtropic.* p. 574

setaceus 'Pyramidalis' (cupressoides), "Cypress asparagus"; a curious form of the "Fern-asparagus" of dense habit, with its needle-like cladodes as well as the stems all erect and striving straight upward like a cupressus. *Subtropic.* p. 574

ASPHODELUS *Liliaceae*
microcarpus (Canary Islands to Asia Minor); perennial herb with fleshy roots; rosette of broad, fleshy foliage; inflorescence on branched stalks to 1 m high, bearing starry white flowers with purple line on each petal. *Subtropic.* p. 585, 603

ASPIDISTRA *Liliaceae*
elatior (lurida) (China). "Cast-iron plant" or "Parlor palm"; old-fashioned tough-leathery foliage plant with thick roots and blackish-green, shining oblong basal leaves to 75 cm long, narrowed to a channeled stalk; purple bell-shaped flowers at the surface of the ground. Ideal for cool, unfavorable locations, very tolerant of neglect. *Subtropic.*

elatior 'Maculata', "Milky Way iron plant"; novelty with soft-leathery, long-lanceolate rich green leaves finely speckled with creamy-white. *Subtropic.* p. 590

elatior 'Variegata' "Variegated cast-iron plant"; attractive variegated form having leaves alternately striped and banded green and white in varied widths. *Subtropic.* p. 577

ASPIDIUM: see Polystichum, Rumohra

ASPLENIUM *Polypodiaceae (Filices)*
bulbiferum (New Zealand, Australia, Malaya), "Mother fern"; with wiry pinnate fronds, having grooved black stem, the pinnae fresh green, and much larger than viviparum and not as deeply lacily cut, the segments becoming linear only when spore-bearing; bulbils or plantlets are produced on upper surface of frond. *Humid-subtropic.* p. 439, 447

lucidum (obtusatum lucidum) (New Zealand, Australia) "Leather fern"; decorative free-growing fern with grayish stems and pendant pinnate fronds of graceful habit and shining-green, reach 1 m in length, the oblong leaflets are leathery, 15 cm long and with toothed and wavy edge. *Humid-subtropic.* p. 438

mayii hort.; a seedling probably with vieillardii from Polynesia; of tufted habit, with short, gracefully arched pinnate fronds, the dark purplish-green pinnae well spaced and deeply notched. *Humid-subtropic.* p. 440

nidus (nidus-avis) (India to Queensland and Japan), the "Birds-nest fern"; great epiphytic rosette of simple oblanceolate, stiffly spreading, shining, friendly green fronds anywhere from 30-90 cm long, of thin leathery texture with prominent blackish midrib and wavy margins, rising from a crown densely clothed with black scales. One of the most interesting and attractive of ferns for pots, keeping surprisingly well but should be kept steadily moist and warm, or new fronds may become deformed; avoid drafts; tolerates poor light to 25 fc. *Humid-tropical.* p. 443, 445

nidus phyllitidis (South China); a robust epiphyte similar to A. nidus, but with many more leaves in a rosette, the leathery leaves narrower and undulate at margins. *Humid-tropical.* p. 443

nidus plicatum; interesting variety in the fern collection of Kew Gardens; from a stout, hairy rhizome rises a semi-rosette of narrow linear stiff leaves to 50 cm long, glossy green, with bold midrib; the blade closely corrugated into continuous folds, margin uneven or wavy. *Humid-tropical.* p. 445

viviparum (Mauritius), "Mother fern"; tufted plant with dark green, finely lacy, arching fronds on firm stems, the little thread-like linear segments giving rise to tiny bulblets from which develop little plants. *Humid-tropical.* p. 439

ASTELIA *Liliaceae*
nervosa (New Zealand); herbaceous, densely tufted, terrestrial perennial, with short, thick rhizome, and numerous linear, strap-like, glossy-green leaves sheathing near base, with prominent elevated parallel ribs; forming orange-red berries on female plants. *Humid-subtropic.* p. 586

solandri (New Zealand), the "Kokaha"; remarkable epiphytic herb with sword-like leaves forming immense tufts often high on lofty trees on North Island, where they store water in the thick, curved bases of their circle of foliage. Linear leaves ½-1 m long, olive green, silky and recurving; pendant panicles of tiny pale yellow flowers, the female plants producing wine red berries. Collospermum hastatum may be a more valid name, but this is supposed to have broader leaves and larger flowers. *Humid-subtropic.* p. 584

ASTER *Compositae*
amellus (S.E. Europe to W. Asia), "Italian aster"; rough, hairy perennial to 60 cm high, branching stems with oblong leaves and violet, yellow centered flowers 5 cm across. *Warm temperate.* p. 314

novi-belgii (E. No. America), "New York Aster"; rhizomatous perennial to 1½ m high, with narrow, smooth leaves; on stiff stems full clusters of bright blue-violet ray flowers 3 cm across, disk flower yellow. *Temperate.* p. 314

ASTER: see also Callistephus

ASTERISCUS *Compositae*
sericeum (Odontospermum) (Canary Islands); handsome dwarf shrub dense with small rosettes of silky-hairy, obovate green leaves 4-5 cm long; in the apex of branches nestle the pretty, thistle-like golden yellow heads 4 to 8 cm across, the large cushions of disk flowers surrounded by the spreading ray florets. *Subtropic.* p. 329

ASTEROSTIGMA *Araceae*
vermicidum (No. Argentina: Tucumán, Salta); robust tuberous herb to 1 m high, the tuber 5-15 cm diameter, with solitary fleshy light green deeply dissected leaf 30-60 cm long, appearing after the inflorescence; the hooded spathe brown purple with small cone-shaped spadix nestling inside. *Subtropic.* p. 102

ASTILBE *Saxifragaceae*
x arendsii (chinensis x japonica), "False spiraea"; vigorous hardy perennial 60-80 cm high; herbaceous, much divided ferny leaves; plume-like airy inflorescence on slender wiry stems, small white, pink or red flowers. Photographed in Gotland, Sweden. *Warm temperate.* p. 881

japonica (Japan), the florists "Spiraea"; robust, hardy, herbaceous perennial easily used for forcing in pots for Easter, when its fluffy, plumy, white or red flower spikes carried well above the fern-like, pinnate foliage, are very showy and attractive. *Warm temperate.* p. 877, 879

ASTROCARYUM *Palmae*
jauari (Amazonas); feather palm with slender trunk, covered with sharp spines, the pinnate fronds rather erect, 3 m long, shiny green above, whitish beneath. *Humid-tropical.* p. 769

ASTROPHYTUM *Cactaceae*
  asterias (No. Mexico), "Sand dollar"; low, dome-shaped, with a star-shaped depression, green tinged coppery and covered with white scales, the areoles tufted with white wool; flowers yellow with red throat. *Arid-subtropic.* p. 265
  capricorne (No. Mexico), "Goat's horn"; hard green globe to 25 cm with some silver marking, 7 or 8 ribs with contorted spines; spreading yellow flower with red throat. *Arid-subtropic.* p. 263
  myriostigma (C. Mexico), "Bishop's cap"; small globe with usually 5 prominent ribs, spines absent, covered with small white spots; flowers yellow. *Arid-subtropic.* p. 264, 265
  myriostigma quadrocostatum, "Bishop's hood"; a variety divided equally by four broad ribs, resembling a real 4-sided parson's cap; the little yellow flower looks pretty in the center of the plump body. *Arid-subtropic.* p. 265
  ornatum (Mexico), "Monk's hood"; small and hard plant subglobose or cylindric with 8 prominent spiral folds, green and beautifully marked with silvery spots, talon-like spines; flowers lemon-yellow. *Arid-subtropic.* p. 264, 265

ATTALEA *Palmae*
  amygdalina (Brazil), "Cohune palm"; typical with short trunk, and gigantic pinnate leaves to 6 m long, the numerous long pinnae bright green; fruit 5 cm long. *Humid-tropical.* p. 768

AUCUBA *Cornaceae*
  japonica 'Crotonifolia'; form with thinner and much broader leaf coarsely serrate, lighter green freely marbled and blotched yellow, as compared with variegata. *Warm-temperate.* p. 357
  japonica 'Dentata'; photographed in Kew Gardens, London; handsome evergreen bush, with glossy green leaves smaller than the species, and irregularly toothed at margins. *Warm temperate.* p. 357
  japonica 'Goldiana', "Gold-leaf bush"; mutant with large leaves almost entirely golden yellow and only the margins light green. *Warm-temperate.* p. 357
  japonica 'Picturata', also known in horticulture as "Goldiana", the "Golden laurel"; mutant with large leaves almost entirely golden yellow and with only the margins light green, and occasionally dotted yellow. A truly showy colorform although more easily subject to leafscorch or spotting from wetness, due to the absence of chlorophyll. *Subtropic.* p. 357
  japonica 'Variegata' (Himalayas to Japan), "Gold dust tree"; evergreen shrub with opposite shining leathery elliptic leaves dark green blotched yellow, and toothed above middle; purple flowers at base; the female plants with scarlet berries. *Warm-temperate.* p. 357

AVERRHOA *Oxalidaceae*
  carambola (India to China), the "Carambola tree"; evergreen 5-10 m, with pinnate leaves of 5-9 leaflets which close when touched or at night; small fragrant flowers white marked with purple, the fleshy drooping 10 cm fruit ribbed, yellowish-brown, quince-scented and edible. *Subtropic.* p. 463, 763

AZALEA (botanically RHODODENDRON) *Ericaceae*
  'Alaska' (1937), (Vervaeneana alba x Snow, with Rhododendron); very satisfactory hybrid of vigorous habit, dark elliptic foliage, and trusses of medium large 6 cm semi-double hose-in-hose pure white flowers; early forcer. (Rutherfordiana). *Subtropic.* p. 400
  'Concinna' ('Queen Elizabeth'); according to Haerens Tuinbouw probably a strong growing erect cultivar of 'phoenicea' which is a variety of Rhododendron pulchrum originally introduced from China. This 'concinna' has replaced 'phoenicea' as understock for grafting of tender Rhodo. simsii hybrids, commonly known as 'Belgian Indica azaleas'; hairy branches with elliptic or oblanceolate leaves 6 cm or over, matte dark green above, glossy beneath, and thinly furnished with appressed brown hairs on both sides; flowers large when grown under glass, 5 to 8 cm diameter, wide-open, single, purplish crimson rose, spotted crimson on upper petals, the margins lightly wavy. *Subtropic.* p. 400
  'Constance'; vigorous lovely variety with large trusses of medium-large single frilled flowers, vivid cerise-pink; early, midseason. (Rutherfordiana) *Subtropic.* p. 400
  'Coral Bells' (Kirin); favorite selection of the 'Kurume' hybrids, small-flowered evergreen azaleas originated on Kyushu Island (So. Japan) during the Meiji era (1868-1912), from species involving Rhododendron obtusum, obtusum 'Kaempferi' and kiusianum; this charming cultivar is of low habit, densely branched, with small shiny, fresh green leaves; early blooming with a multitude of bell-shaped 3 cm hose-in-hose flowers of dainty silver-pink, shading to coral pink in center. *Warm-temperate.* p. 401
  'Delaware Valley White', form of Rhod. mucronatum; a woody semi-evergreen plant of robust, somewhat sparry, tall habit, with shiny obovate light green leaves and large single, pure white 6-8 cm flowers, late blooming; may be held for Mother's Day. Partially deciduous; hardy. *Warm temperate.* p. 401
  'Double Coral Bells', (Roehrs 1950) (Keller No. 27 x Coral Bells), vigorous seedling with shining fresh green foliage and cupped flowers larger than 'Coral Bells', having three rows of rounded petals, vivid salmon pink. *Warm temperate.* p. 400
  'Hino-crimson'; Kurume type evergreen similar to Hinodegiri but while its growth is more compact, it is an improvement over the Hinodegiri in that its single flowers are a clear crimson red without bluish overtones; late bloomer, fairly hardy. *Temperate.* p. 401
  'Hinodegiri' (Mist-of-the-Rising-Sun); an old favorite bushy 'Kurume', fairly hardy with glossy foliage and small 3 cm single flowers vivid carmine-red; late blooming. *Temperate.* p. 401
  'Kingfisher' (Whitewater hyb. Washington); grown by Yoder; frost-tender; habit shapely and spreading; the glossy foliage sets off clusters of large 8-9 cm carmine-rose, hose-in-hose flowers; keeping quality outstanding; early to midseason bloom. *Subtropic.* p. 400
  'Louise Gable', Gable (Stewartstown, Pa.); indicum-kaempferi hybrid of spreading, sparry habit with medium small semi-double, broad petaled flowers of fine lavender-pink; somewhat deciduous; late blooming. *Warm temperate.* p. 401
  'Madonna' (Rhododendron) (Brooks California hybrid); originated on the Pacific coast, with simsii as one parent; compact plant suitable for growing as pot plant; dark green foliage and medium large double flowers pure white. *Subtropic.* p. 401
  'Marjorie Ann' (Pericat hyb.) (Belgian indica x Kurume); spreading, somewhat straggly hybrid with dark foliage and medium small semi-double to double rose-bud flowers with broad petals, a pretty shade of salmon; midseason. *Warm temperate.* p. 401
  'Peggy Ann' (Roehrs-Bauman hyb.); one of the loveliest creations the bees have brought about, between Kurume hybrids and kaempferi; small 4 cm wide open hose-in-hose flowers, with two rows of white petals edged in rosy-pink like apple blossoms; late blooming; fairly hardy. *Warm temperate.* p. 401
  'Polar Bear' (Firefly x Snow); a 'Beltsville' hardy type azalea suitable for late blooming, small-flowered, 4 cm, floriferous, pure white, hose-in-hose with flaring petals; shining green foliage. *Temperate.* p. 400
  'Pride of Detroit'; sport of 'Mad. Pericat', an outstanding variety noted for its color and keeping quality, of spreading growth habit, with large round flaring flowers vivid fiery red, having a double row of petals but somewhat shy; midseason to late. *Warm temperate.* p. 400
  'Purple Heart' (R. obtusum hybrid); hardy, wiry bush with evergreen leaves, and spreading star-like single flowers bright purple. *Warm temperate.* p. 401
  'Redwing' (Pericat hyb.); rapidly growing evergreen, requiring twice as much pinching as other Azaleas if grown for pot culture or else they will grow loose, straggly and floppy; large, wide open showy flowers 8 cm or more, with two rows of petals, ruffled at the margins, cerise-red. Flowers larger and more frilled than Lentengroet, but buds normally best in older specimen. *Warm temperate* p. 400
  'Rose Pericat' (1931); one of the best "Pericat" hybrids, of robust bushy growth, dark foliage and 6-7 cm pleasing clear pink double flowers with salmon sheen, center lacy and with red lines in throat, outer petals form a star; early, *Warm temperate.* p. 401
  'Salmon Beauty'; free growing "Kurume" with medium small 4½ cm hose-in-hose flowers of an even light salmon shade, with spreading, recurved petals; midseason. *Warm temperate.* p. 401
  'Snow'; popular 'Kurume', willing grower with rich green, glossy foliage covered by masses of small 3 cm pure white, hose-in-hose flowers with freely spreading petals; early and midseason. *Warm temperate.* p. 401

'SIMSII HYBRIDS' or 'Belgian Indica'; descendents of Rhododendron simsii native in moist-warm subtropical Central China, Yunnan and along the Yangtse; deep rosy-red, to 4 m high. These hybrids were developed primarily for their large and showy double flowers, principally in Belgium, for forcing and flowering in pots; they are usually grafted on understock of 'Phoenicea concinna' for better growth and longer life; not winter hardy in cold climate. *Subtropic.*
  'Adventgloeckchen' (Chimes) (1929), (Paul Schaeme x Fred Sander); bushy plant with bell-shaped medium flowers glowing carmine-rose, semi-double, and of good texture; early to mid season. p. 403
  'Albert-Elizabeth' (1921); loosely growing variety, a unique bud sport of Vervaeneana, with large double flowers to 9 cm dia., not unlike rhododendron in form, glistening white and decorated with a salmon-red border and marked green on lower petals; p. 403
  'Ambrosiana' ('Mad. Petrick' x 'Reinhold Ambrosius'); superb early flowering, fairly new German hybrid easily forced for

Christmas; vigorous grower with obovate glossy leaves and 7½ cm double flowers like little roses, glowing crimson red. p. 403

'Anytime'; a simsii "Belgian indica" hybrid, sport of 'Jean Haerens' with 6 cm double flowers soft salmon-rose; willing to bloom over a long period; originated by Hahn's Pittsburgh, Pennsylvania. p. 400

'Brilliant'; a Southern sun hybrid, cultivar of the group known and grown in California and the Deep South as the "Southern Sun azaleas"; a phoeniceum hybrid of vigorous habit, evergreen with obovate, olive green leaves, and large single, wide open funnel-flowers 6-8 cm across, carmine-red with crimson spots and lightly frilled; slightly hardy. p. 400, 402

'Chimes'; grown under this name on the Pacific Coast but appears to be the same as 'Adventgloeckchen'; of full shapely habit, with leathery, glossy obovate foliage and firm semi-double flowers with rounded petals 8 cm across bright rosy carmine; early. p. 403

'Eclaireur' (1914), (Eggebrechtii x Etoile de Noel); a sturdy grower with leathery leaves and medium-size flowers rather round and smooth, semi-double to double, rich carmine or wine-red; tends to burn; midseason. p. 403

'Ernest Thiers' (1890); sturdy plant of vigorous, even growth with leathery leaves, and medium large semi-double flowers bold crimson-red, petals of good texture; late. p. 402

'Euratom'; a dwarf simsii hybrid similar to 'Hexe' but with frilled flowers, vivid crimson with glistening hose-in-hose funnels, on plant of compact habit with broad, dark green foliage. p. 400

'Haerens alba' (1921); sport of 'Avenir', strong grower very floriferous with double white, frilled flowers, with a touch of pink in the center; midseason. p. 402

'Haerens Beauty' (1950); very attractive 'Paul Schaeme' seedling, similar to 'Roberta', with double flowers of good substance and frilled petals soft shell pink with outer areas silver pink; midseason. p. 402

'Hexe' (1888) (simsii x obtusum amoenum); excellent late season potplant, very freeblooming and of dwarf but even habit, with dark green leaves, and long lasting flowers a glowing crimson coming alive under light, the hose-in-hose petals frilled. p. 402

'Lentegroet' (Easter Greetings) (1909), (Camille Vervaene x Hexe); similar to Hexe but with larger flowers, 2-3 times the size, crimson-red semi-double; midseason. p. 402

'Leopold-Astrid' ('Picotee'); simsii hybrid 1933; sport of 'Vervaeneana'; a lovely improvement over 'Albert-Elizabeth', with very pretty, 8 cm double flowers white with pink, and bordered in a rich salmon-red, the margins daintily frilled; mid-season. p. 402

'Madame John Haerens' (1907); an excellent and free-blooming variety of vigorous habit and even growth, the large double flowers are elegant and firm, of a pleasing shade of deep rose, (midseason or late). p. 402

'Madame Petrick' (1901); not a strong grower but is known for its dependably early bloom and forces readily for Christmas, double flowers medium rose and a bright carmine rose. p. 402

'Madame Van der Cruyssen' (1867); known for its exceptionally vigorous growth and free blooming; with large single and semi-double flowers of good substance, carmine rose with crimson markings in upper petals; midseason. p. 403

'Niobe' (1879); robust grower with blackish green leaves, and large double frilled flowers, pure white with green chartreuse tint at base of petals; midseason to late. p. 403

'Paul Schaeme' (1912), (Vervaeneana x Kerckhove); well-known as an early forcer for Christmas, combined with robust growth; large double flowers light salmon-red. p. 403

'Petrick alba' (Kerstperel) (1920); sport of 'Mad. Petrick', not very vigorous but early and a dependable and willing forcer for Christmas, double flowers of good substance white with hint of green in base. p. 403

'Reinhold Ambrosius" beautiful German simsii hybrid for early forcing in pots to bloom during winter-time; silky, soft rosy-red, wide open double flowers 7 cm across, with silky sheen, for midseason January, February bloom, and long-lasting. p. 403

'Rose Queen' (New Zealand); compact simsii hybrid, with beautiful large, single flowers 8 cm across, white in center, carmine-rose toward margins. p. 402

'Simon Mardner' (1878); robust old variety with beautiful large double, clear rose flowers of good texture; early blooming. p. 403

'Southern Charm'; a typical, excellent "Southern indica" azalea, sport of 'Formosa', a phoeniceum hybrid; tall evergreen shrub with large 9 cm single flowers carmine rose with red spots, the petals rounded and half free; rough-hairy 6 cm elliptic leaves. Sun tolerant and vigorous; midseason bloom. This group of azaleas may be trimmed and trained into standard or poodle shapes. p. 402

'Triumph' (1923) (Mad. Aug. Haerens x Lentegroet); strong grower, budding readily and early blooming; the double flowers are crimson red and beautifully frilled, holding their fresh color for a long time. p. 403

'Vervaeneana' (1886); very large, rounded flowers, vivid rose with broad white margin, and spotted crimson at base of petals, nicely double; good growth of loose habit, with large green leaves; midseason. Cultivar of Rhod. simsii var. vittatum. p. 403

'Sweetheart Supreme' (1931); probably the best liked of 'Pericat' hybrids, of strong spreading, irregular growth; the buds are rosy-pink unfolding like a dainty sweetheart rose, opening flat to starlike 5 cm flowers with successively smaller circles of delicate pink petals toward center; midseason *Warm temperate.* p. 400

AZALEA: see also Rhododendron

BABIANA   *Iridaceae*
 purpurea (Cape Province), "Baboon flower"; cormous plant with strongly ribbed leaves; long stalks with spike of purple flowers. *Subtropic.* p. 527
 stricta (So. Africa), "Baboon flower"; low cormous herb with hairy sword-shaped pleated leaves and freesia-like flowers blue and white; winter blooming. *Subtropic.* p. 526

BACCHARIS   *Compositae*
 pilularis (California), "Coyote bush", "Chapparal broom"; evergreen shrub; good ground cover in dry climate; forming dense mat; small 2 cm toothed leathery leaves; small yellowish flower heads. *Arid-subtropic,* p. 322

BACTRIS   *Palmae*
 gasipaes (Guilielma utilis), (Costa Rica, Amazon), "Peach palm"; feather-leaf palm with slender ringed trunk armed with very sharp needle-like spines; graceful fronds pinnate 5-10 cm long, deep green, hairy and spiny; fruit resembling a peach, in pendulous bunches, and used as a vegetable. *Tropical.* p. 471, 766
 vulgaris (Brazil), "Spiny-club"; small clustering palm with slender, spiny stems, and pinnate fronds loosely set with narrow pinnae. *Humid-tropical.* p. 766

BAIKIAEA   *Leguminosae*
 insignis (W. Africa); large tree to 25 m, with pinnate leaves, 3-8 leathery leaflets to 25 cm long; large flowers with 5 separate white petals 15-20 cm long, the largest marked yellow. *Tropical.* p. 543

BAMBUSA   *Gramineae*
 multiplex (Vietnam), "Fernleaf bamboo"; a variable woody grass becoming 4 m high, the green reedlike hollow stems bearing graceful twigs with many very small leaves, fern-like, 3-15 cm long, glabrous deep green, silver-blue beneath, and with a ring of hair at base of leaf; the older culms become 3 cm or more thick. *Subtropic.* p. 518
 oldhamii (Sinocalamus); in nurseries as Dendrocalamus latiflorus; (China), "Oldham bamboo"; large bamboo to 18 m high, forming open clumps; erect canes 9 cm dia., pale green, with dense foliage on pendant branchlets. *Subtropic.* p. 515
 textilis, (So. China; Kwangsi, Kwangtung), "Wong Chuk"; a handsome subtropical species forming compact clumps of thin-walled but tough culms with long internodes, very straight to 15 or more m tall and 5 cm thick, with nodding tips; 6-10 branches at nodes, with attractive foliage 15-20 cm long, and to 2½ cm wide. These straight, light and tough canes are split and used extensively for weaving of baskets; hardy to -10°C. *Subtropic.* p. 519
 tuldoides (S.E. China), "Punting pole bamboo"; used to push junks in Chinese rivers; handsome semi-hardy clump bamboo with straight culms to 18 m high and 8 cm dia., with broad, rich green leaves to 15 cm long, and green reverse. *Subtropic.* p. 514
 ventricosa (China), "Buddha's belly bamboo"; so called because of the characteristic dark olive culms swollen between internodes, especially in potgrown clumps; otherwise may grow 15 m high; leaves to 18 cm long. *Subtropic.* p. 514, 517, 518
 vulgaris (Java, and wild in tropical regions of E. and W. Indies, Africa, C. and So. America); "Feathery bamboo"; widely grown ornamental bamboo because of its attractiveness and large size; forms open clump with spreading rhizomes, arching culms 15-24 m high and 10-12 cm thick, hollow with thin walls, green at first later yellow banded lengthwise or all yellow, the joints covered with deciduous brown hairs; dark green leaves 15-22 cm long borne on graceful branches on the upper part of the stem. *Tropical.* p. 509, 516, 518, 519

BAMBUSA: see also Dendrocalamus, Phyllostachys, Pseudosasa, Sasa

BANKSIA   *Proteaceae*
 ashbeyi (W. Australia); woody evergreen shrub to 5 m tall and wide, with vicious, hard, bluish leaves 25 cm long and 3-5 cm wide,

cut each side into triangular lobes; the showy cones silky gray and gold, 12-15 cm long. *Subtropic.* p. 835

grandis (Western Australia), known as "Bull banksia"; handsome and curious tree 3 to 10 m high, with sparry branches, and unbelievable leaves 30 cm long, green to grayish and stiff as metal, deeply cut out to the midrib by alternating teeth like a big-edged saw, richly brown-felted at first; yellow bottle brush flowers 20-30 cm long. My photographs are of typical leaves from trees near Perth. *Subtropic.* p. 834, 835

marginata (South Australia, Victoria, to Tasmania), "Silver banksia"; dense evergreen shrub to 5 m, with hard, lanceolate leaves 5 to 8 cm long, silvery beneath and with margins recurved; greenish yellow floral cones 8-12 cm long. *Subtropic.* p. 835

media (W. Australia); bushy, woody shrub, with spatulate, stiff leaves deeply serrate at margins, cylindrical and grayish beneath; inflorescence cone-like, dense of styles light yellow to red-brownish. *Subtropic.* p. 834, 835

speciosa (Western Australia); spectacular spreading woody shrub 2-3 m tall, thick woody branches with long narrow, green to grayish stiff-leathery leaves to 30 cm long, cut to the midrib into triangular saw teeth; showy inflorescence of curving yellow styles tipped green, silvery in bud. *Subtropic.* p. 836

BARBOSA *Palmae*
pseudococos (Brazil); tall palm to 15 m with slender trunk to 20 cm thick, topped by crown of arching pinnate fronds, the leaflets long and wide-spreading. *Humid-tropical.* p. 767

BARLERIA *Acanthaceae*
albostellata (Rhodesia, Transvaal); subshrub with gray-pubescent stems, ovate leaves to 8 cm long; white, irregular flowers 6 cm dia., in head with large, leafy bracts. *Subtropic.* p. 34

cristata (India), "Philippine violet"; shrub with oblong leaves covered with stiff hairs; spikes with funnel-shaped, five-lobed blue flowers. *Tropical.* p. 34

lupulina (Mauritius), "Hop-headed barleria"; evergreen spiny shrub 60 cm high with narrow lanceolate leaves green with pink or red mid-ribs; two pairs of down-turned thorns at each node; 5-lobed tubular flowers peach-yellow from terminal hops-like spikes. *Tropical.* p. 34

BARLIA *Orchidaceae*
longibracteata (Himantoglossum) (Spain to Greece, Algeria, Canary Islands), tuberous terrestrial rosette with broad fleshy green leaves to 20 cm; inflorescence a spike hyacinth-like with poreclain purple 3 cm flowers with pale spotted lip, somehow reminding of lizards; to 60 cm high (spring). *Subtropic.* p. 711

BARRINGTONIA *Lecythaceae*
asiatica (South Pacific islands, west to India); "Fish poison tree" with large obovate leathery leaves to 60 cm long, fragrant 18 cm white flowers with brush-like, long white stamens tipped crimson, open in the evening, and fall in the morning; yellow-brown quadrangular fruit 10 cm long containing one large poisonous seed. *Tropical.* p. 536

BAUHINIA *Leguminosae*
blakeana, "Hongkong orchid tree"; apparently a sterile hybrid, to 5 m high, with twisted stem, leaves of two jointed leaflets and showy large orchid-like flowers of 5 spreading unequal petals carmine-rose to burgundy, with the fifth petal striped purple. As seen in Queensland, these orchid trees provide a gorgeous show for at least 5 months. *Subtropic.* p. 541

corniculata (forficata) (Brazil), "White camels-foot"; tree to 6 m high, with stems leaning at an angle, the top flattened; leaves deeply lobed; flowers white with thin petals. *Subtropic.* p. 541

corymbosa (Vietnam), "Phanera"; woody vine climbing by tendrils; leaves 3-5 cm long, lobed to middle; pinkish flowers in loose raceme, 3 cm across. *Tropical.* p. 541

divaricata (lamarckiana) (South America); scandent shrub with leaves lightly lobed at apex; flowers with narrow white petals, and long thread-like white stamens. *Subtropic.* p. 541

galpinii (Tropical Africa to Transvaal), "Pride of the Cape"; a rambling shrub clambering to 10 m, 3-8 cm leaves two-lobed, pale green; their bright flowers nasturtium-like salmon-red, with spoon-shaped petals. *Tropical.* p. 541

monandra (Burma), "Pink orchid"; small ornamental tree sometimes called "Jerusalem date" or "Butterfly flower"; deciduous in winter; produces great quantities of big flowers and seldom without blossoms; flower orchid-like, white flushed rose, lip yellow marked red, changing to pink after 24 hours; leaves notched at apex. *Tropical.* p. 541

purpurea (India, Burma, Vietnam), the "Butterfly tree"; small tree 2-6 m high, with thin-leathery leaves broad cleft about one third; fragrant flowers reddish-purple, marked and shaded in other tones. *Tropical.* p. 541

variegata; in horticulture sometimes as B. purpurea (India, China), the purple "Orchid tree" or "Mountain ebony"; small semi-deciduous tropical tree 4 to 8 m high, also grown as a shrub with multiple stems, with curious foliage; the thin-leathery leaves connate or cleft beyond the middle 8-10 cm long, dull green; gorgeous 9 cm axillary flowers like a cattleya orchid, carmine-rose, the petals with dark purple center stripe, the broad lip lined with crimson, fan-like; blooming January to May. *Subtropic.* p. 541

variegata 'Candida', "White orchid tree"; tree to 12 m, with leaves lobed from apex about one-third; flowers pure white with greenish veins. *Subtropic.* p. 541

BEAUCARNEA *Liliaceae*
gracilis (Nolina) (Tehuacán, Mexico), "Bottle palm" or "Pony tail"; curious, stout xerophytic tree to 10 m, with much swollen base and branching; at apex each with a rosette of narrow, grayish-glaucous leaves ½ to 1 m long and 1 cm wide; clusters of small, off-white to reddish flowers. *Arid-tropical.* p. 612

recurvata (Nolina) (Mexico), "Pony-tail"; tree-like plant with tall trunks to 10 m high, swollen at base, topped by a rosette of linear, pendulous, concave, green leaves to 2 m long, rough and thin but not spiny-toothed; panicles of small whitish flowers. A succulent which can store a year-long water supply; related to Yuccas. *Arid-tropical.* p. 582, 585, 586

BEAUFORTIA *Myrtaceae*
sparsa, (Western Australia), "Swamp bottlebrush"; woody shrub with small ovate, hard leaves; the inflorescence a large powder puff of bright scarlet stamens. *Subtropic.* p. 676

BEAUMONTIA *Apocynaceae*
grandiflora (India: Himalayas), "Heralds-trumpet"; woody twiner or small tree with opposite, ovate leaves to 20 cm long, young shoots tinted rose and rusty-haired; large, fragrant, showy trumpet-shaped 12 cm flowers with 5 twisted lobes, white, with dark throat, in terminal cymes. *Subtropic.* p. 73, 80, 82

BEGONIA *Begoniaceae*
acida (Brazil); rhizomatous species with large, roundish leaves to 30 cm, bright green, roughly puckered; tall inflorescence with white flowers. *Tropical.* p. 160

aconitifolia (sceptrum) (Brazil); tall cane-stemmed species with large leaves palmately deep-lobed, green with silver streaks and reddish veins; large, waxy-white flowers. *Tropical.* p. 179

albo-picta (Brazil), "Guinea-wing begonia"; small, sturdy angel-wing type plant with branching, drooping stems carrying narrow, glossy, green leaves tapered and silver-spotted. Greenish-white flowers. *Humid-tropical.* p. 162

'Alto-Scharff' (bradei x scharffiana); beautiful, hirsute bushy plant with white-hairy, wing-shaped leaves with red, depressed veins, and red beneath; hairy pink blooms. *Humid-tropical.* p. 162

'Alzasco' (Lucerna seedling); tall, cane-stemmed angel-wing type with blackish green, satin, ruffled leaves with silver spots, deep purple beneath; bronzy-red flowers in drooping clusters. *Humid-subtropic.* p. 166

x argenteo-guttata (albo-picta x olbia), "Trout begonia"; bushy canestem with "angelwing" leaves growing at an angle, waxy, olive green heavily spotted silver, red beneath; flowers cream with pink. *Humid-tropical.* p. 160

'Beatrice Haddrell' (boweri x sunderbruchii); magnificent rhizomatous plant with star-shaped leaves boldly marked chartreuse on brown; flowers in tall pink showers. *Humid-tropical.* p. 166

'Black Raspberry' (acetosa x imperialis); rhizomatous, leaves green and pebbly, 7-10 cm, red veins, flowers pink; by Lowe Nursery, Gould, Florida, 1970. *Tropical.* p. 174

bogneri (Malagasy Republic); a most unusual new species discovered on the Isle of Masoala off the coast of Madagascar growing on mossy granite cliffs with annual rainfall of 350 cm, by J. Bogner of the Munich Botanical Garden in 1969; grass-like narrow linear leaves shiny green, 10-15 cm long and only 2 mm wide, on edges with minute teeth, on short erect stem from semi-tuberous roots; new leaves reddish; flowers are pale pink, on pinkish stalk to 15 cm high, with few male and fewer female blooms, about 2 cm across. R. Ziesenhenne reports (Begonian April 1973) that this species is an excellent terrarium plant and with artificial light remains evergreen. *Humid-tropical.* p. 174, 179

'Bow-arriola' (boweri x C42); rhizomatous miniature starleaf; satiny green with intermittent purple markings at edge of leaf or along veins; flowers blush pink. *Humid-tropical.* p. 160

bowerae (Chiapas and Oaxaca, So. Mexico); miniature "Eyelash" begonia, rhizomatous, bushy with small, waxy ovate leaves vivid green and blackish-brown patches and erect hairs along margin; flowers pink. *Humid-tropical.* p. 160

bracteosa (roezlii) (Peru), "Machu Picchu begonia"; fibrous-

rooted, bushy plant, with roundish ovate, waxy leaves, bronzy and lightly toothed; flowers pale pink to lavender. *Tropical.* *p. 163*
  bradei (laetevirides, laetevirens, macrocarpa pubescens), "Alto da Serra" hort.; variety with branches soft-hairy and more upright than secreta, velvety olive green leaves with red margins, reddish underneath; flowers pinkish white. *Tropical.* *p. 169*
  'Braemar' (scharffii x metallica); hirsute, leggy plant with white-hairy stems and large, orbicular-pointed, dark, lustrous green leaves, glossy red beneath; flowers white with bright pink beard. *Tropical.* *p. 174*
  bunchii: see x erythrophylla 'Bunchii'
  'Calla lily': see under semperflorens
  cathayana (China: Yunnan); exquisite fibrous-rooted species to 45 cm, with obliquely cordate leaves plush-velvet resembling rex, purplish olive-green with zone of silver and smooth crimson veins; crimson velvet beneath; flowers orange-vermilion. *Subtropic.* *p. 160, 167*
  x cheimantha 'Lady Mac', a popular "Christmas begonia"; the bushy plant covered in winter with masses of clear pink and smallish, but long-lasting flowers; popularly called "Busy Lizzie". B. x cheimantha is an old hybrid group of socotrana x dregei introduced in France 1891 as 'Gloire de Lorraine'. *Humid-tropical.* *p. 164, 165*
  x cheimantha 'Marina', "Scandinavian winter begonia"; Danish sport of Solbakken, with deep rose, durable flowers larger than 'Solfheim' but not quite as dark, and contrasting nicely with the yellow anthers; of medium growth habit with rich green, medium leaves; flowers October on to December and later. *Humid-tropical.* *p. 164*
  x cheimantha 'White Christmas', "Improved Turnford Hall begonia"; a freer growing form of 'Turnford Hall', which was a rather delicate mutant of 'Lorraine'; during winter, the foliage is literally snowed under by buds flushed apple blossom and opening to pure white. *Humid-tropical.* *p. 164*
  chimborazo (Guatemala); clumpy, semi-trailing rhizomatous species with thickish, glossy-green, ovate pointed peltate leaves; flowers soft pink. *Humid-tropical.* *p. 180*
  'China Doll' (bow-arriola x boweri); rhizomatous miniature, with small, obliquely pointed, irregularly lobed, chartreuse leaves with wide brown veins and wavy edges on hairy petioles; clusters of pink flowers in winter. *Humid-tropical.* *p. 175*
  circumlobata (China); charming rhizomatous plant of upright and bushy habit, with palmately compound leaves of broad forest-green leaflets; showy pink flowers. *Subtropic.* *p. 162*
  'Cleopatra' (Maphil x Black Beauty), "Mapleleaf begonia"; translucent maple-leaf nile-green with chocolate-red areas toward margin; rhizomatous; clusters of perfumed, pink flowers. *Tropical.* *p. 160, 161*
  coccinea (rubra) (Orgaos-Brazil), "Angelwing begonia"; tall, bamboo-like canestem to 4 m high with thick ovate leaves glossy green spotted silver, and edged red; drooping clusters of coral-red flowers blooming constantly. *Tropical.* *p. 179*
  coccinea 'Pink'; pendant stems with glossy green angelwing leaves; large clusters of rose-pink flowers. *Tropical.* *p. 179*
  compta (Brazil); tall and slender cane-stem; sharply pointed leaves green with silver stripes outlining the nerves, red beneath; bearded white flowers. *Tropical.* *p. 160*
  conchifolia rubrimacula (Costa Rica), "Zip begonia"; miniature rhizomatous species, with petioles brown-fuzzy; waxy green, rounded, cupped peltate leaves, having red sinus spot in center; the margins slightly scalloped; veins brownish beneath; clusters of small whitish flowers. *Tropical.* *p. 171*
  cooperi; (Costa Rica); fibrous-rooted, shrubby plant with small ribbed, elm-like leaves, and terminal clusters of small white flowers. *Humid-tropical.* *p. 160*
  'Corallina de Lucerna' (Lucerna) (teuscheri x coccinea); tall, branching cane-stem with large green leaves silver-spotted; flowers droop in giant clusters. *Tropical.* *p. 163, 175*
  x crestabruchii (manicata crispa x sunderbruchii), "Lettuce-leaf begonia"; rhizomatous plant with large, heavy, roundish leaves bronzy-green and very much crested, rosy petioles and nerves; flowers pink. *Tropical.* *p. 162*
  'Dancing Girl'; low, fibrous, angelwing variety with whirling growth of twisted leaves no two alike, olive green with silver spotting, and metallic pink along veins and margin; red flowers. *Tropical.* *p. 160*
  decora (Assam, Vietnam, Malaya); lovely little plant with a creeping, hairy rhizome and oblique-ovate, plushy-brown leaves sharply etched with chartreuse veins, underside red with green veins, edges finely toothed; large rose-pink flowers borne high over foliage. *Tropical.* *p. 172*
  deliciosa (Borneo); attractive, fibrous-rooted, smooth plant with thick red joints, palmately lobed leaves dark olive green with spotted silver gray, red beneath; pink flowers. *Tropical.* *p. 167*
  diadema (Borneo); fibrous-rooted, erect, fleshy plant with maple-like, deeply parted, dentate leaves, satiny green, with streaks of silver along ribs; small pink flowers. *Humid-tropical.* *p. 178*
  'Dianna' (Annie Laurie x dichroa); erect, bushy, angel-wing type with pleated, ruffled, bright green, heavily silver-spotted leaves; salmon-pink flowers in pendulous clusters. *Tropical.* *p. 170*
  dichotoma (vitifolia) (Venezuela), "Kidney begonia"; upright rhizomatous, branched plant with oblique-ovate, smooth, green leaves with irregularly toothed edges; clusters of many small white flowers above the foliage. *Tropical.* *p. 162*
  x drostii (scharffii hybrid); tall plant with large dark green leaves deep purple beneath, overall hairy like an autumn haze; flowers pink, red-bearded or hirsute without. *Tropical.* *p. 180*
  echinosepala (Brazil); species with red cane stems dotted with bright green spots, carrying small, obliquely lance-shaped leaves, dull satiny green on top, younger leaves bright green, shiny, reddish maroon and covered with short hairs below, irregularly crenate to serrate margins. *Tropical.* *p. 168*
  egregia (quadrilocularis) (Brazil); fibrous-rooted species with long lanceolate, light green, puckered leaves somewhat cupped and brittle; white flowers with 4-winged seed pods. *Tropical.* *p. 170*
  epipsila (Brazil); low-growing, fibrous-rooted plant with fleshy, roundish, enamel-green leaves, reddish beneath and covered with felt; white flowers. *Tropical.* *p. 162*
  x erythrophylla (feastii) (manicata x hydrocotylifolia), "Beefsteak begonia"; creeping rhizome, and leathery, glossy, rounded leaves dark olive green, red beneath; flowers pink. *Subtropic.* *p. 160, 166*
  x erythrophylla 'Bunchii', "Curly kidney-begonia"; rhizomatous, decorative mutant, with the fleshy leaves lighter green, red-tinged and ruffled and crested at the margins, the lobes rise and meet. *Subtropic.* *p. 166*
  evansiana: see grandis
  "Exotica" (Baiyer River, New Guinea); magnificent, fibrous-rooted begonia with erect, reddish canes and large, oblique-ovate, corrugated leaves of shiny, iridescent taffeta in crimson on deep green, pale green beneath, both sides sparsely covered with short, red bristles, margins densely and irregularly toothed; red hairs on petioles and brown hairs on stems. Semi-dormant in winter. *Tropical.* *p. 161, 166*
  feastii: see erythrophylla.
  fernandoi-costa (Brazil); fleshy, low, fibrous-rooted plant with large, roundish, soft-hairy, bright green leaves, rose-pink beneath; erect clusters of white flowers, each forming a perfect cross. *Tropical.* *p. 168*
  'Fischer's ricinifolia'; pink-flowered ricinifolia seedling with low, creeping rhizome, leaves somewhat smaller and less deeply lobed, red-tinged underneath; habit bushier. *Subtropic.* *p. 170*
  floccifera (India); small clustered, rhizomatous plant with thick oval leaves 8 cm or more across, dark green; covered with yellowish hair when young; flowers white. *Tropical.* *p. 168*
  foliosa (Colombia), "Fernleaf begonia"; willowy plant with drooping branches; fibrous; tiny, waxy, bronzy-green, oval leaves notched toward tip; tiny blush-white flowers. *Tropical.* *p. 180*
  froebelii (Perú); low tuberous species with roundish-cordate leaves bright green with pale veining and covered with soft purplish hairs; thick flowers scarlet; dormant in summer. *Tropical.* *p. 164*
  x 'Frutescans' (fruticosa seedling); low, fibrous-rooted, foliage begonia with small, thin-leathery, olive-green leaves, wavy and cupped to show red underneath; flowers white. *Tropical.* *p. 166*
  'Frutescaria' (fruticosa seedling); fibrous-rooted plant of low habit similar to Frutescans, but with lighter, grayish leaves, more prominent veins and without the wavy margins. Infrequent, white flowers borne high over the foliage. *Tropical.* *p. 168*
  fuchsioides (Mexico), "Fuchsia begonia"; fibrous-rooted plant with slender arching stems 60-90 cm high, very small oblique-ovate toothed leaves to 4 cm long, waxy dark green; nodding fuchsia-like red flowers. *Tropical.* *p. 168*
  x fuscomaculata (rubellina) (heracleifolia x daedalea); small rhizomatous, hairy plant with lobed leaves smooth, bronze-green with pale veins and chocolate spots; flowers pink. *Tropical.* *p. 166*
  'Gehrtii' (Brazil); giant rhizomatous plant with distinct foliage; quilted fresh-green leaves and pale green raised veins with sand colored hair on petioles (similar to paulensis). *Tropical.* *p. 179*
  glabra (scandens) (Brazil); attractive, fibrous-rooted species with much branched, trailing stems, and light green, ribbed, waxy leaves, cordate and nearly round, shallowly toothed; small white flowers. *Tropical.* *p. 169*
  glaucophylla: see limmingheiana
  'Gloire de Jouy' (rex x incarnata); small fibrous-rooted plant

with broad bronzy-green leaves marked with small silver spots with white hairs; red beneath. *Tropical.* *p. 178*

goegoensis (Sumatra), "Fire-king begonia"; low growing gem on creeping rhizome, silky leaves round peltate, puckered, dark olive green with lighter veins, reddish beneath; pink flowers. *Humid-tropical.* *p. 174*

grandis (evansiana) (China, Japan), "Hardy begonia"; tuberous species, with bulbils forming in leaf axils; stands some frost; olive green, ovate leaves, veined purple beneath; flowers pink. *Warm temperate.* *p. 167*

'Grey Feather' (angular compta); cane-stemmed; narrow gray leaves, wavy-toothed and shallowly lobed near base, with broad, silver veins; flowers white. *Tropical.* *p. 178*

'Gwen Lowell' (olsoniae x obscura); bushy plant to 50 cm high, with oblique-ovate, velvety leaves iridescent forest green with pale veins and toothed margins; flowers pure white. *Tropical.* *p. 168*

haageana in hort: see scharffii

x heracleicotyle ('Mrs. Townsend') (heracleifolia x hydrocotylifolia), rhizomatous; leaves small, fleshy, shallowly lobed, waxy green and dark edged; flowers pink. *Tropical.* *p. 160*

heracleifolia (Mexico), "Star begonia"; with creeping rhizome, large, palmately lobed leaves bristly-hairy, bronzy green with green ribs, flowers pink. *Tropical.* *p. 170*

heracleifolia nigricans (Mexico), "Star begonia"; handsome, rhizomatous plant smaller than sunderbruchii with palmately lobed leaves blackish-green with contrasting pale green area along main veins. *Tropical.* *p. 166*

x hiemalis; a class of autumn and early winter-blooming hybrids of B. socotrana with summer-flowering tuberous types, originated by Veitch (1883) and Clibrans (1908) in England, followed by Dutch specialists in Holland about 1946 as the 'Elatior' begonia; semi-tuberous plants preferring moderate warmth. Unfortunately quickly dropping flowers in superheated rooms. *Humid-subtropic.* *p. 165*

x hiemalis 'Emily Clibran'; hybrid of colorful tuberous species combined with the winter-flowering habit of B. socotrana; showy double flowers rose-pink, flushed orange, and waxy, rounded leaves. *Humid-subtropic.* *p. 164*

x hiemalis 'Rieger's Schwabenland' (1964 Pat.). Robust plant with pointed, oblique leaves metallic deep green, and large, glowing scarlet flowers to 5 cm dia., with contrasting yellow stamens. The new German Rieger "Elatior" strain is distinguished by its bushy, free-growing and floriferous habit, with flowers in vivid new colors of superior keeping quality, blooming for 8-10 weeks even in the home, normally in fall-winter. *Humid-subtropic.* *p. 163, 164*

x hiemalis 'The President', one of the most desirable commercial varieties, a good keeper, but subject to mildew; double flowers brilliant crimson-red to 8 cm across, a beauty for Christmas. *Humid-subtropic.* *p. 164*

hispida cucullifera (Sao Paulo), "Piggyback begonia"; interesting, fibrous-rooted species with large, velvety, maple-like, pale green leaves producing adventitious leaflets on top surface of the leaf along the veins; red petioles covered with white hairs; white flowers. *Humid-subtropic.* *p. 166, 171*

hispidavillosa (Mexico); small creeping rhizomatous species with large, variably round, thick, kidney-shaped, light green leaves with prominent veins; leaves and stems coarse-hairy; drooping flowers rosy pink. *Subtropic.* *p. 168*

hydrocotylifolia (Mexico), "Pennywort begonia"; hairy, rhizomatous plant with rounded, thick, waxy leaves, light olive green with dark veins; small rosy flowers. Called "Pennywort begonia" because of its miniature size. *Subtropic.* *p. 170*

'Immense'; rhizomatous ricinifolia seedling, with large star-like, acutely lobed, light green leaves, waxy and flat, some short bristles on surface, edge red; leafstalks with red scale-like hairs; flowers pink. *Subtropic.* *p. 179*

imperialis (Mexico), "Carpet begonia"; choice, small, rhizomatous, hairy plant with small, heart-shaped, velvety, pebbly leaves in dark green, small white flowers. *Humid-tropical.* *p. 170*

imperialis smaragdina (Mexico), "Green carpet begonia"; differs from imperialis in its lighter, emerald-green leaf color. *Humid-tropical.* *p. 170*

incana (peltata) (Mexico); thick-stemmed plant woolly white with scurf; fleshy peltate, white-tomentose leaves; drooping white flowers. *Tropical.* *p. 167*

'Interlaken' (Lucerna seedling); distinctive, tall, few-branched, cane-stemmed begonia with ovate-pointed, deep green leaves, sometimes faintly silverspotted, one basal lobe sharply pointed, margins toothed and wavy; large, coppery-red flowers in drooping clusters. *Subtropic.* *p. 178*

involucrata (Costa Rica); bushy plant with brown stems; fibrous; green leaves lobed and toothed, lightly white-tomentose; flowers white and fragrant. *Tropical.* *p. 168*

'Iron Cross': see masoniana

'Jinnie May' (coccinea seedling); small, compact plant if kept pruned, with small, green, angelwing leaves crowded on well branched stems; bright red flowers all through the year, in short, drooping clusters. *Tropical.* *p. 170*

kellermannii (Guatemala), "Lily-pad begonia"; dainty, fibrous-rooted species with peltate, cupped, oval leaves yellow-green to rose-colored, and felt-like white-scurfy beneath; flowers pink. *Tropical.* *p. 167*

kenworthyae (Mexico: Chiapas); distinctive slow-growing, upright rhizomatous species with smooth, fleshy, slate-colored or bluish green, 5-lobed leaves with lighter veins, looking very much like an ivy leaf, up to 25 cm in size; flowers faintly pink in winter. *Tropical.* *p. 167*

laetevirens, laetevirides: see bradei

leptotricha (Paraguay), "Woolly bear"; dwarf plant with stout, felted, succulent stems and dark green, oval, thickish leaves, the underside thickly felted with brown; flowers ivory; everblooming. *Subtropic.* *p. 169*

limmingheiana (glaucophylla) (Brazil), "Shrimp begonia"; trailing with slender stems, glossy light green leaves with wavy margin; coral red flowers; good for baskets. *Tropical.* *p. 164, 171*

listida (So. Brazil); attractive fibrous-rooted miniature angelwing, with narrow oblique-ovate pointed waxy leaves to 15 cm long; olive green with bronzy cast; chartreuse with cream along center; red reverse, on hairy red stalks. *Tropical.* *p. 161, 175*

'Lucerna': see 'Corallina de Lucerna'

'Lulu Bower'; everblooming cane-stem, with oblique angelwing leaves glossy coppery-green; clusters of salmon-rose flowers. *Tropical.* *p. 164*

luxurians (Chile), "Palm-leaf begonia"; ornamental, tall fibrous rooted plant sparsely rough-hairy; leaves peltate-palmate, with up to 17 narrow, serrate leaflets, reddish above; small cream, hirsute flowers. *Humid-tropical.* *p. 170*

macdougallii (Chiapas); creeping rhizomatous stem with long, reddish petioles; palmately compound, waxy leaves of 7-10 stalked segments, the outer ones sickle-shaped, bronzy green, red beneath, and with toothed margins. *Tropical.* *p. 169*

macrocarpa (secreta) (Trop. W. Africa); fibrous-rooted species with arching or drooping branches, elliptic pointed leaves glossy olive-green, sparsely hairy and toothed; flowers white tinged pink. *Tropical.* *p. 174*

'Manda's Eyelash'; a pretty bowerae cultivar, small rhizomatous plant more vigorous than the species; small waxy oblique-ovate leaves almost peltate, waxy, boldly marked with black around the margins, scalloped and set with eye-lash bristles; sprays of salmon-pink flowers. *Tropical.* *p. 170*

manicata 'Aureo-maculata', "Leopard begonia"; the fleshy, green leaves are blotched yellow or ivory, occasionally rose-red. *Tropical.* *p. 162*

manicata 'Aureo-maculata crispa'; difficult but beautiful form, the waxy, fresh-green leaves blotched yellow and occasionally rose, the margin thickly crested. *Tropical.* *p. 169*

mannii (eminii) (W. Africa), fibrous-rooted "Roseleaf begonia" with toothed, ovate leaves glossy green, pale beneath and tinged red, on long arching stems; flowers white, streaked with red, the ovaries without wings. *Tropical.* *p. 174*

'Maphil' (boweri seedling); colorful dwarf plant with creeping rhizome and red-scaled petioles; small star or maple-shaped leaves chocolate-brown with chartreuse markings, the veins red, margins with eye-lashes. *Tropical.* *p. 174*

x margaritacea (Arthur Mallet x coccinea); distinctive, fibrous-rooted, delicate hybrid with metallic, purplish-pink leaves overlaid with silver; flowers pink. *Tropical.* *p. 167*

x margaritae (echinosepala x metallica); bushy fibrous plant with soft-hairy, small, ovate-pointed and toothed, quilted leaves, faded bronze green, veins purple beneath; fl. pink, bearded. *Tropical.* *p. 167*

masoniana (Indochina or China, introduced from Singapore), the "Iron cross begonia"; one of the most beautiful begonias in cultivation, spectacular rhizomatous plant of robust habit, with white-hairy, reddish stems and large roundish, firm, puckered leaves nile-green, marked with contrasting, bold pattern of brown-red, the older leaves overlaid with silver, and covered with bristly red hair and red-ciliate, 15-18 cm across; waxy flowers greenish-white with maroon bristles on back; flowering March-May. *Tropical.* *p. 169, 171*

'Maybelle'; Robinson 1934; cane stem; small glossy pointed leaves with silver spots; large pendant clusters of white flowers. *Tropical.* *p. 168*

mazae viridis, "Stitchleaf begonia"; small rhizomatous plant with satiny, heartshaped, light green leaves with purple marks like stitches along margin; flowers pink. *Tropical.* *p. 169*

metallica (Bahia), "Metallic-leaf begonia"; bushy, fibrous-rooted, hairy plant with broad-ovate pointed leaves metallic olive-green and depressed purple veins, red-veined beneath; showy pink hirsute flowers. *Tropical.* p. 169, 171

micranthera (Argentina, Bolivia); summer-blooming tuberous plant with short, succulent stems, carrying small softly hairy, oblique cordate, near round green leaves, shallowly lobed and with toothed, eyelashed margins; bright orange-rose flowers with masses of yellow stamens. *Subtropic.* p. 161

'Mrs. Fred Scripps'; scharffiana x luxurians characters, fibrous rooted, vigorous, hairy; leaves velvety forest-green with red veins, deeply lobed, red beneath; basal lobe sometimes forms separate leaflet, like a finger; small pinkish, hirsute flowers. *Tropical.* p. 168

'Mrs. Townsend': see x heracleicotyle

'Mrs. W.A. Wallow'; tall, cane-stemmed, bushy plant; leaves ovate-lanceshaped, slate green, bristly-hairy, with toothed and wavy margin; glowing-red beneath; flowers blush pink. *Tropical.* p. 178

'Nelly Bly' (cyprea seedling); bushy fibrous plant, hairy; quilted leaves ovate, dark green and toothed, red beneath; flowers pink, hirsute, or bearded blossoms. *Tropical.* p. 174

nelumbiifolia (macrophylla) (Mexico), "Pond-lily begonia"; rhizomatous species with large, peltate, round, green leaves, hairy on nerves; flowers pinkish. *Tropical.* p. 162, 169

nitida (minor) (Jamaica); medium-sized begonia having cane-like shoots with weaker branches; thick, glossy leaves, ovate with irregular lobes, olive green faintly spotted silver; flowers in dull, dark pink, tinged yellow. *Tropical.* p. 180

odorata (syn. nitida odorata) (Jamaica); fibrous, bushy, to 2 m high, with fleshy, reniform, glossy green leaves; rosy-pink flower clusters, fragrant. *Tropical.* p. 180

olsoniae (vellozoana hort.) (Brazil); lush, low, rhizomatous species, with large fleshy leaves in changeable colors bronze to green, contrasting with striking ivory veins; flowers whitish, outer edge seemless rosy. *Tropical.* p. 171

'Orange-rubra' (dichroa x Coral Rubra); cane-stemmed; large leaves obliquely ovate, of the angelwing type, smooth green, silver spotted when young; lacquered orange-red flowers. *Tropical.* p. 166

parilis (Brazil), "Zig-zag begonia"; fibrous-rooted, stiff, upright, well branched grower to about 1 m, branches zig-zagging between nodes, older branches covered with brown scurf; leaves long, narrow, shining green on top, red under; long, arching stems with flowers in white with predominant yellow stamens; when in bud, so tight that the cluster resembles a snow ball. *Tropical.* p. 168

paulensis (Brazil); large orbicular, peltate, hairy, quilted leaves waxy, fresh-green with pale veins which are red beneath; rhizomatous; bright red, bearded flowers. *Tropical.* p. 174

'Piccolo' (imperialis x aridicaulis); small rhizomatous plant with white-hairy, brown petioles; broad, oblique-cordate or peltate leaves 10 cm long, puckered and textured like brocade, silvery green with purple tint, and green primary veins; small white flowers. *Humid-tropical.* p. 166

'Picta rosea'; probably coccinea form, cane-stem; narrow waxy, angelwing leaves light olive green with silver spots; pendant clusters of salmon-rose flowers. *Tropical.* p. 174

'Pinafore'; seedling of Elaine by Logee; low-growing, fibrous-rooted beauty with bronzy-green, angelwing leaves spotted silver, curly edged, beet-red beneath; bright salmon flowers; cane-stem. *Tropical.* p. 170

pustulata (Mexico), "Blister begonia"; beautiful rhizomatous species of rambling, hanging tendencies, with pebbly leaves off-emerald green with silver veins, underside deep rose. Similar to imperialis, with fewer hairs. Likes moss-lined wire basket, with warmth and humidity. *Humid-tropical.* p. 169

quadricularis; see egregia

rajah (Malaya); charming, dwarf, rhizomatous species with roundish, rich reddish green, bullate, silky leaves and contrasting veins of yellow green; under surface dull red; flowers pink. *Humid-tropical.* p. 174

repens (Vellozo) (Brazil); low rhizomatous plant with lanceolate slender pointed leaves, waxy metallic green dotted all over with silver spots, margins wavy; white flowers low in center. Photo R. Ziesenhenne, Santa Barbara. *Tropical.* p. 161

rex 'Baby Rainbow'; a miniature rex with green leaves and purple center, margins purplish brown suffused with carmine-red and silver spotted. *Humid-tropical.* p. 173

rex 'Black Knight' ('Midnight'); one of the darkest; thin velvety leaves red-black with stippled band of pink dots; petioles red-hairy. *Humid-tropical.* p. 173

rex 'Can-Can'; shapely diadema hybrid called "Oakleaf" because of the uniquely lobed and ruffled leaves, glossy silver-green overlaid with blush-pink, and in younger leaf with purplish-red in center. Shimmering like crisp taffeta. *Humid-tropical.* p. 172

rex 'Comtesse Louise Erdoedy' (rex Alex. von Humboldt x argenteo-cupreata); first spiralled form, known as "Corkscrew begonia"; large oblique-ovate, red hairy leaves light olive green zoned with silvery rose; one or both basal lobes spirally curled. *Tropical.* p. 173

rex 'Curly Fireflush', "Spiral begonia"; a beautiful curly, erect mutant with the same striking red velvet leaves as 'Fireflush', the basal lobe twisted into a spiral. *Humid-tropical.* p. 173

rex 'Fairy'; diadema hybrid, shapely, upright habit, strong; pointed leaf with silver dominant over the puckered area and tinted pink, dark gray-green along veins. *Humid-tropical.* p. 173

rex 'Filigree'; upright stem with deeply lobed diadema-type leaves coppery green to deep red, veins red, areas between with silver blotching, overcast with red. *Humid-tropical.* p. 173

rex 'Fireflush' ('Bettina Rothschild'); beautiful plant with large heartshaped leaves metallic olive-green with darker center and edge, and entirely covered by plush-like crimson hairs glistening in the light; flowers waxy-white. *Humid-tropical.* p. 172

rex 'Frosty Dwarf' (Curly King Edward x Silver Queen); bushy and sturdy Brooklyn Botanic Garden hybrid of dwarf habit, with firm, oblique cordate leaves silvery, and outer edge olive green; pinkish when young; brown-red beneath. *Humid-tropical.* p. 173

rex 'Glory of St. Albans'; brilliantly colored metallic leaves of glistening, bright rose-purple overlaid with silver, but of weak growth. *Humid-tropical.* p. 172

rex 'Happy New Year'; same leaf pattern as 'Merry Christmas' but with bluish cast over the whole leaf, and silvery middle zone where 'Merry Christmas' is bright ruby red; the silver zone has maroon blush, stronger toward center. *Tropical.* p. 173

rex 'Helen Teupel'; diadema hybrid; bushy plant with long leaves sharply lobed, center and margin dark fuchsia-red, metallic green along veins, silvery-pink areas between. *Tropical.* p. 159, 173

rex 'Her Majesty'; broad, tapering leaves a blackish, satiny, reddish purple with olive-green zone overlaid with rosy silver blotching. *Humid-tropical.* p. 173

rex 'His Majesty', (English hybrid 1903); long-pointed and lightly notched leaves maroon with a broad silver zone, overlaid with pinkish purple. *Tropical.* p. 173

rex 'Merry Christmas' (Ruhrthal, Reiga, or Gundi Busch); beautiful, smooth leaf with well defined color zones; center velvety blood-red, broad adjoining area silver and pink, outer zone nile-green, edged fuchsia-red. *Tropical.* p. 173

rex 'Mikado', "Purple fan-leaf"; good American hybrid of bushy habit and enduring the heat of summer; center of leaf purple spilling into metallic silver and edged silvery purple. *Humid-tropical.* p. 172

rex 'President' (Pres. Carnot); easy grower; quilted leaf with base deep moss-green; raised areas a clear silver. *Tropical.* p. 172

rex 'Salamander'; diadema hybrid, bushy plant with erect slender leaves olive-green, blotched and pearled with silver. *Tropical.* p. 172

rex 'Silver Queen'; flat leaf a soft silver-gray excepting along veins in center and the margin a metallic green. *Tropical.* p. 172

rex 'Thrush' (rex x dregei); distinctly different, upright branching plant, with small maple-shaped leaves crimson-red sprinkled with fine silver; prolific bloomer. *Tropical.* p. 172

rex 'Yuletide'; spectacularly colored leaf with reddish-black center, a well defined middle zone vivid ruby-red, outer zone with spots of red, forest green and silver, leaf edge blackish; to 20 cm long. *Tropical.* p. 172

x richmondensis; floriferous, fibrous-rooted plant with red stem and medium oblique-ovate, waxy green leaves with bronzy overtone and dentate edge, reddish beneath; flowers dainty pink. *Tropical.* p. 175

x ricinifolia (heracleifolia x peponifolia), "Castor-bean begonia"; robust, old hybrid with erect rhizome and large, roundish leaves, deeply lobed like a maple leaf, fresh moss green with a taffeta sheen on top, reverse pale green, pale veins with red scales; petioles ringed by red, broad-scaly hairs, especially near apex; pink flowers. *Humid-subtropic.* p. 162

'Ricky Minter' (manicata crispa x mazae); robust rhizomatous plant with large round bronze to mahogany leaves with prominent pale veins, the margins crisped and frilled; spires of rose-pink flowers in winter. *Humid-subtropic.* p. 170

roezlii: see bracteosa

'Rogeri'; small form of thurstonii, fibrous-rooted; glossy olive-green, cupped leaves occasionally indented at the margin; free-flowering, with hirsute pink flowers. *Humid-subtropic.* p. 175

roxburghii (inflata) (Himalayas); slender fibrous-rooted species sparsely hairy with large, obliquely heartshaped leaves bright glossy green; fragrant white flowers. *Humid-subtropic.* p. 167

rubra (Java); bushy cane-type species with small shiny oblique-ovate, angelwing leaves 8 cm long, shimmering nile-green with purplish sheen; small pinkish flowers; photographed at the Los Angeles Arboretum, Arcadia, California. *Tropical.* p. 180

'Sachsen'; low, spreading, fibrous-rooted; small, angelwing leaves dark bronzy green, underside red; free flowering rosy red. *Humid-subtropic.* p. 164

sanguinea (Brazil); fibrous rooted, smooth and shining plant with many reddish stems, fleshy, oval leaves olive-green, blood-red beneath; small, white flowers. *Tropical.* p. 175

'scandens': see glabra

scharffiana (Brazil); compact, spreading, densely white-hairy, fibrous-rooted species with red stems, broad ovate leaves, olive-green, red beneath; pale pink hirsute flowers. Of smaller habit than scharffii. *Subtropic.* p. 179

scharffii (haageana) (Brazil), "Elephant-ear begonia"; lovely, rugged, white-hairy plant wth fibrous roots, and larger 25 cm ovate leaves brownish yellow-green with red veins, and red beneath; large pinkish flowers with beards (hirsute) in clusters. Parent of many velvety beauties. *Subtropic.* p. 166, 179

schmidtiana (Brazil); small, fibrous-rooted begonia with reddish, hairy stems and oblique, heart-shaped, olive-green, hairy, toothed leaves, beneath reddish with green margin; hirsute flowers in pale pink. *Tropical.* p. 175

schulziana (kraussiantha) (Haiti); small, rhizomatous miniature creeper with thick, smooth, maple-shaped, dentate leaves with about seven lobes; erect inflorescence with pink flowers. *Humid-tropical.* p. 160

secreta: see macrocarpa

semperflorens (cucullata), "Wax begonia"; originally from Brazil, but the derivatives in cultivation are grouped under B. semperflorens cultorum, or "Wax begonias"; fibrous-rooted, glabrous and more or less succulent plants with fleshy, oval leaves, and usually rose-red flowers in axillary clusters. *Subtropic.* p. 159, 165

semp. albo-foliis, "Calla lily begonia"; dainty mutant with glossy light green leaves marbled white and flushed bronzy-red toward margins, and the terminal growth a glistening white suggesting miniature calla lilies; flowers pink. *Tropical.* p. 176

semp. 'Cinderella'; bright green leaves and large, single flowers with prominent tufts of yellow stamens. *Subtropic.* p. 176

semp. 'Flamingo'; sturdy green-leaved plant with single bicolor flowers, white edged with crimson-red. *Subtropic.* p. 176

semp. fl. pl. 'Curlilocks'; eye-catching "Thimble-type" wax begonia with deep coppery-bronze leaves and somewhat pendant clusters of large flowers, the pronounced back petals rosy red, and the center with a high crest of stamen turned into petals, first green then bright yellow. *Subtropic.* p. 176

semp. fl. pl. 'Firefly'; dark coppery-black leaved wax begonia covered by high-crested, thimble-type red flowers with distinct back petals. *Subtropic.* p. 176

semp. fl. pl. 'Lady Frances', "Rose begonia"; endearing and enduring plant, popular in Milwaukee, with waxy, mahogany-red leaves and a profusion of ruffly camellia-type flowers fully double bright pink, with white ovary but without prominent back petals, in constant bloom. *Subtropic.* p. 176

semp. 'Luminosa'; compact variety with flowers soft scarlet, and light green waxy leaves with ciliate red border, turning deep red-brown if grown in full sun. *Subtropic.* p. 176

semp. 'Pink Pearl'; gracilis type, a good compact growing "bedding" variety with fresh-green waxy leaves and covered with single flowers bright rose-pink. *Subtropic.* p. 176

semp. 'Red Pearl'; green-leaved variety with glowing, red flowers. *Subtropic.* p. 176

semp. 'Scandinavia Pink'; vigorous bushy plant of intermediate habit 20-25 cm high, with glossy-green, wavy leaves, tinged bronze, producing masses of bright rose-pink flowers. *Subtropic.* p. 176

semp. 'Scandinavia White'; similar to Scandinavia Pink, but with white flowers; light green foliage with reddish stems. *Subtropic.* p. 176

semp. 'White Comet' (F-1 hybrid); bushy plant 15-25 cm high; the white flowers contrasting well with the bronze foliage. Good pot plant. *Tropical.* p. 176

serratipetala (New Guinea), "Pink spot"; beautiful, ornamental species, fibrous-rooted, freely branching with arching stems carrying small, shiny, frilled, pleated, deeply lobed and doubly toothed leaves in dark olive green with iridescent, deep pink to red spots; deep pink flowers. Name refers to the red-toothed petals of the female flower. *Tropical.* p. 174

'Silver Jewel' (imperialis x pustulata); luxurious as pustulata argentea but sturdier; cordate leaves with silvery blisters, streaked with emerald green. *Tropical.* p. 175

'Silver Star' (caroliniaefolia x liebmannii); rhizomatous; star-shaped leaves, deeply lobed, olive green, largely overlaid with silver and puckered, margins toothed, red beneath; flowers white. *Subtropic.* p. 172

'Sir Percy' (Silver star x speculata); unique rhizomatous plant with small leaves of metallic sheen which sometimes look like silver, sometimes green, dark green edge; brick-red underneath. *Subtropic.* p. 172

'Skeezar' (dayii x liebmannii); rhizomatous, a favorite because of its strong bushy growth and attractive, smallish, rippled, silver-splashed leaves; flowers greenish-white. *Tropical.* p. 168

sparsipila (barkeri) (Central America); fibrous-rooted plant of medium growth, completely covered with brown scurf, with lobed, toothed, green leaves usually depressed in center; large, pink flowers. *Tropical.* p. 178

x speculata (hybrid with rex), "Grapeleaf begonia"; low, hairy, rhizomatous plant with leaves nearly round and shallowly lobed, dull gray-green, speckled gray, reddish beneath; flowers white. *Subtropic.* p. 164

suffruticosa (Natal), "Maple leaf begonia"; semi-tuberous plant similar to dregei but smaller; small dark green leaves palmately lobed and toothed, with red spot at sinuses of lobes; flowers white or pale pink. *Subtropical.* p. 180

x sunderbruchii (ricinifolia x heracleifolia), "Finger leaf begonia"; a rhizomatous "star" begonia with palmately lobed, bright and bronze-green leaves, and silver bands along nerves; flowers pinkish. *Subtropic.* p. 169

sutherlandii (Natal); slender tuberous plant with drooping branches, lance-shaped, serrate, crisped leaves bright green and red veins; flowers orange. *Subtropic.* p. 168

'Templinii', "Crazy-leaf begonia"; mutant of phyllomaniaca, fibrous-rooted, self-branching; leaves ovate-pointed, glossy green and blotched yellow, margins ruffled, forms adventitious leaves all over foliage; flowers pink. *Tropical.* p. 178

'Thurstonii' (metallica x sanguinea); fibrous-rooted, bushy plant with white-haired red stems and cupped ovate leaves glossy, bronzy green, red beneath; flowers pink. *Tropical.* p. 171

'Tingley Mallet' (rex Eldorado x incarnata purpurea); handsome erect shrubby plant, brighter than 'Arthur Mallet' and less difficult; oblique-ovate leaves metallic maroon and red-hairy, margins toothed; flowers pink. *Tropical.* p. 164

tomentosa (Brazil); bushy, fibrous-rooted species with smallish, cordate-ovate, somewhat cupped leaves beautifully covered with white felt; white flowering. *Tropical.* p. 178

'Tom Ment'; large angelwing begonia with waxy, oblique-ovate leaves, dark olive-green and blotched with silver; flowers pinkish white and rose. *Tropical.* p. 179

x tuberhybrida, or "Tuberous begonia"; sturdy hybrids of Andean species, with watery stems and brittle, pointed leaves; large, single and double, waxy flowers, white, yellow, orange or red, during summer. *Humid-subtropic.* p. 159, 165, 177

tub. 'Camellia type'; giant double camellia-like flowers; crimson, pink, white, scarlet, orange, yellow. *Humid-subtropic.* p. 177

tuberhybrida 'Crispa'; carnation-form; double flowers lemon-white with fine red crisped margins. *Humid-subtropic.* p. 177

tuberhybrida 'Elegans'; fine fully double cream-yellow flowers with thin rose borders. *Humid-subtropic.* p. 177

tub. 'Fimbriata plena' (carnation type); very double flowers with petals serrated and frilled at edges, very much resembling a carnation; red, rose, white, orange. *Humid-subtropic.* p. 177

tuberhybrida 'Picotee' (bicolor); large flowers creamy-white, semi-double with contrasting red margins. *Humid-subtropic.* p. 177

tuberhybrida 'Rubro-marginata'; magnificent color combination of fully double flowers 12 cm across, white edged or variegated with crimson-red. *Humid subtropic.* p. 177

tub. 'Triumph' (Rosebud); double flowers white in center, and margins red. *Humid-subtropic.* p. 177

tuberhybrida multiflora 'Helene Harms'; miniature tuberous begonia of compact bushy habit, with smaller, canary yellow double flowers, blooming profusely through the summer; originated from the dwarf Andean species B. davisii, and the velvety marbled foliage of B. pearcei. *Humid-subtropic.* p. 177

tub. multiflora 'Maxima', "Dwarf tuberous begonia"; strain developed by crossing the large camellia-flowered type with multiflora; compact plants with rigid stems and dark foliage, covered with many medium-sized double blooms, red, rose, yellow, white. Ideal for bedding. *Humid-subtropic.* p. 165, 177

tuberhybrida multiflora 'Tasso'; bushy, compact, tuberous begonia derived from B. davisii, with small double flowers of bright red. *Humid-subtropic.* p. 177

tuberhybrida pendula 'Lloydii' (plena), the "Basket begonia"; drooping stems with narrow leaves with numerous small double pendulous flowers produced from a single tuber; red, rose, salmon, white, yellow; derives its pendant habit from B. boliviensis. *Humid-subtropic.* p. 163, 165, 177

ulmifolia (Colombia, Venezuela), "Elm leaf begonia"; tall, rapid-growing, fibrous-rooted species with elm-like ovate leaves rough brown-hairy; small white flowers. *Subtropic.* p. 180

velloziana (Brazil); small, rhizomatous epiphyte; white dotted leaf (in photo 10 cm long), with tapering base, and fine thickly set red cilia (hairs) along margins and petiole. *Tropical.* p. 171

vellozoana hort.: see olsoniae

venosa (Brazil); fibrous rooted plant covered with white scurf; stem thick; succulent cupped leaves reniform and appearing frosted; flowers white. *Tropical.* p. 175

versicolor (China), "Fairy carpet begonia"; with thick, corrugated, roundish leaves radiating a design of silver, emerald-green and bronze, and covered with red hairs, small, rhizomatous plant. Needs high humidity; good for terrarium. *Tropical.* p. 169

x weltonensis (sutherlandii x dregei), "Mapleleaf begonia"; semi-tuberous bushy plant with light green leaves shallowly lobed and toothed, with purple veins, on wine-red petioles and stems; flowers pink. There is also a white form, 'alba'. *Tropical.* n. 178

BELLIS *Compositae*
perennis fl. pl. (W. Europe, Asia Minor), "English daisy"; low perennial with obovate fleshy leaves in tufts, the short flower heads with several layers of white to rosy florets. *Warm temperate.* p. 308

BELOPERONE *Acanthaceae*
guttata (Mexico), "Shrimp plant"; wiry stems with ovate, hairy leaves; white flowers beneath showy, overlapping, reddish-brown bracts, in drooping terminal spikes. *Tropical.* p. 35

BENINCASA *Cucurbitaceae*
hispida (Indo-Malaysia: Japan to Polynesia), "Wax-gourd" or "Chinese watermelon"; annual, tendril-bearing, pumpkin-like vine with brown-hairy stem and broad lobed leaves 15-22 cm; male and female flowers yellow; fruit melon-like, cylindrical 20-40 cm long, hairy, and white-waxy, flesh white. Grown as ornamental curiosity, but in China is the source of the preserving melon and sweet pickles. *Tropical.* p. 383

BERBERIS: see Mahonia

BERGENIA *Saxifragaceae*
crassifolia (Siberia, Mongolia), "Winter begonia" or "Siberian tea"; ornamental perennial herb more or less evergreen with thick, woody rootstock, and large fleshy obovate leaves to 20 cm long, shining green and with the blade extending down the leafstalk; flowers rose-pink or lilac, in clusters above the foliage in spring. *Warm temperate.* p. 881

purpurascens (Himalayas to W. China, Burma); perennial herb with large fleshy, broad oval leaves to 25 cm long, glossy green with depressed yellow ribs, and suffused with purple; carmine-red nodding flowers. *Subtropic.* p. 881

BERTOLONIA *Melastomataceae*
maculata (Brazil); dwarf species with coarsely hairy, dark green leaves shaded light moss-green with reddish margin and lightly marked silver-green along center; deep red beneath. *Humid-tropical.* p. 644

maculata 'Wentii'; an exquisite small tropical plant with corrugated leaves, velvety green with silver zones along parallel veins. *Tropical.* p. 644

marmorata (Ecuador); beautiful herbaceous plant with quilted ovate leaves velvety moss-green, painted silvery-white along the parallel veins, purple beneath, and purple flowers. *Humid-tropical.* p. 644

marmorata 'Sanderiana'; attractive cultivar with corrugated leaves moss-green or coppery, heavily splashed with silver gray, especially along the vein areas; small light purple flowers. *Tropical.* p. 644

'Mosaica'; colorful German cultivar of low habit with beautiful velvety green leaves with white longitudinal bands broadest along the midrib, iridescent in several colors; fairly durable. *Humid-tropical.* p. 644

x BERTONERILA *Melastomataceae*
houtteana (Bertolonia); intergeneric hybrid of Bertolonia, Gravesia and Sonerila, happiest in a warm conservatory; with large, moss-green, herbaceous leaves veined and dotted clear carmine-red, purple underneath. *Humid-tropical.* p. 644

BESCHORNERIA *Amaryllidaceae*
yuccoides (Mexico); succulent rosette of about 20 leaves 45 cm long x 5 cm wide, glaucous gray-green, rough on margins and beneath; flower stems coral red, to 1 m, with numerous nodding bright green flowers subtended by bright red bracts. *Tropical.* p. 59

BETULA *Betulaceae*
papyrifera (Labrador to Minnesota), "Paper birch", "Canoe birch" or "Cluster birch"; the most wide-spread native birch in North America; tall tree to 30 m high, with bark a vivid warm white, flaking and papery; ovate leaves to 10 cm long, coarsely serrate; male catkins to 10 cm long. Occasionally forming multistem clusters. *Temperate.* p. 182

pendula (verrucosa) (Europe and Asia Minor), "White birch"; stately deciduous tree to 20 m tall; bark white and flaking off in layers; branches usually pendulous; leaves angular-ovate to 6 cm long; tiny flowers in catkins. As grown by nurseries often sold in cluster-form. *Temperate.* p. 182

pendula dalecarlica (Europe and Asia Minor), "European white birch"; deciduous tree with weeping branches; lively-green leaves deeply lobed and cut into narrow, pointed segments. Photographed in Germany. *Temperate.* p. 182

populifolia (Nova Scotia to Delaware), "Gray birch"; small graceful tree to 9 m, with chalky gray-white trunk and narrow crown; leaves glossy green on both sides, triangular-ovate to 8 cm long; male catkins 5-8 cm. *Temperate.* p. 182

BIFRENARIA *Orchidaceae*
harrisoniae (Brazil); sturdy epiphyte, 4-angled pseudobulbs with a solitary leathery leaf, large fleshy, 8 cm flowers, the sepals and petals yellowish, tinged with red, the lip violet-red with yellow hairy callus, (March-May). *Subtropic.* p. 709

tyrianthina (Lycaste) (Brazil); epiphyte with clustered pseudobulbs, solitary leaf, and waxy, heavily fragrant flowers 9 cm dia., purple, in spring to summer. *Subtropic.* p. 709

tyrianthina alba (Brazil); attractive epiphyte with ovoid pseudobulbs and solitary oblong leaf; very waxy flowers 9 cm across, heavily fragrant and long-lasting, purple with white throat in the species, all white in form alba. *Subtropic.* p. 722

BIGNONIA *Bignoniaceae*
capreolata (Virginia, Florida, Louisiana), "Crossvine" or "Trumpet flower"; evergreen climbing shrub to 15 m with leaves of 2 ovate leaflets, each to 15 cm long; funnel-form flowers 5 cm across, yellow and brown-red, the flaring limb golden yellow, anthers cream with reddish line on each. *Warm temperate.* p. 187

BIGNONIA: see also Anemopaegma, Campsis, Clytostoma, Cydista, Distictis, Pandorea, Pyrostegia, Saritaea, Tabebuia, Tecoma, Tecomaria

BILLBERGIA *Bromeliaceae*
x 'Albertii' (distachia x nutans), "Friendship plant"; tall clustering tubular hybrid, the offshoots with narrow tapering leaves, dark green and gray-scurfy; inflorescence rosy-red with pendulous green flowers edged blue. *Subtropic.* p. 208

amoena (So. Brazil); fluted rosette with stiff gray-green leaves with pronounced silver cross bands; inflorescence on arching spike with rose bracts and green flowers edged blue. *Tropical.* p. 208

amoena viridis, a "Vase plant"; fluted rosette more richly silver-banded and spotted ivory; rose bract leaves and petals entirely green. *Tropical.* p. 208

x 'Bob Manda', "Manda's urn plant"; a colorful, compact pyramidalis hybrid with broad leathery leaves in a bold rosette, highly variegated and blotched cream over copper to olive and even purple. *Tropical.* p. 208

buchholtzii (Brazil); fluted rosette, coppery and gray with cream spots; inflorescence with red bracts, purple flower. *Tropical.* p. 211

euphemiae (So. Brazil); stiff tubular plant with gray-scurfy, green leaves and gray crossbands; rosy bracts and pendant blue flowers. *Tropical.* p. 208

'Fantasia' (saundersii x pyramidalis), "Marbled rainbow plant"; urnshaped hybrid with coppery green leaves highly blotched and variegated creamy white and pink; rose bracts and blue flowers. *Tropical.* p. 208, 211

'Fascination'; slender rosette, highly variegated cream; pendant inflorescence with rosy bracts. *Tropical.* p. 210

nutans (So. Brazil, Uruguay, Argentina). "Queen's tears" or "Indoor oats"; slender rosette of narrow, silvery bronze foliage, forming clusters. An arching flowerstalk bears the nodding inflorescence of rosy bracts and green flowers edged violet—a teardrop forming on the stigma. Very tolerant. *Subtropic.* p. 208, 211

pyramidalis, "Fool proof plant"; elegant rosette of coppery leaves; erect inflorescence with rose-pink bracts, purple flower. Summer-blooming. *Tropical.* p. 211

pyramidalis concolor (thyrsoidea) (Brazil), "Summer torch"; rosette shaped like a birdsnest, broad, glossy, apple-green leaves; showy but short-lived inflorescence, stem mealy-white, and head of large red bracts and crimson red flowers tipped purplish, with blue stigma. *Tropical.* p. 208, 211, 214

pyramidalis 'Striata'; an attractive seedling clone raised by M. Foster 1950, from the species collected in Brazil; has broad tomentose blue-green leaves, not glabrous yellow-green as in the type and are striated and variegated at margins with cream; flowering in winter. *Tropical.* p. 208

saundersii (Bahia), "Rainbow plant"; smallish tubular species with stiff olive to bronzy foliage blotched ivory and tinted pink; inflorescence bracts red and pendant blue flowers. *Tropical.* p. 208

seidelii (Estado do Rio); gray-striped fluted rosette; inflorescence pendant on threadlike axis; bracts rose pink, petals blue-purple. *Tropical.* p. 208

tessmanniana (Perú), "Showy billbergia"; tall urn-shaped plant with bronzy leaves mottled gray and white; possibly the showiest of this genus for its inflorescence, which is a giant pendulous spike with red and pink flowers, the rosy bract leaves flaring out into a fabulous show. *Tropical.* p. 210, 211

zebrina (Minas Geraes to Rio Grande do Sul), "Zebra urn"; attractive species with long fluted leaves purplish bronze in strong light, heavily crossbanded silvery white, and armed with thorns; inflorescence with large rosy-red bract leaves and nodding violet flowers, the salmon calyx leaves tipped violet. *Tropical.* p. 208

BIOPHYTUM Oxalidaceae
zenkeri (Zaire), "Sensitive life plant"; a small sensitive plant with pinnate leaves 5-8 cm long, occasionally forking, the pinnae short and roundish, dark glossy-green, and which fold back when touched; small yellow flowers. *Tropical.* p. 762

BISCHOFIA Euphorbiaceae
javanica (trifoliata) (Trop. Asia: Java), "Toog tree"; ornamental evergreen tree to 25 m high, somewhat deciduous, with large alternate compound leaves of 3 ovate, fleshy leaflets deep green to bronzy, minutely toothed; small greenish flowers without petals, and reddish pea-size fruit on female flowers. *Tropical.* p. 407

BISMARCKIA Palmae
nobilis (Madagascar); majestic, dioecious fan-palm with great trunks to 60 m high, large crown with gray-green palmate, rigid leaves 3 m across, the stalk streaked with scurfy white; 3 cm plum-like brown fruit on female trees, in huge clusters. *Tropical.* p. 767

BIXA Bixaceae
orellana (West Indies), "Lipstick-tree" or "Annatto-tree"; a showy small evergreen tree 6-10 m high, with broad, smooth, heart-shaped ovate leaves to 18 cm long; 5 cm regular flowers pink, the fleshy fruit spiny and source of orange henna or Annatto dye made from its pulp. The foliage was reddish where I have seen them growing in East Africa. *Tropical.* p. 181

BIOTA: see Thuja orientalis

BLAKEA Melastomataceae
sanguinea (Colombia); found at 3,000 m, in the Andes above Bogotá; small shrub with rugose, ovate leaves on red branches; large cup-shaped flowers crimson-red. *Subtropic.* p. 870

BLANDFORDIA Liliaceae
nobilis (New South Wales), "Christmas bells"; fibrous-rooted perennial to 1 m high; basal two-ranked linear leaves 40 cm long; bell-flowers 4 cm long in large clusters, bronzy-red with yellow lobes. *Subtropic.* p. 602

BLECHNUM Polypodiaceae (Filices)
brasiliense (Brazil, Peru); coarse rosette growing on a scaly trunk to 1 m high, the leathery green fronds deeply pinnatifid, widest in the upper third, the midrib broad, with pinnae overlapping and wavy, and coppery when young. *Humid-tropical.* p. 441

gibbum (Lomaria) (New Caledonia); graceful, symmetrical rosette, developing a trunk to 1½ m high, with broad, thin-leathery, arching pinnate fronds, the shining green pinnae are long and narrow, and almost threadlike on the fertile fronds. *Humid-tropical.* p. 441

serrulatum (No and So. America), "Saw fern"; coarse swamp fern with creeping underground rhizomes, with pinnate fronds to 75 cm long, the pinnae pointed at apex. *Humid-Tropical.* p. 441

BLETILLA (Orchidaceae)
striata, better known in horticulture as Bletia hyacinthina (Vietnam, So. China, Japan); handsome terrestrial orchid 30-60 cm high, growing leafy stems from tuberous rhizomes, bearing 3-5 rather thin plaited leaves; light purple flowers with trilobed lip lined by deep purple ridges; the blooms are 3 to 5 cm across, usually not fully opening, in terminal clusters on erect leafless stalk, rising from the center of the new shoots, mostly in June-July. A good orchid for summer on the patio. Easy to grow and fairly winter-hardy with protection. *Warm temperate.* p. 711

BLIGHIA Sapindaceae
sapida (West Africa: Guinea), the "Aki"; large ornamental, tropical fruit tree from West Africa but very common in the West Indies where it was introduced in slave days; 10-12 m high, with compound leaves of 3 to 5 pairs of glossy green oblong leaflets each about 15 cm; greenish-white fragrant flowers; the attractive oblong, ribbed fruit is orange with red cheeks, 8-10 cm long, triangular in cross-section with leathery red shell which opens when ripe, exposing firm white, nut-flavored pulp; wholesome food when ripe, eaten raw, fried or boiled; the seed coat is poisonous. *Tropical.* p. 455, 876

BOCCONIA Papaveraceae
frutescens (W. Indies, Mexico), the "Plume poppy", or "Tree celandine"; handsome woody shrub to 2 m high, dense with pinnately lobed leaves, 15-35 cm long, grayish-green, smooth above, densely pubescent underneath, on slender branches; small greenish-purple flowers, in panicles 30 cm long. *Tropical.* p. 648

BOEA Gesneriaceae
hygroscopica (East Asia), "Oriental streptocarpus"; lovely small, fibrous-rooted perennial herb about 15 cm high, close to streptocarpus; fleshy oval, bullate fresh green leaves 5-8 cm long, deeply veined and ribbed and covered with short white hairs, the margins pronounced crenate; small 1 cm 5-lobed flowers purplish-violet and wheel-like, showing yellow pollen; if dry, leaves will shrink and curl but spread again when watered. *Humid-tropical.* p. 493

BOERLAGIODENDRON Araliaceae
eminens (Philippines: Mindanao); evergreen tree 5-10 m high, glabrous throughout, large leaves to 60 cm long, palmately 10-14 lobed, lobes reaching nearly to base, irregularly toothed and coarsely incised, shining green on both sides. *Tropical.* p. 133

novo-guineense (Papua); small willowy tree to 6 m high with sparry branches set with large palmate 60-90 cm leaves, on long petioles having thorns near base, the leaflets irregularly lobed and glossy green; the showy inflorescence caught my eye with its enormous panicle like a gigantic pincushion, of bright orange flowers and purplish fruits. *Tropical.* p. 132

BOLIVICEREUS Cactaceae
samaipatanus (Borzicactus) (Bolivia); slender oblique columnar cactus light olive green deeply ribbed and characterized by a hair-ring, the ribs set with long straw-colored needle spines tinged with red, and shorter radials; lateral tubular flowers with series of petals spreading 3 cm, velvety crimson edged in gold, from pale tube. *Arid-subtropic.* p. 248, 250

BOLUSANTHUS Leguminosae
speciosus, (No. Transvaal, Swaziland, Mozambique), the "Tree wisteria"; a slow-growing small tree with gray trunk, about 5 m high as I saw it in the Northern Transvaal, but may reach 12 m;with deciduous, glossy green pinnate leaves usually appearing after the flowers; these pea-shaped bright purple in pendulous 15 cm racemes resembling wisteria. *Tropical.* p. 542

BOMAREA Amaryllidaceae
carderi (Colombia); tropical twining herb with oblong parallel-veined leaves 16 cm long, and 6 cm bell-flowers in pendulous clusters, outer segments rose, inner crenulated, spotted purplish-brown. *Tropical.* p. 60

multiflora (Colombia, Venezuela); attractive vine with oblong leaves 10 cm long, and dense many-flowered umbels, corolla about 3 cm long with outer segments tinged red, inner reddish-yellow spotted claret-brown. *Tropical.* p. 61

shuttleworthii (So. America to Pacific); gorgeous tuberous plant with twining stems; parallel-veined leaves, and large and showy tubular flowers 6-8 cm long, in pendant bracted clusters, bright crimson-red with yellow tips. *Humid subtropic.* p. 53

BOMBAX Bombacaceae
ceiba (malabaricum) (India to Indochina), "Red silk-cotton" or "Red kapok tree"; big soft wooded tree 25 to 30 m high, with buttressed, spiny trunk 2½ m dia.; palmate leaves with 5-7 leaflets; brilliant crimson 10 cm waxy flowers, clustered near ends of the whorled branches. *Tropical.* p. 194, 195

costatum (Sénégal, West Africa); bulky tree flowering leafless with brilliant salmon flowers. *Tropical.* p. 193

ellipticum (Mexico: Jalisco, Veracruz to Yucatán), the "Shaving brush"; a large soft-wooded tree with green trunk with large deciduous leaves of 5 leaflets finger-like, wine-red when young, later dark green; the flower buds composed of 5 purplish petals stuck together, later bursting open and revealing a bundle of delicate, silky rose-pink stamens 8 cm long, tipped with golden anthers. *Tropical.* p. 193, 195

ellipticum 'Album' (Pseudobombax) (Mexico to Guatemala), "White shaving-brush tree"; charming colorform with the bundles of long bursting stamens pure white, tipped with yellow pollen. *Tropical.* p. 193

BOMBYCIDENDRON   Malvaceae
vidalianum (Philippines), "Tree hibiscus"; tropical tree with pendant branches, closely set with ovate leaves in two ranks; at end the hibiscus-like flowers 20 cm across, opening white and turning pink. *Tropical.* p. 620

BORASSUS   Palmae
aethiopum (Africa: No. Transvaal to Kenya, Uganda, Sudan, to Guinea coast), "African fanpalm"; stately palm to 25 m high, with spindle-shaped trunk swollen above the middle, the bark gray and smooth; leaves fan-shaped, to 4 m long including petiole, divided to middle; 15 cm orange fruit. *Tropical.* p. 767, 771

flabellifer (India, Burma, Malaya), the widely cultivated and fabled "Palmyra palm" or "Toddy palm" of India where it finds hundreds of uses, for food, black timber, sugar and toddy; trunk 20-30 m tall sometimes swollen above the middle, to 1 m thick; rounded head of palmate gray-green leaves to 3 m across rather folded, rigid and with stiff tips, the stalk with horny thorns. Female trees with big 12-20 cm edible fruit yellow to brown and black. The intoxicating toddy is taken from the dense flower spikes. *Tropical.* p. 474, 766, 769, 771

BORONIA   Rutaceae
elatior (Western Australia), "Scented tall boronia"; leafy evergreen dense shrub to 1-1½ m high, the shoots thickly furnished with spreading hairs, lacy, fresh-green pinnate foliage with needle-like leaflets, and numerous attractive nodding globose 6 mm fragrant flowers along the stem carmine-red and often of different colors on the same plant, in May. *Subtropic.* p. 873

heterophylla (W. Australia), "Red boronia"; much branched, smooth evergreen shrub to 2 m, with linear leaves 5 cm long, simple or pinnate; small bell-flowers rosy red, 1 cm dia., not opening widely. *Subtropic.* p. 872

megastigma (Western Australia), "Scented boronia"; slender, hardwooded, evergreen 60 cm shrub with downy shoots; sessile, fresh-green, needle-like leaflets having transparent dots; small 1 cm, sweetly scented, axillary flowers with globose brown-purple corolla, yellow inside, in spring. *Subtropic.* p. 873

BORZICACTUS: see Bolivicereus, Matucana

BOUGAINVILLEA   Nyctaginaceae
'Alba'; charming spectabilis cultivar; scandent woody branches with glossy green, slender ovate leaves, and clusters of showy pure white bracts. *Tropical.* p. 689

braziliensis; in the California nursery trade and elsewhere; usually a plant with bracts in shades of purple, from rose-purple to distinct deep blue-purple. *Tropical.* p. 694

x buttiana (peruviana x glabra), the clone 'Mrs. Butt' is widely popular as 'Crimson Lake'; a vigorous, pubescent woody clamberer with large recurved spines, broader and thicker ovate leaves than glabra; blooming in big panicles, with the insignificant flowers subtended by corolla-like, showy cordate bracts of bright crimson, cascading in masses of color from late winter to summer. Requiring sun; tolerates drought. Keep cool in winter. Propagate by cuttings from ripened wood. *Tropical.* p. 694

x buttiana 'Barbara Karst', (peruviana x glabra clone); rather bushy grower with cascading masses of large brilliant red floral bracts borne almost continually, blooming at an early age. *Tropical.* p. 689

x buttiana 'Louis Wathen'; grown in Peradeniya, Ceylon, with magnificent masses of double-bracted flowers in shades of soft salmon-orange; flowers lacking star-like limb. *Tropical.* p. 692

x buttiana 'Madonna'; somewhat tender cultivar with crisped bracts pale lavender pink to blush-white. *Tropical.* p. 689

x buttiana 'Mrs. McLean'; large bracts daintily colored soft salmon peach to light pink. *Tropical.* p. 690

x buttiana 'Praetoria'; bracts peach yellow turning to golden salmon. *Tropical.* p. 694

x buttiana 'San Diego Red'; very vigorous cultivar also known as 'Scarlett O'Hara'; bold branches with green leaves carry heavy masses of large bracts of glowing crimson. *Subtropic.* p. 690

formosa (Brazil); robust spiny shrub with ovate leaves, and inflorescence of creamy flowers subtended by large orchid pink bracts. *Tropical.* p. 689

glabra (Brazil), "Paper flower"; strong woody rambler with bright green, smooth leaves slender pointed and with narrow base; flower clusters smaller than spectabilis, the branchlets with acute purplish-pink bracts in threes, blooming in summer; more compact than spectabilis. *Tropical.* p. 689, 691, 694

glabra 'Salmonea'; variety of somewhat weak habit, with bracts soft peachy salmon. *Tropical.* p. 691

glabra 'Variegata'; grown for the ornamental foliage, grayish green with broad cream-colored borders. *Tropical.* p. 692

'Harrisii'; a strikingly beautiful, though delicate, variegated foliage plant, probably a form of glabra, of low bushy habit, freely branching, with dense ovate, thin leaves, friendly green with cream when young, grayish green with age, liberally splashed with glistening white inward from the margins. Flowers white, subtended by purple bracts. *Humid-tropical.* p. 689, 694

poultonii; in the collection of Botanic Garden Singapore; spiny branches covered toward tips with dense inflorescence of cream flowers and crimson red bracts of medium size. *Tropical.* p. 689

spectabilis (Brazil); free ranging canes often with stout spines, not as bushy and compact as B. glabra; leaves more or less ovate, velvety hairy beneath; bracts usually light brick-red or purple 5 cm long; calyx with many spreading short hairs. *Tropical.* p. 690, 691

spectabilis 'Alba plena'; at the end of branches the heavy elongate clusters of flowers with fully double pure white bracts. *Tropical.* p. 693

spectabilis 'Arborea'; form in the collection of Rio de Janeiro Botanic Garden, of tree-like habit with stout trunk; the bracts light purple. *Tropical.* p. 691

spectabilis 'Carmencita'; robust branches carry flowers with large semi-double bracts of carmine-red. *Tropical.* p. 693

spectabilis 'Lateritia'; as seen in the collection of the Botanic Garden in Rio de Janeiro; robust clamberer with showy bracts a glowing rosy-red, fading to orange and creating a two-color effect. *Tropical.* p. 693

spectabilis 'Manila Magic Pink'; (Pat. Monrovia Nurs., Calif.); one of the showiest cultivars brought from the Philippines; handsome huge clusters of frilly pink double flowers with carmine rose bracts. *Tropical.* p. 693

spectabilis 'Mary Palmer'; unique cultivar observed in Nairobi, Kenya and Bangalore, Mysore, India; bushy plants with scandent branches and flowers some with carmine-red bracts, and others with white bracts on the same plant. *Tropical.* p. 690, 691, 693

spectabilis 'Milflores'; (Philippines); from Ilo-Ilo Visayas Province; a spectacularly beautiful cultivar profusely blooming, with double flower bracts a glowing crimson-red, in large dense clusters; on vigorous woody clambering stems. *Tropical.* p. 691

spectabilis 'Rubra plena'; floriferous cultivar with branches loaded down by the weight of fully double, carmine red bracts. *Tropical.* p. 689, 693

spectabilis 'Tahitian Gold', Monrovia Nursery (California) introduction; cultivar with large clusters of double-bracted flowers rosy-red with salmon sheen. *Tropical.* p. 693

spectabilis 'Tahitian Maid' (Pat. Monrovia Nursery, Calif.); bountiful double floral bracts of blush pink; darker rose when opening. *Tropical.* p. 693

spectabilis 'Variegata'; foliage prettily bordered with creamy white; inflorescence with red-purple bracts. *Tropical.* p. 690

x spectoglabra; spectacular, interspecific hybrid of glabra x spectabilis of bushy habit, and with giant bracts spreading 10 cm across, soft lavender rose. *Tropical.* p. 693

BOURRERIA   Boraginaceae
ovata (So. Florida, West Indies), "Strongback"; evergreen dense shrub or small tree; obovate or spoon-shaped leathery leaves to 8 cm long; small fragrant white flowers in large clusters, followed by scarlet-red berries 1 cm dia. *Tropical.* p. 196

BOUVARDIA   Rubiaceae
longiflora humboldtii (Mexico), "Sweet bouvardia"; beautiful fall-to-winter-flowering shrub with woody and flexible branches, opposite, smooth, fresh-green leaves, and bearing toward the tip several waxy, salver-form, fragrant flowers having long, slender, 6 cm tubes opening into lobes of purest white. *Subtropic.* p. 860

ternifolia (triphylla) (Mexico, Texas), "Scarlet trompetilla"; straggling shrub with thin, woody branches with whorled, ovate, pubescent leaves, opposite on the branchlets, which are terminated by clusters of 3 cm, tubular flowers of fiery scarlet-red, blooming most of the year; horticultural varieties are 'Christmas Red' and 'Fire Chief'. *Subtropic.* p. 860

ternifolia 'Giant Pink'; a favorite florist's form for cut flowers, with clusters of larger, rose-pink flowers with salmon sheen. *Subtropic.* p. 860

ternifolia 'White Joy'; pretty albino form with clusters of numerous, small, pure white flowers, and with pubescent foliage. *Subtropic.* p. 860

versicolor (So. America); small shrub 50 to 80 cm high, with thin-wiry branches becoming pendant; small narrow ovate dark green leaves 4 cm long; terminal clusters of charming slender tubular flowers 3 cm long, orange-red and prettily tipped yellow; blooming from July to September. *Tropical.* *p. 860*

BOWENIA *Cycadaceae*
spectabilis (North Queensland); remarkable, very attractive small cycad, with its unusual bipinnate foliage; forming a clump, with edible tuberous base, and a long wiry stalk bearing a large palmately compound leaf about 60 cm across, the pinnate segments with sickle-shaped, leathery, glossy-green leaflets finely toothed and slender pointed; used for cut. *Tropical.* *p. 384*

BOWIEA *Liliaceae*
volubilis (Schizobasopsis) (So. Africa), "Climbing onion"; grown as a curiosity; succulent, light green bulb, to 20 cm diam., above ground, sending up a twining fresh-green branched stem with few linear deciduous leaves; small greenish-white flowers. *Subtropic.* *p. 603*

BOWKERIA *Scrophulariaceae*
gerrardiana (So. Africa); evergreen shrub 3 m high, with bronzy rugose, ovate leaves 15 cm long, finely toothed at margins; 2-lipped flowers pure white, lower lip inflated. *Subtropic.* *p. 884*

BRACHYCHILUM *Zingiberaceae*
horsfieldii (Alpinia calcarata) (Java); a slender ginger with leafy stems 1-1½ m high, furnished with sessile, long-sheathed, lanceolate leaves to 30 cm long, glossy green and leathery; flowers with greenish-white corolla and white lip; the fleshy fruit a 3-celled red capsule, showy orange when open, with crimson-red seed masses. *Tropical.* *p. 924*

BRACHYCHITON *Sterculiaceae*
acerifolium (New S. Wales, Queensland), "Flame tree"; to 30 m high, resembling Trevesia but without thorns; the attractive, glossy green, palmate leaves deeply 5-7 lobed, to 30 cm across, on long petioles; flowers without corolla, but a rich-red, bellshaped calyx. *Subtropic.* *p. 901*
rupestris (Sterculia) (Queensland), the peculiar "Queensland bottle-tree"; semi-deciduous, 6-15 m high, with a huge bottle-shaped trunk 3½ m in dia., and storing water; the spreading branches with variable blackish-green leaves lanceolate or palmately divided; tomentose bell-shaped flowers. *Tropical.* *p. 901*

BRACHYCORYTHIS *Orchidaceae*
kalbreyeri (Sierra Leone, Liberia to Cameroun, Zaire to W. Kenya); small terrestrial with potato-like tuber, sending up a weak leafy stem to 30 cm high, the well-spaced leaves narrow-elliptic, and terminal raceme of lilac-mauve 5 cm flowers with royal purple lip, scented like rich spice. *Tropical.* *p. 712, 714*

BRAHEA *Palmae*
armata (Erythea) (Mexico: Baja California), "Mexican blue palm"; stout fan-palm to 12 m; the trunk naturally covered by a dense petticoat of dead leaves; stiff, palmate fronds waxy-blue in heavy crown, deeply cut into many segments, the petiole armed with curved white spines; handsome arching spadices to 5 m long. *Tropical.* *p. 780*
brandegei (Erythea), (Mexico: Baja California), "Daughter of the West", or "Hesper palm"; tall, sturdy fan palm to 30 m or more tall, comparatively slender trunk with persistent leaf bases; heavy crown of green, palmate leaves 1-1½ m dia., on long petioles heavily armed with recurving spines; blade divided in center, segments deeply split, waxy white beneath; shining brown, 1 cm fruit. *Tropical.* *p. 781*
dulcis (Erythea) (So. Mexico), "Rock palm" or "Palma dulce"; a true Brahea: medium-sized, variable fan palm, forming erect trunks to 6 m tall, or procumbent on the ground, sometimes in clusters; upper part covered with old leaf-bases; palmate fronds very stiff, 1 m or more across, deeply cut into about 50 segments, green above, bluish glaucescent beneath; petiole edged with small teeth; small yellow fruit in large pendant clusters. *Tropical.* *p. 780*
edulis (Erythea) (Mexico: Guadalupe Is., Baja California), "Guadalupe palm"; robust, squat-growing, about 10-12 m high, trunk to 40 cm thick and ringed with scars; palmate, rigid leaves 2 m across, green on both sides, the numerous segments deeply cut and cleft or torn at apex. *Tropical.* *p. 780*

BRASSAIA *Araliaceae*
actinophylla (Schefflera macrostachya) (Queensland, New Guinea, Java), the "Queensland umbrella tree"; an attractive ornamental tub plant in U.S.A. where it is generally known as "Schefflera actinophylla"; in Queensland becoming a tall tree 30 m or more high; in New Guinea I have seen them growing epiphytic 30 m up on rainforest trees 5 m high; the sparry, willowy brownish, later woody branches each end at their top with a rosette of successively larger, palmately compound leaves forming umbrella-like symmetrical heads; lacquered-green, soft-leathery oblong, stalked leaflets, the young leaves with 3 later 5, in older plants 7 to 12 and 16, to 30 cm long; with pronounced teeth irregularly spaced, more so in mature stage leaves. Terminal inflorescence of several 1-1½ m straight spikes set with sessile round, dense clusters of honey-laden flowers with fleshy wine-red petals, and which form into the purple fruit. Very satisfactory decorator if kept warm and dry. *Tropical.* *p. 132, 137, 139, 144*

BRASSAIA: see also Schefflera

BRASSAVOLA *Orchidaceae*
acaulis (Guatemala, Costa Rica, Panama); pendulous epiphyte with 1-leaved cylindrical stems; leaf terete to 70 cm long; basal inflorescence of fragrant, heavy flowers 10 cm across, greenish white and purplish brown reverse, white lip. *Tropical.* *p. 708, 711*
cordata (subulifolia) (West Indies); small epiphyte with long linear, pendant, channeled leaves on cane-like pseudobulbs, fragrant flowers white, the porcelain-white heart-shaped lip with its tube-like base marked purple; summer. *Tropical.* *p. 708*
cucculata (Trop. America); epiphyte with cylindric pseudobulbs, and fleshy cylindric leaves; flowers with linear white sepals and petals, to 8 cm long, and white lip. *Humid-tropical.* *p. 713*
digbyana (Laelia) (Honduras); white-mealy epiphyte, in habit resembling cattleya, the pseudobulb bearing a solitary, rigid, glaucous leaf; large, showy, fragrant flower, the narrow petals and sepals greenish-white, tinged purple, the rounded lip cream-white with green throat, fringed at the margin with a long beard, (May-July). *Tropical.* *p. 713*
glauca (Laelia) (Mexico, Guatemala); small, bluish-gray epiphyte with slender pseudobulbs bearing a stiff, glaucous leaf; fragrant flowers with linear sepals and petals white, often tinted green, the large lip cream-white sparsely marked with pink, (Dec.-March). *Tropical.* *p. 713*
nodosa (Jamaica, Costa Rica, Colombia to Surinam), "Lady of the night"; epiphyte with stemlike pseudobulbs bearing a solitary, stout, channeled leaf; flowers solitary or in clusters, linear sepals and petals greenish-yellow and broad, pointed, white lip, (Sept.-Dec.). *Tropical.* *p. 713*

BRASSIA *Orchidaceae*
allenii (Panama); pseudobulbless epiphyte with stiff vandaceous, overlapping foliage and showy flowers having long narrow tapering sepals and petals cinnabar red, greenish-yellow toward base, and a broad, almost square yellow lip and spotted brown in front of the white callus. (Spring). *Tropical.* *p. 709*
antherotes (Ecuador); very ornamental epiphyte to 45 cm high, with oblong 1-leaved pseudobulbs; basal racemes of large firm-textured flowers, the long narrow, threadlike sepals and petals deep yellow, with purple-brown blotches at base, the lip brighter yellow (early summer). *Tropical.* *p. 711*
maculata (W. Indies, Guatemala); epiphyte with single-leaved pseudobulbs, and erect spike with waxy flowers, the narrow sepals and petals pale greenish-yellow barred with brown, the white lip marked red-brown, (April-Oct.). *Tropical.* *p. 709*
verrucosa (Mexico, Guatemala, Honduras, Venezuela), "Queen's umbrella"; vigorous epiphyte with pseudobulbs bearing 2 oblong leaves; arching, wiry flower stems with long threadlike, 12 cm sepals greenish-yellow, spotted at base with purple, as are the petals; the finely brown-warty lip white. *Tropical.* *p. 709*

BRASSICA *Cruciferae*
oleracea acephala, grown in pots for ornament, and known as "Flowering cabbage"; thick-leaved, glaucous perennial, a form descended from the European type; with spreading leaves forming a green rosette which with the advent of colder weather develops in its center striking shades of ivory to rose, imitating a showy flower. Good commercial cultivars are 'Red Crown' (Kokan) deep green with purplish center and purple ribs; and 'White Ripple' (Ginpa) ivory-white on green. *Warm-temperate.* *p. 379, 381*
oleracea acephala crispa, the "Miniature flowering kale"; grown as an ornamental pot plant, reported to be a hybrid between "Flowering cabbage" and ordinary Kale, forms a big rosette of large frilled and fringed foliage, glaucous-green in the younger stage, but in autumn as the season advances and they are kept cold, remarkably beautiful colors begin to appear with the center blossoming forth in shades of ivory, yellow, rose and purple. These plants are used as Christmas and New Year's pot plants in Japan, and have equally been used in flower arrangements in New York. *Warm-temperate.* *p. 379, 381*

x BRASSOCATTLEYA *Orchidacae*
'Brittisa' (Brassavola x Cattleya); charming flower with pink sepals and petals; the large lip with undulate frilling, and golden throat. *Humid-subtropic.* *p. 713*

'Carmen' (Brassavola digbyana x 'Mrs. Myra Peters', 1918); large pink flower with fimbriate lip, yellow in throat. *Humid-subtropic.* p. 713

'Heatonensis' (B. digbyana x C. hardyana); very handsome bigeneric hybrid with large flowers to 16 cm across; pure white with tinge of green on outer sepals, lip beautifully fringed, and delicate yellow-green inside. *Tropical.* p. 713

'Illusion' ('Rosa Bonheur' x C. 'Admiration'); beautiful large flower with pure white sepals and petals, and fringed purple lip. *Humid-subtropic.* p. 713

lehmannii; hybrid with smallish pink sepals and petals, but an exceptional large bearded lip, light purple, deeper purple in throat and with yellow lines. *Humid-subtropic.* p. 713

veitchii (Brassavola digbyana x Cattleya mossiae); first bigeneric hybrid between these genera, with large delicate rose-pink flowers having a large, beautifully bearded lip and crimson in throat changing to yellow and fading out to cream. *Tropical.* p. 713

BRAVOA   *Amaryllidaceae*

geminiflora (C. Mexico), "Twin-flower"; deciduous tuberous herb with linear-sword-shaped basal leaves pale green, and 60 cm long racemes of nodding tubular rich orange-red flowers 3 cm long, in July, in pairs. *Tropical.* p. 60

BREVOORTIA: see Brodiaea

BREYNIA   *Euphorbiaceae*

nivosa (disticha, Phyllanthus) (South Sea Isl.), "Snow-bush"; loose growing, green-stemmed pendant branches with alternate, small, fern-like leaves richly mottled and variegated white. *Tropical.* p. 407

nivosa roseo-picta, a form known as "Leaf-flower" because the little oval, papery leaves are attractively mottled or variegated green, white and pink, looking like flowers, with red stems and petioles. *Tropical.* p. 407, 419

BRODIAEA   *Amaryllidaceae*

coccinea (Brevoortia) (Hortus 3: Dichelostemma ida-maia) (Oregon, No. California), "Firecracker flower"; perennial with brown corm; 2-5 flattened linear leaves keeled beneath; inflorescence on slender stalk to 1 m high; pendant tubular scarlet 4 cm flowers tipped with yellow green. Rest in dry season. *Subtropic.* p. 60

BROMELIA   *Bromeliaceae*

antiacantha (fastuosa) (So. Brazil, Uruguay); formidable terrestrial rosette with vicious hooked spines on narrow leaves, inner leaves along inflorescence stalk colored orange-red; showy spikes with bracts mealy-white and lavender flowers. *Subtropic.* p. 207

balansae (Brazil, Argentina) "Pinuela"; large and vicious terrestrial rosette used for fencing; stiff green leaves with dangerous hook spines facing both directions; center turning red before bloom; flowers white, in paniculate inflorescence forming branches of small, ovoid orange-yellow fruit with pineapple flavor. Easily confused with B. pinguin, but balansae has broad sepal tips, pinguin needle-like tips and loose inflorescence. *Subtropic.* p. 207, 210

fastuosa (Brazil); robust basal rosette of numerous thick-leathery, recurving, concave leaves to 1½ m long, 4-5 cm wide, deep green with gray scurf, the margins with sharp yellowish spines; erect inflorescence 60 cm high, with numerous purple flowers and rosy-red bracts. Possibly identical with B. antiacantha. *Subtropic.* p. 206

pinguin (W. Indies, So. America), "Pinguin"; bold basal rosette of many spreading rigid, light green leaves 1½-2 m long, 3 cm wide, armed with large hooked brown spines; flowers with reddish petals, needle-like sepal tips and white-tomentose at apex, in erect mealy panicles shorter than leaves, and more loosely arranged than B. balansae. *Tropical.* p. 212

plumieri (Mexico to Ecuador); vicious plant with heavily armed stiff leaves; inflorescence club-like with pale bracts and purple flowers. *Arid-subtropic.* p. 210

serra (Bolivia, Paraguay and Brazil to Argentina); open rosette of narrow stiff, viciously spiny leaves 1 m or more long; a smaller version of B. balansae; at flowering time, the inner leaves turn bright red; the short inflorescence with white cottony hair; flowers white or purple, 5 cm long. *Subtropic.* p. 207, 210

serra variegata, known as "Heart of flame"; large showy plant but dangerously spiny, the spreading leaves grayish-green with broad ivory margins; when flowering, center turns bright red, and spike with red bract-leaves and maroon flowers. Produces orange-colored fruits. Inflorescence globose, unlike the elongate inflorescence of balansae. *Tropical.* p. 212

BROSIMUM   *Moraceae*

galactodendron, (Tropical America); "Cow-tree", "Milk-tree" or "Palo de Vaca"; large evergreen tree 30 m high with a trunk 2-2½ m dia.; the leaves leathery, oblanceolate; inflorescence curious with one female and many male flowers in a spherical head, the female sunk in center. The natives make incisions in the trunk and drink the abundant milky sap as a substitute for animal milk. *Tropical.* p. 648

BROWALLIA   *Solanaceae*

speciosa (Colombia), "Amethyst flower" or "Sapphire flower"; attractive herbaceous plant with shrubby base, and sprawling slender branches; small glossy, fresh green leaves; profusely blooming with dark purple flowers to 5 cm across, pale lilac beneath. Graceful in hanging baskets. *Subtropic.* p. 890

speciosa 'Blue Troll'; dwarf cultivar with smaller leaves, and flowers purplish blue with white eye. *Humid-subtropic.* p. 890

BROWNEA   *Leguminosae*

coccinea (Venezuela), "Scarlet flame-bean"; beautiful tropical evergreen shrub 2-3 m high with oblong leathery pinnate leaves, and very showy scarlet flowers with stamens protruding, in dense clustering heads; July-August. *Tropical.* p. 538

grandiceps (Venezuela), "Rose of Venezuela"; handsome tropical tree to 20 m high, with stout, hairy branchlets; even-pinnate, leathery leaves to 11 pairs of leaflets, reddish when young; dense flower heads nearly globose, 20-25 cm across, scarlet red. *Tropical.* p. 542, 543, 545

macrophylla (Panama, Colombia); small tree to 10 m, branches and petioles brown-hairy, pinnate leaves with 3-6 pairs of leaflets to 30 cm long; flower heads on trunks and branches 20-25 cm across, orange-scarlet. *Tropical.* p. 538, 543

BRUGMANSIA: see Datura

BRUGUIERA   *Rhizophoraceae*

conjugata (gymnorhiza), (Africa, Malaya and India to So. China and Pacific), "Oriental mangrove" or "Many-petaled mangrove"; spreading shrub with jointed branches and sending out aerial stiltroots into shallow water of tropical tidal shores and lagoons, forming dense interconnected tangles; large shiny green, thick-leathery pointed leaves on erect brown branches; pink, yellow or red flowers with 10 or more calyx lobes and short woolly petals. *Tropical.* p. 841

BRUNFELSIA   *Solanaceae*

americana (West Indies), "Lady of the night"; evergreen shrub to 2 m high, the leathery leaves oval to obovate, 8-10 cm long; exquisite flowers white, fading to pale lemon yellow with age, usually solitary, with a slender tube to 10 cm long and spreading limb to 6 cm across; very fragrant especially at night; bl. in spring and summer. *Tropical.* p. 890, 900

latifolia (Franciscea) (Trop. America), "Kiss-me-quick"; shrub 60-90 cm high, with broad, leathery, obovate leaves to 10 cm long, slightly pubescent beneath; flowers pale violet with white eye at first, changing in a day or so to white, and very fragrant, blooming freely in winter and early spring. *Tropical.* p. 900

pauciflora calycina (Franciscea) (Brazil), "Morning-noon-and-night"; handsome spreading shrub with long lanceolate, shining livid-green leaves and large 6 cm salver-shaped flowers with wavy margins, rich dark purple, blooming successively throughout the year. *Tropical.* p. 890, 900

pauciflora 'Floribunda', "Yesterday, today and tomorrow"; floriferous evergreen shrub with spreading branches, dark green, long elliptic, leathery leaves and large, 5 cm rich violet to lavender flowers with small white eye, quickly fading to white, blooming from January to July. *Tropical.* p. 890, 900

pauciflora 'Macrantha' (Franciscea), "Yesterday-and-today"; cultivar with light purple flowers very large, 4-6 cm across. *Humid-tropical.* p. 890

BRUNSVIGIA   *Amaryllidaceae*

josephinae (So. Africa: Cape Prov.); "Josephines-lily"; showy bulbous plant with strap-shaped, glaucous leaves 60-90 cm long, from bulb 15 cm thick; floral stalk 45 cm high, topped by a large umbel of long-stalked trumpet flowers 6 cm long, in vivid scarlet red, the filaments purplish, anthers white; late summer blooming. *Subtropic.* p. 54

BRUNSVIGIA: see also Amaryllis
BRYOPHYLLUM; see Kalanchoe
BUCKINGHAMIA   *Proteaceae*

celsissima (Queensland), "Ivory curl-flower"; dense evergreen shrub or tree 6-20 m high, with dark green oblanceolate leaves silvery beneath; bottle brushes of curly cream flowers 10-20 cm long. *Tropical.* p. 834

BUDDLEJA   *Loganiaceae*

cordata (Mexico); large shrub 2 m or more, with woody branches, and broad, leathery leaves covered with hairs; small pinkish flowers in large racemes. *Subtropic.* p. 616

madagascariensis (Madagascar); evergreen shrub of tall, lax habit to 6 m high, with branches downy, 12 cm lanceolate leaves dark green above, pale-woolly beneath; orange-yellow flowers in slender terminal panicles, in winter. *Subtropic.* p. 543

BULBINE   *Liliaceae*
natalensis (Natal); low but spreading succulent rosette which I found growing at the old fort at Durban; few concave, very soft-fleshy, fresh yellow-green leaves broad at the base and gradually tapering to a long point and recurving; small bright yellow flowers in a dense raceme on a long snaking wiry stalk. *Subtropic.* p. 584

BULBINELLA   *Liliaceae*
floribunda (So. Africa); perennial with fleshy roots, linear basal leaves, and tall spikes of small flowers a shining daffodil yellow or orange; resembling miniature kniphofia; dormant in summer. *Subtropic.* p. 603

BULBOPHYLLUM   *Orchidaceae*
dayanum (Burma); miniature epiphyte with egg-shaped pseudobulbs, solitary fleshy leaf to 10 cm long; flowers to 4 cm across, evil-smelling, the greenish sepals with long hairs, petals blood-red and green lip. *Humid-tropical.* p. 711
grandiflorum (New Guinea); probably the most remarkable orchid it has been my privilege to see in New Guinea; the large flowers solitary, olive-green with pale chartreuse spots and netted with brown, the sepals 10-12 cm long, strongly arching over at base with the dorsal hanging down in front hood-like, petals and lip very small and almost hidden, the lip greenish dotted with red-brown; flowering in autumn in the north, in New Guinea in May-June; short 4-angled pseudobulbs with solitary leaves, on creeping rhizome. *Humid-tropical.* p. 711
lobbii (Indonesia to Burma); epiphyte with creeping rhizome and small, single-leaved pseudobulbs; large solitary flower sepals and petals buff-yellow, marked with red, the lip golden-yellow, (May-June, Nov.). *Humid-tropical.* p. 716
makoyanum (Cirrhopetalum) (Malaya); beautiful little epiphyte with 3 cm pseudobulb and solitary leaf, the unusual 3 cm flowers 12-14 in a radiating, wheel-like umbel, pale yellow shading to deep crimson toward base, the whole inflorescence looking like a pretty daisy (winter). *Humid-tropical.* p. 713, 715

BURCHELLIA   *Rubiaceae*
bubalina (So. Africa), "Wild pomegranate"; evergreen shrub 2-3 m high, with leathery, dark green, glossy leaves; tubular-inflated flowers 2-3 cm long, coral red; as buds, they are covered with silky hairs glistening in the sun. *Subtropic.* p. 866

BUTEA   *Leguminosae*
frondosa (India, Burma), "Flame of the forest"; showy tree to 15 m, bearing masses of scarlet-red flowers 5 cm long; appearing before the leaves, these consisting of 3 leaflets. *Tropical.* p. 542
monosperma (frondosa) (India); "Flame of the forest"; tropical tree to 15 m, with crooked trunk, twigs pubescent; leaves of 3 leathery leaflets, silky beneath; flowers bright orange red 3 cm across. *Tropical.* p. 543

BUTIA   *Palmae*
bonnetii (So. Brazil), "Acorn butia"; a robust feather palm of smaller habit, trunk short and thick; very arching leaves with leaflets equally distant on the stalk; narrow inflorescence to 60 cm long; orange fruit egg-shaped 2 cm long, edible. *Tropical.* p. 766
capitata (Cocos australis) (Brazil, Uruguay, Argentina), "Jelly palm"; short, stocky, rather coarse palm, slowly reaching 5 m, with thick trunk, covered with persistent leafbases; the long pinnate bluish-gray leaves stiffly recurving, whitish underneath, spiny at base, tough pinnae often 2-3 together; orange fruit with edible pulp. *Subtropic.* p. 766, 771
yatay (Argentina, Uruguay); "Yatay palm"; stout palm to 8 m tall; trunk with leaf bases long persistent, to 45 cm thick; stiff pinnate fronds 3 m long, arching forward and recurving, blue glaucescent; petiole armed; flowers yellowish; large clusters of ovoid fruit 4 cm long, dark yellow or orange, edible. *Subtropic.* p. 766

BUXUS   *(Buxaceae)*
microphylla japonica (Japan), "California boxwood" or "Japanese little-leaf boxwood"; evergreen densely branching shrub to about 2 m high, the wiry shoots with small 1 to 2 cm obovate leathery leaves closely set, and a glossy bright green to deep green. Clipped into hedges, or globes, pyramids or other topiary shapes in containers. Tolerates the dry heat of a California patio, and alkaline soil, but not fully frost-resistant. *Warm temperate.* p. 237
sempervirens (Europe, No. Africa, Asia Minor), "Common boxwood" or "English boxwood"; winter-hardy evergreen shrub or tree 2 to 5 m high, much used for sheared hedges or topiary; of dense habit, with quadrangular branches, small leathery obovate leaves lustrous dark green, 2-3 cm long; small axillary flowers without petals. *Temperate.* p. 237
sempervirens 'Bullata'; dense winter-hardy evergreen shrub, with opposite, relatively large oval, thick-leathery leaves to 3 cm long, deep green and bullate (convex). *Temperate.* p. 237

BYBLIS   *Byblidaceae (Carnivorous)*
gigantea, (Western Australia); from swampy, sandy places of the Swan River near Perth; semi-shrubby plant 30-50 cm high, with woody rhizome, a stem with several branches with long filament-like yellow green leaves 10-20 cm long, clothed with numerous mucilage glands, secreting viscous glue to hold and eventually digest insects; large rosy or violet flowers. *Subtropic.* p. 292

CAESALPINIA   *Leguminosae*
coriaria (Aruba), "Divi-Divi"; thorny tree generally with wind-blown branches; numerous small leaflets; flowers yellow; curved 3 cm pods. *Arid-tropical.* p. 547
echinata (Trop. America), "Brazilwood"; tropical tree with prickly trunk, pinnate leaves with many rhombic leaflets; yellow flowers. *Tropical.* p. 548
gilliesii (Argentina, Uruguay), "Bird-of-paradise shrub"; straggling shrub with bipinnate leaves, the leaflets very small; terminal inflorescence with pale yellow flowers and long red stamens. *Subtropic.* p. 548
pulcherrima (Poinciana) (W. Indies and other tropics) "Dwarf poinciana" or "Pride of Barbados"; prickly, glabrous shrub with delicate bipinnate, mimosa-like leaves, and very gaudy orange-red flowers with crisped, golden-edged petals and long red stamens. *Tropical.* p. 548, 561
pulcherrima flava (Poinciana) (East Africa, widespread in Tropics), "Pride of Barbados" or "Peacock flower"; spiny, handsome-flowered bush to 5 m; with bipinnate fern-like leaves and bright yellow flowers. *Tropical.* p. 561
spinosa (Western So. America), "Tara"; attractive spiny shrub or small tree from desert regions; thorny branches with bipinnate leaves, the oval leaflets glossy green and leathery, on thorny rachis; pyramidal inflorescence with small yellow flowers turning rose, and sweetly fragrant. *Subtropic.*
vesicaria (Cuba to Brazil); smooth shrub with pinnate leaves, the leaflets leathery, broad obovate and with yellow ribs; flowers in large pyramids light yellow. *Tropical.* p. 548

CALADIUM   *Araceae*
bicolor (Trinidad, Guayana, Brazil), "Heart of Jesus"; tuberous, stocky plant with sagittate-ovate, firm leaves mostly green, red veins and scattered white spots; variable; spathe green, top white to yellow. The most imortant parent of most of our present horticultural forms. *Tropical.* p. 96
hortulanum, "Fancy-leaved Caladium"; tuberous herbs with membranous leaves mostly beautifully marked in many colors and patterns, on slender petioles; widely hybridized; the larger peltate-heartshaped leaves have C. bicolor blood; the lanceolate strap-leaved hybrids go back to C. picturatum. *Tropical.*

'A.B. Graf'; an improved 'Scarlet Pimpernelle', with more leaves, sturdy, and more crimson, with a broad margin of contrasting lemon-yellow or nile green. p. 99

'Ace of Hearts'; beautiful leaf, center area transparent-rose to crimson, moss-green toward margin; strong, blood-red ribs. p. 99

'Attala'; large colorful leaf irregularly variegated pink and dark green, with clear red main ribs. p. 99

'Bleeding Heart'; heartshaped, pointed leaf transparent rose in center, followed by zone of white marbling and green border veins boldly crimson. p. 98

'Candidum', "White caladium"; showy, white leaf traced with dark green veins and green border; delicate, yet a good keeper. Ideal for Easter; 20-30 cm foliage. p. 98, 99

'Citation'; shapely variety, the long lanceolate leaves sturdy, pink to milky white, dark green at undulate margin, a touch of rose at base. p. 99

'Cleo' (Texas Wonder); tall, sturdy variety, red in center, background green with pink marbling and spots; veins contrasting red. p. 98

'Debutante'; bright, colorful leaf regularly marbled white on green, the center and vein area rosy to red. p. 99

'Edith Mead'; sturdy plant with medium-small leaves glistening white in center and dark green toward margins, thin red veins. p. 99

'Elizabeth Dixon'; showy leaves basically deep green but more or less variegated white, more so in center where there is a flush of transparent pink; main ribs crimson. p. 99

'E.O. Orpet'; lanceolate strapleaf, almost leathery; a glossy, bright red, shading to blood-red with narow green border; low habit, and good keeper. p. 99

'Frieda Hemple'; top red variety, bushy, compact and sturdy; medium size leaves clear bright red with primary ribs scarlet, the outer edge deep green. *p. 101*

'Hortulania'; entire leaf transparent, almost shimmering, pink or rose, with narrow green border; veins scarlet; very showy. *p. 99*

'Itacapus'; early forcer with large leaves almost entirely metallic-red with a netting of green between, and dark green border. *p. 99*

'Jessie Thayer'; large leaves with unusually bold, scarlet ribs which stand out conspicuously against metallic green, which changes to bronzy-white with pink except for border. *p. 101*

'Jody'; thrifty, bushy plant with smallish arrow-shaped leaves; center of blade and ribs white, splashed red between veins, moss-green along margins. *Tropical.* *p. 99*

'John Peed'; sturdy variety with firm, waxy, warm moss-green leaves with metallic orange-red center area and scarlet-red veins; purplish stem. *p. 98*

'Keystone'; large leaf deep forest-green with white spotting; bold orange-red veins. *p. 98*

'Lady Chrys'; very showy leaves with wavy blade creamy white, the main ribs green, and with blood-red blotches, edged with a fine deep green border line. *p. 101*

'L.L. Holmes'; an improved 'Marie Moir' by Joyner with more but narrower leaves and the corrugated leaves more white, traced by blackish main ribs, and scattered red blotches. *p. 101*

'Lord Derby'; bushy plant with transparent, quilted leaf with wavy edge, delicate rose-pink with network of thin, green veins and green margin. *p. 98*

'Luella Whorton'; large pink leaf with a fine network of dark green and green edge, main ribs carmine-red. *p. 101*

'Macahyba'; large, showy leaves, the center area transparent with marbling of white to rose, scattering out into the moss-green blade reticulated with occasional yellow green; primary ribs crimson to purple. *p. 101*

'Maid of Orleans'; colorful variety deep green richly marbled with rosy-pink and some white; ribs bright red, on red stems. *p. 101*

'Marie Moir', "Blood spots"; a beautiful variety because of its simplicity, pure white with a network of green tracings, and dark green ribs, and a sprinkling of showy blood-red blotches. *p. 101*

'Miss Muffet' (Midget Princess), "Baby elephant-ear"; wonderful little miniature with intriguing leaf, sturdy, compact, nile green or chartreuse, with white primary ribs and deep blood-red blotches and heart. *p. 101*

'Mrs. Arno Nehrling'; elegant large white leaf traced with a network of fine dark green and the distinct main ribs deep crimson, border green. *p. 101*

'Pink Blush'; glossy, crinkled leaves of translucent carmine-rose, occasionally marbled with cream, the main ribs blackish crimson, and a very narrow edge of dark green; dark brown stems. *p. 101*

'Pink Festivia'; delicate translucent leaves soft pink to rose, with crimson to deep green ribs and narrow green border. *Tropical. p. 98*

'Poecile Anglais'; bushy plant with smaller leaf, crimson-red, with broad, moss-green border; veins scarlet. *p. 98*

'Porto Novo'; small-leaved, bushy variety with milk-white star in leaf middle, the center and ribs intense carmine, with carmine spots scattered in outer area. *p. 98*

'Postman Joyner'; bushy, compact variety with medium leaves light red with darker scarlet veins and a narrow margin of forest green. *p. 100*

'Pothos'; sturdy lance leaf with thick milk-white leaves marbled with dark green; tracings of purple in veins. *p. 100*

'Red Flare'; interesting leaf rosy to purplish in center, outer zone white, or greenish variegated green, ribs crimson. *p. 100*

'Red Frill'; small, sturdy strapleaf, rosy-red with main ribs scarlet fanning out toward deep green, wavy margin. *p. 100*

'Rio' ('Rio de Janeiro'); tallish variety, transparent rose, darker ribs; green spots and splashes in center, the narrow green margin dotted pink. *Tropical.* *p. 97*

'Roehrs Dawn'; a great caladium with large, dainty, quilted, transparent leaves of shimmering creamish-white contrasting with the rosy red veins and a narrow green edge. *p. 100*

'Scarlet Beauty'; bushy plant with large leaves mainly transparent rosy-red, main veins red, outer edge deep green. *p. 100*

'Scarlet Pimpernelle'; large leaves bright carmine-rose, mottled grayish green, outer area straw colored, heavy red ribs. *p. 100*

'Spangled Banner'; bushy variety with solid leaves red in center, the carmine main ribs radiating star-like into the dark green leaf which is marbled lavender-pink. *p. 100*

'The Thing'; tall, vigorous plant with large stiff leaves to 45 cm on light stalks, with ivory-white center, soft green outer area, and blotched salmon-red. *p. 100*

'Thomas Tomlinson'; compact, bushy plant with shiny, moss-green leaves transparent lacquer-red in center; green ribs. *p. 100*

'Triomphe de l'Exposition'; bushy plant with smaller leaf yellowish-green with scarlet veins shooting star-like from the red center; blue stem. *p. 100*

'W.B. Halderman', "Coloradium"; bushy variety with leaves somewhat smaller and pointed; transparent rose, with zone of white marbling and dark green border. *p. 98, 101*

'White Princess'; bushy, small growing variety with slender, membranous leaves pure white, with main veins deep green as well as green toward border. *p. 98*

'White Queen'; stunning introduction with pure white, lightly crinkly leaves contrasted by clear crimson primary veins, with a tracing of green laterals and a fine deep green edge. *p. 100*

humboldtii (argyrites) (Pará, Brazil), "Miniature caladium"; smallest and daintiest of genus; the tiny leaves light green with transparent white areas between veins. *p. 96, 98*

CALAMUS *Palmae*

ciliaris (Borneo, Java, Sumatra), "Rattan palm"; very long thin and tough stems with vivid green leaves pinnate with numerous narrow, hairy segments, climbing by leafless branches with hooked spines appearing under each leaf. *Tropical.* *p. 770*

dealbatus (Brazil), a "Rattan palm"; slender tropical fan palm with scandent stems thickly armed with brown spines, and climbing over nearby trees and shrubs, forming thickets; the leaflets spread out flat, the petioles and midrib of the leaf set with dangerous hooks. *Tropical.* *p. 770, 772*

rotang (India, Sri Lanka, Burma, New Guinea), "Rattan palm" or "Rattan cane"; tall-climbing, spiny palm with slender flexible stems, rambling over great areas; glossy-green pinnate leaves 60 to 80 cm long, with linear-lanceolate leaflets; long spiny projections form from the leaves as the palm reaches maturity. The strong, flexible stems are used as canes and rattans in commerce. *Humid-tropical.* *p. 772*

scipionum (Malaysia), "Malacca rattan"; non-climbing "Malacca canes" forming thickets; very spiny, twisting cane-stems, with very long internodes as much as 1½ m apart, making it suitable for walking-sticks and furniture use; large leathery, glossy green pinnate leaves. *Humid-tropical.* *p. 772*

CALANTHE *Orchidaceae*

'Baron Schroeder' (regnieri x vestita); well known terrestrial (1894), with deciduous pseudobulbs, and long arching spray of flowers with orchid pink petals and sepals, the lip winged and with dark purple. *Tropical.* *p. 714*

discolor sieboldii (Japan); terrestrial with 2-3 lanceolate leaves to 30 cm long, strongly striate; erect raceme with 6-12 flowers, sepals and petals wine-red, lip white, diffused with pink in the species; clear yellow flowers in var. sieboldii. *Subtropic.* *p. 712*

furcata (Philippines); evergreen terrestrial, with plaited green leaves in a rosette; the flowers in dense raceme on 1 m, erect spike, white turning blue; the lip resembles a dancing girl, (June-Aug.). *Tropical.* *p. 714*

silvatica (Madagascar); evergreen species with folded leaves to 60 cm long; the erect stalk with flowers 3 cm across, dark purple, the lip with orange warts. *Tropical.* *p. 714*

'William Murray' (vestita x williamsii); terrestrial hybrid with graceful spray of large flowers from alongside the pseudobulb, petals white and lip carmine-red with darker throat, (winter fl.). *Tropical.* *p. 712*

CALATHEA *Marantaceae*

aemula (Brazil); sturdy plant with oblique-oval 15 cm leaf on short petiole, glossy dark grayish-green with broad flame-like irregular band of greenish yellow through the center; reverse gray with purple tinge and soft pubescent. *Tropical.* *p. 632*

argyraea (Brazil), "Silver calathea"; sturdy plant with foliage borne on stiff-erect, winged stalks; leaves oblanceolate, unequal-sided, glossy, the feathered vein areas grass-green, silvery-gray between; reverse wine-red; from the collection at Paris Museum. *Tropical.* *p. 640*

aurantiaca (Brazil); sturdy, showy plant with large leathery, oblique-ovate leaves glossy gray-green to yellowish, with dark green herringbone feathering. *Tropical.* *p. 631*

bachemiana (trifasciata) (Brazil); small plant with slender stalks and oblique, narrow tapering, leathery leaves pale greenish-gray, with dark green, almond-shaped blotches alternately fanning out from midrib, and a narrow border of the same deep color; green beneath. *Tropical.* *p. 639*

bella (Brazil); very ornamental plant with glossy leathery, broad ovate leaves on wiry petioles, the coloring mainly gray with a feather design of dark green radiating from the midrib. *Humid-tropical.* *p. 634*

carlina (Brazil); species resembling kegeliana but the leathery leaves are shorter and broader, oblique ovate 18 cm by 12 cm, glossy gray with dense net of yellow reticulations or cross veinlets, and a fish bone design of grass green and regular thin green veins between as well as a narrow green border; grayish green beneath reticulated green. *Tropical.* p. 631, 632

clossonii (Maranta) (Brazil); slender bushy plant with thin-leathery narrow-lanceolate leaves, the lightly quilted surface yellow-green with short almond-shaped medium-green blotches on each side of midrib. *Tropical.* p. 632

concinna (Acre, W. Brazil); bushy plant with very pretty, glossy, ovate leaves larger than leopardina, the yellow-green blade marked with not as bold but more numerous, usually opposite, narrower bands of dark green, tapering to a narrow line and reaching to the marginal deep green border, dark green lateral veins between; purple reverse. *Tropical.* p. 632

crocata; clustering tropical species with leathery, ovate leaves dark green with gray feather design; inflorescence a globular spike with orange-salmon concave bracts. *Humid-tropical.* p. 631

grandiflora (Brazil); robust plant to 75 cm high, with large 30 cm oblique-oblong leaves shiny grass-green, closely corrugated by its prominent veins, on winged stalks, brown at base of leaf; grayish beneath; flowers yellow. *Tropical.* p. 634

insignis Bull (Brazil), "Rattlesnake plant"; very pretty, bushy species with narrow, tapering, almost linear stiffly erect foliage wavy at the margins, yellow-green with lateral ovals alternately large and small of dark green; underside a showy maroon red; while slow growing at first, I have seen it measure 60 cm or more as growing in Brazil. *Humid-tropical.* p. 631, 632, 634

kegeliana (Brazil); robust, durable plant to 45 cm high with wiry stalks bearing heavy, thick-leathery oblique-ovate leaves having cordate base, laxly pendant; the surface silvery gray attractively patterned with lateral veins and lance-shaped bands of yellowish to dark green; grayish green beneath with midrib showing yellow. *Tropical.* p. 632

leonii (Ecuador); showy plant with elongate ovate leaves on long-stalked petioles, the blade velvety moss green, except for bands of greenish silver along midrib and lateral veins; reverse red; flowers white within green bracts. *Humid-tropical.* p. 634

leopardina (Brazil); attractive species to 40 cm high with beautiful oblique-lanceshaped, waxy leaves on slender stalks, the blade is nile-green marked with bold dark green triangles, alternately on both sides of the midrib, and not reaching the margin; underside tinted bronzy purple. *Tropical.* p. 633

lietzei (Brazil); pretty plant with narrow-ovate, thin-leathery leaves deep green with sharply contrasting, alternate lateral bands in silvery green to margin, and purple beneath; sending up erect runners bearing young plantlets. *Tropical.* p. 632

lindeniana (Brazil); vigorous plant to 1 m high, broad oval leaves deep green, with a feathery, pale olive zone either side of midrib and near the border, a darker zone between, purple underneath except near midrib. *Humid-tropical.* p. 632

loeseneri; tropical ornamental plant of bushy habit; long lanceolate leaves on stiff petioles, the blades corrugated and olive green except for gray feathering along midrib; long stalked inflorescence with showy heads of pointed pinkish bracts. *Humid-tropical.* p. 635

louisae (albertii) (Brazil); tufted plant somewhat similar to lietzei but makes no runners, leaves are larger and more broadly ovate, dark green with light yellowish-green, irregular feathering along the midrib; purple underneath; flowers white out of pale green bract head. *Tropical.* p. 633

makoyana (Minas Geraes), "Peacock plant"; bushy plant stunningly beautiful on both sides of its oval leaves, the surface with a feathery design of opaque, olive green lines and ovals alternately short and long in a translucent field of pale yellow-green; this pattern of lines and ovals is purplish-red beneath. *Humid-Tropical.* p. 632, 633, 635, 639

metallica 'Undulata' (Brazil); large plant with oblique-oblong pointed leaves 40 cm x 16 cm, corrugated undulate throughout the blade to the margins, the surface iridescent satiny-silky forest green with paler midrib, royal purple beneath. *Tropical.* p. 632

micans (Brazil, Peru); perhaps the tiniest of Calatheas; narrow, pointed leaves, shining medium-green with a flame-shaped silvery-white center band above, gray-green beneath. *Tropical.* p. 632

musaica (Espirito Santo); a beautiful species which I have found in Brazil; of low habit, the broad arrowshaped leaves are leathery, glossy yellow green with a network of deep-green bars at right angles to the likewise dark green lateral veins; underside green. *Humid-tropical.* p. 636, 639

ornata 'Roseo-lineata'; long narrow-ovate, unequal-sided, metallic olive-green leaves, in the juvenile stage nicely marked with closely-set pairs of rosy-red lateral stripes, later turning white in older leaves; purple beneath. *Tropical.* p. 636

ornata 'Sanderiana'; showy form with erect, leathery, rather broader leaves a glossy dark olive green with boldly contrasting pink or white stripes curving across the blade in groups of two or more; reverse rich purple-red. *Tropical.* p. 636, 639

picturata 'Argentea' (Venezuela); dwarf plant similar in habit to vandenheckei but with short-stalked leaves almost entirely shining silver except for a border of dark green; wine-red beneath. Cv. 'Wendlinger' may be the same plant. *Tropical.* p. 636

picturata vandenheckei (Amazonas, Colombia); juvenile form of metallica (picturata); dwarf plant with oblique, glossy leathery leaves, deep olive to blackish green, with feathered silver white band in center and two-thirds toward margin; wine-red beneath. *Tropical.* p. 636

princeps (Amazonas); large oblong, showy leaves of yellow-green, and a feathery center band of blackish-green, with dark veins running to the dark border, violet-purple beneath; I was surprised to see it grow in a rock-hard floor of the primeval forest north of Manaus in Brazil. According to Kyburz, princeps is the juvenile form of altissima, which grows to 2 m tall and becomes plain green. *Tropical.* p. 631, 633, 636

roseo-picta (Brazil); low plant with short-stalked, large 22 cm rounded, unequal-sided leaves dark olive-green above with red midrib and narrow zone of bright red near margin, changing to silver pink when old; underside purple. *Tropical.* p. 631, 636

rotundifolia 'Fasciata' (Bahia); showy variety having low habit with large thick leathery leaves 22 cm long and almost as wide, on thick brownish stalks; blade glossy green with broad lateral bands of silvery gray, transversed throughout by dark lines; underside gray tinged purple. *Tropical.* p. 633, 636, 640

sp. 'Burle Marx'; (Brazil); grown by Fantastic Gardens, Miami; erect growing, sturdy plant with oblique broad-ovate, thin-leathery leaves 20 cm long, shining dark green with lighter feathering; reverse in part purple. *Tropical.* p. 631

stromata (Brazil); attractive low bushy plant of stiffish, thin-leathery foliage, leaves broad oblong, 10-12 cm long, with squared-off apex, silver-gray with dark-green herring-bone pattern; reverse deep purple. *Tropical.* p. 631, 640

tigrina hort.; exquisite foliage plant of compact habit, with broad-oval leaves deep olive green, the vein areas silvery gray, underside and petioles cardinal red. *Humid-tropical.* p. 631

'Tuxtla' (So. Mexico); nice species with glossy, obliquely broad-ovate leaves, medium green with a silvery zone along the midrib, and a second silver zone just inside the margin; gray-green underneath. *Tropical.* p. 636

undulata (Perú); tufted small plant 20 cm high, with thin-leathery, oblique-ovate leaves on short petioles with waxy undulate surface following lateral veins to margins, metallic dark green with bluish sheen when young, and a jagged band of greenish silver along midrib, purple beneath over green; flowers white with green bracts margined and dotted white. *Tropical.* p. 636

vandenheckei: see picturata

veitchiana (Ecuador, Perú), "Peacock plant"; strikingly beautiful plant from the Jivaro country with large, stiff-leathery, glossy leaves to 30 cm long, obliquely broad-ovate, in four different shades of green with a peacock-feather design outlined in yellow green, encircling brownish-green halfmoons which adjoin the pale bluish green feathered center zone; marginal area bright green; the peacock-feather design outlined in red over bluish green underneath also. *Humid-tropical.* p. 636, 639

warscewiczii (Costa Rica); heavy plant to 75 cm high, with short-stalked, oblong leaves undulate at margins, velvety deep green except for yellow-green midrib feathering into pale green lateral veins; underside wine-red. Inflorescence a short spike with tubular, leathery white bracts, white flowers. *Tropical.* p. 635

wiotii (Phrynium brachystachyum) (Brazil); dwarf plant with thin-leathery short-ovate leaves 10 cm long, glossy light green with elevated oval blotches of dark green on either side of midrib; reverse mostly gray-green tinted purple. *Tropical.* p. 632

zebrina (Rio, Sao Paulo), "Zebra plant"; bold, vigorous plant to 1 m high, with magnificent, thin leaves deep velvety green, the midrib and lateral veins pale or yellow-green, purplish beneath. *Tropical.* p. 635, 636, 639, 640

CALATHEA: see also Maranta, Ctenanthe, Stromanthe

CALCEOLARIA Scrophulariaceae

herbeohybrida 'Grandiflora', "Pocketbook plant"; herbaceous cultivar derived principally from C. crenatiflora which grows in the cool Andes of Chile and north, developed into beautiful potplants with thin, fresh-green leaves and clusters of showy, membranous flowers distinguished by a large inflated lower lip resembling a handbag, in shades of yellow or red, often brilliantly spotted or tigered orange-red to maroon, blooming in spring. *Humid-subtropic.* p. 882, 884

herbeohybrida 'Multiflora nana'; a herbaceous crenatiflora form of dwarf, bushy habit, ideal as a potplant, with thin ovate, toothed and quilted leaves, and clusters of numerous, smallish pouch-like flowers, most often yellow or orange or red, usually spotted or tigered with crimson. *Humid-subtropic.* p. 882, 883, 884

integrifolia (rugosa) (Chile), "Chilean pouch flower"; shrubby perennial of bushy habit, with woody, hairy stems, small, wrinkled, toothed leaves, and clusters with masses of small 1 cm pouch-flowers, yellow to red-brown. *Humid-subtropic.* p. 884

pavonii (Perú); shrubby perennial we found on the slopes to Machu Picchu, with rough ovate leaves, and 3-4 cm flowers rich yellow, with inflated pouch. *Humid-subtropic.* p. 884

CALENDULA *(Compositae)*
officinalis (So. Europe), "Pot-marigold"; popular annual herb 30-40 cm high originating in So. Europe; with fleshy entire, brittle and clammy leaves not scented, on sticky stalks; the large double flowers 8 cm or more across, closing at night, in apricot, also shades of orange, lemon or golden yellow in other named varieties. Excellent, long-lasting cutflower developing largest blooms if disbudded and grown cool. Their normal outdoor season is from June to frost, however, with a cool greenhouse or wintergarden, pot-marigolds may be had in bloom in the dark of winter; for Christmas from August sowings. p. 308

CALICOTOME *Leguminosae*
villosa (E. Mediterranean), "Thorny broom"; spiny broom-like shrub 1-2 m high, with numerous small yellow flowers; small leaves of 3 leaflets, silvery-hairy below. *Arid-subtropic.* p. 548

CALLA *Araceae*
palustris (Atlantic No. America, No. Europe, Lapland, Siberia), the "Water-arum", or "Wild calla"; bog-aquatic with creeping or floating stems or rhizomes with heart-shaped 15 cm leaves; the inflorescence with 5 cm spathe green outside, white inside and thread-like stamens; forming clusters of red berries. *Temperate.* p. 102

CALLA: see also ZANTEDESCHIA

CALLIANDRA *Leguminosae*
emarginata ('Minima' in hort.) (Mexico, Guatemala), "Miniature powder-puff"; small shrub with leaves compound of several uneven leaflets; small inflorescence a globular head of red stamens. *Subtropic.* p. 548

haematocephala (Inga pulcherrima) (So. Brazil), "Pink powder puff"; rambling shrub with branched pinnate, silky leaves and powder-puff-like balls of conspicuous dark. crimson stamens. *Subtropic.* p. 547

"inaequilatera" in hort. (bot. prob. haematocephala) (Ecuador, Bolivia), "Red powder puff"; foliage pinnate; globose inflorescence larger and more dense than tweedii, with showy, long silky bright red stamens. *Subtropic.* p. 547

'Minima' hort. see emarginata

portoricensis (caracasana) (Central America); shrub to 2½ m, shoots grooved, downy; leaves bipinnate; flowers white in globose heads, the long stamens spreading like a powder puff 5 cm dia. *Tropical.* p. 548

surinamensis (Surinam); low, spreading shrub or woody climber resembling an inverted umbrella, with pinnate leaves always in forked pairs, smothered with pretty, pink brush-like flower heads composed of numerous long, thread-like, silky stamens reddish in the upper part and white below. *Tropical.* p. 548

tweedii (Inga pulcherrima) (S. Brazil), "Mexican flame-bush"; shrub with bipinnate leaves, the numerous leaflets almost linear, 4-6 cm long, silky when young; the long stamens red-purple; somewhat cold-resistant. *Subtropic.* p. 548

CALLIOPSIS see Coreopsis

CALLISIA *Commelinaceae*
repens (Tradescantia minima in hort.) (Trop. America), mini "Turtle vine"; new in commercial cultivation (1977); vigorous grower with long pendant slender vines, closely shingled with waxy, fresh-green, ovate leaves 1-2 cm long; insignificant small axillary flowers with 3 sepals and 3 tiny white petals, and prominent stamens, toward end of stems. Popular in hanging baskets in California. *Tropical.* p. 307

CALLISTEMON *Myrtaceae*
citrinus (lanceolatus) (Metrosideros floribunda) (New South Wales) "Bottle-brush"; tree-like shrub of sparry, bare habit, with hard, heavy wood, sun-loving and drought resistant; the silky twigs with rigid, long-linear 8 cm leaves and cylindrical flower spikes with masses of brilliant crimson brush-like stamens with dark yellow anthers, out of a grayish, felted calyx. *Subtropic.* p. 675, 676, 688

montanus (Western Australia); woody shrub of sparry habit; short, dark green, sharp-pointed sessile leaves; the showy inflorescence axillary. a dense bottle brush of carmine-red filaments; photographed near Perth, W.A. *Subtropic.* p. 676

phoeniceus (W. Australia), "Fiery bottle-brush"; bushy shrub to 2½ m; grayish, thick and smooth, narrow oblanceolate leaves to 10 cm long; inflorescence in rich scarlet bottle brushes with stamen bundles tipped dark red. *Subtropic.* p. 676

rigidus (New South Wales), "Stiff bottle brush"; erect, sparse, rigid shrub or small tree to 6 m; leaves sharp-pointed, gray green or purplish, red flower brushes to 10 cm long; very drought tolerant. *Subtropic.* p. 676

viminalis (New South Wales), "Weeping bottle brush"; shrub or small tree with pendulous branches, 6 to 10 m high; leaves narrow linear, light green 15 cm long; dense spikes of scarlet red stamens. *Subtropic.* p. 676

CALLISTEPHUS *Compositae*
chinensis (China), "China aster"; annual herbaceous plant with corrugated deeply toothed leaves; solitary showy flower heads 12 cm across, with ray petals usually light purple, disk yellow. *Subtropic.* p. 314

chinensis 'Dwarf Double'; a very dwarf "Aster" suitable for pot culture, with fully double flowers 5-8 cm dia., in various colors pink, rose, red to blue. *Subtropic.* p. 314

CALLITRIS *Pinaceae (Coniferae)*
arenosa (Australia and Tasmania), "Cypress pine"; shapely flask-like pyramid, dense branches, dark green. *Subtropic.* p. 342

CALLOPSIS *Araceae*
volkensis (Tanzania), "Miniature calla"; short-branching little plant with underground rhizome, resembling a calla-lily, but only 10 cm high, from Mt. Msasa near the Sigi river; cute, cupped spathe waxy-white, with an attractive yellow spadix; quilted, leathery, heart-shaped leaves; winter flowering. *Tropical.* p. 128

CALOCEPHALUS *Compositae*
brownii (Australia), the "Cushion-bush"; rigid, white-woolly dwarf shrub about 30 cm high, with tiny alternate, linear silvery-gray leaves, and round terminal clusters of yellow composite tubular flowers. *Subtropic.* p. 310

CALODENDRUM *Rutaceae*
capense (So. Africa to Rhodesia), "Cape chestnut"; tall, semi-evergreen tree to 18 m, with oval 15 cm, ruffled leaves and very conspicuous when it bears its large clusters of dainty pinky-mauve flowers, the five slender and recurving petals marked with crimson, measuring 8-10 cm across; of 10 pale pink stamens, five resemble thin petals and are spotted with red; the brown fruits are knobby and split to release the seed. *Subtropic.* p. 873

CALONYCTION *Convolvulaceae*
aculeatum (American and other Tropics), the legendary tropical "Moon-flower"; perennial twiner related to morning-glories (Ipomoea), but blooming at night instead of in sunshine; stems more or less prickly, with milky juice, climbing to 5 m high; cordate, smooth leaves 15 cm long, and large salver-shaped flowers 12-15 cm wide, pure white, and fragrant, opening toward evening and closing in the morning. *Tropical.* p. 360

CALOPHYLLUM *Guttiferae*
inophyllum (native on shores of the Indian and Western Pacific Oceans), the "Beauty leaf", "Kamani" or "Alexandrian laurel"; handsome, low-branching crooked or leaning tree such as the one I photographed on Moorea in French Polynesia; to 20 m high, with rough gray bark, and shiny green, leathery, oblong leaves 8-20 cm long and with yellow midrib; the white flowers suggest orange blossoms and are very fragrant; the 3 cm green fruit in pendulous clusters; thin leathery skin covers a bony shell, containing an oily kernel yielding "dilo oil", used medicinally and for lights. *Tropical.* p. 520, 521

CALOSTEMMA *Amaryllidaceae*
purpureum (New So. Wales to So. Australia); bulbous plant resembling a small Belladonna lily, with several strap-shaped leaves which die down annually; the bell-shaped flowers 2 cm long, purplish crimson, in a cluster topping a fleshy stalk 30-50 cm tall. *Subtropic.* p. 60

CALOTHAMNUS *Myrtaceae*
chrysantheros (W. Australia), "Craw flower"; dense shrub to 1½ m, with thick brown stems often corky; long pencil-like thin dark green leaves 5-10 cm long; the inflorescence an axillary group of stout crimson stamen bundles feathering at apex into slender filaments; seen at King's Park, Perth. *Subtropic.* p. 676, 678

coccineus (W. Australia), "One-sided bottle brush"; interesting shrub with thick branches, set with leathery, linear leaves; the one-sided inflorescence of bundles of united stamens crimson red at base, and feathery toward apex. *Subtropic.* p. 676

CALOTROPIS *Asclepiadaceae*
gigantea (Iran, Tibet, India, China), the "Crown plant"; lush shrub with white glaucous stem containing milky juice, large obovate leaves 15-20 cm long covered with white down, and with pale veins; axillary inflorescence clusters of star-like flowers lavender to purple, with a crown of 5 narrow, fleshy scales. Sacred to Hanuman, the Monkey-god, in India. *Arid-tropical.* p. 146, 543

procera (India, Syria, No. Africa to Morocco), "Felt plant"; handsome hairy shrub to 5 m high, with large opposite, almost sessile leaves 15 cm or more long, white-woolly and with prominent pale ribs; the fragrant crown-like flowers white to lavender with purple spot at base, in clusters. Fiber from the stems is used for rope or fishing lines. The milky juice is poisonous and may irritate the skin. *Arid-tropical.* p. 146

CALVOA *Melastomaceae*
orientalis (W. Africa); small tropical herbaceous plant, the branches with thickened nodes; three-nerved ovate, ornamental leaves, the surface rough-textured, metallic bronze with silver-pink spots; 2 cm rose-purple flowers. *Tropical.* p. 644

CALYCOPHYLLUM *Rubiaceae*
spruceanum (Perú, Bolivia, Brazil), "Pan-mulato"; interesting tall tree with greenish bark polished like a candle, to 25 m high; oblong leaves to 18 cm long; small white flowers in terminal clusters. The wood is useful as timber. *Tropical.* p. 863, 870

CALYMANTHIUM *Cactaceae*
substerile (Northern Perú); tree-like with few-ribbed stems and long straight spines; brown-red flowers. *Arid-tropical.* p. 248

CAMASSIA *Liliaceae*
esculenta (Pennsylvania to Georgia and Texas); bulbous herb with linear basal leaves, and floral stalks to 60 cm high, bearing starry flowers in shades from light to deep blue. *Warm temperate.* p. 602

CAMELLIA *Theaceae*
japonica (Mountains of Japan and Korea), the "Common camellia"; a fine ornamental evergreen tree to 12 m high, well-shaped with woody branches and dark green, glossy, leathery, ovate leaves to 10 cm long, finely toothed; axillary flowers to 12 cm across, variably single to double, white, pink, red, or variegated. *Subtropic.* p. 904

japonica 'Alba plena'; fine cultivar; evergreen tree to 10 m high, with rather erect, woody branches, handsome, alternate, leathery, ovate, slender-pointed leaves deep shining green, with finely toothed margins; large round, sessile flowers of the formal type, snow-white, double, early blooming and long lasting, Oct.-Jan. *Subtropic.* p. 904

japonica 'Colonel Firey' (C.M. Hovey); compact grower with twisted, narrow, dark green leaves, and perfectly formal, very double, large flowers of brilliant crimson, blooming midseason to late. *Subtropic.* p. 904, 905

japonica 'Covina'; semi-double to rose-form, rose-red flowers, midseason to late; compact grower. *Subtropic.* p. 904

japonica 'Daikagura'; an early variety from California, slow-growing, of open, willowy habit, but blooming freely when quite young; dark green foliage and large, peony-type flowers rosy-red splashed with white, Oct. through December. *Subtropic. p. 904, 905*

japonica 'Debutante' (Sara Hastie); originating in South Carolina, this is a shapely, fast-growing beauty, ideal pot plant, early and free-flowering, with exquisite, full peony-form, firm, small 4-5 cm flowers of clear rose-pink, from October to January and later. *Subtropic.* p. 904, 905

japonica 'Delfosse'; medium-large rose-form double, glowing red. *Subtropic.* p. 904

japonica 'Elegans' (chandleri); an old variety; wide-spreading tree with long, glossy foliage and very large flowers of the peony type, variegated cherry-red with white, borne in profusion in late January. *Subtropic.* p. 905

japonica 'Jordan's Pride' (Herma); free-blooming, dependable variety with large formal semi-double 8 to 10 cm flowers of soft shell pink, irregularly streaked deep pink and bordered white, occasionally altogether pink (midseason to late; January to March). *Subtropic.* p. 905

japonica 'Mad.LeBois'; medium large, rose-form, double, salmon-red; vigorous, compact, upright growth. *Subtropic.* p. 904

japonica 'Noblissima'; early blooming pure white peony-form with yellow stamens; fresh green foliage; tall upright growth. *Subtropic.* p. 904

japonica 'Pearl Maxwell'; large formal, double flowers soft shell pink; looks like a glorified 'Pink Perfection' (midseason to late). *Subtropic.* p. 905

japonica 'Pink Perfection'; popular variety of symmetrical, vigorous growth; small, well-formed double, rather flat flowers of delicate shell-pink, free flowering from November-April. Earlier name: 'Frau Minna Seidel.' *Subtropic.* p. 905

japonica 'Princesse Baciocchi'; a late-flowering, tall-growing beauty from Savannah, with full double flowers of the rose type, dark red marbled with white; blooms freely from February into May. *Subtropic.* p. 904

japonica 'Purity'; well-known variety of slender habit, free-blooming while quite young, with glossy, pointed leaves, and exquisite white, double flowers of the formal type, of porcelain texture and long lasting, Nov.-April. *Subtropic.* p. 904, 905

japonica 'William S. Hastie'; outstanding variety of slender habit with long pointed, dull green foliage, and freely bearing large, perfect, rose-form flowers of brilliant crimson, from February-April. *Subtropic.* p. 904, 905

reticulata (China); "Temple flower"; sparry shrub to 6 m high, with scattered rigid foliage having prominent veins; large single flowers 8 to 15 cm across, vivid rosy-red, with wavy petals, blooming in February. *Warm temperate.* p. 906

sasanqua (China, Japan), "Sasanqua camellia"; evergreen shrub of loose habit and with branches pubescent when young; small, thin-leathery leaves elliptic and toothed, shining dark green and hairy on the midrib above; very floriferous with small, 4-5 cm single, slightly fragrant blossoms white in the type, with 5 more petals and with yellow stamens, but running into many forms semi-double, white, pink to cherry-red; more hardy than japonica, but the flowers shed easily; Oct.-April. Winter-hardy in So. N.J. (Zone 7). *Warm temperate.* p. 904, 905

sasanqua 'Yuletide'; striking flowering plant for the winter season; erect bushy plant with dark green small foliage finely crenate; brilliant fiery red to maroon single blooms shaped like cups, in the center a nest of bright yellow stamens in charming contrast; flowering from autumn on into the Christmas season. *Warm temperate.* p. 906

sinensis (Thea) (China, India), "Tea plant"; evergreen, tree-like shrub to 10 m high, with alternate waxy, thin-leathery, elliptic leaves finely serrate, and small, white, fragrant flowers; makes a handsome plant for tubs. Where I have seen tea cultivated in Sikkim, Ceylon, Java or Japan, the bushes are kept low for women to pick each tender tip every 3 weeks, and gentle heating, rolling and fermenting produces the aromatic tea of commerce; fruit a woody capsule. *Subtropic.* p. 904, 905

CAMOENSIA *Leguminosae*
maxima (Angola); shrubby climber; shiny leaflets obovate pointed; flowers pale cream, edges of petals frilled; standard projecting about 10 cm. *Tropical.* p. 544

CAMPANULA *Campanulaceae*
isophylla (Italy); slender trailer with thin stems and small cordate leaves, literally covered by star-shaped, pale violet flowers 3 to 4 cm dia. *Subtropic.* p. 287

isophylla alba (Italy), "Star of Bethlehem"; dainty trailer for hanging basket; small green ovate toothed leaves on thin stems and white star-like, saucer-shaped flowers. *Subtropic.* p. 287

isophylla mayii, "Italian bellflower"; variety with larger, heavier, white-hairy leaves, and pale blue flowers larger than isophylla. *Subtropic.* p. 287

medium (So. Europe, France), "Canterbury bells"; hairy biennial with tall branched stem bearing numerous large bell-shaped flowers violet-blue. *Warm temperate.* p. 287

medium 'Dean hybrid'; a hardy perennial with showy bell-shaped flowers of exceptional size, some measuring to nearly 12 cm, in shades from pale lavender to dark blue, and pink to carmine-rose. *Warm temperate.* p. 287

CAMPELIA *(Commelinaceae)*
zanonia 'Mexican Flag', (long known as Dichorisandra albo-lineata), a striking cultivar of the green species (Mexico to Brazil); attractive and vigorous foliage plant, with erect, thick-fleshy stalk to 90 cm high, and large thin-fleshy, lance-shaped, light green leaves 20-30 cm long in a rosette, striped white lengthwise throughout center and with broad, cream-white marginal bands edged red; flowers purple and white, in dense cluster, from leaf axils. *Tropical.* p. 306

CAMPSIS *Bignoniaceae*
grandiflora (chinensis) (China, Japan), "Chinese trumpet-creeper"; climbing by aerial rootlets, related to Bignonia but without tendrils and with opposite leaves, pinnate, deciduous; the 7-9 ovate leaflets toothed; terminal clusters of large funnel-shaped orange and scarlet flowers 8 cm across. *Subtropic.* p. 184

radicans (Pennsylvania to Florida and Texas), "Trumpet-vine"; high climber to 12 m, by means of aerial rootlets; pinnate leaves with 9-11 oval leaflets partially pubescent; trumpet-shaped flowers in terminal clusters, orange with scarlet limb, 5 cm across. *Warm temperate.* p. 184, 191

CAMPTOTHECA

CAMPTOTHECA *Nyassaceae*
  acuminata (China); quick-growing deciduous tree to 25 m, with ovate, glossy green, corrugated leaves to 15 cm long; small flowers with long white stamens, in clusters, developing into the curious pale yellow winged fruit. *Warm temperate.* p. 673

CANANGA *Annonaceae*
  odorata (Burma to Malaya and Australia), "Ylang-ylang" or "Perfume tree"; charming tropical tree with crooked trunk, to 25 m tall, and pendant brittle branches, oblong pointed, soft-leathery foliage 12-20 cm long; large axillary clusters of peculiar greenish-yellow flowers with long lanceolate drooping, twisted petals of soft-leathery texture 5-10 cm long, and intensely fragrant, strongest in early morning. The flowers are worn for their perfume or used in leis as in Tonga; and distilled for their fragrant essential oil in perfume. *Tropical.* p. 72

CANARINA *Campanulaceae*
  campanulata (Canary Islands), "Canary bellflower"; beautiful herbaceous perennial arising from a tuber, the branches with ovate, glaucous leaves irregularly toothed; open bell-shaped flowers 5 cm long, yellowish purple or orange with red lines. *Subtropic.* p. 287

CANARIUM *Burseraceae*
  vulgare (commune) (Malaysia, Java), "Java almond"; large buttressed tree with gray bark; leaves 30 cm or more long, have 3-5 pairs of oval leaflets and one at tip; small yellowish-white flowers with separate sexes; the fruits are ovoid, somewhat 3-sided blue-black nuts each with one kernel which is edible raw or roasted with almond flavor. *Tropical.* p. 237, 304, 474

CANAVALLIA *Leguminosae*
  ensiformis (Tropics), "Jack-bean" or "Chickasaw Lima-bean"; bush or climber; leaflets ovate, 12-20 cm long, strongly veined, dull green; flower small, light purple, keel curved; long pendant green beans; ripe seeds used as substitute for coffee. *Tropical.* p. 544

CANISTRUM *Bromeliaceae*
  fosterianum (Brazil: Bahia); tubular rosette partially covered with scurfy scales, and heavily mottled with chocolate brown, margins with tiny spines; long slender stalk with clustered inflorescence of red bracts and white flower petals. *Tropical.* p. 211
  lindenii (Brazil); large epiphytic open rosette to 1 m across, with broad apple-green leaves mottled dark green, the margins with teeth; the dense inflorescence of pure white flowers charmingly rests deeply in a nest of the surrounding inner leaves, turned creamy-white at blooming time. *Tropical.* p. 210, 211

CANNA *Cannaceae*
  generalis 'Brilliant'; green-leaved, with large cluster of scarlet red flowers to 60 cm high, free blooming. *Tropical.* p. 288
  generalis 'City of Portland'; luxuriant, compact hybrid with green foliage, and heavy rose-pink flowers. *Tropical.* p. 289
  generalis 'Cleopatra'; a fancy cultivar from Texas, with flower heads 20 to 25 cm across, huge yellow petals with contrasting red spots and others solid red; foliage green with purple streaks and also leaves solid purple on same spike. *Tropical.* p. 289
  generalis 'Confetti'; a Dwarf Pfitzer hybrid of compact habit, green foliage, and huge clusters of deep yellow flowers covered with red dots. *Tropical.* p. 289
  generalis 'Felix Ragout'; robust cultivar with long green foliage; clusters of large, pure yellow flowers. *Tropical.* p. 289
  generalis 'King Humbert', "Bronze garden canna"; striking perennial tropical herb with thick branching rootstocks and large bronze-red leaves, with a showy truss of large orange-red flowers; to 1 m high. *Tropical.* p. 288
  generalis 'Lucifer'; spectacular cultivar of dwarf habit to 60 cm high, with glossy deep green foliage, and from the center a stout stalk bearing a cluster of numerous bicolored flowers Chinese-red, with bright yellow along margins; ideal for growing in planters (Park Seed, S.C.). *Tropical.* p. 289
  generalis 'President', an old "Garden canna"; variety with large green leaves and a terminal cluster of large and showy scarlet flowers; to 1 m high. *Tropical.* p. 289
  generalis 'Red King Humbert'; striking perennial tropical herb with thick branching rootstocks and large bronze leaves, and large scarlet red flowers, 1 m high. *Tropical.* p. 288
  generalis 'Richard Wallace'; vigorous giant hybrid with fresh green foliage, and clusters of large yellow flowers, covered with red spots. *Tropical.* p. 289
  generalis 'Rigoletto'; "Grand opera" hybrid with green foliage and large canary-yellow flower, becoming darker. *Tropical.* p. 288
  generalis 'Striatus'; foliage green and beautifully striped with cream in feather fashion and midrib; margins red; flowers orange-red. *Tropical.* p. 289
  indica 'Moorea' (Polynesia), "Indian shot"; slender species with bamboo-like canes, long pointed glaucous green leaves, and orange-scarlet flowers with few large petals. *Tropical.* p. 288
  iridiflora (Perú) (photo possibly humilis); to 3 m tall, lanceolate leaves ½-1 m long, woolly beneath at first; flowers rose, in racemes, the nodding corolla 6-12 cm long. *Tropical.* p. 288
  limbata (Brazil); pretty species to 1 m high, with lanceolate, pointed leaves glossy grass green, 30 cm long, on slender canes; flowers in branched inflorescence, green sepals and lanceolate petals red with throat and base yellow. *Tropical.* p. 288

CANTUA *Polemoniaceae*
  buxifolia (Perú, Bolivia, N. Chile), "Sacred flower of the Incas"; beautiful flowering, dense bush which I saw growing 2½ m high in Puna climate at 4,000 m in the Cordilleras of Bolivia, the woody shrubs with small leaves; colorful, pendant, 8 cm tubular flowers, pale green calyx, yellow tube striped red, the lobes inside rosy-red, outside crimson. *Subtropic.* p. 825, 826

CAPSICUM *Solanaceae*
  annuum conoides 'Fiesta', an ornamental "Christmas pepper"; small tropical shrub originally from South America, but grown as an herbaceous annual pot plant; much branching, the twigs dense with narrow lanceolate, fresh green leaves, and bearing a multitude of small white flowers; followed by an abundance of slender, waxy, tapered miniature peppers, first cream then scarlet red, 5-8 cm long, beginning in September, to Christmas. *Tropical.* p. 894
  annuum conoides 'Red Chile'; actually an edible vegetable with its sweet-spicy, large conical, 5 cm fruit, but also an attractive annual potplant with white star-like flowers and fruits beautifully lacquered cardinal red, born stiffly above the rich green, scattered foliage on few, wiry branches. *Tropical.* p. 900
  annuum grossum, the "Bell-pepper", also known as "Sweet-pepper"; stout plant about 60 cm high, little branched, with large turgid leaves 10-12 cm long; white flowers, and producing large oblong, puffy fruit wrinkled on the surface; the "Green Bell-pepper" is waxy dark green, and of a mild flavor. *Tropical.* p. 894
  annuum 'Longum', "Cayenne pepper"; low herbaceous bush with narrow, fresh green leaves; white flowers; fruit slender pointed, dark green, maturing to red, 15 cm long; used for drying or sauce. *Tropical.* p. 472

CARAGUATA: see Guzmania

CARALLUMA *Asclepiadaceae*
  europaea (Mediterranean); succulent, four-angled, leafless stems with blunt teeth, gray-green and branching; flowers greenish with brown lines. *Subtropic.* p. 147
  mammillaris (South Africa: Great Karroo, Namaqualand); mat-forming stems with 5-6 angled columns 3 cm thick, the angles with spreading teeth, bronzy green overcast with glaucous purple; flowers inside black-purple, tube whitish, dark spotted. *Arid-subtropic.* p. 154
  praegracilis (Transvaal); clustering succulent with stems to 10 cm high, bluish green with long spine-like knobs; ochre yellow star-shaped flowers with red-brown spots and long tails. *Arid-subtropic.* p. 147

CARDIOSPERMUM *Sapindaceae*
  halicacabum (Pantropic to India, Africa, America), the "Balloon vine" or "Heartseed"; tendril-climbing herb to 3 m, with twice 3-parted thin leaves 10 cm long, and small white flowers; fruit a brown, thin-shelled inflated, angled capsule 3 cm dia. containing 3 black seeds each with a white, heart-shaped scar. *Tropical.* p. 876

CAREX *Cyperaceae*
  elegantissima: see foliosissima albo-mediana
  foliosissima albo-mediana (elegantissima) (Sakhalin, to Japan), grass-like tufted plant with elegant, flat leaves bright green with a white stripe near each margin, of stiff-erect habit. *Warm-temperate.* p. 390
  morrowii variegata (Japan), "Variegated sedge grass"; attractive, tufting, grass-like plant with narrow-linear, 6 mm wide concave leaves recurving, very colorful ivory-white to pale yellow, edged with green margins. *Subtropic.* p. 390

CARICA *Caricaceae*
  papaya (Colombia), "Melon tree"; dioecious succulent tree to 8 m with leaves digitately 7-lobed; flowers greenish-yellow, the hanging fruits to 30 cm long resembling a melon. *Tropical.* p. 291, 459, 475
  papaya 'Oblonga', "Papaya"; variety with oblong, gourd-like fruit to 30 cm long; the fine-flavored fruit is eaten like a melon, or may be baked like squash, stewed or candied. *Tropical.* p. 459

CARISSA *Apocynaceae*
grandiflora (macrophylla) (Natal), "Natal plum"; woody shrub to 9 m, armed with massive forked spines, lustrous green, ovate leaves, and large 5 cm white, fragrant flowers; big plum-like scarlet fruit to 5 cm long, and edible. *Subtropic.* p. 74, 462

humphreyi variegata; very decorative little bush dense with oval, leathery leaves 2 cm long, tipped with short spine, rich green variegated and edged with ivory-white; holding variegation well. *Subtropic.* p. 74

CARLEPHYTON *Araceae*
glaucophyllum (Madagascar); tuberous plant of Diego Suarez Prov.; tubular to 6 cm dia.; after a period of rest appears a glaucous green leaf, usually solitary, broadly heart-shaped on reddish petiole 25-50 cm long; boat-shaped 8-10 cm spathe, pale lemon-green inside, white outside on upper part; long slender spadix. *Tropical.* p. 102

CARLINA *Compositae*
acaulis (Europe); stemless perennial thistle with spiny leaves in a rosette, very deeply incised; solitary heads silvery, to 12 c m across. *Temperate.* p. 322

CARLUDOVICA *Cyclanthaceae*
palmata (Colombia, Ecuador, Perú), "Panama hat plant"; stemless plant with friendly green, fan-shaped, palm-like leaves usually cut into 4 parts, the lobes cut again; used for making of hats. *Tropical.* p. 391

CARNATION: see Dianthus

CARNEGIEA *Cactaceae*
gigantea (Cereus) (So. Arizona, No. Mexico), "Giant saguaro"; tall tapering column to almost 18 m, branching with age, close-ribbed, strong light brown spines; white flowers open in daylight. One of largest tree-like cacti known. *Arid-subtropic.* p. 245, 247, 253

gigantea 'Monstrosa'; a spectacular crested form of the giant saguaro, photographed in Saguaro National Forest south of Tucson, Arizona; the crest fan-like on top of a 6 m column. *Arid-subtropic.* p. 247

CARPENTERIA *Saxifragaceae*
californica (No. California), "Tree anemone"; handsome evergreen flowering shrub of the foothills of the Sierra Nevada; slow-growing to 2 m or more; the new branches purplish, with long-elliptic, thick-leathery leaves to 10 cm long, glossy dark green above, whitish beneath; large anemone-like pure white flowers 4-8 cm across, and pleasantly fragrant. *Warm-temperate.* p. 881

CARPOBROTUS *Aizoaceae*
edulis (So. Africa), "Hottentot-fig"; creeping, freely growing, branching succulent, often planted as a sand-binder along subtropical seashores; branches angled, to 1 m long, with long linear 3-angled fleshy, grass green or glaucous leaves 8-12 cm long, keel minutely serrate; large flowers 8-10 cm across with narrow, silky rays opening with the warming sun, yellow or purple; large fruits edible. In South Africa, I have observed the purple form primarily on the cool Cape Peninsula, the yellow in the glaring hot Karroo. *Arid-subtropic.* p. 45, 46

CARYA *Juglandaceae*
pecan (olivaeformis) (Indiana to Alabama, Texas and Mexico), the "Pecan nut"; handsome partially deciduous hard-wooded tree to 25 m high, planted by the millions in the South for their delicious nuts; Thomas Jefferson brought the first pecans from the Mississippi Valley to George Washington where he planted them at Mt. Vernon, Virginia in 1775. Wide-spreading, with grayish, deeply furrowed bark and graceful pinnate foliage; thin-leathery fresh green, crenate leaflets with yellow midrib 10 to 15 cm long; the male flowers in pendulous catkins, the female flowers in terminal spikes, producing clusters of oblong, angled green husks 4-8 cm long which split into 4 sections, and holding the edible, and sweet-tasting kernel or nut; ripening in autumn. *Warm-temperate.* p. 467

CARYOTA *Palmae*
mitis (Burma, Malaya, Indonesia), "Clustered fishtail palm"; with numerous suckers growing up in clusters with green-gray trunks 7 m high topped by dense tufts of irregularly bipinnate dull-green leaves, the pinnae fan-shaped, jagged at apex and many-veined, nodding at the tips; fruit blackish-red. *Tropical.* p. 770, 773

urens (Himalayas, India, Burma, Ceylon, Malaya), "Wine palm" or "Fishtail"; I remember these majestic, solitary palms giving special character to the beautiful, verdant landscape of the mountains of Ceylon; glossy gray trunk more than 30 m high, topped by bipinnate, arching leaves with thick wedge-shaped, loosely spaced segments; yields wine. *Tropical.* p. 773

CASASIA *Rubiaceae*
clusiaefolia (So. Florida, Bahamas, W. Indies), "Seven-year apple"; evergreen shrub to 3 m, bushy or tree-like; oblong, glossy leaves to 15 cm long, in clusters at branch tips; star-like white flowers to 4 cm across; very fragrant; ovoid fruit 8 cm long, green to brown to blackish when ripe; licorice-flavored, edible, but not tasty; fruit almost always present on the plant. *Tropical.* p. 864

CASIMIROA *Rutaceae*
edulis (Mexico, C. America), the "White sapote" or "Mexican apple"; fast-growing tropical fruit tree to 15 m high, treasured by the Aztecs; shiny green leathery leaves, palmately compound of 5-7 leaflets 12 cm long; small greenish or whitish flowers; the handsome apple-shaped fruit to 10 cm dia. pendant on long stalks, yellow-green with thin waxy glaucous cover; the edible soft and juicy, cream-colored flesh very sweet resembling mango, and with fragrant aroma, but may have a bitter after-taste. The seeds are used medicinally in Mexico to induce sleep. Planted in South Florida and So. California in several good-flavored varieties; eaten fresh or in sherbets. The selected cultivar 'Vernon' (shown on pg. 463) is usually grafted, with lucious golden yellow fruit of better, sweeter flavor than the species. *Subtropical.* p. 463, 872

edulis 'Suebelle', "White sapote"; a favored cultivar of medium size for the home garden, because of its sweet, rich, soft pulp and small seeds; ripening most in November, turning yellow; fruit has musky sweet flavor and does not keep long. *Subtropic.* p. 461

CASSIA *Leguminosae*
alata (Trop. America), "Candle-bush"; short-lived big bush 1 to 4 m high, with pinnate foliage; 6 cm leaflets; erect candle-like spikes of golden yellow flowers. *Tropical.* p. 550

angolensis (Uganda); robust bush of Central Africa, with long, pinnate leaves, the leaflets narrow lanceolate, in numerous pairs. *Arid-tropical.* p. 550

artemisioides (Australia), "Feathery senna"; bushy shrub 2-3 m high, covered with silky-gray pubescence, with leaves of 6-8 stiff linear needle-like leaflets, and bright yellow, showy flowers with black anthers. *Subtropic.* p. 538, 550

bicapsularis (bot. pendula) (bicuspidata in hort.) (Trop. America); showy, robust bush to 3 m high, with pinnate leaves, the oval leaflets leathery; erect terminal inflorescence with large cupped golden yellow flowers 4 cm across. *Tropical.* p. 550

corymbosa (Argentina), "Flowering senna"; sunloving shrub with pinnate foliage of 3 pairs of leathery leaflets, and showy clusters of nearly regular, rather cupped flowers of golden yellow. Sometimes listed as "marilandica" in horticulture. *Subtropic.* p. 550

didymobotrya (nairobensis), (Trop. Africa), "Popcorn-bush"; bushy shrub or small tree to 3 m, young shoots and leaves finely downy; leaves 35 cm long; leaflets in 4-18 pairs; flowers in erect, showy racemes 15-30 cm long, several borne near end of shoots; large cupped golden yellow flowers. *Tropical.* p. 549, 552

excelsa floribunda (Brazil); tree with pendant pinnate foliage, and long racemes with fragrant yellow flowers. *Tropical.* p. 550

fistula (India, Sri Lanka); "Golden shower" or "Pudding pipe tree"; beautiful tree of moderate size; leaves with leaflets in 4 to 8 pairs, 5-15 cm long; fragrant flowers golden yellow, in drooping racemes to 30 cm long, resembling huge bunches of grapes. *Tropical.* p. 547, 550, 551

grandis (Trop. America), "Pink shower tree"; beautiful flowering tree to 15 m tall, with spreading branches; leaves divided into 8 to 20 pairs of leaflets, each 6 cm long, pink when young; pinkish-lavender buds open into rich coral-pink flowers massed in thick clusters. *Tropical.* p. 551

javanica (nodosa) (Indonesia), "Appleblossom shower"; low tree with wide-spreading crown; pinnate leaves with obtuse, 5 cm leaflets; flowers of short duration, with petals at first pale red, changing to dark red, then paling again. *Tropical.* p. 551

moschata (Panama, Colombia), "Bronze shower tree"; showy flowering tree to 10 m, leaves pinnate with 10-18 pairs of leaflets 3-5 cm long; pendant racemes to 25 cm long, of yellow flowers becoming brick-red with age. *Tropical.* p. 549

multijuga (W. Indies, So. America), "November shower"; small tree to 6 m high, pinnate foliage; golden yellow flowers 5 cm dia., in erect clusters. *Tropical.* p. 552

nodosa (E. Himalayas to Malay Peninsula), "Pink and white shower"; tropical tree to 15 m high, with pinnate leaves, 5 to 12 pairs of elliptic leaflets; often flowering leafless; corolla 2 cm long, pale, becoming bright pink. *Tropical.* p. 549

peltophora (Sri Lanka); large spreading tree with long pinnate leaves of numerous narrow leaflets; large erect panicles of golden yellow small flowers. *Tropical.* p. 547

'Rainbow Shower' (fistula x javanica); beautiful hybrid tree and the landmark of Honolulu; with showy masses of flowers from

cream to orange and red, from March to August. *Tropical.*
p. 547, 550

roxburghii (marginata) (Sri Lanka), "Red cassia" or "Ceylon senna"; large, showy, spreading tree with dense ranks of pinnate leaflets along a willowy axis; the arching branches beautifully set with masses of axillary flowers, salmon rose. *Tropical.* *p. 551*

siamea (Malaysia, naturalized in W. Africa), "Kassod tree"; tropical tree to 12 m tall; pinnate leaves with leathery leaflets to 8 cm long; inflorescence forming large terminal clusters of yellow flowers 3 cm across. Planted in Africa as shade trees for coffee plantations, and known there as "Coffee senna". *Tropical.* *p. 552*

speciosa (Brazil); large shrub with leaflets in 2 pairs, obliquely ovate, soft-hairy beneath; inflorescence in wide, terminal clusters to 50 cm long, flowers deep yellow, 4-6 cm across. *Tropical.* *p. 552*

spectabilis (Trop. America), "Popcorn bush"; showy spreading tree to 12 m high, with long pinnate leaves, the leaflets lanceolate and bright green; branches bearing spectacular erect 30-60 cm racemes of large beautiful, cupped golden-yellow 4 cm flowers like shining lamps. *Tropical.* *p. 552*

surattensis (glauca) (Marquesas, India,' Sri Lanka); "Scrambled eggs"; tall shrub or tree, shoots angular; leaves 15-22 cm long; leaflets in 6-9 pairs, very glaucous beneath, 4-6 cm long; flowers in a fine, erect cluster of axillary and terminal racemes 10 cm long; corolla 3 cm across, yellow; pod 15-20 cm long, wide, flat. *Tropical.* *p. 550*

sylvestris (Brazil); tall tropical bush with robust canes; leaves pinnate with big lanceolate leaflets; inflorescence in tall panicles of golden yellow flowers. *Tropical.* *p. 552*

CASSINE *Celastraceae*
orientalis (Madagascar, Mauritius), "False olive"; handsome tropical shrub or small tree with ornamental waxy, bluish green foliage on reddish stems, the opposite leaves long-linear in the juvenile stage; in the maturity stage shorter obovate 5-8 cm long with scalloped edges; small greenish and white flowers in axillary clusters, followed by the olive-like fruit. *Tropical.* *p. 302*

CASTILLEJA *Scrophulariaceae*
coccinea (No. America), "Scarlet paintbrush"; hairy annual or biennial to 60 cm high; basal leaves obovate, stem leaves 3-5 lobed; floral bracts scarlet, corolla pale yellow. *Warm temperate.* *p. 886*

lanata (Texas, Arizona, Mexico), "Indian paint brush"; perennial covered with white wool, with narrow leaves; small greenish flowers with showy scarlet bracts and calyx. *Subtropic.* *p. 883*

CASUARINA *Casuarinaceae*
decussata (Western Australia), "Karri oak"; an arid-climate tree with sparry branches, and long spreading roots; very unique foliage, very tiny and like scales appressed to branches, looking like pines or horsetails, grayish green; flowers reddish and insignificant, and seed-vessels like globular cones. *Arid-subtropic.* *p. 300*

equisetifolia (No. Australia, Queensland), the "Horse-tail tree" or "Australian pine"; hardwooded tree to 25 m high, with pendulous branches swaying in the breeze; wirelike branchlets with apparently leafless twigs, the leaves reduced to scales and suggesting the horsetail (Equisetum). Adaptable as a patio tree or the sunny, airy greenhouse, when sheared into compact shapes or imaginative topiary forms. *Tropical.* *p. 300*

sumatrana (Malaya), "She-oak"; dense tree with scale-like, appressed leaves, similar to pines, and with small cone-like fruit. *Tropical.* *p. 300*

CATALPA *Bignoniaceae*
bignonioides (Georgia, Florida, Mississippi), the "Indian-bean"; tree 8-18 m high, of rounded, spreading habit, with ovate leaves 15-25 cm long, pale green, downy beneath; inflorescence a broad erect panicle with bell-shaped flowers, and spreading, frilled lobes, white, with two yellow stripes and purple spots, 5 cm across. *Warm-temperate.* *p. 184*

CATASETUM *Orchidaceae*
roseum (Mexico: Oaxaca); a small, very pretty epiphyte with short pseudobulbs and deciduous leaves, numerous showy and long-lasting, cupped star-like waxy flowers on short, pendant raceme cream flushed rose, petals and lip fringed with rosy hairs; lip greenish inside (early spring). *Tropical.* *p. 712*

trulla (Brazil); pseudobulbs bearing broad leaves, the inflorescence on arching racemes with 5 cm flowers having greenish sepals and petals, the lip spoon-shaped, concave, fringed at margins greenish-white with brown tip; September. *Tropical.* *p. 712*

viridiflavum (Cent. America to Perú); terrestrial with plaited leaves 30 cm long, the inflorescence shorter than the foliage, with up to 12 flowers, petals wider than sepals all yellow-green and concave, lip marked with yellow inside, conical-saccate and hooded (summer). *Tropical.* *p. 712*

CATESBAEA *Rubiaceae*
spinosa (Cuba, Bahamas), "Lily-thorn"; spiny evergreen shrub to 4m high, with small leathery leaves in clusters; the spines above the leaf axils; flowers funnel-shaped with a very long tube, 8-15cm long, gradually widening; the corolla pale yellow and drooping; orange-yellow fruit. *Tropical* *p. 462, 864*

CATHARANTHUS *Apocynaceae*
roseus (Vinca rosea) (Java to Brazil), "Madagascar periwinkle"; erect, fleshy plant with oblong leaves glossy green with white center vein; showy flowers rosy-red with purple throat. *Tropical* *p. 74*

roseus 'Bright Eyes' (Vinca rosea); a "Rose periwinkle" of compact habit, very prolific bloomer with pretty white flowers having a bright red spot in center. *Tropical* *p. 74*

CATOPHRACTES *Bignoniaceae*
alexandrii (Namib Desert, S. W. Africa); spiny shrub with small lobed, felty leaves; large pinkish white flowers with yellow throat. *Arid-Tropical* *p. 184*

CATTLEYA *Orchidaceae*
aclandiae (Northern Brazil); epiphyte with cylindric pseudobulbs, 2-leafed, to 10cm long; waxy flowers to 12cm across, sepals and petals green with purple blotches, lip magenta; very fragrant. *Humid-Tropical* *p. 712*

amethystoglossa (Brazil); tall plant 60-90 cm high with long terete stems topped by 2 leaves, and clusters of 3-20 flowers 8-10 cm across, sepals and petals bright rose and spotted with purple lip with whitish side lobes and violet middle lobe (Nov.-July). *Tropical* *p. 717*

bicolor (Brazil); epiphyte with long cane-like pseudobulbs bearing 2 leaves, flowers with linear petals and sepals brownish green, the narrow lip rosy-purple sometimes edged with white. (Sept.-Nov.). *Tropical* *p. 717*

citrina (Mexico), "Tulip cattleya"; beautiful epiphyte with silvery, globular pseudobulbs and strapshaped leaves; the fragrant, bell-like flowers are borne singly and pendant, bright lemon-yellow, lip edged white and crisped, (April-June). *Tropical* *p.717*

dowiana (Costa Rica), "Queen cattleya"; epiphyte with stout pseudobulbs bearing solitary leaves, and beautifully colored, fragrant flowers widely used for hybridizing, the crisped petals golden yellow, the sepals yellow shaded buff, and a great, frilled lip deep crimson-red streaked with old gold, (May-Aug.). *Humid-Tropical* *p. 720*

dowiana aurea (Colombia), "Queen cattleya"; a superb epiphyte with stout pseudobulbs bearing solitary leaves, and beautifully colored, highly fragrant flowers to 15cm across, the crisped petals and sepals golden yellow, and a great, frilled lip deep crimson-red distinctly streaked or banded with old gold; blooming from May to Autumn. *Tropical* *p. 720*

'Enid' (gigas x mossiae); robust, elegant hybrid with big, purplish-mauve flowers of firm texture, the lip overlaid with purple, yellow in the throat; blooming at various times, often twice a year. *Tropical* *p. 717*

forbesii (Brazil), a "Cocktail orchid"; clustering epiphyte, and a sight to see it nesting in small trees, loaded with medium-size flowers, the sepals and petals yellow-green, lip white outside, lightly streaked with red inside, borne on stem-like, 2-leaved pseudobulbs, (May-Oct., and various). *Tropical* *p. 716*

gaskelliana (Venezuela, Brazil), "Summer cattleya"; epiphyte with 1-leaved pseudobulbs; fragrant flowers in clusters of 3-5, petals and sepals lavender, the fringed lip purple with yellow throat, (May-July). *Tropical* *p. 716*

granulosa (Guatemala and Brazil); free-growing species with slender terete 2-leaved stems 40-50 cm high, and olive green, fleshy 10 cm flowers spotted with brown, the spreading lip white dotted with purple (Aug.-Sept.). *Tropical* *p. 717*

granulosa schofieldiana; flowers with large spreading sepals and petals lemon-yellow spotted with purple, midlobe of lip covered with magenta warts. *Tropical* *p. 717*

intermedia (Brazil); epiphyte with 2-leaved, slender, cane-like pseudobulbs; slender flowers to 12 cm, usually 3-5 together in clusters, pale rose, with middle lobe of lip purple and crisped, (April-Nov.). *Tropical* *p. 717*

intermedia alba (parthenia) (Brazil); variety with pure white flowers on cane-like pseudobulbs much longer than the species; summer. *Tropical* *p. 712*

intermedia var. aquinii; a curious variety in which the normally irregular flower becomes regular because petals imitate the lip; sepals and petals soft rose, the midlobe of lip and tip of petals purple. *Humid-subtropic.* *p. 717*

intermedia superba (Brazil); variety with larger flowers, sepals and petals delicate rose, tips of petals purple. *Tropical* *p. 712*

labiata alba (Trinidad, Brazil), "Autumn cattleya"; epiphyte with pseudobulbs bearing a solitary leathery leaf, the showy flowers 15-18 cm across, 3-7 in a cluster, in the species with sepals and petals bright rose, the latter broad and wavy, large lip deep velvety-crimson, throat marked yellow; the var. alba pure white flowers and deep yellow throat. *Tropical* p. 716

leopoldii (guttata var.) (Brazil: Bahia); interesting variety with terete pseudobulbs 50 cm long and a pair of short leaves; bold inflorescence carrying as many as 30 fragrant, waxy 8-10 cm flowers, sepals and petals bronze spotted crimson, small lip rich crimson-purple; free-blooming (summer). *Tropical* p. 712

luteola (Brazil, Ecuador, Perú, Bolivia); dwarf epiphyte, with solitary leaves on clustered pseudobulbs, small flowers in clusters, lemon-yellow, the middle lobe of lip white and wavy, the side lobes sometimes streaked with purple, (Nov.-Aug.). *Tropical* p. 712

mossiae (Venezuela); "Easter orchid"; epiphyte with pseudobulbs bearing solitary leaves; the handsome flowers 12-15 cm dia., 3-5 or more in a cluster, blush or light rose, large frilled lip crimson and rose with golden yellow markings, often on a suffused white ground, (March-June). *Tropical* p. 716, 717, 719

'Priscilla alba' ('Enid' x lueddemaniana); old English hybrid 1926, with elegant flowers, sepals and petals pure white, the lip frilled and with purple, in rich golden yellow. *Tropical* p. 717

'Rex' (Perú); distinctive Andean epiphyte with 1-leaved pseudobulbs, and striking flowers 12-15 cm across, sepals white tinged primrose, petals creamy-white and wavy, lip crimson with yellow in upper throat and frilled (summer). *Tropical* p. 717

skinneri (Guatemala); epiphyte with long, twin-leaved pseudobulbs; the small 5-8 cm flowers in clusters, sepals and petals rose-purple with lightly pointed dark lip, yellowish-white in throat, (April-June). *Tropical* p. 717

CAVENDISHIA *Ericaceae*
acuminata (Colombia, Ecuador); evergreen shrub at home in the Andes, with pendulous branches; leathery leaves 5-8 cm long; showy inflated flowers 2 cm long, bright crimson red with pale green tips, covered in bud by large scarlet bracts. *Subtropic.* p. 397

CECROPIA *Moraceae*
palmata (West Indies to So. America), "Snakewood tree"; fast growing tree with soft wood and hollow stems, having milky sap; large corrugated leaves 7 to 11-lobed to the middle or more, 30 cm or more across, underside white tomentose; small flowers in dense spikes. *Tropical.* p. 648

peltata (Puerto Rico to Jamaica), "Guarumo"; noted tropical tree to 18 m high, of sparry habit and with milky juice, its hollow branches often inhabited by vicious ants; the large decorative, peltate, long stalked leaves are palmately 7-9 lobed, rough hairy above and shimmering white-downy beneath, flashing like silvery mirrors when turned up by a breeze. *Tropical.* p. 648

CEDRELA *Meliaceae*
odorata (West Indies to So. America), "West Indian cedar"; tall aromatic timber tree 30 m high with buttressed base and smooth trunk 1-2 m dia.; pinnate leaves with 8 pairs of lanceolate leaflets; small yellowish flowers, followed by leathery fruit 4 cm long. The reddish heartwood is used in making cigar boxes. *Tropical.* p. 641

CEDRUS *Coniferae: Pinaceae*
atlantica (Morocco, Algeria), "Atlas cedar"; slow-growing, large evergreen tree to 30 m high, with wide-spreading branches; stiff, bluish-green needle-shaped, 4-angled leaves 2 cm long, clustered on short stout, lateral spurs; ovoid male cones erect, 8 cm long; female cones smaller. *Warm Temperate.* p. 342

atlantica 'Glauca' (No. Africa) "Blue Atlas cedar"; unusually attractive tree of open habit with spreading branches having upright leading shoots with rigid needles in whorls on lateral branches, glaucous bluish silver. *Temperate.* p. 342, 343

atlantica 'Glauca pendula', "Weeping blue cedar"; a variety of weeping habit, with the lateral pendulous branches drooping like a bluish curtain 3 m or more from limbs. *Temperate.* p. 343

deodara (Himalayas) "Deodar cedar"; outstandingly graceful, tall tree of pyramidal habit, with branches and leading shoots pendulous, needles to 5 cm long, dark bluish green. *Warm Temperate.* p. 343, 344, 346

deodara 'Compacta' (Himalayas), "California Christmas tree"; dense pyramid with stiff, sharp spiny blue needle bundles. *Subtropical.* p. 346

libani (Syria, Lebanon), "Cedar of Lebanon", aristocratic conifer; when young with branches erect, tufted needles on short lateral shoots, to 3 cm long, green or glaucous; with age trees will branch and grow from 15 to 30 m high, with branches horizontal and fan-like. *Temperate.* p. 341, 344

CEIBA *Bombacaceae*
pentandra (Tropics of America, Africa, Asia), the "Kapok tree" or "Silk-cotton"; large deciduous tree to 40 m high, with widely spreading branches, trunk to 3 m thick and often prickly, and forming board-like buttresses; digitately compound leaves, and large fragrant flowers silky outside, yellowish inside; the fruit a leathery capsule with the seeds embedded in long cotton-like fiber or kapok. *Tropical.* p. 195

CELMISIA *Compositae*
coriacea (New Zealand), the "Mountain daisy"; noble rosette, at home along the Southern Alps and to the Fiordland, with stiff-leathery sword-like leaves 30-60 cm long, beautifully covered with silver skin above when young, later dark green and shining; beneath dense with silvery wool; large daisy-like rayflowers 8-10 cm across, white with golden center, on woolly stalks. *Warm-temperate.* p. 308, 309

spectabilis (New Zealand); handsome stout herb in tufts of lanceolate, leathery leaves 10-20 cm long, covered with white soft felt; flower heads yellow. *Warm temperate.* p. 308

CELOSIA *Amaranthaceae*
argentea 'Castle Gould'; a cultivar of C. arg. childsii from Johore, with immense heads of loosely twisted, pendant, feathery plumes in shades of red or yellow. *Tropical.* p. 54

argentea 'Golden Spire', "Chinese wool-flower"; showy herbaceous tropical annual 30-50 cm high, succulent angled stems with narrow ovate fresh green leaves; toward apex the branching inflorescence of numerous plume-like spikes a glistening vivid yellow. *Tropical.* p. 52

argentea plumosa, the "Feather celosia" or "Burnt plume"; herbaceous annual about 1/3 m high in bushy types, with fleshy stem and fresh green ovate leaves; the branches terminated by dense erect spikes of feathery plumes, highly colored fiery red as in 'Fiery Feather', or yellow such as in 'Golden Feather'; developing best in high heat with plenty of sunshine. *Tropical.* p. 52

argentea plumosa 'Flame', "Burnt plume"; herbaceous annual from Tropical Asia with fresh-green ovate leaves, and with stem and branches terminated by dense chaffy spikes highly colored as fiery red plumes. *Tropical.* p. 51

argentea pyramidalis (Trop. Asia), "Plume celosia"; a taller growing annual ½-1 m high, with erect stems bearing spectacular pyramidal plumes with silky threads, and looking like the feathery inflorescence of Pampas grass, but in brilliant colors of luminous crimson-red, bronze, copper or golden yellow, carried well above the fresh-green foliage. Useful cut flowers in late summer; may also be dried for winter bouquets. *Tropical.* p. 52

cristata (E. Indies) "Cockscomb", one of the most spectacular summer-flowering annuals; a tropical fleshy herb 30 cm or more high with succulent stem and big fresh-green or sometimes bronzy lanceolate leaves; the inflorescence in showy stiff, cockscomb-like velvety fans 15 to 25 cm wide, often very contorted and fluted; most effective in glowing crimson, but available also in yellow or orange color forms. Large flower heads develop best during hot summer. *Tropical.* p. 51, 52

CENTAUREA *Compositae*
candissima (Sicily), "Dusty miller"; small perennial densely covered with white matted wool, leaves pinnately parted into broad lobes in maturity; inflorescence golden yellow. *Subtropic.* p. 310

cyanus (Europe, Near East), "Cornflower"; herbaceous winter annual with stiff stems, narrow gray green leaves and flower heads 3-4 cm across, usually blue or purple. *Subtropic.* p. 308

gymnocarpa (Italy: Capri), "Velvet centaurea"; perennial 50-80 cm high, with white felt-like leaves more finely divided than C. cineraria; purple flower heads. *Subtropic.* p. 310

CENTAUREA: see also Senecio

CENTRANTHUS *Valerianaceae*
ruber (Europe, No. Africa), "Red valerian"; compact perennial to 1 m, with ovate, glaucous leaves 10 cm long; small fragrant, carmine-red flowers in dense terminal clusters. *Subtropic.* p. 916

CENTROPOGON *Lobeliaceae*
x lucyanus; tropical herbaceous plant raised in Marseilles 1856, said to be a hybrid of Centropogon fastuosus of Mexico x Siphocamphylus betulaefolius from Brazil; a good basket plant; prostrate herbaceous branches with ovate, toothed, green leaves, and long tubular, bilabiate flowers rosy-carmine or burnt red, 4 cm long; winter-blooming. *Tropical.* p. 616

CEPHAELIS *Rubiaceae*
tomentosa (Mexico to Bolivia); evergreen tropical shrub to 5 m, with elliptic leaves to 25 cm long, white tomentose beneath; small yellow flowers subtended by bright red bracts, to 5 cm across; berries blue. *Tropical.* p. 867

CEPHALOCEREUS *Cactaceae*
   fluminensis (S.E. Brazil); sprawling columns branching from the base, often coiling like a snake but branches erect, ascending or pendant, to 10 cm thick, with 10-17 ribs, the areoles with yellow spines and abundant white wool; flowers pink outside, white within. *Arid-tropical.* *p. 246*
   leucostele (Stephanocereus) (Brazil), from the deserts of Bahia, normally single columns to 5 m tall and with 12-18 shallow ribs, and dense wool covering the stem; cephalium densely white-woolly with a peculiar collar of yellow bristles 8 cm long; flowers white. *Arid-tropical.* *p. 254*
   palmeri (E. Mexico), "Woolly torch cactus"; tree type to 8 m, slender columns, dark green, glaucous and bluish when young; 7-9 rounded ribs, white-hairy at top; flowers purplish. *Arid-tropical.* *p. 247*
   polylophus (Neobuxbaumia) (Mexico), fat column to 12 m; close-ribbed, deep green, yellow-spined; dark red flowers to 8 cm long. *Arid-tropical.* *p. 247, 248*
   royenii (West Indies); strong columns to 6 m high branching near base, to 8 cm thick, bluish green, with 7-11 ribs, the areoles with white hair tufts, and long light brown spines; small shell-pink tubular flowers brownish outside. *Arid-tropical.* *p. 247, 248, 571*
   senilis (Mexico), "Old man cactus"; slender column to 6 m or more high, and to 30-40 cm thick, closely ribbed, and covered with long gray hairs; nocturnal flowers rose-colored, to 9 cm long. *Arid-tropical.* *p. 247, 248, 249*

CEPHALOPHYLLUM *Aizoaceae*
   'Red Spike' (Cape Prov.: Karroo); striking succulent forming dense cushions of erect cylindrical glaucous green leaves 7 to 10 cm long; showy flowers ruby-red. According to Herre of Stellenbosch, the "Diamond" among the Mesembs. *Arid-subtropic.* *p. 44*

CEPHALOTAXUS *Cephalotaxaceae (Coniferae)*
   harringtonia (Japan), "Harrington plum yew"; shapely evergreen, long narrow needles, dark green 3-5 cm long. *Warm temperate.* *p. 352*

CERATONIA *Leguminosae*
   siliqua (Eastern Mediterranean Reg.), "St. John's bread" or "Carob"; evergreen tree 12-15 m, with stout trunk and rounded head of branches; leaves pinnate, 15-30 cm long, leathery; flowers in cylindrical racemes to 15 cm long; fruit a brown pod 12-30 cm long, the pulp of which is sweet and much valued in S. Europe as animal food, or for chewing. *Subtropic.* *p. 474, 544, 561*

CERATOPTERIS *Ceratopteridaceae (Filices)*
   thalictroides (in quiet waters of the Tropics), "Water fern"; a true aquatic fern with rosettes of light green, succulent fronds, barren ones floating and forming young plantlets along the lobed margins, the fertile fronds divided into linear segments. *Humid-tropical.* *p. 438*

CERATOSTIGMA *Plumbaginaceae*
   willmottianum (W. China), "Chinese plumbago"; low spreading shrub to 1½ m high, with angled branches, obovate dull green leaves 5 cm long, bristly on both sides; salver-shaped flowers 2 cm across; limb sky-blue with white eye and with rosy tube; summer-blooming. *Subtropic.* *p. 811*

CERATOSTYLIS *Orchidaceae*
   rubra (Philippines); charming small epiphyte; clustered stems with glossy green, fleshy semi-cylindrical, grooved leaves to 12 cm long; showy short-stalked flowers opening flat, brilliant salmon-red, 3 cm across. *Humid-tropical.* *p. 715*

CERATOZAMIA *Cycadaceae*
   mexicana (Mexico), "Mexican horncone"; palm-like cycad with stiff pinnate glaucous and hairy leaves borne in a whorl on a short hairy trunk, to 2 m high, with up to 150 leathery leaflets, the stalks spiny; the flowers in cones. *Tropical.* *p. 384*
   sp. 'Hilda' (Mexico), "Bamboo zamia"; palm-like cycad with only short trunk, bearing stiff pinnate fronds 1-1½ m long, in shade even 2 to 3 m tall; the bamboo-like axis beige and lightly armed with prickles; thin-leathery, shiny grass green leaflets 15 cm long, whorled in pairs nearly around stem. Quite unusual in its type. *Tropical.* *p. 384*

CERBERA *Apocynaceae*
   manghas, (Polynesia, Hongkong, Malaya, India); evergreen shrub or small tree to 6 m high, with milky latex; alternate leathery ovate leaves 15-30 cm long crowded on young branches; fragrant white tubular flowers with 5 spreading lobes and rose-pink center, 3-8 cm across; the pink to black roundish fruit 5-10 cm long. *Tropical.* *p. 74*
   odollan (India to Pacific Is.), "Reva"; shrub or small tree with milky sap; deep green obovate, leathery leaves to 30 cm long, crowded at branch ends; fragrant white waxy flowers with spreading lobes, to 6 cm across, ovoid fruit 5-10 cm long, becoming red and with poisonous seeds. *Tropical.* *p. 74*

CERCIDIUM *Leguminosae*
   floridum (torreyanum) (Arizona to Mexico), "Blue palo verde" of the desert; small bush or tree to 10 m; spiny blue green branches with finely divided, bipinnate foliage which drops early; in spring masses of small bright yellow 2 cm flowers. *Arid-subtropic.* *p. 562*

CERCIS *Leguminosae*
   canadensis (Eastern U.S.), "Eastern red bud"; deciduous tree to 12 m, with leaves rich green, broadly ovate to 10 cm across; small 2 cm rose-pink flowers grown in great profusion on bare twigs and branches. *Warm temperate.* *p. 552*
   siliquastrum (S. Europe and the Orient), "Judas tree" or "Red-bud"; tree to 12 m, usually smaller and more shrubby, leaves deeply cordate, 6-10 cm wide, glaucous green; flowers bright purplish-rose 1-1½ cm long, in clusters of 3 to 6, produced in immense quantities directly from branches young and old, often appearing before the leaves. *Warm temperate.* *p. 552*
   siliquastrum alba (So Europe, W. Asia), "Judas tree", also known as "White redbud"; deciduous to 12 m high, with reddish branches; leaves round-cordate 12 cm long, small white flowers. *Warm temperate.* *p. 430*

CEREUS *Cactaceae*
   hexagonus (Colombia), "South American blue column"; tall tree type to 15 m, branching near base; glaucous blue-green fleshy columns, with 4-6 high ribs, and few spines; white flowers 20-25 cm long; reddish fruit to 12 cm. *Tropical.* *p. 247, 249, 253*
   hildmannianus (Brazil: Guanabara); treelike column cactus, freely branching, to 5 m high, with stems 10-12 cm dia, at first bluish then passing to grayish green, with 5 or 6 winglike acute ribs, notched at 2 to 3 cm, nearly spineless; funnel-flowers to over 20 cm long, greenish outside, white inside; large red, edible fruit. *Tropical.* *p. 247, 248*
   jamacaru (Brazil); branching tree type to 10 m, slender column quite blue when young, deeply 4-6 ribbed, spines yellow to brown; flowers white. *Tropical.* *p. 250*
   peruvianus hort.; the commonly cultivated tree-cereus in the trade, possibly a hybrid; fleshy columns 6-9 ribbed, glaucous bluish-green, free branching; few brown spines; whitish flowers; crimson red fruit is edible. *Tropical.* *p. 248, 250, 459, 464*
   peruvianus monstrosus (Mexico), "Curiosity plant"; an irregular growing monstrose form constantly forming new heads, glaucous bluish-green, slow growing but always retaining its habit. *Tropical.* *p. 250*

CEREUS: (see also Carnegiea, Cephalocereus, Cryptocereus, Echinocereus, Espostoa, Hylocereus, Lemaireocereus, Lophocereus, Myrtillocactus, Nyctocereus, Oreocereus, Pachycereus, Selenicereus, Trichocereus)

CEROPEGIA *Asclepiadaceae*
   aristolochioides (Trop. Africa: Sénégal to Ethiopia); slender stems slightly succulent, with ovate glossy green leaves; flowers a whitish curved tube to 15 mm long, the lobes inside yellow, outside burnt crimson. *Tropical.* *p. 148*
   armandii (S.W. Madagascar); succulent stems stapelia-like to 2 cm thick, sprawling 10-15 cm long, the flowering stem sections much prolonged and twining to over 1 m long; 7 mm ovate leaves; flowers lantern-like, beige with greenish cage of 12 mm lobes. *Tropical.* *p. 148*
   ballyana (Kenya); robust, twining pencil-like stems, with fleshy leaves, and canary-cream flowers with red-brown spots 6-7 cm long, the blackish lobes twisted together. *Tropical.* *p. 148*
   fusca (Canary Isl.), "Carnival lantern"; succulent shrub with upright, forked, cylindrical columns constricted at joints, gray to purplish; flowers brown and yellow. *Subtropic.* *p. 148*
   haygarthii (Natal), "Wine-glass vine"; twining, succulent stem with small ovate leaves; flower like a fluted wineglass with bent stem, cream with specks of maroon, covered by maroon parachute; from center rises a maroon stalk topped by red knob, sparsely covered by white hair. *Subtropic.* *p. 148*
   krainzii (Canary Islands); stiffly erect succulent 40-60 cm high branching from base, with stems olive grey to whitish; tubular flowers in large clusters, lemon-yellow, 2 cm long, white-hairy inside. *Subtropic.* *p. 148*
   sandersonii (Natal), "Parachute plant"; twining succulent with small ovate, fresh green leaves on younger tips; flowers like a parachute mottled light and dark green. *Subtropic.* *p. 148*
   serpentina (So. Africa: Transvaal); succulent with fleshy stem 1 cm dia, with basal shoots twining, slightly 3-angled, olive green tinged purple; tiny deciduous fleshy leaves; flask-shaped flowers 6-9

cm long, with thin whitish tube opening into flaring mouth, the pointed lobes spotted chocolate. *Arid subtropic.* p. 148
woodii (Natal), "String of hearts"; because of the trailing, threadlike stems set with pairs of succulent, heartshaped, bluish leaves 1-2 cm across, marbled with silver, purple beneath; corm-like roots; flowers purplish. *Subtropic.* p. 148, 149

CEROXYLON   *Palmae*
andicola (Colombia); "Waxpalm" of the Andes, found at altitudes above 3000 m; tallest amongst palms, with straight trunks to 65 m tall, covered with a layer of wax and looking like marble columns to 60 cm thick, slightly swollen in the middle; crown of pinnate fronds to 6 m long; linear dark green leaflets split at end, whitish powder underneath, and drooping; berry-like 2 cm fruit. *Subtropic.* p. 774

CESTRUM   *Solanaceae*
aurantiacum (Guatemala); rambling or half-climbing shrub with matte green ovate leaves 5-8 cm long, and leafy pyramidal clusters of bright orange-yellow 2½ cm flowers with reflexed lobes. *Subtropic.* p. 895
diurnum (W. Indies), "Day jessamine"; tropical evergreen shrub or small tree to 5 m, sometimes scandent; leathery oblong leaves 10-12 cm long; flowers in clusters, fragrant by day, shaped like miniature trumpets, 2-3 cm long, white with a touch of green; glossy black fruit. *Tropical.* p. 895
elegans (purpureum) (Mexico); rambling evergreen shrub to 3 or more metres, with soft-hairy branches and ovate leaves to 10 cm long; the flowers in nodding clusters, slender tubular corolla with spreading lobes, glowing wine-red. *Tropical.* p. 891, 900
nocturnum (West Indies), "Night jessamine" or "Queen of the Night"; evergreen shrub to 2 or 3 m high with wiry, brownish branches; shining, thin-leathery, 8 to 15 cm ovate leaves, and small, greenish or creamy-white, slender tubular flowers with pointed lobes, to 3 cm long, axillary in profuse clusters, intensely perfumed at night, and blooming mostly from July to September, but at intervals throughout the year. *Tropical.* p. 895
parqui spurium (Chile), "Willow-jasmine"; partially evergreen shrub with willowy, erect stems, to 2 m high, small ovate fresh-green leaves 5-12 cm long; at branch ends the heavy clusters of slender tubular flowers 3-4 cm long, yellow, and strongly perfumed at night. *Subtropic.* p. 895

CETERACH   *Polypodiaceae (Filices)*
officinarum (C. and So. Europe to N. Asia and Africa), the "Scale-fern"; small tufted xerophyte, rhizome covered with black hairs, the pinnate or lobed fronds 15 cm long, thick-leathery, dark green and smooth above, underneath dense with yellow-brown scales. *Warm temperate.* p. 443

CHAENOMELES   *Rosaceae*
japonica (Japan), "Japanese quince"; dwarf evergreen shrub of wide-spreading habit, to 1 m high, spread to 2 m, with short spines; glossy green, leathery, spoon-shaped leaves, crenate at margins; orange-red apple-blossom flowers 3 cm dia., on previous year's growth; followed sometimes by fragrant, apple-shaped yellowish fruit, gnarled, to 4 cm dia.; used for making preserves. *Warm temperate.* p. 470, 848
speciosa (lagenaria) (Cydonia japonica hort.) (China; cult. in Japan), "Japanese quince"; deciduous shrub to 3 m high, of dense habit, more or less spiny, with sharply toothed, glossy green, ovate leaves; the showy, waxy flowers red, to 5 cm across, in spring before the leaves; roundish fruit greenish-yellow and fragrant. *Warm temperate.* p. 848

CHAENOMELES: see also Cydonia

CHAMAECEREUS   *Cactaceae*
sylvestri (Argentina), "Peanut cactus"; little clusters of short cylindrical, fresh green branches with soft white spines; flowers orange-scarlet. *Subtropic.* p. 257
sylvestri 'Aureus' (Argentina); the little column is lacking chlorophyll and consequently entirely yellow with white hairs. *Arid-subtropic.* p. 257
sylvestri 'Monstrosus' (Argentina); monstrose form with columns fused into fan-like pads. *Arid-subtropic.* p. 257

CHAMAECYPARIS   *Cupressaceae (Coniferae)*
nootkatensis 'Pendula' (Pacific coast to Alaska), "Alaska cedar" or "False cypress"; beautiful conifer to 30 m high, with fissured bark, spreading branches; leaves closely appressed on pendulous branchlets; cones 1 cm across, reddish brown. *Warm temperate.* p. 337

CHAMAEDOREA   *Palmae*
elegans (Collinia) (Mexico), "Parlor palm"; small graceful, relatively fast grower with thin stem to 3 m, 3 times as high as Neanthe bella and eventually forming clusters; pinnate leaves loosely spirally arranged, broadly lanceolate, thin-leathery segments dark green; good keeper in shady locations; fruit yellow to white. *Tropical.* p. 770, 777
elegans 'Bella' (Neanthe bella) (E. Guatemala); a tree palm in miniature, from mountain forests of Mayaland, more dwarf in habit than Chamaedorea elegans and slow-growing, thin stems bearing near the top a graceful rosette of small, pinnate fronds, with narrow, hard-leathery dark green pinnae; will flower as a youthful plant in pots. *Tropical.* p. 777
erumpens (Honduras), "Bamboo palm"; suckering dwarf palm, forming bushy, erect clusters of thin, bamboo-like reed-stems, loosely furnished down to the base with short pinnate, drooping leaves, the segments are broad, almost papery, dark green and recurved; a good keeper when well established. *Tropical.* p. 770
metallica, introduced in hort. as tenella (Mexico), the "Miniature fishtail"; one of the smallest of palms, slowly developing a stiff stem to 1 m high, and which bears a rosette of broad, leathery, shining green entire leaves forked only toward apex, and distantly toothed; blooming when quite young with simple, short spadix bearing minute yellow flowers. A tough and unusual small decorative plant. *Tropical.* p. 770
seifrizii (graminifolia) (Mexico: Yucatán), "Reed palm"; small stoloniferous palm with clustering slender cane-like stems alternately furnished near the top with broadly spreading pinnate fronds, their pinnae long and narrow and spaced apart, giving the plant a lacy appearance; withstands some cold. *Tropical.* p. 770
stolonifera (Guatemala), "Climbing fishtail palm"; stoloniferous small palm to 2 m high, covering, in time, large areas by running stolons; the plaited leaves are satiny green and slightly glaucous and remain simple or bifid, on slender rattan-like cane 1 cm thick. *Tropical.* p. 770

x CHAMAELOPSIS   *Cactaceae*
'Firechief' (Chamaecereus x Lobivia); striking bigeneric hybrid by Harry Johnson of California; small olive-green finger-like, 9 to 10 ribbed columns bearing needle spines and white radials, forming clusters; free-blooming with brilliant flame-red flowers to 8 cm across. *Arid-subtropic.* p. 258

CHAMAERANTHEMUM   *Acanthaceae*
igneum (Xantheranthemum) (Perú); low, spreading herb with beautiful, dark brownish-green, oblanceolate, veloured leaves, and vein pattern in red to autumn-yellow; bright yellow flowers. *Tropical.* p. 34
venosum (Brazil); dwarf plant with wiry stems spreading close to the ground, its small, hard-leathery 8-10 cm ovate leaves grayish, and attractively marked with a network of silver along veins. *Tropical.* p. 35, 38

CHAMAEROPS   *Palmae*
humilis, (Mediterranean, So. Europe, No. Africa), "European fan palm"; usually dwarf, clusterforming, sometimes to 6 m in arborescent forms, trunk rough, clothed with fiber; tough leaves relatively small, stiff and folded, with many narrow segments nearly to the base, on spiny, flat stalks; small red-brown fruit. *Subtropic.* p. 769, 770, 772, 777

CHAMAEROPS: see also Trachycarpus

CHAMBEYRONIA   *Palmae*
macrocarpa (Australia, New Caledonia); beautiful, unarmed feather palm with trunk to 20 m tall, the pinnate fronds to 1 m or more long; alternate dark green leaflets, prominently ribbed, on yellow or brown petiole; attractive with young unfolding leaves deep blood-red; ovoid fruit 3 cm long. *Tropical.* p. 774

CHEIRANTHUS   *Cruciferae*
x allionii, "Siberian wallflower"; properly Erysinum; subshrubby plant to 40 cm high, with mid-green narrow lanceolate leaves; orange flowers cross-shaped 2 cm across. *Warm temperate.* p. 380
cheiri (So. Europe), "Wallflower"; woody perennial, but grown as a biennial, blooming in spring; sweetscented flowers on terminal racemes like stocks, but in yellow-brown shades; narrow leaves slightly pubescent. *Subtropic.* p. 380

CHIMONANTHUS   *Calycanthaceae*
praecox (fragrans), (China), "Wintersweet"; deciduous shrub to 3 m high, blooming early winter from November on to March in the Northern hemisphere, long preceding the foliage; waxy 2 cm flowers with outer sepals and petals yellow, inner petals smaller and brown-purple; followed by the lustrous green lanceolate leaves 8-15 cm long. *Warm temperate.* p. 287

CHIONANTHUS   *Oleaceae*
virginicus (Pennsylvania to Florida and Texas), the "American fringe-tree"; deciduous tree to 10 m, with oblong leaves to 20 cm,

veins downy beneath; white flowers with usually 4, sometimes 5-6 narrow linear 2½ cm petals, in drooping panicles; dark blue fruit. *Warm temperate.* p. 696

CHIONODOXA *Liliaceae*
  sardensis (Asia Minor), "Glory-of-the-snow"; bulbous herb with rolled channeled leaves, and all-blue flowers with long spreading lobes on nodding stemlets, 3 to 6 on each short stalk, in early Spring. *Subtropic.* p. 602

CHIRANTHODENDRON *Bombacaceae*
  pentadactylon (syn. Cheirostemon platanoides) (Mexico, Guatemala), "Mexican hand-flower", "Monkey-hand tree" or "Manitos"; tree to 30 m high, with large cordate, 3 to 7 lobed leaves to 30 cm long, green above, tomentose beneath; strange flowers with 5 cm deep red waxy calyx, from the center of which extends a handlike appendage, complete with "finger-nails". *Subtropic.* p. 193

CHIRITA *Gesneriaceae*
  lavandulacea (Malaya), "Hindustan gentian"; erect, branching, rather succulent plant with large ovate, soft hairy, opposite leaves with toothed margins, and whorls of axillary flowers with white pouchlike corolla tube and spreading limb of pale lavender blue, marked yellow in throat. *Humid-tropical.* p. 491
  sinensis (So. China), "Silver chirita"; attractive stemless plant to 15 cm high, with tuberous roots, and large quilted thick-fleshy ovate leaves set crosswise, white-hairy over deep green and silver variegation, crenate margins, on flat petioles; lavender-lilac flowers gloxinia-like, in stalked clusters. *Humid-tropical.* p. 491, 498

CHLIDANTHUS *Amaryllidaceae*
  fragrans (Andes of Perú to N.E. Argentina), "Perfumed fairy lily"; bulbous plant with basal strap-shaped sheathing leaves and fragrant yellow 8 cm trumpet flowers, appearing before the leaves, in umbels on solid stalks; summer. *Subtropic.* p. 60

CHLOROPHYTUM *Liliaceae*
  bichetii (Trop. W. Africa), "St. Bernard's lily"; small tropical herb with broad, grass-like but thin-leathery leaves fresh-green with yellowish white stripes, forming bushy tufts. *Tropical.* p. 576
  bracteatum (Zaire); low rosettes seen at Kew Gardens, of broad lanceolate, glossy green leaves; white flowers. *Tropical.* p. 590
  comosum 'Milky Way'; a beautiful cultivar with broad-linear recurving foliage; cream-white and with friendly green edging; very attractive in a hanging basket. *Subtropic.* p. 590
  comosum 'Variegatum', "Ribbon plant"; large rosettes of arching, fresh green, linear leaves 25-40 cm long, 2-3 cm wide, having margins edged in white; long racemes appearing from the center will first flower, then develop tufts of leaves with aerial roots. *Subtropic.* p. 584
  comosum 'Vittatum' (C. elatum vittatum) (Cape of Good Hope), "Spider plant"; low clustering rosette with channeled narrow-linear, recurving leaves, dark green banded white in center; successive plantlets develop from the long flowering racemes which become pendant if used in baskets; leaves only 10-20 cm long, 1 cm wide. *Subtropic.* p. 584, 590
  ignoratum (Trop. Africa); fleshy lanceolate green leaves 30 cm long; inflorescence on erect stalks with small green flowers salmon-pink in center. *Subtropic.* p. 584
  laxum (Trop. Africa); ornamental, bushy plant with glossy green, strap-shaped leaves prettily bordered with white; flowers whitish. *Tropical.* p. 590

CHORDOSPARTIUM *Leguminosae*
  stevensonii (New Zealand); curious leafless tree to 8 m tall; with thick trunk, and pendulous switch-like pendulous branches; tiny lavender flowers. *Warm temperate.* p. 549

CHORISIA *Bombacaceae*
  insignis (Perú, N.E. Argentina), "White floss-silk tree" or "Drunken tree"; big tree to 15 m high, with fat, flask-shaped green trunk to 2 m thick, big spines on the younger trunk and branches, the sparry branches with palmately compound leaves of 5-7 obovate, waxy green leaflets 6-8 cm long and usually toothed near apex; beautiful, lily-like, large waxy flowers to 14 cm long, with undulate petals at first yellow with purple markings toward base, later fading to cream or white. *Subtropic.* p. 193, 197
  speciosa (Brazil, No. Argentina), "Floss-silk tree"; large tree about 15 m high, its trunk usually studded with stout sharp thorns, digitately compound toothed leaves and showy, variable, usually pinkish 5-petaled flowers; pear shaped fruits containing silky floss on the seeds. *Tropical.* p. 193, 197
  speciosa 'Majestic Beauty'; a medium size Floss-silk tree, offered as grafted cultivar in California nurseries (1976), more evergreen than the species which blooms usually during the brief leafless stage in winter; bright green thornless trunk; in autumn the spectacular, showy flowers with flaring, crisped petals 10 to 13 cm across, rosy-carmine with white center. *Subtropic.* p. 193

CHORIZEMA *Leguminosae*
  cordatum (ilicifolium) (Western Australia), "Australian flame pea"; shrub with weak, wiry branches, and small leathery, ovate or cordate leaves sharply toothed and spine-tipped; small flowers red-orange with a yellow blotch, and purplish keel. *Subtropic.* p. 544

CHOYSYA *Rutaceae*
  ternata (Mexico), "Mexican-orange"; evergreen shrub to 3 m high, with lustrous, yellow-green trifoliate leaves, held toward the end of the branches, each leaflet to 8 cm long, and forming fans giving the plant a dense, massive look; fragrant white flowers 3 cm across, like small orange blossoms, in clusters above the foliage; fruit of 8-10 cm dia.; leaves are strongly scented when bruised. *Tropical.* p. 872

CHRISTIA *Leguminosae*
  vespertilionis (Lourea) (Brazil); evergreen bush with scandent, rambling, wiry branches, and leaves with two long lobes, set at an angle like swallow-tails spreading 15 cm, green with silvery-gray or ivory ribs; small white flowers. *Tropical* p. 546

CHRYSALIDOCARPUS *Palmae*
  cabadae (Madagascar), "Cabada palm"; fast-growing clustering graceful feather palm to 10m tall; slender trunks with swollen base; the leaves are arranged in vertical rows of three as seen from below; slender, glossy green leaflets in one plane held at an open angle, on thin wiry petioles; small 2cm red fruit. *Tropical* p. 770
  lutescens (Areca) (Madagascar), "Butterfly palm"; with slender, graceful, yellowish stems, forming an attractive clump to 8 m tall, with pinnate foliage nearly to the base, narrow, papery, the pinnae glossy yellow-green and well spaced, on yellow, willowy, furrowed stalks; fruit violet-black. *Tropical* p. 773, 774

CHRYSANTHEMUM *Compositae*
  frutescens (Canary Islands), "White marguerite"; bushy branching herbaceous plant with lacy, divided grayish-green leaves and slender single white flowers with yellow disk. *Subtropic* p. 314
  frutescens chrysaster, "Boston yellow daisy"; form with fleshier stem, fresh green, divided leaves and lemon-yellow rays heavier and not as free flowering as the type. *Subtropic* p.314
  frutescens 'Mary Wootten'; a famous "Tree-marguerite" named after a lovely Miss New Zealand; (Pink Beauty x Mrs. Sanders); 8-10 cm flowers, pleasing rose-pink, the high cushion center with the outer fringe darker, center more soft pink; very desirable in this color range; blooming over a long period. *Subtropic* p. 314
  frutescens 'Mrs Sanders'; herbaceous shrub of the group grown in New Zealand as "Tree-marguerites", wide-branching plant with lacy-pinnately cut bright green leaves and double flowers with large pincushion centers, pure white; widely used by florists for cut and grown in gardens. *Subtropic* p. 314
  frutescens 'Roseum', "Pink marguerite"; bushy, branching herbaceous plant with lacy, divided grayish-green leaves and many slender, single flowers with ray petals light pink. *Subtropic* p. 314
  frutescens 'Snowflake', "New Zealand tree-marguerite"; rambling subshrub with pure white somewhat shaggy, double flowers glistening white, about 6 cm dia. *Subtropic* p. 314
  frutescens 'Wellpark Beauty'; an English "Tree-marguerite," with waxy green, irregularly cut bipinnatisect foliage, and soft lilac-pink flowers with anemone type double cushion soft pink, darker red in center. *Subtropic* p. 314
  maximum 'Wirral Supreme'; a large-flowered "Shasta daisy", originally from the Pyrenees of Spain; robust, hardy perennial to 75cm and more high, often grown as biennial, with lanceolate, serrate, deep green foliage, the stems topped by large 5 to 10 cm marguerite-like white flowers with a double row of ray-flowers, and a white and yellow anemone center cushion; late summer blooming. *Warm temperate* p. 308
  morifolium (hortorum), the "Florists' chrysanthemum", known in slang as "Mum"; cultigen of Chinese origin and cultivated in China and Japan for 3000 years; perennial herb ½-1 m or more high, and much-branched, with succulent stems becoming woody, and strong-scented lobed leaves gray pubescent; flowers in showy terminal heads of various colors except blue, the florets modified by long cultivation into petal-like marginal ray-flowers and small, usually tubular disk-flowers forming center cushions; developed into several groups such as single, cascade, anemone (with center cushion), pompon (globular), decorative (aster-like), spider, incurved, and large exhibition; their growth habit is regulated by pinching. *Warm temperate* p. 309, 316
  mor. 'Beautiful Lady', "Anemone-mum"; large "anemone" of bushy habit with strong stems and good foliage, flowers semi-

double lavender pink lighter toward center, and a yellow cushion (Nov. 25). *p. 313*

mor. **'Bonaffon Deluxe'**, "Incurved", well-known pot plant variety, with numerous medium-large deep lemon yellow flowers of good lasting quality, on wiry stems, and with foliage of good substance (Nov. 20). *p.313*

mor. **'Cathay'**, a "Spider-spoon" type with large flowers of exquisite shape and color; the slender ray-flowers in a burst of rose-pink splendor curled inward at tips like little spoons, and red inside. *Warm-temperate* *p. 312*

mor. **'Decorative incurved'**; typical of the best of florists' "Mums", grown for cutting or in pots, with firm heads of incurved ray petals. *Warm temperate* *p. 311*

mor. **'Delaware'**, "Incurved type"; a typical good commercial, free branching pot plant with firm foliage and semi-incurved, deep amaranth-red bleaching to bronze; reverse yellow. *p. 313*

mor. **'Golden Cascade'**, "Daisy-cascade"; commercial "cascade" variety of the single daisy type very floriferous with light yellow ray-petals around a golden yellow center disk; earlier than 'Jane Harte'; its dense habit with long wiry branches is equally adaptable to training to forms as well as broad bushy globes. (Oct. 30). *p. 312*

mor. **'Golden Lace'**, "Fuji-mum"; morifolium cultivar of the spider type, large flowers with graceful threadlike, golden yellow quilled florets hanging from the double center, the tips are open and curled (Oct. 25). *p. 312*

mor. **'Hawaii'**, "Incurved Exhibition"; beautiful large incurved type, soft pink and of firm texture. *Warm temperate* *p. 313*

mor. **'Illini Bonbon'**, "Cushion anemone"; small but attractive "anemone" of short, bushy habit ideal for pots, the stiff stems with massive clusters of 4cm flowers of daisy-like flat petals lemon yellow, surrounding a high center cushion of golden yellow tubular petals (Nov. 18). *p. 312*

mor. **'Imperial'**; a "decorative" form of 10cm aster-like flowers with boat-shaped petals carmine pink with darker streaks and pale reverse, the double center attractively tipped with pale gold; free-blooming, medium-tall pot plant (Nov. 20.) *p. 313*

mor. **'Indianapolis Yellow'**; large "incurved", dependable type, assuring good buds and elegant flowers, with rich yellow petals in part tubular, bleaching to pale yellow with age, but of long-lasting quality; a rather tall variety for pots, very vigorous (Nov.). *p. 311*

mor. **'Lorraine'**; a good true "Spider mum"; very floriferous with thin but wiry stems; large 15 cm flowers as growing in pots; long tubular petals with a quilled spoon at tips, clear lemon yellow (Nov. 15). *p. 312*

mor. **'Oregon'**; "Semi-incurved" variety exclusively for pots; stronger habit and larger flower than 'Wilson's White', compact stiff growth with healthy foliage, and double white flowers incurved at first but becoming shaggy (Nov. 5). *p. 313*

mor. **'Peggy Ann Hoover'**; attractive "spider" mum of the "Fuji" type, large fine mauve pink flowers with thread-like tubular petals, a tint of pale yellow in center; thin-necked but wiry stems and good in pots (Oct. 20). *p. 312*

mor. **'Precose'**; "White anemone"; large "anemone" type spreading flower 11cm across lemon-yellow in bud opening to white with greenish center, with several rows of petals; bushy pot plant (Nov. 10). *p. 312*

mor. **'Princess Ann'**, "decorative" type, and excellent pot plant of compact, shapely growth habit with firm foliage and substantial flowers 10cm across, broad-petalled, peach-pink slightly darker in center, paling to light pink at edges. (Nov. 10). *p. 313*

mor. **'Real Mackay'**, "Cascade-mum"; a dependable "cascade" type branching bush, with flexuous branches bearing the numerous large daisy-like 5cm flowers fine rose to pale pink, with yellow cushion center. *p. 312*

mor. **'Red King'**; large "anemone" type 9cm flowers with half dozen layers of recurved boat-shaped petals rusty-red, and contrasting yellow beneath; a robust pot plant (Nov. 13). *p.312*

mor. **'Red Star'**; compact plant with shiny green foliage, and decorative, aster-type double flowers with flat petals yellow to bronze, cinnamon beneath. *Warm temperate* *p. 313*

mor. **'Sungold'** with vigorous "incurved" type with large flowers deep lemon yellow, fully double and of good keeping qualities, some petals tubular; strong stems and good commercial pot plant for Thanksgiving (Nov. 25). *p. 311*

mor. **'Sweepstake'**, "Pompon-mum"; typical small-flowered "pompon" with full round golden-yellow 4½cm blooms, of dense, bushy habit with small, dark green foliage; used for cut, pots or garden (Oct. 7). *p. 312*

mor. **'Turner'**; well-known, old variety with large greenish golden yellow flowers on long stiff stems and used as a cut flower. *Warm temperate* *p. 311*

mor. **'Venoya'**; listed as a "large anemone" but with its thick layers of shaggy petals appearing like a decorative type, except for its green-eyed cushion, framed by blush to lavendar pink petals; a tall pot plant (Nov. 22). *p. 312*

mor. **'Warhawk'**; striking pot plant variety with large and showy incurved, double flowers 12cm across, burnt brown-red with amber-yellow reverse; firm leathery foliage. *Warm temperate p. 313*

mor. **'Wilson's White'**; intermediate "Semi-incurved" pot variety with small foliage and broad-petalled pure white flowers tinted greenish in the center (Nov. 1). *p. 313*

mor. **'Yellow Delaware'**, intermediate "semi-incurved" sport of 'Delaware', rich yellow 10cm flowers (pot-grown), the lower petals turning outward and center opens; (Nov. 8). *p. 313*

mor. **'Yellow Princess Ann'**, favorite "Decorative"; mutation of 'Princess Ann' with deep yellow decorative flowers, of good keeping quality. *Warm temperate* *p. 313*

mor. **'Yellow Spoon'**, "Spoon-mum"; large single flowers with spreading ray-petals, tubular and flaring into a cupped spoon; golden yellow. *Warm temperate* *p. 312*

CHRYSOLARIX: see Pseudolarix

CHRYSOPHYLLUM   *Sapotaceae*

cainito (C. America), "Star apple" or "Caimito"; handsome ornamental evergreen tree and also a fruit tree, to 15 m high, with gracefully pendant branches; elliptic leaves 15-18 cm long, shining rich green above, and silky coppery-brown underneath; small purplish-white flowers, followed by the smooth roundish fruit 5-8 cm dia, with green or black-purple skin and sweet white flesh, star-shaped in transverse sections. *Tropical* *p. 463, 880*

CHRYSOTHEMIS   *Gesneriaceae*

friedrichsthaliana (Alloplectus lynchii, Tussiaca) (Guatemala, Trinidad, W. Indies); erect, tuberous-rooted, succulent plant with waxy green, hairy, lanceshaped, crenate leaves to 30cm long; axillary flower clusters, the short orange corolla with dark-striped lobes peeking out of large greenish-yellow calyx. *Humid-Tropical.* *p. 497*

pulchella (Trinidad), erect succulent herb producing tubers with age; large fleshy opposite, rough-bristly, puckered 15 cm leaves shiny bright green, with crenate margins; the flowers in the leaf axils are small, bristly 1 cm tubes buttercup-yellow with red stripes and markings inside, but curiously interesting for the cluster of large flame-red calyces which appear well before the blossoms and keep long afterwards, giving the plant a colorful appearance over a long period. Blooms in spring; allow to go dormant when too tall. *Humid-Tropical* *p. 491, 493*

CHYSIS   *Orchidaceae*

aurea (Mexico, Colombia, Venezuela); bright colored epiphyte with spindle-like stems bearing 3-5 deciduous leaves, the thick-waxy 5 cm flowers closely clustered in short raceme, petals reddish-buff with golden yellow edges and base, throat yellow marked red. (March-Aug.). *Humid-tropical* *p. 712*

bractescens (Mexico, Guatemala); epiphyte with compact pseudobulbs and deciduous leaves, the fragrant waxy, large 8cm flowers in arching raceme, ivory-white with yellow lip marked red. (Feb.-May).*Humid-Tropical* *p. 715*

CIBOTIUM   *Dicksoniaceae (Filices)*

barometz (So. China, Assam, Malaya), "Lamb of Tartary"; from a hairy rhizome rise the handsome, fragrant light green fronds of leathery texture, bipinnately toothed, and bluish beneath, on wiry stalks covered with light brown hairs. *Humid-Tropical.* *p. 436*

chamissoi (menziesii in hort.) (Hawaii), "Man fern" of the islanders, with a short, stout, fibrous trunk, and handsome, massive, tripinnate fronds, delicate, glossy nile-green and of a smooth leathery-hard texture, on stalks covered with blackish woolly hair. *Humid-tropical.* *p. 435*

glaucum (chamissoi in hort.) (Hawaii), "Blonde treefern" of the rainforest, with stout, fibrous trunks, consisting of a mass of aerial roots around a core of starch, and capable of storing moisture, with heads of large, soft-textured, tripinnate crinkled fronds luxuriously green, the stalks covered with soft pale hair. *Humid-tropical.* *p. 435*

schiedei (Mountains of Mexico and Guatemala), "Mexican treefern"; favorite decorator of florists, with shapely crown of graceful, light green fronds, thin-leathery and dainty yet durable, lacy tripinnate, and glaucous beneath; will eventually, over many years, form a fibrous trunk to 5 m high. *Humid-tropical.* *p. 436*

CINCHONA   *Rubiaceae*

officinalis (Colombia to Perú), "Lojas quinine", formerly known as "Jesuits bark"; small tree with rough bark; lanceolate leathery leaves to 10 cm long; clusters of 2 cm salverform flowers, deep pink; the fruit a capsule opening from base. The bark is stripped on young trees, and yields the quinine of commerce, used in anti-malarial medicine. *Tropical.* p. 861

CINERARIA: see Senecio

CINNAMOMUM   *Lauraceae*
   camphora (China, Japan), "Camphor tree"; handsome tree to 12 m high or more, with dense crown of shiny, dark green leaves 5-12 cm long; when crushed, the leaves have the odor of camphor; small yellow flowers; small black fruit. *Subtropic.*   p. 534
   iners (Tropical Asia); small tree with wiry branches, the twigs with opposite obovate leaves 10 cm long, coppery glossy green and with parallel veining. *Tropical.*   p. 534
   zeylanicum (Ceylon, S.W. India), "Cinnamon tree"; small evergreen tree to 10 m high; glossy green, leathery ovate leaves to 18 cm long, 3-nerved from base; the young growth with foliage strikingly crimson-red with white veins; small yellowish flowers. Dried bark yields the commercial spice cinnamon. *Tropical.*   p. 534

CIRRHOPETALUM: see Bulbophyllum

CISSUS   *Vitaceae*
   amazonica (Amazonas); very attractive tropical climber with string-like vines and closely set, alternate, narrow-lanceolate, somewhat fleshy leaves in two ranks, becoming broader with age, the glossy surface deep metal-gray with contrasting silvery vein area and brown midrib, the margins remotely toothed, and maroon beneath. *Tropical.*   p. 920
   antarctica (Rhoicissus) (New So. Wales), "Kangaroo vine"; attractive plant with shrubby base and flexuous, slowly climbing, tomentose branches, the elegant, firm leaves almost leathery, bright shining green to a deep metallic shade, and light brown veins, the base cordate, and margins sinuately toothed, becoming 15 cm long; a good keeper. *Subtropic.*   p. 922
   discolor (Java, Cambodia), "Rex begonia vine"; beautiful tendril-climber, with thin angled, dark-red veins and petioles, the strikingly colored, oblong-cordate, quilted leaves to 15 cm long, the sunken network of veins moss-green, with elevated ridges painted shimmering silver, and the center variegated violet and red-purple with velvet sheen; toothed margin and reverse glowing maroon. *Tropical.*   p. 919, 922
   gongylodes (Brazil, Paraguay, Perú); robust liane for the large conservatory, rapid climber with 4-angled stems, large trifoliolate rugose leaves to 20 cm long; hairy at the nerves and at the margins; the tendrils with suction cups. Most attractive are the long pendant, red aerial roots looking like a bead curtain. At end of growing season the ends of branches produce a terminal fleshy tuber which drops and develops into a new plant. *Tropical.*   p. 919, 922
   hypoglauca (Victoria, New South Wales); climbing shrub without tendrils, to 5 m or more; leaves divided into 5 oblong, leathery leaflets with some teeth along margins, polished bronzy green. *Subtropic.*   p. 922
   juttae (S. W. Africa); queer-looking succulent with thick-fleshy, stubby, pale gray, smooth trunk 1-1½ m high, dividing into a few thinner branches toward apex, sessile oval, serrate leaves 10-15 cm, shining waxy green above, translucent hairs beneath, soon falling; panicles of small flowers followed by berries. *Arid-subtropic.*   p. 920, 921
   quadrangularis (Vitis) (Trop. Africa, Arabia, India, Moluccas); odd succulent with thick, fleshy, rich green, 4-winged stems much constricted at the nodes, climbing, mostly glabrous, often nearly leafless, looking like a spineless cactus; small herbaceous, 3-lobed, green leaves; green flowers and berried red. I have seen this plant clambering over euphorbias and various scrub on the steppes in Uganda, Tanzania and northern Transvaal. *Arid-tropical.*   p. 921
   rhombifolia (Vitis, Rhoicissus) (West Indies, N. So. America) "Grape-ivy"; scandent, herbaceous plant with vine-like, flexuous, brown-hairy branches having coiling tendrils, and compound leaves of 3 rhombic-ovate, stalked, thin-fleshy leaflets wavy-toothed, the glabrous surface fresh green to metallic deep green and with brownish veins and petioles, pubescent underneath, the young growth covered with white hairs. Small flowers with 4 petals, unlike Rhoicissus which have 5-7 petals. *Tropical.*   p. 922
   rhombifolia 'Ellen Danica'; excellent, vigorously growing grape-ivy; mutation of 'Jubilee' (Nielsen-Denmark) found in Odense about 1965; of bushy habit with foliage divided and larger than the species, the leaflets like deeply lobed and incised oakleaves, rich glossy green, on brownish branches. Suitable for larger containers or for pyramids. *Subtropic.*   p. 922
   rhombifolia 'Mandaiana' (Vitis); meritorious sport introduced by Manda in 1935, of more upright, compact habit, thicker stems and substantial, fleshy, deep green leaves, the leaflets broad and firm, almost leathery, lightly quilted and recurved, shining like polished wax; very attractive and a good keeper. *Tropical.*   p. 922
   rotundifolia (E. Africa: Tanzania, to Yemen), "Arabian wax cissus"; climbing plant with green, 4-angled stems having sharp corky edges in older specimen; rounded waxy-fleshy leaves 6-8 cm dia., distantly toothed at margins, glossy fresh green to olive; inconspicuous green flowers, followed by red berries. *Arid-tropical.*   p. 922

CISSUS: see also Piper, Parthenocissus, Rhoicissus, Tetrastigma

CISTUS   *Cistaceae*
   incanus (villosus) (Corsica to Crimea), "Rock rose"; low shrub to 1 m, with small, rugose, ovate leaves, and large crepy rose-pink flowers 6 cm dia. *Subtropic.*   p. 302
   ladanifer (Portugal), "Laudanum"; compact bush 1 m high, with rugose, fragrant leaves to 10 cm; large 3 cm flowers, carmine pink. *Subtropic.*   p. 302
   salvifolius (S.E. Europe), "Sageleaf-rock-rose"; low spreading shrub to 60 cm high, photographed amongst ancient Greek ruins on the island of Rhodes; with small gray-green rugose leaves 3 cm long, and flowers like wild roses, with white or pinkish petals of crepy texture, with yellow spots in base, 4 cm across; blooming very profusely in spring. *Subtropic.*   p. 302

CITRULLUS   *Cucurbitaceae*
   lanatus 'Dixie Queen' (Trop. Africa), "Watermelon"; coarse annual vine with branched tendrils; leaves pinnately divided; large 4 cm yellow flowers; fruit globose and banded silver, 25 cm long; deep red flesh with sweet flavor. *Tropical.*   p. 382
   lanatus 'Family Fun', "Watermelon"; early ripening large fruit 6 to 7 kg.; oblong with rind medium green striped dark green; crisp red flesh is sweet and tasty. *Tropical.*   p. 472

CITRUS   *Rutaceae*
   aurantifolia (Malaya, India), the "Lime" or "Key lime"; small thorny tree to 5 m tall, very tender; with aromatic small 6-8 cm glossy green, oval leaves; waxy-white, fragrant, axillary flowers; small oval, green to greenish yellow fruit 3 to 5 cm dia. usually in clusters; smooth, thin-skinned with very acid seedy pulp; used in drinks, on seafood or in preserves. *Subtropic.*   p. 479
   aurantium (Vietnam), "Sour Seville orange"; the pictured variety 'Temple' bears high quality, delicious, juicy fruit of large size with deep orange, glossy skin, a vigorous grower, bearing freely when very young, and lending itself to culture in tubs; large, fragrant, white, waxy flowers, the slender-pointed leaves with broadly winged petioles. *Subtropic.*   p. 479
   aurantium 'Marginata'; ornamental variety cultivated in Japan, with leaves prettily edged in white; few fruit. *Subtropic.*   p. 874
   aurantium 'Myrtifolia' (China), "Myrtleleaf orange"; dwarfish, thornless tree with dense, deep green, stiff, little 2 cm leaves; waxy flowers and small, sour, bright orange fruits; while blooming in April in the northern hemisphere, I remember the strikingly impressive sight of these trees in Uruguay and Argentina in October, when literally covered with fragrant white flowers. *Subtropic.*   p. 479, 482, 874
   'Dweet' (Tangor) (Orange x Tangerine) (reticulata x sinensis); dwarf tree to 2½ m of open habit; fruit globe-shaped with orange skin; seedy, but packed with sweet juice. *Subtropic.*   p. 480
   limon 'Eureka', "Acid lemon"; medium size tree, open and spreading, almost thornless; large dark green leaves; fragrant waxy white flowers; oblong fruit with protruding apex, lemon-yellow, 8 cm or more long; very juicy and acid. *Subtropic.*   p. 479, 874
   limon 'Meyer' (hybrid of lemon with sweet orange), "Meyer lemon" or "Dwarf Chinese"; semi-dwarf, spreading tree, nearly thornless, with small leaves, often grown in pots as it produces sweetly-scented, lavender to white flowers while young, and bears good-looking, oval lemons of table quality, furnishing excellent acid juice almost the year round. *Subtropic.*   p. 479, 874
   limon 'Ponderosa' (Maryland hybrid 1887), "American wonder-lemon"; ornamental tree 2½-3 m high, often grown in tubs, with irregular branches and short, stout spines, large oblong leaves on short petioles nearly wingless, the large waxy flowers white, the somewhat pear-shaped, big fruit lemon-yellow, almost 12 cm long, weighing 1 kg, and while edible, its juice is sour. *Subtropic.*   p. 479, 873
   maxima (grandis), the "Shaddock" or "Pummelo", probably from So. China, Malaya and Polynesia; tender tropical spiny tree to 9 m, with angular twigs and large oval 10-20 cm leaves slightly pubescent beneath, the petiole widely winged; large white flowers; fruit 15 cm large and pale yellow, broadly pear-shaped, the acid flesh coarse-grained and separating readily. *Subtropic.*   p. 479
   microcarpa (South China), "Four-season tangerine"; possibly hybrid of Mandarin x Fortunella; depressed globose fruit 3 cm dia., with thin peel and with intensely acid pulp. *Subtropic.*   p. 482

'Minneola' (Tangelo) (Grapefruit x Tangerine) (reticulata x paradisi); reddish-orange fruit with smooth skin, few seeds, unusual exotic tangy flavor and very juicy. *Subtropic.* p. 480

mitis (Philippines), "Calamondin"; a small, spineless tree with upright slender branches dense with broad-oval, leathery leaves on narrowly-winged petiole; small white, fragrant flowers borne singly at the tips of twigs; small 4 cm round fruit somewhat flattened, deep orange-yellow, loose skinned. Probably a hybrid of lime with kumquat; strongly acid, pleasant flavor like lime; season Nov.-Dec., but occasionally everbearing. *Subtropic.* p. 481, 482, 874

x paradisi (West Indies), "Grapefruit"; spreading tree with big leaves on broadly winged petioles, and large white flowers, the smooth branches bending under the weight of the large and heavy, pale lemon-colored, round fruit, much grown for its juice. C. paradisi of the West Indies is a satellite species rising from the tropical C. grandis (maxima) of S.E. Asia. *Subtropic.* p. 481

x paradisi 'Marsh Seedless', "Seedless grapefruit"; important commercial grapefruit of the West; large light yellow fruit with few or no seeds, and big glossy leaves. *Subtropic.* p. 480

x paradisi 'Ruby Blush', "Pink grapefruit"; large fruit with orange skin blushing pink; deep pink flesh and quite sweet tasting in flavor. *Subtropic.* p. 480

reticulata (Southeast Asia), "Mandarin orange"; small spiny tree with slender branches; lanceolate leaves; flowers white and fragrant; fruit deep orange-yellow depressed globose, to 8 cm dia.; loose thin skin, sweet pulp. *Subtropic.* p. 481

reticulata 'Clementine', "Tangerine"; tree to 4 m high, of open habit, with slender branches hanging under weight of fruit; fruit deep orange-red with glossy surface, 5-6 cm dia., apex depressed, skin loose; tender melting flesh, Nov.-Dec. *Subtropic.* p. 480

reticulata 'Dancy', "Dancy tangerine"; small tree of spreading growth with thin branches and twig-like shoots becoming pendulous; narrow dark green leaves, and bright orange colored fruits to 8 cm with zipper-skin, peeling easily and segments easy to separate. *Subtropic.* p. 480, 874

reticulata 'Satsuma' (Japan), "Mandarin orange" or "Unshin"; early ripening fruit October-December, medium size 5-6 cm, depressed globose, smooth, loose thin skin, pulp acid-sweet. *Subtropic.* p. 480

sinensis 'Valencia' (Spain), "Valencia orange"; large vigorous tree to 10 m high, fuller growing than Washington Navel, with regular branches and rounded top; flexible slender spines sometimes absent; dark shining green, ovate, leaves narrowly winged; waxy white, sweetly fragrant flowers; fruit nearly globose, golden yellow 6-8 cm dia., and with sweet pulp, ripening March to July and lasting on the tree for months, improving in sweetness. It is the most widely planted juice orange in the world. Known in California since 1876, it had its origin in Spain under the name "Naranja tarde de Valencia". *Subtropic.* p. 480, 481, 874

sinensis 'Washington Navel' or 'Bahia', from Brazil, "Winter orange"; excellent, large commercial sweet orange, early-ripening, and seedless, from November to April (Valencia from March to Nov.); best adapted to California and Arizona climate and produces low yield and poor fruit in tropical regions; it derives its name from the characteristic rudimentary secondary fruit imbedded in the apex of the primary fruit, and which serves as a trademark; deep orange in color, and easy peeling; flowers in large clusters, waxy-white and sweetly fragrant. *Subtropic.* p. 480, 873

taitensis (otaitense), "Otaheite orange"; dwarf tree grown as an ornamental pot plant, with glossy-green, dense foliage and few spines, the waxy-white, very fragrant flowers tinged with pink appearing in January, followed toward Christmas by golden orange fruits; a miniature version of the sweet orange, except that the juice is tart, more resembling a lime; the best for pots. Probably a lemon (or lime) x mandarin (aurantifolium x reticulatum) hybrid, possibly of garden origin in South China. *Subtropic.* p. 873

CLARKIA *Onagraceae*

amoena whitneyi (Godetia grandiflora), "Satin flower"; of California origin; sprawling, low, leafy, summer-blooming annual to 30 cm or more; with stout stem and lanceolate leaves to 5 cm long; satiny 8-12 cm flowers looking up, petals pinkish with central carmine spots; also in other colors, rosy red to white. *Subtropic.* p. 703

elegans 'Rosea plena' (unguiculata); species from California, "Farewell-to-spring"; tall garden annual to 1 m high, ovate to lanceolate leaves, and showy flowers on erect, reddish-glaucous stems; single in the species, double carmine-rose in this form of plena, 5 cm across, from July to October. *Subtropic.* p. 702

elegans 'Rubra plena'; with double flowers rich ruby-red. *Subtropic.* p. 702

CLAVIJA *Theophrastaceae*

tarapotana (Perú, Bolivia); evergreen palm-like subtropical tree with unbranched stem, bearing a rosette of large obovate, leathery leaves blunt at the tip, about 40 cm long; axillary from near the apex the pendant inflorescence; numerous racemes of small very fragrant orange-red flowers, followed by grape-like clusters of red berries. *Subtropic.* p. 907

CLEISTOCACTUS *Cactaceae*

baumannii (Argentina, Paraguay), "Firecracker"; thin, ribbed column clambering and branching at base, covered with white or brown spines; flowers scarlet, 6-7 cm long. *Arid-subtropic.* p. 246

candelilla (Bolivia: Cochabamba), aptly named the "Small candle in the Andes" because of the flame-like tubular red flowers, tipped with yellow, that appear near the apex of the slender ribbed columns of this natural fence or thicket-forming cactus, branching from the base, and ½-1 m tall; spreading radial spines yellowish, and 3-4 needle-like central spines. *Arid-subtropic.* p. 250

smaragdiflorus (Argentina, Uruguay, Paraguay), "Firecracker cactus"; slender, erect stems with 12-14 ribs, 5 cm thick, 2 m high, later leaning; closely set yellow areoles and numerous thin yellow to brown spines very sharp; flowers orange red inside but remaining closed. *Arid-subtropic.* p. 250

strausii (Bolivia), "Silver torch"; slender, light green, clustering, many-ribbed column 6 cm thick, covered with bristle-like white spines, central spine pale yellow; flowers red. *Arid-subtropic.* p. 248, 252

strausii jujuyensis, (Argentina), "Silver torch"; form with areoles bearing white bristly spines and one brown central spine. *Arid-subtropic.* p. 250

wendlandiorum (Andes of Bolivia); slender column with many shallow ribs and dense with soft golden spines; inflated flowers, orange and red. *Arid-subtropic.* p. 250

CLEMATIS *Ranunculaceae*

cirrhosa (E. Mediterranean), "Virgin's bower"; evergreen climber with shining leathery green rounded leaves, and large bell-shaped cream-colored nodding flowers. *Subtropic.* p. 847

x jackmanii; the jackmanii group (lanuginosa x hendersonii) is one of the best of "Virgin's bowers" for size of flower, vigorous in growth, and profuse in bloom; climber to 3 m, with pinnate or 3-foliolate leaves; flowers 10-15 cm across, violet-purple, and long-lasting, in July to October; fairly hardy. *Warm temperate.* p. 842, 845

lanuginosa 'Nellie Moser'; a free-blooming cultivar with flowers pale mauve-pink, and with a carmine-red bar down the middle of sepals; June-Sept. *Warm temperate.* p. 843

x lawsoniana 'Henryi' (lanuginosa x patens); climbing hybrid with leaves in threes, tremendous size white flowers 15-20 cm across. *Warm temperate.* p. 842

x lawsoniana 'Ramona'; a "Virgin's bower" vine of the lanuginosa (China) x patens (Japan) hybrid complex; shrubby deciduous winter-hardy climber with leaves of 3 leaflets, and striking flowers with spreading petals 15 cm or more across; in the clone 'Ramona' lavender blue with purple stamens, blooming during summer. One of the most beautiful of flowering hardy vines. *Temperate.* p. 845

vitalba (Europe, N. Africa, S.W. Asia), "Travellers-joy" or "Old-man's beard"; climbing to 10 m; leaves of 5 entire toothed or 3-lobed leaflets; fragrant flowers greenish-white, to 2½ cm across; very floriferous, in massed clusters. *Subtropic.* p. 842

CLEOME *Capparidaceae*

spinosa (pungens) (W. Indies), "Spider flower"; pubescent strong-scented annual with spiny stem, palmately compound leaves and large rose-purple flower clusters having long spidery stamens. *Tropical.* p. 290

CLERODENDRUM *Verbenaceae*

buchananii (East Indies); large palmately lobed leaves, and showy erect terminal clusters of bright red flowers similar to C. fallax. *Tropical.* p. 913

bungei (China); erect-growing shrub with wiry branches to 2 m high, with large, broadly ovate, quilted leaves to 30 cm long and coarsely toothed, dark green on the surface; fragrant flowers rosy-red in a head-like cluster 10-20 cm across, in June-Sept. *Subtropic.* p. 914

foxii hort.; magnificent shrub, in the Florida nursery trade, resembling C. paniculatum; glossy green, corrugated foliage of various shapes, in the center a tall erect inflorescence fountain-like, with masses of slender tubular scarlet flowers. *Subtropic.* p. 915

fragrans pleniflorum (China, Japan), "Glory tree"; shrubby plant with stiff stems covered with white hairs, broad ovate, opposite leaves light green and hairy on the surface, green beneath, with margins toothed; the fragrant, 2 cm flowers in a dense terminal cluster, delicate peach-white and fully double, calyx purplish-red. *Subtropic.* p. 913

inerme (India: Bombay), the "Indian privet"; evergreen branching shrub with hard-leathery, small boxwood-like shiny deep green leaves, and small white flowers with purple stamens; indigenous to mangrove areas near Bombay, and planted at the Hanging Gardens of Bombay, trained over wire frames, into topiary shapes of animals; sheared every 2 weeks. *Tropical.* p. 913, 914

myricoides (Trop. and So. Africa); dense shrub to 2 m, with spoon-shaped oval, matte green leaves 6-9 cm long, crenate at margins; the pretty flowers light lavender-blue with violet on lip, and long curling stamens. *Subtropic.* p. 914

paniculatum (E. Trop. Asia), "Kashmir bouquet"; robust bush with broad, shining green, 5-lobed leaves, and showy terminal panicles to 30 cm high, of scarlet red flowers. *Tropical.* p. 913, 917

sahelangii; in col. Andromeda Garden, Barbados; shrubby, scandent bush with glossy green, corrugated ovate leaves 10-15 cm long, the inflorescence at end of branches with clustered white flowers having long slender tubes and 5 linear, spreading lobes. *Tropical.* p. 914

speciosissimum (fallax) (Java, Ceylon), "Glory bower"; a magnificent shrub I remember from the mountains of Java and Ceylon, the 30 cm flower clusters a blaze of fiery scarlet above white-pubescent, heart-shaped foliage, on white-hairy, angled stems to 1 m high. *Tropical.* p. 915, 917

x speciosum (splendens x thomsonae); shrubby plant of semi-erect habit with scandent branches, dark green glossy leaves, and terminal clusters of flowers having a pinkish calyx and deep crimson corolla shaded violet, blooming in summer. *Tropical.* p. 917

splendens (W. Africa: Senegambia to Angola); twining shrub with large, oblong or elliptic, corrugated, leathery leaves to 15 cm between which show large clusters of brilliant scarlet 2 cm flowers. *Tropical.* p. 914

thomsonae (balfouri has larger fl.) (Trop. West Africa), "Bleeding heart vine"; twining evergreen shrub, climbing to 4 m, with ovate, quilted, glossy, papery leaves to 12 cm long, deep green; showy flowers in forking clusters, the inflated calyx pure white changing to pink, corolla deep crimson; spring-blooming. *Tropical.* p. 917

trichotomum (Japan); hardy shrub or tree 3 to 6 m high, deciduous in colder regions, with pithy, downy branches and dark green ovate leaves to 20 cm long, downy beneath; in upper leaf axils the clusters of small white, fragrant flowers with spreading petals 4 cm across, from inflated, fleshy crimson red calyx; small blue berries. *Warm temperate.* p. 917

ugandense (Uganda to Rhodesia); smooth climbing shrub to 3 m high, bright green elliptic or obovate leaves 10-12 cm long, toothed at margins; flowers to 3 cm long, calyx lobes crimson, corolla with 3 lobes pale blue and 1 lobe violet-blue, filaments purple and anthers blue. *Tropical.* p. 914

wallichii (India); handsome evergreen shrub of dense habit, 2-3 m high; the narrow lanceolate leaves 12-18 cm long, glossy rich green with depressed veins and wavy at margins; the inflorescence a gracefully pendant cluster on thin-wiry red stalks, flowers white with red calyx and prominent, curving stamens. *Subtropic.* p. 914

CLETHRA Clethraceae
arborea (Madeira), "Lily-of-the-Valley tree"; beautiful small evergreen tree to 8 m, with rusty-pubescent young shoots; glossy bronzy green elliptic leaves 8 to 10 cm long, at tips of branches the lovely erect inflorescence in spikes of ranked nodding, waxy bell-flowers creamy white and sweetly fragrant. *Subtropic.* p. 303

CLEYERA: see Eurya

CLIANTHUS (Leguminosae)
formosus (dampieri) (Australia), "Glory pea"; a striking clambering shrub to 1 m high covered with long white, silky hairs; with glaucous-gray pinnate leaves feather-fashion, and pendant, showy clusters of scarlet, pea-shaped flowers to 8 cm long, and a parrot-bill-shaped keel-petal, blooming from March on. Young plants may flower from August into winter; silky-hairy pods follow blooms. *Tropical.* p. 544, 545

puniceus, (New Zealand), "Parrot's-bill" or "Red kowhai"; evergreen scandent shrub to 3½ m; leaves pinnate; flowers in axillary clusters on a pendulous stalk wholly brilliant red; keel canoe-shaped, 6 cm long. *Subtropic.* p. 544, 553

CLIDEMIA Melastomataceae
vittata (Perú); tropical foliage plant with stem and branches reddish-hairy; elliptic leaves 20-30 cm long with puckered surface, intense green above with central stripe of silver, reddish hairy beneath; flowers rosy. *Tropical.* p. 647

CLITOREA (Leguminosae)
ternatea (Pantropic: Panama to India and Moluccas), "Butterfly pea"; tropical annual or biennial slender twiner with pinnate leaves, and pea-like solitary, showy flowers 5 cm long with broad, fan-like lip narrowing to the base, bright blue with beautiful markings in the throat, blooming all summer. *Tropical.* p. 544

CLIVIA Amaryllidaceae
miniata (Imantophyllum) (Natal), "Kafir lily"; bulb-like plant with fleshy roots, with long, waxy, straplike, arching leaves; broad, bell-shaped, erect, orange-red flowers, yellow toward center, in stiff umbels. *Subtropic.* p. 64

nobilis (So. Africa), "Greentip Kafir lily"; inflorescence with many drooping, narrow, funnel-shaped flowers salmon red with green tips. *Subtropic.* p. 64

CLUSIA Guttiferae
rosea (W. Indies, Panama, Venezuela), "Fat pork tree"; evergreen tree 6 m or more high, growing naturally on rocks or epiphytic on other trees, with obovate thick leathery, deep green, opposite leaves to 20 cm long, without lateral veins; large rose flowers. *Tropical.* p. 520, 521

rosea 'Aureo-variegata', "Variegated balsam apple"; unusually showy variety having the broad obovate leaves beautifully variegated dark green, light green and creamy to golden yellow. *Tropical.* p. 523

rosea 'Marginata', "Monkey-apple"; the large fleshy leaves rich green and bordered with white. *Tropical.* p. 523

CLYTOSTOMA Bignoniaceae
callistegioides (Bignonia) (Brazil, Argentina), "Argentine trumpet vine"; climbing shrub with opposite oval leaves and simple tendril, large funnelshaped flowers in pairs, with spreading lobes, lavender and streaked. *Tropical.* p. 184

CNIDOSCOLUS Euphorbiaceae
phyllacanthus (So. America), "Tread softly"; shrub with milky juice, spiny stems and petioles; leaves with spiny lobes and stinging bristles; flowers with showy white calyx, fruit an ovoid capsule. *Tropical.* p. 407

COBAEA Polemoniaceae
scandens (Mexico), "Monastery-bells"; attractive climber with pinnately compound leaflets terminating in a branched tendril used for climbing, and with large bell-shaped flowers having a violet calyx, and a corolla first greenish, then violet. *Subtropic.* p. 826

COCCOLOBA Polygonaceae
pubescens (grandifolia) (Trop. America), "Moralin"; great tree to 25 m high, with woody stems and giant round, leathery, sessile leaves to 1 m across, fresh green and with sunken veins, the margins turned down, with rusty hairs beneath, especially on the prominent ribs. *Tropical.* p. 824, 827

uvifera (South Florida, W. Indies), the familiar "Sea-grape"; along sandy tropical shores, a good decorative shrub or tree to 6 m, with flexuous branches and stiff-leathery, rounded leaves to 20 cm across, glossy yellowish to olive green with prominent crimson-red veins on young growth, later changing to ivory; the flowers white; fruits purple, resembling bunches of grapes, used for jelly. *Tropical.* p. 467, 824, 827

COCCOTHRINAX Palmae
argentata (So. Florida, Bahamas), a small "Silver palm" to 6 m, with small, very deeply divided star-shaped leaves which appear peltate, the segments narrow and more or less pendant, glossy green above and silvery-white beneath. *Tropical.* p. 774

fragrans (Cuba, Haiti, Jamaica), "Cuban silver palm"; slender palm to 5 m, with solitary, smooth trunk, and near circular 1 m palmate fronds with narrow segments deeply cut and recurving, glossy green above and bluish beneath; the spadix with yellow, very fragrant flowers. *Tropical.* p. 774

COCCULUS Menispermaceae
laurifolius (Himalayas), "Platter-leaf"; decorative sparry shrub to 5 m high, with wiry green branches and alternate stiff-leathery, ovate to narrow-elliptic leaves 12-15 cm long, shining forest green, concave with prominently raised yellow-green parallel veins. *Subtropic.* p. 642, 646

COCHEMIEA

COCHEMIEA *Cactaceae*
poselgeri (Baja California), "Long hooks"; sprawling cylindric plant with numerous stems, to 2 m long and 4 cm thick, bluish green, with conical warts and white-woolly in axils, reddish radials and solitary hooked central spine; small, glossy scarlet flowers. Arid-Tropical. *p. 274*

COCHLEANTHUS *Orchidaceae*
discolor (Chondrorhyncha, Warscewiczella) (Cuba, Costa Rica, Panama); tufted epiphyte with lanceolate leaves to 60 cm long and flowers with pale yellowish sepals, petals pale yellowish at base and pale violet on their upper half and a yellow spot at apex, lip deep violet with whitish front; (spring-early summer). Tropical. *p. 758*

COCHLIODA *Orchidaceae*
densiflora (Perú, Bolivia); miniature epiphyte with 5 cm pseudobulbs, linear leaves, and orange-scarlet flowers 3 cm across, lip recurved. Humid-tropical. *p. 711*
rosea (Andes of Perú); epiphyte with flattened pseudobulb 8 cm long, solitary linear soft leaf; inflorescence gracefully arching, to 40 cm long, 4 cm flowers dark rose-red. Subtropic. *p. 715*

COCHLIOSTEMA *Commelinaceae*
odoratissimum (jacobianum) (Brazil and Ecuador); large epiphyte of the habit of a billbergia, 30-75 cm high; oblong lanceolate leaves to more than 1 m long, and 10 cm wide, rich green above and purplish red with violet lines beneath, sheathing at the base and forming a rosette; numerous beautiful, deep violet-blue flowers of three petals and large white claw, and leaf-like hairy yellow-green sepals with reddish tip. Tropical. *p. 307*

COCHLOSPERMUM *Cochlospermaceae*
vitifolium (Mexico, Central America, So. America), "Buttercup tree"; highly ornamental deciduous tree, with 5-lobed, toothed leaves to 30 cm dia.; attractive golden yellow flowers to 8 cm across appear before the leaves. Tropical. *p. 303, 304*
vitifolium plenum (Maximilianea) (Mexico, C. America, Puerto Rico, So. America), "Brazilian-rose" or "Rosa amarilla"; magnificent tropical tree to 12 m high, deciduous during the dry season; with palmately lobed leaves to 30 cm across, and showy yellow, fully double rose-like flowers of 8 cm dia., appearing before the foliage. The velvety seed capsules bear an abundance of silky white cotton. Tropical. *p. 303*

COCOS *Palmae*
nucifera (Indian Archipelago), "Coconut palm", widely spread into all tropical regions, especially along the sea where its curving-erect swaying trunks, to 30 m high, leaning toward the water, and topped by majestic crowns of glossy, feathery fronds are a feature of tropical shores; producing its edible nuts inside a large yellow-brown husk. Tropical. *p. 465, 476, 769, 775, 802*
nucifera 'Dwarf Green', (Malaya), "Green Malay coconut"; a strain brought in from Malaysia having shorter trunk and starting to produce its green fruit earlier (3 years on) than the larger species; also has been found more resistant to the "lethal yellowing", a parasitic fungus, and an epidemic disease that has killed 200,000 coconuts in Florida and the Caribbean since 1970. Tropical. *p. 476*
nucifera 'Dwarf Samoan' (Marquesas), a dwarf growing coconut forming stout but short trunk, to 8 m high, and bearing fruit larger and more rounded than that of C. nucifera typica; begins to bear at 5 years from planting seed while var. typica takes 8-10 years; husk color either green or red, with average 550 gm of fresh meat per nut against the normal 350 gm in the type coconut. Tropical. *p. 476, 775*
nucifera 'Golden Malay', "Dwarf coconut"; originated from seedlings of the dwarf King-coconut from the Andaman Islands; more diminutive and more slender-trunked than 'Dwarf Samoan', and smaller-fruited (yellow, green, or red); starts producing in about three years, in time 100 or more nuts per year as against 30 or 40 of the common coconut palm, and has a trunk so low that the clusters may need props. Tropical. *p. 474, 775*

COCOS: see also Syagrus

CODIAEUM *Euphorbiaceae*
variegatum pictum (So. India, Ceylon, Malaya, Sunda Isl.); "Crotons" are beautiful tropical shrubs with highly ornamental leaves, thick leathery, glabrous, ovate to linear, entire or lobed, or spirally twisted; variegatd in beautiful colors, the greens and yellows of the leaves later often change to shades of red; small female and male flowers, the latter with white petals; developed into many varieties and hybrids; some of the better known cultivars are listed below. Tropical.
'America'; fast growing plant with large leathery leaves broadening from the base but constricted near apex, strikingly colored copper to maroon and a network of veins shading from pale yellow to red. Tropical. *p. 413*
'Appleleaf'; bushy plant with small ovate, colorful leaves 8-10 cm long, vari-colored with yellows, salmon shades and red. *p. 410*
aucubaefolium (Polynesia); bushy plant with small elliptic thin leathery leaves bright glossy green blotched and spotted yellow, resembling a miniature Aucuba. Tropical. *p. 410, 414*
'Baronne de Rothschild'; compact variety with ovate or lanceolate leaves wavy at margins; young foliage green and yellow, older leaves rich in reds. Tropical. *p. 413*
'Clipper'; an elegant and vigorous new Cutler hybrid descended from 'Monarch', with large lanceolate waxy leaves to 45 cm long, shining green with prominent lemon-yellow veins and margins turning rosy with age. Tropical. *p. 413*
'Cornutum' (Polynesia); compact grower with horn-like tips at apex; green and yellow. Tropical. *p. 414*
'Craigii' (1910); attractive plant with stiff, deeply trilobed leaves, bright green with yellow veins. Tropical. *p. 410*
'Crispum'; long linear recurving leaves glowing crimson or orange red, predominating over the dark green. Tropical. *p. 413*
'Cronstadtii'; remarkable variety with narrow, long strap-shaped leaves variously flat, or spirally twisted and crisped at margins, sometimes with interruptions in the leaf blade, variegated green and yellow; listed by Dreer 1884. Tropical. *p. 416*
'Delaruye' ('Mons. Ernest Delarue'); trilobed leaves light green, attractively colored by golden yellow blotching and broad yellow veins. Tropical. *p. 410*
'Delaware' ("Croton"); large obovate leaves with prominent yellow ribs; largely yellow variegation toward apex. Tropical. *p. 412*
'Dormannianum'; small curved leaves dark red and bronze green. Tropical. *p. 414*
'Eburneum'; large ovate leaves contracted or indented toward apex, silvery-green with yellow border areas, sometimes creamy-yellow over large portions of the blade; photo by G. Roof of Miami. Tropical. *p. 416*
'Elaine' (1910), "Lance-leaf croton"; a good variety with stiff upright lanceolate leaves, shallowly lobed, smaller than 'Reidii', fresh green with bright yellow veins tinted into pink and red. Tropical. *p. 410*
'Excurrens'; leaves very variable from narrow-linear to small oval or obovate; waxy leaves fresh green with part of blade creamy yellow and midrib yellow or red; with age shading more into reds. Known in England since 1884. Glenn Roof of Miami tells me that this variety in cultivation can have leaves both broad and oval as well as narrow-strap-shaped, twisted or with interrupted sections of leaf blade, all on the same bush. Tropical. *p. 416*
'Exotica'; a beautiful Florida (Cutler) hybrid with giant halberd-shaped leaves to 35 cm long; blackish-green with cherry-red vein areas, later autumn-colored in bronzes and salmons. *p. 410*
'Fred Sander' (1910); yellow stems with leaves distinctly 3-lobed, yellow central area and fresh green margins, soft stem often yellow; delicate and susceptible to checks. Tropical. *p. 415*
'Gen. Paget'; distinct variety with broad, large and wavy, oblanceolate leaves deep green with bright yellow ribs and marbling, some entirely creamy-yellow. Tropical. *p. 410*
'Gloriosa'; an excellent hybrid, fast growing yet tough, and colors well; the large oval glossy leaves are rich green with yellow veins when young, later olive green to maroon with veins golden yellow to red. Tropical. *p. 410, 413, 415*
'Gloriosum superbum', "Autumn croton"; dependable variety with long leathery leaves lobed near the apex, nicely variegated according to age of leaf with yellow veins and margins changing to autumn and crimson, the blade fresh green to almost red-black; good keeper. Tropical. *p. 410*
'Illustris'; odd, trilobed green and yellow leaves with protruding midrib. Tropical. *p. 414*
'Imperialis' (Appleleaf); small, compact plant and good keeper, with simple, elliptic leaves almost entirely colored yellow, shading to peach and turning rose and red, green midrib and apex turning metallic purple. Tropical. *p. 412, 415*
'Interruptum' (Polynesia), a "Garden croton"; long linear leaves red and yellow, with midrib extending beyond the blade. Tropical. *p. 415*
'Jan Bier' ("Croton"); European hybrid with showy leaves rich yellow in center, the ribs spreading out yellow in herringbone pattern. Tropical. *p. 412*
'Lineatum'; long narrow leaves recurving, dark green with boldly contrasting yellow midrib; flower a raceme of small whitish flowers and stamen bundles. Tropical. *p. 415*
'Lord Belhaven' ("Croton"); large elliptic leaves coppery-green with yellow and yellow midrib. Tropical. *p. 412*
'MacArthurii', also known as 'Sir William'; vigorous bush with large and heavy long-ovate leaves, light moss-green richly variegated with yellow. Tropical. *p. 412*

'Mad. Blanc' ("Croton"); long linear leaves with broad yellow band feathered along the middle, the sides coppery green. *Tropical.* p. 412

'Majesticum'; a Polynesian variety of drooping habit, with very long, narrow-linear recurving leaves to 45 cm or more long, deep green with yellow ribs when young, the green changing to olive with age, and the yellow becoming crimson in sunlight. *Tropical.* p. 411, 414

'Martha Brooks'; large and showy leaves of the oakleaf type, green dominated by orange and reds mainly in the vein areas. *Tropical.* p. 416

'Mona Lisa', "White croton"; outstanding Bachmann cultivar with large, broad ovate leaves lobed or constricted toward apex, the young growth almost entirely a waxy creamy-white edged in green, later the center area and ribs with reddish tints, the green turning into olive or coppery. *Tropical.* p. 410

'Mons. Florin'; compact grower with hard, medium large, broad elliptic or obovate leaves, apple green with contrasting yellow veining traced in lacy network pattern. *Tropical.* p. 416

'Mortimer'; generally known as 'Piecrust'; narrow-lanceolate leaves 15-20 cm long shiny green, brilliantly blotched yellow and cream, turning crimson-rose to dark blood-red; unusual because the margins are undulate and frilly (Royal Palm Nurs., Oneco, Florida). *Tropical.* p. 412

'Mrs. Duncan Macaw'; long semi-oakleaf type, orange red veins and midrib through the green. *Tropical.* p. 414

'Nobile'; shapely plant with narrow lanceolate leaves green with yellow midrib, in older foliage purple with maroons. *Tropical.* p. 413

'Norwood Beauty' (1924); small and tough 3-lobed leaves of the oakleaf type, dark green with bronze to brown-red and rosy edge, yellow in vein areas. *Tropical.* p. 415

'Ovalifolium'; small ovalish leaf rounded at apex; dark green is dominated by the yellow midrib, veins and margins. *Tropical.* p. 416

'Picturatum' (New Hebrides); long narrow linear leaves with yellow midrib, bright red with age. *Tropical.* p. 414

'Punctatum aureum'; densely bushy, free branching shrub with narrow linear, dark glossy green leathery leaves freely spotted yellow; good as miniature plant. *Tropical.* p. 414, 415

'Rubrum'; large and showy semi-oakleaf with the center totally a glowing salmon-rose, and extending out between the coppery green. *Tropical.* p. 415

'Recurvifolium'; long lanceolate leaves recurving and folding in; green with yellow, in older leaves red. *Tropical.* p. 416

'Stewartii'; an old well-known variety from New Guinea, with leathery, obovate leaves deep bronzy-green with contrasting veins and blotching of yellows, the midrib and petiole red; tolerates cooler temperature. *Tropical.* p. 416

'Stoplight'; bushy plant of broad linear, leathery leaves a blackish-red, the center and undulate margins yellow to salmon-red. *Tropical.* p. 413

taeniosum (Java); long ribbon-like narrow leaves 30 cm and more long, highly variegated with greens, yellow, pinks and reds. *Tropical.* p. 411

'Tapestry'; strikingly colorful plant of compact habit, with long narrow, barely lobed leaves nearly obtuse at apex; new growth intermingled green and yellow with red veins, with age becoming coppery-green and lovely rose, with lavender tinting. *Tropical.* p. 412

'Van Houtte' ("Croton"); showy hybrid with large undulate leaves, the veins yellow to red, the green turning red and purple. *Tropical.* p. 412

'Volutum' (Polynesia), "Ramshorn"; short recurved, linear leaves, veins and midrib yellow. *Tropical.* p. 414

'Vulcan'; magnificent hybrid cultivar with obovate pointed leaves, similar to 'Apple leaf' but more colorful; the younger foliage creamy-yellow with pink fringe, green at tip and midrib; older leaves almost bright wine-red. Photographed at Rochford Nurseries near London. *Tropical.* p. 416

'Warrenii' (Polynesia); robust tall growing stem topped by a crown of long linear, pendant leaves spirally twisted or undulate, red in center with yellow blotches, margins coppery green. *Tropical.* p. 410, 415

CODONANTHE   *Gesneriaceae*
crassifolia (Costa Rica, Panama, N. So. America); branched creeping stems rooting below nodes, with stiff-waxy elliptic leaves light olive green, underside red-spotted and with red midrib; slipper-shaped flowers creamy-white spotted orange in throat. *Tropical.* p. 494

gracilis (Brazil); small epiphytic clamberer found in the wooded Organ Mountains, with thin purplish branches, rooting below nodes, with pairs of glossy, deep green elliptic leaves, reddish beneath; tubular flowers with spreading lobes, creamy-white spotted with purple in throat. *Tropical.* p. 491

COELOGYNE   *Orchidaceae*
cristata (Nepal, Himalayas); small easy-growing epiphyte with branching rhizome forming mats of fleshy pseudobulbs with thin twin leaves; large 8-10 cm fragrant flowers in drooping clusters, snow-white with fringed keels on lip. (Feb.-April). *Humid-subtropic.* p. 721

dayana (Borneo, Sumatra), "Necklace orchid"; decorative epiphyte with pseudobulbs having 1-2 narrow leaves, 1 m long pendulous racemes, loosely many-flowered, 6 cm flowers with narrow lemon-yellow sepals and petals, and broad, lobed lip blotched with chocolate and with white keels (May-Aug.). *Tropical.* p. 715

flaccida (Nepal, Himalayas); epiphyte with dark green thin leaves, the small flowers in graceful pendulous racemes and of heavy odor, cream-white with lip having brownish streaked side lobes and yellow in the middle. *Tropical.* p. 715

massangeana (Assam to Java); epiphyte with oval 2-leaved pseudobulbs; numerous 6 cm flowers loosely spaced on long pendulous racemes, sepals and petals pale yellow, chocolate lip veined yellow and edged with white. (March-July). *Tropical.* p. 715

pandurata (Borneo, Sumatra), "Black orchid"; straggling epiphyte with large pseudobulbs and shining leaves; the large flowers to 10 cm, in pendulous sprays 60 cm long and very fragrant, sepals and petals lovely green, lip greenish-yellow with black raised ridges. (May-July). *Tropical.* p. 715, 716

parishii (So. Burma: Moulmein); small epiphyte with squarish pseudobulbs and several 8 cm flowers resembling small pandurata, with slender pointed sepals and petals pale nile-green, the crested lip marked with black (spring). *Tropical.* p. 715

sparsa (Philippines); dwarf epiphyte with tapering pseudobulbs and 8-10 cm glaucous leaves; short racemes of small 2½ cm fragrant flowers white tinted green, and white, lobed lip with sides dotted purple and brown cross-bar on front lobe (Jan.-May, Nov.). *Tropical.* p. 715

COFFEA   *Rubiaceae*
arabica (Ethiopia, Angola), "Arabian coffee"; evergreen shrub to 5 m high, with willowy branches bearing shining dark green, ornamental, elliptic foliage to 15 cm long, and wavy at the margins; the pure white, fragrant flowers, in axillary clusters at the base of leaves, are followed by brilliant crimson, pulpy, 1½ cm berries containing the two seeds or "beans", which are roasted into coffee. Superior type of 'Blue Mountain' coffee, grown at higher altitudes or cool climates. *Tropical.* p. 466, 471, 860, 861

laurifolia (Trop. Africa); in collection of Marnier-Lapostolle; a species of very robust habit with wiry branches bearing pairs of opposite leaves, and between them red berries of large size - 2 cm dia. *Tropical.* p. 466, 864

liberica (West Africa: Liberia), "Liberian coffee"; more robust than arabica, with longer, wider obovate, shining leaves to 30 cm long, and with a shorter point; flowers in a dense cluster of 15 or more; fruit black 2 cm long; a chief source of coffee, thriving in hot climates where C. arabica will not grow well. *Tropical.* p. 466, 864

robusta (Zaire), "Robusta coffee"; vigorous, small evergreen tree of luxuriant habit, with willowy branches furnished with large ovate-elliptic, sharp-pointed leaves shining dark green; the axillary flowers pure white, later developing large brownish-red berries and containing a pair of coffee beans; not as valuable commercially as C. arabica but a better producer in the tropics. *Tropical.* p. 471, 861, 864

COLA   *Sterculiaceae*
nitida (Trop. W. Africa), "Cola nut"; evergreen tree to 20 m high, with glossy green, obovate leaves 15-20 cm long; small flowers with yellow calyx; orange-colored leathery fruit 2 cm dia., enclosing the caffeine-holding seed or cola nut. *Tropical.* p. 466, 902

COLCHICUM   *Liliaceae*
autumnale (Europe, No. Africa), "Autumn crocus"; with long tubed lilac-violet flowers resembling slender-stemmed goblets produced from the corms without foliage in August-September, the broad basal leaves appear the following spring. *Warm temperate.* p. 526, 575

COLEUS *Labiatae*

blumei 'Beckwith Gem'; attractive cultivar with leaves blackish-red at the base and center, bronzy scarlet alongside, and yellow along the crenate margins. *Tropical.* p. 530

blumei 'Brilliancy'; horticultural cultivar of the "Painted nettle" from Java, a soft but showy-leaved herb with square, succulent stems and blue flowers; the leaves of 'Brilliancy' are boldly crimson-red, marked gold on finely crenate edge. *Tropical.* p. 530, 531

blumei 'Klondyke', "Butterfly coleus"; robust variety wildly streaked and variegated brownish-red and yellow. *Tropical.* p. 530

blumei 'Pride of Autumn'; most attractive cultivar with ivory-white stem and medium-sized ovate leaves having bright carmine center, outer area chocolate brown, and a contrasting margin lightly crenate; underside pale green with carmine center showing through. *Tropical.* p. 530

blumei 'Red Verschaffeltii'; vigorous plant with showy leaves almost entirely crimson red, the outer zone with blackish red. *Tropical.* p. 530

blumei 'Verschaffeltii' (Java), "Painted nettle"; most beautiful variety with large, glowing crimson-red leaves, purple in center, and narrow, nile-green border, underside purplish-red also, stems and petioles white to purple; flowers pale blue. *Tropical.* p. 531

rehneltianus 'Red Trailing Queen'; robust creeper with small, broad-ovate leaves, the center purple carmine, outer zone deep blood red, broad border of dull green into lobes. *Tropical.* p. 530

rehneltianus 'Trailing Queen'; a popular commercial cultivar densely branching especially if pinched, with somewhat larger, ovate leaves chartreuse to emerald green along margins, prominently overlaid and marbled with brownish purple, carmine center near base. *Tropical.* p. 530

COLLETIA *Rhamnaceae*

cruciata (So. Brazil, Uruguay), "Anchor plant"; curious rigid glossy green spiny shrub with flattened branches armed with flat triangular spines which are really modified branches, borne in pairs, each pair set at right angles to its neighbor, and sharply pointed; tiny deciduous leaves, and small tubular yellowish-white flowers, in autumn. *Subtropic.* p. 832, 857

COLLETOGYNE *Araceae*

perrieri (Madagascar); curious small tropical plant with short-stalked inflorescence, the waxy cupped spathe pale foam green spotted black inside, and forming a shield for the short, brownish-purple spadix with white flowers ringed purple-brown; the broad hastate, horizontal leaves standing like an umbrella over the clustered flowers. *Tropical.* p. 102

COLOCASIA *Araceae*

antiquorum (India), "Egyptian taro"; aquatic or swamp plant with slender stalks carrying the thin-textured, peltate leaves, shimmering green with bronze-purple shading between veins; long green to purplish, expanded spathe. *Tropical.* p. 103

antiquorum illustris (E. Indies), "Black caladium"; the soft, peltate, heartshaped leaves are fresh, spring-green, especially in vein areas, the balance brownish-purple which shows through to grayish reverse; stems green or violet. *Tropical.* p. 103

esculenta (Hawaii and Fiji), "Elephant's ear" or "Taro"; soft, fleshy herb with edible tuber and large, quilted leaf to 1 m, bright, satiny green. *Tropical.* p. 103, 105

fallax (Sikkim); shield-shaped leaf metallic, fresh-green with lighter veins. *Subtropic.* p. 103

gigantea (indica) (Malay Penin., Java), an "Elephant's ear"; large herb forming trunk-like stems to 60 cm or more; the big sagittate-peltate leaves fresh-green above, white-glaucous beneath, 60 cm or more long, undulate at margins; hooded white spathe to 20 cm long. Photo by J. Bogner, Munich Bot. G. *Tropical.* p. 85

COLUMNEA *Gesneriaceae*

campanulata (Colombia); shrubby plant to 60 cm high, with corrugated ovate leaves; flowers inflated tubular with flaring mouth, golden yellow with red hairs. *Tropical.* p. 493, 507

crassifolia (Mexico, Guatemala); erect, succulent plant with long linear, fleshy, glossy-green leaves and nearly erect, orange-red, spread-open, tubular flowers, covered with erect red hairs. *Humid-tropical.* p. 494

gloriosa (Costa Rica); epiphytic trailer with rooting and pendulous stems, small oblong leaves covered with brown-red hairs, and large, solitary, bilabiate, fiery red flowers to 8 cm long, with wide open yellow throat, and broad, helmet-like upper lip; good in baskets. *Humid-tropical.* p. 491

gloriosa 'Variegata' (Costa Rica); long pendant brown wiry stems, dense with opposite small elliptic leaves 3 cm long, green with white margins. *Tropical.* p. 491

hirta (Costa Rica), "Goldfish plant"; epiphytic creeper with rooting stems covered with red hairs, small ovate, red-satiny leaves; vermillion red, bilabiate flowers marked with orange, solitary from axils, smaller than gloriosa but more floriferous. *Humid-tropical.* p. 491, 494

linearis (Costa Rica), "Pink goldfish plant"; bushy trailing epiphyte with long narrow, shiny leaves and axillary rose-pink, bilabiate flowers with flattened throat, covered with silky-white hair. *Humid-tropical.* p. 494

microphylla (diminutifolia) (Costa Rica), "Goldfish vine"; soft trailing plant with tiny rounded or broad-elliptic, coppery hairy leaves, and relatively very large, spread-open bilabiate flowers burnt-red with yellow along bottom of tube, similar to gloriosa but smaller. *Humid-tropical.* p. 495

nicaraguensis (localis) (C. America); profusely blooming, stiff-stemmed, epiphytic trailer with large lanceolate, corrugated, satiny green leaves red beneath, and axillary, brilliant red flowers, often yellow at the throat. *Humid-tropical.* p. 494

panamensis (Panama); scandent stems with small pubescent oval leaves; large erect tubular, bilabiate flowers 6 cm long, crimson red and bristly hairy. *Tropical.* p. 494

schiedeana (E. Mexico); large upright or climbing shrub with long lanceolate green leaves covered with white plush, burgundy-red beneath; bilabiate flowers orange-yellow, striped and spotted crimson. *Humid-tropical.* p. 494

teuscheri (Trichantha minor in horticulture) (Ecuador); rambling creeper with long thread-like vines set with lance-shaped blackish-green leaves shiny above and hairy beneath; from the leaf axils the very unusual tubular 5 cm flowers dark purple, with 4 stamens, the calyx a feathery cluster of carmine bristles, and the face of the corolla contrasting yellow and black, with 5 curious yellow horns at the mouth; blooming for 4 months, starting in summer. *Humid-tropical.* p. 493, 498, 502

tulae 'Flava' (Puerto Rico, Haiti); climbing and trailing plant from mountainous areas, with aerial roots and pubescent, soft green leaves; hairy, bilabiate flowers bright yellow, with long tube and spreading lobes. *Humid-tropical.* p. 494

x vedrariensis (magnifica x schiedeana); erect, vigorous shrub with dark waxy, elliptic leaves, and downy tubular flowers widespread at opening, bright scarlet with network of pale yellow veining. *Humid-tropical.* p. 494

'Vega' (C. Stavanger x vedrariensis), "Goldfish bush"; 1956 diploid bush with erect or arching wiry branches first light green, later reddish, fresh green round-oval, smooth leaves 3 cm long; the erect bilabiate flowers 6-8 cm long and evenly cox-comb red; of stiffer habit than Stavanger, flowering 3 weeks earlier; set buds by chilling, cool and dry in Sept., after 6 weeks give warm 15°C. *Humid-tropical.* p. 494

verecunda (Costa Rica); compact, shrubby, erect plant with waxy, deep olive-green, lanceolate, convex leaves red beneath, shielding the axillary, flattened, bilabiate, lemon to golden-yellow flowers, beautifully wine-colored below. *Humid-tropical.* p. 494

COLVILLEA *Leguminosae*

racemosa (Madagascar); spreading, showy tree to 15 m, withfern-like bipin ate foliage and robust, curving whips of branches bearing masses of pendant racemes to 60 cm long, brilliantly fiery-red flowers with yellow stamens. *Tropical.* p. 540, 545

COMBRETUM *Combretaceae*

fruticosum (Trop. America), "Burning bush"; large woody climber with corrugated elliptic leaves, and showy terminal one-sided spikes, orange-scarlet and long, orange stamens. *Tropical.* p. 302, 303

platypterium (Tanzania to Nigeria); climbing parasitic shrub with large glossy green leaves; the long inflorescence with tubular flowers, yellow to red. *Tropical.* p. 303

CONGEA *Verbenaceae*

tomentosa (India), "Shower orchid"; climbing shrub with opposite ovate entire 8 cm leaves hairy beneath; white flowers subtended by lilac leaf-like, hairy bracts. *Tropical.* p. 918

CONICOSIA *Aizoaceae*

communis (South Africa); ground-covering spreading succulent, with long-linear, 3-angled soft leaves 10-15 cm long without dots, arranged in spirals on the stem; yellow ray flowers on pedicels which die after fruiting. *Arid-subtropic.* p. 46

CONOCARPUS *Combretaceae*

erectus (Seashores, Trop. America, and W. Africa), "Button mangrove"; prostrate shrub or tree from mangrove swamps, with leathery ovate leaves to 10 cm long; flowers greenish, followed by purplish fruit clusters. *Humid-tropical.* p. 303

CONOPHYLLUM *Aizoaceae*
   tenuifolium (So. Africa); branching succulent with long linear, fleshy, angled, sterile leaves glossy green, splitting into forks from the middle up. *Arid-subtropic.* p. 42

CONOPHYTUM *Aizoaceae*
   auriflorum (So. Africa: Cape Prov.); mat-forming low succulent with thick conical bodies ½ cm across, dark green with reddish on sides; small golden-yellow flowers. *Arid subtropic.* p. 48
   elishae (So. Africa); small, clustering succulent with thick, club-like, two-lobed leaves grayish-green with purple; flowers yellow. *Arid-subtropic.* p. 48
   griseum (Southwest Africa), "Silver stars"; clusters of tiny flattened, split clubs coppery-rusty with surface windows and blackish dots; masses of small white, daisy-like flowers showing yellow anthers. Prob. form of fenestratum. *Arid-subtropic.* p. 48
   luisae (So. Africa: Namaqualand); low succulent forming mats; bodies heart-shaped, 1 cm across, gray-green with darker spots; glossy yellow 2 cm flowers. *Arid subtropic.* p. 48
   pearsonii (So. Africa: Cape Prov.); cushion-forming succulent with obconical bodies to 2 cm dia., the upper surface flat convex, grayish with bluish dots; glossy mauve pink flowers 2 cm across. *Arid subtropic.* p. 48
   taylorianum (Southwest Africa); small succulent forming dense, flat cushions; bodies somewhat cordate, to 1½ cm across, chalky green with crimson borders; shiny flowers pink to magenta, to 2½ cm across. *Arid subtropic.* p. 48

CONVALLARIA *Liliaceae*
   majalis (Europe, temp. Asia), "Lily-of-the-valley"; small perennial herb with horizontal slender creeping rootstock, the upright parts called pips, with 2 basal, fresh green lanceolate leaves and stalk with nodding little bells of fragrant, waxy-white flowers. *Temperate.* p. 592, 595

COPERNICIA *Palmae*
   baileyana (Cuba); robust fan-palm, to 15 m, with smooth, thick trunk, and numerous large 1½ m, glossy green hard-leathery leaves in a dense, roundish crown, on thorny stalks. *Tropical.* p. 776
   ekmanii (Haiti); straight erect fan palm, the trunk covered with leaf bases; the stiffly erect palmate grayish foliage; the two outside segments of each leaf armed with strong teeth. *Tropical.* p. 776
   macroglossa ("torreana" of hort.) (Cuba), the "Cuban petticoat palm", so called because of the dense cylindrical hanging growth of dead persistent leaves covering the trunk of this fan palm; growing to about 8 m, the stiff, folded, lustrous deep green leaves 1½-2 m across, on short thorny stalk. *Tropical.* p. 776
   rigida (Cuba); curious-looking fan palm with long and narrow wedge-shaped palmate, plaited leaves arranged in densely packed, erect layers as if tied together, the long segments stiff-erect, on short 30 cm petioles. *Tropical.* p. 776
   sueroana (Cuba); strange-looking palm with densely shingled, palmate, fan-like leaves, deeply cut; erect petioles covering and hiding base of trunk; in appearance like a bursting fountain. *Tropical.* p. 778

COPROSMA *Rubiaceae*
   repens 'Marginata' (baueri) (New Zealand), "Mirror plant"; flexible densely leafy dioecious shrub with opposite, roundish, soft-leathery leaves 6-8 cm long, bright green and glossy as if varnished; small flowers greenish, the berries on female trees orange-yellow; planted in Southern California along the seashore because of its resistance to salt spray. *Subtropic.* p. 864
   robusta (New Zealand), "Kakaramu" of the Maoris; stout shrub to 4 m high, with pale brown bark, shining leaves 3-8 cm long; insignificant greenish flowers, followed by masses of yellow to red berries 1 cm dia. *Subtropic.* p. 864
   williamsii (New Zealand); small evergreen with narrow elliptic leaves variegated with cream, center green. *Subtropic.* p. 864

CORDIA *Boraginaceae*
   alba (dentata) (Trop. America), "Jackwood"; large bushy shrub or small evergreen tree to 8 m; broad ovate, rich green leaves with toothed margins to 10 cm long; small white or pale yellow flowers to 2 cm across, in large clusters. *Tropical.* p. 196
   dodecandra (Mexico: Veracruz to Chiapas; Guatemala), "Anacahuite tree"; evergreen tree to 30 m; broadly ovate rough leaves 6-12 cm long; crinkled flowers reddish-yellow or orange-scarlet, with 12 to 17 lobes, 5 cm long. *Tropical.* p. 196
   lutea (Polynesia: Marquesas, Galapagos, W. Perú); shrub or small tree 2-3 m high; broad-ovate, dark green rugose leaves to 10 cm long; pretty, bright yellow crisped 3 cm bell flowers with more than 5 lobes, in large clusters. *Dry-tropical.* p. 196
   sebestena (Florida Keys, Bahamas, West Indies to Barbados), "Geiger tree"; small evergreen tropical tree to 8 m, with rough brown bark; dark green, oval leaves 8 to 20 cm long, with pale veins, in clusters at ends of branches; flowers to 5 cm across, brilliant orange-scarlet, quilted and ruffled and of a crepy texture; small white 2 cm fruit sweet and edible. *Tropical.* p. 196
   sebestena 'Aurea', "Geiger tree"; variety with orange-yellow flowers; edible whitish 2 cm fruit. *Tropical.* p. 459
   superba (Brazil); large tropical shrub with stout young shoots; long obovate, rough leaves 15-20 cm long, clustered at ends of branches; white, crepy, bell-shaped flowers 5 cm wide, in long-stalked clusters. *Tropical.* p. 196

CORDYLINE *Liliaceae*
   australis (New Zealand), "Dracaena indivisa" of florists, tree to 12 m high, with single erect stem crowned by a dense cluster of flat and narrow, arching leaves tough-leathery, bronzy-green with fresh-green midrib; fragrant white flowers, following which the head will fork. Known as "Grass palm", or in New Zealand as "Cabbage tree"; an ancient monocot. Roots are typically white in Cordyline (yellow in Dracaena) *Subtropic.* p. 582, 587
   australis 'Atropurpurea' (cuprea); handsome color variation; dark-leaved form with the slender leaves entirely a coppery reddish brown. *Subtropic.* p. 578
   australis 'Variegata', "Tikouka" in New Zealand; elegant small tree with slender trunk and a crown of narrow-linear, leathery, graceful leaves olive green, variegated with broad longitudinal bands and margins of creamy-yellow. *Subtropical.* p. 578
   banksii (New Zealand); slender stems to 3 m high, often clustering and sometimes branched, dense with 1 m long recurving strap-shaped leaves dark green with pale yellow midrib, to 8 cm wide, veins often red; flowers in pendant panicle. *Subtropic.* p. 578
   fruticosa (Polynesia); similar to Cordyline terminalis 'Ti', but of stockier habit, thicker cane and broader, oblanceolate leaves glossy green to bronze. *Tropical.* p. 578
   indivisa (New Zealand); slender tree to 8 m high, usually with single, flexible stem, with a large head of sessile, flat, tough leaves, matte green with raised orange midrib, glaucous blue beneath, to 20 cm, or twice as wide as australis; not branching after flowering. *Subtropic.* p. 582, 587, 611
   indivisa 'Purpurea'; found in California, tree-forming with woody trunk usually clothed with foliage, toward apex a dense crown of broad, flexuous leathery leaves entirely suffused with bronzy purple; young growth sprouting from base has foliage much narrower. *Subtropic.* p. 578
   rubra 'Bruantii'; form with gracefully recurved, coppery-red leaves, of this Australian, slender tree-like species, intermediate between terminalis and stricta, growing to 5 m, leaves continually ascending, closely set, narrow oblanceolate and thick, normally dull green, reddish reverse when young, prominent midrib, on broad grooved stalk. *Subtropic.* p. 578
   stricta (Drac. congesta) (Queensland, New So. Wales); treelike, with slender stem to 4 m high, occasionally branched, narrow clasping leaves swordshaped, leathery, matte green with rough edges, inconspicuously toothed, narrowing toward a constricted base; young growth reddish. *Subtropic.* p. 578
   terminalis (India, Malaysia to Polynesia), "Tree of kings"; a commercial "Red dracaena" with rather slender, leathery, sword-shaped leaves in a cluster of erect habit, on a cane to 3 m high; the normal leaves are copper-green shading into red, the young winter growth intense rosy crimson; flowers lilac-tinted, in panicles, followed by red berries. *Tropical.* p. 579
   terminalis 'Amabilis', "Pink dracaena"; strong growing form with broad, shining green to bronze foliage prettily variegated or edged with cream suffused with pink. *Tropical.* p. 579
   terminalis 'Baby Ti' (Hawaii); one of the tiniest of "Red dracaenas", miniature rosette with narrow, concave, gracefully recurved leaves deep metallic green suffused with copper, and a red border. *Tropical.* p. 579
   terminalis 'Calypso Queen'; very compact plant, with small leaves deep bronzy red; the center purplish. *Tropical.* p. 576
   terminalis 'Firebrand', excellent commercial "Red dracaena", which in addition to being a compact rosette, also holds its foliage well; slender stiff, gracefully recurving leaves a satiny purplish-red with glaucous sheen, the younger foliage mahogany red with crimson midrib. *Tropical.* p. 576
   terminalis 'Hawaiian Bonsai'; dwarf form of C. terminalis, with red petioles, and leaves rather narrow and stiff, bronzy green with red and pink variegation. *Tropical.* p. 576
   terminalis 'Hawaiian Flag'; Hawaiian cultivar with magnificent colors in broad, decorative leaves, in greens, ivory, pinks and crimson red. *Tropical.* p. 579

terminalis 'Liliput'; short, compact cultivar with broad leaves, bronzy green edged in rose, the upper leaves primarily carmine pink. *Tropical.* p. 579

terminalis 'Mad. Eugene André', "Flaming dragon tree"; showy commercial variety of spreading habit, with broad leaves deep green to coppery, and red border; the new growth in winter a beautiful shocking pink, indicating the flowering season, after which the lower leaves tend to drop if disturbed. *Tropical.* p. 576

terminalis 'Negri', known in horticulture as Dracaena Negri, the "Black dracaena"; a striking cultivar from Louisiana, outstanding for its big leathery, glossy coppery maroon leaves almost black, $\frac{1}{4}$ to $1\frac{1}{2}$ m long and 12 cm wide, in graceful rosette on slender canes. *Tropical.* p. 579

terminalis 'Schubertii'; robust variety with long woody canes and long, sword-shaped leaves deep red. *Tropical.* p. 579

terminalis var. 'Ti' (Taetsia fruticosa) (Hawaii to New Guinea) "Tree of kings" or "Hawaiin good luck plant"; palm-like, robust, with slender cane to 4 m high, topped by a cluster of oblong leaves to 60 cm long spirally arranged, smooth, flexible and plain green; this foliage is used for hula skirts; sections of cane will sprout young plants. *Tropical.* p. 578, 587

terminalis 'Tricolor', "Tricolored dracaena"; very colorful rosette of broad leaves beautifully variegated red, pink, and cream over a base of fresh green and a good keeper. *Tropical.* p. 576, 587

'Volckaertii'; a Belgian terminalis cultivar 1929, with broader, somewhat spatulate leaves light and dark green, grayish beneath; in habit more vigorous and more tough and durable than terminalis. *Subtropic.* p. 578

COREOPSIS *Compositae*
tinctoria, "Calliopsis" in hort. (Minnesota to Arizona); herbaceous summer-flowering biennial with divided leaves and long-stalked flowers with toothed yellow florets and crimson base, disk brownish-purple. *Temperate.* p. 326

CORNUS *Cornaceae*
alba 'Argenteo-marginata' (Siberia, No. China, No. Korea), "Tatarian dogwood"; red stems; elliptic leaves, milky green heavily bordered and variegated with cream. *Temperate.* p. 356

capitata (Himalayas); attractive evergreen tree dense with leathery 10 cm elliptic leaves, reverse grayish and finely pubescent, on willowy branches; inconspicuous flowers subtended by fleshy, 4 cm sulphur-yellow bracts; fruit puckered strawberry-like, salmon-red to crimson. *Subtropic.* p. 357, 456

florida (Maine to Florida and Texas), "Flowering dogwood"; popular deciduous flowering tree to 12 m high, with spreading branches, and blooming on bare wood, the small inconspicuous yellow flower cluster surrounded by 4 large, gleaming petal-like white bracts 4-5 cm long, notched at apex, turning pink with age; the 8-12 cm ovate green leaves will turn red in late autumn; berry-like fruit scarlet. *Temperate.* p. 356

florida rubra, (Maine to Florida and Texas); deciduous tree to about 10 m high, with ovate leaves 10-15 cm long and small greenish flowers in dense heads, subtended by 4 large showy, notched rosy, petal-like bracts; white in the species florida; fruit red. *Warm temperate.* p. 356

florida 'Welchii', "Variegated dogwood"; variety with ornamental leaves green, variegated or edged yellow and red. *Warm temperate.* p. 356

florida 'White Cloud' ('Cloud Nine'); a dwarf "Dogwood" of shrubby dense habit without a leader, full of flowers, with 4 white bracts much more rounded in form and borne more horizontally than those of florida, of which this is a cultivar; flowering at an early age. Photo. at North Haven Gardens, Dallas, Texas in spring 1971. *Temperate.* p. 356

kousa (Japan, Korea), "Kousa dogwood"; shrub or small tree to 6 m; leathery oval leaves 10 cm long, with yellow veins and glaucous beneath; inflorescence with globular green head, subtended by 4 petal-like narrow bracts creamy white. *Temperate.* p. 356

nuttallii (Brit. Columbia to No. California), "Pacific dogwood"; spectacular flowering tree 5 to 25 m high, often encountered in Sequoia National Park and the Sierra Nevada, blooming on bare branches, the flowers with 4 to 6 or 8 glistening white bracts 4-6 cm long, later turning pink, around a center cluster of tiny flowers; the downy foliage 8-12 cm long and glaucous beneath. *Warm temperate.* p. 356

CORONILLA *Leguminosae*
emerus, (Central and So. Europe), "Scorpion senna"; deciduous shrub with angled shoots; unequally pinnate leaves; yellow flowers with long claws. *Warm temperate.* p. 546

CORREA *Rutaceae*
alba (S.E. Australia), "White correa"; evergreen shrub 1 m or more, with leathery oval leaves buxus-like, 3 cm long, woolly beneath; small white 2 cm bell-flowers split into 4 petals; resistant to salt-spray. *Subtropic.* p. 872

CORTADERIA *Gramineae*
selloana (argentea) (Argentina), the "Pampas-grass"; gigantic tufted grass 2-3 or even 6 m high, forming great clumps, the leaves 1-3 m long and 2 cm wide, with rough edges; female spikelets borne in panicles, beautiful silvery-white, silky plumes to 1 m long. The plumes can be dried for indoor decoration. *Subtropic.* p. 511

CORYANTHES *Orchidaceae*
bicalcerata (Perú), "Bucket orchid"; curious epiphytic orchid related to Stanhopea; conical pseudobulbs topped by thin-leathery leaves; from the base a pendulous stalk with 2 or 3 very complicated hooded flowers, white with purple markings and with strong, penetrating odor; the lip is thick-fleshy, wax-like in texture. *Humid-tropical.* p. 720

leucocorys (Amazonian Perú), distinct species with large flowers, waxy-white with deep red spots; the lateral sepals 10 cm long and 5 cm wide, petals narrow; sepals greenish yellow marked with red spots, petals with fainter purple markings. *Humid-tropical.* p. 722

macrantha (Trinidad, Venezuela, Guayana), "Helmet orchid"; epiphyte of the habit of Stanhopea, pushing out from the base of its pseudobulbs a pendulous stem with large odorous flowers, the delicate sepals and petals are yellow spotted purple, the helmet-shaped wax-like lip orange spotted with red. *Tropical.* p. 722

macrocorys (Perú), "Helmet orchid"; from the upper Amazon; flowers of medium size, the lateral sepals concave and arranged to form a shield to the lip; sepals and petals greenish white marked with cinnamon-red; purplish hood with large horns and suffused with purple (summer). *Humid-tropical.* p. 720

maculata punctata (Guayana); colorful form with sepals and petals yellowish and spotted red, hypochil (basal part of lip) flushed with orange-red and spotted purple. *Humid-tropical.* p. 722

CORYPHA *Palmae*
elata (Bengal, Burma, Pacific Is.); robust fan palm to 20 m high; trunk more slender than the 45 cm dia. related Talipot palm, and covered with leaf bases; the bright green fan-leaves 2-3 m across, segments divided halfway, end of leafstalk extended into the leaf forming a rib; at maturity the 6 m monocarpic inflorescence appears above the crown, then dying. *Tropical.* p. 769, 778

umbraculifera (Ceylon, So. India), the "Talipot palm"; slow-growing to 30 m high, with straight trunk and a great crown of fan-shaped, plaited bright green leaves about 4 m wide, on spiny stalks; pyramidal terminal inflorescence of creamy-white flowers, and small olive-colored fruit; after fruiting the tree will die. *Tropical.* p. 778

CORYPHANTHA *Cactaceae*
elephantidens (Mexico: Michoacan) "Elephant-tooth cactus"; beautiful globose plant to 15 cm high and broader to 20 cm, blue-green obtuse knobs 5 cm long, densely white-woolly in axils; 6-8 brownish spines and all radial; large showy flowers about 10 cm across, and yellowish to salmon-rose colored with crimson-red center and red tips. When old, freely sprouting from the warts. *Arid-tropical.* p. 265

palmeri (Coahuila to Durango, Mexico); small pin-cushion with tubercles in 13 rows, tipped by tan radials; flowers pale yellow. *Subtropic.* p. 276

vivipara (from Manitoba and Alberta to Kansas, Colorado, Texas), "Spiny stars"; olive green globes 5 cm thick, clustering, dense with high woolly nipples covered with spreading white radials and central spines; flowers flushed rose. *Temperate.* p. 274, 276

vivipara arizonica (Arizona); ovoid body to 10 cm high, forming mounds covered by white radial spines; 5 cm flowers, deep pink. *Subtropic.* p. 276

CORYPHANTHA: see also Escobaria

COSMOS *Compositae*
bipinnatus (Mexico), "Garden cosmos"; showy summer and fall-blooming annual that may grow to 2 m and more, of open branching habit, with light green bipinnate leaves finely cut into lacy segments, and daisy-like heads 5 to 8 cm across, with broad petal-like ray flowers in white, shades of pink, rose, lavender, purple or crimson, with tufted yellow centers. *Tropical.* p. 315, 326

COSTUS *Zingiberaceae*
igneus (Brazil), "Fiery costus"; stout leafy stems maroon with 15 cm leaves green above, reddish beneath, spirally arranged; large 6 cm flowers deep orange. *Tropical.* p. 924

malortieanus (zebrinus) (Costa Rica), "Stepladder plant"; showy, suckering plant with stout stalks to 1 m high in spirals, with broad, recurved, fleshy leaves to 30 cm long, bright emerald green, banded lengthwise with dark zones and covered with glistening hair; flowers yellow marked with red. *Tropical.* p. 926

sanguineus (C. America), "Violet spiral flag"; beautiful "Spiral flag" with coppery green stems clasped by wine-red petioles bearing in spiral order, the gracefully recurving, oblique-elliptic, fleshy leaves of shimmering velvet-bluish-green, marked with a central band of silver, thin gray lines, and a zone of yellow-green toward the margin; deep blood-red beneath. *Tropical.* p. 924

speciosus (India), "Spiral ginger"; perennial herb with heavy rootstock, clustering, slender reed-like, green, leafy stems to 3 m long, growing upward in loose spirals or drooping in graceful curves, set in spiral order with fresh-green, oblanceolate, glossy, succulent, slender-pointed leaves, silky beneath; flowers in dense spike, white with yellow center and red bracts. *Tropical.* p. 926

spicatus (West Indies); "Indian head ginger"; stout perennial herb with leafy stems to 2½ m tall; smooth, thick ovate leaves to 15 cm; club-like cylindrical inflorescence of overlapping red bracts, and yellow flowers. *Tropical.* p. 926

spiralis (Colombia to Guayana and Brazil), "Scarlet spiral flag"; rhizomatous plant 1 to 3 m tall with canes set with fleshy green, lanceolate leaves up in spiralling fashion; at apex a tight cone of appressed crimson bracts and with partially enclosed reddish flowers. *Tropical.* p. 926

COTINUS Anacardiaceae

coggygria (Rhus cotinus) (So. Europe to Asia), "Smoke tree" or "Venetian sumac"; deciduous bushy shrub to 5 m high and wide; roundish 4-8 cm leaves, bluish green in summer, turn yellow to orange-red in autumn; the inflorescence in large puffs of tiny greenish blossoms. *Warm temperate.* p. 71

COTONEASTER Rosaceae

dammeri radicans (China), "Bearberry"; evergreen prostrate and creeping shrub, with ovate, corrugated, leathery leaves 2-3 cm long; small white flowers, wax-red berries. *Subtropic.* p. 848

franchetii (W. China, Burma); evergreen shrub to 3 m high, fountain-like, with arching branches; thickish, ovate leaves 3 cm long, dull green above, pink flowers, followed by 3 cm scarlet fruit in clusters. *Subtropic.* p. 848

glaucophyllus (China), "Bright-bead cotoneaster: evergreen erect arching shrub to 2 m, with oval 5 cm gray-green leaves, closely set; flowers pinkish followed by 1 cm dull-red fruits. *Warm temperate.* p. 848

lacteus (parneyi) (W. China), "Red cluster-berry"; evergreen to 2 m or more, with arching branches, leathery, corrugated leaves to 5 cm long, deep green above, white-hairy beneath; small white flowers, followed by long-lasting red fruit. *Warm temperate.* p 848

x watereri 'Cornubia' (frigidus x henryana); tall evergreen shrub to 6 m high and wide; oval leaves 8-12 cm long, corrugated and with red veins; white flowers; large red fruit in profuse clusters. *Warm temperate.* p. 848

x watereri 'Pendulus'; form more compact and with branches arching downward. *Warm temperate.* p. 848

COTYLEDON Crassulaceae

heterophylla (So. Africa); branching succulent with woody stems bearing short fat spatulate light green leaves thickly covered with white felt, the apex toothed and with brown tips. *Arid-subtropic.* p. 363

ladismithiensis (Cape Prov.), "Cub's paws"; attractive branching succulent with small and thick-fleshy leaves yellowish green and covered with white hair, apex claw-like and maroon; shiny flowers pale yellow tinged apricot, darker outside. *Subtropic.* p. 363

macrantha (E. Cape Prov.); showy and beautiful succulent to 1 m high much grown on the Riviera; stout stem with erect branches, dense with fleshy, fresh green obovate, concave leaves edged in red, to 10 cm long; inflorescence on stiff stalk with inflated tubular, nodding flowers intensely red, inside greenish. *Subtropic.* p. 364

orbiculata (Namibia), large robust succulent with thick, obovate leaves covered with waxy silver-white bloom and with red margin; flowers yellowish red. *Arid-subtropic.* p. 361, 363, 364, 365

orbiculata oophylla (Namibia); elongate rosette of thick-fleshy obovate gray-green leaves covered with silvery blue powder, the apex edged in brown-purple; nodding bell-shaped flowers orange red. *Arid-subtropic.* p. 361

paniculata (Namibia to Karroo and to Cape of Good Hope), the "Botterboom" of the Boers; thick-stemmed succulent tree-like to 3 m high, with swollen trunk covered with papery yellow-brown skin, few branches with obovate leaves 5-10 cm long, gray-green with yellow margin, deciduous during resting time. Red flowers with green stripes. *Arid-subtropic.* p. 366

undulata (Cape Prov.), "Silver crown"; interesting succulent with opposite, broad wedge-shaped leaves covered with silver-gray bloom, the crimped apex pure white; flowers orange. *Arid-subtropic.* p. 361

wallichii (South Africa); branching succulent with pencil-like, sub-cylindrical leaves grooved on top, and incurving, glaucous gray, bronzing in the fierce sun, and deciduous; flowers green, dotted red; very poisonous. *Arid-subtropic.* p. 365

COUROUPITA Lecythidaceae

guianensis (Guayana), the "Cannonball-tree"; tall, soft-wooded, deciduous tree with armed branches, obovate, serrate 30 cm leaves, and fragrant 12 cm waxy flowers rose-colored inside, orange-yellow outside, borne on the tree-trunk on tangled stems, later followed by the large round 15-20 cm brownish fruits. *Tropical.* p. 535

COWANIA Rosaceae

mexicana (California to Mexico), "Cliffrose" or "Quinine bush"; dense shrub to 2 m, with leathery, pinnatifid leaves and sticky; cream-colored flowers 2-3 cm across. *Subtropic.* p. 546

CRASSOCEPHALUM Compositae

mannii (Gynura) (Uganda); branching small tree to 8 m, with green to purple stems and rosette of large oblanceolate leaves 45 to 75 cm long, prominently toothed at margins; yellow flowers in terminal inflorescence. *Tropical.* p. 316

CRASSULA Crassulaceae

acinaciformis (Zimbabwe, Transvaal); woody root stock bearing annual shoots 50-70 cm high, forming aloe-like rosette of soft-fleshy leaves a beautiful wine-red; inflorescence in large clusters of whitish flowers. *Arid-subtropic.* p. 374, 375

arborescens (So. Africa: Namaqualand to Natal), known commercially as argentea or "Silver dollar" or "Silver jade plant"; heavy, branching succulent growing tree-like with thick trunk to 4 m high, the robust branches with boldly fleshy, broad obovate, opposite leaves 4-8 cm long, united at the base, silver gray with reddish dotting and contrasting red margin; starry flowers white, turning pink. Slow growing and rarely blooming; excellent in pots, *Arid-subtropic.*

"arborescens" of horticulture: see argentea

argentea (Cape Prov., Natal), known as C. arborescens, the "Jade plant", in the trade, and so listed in Loudon's Encyclopedia of Plants (England 1836); freely-branching or forking succulent growing to 3 m; leaves spatulate, thick-fleshy, upper surface rounded, lower flat, glossy jade-green, turning reddish in the sun, edged red; an old house plant; flowering pinkish-white, with age. *Subtropic.* p. 364, 367

argentea 'Sunset'; a California cultivar grown by Hummel, with three-tone leaves in glossy green variegated with yellow, and contrasting red margin; best under cool conditions. *Subtropic.* p. 367

corymbulosa (S.E. Cape Prov.); erect-growing plant with long gray green, triangular leaves in 4 rows, to 3.5 cm long, diminishing toward the apex; inflorescence 30 cm long with starry white flowers. *Arid-subtropic.* p. 374

falcata (Rochea) (Cape Prov.), "Scarlet paintbrush"; handsome succulent with wide, flattened and sickle-shaped curved leaves arranged like parallel shingles, rough gray-green, and sending up fleshy stalk with showy terminal cluster of bright crimson flowers. *Subtropic.* p. 367

'Flame'; possible perfoliata hybrid; attractive succulent pyramid with concave, ovate leaves 4-6 cm long, light olive-green in younger ones, older leaves and all margins a glowing coppery rose. *Arid-subtropic.* p. 364

hemispherica (Namibia), "Arab's turban"; small round plant forming cushions, the rounded leaves are close together, overlapping and curving downwards, dark gray green and finely fimbriate at edges; flowers white. *Arid-subtropic.* p. 367

hottentotta (Namibia); small clustering succulent with columnar branches dense with short pointed, fleshy leaves in four ranks, silver gray with brownish edges and green spots; sometimes confused with C. cornuta. *Arid-subtropic.* p. 367

mesembryanthemopsis (Namibia); clustering rosettes of 4-5 succulent pairs of squared-off, oblique angular leaves, flat on top, grayish green with white etching; white flowers in sessile head. *Arid-subtropic.* p. 367

'Morgan's Pink' (C. mesembryanthemopsis x falcata); pretty little clustering plant dense with irregularly wedge-shaped thick leaves covered with pearly gray puckers, bearing a tight cluster of rosy-salmon fragrant flowers with yellow anthers, lasting a long time. *Subtropic.* p. 367

multicava 'Variegata'; Natal species, freely branching succulent with broad spatulate, fleshy leaves, 3-8 cm long, glossy gray green with white variegation; flowers pinkish white and carrying bulbils. *Arid-subtropic.* p. 364

obliqua (argentea) (Cape Province, Natal), "Jade plant"; succulent leaves, fresh green and spatulate, 3-4 cm long, the tips pointed. *Arid-subtropic.* p. 364

perfoliata (Cape Prov., Natal); elegant erect plant with tapering, fleshy leaves in 4 rows, gradually broadening and becoming lanceolate, deep grayish green covered with a silvery roughness; flower heads brilliant scarlet. *Subtropic.* p. 366

x portulacea (argentea x lactea), "Baby jade"; type of "Jade plant" with fleshy obovate pointed leaves somewhat smaller, and entirely green; upper surface flat, rounded beneath. *Subtropic.* p. 366

x portulacea 'Convoluta'; very compact succulent with small obovate pointed leaves with red margins, 2-3 cm long, the upper surface convex and turned down. *Arid-subtropic.* p. 364

pyramidalis (Cape Prov.: Namaqualand); very attractive little succulent, grass green with tricornered flat leaves of equal size arranged in four ranks close above each other as in a perfect pyramid, over 2 cm dia., apex tinted red in the sun; flowers white. *Subtropic.* p. 367

rosularis (Cape Prov., Natal); succulent perennial with flat, basal rosettes producing branches; thin-fleshy leaves linear-spoon-shaped to 8 cm long, glossy green, edges white-horny ciliate; flowers white. *Subtropic.* p. 374

rubicunda (S.E. Africa to Mozambique); showy succulent rosette of concave, lanceolate tapering leaves, silvery gray tinted with purple, the lower ones 5 to 20 cm long; leafy stalk with terminal inflorescence in flat clusters of tiny flowers brilliant red. *Arid-subtropic.* p. 364

rubicunda in hort: see Rochea coccinea

rupestris (monticola) (Cape Prov.), "Rosary vine"; shrubby succulent with scandent branches, the thin wiry stems set with opposite, sessile fat triangular leaves united at the base, to 2½ cm long, glaucous light gray-green, the margins tinged red; clusters of yellowish flowers. *Arid-subtropic.* p. 366

sladenii (Cape Prov.); attractive succulent with elongate stems, roundish fleshy grayish-green leaves with flat surface, bottom rounded; small white flowers in inflorescence. *Subtropic.* p. 367

teres (So. Africa), "Rattlesnake"; dwarf clustering plant with roundish columns, closely appressed, pale green leaves, having a pale-translucent margin, which gives it a glazed appearance; flowers white. *Subtropic.* p. 362

tetragona (Cape. Prov.), "Miniature pine tree"; erect shrub with opposite, thin spindle-shaped, glossy green leaves; small white flowers. *Subtropic.* p. 364

'Tricolor Jade' (argentea x lactea); habit upright, leaves somewhat pointed and a true green; the glossy leaves are beautifully variegated rich green, gray, white and pink, shading to purplish towards margin, surface flat, round beneath. Has the typical needlework patch of lactea. *Subtropic.* p. 367

CRASSULA: see also Rochea

CRESCENTIA *Bignoniaceae*

alata (Parmentiera) (Sonora to Guerrero and Guatemala), "Gourd tree" or "Mexican calabash"; sparry tree to 8 m, with gray bark and wide-spreading or drooping branches; leaves divided into 3 leaflets, on winged petioles; bell-shaped greenish and purple flowers; globular green to brown fruit to 12 cm dia., from main branches, and used for household utensils. *Tropical.* p. 186

cujete (Trop. America), the "Calabash tree"; small tree to 12 m high, with sparry, spreading branches, simple oblanceolate usually clustered leaves to 15 cm long, borne in groups of 2 or more on spurs from axillary buds on thick gray branchlets; flowers bell-shaped 5-8 cm long, yellow with brownish markings; interesting are the melon-like smooth, yellow to black fruit 12-30 cm or more dia., like a hard-shelled gourd, on its trunk and older branches; like gourds they can be tied around their girth as they mature to produce unusual forms, for making into cups or other vessels in tropical lands. *Tropical.* p. 185, 187, 456, 457

CRINUM *Amaryllidaceae*

amabile (Sumatra), "Giant spider lily"; large bulbous plant with neck to 30 cm long, crowned by rosette of fleshy leaves shaded purple, 10 cm wide and 1 m long; robust floral stalk bearing fragrant flowers with red tube, and red down center of the white segments 12 cm long. *Tropical.* p. 63

asiaticum (Trop. So. Asia, Melanesia), known as "Poison-bulb"; showy, large rosette with bulb-like roots, very decorative alone for its numerous broad, sword-like, fleshy, yellow-green leaves 8-12 cm wide, with broad clasping base, center rib depressed 1-1½ m long; the white waxy flowers between foliage, with narrow, pointed petals inside purple-red; Melanesian maidens wear these fragrant blooms in their black hair. *Tropical.* p. 60, 63

augustum (amabile hort.) (Sumatra), a "Crinum lily"; sturdy plant with bulb sometimes 15 cm thick and 30 cm neck; many leaves 8-10 cm broad; numerous fragrant flowers on 1-1½ m stalk, the lanceolate segments deep wine-red outside, lighter inside. *Tropical.* p. 54, 62

japonicum 'Variegatum' (asiaticum var.) (Kanto to Kyushu), "Hama Omotu"; subcylindric bulbs 30-50 cm long; lustrous green, fleshy leaves to 70 cm, prettily banded and variegated creamy-white; flowers white. *Subtropic.* p. 68

moorei (So. Africa: Natal), "Longneck swamp lily"; herbaceous plant from a large bulb with stem-like neck to 30 cm; broad and thin, smooth-edged, somewhat wavy, bright green leaves 60-90 cm long, and showy, lily-like, soft rose bell-shaped fragrant flowers 12 cm or more across, very attractive with pink filaments, 6 to 8 blooms to a cluster, on stout stalk, during summer; free bloomer. *Subtropic.* p. 62

moorei album (So. Africa); bulbous plant with strap-shaped leaves and showy, lily-like flowers in umbels. *Subtropic.* p. 62

x powellii (bulbispermum x moorei). "Powell's swamp lily"; a spectacular old English hybrid (1732), with globose bulb, carrying abundant, decorative foliage, about 20 sword-shaped leaves to 120 cm long; a 60-90 cm firm floral stalk appears in summer, crowned by a cluster of up to 10 trumpet-shaped flowers 10 cm long, generally deep rose, and opening in succession; blooming at a time where there are few other bulbous plants of its stature or beauty in flower. Slightly hardy. *Subtropic.* p. 62

x powellii 'Album'; chaste form of pure white, large lily-like flowers 10-12 cm long, on stiff-erect stalks over shining green foliage. *Subtropic.* p. 62

x powellii 'Roseum'; handsome evergreen variety with long recurving, glossy green leaves; stiff erect stalks carrying clusters of soft pink trumpet flowers. *Subtropic.* p. 62, 63

procerum var. splendens; a form closely allied to C. asiaticum but with a massive bulb to 30 cm diameter and 50 cm high; like amabile it has flowers bright red outside, lighter inside, carried in a large rounded umbel, with a compact group of buds in the central area. *Tropical.* p. 62

yemense (latifolium), (South Arabia); bulbous plant with numerous shining green strap-shaped leaves 60-80 cm long; trumpet-shaped pure white flowers, the tube 8-10 cm long, with petals reflexed at tip; 10 to 20 blooms on long 30 to 60 cm stalk. *Arid-tropical.* p. 62

CROCOSMIA *Iridaceae*

masonorum (Tritonia) (So. Africa: Transvaal), "Golden swan tritonia"; magnificent cormous herb forming clumps of ribbed, sword-shaped basal leaves 75 cm long and 5 cm wide; branched arching stems bearing two-tiered spike-like clusters of starry flowers glowing orange-scarlet, 4 cm across, and blooming July or August; similar to Tritonia crocosmaeflora, the "Montbretia" of gardens, but with broader foliage, broader flower segments of deeper red, and inflorescence more pendant. Goes dormant during winter. Not winter-hardy. *Subtropic.* p. 525, 527

CROSOSMIA: see also Tritonia

CROCUS *Iridaceae*

speciosus (S.E. Europe to Turkey and Iran), "Autumn-flowering crocus"; cormous plant with 4-5 grass-like, linear leaves, appearing after blooming; the short-stalked flowers with spreading segments 5-6 cm long, the petals bluish-lilac feathered with purple veins inside, blooming September-October. *Warm-temperate.* p. 526

CROSSANDRA *Acanthaceae*

infundibuliformis (undulaefolia) (India), "Firecracker-flower"; shrubby plant with glossy, ovate leaves; showy salmon-red tubular flowers with split limb, in angled, bracted spike. While grown in our greenhouses as a small 10 cm potplant, I have seen these in India and Ceylon as an ever-blooming 1 m bush. *Tropical.* p. 35

infundibuliformis 'Mona Wallhead'; a Swedish cultivar of compact habit about ½ m high, with shining, black-green leaves and flowers salmon-rose. *Tropical.* p. 34

pungens (Tanzania: Usambaras); attractive, bushy plant to 60 cm high, with oblong leaves olive-green, with creamy-white veins; flowers orange, set in bracts hairy at margins. *Tropical.* p. 34

CROTALARIA *Leguminosae*

laburnifolia (Australia), "Queensland bird-flower"; shrub 2 m or more high with leaves of 3 leaflets; bird-like greenish-yellow pea-flowers, hanging by their beaks to the stems. *Subtropic.* p. 546

retusa (Tropics, Asia), "Rattle-box"; stout pubescent annual 1 m or more, leaves with one oblanceolate leaflet; terminal inflorescence with yellow flowers. *Tropical.* p. 546

CROTON *Euphorbiaceae*

gossypifolius (Trinidad); sparry tree with large cotton-like lobed, green leaves; yielding croton oil used as a purgative. *Tropical.* p. 411

poilanii (Vietnam: Trang Bom); small tree to 5 m high with slender gray trunk and willowy branches, dense with small 8-10 cm elliptic, plain dark green, leathery leaves crenate at margins; the female flowers usually without petals, the male flowers with 5 petals and inconspicuous; croton-oil is made from seeds. I found this species in ironwood forest 51 km northeast of Saigon. *Tropical.* p. 411

CROTON: see also Codiaeum

CRYPTANTHUS   *Bromeliaceae*
   beuckeri (Brazil), "Marbled spoon"; irregular rosette of flat, thin leathery spoon-shaped leaves with slender point, rich green marbling on pale green; flowers whitish. *Tropical.* p. 212
   bivittatus minor (roseus pictus) (Brazil), "Rose-stripe star"; flattened, small, star-like, terrestrial rosette satiny olive-green with two pale bands, overcast with salmon rose, turning coppery red in strong light; finely toothed. *Tropical.* p. 212
   bivittatus 'Pink Starlite'; beautiful small, flat rosette, patented in Florida, ave. 6-8 cm across and nearly a perfect star in shape, the almost triangular leaves moss-green along center, toward margins a contrasting cream in wide bands, the finely toothed edges a vivid, pleasing rosy-red; probably a bivittatus cultivar; ideal for small dishgardens and terrarium planting. *Tropical.* p. 213
   bivittatus 'Tricolor'; handsome large rosette of narrow lanceolate, stiff leathery leaves to 20 cm long, olive green with cream-white marginal bands tinged rose, crisped and wavy along sides. *Humid-tropical.* p. 213
   bromelioides 'Racinae'; a charming hybrid by Mulford Foster of Orlando, Florida, named after his wife Racine; attractive star with wide-spreading, stiffly lanceolate leaves 12-18 cm long, coppery bronze with irregular, silvery crossbands, the margins undulate or crisped; pure white flowers from the center cup; photographed in Palmengarten collection, Frankfurt, Germany. *Tropical.* p. 213
   bromelioides 'Tricolor', "Rainbow star"; strikingly variegated sport with the fleshy, fresh-green leaves edged and striped ivory-white, the margins and base tinted carmine-rose. *Tropical.* p. 212, 213
   diversifolius (Brazil), "Vary-leaf star"; larger rosette with arching, wavy-margined leathery leaves of diverse shape: broad-ovate, long-oblanceolate and spoon-shaped; dark bronze-purple covered with silvery scales. *Tropical.* p. 212
   fosterianus (Pernambuco), "Stiff pheasant-leaf"; large rosette to 85 cm across, marked similar to zonatus but leaves much thicker and very stiff, habit very flat, coppery green to purplish brown, with tan zebra banding. *Tropical.* p. 212
   x 'It'; aptly known as "Color-band"; spectacular mutant developed from a Foster hybrid probably involving C. bivittatus lueddemannii. Terrestrial rosette with strap-like leaves to 45 cm long and 4 cm wide, coppery green with longitudinal striping and broad outer bands of ivory, the margins a glowing rosy-red intensified by good light. *Tropical.* p. 212, 213
   x mirabilis (beuckeri x osyanus); handsome Richter hybrid with broad-ovate, recurving, undulate leaves in various shadings light green, olive and copper-brown, more pronounced in blotches or cross bands in cream, and shading into glowing rose. Photographed in the collection of Leningrad Botanic Garden, U.S.S.R. *Tropical.* p. 212
   zonatus zebrinus (fuscus), "Zebra plant"; strikingly beautiful form with bronzy-purple long wavy leaves, the pronounced silvery to beige crossbanding resembling those of a zebra. C. zonatus is from Pernambuco, a rosette to 45 cm across. *Tropical.* p. 212, 213

CRYPTOCEREUS   *Cactaceae*
   anthonyanus (Chiapas), "Anthony's rick-rack"; nightblooming climber of the rain forest, using aerial roots; unusual deeply lobed stems looking like fish bones; flowers intensely fragrant, lasting but a single night, beautifully colored burning red with cream-yellow petals. *Tropical.* p. 255, 256

CRYPTOCORYNE   *Araceae*
   affinis (Malay Peninsula), "Water trumpet"; rhizomatous aquatic herb, with lanceolate leaf blades, puckered dark green with pale veins, wine-red beneath; spathe 12-20 cm long, pale green outside and with purple lines, the blade spiralled and black-purple inside. *Tropical.* p. 103
   ciliata (Monsoon India: Bengal to Malaya), "Fragrant tape grass"; tropical aquatic herb or bog plant to 40 cm high, with creeping rhizome and stalked lance-shaped leaves of firm texture, usually growing under water, fresh to yellowish-green with undulate margins, the depressed midrib dark green. Inflorescence submerged with the enclosed spadix bearing female flowers on the lower part, male flowers on the upper, only the extended spathe above water surface. *Humid-tropical.* p. 104
   petchii (Sri Lanka: Ratnapura); aquatic tropical plant with robust rootstock, forming a rosette of long-stalked lance-shaped green leaves 15-20 cm long, 2 cm or more wide; attractive tubular spathe 7-10 cm long, the 4 cm tube opening into a long pointed, fleshy and warty limb, olive green to brown, the inside of tube rich purplish-brown. *Tropical.* p. 103
   wendtii (Thailand); tropical aquarium plant for both under water and above surface culture, with sturdy roots and branching; stalked leaves spear-shaped 8-11 cm long, and 3 cm wide, glossy green with brownish purple in habitat; spathe 8 cm long, tubular with twisted mouth and tail on top, brown purple. *Tropical.* p. 104
   willisii (Trop. S.E. Asia), "Water trumpet"; aquatic plant with narrow lanceolate leaves, when young reddish-brown marked with greenish-black; curly margin. *Tropical.* p. 103

CRYPTOCORYNE: see also Lagenandra

CRYPTOMERIA   *Coniferae: Taxodiaceae*
   japonica (Japan and C. China), "Japanese cedar"; graceful evergreen tree of pyramidal habit and somber appearance to 50 m tall, with reddish-brown bark peeling in strips; the branches spreading and slightly pendulous, clothed with short 1-2 cm needle-like green leaves turning brownish in winter; roundish female cones globose 2 cm dia. *Warm-temperate.* p. 352, 354
   japonica 'Lobbii' (Japan: Kyushu); compact pyramidal tree 20-30 m high, densely branched, branches shorter than the species, slightly decurving; longer, pointed branchlets with leaves sickle-shaped and densely set, directed forward, 1 cm long, flat and glossy dark green, not changing color in winter. *Subtropic.* p. 354

CRYPTOPHORANTHUS   *Orchidaceae*
   nigriflorus (Trop. America) "Window orchid"; diminutive epiphyte with oval leaves; queer flower from leaf base; the sepals are joined to form a sac, in which the small petals and lip are enclosed, black maroon and covered with translucent bristles. *Tropical.* p. 722

CRYPTOSTEGIA   *Asclepiadaceae*
   grandiflora (Mascarene Islands), "India rubber-vine"; scandent tropical shrub with willowy, climbing branches, bearing opposite, glossy green soft-leathery leaves with ivory midrib, 10 cm long; large bell-shaped flowers 5 cm across, white inside, and reddish-purple outside, changing with age to pink. Cultivated in India for rubber-yielding latex; plants however, are poisonous. *Tropical.* p. 80, 145, 146
   grandiflora alba, "White Rubber vine"; strong-growing willowy vine with milky juice; fleshy lanceolate leaves 10 cm long; waxy white funnel-shaped flowers. Photographed in Western Australia. *Tropical.* p. 146
   madagascariensis (Madagascar); woody tropical evergreen vining plant with dark green shining foliage, pubescent underneath; pink or whitish trumpet flowers having corolla segments divided. *Tropical.* p. 145

CTENANTHE   *Marantaceae*
   compressa (pilosa) (Pernambuco, Brazil); erroneously called "Bamburanta" in California nurseries; waxy-green, unequal-sided, leathery leaves oblong, grayish-green beneath, borne at an angle on wiry petioles, the plant is bushy but throws up bare stems bearing 2 to 4 small plants at the top forming a heavy tuft of foliage; flattened, hairy bracts. *Tropical.* p. 640
   kummeriana (Brazil); clustering leafy bush becoming 75 cm high and sending out stolons; lanceolate leaves papery when young and dark green with attractive lateral bands of silver along the nerves, shimmering purple beneath; the silvery variegation disappearing with age. *Tropical.* p. 640
   lubbersiana (Phrynium) (Brazil); spreading herb with forking stems, the oblique linear-oblong leaves are firm, green, variegated and mottled with yellow above, paler beneath. *Tropical.* p. 637
   oppenheimiana (Brazil); strong, compact, branching plant forming a dense, broad bush up to 2 m high, the narrow lanceolate leaves are leathery and firm, dark green with lateral bands of silver, and wine-red beneath, attached at an angle to downy stalks. *Tropical.* p. 637
   oppenheimiana 'Tricolor', "Never-never plant"; very colorful, tufted variety with narrow leaves, highly variegated white over green and silver gray, their wine-red underside in vivid contrast, and glowing through the surface. *Tropical.* p. 637, 638, 639
   setosa (Brazil); slender plant with thin, narrow lanceolate leaves attached at an angle to reed-like, hairy stalks; blade almost all yellowish-silver with broad lateral bands on metallic green base; reverse purple. *Tropical.* p. 637

CUCUMIS   *Cucurbitaceae*
   anguria (So. U.S.: Florida and Texas), "West Indian gherkin" or "Gooseberry gourd"; cucumber-like vine with slender angled, rough stems and palmately lobed 10 cm leaves; yellow flowers; the females producing spiny ovoid, yellowish-green fruits 5-8 cm long, and which can be used for pickles. Interesting for the conservatory. *Subtropic.* p. 382

melo 'Reticulatus' (W. Africa), "Cantaloupe"; tender, hairy annual trailer, angled leaves, yellow flowers 3 cm across; medium-size fruit 15-18 cm, with hard, rough warty rind, and orange-salmon flesh of sweet, delicious flavor. *Tropical.* p. 382

sativus (W. Trop. Asia), the "Cucumber"; rough-hairy annual trailing or tendril-climbing vine with herbaceous, lobed leaves; male and female yellow flowers 2-3 cm across; cylindric oblong edible fruit; smooth or prickly, green with pale marbling; used for salads and pickles. The pictured cultivar 'Patio Pik' (Geo. Ball, W. Chicago 1973) is adapted for growing on trellis in containers on the patio, for both ornamental and culinary purpose. *Tropical.* p. 383

CUCURBITA   *Cucurbitaceae*

maxima (So. America origin), "Squash"; tendril-bearing herbaceous, running vine, slightly prickly, with large, orbicular, lobed leaves, rough to the touch; yellow flowers with soft corolla, the female flowers producing the vegetable squash. The "Winter butternut squash" pictured on p. 383 is of very curious, oblong flask-shape, 25 cm long, the skin cream-colored, flesh deep orange, and of excellent flavor. Winter squash is commonly eaten when baked. *Humid tropical.* p. 383

pepo (Trop. America), "Winter squash"; herbaceous annual with prickly vines and harsh, rugose lobed leaves; producing hard-shelled, so-called winter squash with woody rinds and firm, close-grained, fine-flavored flesh, and used for baking and pies. They come in various shapes, turban, warted and banana, as in the var. melopepo or the "Banana squash" shown on pg 383, with curved cylindric, pale yellow fruit 30-40 cm long, photographed in Vista, California. *Tropical.* p. 383

pepo 'Lady Godiva', a "Pumpkin"; annual prostrate vine with prickly stems, leaves triangular oval, deeply lobed; large yellow flowers; yellow fruit 30 cm long. *Tropical.* p. 382

pepo 'Little Dumpling', a "Vegetable gourd"; of medium size, grown on stakes in England during summertime, both as an ornamental, but also used for cooking; may be stuffed with meat and baked like pepper. *Tropical.* p. 383

CULCASIA   *Araceae*

rotundifolia (W. Africa: Gabon); rainforest climber with slender wiry stems forming aerial roots; ornamental broad-ovate leaves with cordate base, coppery deep green with silky surface, 6-10 cm long. *Tropical.* p. 97

CUNNINGHAMIA   *Taxodiaceae (Coniferae)*

lanceolata (China), "China fir"; when visiting China I confused this handsome tree with Araucaria bidwillii which it closely resembles, but the glossy dark green needles are narrower, more close, and finely serrate. *Subtropic.* p. 352, 353

CUNONIA   *Cunoniaceae*

capensis (So. Africa), "African red alder"; small evergreen tree with pinnate, leathery glossy green leaves with toothed margins; on reddish petioles; dense feathery axillary racemes of white flowers; attractive cool house decorator. *Subtropic.* p. 391

CUPANIOPSIS   *Sapindaceae*

anacardioides (Australia: N.S.W.), "Tuckeroo" or "Carrotwood"; evergreen tree to 10 m high; pinnate leaves with 5-10 obovate, leathery leaflets to 15 cm long; small white flowers in axillary clusters; fruit a 3-lobed capsule 2 cm dia. *Subtropic.* p. 876

CUPHEA   *Lythraceae*

melvilla (Guayana), "False heather"; heather-like perennial herb, to 1 m high, with small sessile, lanceolate leaves; slender flowers 2-3 cm long, red at base, green at apex. *Tropical.* p. 621

platycentra (ignea) (Mexico), "Cigar flower"; low herbaceous plant with slender stems, lanceolate leaves and solitary axillary flowers having a slender, bright scarlet calyx with white mouth and dark ring at end, without petals. *Tropical.* p. 620

CUPRESSOCYPARIS   *Cupressaceae (Coniferae)*

'Leylandii' (Chamaecyparis nootkatensis x Cupressus macrocarpa); very ornamental with red stems, and striking green, flattened branchlets, cones 1.5-2 cm dia. *Warm temperate.* p. 338

CUPRESSUS   *Cupressaceae (Coniferae)*

funebris (China), "Mourning cypress"; from the Yangtse Valley, tree to 20 m, with pendulous branches, crown loosely spreading; branchlets on a flat plane; leaves 3-angled and densely appressed, bluish or light green; cones globose, 1-2 cm dia. *Subtropic.* p. 337

lusitanica (Mexico, Guatemala); "Portuguese cypress" or "Mexican cypress"; evergreen tree to 30 m, with red-brown bark splitting in long strands; branches widely spreading, with branchlets more or less pendulous; small appressed, pointed needles glaucous green; cones 1 cm dia. and glaucous. *Subtropic.* p. 338

macrocarpa (California) (Cupressaceae), "Monterey cypress"; similar to arizonica but the scale-like leaves are bright to dark green, branches spreading. *Subtropic.* p. 341

macrocarpa 'Aurea', "Golden Monterey cypress"; New Zealand cultivar, of horizontal habit and golden foliaged. *Subtropic.* p. 338

sempervirens (So. Europe, W. Asia, No. India), "Italian cypress"; the classic conical cypress of the Greek and Roman writers, with very short branches, forming a dense, narrow column slowly to 20 m or more high; the stout branchlets with scale-like leaves dark green with grayish cast in 4 ranks. Esteemed for formal effect because of its stiff, picturesque outline. Suitable in containers for warm, sunny places; cool in winter. *Subtropic.* p. 337, 338

sempervirens 'Stricta'; branches densely erect, forming a pyramid, to 20 m high. *Subtropic.* p. 338

CURCUMA   *Zingiberaceae*

domestica (India), "Turmeric"; herbaceous stemless tropical plant from short, tuberous rhizomes, with yellow flesh; rosette of leaves 45 cm long; from the center a cylindric tuft 12 cm long, of greenish white pouch-like bracts, the topmost ones pink; flowers creamy yellow. Turmeric is used as curry spice and dye, obtained from the dried and ground rhizomes. *Tropical.* p. 926

inodora (India: Maharashtra); showy plant which I collected in rainforest east of Bombay, with large, fresh green, ribbed leaves, and erect cane-like inflorescence with cup-like green lower bracts tipped pink, topped by a tuft of colored floral bracts deep rose. *Tropical.* p. 925

roscoeana (Burma), "Hidden lily"; robust, perennial herb with tuberous roots, sending up 6-8 long-stalked, lanceolate, ribbed, handsome leaves with dark green nerves, the inflorescence a splendid spike about 20 cm long, with cone-like, showy bracts gradually changing from green to vivid scarlet-orange, corolla yellow, with rich golden lip. *Tropical.* p. 923

siamensis (Thailand), a "Hidden lily" of compact habit; rosette of broad, elongate leaves 30-40 cm long, corrugated in feather fashion and with depressed yellowish midrib; in the center, the attractive inflorescence with green-cupped lower bracts, topped by concave bracts of creamy pink. *Tropical.* p. 925

CUSSONIA   *Araliaceae*

holstii (E. Africa: Kenya), "Cabbage tree"; small shapely tree with slender stem and handsome leathery foliage deep glossy-green, stalked ovate leaflets with crenate margins, palmately compound and carried on long wiry petioles. *Tropical.* p. 134

paniculata (So. Africa), "Cabbage tree"; evergreen small tree forming rosette of heavy fleshy leaves palmately compound, the grayish leaflets with serrate and lobed margins, and covered with silvery bloom; on single trunk to 3 m high. *Subtropic.* p. 132, 133, 134

spicata (Transvaal), "Spiked cabbage tree"; evergreen tree with handsome leathery leaves palmately compound, the smooth, grayish-green segments dentately cut or lobed again, and arranged in flat rosettes. *Subtropic.* p. 134

CYATHEA   *Cyatheaceae (Filices)*

arborea, (mountains of Puerto Rico to Jamaica), "West Indian treefern"; slender treefern to 15 m high, with mostly bare brown trunk, the upper part covered with pale brown scales, and crowned by bipinnate, finely toothed, ample fronds soft-textured, fresh green, paler below, and without spines. Very graceful in appearance. *Humid-tropical.* p. 431, 433, 434

deckenii (Tanzania: Usambara Mts.), "Amani treefern"; short, straggling treefern with slender trunk, entirely covered with long brown fiber; the fronds bipinnate on long rough-hairy petioles. *Humid-tropical.* p. 436

dregei (So. Africa); robust treefern seen at Kirstenbosch, with blackish trunk and stiff-erect, smooth green fronds on yellow axis, bipinnate into fine serrate leaflets. *Humid-subtropic.* p. 436

lunulata (Fiji); stately, palm-like treefern with slender trunk covered with brown fiber, except for leaf scars; fronds bipinnate, of firm texture. *Humid-tropical.* p. 433

medullaris (New Zealand, S.E. Australia), the "Black tree-fern" or "Mamaku"; tallest of N.Z. tree ferns to 20 m or more, slender black trunk, at base covered with matted aerial roots, on top a great crown of spreading, curving, feathery fronds 2-6 m long, firm, tripinnate, deep green above, paler beneath; the apex and leaf bases clothed with black scales. *Humid-subtropic.* p. 432, 434, 436

spinulosa (China: Kwangtung Prov.); graceful small treefern with slender trunk covered with black-brown fiber; fine-feathered rich green fronds of firm texture. *Subtropic.* p. 431

usambarensis (Tanzania); robust treefern from the moist forests of the Usambara Mountains; trunk of medium height, to 15 cm thick and covered by dark brown fibers; wide-spreading bipinnate fronds with pinnae widely separated. *Humid-tropical.* p. 432

CYBISTAX: see Tabebuia

CYCAS *Cycadaceae*
circinalis (India, Madagascar, New Guinea), "Fern palm"; palm-like tree with stout trunk to 3½ m or more high, topped by a graceful rosette of stiff-glossy leaves pinnately divided, the leaflets flat on edges; male and female inflorescence on separate plants. I have seen colonies of this species in the Bulolo highlands in New Guinea, at 1,200 m in tall Kunai grass. *Tropical.* p. 385, 387
media (No. Australia, Queensland), "Australian nut palm"; trunk to 5 m high, leaves shorter than circinalis, with numerous leaflets, narrowed to a spine, mostly flat. *Tropical.* p. 384, 385
neo-caledonica (New Caledonia), "Bread palm"; graceful large cycas with the trunk 18 cm thick, the arching fronds 2 m long, the narrow, closely set leaflets shining green. *Tropical.* p. 384
revoluta (So. Japan to Java), "Sago palm"; this palm-like tree has stout trunk to 3 m high, sometimes branched, the deep green stiff pinnate foliage keeps like iron; segments are rolled down at margins. *Subtropic.* p. 384, 385, 387
rumphii (No. Australia, Malay Archipelago), "Bread palm"; palm-like, with trunk to 6 m, leaves similar to media, but shorter and with fewer leaflets, thinner than circinalis. *Tropical.* p. 388

CYCLAMEN *Primulaceae*
persicum (indicum) (Greece and Mediterranean islands to Syria); charming low, fleshy herb with large, hard tuberous roots, heart-shaped basal leaves in a rosette, prettily patterned with silver; and long-stalked solitary, 4 cm flowers nodding, fragrant flowers; with purplish-rose flaring corolla lobes elegantly reflexed. *Humid-subtropic.* p.829, 832, 833
persicum 'Bonfire', florists "Alpine violet"; elegant clone, of compact habit and somewhat slower, but sturdy growth, the noble flowers borne well above the foliage, in brilliant salmon-scarlet, petals erect except one pendant. *Subtropic.* p. 833
persicum 'Fimbriatum'; a cultivated form with richly patterned leaves, and large flowers with spreading petals fringed at margins, soft pink with carmine-red eye. *Subtropic.* p. 833
persicum 'Pearl of Zehlendorf'; the favorite color form, with elegant flowers vivid salmon-red, deeper toward eye, shading to lighter salmon at margins, free-blooming. *Subtropic.* p. 833
persicum 'Rococo'; curious fleshy flowers rather flat and wheel-round with margins beautifully frilled, usually light salmon pink with darker eye; not free-blooming. *Subtropic.* p. 833
persicum 'Rose of Aalsmeer'; an excellent Dutch horticultural cultivar of robust habit, dense with fleshy leaves, and large flowers with broad petals a lovely rich rose-pink. *Subtropic.* p. 832
persicum 'Rose von Zehlendorf'; old German cultivar with fine flowers soft salmon pink with red-purple eye. *Subtropic.* p. 833
persicum 'Swiss dwarf'; diminutive cultivar of persicum, with attractive foliage marked silver; flowers on slender stems with petals 3 cm long; developed in Switzerland for lilliputian effect as house plant. *Subtropic.* p. 833
persicum 'Vogt's Double'; beautiful, double-flowering, American strain developed since 1920, with heavy blooms with numerous petals, in shades of salmon-pink. *Subtropic.* p. 833

CYCNOCHES *Orchidaceae*
chlorochilum (C. America, Colombia, Venezuela); epiphyte with tall pseudobulbs and plaited leaves, and clusters of large waxy flowers greenish yellow, lip creamy-white with a blotch of black-green at base. (Oct.-April). *Humid-tropical.* p. 724
pentadactylum (Brazil); curious species with spindle-shaped leafy pseudobulbs and plaited leaves; large flowers yellowish-green, banded throughout with brown, the lip whitish spotted red, and is divided into five parts like a hand; summer. *Humid-tropical.* p. 724
ventricosum (Guatemala), "Swan orchid"; striking epiphyte; stout pseudobulb with arching raceme of sweet-scented, waxy flowers greenish-yellow, a white lip, and an arched slender column resembling a swan's neck. (July-Nov.). *Humid-tropical.* p. 724

CYDISTA *Bignoniaceae*
aequinoctialis (W. Indies to C. America and Brazil), "Garlic vine"; scandent shrub climbing by tendrils, with shiny leathery, obovate, dual leaves to 15 cm long, and with clusters of funnelform flowers a pretty rose with white throat. Erroneously called "Garlic vine" in Florida, its leaves are not garlic-scented; the true "Garlic-scented vine" is Pseudocalymma alliaceum. *Tropical.* p. 187

CYDONIA *Rosaceae*
oblonga (Turkestan, Iran), "Fruiting quince"; sparry, thornless tree to 6 m high, deciduous in colder regions; with attractive ovate 5-10 cm leaves of rough parchment texture, dark green above and whitish tomentose beneath; yellow in autumn; 5 cm flowers at young branch tips; the fruit apple-like, greenish-yellow, somewhat wrinkled, and covered with light brown felt, 6 cm or more across and showing a woody calyx. The large, fragrant fruits are inedible when raw, but used for making jams and jellies. For flowering quince, see Chaenomeles. *Subtropic.* p. 482, 849
oblonga cv. 'Pineapple', "Pineapple quince"; roundish, light golden fruit, covered with white felt; tender white flesh of pineapple flavor. *Subtropic.* p. 470, 849
sinensis (Chaenomeles) (China), "Chinese quince"; small tree to 6 m, ovate leaves; 4 cm flowers with pink petals and white base; dark yellow ovoid fruit, 6-10 cm long, used for stewing, marmalade or jelly. *Warm temperate.* p. 470

CYDONIA see also Chaenomeles

CYLINDROPUNTIA: see Opuntia

CYMBIDIELLA *Orchidaceae*
humblotii (Madagascar); remarkable cymbidium-like species, with cylindric pseudobulbs and numerous narrow, strap-shaped leaves arranged fan-like; erect inflorescence to 80 cm tall, with cluster of long-lasting waxy, 8 cm flowers yellow-green, petals and lip bordered with black (winter). *Tropical.* p. 720
rhodochila (Madagascar); epiphyte often classified as Cymbidium; almost always growing together with Platycerium; the folded leaves grow from the axils of the older ones creating a braided effect; floral stalks from base of pseudobulbs, to 1 m high, bearing 20 or more yellowish green flowers, with brownish spots on petals, and a glossy crimson lip with yellow stripe (autumn). *Tropical.* p. 718, 722

CYMBIDIUM *Orchidaceae*
x alexanderi (eburno-lowianum x insigne); well-known commercial white with pink hybrid of robust habit; terrestrial with bold, arching raceme of waxy, long-lasting flowers flushed pink, with a purple horseshoe on the lip; the form alba has pure white flowers. *Subtropic.* p. 723
canaliculatum (Queensland); distinctive small terrestrial almost pseudobulbless with rigid, gray keeled leaves to 30 cm long, and stiff, arching raceme with numerous small waxy flowers intensely dark maroon-brown, almost black, edged lemon-yellow; lip pink and green spotted crimson. (April-May). *Tropical.* p. 724
'Ceres' (l'Ansonii x insigne sanderi) (1919); waxy flowers bronzy-red with faint yellow border around petals; lip marked purple with yellow lines. *Tropical.* p. 723
'Doris' (insigne x tracyanum); well-known commercial primary hybrid 1912, with large, fleshy, long-lasting flowers; light yellow sepals, petals closely veined with red, the lip spotted with red-brown on front lobe and veined in same color on side lobes. *Subtropic.* p. 724
eburneum (No. India, Burma); handsome terrestrial from the Himalayas and the Khasia Hills at 1,500-1,800 m, of compact habit forming tufts of 2-ranked linear foliage, and short racemes of 1-3 large waxy, fragrant flowers 10-12 cm across, blush-white, the lip yellow inside and with hairy keels (March-June). *Subtropic.* p. 724
'Flirtation' (pumilum x 'Zebra'); attractive, variable semi-miniature recently developed, ideal as a shapely, compact flowering pot plant, blooming successively twice a year, usually Nov. through April; arching racemes with flowers spreading 5 cm, pale pink to greenish ivory, shaded to coppery purple and striped orchid, lip white in throat spotted and blotched maroon. *Subtropic.* p. 724
lancifolium (Himalayas at 2000 m; India to Japan, south to Malaysia); charming, terrestrial forest dweller of low habit; spindle-shaped pseudobulbs clothed with 3 to 5 ribbed thin-leathery leaves 20 cm long; stiff-erect stalks bearing inflorescence 30 cm high of scattered, fragrant and long-lasting flowers 4 to 5 cm across, greenish-cream with purple spots, the lip white boldly marked with maroon, blooming in spring and early summer. *Subtropic.* p. 724
'Louis Sander' (alexanderi x 'Ceres') (1924); large substantial flower yellow overlaid with beige-brown; front of lip purple. *Tropical.* p. 723
'Madonna' (alexanderi 'Westonbirt' x Mem. P.W. Janssen); magnificent, robust plant of splendid habit, with stiff erect spikes carrying numerous large 10 cm open flowers, glistening white with unique green overcast, the blush pink lip spotted bright maroon-purple along margin; free-blooming in Feb.-March every year. *Subtropic.* p. 714
'Pixie' (Ceres x Landrail) (1955); large-flowered seedling with petals and sepals buff greenish-yellow lined with purplish brown, the lip cream-white with attractively contrasting band of bold crimson along pointed lip, lemon in throat. *Subtropic.* p. 714
'Priscilla' (insigne x 'Yellow Hammer'); good standard bronze type with large, durable flowers coppery yellow, the lip yellow, heavily marked with crimson inside margin. *Subtropic.* p. 723

pumilum (China: Yunnan); dwarf terrestrial with linear leaves to 30 cm long; small waxy flowers 3 cm across, reddish-brown and yellow. *Subtropic.* p. 714

'Swallow var. Broadview' (alexanderi x pauwelsii) (1916); elegant arching spray of lemon-yellow flowers with crimson lip. *Tropical.* p. 723

virescens angustifolium (Japan); tufted, grass-like plant with linear leaves to 25 cm long; stalked flowers 5 cm dia., vivid green with faint purple stripes, two petals join over lip to form a hood, lip curved under, creamy-white with yellow throat and purple markings; winter. *Subtropic.* p. 724

CYNARA *Compositae*
scolymus, (So. Mediterranean Reg., Canary Islands), the "Artichoke" or "Globe artichoke"; robust ferny-looking perennial of irregular, fountain-like form, to 1 m or more high; silvery green leaves deeply dissected and somewhat spiny, white cottony beneath; heads with purple disk flowers resembling a thistle, 15 cm across with a nest of fleshy scales beneath. The fleshy buds are cut before they open, and are edible when cooked. Flowers may be cut for fresh or dried arrangements. *Subtropic.* p. 331

CYNODON *Gramineae*
dactylon (Warm regions of both Hemispheres), "Bermuda grass" or "Devil's grass"; perennial grass with creeping stolons or rhizomes, spreading aggressively; the green or grayish-green leafblades are flat and short, 4 mm wide, on basal shoots often 2-ranked; very heat tolerant; can be grown from seed, rhizomes or sprigs. Several excellent hybrid strains with unusually fine, thread-like foliage merely 2 mm wide, have been developed for lawn-planting in Florida and California. *Tropical.* p. 510

CYPELLA *Iridaceae*
herbertii (Brazil, Uruguay, Argentina); bulbous plant about 30 cm high, with linear, plaited basal as well as stem-leaves; few flowers iris-like with spreading perianth segments, yellow varying to deep chrome, with spurs in center barred deep yellow; late summer. *Subtropic.* p. 527

CYPERUS *Cyperaceae*
alternifolius (Madagascar), "Umbrella plant"; clustering perennial bogplant with ribbed stalks to 1 m high, bearing a crown of bright-green, grass-like leaves around a head of small green flowers. I have seen this also along the Athi River in Kenya. *Tropical.* p. 390

alternifolius 'Variegatus', "Variegated umbrella plant"; a fugitive variety with stems and leaves striped and banded lengthwise with creamy-white contrasting with shiny green. *Tropical.* p. 390

diffusus (laxus) (Mauritius), "Broadleaf umbrella plant"; compact bushy plant sending out runners with suckers which are used for propagation; sturdy, 3-angled stalks with a crown of broad, matte green, rather rough leaves and long, pale brown spikelets. *Tropical.* p. 390

diffusus 'Variegatus', "Striped umbrella palm"; compact-growing variety having its broad green leaflets striped and banded pale yellow or cream. *Tropical.* p. 390

papyrus (Egypt), "Egyptian paper plant"; a stately plant for pools, with stout dark green stalks to 2 m high, topped by brush-like umbel of drooping, threadlike leaves. Along the shores of Lake Victoria in Africa I have seen them 5 m tall. Used in Egypt for making "papyrus" since 2750 B.C. *Tropical.* p. 389, 390

CYPHOMANDRA *Solanaceae*
betacea (So. Brazil), the "Tree-tomato" or "Tamarillo"; tree-like shrub somewhat woody, to 3 m, with heartshaped ovate 30 cm soft-hairy, fleshy leaves, and fragrant 1 cm flowers in pendulous raceme, the 5-lobed corolla greenish pink with a dark stripe on back of each segment; the egg-shaped, edible, reddish 8 cm fruit resembling a tomato in looks and flavor. *Subtropic.* p. 473, 894

CYPHOSTEMMA *Vitidaceae*
laza (Madagascar) peculiar succulent forming fleshy flask-like swollen trunk to 1 m dia., and 2 m tall, with parchment-like bark, the branches prostrate to clambering 3-5 m long, initially white-felted; fleshy pinnatifid leaves to 16 cm long; flowers green. *Arid-tropical.* p. 919

CYPRIPEDIUM *Orchidaceae*
acaule (Newfoundland to N. Carolina), "Pink ladyslipper"; terrestrial of the temperate zone, with two light green plaited leaves and large solitary flowers with hairy sepals and petals greenish-brown, the divided, pouch-like lip rose veined with crimson. (May-June). *Temperate.* p. 714

calceolus pubescens (No. America), "Yellow ladyslipper"; terrestrial orchid with leafy stem bearing 1-2 flowers, the broad leaves soft-hairy, sepals and petals purplish-brown to green, petals more or less twisted, the rounded pouch yellow veined with red. (April bl. in south, Aug. in north). *Temperate.* p. 714

CYPRIPEDIUM: see also Paphiopedilum

CYRTANDRA *Gesneriaceae*
schraderi ('Oomsis') (Papua); clamberer with semi-woody stems and long thin-leathery, 15 cm ovate to oblanceolate leaves; axillary tubular-inflated flowers salmon-red tipped yellow. *Tropical.* p. 495

CYRTANTHUS *Amaryllidaceae*
herrei (So. Africa), a "Fire lily"; attractive bulbous plant with fleshy, vivid green, strap-shaped leaves; the swollen floral stalk with dense clusters of pendant tubular red flowers. *Subtropic.* p. 67

mackenii (Natal), "Ifafa lily"; bulbous herb with linear leaves 30 cm long, and umbels on 30 cm stalks with pure white fragrant flowers having a 5 cm long curved tube and 6 spreading lobes. *Subtropic.* p. 53

CYRTOMIUM *Polypodiaceae (Filices)*
falcatum (Japan, China, India, Celebes, Hawaii) the "Holly-fern"; with handsome pinnate fronds on brown scaly stalks, the leathery, shining dark green leaflets are ovate, slender pointed and very durable under adverse conditions. *Subtropic.* p. 445

CYRTOPODIUM *Orchidaceae*
andersonii (Brazil, W. Indies); bold terrestrial and on rocks, with long pseudobulbs and tall 1 m branching spikes of yellow flowers tinged with green, and rich yellow lip. (spring bl. but saw it in bloom near Rio, in September.) *Tropical.* p. 731

paranaensis (Brazil: Paraná); robust terrestrial with spindle-shaped pseudobulb, deciduous prominently nerved leaves to 1 m long; erect tall stalk with fragrant flowers to 5 cm across, with greenish yellow sepals and yellow petals and lip. *Subtropic.* p. 715

CYRTORCHIS *Orchidaceae*
praetermissa (West and S.E. Africa); small epiphyte to 20 cm high, of vandaceous habit; inflorescence from leaf axils, pendant racemes of white waxy flowers with 3 cm spurs. *Tropical.* p. 720

CYRTOSPERMA *Araceae*
johnstonii (Alocasia) (Solomon Isl.); firm, sagittate leaves with long, widely spread basal lobes fresh green with crimson veins; spiny stem with dark purple zebra bands. *Tropical.* p. 108

senegalense (Trop. West Africa); from swampy forest areas, tall rhizomatous plant with leaf-stalks to 2 m long, covered with small spines; green leaves oblong arrow-shaped 30-60 cm long; the spathe greenish-yellow outside, purplish-red inside and striped with yellow, 20-50 cm long at apex with a long slender, twisted point. *Humid-tropical.* p. 102

CYRTOSTACHYS *Palmae*
lakka (Malaya, Borneo, Pacific Islands), "Sealing wax palm"; beautiful feather palms, clustering with slender, glossy green trunks to 5 m high and 8 cm thick; very handsome with their 60 cm long scarlet red leaf bases sheathing the base of the crown of fronds; the leaves pinnate on scarlet petioles, the numerous leaflets dark green above, glaucous beneath; small 1 cm black fruit from red stalk. *Tropical.* p. 779

renda (Sumatra), the "Sealing wax palm"; clustering feather palm with slender trunks to 10 m high; pinnate fronds; linear leaflets gray beneath, the leaf bases and petioles a showy red. *Tropical.* p. 779

CYTISUS *Leguminosae*
x racemosus (canariensis x maderensis magnifoliosus), "Florists' genista"; a hybrid favored over canariensis because of its smaller, darker green leaflets, and shorter, more numerous racemes with flowers deep yellow, set closely on terminal racemes; longer-lasting blooms than canariensis. *Subtropic.* p. 538, 546, 562

scoparius 'Windlesham Ruby'; a form of the "Scotch broom", to 2 m high, with flowers rose and red, instead of yellow, as in the species; small leaflets 1 to 2 cm long. *Warm temperate.* p. 546

DACRYDIUM *Podocarpaceae (Coniferae)*
cupressinum (New Zealand), known as "Rimu" by the Maori, or "Red pine"; tall forest tree to 30 m, with cypress-like branchlets and overlapping scale-like leaves when mature; very interesting when juvenile when the branchlets are like Lycopodium and pendulous, the needles soft and awl-shaped. *Subtropic.* p. 348

DAHLIA *Compositae*
imperialis (Mexico), a "Tree-dahlia"; tree-like 2-6 m tall, with tuberous roots, few woody stalks mostly solid and 4-grooved; compound leaves lacy; florets white, tinged purple near base, disk yellow. *Subtropic.* p. 317

pinnata (variabilis) hybrid, the "Garden-dahlias", originally from Mexico, but now in about 14 groups, with thousands of named clones; herbaceous bush 1½-2 m tall, with tuberous roots, stout semi-woody stalks, turgid leaves 5-foliolate, fresh-green above, grayish beneath; showy flower heads in many horticultural forms, sizes and colors, originally purple, but now also in white, yellow, salmon, bronze, red, pink, violet. *Subtropic.* p. 316

pinnata 'Apex'; elegant straight cactus-flowered type of tall habit, flaming orange, suffused with yellow. *Subtropic.* p. 317

pinnata 'Collarette'; open-centered heads, with one row of broad, flat floral rays, and in addition one or more rows of shorter petal-like florets, usually of a different color, forming a collar around the central disk. *Subtropic.* p. 317

pinnata 'Coltness Monarch' (single); bushy, compact plant with large single flowers in bronze, red or apricot. *Subtropic.* p. 317

pinnata 'Hit Parade'; a semi-cactus type, brilliant scarlet red with tips shaded yellow. *Subtropic.* p. 317, 321

pinnata 'Jugendliebe' (formal decorative); broad-petalled, crimson red in center, tips a contrasting pure white. *Subtropic.* p. 317

pinnata 'Pompon'; attractive usually small, fully double flower heads round, ball-shaped or slightly flattened, the floral rays cupped or tubular and blunt, arranged spirally. *Subtropic.* p. 317

pinnata 'Schweizerland', "Garden dahlia"; formal decorative type, striking color combination with outer petals white, toward center blood-red. *Subtropic.* p. 316

pinnata 'Siegerland'; cactus-flowered type, yellow at base, but heavily overlaid with orange scarlet. *Subtropic.* p. 317

pinnata 'Unwin's Dwarf' ("Mignon"); dwarf growing strain suitable for growing as a compact pot plant, producing 6 cm and 8 cm double and semi-double blooms in colors red, salmon, yellow, and lavender; usually grown from seed but forming small tubers; blooming first year in cool climate. *Subtropic.* p. 317

DAHLSTEDTIA *Leguminosae*
pinnata (Brazil); sparry tree with grayish wood; flowers in clusters, pink and tipped with red, blooming on bare branches. *Tropical.* p. 543

DAIS *Thymelaeaceae*
cotinifolia (So. Africa), "Pompon tree"; delightful small tree to 6 m; obovate leaves 8 cm long; the lovely pompon clusters of pink tubular flowers measure 8 cm in dia., and cover the tree in profusion in spring. *Subtropic.* p. 909

DALECHAMPIA *Euphorbiaceae*
spathulata (Mexico); low shrub to 1 m high, leaves oblanceolate 15-20 cm long; grown for its showy bracts, rose-red and very durable; flowers without petals. *Tropical.* p. 407

DAMMAROPSIS *Moraceae*
kingiana (New Guinea); extraordinarily interesting small branching tree about 3 to 5 m high which I collected in the Finisterre Mountains; very decorative with great leaves at first ovate, later much broader and 60 cm long, turgid, leathery, deeply corrugated, the margins undulate, deep glabrous green above, with red ribs when young, later changing to ivory, and pale beneath, petioles red; large pineapple-like brown fruit. *Tropical.* p. 648, 651

DAPHNE *Thymelaeaceae*
odora (China, Japan), "Winter daphne"; ornamental evergreen shrub to 1 m high, with glossy dark green, leathery, 8 cm leaves; the flowers are creamy-white to purple and very fragrant, in dense terminal clusters, blooming Jan. to March. *Subtropic.* p. 909

odora 'Marginata', "Variegated daphne"; attractive, slow-growing, compact bush having its glossy leaves prettily edged with ivory, and the waxy pink flowers radiating an intense sweet perfume; slower than odora. *Subtropic.* p. 909

DARLINGTONIA *Sarraceniaceae (Carnivorous Plants)*
californica (California, Oregon), "Cobra plant"; carnivorous bog plant with hollow, light green pitchers, and equipped with downward pointed hairs entrapping insects; the top is hooded and has translucent windows, and a cobra-like, purple-spotted forked tongue; nodding flowers purple. *Subtropic.* p. 293, 297

DASYLIRION *Liliaceae*
glaucophyllum (E. Mexico); stem-forming dense formal rosette of hundreds of linear leathery leaves 1 m long and 2 cm wide, lightly convex and with prominent rib off-center; olive green with faint glaucescence, the margins closely set with small hookspines in both directions, tips usually straw-colored and splitting; inflorescence to 5 m high, with white flowers. *Tropical.* p. 588

longissimum (Mexico), "Mexican grass-tree" or "Junquillo"; stemless xerophyte forming dense bundle of narrow-linear, stiffly erect olive green leaves without ribs, 1 m or more long; inflorescence a narrow cluster of small whitish flowers. *Arid-subtropic.* p. 586

serratifolium (S.E. Mexico); dense rosette with stout stem, narrow wiry, grass-like, green leaves to 1 m long, 2-3 cm wide; margins set with vicious spines; apex divided and with feathery tufts of hairs; white flowers on dense clusters. *Tropical.* p. 586

wheeleri (Arizona, Texas, Mexico), "Desert spoon"; dense rosette of narrow, flexuous, ribbed leaves to 1 m long and 2 cm wide, glaucous silvery, and with marginal spines tipped brown. *Arid-subtropic.* p. 586

DATURA *Solanaceae*
arborea (Peruvian Andes), "Angel's trumpet"; small, pubescent tree to 3 m high, with soft-hairy, ovate leaves; the nodding, trumpet-shaped flowers not over 18 cm long, white with green nerves and recurving pointed lobes, the long green calyx spathe-like, tapering to one tip. *Tropical.* p. 891, 893

aurea (Perú); sparry shrub with brittle, woody branches; large corrugated leaves 30-40 cm long; handsome pendant flowers 15 cm long, calyx with 2 to 4 lobes, and corolla a rich apricot yellow. Seen at Botanical Gardens in Ghent (Belgium), Dortmund and Tubingen (Germany). *Tropical.* p. 892

candida (So. America), "Tree datura"; with large pendant white flowers 20 cm or more long; commonly in cultivation as D. arborea. *Tropical.* p. 892

chlorantha (Peruvian Andes); free-blooming shrub with broad ovate, wavy leaves, and fragrant yellow, pendulous flowers, tubular calyx with 5 teeth, corolla funnel-shaped; August-October; prickly fruit. *Tropical.* p. 892

mollis (Ecuador); extremely showy tree-like shrub with herbaceous, ovate leaves and delicately tinted, large nodding trumpet-like flowers, vivid salmon-pink suffused with orange, and a long spathe-like calyx. *Tropical.* p. 892

rosei (sanguinea hort.) (Ecuador); small tree-like shrub with rough hairy, angular foliage; pendant trumpet flowers 18 cm long deep crimson red with white base, and the green calyx with single lobe. Photographed at Cap-Ferrat, France. *Tropical.* p. 894

sanguinea (Perú); showy tropical shrub becoming 4 to 5 m high, the brittle branches with clustered, long-ovate soft-hairy foliage 16 cm long, and carrying the weight of large pendulous 25 cm trumpets like so many bells, flesh pink to orange red toward apex, with yellow veins, calyx with two or more pointed lobes. Photo at Nassau Botanic Garden, Bahamas. *Tropical.* p. 893

suaveolens (Brazil), "Angel's trumpet"; tree-like shrub to 5 m high, with heavy canes, large, lanceolate, 30 cm leaves, oblique at the base, glabrous green, thin and quickly wilting, the large, nodding, funnel-shaped flowers with a tubular, 5-toothed calyx and a showy, white corolla to 30 cm long and fragrant. *Tropical.* p. 891, 893

versicolor (Brugmansia) (So. Ecuador: Guayaquil); large shrub or small tree 2-4 m high, with sparry, brittle branches, and big oblong-elliptic, soft pubescent leaves; showy flowers like large hanging trumpets 30 cm or more long, with spathe-like calyx; the thin constricted tube expanding to the reflexed petal-lobes, pinkish and turning apricot-peach with age, fragrant in the evening. *Tropical.* p. 893

DAVALLIA *Polypodiaceae (Filices)*
canariensis (Canary Islands, Spain, No. Africa), the "Hare's-foot fern"; sturdy epiphytic fern; the creeping rhizome covered with brown hair-like scales tipped with white, like a rabbit's foot; wiry stalks bear the leathery, light green fronds 20-50 cm long, finely cut in tripinnate fashion into broad ultimate toothed leaflets. A wonderful keeper in baskets. *Humid-subtropic.* p. 444

fejeensis (Fiji Islands), "Rabbit's-foot fern"; a name referring to its brown woolly, creeping rhizomes, from which rise the graceful durable fronds on wiry stems, more finely cut than solida. *Humid-tropical.* p. 444

fejeensis 'Plumosa' (Polynesia), a dainty, dwarf variety with fresh-green, more finely cut, 4-pinnatifid, plume-like, pendulous fronds of long-lasting character. *Humid-tropical.* p. 444

griffithiana (India, So. China), epiphytic fern with creeping rhizomes, covered with glistening white scales, wiry fronds 3-4 pinnatifid into well spaced segments, with the ultimate leaflets deeply toothed; not deciduous. *Humid-tropical.* p. 440

mariesii (bullata) (Japan), "Ball fern"; long creeping, flexible, slender, light brown, hairy rhizomes with uniformly small, 4-pinnate, finely lacy yet tough fronds; I saw them in Japan trained into many shapes, such as balls, pillars, bells, animals, monkeys and dolls; deciduous in cool climates. *Humid-tropical.* p. 440

trichomanioides (Malaya), D. "canariensis" of horticulture; "Squirrel's-foot fern"; small creeping rhizome covered with pale brown scales; wiry gray stalks bearing rather leathery fronds 15-22 cm long, 4 times pinnate, with leaflets overlapping, the final pinnae cut into strap-shaped segments. More durable than fejeensis, seldom fertile. *Humid-tropical.* p. 444

DAVIDIA *Nyassaceae*
involucrata (Western China), "Dove tree"; deciduous tree to 15 or 20 m, with rounded crown; cordate leaves vivid green 8-15 cm long; small clustered, red anthered flowers are carried by two large unequal, creamy-white bracts, the larger 15 cm long. When in bloom, the tree gives the effect of white doves resting among green leaves. *Warm temperate.* p. 673

DEAMIA *Cactaceae*
testudo (So. Mexico to Colombia), "Tortoise cactus"; epiphytic climber with aerial roots and glaucous stems very variable in shape, triangular, flat like a tortoise, or to 8-ribbed; large white diurnal flowers, flushed copper outside. *Arid-tropical.* p. 256

DELONIX *Leguminosae*
regia (Poinciana) (Madagascar), the "Flamboyant", "Royal poinciana" or "Mohur tree"; a regal flowering tree and one of the showiest; widely planted in the tropics; wide-spreading and to 15 m high; finely subdivided, ferney leaves often fall before blooming time; brilliant 8-10 cm flowers scarlet-red with broad cream lip marked red; followed by flat, woody black pods to 45 cm long. *Tropical.* p. 537, 540, 543

DELOSPERMA *Aizoaceae*
'Alba' (So. Africa), "White iceplant" or "Disneyland iceplant"; dwarf, spreading succulent rooting freely from stems; small fleshy, angled leaves forming a dense green carpet; the little white 1 cm flowers not showy. *Dry subtropic.* p. 44
brunnthaleri (So. Africa: Natal); wide spreading succulent, with long slender, rich green leaves to 4 cm long, flat or channeled on upper side; flowers with petals rose-pink. *Arid-subtropic.* p. 46

DELPHINIUM *Ranunculaceae*
ajacis (Switzerland to Taurus), "Larkspur"; erect branching annual to 1½ m high, with lacily divided leaves and long, showy spikes of flowers, often double, with a prominent spur, and sepals in shades of blue, violet, pink to white. *Warm temperate.* p. 844
elatum (Pyrenees to Siberia), "Candle larkspur"; winter-hardy perennial herb to 2 m high, the branches upright, large leaves palmately 5-7 parted; inflorescence in dense spike-like racemes 30 cm long, of 2-3 cm flowers sky blue or purple, with curved spurs. *Temperate.* p. 845
elatum 'Pacific hybrid'; a race of mainly seed-grown varieties with candle-like spikes of semi or double flowers on strong stems 1 to 2 m tall; they range in color from white through pink and purple to blue. *Warm temperate.* p. 844, 845

DENDROBIUM *Orchidaceae*
acuminatum: see Epigeneium lyonii
chrysotoxum (Burma, Yunnan), "Fried egg orchid"; strong epiphyte with cane-like pseudobulbs, producing from their leafy top arching racemes of golden yellow flowers with fringed lip having a deep orange throat. (March-July). *Tropical.* p. 727
d'albertisii (Papua), a "Rabbit's-ear orchid"; pretty epiphyte with square, slender tapering stems set alternately with narrow, fleshy leaves, the axillary erect racemes of fragrant, spurred flowers with white sepals, the narrow greenish twisted petals standing erect resembling rabbit's ears, the lip marked purple. *Tropical.* p. 721
densiflorum (Himalayas, Burma); beautiful epiphyte with tall, stout cane-like 4-sided pseudobulbs bearing 3-5 leathery leaves; the flowers golden yellow with velvety orange-yellow lip, in dense, pendant trusses, from near the top of either old or young growth. (March-May). *Tropical.* p. 727
falcorostrum (Australia); showy epiphyte with slender pseudobulbs to 20 cm high, and leathery leaves; topped by terminal inflorescence with up to 20 fleshy, creamy white flowers having narrow sepals and 2 cm petals, lip marked yellow and striped crimson; free-flowering, spring on. *Subtropic.* p. 730
farmeri var. albiflorum; variety with sepals and petals creamy-white, and orange-yellow disk on lip. D. farmeri is from the Himalayan region; a compact epiphyte with club-shaped pseudobulbs tipped by leathery leaves; pendulous clusters of 5 cm flowers with rose petals. *Tropical.* p. 731
fimbriatum (Himalayas, Burma); tall epiphyte with cane-like stems with numerous leaves topped by loose clusters of flowers, deep yellow with fringed lip rich orange and velvet-like, a large blood-red blotch at base of lip, in the variety oculatum (March-April). *Tropical.* p. 727
formosum giganteum (Upper Burma); showy variety with shorter but stouter stems, the fragrant flowers much larger, 12 or even 15 cm across, pure white with bright golden blotch in throat. (spring). *Tropical.* p. 727
gracilicaule (Australia); slender cane-like pseudobulbs to 50 cm high, topped by several leaves, and short clusters of small variable 2 cm flowers curiously hooded, creamy-yellow, spotted with brown-red. *Subtropic.* p. 727

infundibulum (So. Burma); an epiphyte of great beauty with black-haired cylindric stems to 60 cm long and 2-4 showy flowers 8-10 cm across from the upper joints, snow-white with orange-yellow stain in the throat (May-Aug.). *Tropical.* p. 726, 727, 730
johnsoniae (macfarlanei) (New Guinea); noble epiphyte with erect, cylindric, 2-3 leaved pseudobulbs bearing terminal racemes of large 10-12 cm flowers pure white except for 3-lobed lip marked purple; summer. *Tropical.* p. 727
macrophyllum, (Philippines, Java, New Guinea); one of the finest of dendrobiums; very variable; epiphyte with slender pseudobulbs becoming stout, to 30 cm tall; persistant leathery leaves glossy green; inflorescence erect with clusters of heavy-textured flowers to 5 cm across, and covered with blackish bristles; sepals and petals cream to greenish and spotted purple, the waxy lip yellow green veined purple (spring). *Tropical.* p. 721, 725
'New Guinea' (macrophyllum x atroviolaceum); a "different", compact hybrid with 3-4 broad leaves and an erect, short raceme of long-lasting long-stalked flowers with pointed greenish ochre-yellow sepals and petals, the cupped lip patterned with crimson over yellow, 5 cm dia. (early spring). *Tropical.* p. 727
nobile (Himalayas, Assam, Yunnan); popular epiphyte with large fragrant 8 cm flowers produced in twos and threes from nodes of 2-yr. pseudobulbs, sepals and petals white, tipped rosy-pink, lip white with rosy tip and deep velvet-crimson throat. (Jan.-June). *Tropical.* p. 726, 727
nobile 'Albiflorum'; flowers white; yellowish lip with black-purple blotch. *Tropical.* p. 727
parishii (Yunnan to Vietnam); pseudobulbs usually prostrate or pendulous, with deciduous leaves; inflorescence with 2-3 flowers from the leafless stems, to 5 cm across, rose-purple, lighter in lip. *Subtropic.* p. 726
phalaenopsis (bigibbum var.) (Queensland, New Guinea, Timor); a most beautiful epiphyte with stiffly slender canes 60-90 cm tall, bearing leaves in 2 ranks, and long, arching chain-like sprays of large, neatly draped 8 cm flowers, the sepals pale magenta with reticulated darker nerves, petals rose, much larger; lip dark purplish-red; blooming variously, mainly in spring, sometimes also in fall to November. *Tropical.* p. 726
phalaenopsis 'Pompadour'; (Louis Bleriot x schroederianum); an old French hybrid (Vacherot 1934), vigorous, compact grower with leafy stems, very popular for cut flower trade and shipping, in Bangkok, Thailand; long, striking sprays of bright magenta-red flowers 8 cm across; sepals and petals strong and thick; except for short resting period, flowering all year. *Tropical.* p. 726
pierardii (Himalayas, Burma); epiphyte from steaming forests, with long cylindric, pendant stems, the flowers from the nodes, sepals and petals delicate pale rose, lip yellowish, veined with rose-purple. (Feb.-May) *Tropical.* p. 725
primulinum (Himalayas, Burma); epiphyte with cylindrical, fleshy stems, erect or pendulous, with flowers in 2 rows from nodes along the leafless pseudobulb, the large blooms with small sepals and petals pale rose, the shell-shaped lip primrose-yellow streaked red. (Feb.-May). *Tropical.* p. 725
secundum (Burma, Vietnam, Pacific Isl.); stout cylindric stems 60-90 cm high, the small somewhat tubular flowers shining rosy purple, with pointed orange-yellow lip, in dense one-sided stiff racemes from the top (winter). *Tropical.* p. 726, 727
senile, (Thailand, Laos), "Cowslip orchid"; pseudobulbs spindle-shaped, covered with white hairs; persistant, leathery leaves; solitary or paired flowers from the upper nodes, waxy, long-lived, 5 cm across, bright yellow; lip darker and with green about disc (spring). *Tropical.* p. 725
superbum (Borneo, Celebes to Philippines); noble, deciduous epiphyte of pendulous habit, with cylindric stems to 1 m long, the 8-10 cm flowers appearing along them on each side, rose-purple with base of pubescent lip blood purple. *Tropical.* p. 726
x thwaitesiae (ainsworthii x wiganiae); cane-stem hybrid with large 10 cm flowers, one of the showiest of the genus; sepals and petals ranging from buttercup-yellow through gold and orange, and a deep brown-red disk on lip. *Tropical.* p. 725
thyrsiflorum (Burma); showy epiphyte with the habit of densiflorum but stems are more round and taller, to 75 cm high, the flowers in longer pendant racemes, sepals and petals white, often flushed pink, the pubescent lip golden orange. (Feb.-May). *Tropical.* p. 727
veratrifolium (New Guinea); robust epiphyte with elongate pseudobulbs and large clusters of spreading, stiff racemes of small flowers with white sepals and erect green-tipped petals looking like rabbit's ears, all marked with violet; Autumn. *Tropical.* p. 725

DENDROCALAMUS *(Gramineae)*
giganteus (Burma, Malaya), "Giant bamboo"; this tropical giant of the tribe, largest of all bamboos, grows to 35 m high with

stems 25 cm or more in diameter, thin-walled, joints to 40 cm apart; later developing branches with large leaves to 50 cm long borne in graceful masses toward the top. The young emerging culms, growing at the rate of 45 cm a day, were an instrument of torture and death to prisoners of war in Ceylon, pushing through their bodies when they were tied down to the ground in a bamboo grove. In temperate climate for the large conservatory only. Widely planted in Southeast Asia and used for building, water-spouting or plant pots. *Tropical.* p. 517, 519

DENDROCHILUM Orchidaceae
glumaceum (Platyclinis) (Philippines); small clustering epiphyte with lovely arching slender spikes of scented, 2-ranked flowers, sepals straw-white and a greenish-yellow lip. *Tropical.* p. 724

uncatum (Philippines); graceful, small epiphyte with tufted, slender conical pseudobulbs and linear leaves; the arching floral racemes short with numerous small pale green flowers with pointed sepals and petals. (Sept.-Jan., May-July). *Tropical.* p. 724

DENDROPANAX Araliaceae
trifidus (Gilibertia) (Japan); small evergreen tree with gray-brown branches and thick green branchlets, shiny green, ovate leaves 7-12 cm long, 3-nerved, sometimes 2 or 3-lobed or 3 to 5-cleft; small yellow-green flowers in clusters, followed by black berries. *Warm temperate.* p. 134

DENDROSICYOS Cucurbitaceae
socotrana (Yemen: Socotra Isl.), "Cucumber-tree"; odd desert tree with conical swollen stem to 6 m tall, often 1 m dia., with white bark, and sparse pendant branches with scattered, lobed leaves 7 cm dia., rough with white bristles; axillary flowers, and hairy fruit. *Arid-tropical.* p. 380

DERRIS Leguminosae
indica (Trop. Asia); woody plant with glossy-green, pinnate foliage; small pale pink flowers. *Tropical.* p. 560

DESMONCUS Palmae
horridus ("major" of hort.) (Trinidad, Guayana, Surinam), "Wait-awhile palm"; a horrible climbing palm, found in thickets along rivers in dense jungle on the Guiana coast, the dark green pinnate fronds very ornamental, but the axis with black spines and its whip-like tips set with pairs of vicious fish-hook spines; I can testify to some bloody slashes while exploring in this area. *Tropical.* p. 774

DEUTEROCOHNIA Bromeliaceae
longipetala (Dyckia) (Peru); dense rosette of spreading spiny leaves to 40 cm long; sessile flowers with yellow petals spotted green on tips. *Subtropic.* p. 214

DEUTZIA (Saxifragaceae)
gracilis (Japan), "Slender deutzia"; attractive floriferous winter-hardy deciduous shrub of low habit, about 1 m high or sometimes more, with slender branches, wide-spreading or arching; lanceolate leaves with star-like hairs above, serrate at margins, to 6 cm long; masses of pure white 2 cm flowers in open simple or compound clusters; blooming outdoors in spring. *Warm temperate.* p. 880

DIANTHUS Caryophyllaceae
'Allwoodii' (caryophyllus x plumarius); perennial forming tufts of firm, blue-gray leaves, 10-50 cm high; flowers with fringed petals, in various colors. *Warm temperate.* p. 298

barbatus (So. and S. Eastern Europe to So. Russia and East), "Sweet William" or "Bunch pink"; popular garden perennial better treated as a biennial; from some seed also annual; smooth herb 45-60 cm high with green flat and broader leaves than most Pinks; the sturdy stem with 3 cm flowers in large bracted heads, the spreading petals fringed and calyx bearded, in colors from white, pink, rose, red, purplish to bi-colored, blooming from June to August; winter-hardy. *Warm temperate.* p. 298

caryophyllus (So. Europe to India), "Carnation"; tufted glaucous plant of stiff habit, with bluish linear leaves, the flowers showy, double, dentate, in many colors, and spicy-fragrant. *Subtropic.* p. 299

caryophyllus 'Apollo'; a diploid form of caryophyllus, typical of the "American Remontant carnation" commercially grown for cut flowers under glass, because of their perpetual, year round growing and flowering habit. 'Apollo' is a clear pink double flower with salmon sheen, and petals crenate, the inner ones incurved, and of good substance on stiff stem. *Subtropic.* p. 298

caryophyllus 'Arthur Sim'; "American Remontant" carnation, cultivar for commercial cut flower production; long, stiff-erect stems carrying large fully double flowers 8 cm across of exquisite coloring, soft pink, the crisped petals edged with crimson red. *Warm temperate.* p. 299

caryophyllus chorus, "Clove pinks"; variety with pendant stems used for hanging baskets, as seen in Hardangerfjord, Norway; deep red, double flowers. *Tropical.* p. 299

caryophyllus minima, "Miniature carnation"; short-stemmed, with semi-double flowers crimson red and fringed white margins. *Warm temperate.* p. 298

caryophyllus 'Mini Queen'; a special race of cut-flower carnations with tall, erect, wiry stems 30-50 cm long, each carrying several medium size 4-5 cm fully double, spicily fragrant flowers deep rose with white edges, the petals fringed; (photographed at the Scandinavian Flower Show, Malmoe, Sweden October 1970). *Warm temperate.* p. 299

caryophyllus 'Pink Littlefield'; large flaring flower 10 cm dia., with toothed petals carmine pink, on wiry stem. *Subtropic.* p. 298

caryophyllus 'Pink Sim'; large flowered florists' carnation, fully double and a dainty baby-pink. *Warm temperate.* p. 299

caryophyllus 'Portrait'; high double flower 10 cm dia., clear rich rose pink with trace of white markings, fringed edge; wiry stem. *Subtropic.* p. 298

caryophyllus 'Red Sim'; popular florists' "Carnation"; sport of 'William Sim', excellent cut flower; brilliant red, fully double. *Warm temperate.* p. 298

caryophyllus 'White Sim'; sport of 'Wm. Sim', high double flower 10 cm across, pure white with creamy base and with lacy edge, spicy fragrant; stems wiry and slender. *Subtropic.* p. 298

chinensis heddewigii (China, Japan), "Rainbow pink"; blue-green tufted biennial; flowers in velvety colors and markings and petals much cut and frilled. *Warm temperate.* p. 299

chinensis 'Queen of Hearts', "Chinese pinks"; biennial, 15-50 cm high, with narrow leaves; stems branching at top, with 3 cm fringed single flowers, scarlet red. *Warm temperate.* p. 298

chinensis 'Snowfire'; bushy annual to 20 cm high, covered with fringed flowers white with red center. *Warm temperate.* p. 299

gratianopolitanus (Great Britain, Central Europe), "Cheddar pinks"; bushy perennial forming mounds, with blue-gray foliage; weak stems to 30 cm long with very fragrant pink blooms, petals toothed. *Warm temperate.* p. 299

DIASTEMA Gesneriaceae
vexans (Colombia); herbaceous gesneriad with hairy branching stems 10 cm high; ovate to lanceolate toothed, hairy green leaves 8 cm long on long petioles; stalked axillary white flowers with tube spotted purple-brown, and each spreading lobe has a purple brown spot at the base. *Tropical.* p. 497

DICENTRA Fumariaceae
spectabilis (Japan), "Bleeding heart"; perennial herb with divided leaves and arching stalks with pendulous, heartshaped, rosy flowers and protruding inner petals of white. *Temperate.* p. 484

DICHONDRA Convolvulaceae
micrantha (repens) (West Indies), known as "Lawn-leaf"; used as a grass substitute in California; low herb creeping close to the ground and rooting, with close-matting stolons; small, rounded or kidney-shaped fresh green, silky leaves to 1 cm across, tightly overlapping; the little flowers greenish-yellow; seeds freely. *Subtropic.* p. 357, 510

DICHORISANDRA Commelinaceae
reginae (Perú), "Queen's spiderwort"; stiffly erect, rather slow growing plant, the waxy, long pointed, dark green leaves banded and spotted on both sides with sparkling silver, the center metallic red-violet, and spilling over into deep purple beneath; flowers lavender. *Tropical.* p. 306

siebertii (Brazil); a bushy spreading plant which I first saw in Sao Paulo where it is used as a garden ornamental; the arching branches have ovate, glossy, metallic green leaves with two silvery length stripes; flowers blue. *Tropical.* p. 307

thyrsiflora (Brazil), "Blue ginger"; tall canes bearing rosettes of broad lance-shaped, glossy green leaves with purplish reverse, umbrella-like, topped by a huge raceme of brilliant, deep electric-blue flowers with yellow anthers. *Tropical.* p. 305, 307

DICHORISANDRA: see also Geogenanthus, Campelia

DICKSONIA Dicksoniaceae (Filices)
antarctica (Australia, Tasmania), "Tasmanian treefern"; tree fern with woody trunk to 15 m high, covered with matted aerial rootlets, fronds to 2m long, 3 pinnate, dark green with hard lanceolate toothed segments which are turned down at the margins, and a network of straw-colored veins. *Humid-subtropic.* p. 434

fibrosa (New Zealand), "Golden treefern"; becoming tree-like, with handsome fronds bright glossy green, lighter and hairy beneath, 2-3 m long, bipinnate; trunk light brown. *Humid-tropical.* p. 432, 436

squarrosa (New Zealand), "Westland treefern"; a medium-sized tree-fern to 6 m high, the slender black trunk clothed with leaf-bases, crown with fronds nearly horizontal, 2-3 pinnate, to 2½ m long, dull, dark green and stiff-leathery, and harsh to the touch; on black-brown stalks clothed when young by long brown, stiff hairs. *Humid-subtropic.* *p. 432, 434*

DICTYOSPERMA *Palmae*
  album (Mauritius), the "Princess-palm"; handsome feather-palm to 10 m, the slender, dark gray trunk with numerous vertical cracks, topped by a crown-shaft and the graceful pinnate fronds to 4 m long; the drooping leaflets shiny green with pale veins; the last undeveloped leaf always stands up in the center and unfolds from its binding as from a needle; fragrant reddish flowers at an early stage. *Tropical.* *p. 779*
  aureum (Réunion), "Yellow Princess palm", or "Hurricane palm", as they withstand high winds; straight, slender, dark gray trunk with swollen base; ring scars forming steps and many vertical cracks, to 8 m tall; long green crownshaft opens to a handsome crown of pinnate fronds 3 m long, with shiny dark green leaflets in one plane; the rachis twists 90 degrees so that leaflets parallel the trunk; reddish-yellow flowers and purplish fruit. *Tropical.* *p. 779*

DIDIEREA *Didiereaceae*
  trollii (Madagascar); weird succulent with thick arms waving in all directions forming tangled twigs, eventually forming trunks with horizontal branches; stems covered with convex tubercles with short shoots whose tips bear 5 thin thorns; fleshy leaves 1-2 cm long; flowers greenish yellow. *Arid-tropical.* *p. 389*

DIEFFENBACHIA *Araceae*
  amoena (Colombia, Costa Rica); sturdy, thick-stemmed species with large, oblong-pointed, glabrous leaves, deep green and marked with cream-white bands and blotches along veins. *Tropical.* *p. 108*
  amoena 'Tropic Snow' (syn. 'Tropical Topaz' or 'Hi-Color'); beautiful Florida (Chaplin) mutation of very compact habit; showy leathery foliage glossy deep green highly variegated from center with cream and nile green. *Tropical.* *p. 106, 107*
  'Arvida': see 'Exotica'
  x bausei (maculata x weiri); beautiful, pointed leaf delicate yellowish-green with dark green and white spots and dark green at edge. *Tropical.* *p. 107, 108*
  'Exotica' ('Arvida') (Costa Rica); attractive mutant, probably of hoffmannii, from the Rio Ysidro, quite compact and shapely; of slender, narrow habit; smallish ovate pointed, firm leaf, deep matte-green and highly variegated cream-white, and of good texture. *Tropical.* *p. 108*
  'Exotica Perfection'; a Florida selection of compact habit, the smallish leaves 15-20 cm long, broadly bordered with deep matte green and of leathery feel; highly variegated greenish ivory. *Tropical.* *p. 106*
  fournieri (Colombia); upright, large, shiny leathery black-green leaves with white spots; slender and very elegant. *Tropical.* *p. 108*
  maculata (picta) (Brazil), "Spotted dumbcane"; glossy, grass-green, oval leaves with cordate base and ivory-white marbling and blotching; petioles dotted pale green; spathe greenish; found growing wild by the writer at the confluence of the Solimoes and Rio Negro in Amazonas. *Tropical.* *p. 107*
  maculata jenmannii (Guayana); long-oblong leaves glossy, fresh green with distinct ivory bars in feather design. *Tropical.* *p. 108*
  maculata 'Roehrs superba', (Roehrs 1950); very handsome compact cultivar of maculata (picta), with thicker, more durable foliage and more highly variegated than the species, with creamy-white over the rich green leaf; a highly decorative indoor foliage plant tolerant of low light and revelling in warm room temperature. Changed from picta to maculata according to findings in Baileya Dec. 1962. *Tropical.* *p. 106*
  maculata 'Rudolph Roehrs' (Roehrs cultivar, named by the author 1937), "Gold dieffenbachia"; a striking mutant having oblong pointed 25-30 cm leaves almost entirely yellow or chartreuse, with ivory-white blotches, and only the midrib and border dark green. Both beautiful and showy, and proven an excellent house plant. *Tropical.* *p. 106*
  maculata 'Superba' (Roehrs 1950); very attractive, compact mutant of maculata, with thicker, more durable foliage dominated by a high degree of creamy-white variegation, over the glossy green leaf. *Tropical.* *p. 107*
  picta: see maculata
  reginae (Estado do Rio); oblong pointed leaves somewhat pendant, glabrous, pale pea-green on dark green and blotched cream-white; bluish-green beneath. *Tropical.* *p. 105, 106*
  'Sao Antonio' (S.E. Brazil); attractive novelty with broad, succulent leaves tapering to a point, rich green with yellow-green marbling near midrib, and freely spotted with white; the petioles white traced with green. *Tropical.* *p. 107*
  seguina (Puerto Rico, W. Indies), "Dumbcane"; robust species with long elliptic, dark green fleshy leaves, base acute, prominent midrib and depressed veins; pleasing green beneath; spathe green. *Tropical.* *p. 106*
  x splendens (leopoldii x maculata); velvet, bronze-green leaf with nile-green, and ivory spots and ivory midrib. *Tropical.* *p. 108*
  'Tropic Snow' see under amoena
  wallisii (No. S. America); tall-growing, robust plant with oblique lanceolate leaves with long tapering apex, matte to semi-glossy grass-green, a broad feathery band of light gray to creamy-silver along midrib, on winged petioles. *Tropical.* *p. 108*
  'Wilson's Delight'; handsome, bold plant; broad, fleshy leaves deep green with broad, cream-white midrib, the cream fanning out into the lateral veins. *Tropical.* *p. 107*

DIERAMA *Iridaceae*
  pulcherrima (So. Africa: Cape to Transvaal), "East London harebell"; pretty, cormous herb about 1½ m high, with basal long-linear, narrow, rigid leaves; blooming on branched, pendant spikes, the 3 cm funnel-shaped flowers bright purple or blood-red, in autumn. *Subtropic.* *p. 527*

DIETES *Iridaceae*
  grandiflora (So. Africa), "Large wild iris"; sword-shaped leaves in flat fan; big flowers with large white to pinkish petals and purple stripes in center. *Subtropic.* *p. 526*
  vegeta (Moraea iridioides) (So. Africa: E. Cape, Natal), "African iris"; lovely, free-blooming rhizomatous, clustering plant with linear, dark green basal leaves arranged fan-like from creeping root stock; on long branching stalks the somewhat fleeting, iris-like flowers 5-8 cm across, white tinged with blue, and yellow on outer petals, the crest of style marked with blue; summer-blooming. Widely planted in California where it seems to be in bloom all year. *Subtropic.* *p. 529*

DIGITALIS *Scrophulariaceae*
  purpurea (W. Europe), "Foxglove"; biennial with wrinkled leaves in rosettes and on the erect ½-1 m showy spike of dense, bell-shaped, bilabiate, nodding flowers, purple, varying to white, with dark purple spots, edged white inside, in summer. *Warm temperate.* *p. 887*
  purpurea 'Alba', "White foxglove"; color variant with a spike of white flowers, yellow in throat. *Warm-temperate.* *p. 887*

DILLENIA *Dilleniaceae*
  indica (India to Java and Philippines), "Elephant apple"; handsome tree to 12 m high, usually evergreen but may lose foliage in dry season; very ornamental, large oblanceolate, thick leaves 30-40 cm long, bright light green above and strongly pinnately depressed-veined; rough beneath, and margins toothed; fragrant magnolia-like white flowers with golden stamens. Globose, acid fruit to 10 cm dia, edible when fresh; used in curries and jellies. *Tropical.* *p. 391, 392, 463*
  suffruticosa (Wormia burbidgei) (Malaysia), "Simpoh"; large shrub to 10 m; broad elliptic leaves to 25 cm long; yellow flowers 10 cm across; fruits open in star shape. *Temperate.* *p. 391*

DIMORPHANTHERA *Ericaceae*
  sp. 'Finisterre'; robust woody clamberer with long elliptic, corrugated leaves; the striking tubular flowers bright scarlet with base and tips canary-yellow, 3 cm long, appearing directly on the branches. *Subtropic.* *p. 398*

DIONAEA *Droseraceae (Carnivorous Plants)*
  muscipula (Carolinas), "Venus fly trap"; carnivorous perennial rosette of damp mossy places; the two halves of a leaf are turned upward and equipped with long teeth which close traplike when entered by flies. *Warm temperate.* *p. 292*
  muscipula 'Coccinea', "Red fly-trap"; form of species with inside of traps blood-red. *Warm temperate.* *p. 292*

DIOON *Cycadaceae*
  edule (Mexico), "Virgin's palm"; palm-like plant from the hot open country, closest to the fossil cycads, with stiff pinnate leaves having spiny tips, the young petioles are covered with white wool; flowers in cones to 30 cm long. *Tropical.* *p. 384*

DIOSCOREA *Dioscoreaceae*
  bulbifera, (Trop. Asia, Philippines), "Air-potato" or true "Yam"; twining plant without the large underground tuber of most yams, but forming aerial, axillary, angular gray-brown tubers to 30 cm long and edible; alternate leaves heartshaped, and small dioecious, greenish flowers. *Tropical.* *p. 393*
  discolor (Surinam), "Ornamental yam"; beautiful broad leaf with long basal lobes, dark olive green marbled light green and silver-gray, veins carmine to silver, reverse purple. *Tropical.* *p. 392, 393*

elephantipes (Testudinaria) (Cape Prov., Natal, Transvaal), "Elephant's foot" or "Hottentot-bread"; curious-looking partially exposed tuber to 90 cm across, the bark divided into corrugated angled knobs and from which rises the annual vining stalk bearing heart- or kidney-shaped leaves and greenish-yellow flowers with dark spots. *Subtropic.* p. 393

macrostachys (Testudinaria) (Mexico), the "Tortoise plant"; large, knobby woody, chocolate brown tuber 20 cm or more across, resting above the soil and looking like a horny turtle, its coat broken into numerous angled plates; sprouting slender, asparagus-like shoots becoming vine-like, 1 m or more high, with alternate leaves and small axillary greenish-yellow flowers; grown as a curiosity in greenhouse. *Arid-tropical.* p. 392, 393

sansibarensis (E. Africa); robust vine with large leaves 15 cm across, corrugated with parallel veins, dark forest green, indented each side and with slender tips; purple beneath. *Tropical.* p. 393

DIOSPYROS Ebenaceae

digyna (ebenaster) (Mexico, C. America), "Black sapote"; evergreen tree to 20 m, with alternate, simple elliptic leaves, glossy dark green; small fragrant, whitish flowers; globular fruit 10 cm dia, olive green becoming dark; soft chocolate brown flesh. Popular fruit in Mexico. Photographed in Borda Gardens, Cuernavaca, Morelos. *Semi-tropical.* p. 463

kaki (Japan, China), "Kaki" or "Japanese persimmon"; a well-known fruit tree of the Orient; bushy, deciduous tree 8 to 12 m high, with brownish branches, broad ovate leathery leaves 8-20 cm long and shining green above, pubescent beneath; 4 cm flowers yellowish-white; on female trees the depressed globose glowing red fruit 8 cm dia., with custard-like orange flesh; especially piquant and delicious when served frozen, with a taste like fruity sherbet. *Subtropic.* p. 392, 463

kaki 'Fuyu', "Japanese persimmon"; with firm-fleshed fruit, like a flattened apple; orange-colored and non-astringent, even when not ripe; ready in late autumn. *Semi-tropical.* p. 461

malabarica (India), "Malabar ebony"; tree to 12 m, with smooth gray bark, leathery ovate leaves 10-20 cm long; flowers ochre-yellow; fruit 4-6 cm, ripening orange to red; pulp edible. *Tropical.* p. 393

DIPLACUS: see Mimulus

DIPLADENIA: see Mandevilla

DIPLOTHEMIUM Palmae

maritimum (Allagoptera arenaria) (S.E. Brazil); small feather palm forming colonies on sea-shore dunes from Bahia to the Restinga south of Rio; trunkless, with pinnate leaves arising directly from the ground, 1½ to 2½ m high, recurving, waxy dark green above, white-fuzzy beneath; inflorescence a spike with reddish spadix, the green fruit edible. *Subtropic.* p. 804

DISA Orchidaceae

uniflora (grandiflora) (So. Africa), "Pride of Table Mountain"; terrestrial rosette of shining, lanceolate leaves with tuberous rootstock, and producing an erect, leafy spike to 60 cm high with a lax cluster of showy flowers to 10 cm across, the helmet-shaped dorsal sepal red outside, inside lighter, veined with crimson and shaded with yellow, the lower sepals vivid scarlet, lip small, (Jan.-April). *Subtropic.* p. 720, 721, 738

DISCHIDIA Asclepiadaceae

bengalense (India); epiphytic trailer, with twisting, wiry stems rooting at the joints; opposite narrow, fleshy two-edged, flattened leaves; small urn-shaped flowers. *Humid-tropical.* p. 146

platyphylla (Philippines); rambling vine with thin stems, opposite fleshy, gray-green leaves rounded or kidney-shaped, and pitcher-shaped, beige flowers 3-4 mm. dia. *Tropical.* p. 149

rafflesiana (Malaya, New Guinea), "Malayan urn vine"; epiphytic climber with oddly thick-fleshy oval leaves formed into pitchers; the mature, 8-10 cm, opposite leaves are pear-like, hollow, purplish inside, in habitat frequented by ants; yellowish flowers in umbels. *Tropical.* p. 149

DISOCACTUS Cactaceae

eichlamii (Guatemala); pendant epiphyte with cylindrical and flattened stems; day-blooming with ascending long-tubed flowers 4 cm long, with bent base, brilliant red. *Tropical.* p. 283

DISTICTIS Bignoniaceae

buccinatoria (Phaedranthus; Bignonia cherere) (Mexico), "Blood trumpet"; vine with showy flowers which I first saw, smothered with blooms, around the Union Buildings in Pretoria, South Africa; rough-leathery oval leaves and axillary clusters of large trumpet-shaped waxy flowers with flaring lobes, crimson-red with scarlet sheen, and yellow throat. *Subtropic.* p. 183, 192

x riversii (Phaedranthus, Tecoma, Bignonia) (buccinatoria x laxiflora); tendril-climbing woody vine, with paired obovate leaves, shining green; showy large flowers with spreading petals 6-8 cm across, rosy purple with white throat. *Tropical.* p. 183

DIZYGOTHECA Araliaceae

elegantissima (Aralia) (New Hebrides), "False aralia"; graceful shrub with palmately compound, leathery leaves, the threadlike, narrow segments metallic, red-brown and lobed; slender stem mottled cream. Forms a small tree to 8 m high. *Tropical.* p. 132

elegantissima 'Castor'; a compact growing, freely branching cultivar seen in Germany, with palmately compound leaves, the leathery leaflets broader than the species, purplish-green with ivory-white midrib. *Tropical.* p. 132

DODECATHEON Primulaceae

meadia (Manitoba to Pennsylvania and Texas), "Shooting-star"; small perennial herb with wavy-margined basal leaves, and nodding cyclamen-like flowers with reflexed corolla lobes, rose with white base and long purple anthers. *Temperate.* p. 830

DODONAEA Sapindaceae

viscosa (Arizona to W. Indies), "Hopseed-bush"; fast-growing evergreen bush to 5 m high, with many upright branches, narrow willow-like 10 cm green leaves; small whitish flowers in short clusters; the straw-like pinkish winged seed capsules were used as hops by early settlers. *Subtropic.* p. 876

viscosa 'Purpurea'; "Purple hop-bush"; introduced from New Zealand; a rich bronze-purple leaved cultivar, the color deepening in winter; best coloring in full sun. *Subtropic.* p. 876

DOLICHANDRONE Bignoniaceae

lutea (Burma); glossy pinnate leaves; large trumpet flowers with flaring lobes, golden yellow with red lines. *Tropical.* p. 183

DOLYCHOS Leguminosae

lablab (Trop. Africa: Egypt), "Hyacinth-bean"; woody perennial climber with purple stems, usually grown as an annual; leaves of 3 ovate, purplish-green leaflets 8 to 15 cm long, and veined with purple; rosy-lilac 2 cm butterfly flowers in loose racemes, standing out from the foliage; followed by velvety, bean-like pods 6 cm long. *Tropical.* p. 557

DOLICHOTHELE Cactaceae

longimamma (C. Mexico), "Finger-mound"; interesting plant of globular shape to 10 cm high but consisting almost entirely of cylindrical knobs to 5 cm long, deep green and soft-fleshy, tipped with scattered soft pale spines; flowers yellow. *Arid-tropical.* p. 276

DOMBEYA Sterculiaceae

burgessiae (Kenya to Natal), "Wedding flower" or "Rose-mound"; palmately veined, broad cordate leaves to 20 cm long, the surface corrugated and with contrasting yellow veins; charming pale pink flowers with red veins, 3 cm across, in small clusters. *Tropical.* p. 903

x cayeuxii (D. mastersii and wallichii), "Mexican rose" or "Pink ball"; a hybrid shrub of tropical African parents, with soft, rough-hairy cordate-kidney-shaped and palmately veined and netted, rich green leaves 15-20 cm across, dentate at margins; the lovely inflorescence a pendant 8-10 cm ball-like cluster of fragrant 4 cm flowers rosy-pink with pale center, and which bloomed for me in January. *Tropical.* p. 903

elegans (Réunion); slender branched, smooth shrub, with broad cordate leaves to 12 cm long, serrate at margins and palmately veined; smallish flowers in attractive dense clusters, soft pink. *Tropical.* p. 903

spectabilis (rotundifolia) (Transvaal), known in Africa as the "Wild pear"; deciduous tree to 8 m high, a gorgeous sight on the High Veld when trees are covered with small cream-white clustered flowers against the bare black branches; small round toothed leaves usually after bloom. *Subtropic.* p. 901

tiliacea (natalensis) (S.E. Africa), "Heartleaf dombeya"; small tree to 8 m, with gray-brown bark; soft leaves cordate-ovate, to 9 cm long, irregularly toothed; pretty white flowers with crimson red center, 3 cm across, in showy clusters though often hidden between the foliage. *Subtropic.* p. 903

wallichii (E. Africa, Madagascar); evergreen tree to 10 m, with large 30 cm, lobed, herbaceous leaves matte-green, densely hairy beneath; showy bell-like flowers with flat, spreading petals salmon-rose to scarlet, in dense hanging heads. *Tropical.* p. 903

DOROTHEANTHUS Aizoaceae

bellidiformis (Cape Prov.), "Livingstone-daisy"; small annual herb branching from the root; short, fleshy obovate leaves, rough-puckered, to 8 cm long; masses of daisy-like flowers in colors red purple, salmon, straw-yellow, yellow, lavender, white or white with red tips, opening to the sun. *Arid-subtropic.* p. 42, 46

DORSTENIA *Moraceae*
yambuyaensis (Zaire); erect, bristly herb with oblanceolate leaves to 15 cm long, shining green above, pale and dull beneath, and irregularly toothed; the green receptacle angularly rounded, almot 2½ cm across, the winged margins with rays to 10 cm long. *Tropical.* p. 650

DORYANTHES *Amaryllidaceae*
excelsa (New South Wales, Queensland), the "Globe spear-lily"; bold rosette of numerous succulent sword-shaped leaves to 2m long; and a leaning inflorescence to 4½ m long, the heavy head of 10 cm flowers with divided perianth segments crimson-red outside, pale inside, and with green bracts. *Subtropic.* p. 61

DOVYALIS *Flacourtiaceae*
caffra (So. Africa), "Kei-apple"; dense thorny shrub to 5 m; wiry branches with one large stiff spine to each node; leaves shiny green; small yellowish flowers, yellow 2½ cm fruit with yellow flesh of apricot flavor. *Subtropic.* p. 458
hebecarpa (Aberia gardneri) (Sri Lanka, India), "Ceylon-gooseberry"; decorative evergreen growing into a small dioecious tree to 6 m high; the alternate ovate, undulate, 10 cm papery leaves, glossy green with metallic sheen, and rosy midrib, grayish beneath; the thin wiry branches set with scattered 1 cm needle-like thorns; small greenish flowers, and 2½ cm purple, edible fruit. *Tropical.* p. 456, 458
x hybrida (abyssinica x hebecarpa), "Tropical apricot"; natural hybrid in Florida; large spreading shrub to 5 m, light green ovate leaves 8-10 cm long, on long drooping branches, some with thorns; small greenish flowers; 3 cm globular fruit with velvety brownish skin and soft melting flesh of apricot flavor. *Subtropic.* p. 458

DOXANTHA *Bignoniaceae*
unguis-cati (West Indies to Argentina), "Cats-claw"; tropical woody vine high-climbing into trees, with opposite leaves each with a pair of leaflets separated by a tendril split into 3 claws; leaflets are thin, oval-pointed, to 8 cm long; funnel-form, allamanda-like lobed flowers rich yellow, to 10 cm wide, borne profusely in clusters in spring, often also in autumn. *Tropical.* p. 183

DRACAENA *Liliaceae*
arborea (North Guinea), "Tree dracaena"; tree-like, to 12 m, with dense head of broad, swordshaped, evenly fresh-green, sessile, wavy leaves to 90 cm long, with sunken veins and prominently raised midrib. Roots of Dracaenas are orange-yellow (white in Cordyline). *Tropical.* p. 586, 589
aurea (Hawaii); much-branched native Hawaiian tree to 6 m high, long narrow, light green, recurving leaves without petioles, clustered at ends of slender branches. *Tropical.* p. 586
cantleyi (Singapore); rosette of very ornamental, long-oblanceolate, leathery leaves, fresh grass-green liberally sprinkled with oblong blotches of chartreuse. *Tropical.* p. 577
deremensis 'Bausei'; similar to warneckei but having the two white marginal bands brought close together in the middle of the leaf, with only a thin milky-green center stripe to separate them. Deremensis is from Trop. Africa. *Tropical.* p. 577
deremensis 'Compacta', "Dwarf bouquet" or "Calypso Queen"; very compacted, dense rosette or bundle of thin-leathery lanceolate, corrugated dark green, shiny leaves; introduced from Puerto Rico and Florida 1973; very attractive. *Tropical.* p. 580
deremensis 'Janet Craig'; green sport of warneckei, a rosette of freer, larger growing habit, similar to fragrans but more erect and stiff, with dark green, lustrous, wavy-margined leaves, corrugated lengthwise; of excellent keeping qualities. *Tropical.* p. 580
deremensis 'Roehrsii'; cultivar with leaves more or less variegated with golden yellow, over the white and green foliage. *Tropical.* p. 583
deremensis 'Roehrs Gold' (1956); a handsome cultivar of warneckei, of shapely erect habit, with the broad center of the leaves creamy yellow, bordered by translucent white lines and edged by rich green borders. *Tropical.* p. 581
deremensis 'Souvenir de Aug. Schryver'; beautiful Belgian cultivar, first seen in Germany 1970; a symmetrical, wide-spreading rosette of soft-leathery, lanceolate leaves, deep glossy green in center with broad creamy-white margins. *Tropical.* p. 581
deremensis 'Warneckei' (Trop. Africa), "Striped dracaena"; attractive house plant for dark locations; branching stout canes to 5 m high, covered with sessile, swordshaped, leathery leaves, fresh green streaked milky green in center, and bordered by a translucent white band on each side inside the narrow bright green edge. *Tropical.* p. 577, 580
deremensis 'Warneckei marginata'; handsome variegated cultivar seen in the Rhineland (Germany), with the broad ivory white bands along the outer margins of the otherwise green leaf. *Tropical.* p. 580

draco (Canary Isl.); "Dragon tree"; forming large trees to 20 m, with monstrous trunks of 4½ m dia., decorative as young plants with crowded rosette of sessile, sword-shaped, thick fleshy leaves, smooth to glaucous green, translucent edges, outlined in red if in the sun; flowers greenish. Superstition believes that the dark red, resinous substance exuding from leaves and trunk is "dragon's blood". *Subtropic.* p. 586, 589, 613
fragrans (Guinea, Nigeria, Ethiopia); tree-like, with cane to 6 m, sometimes branched, with broad, laxly recurved, oblanceolate, soft-leathery, sessile leaves continually ascending into a shining green, showy rosette, fragrant yellowish flowers in clusters. *Tropical.* p. 577, 580, 613
fragrans 'Lindenii'; variety with the wide, limply pendant leaves having broad, often indistinct, greenish-yellow bands at margins, center green. *Tropical.* p. 580
fragrans 'Massangeana', "Cornstalk plant"; an old-fashioned robust house plant with its rosette of rich green, laxly arching leaves broadly striped and banded light green and yellow down the center. *Tropical.* p. 580, 581
fragrans 'Victoriae', "Painted dragon lily"; beautiful, slow-growing conservatory plant with gracefully pendant, wide, soft-leathery leaves green streaked silvery-gray in center, bordered by contrasting broad margins of cream to clear golden yellow. *Humid-tropical.* p. 577, 583, 585
godseffiana (Zaire, Guinea), "Gold-dust dracaena"; small shrubby plant with spreading, wiry stems bearing thin-leathery, elliptic leaves in pairs or whorls of three, glossy deep green, irregularly spotted yellow, maturing to white; greenish-yellow flowers followed by red berries. *Tropical.* p. 583
godseffiana 'Florida Beauty'; a striking seedling of kelleri, with the same characteristic thick-leathery leaf but almost entirely covered with creamy-white blotching. *Tropical.* p. 577
godseffiana 'Friedman' ('Milky Way'); charming small branching plant; thin-wiry stems with pairs of elliptic, thin-leathery leaves to 10 cm long, glossy olive green, splotched and spotted cream, and a broad ivory band down center. *Tropical.* p. 583
goldieana (So. Nigeria, 900 m), "Queen of dracaenas"; the most spectacular dracaena in cultivation, but requiring humid-warm conditions to thrive; I remember a branched specimen 3 m high in Pará, its slender canes furnished with leathery foliage from bottom to top; the ovate, glossy deep green, long-stalked leaves strikingly marked with crossbands of pale green, maturing to almost white. Flowers in a dense spiral head, fragrant white, opening at night. *Humid-tropical.* p. 577, 581, 583
hookeriana (rumphii) (Cape of Good Hope, Natal), "Leather dracaena"; heavy plant with woody trunk to 2 m high, occasionally branching, topped by a crowded rosette of spreading sword-shaped, very thick leathery leaves dark green and glossy, to 3-5 cm wide and with translucent edges; slow-growing; panicle of greenish flowers. *Subtropic.* p. 583
hookeriana 'Rothiana'; horticultural form widely grown in the San Francisco area and a favorite because of its good keeping qualities; the thick, glossy dark green leaves 5-8 cm wide. *Subtropic.* p. 583
marginata (gracilis) (Madagascar), "Madagascar dragon tree"; a favored decorator; tree-like, with branching slender trunk, growing to 5 m high, each cane topped by a dense terminal rosette of thick-fleshy, narrow linear clasping leaves 40 cm long, rigidly spreading horizontally, shiny deep olive green prettily edged in red. The flexuous canes have the tendency to grow snaky and twisted, giving a branched specimen very artistic appearance. Slow and durable. *Tropical.* p. 580, 582, 589
marginata 'Tricolor', "Rainbow tree"; magnificent cultivar found in Japan, developed in Puerto Rico and Florida (1973); multi-colored fountain of leaves striped cream between green and with rosy-red margins; bright light for most intense glow. *Tropical.* p. 582, 583
sanderiana (Cameroons, Zaire), "Ribbon plant"; neat, very durable, little rosette, erect on slender cane until becoming too heavy, with narrow lanceolate, elegantly twisted leaves, deep green somewhat milky, and with broad marginal bands of white. *Tropical.* p. 577, 583
sanderiana 'Celes' (Pat.); handsome Puerto Rico mutation of stiffly erect habit; oblanceolate linear, leathery leaves 1-2 cm wide, dark green with white margins; for dishgardens. *Tropical.* p. 583
sanderiana 'Margaret Berkery' (1958); attractive, robust Puerto Rico sport with bamboo-like stem, clasping leafbases, short 12 cm lanceolate, leathery deep green leaves contrasting against the broad cream-white center band; translucent margins. *Tropical.* p. 577

DRACAENA: see also Pleomele, Cordyline

DRACONTIUM   Araceae
asperum (N. Brazil, Venezuela: Orinoco); giant tuberous herb with short 10-15 cm leathery, hooded spathe brownish-green, violet inside; followed by a solitary much divided leaf, 90 cm across, 3-parted and cut again, on stalk, to 3 m high, mottled with green and brown or lilac. *Tropical.* p. 103

DRACOPHYLLUM   Epacridaceae
traversii (New Zealand), "New Zealand grasstree"; handsome small tree found on South Island, of candelabrum-like habit, sparry very hard woody branches, and thick-leathery, stiff-recurving linear concave leaves deep bronzy, glossy green, with bases clasping and tips curling, in rosettes at branch-ends; the inflorescence in brownish terminal clusters. *Subtropic.* p. 394, 395

DRACUNCULUS   Araceae
vulgaris (Arum dracunculus) (Mediterranean); tuberous herb with pedately dissected, bright green leaves with 13-15 segments; spathe tube purplish-white striped purple, the purple blade 30 cm long, deeper at margin; spadix tipped purple also; offensive odor in bloom. *Subtropic.* p. 102

DRIMIOPSIS   Liliaceae
kirkii (Zanzibar); bulbous plant with lax, strapshaped, keeled, fleshy leaves narrowing toward base, pale blue-green with dark blotches; white flowers on short spike in July. *Tropical.* p. 606

saundersiae (South Africa: Transvaal); low bulbous plant with 2-4 broad-lanceolate, fleshy leaves grayish-green blotched with deep green; spike of greenish-white, small flowers. *Subtropic.* p. 606

DROSANTHEMUM   Aizoaceae
hispidum (S.W. Africa: Namibia, Cape); succulent shrub to 60 cm high; the rooting branches covered with rough white hairs; cylindrical light green to reddish leaves 2 cm long, covered by translucent pimples; purple flowers 3 cm dia. *Arid subtropic.* p. 44

speciosum (Cape Prov.); shrubby succulent with inclined branches, tiny 16 mm fleshy leaves curved upwards, semi-cylindrical, blunt, fresh green, set with crystalline blisters; masses of beautiful daisy-like flowers glowing brownish to orange-red, greenish in center. *Subtropic.* p. 46

DROSERA   Droseraceae (Carnivorous Plants)
binata (Australia), "Twin-leaved sundew"; small perennial 15 cm high, with long-stalked leaves deeply divided into two linear lobes with sensitive glandular hairs discharging sticky fluid to hold and digest insects. *Subtropic.* p. 292

capensis (So. Africa); perennial 15 cm high with linear oblong, blunt leaves tapered toward base, and densely furnished with hairs giving off sticky sap attracting insects, whereupon the sensitive hairs curve inward acting as a trap until the fluid digests their body. *Subtropic.* p. 292

dichotoma (E. Australia, New Zealand); a subtropical "Sundew"; robust, stemless species to 45 cm high, with smooth, linear leaves simple or bi-forked, each division linear 5-15 cm long, 1½-2 cm wide, densely covered by red-brown gland-hairs, from their apex secretes the mucilous sap which entraps insects; strong floral stalks with numerous pure white flowers 3 cm across. *Subtropic.* p. 292

schizandra, a "Sundew"; rosette of obovate leaves in complete circle, hugging the ground, covered by tiny red-brown tentacles, each tipped with a droplet of perfumed, sticky liquid, attracting insects. *Subtropic.* p. 296

DRYMOGLOSSUM   Polypodiaceae (Filices)
microphyllum (carnosum) (Himalaya, Nepal, Japan, China, Réunion), the "Bean-leaf fern"; wide-creeping, rooting, thread-like rhizome with small sterile waxy-glossy leaves 2-5 cm long, oval or elliptical, arranged shingle-like, the leathery fertile leaves linear-spoon-shaped 5-8 cm long. *Humid-subtropic.* p. 442

niphoboloides (Madagascar); creeping epiphytic fern forming dense masses; thin rhizomes with long obovate, leathery leaves 5-8 cm, covered with white felt. *Humid-tropical.* p. 444

DRYMONIA   Gesneriaceae
turrialvae (Costa Rica to Panama); square succulent stems flushed with red; large ovate leaves on red petioles, deep glossy green blades to 30 cm long, with sunken veins, and scalloped margins; flowers white, tinged pink, 4 cm long. *Tropical.* p. 492

DRYNARIA   Polypodiaceae (Filices)
fortunei (South China); found and photographed in Kwangtung, China, growing on rocks but also epiphytic; short fleshy rhizome, with leaves of two kinds, the shield-like sterile fronds 5-8 cm long, shielding the base, and the long fertile, deeply lobed fronds 30 to 80 cm tall, fresh green and of leathery texture, the triangular segments alternating along the rachis. *Humid-subtropic.* p. 441

quercifolia (Polypodium), (So. India, Queensland, Fiji), "Oak-leaved fern"; growing on trees and rocks, with a thick, brown, woody rhizome, having stalkless brown, bluntly lobed barren fronds and long-stalked deeply pinnate rigid fertile fronds with brown veins, the pinnae cut to center. *Humid-tropical.* p. 440, 441

rigidula (Malaysia, Queensland, Fiji); epiphytic fern with thick creeping rhizome looking like a rabbit's foot covered with long brown hair-like scales, the fronds in two forms: the sterile ones stalkless to 22 cm long and merely lobed, the fertile fronds pinnate, leathery and rich green, on brown rachis first erect, later pendant, ½-1 m long. *Humid-tropical.* p. 441, 443

rigidula var. whitei; attractive variety with lateral pinnae finely cut into pointed lobes. *Humid-tropical.* p. 441

DUCHESNEA   Rosaceae
indica (India), "Mock strawberry"; perennial herb with trailing branches rooting along the ground, bright green three-parted leaves; 2 cm yellow flowers followed by strawberry-like 2 cm red fruits of insipid taste; used as ground cover in California, and for hanging baskets. *Subtropic.* p. 849

DUDLEYA   Crassulaceae
brittonii (Mexico); low rosette of short stems, with few linear leaves, rounded at base, glaucous silver gray, and tipped red, later bluish; flowers whitish. *Arid-tropical.* p. 366

farinosa (Coast of N. California); attractive rosette to 10 cm across, succulent, narrow-oblong, green or white-mealy leaves, and short-pointed; double, starry, pale yellow flowers. *Arid-subtropic.* p. 365, 366

traskiae (So. California); clustering rosette, silvery glaucous blue with red on outside; leaves narrow, obovate, crowded. *Arid-subtropic.* p. 362

DURANTA   Verbenaceae
lorentzii (Argentina); bushy shrub or small tree to 3 m, with 4-angled pendant branches; small ovate, leathery leaves finely crenate; short tubular flowers with flaring limb white, in terminal interrupted clusters toward ends of willowy branches; small juicy, orange fruit. *Subtropic.* p. 916

repens (Florida, West Indies, Mexico, to Brazil), "Pigeon-berry"; small tree to 6 m, occasionally spiny and with drooping 4-angled branches; ovate 10 cm leaves; small flowers with cylindrical corolla and spreading limb lilac-blue, followed by orange-yellow berries. *Tropical.* p. 916

repens 'Variegata', "Sky flower"; spreading willowy branches with thin-leathery, opposite leaves 8 cm long, glossy green and attractively bordered with creamy white; clusters of lilac flowers, followed by the 6-10 mm yellow berries. *Subtropic.* p. 916

DURIO   Bombacaceae
testudinarum (Malaysia: Sabah, Malaya), the "Durian tanah"; fantastic tropical forest tree with more or less buttressed trunk at base of which are produced the numerous large globular fruit, hanging on short stalks directly from the trunk and elevated roots; the fruit is a thorny capsule 15-20 cm long with thick walls, splitting into five parts, enclosing the custard-like edible pulp; flowers red, with prominent stamens; leathery, lanceolate leaves covered with silvery scales underneath. *Tropical.* p. 194

zibethinus (Borneo, and widespread in Malaysia), the "Durian"; medium to large tropical evergreen tree to 25 m high, famous for its edible fruit; long ovate, shiny dark green, leathery leaves 10-16 cm long; white flowers with prominent stamens 5 cm long, are clustered in the branches; the green ovoid fruit is 20-30 cm long, with a spiny covering enclosing a creamy delicious, but unpleasant-smelling pulp, and is produced from the branches and old wood. *Tropical.* p. 194, 461

DYCKIA   Bromeliaceae
brevifolia (sulphurea) (So. Brazil), "Miniature agave"; dwarf clustering rosette of stiff succulent glossy green sharp-pointed leaves with silver lines beneath; inflorescence off center, a long stalk with bright orange flowers. *Tropical.* p. 212

fosteriana (Paraná), "Silver and gold dyckia"; very ornamental small rosette in a dense whorl of stiff silvery-purple arched leaves with silver spines; inflorescence a spike of rich orange flowers. *Subtropic.* p. 212

ECCREMOCARPUS   Bignoniaceae
scaber (Chile), "Glory flower"; scrambling bush with lobed leaves; inflated tubular flowers, orange-scarlet. *Subtropic.* p. 184

ECHEVERIA   Crassulaceae
agavoides (Urbinia) (San Luis Potosi), "Moulded wax"; solid starlike rosette with rigid fleshy leaves triangular pointed, glossy pale apple-green, margins frequently reddish, spinetipped; flowers reddish tipped yellow. *Tropical.* p. 371

agavoides corderoyi, "Urbinia red" in hort.; star-like succulent rosette of pointed boat-shaped leaves, more numerous than species, apple green, with margins and tips red-brown. *Tropical.* p. 371

ECHEVERIA

albicans (Mexico); charming rosette related to elegans, but with more leaves, to 20 cm across, glaucous-blue; flowers pink. *Tropical.* p. 371

x carnitricha (carnicolor x leucotricha); open rosette of fleshy, incurving spatulate or obovate smooth leaves 5 cm long; the stems elongated into the leafy inflorescence with orange-red flowers. *Tropical.* p. 368

'Chantilly Lace'; giant crenulata hybrid, 20-30 cm across, with wide, glaucous leaves suffused with copper, the margins lacily wavy. *Tropical.* p. 371

chihuahuaensis (Mexico); large formal rosette densely composed of thick-fleshy obovate short-pointed keeled leaves, glaucous silvery flushed purplish-red toward apex. *Subtropic.* p. 361

ciliata (Oaxaca, Mexico); solitary flat rosette of deep green, short leaves, ciliate at margins and with red cusp; flowers green to scarlet. *Tropical.* p. 368

crenulata (hoffmannii) (Morelos), "Scallop echeveria"; showy stemforming large rosette with broad, obovate leaves to 30 cm long, and tapering to narrow petiole, pale green and glaucous bluish-gray, the margins undulate and red; flowers yellowish red. *Tropical.* p. 369, 370

crenulata 'Roseo-grandis'; a very showy and decorative large rosette 25 cm diameter, with always rich green, waxy leaves beautifully undulated at the maroon-red margins; some light bluish glaucescence on young inside leaves. *Tropical.* p. 370

derenbergii (Oaxaca), "Painted lady"; small, globe-shaped clustering rosettes of numerous thick leaves pale green, glaucous with waxy silvery blue bloom and tipped red; flowering freely on short stems, golden yellow with orange. *Tropical.* p. 370, 371

'Desmetiana' hort. (So. Mexico); an old-time carpet-bedding plant in Europe, close to peacockii or a synonym for it; elegant blue-glaucous rosette 10-15 cm dia., with incurving leaves tipped with rose. *Subtropic.* p. 368, 369

'Doris Taylor' (setosa x pulvinata), "Woolly rose"; tall robust rosette with glossy deep green, obovate leaves tipped red, and beautifully covered with white hair. *Tropical.* p. 370

elegans (Hidalgo), "Mexican snowball"; like small balls of ice opening into beautiful clustering rosette of spoonshaped leaves, waxy glaucous pale blue, with white translucent margins; flowers coral pink on pink stem. *Tropical.* p. 368, 369, 370

elegans 'Kesselringii', (Mexico); lovely globular, tight rosette 8-10 cm across, with glaucous-bluish incurving, succulent leaves, concave spoon-shaped above and rounded beneath; photographed at Gruga Botanical Garden, Essen, Germany. *Subtropic.* p. 370

gibbiflora 'Carniculata'; strange looking succulent I photographed in France, with erect stout stem, spoon-shaped leaves glaucous blue, tinted copper-purple; at crest with warty outgrowth and recurving. *Tropical.* p. 368

gibbiflora 'Flammea'; showy large rosette with broad, concave glaucous leaves overlaid with bronzy red, and occasional leaves variegated with cream. *Tropical.* p. 371

gibbiflora metallica; handsome smooth rosette 30 cm or more across, with large spoonshaped leaves bronzy amethyst with metallic luster and glaucous purple, margins translucent. *Tropical.* p. 369

x gilva (agavoides x elegans) (Mexico), "Wax rosette"; shapely open rosette of pale green, broad obovate, spoon-like leaves, slightly silvery glaucous, with tip flushed orange red. *Tropical.* p. 371

'Hoveyi variegata'; striking large rosette of spoon-shaped, obovate leaves undulate at apex, glaucous blue with darker areas and variegated with pink. *Tropical.* p. 371

hyalina (Mexico: San Luis Potosi); handsome rosette of densely shingled, succulent leaves to 6 cm long, with sharp tip, delicate glaucous blue green in color, flushed with red and with translucent margin; tall inflor. of flesh-colored flowers. *Tropical.* p. 370

leucotricha (Mexico: Puebla), "White-plush plant"; beautiful loose rosette with thick boat-shaped, oblanceolate leaves densely covered with glistening white hair, edged with brown hair toward apex; flowers cinnabar-red. *Tropical.* p. 370

lindseyana (Mexico); beautiful stemless rosette rather flat, of obovate, thick-succulent leaves 5-9 cm long, keeled beneath, glaucous grayish green with red pointed tips; inflorescence on 50 cm stalks, with crimson-rose flowers, yellow at tips and inside. Described by E. Walther, San Francisco 1959. *Tropical.* p. 370

'Lipstick'; open star-like rosette with broad ovate sharp pointed leaves, apple green with red toward apex; similar to agavoides and corderoyi, but more flat in habit. *Tropical.* p. 371

pallida (Argentina), "Argentine echeveria"; spreading rosette of large obovate, spoon-shaped leaves pale green with edge lightly glaucous; bellshaped pink flowers. *Subtropic.* p. 370

peacockii (desmetiana) (Mexico); lovely symmetrical, dense rosette of numerous cupped, oblanceolate fleshy leaves richly covered with silvery-blue glaucescence, edged and tipped with red; red flowers glaucous outside; a favorite for carpet-bedding in Europe. *Tropical.* p. 368

'Perle von Nuernberg' (potosina x gibbiflora metallica); large open rosette to 15 cm across, with broad, almost flat, obovate leaves deep bluish and glaucous, edged in red, with fine patina. *Tropical.* p. 368

potosina, (Mexico: San Luis Potosi); charming blue-glaucous rosette, with broad leaves 4-6 cm long, and with purple tips; cup-shaped pink flowers. *Tropical.* p. 370

pulvinata (Oaxaca), "Chenille plant"; beautiful stemforming rosette of thick fleshy, broad obovate ashy-green, red edged leaves thickly covered with silver white felty hair, and forming a shaggy, rustcolor stem; spiked red flowers. *Tropical.* p. 368

'Pulv-oliver' (pulvinata x harmsii), "Plush plant"; fine Oliveranthus hybrid with oblancelate, broad medium-green leaves edged maroon only at apex, covered with glistening white velvety hairs; stem-forming; flowers orange and red. *Tropical.* p. 372

runyonii (Mexico); clustering succulent rosette of numerous rather flat, obovate 8 cm leaves upcurving, very glaucous-blue; flowers pink. *Tropical.* p. 372

secunda (So. America), "Blue hen-and-chickens"; small, saucer-shaped rosette of obovate leaves which stay narrow, always bluish-green, not metallic, apex with short tip red; flowers reddish. *Tropical.* p. 368

shaviana (Mexico); beautiful rosette 7 to 10 cm across, spatulate, rather thin succulent leaves with upturned tips, glaucous-blue with bright red or white wavy edges; tall glaucous inflorescence with flowers pink outside, orange inside. *Tropical.* p. 369

simulans (gilva) (Northern Mexico), "True Mexican rose"; dense rosette of many comparatively narrow pointed, fleshy, concave leaves rather incurving in center, light green and lightly glaucescent, the upper half of leaf rich crimson red, particularly the short cusps; flowers reddish-yellow. *Subtropic.* p. 372

zahnii 'Hoveyi'; colorful large loose rosette of long spatulate, fleshy leaves, furrowed lengthwise and variegated in delicate blended stripes of soft pale bluish-green, mauve, pink and cream, and glaucous with waxy silvery bloom, apex raggedly crenulated; flowers pinkish orange. *Tropical.* p. 370

ECHEVERIA: see also Oliveranthus

ECHIDNOPSIS *Asclepiadaceae*

archeri (Kenya); succulent with 8-angled, knobby gray stem 2 cm dia., first erect, later creeping to 35 cm long; scale-like persistent leaves; small cup-shaped crimson flowers. *Arid-tropical.* p. 147

ECHINOCACTUS *Cactaceae*

conothelos (Thelocactus) (Tamaulipas, Mexico); globe to 10 cm high, ribs cut into high tubercles, these with strong spines; flowers purplish-red. *Arid-tropical.* p. 264

grandis (So. Mexico: Tehuacan), "Giant Visnaga"; monstrous barrel cactus to 2 m tall and 1 m thick; with 30-50 ribs; 5-6 yellowish radial spines, and straight central spine to 5 cm long; yellow funnel-form flowers 5 cm long, and 5 cm woolly fruit. *Arid-tropical.* p. 249

grusonii (C. Mexico), "Golden barrel"; growing into a giant globe of 90 cm dia., light green, closely ribbed, covered with golden spines; yellow flowers imbedded around top, opening in sun. *Arid-tropical.* p. 263, 266, 267, 275

texensis (Texas, No. Mexico); small depressed globe to 30 cm high, with high ribs set with sharp, curved spines; flowers pink with red center. *Arid-subtropic.* p. 266

ECHINOCEREUS *Cactaceae*

albispinus (Oklahoma), "Lace cactus"; clusters of small green elongate-globular plants, their numerous ribs set with short white spines; large sweetly scented flowers with narrow petals dainty mauve-pink, throat old rose. *Subtropic.* p. 259

blanckii (So. Texas, Mexico), "Hedgehog"; branching plant with ascending or sprawling stems to 30 cm long and 3 cm thick, dark green, with 5-6 warty ribs set with glossy black central spines; beautiful purplish-pink flowers 9 cm long. *Subtropic.* p. 259

chloranthus (W. Texas), "Brown-flowered pitaya"; globular, spiny cactus with bundles of white needle spines, small red-brown flowers. *Arid-subtropic.* p. 259

dasyacanthus (W. Texas, Mexico), "Rainbow cactus"; small cylindric, hard and closely ribbed xerophyte densely stiff-spined in zones of straw-yellow, purple, tan, red-brown and amethyst, spines pointing out and downward; flowers yellow. *Arid-subtropic.* p. 257, 259

ehrenbergii (Mexico); freely suckering plant of erect green stems 15 cm high and 3 cm thick, with 6 notched ribs and glassy-white needle-spines; flowers purplish to violet-red. *Tropical.* p. 257

engelmannii (No. Mexico, U.S.: California, Arizona, Nevada, Utah), "Hedgehog cactus"; cylindric column 10 to 45 cm high, and 4-6 cm thick with 10-13 warty ribs, branching from base and forming clumps, with long needle spines; funnel-shaped day-blooming purple flowers to 10 cm long. Photo shown on p. 259 is of a plant carefully copied from nature and made of glass, in the collection of Harvard University. *Arid-subtropic.* p. 258, 259

enneacanthus (Sonora, Mexico); low plant with attractive, large flowers carmine-rose with deep red eye. *Arid-subtropic.* p. 258

longispinus (probably Mexico), a "Pitaya"; elongate plant, a variation of E. pectinatus, similar to baileyi but the cream spines appear longer and are red-tipped toward apex. *Tropical.* p. 258

papillosus (West Texas), "Hedgehog cactus"; sprawling and branching green stems 3 cm thick with 6-10 ribs, notched into prominent warts; yellowish white to brown radials and one yellow central spine; flowers yellow, reddish at base. *Arid-subtropic.* p. 259

pectinatus (Texas and Mexico), "Pitaya"; small globe to 30 cm high, densely covered with flattened spine bundles; large carmine-rose flowers to 12 cm across. *Arid-subtropic.* p. 259

pentalophus (So. Texas, Mexico); cylindric branches sprawling and ascending to 12 cm long and 3 cm thick, green, with 4-6 warty ribs, and white areoles; pretty round flowers lilac or pink with reddish violet. *Arid-subtropic.* p. 259

perbellus (Texas), a "Lace cactus"; colorful globe or short cylinder to 10 cm, closely ribbed, radial spines shaded from reddish to nearly white; flowers purple. *Arid-subtropic.* p. 260

reichenbachii (Texas, Mexico), "Lace cactus"; small globe to 20 cm, close-ribbed, attractively covered by flat rosettes of stiff spines in a range of colors from white to red-brown; flowers iridescent light purple. *Arid-subtropic.* p. 257

rigidissimus (pectinatus) (Arizona, Sonora), attractive, short-cylindric plant 10-20 cm high, known as a "Rainbow cactus"; gray-green covered with small, stiff radial spines interlocking in older plants, alternate zones of cream, pink, reddish and brownish, the apex beautifully tinted purplish-red; purple flowers. Marshall-Bock suggests that this purple-flowered type should be listed a variety of E. dasyacanthus, and report a plant of dasyacanthus seen in New Mexico with a purplish flower from one side of the stem and a yellow flower opposite. *Arid-subtropic.* p. 257

sciurus (So. Baja California); low tufting plant with green stems 20 cm high and to 5 cm thick, 12-17 shallow, notched ribs covered with short bristle-like yellow spines; large and beautiful 12 cm flowers rosy violet. *Arid-tropical.* p. 259

triglochidiatus (W. Texas, New Mexico, Colorado), "Hedgehog cereus"; stems to 60 cm long, with 5-8 notched ribs in spirals, small reddish to fawn spines, with attractive scarlet flowers with yellow center. *Warm temperate.* p. 258

triglochidiatus neo-mexicanus (New Mexico); from West of the Organ Mountains, forming clusters; bluish-green stems to 7 cm dia., with 11-12 ribs; charming flowers 5 cm long a striking salmon-orange. *Warm temperate.* p. 258

viridiflorus (So. Dakota, Wyoming to Texas), "Green-flowered hedgehog"; small, nearly globular, rigid, dark green plant to 20 cm high, with 14 ribs, long stiff areoles and spines, in zones of white or brownish; flowers greenish. *Temperate.* p. 257, 258

ECHINOMASTUS Cactaceae
macdowellii (No. Mexico); small globular plant 8 cm high, with 20-25 ribs, white radials and long white needle spines; large flowers rose-pink. *Arid-subtropic.* p. 266

ECHINOPSIS Cactaceae
hamatacantha (Argentina), "Lily cactus"; globular, flattened plant depressed at top, 15 cm thick, dark dull-green with some 15 notched ribs, straight and recurved spines; flowers white, night-blooming. *Arid-subtropic.* p. 260

multiplex (S. Brazil), "Easter lily cactus"; small barrel to 15 cm high, forming clusters, dark green, close-ribbed, with sharp brown spines; long, erect funnelformed pink or rosy flowers to 25 cm long. *Arid-subtropic.* p. 261, 263

multiplex hybrida; according to Britton-Rose possibly with the white-flowering E. turbinata; a freely suckering, globular, "Lily-cactus" with fresh-green bodies many-ribbed and with small sharp spines, the large, showy funnel-form flowers 12-20 cm long, white inside, flushed pink to purplish outside, open in the morning; widely grown in California nurseries under this name for dishgarden planting. *Arid-subtropic.* p. 258

ECHITES Apocynaceae
rubrovenosa (Prestonia), (So. America: Rio Negro); tropical wiry climber with striking foliage on slender twining stems; the large elliptic emerald green leaves to 15 cm long, netted with red veins. The beautiful coloring of the leaves is brought out to perfection only under moist-warm conditions. *Tropical.* p. 78

ECHIUM Boraginaceae
fastuosum (No. Europe, Canary Isl.), "Pride of Madeira"; gray-hairy shrubby perennial ½ to 2 m high, with narrow lanceolate leaves covered with soft white hair, and deep blue, bell-shaped flowers with protruding red stamens, on one-sided spikelets formed in cylindrical inflorescence. *Subtropic.* p. 196

EDITHCOLEA Asclepiadaceae
grandis (Somaliland, Kenya, Tanzania); succulent close to Caralluma but with larger 10 cm flowers, pale yellow, with red-brown spots on the 5 lobes, and ciliate margins; the 5-angled grayish-green stems to 30 cm, with hard, brown, thornlike teeth. *Arid-tropical.* p. 147

EICHHORNIA Pontederiaceae
azurea (Brazil), "Peacock hyacinth"; aquatic herb rooting in shallow water, and sending flexuous runners across its surface, with fleshy stalks not inflated, fresh green, waxy leaves rounded or obovate; terminal spikes of showy flowers, lilac-blue with purple center, the two lower petals fringed. *Humid-tropical.* p. 825, 828

crassipes (speciosa) (Trop. and Subtrop. America), "Water hyacinth"; small aquatic plant usually floating, with bluish, feathery roots, and runners forming clumps, the roundish, glossy green leaves in rosettes, on petioles much inflated at base causing them to stay afloat; clusters of large, lilac flowers with yellow center surrounded by blue blotch and purple stripes. *Humid-tropical.* p. 825, 827, 828

ELAEAGNUS Elaeagnaceae
philippinensis (Philippine Islands), "Lingaros"; evergreen shrub to 3 m or more, with drooping branches, long ovate leaves 10 cm long, silvery underneath and with undulate margins; small axillary flowers; glossy red, edible berries 1 cm dia. *Tropical.* p. 462

pungens (Japan, China), the "Silver berry"; slow-growing brown-stemmed evergreen shrub of rather rigid, angular sprawling habit of growth, to 2-5 m high; leathery oval grayish green leaves with rusty-brown tinting, 3-8 cm long and with wavy edges and silvery beneath; the surface is covered with silvery scales reflecting sunlight which give the plant a special sparkle; white, fragrant flowers. *Warm temperate.* p. 393

ELAEIS Palmae
guineensis (Trop. West and C. Africa), "African oil-palm"; feather-palm to 18 m or more tall, with ringed trunk crowned by graceful pinnate fronds to 5 m long, the ridged narrow leaflets green on both sides; male and female flowers in separate clusters on the same tree; the ovoid black 4 cm fruit in large clusters of 200-300; source of palm-oil. *Tropical.* p. 778, 802

ELAPHOGLOSSUM Polypodiaceae (Filices)
pachycraspedon (Eastern Perú); tropical epiphytic fern with creeping rhizome, and long-stalked tongue-like leaves, hard-leathery and metallic glossy blue-green, 25 cm long and 5 cm wide, with bold midrib; underside dense-scaly when young; new growth hairy. *Humid-tropical.* p. 446

ELLEANTHUS Orchidaceae
capitatus (West Indies, Mexico to Perú); terrestrial of the habit of Sobralia, with leafy, slender reed-stems 30 cm or more; thin but stiff leaves prominently length-ribbed ending in two short points; terminal heads of rose-purple flowers; April-June. *Tropical.* p. 720

EMBOTHRIUM Proteaceae
wickhamii (New Caledonia); magnificent evergreen shrub dense with long oval, leathery concave, waxy leaves, the inflorescence a striking flat brush of brilliant orange-red filaments tipped with yellow. I photographed this tropical bush, 2 m high, near Noumea. *Tropical.* p. 838

EMILIA Compositae
javanica (sagittata) (Pantropic), "Flora's paintbrush"; showy erect annual to 60 cm or more with elliptic leaves with winged stalks; flower heads loosely clustered, 2 cm dia., without rays, scarlet red. *Tropical.* p. 315

EMMENOSPERMA Rhamnaceae
alphitonioides (Australia); sparry evergreen shrub photographed at Taronga, near Sydney; stiff branches with long ovate, concave, leathery green leaves, and with terminal clusters of bright orange berries; very handsome. *Subtropic.* p. 832

ENCEPHALARTOS Cycadaceae
altensteinii (South Africa: E. Cape), "Prickly cycad"; stout trunk to 2 m high bearing crown of pinnate leaves 1-2 m long, the numerous leaflets thick, and with 3-5 spiny teeth on each margin, shining deep green; cylindrical male cones 30-40 cm long; female cones broadly oval to 45 cm tall. *Subtropic.* p. 384

arenarius (So. Africa); handsome rosette of twisted leaves, glaucous blue with recurved apex, tips divided into spine-like sharp points, the margins of the rigid leaflets toothed. *Subtropic.* p. 386

ferox (South Africa); palm-like small tree forming trunk to 2 m high, crowned by a rosette of glaucous green, stiff pinnate fronds 1-2 m long, the oblong pinnae 15 cm long with 2 to 4 teeth on either margin, apex spiny; colorful cones 1-3 together, the male cylindrical to 40 cm long, the females ovoid, salmon pink or red. *Subtropic.* p. 386

hildebrandtii (Trop. E. Africa); splendid plant growing to 6 m high, pinnate leaves light green, somewhat rigid, to 3 m long, leaflets shorter near base, and are spiny along margin; the male cone to 60 cm long. *Tropical.* p. 384

horridus (So. Africa), "Ferocious blue cycad"; wide spreading ornamental cycad with short stout trunk crowned by a cluster of stiff pinnate leaves, glaucous green, 60 cm long, leaflets oblique lanceolate and spinily lobed; male cone cylindric and stalked, 30 cm long, female cone broad ovoid 40 cm long. *Subtropic.* p. 386

kosiensis (So. Africa: Zululand); short trunk, bearing rosette of pinnate, vicious leaves about 1 m long; the broad leaflets 8-15 cm long formidably armed with large triangular teeth and spines; cones brilliant orange-red. *Subtropic.* p. 386

latifrons (horridus latifrons) (So. Africa: S.E. Cape); "Spiny kaffir bread"; stout, globular trunk to 2 m high, with recurved, vicious pinnate leaves 60-90 cm long, reflexed toward apex, the broad, thick leaflets overlapping and with coarse spine-tipped lobes; the cones brownish-yellow. *Subtropic.* p. 386

laurentianum (Trop. Africa: Uganda); palm-like tree with trunk to 10 m high and 75 cm thick; stiff, straight, pinnate leaves to more than 6 m long, the narrow-linear, leathery pinnae dark green, and with spine-tipped marginal teeth; cones red. *Tropical.* p. 387

lehmannii (So. Africa), "Blue-leaved cycad"; thick trunk to 1 m or more high, crowned by a mighty rosette of elegant, recurving fronds about 1 m long, the glaucous blue-green leaflets 16-18 cm long, with spiny tip and marginal spines. *Subtropic.* p. 386

transvenosus (South Africa: North Transvaal), the "Mujaji-palm"; forming tall, stout trunks crowned by stiff pinnate leaves; the glaucous dark green, leathery pinnae lightly hairy and curved back from the midrib. By special permission I was privileged to visit Mujaji village in the mountains of northern Transvaal, their habitat, where they are venerated as dedicated to the Rain Queen, with specimen, occasionally branched, 5 m tall; propagation from bulbils forming along trunk. *Subtropic.* p. 387, 388

ENSETE *Musaceae*
   maurelii (Musa) (Ethiopia), "Black banana"; symmetrical plant with the decorative habit of ventricosum; rosette of sturdy, broad, dark-complexioned leaves, deep green suffused with blackish-red above, with brown-purple midrib and margins, purplish-brown beneath. *Tropical.* p. 662

ventricosum (Musa ensete) (Moist Central and E. Africa), "Abyssinian banana"; to 12 m, stout solitary pseudostems, not stoloniferous, conspicuously swollen at the base, bearing erect broad leaves to 6 m long in a cluster, banana-like, and dying after fruiting. Leaves bright green with red midrib and purple edge, on short red stalks. Erect inflorescence with dark red bracts and whitish flowers. Used as shapely centerpiece in summer beds. Have seen this at alt. 1,500 m on Mt. Kilimanjaro (Tanzania), and at 2,300 m in Kikuyu country of Kenya. *Subtropic.* p. 662, 666

ventricosum 'Montbeliardii'; a form very tall and slender, with leaves narrow-lanceolate, having blackish petioles and midrib. *Tropical.* p. 668

EPACRIS *Epacridaceae*
   impressa (Tasmania, Australia); evergreen shrub 1-1.5 m high with erect, downy shoots, small leaves 2 cm long; tubular flowers pale yellow with red hairs at base, 2-3 cm long. *Subtropic.* p. 394

longiflora (New South Wales), "Australian fuchsia"; heath-like shrub 1 m or more, with downy shoots and closely-set, sharply pointed leaves 1-2 cm long; pendant slender tubular flowers 3-4 cm long, rosy crimson with white tips. *Subtropic.* p. 398

EPERUA *Leguminosae*
   jenmanii (Trinidad); evergreen tree with pinnate leaves of 3-4 pairs of leathery 15 cm leaflets; purple trumpet flowers in pairs 10 cm long; inside white with purple spots. *Tropical.* p. 544

EPIDENDRUM *Orchidaceae*
   altissimum (hodgeanum) (Bahamas); slender 8 cm pseudobulbs with 2-3 leathery leaves; branched inflorescence to 1 m tall, with waxy, greenish-yellow 3 cm flowers flushed with brown, the lobed lip white; fragrant and long-lasting (winter). *Tropical.* p. 724

atropurpureum (Mexico to Panama, W. Indies, Brazil), "Spice orchid"; handsome epiphyte with pear-shaped pseudobulbs bearing 2-3 long leaves, stout racemes of 5-8 cm flowers having sepals and petals chocolate with incurved green tips, often tinged with purple, large white lip with crimson center. (Dec.-June). *Tropical.* p. 729

ciliare (W. Indies, C. America, Brazil); strong epiphyte with oblong, compressed pseudobulbs with 1-2 leathery leaves; the inflorescence an erect cluster of waxy, fragrant flowers with long linear, greenish-yellow sepals and petals and a lobed and fringed white lip spotted yellow. (Dec.-Jan., Sept.) *Tropical.* p. 729

dichromum (Brazil: Bahia); handsome epiphyte with 2-3 leaved ovate pseudobulbs and stalks 60-90 cm high bearing bold clusters of 5 cm flowers with pure white sepals and petals, the 3-lobed lip rosy-crimson, yellow at base, and margined white (early summer). *Tropical.* p. 732

erubescens (Oaxaca); desirable mountain epiphyte found at 2,950 m, the limit of orchids there; with spindle-shaped, 2-leaved pseudobulbs; the waxy flowers in large panicles, sepals and petals delicate mauve, the lip yellow at base. (winter). *Subtropic.* p. 728

falcatum (Mexico); pendulous epiphyte with tiny cylindrical pseudobulb and a solitary thick, linear leaf, 1-2 spidery, waxy flowers, the narrow sepals and petals greenish-yellow, lip trilobed, side lobes wing-like, midlobe narrow, white with yellow throat. (June-Aug.) *Tropical.* p. 728

fragrans (W. Indies, C. America, Brazil); stocky epiphyte with compressed pseudobulbs with solitary leaf, the waxy, fragrant flowers in short clusters, sepals and petals creamy-white, narrow linear and curled, the heart-shaped pointed lip white, lined red-purple. (Feb.-Aug.) *Tropical.* p. 729

ibaguense (radicans) (Perú, Colombia, Guatemala, Mexico), "Fiery reed orchid"; reed-type terrestrial with slender leafy stems to 1 m high, rooting from nodes, leaves very fleshy; small flowers on a long wiry stem in dense cluster, orange-scarlet or cinnabar-red, with yellow in fringed lip. *Tropical.* p. 726, 728, 729

ibaguense purpurea, "Purple reed orchid"; color form with purple flowers. *Tropical.* p. 728

ibaguense schomburgkia (Guayana, Brazil, Perú); slender stems with 2-ranked leaves, and large 15 cm terminal cluster of brilliant flowers orange to vermillion, lip yellow at base. *Tropical.* p. 728

lanipes (Brazil, Venezuela, Perú); pretty species with leafy, slender reed-stems topped by a branching raceme of numerous little 1 cm yellowish flowers with narrow sepals, the ovaries and stemlets hairy, and with heliotrope fragrance (spring-summer). *Tropical.* p. 726

lemorea (Perú); handsome epiphyte found by Lee Moore on the upper Amazon; waxy, long-lasting flowers with dark brown sepals and petals, the lip white with red veins. *Tropical.* p. 730

mariae (So. Mexico); excellent little epiphyte, small pear-shaped pseudobulbs with 1-2 leaves, topped by clusters of several large waxy flowers with greenish yellow sepals and petals, and very broad, wavy white lip with green throat (summer). *Tropical.* p. 729

x o'brienianum (evectum x radicans), "Butterfly orchid"; slender reed-stems with aerial roots and alternate leaves like in ibaguense, and extending into long stalks with showy clusters of brilliant carmine-red flowers, with bright yellow on the crested lip (spring-summer). *Tropical.* p. 728, 729

oncidioides (C. America to Brazil); stately epiphyte with spindle-shaped bulbs bearing 2-3 straplike leaves; arching inflorescence to 1 m long with many 4 cm flowers bronzy-yellow, lip marked red (June). *Tropical.* p. 728, 729

paniculatum (Bolivia, Perú, Ecuador, Colombia); a noble plant 1 m or more high, the dark green leaves arranged in two ranks along the stem, and on top a pendant inflorescence of nodding clusters with a profusion of lovely 2 cm rosy-lilac fragrant blossoms, the column tipped yellow (April). *Subtropic.* p. 728

pentotis (C. America, Brazil); stiff epiphyte with long spindle-shaped pseudobulbs and 2 leaves, the large waxy, very fragrant flowers back to back in pairs on short stem, tapering fleshy sepals and petals greenish-white, lip white with purple (Mar.-July). *Tropical.* p. 729

prismatocarpum (Costa Rica, Panama), "Rainbow orchid"; beautiful robust epiphyte with flask-shaped pseudobulbs bearing 2 leaves; and erect spike of brightly colored, waxy, long-lasting flowers, sulphur-yellow blotched maroon, lip marked rosy-red (May-June). *Tropical.* p. 729

pseudepidendrum (Costa Rica, Panama); excellent species with leafy, stem-like pseudobulbs to 1 m high; arching inflorescence with 8 cm flowers having narrow spoon-like green sepals and petals; and orange-red, fringed lip edged yellow. *Tropical.* p. 728, 730

radicans: see ibaguense

radiatum (lancifolium) (Mexico); epiphyte with short spindle-shaped pseudobulbs and 2-3 stiff leaves; compact clusters of fleshy cream-white flowers, broad white lip with radiating purple lines (May-June). *Tropical.* p. 729

stamfordianum (Honduras, Guatemala, Venezuela, Colombia); beautiful epiphyte with spindle-shaped pseudobulbs with 2-4 leaves, the inflorescence a lax panicle to 60 cm long from the base of the bulb, with fragrant 3 cm flowers yellow spotted bright red, lip fringed. (Feb.-May). *Tropical.* p. 729

vespa (variegatum) (Brazil); stout epiphyte with spindle-shaped, 2-3 leaved pseudobulbs, and stiff spike of small waxy flowers, the sepals and petals greenish-yellow spotted red-brown, small white lip with purple center. (April-July). *Tropical.* p. 729

EPIGENEIUM *Orchidaceae*
lyonii (Sarcopodium, Dendrobium) (Philippines); handsome small epiphyte from the mountains of Luzon, with angular pseudobulbs each with two leaves; pendant inflorescence with beautiful star-like flowers 8 cm across, crimson-red in center fading to a delicate pink near tips, and very fragrant. *Tropical.* p. 732, 750

EPILOBIUM *Oenotheraceae*
angustifolium (Chamaenerion), (No. Temp. hemisphere), "Fireweed" or "French willow"; shrubby perennial to 2 m, with lanceolate, grass-green, 12 cm leaves; with undulate margins; 3 cm flowers rose to purple, or white in var. album. Seen at Rochford's Nurseries in England. *Warm temperate.* p. 702

x EPIPHRONITIS *Orchidaceae*
veitchii (Epid. radicans x Sophronitis grandiflora); attractive small bigeneric hybrid with reed-type stems with alternate leaves, topped by clusters of fiery red 2½ cm flowers with a golden yellow center on the lip (spring to fall). *Subtropic.* p. 732

EPIPHYLLANTHUS *Cactaceae*
obovatus (E. Brazil); branching cactus of the habit of Zygocactus but with flat joints resembling a miniature Opuntia, waxy green tinted purple in the sun, with areoles scattered over the surface set with tiny whitish spines; small zygomorphic flowers purplish rose. *Tropical.* p. 281

EPIPHYLLOPSIS: see Rhipsalidopsis

EPIPHYLLUM *Cactaceae*
ackermannii: see Nopalxochia
chrysocardium (Mexico: Chiapas), "Golden heart"; an interesting species growing at the base of trees, with bright green, flat stems cut into narrow segments like fishbones, and looking more like a fern; very large white flowers noted for their beautiful golden filaments; night-blooming, with unpleasant odor. *Tropical.* p. 282

hybridum, "Orchid cactus" or "Pond-lily-cactus"; there are literally thousands of named "Orchid cacti", with large flowers of great beauty in iridescent colors from nearly blue through red, pink, yellowish to white. Recorded genera which were used in breeding were such epiphytes as Heliocereus, Nyctocereus, Hylocereus, Selenicereus. *Humid-tropical.* p. 283

hyb. 'Argus'; a "Pond-lily-cactus" with flattened stems, and exquisite large broad-petalled flowers 12 cm across, soft flesh-pink suffused with red in center. *Humid tropical.* p. 282

hyb. 'Brilliant' (Phyllocactus); large growing plant with broad flattened, fresh green branches and very large bright red double flowers. *Tropical.* p. 282

hyb. 'Chas. Preston'; an "Orchid cactus" with flattened green branches; large crimson red flowers of waxy texture; shading to pink toward tips. *Humid-tropical.* p. 284

hyb. 'Elegantissimum', "Dwarf orchid cactus"; superb, dwarf basket type; large, perfect flower with broad petals bright crimson, the outer petals flushed scarlet, and green throat, on glossy green flattened branches. *Tropical.* p. 282

hyb. 'Flamingo'; an "Orchid cactus" with uniquely colored flowers an iridescent coppery red, edged with pink and crimson-red deep inside, contrasting vividly with the bold, cream-colored stamens. *Humid tropical.* p. 282

hyb. 'Grace-Ann'; free-flowering with showy flower, outside crimson-red, inner petals purplish-rose. *Humid-tropical.* p. 284

hyb. 'Hermosissimus' (E. ackermannii x Nopalxochia); one of the brightest and best of "orchid" cactus, a vivid example of the bicolor type with deep crimson outer petals, while the inner petals and throat are a glowing electric violet painted orange through the center; free bloomer of vigorous growth. *Tropical.* p. 283

hyb. 'Hermosus', "Orchid cactus"; beautiful ackermannii cultivar with flat stems and large 15-18 cm showy tubular-trumpet flowers carmine to fuchsia-red, changing to rich purple in its center. *Tropical.* p. 281

hyb. 'Matador'; a charming "Orchid cactus" with scandent flattened stems; large salmon rose flowers with pink sheen. *Humid-tropical.* p. 281

hyb. 'Peacockii'; Heliocereus hybrid, with three-angled and flat branches; flowers large red and an overtone of iridescent purple derived from Heliocereus. *Humid-tropical.* p. 284

oxypetalum (latifrons) (Mexico to Brazil), "Queen of the night"; the best nightblooming cactus for the home, grows large, branches usually flat and thin, waxy green and deeply crenate; large fragrant white flowers, reddish outside. *Tropical.* p. 281

stenopetalum (Mexico: Veracruz); epiphytic clamberer with woody stem and angular leaf-like branches obliquely scalloped; showy, fragrant, nocturnal flowers with tube 25 cm long, and narrow white petals; lasting only one night. *Tropical.* p. 281

EPIPREMNOPSIS *Araceae*
media (Borneo); wiry tree climber with palmately 3-lobed parchment-like leaves deep green with faint mottling of yellow-green on puckered surface. Baileya (March 62) puts this under Epipremnum medium. *Tropical.* p. 104

EPIPREMNUM *Araceae*
elegans (New Guinea); clambering epiphyte or tree-climber with thick, almost woody stems, and large pinnately lobed leaves to 1 m long, shining fresh green and leathery, borne on long angular, stiff petioles; spectacularly showy cone-like fruit 30 cm long, brilliant shiny orange-red. We collected this species in the lower Markham Valley Oomsis rain-forest. *Humid-tropical.* p. 105

pinnatum (Malaya through Java to New Guinea); prolific tropical climber with large, oblong leaves pinnately parted into regular segments, and tiny pinholes appearing as silvery dots along midrib. The juvenile leaves are entire, oblique-ovate. *Humid-tropical.* p. 105, 109

EPIPREMNUM: see also Scindapsus

EPISCIA *Gesneriaceae*
cupreata (Colombia), "Carpet plant"; tropical creeper, rooting at joints, with soft-hairy, oval, wrinkled leaves almost entirely a metallic copper, faintly marked silver; small solitary flowers with short corolla tube yellow-tinged below, the lobes orange-scarlet, and yellow with red spots inside. *Humid-tropical.* p. 496

cupreata 'Acajou'; an exquisite hybrid of cupreata x variegata, leaves dark mahogany except for the contrasting network of veins and the central area of bright metallic silver-green; very excellent and a good keeper, more pubescent than 'Harlequin'. *Humid-tropical.* p. 497

cupreata 'Cleopatra' sport of 'Frosty', registered 1964, similar to E. 'Pink Brocade'; gorgeous leaves green and milky green in center, toward margins variegated cream, pink and glowing crimson red. *Humid-tropical.* p. 497

cupreata 'Frosty', a silverleaf "Flame violet"; lustrous, robust growing cultivar with downy emerald-green leaves tinted copper at margins, the center area silvery white splashing over through the lateral veins. *Humid-tropical.* p. 497

cupreata 'Metallica', "Kitty episcia"; variety with large quilted coppery leaves having a band of pale silvery green through the center, margined with metallic pink; similar to the cultivar 'Kitty'. *Humid-tropical.* p. 497

cupreata 'Musaica' (Panama); seedling from the Canal Zone with coppery leaves attractively painted with an even network of silver veins. *Humid-tropical.* p. 497

cupreata 'Shimmer'; low rosette of corrugated green leaves netted and splashed with carmine-rose; flowers scarlet red. *Humid-tropical.* p. 496

cupreata 'Tropical Topaz' (Panama Canal Zone), a mutant form of E. cupreata once known as "Canal Zone yellow", this clone is distinguished by having bright yellow flowers instead of scarlet as in cupreata viridifolia; plain pale green leaves with erect hairs instead of appressed. *Humid-tropical.* p. 496, 497

dianthiflora (C. Mexico), "Lace-flower vine"; small elliptic, pubescent, vivid to dark green, crenate leaves with purple midrib, in a miniature clustering rosette, sending out prolific rooting branches; free-blooming with glistening white, deeply fringed flowers, appearing singly in the leaf axils. *Humid-tropical.* p. 497

lilacina (Costa Rica), "Blue flowered teddy-bear"; a beautiful blue-flowered species also known as "Fannie Haage" and 'Variegata'; pubescent creeper with dark coppery, rough-puckered 5 to 10 cm leaves decorated with a prominent fishbone pattern of silvery green; large flowers of lavender-blue, 3 to 4 cm long, good grower, but sensitive to cold. *Humid-tropical.* p. 496, 497

lilacina 'Cuprea' (Costa Rica); robust leaf variation from the Rio Reventazon with deep coppery velvet leaves only faintly marked with silver-green along center; free flowering lavender-blue with pale eye. *Humid-tropical.* p. 496

lilacina 'Viridis' (fendleriana) (Costa Rica); attractive variety having bright emerald-green leaves faintly traced with glistening silver from the midrib; foliage feels plushy; flowers light blue. *Humid-tropical.* p. 496

punctata (Guatemala, Mexico); rank creeping, almost smooth species with leathery, ovate, crenate leaves green except for red-purple midrib, on lightly erect branches; tubular flowers solitary with spreading fringed lobes creamy-white and spotted purple into the throat. *Tropical.* p. 497

reptans (fulgida, coccinea) (Brazil, Guyana, Surinam, Colombia), "Flame violet"; tropical pubescent creeper with broad ovate, quilted, brown-green leaves and bright silvery-green veins, margins crenate; flowers with long blood-red corolla tube with fringed edges and pink inside. *Humid-tropical.* p. 496

reptans (fulgida) **'Lady Lou'**, a handsome sport having its bronzy-green leaves with narrow silver veining beautifully variegated clear rose-pink and occasional white or gray; flowers red. *Humid-tropical.* p. 497

x wilsonii 'Pinkiscia' (cupreata x lilacina); lovely hybrid with leaves of thick texture, metallic chocolate-bronze with a touch of silver along the green veins, and covered with pink hair; flowers with broad rose-pink limb, and orange-yellow dotted rose inside the long tubes. *Humid-tropical.* p. 496

EPITHELANTHA Cactaceae
micromeris (Texas, N. Mexico), "Button-cactus"; tiny globes usually 1-2 cm dia., but may grow to over 5 cm; depressed on top; small tubercles in spirals almost entirely covered by flattened white spines; pinkish flowers. *Arid-subtropic.* p. 263, 265

EPITHEMA Gesneriaceae
involucratum (Celebes); herbaceous plant with large ovate leaves 12 cm long, glossy green with depressed veins; clusters of small white flowers on long wiry petioles. *Humid-tropical.* p. 504

ERANTHEMUM Acanthaceae
nervosum (pulchellum) (India), "Blue sage"; shrubby plant with rough, ovate, dark-green leaves and depressed veins; blue flowers from pointed bracts. (syn. Daedalacanthus). *Tropical.* p. 38

ERANTHEMUM: see also Pseuderanthemum, Chamaeranthemum

EREMURUS Liliaceae
elwesii (Turkestan), "Desert candle"; stately perennial from the steppes of Central Asia, with narrow basal leaves and strong spike 2 m or more, with numerous stalked flowers having spreading segments, pink with deeper center stripe; May-blooming. *Warm temperate.* p. 603

ERIA Orchidaceae
merrillii (Philippines); epiphyte with angular pseudobulbs grayish green, borne at intervals on the creeping rhizome, paired fleshy leaves; inflorescence from near apex, a raceme dense with starry white flowers, the center disc light yellow. *Tropical.* p. 723

spicata (Himalayan Region to Vietnam,) "Lily-of-the-valley orchid"; swollen pseudobulbs bearing 3-4 shiny, leathery leaves; the small inflorescence to 15 cm high, with nodding flowers very fragrant, 2 cm across, not opening fully, translucent waxy-white or straw-yellow, the lip golden-yellow at front. *Subtropic.* p. 732

ERICA Ericaceae
baccans (So. Africa), "Berry heath"; erect shrub 1-1½ m high, tiny linear bluish-green leaves; the globular bell-flowers, mostly in terminal fours, deep rose, the 6 mm corolla narrowed at throat; winter-spring. *Subtropic.* p. 396

bauera (bowieana) (So. Africa), "Bridal heath" or "Albertina heath"; charming flowering shrub to 1½ m high, with bluish-green leaves in fours, standing straight out; waxy flowers in leafy raceme, with tubular inflated corolla 2 cm long, translucent pink, in spring (August to October in So. Africa). *Subtropic.* p. 397

'Felix Faure' (hyemalis hybrid), "French heather"; branching evergreen of low, compact habit to 30 cm high; bright green needles; tubular flowers 3 cm long, purplish-orange with whitish tips. *Subtropic.* p. 397

gracilis (So. Africa), "Rose heath"; bushy shrub with small light green linear leaves, the side shoots loaded with terminal clusters of tiny globose rosy flowers (winter). *Subtropic.* p. 396

hyemalis (So. Africa), "French heather"; spreading bush producing tall, tapering racemes of tubular rose-pink flowers 2-2½ cm long, and tipped with white, winter blooming. Possibly form of summer-fl. perspicua. *Subtropic.* p. 397

mammosa (So. Africa); tall shrubby plant to 1 m, the little needle-like leaves in fours, or scattered; tubular flowers pendulous in a dense cluster, corolla 2 cm long, reddish-purple, blooming July to October (Autumn). *Subtropic.* p. 397

melanthera (So. Africa), "Christmas heather"; compact shrub with downy shoots and tiny linear leaves; the small globular flowers delicate pale rose with prominent black anthers, blooming in winter. *Subtropic.* p. 396

regia (Cape Prov.), the "Royal heath"; striking heath of straggling habit 60-90 cm high but bearing lovely glossy sticky flowers with inflated tubular, waxy corolla 2 cm long, rich crimson-red; the small appressed needle-leaves fresh green. The photos shown are probably E. regia var. variegata, their bicolored corolla crimson with white base, and popular in cultivation. *Subtropic.* p. 397, 398

speciosa (So. Africa); sparry heath 60-120 cm high, the short, plump needles close to stems; long tubular curved flowers to 3 cm long, and sticky, crimson red with yellow tips. *Subtropic.* p. 397

taxifolia (Cape Prov.); small shrub to 30 cm high, known as the "Double pink heath" because its small flowers appear to be double; the jar-shaped deep pink corolla, tipped red, is half concealed, like petticoats, by broad pink sepals; in clusters on the branchlets; linear leaves 3-sided. *Subtropic.* p. 396

vagans 'Rubra' (W. Europe), "Cornish heath"; low spreading shrub to 30 cm high, small needle-like leaves and urn-like flowers 5 mm long, rosy-purple. *Warm temperate.* p. 396

'Wilmorei', "Prince of Wales heather"; a grand old large-flowered hybrid well-known since 1835, probably of E. perspicua but with swollen tubes and more bushy; showy spikes with stiff linear leaves in threes, and tubular rosy flowers 4 cm long, prettily tipped with white; blooming winter to spring; very showy and with long-lasting flowers. *Subtropic.* p. 397

ERIGERON Compositae
hybridus; involved hybrid; bushy perennial; leafy-stemmed to 60 cm high; many large heads with lilac blue ray petals and yellow disk. *Tropical.* p. 315

speciosus (Pacific coast), "Fleabane"; free-blooming perennial to 75 cm high, with narrow, smooth leaves; purple daisy-like flowers, 4 cm across, the ray petals thread-like in 2 or more rows. *Warm temperate.* p. 315

ERIOBOTRYA Rosaceae
deflexa (Taiwan), "Bronze loquat"; woody evergreen 2 m or more high, with obovate wrinkled leaves 12-25 cm long, bright coppery when young and rusty hairy, serrate at margins; small white flowers; 2 cm fruit not edible. *Subtropic.* p. 849

japonica (China), "Chinese loquat" or "Japan plum"; symmetrical evergreen tree with noble decorative thick foliage, 15-30 cm long obovate, glossy-green on the surface, strongly ribbed, and toothed, the underside and stalk rusty-woolly; fragrant white flowers, and pear-shaped yellow to orange edible fruit 3 to 8 cm long, of pleasant, sprightly flavor. *Subtropic.* p. 456, 462, 849

ERIOPSIS Orchidaceae
biloba (Perú, Guianas, Brazil); epiphyte with ovoid pseudobulbs bearing 2 leaves 40 cm long; inflorescence densely many-flowered, almost cylindrical; waxy, fragrant blooms 3 cm across, golden yellow with brown-red margins, lip deep yellow spotted with red (autumn). *Tropical.* p. 734

ERODIUM Geraniaceae
chamaedryoides (Mallorca, Corsica), "Alpine geranium"; dainty herbaceous perennial to 8 cm high, small cordate leaves crenate at margins; flowers white with red veins. *Subtropic.* p. 490

ERVATAMIA: see Tabernaemontana

ERYCINA Orchidaceae
echinata (Mexico); small epiphytic oncidium-like orchid, with ascending rhizome producing small 3 cm pseudobulbs at intervals, with solitary leaves; wiry stalks bear bird-like 2 cm flowers greenish-yellow or pinkish, the large lip golden yellow (spring). *Tropical.* p. 732

ERYNGIUM Umbelliferae
alpinum (Europe: Maritime Alps to Balkans), "Alpine thistle"; handsome perennial 60 to 80 cm high, with rich green leaves lobed or palmately cut; the inflorescence with blue or white flowers in cone-like heads, subtended by ray-like silvery bracts. *Warm temperate.* p. 911

bromeliifolium (Mexico); perennial rosette of fleshy olive green, channelled leaves edged by large thorny teeth; tall stalk to 2 m, with thistle-like inflorescence of bluish flowers. Photographed outdoors in the Botanic Garden, Strasbourg, France. *Subtropic.* p. 908

ERYTHEA: see Brahea

ERYTHRINA Leguminosae
abyssinica (Ethiopia, Kenya, Uganda), "Red hot poker tree" or "Kaffir-boom"; deciduous tree to 12 m, with corky, yellow-brown bark with woody spines; the branchlets armed with strong recurved thorns; the leaves trifoliolate, leaflets ovate, and gray-hairy beneath; showy flowers with narrow lobes brilliant scarlet, in erect poker-like inflorescence. *Tropical.* p. 555

x bidwillii (crista-galli x herbacea), "Florida coral-bean"; handsome plant 1-2 m high, with leaves of 3 leaflets, and erect inflorescence of crimson-red curved flowers, blooming April to November. *Subtropic.* p. 545

caffra (humeana) (E. Cape Prov., Natal), "Kaffirboom coral tree"; large spreading, deciduous tree to 25 m, trunk and branches with hooked thorns; pointed leaflets in three's, spade-shaped; tubular flowers more than 3 cm long, in a spreading spike, brilliant cinnabar-scarlet fading to purple; protruding stamens give a whiskery look. *Subtropic.* p. 554

coralloides (corallodendrum) (W. Indies) "Coral tree"; tree to 6 m, with woody stems prickly or unarmed; leaves of 3 broad leaflets; flowers with standard never opening, deep scarlet, in long racemes; appearing after leaves fall. *Tropical.* p. 545

crista-galli (Brazil), "Coral tree"; small tree with trifoliate leaves on spiny petioles, and showy butterfly flowers deep scarlet-red, in dense racemes, usually produced before the leaves. *Tropical.* p. 554, 555, 564

falcata (So. Brazil); tropical tree with bright coral scarlet flowers; I will never forget the breathtaking sight of these leafless trees in red, blazing against the azure blue sky of the Organ mountains in the dry landscape of Minas Gerais in Brazil. *Tropical.* p. 554, 555

humeana (So. Africa), "Natal coral tree"; semi-evergreen shrub or tree; may grow to 10 m, but begins to bear its bright crimson-red blooms when 1 m high; flowers in long-stalked clusters at branch ends and inclined to be pendant or arching; foliage with 3 broad leaflets. *Subtropic.* p. 554

indica picta (India, Malaya, Australia); a beautiful variegated-leaf form of E. indica; bushy tree with black spines, the large leaves with three broad leaflets fresh glossy-green, strikingly variegated cream to golden yellow along the primary veins; photographed in Queensland and Java, *Tropical.* p. 554, 555

lysistemon (Transvaal, Zululand, Natal), "Kaffirboom"; spectacularly showy deciduous, thorny tree to 12 m with gray branches; leaves after bloom, the obovate leaflets arranged in three's; large tube-shaped, bright salmon-scarlet flowers, in large compact clusters reminding of a coxcomb. *Subtropic.* p. 554

pallida (Trop. America), "Pink coral tree"; straggling tree with willowy branches, leaves of 3 leaflets; erect spike of waxy, sickle-shaped flowers soft salmon-pink. *Tropical.* p. 554

vespertilio (E. Australia), "Batswing coral tree" or "Gray corkwood"; low tree with trunk to 1 m dia.; slender, unique, stalked leaves of 3 broad winged pointed leaflets; open salmon-red pea-flowers 4 cm long, in pendant clusters to 25 cm long; fruits an elongate legume constricted between seeds. *Subtropic.* p. 554

ERYTHRONIUM   *Liliaceae*
grandiflorum (Brit. Colombia, Montana, Oregon, Utah), "Avalanche lily" or "Glacier lily"; beautiful cormous perennial, with green-elliptic leaves to 15 cm long; lily-like flowers variable, usually bright golden yellow with recurved petals 4 cm long. I photographed this plant in July, breaking through the melting snow, in Glacier National Park, Montana. *Temperate.* p. 594

ERYTHROXYLUM   *Erythroxylaceae*
coca (Eastern Andes), "Cocaine plant" densely leafy shrub to 4 m, obovate leathery leaves to 6 cm; small 1 cm yellowish flowers, followed by small reddish berries. Coca leaves mixed with lime are chewed daily by the Indians of the Andes. *Subtropic.* p. 407

ESCALLONIA   *Saxifragaceae*
laevis (organensis) (So. Brazil: Organ Mts.); sturdy ornamental evergreen, bushy shrub to 2 m, with angled branchlets; obovate firm bronzy-green, toothed leaves to 8 cm long, often red-margined; small 1 cm fragrant, rosy flowers apple-blossom-like, in dense terminal clusters; autumn. *Subtropic.* p. 880

ESCHSCHOLTZIA   *Papaveraceae*
californica (California, Oregon), "California poppy"; soft herb with gray-green, 3-pinnatifid leaves and large showy, poppy-like flowers, bright yellow with orange-red base, opening in the sun; widely naturalized in southern Chile where I noted them climbing even into the Andes. *Subtropic.* p. 812, 813

californica 'Rubra', "Red California poppy"; color form glowing scarlet red. *Subtropic.* p. 812

mexicana (Arizona, Sonora); smaller version of the "California poppy", with flowers dark orange. *Warm-temperate.* p. 813

ESCOBARIA   *Cactaceae*
chaffeyi (Coryphantha) (Zacatecas, Mexico); column to 12 m, 5-6 cm dia., almost completely covered by white spines; large flowers creamy-red with brown central band. *Arid-tropical.* p. 278

tuberculosa (chisoensis) (Texas); also known as albicolumnaris; beautiful fingers to 18 cm high and 6 cm dia., entirely covered with white hair-like spines; flowers purplish-rose. *Arid-subtropic.* p. 277

ESCONTRIA   *Cactaceae*
chiotilla (S. Mexico), "Little star big-spine"; slender columns branching tree-like, sparry-branched to 6 m; seedlings stocky, bright green, 7-8 sharp ribs, light colored radial spines, central spines to 8 cm long, apparently dropping in old specimens; small 3 cm yellow flowers; edible purple fruit. *Arid-tropical.* p. 246, 250

ESPOSTOA   *Cactaceae*
lanata (Perú, Ecuador), "Peruvian old man"; small column, 20-25 ribs, beautifully covered with cottony, snow-white hair; flowers pinkish. Becoming 1 m high with age. *Arid-subtropic.* p. 248

EUCALYPTUS   *Myrtaceae*
caesia (Western Australia), the "Gungurru"; handsome small, graceful tree of weeping habit 5 to 7 m high; wood, leaves and seed pots beautifully coated with waxy, silvery white powder; gray-green lanceolate leaves contrast with the red stems; bark white and mottled, curling when older; dusty rose-pink flowers; seed capsules shaped like silver bells. *Subtropic.* p. 678, 679

calophylla (W. Australia), "Marri" or "Red gum"; round headed large tree 30-50 m high, with rough, flaky bark; broad lanceolate, beautiful leaves 10-18 cm long, deep green; large showy heads of white stamen flowers, sometimes rose or red; the large urn-shaped capsules to 3 cm dia. *Arid-subtropic.* p. 679

camaldulensis (rostrata) (Australia), "River red gum"; wide-spreading tree 25-40 m high; smooth bark, red when young, later grayish and mottled; gracefully weeping branches with long-slender leaves; unimportant white to pale yellow flowers in drooping clusters, followed by pea-sized seed capsules. *Arid-subtropic.* p. 680, 682

cinerea (cephalophora) (N.S. Wales, Victoria), "Silver dollar tree"; small glaucous tree, with brown-red, willowy branches bearing pairs of sessile leaves rigid, stiff-leathery and silvery glaucous, rounded like a silver coin, to 8 cm across, in the juvenile stage; later when 2 m or more high the leaves become ovate to lanceolate, with yellow midrib; used by florists as cut branches; often preserved with glycerine. *Subtropic.* p. 679, 688

citriodora (Queensland), "Lemon-gum"; one of the most graceful of trees to 25 m, with smooth powder-white to pinkish trunk; branches pendant with long 8-20 cm sickle-shaped leaves, golden green and lemon-scented; white flowers in clusters. Older specimen are designer's trees, very picturesque with bare trunks. Oil has been distilled from leaves. *Tropical.* p. 680, 681

deglupta (New Guinea, Indonesia, Philippines), "New Guinea gum" or "Mindanao gum"; noble tall tree, with trunk displaying a striking coloring as the bark peels, revealing a beautiful spectrum of browns, yellows, grays, greens, pinks and reds; handsome dark olive, lanceolate foliage on pendulous branches. *Tropical.* p. 680

diversicolor (W. Australia), "Karri gum"; magnificent tree to 60 m tall, important for timber; bark smooth, gray to orange-yellow; lanceolate leaves not sickle-shaped, 15 cm long, and of different color each side; canary yellow flowers in showy clusters. *Subtropic.* p. 678

erythrocorys (W. Australia), "Red-cap gum" or "Illyarie"; small tree 3 to 10 m high; smooth bark, peeling in thin flakes; thick, shiny green narrow sickle-shaped leaves 10-18 cm long; the inflorescence with bright scarlet caps which drop off to reveal yellow stamen flowers. *Subtropic.* p. 675, 679, 688

ficifolia (S.W. Australia), "Scarlet flowering gum"; slower growing, ornamental tree to 15 m high, with dark and furrowed bark, thick lanceolate green leaves with yellow midrib; well known for its showy masses of flowers with bright scarlet stamens and dark red anthers. *Subtropic.* p. 675, 678, 679, 688

globulus (Tasmania, Victoria), "Tasmanian blue gum"; a pretty plant when small, with shoots and clasping cordate leaves a glaucous blue, but rapidly growing into a gigantic tree to 100 m high, with leaves becoming long and dark green; much used for windbreaks in California; its foliage contains aromatic oil used for medicinal purposes. *Subtropic.* p. 679

globulus 'Compacta', the "Bushy blue gum"; a densely branched lower, more compact tree with grayish trunk, very glaucous-blue ovate leaves when young; with age leaves become 30 cm long, sickle-shaped, dark shining green. Flowers white. *Subtropic.* p. 679

lehmannii (W. Australia), "Bushy yate"; dense, shrub-like small tree 6-10 m high, with smooth, gray-brown bark; light green, long oval 5 cm leaves; dense clusters of apple-green flowers open from striking yellowish horn-shaped buds; followed by pointed seed pods which are fused into a globular mass. *Subtropic.* p. 679

macrocarpa (Western Australia), "Mottlecah" or "Blue-bush"; curious sprawling shrub to 5 m; the stiff stems closely set with broad ovate, silvery blue, clasping leaves in ranks, 5-12 cm long and almost as broad; flat-topped fluffy flowers with filaments usually red or pink, directly from leaf axils. *Subtropic.* p. 678, 679, 688

maculata (Queensland to Victoria), "Spotted gum"; dignified tall tree with single trunk branching to make wide head, 20-30 m

high; pearl gray bark mottled dark red to violet; glossy dark green lanceolate leaves; whitish puffy flowers in clusters to 8 cm wide; small urn-shaped capsules. Subtropic. p. 681
nicholii (New So. Wales), "Willow peppermint"; graceful, willowy tree to 15 m, with soft brown, persistent bark; narrow light grayish green foliage 8-12 cm long, purplish in spring; small whitish flowers; tiny 1 cm peaked pods. Subtropic. p. 680
polyanthemos (Victoria, New So. Wales), "Red box" or "Silver dollar tree"; tree to 45 m high, with persistent bark, the slender, sparry, spreading branches with distantly spaced attractive leathery leaves almost round, glaucous bluish gray and edged all around with purple. Subtropic. p. 680
rhodantha (W. Australia), "Rose Mallee"; ornamental sprawling shrub to 4 m, with horizontal branches densely set with clasping, sessile leaves in lovely glaucous-blue, 5-10 cm long and nearly round; showy inflorescence of long bright crimson-red stamens tipped yellow, looking like small brushes. Subtropic. p. 680, 681, 688
rudis (W. Australia), "Desert gum"; robust upright, spreading, often weeping tree 15-20 m high; with rough, dark gray, persistent bark; juvenile leaves broad ovate and silvery, mature leaves lance-shaped 10-15 cm long and more green; white flowers in clusters, not showy; small 1 cm capsules. Subtropic. p. 680, 681
salubris (Western Australia), "The Gimlet"; attractive tree to 15 m, with rich red-brown trunk and smooth with prominent spiralled flutings; branches are willowy, with shining dark green, narrow leaves. My photo shows a tree sheared for compactness in the streets of Perth. Subtropic. p. 682
sideroxylon (Queensland to Victoria), "Red ironbark"; graceful medium-sized tree to 25 m, with furrowed, non-shedding, red to nearly black trunk; slim blue-green leaves turn bronze in winter; fluffy flowers light pink in pendulous clusters; seed capsules goblet-shaped. Subtropic. p. 681
sideroxylon rubra; variety with showy stamen flowers deep ruby-red; medium sized tree 6 to 15 m high, with hard blackish bark, deeply furrowed and persistent; slim blue-green leaves turn bronze in winter; goblet-shaped seed capsules. Subtropic. p. 681
torquata (Western Australia), "Coral gum"; small aromatic tree to 4 m high, with willowy light brown stems, and blue-grayish-green leathery, lanceolate leaves; masses of flowers at an early age, in umbels, with vivid coral-red stamens, the base of calyx dilated into a ring. Subtropic. p. 681, 688
viminalis (S.E. Australia), "Manna gum"; picturesque tall spreading tree to 50 m high, with beautiful smooth whitish trunk, the rough cover peeling in ribbons; drooping willow-like branches with narrow 10-15 cm leaves; little white flowers in long, open clusters; small 1 cm seed capsules. Bark yields manna, eaten by aboriginals; the leaves form the principal food of Koala bears. Subtropic. p. 681, 682
woodwardii (W. Australia), "Yellow-flowered gum" or "Lemon-flowered gum"; smallish tree to 12 m high, very glaucous, with smooth whitish gray bark; thick and rigid ovate to lanceolate gray leaves, and bright yellow flowers in clusters from powdery white buds. Subtropic. p 681

EUCHARIS Amaryllidaceae
grandiflora (amazonica) (Andes of Colombia), "Amazon lily"; bulbous plant with broad, basal leaves narrowed into petioles; umbels of fragrant white flowers with spreading segments, 5 to 8 cm across. Tropical. p. 64, 66, 67
korsakoffii (Perú); a "Miniature Amazon lily", found near Moyobamba, Dept. San Martin; with globose 3 cm bulb, long-stalked fleshy glossy deep green, lance shaped leaves 15-18 cm long; the floral stalk to 28 cm with cluster of small, pure white, waxy flowers, the flaring segments 2 cm long. Subtropic. p. 64

EUCOMIS Liliaceae
pole-evansii (South Africa), "Giant pineapple lily"; bulbous plant to 1 m or more high; leaves in basal rosette; spike-like raceme of small greenish white flowers; blooming in South Africa in December. Subtropic. p. 603
undulata (So. Africa), "Pineapple flower"; bulbous plant with broad, undulated bright green leaves to 8 cm wide, with tough margins and keeled beneath, thin and spreading, plain green; the inflorescence on cylindric stalk with dense head of bright green flowers crowned by a tuft of small leaves; August. Subtropic. p. 603

EUGENIA Myrtaceae
megacarpa (E. Indies, Malaya); evergreen tropical shrub or tree with quadrangular twigs; leaves ovate, 12-25 cm long; flowers white 2½ cm across; the oblong, constricted fruit 8 cm long, glossy crimson red, not edible. Tropical. p. 688
myrtifolia 'Gracilis'; charming and shapely small evergreen bush densely branched, and with light glossy green, needle-like linear leaves 2 cm long. Seen at Fabricius, Leverkusen, Germany. Subtropic. p. 688

uniflora (Brazil, Guayana), "Surinam cherry"; small glabrous tree to 8 m high, with glossy, ovate, 5 cm leaves, and fragrant white flowers; producing distinctive grooved, round fleshy fruit of deep crimson, edible and of spicy flavor. Tropical. p. 460, 677

EUGENIA: see also Syzygium

EULOPHIA Orchidaceae
paiveana (Kenya to Southern Africa); xerophytic terrestrial with conical pseudobulbs bearing narrow lanceolate ribbed thin but tough leaves; basal tall, erect raceme 75 cm high, with flowers having green sepals marked brown, petals golden yellow, lip yellow with purplish veins and a brown spur. Tropical. p. 731
porphyroglossa (Lissochilus) (Kenya, Uganda, Zaire, Sudan), "Swamp orchid"; very large terrestrial, with indistinct pseudobulbs, sword-shaped plaited leaves to 75 cm long; the tall rigid erect raceme to 1³/₄ m high, with solid textured, 6 cm flowers with bronze-purple sepals, petals magenta outside, white within, lip yellow and purple, the spur bronze. Tropical. p. 720

EULOPHIELLA Orchidaceae
rolfei (roempleriana x elisabethae from Madagascar); lovely epiphyte, whose parents typically grow on pandanus trees; pseudobulbous plant with creeping rhizome, and long plicate foliage; basal flowers 8 to 10 cm across on stalk 30 cm or more, fragrant, with rose-carmine, waxy petals and sepals, and white lip. Tropical. p. 735

EULYCHNIA Cactaceae
floresii (Chile), "White fluff-post"; very pretty column, olive green, with many close ribs, attractively and regularly dressed with large white-woolly tufts from the areoles, long brown-tipped central spines; funnel-shaped white flowers. Arid-subtropic. p. 251

EUONYMUS Celastraceae
fortunei colorata, "Purple-leaf winter creeper"; from China; evergreen sprawling shrub with elliptic leaves, deep green or milky green and cream, turning deep purple in fall. Temperate. p. 302
fortunei gracilis (Japan, Korea), "Silver-edge creeper"; small evergreen shrub climbing by rootlets, small oval, serrate leaves gray green, variegated and edged white, tinted pink. Temperate. p. 301
fortunei radicans (Cent. Japan, So. Koera), "Creeping euonymus"; evergreen shrub creeping or climbing by rootlets, with small, leathery, deep green leaves 2-3 cm long, margins serrate. Temperate. p. 301
fortunei vegeta (Central and Western China), "Big-leaf winter creeper"; low, spreading shrub to 1½ m high, or climbing by rootlets; leaves ovate to 5 cm long; orange seeds in little hat-boxes. Temperate. p. 301
japonica 'Albo-marginata' (So. Japan), "Japanese spindle-tree"; dense evergreen shrub to 5 m high, with erect willowy branches, closely set with opposite leathery glossy, small oval leaves 3-6 cm long, obscurely toothed, dark green in the species, and with small greenish-white flowers. This "Silver-leaf euonymus" has somewhat smaller, narrower leaves milky green bordered with a narrow white edge. Warm temperate. p. 301
japonica 'Argenteo-variegata' (Japan), known as "Silver Queen"; leaves oval, glossy fresh green with broad white marginal variegation. Warm temperate. p. 301
japonica 'Aureo-marginata; compact bush with colorful oval leaves, deep green with yellow margins. Warm temperate. p. 301
japonica 'Aureo-variegata Gold Spot'; as grown in So. California, with glossy leaves, deep green and all centers creamy-yellow. Warm temperate. p. 301
japonica 'Aureo-variegata Yellow Queen'; the waxy green oval leaves are variegated and margined yellow toward edge. Warm temperate. p. 302
japonica medio-picta, "Goldheart euonymus"; waxy oval leaves fresh green at margin, golden yellow in the center and down the petiole and stem. Warm temperate. p. 301

EUPATORIUM Compositae
sordidum (ianthinum) (Mexico), "Boneset"; shrubby plant with stems densely red-haired; corrugated leaves broad ovate, 10 cm long, with toothed margins; the clustered disk-flower heads violet and sweetly fragrant. Subtropic. p. 315

EUPHORBIA Euphorbiaceae
atropurpurea (Canary Islands: Tenerife); succulent shrub branching in ranks, with thickened fleshy stems 2 cm dia., becoming woody, with leaf scars, young branches densely leafy, with glaucous bluish narrow foliage 5-10 cm long; the inflorescence in clusters of flowers with deep maroon bracts. Subtropic. p. 420
caerulescens (Cape Prov.); thorny shrub, branching underground and forming clusters of upright columns 5 cm thick, 4-6-angled, constricted to joints and with hollow sides, bluish green, margins toothed and horny. Subtropic. p. 422

canariensis (Canary Islands), "Hercules club"; large cactus-like succulent branching from the base, to 10 m high; the stems 5, 4 or 6 angled, brownish fresh green, the sinuate angles set with black spine-pairs; small yellow flowers; poisonous. *Subtropic.* p. 422, 425

candelabrum (Sudan, Uganda); massive candelabra-like tree to 10 m high, at home in the Nile basin, with 4-angled yellowish green branches constricted to joints, the angles wavy-toothed and set with spines; commonly used as living fences. *Arid-tropical.* p. 417, 422

caput-medusae (So. Africa), "Medusa's head"; short globose stem with many snake-like gray-green branches densely knobbed and with tiny leaves at the growing tips. *Arid-subtropic.* p. 424

clandestina (So. Africa); thornless succulent to 60 cm high, the stem cylindric club-shaped, covered shingle-like with wart-like knobs, the apex crowned with willow-like 6 cm linear channeled glaucous leaves edged purple, almost hiding the sessile green flowers; the gray-green knobs with milky streaks. *Arid-subtropic.* p. 421

cooperi (So. Africa); thorny tree to 5 m high, with round trunk and arching, upright branches divided into 4-6 angled joints, the angles winglike, green and marked with darker crossbands; poisonous. *Arid-subtropic.* p. 417, 422, 425

cotinifolia (Mexico to So. America), "Hierba mala"; ornamental shrub or small tree 3-6 m high, coppery-red leaves usually 3 at a node, rounded ovate, 5-12 cm long; used for fish poison; whitish inflorescence. *Tropical.* p. 408

fulgens (jacquinaeflora) (Mexico), "Scarlet plume"; leafy shrub with slender thin-wiry branches gracefully arching, the small leaves narrow lanceolate, and the flowers in brilliant terminal sprays of small petal-like orange-scarlet bracts. *Tropical.* p. 419

x 'Giant Christ-thorn', "Giant California Christ-thorn"; a phenomenal flowering plant; the result of 25 years of hybridizing by Ed Hummel of Carlsbad, California, involving E. milii bojeri, breonii and lophogona; a swollen, stout grayish stem to 6 cm thick, with long brown thorns; toward the apex with bluish-green leathery leaves 20-25 cm long and, arranged as a giant bouquet, strong stalks bearing the clustered inflorescence with firm, large bracts 2 to 5 cm across, larger than a silver dollar, in glowing cerise pink with salmon sheen. Slow-growing. *Tropical.* p. 420

grandicornis (Natal to Kenya), "Cow horn euphorbia"; thorny branching succulent to 2 m, of interesting shape, the green to gray-green branches 3-angled, to 15 cm thick, irregularly constricted, angles winglike with horny margins. *Arid-tropical.* p. 421

heterophylla (Illinois to Perú) "Annual poinsettia", or "Mexican fire plant"; noted annual with variable green leaves linear, ovate or even fiddle-shaped, the upper ones red at the base, bractlike, surrounding the little flowers. *Warm temperate.* p. 426

ingens (Natal, Transvaal, Mozambique, Rhodesia), a giant "Candelabra tree" to 10 m high, stems to 5-angled, the leafless branches succulent, constricted into joints, dark green, 4-angled, winged and wavy. In habitat in the bush-veld of northern Transvaal, I noted that seedlings in their juvenile stage up to 1½ m are beautifully marked with silver. *Arid-tropical.* p. 417, 425

x keysii (lophogona x milii), a "Giant crown of thorns", also known as "Flamingo"; resembling milii in habit but with branches much more stout, 15 cm long ovate fleshy leaves that turn coppery-red in the sun, and stalked clusters of large showy flowers rosy-carmine, or red in saline soil. *Tropical.* p. 420, 421

lactea (India, Ceylon), "Dragon bone-tree"; cactus-like plant growing like a candelabra, the branches 3-4 angled, distantly but deeply scalloped and black-spined, dark green with greenish-white band down the center. Used in India medicinally, as a hot jam, for rheumatism. *Tropical.* p 417, 421, 422

lactea cristata, "Elkhorn"; an intricately monstrose form with fanshaped crested branches forming a snaky ridge or crowded cluster, attractively green marked silver-gray. *Tropical.* p. 421

leucocephala (Kenya), "Flor de Nino"; attractive shrub seen and photographed near Nairobi, with pinkish stems bearing narrow oblanceolate, channeled leaves, and clusters of small yellow flowers with white bracts. *Tropical.* p. 408, 419, 423

lophogona (Madagascar); branching shrub with clubshaped branches, angled with the ridges reddish and fuzzy-hairy, spatulate leaves with pale veins, stalked inflorescence with bracts light pink. *Tropical.* p. 420

mammillaris 'Variegata', "Indian corn cob"; a beautiful cultivar which I photographed in California at Los Angeles Plant Co., Manhattan Beach; the notched columns are largely greenish-white marked fresh-green and tinted pink, with buff spines. *Subtropic.* p. 421, 424

marginata (variegata) (Plains of Minnesota, Dakota to Texas), "Snow-on-the-mountain" or "Ghost-weed"; hardy pubescent annual to 60 cm high, numerous branches with glaucous gray-green ovate to oval soft-fleshy leaves, 3-8 cm long with milky sap, the upper ones white-margined; flowers in umbels with bract-leaves having showy white border, and small white flowers in September; long lasting. *Warm temperate.* p. 408

milii (splendens) (W. Madagascar), "Crown of thorns"; xerophytic spiny shrub with slender scandent woody stems to 2 m long, the spreading branches about 1 cm dia., grooved and armed with spines; obovate 4 cm leaves dull green, deciduous, and soon falling if disturbed or too dry; flower bracts soft salmon-red with pale center. May be trained against trellis or wire frame; very cheerful when in bloom at Easter time; good house plant for warm location. From stem cuttings. *Tropical.* p. 420

milii breonii (Madagascar); fleshy, spiny shrub with stout circular, shallowly ribbed, thorny branches about 3 cm thick; bronzy, lanceolate leaves 10-15 cm long in terminal rosettes; the cup-flowers brilliant crimson-red. *Tropical.* p. 420

milii hislopii (Madagascar); branching shrub with stout gray stem, furnished with heavy brown spines, later gray, the satiny bright green leaves are long and thin, the saucer-type bracts are large and soft salmon red with pale center. *Tropical.* p. 420

milii var. imperatae (Madagascar), "Mini-Christ-thorn"; small, very bushy shrub to 50 cm high, with slender woody, brown stems not swollen at the base, the spines rather distant; leathery roundish or obovate leaves 1-2 cm long; the flower bracts red or yellow. *Tropical.* p. 420

myrsinites (So. Europe); branching succulent with prostrate then ascending yellow stems densely shingled with sessile ovate leaves bluish-glaucous, in whorls or spirals; umbels of yellow flowers. *Subtropic.* p. 424

neriifolia 'Variegata cristata', "Variegated oleander cactus"; weird fasciation with stems crested fan-like, grass-green with milk-white variegation, fleshy light green obovate leaves and short brown spines. *Tropical.* p. 421

neohumbertii (Madagascar); small succulent column with 5 angles, densely set with thin brown spines, the fleshy moss-green obovate leaves 15-25 cm long, in a rosette on top; small dark reddish cupped flower bracts. *Arid-tropical.* p. 408

obesa (So. Africa), "Baseball plant"; beautiful small globes to 12 cm, in male and female plants, gray-green marked with reddish length and cross stripes, and rows of small knobs along ridges. *Subtropic.* p. 421

oncoclada (Madagascar); thornless succulent shrub with cylindrical gray-green stems 2 cm thick, the segments tapering to both ends; tiny leaves and minute yellowish flowers. *Arid-tropical.* p. 422

polyacantha (Ethiopia), "Fish bone cactus"; sparry branching plant to 1½ m, with 4-5 angled leafless stems grayish green, horny angles crenately toothed and spiny. *Subtropic.* p. 424

pulcherrima (So. Mexico), "Poinsettia" or "Flor de Noche Buena"; branching shrub to 3 m high, with woody trunk and milky juice, ovate leaves, deciduous when disturbed or resting, the terminal shoots forming dark red-velvet, lanceolate bract leaves, surrounding the tiny yellowish flowers during the short-day period of year. Buds will be initiated about Sept. 21-25 at latitude 40-45 deg. north, provided the night temperature is below 18°C. *Tropical.* p. 423, 426

pulcherrima 'Annette Hegg' (Norway 1964), "Christmas poinsettia"; durable variety of compact habit; freely branching; wiry stems; ovate leaves long lasting; blooms early; does well at temperate 16°C; develops broad, smooth bracts vivid medium red, even under low winter light intensity. *Tropical.* p. 426, 427

pulcherrima 'Annette Hegg Supreme'; introduced 1970, a bright red sport of 'Dark Red Hegg', with the most brilliant crimson-red color of the multiflower types, especially under artificial light; the bracts very substantial, and of larger size than 'Annette Hegg', forming stars with close-in center, according to my measuring in our green houses, of 40 cm across at Christmas time. *Tropical.* p. 426

pulcherrima 'Eckespoint' (Paul Ecke 1967); strain of striking diploid hybrids of medium height, characterized by stocky growth, strong self-supporting stems, vigorous oakleaf foliage and a close-in circle of lush bracts lightly crinkled. The variety 'C-1 Red' has bracts intense scarlet-red with rosy sheen, and is ideal for Christmas bloom. Needs some warmth, feeding, and attention to moisture to best hold foliage. *Tropical.* p. 427

pulcherrima 'Ecke's White' (E. pulch. alba cultivar), a "White poinsettia"; shapely, freely branching and late flowering seedling with bright green, ovate leaves, the flexible, wiry stems producing perfect close disks of creamy-white, membranous bracts; if kept cool, will keep both leaves and bracts from Christmas to Easter. *Tropical.* p. 426

pulcherrima 'Henriette Ecke', "Double poinsettia"; sported in 1927 in Hollywood, blooming early December, and with broader vermilion horizontal bracts and "double" crown. *Tropical.* p. 427

pulcherrima 'Mrs. Paul Ecke', a sport of 'Oakleaf' (1926); with shorter and heavier stems, not as freely branching, thicker leaves and wider fleshier bracts of blood red, brighter than 'Oakleaf'. *Tropical.* p. 427

pulcherrima 'Oakleaf' (1918 seedling); vigorous growing tall and slender, freely branching seedling with oakleaf-type lobed leaves and somewhat long, narrow, stalked bracts of darkish red. *Tropical.* p. 426

pulcherrima plenissima, "Double poinsettia"; wild in Mexico; large shrub with flexible branches and grayish-green leaves producing a striking inflorescence composed of a circle of vermillion red lanceolate bracts, as well as a bushy central head of smaller bracts, transformed from flowers; normally blooms in late winter. *Tropical.* p. 426

pulcherrima 'Praecox'; variety of very compact habit, used for outdoor planting in Spain and the Canary Islands. *Tropical.* p. 427

pulcherrima 'Rosea', "Pink poinsettia"; sparry plant with pale green ovate leaves, green petioles and veins, and full, but smallish heads of fleshy, obovate bracts of delicate rose, with darker veining; flowering early. The cultivar 'St. Louis' is a sport of the pink poinsettia. 'Ecke Pink' is a clone of rosea. *Arid-tropical.* p. 426

pulcherrima fol. variegata, the "Variegated-leaf poinsettia", interesting variety with delicate, pale gray-green, lobed leaves variegated white along margins, flowering bracts carmine red. *Tropical.* p. 426

pulvinata (So. Africa), "Cushion euphorbia"; low, clustering succulent with many even branches 5 cm high, mostly 7-ribbed and lightly notched, dark green, numerous wine-red blunt spines to 15 cm long. *Subtropic.* p. 424

punicea (West Indies: Jamaica, Cuba, Bahamas); tree-like to 9 m high, the branches with leaf-scars and with obovate, sharp-pointed leaves toward the ends, glaucous beneath; terminal clusters of flowers subtended by bright crimson-red bract-leaves 8 cm long. *Tropical.* p. 418

quadrialata (Tanzania); symmetrical tree with straight, angled maintrunk, branching laterally more or less in whorls like a Christmas tree; the green branches 3-angled. *Tropical.* p. 422

robecchii (Kenya); large tree looking like telegraph poles, as seen on the steppe in Tsavo National Park, in candelabra shape to 14 m high, the horizontal branches in distinct groups, 3-angled and twisted, 3 cm or more thick; leaves scale-like, the spines in pairs. *Arid-tropical.* p. 425

splendens: see milii

stenoclada (Madagascar), "Silver thicket"; beautiful leafless but forbidding shrub 2 m or more high, densely bushy and flattened silvery gray limbs forking into vicious long-pointed spikes and thorns. *Tropical.* p. 422, 425

tirucalli (Uganda, Zaire, Zanzibar), "Milk bush"; forming tree to 10 m high, branches cylindrical and pencil-thick, glossy green and bursting with poisonous milk; narrow deciduous leaves. *Tropical.* p. 422, 425

tuberosa (W. Cape Prov.); perennial, fleshy herb with large tuberous root, and short basal stem set with rosettes of concave-folded or undulate lanceolate, soft-leathery grayish 5 cm leaves; forming large globular capsules at first reddish, then greenish-gray. *Subtropic.* p. 424

vigueri (W. Madagascar); succulent shrub 30 cm to more than 1 m high, the dark green stem thickened and 2-3 cm dia., with 6 angles, set with light brown spines; large obovate leaves tufted at apex of branches 8-10 cm long, emerald green with whitish midrib, the petiole red; bract flowers scarlet. *Arid-tropical.* p. 418

x 'Zigzag' (grandicornis x pseudocactus) x pseudocactus; "Zigzag cactus"; erect 3-angled, rather compressed column growing more or less upright in hither-thither style; 3 cm thick, yellow-green and prettily painted dark green in herringbone pattern, the wavy edges meandering upward in zigzag fashion, corky gray-brown and set with pairs of sharp spines. A worthy California (Hummel) hybrid involving ³/₄ E. pseudocactus and ¹/₄ grandicornis; quite tough and durable; slow-growing. *Subtropic.* p. 421

EUPRITCHARDIA: see Pritchardia

EURYA   *Theaceae*
japonica (Cleyera) (Temp. E. Asia, Himalayas); small evergreen shrub or tree, with elliptic, smooth, leathery leaves to 6 cm long; small axillary flowers a greenish white, less than 1 cm across, with unpleasant odor; small berries. Photo taken in Strybing Arboretum, San Francisco; not to be confused with Cleyera japonica (see Hortus III). *Warm temperate.* p. 906

japonica 'Variegata' (Japan, Korea); very neat evergreen bush of slow growth, the glossy, oblique-elliptic or obovate-pointed, leathery leaves remotely toothed, dark and milky-green, beautifully variegated creamy-white from the margin toward the center, decoratively arranged and densely overlapping on willowy, spreading branches and woody stems; an excellent keeper. *Subtropic.* p. 904

EURYALE   *Nymphaeaceae*
ferox (Tropical and Subtropical India to China and Japan); a widespread aquatic perennial with large circular floating bright glossy green leaves 1 to 2 m across, very spiny and with inflated bubbles, and spiny-ribbed beneath; small 5 cm fl. green outside, purplish red within, barely above the water and short-lived; the seeds are edible when roasted. Much at home in rice paddies in the Orient; stunning in the conservatory pool. *Humid-tropical.* p. 705

EURYOPS   *Compositae*
pectinatus (So. Africa), the "Resin bush"; shrubby, free blooming perennial with thick green leaves covered by white pubescence, pinnatifid and cut into narrow segments, strongly scented when bruised; daisy-like flowers with firm, pretty ray flowers sky blue, surrounding a yellow disk, 5 cm across; blooming nearly year-round. *Subtropic.* p. 315

EUTERPE   *Palmae*
edulis (Brazil), "Assai palm"; slender feather palm with graceful trunk to 30 m high, topped by prominent crown shaft and the crown of arching pinnate fronds 2 to 3 m long, dense with drooping, thin, plaited leaflets, on scaly rachis; small round fruit. Usually forming multistemmed clusters. *Tropical.* p. 774, 781

globosa (West Indies: Cuba, Puerto Rico, to Grenada), the "Mountain palm" or "Palma de Sierra" as it is known in the Luquillo mountain region of Puerto Rico; handsome palm to 12 m high, with ringed trunk bearing rather few, broad pinnate fronds to 2½ m. *Tropical.* p. 781

oleracea (Brazil), "Assai palm"; slender single trunk to 20 m tall, 12 cm in dia.; crownshaft red-green, to 1 m long or more; graceful spreading fronds pendulous, 50-80 on each side; white inflorescence; purple-black fruit, 1 cm in dia. *Humid-tropical.* p. 781

EXACUM   *Gentianaceae*
affine (Socotra Is.), "Mexican violet"; bushy little herbaceous, free-flowering biennial with waxy, stalked, ovate leaves and tiny, wide open, bluish-lilac, star-like, fragrant flowers with pretty eye of deep yellow stamen. *Tropical.* p. 484

EXCOECARIA   *Euphorbiaceae*
bicolor (Vietnam), "Picara"; smooth shrub with milky sap used as fish poison; shiny green, leathery leaves 12 cm long, and red beneath; small flowers in narrow spikes. *Tropical.* p. 421

FAGRAEA   *Loganiaceae*
fragrans (India to Malaya), "Tembusu"; tropical tree to 25 m, with elliptic leaves 15 cm long, and showy white flowers, fragrant and blooming at night. *Tropical.* p. 616

x FATSHEDERA   *Araliaceae*
lizei, "Ivy tree". "Miracle plant" or "Botanical wonder"; bigeneric hybrid between Fatsia japonica 'Moseri' and Hedera helix hibernica, the "Irish ivy". Evergreen shrub combining characters of both parents, growing erect over 2 m high if the woody stem has support; leathery leaves 5-lobed, similar to English ivy but larger, often 12-20 cm wide, dark lustrous green. Small light green flowers in dense clusters. *Subtropic.* p. 138

lizei 'Variegata'; "Variegated pagoda tree"; horticultural form with the fresh green leaves prettily variegated and edged cream-white. Prefers moist-temperate climate. *Subtropic.* p. 138

FATSIA   *Araliaceae*
japonica (Aralia sieboldii) (Japan), "Japanese aralia"; evergreen shrub with leathery, palmately lobed leaves dark, shining green, the broad lobes pointed and toothed; flower milky-white. *Subtropic.* p. 132, 133

japonica 'Moseri'; a robust French seedling of lower, more compact habit, and with more and larger, rich glossy green, deeply lobed, crenate leaves having yellow veins; pubescent beneath and on stem, turning woody. *Subtropic.* p. 137

FAUCARIA   *Aizoaceae*
tigrina (So. Africa: Cape Prov.), "Tiger jaws"; ferocious-looking small succulent with opposite, thick, keeled or boat-shaped, gray-green leaves 3-5 cm long, and marked with numerous white dots; margins armed with stout recurved teeth so that the leaf-pairs resemble a gaping jaw. Large golden yellow sessile flowers. *Arid-subtropic.* p. 48

tuberculosa (So. Africa: Cape Prov.), "Pebbled tiger jaws"; striking little succulent with wide-open jaws of dark green thick leaves about 2 cm long, upper side covered with white tubercles, the edges armed with stout teeth, underside keeled; 4 cm flowers yellow, opening in the afternoon. *Arid-subtropic.* p. 48

FEIJOA *Myrtaceae*

deflexa (So. America); tall erect evergreen shrub, in the California nursery trade; woody brown stems with large lanceolate leathery leaves 15-20 cm long, depressed veins and jagged saw teeth along margins. *Subtropic.* p. 686

sellowiana (So. Brazil, Paraguay, Uruguay, No. Argentina), "Pineapple guava"; small tree to 5 m high, dense with whitish-felted branches and small thick 6 cm oval, opposite leaves waxy dark green above and lightly rugose, midrib white, white to brown-tomentose beneath; flowers with fleshy petals white tomentose outside, purplish inside, and dark red stamen; green edible fruit tinged red, with guava flavor. *Subtropic.* p. 686

sellowiana 'Nazemetz', "Pineapple feijoa"; selected cultivar grown in California, with larger, thicker fruit 8-10 cm long, and of a special sweet flavor. *Subtropic.* p. 463

FELICIA *Compositae*

amelloides 'Astrid Thomas'; a "Blue marguerite" of compact habit, a charming mutation of the So. African species, in the California nursery trade, with fresh green, thick leaves; the flowers with firm, broad ray florets a pretty sky-blue, surrounding the yellow center cushion. *Subtropic.* p. 315

amelloides 'Variegata', variegated "Blue daisy"; low spreading, shrubby plant with small green leaves prettily bordered with white; flowers sky blue with yellow center; an attractive ground cover in climates such as California. *Subtropic.* p. 315

fruticosa (So. Africa), "Aster bush"; much branched shrub to 75 cm high, with crowded, small linear leaves; 3 cm aster-like heads with purplish-blue ray flowers and yellow disk. *Subtropic.* p. 315

FENESTRARIA *Aizoaceae*

aurantiaca (Cape Province), "Window plant"; clustering succulent with club-like gray-green glaucous leaves 6 cm tall, the pearly flat top whitish and with translucent windows; flowers golden yellow. *Arid-subtropic.* p. 42, 48, 49

rhopalophylla (S.W. Africa), "Baby toes"; tufts of cylindrical, succulent leaves thicker above and bearing a translucent "window" at top; flowers white; wants dry, in sandy soil with lots of lime. *Arid-subtropic.* p. 44

FERNS: see FILICES in Pictorial Section

FEROCACTUS *Cactaceae*

acanthodes (S. Nevada, Calif., Baja California), "Fire barrel"; globular plant glaucous green, becoming cylindric to 3 m tall, very spiny, ribs up to 27, the 12 cm long curving awl-shaped spines pinkish to red; flowers yellow to orange. *Arid-subtropic.* p. 264, 265, 267, 268

covillei (Mexico); globular to barrel type to 1½ m, dull green, 22-32 ribs, strong spines variable in color, red to white; flowers red, tipped yellow. *Arid-tropical.* p. 265

fordii (Baja California); first globular, later elongate body to 2.5 m high, covered with masses of stiff spines; flowers yellow to red. *Arid-tropical.* p. 263

horridus (peninsulae) (Baja California), "Fishhook cactus"; globular to elongate, to 1 m high, with 13 fish ribs, glaucous gray; long fish-hook spines and flattened radial bundle; flowers yellow to red. *Arid-tropical.* p. 263

latispinus (Mexico), "Fish-hook barrel"; depressed globe to 40 cm high, dull grayish-green, with many dented folds, the central spines very broad, curving and crimson-red; bellshaped flowers rose to purple. *Arid-tropical.* p. 268

orcuttii (Baja California); viciously spiny barrel cactus to 1.3 m high with long, stout spines; from center a cluster of brownish flowers, yellow inside. *Arid-tropical.* p. 263

FESTUCA *Gramineae*

ovina glauca (No. Temp. Europe and Asia), "Blue fescue"; low tufted ornamental grass with thread-like, linear leaves rolled inward or circular, 15-20 cm long, silvery-blue-glaucous and almost wiry-stiff. *Warm temperate.* p. 510

FICUS *Moraceae*

aspera (parcellii) (South Pacific Islands), "Clown fig"; small shrub with sparry branches bearing large oblong, slender pointed leaves to 20 cm long, toothed, rough hairy, grass-green and milky-green, wildly variegated and marbled ivory white; even the pear-like fruits are variegated white. *Tropical.* p. 648, 650

aurea (So. Florida), "Strangler fig"; epiphytic at first, growing into a tree 20 m high, leaves oblong, to 10 cm, and narrowed at both ends, but round-tipped, leathery, green, on buff branches; fruits orange. *Subtropic.* p. 648

auriculata (roxburghii) (India), "Ornamental fig"; low, spreading tree to 6 m high, very showy for its large foliage on slender woody branches, the light brown petioles covered with white hairs; big papery, rounded and toothed, slightly glossy leaves to 40 cm long, their surface covered with a fine pubescence, and with depressed veins; when young an attractive mahogany-red. Big flat-globose figs are borne on the stems. Not a house plant as foliage easily catches red spider and drops. *Tropical.* p. 653, 660

barteri (Nigeria); interesting as ornamental, tree to 7 m high, with unusually long narrow, oblanceolate leaves 30-50 cm long; orange colored edible, axillary fruit 2½ cm dia. *Tropical.* p. 653

benghalensis (India, Ceylon), "Banyan tree"; large tree, the top spreading by aerial roots which become secondary trunks; leathery leaves ovate or elliptic, to 20 cm long, dark green, with yellow-green veins, on pubescent stem; round red fruits. *Tropical.* p. 651

benghalensis krishnae (India, Pakistan), the "Sacred fig tree", probably a form of benghalensis; small tree sacred to the Hindu god Krishna, said to have folded its leaves into cups to collect drops of dew, when the god was thirsty in the desert; leathery, irregularly cupped, deep green leaves with raised ivory ribs, finely pubescent inside, on grayish branches. Very interesting collection plant because of its curious leaves. *Tropical.* p. 648, 656

benjamina (India, Malaya), the graceful "Weeping fig"; a beautiful tropical tree of dense growth, forming aerial roots, and with branches of somewhat pendant habit; the pendulous, shining deep green leaves long ovate, slender pointed 8-10 cm long; small round fruit blood-red when ripe. One of the most attractive tubbed decorators for tropical effect, preferring smallish containers, and rejoicing in warmth and good light, 40 fc. and up to 500 fc., the more light the more leaves. Allow surface to become dry, then water; tolerates air-conditioning. *Tropical.* p. 649, 651, 654, 655, 657

benjamina comosa (glomerata) (Burma, Philippines); large tree with small ovate leaves; fruit larger than benjamina, 2 cm long, orange yellow. The tree photographed in Fairchild Gardens labeled comosa has cauliflorous fruit on main branches. *Tropical.* p. 653

benjamina 'Exotica' (Java, Bali), "Java fig"; especially graceful, weeping form with slender, arching branches, and smaller, rather narrow and pendulous glossy green, leathery, oblique 9-12 cm leaves, which are given special charm by the coquette twist of their slender tip; small red berries; thrip-proof. I have seen large trees of this variety quite common in Bali. Good indoor decorator tub plant but needs at least 50 fc of light; some leaves may drop at first when temperature changes, or a tree is moved to air-conditioning. *Tropical.* p. 650, 651, 655

benjamina 'Variegata', "Variegated mini-rubber"; handsome cultivar grown in Florida and California; charming as bushy little tree, with slender brownish stem and small elliptic leaves 6-8 cm long, glossy light green, prettily margined and variegated with ivory-white. *Tropical.* p. 656

brandegeei (Mexico); a curious xerophytic tree from Isla Ildefonso, Baja California, with beige trunk from a swollen base bottle-like; grayish green, leathery leaves, broad ovate with cordate base; young plants may be dwarfed by retarding growth and kept as bonsai. *Arid-tropical.* p. 656

capensis (Cape Prov. to Ivory Coast and Sudan), "Cape fig"; wide spread cauliflorous tree to 12 m, with broad ovate, smooth, thin-leathery leaves to 22 cm long, larger than sycomorus, prominent pale ribs, margins coarsely serrate; small 2 cm edible green figs are dangling profusely on leafless stalks from the trunk and larger branches. *Tropical.* p. 653

carica (Mediterranean region, Asia Minor), "Common fig"; broad, irregular, deciduous tree to 10 m high, much planted for its sweet, pear-shaped fruit; leaves thick, rough above, deeply 3-5 lobed and with palmate veining. *Subtropic.* p. 456, 651, 660

carica 'Mission', "Black Mission fig"; medium large pear-shaped fruit with purplish black skin and light strawberry-colored flesh; it has excellent flavor and good for fresh and dried fruit. *Subtropic.* p. 459, 660

cordifolia (South India: Madras), "Strangler fig"; large tropical fig-tree with milky latex, gray bark, and leathery, heart-shaped leaves. Remarkable because, like several other species, it has its start as an epiphytic seedling on a convenient tree, and over the years growing larger and more bold, its roots embracing the host tree, eventually strangling and killing it. *Tropical.* p. 657

diversifolia (lutescens) (India, Malaya, Java), "Mistletoe fig"; woody shrub with small obovate, hard leaves 5 cm long, dark green with brown specks above, pale beneath; liberally bearing small yellowish fruit lined with gray. *Tropical.* p. 653

dryepondtiana (Zaire), "Congo fig"; brown woody stem with stiff-oval, quilted wavy leaves metallic deep olive green, the underside red-purple. A very attractive colorful species. *Tropical.* p. 650

elastica (India, Malaya), "India rubber plant"; durable old houseplant, but grows into large trees to 30 m high, and rooting from the branches; young plants erect, with brown woody stem and thick-leathery oblong leaves glossy deep green, the young leaves enclosed in a rosy sheath; yielding latex-bearing milky sap. *Tropical.* p. 650, 652

elastica 'Decora' (Indonesia), "Wideleaf rubber plant"; superb commercial seedling, of bold habit, larger and much broader, heavier leaves, 25-35 cm long, deep glossy green, with prominent lateral depressed veins, at right angles to the ivory midrib which is red beneath, the sheath at the growing tips is red. *Tropical.*
p. 652, 658

elastica 'Doescheri' (New Orleans 1925), "Variegated rubber plant"; outstanding variegated cultivar having a striking range of colors, from green with gray into white and cream-yellow, while midrib and leafstalks are pink; coloring is stable and not reverting to green. *Tropical.* p. 650

elastica 'Schryveriana' ('Schrijvereana'); robust Belgian cultivar (1959), of the broad-leaved 'Decora', with glossy, deep-green leaves irregularly variegated gray-green in the center, light green, sea green and cream or light yellow toward the marginal area, the midrib cream, the petioles red. *Tropical.* p. 652

elastica 'Variegata', the common "Rubber plant", in variegated form, the leathery leaves are usually variegated gray and edged in creamy-yellow. *Tropical.* p. 652

gibbosa (parasitica) (Indochina: Cambodia; Malaya, Ceylon); an epiphyte in young state, growing in the forks of trees, or as I have observed in Cambodia, on the stones of the ruins of Angkor, surrounding its host with strangling roots and becoming a large tree with spreading branches, bearing the smallish narrow elliptic, oblique leaves 4-15 cm long, often yellowish beneath; pear-shaped 8 cm figs reddish orange. *Tropical.* p. 657

gibbosa batlinii; variety with twigs bearing glossy green, leathery leaves 10-12 cm long, some with one side having one pointed lobe. *Tropical.* p. 656

gilletii obovata (Zaire); big forest tree with long-stalked, obovate leaves 10-12 cm long, thin-leathery, rich green. *Tropical.* p. 656

heteropoda (Moluccas, New Guinea); large tree with gray bark, twigs hollow when young; opposite broad oval leaves with dentate margins; pear-like fruit borne from tubercles on the main trunk. *Tropical.* p. 653

lyrata (pandurata) (Trop. West Africa), "Fiddleleaf plant"; close headed tree to 12 m high, with large, thick-leathery leaves 30 to 60 cm long, fiddle-shaped, wide rounded apex, deep waxy green, quilted and wavy, with attractive yellow-green veins, on woody stem; fruits with white dots. *Tropical.* p. 652, 654

macrophylla (Queensland, New South Wales), the "Australian banyan" or "Moreton Bay fig"; large tree with ovate to broad oblong leathery leaves blunt at apex, to 25 cm long and 10 cm wide, cordate at base, glossy green with pronounced ivory veins, shiny light green underneath, netted. *Subtropic.* p. 652

mysorensis (So. India, Burma); large tree with twigs tomentose becoming glabrous, large leathery, rich green, ovate leaves to 30 cm long, glossy under the loose pubescence, and with pale yellowish veins; fruit orange-red; tolerates some frost. *Tropical.* p. 656

nekbudu (utilis) (Trop. Africa), "Zulu fig"; large forest tree, young growth pubescent, fresh green leathery leaves thick, oval to obovate, 15-40 cm long, rounded at both ends and pretty, yellow ribs; good examples of their shapely form are the old specimen in front of the Capitol in San Juan, Puerto Rico. *Tropical.*
p. 653, 657

nitida (Malaya), "Indian laurel"; attractive, glabrous thick-topped banyan tree forming aerial roots and buttressed trunk in habitat; with erect banches and small, rubbery, elliptic leaves 10 cm long, waxy, nice green and very smooth; can be shaped into pyramids and standards resembling a true laurel. Good tub plant; also widely used as a containerized street tree in our Southwest. In the interior of buildings this sun-lover requires 50 fc. to 500 fc.; the more light, the more leaves it can hold. *Subtropic.* p. 654, 658

palmeri (Baja California, Mexico); a magnificent xerophytic, desert-climate tree to 20 m tall, growing on rocky cliffs as a strangler fig; base swollen and bulbous; cordate 15 cm leaves soft-leathery grayish green with ivory ribs, velvety beneath; small yellowish figs. *Arid-tropical.* p. 656

petiolaris (Mexico), "Blue Mexican fig"; small tree to 1³⁄₄ m high, developing very wide, swollen base; attractive, heartshaped, leathery, wavy leaves with long pointed tip, metallic blue-green, with ivory pink to showy red veins and petioles. (syn. F. jaliscana). *Arid-tropical.* p. 652

pseudopalma (Philippines), "Dracaena fig"; showy tree to 10 m high, with thin-leathery palm-like leaves 60-90 cm long, the stiffish foliage clustered near the top, acuminate and long-tapering toward the base, and coarsely notched, the surface somewhat glossy. *Tropical.* p. 661

pumila (repens, stipulata) (China, Japan, Australia), "Creeping fig"; freely branching creeper with small, obliquely cordate, dark green leaves less than 3 cm long, clinging to walls by roots like ivy and then flattened; fruiting branches erect, with stiff, much larger, oblong leaves to 10 cm long. *Subtropic.* p. 650, 655, 658, 660

racemosa (glomerata) (South India); large tree with ovate, glossy green leaves 15-20 cm long; the yellow-orange, edible fruit cauliflorous from main trunk and branches. *Tropical.* p. 661

radicans 'Variegata', "Variegated rooting fig"; attractive variegated creeper much grown under glass, with lanceolate, grayish green leaves irregularly marked with creamy-white, the variegation beginning at the margin. *Tropical.* p. 650, 660

religiosa (India), "Bo-tree"; sacred to Hindus and Buddhists, and under which Buddha received enlightenment; large glabrous tree to 30 m high, with gracefully pendant, thin, bluish-green leaves, heartshaped with tail-like driptip, ivory or pinkish veins, on slender branches; purple fruit. *Tropical.* p. 650, 651, 659, 660

retusa (So. China, Macao, Philippines), "Chinese banyan"; shapely tree with dense foliage, the branches first ascending but becoming pendulous, with small leathery leaves broadly obovate and waxy; ideal for shaped standards. *Subtropic.* p. 654, 658

roxburghii: see auriculata

rubiginosa (australis) (New South Wales, Queensland), "Rusty fig"; broad tree, spreading by means of aerial roots like the banyan; small leathery 8-15 cm leaves oval or elliptic, rusty-pubescent beneath. *Subtropic.* p. 649, 652

rubiginosa fa. australis (New South Wales); evergreen tree with spreading willowy branches, in this form lacking the densely pubescent, rusty-downy character of the foliage; soft-leathery glossy dark green, oval to obovate leaves, with pale midrib, about 8 cm long in container-grown or smaller plants, to 20 cm long as trees in habitat under favorable moist conditions. *Subtropic.* p. 654, 655

rubiginosa 'Variegata'; a pretty miniature rubber plant with graceful branches and small, egg-shaped, deep green, leathery leaves richly marbled and edged yellowish-cream; tomentose during juvenile stage only. *Tropical.* p. 650, 652, 654

saussureana (Brazil, Guayana); sparry tree with stems, young branches, and reverse of foliage cinnamon-brown; long obovate leaves to 30 cm long by 10 cm wide, with rough surface, dark-green with light green veins. *Tropical.* p. 656

sycomorus (Syria, Egypt, Sudan, south to the Transvaal), "Sycamore fig"; round headed, freely branching tree to 12 m high, partially deciduous. In Africa, one sees them often along rivers, very attractive with their pale fawn-yellowish trunk and shapely head; ovate, rugose leaves with hairy ribs, 8-10 cm long, soft when young, almost bluish-green; when mature with leathery "feel" but quickly desiccating, fresh green to deep olive, blunt apex; greenish flowers and small, abundant edible fruit produced from the trunk and branches. *Arid-tropical.* p. 659

triangularis (Trop. Africa); small evergreen ornamental tree with curious triangular, thick-fleshy, dark green leaves 5 cm long, the margins rolled down; masses of small 1 cm greenish beige berries are growing cauliflorous directly from the woody branches. *Tropical.* p. 653, 659

variolosa (Hongkong); sparry tree inhabiting the mountains of Hongkong, with thick-leathery oblong leaves 15-20 cm long, the globular 3-4 cm fruit cauliflorous on old wood. *Subtropic.* p. 653

wildemanniana (cyathistipula) (Trop. Africa); introduced by Vosters as 'panduriformis'; sturdy plant small enough to look good in a 10 cm pot; leaves are glossy dark forest green and very leathery oblique elliptic, and with their ivory midrib and lateral veins look very attractive. *Tropical.* p. 650, 654

FILIPENDULA   *Rosaceae*
vulgaris (Spiraea filipendula) (Europe, Asia), "Meadow sweet" or "Dropwort"; hardy perennial with tuberous rootstock; pinnatifid, fern-like leaves; flowers white tinged with red, in large panicles. *Temperate.* p. 849

FIRMIANA   *Sterculiaceae*
simplex (platanifolia) (E. Asia: Okinawa to Vietnam), "Phoenix tree" or "Chinese parasol tree"; large semi-deciduous tree to 20 cm high, with smooth trunk; decorative leaves palmately 3-5 lobed to 30 cm across, with contrasting yellow ribs; flowers without petals, but with showy yellow calyx 2 cm across; fruit opening into 5 papery sections displaying seeds. *Subtropic.* p. 902

FITTONIA   *Acanthaceae*
verschaffeltii (Perú), "Mosaic plant"; a pretty tropical ground cover; low, creeping herb with hairy, rooting stems and colorful oval 8-10 cm leaves dark olive green and entirely covered with a network of deep-red veins. Although the leaves are somewhat papery they are fairly sturdy. Small yellow, green-bracted flowers. *Tropical.* p. 34

verschaffeltii argyroneura (Perú), "Nerve plant"; charming tropical herb, spreading and hugging the ground, with flat, papery, oval leaves vivid green and beautifully netted with white veins. Likes warmth and high humidity. *Tropical.* p. 34

FLACOURTIA *Flacourtiaceae*
indica (Trop. Asia, Madagascar), "Governor's plum" or "Ramontchi"; fruit-bearing shrub or tree to 8 m high; the branches with long slender spines and ovate, glossy green, toothed leaves to 8 cm long, partially deciduous; small yellowish flowers in leaf axils, followed by the globose maroon-red fleshy fruit to 2 cm dia., tipped with short radiating styles; the juicy, edible pulp surrounding several flattened stones. *Tropical.* p. 458, 484

FOCKEA *Asclepiadaceae*
capensis (So. Africa); curious dioecious succulent with turnip-shaped root, forming with age, a tremendously thick, swollen caudex, from the apex of which rise thin woody, twining branches, with tiny opposite holly-shaped but fleshy, undulate leaves grayish green. *Arid-subtropic.* p. 149

FORTUNELLA *Rutaceae*
hindsii (Hong Kong, Kwangtung), the "Dwarf kumquat"; ornamental, small spiny tree, with leathery green, oval-obovate leaves, and nearly round, small 1 cm orange-red fruit of 3-4 cells, almost without juice. *Subtropic.* p. 482
japonica (South China), "Marumi kumquat"; small shrub usually spiny, with elliptic leaves; white flowers; small round fruit 3 cm dia., deep orange; edible with sweet rind and acid juice. *Subtropic.* p. 482, 874
margarita (Kwangtung), "Nagami kumquat"; small, vigorous, bushy citrus, often grown as a standard in tubs, with slender, erect, angled branches, long-pointed, shining dark green leaves, fragrant white flowers, and small, rather persistent, oblong, golden orange fruit of 2 cm dia., with finely-flavored pulp, produced prolifically in Oct.-Jan. and longer. The sweet fruit is used for preserves. *Subtropic.* p. 482, 873

FOUQUIERIA *Fouquieriaceae*
splendens (N. Mexico, Arizona, So. California), the "Ocotillo" of the desert; a xerophytic shrub with many whip-like gray, furrowed basal stems, to 6 m; rigid spines bear clusters of small oval, deciduous leaves in their axils; and at the very tips of branches the brilliant-red flowers in showy racemes to 25 cm long. *Arid-subtropic.* p. 430

FRAGARIA *Rosaceae*
chiloensis (Coastal Alaska to California and So. America), "Sand strawberry"; ground cover forming low mats, with rooting runners; leaves of 3 leaflets, with toothed margins; 2 cm white flowers followed by bright red, seedy fruit 2 cm dia. *Warm temperate to subtropic.* p. 849

FRANCISCEA: see Brunfelsia

FRAXINUS *Oleaceae*
uhdei (Mexico), "Evergreen ash"; charming evergreen to semi-evergreen tree to 15 m high, with widely spreading crown and pendulous branches; pinnate leaves with 5-9 lanceolate leaflets glossy deep green, 10 cm long, the margins with fine teeth; very popular in California; also planted along the Paseo de la Reforma, Mexico City. *Subtropic.* p. 696

FREESIA *Iridaceae*
x hybrida; large-flowered strain of such South African species as refracta and armstrongii; bulb-like corms with linear leaves and very fragrant, firm, funnelform flowers on bent wiry spikes, creamy-white through orange to lilac-blue; August blooming unless forced. *Subtropic.* p. 526

FREMONTODENDRON *Sterculiaceae*
mexicanum (Fremontia) (San Diego and Baja California), "Southern flannel bush"; evergreen shrub or small tree to 6 m, with spur-like branches bearing 5-lobed leaves, dark green above and covered with felt beneath, 4-8 cm long; flowers with showy orange-yellow calyx to 9 cm across, but without petals; black seeds. *Subtropic.* p. 903

FREYCINETIA *Pandanaceae*
arborea (Hawaii), "Climbing screw pine"; woody climber with brittle, rooting stems twining into treetops; every metre new branches will form, ending in tufts of narrow, spiny leaves tapering to a point, their center produces a curious, ornamental inflorescence of fleshy rosy-red bracts surrounding the purplish-red spadix. *Tropical.* p. 808
funicularis (insignis), (Indonesia: Java); a beautiful climbing screw-pine with long flexuous stems; sword-shaped leathery leaves glossy green with very few minute marginal spines; the terminal inflorescence very showy, with large triangular bright red bracts and cylindric yellow spadices in center. *Tropical.* p. 810
multiflora (Philippines), "Climbing pandanus"; climbing tropical shrub with slender rooting stems becoming pendant from trees; sessile leathery, black-green, linear lanceolate keeled leaves, and showy terminal inflorescence with glowing salmon-rose bracts; and with rosy spadix. *Tropical.* p. 810
rigidifolia (Thailand, Malaya, Borneo); climbing dioecious screw-pine with leathery leaves to 25 cm long and 14 mm wide; at apex the attractive inflorescence with 6-8 cm red bracts, the cylindric spadix pale yellow. *Tropical.* p. 810
stenophylla (New Guinea); woody climber we collected in the mountains of N.E. New Guinea, with dense, glossy green leathery leaves; at blooming time, the terminal growth turns a vivid orange. *Tropical.* p. 810

FRITHIA *Aizoaceae*
pulchra (Transvaal), "Purple baby toes"; pretty, clustering rosette similar to Fenestraria, with club-like succulent gray leaves under 3 cm high, the flat top with translucent windows; large carmine-purple flowers with white center. *Subtropic.* p. 48

FRITILLARIA *Liliaceae*
imperialis (Iran, W. Himalaya), "Crown Imperial"; simple-stemmed stout, strong-smelling herb to 1 m, with scaly bulb, lanceolate leaves, with some in a terminal whorl above the nodding 5 cm flowers, with 6 veined segments purplish, terracotta or yellow, borne on a stiff-erect, purple-spotted stalk. *Subtropic.* p. 602
meleagris (Britain, Norway, C. Europe to Caucasus), "Checkered-lily" or "Snake's head"; bulbous perennial to 45 cm high, with scattered linear to oblanceolate glaucous leaves along middle of stem; solitary bell-shaped 5-8 cm flowers, purple with white checkering, or white with green network of veins. *Warm temperate.* p. 602
pudica (Brit. Colombia to Utah), "Yellow bell"; bulbous perennial with stem to 30 cm; linear leaves to 20 cm long; orange-yellow bell flowers. *Warm temperate.* p. 598

FUCHSIA *Onagraceae*
excorticata (New Zealand), called "Tree fuchsia", aptly descriptive, becoming a tree to 15 m high, with woody trunk to 60 cm thick, having loose, papery bark and brittle branches, thin dark green leaves lanceolate 5-12 cm long, silvery beneath; small waxy flowers with spreading calyx lobes first yellowish then deep dull red, the petals deep purple with blue stamens. *Subtropic.* p. 701
x hybrida (speciosa), a group name for most of our cultivated fuchsias, nick-named "Lady's eardrops"; and derived from such species as fulgens, magellanica and corymbiflora, having sturdier twigs, broader ovate leaves to 10 cm long, and large gay flowers, always pendant, with long showy stamens. Most fuchsias can be had in flower the year round, but most profusely during the season of warm long days from March to November, and they generally perfer a cool, moist climate. *Humid-subtropic.*
hyb. 'Caledonia'; an almost hardy variety of lax habit and free blooming, small single flowers with long tube and sepals red-cerise, the corolla reddish-violet. *Subtropic.* p. 701
hyb. 'Cascade'; ideal basket plant with trailing branches, and single flowers with deep carmine-red petals and well recurving white sepals flushed carmine. p. 700
hyb. 'Checkerboard' (1948); single flowers with long red tube, the spreading sepals white, corolla red; ideal for pyramids with myriads of pendulous blooms. *Subtropic.* p. 702
hyb. 'Dollar Princess'; an old floriferous pot and basket variety still in cultivation after 50 years, medium small but full plant and small, compact flowers with densely double violet-purple corolla and carmine sepals. p. 700
hyb. 'Golondrina' (1941); vigorous grower and ideally suited to training as standard; masses of long slender, pendulous single flowers, the long reflexing sepals light red, the corolla deeper carmine-red. *Subtropic.* p. 703
hyb. 'Lace Petticoats' (1952); intriguing large flowers with broad white recurving sepals and double white corolla with center petals serrated and fringed like lace with undertone of blush pink. *Subtropic.* p. 700
hyb. 'Mrs. Victor Reiter'; early blooming, good variety with lovely single flowers with slender white, spreading sepals and small carmine red corolla. *Subtropic.* p. 702
hyb. 'Pink Cloud'; free-blooming upright bush, suitable for espalier or pillar; large single flowers with short white tube, long pink sepals, and clear pink, flaring corolla. *Subtropic.* p. 700
hyb. 'Strawberry Queen' (Haag 1937); luxuriant bloomer, somewhat hardy; flowers small but profuse, single with red sepals and strawberry-red corolla. *Subtropic.* p. 701
hyb. 'Winston Churchill'; excellent commercial potplant of compact habit and a willing bloomer, with dark green leaves and

large round flowers with salmon-red sepals, and deep purplish-blue, fully double petals, on stiff stem, (spring). *Subtropic.* p. 700

lycioides (Chile); evergreen shrub to 3 m high, with stout woody branches, ovate leaves to 3 cm long; during summer with a great profusion of flowers of spreading red sepals and purplish petals. *Subtropic.* p. 701

magellanica (Perú, Chile, to Tierra del Fuego), known as the "Hardy fuchsia"; bushy, herbaceous plant growing into a shrub 2 m high, and with support on walls to 6 m, arching branches with small ovate leaves to 5 cm long, wavy-toothed at margins, and pendulous slender flowers with purplish-red calyx lobes longer than the purple-blue petals. *Warm temperate.* p. 703

procumbens (New Zealand), "Trailing fuchsia"; creeping, branching plant with slender stems, roundish 2 cm leaves, forming mats; tiny erect, 2 cm flowers with calyx tube pale orange and reflexed purple calyx lobes, petals absent; followed by bright red small berries. Good basket plant. *Subtropic.* p. 702

splendens (Mexico, Guatemala); much branched shrub to 2 m, shiny 12 cm, corrugated leaves ovate-cordate, pale green, toothed; flowers drooping, rather short, about 3 cm long, crimson red, tipped green, base swollen, compressed above, and with prominent stamens. *Subtropic.* p. 702

triphylla (West Indies: Haiti); bushy shrub to 60 cm high, with downy shoots, ovate to obovate, bronzy, toothed leaves to 8 cm long, purplish beneath; nodding 3 cm flowers in terminal racemes, the spreading tube cinnabar-red, the short corolla vivid coral-red. *Tropical.* p. 701

triphylla 'Gartenmeister Bohnstedt', the "Honeysuckle fuchsia"; a pretty descendant of this little West Indian shrub, with dark metallic green, purple veined leaves, red-purple beneath, and lustrous, slender tubular flowers of salmon-rose with inner petals orange-scarlet, on purple stems, in terminal, nodding clusters, blooming during winter over a long period of time. (1906). *Humid-subtropic.* p. 700, 702, 703

FUIRENA Cyperaceae
simplex (South America); tropical aquatic or bog plant with reed-like stems, set along their length and at the top with long-linear, glossy dark green leaves, showing their paralleled veining. *Humid-tropical.* p. 390

FURCRAEA Amaryllidaceae
foetida 'Medio-picta' (gigantea) (So. Brazil), "Mauritius hemp"; giant succulent rosette of up to 50 erect oblanceolate leaves to 2½ m long and 20 cm across, shining green and with creamy-white bands down center; tall inflorescence to 8 m long. Cultivated in Mauritius etc. for its fiber. *Tropical.* p. 61

gigantea (foetida) (So. Brazil), "Giant false agave"; giant rosette of up to 50 fleshy, sword-shaped leaves 1-1½ m long, flat, shining green and with negligible spines; forming 1-1½ m stems; stalk 9-12 m high, bearing branched inflorescence with flowers milk-white inside, greenish outside. *Arid-tropical.* p. 55, 58

gigantea 'Striata'; spectacularly showy, variegated, open rosette of broad, sword-shaped fleshy leaves 45-60 cm long, sharp-pointed and with occasional marginal spines; highly colored with stripes and broad bands of ivory and milky green over rich green. *Tropical.* p. 56

selloa marginata (Colombia), "Variegated false agave"; succulent rosette, forming trunk. Leaves sword-like, narrowed toward base, thin and flexible, glossy-green with broad cream margins, armed with vicious curved brown teeth. *Tropical.* p. 56

GAILLARDIA Compositae
aristata (Minnesota to Brit. Colombia); the "Blanket flower"; flower garden hardy perennial 60-90 cm high, more or less rough-hairy, with narrow, gray-green foliage to 12 cm long; colorful flower heads 8-10 cm across, the ray florets red with yellow tips. The plants in cultivation are called G. 'Grandiflora', which may vary in colors, in warm shades of red and yellow with orange or maroon, from June until frost. *Temperate.* p. 319, 321

x grandiflora; hybrid of aristata x pulchella; perennial with gray-green foliage more vigorous than the native species and blooming early from seed; flower heads 8-10 cm across, orange-yellow with blood-red toward blackish center cushion. *Warm temperate.* p. 321

x grandiflora 'Mandarin' (aristata x pulchella), "Blanket flower"; gorgeous perennial similar to aristata but more vigorous and easier to grow, often blooming first year from seed; to 1 m high, with rough, gray-green foliage, and showy flowerheads 8-10 cm across, the ray petals maroon-red tipped with yellow, blooming in summer. *Warm temperate.* p. 319

GALANTHUS Amaryllidaceae
nivalis (France east to Caucasus,) "Common snowdrop"; early spring-blooming bulb with 2-3 narrow basal leaves and solitary white 2 cm flowers, the short inner segments green on sinus. *Warm temperate.* p. 70

GALPHIMIA Malpighiaceae
glauca (Thryallis) (Mexico to Panama), "Gold shower thryallis"; handsome bush of open habit, 1 to 1½ m high, with scandent branches, opposite thin-leathery oblong bluish glaucous leaves to 5 cm long, and small 2 cm yellow flowers in showy, elongate clusters. Very floriferous evergreen for the warm greenhouse or conservatory. *Tropical.*

GALTONIA Liliaceae
candicans (So. Africa), "Giant summer hyacinth"; beautiful bulbous plant to 1 m high, with flat, linear leaves and showy white fragrant flowers, nodding like little bells in tall, loose clusters; summer. *Subtropic.* p. 603

GARCINIA Guttiferae
mangostana (Malaya), the "Mangosteen"; important tropical fruit tree 6 to 10 m; in Peradeniya, Sri Lanka are magnificent specimens to 30 m high; the opposite, dark green and leathery elliptic foliage is 15 to 25 cm long; purple or yellow-red male flowers with fleshy petals are 4 cm across; the brilliant purplish-crimson fruit, measuring 6 to 8 cm dia., with rind and edible segments like an orange, yellow juice and with large flat seeds; at the base with 4 persistent sepals. The white delicious pulp has a flavor between a grape and peach. Fruit usually ripens in Peradeniya in July or August, but differs according to season. *Tropical.* p. 457, 471, 520

xanthochymus (India), the "Camboge tree"; tall tropical evergreen shrub or tree to 12 m high, with handsome dark green leathery lanceolate leaves 25-45 cm long; small 2 cm white flowers in summer, followed by lemon-yellow, pointed fruit 6 cm long, the thick rind containing pith-like flesh and one large seed. The slightly acid fruit yields cathartic. *Tropical.* p. 520, 521

GARDENIA Rubiaceae
carinata (Malaya), "Kedah gardenia"; small tree to 12 m high, twigs and underside of foliage finely hairy; obovate, thin leaves 10-30 cm long, with depressed veins; flowers 5-10 cm across, cream-yellow deepening to rich egg-yellow, with 6-9 petals; oblong fruit to 4 cm long, green, ripening yellow. *Tropical.* p. 860

jasminoides 'Fortuniana' (syn. florida) (So. China), "Cape jasmine"; robust shrub with strong woody branches rather sparry, large, shining dark green, leathery, quilted leaves, and big 10 cm, fragrant, waxy-white flowers of heavy substance, turning creamy-yellow with age, spring and summer; the yellow fruit is eaten in China. *Tropical.* p. 860

jasminoides 'Veitchii', "Everblooming gardenia"; double-flowering form of the Chinese species, of compact, bushy habit, dense with smaller, shining green leaves, and while the sweetly fragrant, pure-white flowers average only 8 cm in dia., they are extra double and willing to bloom from January to May; often laden with buds and flowers. *Tropical.* p. 860, 861, 866

radicans florepleno (Japan), "Miniature gardenia"; dwarf, shrubby plant with spreading, rooting branches, narrow, 5-8 cm pointed leaves and small, solitary, irregularly double, white flowers, very fragrant. *Subtropic.* p. 860

taitensis (Society Islands: Tahiti), the charming "Symbol flower" or "Tiare tahiti"; considered the most beautiful flower of the islands; evergreen bush some 2 m high, with dark and glossy obovate 10 cm foliage with creamy midrib; the fleshy salver form waxy flowers with 6-8 linear petals arranged like a pin-wheel, 4-5 cm across, snow-white, incomparably scented, blooming in the evening and lasting several days; Tahitians put a blossom behind the ear after bathing, collecting it in the bud. *Tropical.* p. 866, 867

thunbergia (So. Africa: Natal); robust bush with woody branches, leathery, glossy green, broad-elliptic 10-15 cm leaves; large 8-14 cm long-tubed, waxy-white single flowers of 8-9 petals overlapping shingle-like, very fragrant. *Subtropic.* p. 866, 867

GASTERIA Liliaceae
batesiana (So. Africa: Zululand); star-like succulent rosette with broad triangular olive green leaves keeled beneath, rough by large deep green dots and smaller silver tubercles. *Arid-subtropic.* p. 591

liliputana (Cape Province), a miniature "Ox-tongue"; attractive, suckering miniature, smallest of the genus, with first two-ranked, later spirally arranged, thick, short-stubby, keeled leaves dark green and glossy, mottled nile-green. *Arid-subtropic.* p. 597

verrucosa (So. Africa), a 'Warty aloe'; attractive succulent suckering from the base, with 2-ranked, fleshy, concave leaves pointed at tip, dull deep green and covered with raised white warts. *Arid-subtropic.* p. 594, 596

GASTORCHIS Orchidaceae
luteus (Madagascar); rare terrestrial similar to Phaius, on occasion growing on pandanus trees; elongate stem-like pseudobulbs set

with folded, ribbed foliage; erect inflorescence bearing a raceme of waxy flowers of peach cream, the lip blotched with crimson. *Tropical.* *p. 730, 732*

x GASTROLEA   Liliaceae
beguinii (Aloe in hort.) (Aloe aristata x Gasteria sp.), "Pearl aloe"; dense rosette of rather narrow, leathery, short-pointed, keeled leaves dark green marked with pale warty dots, and warty teeth at margins. *Arid-subtropic.* *p. 597*

'Spotted Beauty' (Gasteria x Aloe); a California hybrid combining the rosette-shape of Aloe with the hard nature of Gasteria in an attractive, trim miniature; deep green leaves, tinged bronze, are covered with white tubercles and have translucent spines on keel and margins. *Arid-subtropic.* *p. 597*

GAZANIA   Compositae
nivea (So. Africa); small, spreading plant from thick, woody rhizome; crowded silvery leaves, much divided and white-hoary; daisy-like flowers golden yellow. *Subtropic.* *p. 319*

pavonia var. hirtella (So. Africa); low plant with small pinnate leaves, silver gray beneath; large daisy-like flowers deep orange, with white and brown peacock design at the base of each floret, forming a ring around center. *Subtropic.* *p. 319*

rigens 'Aztec', "Treasure flower"; charming flowers with yellow disk, framed by green ring; the ray flowers white and pink flaring with yellow and maroon. *Subtropic.* *p. 319*

rigens 'Copper King'; clumping plant with gray leaves and exceptionally large flowers orange with chestnut brown and violet. *Subtropic.* *p. 319*

rigens 'Fire Emerald'; fancy hybrid of the clumping type, with large flowers 5-6 cm, orange-yellow or bronzy red with green center ring. *Subtropic.* *p. 318*

rigens 'Gold Nuggets'; beautiful hybrid involving G. rigens and others; with small foliage, forming clumps; short-stalked large, showy daisy-like flowers 5 to 8 cm across, rich yellow with black at base of each ray; a dazzling color display during peak of bloom in late spring and early summer, in mild climate. *Subtropic.* *p. 319*

rigens hybrida, small perennial dense with narrow obovate leaves white-woolly beneath, the pretty flower heads golden yellow in the So. African species, brown-black in center; hybrids in various colors red, cream, orange, buff, with pretty markings. *Subtropic.* *p. 319*

uniflora (Natal), "Trailing gazania"; grows to same height as rigens, but spreads rapidly by long, trailing stems; foliage is clear silvery gray. The yellow, orange, white or bronze 6 cm flowers bloom in profusion. *Subtropic.* *p. 319*

GEISSORHIZA   Iridaceae
rochensis (Cape Prov.), "Wine-cups"; a very pretty small cormous herb 15-22 cm high, with long-linear, ribbed, basal leaves; wiry stalks bear several wide cup-like flowers 3 cm across, the outer area satiny violet-blue, the base crimson, both colors separated by a thin white line. *Subtropic.* *p. 525*

GELSEMIUM   Loganiaceae
sempervirens (Virginia to Florida, Texas and Central America), "Carolina yellow jessamine" or "False jasmine"; sparry evergreen shrub with scandent or clambering brown branches reaching to 6 m; leathery glossy-green ovate-lanceolate, opposite leaves 4 to 10 cm long; small flaring trumpet flowers 3 cm dia., yellow and sweetly fragrant like a tea rose blooming from December into summer and later. *Warm temperate.* *p. 616*

GENISTA: see Cytisus

GENTIANA   Gentianaceae
acaulis (Europe: Alps and Pyrenees), "Blue gentian"; very pretty low, tufted perennial to 10 cm high with narrow elliptic leaves and solitary 5 cm, funnel shaped flowers deep blue, in spring. *Temperate.* *p. 484*

lutea (Pyrenees, Alps, Balkans, Asia Minor), "Yellow gentian"; leafy perennial 1-1½ m tall, ovate leaves blue-green; 2½ cm flowers in whorls on erect stalks, yellow, veined or spotted. *Temperate.* *p. 483*

GEOFFROEA   Leguminosae
inermis (Panama); spreading tree with large pinnate leaves on arching branches, terminated by racemes of waxy small flowers, lilac-pink with purple center. Seen along the streets in Balboa. *Tropical.* *p. 545*

GEOGENANTHUS   Commelinaceae
undatus (Perú), "Seersucker plant"; compact, low growing suckering plant with stiff-fleshy broad ovate, quilted leaves dark metallic green with parallel bands of pale gray; wine-red beneath. *Tropical.* *p. 305, 306*

GERANIUM   Geraniaceae
grandiflorum (Sikkim); branching perennial 30 cm high, with roundish long-stalked leaves deeply 5-lobed and toothed; the 4 cm flowers characteristic with 5 equal petals and 10 stamens, lilac with purple veins. *Subtropic.* *p. 490*

GERANIUM see also PELARGONIUM

GERBERA   Compositae
jamesonii (Transvaal), "African daisy"; herbaceous tufted perennial with pinnately lobed leaves very woolly beneath; long-lasting flowers daisy-like with slender, usually orange-flame colored florets, but also in shades from brick-red to yellow and white. *Subtropic.* *p. 318, 331*

jamesonii plena, "Double Barberton daisy"; double flowered with several rows of ray flowers. *Subtropic.* *p. 318*

jamesonii 'Rosea'; suberb flowers with rose-pink ray petals; the inflorescence spreading 8-10 cm across in fancy strains. *Subtropic.* *p. 318*

viridifolia (South Africa); tufted perennial with crown of rootstock silky; obovate leaves, and long-stalked daisy-like flowers variable in colors ranging from white to pink above, pink, purplish or brown beneath. *Subtropic.* *p. 318*

GERRARDANTHUS   Cucurbitaceae
macrorhizus (E. and So. African deserts); monstrous disk-shaped caudex 60 cm or more dia., covered with hard, woody bark; branches with triangular lobed leaves, and orange-brown flowers. *Arid-subtropic.* *p. 381*

GESNERIA   Gesneriaceae
cuneifolia (Pentarhaphia reticulata) (Cuba, Puerto Rico, Hispaniola), "Firecracker"; low-growing rosette of leathery, glossy grass-green, long wedge-shaped leaves with toothed margins; and tubular somewhat bottle-shaped flowers burning red, yellow inside, borne singly on short stalks. *Humid-tropical.* *p. 498, 502*

GIBASIS   Commelinaceae
geniculata (Jamaica); in hort. trade as Tradescantia multiflora, the "Tahitian bridal veil"; small free-branching creeper with string-like stem, narrow shining olive-green 2½ cm leaves purplish beneath; tiny white flowers. *Tropical.* *p. 307*

GIBBAEUM   Aizoaceae
petrense (So. Africa: Cape Prov.), "Flowering quartz" or "Shark's heads"; minute, nearly stemless succulent with fleshy roots forming clumps; growths with 1 or 2 pairs of modified fat leaves, united at base and only 1 cm long, whitish gray-green, flat on top, keeled below; small reddish flowers. As in Lithops, water must be withheld when growth is completed and the plant rests. *Arid-subtropic.* *p. 44*

GILIBERTIA: see Dendropanax

GINKGO   Ginkgoaceae
biloba (China), the "Maidenhair-tree"; deciduous resinous tree to 40 m high; perhaps the most ancient existing flowering plant, sole survivor of an extinct race of the carbon age. Woody trunk with gray, corky bark, and broad fan-like leathery grayish green leaves with parallel veining; flowers without petals in catkins, male and female on separate trees; grown in pots as dwarfed bonsai in Japan. *Temperate.* *p. 508*

GLADIOLUS   Iridaceae
x hortulanus (South Africa), "Painted lady" or "Garden gladiolus"; strong floriferous plant, descendent of numerous So. African species including natalensis and primulinus; bulb-like corm with basal flattened, sword-shaped leaves in ranks; showy wide-open funnel-flowers with rounded petals, 10 cm or more across, in wide color range white, buff, yellow, orange, salmon, reds, rose, purple; on stiff fleshy spike 60 to 90 cm long, the buds opening flower after flower, and an excellent long-lasting cut flower. *Subtropic.* *p. 528*

GLECHOMA   Labiatae
hederacea variegata (Nepeta) (Europe, Asia), "Gill-over-the-ground"; small, lively creeper useful as ground cover or for baskets, the hairy, kidney-shaped leaves light green, variegated white at the crenate margins, to 2 cm across; small light blue flowers. *Warm temperate.* *p. 532*

GLEDITSIA   Leguminosae
triacanthos (Eastern U.S.), "Honey locust"; deciduous thorny tree with spreading branches, to 20 m or more high; pinnate leaves with oval leaflets in 10-14 pairs, some bipinnate; foliage turns yellow in autumn; inconspicuous flowers. *Temperate.* *p. 558*

GLEICHENIA   Gleicheniaceae (Filices)
linearis (dichotoma) (Malaya to Borneo), "Savannah fern"; thicket-forming tropical fern with long creeping rhizome; leathery

leaves on zigzag rachis, repeatedly 2 or 3 branched, ultimate branches bearing a pair of forked pinnae. *Humid-tropical.* p. 437

GLIRICIDIA *Leguminosae*
sepium (Central America, Colombia), "Madre de cacao"; ornamental tropical tree to 9 m, with pinnate leaves; clusters of flowers pinkish-lilac with white in profusion before the leaves; pods to 12 cm long; a favorite shade tree for coffee and cacao plantations. *Tropical.* p. 553, 558

GLOBBA *Zingiberaceae*
winitii (Thailand); perennial herb to 1 m high, with slender rhizome and fibrous roots, large lanceolate leaves, the base sheathing the stem, and hairy underneath; flowers in a loose, drooping terminal panicle to 15 cm long, with rose-purple bracts, and corolla yellow with a curved tube; autumn. *Tropical.* p. 925

GLORIOSA *Liliaceae*
carsonii (Central Africa); tuberous climber with shiny wiry stems set with clasping 10-12 cm thin leaves ending in a grasping tendril; magnificent flowers with broad recurving 6 cm segments wine-purple, edged with lemon-yellow especially toward base. *Tropical.* p. 592, 594
'Greeneae' hort.; canary-yellow, smooth, straight petals, lightly tinted copper. *Tropical.* p. 594
rothschildiana (Kenya, Uganda), "Glory lily"; a climbing "lily" with tuberous roots, the fresh-green, lanceolate leaves prolonged into tendrils; striking flowers with broad, recurved petals crimson-scarlet, golden-yellow toward base; blooms early spring on, but can be flowered anytime. I remember this growing north of Mombasa, clambering over coastal jungle. *Tropical.* p. 592, 593, 594
simplex (virescens, plantii, greenei) (No. Uganda, Mozambique, Zaire, Fernando Po), a dwarf "Glory lily"; with broader petals, not crisped, and which I first saw in Surinam where, in the shade of trees, it is a clear yellow, while in sunlight, the petals turn orange; summer-blooming. *Tropical.* p. 593
superba (India, Ceylon), "Glory lily"; tall vining herb blooming late summer to fall only, flowers are smaller, with narrow but crisped petals, first green, then yellow, changing to orange red; at its best into autumn. *Tropical.* p. 591, 592, 593
superba 'Aurea', "Golden glory lily"; flowers with gracefully crisped petals all orange-salmon. *Tropical.* p. 591
verschuuri (East Africa); tuberous vine rather compact-growing to 2 m, leaves broader than rothschildiana, shorter pedicels, 9 cm flowers with broad, reflexed segments of substantial texture, deep vivid crimson, yellow at base and margins. *Tropical.* p. 591, 592

GLOTTIPHYLLUM *Aizoaceae*
depressum (So. Africa: Cape Prov.); "Tongue-leaf"; robust succulent creeper with nearly flat, obliquely tongue-shaped, thick-fleshy green leaves to 10 cm long; flowers to 6 cm across, with yellow petals. *Arid subtropic.* p. 44
fragrans (Cape Prov.); low succulent with fleshy, twisted, tongue-shaped leaves flat on top, keeled beneath, 6-8 cm long; flowers 8-10 cm across, shining golden yellow and sweetly scented. *Arid-subtropic.* p. 44

GLOXINIA *Gesneriaceae*
perennis (maculata) (Colombia, Brazil, Perú), "Canterbury bells"; fleshy, spotted stem to 45 cm high, on scaly rhizome but no tuber, bearing large downy, bell-shaped fragrant flowers purplish blue with darker throat; basal leaves heartshaped, crenate, waxy above and reddish beneath. *Tropical.* p. 501

GLOXINIA: see also Sinningia

GLYPTOSTROBUS *Taxodiaceae (Coniferae)*
pensilis (lineatus) (S.E. China), "Chinese water pine"; small deciduous buttressed tree, usually at water's edge, with brown, split-barked trunk; coniferous needles linear and flat on sterile branches, scale-like on fertile ones; 2 cm ovoid cones. *Subtropic.* p. 354

GMELINA *Verbenaceae*
hystrix (Philippines), "Hedgehog"; spiny climbing shrub with habit of bougainvillea; elliptic leaves 8 cm long, glaucous beneath; large yellow flowers obliquely 2-lipped bell-shaped, 8 cm long, with large reddish purple bracts. *Tropical.* p. 917

GOETHEA *Malvaceae*
strictiflora (Brazil); small shrub with long-ovate, crenate leaves colored red; the inflorescence axillary or directly out of woody stems, with bracts beautifully glowing crimson. *Tropical.* p. 620

GOMPHRENA *Amaranthaceae*
globosa (India), "Globe amaranth"; erect annual branching herb with elliptic leaves ciliate at edge, and strawy clover-like flower heads which if cut just before maturity are "everlasting", retaining their color, usually white, to purple, for a long time. *Tropical.* p. 52

GONATANTHUS *Araceae*
pumilus (sarmentosus) (Sikkim); small, tuberous plant of the Himalayas with peltate, waxy, green leaves, some marked brown; thickened veins and gray-marked stem; plants I dug out of rock crevices in the Himalayas, kept bare root in good condition for weeks without wilting. *Subtropic.* p. 102, 109
rhizomatosus (Transvaal, Natal, Mozambique); very curious rhizomatous aroid with leaves dividing into several dissected branches, the leaflets glossy green, lanceolate and slender pointed; inflorescence on 10 cm stalk with 6-7 cm greenish-yellow spathe closely wrapped around spadix, slit open on one side and coming to an awl-shaped point. *Subtropic.* p. 104

GONATOPUS *Araceae*
boivinii (Trop. E. Africa: Zanzibar); from a flattish tuber 15 cm dia. rises the solitary, robust leaf, several times divided and distantly pinnate, the ovate or elliptic leaflets 10 cm long and quilted, on 75 cm stalk marbled with blackish brown; the inflorescence on 30 cm spotted stem, with lanceolate spathe yellow-green outside marbled with brown, spadix yellowish white. *Tropical.* p. 102

GONGORA *Orchidaceae*
quinquenervis (maculata) (Brazil); epiphyte; bulbs and 2 ribbed leaves both pale green; nodding flowers yellowish-white spotted with dull wine-red, in pendulous raceme. *Tropical.* p. 734, 735

GORDONIA *Theaceae*
lasianthus (No. Carolina to Mississippi), "Black laurel"; tall evergreen tree to 30 m high; with leathery elliptic, dark green and glossy leaves to 15 cm long, serrate in upper half; axillary, solitary flowers white, 6 cm across and with prominent yellow anthers. *Warm temperate.* p. 905

GOSSYPIUM *Malvaceae*
herbaceum (Arabia, Asia Minor), "Levant-cotton"; annual, little-branched herb to 1 m, with thin-leathery, palmately 5-7 lobed leaves, and showy yellow flowers with purple center; capsular fruits or "bolls" whose seeds bear fleece or lint, furnishing cotton. *Tropical.* p. 620

GOVENIA *Orchidaceae*
utriculata (Mexico, C. America to Argentina, W. Indies); terrestrial with subterranean tuber-like pseudobulbs from fibrous rhizome; 1 or 2 large plicate, stalked leathery leaves to 40 cm; wiry stalk supports a many-flowered inflorescence, to 1 m long; hooded flowers 4 cm across, creamy-white tinged with purple, spotted and striped brown within, lip marked red (autumn). *Subtropic.* p. 734

GRAMMATOPHYLLUM *Orchidaceae*
scriptum (measuresianum) (Moluccas); attractive epiphyte of relative small size in the genus, with stem-like ribbed pseudobulb some 60 cm high bearing strap-shaped leathery leaves, and basal raceme 60-90 cm long with many 8 cm waxy flowers, the spreading oblong sepals and petals canary yellow blotched all over with brown and gold (summer). *Tropical.* p. 732, 734
speciosum (Java, Malaya, Vietnam), magnificent epiphyte, meriting the title of "Queen of orchids"; with stout stems 1½-3 m long, bearing numerous 2 ranked strap leaves, and basal floral stalks to 1½ m long with as many as 100 showy flowers 8-15 cm across, the undulated sepals and petals rich yellow, blotched with reddish-brown (winter). *Tropical.* p. 730, 731

GRAPHORKIS *Orchidaceae*
lurida (Trop. West Africa); clustering epiphyte resembling a small cymbidium; small pseudobulbs with pleated, deciduous leaves 30 cm long, and with aerial roots; wiry, branched raceme with many small 1½ cm flowers brownish-purple, lobed lip white with yellow midlobe (winter). *Tropical.* p. 732

GRAPTOPETALUM *Crassulaceae*
filiferum (Mexico: Chihuahua); attractive stemless rosette 5-6 cm across, with 75-100 spatulate fleshy leaves, densely shingled and recurving downward, forming a symmetric mound, convex and shining green to silver-gray, with wing-like white margins and ending in a brown bristle; branched floral stalk with flower spotted red inside, white outside. *Arid-subtropic.* p. 369
paraguayense, also known by its synonym Sedum weinbergii (Mexico), "Ghost plant" or "Mother-of-pearl plant"; branching succulent loose rosette forming thick stem and fleshy, broad 5-8 cm leaves flat and somewhat recurved on surface, keeled beneath, amethyst-gray with silvery bloom; flowers white. The foliage has a subtle opalescent blending of colors, but the leaves are brittle and drop easily when handled. *Arid-subtropic.* p. 374

GRAPTOPHYLLUM *Acanthaceae*
pictum (hortense) (South East Asia), "Caricature plant"; dense evergreen shrub to 2½ m high, with leathery elliptic leaves 10-15 cm long, deep green variegated creamy-white along central area; crimson red flowers. *Tropical.* p. 36

pictum 'Tricolor' (New Guinea); beautiful shrub with oval, pointed, leathery leaves, purplish green variegated with yellow and pink, center vein and stem red. *Tropical.* p. 36

GRAVISIA: see Aechmea

GREENOVIA   Crassulaceae
aurea (Sempervivum) (Canary Islands); beautiful cushion-forming succulent shrub, in habitat growing to altitudes of 1700 m, cup-shaped rosettes to 40 cm across, of thin, erect, obovate leaves waxy blue-green, and with narrow white margin; flowers golden yellow with spreading linear petals, in showy terminal clusters. *Subtropic.* p. 374

GREIGIA   Bromeliaceae
sphacelata (Chile); terrestrial rosette to 1 m, spreading soft-leathery glossy-green leaves, spiny-margined; flowers rose, in heads with spiny bracts. *Subtropic.* p. 215

GREVILLEA   Proteaceae
asplenifolia (New So. Wales); tall shrub to 5 m, linear, saw-edged leaves 10-25 cm long, silky beneath; red flowers arranged like a tooth-brush on the underside of axillary spikes. *Subtropic.* p. 834

banksii (Queensland), "Scarlet grevillea"; small tree 4½-6 m high, hairy branches with pinnate leaves to 25 cm long, segments linear and white-silky beneath; bright coral red flowers downy outside, in dense, one-sided clusters. *Tropical.* p. 834

robusta (Queensland, New S. Wales), "Silk oak"; daintily lacy, ornamental plant while small, but growing into a mighty tree 45 m high; silvery downy shoots with fern-like, green leaves, 2-pinnate into finely lobed segments silky-haired, giving them a grayish appearance; flowers golden yellow in one-sided racemes to 10 cm long. *Subtropic.* p. 834, 840

thelemanniana (W. Australia), "Orange spider-flower"; showy shrub with bipinnate silky, bluish leaves divided into linear segments; flowers with rosy-red tube and yellowish-green recurved lobes, and long, curving 2½ cm red style. *Subtropic.* p. 834

GREWIA   Tiliaceae
caffra (So. Africa), "Lavender star-flower"; spreading evergreen shrub to 2-3 m high, dense with small oval waxy, moss-green leaves with finely crenate margins, 6-8 cm long; charming starry flowers soft lavender pink with prominent yellow stamens in center, 4 cm across; larger and more perfectly shaped flowers than the similar G. occidentalis. *Subtropic.* p. 909

GREYIA   Melianthaceae
radlkoferi (Transvaal), "Mountain bottle brush"; bushy shrub to 12 m; with big, fleshy-herbaceous, downy, bronzed foliage, deeply toothed; terminal erect raceme of glossy scarlet red, cup-shaped flowers. *Subtropic.* p. 646

sutherlandii (Natal, Transvaal), "Mountain bottle-brush"; small tree with deciduous, broad, deeply toothed, smooth leaves to 8 cm; inflorescence brush-like, bright scarlet, with conspicuous purplish-red filaments. *Subtropic.* p. 643

GRISELINIA   Cornaceae
lucida 'Variegata' (New Zealand); lovely ornamental, dense bush or tree to 8 m, often epiphytic and forming aerial roots; the oblique oval thick-leathery leaves normally glossy green, 6-10 cm long; in this cultivar variegated or edged in ivory white; minute male flowers with purplish-green petals. *Subtropic.* p. 357

GUADUA   Gramineae
angustifolia (Panama, Jamaica, Colombia, Venezuela, Brazil to Argentina); a giant non-hardy timber bamboo with culms to 20 m or even 30 m tall, and 10-20 cm thick, widely arched above; the internodes hollow with wood 2 cm or more thick, higher up the branches usually thorny, the leaves oblong lanceolate 16 up to 20 cm long. Used much for timber in South America, also by opening large culms out flat in place of sawn boards of wood. Locally cut into sections to use instead of flower pots. *Tropical.* p. 518, 519

GUAIACUM   Zygophyllaceae
sanctum (So. Florida, W. Indies, Mexico, to No. South America), "Lignum vitae"; ornamental evergreen shrub or small tree, sometimes to 8 m high, of very hard, resinous wood; handsome glossy green, opposite, pinnate leaves to 10 cm long, the leathery leaflets oval, and smaller than in G. officinale; at branchtips large clusters of felty blue flowers, followed by angled 2 cm orange-yellow fruit. The commercially valuable wood is dark brown with black streaks, and is so heavy that it sinks in water. *Arid-tropical.* p. 928

GUNNERA   Haloragidaceae
chilensis (scabra) (Chile, Ecuador, Colombia); large perennial herb with creeping rhizome and dark green hard, puckered leaves to 1½ m diameter, palmately and deeply cut into pointed and toothed lobes, on reddish stalks with fleshy green spines; the inflorescence spikes with branches short and thick, while they are long in manicata. *Subtropic.* p. 508

manicata (So. Brazil); the largest species with huge leaves to 2 m across, palmately lobed, more kidney-shaped than chilensis; hard, rough puckered, light green with buff veining, on thick light brown stalks with thorn-like prickly hairs. *Subtropic.* p. 508

GURANIA   Cucurbitaceae
malacophylla (Upper Amazon); tall tendril climber with usually 3 lobed, hairy leaves; male flowers reddish, in long-stalked clusters; the female flowers later, axillary. *Tropical.* p. 382

GUSSONAEA   Orchidaceae
physophora (Microcoelia) (Madagascar, Zanzibar); remarkable leafless epiphyte from Tsarantanana forest; an interesting orchid reduced to a cluster of fleshy, grayish-green, flattened roots 1 cm wide, clinging to the trunks of trees; small white flowers rising from a tiny central stem, the lip with a long spur. *Tropical.* p. 736

GUZMANIA   Bromeliaceae
berteroniana (Puerto Rico), "Flaming torch"; formal rosette of wine-red or sometimes fresh green, thin leathery leaves with showy inflorescence in form of a tight cylindrical head of scarlet bracts with yellow flowers. *Tropical.* p. 216, 217

cateriflora (sanguinea) (Costa Rica, Trinidad, Ecuador); handsome epiphyte from tropical forests to 1000 m elevation; wide open rosette of bronzy green, leathery leaves to 40 cm long, recurved at apex, the center nest turning yellow and brilliant red-maroon toward tips before blooming; flowers straw-yellow deep in center cup. This name should probably be referred to sanguinea or a form of it. *Tropical.* p. 215

'Fantasia'; beautiful hybrid, photographed in the Palmengarten Frankfurt, Germany; rosette of narrow, soft-leathery, glossy green leaves laxly pendant; showy inflorescence with branched heads of yellow flowers as in a bouquet, subtended by coppery red bracts. *Tropical.* p. 215

insignis (Colombia); rosette of leaves 60 cm long and 5 cm wide, reddish underneath; showy branched inflorescence with light amaranth-red floral bracts and lemon-yellow flowers, on leafy stalk. Dr. L.B. Smith (Smithsonian) refers this to Tillandsia insignis. *Tropical.* p. 215

lindenii (Northeastern Perú), "Large snake vase"; showy large rosette of glossy-leathery leaves beautifully mottled moss-green with jagged ivory or pinkish cross-bands, made up of parallel lines across the leaf; these lines show brownish beneath. Inflorescence with numerous white flowers. *Tropical.* p. 217

lingulata (C. America to Guayana, Pará, Mato Grosso, Ecuador, Bolivia); striking epiphytical rosette from the rainforest, with smooth metallic green leaves, forming a showy, raised head of leathery, brilliant fiery-red bracts, with a contrasting center of hooded, waxy orange-red inner floral bracts tipped yellow to white, and with white flowers. *Tropical.* p. 216

lingulata 'Major' (Broadview), "Scarlet star"; magnificent clone of a plant collected in Ecuador, epiphytic rosette of smooth, metallic green leaves, from the center of which rises the bold inflorescence of recurving, glossy-leathery bracts in vivid scarlet red, its center with short waxy, incurved red bracts tipped yellow, and white flowers; the scape leaves typically red at base. *Tropical.* p. 216, 217

lingulata minor, "Orange star"; small, clustering rosette of strap-like, thin-leathery, yellowish green leaves, with maroon pencil lines starting at base and diminishing toward tip; long floral bracts bright orange-red, and small white flowers. *Tropical.* p. 216

x magnifica (lingulata cardinalis x minor); a magnificent epiphyte when the inflorescence develops its stalked star-like, raised 15 cm head of brilliant scarlet, leathery bracts lasting from November to April, flowers white; soft leaves green with reddish lines. *Tropical.* p. 216

melinonis quitense (Caraguata) (Fr. Guyana, Colombia, Ecuador, Amazonian Brazil); epiphytic small rosette, with strap-shaped soft-leathery leaves 30 cm long, glossy green above, red-brown beneath; slender, club-shaped inflorescence with shingled bracts magenta-red in the species, rose-pink in the var. quitense; flower petals lemon-white. *Tropical.* p. 217

monostachya (tricolor) (W. Indies, C. America, to Brazil), "Striped torch"; formal rosette of thin-leathery bayonet shaped yellow-green leaves; inflorescence a stiff spike with bracts salmon-red striped brown, and white flowers. *Tropical.* p. 217

'Naranja' ('Hummel's Memoir' x lingulata 'Major'), a lovely hybrid rosette with shiny, fresh green recurving leaves, from the center a raised nest of scarlet-red bracts, the inner bracts tipped yellow, and white flowers. *Tropical.* p. 216

'Omar Morobe'; beautiful plant with flaccid band-like leaves, primarily white or cream with pink blush, and olive-green margins. *Humid-tropical.* p. 217

sanguinea (C. America, Trinidad, Colombia, Ecuador); stout, compact, rather flat rosette with broad, thick leaves 30 cm long, scaly at base, the inner leaves ruby-red from the middle up to the apex, the lower part yellow and chartreuse at flowering time; the flowers a slender yellow tube with spreading white lobes, in a center cup. *Tropical.* p. 215

sanguinea brevipedunculata; glowing red rosette with spreading leaves; yellow flowers on short stalks. *Humid-tropical.* p. 217

'Symphonie' (zahnii x peacockii); rosette of dark coppery foliage; from the center rises the showy, starry head of pointed, lacquered crimson bracts with yellow flowers. *Humid-tropical.p. 127*

zahnii (Colombia, Panama); very ornamental plant with strap-like, papery, olive green leaves pencil-striped maroon-red, the center tinted pink to coppery red; strong branched inflorescence with pink to yellow bracts and white flowers. *Tropical.* p. 217

GYMNOCALYCIUM   *Cactaceae*

baldianum (Argentina); small globes with large tubercles topped by curved radial spines; flowers salmon-red. *Arid-subtropic.* p. 269

bruchii (Argentina), "Chin cactus"; depressed little globe, bearing many offshoots, shallowly ribbed, and with curved bristle-like white spines; pinkish flowers; will grow faster and bloom earlier if grafted. *Arid-subtropic.* p. 266

damsii (Paraguay), a "Chin cactus"; small globular cactus, flattened on top, with 10-12 wide, tubercled ribs, spines white tipped brown; flowers white, outer segments green tipped red. *Arid-subtropic.* p. 266

denudatum (So. Brazil to Argentina), "Spider cactus"; broad globe to 15 cm dia., deep green, with 5-8 very broad, notched ribs, spidery, curved yellowish spines; flowers white or pale rose. *Arid-subtropic.* p. 268

fleischerianum (Paraguay); small coppery globe to 7 cm high, covered with brown, bristle-like radial spines; flowers rosy, white inside. *Arid-subtropic.* p. 268

leeanum (Argentina, Uruguay), "Yellow chin-cactus"; small depressed globe, glaucous green, with 11-14 indistinct ribs, covered with slender spines; large flowers yellow, the outer perianth segments purplish. *Arid-subtropic.* p. 268

mihanovichii (Paraguay), "Plain chin cactus"; grayish green, depressed little globe to 6 cm thick, with 8 triangular, notched ribs and banded with maroon, straw-colored spines; free-flowering chartreuse. *Arid-subtropic.* p. 268

mihanovichii friedrichii (Uruguay, Argentina), "Rose plaid cactus"; little depressed globe, coppery green with triangular ribs, banded cream and yellow spines; flowers when young, pink. *Arid-subtropic.* p. 268

mihanovichii friedrichii 'Rubra'; the novel "Red cap", "Oriental moon", "Hibotan"; strikingly colorful red (or yellow) small globe; variant with chlorophyll-poor body; for better growth and survival usually grafted on night-blooming Hylocereus undatus. *Arid-subtropic.* p. 267, 268, 269

mihanovichii 'Multicolor'; striking cultivar with the little globes in various colorings from pink to red, purple and green. In the collection of Roberto Seidel, Corupá, Brazil. *Arid-subtropic.* p. 266

multiflorum (So. Brazil to Argentina); globe to 10 cm high, glaucous bluish green, 10-15 ribs with narrow furrows, cut into prominent knobs with a chin below, thick radials radiating comb-like; flowers pinkish. *Arid-subtropic.* p. 268

saglione (Argentina: Tucuman); broad depressed globe, to 30 cm dia., when young the top surface velvety grayish green, wrinkled into an attractive design, later with 13-32 ribs with large low tubercles with curved spines; flowers pinkish. *Arid-subtropic.* p. 268

venturianum (Uruguay: Montevideo); globe lightly depressed, 5 cm thick, pale bluish green, with 9 broad rounded ribs cut into knobs with chins below; yellow radials; flowers bright carmine. *Arid-subtropic.* p. 268

GYNURA   *Compositae*

aurantiaca (Java), the popular "Velvet plant"; a beautiful tropical herbaceous plant with stout stems and fleshy, broad-ovate, serrate 12 cm leaves densely velvety with violet or purple hairs and deeper purple veins; orange disk-flowers. *Tropical.* p. 324, 327

bicolor (Moluccas), "Oakleaved velvet plant"; tropical branching foliage plant with heavy, fleshy, slightly clammy leaves, 12-15 cm long, deeply and irregularly lobed on sides, metallic green with purple cast and short-hairy on top, midrib light purple and glossy, rich purple beneath; terminal flower heads with orange florets. *Tropical.* p. 327

x sarmentosa (aurantiaca x bicolor), "Purple passion vine"; twiner with reddish stem, lanceolate leaves with wavy-toothed or shallowly lobed margins, and covered by purple hairs; wine-red beneath; small orange flower heads in clusters at ends of branches. In Europe (Hay-Synge Blumenbuch 1973) as G. procumbens (Lour.) Merr. *Tropical.* p. 324, 327

GYPSOPHILA   *Caryophyllaceae*

paniculata (Central and Eastern Europe to Central Asia), "Baby's breath"; perennial, much branched with thin, wiry stems to 1 m high; narrow leaves, and with hundreds of tiny white flowers. Used in flower arrangements. *Temperate.* p. 298

HABENARIA   *Orchidaceae*

glazioviana (Brazil), "Fringed orchid"; large terrestrial with fleshy roots, erect simple stems to 1 m high, basal oblanceolate leaves, and a raceme of greenish-yellow flowers, the lip with several narrow or broader lobes. *Subtropic.* p. 734

HABRANTHUS   *Amaryllidaceae*

bagnoldii (Rhodophiala chilensis, Hippeastrum) (Chile); small bulbous plant to 30 cm high; narrow linear, glaucous leaves; summer-blooming with yellow trumpet flowers tinged with red, 5 cm long. *Subtropic.* p. 67

tubispathus (Zephyranthes robusta) (Argentina); small bulbous herb to 22 cm high, with recurved linear leaves, appearing after blooms; flowers rosy-red, to 8 cm long, with short greenish tube. *Subtropic.* p. 67

HADRODEMAS   *Commelinaceae*

warscewiczianum (Tripogandra), (Guatemala); very thick-fleshy rosette with stout stem, resembling Dracaena or Agave; the clasping leaves broad, long pointed, to 30 cm long, pale green, ciliate at edge and recurved; clusters of small pale purple flowers on long stalk. *Tropical.* p. 305

HAEMANTHUS   *Amaryllidaceae*

albiflos (So. Africa), "White paint brush"; bulbous plant to 30 cm high, fleshy, wide evergreen leaves with ciliate margins, the inflorescence of white flowers in heads 5 cm across. *Subtropic.* p. 66

katherinae (Natal), "Blood flower"; robust bulbous plant branching from offsets, the soft-fleshy fresh green sword-shaped leaves with channeled midrib running into channeled petiole, separate solid stalk bearing umbrella-shaped head of star-like flowers with salmon petals and long red stamens. *Subtropic.* p. 63

multiflorus (Trop. Africa), "Salmon blood lily"; bulbous plant with leaves on short spotted petioles; the showy inflorescence forming a perfect ball with up to 100 flowers, separate from leaves, coral pink to red, crimson at base of narrow petals with long extended stamens tipped yellow. *Subtropic.* p. 66, 69

natalensis (So. Africa: Natal), "Natal paintbrush"; bulb bearing 7-9 glossy green, strong-nerved lanceolate leaves; dense globose head of pale green 3 cm flowers, showing orange-salmon styles. *Subtropic.* p. 69

puniceus (So. Africa), "Pink paintbrush"; bulbous plant with many wavy leaves on reddish stems; the summer flowers in dense 10cm heads pale red to white and not too showy, but they are followed by a nestful of scarlet berries. *Subtropic.* p. 69

HAEMARIA   *Orchidaceae*

discolor dawsoniana (Anoectochilus) (Malaya), "Jewel orchid"; vigorous terrestrial with creeping, branching rootstock and fleshy, ovate, gorgeous leaves of blackish red-green velvet with a network of coppery-red veins, wine-red beneath; small, waxy-white flowers, with yellow center, in terminal raceme, (Oct.-Feb.). *Tropical.* p. 732, 741, 745

HAEMATOXYLUM   *Leguminosae*

campechianum (Yucatán, W. Indies), "Bloodwood tree"; fast-growing spiny shrub or tree to 8 m, with gnarled trunk; leaflets in 2 to 4 pairs, 3 cm long; narrow clusters of small yellowish flowers. *Tropical.* p. 558

HAKEA   *Proteaceae*

saligna (Queensland, N.So. Wales), "Willow-leaf hakea"; fast-growing upright shrub to 3 m; brown-woody branches with scattered, thick-leathery narrow grayish-green leaves 8-15 cm long; from the leaf axils dense masses of small white flowers with long twisted styles. *Subtropic.* p. 834

victoriae (Western Australia), "Royal hakea"; magnificent, colorful sparry tree 3-4 m high, with broad, rigid leaves 5 to 10 cm across, densely shingled, and with sharp spines along margins, light glossy green beautifully veined and variegated with yellow and into red, the base golden yellow and looking like big flowers; pinkish blooms among the foliage. *Subtropic.* p. 836

HAMATOCACTUS   *Cactaceae*

setispinus (No. Mexico, So. Texas), "Strawberry cactus"; globose plant to 15 cm high, with 13 thick ribs arranged spirally, grayish matte green, the areoles with tufts of white wool, radials gray-brown, the long central spines with fish hook; flowers yellow with red throat. *Arid-subtropic.* p. 266, 268, 269

HAMELIA   *Rubiaceae*

patens (Florida, W. Indies to Bolivia), "Fire bush"; evergreen shrub or small tree to 8 m, gray pubescent; elliptic leaves 15 cm long; tubular flowers scarlet red, 2-3 cm long, in one-sided raceme. *Subtropic.* p. 867

HAPALINE *Araceae*
brownii (Malaysia); attractive small tuberous aroid with waxy, heart-haped leaves green lightly spotted with white, 7-9 cm long, 6 cm wide; ovate spathe cream on slender stalk. *Tropical.* p. 130

HARDENBERGIA *Leguminosae*
violacea (Australia, Tasmania), "Purple coral pea"; evergreen twining shrub found wild growing over low bushes. Leaves simple, cordate-ovate 4-12 cm long; axillary clusters of small flowers, often in pairs, purple with yellow basal spot, notched at the top. *Subtropic.* p. 560

HARPULLIA *Sapindaceae*
arborea (Philippines), "Puas"; tropical tree with thin, green pinnate foliage; green-petaled flowers; the fruit in coppery red capsules 4 cm dia., with large black seeds inside. *Tropical.* p. 875
pendula (New South Wales, Queensland), "Tulipwood"; evergreen tree to 15 m, with pinnate leaves, lanceolate leaflets glossy green, 8-12 cm long; long sprays of small dull yellow flowers, followed by pairs of orange berries 3 cm dia., with black shiny seeds. Wood is used in carpentry. *Subtropic.* p. 875
zanguebarica (Kenya: Lamu), "Mgambo" in Swahili; tropical timber tree with rough bark; leaves pinnate with up to 10 pairs of glossy, elliptic leaflets to 5 cm long; small green flowers in dense clusters; seed capsule 3-lobed 3 cm long, crimson-red inside with velvety black seed. *Tropical.* p. 875

HARRISIA *Cactaceae*
bonplandii (Brazil, Paraguay, Argentina); night-blooming cactus erect at first, later arching or clambering, stems 5-8 cm thick and strongly 4-5 angled, velvety dark green, with a few gray spines; white flowers closing soon after sunrise. *Subtropic.* p. 251
jusbertii (Eriocereus) (Paraguay), "Moon cactus"; clambering or climbing cereus; dark green slender stems 4-6 cm dia, with usually 5 ribs and concave sinus between, set with straw-colored spines; large white funnel flowers 18 cm long, brownish outside, blooming freely at night and fragrant. *Tropical.* p. 251

HATIORA *Cactaceae*
salicornioides (S.E. Brazil), "Drunkard's dream"; so called in Brazil because of the bottle-shaped branchlets of this epiphytic tree dweller; plant much branched, green or purplish; salmon flowers tipped yellow. *Tropical.* p. 285

HAWORTHIA *Liliaceae*
armstrongii (Cape Prov.), "Wart plant"; suckering erect rosette densely spirally leafy; blackish-green sharply pointed and obliquely keeled succulent leaves, with pale tubercles. *Subtropic.* p. 607
chalwinii (reinwardtii var.) (So. Africa), "Aristocrat plant"; attractive columnar succulent to 15 cm high, the leafy stem with ovate-triangular, fat leaves to 5 cm long, arranged in tight spiral, purplish-bronze with pearly-white tubercles; flowers white. *Arid-subtropic.* p. 597
cooperi (So. Africa), "Window haworthia"; rosette similar to pilifera, with longer, fleshy, boat-shaped leaves purplish-brown, with translucent ciliate edges and window-like, translucent point. *Subtropic.* p. 597
fasciata (So. Africa), "Zebra haworthia"; very attractive commercial, small erect rosette, making offsets, of slender tapering somewhat incurved leaves dark green, with large white warts in neat connected cross bands. *Subtropic.* p. 597
limifolia (Swaziland), "Fairy washboard"; exquisite small rosette 8-10 cm across; leaves triangular pointed, the upper surface concave, dark green-brown, on both sides with 15-20 transverse ridges. *Arid-subtropic.* p. 596
"Margaritifera" hort.; widely cultivated as a durable dishgarden plant; low suckering rosette of numerous spreading, slender-tapering leaves, matte deep green with evenly scattered pearl-white tubercles; small white bilabiate flowers in racemes. *Subtropic.* p. 597
papillosa (Cape Prov.), "Pearly dots"; beautiful, robust rosette with erect, lanceolate leaves rich deep green, adorned with rows of greenish-white raised, often hollow warts, rounded beneath. *Subtropic.* p. 597
retusa (Cape Province), "Star cactus"; stemless rosette forming clusters, with fat and stubby deltoid pale green leaves to 4 cm long, the upper half recurved with a smooth triangular face almost translucent and with pale lines. *Subtropic.* p. 597
truncata (Cape Prov.), "Clipped window plant"; most unusual "Window plant" with succulent leaves oval in cross-section, arranged in 2 ranks, dark green-brownish and rough-warty, apex of each leaf flat as if cut off and translucent. *Subtropic.* p. 597

HEBE *Scrophulariaceae*
buxifolia 'Variegata' (Veronica) (New Zealand); dense evergreen bush resembling euonymus, with small, soft, 2 cm obovate leaves, opposite and overlapping, dark, waxy green with gray, variegated cream-white from the margin inward; small white flowers in closely packed spikes. *Subtropic.* p. 882
x imperialis (speciosa x salicifolia); luxuriant bushy evergreen with shiny green to reddish oval leaves in opposite ranks; large cylindrical heads of magenta purple flowers in summer. Resembles H. speciosa. *Subtropic.* p. 887
speciosa (Veronica) (New Zealand), "Showy veronica"; robust, evergreen shrub with spreading, angled branches dense with opposite, thick, oblong leaves glossy dark green and downy on the midrib above; the small purple-crimson flowers in dense axillary racemes opposite near tips of branches, summer blooming. *Subtropic.* p. 887

HECHTIA *Bromeliaceae*
argentea (Mexico), "Vicious hechtia"; attractive terrestrial rosette of stiff dagger-shaped, dark glossy green leaves 30 cm or more long, with prominent sawtooth spines; gray pencil lines beneath; tall inflorescence with orange flowers. *Subtropic.* p. 215
stenopetala in hort.; correct name may be ghiesbreghtii (So. Mexico); handsome if viciously spiny rosette to 30 cm across, found growing in the lava rock of the Pedregal region near Mexico City; stiff, concave leaves light green and heavily splashed with blood-red when growing in sunny exposure; the margins with strong hook thorns; flowers white, in cylindric inflorescence. From the base of the mother plant, long runners form to produce new plantlets. Bot. Mag. shows stenopetala to have narrower leaves; plant shown could be glomerata exposed to sun. *Tropical.* p. 215

HEDERA *Araliaceae*
canariensis (Azores, Canaries, Morocco), "Algerian ivy"; burgundy-red twigs and petioles with glossy, fresh green, leathery leaves broadly ovate and quite flat, even recurved, shallowly lobed and usually with slender main lobe; covered with grayish-white scales. *Subtropic.* p. 136
canariensis arborescens 'Variegata', the "Ghost tree ivy"; so called because of the ghostly trembling of the pendant foliage in a breeze; arborescent or fruiting form of the variegated Algerian ivy, its ovate, hard leaves with cream variegation on light green or gray; fruit black. An excellent, very durable, but slow-growing decorative plant in containers, for the sunny window or the patio, the beautiful variegated foliage stiff as if varnished. From half-hard cuttings. *Subtropic.* p. 135, 136
canariensis 'Gloire de Marengo'; long cultivated in Europe under this name, this is probably identical with H. canariensis 'Variegata' or one of its color forms. The relatively large leaves are usually 3-lobed, green in the center changing to milky-gray, then white or cream to yellowish in irregularly variegated areas, mostly along the margins; stems and petioles red. *Subtropic.* p. 136
canariensis 'Variegata' (maderensis), known as 'Gloire de Marengo' and "Hagenburger's" commercially, very colorful with the thin-leathery leaves in the center fresh-green to slate-green, joined by a zone of blue or gray-green, and marginal variegation of creamy white. *Subtropic.* p. 136
colchica (Caucasus to Iran), "Persian ivy"; a bold woody climber with juvenile leaves thick-leathery, somewhat heart-shaped, dark dull green, to 25 cm long, occasionally lightly lobed. *Warm-temperate.* p. 135
colchica 'Dentato-variegata' (Caucasus, Iran), "Variegated Persian ivy"; twigs pea-green with large, leathery, green leaves broadly ovate to 25 cm long, lightly lobed and remotely toothed, with dense, scaly pubescence, and broad cream-white margin. *Warm temperate.* p. 138
helix (Europe, Asia, N. Africa), "English ivy"; root climbing vine with juvenile leaves 5 lobed, glossy forest green, with creamy veins, to 6 cm, and somewhat cupped; this species has 4-12 stellate hairs covering the foliage, mostly underneath, seen through a magnifying lens; in the arborescent stage leaves are unlobed, the fruit black. *Temperate.* p. 135, 136, 137
helix 'Abundance'; slow trailer, and bushy, with broad and large 10 cm 4-5 to 7 lobed variable leaves dark green with pale veins, some are wavy in the sinus, some have 2-pointed apex. *Temperate.* p. 138
helix baltica (Latvia), "Baltic ivy"; clinging vine similar to English ivy but the leathery foliage not so large and more cut, whitish veins; very hardy. *Temperate.* p. 137
helix cristata 'Curlilocks'; sport with young growths sprouting from every axil of the crested leaves resulting in densely bushy vines. *Temperate.* p. 138

helix 'Glacier'; good vining growth combined with nicely variegated, small, triangular, leathery leaves of several shades of green down to gray, with white marginal areas and pink edge. *Warm temperate.* p. 136

helix 'Harald', "White and green ivy" or "Improved Chicago variegata"; a medium small-leaved yet robust, variegated clone favored in Europe because its 4 to 7 cm 3 to 5 shallowly lobed leaves, somewhat rounded, are mostly green but not gray in the center, broadly margined creamy white; also quite durable as a house ivy. *Warm temperate.* p. 136, 138

helix hibernica 'Variegata'; creeping vine with firm broad leaves similar to English ivy, shallowly 3-5 lobed, dark green with irregular cream-white variegation and mottling; veins milk-white. *Temperate.* p. 138

helix 'Jubilee'; tiniest of the variegated leaf forms, self-branching, the little snubnosed leaves friendly light green, gray and white, and quite irregular. *Warm temperate.* p. 138

helix 'Maculata'; may be a slow-growing, variegated form of the Irish ivy; the leaves are roundish, shallowly 5-lobed, rather flat and fleshy, and the yellow-green mixed with dark green beautifully variegated white or cream. *Temperate.* p. 138

helix 'Manda's Crested'; attractive plant with star-shaped, jade-green leaves with rosy edge, the long lobes fluted and undulate, on straight, upright, reddish stalks. *Warm temperate.* p. 138

helix 'Minima'; form with flexuous twigs, green petioles and small, 3-5 lobed, thin leathery leaves, somewhat undulate, base cordate. *Temperate.* p. 137

helix 'Patricia'; sport of 'Pittsburgh', an excellent Philadelphia cultivar; dense self-branching ivy with medium-sized leathery leaves to 3-6 cm wide and long, usually remotely 5-lobed, and prettily curled in at sinuses. A good keeper and favorite in a 13 cm pot, with its neatly draped reddish branches. Better keeper indoors than the "Pittsburgh ivy". *Warm temperate.* p. 136

helix 'Ripples'; heavily loaded pendant branches with medium-large roundish foliage irregularly lobed, the margins more or less curled or crested. *Temperate.* p. 138

helix 'Star'; shapely selfbranching, bushy variety, leaves 5-pointed, star-shaped with slender finger-like lobes. *Temperate.* p. 138

helix '238th Street'; this ivy produces, in addition to flowering twigs, stiff, viny, green shoots with unlobed, waxy green leaves, spreading horizontally, and which stay green in winter; flower greenish-cream, fruit green to black. *Warm temperate.* p. 136

helix 'Williamsiana'; shapely and vigorous variety with 3-5 lobed leaves, the long tips curled downward while edges of leaves are wavy; greenish ivory border around apple-green or gray-green center. *Warm temperate.* p. 138

HEDYCHIUM *Zingiberaceae*

coccineum (Himalaya, Burma, Ceylon), "Scarlet ginger-lily"; robust perennial herb with suckering rootstock and leafy stems to 2 m high, long, stiff leaves to 50 cm long, smooth green above, bluish beneath; the scarlet red flowers with long corolla tube and pink filament, dense on a stout spike. *Tropical.* p. 924

coronarium (Himalayas into China), "Butterfly lily"; also kown as "Garland flower", this "White ginger" is most popular in Hawaii and used for leis because of the sweet perfume of its broad-petaled, pure white flowers, showing a yellow heart on their lip, and appearing from behind a green, waxen bulb of scale-like bracts in terminal clusters, on robust, leafy canes to 2 m long, the leaves silvery-haired beneath. *Subtropic.* p. 927, 928

flavescens (coronarium flavescens) (Indian Himalayas), "Cream ginger"; robust perennial herb with leafy stems to 2 m high, from stout rhizomes; leathery, lanceolate leaves 30 cm or more long; at apex a cluster of fragrant creamy flowers with yellow in base, and long red stamens. (see Neal: In Gardens of Hawaii, p. 252). *Tropical.* p. 928

flavum (Indian Himalaya), "Yellow ginger"; luxuriant green herb with leafy canes to 1½ m high; long, glabrous, slender pointed leaves almost 60 cm long and sheathing, alternate along the stem in 2 rows, bearing on its summit a head of shingled green bracts from which appears a broad cluster of heavily perfumed flowers, yellow with orange patch and with cream stamens; threaded into leis in Hawaii. *Tropical.* p. 928

gardnerianum (No.India), "Kahili ginger"; beautiful, desirable species growing in the Himalayas to 2500 m, of stiff habit, canes to 1½ m high, leaves to 45 cm and powdery-white beneath when young; the delightfully fragrant flowers in elongate, open terminal spikes to 45 cm long, and from a cylindrical cone of green bracts appear the yellow flowers having long, conspicuous, bright red filaments. *Tropical.* p. 927, 928

longicornutum (Malaya); tropical epiphyte with fleshy roots forming stiff canes, dressed along their length with glossy green lanceolate leaves; at apex a cluster of large, showy salmon-red flowers having one broad fan-like corolla lobe, the others narrow, thread-like. *Humid-tropical.* p. 925

HEIMERLIODENDRON *Nyctaginaceae*

brunonianum 'Variegatum' (Pisonia) (New Zealand), "Birdcatcher-tree"; showy sport of the "Para-Para" tree 6 to 15 m high and native from Tahiti and Marquesas to New Zealand and Australia, with oblong foliage to 40 cm long, having very sticky ribs, on short, robust petioles, and slightly angled stem; the glossy leaves are marbled in two shades of green, edged with warm cream to almost white; unfolding young growth has a tinge of red at edge and midrib; clusters of inconspicuous greenish flowers. The seed pods are covered with a sweet gum which attracts birds. *Subtropic.* p. 692, 694, 698

HELENIUM *Compositae*

bigelovii (California), "Sneezeweed"; perennial to 1 m, with unbranched stems forming a clump; lanceolate leaves sparsely hairy; disk cushion brownish, ray flowers yellow. *Subtropic.* p. 320

HELIAMPHORA *Sarraceniaceae (Carnivorous Plants)*

nutans (Guayana), "Sun pitcher", carnivorous rosette from the moist Roraima savannah, with funnel-shaped pitchers hairy inside, forming an insect trap of the pitfall type; leaves are red-veined; the delicate nodding flowers white. *Humid-subtropic.* p. 297

HELIANTHUS *Compositae*

annuus (Western United States), the "Common sunflower"; coarse stiff-hairy annual with stout straight stalk to 5 m tall, rarely branched and leafy with large ovate, rough hairy, toothed foliage to 30 cm long, and topped by tilted, immense heads sunlike to 50 cm across in cultivation, large disk of yellow tubular rays and blackish bracts, the large outer ring of florets orange-yellow. The white seed is used as food, as birdfeed, and furnishes oil. Wild on the prairies the flowers are only 8-15 cm across. *Temperate.* p. 320

annuus nanus, "Dwarf sunflower"; of smaller habit than the species, smaller leaves and flowers. *Temperate.* p. 320

grosseserratus (Prairies from Ohio to Dakotas and Texas); tall perennial, with toothed leaves; flowers yellow 5 cm across, in succession above each other on stems to 3 m high. *Temperate.* p. 320

laetiflorus (Kansas, Nebraska, Colorado), "Showy sunflower"; perennial of the high prairie; 15 cm to 2 m tall, with usually paired leaves; variable flowers, dark brown or yellow disk and yellow ray flowers, 4-6 cm across. *Temperate.* p. 320

HELICHRYSUM *Compositae*

belloides (bellidioides) (Gnaphalium) (New Zealand), "Immortelle"; much branched perennial with trailing stems to 50 cm; obovate leaves white-woolly beneath; flowers yellow with brown disk. *Subtropic.* p. 320

bracteatum (Australia), "Strawflower"; annual herb with narrow leaves and stalks of solitary flower disks enclosed by strawlike colored bracts in shades from white through yellow to red, and suitable for drying. *Subtropic.* p. 320

petiolatum (Gnaphalium lanatum) (So. Africa), an "Immortelle"; shrubby perennial with woolly stems and vine-like shoots; oval white-woolly leaves; flower heads with yellow disk flowers and cream-white bracts. *Subtropic.* p. 310, 321

HELICONIA *Musaceae*

acuminata 'Espiritu Santo' (Venezuela); clustering tropical plant 2-3 m tall, with dull green, leathery, lanceolate 50 cm leaves on long, slender petioles; erect inflorescence on wiry stalks of 2-ranked slender bracts orange shading to yellow at edge, 12 cm long; the small flowers orange yellow. Grown in Florida. *Tropical.* p. 664

aurantiaca (So. Mexico); reed-like stems 75 cm high, bearing oblong, smooth-leathery leaves, topped by a showy, erect inflorescence with lower bracts orange tipped green, the upper bracts yellowish-red, flowers greenish-red stalked red. *Tropical.* p. 665

bihai (W. Indies, Brazil, New Caledonia, Samoa), "Wild plantain" or "Firebird"; large perennial to 5 m high, long-stalked, oblong, smooth-textured, pointed, green leaves having a pale midrib and raised lateral veins; greenish-yellow flowers clustered in the axils of large, stiff boat-shaped, crimson-red, flattened bracts with pointed tip, and arranged in two ranks, on erect inflorescence. *Tropical.* p. 663

bourgaeana (Mexico: Veracruz); robust evergreen herb of the rainforest where I have seen it to 3 m high; with dark green long-stalked elongate leaves; hidden between the foliage the bold inflorescence erect with large stubby, wide-open boat-shaped bracts in two ranks, glossy crimson red suffused with rose; greenish flowers tipped white. *Humid-tropical.* p. 667

brasiliensis (Brazil), "False bird-of-paradise"; clustering perennial with long-stalked leathery leaves green above, pale beneath, to 60 cm long; lanceolate scarlet bracts, 6-10 cm long; flowers pink to purple. *Tropical.* p. 665

HELICONIA

caribaea (West Indies: Martinique), "Wild plantain"; huge perennial with large leaves 90 cm or more long and rounded at base resembling bananas but arranged in two ranks, the striking inflorescence is carried erect between the foliage, being a series of large, fleshy boat-shaped, pointed, stiff bracts holding water, and compacted shingle-like on two alternate series, waxy golden-yellow with keel and tip greenish. *Tropical.* p. 663, 664

caribaea var. purpurea; of similar habit as the species but the alternate, rigid, clasping spathes crimson red with a touch of yellow at the upper edges. *Tropical.* p. 663, 665

collinsiana (pendula in Hortus) (Guatemala); robust tropical perennial with lush growth to 3 m tall, big banana like foliage dull green, on thick petioles; the striking inflorescence pendant below the leaves, with spreading bracts crimson-red and covered with waxy powder, yellowish toward tips; cream flowers. *Tropical.* p. 666

distans (Trop. America); long-stalked, clustering plant with fresh green foliage; slender bracts lemon-yellow at base, crimson on top and to apex. *Tropical.* p. 663

humilis (Trinidad, Brazil), "Lobster's claw"; related to H. bihai, but with leaves shiny green, and smaller, salmon-red, boat-shaped bracts changing into green toward tip, and ridge of greenish-yellow; flowers yellowish white. *Tropical.* p. 663, 667

illustris 'Aureo-striata'; striking foliage plant with broad, recurving leaves shining fresh-green with contrasting ivory-yellow and pink midrib and closely parallel lateral ivory veins showing on both surfaces; sheathing petioles pinkish striped or mottled green. *Tropical.* p. 668

illustris 'Rubra', as grown at Hummel's, Carlsbad, California; colorful tropical plant with large 40 cm, pointed leaves first green with red midrib, turning brown-red; glossy wine-red beneath, petiole salmon-red mottled with black. *Tropical.* p. 664, 668

illustris rubricaulis; has beautiful foliage light green with midrib and the dense lateral veins clear rose pink, on clasping petioles vermillion red, and red underneath; rather delicate. *Tropical.* p. 668

indica; very decorative plant 1½-2 m high, with erect foliage broadly lanceolate, green and heavily overlaid with crimson red, on stiff red petioles. *Tropical.* p. 668

jacquinii (Fiji), "Lobster's claw"; large banana-like plant, tall, stout stalks with broad leaves, bearing between the foliage the erect, showy inflorescence of alternating, boat-shaped slender pointed leathery bracts salmon shaded to crimson, the upper edge and tips green. *Tropical.* p. 663

latispatha (Cent. America, No. S. America); showy plant with green broad-oblong leaves 1 m long and 30 cm wide, and an erect flexuous inflorescence with well separated boat-shaped bracts 15 cm long, orange yellow at the base near axis, lacquer red toward slender-pointed tip; flowers greenish. Occasional forms with colors reversed, or entirely red and entirely yellow. *Tropical.* p. 663, 664

marginata (Venezuela); plant to 2 m tall, with erect, narrow lanceolate leaves; inflorescence pendulous 30-40 cm long, the red bracts bordered with yellow; flowers yellow. *Tropical.* p. 665

mariae (C. America: Venezuela, Colombia), the "Beefsteak heliconia"; big plant to 5 m tall, with hanging inflorescence 25 to 80 cm long, having overlapping scarlet-red bracts with pink flowers. *Tropical.* p. 664, 666

psittacorum (coast of Guayana, Brazil), "Parrot flower"; tufted perennial with long-stemmed, narrow lanceolate, leathery, rich green leaves and a stalked inflorescence of shining orange, long-pointed bracts tipped red, and greenish-yellow flowers with black spots near apex. *Tropical.* p. 664, 666

psittacorum rhizomatosa (C. Venezuela), "Parakeet flower"; variety with thick, swollen rhizome; leaves narrow lanceolate; inflorescence with rose-pink bracts and red flowers tipped purple. *Tropical.* p. 666

psittacorum 'Rubra'; cultivar with bracts scarlet red, and flowers tipped blue. *Tropical.* p. 666

revoluta (Venezuela); giant plant 2-3 m high, with great leaves long lanceolate; inflorescence pendulous 45-60 cm long, 10 or more boat-shaped bracts scarlet-red; flowers cream. *Tropical.* p. 663

rostrata (Perú) "Hanging lobster-claws"; beautiful tropical herb 1 to 3 m high, with banana-like, leathery, green leaves; magnificent pendant inflorescence of alternating bracts scarlet red tipped with cream to yellow. *Tropical.* p. 666, 667

schiedeana (Mexico: Veracruz); tall clump-forming perennial to 2 m high, with long-stalked narrow-oblong leaves, usually slashed from the margins; the erect tall inflorescence on a red floral stalk with folded, sharp-pointed bracts red with greenish-yellow tips, well spaced along the rachis; out of the little boats peek yellow flowers, followed by the prominent clusters of fruit, first green then blue. *Humid-tropical.* p. 667

sharonii (Ecuador); handsome ornamental plant of compact habit; short, stiff red petioles carry broadly lanceolate leaves, deep green with red midrib and deep red reverse. *Tropical.* p. 668

spectabilis (South Pacific: Melanesia); showy clustering plant with large, succulent, pointed leaves bronzy-green having a brown-red midrib, wine-red beneath, on winged, deep red petioles; the several overlapping reddish bracts are deeply boatshaped in a short-stalked inflorescence not higher than the foliage. *Tropical.* p. 664

spectabilis 'Edwardus Rex'; richly colored, robust plant with large showy leaves of an intense, deep red with coppery sheen, more so underneath, especially the veins and stalks. *Tropical.* p. 668

wagneriana (Costa Rica, Panama); clump-forming tropical perennial, with stems to more than 1 m and long banana-like, but leathery leaves 1-2 m long, 30 cm wide; erect inflorescence with up to 20 folded bracts in 2 ranks and overlapping, light red shading to dark crimson, tips green; white flowers. p. 663, 666, 673

HELIOCEREUS *Cactaceae*

speciosus (Central Mexico), the "Sun cactus" because of its large and showy, day-flowering, bright scarlet flowers with a lovely steel-blue sheen; 4-angled stems erect or clambering, freely branching; the spines all alike. *Tropical.* p. 281

HELIOPSIS *Compositae*

helianthoides (Eastern United States), "Oxeye"; bushy perennial to 1½ m, smooth ovate leaves; flowers 6-8 cm across; disk brownish yellow, ray flowers pale yellow. *Temperate.* p. 320

helianthoides scabra (Eastern U.S.), "Ox-eyes"; perennial 1 m high, with rough-textured ovate foliage, and 8 cm flowers with brownish disk and orange-yellow ray-flowers. *Temperate.* p. 329

HELIOTROPIUM *Boraginaceae*

arborescens (Perú), "Heliotrope"; fleshy herb with small ovate wrinkled leaves and large clusters of small purple flowers very fragrant like vanilla. *Tropical.* p. 196

HELIPTERUM *Compositae*

roseum (Acroclinium) (W. Australia), "Everlasting"; annual strawflower to 60 cm, smooth linear leaves; solitary heads 5 cm across, usually rose colored. *Subtropic.* p. 320

HELLEBORUS *Ranunculaceae*

niger (Europe: Alps, Italy, Balkans), the "Christmas-rose"; known since the 16th century; a wintergreen herb with palmately divided leathery, olive green leaves; solitary flowers white or purplish, with green petals 4 to 6 cm across. *Temperate.* p. 843

orientalis 'Atropurpureus' (W. Asia: Black Sea area, Armenia), the "Lenten-rose", with basal leaves palmately divided; 4½ cm flowers dark purple outside, greenish inside. *Warm temperate.* p. 843

HELXINE *Urticaceae*

soleirolii (Corsica, Sardinia), popularly known as "Baby's tears", also as "Mind-your-own-business", "Irish moss", "Corsican curse" and "Japanese moss"; low moss-like creeping herb hugging the ground and forming dense mats or cushions as ground cover or over pots in subtropical plantings; tiny roundish lush-green leaves 6 mm or less across on threadlike intertwining branches, with minute greenish flowers in leaf-axils. *Subtropic.* p. 907, 911

HEMEROCALLIS *Liliaceae*

aurantiaca (China), "Golden summer daylily"; perennial with fleshy, tuberous roots; evergreen; sturdy, early-blooming; strap-shaped leaves 60-90 cm long, 3 cm wide; fragrant, bright orange flowers 10 cm long. *Warm temperate.* p. 575

HEMIGRAPHIS *Acanthaceae*

colorata (Java), "Red ivy"; prostrate tropical herb with stringy, rooting branches and opposite, broad-cordate, puckered and toothed leaves 6 to 10 cm long, shimmering silvery violet, underneath red-purple; terminal heads of small white flowers between large bracts. *Tropical.* p. 36

'Exotica', (New Guinea), "Purple waffle plant"; trailing plant of robust habit, with flexible pubescent reddish branches, opposite oval to ovate 8 cm leaves irregularly depressed-puckered, metallic purplish-green surface, margins crenate, reverse wine-red. A showy species we found in N.E. New Guinea. *Tropical.* p. 38

repanda (Malaysia); prostrate tropical herbaceous plant spreading by rooting stems, the red branches with narrow linear crenate leaves 6 cm long, satiny purplish green, and red beneath; small white 15 mm flowers in cone-like spikes; good in hanging baskets. *Tropical.* p. 36

HEMIONITIS *Polypodiaceae (Filices)*

palmata (W. Indies, Mexico, So. America), "Strawberry fern"; small tropical fern with palmate pubescent fronds to 15 cm long and wide, with 5 triangular divisions; the fertile leaves stiff and long-stalked, the sterile ones on short stalks. *Humid-tropical.* p. 448

HEMITELIA *Cyatheaceae (Filices)*

horrida (Puerto Rico); small treefern seen on the Toro Negro in the Cordillera Central; spineless dark brown, slender trunk, and narrow bipinnate fronds of firm texture. *Tropical.* p. 436

HEPTAPLEURUM: see Schefflera

HERACLEUM *Umbelliferae*
mantegazzianum (Caucasus); "Giant hogweed"; coarse biennial or perennial herb to 3 m, leaves to 1 m long, twice divided into threes; white flowers in giant clusters. *Warm temperate.* p. 908

HEREROA *Aizoaceae*
granulata (So. Africa); succulent dark green leaves 7 cm long, keeled toward apex, roughened by transparent dots. *Arid-subtropic.* p. 49
muirii (Cape Prov.: Karroo); distinctive cluster forming smooth succulent rosette to 7 cm high, with spreading, soft-fleshy, boat-shaped, semicylindrical, silver-gray leaves rough above and keeled, 5 cm long; yellow flowers, on leafy stalk. *Arid-subtropic.* p. 49

HERITIERA *Sterculiaceae*
fischeri (macrophylla) (E. Indies), "Glass-tree" or "Looking glass tree"; handsome tropical evergreen tree with large oblong, quilted leaves 25 cm or more long, shining yellow green above, opaque silvery beneath, reflecting the sun like so many mirrors; reddish when young; small greenish flowers in clusters. Photographed as H. fischeri at the Botanic Garden in Leningrad. *Tropical.* p. 901

HERNANDIA *Hernandiaceae*
ovigera (peltata) (Queensland, Trop. Asia, Trop. Africa), "Sea cups"; handsome evergreen tree to 12 m, with broadly ovate peltate decorative leaves to 20 cm long, leathery-glossy olive to dark green; small greenish-yellow flowers. *Tropical.* p. 523
sonora (East Indies); evergreen shore tree with buttressed roots; leaves arranged spirally, ovate, leathery; small yellowish flowers; strange fruit with black oily nut set inside grayish cup 5 cm wide. In India the seeds are used medicinally. *Tropical.* p. 523

HERNIARIA *Caryophyllaceae*
glabra (So. Europe, No. Africa, Turkey to C. Asia), "Green carpet"; low, dense groundcover forming thick carpets and hugging the ground; hairy perennial with gray green to bright green oval leaves 4-12 mm long, ciliate at margins; tiny insignificant greenish-white flowers. Used in California for rockeries and between stepping stones; turns bronzy in winter. *Warm-temperate.* p. 510

HETERANTHERA *Pontederiaceae*
dubia (graminea) (U.S.A., Mexico, Cuba), "Water star-grass"; aquatic plant with stems submerged in water, or floating on the water surface, or creeping on wet ground; linear green leaves 6-12 cm long; spidery yellow flowers carried above the water. *Humid-subtropic.* p. 828
osteniana (zosterifolia) (So. Brazil, Paraguay, Uruguay, Bolivia, Argentina); aquatic plant with stems both submerged or floating, or creeping on boggy ground; fresh-green linear leaves 2-4 cm long; starry blue flowers on terminal stalks. *Humid-subtropic.* p. 828
reniformis (Trop. and Subtrop. America); aquatic plant with long-stemmed kidney-shaped small leaves, rich green with yellow base; carried above the water; sparse flowers, white or pale blue. *Humid-subtropic.* p. 828

HETEROCENTRON *Melastomataceae*
roseum (Lasiandra) (Mexico), "Pearl flower"; small evergren shrub to 50 cm, with 4-angled branches; opposite elliptic 5 cm leaves, and clusters of bright rose flowers of 4 spreading petals. *Tropical.* p. 647

HETEROMELES *Rosaceae*
arbutifolia (Photinia) (California, Baja California), "Toyon" or "California holly"; also known as "Christmas-berry"; evergreen drought-resistant shrub or multiple trunk tree of the Sierra Nevada foothills, 3 to 8 m high, with leathery, recurved, shining leaves to 10 cm long; small white 1 cm flowers, followed by masses of bright red, long persistant berries through winter; a striking ornamental. *Subtropic.* p. 849, 852

HETEROPTERYS *Malpighiaceae*
beecheyana (Mexico to Bolivia), "Sobach"; climbing plant with ovate leaves 3-8 cm long, downy beneath; small 2 cm lilac flowers in large clusters; fruit 4 cm long and reddish, with 1-3 wings. *Subtropic.* p. 620

HEUCHERA *Saxifragaceae*
sanguinea (Arizona, Mexico), "Coral bells"; attractive herbaceous perennial with a tuft of rounded, lobed leaves with silver markings; slender stalks bear airy racemes of small bell-shaped flowers, bright rose-red; winter-hardy. Seen in the Swiss Alps as a potted plant. *Subtropic.* p. 879, 881

HEVEA *Euphorbiaceae*
brasiliensis (Amazonas), "Para rubber"; tropical tree, growing to 18 m, with milky juice containing latex which when coagulated becomes rubber; large palmate leaves of 3 stalked, quilted, papery segments bluish beneath; flowers greenish-white. *Tropical.* p. 429

HIBBERTIA *Dilleniaceae*
scandens (Australia: N.S.W., Queensland), "Guinea goldvine"; shrub with trailing stems, and ovate leaves 8 cm long, silky beneath; flowers yellow, to 5 cm across. *Subtropic.* p. 391

HIBISCUS *Malvaceae*
arnottianus (Kauai, Hawaii); dense bush with dull green thin-leathery ovate leaves, flowers white with protruding style carmine-red. *Tropical.* p. 625, 627
brackenridgei (Hawaii), "Native yellow hibiscus"; shrub or small tree to 10 m, the branch tips woolly and sometimes thorny; roundish lobed, toothed leaves 10 cm across; flowers yellow with petals to 8 cm long, sometimes with maroon base. *Tropical.* p. 628
brevipedunculata (Islands of the Caribbean); hairy bush with deeply lobed leaves; calyx lobes linear and spreading; cupped flowers carmine-pink with violet center. *Tropical.* p. 629
diversifolius (Trop. E. Africa to Pacific); a tropical "Mallow"; prickly shrub 2 m high, with leaves varying from round to angular to deeply three to five-lobed; flowers 15 cm dia., yellow to purplish with blackish-red center. *Tropical.* p. 629
elatus (Jamaica, Cuba), "Cuban bast"; ornamental tree to 25 m, also yielding fiber; heart-shaped leathery leaves 12-20 cm aross; flowers opening yellow and changing to orange and red; petals 8-12 cm long not overlapping; larger than H. tiliaceus. *Tropical.* p. 628
huegelii (Australia), "Blue hibiscus"; erect, tomentose shrub with small, deeply 3 to 5 lobed, dark green, coarse-hairy leaves, and numerous small, single, rosy-purple flowers with spreading petals. *Subtropic.* p. 628
moscheutos 'Southern Belle'; a "Rose-mallow"; garden hybrid to 80 cm high, serrate leaves, white-downy beneath; flowers crimson-red or rose with red center; blooms extremely large. *Warm temperate.* p. 630
mutabilis (South China), "Confederate rose" or "Cotton-rose"; fast-growing shrubby bush becoming tree-like where planted in the tropics and subtropics; green stems becoming woody, the large 3 to 5-lobed leaves 10-20 cm wide, dull green and rough pubescent; toward branch ends the showy axillary flowers 10-12 cm across, opening white or rose in the morning with crimson center and a divided maroon column; by evening the flower becomes deep red. In colder areas the plant is deciduous. *Subtropic.* p. 628, 629
rosa-sinensis (Trop. Asia), "Chinese hibiscus"; magnificent flowering shrub of vigorous habit, to 3 m high; glossy green, serrate leaves, and mostly very large 5-petaled flowers from white to yellow and red to magenta. Hibiscus are among the most showy flowers of the tropics, especially the Chinese hibiscus, which is the state flower of Hawaii where some 5000 varieties are known. Raw flowers are eaten there to aid digestion. In Tahiti, a flower worn over the right ear shows that one is looking for a mate. Blossoms of most varieties remain open for one d y only, unfolding early in the morning and dying after closing near sunset, but in good sunlight are almost everblooming, especially if given liberal watering and nourishment. Much used as flowering shrubs and hedges in tropical gardens. Chinese hibiscus are also wonderful container plants for pots or tubs indoors, or on the summer patio, preferring some rest, and cutting back in winter; need lots of food and water when growing, to support adequate foliage and prodigious bloom. *Tropical.* p. 626, 627, 629
rosa-sinensis albus; free-flowering pure white blooms of medium size. *Tropical.* p. 626
rosa-sinensis 'Aurantiacus'; in California nursery trade; salver-shaped single flowers of waxy texture, vivid salmon with purple center, the margins frilled. *Tropical.* p. 627
rosa-sinensis 'California Gold'; shapely bush of robust habit, with fresh-green, broad ovate leaves, and medium large, firm, single flowers golden yellow, with crimson-red center, blooming with the sun. Very popular in California gardens. *Subtropic.* p. 626
rosa-sinensis 'Cheerful'; single flowers deep rose with white center. *Tropical.* p. 626
rosa-sinensis 'Cooperi' (E. Indies), "Checkered hibiscus"; ornamental shrub mainly grown for its colorful foliage, the narrow lanceolate leaves are metallic green and brightly variegated and marbled with dark olive, white, pink and crimson; small scarlet flowers. *Tropical.* p. 626, 629, 630
rosa-sinensis 'Crown of Bohemia'; bushy shrub to 3 m, double golden yellow flowers; petals change to orange and carmine toward base. *Tropical.* p. 629
rosa-sinensis 'Hula Girl'; large single flowers 15 cm across; yellow to salmon with bold crimson center and purple lines; prolific blooms. *Tropical.* p. 626
rosa-sinensis 'Lateritia'; an excellent cultivar of compact, bushy habit, with dark green, lobed leaves; charming 10 cm flowers bright

orange yellow with blackish-red eye in center; a favorite pot plant for window sills in Germany. *Tropical.* p. 630
  rosa-sinensis 'Lutea'; single flowers a delicate shade of salmon-yellow. *Tropical.* p. 626
  rosa-sinensis 'Matensis', "Snowflake hibiscus"; probably a clone or colorform of H. rosa-sin. Cooperi; vigorously branching shrub with willowy reddish stems dense with rough ovate, toothed leaves grayish-green, variegated mainly toward margins with creamy-white; single 8 cm flowers carmine-red with crimson veins and center. *Tropical.* p. 627
  rosa-sinensis 'Mist'; large yellow flowers fully double with fluffy inner petals, and quilted foliage. *Tropical.* p. 626
  rosa-sinensis cv. 'Natal' (So. Africa); free-blooming bush, dense with shiny foliage, and white flowers flushed with pink, and deep red center. *Subtropic.* p. 627
  rosa-sinensis plenus (India to China), "Double Rose of China"; magnificent robust flowering bush of somewhat sparry habit, with fresh green foliage and large showy blooms, 10-15 cm across, typically carmine-rose with dark center, but with flowers fully double; in bloom whenever the sunlight is intense enough, best during the warm summer season but also in winter. Other cultivars are in shades of yellow to deep red. *Tropical.* p. 629, 630
  rosa-sinensis 'Scarlet', also known as 'Brilliant'; a cultivar very floriferous with single, medium-large flowers in blazing scarlet-crimson, blooming with the sun from early spring into late fall and through the winter if kept light and warm; its strong habit with its glossy foliage, lends itself to training into attractive tree-like semi-standards. *Tropical.* p. 626, 629
  rosa-sinensis 'Toreador'; vigorous, erect grower with pubescent, serrate leaves; free-blooming with single 12-15 cm flowers maize-yellow, in the center a contrasting, definite eye of ruby-red with striking effect. *Tropical.* p. 629
  rosa-sinensis 'White Wings'; large flowers with linear, spreading petals 9 cm long, freely separated to base, white with contrasting feathery crimson base, and long red staminal column. Willing grower of spreading habit, and profuse with blooms. *Tropical.p.* 629
  schizopetalus (Trop. E. Africa), "Japanese lantern"; glabrous shrub with slender drooping branches, and smooth, ovate-elliptic, toothed leaves; the showy orange-red flowers hanging from slender stalks, petals deeply slit and recurved, and a long projecting, pendulous staminal column. *Tropical.* p. 623, 630
  schizopetalus 'Pagoda' (Colombia), "Flora en flora"; curious mutant from Colombia, with bright red flowers, the tubular column of united stamens white, the style developed into a petaloid secondary cluster of frilled petals resembling a miniature rose. *Tropical.* p. 630
  syriacus (East Asia), "Rose of Sharon" or "Shrub althaea"; shrub or small tree to 3 m; leaves to 8 cm long, triangular or three-lobed and coarsely toothed; flowers single or double, 6-8 cm across, rosy-lavender with purple center, but also other shades including white. *Subtropic.* p. 623, 630
  tiliaceus (Trop. Asia and Polynesia); crooked shrub or much branched tree to 4 m or more, with large cordate, matte-green leaves, white hairy beneath; 8 cm flowers open lemon-yellow with or without brown-red base inside, later in the day they change to orange and by night to red. In Polynesia, the fiber of its inner bark is used for ropes or tapa cloth. *Tropical.* p. 628
  trionum (So. Europe, W. and C. Asia, E. and So. Africa), "Flower-of-an-hour"; bushy annual with leaves deeply divided into coarsely toothed lobes; cupped 8 cm flowers sulphur-yellow or white with velvety purple center. *Subtropic.* p. 630

HIMANTOGLOSSUM: see Barlia

HIPPEASTRUM *Amaryllidaceae*
  'Giant White' (leopoldii hyb).; typical of the high perfection of breeding, with large, 20 cm, full round flowers glistening white and faintly green in the throat; thick petals of good lasting quality. *Tropical.* p. 66
  'Leopoldii hybrid' (Amaryllis), a Dutch "Amaryllis"; the leopoldii hybrids are considered the finest class of fancy flowering amaryllis in pots, the result of breeding H. leopoldii from Perú with reginae and other species; large flat, open-faced flowers generally 20 cm but even to 30 cm across, with roundish, overlapping segments, and short tubes. Hybrid clones come in a wide range of colors, from deep glowing red, clear scarlet, orange-red, light rose, and white with stripes or finally, all white. *Tropical.* p. 67
  procerum (Worsleya rayneri) (Brazil), "Blue amaryllis"; bulb with neck to 30 cm long, from which extend the drooping leaves; lilac blue flowers with pointed segments to 15 cm long. *Tropical.* p. 65
  puniceum (H. equestre) (West Indies and Mexico to Brazil), popularly known as the "American belladonna" or "Barbados lily"; a smaller-flowered amaryllis with bulb having brown scales and short neck, narrow strap-shaped, pointed waxy green leaves 2 to 3 cm wide, and long round stalk bearing 2 or more obliquely trumpet-shaped flowers of 12 cm dia. and 10 cm long, salmon-red, with center creamy-whitish with greenish bands, an oblique feathery corona in base; stamens when straight larger than stigma; an old-fashioned flowering houseplant almost continuously in bloom. *Tropical.* p. 66
  reginae (Trop. America, Antilles, W. Cent Africa), "Royal amaryllis" or "Mexican lily"; colorful species with large scarlet trumpets 12 cm dia. and greenish-white star in throat, on 30 cm stalks, blooming in summer. When flowering is over, the bulb produces the strap-like leaves. *Tropical.* p. 66
  reticulatum striatifolium, "Stripe-leaf amaryllis"; an interesting variety of this bulbous Brazilian plant because of its strapshaped dark green leaves having a prominent ivory-white midrib; the lily-like flowers are rose-pink lined with darker rose. *Tropical.* p. 65
  vittatum 'Equestre Red' (Amaryllis); handsome hybrid as grown in New Zealand; prolific bloomer with stiff erect, hollow stalks topped by clusters of large trumpet flowers brilliant scarlet, with white star in center. *Tropical.* p. 67
  'Vittatum hybrid', commercial "Amaryllis", developed from species of the Peruvian Andes; large bulb producing spectacular trumpet-shaped flowers to 15 cm across, several on a hollow stalk, most often scarlet red with white; strap-shaped leaves appearing with or after bloom, in late winter; parentage includes H. leopoldii, reginae, aulicum, solandriflorum, reticulatum. *Tropical.* p. 65, 66

HIPPOBROMA *Lobeliaceae*
  longiflora (Indonesia); low growing, pretty plant seen around temples and kampongs in Bali; soft-fleshy lanceolate, dark green leaves with toothed margins and pale midrib; white flowers with spreading petals and long slender tube. *Tropical.* p. 616

HIPPOMANE *Euphorbiaceae*
  mancinella (W. Indies), "Manzanillo"; large spreading shrub with milky, poisonous sap, leaves shining green, ovate, 8 cm long; flowers inconspicuous; the fruit like little green apples; very ornamental and prominent along the shores of Curacao. *Tropical.* p. 407

HIPPOPHAE *Elaeagnaceae*
  rhamnoides (Europe; Caucasus, Iran, Siberia), "Sea-buckthorn"; spiny shrub, very hardy, to 10 m high, covered with silvery scales; narrow willow-like, thin-leathery leaves; inconspicuous yellow flowers, and masses of bright orange-yellow 1 cm fruit, staying on all winter. *Temperate.* p. 393

HOFFMANNIA *Rubiaceae*
  ghiesbreghtii (Higginsia) (Mexico), "Tall taffeta plant"; erect, herbaceous ornamental with 4-cornered, winged, green stems to 1 m tall, the long lanceolate leaves soft and shimmering-velvety, bronzy to moss-green, quilted between the silvery green and pink ribs, rosy-red beneath; small, tubular flowers yellow with red spot. *Tropical.* p. 862
  ghiesbreghtii 'Variegata', "Variegated taffeta plant"; delicate form with its long perfoliate leaves variegated and mottled on the quilted surface with cream and pink over milky-green and bronze; red beneath. *Tropical.* p. 862
  refulgens (Chiapas), "Quilted taffeta plant"; beautiful, low, herbaceous plant with the short-jointed, red stem hidden by the broadly obovate, heavily quilted leaves almost fleshy, iridescent-velvety coppery-purple, shading to greenish, with bluish or light green ribs and red-hairy, red margin; wine-red beneath; flowers pale red. *Humid-tropical.* p. 862

HOHENBERGIA *Bromeliaceae*
  stellata (Brazil, Venezuela); loose rosette of green leathery leaves, finely toothed; inflorescence a long leaning spike with alternate clusters of red bracts with purple flowers. *Tropical.* p. 219

HOHERIA *Malvaceae*
  populnea 'Argentea' (New Zealand), "Variegated lace-bark"; evergreen small tree, forming trunk covered with tough, lacy, perforated inner bark, willowy deep brown branches, and alternate, ovate, rugose 12 cm leaves with irregularly double sawed margins, matte, rich deep green beautifully variegated creamy-white toward edge; white, starry axillary flowers in profusion. *Subtropic.* p. 622
  populnea 'Aureo-variegata' (New Zealand); dense evergreen bush with elliptic leaves primarily bronzy-yellow, the center grayish green. *Subtropic.* p. 622

HOLARRHENA *Apocynaceae*
  antidysenterica (Wrightia) (Trop. Asia), the "Easter tree"; tropical evergreen shrub with cord-like branches, and opposite, oval leaves; charming tubular pure white flowers with 5 spreading,

ovate petals, 4 cm across, in center a small white corona; blooming over a long period. *Tropical.* p. 76, 80

pubescens (antidysenterica var.) (Bengal, Trop. Africa), "Jasmine tree" in Africa, "Ivory tree" in India; bushy small tree, with shiny green, ovate leaves 15 cm long; the numerous small waxy, fragrant white 3 cm flowers in clusters along branches; fruit as a pair of long erect pods. *Tropical.* p. 78

HOLCOGLOSSUM *Orchidaceae*
quasipinifolium (Taiwan); small epiphyte with slender cylindrical foliage, and small greenish-white, complex flowers, with large lobed lip prolonged into a horn curved forward, the side lobes red-brown. *Subtropic.* p. 761

HOLMSKIOLDIA *Verbenaceae*
sanguinea (Himalayas), "Chinese-hat-plant"; subtropical, straggling shrub attaining 10 m, with slender-pointed ovate leaves to 10 cm long, and curious flowers having scarlet tubular corolla 2 cm long, and a spreading bell-shaped, orange calyx. Widely grown for its odd and beautiful blossoms. *Subtropic.* p. 914, 917

sanguinea 'Aurea', "Mandarin-hat plant"; originally from subtropic Himalaya and Malaysian region, now widely planted or naturalized in Tropical America; handsome evergreen wide-ranging rambler with wiry branches and opposite broad green leaves; toward the end the showy racemes of flowers with both hat-like calyx and tubular corolla vivid orange yellow, blooming all summer or longer. *Tropical.* p. 917

HOMALOCLADIUM *Polygonaceae*
platycladum (Muehlenbeckia) (Solomon Isl.), "Ribbon bush", or "Tapeworm plant"; odd curiosity plant with perfectly flat, jointed, fresh-green stems and small, lanceolate leaves; leafless in the blooming stage, with small greenish flowers at alternate joints; in the tropics making round canes to 4 m long. *Tropical.* p. 824

HOMALOMENA *Araceae*
picturata (Colombia); small, pretty herb with cordate-ovate dark green leaves variegated silvery-white along each side of midrib, on sheathing slender petioles; spathe green, spadix white. *Tropical.* p. 108

pygmaea (Malaysia: Java, Sumatra, Borneo, Celebes, Philippines); small, clustering herb with short rootstock, leaves elliptic-lanceolate, dark green, reverse purplish, 5-10 cm long, with undulate margins, on short, partly vaginate petioles; oblong spathe yellowish green. The var. purpurescens has purple spathe. *Humid-tropical.* p. 96

sulcata (Borneo); dwarf plant with long, heartshaped, metallic coppery green leaves on slender brown petioles in rosette. *Humid-tropical.* p. 106

wallisii (Curmeria) (Colombia), "Silver shield"; low, compact, leathery plant with broad, oval, reflexed leaves dark olive-green and beautifully blotched with yellowish-silver, translucent-silvery edge. *Tropical.* p. 106, 108

wendlandii (crinipes) (Costa Rica); hastate leaf dark green with pale center vein and reddish margins, glabrous; pale beneath. *Tropical.* p. 96

HOODIA *Asclepiadaceae*
gordonii (So. Africa); succulent with thick, cylindrical, angled stems covered with warty teeth; cupshaped flowers yellowish-brown. *Arid-subtropic.* p. 154

lugardii (Botswana to Namibia); handsome succulent gray columns to 45 cm tall; with about 16 deep angles; the ridges with small spiny tubercles; rather flat, roundish flowers lightly 5-lobed, brick red, lobes 1 cm long. *Arid subtropic.* p. 154

rosea (So. Africa, Botswana), "African hat plant"; clustering cactus-like succulent to 30 cm high with leafless circular grayish-green stems branching from the base, ribs about 14, dense with tubercles armed with pale brown spines; the spectacular flowers have a saucer-shaped flat corolla 8 cm across, rosy to cinnamon and covered with thin hairs. Keep dry and warm in winter. *Arid-subtropic.* p. 147

HOSTA *Liliaceae*
crispula (Japan) (syn. H. fortunei var. marginato-alba, H. coerulea); elegant plant densely tufted, slow growing but forming large groups; elliptic long-pointed leaves with 7-9 veins, matte green above, bordered white, and undulate; glossy beneath; lavender flowers, on long lax stalk. *Warm temperate.* p. 590

sieboldiana (Japan), the "Seersucker-plantain-lily"; giant roundish to ovate 25-40 cm leaves with many deeply depressed veins, deep green on surface, glaucous-blue beneath; nodding flowers white tinged lilac, in cluster above foliage. *Warm temperate.* p. 590

HOULLETIA *Orchidaceae*
lansbergii (Guatemala to Venezuela and Brazil); attractive species with small pseudobulbs carrying a 30 cm solitary, rigid ribbed leaf; pendulous inflorescence of fleshy, fragrant, waxy flowers 4 cm long, sepals orange with red spots, the smaller petals similar but darker, lip white with red stripes and purple warts (autumn). *Tropical.* p. 734

HOWEIA (KENTIA) *Palmae*
belmoreana (Kentia) (Lord Howe Island), "Sentry palm"; handsome feather palm to 8 m, with pinnate leaves erect and then arching downward, the segments crowded, narrower than forsteriana, first upward then gracefully pendant and slender-pointed, on reddish stalks; fruit yellow green. *Subtropic.* p. 777

forsteriana (Kentia) (Lord Howe Isl., near Australia), "Paradise palm"; elegant, sturdy decorator, widely used by florists because of its good keeping qualities; graceful pinnate fronds growing successively larger on slender stalks, the well-spaced, waxy deep green pinnae leathery and durable; with age forming a robust trunk 20 m high; yellow-green, olive-shaped fruit, in heavy clusters, in successive 4-strand racemes, ripe about 4 years after first flowering. *Subtropic.* p. 774, 780, 783

HOYA *Asclepiadaceae*
australis (Queensland), "Porcelain flower"; robust vine with broad oval, pointed leaves, thick, waxy, green with occasional silver spots, fuzzy beneath; dainty pink, waxy flowers with red crown, fragrant. *Tropical.* p. 153

bandaensis (Java, Moluccas); robust climber with large, oval-pointed, fleshy leaves dark green and glossy; blooms greenish-white with scarlet centers. *Tropical.* p. 150

bella (paxtonii) (India), "Miniature wax plant"; dwarf, shrubby plant with flexuous branches first upright, later drooping, the small, thick leaves ovate, deep green; flowers waxy-white with purple center. *Tropical.* p. 150, 153

carnosa (Queensland, So. China), called "Wax plant" because of the waxy, wheel-shaped, fragrant, pinkish-white flowers with a red, star-shaped crown in pendant umbels; root climbing vines with elliptic fleshy-waxy leaves. *Subtropic.* p. 153

carnosa 'Compacta', "Hindu rope plant" or "Honey plant"; contorted cultivar of Hummel's H. carnosa 'Exotica' in California, a tortuous wax vine densely compacted with folded and cupped foliage oddly twisted, the 4 to 5 cm leaves waxy green with occasional silver spots, grayish beneath; the waxy flowers in a dense cluster, old ivory tinted pink petals and starry crown, on brown base. Slow-growing and ideal for training on trellis for near the window. *Tropical.* p. 153

carnosa 'Compacta Regalis', "Variegated Hindu rope"; curiously interesting Florida plant, slow-growing and rather erect for a long time, dense with waxy, thick leaves, twisted and crinkled, green and prettily bordered with ivory to rosy margins. *Subtropic.* p. 150

carnosa 'Hummel's compacta', "Hindu rope"; tortuous vine densely compacted with twisted foliage; blooms when small. *Tropical.* p. 153

carnosa 'Marginata'; charming twiner with slender flexible stems, and waxy, thick-leathery leaves 6-8 cm long, olive green with pronounced ivory white margins turning pink in strong light. *Tropical.* p. 153

carnosa 'Rubra', "Krimson Princess"; brilliantly colored Florida vining cultivar of H. carnosa with younger leaves rosy crimson over ivory center, the margins rich olive green; developed and patented by Cobia's of Winter Garden, Florida. *Subtropic.* p. 150

carnosa 'Silver Princess' (Argentea picta); attractive cultivar with leaves painted milky silver, variegated white to rosy-red toward margin. *Subtropic.* p. 150

carnosa 'Tricolor', "Krimson queen"; (Pat. Cobia, Florida); foliage beautifully coppery with salmon-rose while young, green edged ivory when mature; blooms pink. *Tropical.* p. 150

carnosa 'Variegata' "Variegated wax plant"; ornamental variety with the fresh green to bluish leaves broadly bordered creamy-white and even pink; more variegated than 'Marginata'. *Tropical* p. 152, 153

carnosa 'Verna Jeanette'; pretty variation of H. carnosa marginata with leafblades covered as if with silver, some dark green mottling showing through; margins creamy-white; young leaf edges tinted pink. *Subtropic.* p. 153

coronaria (Java); long vining climber with waxy, oval, fresh-green leaves with margins recurved, hairy beneath; waxy flowers pale yellow with 5 red spots at base of crown. *Tropical.* p. 150

darwinii (Philippines); clambering vine with thin-wiry stems, and distant internodes bearing oblong-ovate recurved leaves; stunning cupped 3 cm flowers rose with purple center. Occasionally with pouched leaves as in Dischidia. *Tropical.* p. 150

gigas (New Guinea); one of the many Hoyas we collected on government expedition in the Finisterre Mountains; a beautiful liane with 4-6 cm waxy flowers reddish brown with light tips. *Tropical.* p. 151

globulosa (Sikkim, Himalayas); slender vine with thick oval-pointed, spoon-like, lightly hairy leaves, and round clusters of cream colored waxy flowers, the corona marked brown-red at base. *Subtropic.* p. 152

imperialis (Borneo), "Honey plant"; tall climber with downy stems and elliptic shiny leaves lightly downy, margins wavy; flowers with cream corona in reddish-brown, waxy flowers to 7 cm across. *Tropical.* p. 151

kerrii (obovata hort.) (Thailand, Vietnam, Fiji), "Sweetheart hoya"; robust twiner; curious, thick succulent, leathery leaves in form of an inverted heart; small flowers cream-white, corona lobes rose-purple. *Tropical.* p. 153

longifolia (Central Himalayas); climber with stout stem; very fleshy oblanceolate, concave leaves grayish green, 15-20 cm long; clusters of starry white flowers 2-3 cm across. *Subtropic.* p. 152

'New Guinea Red' (New Guinea); a waxvine collected in the Eastern Highlands, and grown by Seaborn in California; robust species with wiry stems, ovate or obovate moss-green leaves 8-10 cm long; rusty-red flowers. *Tropical.* p. 145

'New Guinea White' (Papua); large glossy oblong leaves and clusters with to 15 large waxy, cupped 3 to 5 cm flowers pure white. *Tropical.* p. 150

polyneura (Himalaya Reg.); glossy leaves greenish-silver with darker veins; waxy flowers like white stars, corona bronzy red. *Subtropic.* p. 152

purpureo-fusca (Java), "Silver-pink vine"; found in Queensland, and introduced as 'Silver Pink' from Hawaii; a beautiful vining plant with raised pinkish silver blotching on waxy-green leaf, petioles red; rusty-red, white-hairy flowers, corona pinkish and purple. *Tropical.* p. 145, 153

rubida (New Guinea); a beautiful climber which we collected in the wild Finisterre Mountains; fleshy glossy green, 10 cm obovate leaves; stunning 3 cm flowers lacquered maroon, with red corona. *Tropical.* p. 151

sikkimensis (Himalaya); dwarf plant with flexible, pendant branches set with small fresh green, pointed leaves; clusters of white-waxy starlike flowers, pink crown with purple tips. *Subtropic.* p. 152

"Silver Pink": see purpureo-fusca.

HUERNIA *Asclepiadaceae*

macrocarpa cerasina (Ethiopia, Sudan), a "Dragon flower"; small leafless succulent with olive-green, 5-angled and toothed stems to 10 cm high, star-like flowers near base velvety blackish-red with wart-like elevations; reverse gray-brown. *Arid-tropical.* p. 147

namaquensis (So. Africa: Cape Prov.); dwarf leafless succulent with 4 to 5-angled erect grayish stems 4-6 cm high; star-shaped, cupped flowers lemon-yellow covered with red spots. *Arid-subtropic.* p. 147

reticulata (So. Africa: Cape Prov.); clustering succulent 5-10 cm high, stems with 5 acute, toothed angles, coppery green mottled with red; plate-like flowers 5 cm dia., yellow reticulated with red; fleshy blackish-red crown. *Arid-subtropic.* p. 147

zebrina (So. Africa), "Owl eyes" or "Zebra flower"; thick clusters of little, fleshy, angled stems, marked reddish, with spreading teeth; small, yellow flowers with purple bands. *Arid-subtropic.* p. 147

HUMATA *Polypodiaceae (Filices)*

tyermannii (Cent. China), "Bears-foot-fern"; related to Davallia, with creeping light brown rhizome covered with silvery-white scales and looking like a rabbit's foot; brown, channeled petiole bears sturdy dark green, leathery, tripinnate 15 cm frond, always with spores; slow growing. *Humid-subtropic.* p. 442

HUMBERTINA: see Arophyton

HUMULUS *Moraceae*

japonicus 'Variegatus' (scandens) (Temp. E. Asia), "Japanese hops"; ornamental rough-stemmed, tall twining perennial, with serrate leaves deeply 5-7 lobed, rugose surface in shades of green, yellow and splashed with white; small flowers in cone-like, bracted spikes, used in the brewing of beer. *Warm temperate.* p. 650, 660

HUNNEMANNIA *Papaveraceae*

fumariaefolia (Mexico), "Mexican tulip-poppy"; erect perennial 60-90 cm high, woody at base, grown as annual; glaucous 3-parted leaves 5-10 cm, with linear segments, and long-stalked yellow flowers 5-8 cm across, of four concave petals. *Subtropic.* p. 812

HUNTLEYA *Orchidaceae*

meleagris (Brazil); epiphyte without pseudobulbs, bluish-green leaves, to 40 cm long, arranged fan-like, in 2 ranks, the solitary 8 cm flowers on long stalks, with sepals and petals whitish at base, passing into yellow, heavily flushed and marked red-brown, fringed lip cordate, white, with front yellowish-brown. (June-July). *Humid-tropical.* p. 734

HYACINTHUS *Liliaceae*

'Anne Marie' (orientalis); delicate soft, clear pink flowers, but smallish heads; fine for earliest forcing. *Subtropic.* p. 599

'Blue Giant'; big heads of large porcelain blue bells, washed with violet. *Subtropic.* p. 599

'Blue Jacket; large heads of single dark blue flowers, with small purple stripe; excellent for Spring forcing. *Subtropic.* p. 599

'Carnegie' (orientalis cv.); an excellent late-blooming Dutch hyacinth of robust, compact habit, with large and full heavy heads of pure white, waxy flowers; a favorite for flowering in pots especially if Easter is late, while the ivory-white 'L'Innocence' is recommended for an early Easter. *Subtropic.* p. 599

'Delft Blue', a "Dutch hyacinth"; excellent for Easter flowering in pans; large 3-4 yr. bulbs will produce a full spike of bright porcelain-blue flowers of rich fragrance. *Subtropic.* p. 599

'Lord Derby'; large head of single flowers vivid rose-pink, for mid-season blooming. *Subtropic.* p. 599

'Ostara' (orientalis cv.); robust Dutch garden hyacinth with large, heavy spikes of medium sized, 3 cm porcelain-blue or violet-purple waxy flowers on pale center, suitable for both early and Easter flowering in pots; an improvement of the old favorite, pale blue 'Bismarck'. *Subtropic.* p. 599

'Pink Pearl'; an excellent, vivid pink hyacinth, often with two spikes per bulb; the trumpet-shaped flowers are more loosely set, the stout stalks are brittle as in all hyacinths, and leaves are linear; the species is native from Greece to Iraq. *Subtropic.* p. 599

HYDNOCARPUS *Flacourtiaceae*

anthelmintica (Thailand), "Chaulmoogra tree"; evergreen tree to 15 m, with thin, leathery narrow leaves; flowers with numerous stamens; fruit 6 cm dia., a hard shell covered with brown velvet; the seed containing oil for treatment of skin diseases or leprosy. *Tropical.* p. 484

HYDRANGEA *Saxifragaceae*

aspera (Himalayas, China); shrub to 2 m or more, of stiff habit, with downy branches; slender-pointed 10-22 cm, toothed leaves with harsh surface, gray-downy underneath; 3-4 cm sterile flowers with 4-6 blue or white petals. *Warm temperate.* p. 879

macrophylla 'Alpengluehn' (1950); robust freely branching plant of even habit and easily budding with firm, shining fresh-green, deeply dentate leaves and medium heads of crimson-rose flowers of good substance, greenish in opening. *Warm temperate.* p. 878

macrophylla 'Chaperon Rouge' (Red Ridinghood), also known as 'Chapeau Rouge' (Red Cap), "French hortensia"; superb French (Mouillere 1951) cultivar of compact habit, with smallish dark green foliage on stiff-erect stem, which bears a large firm, rounded head dense with flowers in vivid, pure rosy-crimson to clear carmine, depending on degree of temperature and pH of soil; ideal pot plant, medium-early. *Warm temperate.* p. 878

macrophylla (hortensis) **'Enziandom'** (1949); excellent, free-branching, tallish potplant for early forcing, with dark green, glossy leaves and large, dense heads of flowers, by nature vivid red but by the use of aluminum sulphate in an acid soil, of pH5 or below, easily coloring a beautiful, deep gentian-blue. *Warm temperate.* p. 878

macrophylla 'Flamboyant'; vigorous grower with smallish leaves on good stems supporting the large heads of carmine 5 cm ruffled flowers, a shade darker than Merveille. *Warm temperate.* p. 878

macrophylla 'Kuhnert' (1926) (Gartendirektor Kuhnert), "Snowball"; an old freely branching variety easily becoming tall, with small toothed foliage and medium size flower-heads normally rose, but mostly grown as a good blue; not as showy as 'Enziandom', but a ready forcer and good keeper, responding willingly to aluminum treatment with a clear sky-blue and cream center, to corn-flower blue. *Warm-temperate.* p. 877, 878

macrophylla 'Merveille' (1927), "French hortensia"; robust, vigorous, midseason variety of excellent keeping qualities, with stout stem not requiring staking, firm foliage and large heads of big round, carmine rose flowers of good texture; lend themselves also to coloring into blue or lilac; the showy calyces hold well after forcing. *Warm temperate.* p. 878, 879

macrophylla 'Otaksa'; formerly well-known old "Hortensia", in cultivation for about 100 years, one of the first to be tried for pot forcing but later used for planting outdoors; a hybrid Japanese clone of Thunberg's original H. macrophylla, forming the foundation for the modern race of pot-flowered hydrangeas. 'Otaksa' is of dwarfer habit than the type species, yet vigorous, with obovate, short-pointed leaves rather thick and smooth, the rounded heads of blooms partly with insignificent fertile flowers, partly showy sterile ones with obovate, entire petals, pink or blue. However, it flowers only on 2nd year wood, and only if buds were not hurt by cold. *Warm temperate.* p. 877

macrophylla 'Regula'; German cultivar of compact habit, with smallish leaves, and solid heads of greenish-white flowers. *Warm temperate.* p. 878

macrophylla 'Rosabelle' (1928); a good variety where extremely large "Snowballs" are desired, of robust growth with rigid, thick stems and tough, dull green, smooth leaves becoming rather tall, supporting fewer but giant heads of carmine-rose flowers 20-25 cm across, the individual showy sepals large and not crenate; late season and Mother's Day; normally extremely durable but does not keep well if forced early. *Warm temperate.* p. 878

macrophylla 'Soeur Therese' (1945); willing grower and early forcer, freely branching, with strong stems and dark foliage which sets off beautifully the noble flower heads of pure white, a great improvement over E. Mouillere. *Warm temperate.* p. 878

macrophylla 'Strafford' (1937) (Mad. Cayeux); favorite commercial, well-proportioned variety with strong erect stem, small, dark green foliage, and medium-large, firm heads of clear rosy-red flowers—actually the enlarged calyx lobes, which are first pale yellow-green, then colored; mid season, free to bud, but flowers have a tendency to drop after forcing. *Warm temperate.* p. 878

macrophylla 'Todi', "German hortensia"; of short habit, with rather weak stem carrying a heavy head of individually large, dark pink flowers with entire sepals; aluminum suplhate treatment results in purple flowers; inclined to blindness on the numerous branches. *Warm temperate.* p. 878

macrophylla 'Variegata'; a variegated form of the "Big leaf hydrangea", with broad leaves silvery green and edged with creamy white; fairly hardy but deciduous outside; inflorescence in terminal panicles, some of the marginal sterile flowers with large petal-like sepals, colored blue in acid soil with available aluminum, pink in sweet soil with calcium. *Warm temperate.* p. 878

quercifolia (Georgia to Florida and Mississippi), "Oakleaf hydrangea"; shrub to 2 m, with stout brown-woolly branches, deeply 3-7 lobed 20 cm leaves, dull green above, pale downy underneath; pyramidal 30 cm panicles of white sterile 2-3 cm flowers turning purplish, and small fertile ones; fairly hardy. *Warm temperate.* p. 880

HYDROCLEIS *Butomaceae*

nymphoides (C. America to Argentina), "Water poppy"; perennial water plant with floating leaves, broadly heart-shaped or almost round, leathery-pulpy, and glossy green; everywhere in the leaves and stalks are airy tissues that causes them to float; beautiful flowers shining yellow with red center, 4-5 cm wide. *Humid-tropical.* p. 237

HYDROSME *Araceae*

rivieri (Amorphophallus) (Vietnam), "Devil's tongue" or "Leopard palm"; curious tuberous plant with a 1 m flower-spike which carries the large, reddish spadix and calla-like green and purple spathe, and with unpleasant odor; the foliage, appearing after flowering, is on a single, rose-marbled stalk with 3 palm-like branches bearing numerous elliptic segments. *Tropical.* p. 91

HYDROSTACHYS *Hydrostachyaceae*

pinnatifolia (Madagascar, E. Africa); aquatic plant with floating foliage in running water, with long ferney leaves, greenish spikes with tiny flowers. *Humid-tropical.* p. 522

HYLOCEREUS *Cactaceae*

lemairei (Trinidad, Tobago); beautiful night-blooming climber with triangular dark green stems rooting on one side; very large and odorous flowers 25 cm long with outer linear segments greenish yellow tipped bronze, the broad inner perianth segments white suffused pink. *Tropical.* p. 256

undatus, known in horticulture as Cereus triangularis (Brazil), "Honolulu queen"; one of the largest night-blooming cereus; epiphytic, deep green 3-angled clamberer of 5 cm dia.; white flowers nearly 30 cm long, blooming one night. Edible red fruit. Used as understock for grafting of Gymnocalycium, Rhipsalis or Zygocactus, for better root system. *Tropical.* p. 256

HYMENOCALLIS *Amaryllidaceae*

caribaea (Lesser Antilles), tropical summer-flowering "Spider lily" with globose bulb, a dozen leaves in several ranks, narrowing at base, and solid scape bearing an umbel of elegant fragrant white flowers with toothed crown and long linear segments, green outside at base. This may be H. pedalis from South America, naturalized in Florida and the West Indies. *Tropical.* p. 70

harrisiana (Mexico), "Mexican spider lily"; bulbous plant with sessile, oblanceolate leaves to 30 cm long; few-flowered inflorescence in clusters on erect stalk, the flowers with slender, greenish tube 10 cm long, the narrow-linear segments shorter and whitish; well-formed corona with green stamens. *Tropical.* p. 66

littoralis (Polynesia: Marquesas; So. America); a "Spider lily" which I photographed on the South Pacific Island of Nuku Hiva; tropical bulbous plant with broad leaves, and very spidery, pure white, waxy flowers, with 8-10 cm tube and long, thread-like segments, in the center a wavy-edged cup or corona. *Tropical.* p. 70

longipetala (Perú), "Crown beauty"; a spider-lily to 1 m tall, strap-shaped basal leaves to 60 cm long; the fragrant white blossoms with linear, curving petal lobes to 10 cm long, and long filaments. *Subtropic.* p. 70

narcissiflora (Ismene calathina) (Andes of Perú, Bolivia), "Peruvian daffodil"; bulbous plant with strap-shaped leaves; umbels of large, fragrant, white flowers, crown funnel-shaped, lobes fringed. *Subtropic.* p. 70

speciosa (W. Indies), "Winter spice"; evergreen flowering plant with big bulb; the thick-fleshy, dark green oblanceolate leaves on tapering, channeled petioles; from the center rises a flattened, glaucous stalk crowned by a cluster of fragrant, spidery pure white flowers with long greenish tube and linear segments 5 cm long, a distinctive inner cup bearing the long anthers. *Tropical.* p. 70

'Sulphur Queen' (narcissiflora x amancaes), a "Basket flower"; handsome fleshy-rooted bulbous plant with decorative strap-shaped basal leaves; blooming in summer with lily-like flowers pale yellow, 10 cm long, the inner petals fringed. *Subtropical.* p. 61

HYMENOCYCLUS *Aizoaceae*

latipetalus (Cape Prov.), "Carmine ice plant"; creeping woody perennial with succulent keeled leaves to 3 cm long, glaucous blue to reddish; buds purplish-red opening into daisy-like flowers of orange-red. *Subtropic.* p. 45, 49

luteola (Malephora) (So. Africa), "Yellow trailing ice plant"; spreading succulent, forming dense cushions to 30 cm high; short gray-green leaves cylindrical or flattened on top; bright yellow flowers 3-4 cm across but blooming sparsely. *Arid-subtropic.* p. 45

HYMENOSPORUM *Pittosporaceae*

flavum (Queensland, N.S. Wales), "Sweetshade"; graceful evergreen shrub or tree to 15 m high, of open habit, with shining light green, leathery leaves 8-15 cm long; clusters of honey-scented bright yellow flowers 3 cm wide, blooming spring into summer. *Subtropic.* p. 823

HYOPHORBE: see Mascarena

HYOSCYAMUS *Solanaceae*

aureus (Crete), "Golden henbane"; spreading semi-woody, white-woolly plant dense with short, serrate leaves; tubular trumpet-like flowers 2½ cm long creamy-white with throat striped purple, the larger, expanded lip yellow. Plant photographed in Heraklion, growing out of an ancient stone wall, the branches hanging downward. *Subtropic.* p. 898

HYPERICUM *Hypericaceae*

calycinum (S.E. Europe, W. Asia Minor), "Rose of Sharon"; stoloniferous evergreen subshrub with large golden yellow flowers, to 5 cm across; short rough leaves, dense. *Warm temperate.* p. 524

formosum scouleri (W. No. America); rhizomatous perennial to 40 cm high, with small, rough 3 cm leaves; small yellow flowers dotted at tip with violet glands. *Warm temperate.* p. 524

leschenaultii (Malaya); ornamental shrub 2-3 m high usually evergreen but deciduous in colder climate; obovate leathery leaves 4-6 cm long, glaucous beneath; large cupped, clear yellow flowers with concave petals, to 8 cm across, displaying the numerous stamens. *Tropical.* p. 524

x moserianum, (calycinum x patulum), "Gold flower"; evergreen shrub or perennial to 1 m tall where winters are mild; grows as hardy perennial in cold-winter areas; forms mounds, with arching, reddish stems; leaves 5 cm long, bluish beneath; large cup-shaped golden yellow blooms 6 cm across. *Warm temperate.* p. 524

patulum henryi (Japan), "St. John's-wort"; evergreen spreading shrub to 1 m or more high, with 2-edged, purplish branches, stiff-leathery oblong leaves to 6 cm, glaucous beneath; the showy flowers bright golden yellow, with long silky stamens; extra large in the variety henryi to 6 cm across; photographed in Leucadia, California. *Warm temperate.* p. 524

x rowallane; evergreen bush of straggly growth to 2 m high, with handsome 8 cm golden flowers, profuse in late summer and fall. *Subtropic.* p. 524

HYPHAENE *Palmae*

thebaica (Upper Egypt, Sudan, Kenya, Tanzania), "Dhoum palm" or "Gingerbread palm"; botanical wonder with its repeatedly forked habit, to 15 m high; the slender trunk and branches smooth like a cordyline, each branch end tipped by a rosette of smallish, stiff, green fan-leaves, the blade 60-75 cm long, deeply cut to the middle, on spiny petiole. The orange edible fruit tastes like gingerbread. Although palms normally are not hosts to epiphytic orchids, I have often seen Ansellias growing in the forks of their branches in East Africa. *Tropical.* p. 777, 782, 783

ventricosa (Zimbabwe, Zaire), "Gingerbread palm"; tall, slender fan-palm 10 m high, usually growing with a single trunk, frequent along the Zambesi river, their deep roots reaching for moisture, their trunks somewhat bulging above the middle, the deeply cut, smallish, palmate, blue-green fronds, on long stalks, forming a large cluster reaching for the burning sun. *Tropical.* p. 782, 783

HYPOCYRTA Gesneriaceae
nummularia (S. Mexico, C. America); tiny fibrous-rooted creeper on mossy tree trunks with branching, red-hairy stems and oval, crenate 2-8 cm leaves; the flowers a vermillion-red corolla tube with pouch and small yellow lobes; suitable for hanging baskets. *Humid-tropical.* p. 498

selloana: see Nematanthus fissus

HYPOESTES Acanthaceae
aristata (So. Africa), "Ribbon bush"; perennial herb of erect habit to 1 m high, with downy ovate leaves 8 cm long, entire; downy 3 cm tubular flowers rose-purple with short lobes striped and spotted purple or white. *Subtropic.* p. 34

sanguinolenta (rotundifolia, phyllostachia) (Madagascar), "Freckleface"; herb with soft, downy, small leaves green with rosy-red marking; flowers lilac. *Tropical.* p. 34

HYPOXIS Hypoxidaceae
rooperi, (South Africa), "Star flower"; stemless herb with corm-like rhizome, forming a rosette of leathery, lanceolate, pubescent leaves; clusters of orange yellow starry flowers 4 cm across. *Tropical.* p. 522

IBERIS Cruciferae
sempervirens (So. Europe), "Edging candytuft"; evergreen subshrub to 30 cm high, with linear leaves and 4 cm clusters of white flowers. *Warm temperate.* p. 379

umbellata (Mediterranean region), "Globe candy-tuft"; hardy annual herb 15-40 cm high, with narrow mid-green leaves and pale purple flowers in 5 cm clusters. *Warm temperate.* p. 379

IBERVILLEA Cucurbitaceae
sonorae (Mexico); climbing plant with heavy, bottle-shaped base, the fissured bark grayish-white and corky; several stems 2-3 m long; 3-lobed leaves 4-10 cm long; small hairy flowers; ovoid fruit 3-5 cm long, red when ripe. *Arid-subtropic.* p. 382

IBOZA Labiatae
riparia (Moschosma) (So. Africa), "Misty plume bush"; stout, musk-scented perennial subshrub to 1½ m, with 4-angled stems, broad ovate, toothed leaves, and numerous showy flowers creamy-white with dark anthers in erect panicles. *Subtropic.* p. 532

IDRIA Fouquieriaceae
columnaris (Mexico, Arizona), "Bojum tree"; bizarre desert tree with soft swollen, often hollow trunk to 15 m high, tapering to apex, spreading spiny branches and obovate, deciduous leaves 2 cm long; small yellow flowers in long panicles. *Arid-subtropic.* p. 430

ILEX Aquifoliaceae
aquifolium (Europe, No. Africa, W. Asia), "English holly"; evergreen tree to 12 m or more; alternate leathery, ovate leaves shining, and with coarse spiny teeth along wavy margins; small unisexual flowers whitish, followed by scarlet-red berries on female trees. *Warm temperate.* p. 84

aquifolium 'Marginata'; attractive horticultural form with black-green waxy leaves having ivory margins, silvery in places, especially on the underside. *Warm temperate.* p. 84

cornuta (Eastern China), "Chinese holly" or "Horned holly"; evergreen shrub of dense bushy growth, to 3 m high; recurved shining green, leathery leaves 8-10 cm long, nearly rectangular with pronounced spines at the 4 corners and at the tip of each leaf; large bright red, long-lasting berries. The cv. 'Dazzler' is of more compact habit. *Warm-temperate.* p. 84

cornuta rotundifolia ('Rotunda'), "Dwarf Chinese holly"; compact low-growing holly with dense branching habit, usually broader than tall; leaves dark shining green, 5-10 cm long, somewhat rectangular, with a large spine at each corner, one at the end and 2 smaller ones, much recurved, in the middle; flowers small, dull white; produces no berries, but its low, globular appearance sets it apart. Limited hardy. *Warm temperate.* p. 84

crenata (Japan), "Japanese holly"; stiff evergreen to 6 m, with smooth, dark green, oval leaves 6 cm long, sharply pointed and sparsely toothed; small black berries. *Warm temperate.* p. 84

opaca (Massachusetts to Florida and Texas), "American holly"; evergreen, spreading tree to 15 m, with stiff, elliptic or obovate leaves dull green above, yellowish beneath, to 9 cm long, and with large spiny teeth; small red fruit, usually solitary. *Temperate.* p. 84

rotunda (Japan, China, Korea, Vietnam, Taiwan), "Kurogane holly"; smooth evergreen tree to 18 m high, forming massive trunk; leathery obovate, dark green, glossy leaves 4-8 cm long and with smooth margins; small white flowers; dense clusters of glossy scarlet-red 1 cm berries. *Warm-temperate.* p. 84

verticillata (No. America: Newfoundland to Texas), "Winterberry"; wide-spread deciduous dense shrub 2 to 5 m high, smooth or slightly hairy; obovate leaves to 10 cm long, dull green above, the margins saw-toothed; 1 cm berries red or yellow. *Temperate.* p. 84

wilsonii (China, Taiwan, No. Burma); small evergreen tree to 8 m; thick-leathery obovate, glossy olive green leaves to 6 cm long and with acute tip, the margins smooth; glossy, crimson-red berries. *Warm-temperate.* p. 84

IMPATIENS Balsaminaceae
balsamina (India to China), "Garden balsam"; annual herb 45-60 cm tall, with succulent knotty stem and soft-fleshy, often reddish branches and lanceolate leaves; charming double camellia-type 5 cm flowers in the leaf-axils, white to red, borne close to succulent stem; leaves lanceolate. *Tropical.* p. 156

balsamina 'Beijo do Frade'; a beautiful cultivar from Brazil, with reddish, succulent stems and fleshy, fresh-green obovate, 10 cm leaves toothed at margins; the showy, waxy flowers crimson-rose with bold white blotching. *Tropical.* p. 156

balsamina plena, "Rose balsam"; charming variety with soft baby pink double flowers shaped like a rose-bud, peeking out between the dark green foliage. *Tropical.* p. 156

glanduligera (roylei) (India); tall herbaceous plant with reddish succulent stem, large ribbed leaves on long petioles; flowers in clusters, toward top of stem, inside pink and outside red and with short spur. *Tropical.* p. 156

hawkeri (New Guinea, Sunda Isl.); herbaceous species with branching, purple stem to 50 cm high; dark green, quilted leaves with red midrib, 10-12 cm long and finely toothed; showy flower scarlet-red with white eye, but variable, 4 cm across and with red spur. *Tropical.* p. 157

hawkeri hybrida, "New Guinea hybrid"; highly hybridized with selected cultivars and the species from the highlands of New Guinea; this has produced a series of named hybrids with broad elliptic leaves variously colored with green, reds and yellows; the everblooming flowers in red, pink and white. *Tropical.* p. 158

kilimanjari (Tanzania), a "Snapweed"; spreading creeper which I collected in the rainforest zone at about 2,100 m on Mt. Kilimanjaro, creeping over fallen, rotting trees and rocks; watery-brittle round stems, with small shiny, fresh-green, crenate leaves with sharp-tipped apex, and beautiful waxy, fiery-scarlet flowers ending in a thick curling yellow spur tipped with green. *Subtropic.* p. 157

linearifolia (New Guinea); a colorful, variable species which I collected in the mountains surrounding the mysterious Chimbu Valley in the Central Highlands; herbaceous, rather succulent, erect branching plant with long elliptic, almost linear, ciliate, smooth leaves, usually with yellow center and dark green outer area, the midrib wine-red; rosy flowers with long spur. *Tropical.* p. 157

linearifolia hybrida; by continuous hybridizing of the New Guinea species, attractive color forms have been produced; the red stem with narrow leaves variegated green and yellow, and red midrib; flowers carmine-red. *Tropical.* p. 157

oliveri (Trop. E. Africa), "Giant touch-me-not"; large growing, fleshy herb with long, oblanceolate, succulent leaves olive-green with prominent, pale midrib, edged with coarse bristles; large spurred flowers a delicate lilac-pink. *Tropical.* p. 158

platypetala aurantiaca (Celebes); fleshy herb with fresh-green, ovate, corrugated leaves and pink midrib; spurred flowers orange-yellow with crimson eye; very attractive. *Tropical.* p. 158

platypetala 'Tangerine'; lovely cultivar offered by Park Seed, Greenwood, So. Carolina; herbaceous plant 35 cm high, prolific with large single flowers bright tangerine with red eye, very showy. Probably a form of I. platypetala aurantiaca from Celebes. *Tropical.* p. 158

repens (Ceylon), "Creeping impatiens"; small trailer with creeping fleshy red branches and alternate small round or kidney-shaped, ciliate leaves 2 cm wide, waxy deep green, purplish beneath, on long red petioles; golden yellow hooded flowers with brownish net-like striping, and curved spur. *Tropical.* p. 156

sultanii: see walleriana

tuberosa (Madagascar); fleshy cane-stem with characteristic thickened base to 40 cm dia.; erect brown stem 40-60 cm tall, with soft ovate waxy, grass-green serrate leaves 10-15 cm long, underside amaranth-red; flowers rosy-red, 3 cm across. *Tropical.* p. 156

'Uganda Red' (Uganda); dense herbaceous bush with waxy, corrugated small ovate green leaves, against which contrast the vivid, bright crimson flowers characterized by two large front lobes. *Tropical.* p. 158

walleriana (sultanii) (Zanzibar), "Patient Lucy"; in continuous bloom, the carmine flowers with an upturned spur; long, tapering, fresh-green, waxy, crenate leaves on watery-succulent stem. *Tropical.* p. 156, 158

walleriana holstii (Trop. E. Africa), "Busy Lizzie"; watery-succulent herb with fleshy stem striped red, and small, coppery, ovate leaves; flowers fiery vermillion-scarlet with spurs pointing downward. *Tropical.* p. 156

walleriana 'Red Ripple'; charming cultivar, grown as an annual from seed; bushy, compact plants 25-30 cm high, with happy-green to bronzy foliage, and prolific blooms of red with white star pattern. *Tropical.* p. 158

walleriana 'Variegata' (sultanii 'Variegata'), "Variegated patient Lucy"; long, tapering leaves gray-green, irregularly bordered white; flowers carmine-red. *Tropical.* p. 156

INDIGOFERA *Leguminosae*
incarnata (decora) (Japan, China), "Indigo"; deciduous shrub, with attractive pinnate leaves of 3-6 pairs of leaflets 6 cm long; inflorescence wisteria-like in pendant racemes of rose-pink flowers with pale standard, showy. *Warm temperate.* p. 563

INGA *Leguminosae*
edulis (C. and So. America), "St. John's-bread"; tropical tree to 15 m, with broad crown and gray bark; pinnate leaves glossy dark green, the leaflets separated by winged axis; flowers with long white stamens and brown hairy corolla; long 4-angled pods contain edible, sweet white pulp, and split open when ripe. *Tropical.* p. 563

IOCHROMA *Solanaceae*
coccineum (Perú); robust shrub with downy shoots; and soft herbaceous green, flaccid leaves 8-12 cm long; clusters of pendant lovely, tubular bell-shaped flowers carmine-red, 5 cm long, near apex of branches. *Tropical.* p. 890

cyaneum (tubulosum) (Colombia), the "Violet bush"; hairy shrub 1 to 2 m high, with soft herbaceous, thinly downy, long elliptic 12 cm green leaves; axillary pendulous clusters of lilac-purple cylindric-tubular flowers 4½ cm long, pale lavender at the toothed mouth, and blooming during summer from June on. Ideal for the summer patio, as they love fresh air. *Tropical.* p. 890, 900

warscewiczii (Trop. America); handsome spreading shrub with pubescent shoots, and large herbaceous, ovate leaves with depressed ribs; from the apex a cluster of charming pendant flowers 5 cm long with balloon-like calyx and lavender-blue tubular corolla striped dark blue; photographed at Royal Botanic Garden, Sydney. *Tropical.* p. 890

IPOMOEA *Convolvulaceae*
arborescens (Mexico: Sonora to Morelos and Veracruz), "Casahuate tree" or "Palo blanco"; large shrub or small tree of dense habit, with smooth white bark; glossy green, heart-shaped leaves, downy beneath; large and showy white flowers 5-8 cm across, with frilled margins and red centers. *Tropical.* p. 358

batatas (E. Indies), "Sweet potato vine"; an economic as well as a hanging basket plant, trailing perennial with deep tuberous edible roots, the long vines are stem-rooting, with milky juice, the variable leaves ovate, angular or digitately lobed; the tubers are often grown in water. *Tropical.* p. 358, 465

cairica (Old World Tropics); perennial high climber with slender, thread-like stems; palmately compound leaves of 5 segments; lilac-pink flowers with purple eye, 5 cm across. *Tropical.* p. 358

crassicaulis (fistulosa) (Brazil); large straggling herbaceous plant, becoming woody, climbing when finding support, the stems with milky juice; leaves heart-shaped to 15 cm long, soft hairy beneath; funnel-shaped pinkish flowers with ruffled margins, 6 cm across, singly or in pairs. *Tropical.* p. 358

digitata (Pantropic); slender climber with leaves palmately divided, the sinus between lobes acute; flowers carmine pink with darker eye. *Tropical.* p. 359

fistulosa (Florida, tropical America); climbing vine with slender pointed leaves, and large funnel flowers deep rose with darker eye. *Tropical.* p. 358

horsfalliae (W. Indies), "Morning glory", or "Princess vine"; twining winter-flowering perennial with palmately lobed leaves, the showy, 6 cm waxy-glossy, bell-shaped flowers deep rich rose or red. *Tropical.* p. 359

learii (acuminata in Hortus 3), (Trop. America), "Blue dawn-flower"; perennial twiner with ovate entire or lobed leaves to 20 cm long; trumpet-shaped flowers to 12 cm across, with white tube and sky-blue limb, turning purple and pink. *Tropical.* p. 360

mauritiana (Mauritius); robust climber with large leaves, deeply cut into narrow fingers, sinus between lobes usually rounded; large showy flowers 6 cm across, pale lavender with contrasting purple eye. *Tropical.* p. 359

murucoides (Mexico: Michoacan to Oaxaca), "Morning-glory tree"; sparry tree or large shrub, with oblong-lanceolate leaves, 15-20 cm long, white-woolly beneath, on tomentose branchlets; large crepy, ruffled flowers, 6-10 cm across, pure white and apparently night-blooming, fading to creamy yellow in the morning. *Tropical.* p. 358

nil (hederacea) (Old World Tropics), "Imperial Japanese morning-glory"; floriferous hairy perennial; rank tendril-climbing vine, with yellow-green, 3-lobed leaves 10-15 cm wide; showy axillary, funnel-form flowers about 10 cm wide, blue purple or rose. Can be grown as a bushy pot plant for the cool, sunny window as in Japan, where plants are pinched back every second or third leaf to prevent climbing, resulting in large blooms all summer and fall; blooms open in morning and fade afternoon. *Tropical.* p. 358

palmata (W. Africa: Guinea coast); deciduous twiner with white tubercular stem and tuberous root; leaves palmately divided into linear lobes; funnel-form flowers with 5-angled limb white with purple center. *Tropical.* p. 360

pes-caprae (Sandy beaches, pantropic), "Beach morning-glory"; creeper on sandy shores, fleshy oval, fresh-green leaves 10 cm long; flowers purplish pink with carmine center. *Tropical.* p. 360

purpurea (Pharbitis) (Trop. America), "Common morning-glory"; widely naturalized; annual twiner with 12 cm cordate-ovate leaves and large funnel-shaped flowers deep purple with pale tube, opening from early morning until about 10 a.m.; garden forms in white, pink, carmine, blue, striped, or double. *Tropical.* p. 360

tricolor 'Heavenly Blue' (rubro-coerulea) (Trop. America); annual climber with cordate leaves and large blue, disk-shaped flowers, to 10 cm dia. *Tropical.* p. 359

tuberosa (India), "Wood-rose" or "Ceylon morning-glory"; perennial viner with 20 cm leaves digitately parted into 5-7 narrow lobes; funnel-form yellow flowers; the globular fruit ultimately a woody pod, when ripe opens and which with its persistent large, rounded leathery sepals form the "Wooden" rose some 8-10 cm across; used for decoration, especially in Hawaii. *Tropical.* p. 360

IRESINE *Amaranthaceae*
herbstii (Achyranthes verschaffeltii) (So. Brazil), a "Beefsteak plant"; bushy tropical herb with waxy 2 to 6 cm leaves almost round, and notched at tip; glowing purplish-red and traced with light red veins. The ornate foliage coloring is brought out best in good sunlight. A bedding or border plant where red is required. Small woolly flowers not showy. *Tropical.* p. 50

herbstii 'Acuminata' (versicolor), "Painted bloodleaf"; variety with ovate, sharply-pointed leaves deep red with veins a showy, light carmine, also the stem. *Tropical.* p. 50

herbstii 'Aureo-reticulata', "Chicken gizzard"; rounded leaf with notched apex, fresh green with yellow veins; stem and petioles red. *Tropical.* p. 50

lindenii formosa (reticulata), the "Yellow blood-leaf"; very attractive, colorful form with broader, pointed 5-8 cm leaves yellow with light green area between veins; stems and petioles red. Charming for carpet bedding and as a window plant. *Tropical.* p. 50, 52

IRIS *Iridaceae*
x germanica (probably Mediterranean), the "Common bearded iris"; rhizomatous, with flat glaucous leaves to 45 cm long; tall stems 60-90 cm, forked, with fragrant flowers, the outer segments bright purple, the claw white with brownish veins; beard yellow, the erect inner segments deep lilac; May. *Warm temperate.* p. 528

'Ideal'; a xyphium x tingitana hybrid for forcing; darker purple than 'Wedgwood'. *Warm temperate.* p. 528

japonica 'Variegata' (Japan, China); attractive cultivar of the Japanese iris, with a fan of flat, gray-green leaves having inner edges broadly banded ivory-white; the many-flowered inflorescence 40 cm high, with blooms 5 to 8 cm across, pale lavender blue with conspicuous yellow crests and orange markings on falls. *Warm temperate.* p. 526

pallida 'Variegata' (South Tyrol), "Orris"; fan of flat leaves light or milky green edged in cream; flowers lavender-blue with brown, beard white, tipped yellow. *Temperate.* p. 528

pseudacorus (W. Europe), "Yellow flag"; long sword-shaped leaves; large golden yellow flowers. *Warm temperate.* p. 528

'Wedgwood', "Forcing iris"; bulbous hybrid of the No. African I. tingitana and the "Spanish iris", xyphium; sword-like leaves and stiff spike with striking violet-blue flowers, the outer segments or "falls" lavender and marked with yellow; much grown under glass. *Subtropic.* p. 525

ISMENE: see Hymenocallis

ISOCHILUS *Orchidaceae*
linearis (Mexico, Cuba, Argentina); curious orchid forming clumps to 75 cm high, of thin, grass-like arching stems clothed by alternating narrow-linear leaves; small 1 cm flowers toward the apex commonly rosy-magenta, with darker lip, blooming over long period. *Subtropic.* p. 735

ISOLEPIS: see Scirpus

ISOLOMA: see Kohleria

ISOTOMA *Lobeliaceae*
fluviatilis (Laurentia in Hortus 3) (S. Australia, New Zealand), "Blue star creeper"; low prostrate perennial forming mats; small rounded 1 cm rich green leaves; masses of little light blue star flowers. Used as groundcover in California. *Sub-tropical.* p. 616

IXIA *Iridaceae*
speciosa (So. Africa), "Corn lily"; cormous herb with short stems 15-30 cm high, basal, grass-like leaves, and large clusters of deep crimson flowers. *Subtropic.* p. 528

IXORA *Rubiaceae*
borbonica (bot. Enterospermum), a beautiful tropical foliage plant somewhat resembling a croton, from the Indian Ocean Isle of Réunion; branching shrub with stiff-leathery narrow-lanceolate leaves about 25 cm long, in bluish or mossy-green, mottled with pale green; the midvein a bold salmon-red; small whitish flowers. *Tropical.* p. 868, 869
chinensis (So. China, Malaya); small evergreen shrub with rich green, firm, obovate leaves 10 cm long; waxy tubular flowers with spreading lobes light orange-red, varying to yellow and white, in dense clusters. *Tropical.* p. 869
chinensis 'Alba'; color form with clusters of pure white flowers. *Tropical.* p. 869
chinensis 'Fraseri'; color form seen in Sydney Botanic Garden, with large clusters of flowers soft salmon pink. *Tropical.* p. 868
coccinea (East Indies), "Flame of the woods"; evergreen, tropical, flowering shrub to 1 m, with sparry branches; opposite, sessile, short, leathery leaves, and clusters of dark scarlet, tubular flowers with spreading lobes. *Tropical.* p. 869
'Herrera's Pink'; compact bush with long elliptic leaves, and large clusters of carmine-pink flowers. *Tropical.* p. 869
javanica (Java), "Jungle geranium"; showy flowering shrub with forking, willowy, red branches; opposite, long slender-pointed, smooth, leathery leaves; and large, terminal clusters of waxy, soft salmon-red flowers having long thin tubes and expanded lobes, blooming in summer. *Tropical.* p. 830, 862
macrothyrsa (duffii) (Sumatra), "King ixora"; large tropical flowering shrub freely branching, and a prolific bloomer, with oblong lanceolate deep green leaves to 30 cm long and slender-pointed, the 2 cm flowers rosy-red, becoming tinged crimson with age, in large striking clusters. *Tropical.* p. 861, 863, 869
macrothyrsa 'Super King'; probably hybrid of I. coccinea var. fraseri x macrothyrsa (duffii); vigorous free-blooming, very showy evergreen shrub with stout, cane-like branches with large leathery leaves, bearing 15 cm ball-shaped clusters of brilliant orange-scarlet flowers tinted cinnamon, 1-3 cm across. *Tropical.* p. 862
odorata (Madagascar); tropical evergreen shrub to 1 m high, dense with leathery green lanceolate leaves to 30 cm long; terminal clusters of ray-like slender tubular flowers 8-10 cm long, pinkish with red base, and very fragrant. *Tropical.* p. 869
undulata (Bengal); evergreen to 1 m high, elliptic, undulate leaves with slender points; flowers coral-red, sometimes white. *Tropical.* p. 869

JACARANDA *Bignoniaceae*
mimosaefolia (acutifolia) (Brazil), "Mimosa-leaved ebony"; semi-evergreen tropical tree 8 to 20 m high, with lacy, bright green, bipinnate foliage to 45 cm long, lightly downy. In early spring the leaves drop, following which appear the big 20-30 cm erect clusters of hanging 5 cm lavender blue flowers, with silky, inflated tubular corolla. In the juvenile stage an attractive house plant because of its fern-like foliage. *Subtropic.* p. 185
obtusifolia (filicifolia) (Venezuela, Guayana), "Green ebony"; graceful tree to 18 m high; feathery leaves to 40 cm long, with many pairs of leathery 2 cm leaflets, shining above and glaucous beneath; bluish-purple flowers to 5 cm long usually borne on old leafless branches. Ovary smooth in rhombifolia. *Tropical.* p. 186

JACOBINIA *Acanthaceae*
carnea (Justicia magnifica) (Brazil), "Flamingo plant"; upright plant with ovate, grayish-green, satiny leaves, reddish stem and terminal head of arched clear rose flowers. *Tropical.* p. 39
velutina (Brazil), "Brazilian plume"; somewhat shrubby plant with long, olive-green leaves, soft hairy on both sides. The beautiful inflorescence a dense, fountain-like terminal head of arched rosy-pink blooms; from early summer. *Tropical.* p. 39

JACQUINIA *Theophrastaceae*
aculeata (ruscifolia) (Cuba); evergreen dense shrub to 3 m; stiff-leathery lanceolate leaves to 3 cm long, in whorls and spine-tipped; small red flowers in terminal clusters, followed by red berries. *Tropical.* p. 674
armillaris arborea (W. Indies); attractive small evergreen shrub or tree to 3 m or more, with thick-leathery, grayish green obovate leaves; terminal clusters of fragrant white inconspicuous flowers, followed by conspicuous glossy bright orange-red berries. *Tropical.* p. 674, 906
pungens (Mexico), "Cudjoe wood"; evergreen shrub with leathery, long-elliptic leaves; small scarlet red cup flowers opening star-like in clusters; small orange fruit. *Tropical.* p. 674, 906

JASMINUM *Oleaceae*
floridum (China, Japan); evergreen or partially deciduous sprawling shrub; glossy green leaves divided into 3 to 5 leaflets; clusters of golden yellow, scentless 2 cm flowers. *Subtropic.* p. 697
gracillimum (No. Borneo), "Star jasmine" or "Pinwheel jasmine"; climbing shrub with thin-leathery, ovate, deep green leaves; showy fragrant, starry flowers 4 cm across, with 7-9 ovate petals pure white. *Humid-tropical.* p. 695
grandiflorum (Kashmir, Himalayas), "Poets jasmine", also known as "Spanish jasmine"; straggling tender bush with slender, angled branches, opposite, pinnate leaves of usually 7 small leaflets, and showy white, fragrant flowers, reddish beneath, commonly in clusters; June-October blooming. *Subtropic.* p. 695
humile revolutum (Himalaya, Afghanistan), "Italian jasmine"; shrub of spreading habit, with strong, angled branches, alternate, dark green, pinnate leaves with 3 to 7 thick leaflets revolute at the edges; fragrant, lemon-yellow flowers in axillary and terminal clusters from June to September. *Subtropic.* p. 697
mesnyi (primulinum) (China), "Primrose jasmine"; free-flowering, evergreen, rambling shrub up to 5 m if trained, with 4-angled, glabrous branches, opposite trifoliate leaves with small, lanceolate shining green leaflets and showy, single or semi-double, solitary yellow flowers with darker center, to 4 cm across, in spring. *Warm temperate.* p. 694, 697
multiflorum (China, India), "Angel-hair jasmine"; popular, wide spread tropical jasmine, prized for centuries in the Orient, especially India; vigorous, freely spreading, downy shrub; cordate, dull green leaves hairy beneath; pure white, star-like, sweetly scented flowers, borne in clusters at the ends and along the often drooping branches. *Subtropic.* p. 697
nitidum (So. Pacific: Admiralty Isl.); known in Florida hort. as J. ilicifolium, in California hort. as magnificum; the "Angelwing jasmine"; semi-vining small evergreen shrub with shiny dark green leaves ovate with tapering tip, 6 cm long; large 4 cm windmill-like, glistening white flowers with lanceolate petals, purplish in bud, and sweetly fragrant. *Tropical.* p. 694, 697
rex (Southwest Thailand), "King jasmine"; glabrous climber with young branches green, round and wiry; simple, rigidly hard opposite leaves broad-ovate, dark green, 10-20 cm long; large pure white, salver-shaped flowers without scent, 5 cm or more across, during winter, usually in 2-3 flowered clusters. Very showy in bloom. *Tropical.* p. 694, 697, 700
sambac (Arabia, India), "Arabian jasmine"; woody shrub clambering to 2 m high, with firm, broad elliptic, dark green leaves, opposite or in threes, to 8 cm long; 3 cm flowers in clusters, gardenia-white but turning purple as they fade, and very fragrant, blooming from early spring to late fall. *Tropical.* p. 699
sambac 'Grand Duke of Tuscany' (trifoliatum), "Gardenia jasmine"; a button-flowered form from Italy with large, tightly double, gardenia-white blooms that won't drop off, of a penetrating sweet fragrance; waxy, quilted, oval leaves in whorls, on stiff pubescent stems; flowers more fully double than 'Maid of Orleans' and slower growing. *Tropical.* p. 697
sambac 'Maid of Orleans', "Arabian jasmine"; woody evergreen shrub of sparry habit and slow growth, with broad ovate, deep green leaves, and waxy-white, semi-double, somewhat cupped 2-3 cm flowers, and intensely fragrant. *Tropical.* p. 694
simplicifolium (South Sea Islands), "Little star jasmine"; rambling evergreen shrub, with leaves privet-like, ovate, 5 cm or less long, dark green and glossy, the sweetly fragrant, tiny star-shaped flowers white, 1½ cm across, with narrow-linear petals, in terminal clusters very free-blooming. Attractive as a little flowering pot plant. *Tropical.* p. 697

undulatum (China, India); scrambling bush with glossy green, leathery, ovate leaves; the flowers white, with thread-like linear petal lobes. *Subtropic.* p. 697

JATROPHA *Euphorbiaceae*
  gossypifolia (Nigeria); leafy bush with milky juice, red stems, red-hairy petioles and broad, deeply palmately lobed foliage, glossy coppery green with red veins; small scarlet flowers. *Tropical.* p. 428
  integerrima (Cuba), "Peregrina"; evergreen shrub to 1 m, with milky sap, more or less 3-lobed leathery, glossy green leaves, and clusters of red flowers. *Tropical.* p. 418, 421
  multifida (Mexico to Brazil), "Coral plant"; growing tree-like, with milky juice, leaves nearly circular in outline but deeply parted, the narrow segments pinnately lobed; flowers scarlet. Seen as a street tree in Nigeria, West Africa. *Tropical.* p. 418, 419
  pandurifolia (West Indies); evergreen shrub to 1 m or more high, with variable ovate or obovate, somewhat fiddle-shaped leaves with an occasional tooth here or there, and small scarlet flowers in branched clusters. *Tropical.* p. 418
  podagrica (West Indies, C. America, Colombia); succulent shrub with short trunk thickened at base, to 60 cm high, knobby branches, and peltate, 3 to 5-lobed, dark green leathery leaves, to 25 cm across; fleshy inflorescence of small scarlet-red or coral flowers in long, red-stalked clusters. *Tropical.* p. 418

JUANULLOA *Solanaceae*
  aurantiaca (Perú); evergreen shrub 1-2 m high, growing epiphytic in habitat; branches covered with felt; oval-pointed leathery matte dark green leaves 5-12 cm long; axillary pendant tubular flowers toward ends of branches, orange corolla in large, fleshy, angled calyx, in forked racemes. *Tropical.* p. 895

JUBAEA *Palmae*
  chilensis (spectabilis) (Chile),"Syrup-palm"; or "Chilean wine-palm"; massive feather-palm with trunk usually swollen and to 1 m thick and 12-24 m high, crowned by spreading pinnate fronds 2-4 m long; the numerous green pinnae in pairs, standing out in different directions, and split at apex; the short thick petioles covered with brown fibers at base. The sap of the trunk yields syrup. *Subtropic.* p. 780, 782

JUGLANS *Juglandaceae*
  regia (S.E. Europe, W. Asia), "English walnut"; deciduous tree to 30 m high, with silvery gray bark; large odd-pinnate leaves; flowers in drooping catkins; edible fruit a furrowed nut within a thick globular husk. *Temperate.* p. 467, 520

JUNIPERUS *Coniferae: Cupressaceae*
  chinensis 'Pfitzeriana', ("Pfitzer juniper"); evergreen dense shrub broadly spreading, to 2 m high and 3 m across; the branches horizontal, with scale-like awl-shaped, gray-green, sharp-needled foliage; dioecious with male catkins and female berry-like cones on separate plants. Widely planted in California. *Temperate.* p. 340
  chinensis 'Sargentii' (Japan) (Cupressaceae); procumbent variety, used for bonsai culture in Japan, with whorled, scale-like leaves, or linear with white bands below. *Temperate.* p. 339
  chinensis 'Torulosa' ('Kaizuka'), (Japan), "Hollywood juniper" or "Twisted Chinese juniper"; very artistic irregular growing juniper to 5 m high, branching with an appealing twisted effect; the tiny imbricated scaly leaves rich green on brown stem, the cordlike branchlets somewhat flattened and contorted; cones berry-like 5-7 mm dia., at first blue-glaucous, dark brown later. Fairly winter-hardy. Very decorative as container plant. *Temperate.* p.337, 338
  communis (No. America, Eurasia), "Common juniper"; dense shrub or tree to 15 m, reddish bark peeling off; branchlets triangular, leaves awl-shaped 1½ cm long, with white band above; cones to 5 mm dia., bluish black. *Temperate.* p. 338
  excelsa 'Stricta' (S.E. Europe to Asia), "Greek juniper"; pyramidal dense conifer to 20 m, with scale-like leaves bluish-green; female cones glaucous, purplish, berry-like. *Subtropic.* p. 338
  sabina 'Tamariscifolia' (So. Europe), "Tamarix juniper"; low spreading evergreen to 50 cm high but 2 m wide, with branches arching out from the center carrying the upward-facing branchlets, with short awl-shaped bluish green needles, scale-like near the top; fruit brownish glaucous blue, fairly hardy in colder climate; a favorite for ground cover. *Temperate.* p. 341
  scopulorum 'Blue Haven' ('Blue Heaven'); excellent commercial cultivar of the "Colorado Red cedar" or "Rocky Mountain juniper"; tall-growing pyramidal trees from Brit. Columbia to Arizona and Texas; this form is of compact habit, to 6 m tall, with appressed scale-like leaves strikingly silver-blue; the cones nearly globular 6 mm dia., glaucous, blue. *Temperate,* but widely planted in California. p. 340
  scopulorum pendula 'Tolleson's Weeping'; remarkable evergreen growing upright but with branches arching, and the silvery-bluish-gray, string-like foliage hanging gracefully as if weeping, with artistic effect, in landscape planting in California. *Temperate.* p. 340
  virginiana 'Tripartita', "Red cedar"; dwarf, dense evergreen 2 m or more across, with stout, irregular, pointed branches ascending from base then spreading out; green, needle-shaped foliage; resembling 'Pfitzeriana' but of smaller habit. *Temperate.* p. 341

JURINEA *Compositae*
  species Wats, (Asia Minor); interesting xerophytic subshrub gray-hairy; notched foliage to 20 cm, on hairy petioles; globe-shaped inflorescence 5 cm dia., filled with small straw-yellow flowers between the long filaments. *Arid-subtropic.* p. 322

JUSTICIA *Acanthaceae*
  aurea (umbrosa) (Mexico, C. America); bushy shrub with angled stems, grayish green, lanceolate leaves 20-30 cm long; bilabiate, curving flowers in dense terminal bouquet; corolla 5 cm long, clear yellow. *Subtropic.* p. 39
  betonica (Jacobinia) (Trop. Asia, South Pacific), "White shrimp plant"; handsome bush photographed in Tonga; opposite, corrugated ovate leaves; the stems terminated by slender bracted spikes, the bracts white with green veining; flowers white to bluish. *Humid-tropical.* p. 39
  pohliana obtusior (Jacobinia) (Brazil); erect, herbaceous plant with smooth, ovate-lanceolate leaves; showy terminal inflorescence with curved-tubular, hooded, 2-lipped flowers rosy-crimson, from rounded bracts. *Tropical.* p. 39
  rizzinii (Jacobinia pauciflora) (Brazil); downy shrub to 60 cm high, with small 2 cm elliptic leaves on round stems and numerous nodding 3 cm solitary tubular flowers with short lips, scarlet tipped with yellow. *Tropical.* p. 39

JUSTICIA: see also Jacobinia, Pachystachys

KAEMPFERIA *Zingiberaceae*
  brachystemon (East Africa), "Dwarf ginger lily"; tropical clustering plant with tuberous roots, forming rosettes of light green, elliptic leaves 10-12 cm long; pretty flowers from base light blue with white eye 5 cm across. *Tropical.* p. 925
  galanga (India); rhizomatous, stemless herb with aromatic, edible tubers and opposite, horizontally spreading, fleshy, roundish, to oblique-ovate leaves shiny green; flowers with two violet bands on lip. *Tropical.* p. 924
  gilbertii (So. Burma), "Variegated ginger-lily"; stemless, fleshy-rooted herb with tufted, oblong-lanceolate soft-fleshy leaves grass-green, the marginal areas prettily variegated with milky-white, and gray beneath; the flowers on separate stalks with white corolla and long white lip with violet stripes, in summer. *Tropical.* p. 924
  grandiflora (Kenya); erect herb with tuberous rhizome bearing a rosette of stalked, shining fresh green, unequal sided, lanceolate leaves with depressed veins, and light blue flowers, appearing before the foliage. *Tropical.* p. 924
  'Pobeda' (K. decora x kirkii elatior); similar to decora but with spikes bearing white flowers; a yellow eye framed by lavender bars; aromatic tuberous plant with large leathery, lanceolate leaves rich green (Korsakoff, Florida 1969). *Tropical.* p. 926
  pulchra (E. Tropical Asia: Burma), "Pretty resurrection lily"; attractive tropical rhizomatous herb with broad, corrugated leaves flat to the ground, a gray band in peacock design over the bronze blade; large 4 cm light purple flowers with broad petals and narrow translucent lip, white eye in center. *Tropical.* p. 924

KALANCHOE *Crassulaceae*
  beharensis (S. Madagascar), "Elephant ear" or "Napoleon's hat"; woody succulent shrub to 3 m high, with large, broadly arrowshaped, lobed leaves, rich green but densely rusty-haired above, silver-haired beneath and on leaf stalks; flowers yellowish, violet inside. *Arid-tropical.* p. 373, 376
  'Maltese Cross', (beharensis x (possibly) tomentosa), "Fernleaf felt bush"; an attractive California hybrid; fleshy succulent entirely covered with short stiff hairs; with features of the giant K. beharensis but much smaller and more graceful, the thick, triangular lobed leaves dentate, gray-felty with brown edge. *Arid-tropical.* p. 373
  blossfeldiana (globulifera coccinea) (Madagascar), "Flaming Katy"; compact branching plant with small obovate, glossy-green leaves, topped during the short-day season of the year with clusters of bright scarlet red flowers. *Tropical.* p. 373
  blossfeldiana 'Tom Thumb', a dwarf and very compact "Christmas kalanchoe"; with bronzy foliage, and covering itself during late winter with masses of bright red flower clusters. To encourage budset, place the plant outside for the summer. If flowers are wanted earlier than usual, daylight should be restricted to between 9 and 10 hours during July and August by covering the plant with black paper or a box; for Christmas flowering short-day treatment from early September until buds show. *Subtropic.* p. 378

daigremontiana (Bryophyllum) (Madagascar), "Devil's backbone"; easy-growing robust, erect plant with fleshy, long tricornered brownish-green leaves nicely arched and producing plantlets from the serrate margins, reverse gray flecked purple; flowers gray-violet. *Tropical.* p. 372

farinacea (Socotra Is.); compact almost compressed succulent with short, mealy-white stem; the thick obovoid, flat or slightly concave leaves 2-3 cm long, pale green and covered with waxy silvery powder, the margins pink; inflorescence a sessile cluster of yellow tubular flowers with red lobes. *Arid-tropical.* p. 373

fedtschenkoi 'Marginata', "Aurora Borealis plant"; an attractive sport which I first photographed in Mr. Orpet's garden in Santa Barbara; the pale bluish gray leaves beautifully margined creamy-white flushed with pink. *Tropical.* p. 372

fedtschenkoi 'Rosy Dawn'; exquisite variegated cultivar with fleshy obovate leaves 5 cm long, and with crenate margins, in pastel coloring of cream streaking through the center, pinkish on midrib and edge, glaucous green along sides. *Subtropic.* p. 372

flammea (Somaliland), succulent plant with few branches, the grayish-green leaves ovate and toothed; inflorescence on rather long, branching stalks with clusters of large scarlet flowers, yellow outside. *Arid-tropical.* p. 372

gastonis-bonnieri (Madagascar), "Life plant"; loose, fleshy plant with large, lanceolate leaves, pale to coppery green with darker spots, glaucous white, especially young growth, brownish margins as if stitched and toothed; forms young plantlets on foliage. Flowers with corolla tube pale pink. *Arid-tropical.* p. 372

marmorata (Ethiopia), "Pen-wiper"; stout plant with fleshy, broad obovate leaves pinkish to bluish green, dusted glaucous blue and blotched purple on both sides, margins scalloped; flowers white. *Arid-tropical.* p. 372

paniculata (So. Africa); shrubby succulent with spoon-shaped obovate, sessile leaves irregularly crenate at margins, waxy gray with pinkish edge. *Arid-subtropic.* p. 372

pinnata (Bryophyllum) (India, and other trop. regions), "Air plant"; also known as "Miracle-leaf" or "Curtain plant", because young plantlets are produced from the leaves, even if broken off; the fleshy foliage 5 to 20 cm long, greyish green and tinged with red, at first undivided oval and notched, in later stages divided into 3 to 5 scalloped leaflets; nodding greenish flowers tinted purple. *Arid-tropical.* p. 373

pumila (C. Madagascar), "Dwarf purple kalanchoe"; bushy plant with closely-set obovate leaves notched on upper margin, purplish brown and covered with white bloom; pitcher-shaped red-violet flowers. *Tropical.* p. 373

rhombopilosa (Madagascar); beautiful small succulent with branching silvery stems mottled brown; rhombic olive green to coppery brown, fan-like leaves 2 cm long, largely covered with silvery-white scurf, the outer margin beige with acute points; tall inflorescence with small yellow-green flowers lined with red. Falling leaves root easily and soon develop plantlets. *Tropical.* p. 373

'Roseleaf' (beharensis x pilosa); an attractive Hummel hybrid with stocky stem, triangular, thick, spatulate-pointed, toothed leaves symmetrically arranged as in a cross, with brown felt above, and silver felt beneath; teeth brown. *Tropical.* p. 373

synsepala (C. Madagascar), "Cup kalanchoe"; attractive short-stemmed succulent with cupped, broad oval, fleshy leaves, pale green with purple band inside of marginal teeth; flowers white or light pink. Unique for its sending out young plants on runners from the leaf axils, and forming plantlets on roots. *Tropical.* p. 365

thyrsiflora (in hort. as "Vertical leaf") (Cape Prov., Transvaal); stem-forming succulent with almost oval leaves flat on the surface, rounded beneath and arranged diagonally, light green with silvery hoary covering, apex flushed red; flowers yellow. *Subtropic.* p. 378

tomentosa (pilosa) (C. Madagascar), "Panda plant"; strikingly beautiful succulent with erect branching stem and soft fleshy spoon-shaped leaves entirely clothed in dense white felt, apex dentate and the teeth marked brown; whitish flowers with light brown stripes. *Arid-tropical.* p. 372, 373

tubiflora (verticillata) (Bryophyllum) (Madagascar), "Chandelier plant"; slender erect succulent with pinkish-brown stem with many almost cylindric pinkish leaves blotched purplish, young plants forming at tips; flowers red. *Arid-tropical.* p. 372, 378

waldheimii (Madagascar), "Ghost plant"; shrubby, branching succulent, with fleshy, obovate, bluish glaucous leaves notched toward apex, 10 cm long; nodding orange or salmon-pink flowers. *Tropical.* p. 374

'Yellow Darling' (blossfeldiana x schumacheri), "White kalanchoe"; compact, short bushy plant with waxy light green, crenate leaves topped by dense clusters of creamy-white flowers with lemon-yellow center. *Tropical.* p. 378

KALMIA *Ericaceae*
angustifolia 'Rubra' (E. No. America), "Sheep laurel"; erect evergreen of thin, open habit, to 1 m; leathery fresh green 6 cm leaves usually opposite; small 1 cm saucer-shaped flowers deep purple, in axillary clusters, along terminal part of previous season's growth. *Temperate.* p. 396

latifolia (E. No. America), "Mountain-laurel" or "Calico bush"; thicket-forming evergreen shrub to 3 m or more high, with leathery elliptic, glossy green leaves and large terminal clusters of beautiful pink saucer-shaped flowers marked inside with purple. *Temperate.* p. 396

KARATAS: see Nidularium

KENNEDIA *Leguminosae*
rubicunda (Australia: Victoria, N.S.W., Queensland), "Coral-pea"; strong-growing bean-like twiner, the young growth covered with silky brown fur; variable palmate-trifoliate leaves, the leaflets ovate, 5-8 cm long; red pea-shaped, pendulous axillary flowers in pairs, a long standard, or upstanding petal, reflexed upward. *Subtropic.* p. 557

KENTIA: see Howeia, Ptychosperma

KENTIOPSIS *Palmae*
macrocarpa (Chambeyronia in Hortus 3) (New Caledonia), "Lucian palm"; handsome feather palm with slender trunk to 20 m; open crown with graceful pinnate fronds that may be from 1 to 2 or 3 m long; dark green leaflets prominently ribbed, on yellow petiole, the young unfolding leaves blood-red. Photographed under this name at Botanical Garden Rio de Janeiro, but this may be same as Chambeyronia macrocarpa. *Tropical.* p. 782

olivaeformis (New Caldeonia); lofty feather palm, with elegant pinnate fronds, gracefully recurving, the broad leaflets glossy green and leathery. *Tropical.* p. 780

KERRIA *Rosaceae*
japonica 'Pleniflora' (China, Japan), "Japanese rose"; deciduous shrub 2 m or more high, ovate, doubly-toothed leaves 5 cm long; flowers golden yellow, fully double in this cultivar, 5 cm across. *Temperate.* p. 850

KIELMEYERA *Guttiferae*
coriacea (So. Brazil); ornamental tree with thick-leathery, obovate leaves 15-20 cm long; white flowers in clusters, with large cushion of yellow stamens. *Tropical.* p. 520

variabilis (So. Brazil); handsome small tree with milky sap, large oval, glossy green, leathery leaves 12-15 cm long; terminal racemes of beautiful white flowers 5 cm across, with crisped petals and center cushion of yellow stamens. *Tropical.* p. 521

KIGELIA *Bignoniaceae*
pinnata (Trop. Africa: Sudan, Uganda, Kenya, Zimbabwe, Transvaal, Mozambique), "Sausage-tree"; spreading tree to 15 m, with pinnate leaves, and pendant racemes of large, showy flowers with curved tubular-flaring corolla orange-yellow at base, the 6 cm corrugated lobes blood-red; on reverse striped yellow; a curiosity mainly because of the cylindric, pale brownish-gray fruit 30-45 cm long, hanging on a cord often to 1 metre long; the Kikuyu make beer with this fruit, adding sugar water and honey. *Tropical.*
p. 185, 187, 454, 459

KLEINIA: see Senecio

KNIPHOFIA *Liliaceae*
uvaria (Tritoma) (So. Africa), "Torch lily"; stout perennial herb wth thick roots, clumps of long grass-like basal leaves, and a showy poker-like 1 m spike of nodding tubular flowers, the upper ones scarlet-red, and lower ones yellow. *Subtropic.* p. 591, 615

uvaria maxima, "Big poker plant"; plant taller, stouter; with larger orange-scarlet flowers in longer spike-like racemes. *Subtropic.* p. 591

uvaria praecox; forming impressive clusters of narrow sword-shaped leaves with masses of orange-red spikes of medium height. *Subtropic.* p. 591, 595

KOCHIA *Chenopodiaceae*
scoparia (S.E. Europe, Temperate Asia), "Summer cypress", "Belvedere cypress" or "Fire bush"; showy, densely branched, ornamental annual, with its formal globe or columnar shape resembling Cupressus, ½ to 1 m high; a subshrub with numerous narrow, partly almost threadlike leaves 5 cm long, fresh green; the foliage in forma trichophylla coloring purplish-red in autumn hence the name "Burning bush". The tiny green axillary flowers are insignificant. *Warm temperate.* p. 302

KOELREUTERIA *Sapindaceae*
elegans (formosana) (Fiji), "Chinese rain tree" or "Shrimp tree" (in Florida); spreading, flat-topped tree 6-15 m high; bipin-

nate leaves to 45 cm long with leaflets narrow ovate and lustrous green, later turning yellow before dropping; small fragrant, yellow flowers; fruit a capsule separating into 3 rosy-salmon papery segments imitating shrimps. *Tropical.* p. 875

KOHLERIA   Gesneriaceae
amabilis (Colombia); low growing species with weak stems and scaly rhizomes; small friendly green, hairy, scalloped leaves with red-brown pattern along veins; free-blooming with small bright pink flowers dotted carmine-red on limb; good for hanging baskets. *Tropical.* p. 498

bogotensis (Isoloma pictum) (Colombia); erect tropical herb with velvety dark green to brownish-tinged, ovate leaves, pale along the veins, not bordered with reddish hairs; nodding flowers red above shading to yellow at the swollen lower side of tube, the lower lobes yellow marked deep crimson. *Tropical.* p. 498

'Bogotensis hybrid', a "Tree gloxinia"; plant prostrate when older, the white-hairy leaves are more slender, bronzy-green and lacking pale veins; corolla tube crimson-red changing to white beneath, throat and limb white spotted and lined with red. *Tropical.* p. 498

eriantha (Isoloma hirsutum) (Colombia); erect, soft-hairy species from scaly rhizome, with deep green, ovate, toothed leaves having a conspicuous border of reddish hairs; orange-red corolla tubes widening toward the throat which is pale yellow and marked blood-red. *Tropical.* p. 498, 502

'Eriantha hybrid' (Isoloma hirsutum multiflorum); robust plant more floriferous than the species and with more prominently inflated orange-red corolla tube and large, spreading limb, throat and lower lobes yellow and all marked red. *Tropical.* p. 498, 502

'Sciadotydaea' (Sciadocalyx x Tydaea sections of Kohleria); a larger growing upright hybrid of stout habit with ovate, white-downy, green leaves and large flowers with hairy purplish-red tube and spreading limb, the lower lobes greenish-yellow dotted and lined with violet-purple. *Tropical.* p. 498

KOPSIA   Apocynaceae
flavida (Ochrosia), (India, Java to Philippines), "Shrub vinca" or "Penang sloe"; bushy shrub or tree to 12 m, with lanceolate, thin, shining green leaves to 20 cm long, yellowish beneath, becoming blood-red when old, and containing latex; short clusters of 4-5 cm white flowers with yellow throat. *Tropical.* p. 76

fruticosa (India, Malaysia to Philippines), the red-eyed "Shrub vinca"; large tropical evergreen shrub, with opposite elliptic-lanceolate, somewhat leathery leaves 10 to 20 cm long; the vinca-like flowers in clusters, white with crimson center, 5 cm across, blooming in late spring. *Tropical.* p. 80

KOPSIA of hort.: see Ochrosia

LABURNUM   Leguminosae
anagyroides (C. and So. Europe), "Golden chain"; poisonous but beautiful tree to 10 m high; leaves of 3 leaflets to 8 cm long; pendant inflorescence 15-25 cm long, of deep yellow 2 cm flowers, in late spring. *Warm temperate.* p. 558

x watereri (alpinum x anagyroides), "Golden chain"; small ornamental deciduous tree from So. Europe, with trifoliate leaves and long pendulous, very showy, 40 cm racemes of rich yellow flowers. *Temperate.* p. 553

LACHENALIA   Liliaceae
aloides (tricolor) (So. Africa), "Tricolor Cape cowslip"; dainty hyacinth-like bulbous plant to 30 cm high, with two broad linear lanceolate spreading, fleshy leaves dark green and spotted purple; bright, colorful nodding waxy tubular flowers 2½ cm long, the outer segments yellow tipped green, inner ones scarlet-red at tip, much exceeding the outer, on erect fleshy stalk; spring blooming. *Subtropic.* p. 591, 599

LAELIA   Orchidaceae
albida (Mexico, Guatemala); small epiphyte with 2 narrow-leaved pseudobulbs, and wiry stalk bearing a cluster of 5 cm fragrant flowers, delicate lemon-white often tinged with rose, lip pink or white streaked in center with yellow lines. (Nov.-Jan.). *Tropical.* p. 733

anceps (Mexico); epiphyte with egg-shaped, angled pseudobulbs to 12 cm high, usually single-leaved, topped by long wiry stems to 1 m high, bearing small clusters of slender, variable 8-10 cm flowers, sepals mostly lilac-rose, the petals slightly darker, lip purplish crimson with yellow throat, (Nov.-Feb.). *Tropical.* p. 733

anceps alba (Mexico); very attractive variety with flowers pure white except for shade of yellow in throat *Subtropic.* p. 732

cinnabarina (Brazil); charming epiphyte of compact habit with cylindric tapered 12-25 cm pseudobulbs and usually single leaved; flower stalks 30-60 cm high, with clusters of 5-8 cm flowers bright orange-red, lasting 6 weeks (March-May, Nov.). *Tropical.* p. 733

flava (Brazil); small epiphyte with clustered, thin, 1-leaved pseudobulbs, and a slender truss of small golden yellow flowers with narrow petals and sepals, and crisped lip, (Jan.-June). *Tropical.* p. 734

gouldiana (Mexico); attractive epiphyte with rounded, tapered pseudobulbs with 2 leathery leaves; the dainty 10 cm flowers clustered on slender stalks, sepals and petals light purple and darker crimson at the tips, lip trilobed, deep crimson with yellow throat, (Dec.-Jan.). *Tropical.* p. 733

grandis (Brazil: Bahia); beautiful epiphyte distinctive for the color of its large 10-18 cm flowers, sepals and petals coppery-brown, the lip white flushed purple, darker in throat; furrowed stems with solitary leaf (May-July). *Tropical.* p. 733

x harpophylla (Brazil); handsome epiphyte with 1-leaved terete stems to 45 cm high, and topped by a cluster of showy 5-8 cm flowers with lanceolate sepals and petals brilliant orange vermillion and the trilobed lip margined with white. *Tropical.* p. 733

x harpophylla 'Aurea' (Brazil); variety with sepals and petals in shades of salmon, lip yellow. *Tropical.* p. 733

lundii (regnellii) (Brazil); small pseudobulbs usually two-leaved; 4-5 cm flowers lilac-rose, with narrow sepals and petals, the lip trilobed; summer. *Tropical.* p. 733

pumila (Brazil); variable dwarf epiphyte with small, one-leaved pseudobulbs from a creeping rhizome; single, short-stalked, relatively large, flat flowers usually orchid-lavender, with broad petals and maroon-purple lip, yellow in the throat. *Subtropic.* p. 733

purpurata (Brazil); grand, robust epiphyte with tall, clubshaped pseudobulbs 30 cm or more, and a long, dark green, solitary leaf; the elegant flowers in clusters, narrow sepals and slender petals glistening white, sometimes flushed pink, large lip crimson-purple, with pale yellow throat striped crimson. *Tropical.* p. 733, 737

rubescens (acuminata) (Mexico); dwarf epiphyte with oval, flattened pseudobulbs with solitary leaf; a wiry stem to 45 cm tall bearing a small cluster of 5 cm lilac-mauve flowers, lip lemon-white with purple throat, and fragrant. (Sept.-March). *Tropical.* p. 733

x LAELIOCATTLEYA   Orchidaceae
bella (C. labiata x L. purpurata) (1884); fine hybrid resembling in habit a cattleya, large flowers with sepals and broader petals pale lilac and the lip with broad, wavy lobe of warm purple, with two zones of yellow in throat (Jan.-Feb.). *Tropical.* p. 715, 737

canhamiana alba (L. purpurata x C. mossiae reineckeana); excellent intergeneric hybrid of robust habit, free blooming with clusters of large elegant flowers, sepals and petals ivory or pure white, with deep violet-purple frilled lip edged white, and a golden throat, (Feb.-June). *Tropical.* p. 737

hassallii alba 'Majestica' (L.C. Britannia x C. warscewiczii); charming bigeneric hybrid with glistening white sepals and broad wavy petals, and bright purplish-violet lip, yellow in the throat and frilled (spring). *Tropical.* p. 737

LAGENANDRA   Araceae
koenigii (Sri Lanka); tropical aquatic plant from Singaraja; robust root-stock carries dark green ribbon-like leaves slender at each end, 40-50 cm long, 1-1½ cm wide; small spathe smooth outside, pale green and red-purple. *Humid-tropical.* p. 104

ovata (Cryptocoryne) (W. India, Sri Lanka); large tropical waterside plant with thick, creeping rootstock, bearing rosette of lanceolate, fleshy leaves matte grass green, nearly 1 m long with stalk; spathe short and thick-fleshy covered with coarse purplish-red scales. *Humid-tropical.* p. 104

LAGENARIA   Cucurbitaceae
siceraria 'Clavata' (Old World Tropics), "White-flowered calabash gourd"; long-running, tender annual vine, with cordate-ovate leaves, white flowers; smooth, pendant fruit, hard-shelled when ripe, 30-40 cm long. *Tropical.* p. 383

siceraria 'Rotunda'; a greenhouse grown gourd with pear-shaped, yellow fruit. *Tropical.* p. 382

LAGERSTROEMIA   Lythraceae
hirsuta (reginae) (India to New Guinea); evergreen tree to 10 m high, with oblong, leathery deep green leaves 20 cm or more long; flowers purplish or white, followed by woody capsules 2-3 cm dia. *Tropical.* p. 622

indica (Japan, Korea, China), the "Crape myrtle"; handsome flowering tree to 10 m high, foliage falling annually, the elliptic leaves 2 to 6 cm long; at branch tips the gorgeous clusters of frilled, though scentless flowers pink or purple and resembling crepe-paper, blooming profusely all summer, and fall from August to October; best in hottest or sunniest season. In older trees the bark flakes off to reveal a smooth pinkish inner bark. *Subtropic.* p. 618

indica 'Rubra', "Red crape-myrtle"; variety with flowers clear crimson-red. *Warm temperate.* p. 618

speciosa (India to Australia), "Rose of India" or "Queen's crape myrtle"; in Hawaii as the "Giant crape myrtle"; a beautiful flowering tree to 20 m high; thin-leathery ovate leaves 25 cm long, and clusters of large, very showy flowers with frilled petals, to 8 cm across; colors usually mauve to purple, but in Tahiti have seen trees with flowers in gorgeous clear rose. *Tropical.*   p. 617, 618

LAGUNARIA   *Malvaceae*
patersonii (New South Wales, Queensland), the "Pyramid-tree"; of symmetrical pyramid shape, evergreen, 6-15 m high; the young growth scurfy; thick, ovate 10 cm leaves white-scaly beneath; 6 cm axillary, bell-shaped flowers rosy-pink, with 5 recurved petals. *Subtropic.*   p. 624

LAGUNCULARIA   *Combretaceae*
racemosa (Trop. America, W. Africa), "White mangrove"; sprawling shrub with opposite leathery gray green, obovate leaves; growing in Fairchild Gardens, Miami, Florida. *Tropical.*   p. 303

LAMIUM   *Labiatae*
galeobdolon 'Variegatum' (Quebec, C. and E. Europe, to Urals), "Yellow archangel"; rampant creeper with square thread-like, rooting stems covered with appressed pale hairs; nettle-like, opposite, crenate leaves 3-5 cm long, deep green and rugose, prettily zoned and painted with silver; two-lipped yellow flowers, the lower lip marked red. *Temperate.*   p. 530, 531

LAMOUROUXIA   *Scrophulariaceae*
rhinanthifolia (Mexico); striking woody shrub 1-1½ m high with stiff, sharply spined leaves; the inflorescence candelabra-like with tubular scarlet, pubescent flowers 4-5 cm long; photographed on Monte Alban, Oaxaca, So. Mexico. *Subtropic.*   p. 885

LAMPRANTHUS   *Aizoaceae*
aurantiacus (So. Africa: Cape Flats), "Golden ice plant"; bush type; succulent shrub forming dense bush of upright habit; olive green or grayish 3-sided leaves 3-5 cm long; blooming profusely in spring with large 3-5 cm glossy flowers an iridescent bronzy orange. *Arid subtropic.*   p. 47
aurantiacus 'Glaucus' (So. Africa: Cape Prov.), "Yellow bush ice plant"; sparsely branching succulent to 45 cm high; gray-green tapering leaves 2-3 cm long, obtusely 3-angled, with red tips; gleaming yellow flowers 4-5 cm across. *Arid-subtropical.*   p. 45
aureus (Mesembryanthemum) (Cape Prov.), "Golden ice plant"; succulent creeper with brown-barked branches, leaves shortly connate, sides convex, 5 cm or more long, narrowing to a point, fresh-green, with transparent dots; large 5 cm flowers deep, shining orange with yellow center. Naturalized in So. California. *Subtropic.*   p. 43, 45, 47
productus (Cape Prov.: Karroo), "Purple ice-plant"; freely branching, shrubby succulent, with bluish to coppery, long linear leaves 3-4 cm long; the large 3 cm flowers glistening purple with long rays; popular for planting on steep banks in California. *Arid-subtropic.*   p. 47
roseus (Cape Prov.: Table Mountain), "Pink ice plant"; spreading succulent, with compressed 3-angled, linear leaves 3 cm long, covered with translucent spots, and large soft pink flowers becoming deeper toward center, and with yellow in center. *Arid-subtropic.*   p. 45, 47
spectabilis (Mesembryanthemum spectabile) (So. Africa: Cape Prov.), "Red ice plant"; somewhat woody perennial, branching with long prostrate reddish flowering stems; succulent keeled leaves 2-5 cm long, glaucous gray olive; the 4-5 cm flowers gleaming in brilliant color, normally purplish, and with longer leaves, the color forms in the California trade as L. spectabilis are probably hybrids such as the pictured "Red ice plant" in shining crimson; other colors available are pink or rose. Beautiful flowering plants in spring and summer. *Arid-subtropic.*   p. 43, 45, 47
spectabilis rosea (So. Africa); a charming color form with large flowers a shining baby-pink of iridescent quality, opening to the warming sun as "Mid-day flower". *Arid-subtropic.*   p. 45
tricolor (Transvaal), "Copper ice plant"; low creeper with small linear, succulent, glaucous leaves; masses of brilliantly colored daisy-like heads of ray-flowers brick-red to crimson toward apex, with whitish button in center. *Arid-subtropic.*   p. 47
zeyheri (So. Africa: Cape Prov.); succulent creeper with curved branches; slender cylindrical, bluntly tapered smooth leaves of soft texture, glossy green and spotted, 4 cm long; purplish-violet to pinkish flowers 5-6 cm across. *Arid-subtropical.*   p. 43, 45

LANTANA   *Verbenaceae*
camara (West Indies), "Shrub verbena"; small hairy shrub with thin-woody, angled branches sometimes prickly, with ovate, toothed, rough-bristly leaves; very floriferous with stiff-erect, small but showy heads of verbena-like flowers, changeable, usually opening pink or yellow, becoming red or orange, and several color combinations may be found on the same plant; summer-blooming. *Tropical.*   p. 918
montevidensis (Uruguay), "Trailing lantana"; small, downy, spreading shrub with weak, vine-like, pendant branches, used as a ground cover or in baskets, small ovate, rough leaves, a profuse bloomer winter and summer, with pretty 3 cm heads of rosy-lilac flowers. *Subtropic.*   p. 918

LAPAGERIA   *Liliaceae*
rosea (Chile), "Chile bells"; showy vine with alternate ovate, leathery leaves, and many large 8 cm pendulous axillary bell-shaped flowers, rich rosy crimson, spotted white inside; summer blooming. *Subtropic.*   p. 593
rosea var. albiflora (Chile), "White Chile bells"; a variety beautiful in contrast, with the hanging waxy flowers pure milky-white; the long-elliptic leathery leaves grayish green with wavy margins. *Subtropic.*   p. 591

LAPEIROUSIA   *Iridaceae*
laxa (cruenta) (Anomatheca) (Transvaal), "Woodland painted petals"; miniature cormous plant with flat narrow basal leaves, the small flowers on a one-sided wiry raceme, long tubes with spreading lobes bright orange-red with blood-red border; summer blooming. *Subtropic.*   p. 525

LASIA   *Araceae*
spinosa (Malaysia); tropical swamp plant having thick stem with long, spiny leafstalk and variable, matte, green leaf with spines beneath, arrowshape, later with lateral lobes; recurved spathe brown. *Tropical.*   p. 106

LASIANDRA: see Heterocentron

LATANIA   *Palmae*
borbonica: see Livistona chinensis
loddigesii (Mauritius), "Blue latan palm"; handsome fan palm 15 m tall, with rough, slender trunk, bearing a large crown of numerous palmate, rigid blue-gray leaves to 1 m across, the deeply cut segments fuzzy beneath, the petioles and leaf veins colored orange. *Tropical.*   p. 780
lontaroides (borbonica, commersonii) (Mauritius), "Red latan"; robust, rapid-growing fan palm to 16 m, with gray trunk swollen at base, bearing a large crown of numerous handsome thick leaves palmate fan-shaped, gray-green, and deeply cut, 2-2½ m across; the segments edged with tiny sawteeth, veins and margins tinged with red, fuzzy beneath; the stalks colored orange, thorny when young, the rachis extending into the leaf for 45 cm or more. Highly ornamental. Loves warmth and moisture with good drainage. *Tropical.*   p. 785
verschaffeltii (Mascarene Isl.), "Yellow latan palm"; robust fan palm more than 15 m high, with rough trunk, palmate fronds light green, white tomentose beneath, with thin yet rigid segments, long-pointed; petiole yellow. *Tropical.*   p. 780

LATANIA: see also LIVISTONA

LATHYRUS   *Leguminosae*
clymeneus articulatus (E. Mediterranean); annual with tendril climbing stems and clusters of 2-4 purple flowers with rose-colored wings; narrow leaflets bluish-gray. *Subtropic.*   p. 557
odoratus (Sicily), "Sweet pea"; annual, tendril-climbing herb with brittle, winged stems, paired oval leaves, and large, sweetly fragrant, butterfly-like flowers originally purple, but now hybridized into many delicate pastel tints. *Subtropic.*   p. 557, 559

LAURENTIA: see Isotoma

LAURUS   *Lauraceae*
nobilis (Asia Minor, naturalized in So. Europe), "Sweet bay"; or true laurel; evergreen, pyramidal aromatic tree, very leafy, with elliptic, stiff thin-leathery leaves dark green and lightly crimped, and small, inconspicuous greenish-yellow flowers; much grown in tubs and clipped into formal shapes, mainly in Belgium. The leaves are used in cooking. *Subtropic.*   p. 534, 535

LAVANDULA   *Labiatae*
officinalis (spica) (Mediterranean reg.), "English lavender"; aromatic subshrub with square stems with narrow linear leaves white tomentose when young; spikes of fragrant purple-blue flowers in late summer. Use: medicinal, in perfumes, soaps, sachets; moth prevention. *Subtropic.*   p. 530

LAVATERA   *Malvaceae*
trimestris (Mediterranean), "Tree-mallow"; rough-hairy annual herb 1-2 m high, with lower leaves nearly round, upper ones angled or trilobed and toothed; showy axillary, solitary saucer-like flowers 6-10 cm across, rosy with darker veins. *Subtropic.*   p. 624

LAWSONIA *Lythraceae*
  inermis (North Africa, Asia), "Mignonette tree" or "Henna"; shrub or tree to 6 m high, with elliptic leaves 2-5 cm long; fragrant rosy 1 cm flowers. The dried ground leaves yield Henna, a fast orange dye. *Subtropic.* p. 620

LECYTHIS *Lecythidaceae*
  ollaria (Guayana), "Monkey-pot tree"; large tropical tree with alternate, leathery, dotted leaves, and curious flowers orchid pink, with calyx lobes overlapping; globular woody fruit, the lid falling off when the fruit is ripe. *Tropical.* p. 536

LEEA *Vitaceae*
  rubra (India, Burma, Malaya); evergreen shrubby herb with dark bronzy, pinnate foliage, and clusters of flowers opposite the leaves, brick-red; when open the petals are pink inside, red outside; the fruit a small berry. *Tropical.* p. 920

LEMAIREOCEREUS *Cactaceae*
  marginatus (Pachycereus) (Central and So. Mexico), "Organ-pipe cactus"; beautiful column cactus tree-like, or branching from base, to 8 m tall; slender, smooth green stems to 12 cm thick, with 5-7 ribs and gray spines; bell-tubular flowers 5 cm long, red outside, greenish-white inside; globular orange fruit 4 cm dia, with reddish flesh and black seed. Often planted in Mexican villages as a natural fence. *Arid-tropical.* p. 246, 252, 253
  stellatus (So. Mexico), in the trade as "treleasei"; pretty, slender columns to 3 m high, branching; pale bluish-green, 8-12 ribs, short white spined; flowers red, borne at the apex; edible fruit known in Mexico as "Joconostle". *Arid-tropical.* p. 253
  thurberi (Arizona, Mexico), "Arizona organ pipe"; forbidding-looking columns with 12-17 acute ribs, dark green to grayish, the dense areoles with black cushions and clusters of stiff spines gray to black; branching from the base and becoming 6 m high; dayblooming, purplish flowers with white margin. *Arid-subtropic.* p. 246, 253
  treleasei (Mexico); stout column to 6 m, scarcely branched, closely ribbed and covered with short yellow spines; flowers pinkish. *Arid-tropical.* p. 251
  weberi (Sierra Madre of Oaxaca and Puebla, Mexico), "Candelabra cactus"; magnificent much branched tree to 10 m high, columns glaucous dark green, 10-ribbed, short and sharp gray to black spines; flowers white with brown hairs, to 10 cm long; orange-red fruit 8 cm long and edible. *Arid-tropical.* p. 249

LEOCHILUS *Orchidaceae*
  labiatus (C. America, W. Indies); dwarf epiphyte with small compressed pseudobulbs, and solitary leaves 6 cm long; inflorescence to 25 cm high with 2 cm fleshy yellow flowers overlaid with reddish brown. *Tropical.* p. 734

LEONOTIS *Labiatae*
  leonorus (So. Africa), "Lion's ear"; tall perennial with light green, soft pubescent, crenate, elliptic leaves; the 2-lipped, orange-red, downy flowers in dense showy whorls, during winter. *Subtropic.* p. 531

LEONTOPODIUM *Compositae*
  alpinum (Alps, Pyrenees, Himalaya), "Edelweiss"; low tufted woolly perennial herb which I found even in Central Asia in view of Mt. Everest, the star-shaped floral bracts very pronounced silvery-white at high altitudes. *Temperate.* p. 323, 331

LEPANTHES *Orchidaceae*
  pulchella (Jamaica); small cluster-forming epiphyte with slender secondary stems supporting a single 2 cm leaf at apex; from the base the flower spikes with minute blossoms less than 1 cm across, complex and colorful but short-lived, greenish-yellow and crimson-red (all year). *Tropical.* p. 735

LEPISMIUM *Cactaceae*
  cruciforme (Pfeiffera) (E. Brazil), "Tree-and-rock cactus"; branching cactus with angled stem waxy green tinted purple, the areoles sunken in margins set with tufts of white hair-spines; small white flowers; red fruit. *Tropical.* p. 285

LEPTOPTERIS *Osmundaceae (Filices)*
  superba (New Zealand), the "Prince of Wales plume"; considered the most beautiful fern in New Zealand; we found this species hiding in deep calcareous sink holes deep in dense mountain forest near Nelson on South Island; from a stout rhizome rise shimmering deep green, plume-like tripinnate fronds from 45 cm to 1 m long, on woolly-hairy stalks; thin and membranous leaflets steeped in moisture and densely arranged and overlapping feather-like. Orchid growers have used the fibrous roots as potting medium, although it takes hundreds of years to form a large caudex. *Humid-subtropic.* p. 437

LEPTOSPERMUM *Myrtaceae*
  laevigatum (S.E. Australia, Tasmania), "Australian tea tree"; large shrub to 10 m, forming picturesquely twisted gray-brown trunks; fine-textured leathery oval leaves 3 cm long; small white flowers 2 cm wide. *Subtropic.* p. 683
  petersenii (citratum) (New South Wales), "Lemon-scented tea tree"; evergreen shrub or tree 6 m high, with linear 3-5 cm lemon-scented leaves, and yielding aromatic oil; small white flowers. *Subtropic.* p. 685
  scoparium (Australia, New Zealand), "Tea-tree"; an attractive flowering shrub which may grow to 6 m high, the young growth silky, dense with tiny rigid sharp pointed leaves under 1 cm, dotted with fragrant oil glands; and numerous small white or rosy axillary flowers amongst the foliage, in summer. *Subtropic.* p. 685, 687
  scoparium 'Album', "White tea tree"; variety with large white flowers, contrasting from the usual pink or red. *Subtropic.* p. 685
  scoparium 'Keatleyi'; largest-flowered and most rangy of the scoparium cultivars, to 3 m high; single pink flowers 2 cm across, with red stamens. *Subtropic.* p. 685
  scoparium 'Rubrum plenum', "Manuka"; tall-growing cultivar with fully double flowers deep crimson red. Popular in New Zealand for cut flower use. *Subtropic.* p. 683
  scoparium 'Ruby Glow', "New Zealand tea tree"; excellent commercial cultivar of compact habit, with age growing to 2 m or more; tiny dark needle-like foliage, and semi-double, crimson red flowers with dark center, 2 cm across in great profusion. *Subtropic.* p. 683

LESCHENAULTIA *Goodeniaceae*
  laricina (Western Australia); ornamental, much branched shrub 30-50 cm high, with heath-like crowded, grayish green needle-like leaves 1 cm long; in the upper leaf axils the 3 cm flowers with split tube and oblique corolla with spreading lobes, in a variety of colors from white or lilac to rich red with yellow base. *Subtropic.* p. 508

LEUCADENDRON *Proteaceae*
  arboreum (So. Africa); large sparry shrub 5 m high, the woody branches densely set with lanceolate, grayish leaves; the globular, feathery inflorescence salmon yellow. *Subtropic.* p. 836
  argenteum (So. Africa), "Silvertree"; beautiful, eye-catching tree to 10 m high, the branches dense with clasping, pointed leaves to 15 cm long, thickly covered with silvery pubescence which glistens like shining silver in the sun; globular flower heads yellow. *Subtropic.* p. 837, 839

LEUCAENA *Leguminosae*
  glauca (Tropical America, naturalized elsewhere in tropics), "Wild tamarind"; tree to 9 m with brown stems; leaves 2-pinnate; leaflets in 10 to 20 pairs, glaucous beneath; flowers lemon-yellow in globose, fluffy heads about 3 cm across, stamens 3 times as long as petals. *Tropical.* p. 556

LEUCHTENBERGIA *Cactaceae*
  principis (Mexico), the "Prism cactus"; to 20 cm high, a very different type of small cactus with a parsnip-like root and the elongated tubercles looking like triangular fleshy grayish-green leaves as in Agave but with the tips cut off and bearing grayish wool and angular papery spines; fragrant yellow flowers near the center, borne at the tip of new tubercles. *Arid-tropical.* p. 269
  principis 'Variegata', the "Agave cactus"; with tall grayish green conical knobs 10-12 cm long, having reddish angles and variegated with ivory-white; the tips with long radiating papery spines. *Arid-tropical.* p. 269

LEUCOJUM *Amaryllidaceae*
  vernum (Cent. Europe), "Spring snowflake"; charming small bulbous herb to 30 cm high, narrow linear basal leaves; the nodding solitary bell-flowers white, tipped with green, 2 cm long, in early spring. *Temperate.* p. 70

LEUCOSPERMUM *Proteaceae*
  catherinae (Cape Prov.), the "Catherine wheel"; handsome shrub with stiff, gray, lanceolate leaves edged red, and terminal head of flowers with spreading flesh-colored filaments reddish near tips. *Subtropic.* p. 836
  conocarpum (Cape Prov.), "Cripple-wood"; showy shrub with gnarled wood, crowded broad leaves, and golden yellow flowering cone. *Subtropic.* p. 838, 839
  nutans (So. Africa: S.W. Cape), "Nodding pincushion"; beautiful bush about 1 m high, of low spreading shape, dense with small 8 cm tough-leathery, concave, gray green sessile leaves, and a showy, symmetrical inflorescence, the flowers curled tightly in the center like a pincushion, 10 cm across, with long curving, waxy styles varying in color from yellow to orange, or pinkish; the heads sometimes nodding to one side, and lasting for about one month. *Subtropic.* p. 836, 839

prostratum (So. Africa), "Creeping pincushion"; low creeping shrub with densely matting, long wiry branches set with needle-like leaves, and masses of small round heads of ball-like flowers looking like strawberries, 4-5 cm across, from lemon to deep orange and red as they mature, resulting in a variety of colors at one time. *Subtropic.* p. 839

reflexum (So. Africa), the "Rocket-pincushion"; strikingly beautiful gray-downy shrub which can become 4 m high; the branches densely and regularly set with small obovate silvery leaves 6 cm long and becoming shorter toward the brilliant terminal head of salmon tubular flowers with yellow base, exceeded by the showy 8 cm long, thread-like glossy scarlet styles and stigmas, which become reflexed and look like a rocket with red streamers. *Subtropic.* p. 836, 838

LIATRIS   *Compositae*
elegans (Texas, Oklahoma, Arkansas, east to S. Carolina), "Blazing star"; thistle-like perennial herb to 1 m and more, sparry branches with spiny, silvery wings; globular heads purple. *Warm temperate.* p. 322

LIBOCEDRUS   *Cupressaceae (Coniferae)*
bidwillii (Calocedrus) (New Zealand), "Incense cedar"; tree to 20 m, with bark falling in narrow strips; aromatic needles arranged in two vertical rows, the juveniles flattened; the older leaves fern-like; female cones ovoid, 1 cm long. *Warm temperate.* p. 339

plumosa, (New Zealand), "Kawaka"; beautiful symmetrical conifer to 30 m, with lush green, scale-like leaves in fern-like arching sprays, becoming very dense with age. *Warm temperate.* p. 339

LICANIA   *Chrysobalanaceae*
tomentosa (No. Brazil), the "Oiti"; large tropical tree which I photographed in Bahia, with thin-leathery ovate, glaucous green leaves 8 cm long, the flowers deep maroon and yellowish; the ovoid fruit 5-6 cm long, with leathery yellow skin covering a large hard seed, containing about 30% of oil and burning readily when ignited; locally strung on sticks for illuminating; oil is used for candles, soap and grease. *Tropical.* p. 303, 463

LICUALA   *Palmae*
grandis (New Britain Isl. near New Guinea), "Ruffled fan palm"; very attractive small fan palm with slim solitary 3 m trunk, topped by plaited bright green leaves almost round, lobed and toothed along the continuous margin, on slender, thorny petioles; glossy crimson fruit. *Tropical.* p. 784, 785

muelleri (Australia); beautiful fan palm of compact habit, slow growing, remarkable for the near perfect circle of its glossy-green fronds approximately 60 cm in dia., the wheel-shaped blade divides itself into numerous sectional folded fans, their apex lobed with each fold, and notched. *Subtropic.* p. 785, 789

peltata (India: Bengal); clustering fan palm with slender trunks to 5 m, leaves orbicular appearing as if peltate, 120-150 cm across, divided into wedge-shaped, ribbed segments. *Tropical.* p. 784

rumphii (Moluccas); dwarf clustering fan palm 1 m high, with leaves divided into irregular pleated sections, the apex as if cut off, and toothed. *Tropical.* p. 784

spinosa (Malaya to Java), "Fan palm"; clustering fan-leaf, densely suckering palm forming compact tufts with a mass of foliage from top to bottom; glossy-green leaves parted to the center into plaited segments ending abruptly in a toothed apex as if cut off, the rigid petioles armed with curved black thorns; fruit lustrous red. *Tropical.* p. 777

LIGULARIA   *Compositae*
przewalskii (Senecio) (No. China); handsome perennial, 100-150 cm tall, with creeping rootstock; purple stems and large, deeply cut leaves; small yellow flowers in long spike. *Temperate.* p. 329

tussilaginea 'Argentea' (kaempferi); solitary cordate-orbicular toothed leaves rising from rhizomatous basal rosette, glaucous fresh to grayish-green, variegated cream or white especially at margin. *Subtropic.* p. 326

tussilaginea 'Aureo-maculata' (L. kaempferi or Farfugium japonicum) (Japan), "Leopard plant"; with large rounded, green, smooth leaves blotched yellow and cream, 10 to 15 cm long; light yellow daisy-like flowers. *Subtropic.* p. 326, 327

LIGUSTRUM   *Oleaceae*
japonicum (Japan, Korea), the "Japanese privet"; fast growing evergreen bush with leathery, rich dark glossy green ovate, short-pointed foliage to 3 or even 6 m high, leaves 4-10 cm long, on minutely downy twigs; small white flowers in terminal clusters, and black berries. *Warm temperate.* p. 695, 696

lucidum (China, Korea, Japan), the "Glossy privet" of the South, or "White-wax tree" in China, planted widely in the southern states; dense leafy evergreen shrub to 10 m high, with glossy, dark green, thick-leathery, ovate leaves 8-15 cm long, on flexible, glabrous branches; white flowers in long panicles followed by black berries. *Subtropic.* p. 698, 699

lucidum 'Silver Star'; handsome compact, dense evergreen bush of slow growth; oval leaves with deep green centers and edged in creamy silver. *Warm temperate.* p. 698

lucidum 'Texanum', the "Wax-leaf privet"; a cultivar of compact habit, with glossy, thick-leathery dark green, ovate-acuminate leaves 6-9 cm long, somewhat wavy; ideally suited to shaping by trimming, and an enduring decorator under unfavorable conditions. *Subtropic.* p. 698

ovalifolium (Japan), widely known as the "California privet"; densely branching half-evergreen shrub of erect habit to 4 m high, and much used for hedging including planter boxes for the sidewalk restaurants of New York, where it is quite hardy; the willowy, fast growing shoots are either evergreen, or if the leaves fall off during cold winters, quickly cover themselves anew with oval, rich green smooth leaves 3-7 cm long; summer blooming, if not trimmed back, with erect panicles of white flowers, followed by shining black berries. *Temperate.* p. 696

sinense 'Variegatum', (China), "Chinese silver privet"; shapely variegated bush; the green species in China to 4 m high and deciduous; densely branched with thin twiggy, woody stems, the small ovate leaves 1 to 2 cm long and faintly dentate, to 8 cm long in older plants, milky green with white borders irregularly outlined; pubescent on midrib beneath; fragrant whitish flowers. Not very winter hardy. *Warm temperate.* p. 700

LILIUM   *Liliaceae*
amabile 'Enterprise', "Korean lily"; stiff erect stems to 1 m or more, with lanceolate leaves, and nodding Turk's cap flowers, with waxy recurved petals brilliant red, spotted with black. *Warm temperate.* p. 600

auratum (Japan), "Goldband-lily"; tall erect, leafy stalk from a scaly bulb, to 2 m, with flaring, fragrant trumpet-flowers 25 cm across, white spotted with crimson, each segment with a central yellow stripe. *Subtropic.* p. 600, 601

x aurelianense (henryi x sargentiae), "Aurelian lily"; hardy tall-stemmed lily, closely set with narrow green leaves; horizontal fragrant flowers to 12 cm across, yellow-orange, the petals tending to recurve. *Warm temperate.* p. 600

cernuum (Korea, No. China); to 1 m high, 1-6 flowers in racemes, 4-6 cm long, purplish pink with wine-purple spots and fragrant; petals strongly reflexed. *Temperate.* p. 601

'Cinnabar' (Mid-century hyb.); bulbous plant producing erect cluster of flowers of medium size, deep scarlet, with brown spots. Not continuously winter-hardy. *Warm temperate.* p. 601

dauricum (Siberia), "Candlestick-lily"; about 1 m high, with erect flowers to 12 cm wide, the spreading elliptic segments orange-red spotted with purplish-black. *Temperate.* p. 600

davidii (Western China); to 1 m high, linear leaves; 1-20 flowers 8 cm long, in clusters, orange red and spotted with black segments reflexed. *Warm temperate.* p. 601

grayii (Virginia to Carolinas) "Orange bell-lily"; stoloniferous with stems to 1 m, elliptic leaves in 3-6 whorls; bell-shaped nodding flowers 5 cm long, reddish orange spotted with brown. *Warm temperate.* p. 601

'Harmony' (tigrinum hybrid); a "Mid-century lily", one of the finest, early-flowering in rich orange, the wide-spread blooms 12 cm across and spotted with brown; numerous flowers and buds on a stiff-erect leafy stem about 50 cm high or more; normally blooming in June. One of the best for forcing in pots; may be planted in the garden after flowering, but leave all the foliage possible to help mature the new bulb. *Warm temeprate.* p. 601

'Imperial Crimson', an "Empress hybrid lily"; fantastic strain of magnificent, vigorous lilies raised by Oregon Bulb Farms, involving blood of L. auratum, speciosum, japonicum and rubellum; tall, leafy, wiry stalks 1½-2 m high carrying giant rather flat, firm flowers 20 cm across, the petals gently curling back, basically white but toward the middle changing to pink with red spotting, the center of each segment intense crimson; blooming in August. *Warm temperate.* p. 594

langkongense (Western China); stoloniferous; stem to 1 m; oblong leaves dark green; nodding, fragrant flowers to 4 cm long, white and tinged rose-purple and spotted purple segments reflexed. *Warm temperate.* p. 601

'Limelight'; vigorous, tall lily to 1½ m, the stems with narrow leaves; fragrant flowers funnel-shaped to 20 cm across, lime yellow. *Warm temperate.* p. 600

longiflorum 'Ace'; Oregon cultivar of the "Croft Easter lily" complex; prolific bloomer of rather short, compact habit, dark shining green foliage on stem about 45 cm above pot, toward the apex of which appear buds and flowers one above the other, unlike

Croft which are all in one cluster; the firm-textured white trumpets are generally shorter (12-14 cm long) and force well from a slow start. *Warm temperate.* p. 594

longiflorum 'Croft', leading "Easter lily", superior commercial pot plant because of its large and elegant, firm-textured, white, trumpet-shaped flowers 12-16 cm long, carried horizontally on stiff, medium-tall, green stems, densely furnished with glossy-green leaves; developed in the Pacific Northwest from a hybrid of the Japanese L. 'Giganteum' x 'Erabu'. *Warm temperate.* p. 599, 601

'Prosperity'; compact plant 50 cm high, slender stalk with papery foliage, 12 cm flowers with firm segments, lemon yellow with brown spots. *Warm temperate.* p. 601

pumilum (China), "Coral lily"; to 50 cm high, with wiry stem, numerous linear leaves; fragrant nodding flowers, 5 cm dia., waxy scarlet, with reflexed segments. *Warm temperate.* p. 601

speciosum (Japan), the "Japanese lily"; well-known, attractive species with scaly bulb and wiry stem to 1 m high with scattered rather broad, lanceolate leaves, bearing 3-10 stalked fragrant flowers 10-12 cm across, with reflexed petals, white tinted rose, and spotted with purplish-red. *Warm temperate.* p. 600

speciosum 'Album', "White Japanese lily"; stems purplish-brown, flowers with reflexed white petals; long white stamens with red anthers. *Warm temperate.* p. 600

speciosum 'Lillian Wallace'; large, wide open flowers with reflexed petals, white and with purplish-red spots and splashes. *Warm temperate.* p. 601

'Susan'; waxy flowers of medium size, with segments split down to calyx and reflexed, flesh-pink with purple spots. *Warm temperate.* p. 601

tigrinum (China), "Tiger lily"; hardy, stem-rooting bulbous plant 1 m or more high; strongly recurved Turk's cap flowers bright orange-red, spotted purple-black, and prominent anthers with red pollen. *Warm temperate.* p. 600

LIMNOBIUM   *Hydrocharitaceae*
stoloniferum (laevigatum) (Mexico, West Indies, to Brazil); perennial floating aquatic plant, with long trailing runners forming rosettes of dark green oval, spongy leaves 2-3½ cm long; flowers greenish-white, arranged in three's. Photographed in the tropical pool in Botanic Garden Munich. *Humid-tropical.* p. 522

LIMONIUM   *Plumbaginaceae*
sinuatum (Mediterranean reg.), "Statice" or "Sea-lavender"; biennial or perennial with tufted, lobed leaves, and panicles of numerous, small, clustered flowers with blue calyx and yellowish-white corolla, on winged branches; used as an "everlasting" when cut. *Subtropic.* p. 811

suworowii (bot. Psylliostachys suworowii) (C. Asia to Iran and Israel), "Sea lavender"; annual herb with oblanceolate leaves 15 cm long; inflorescence in spikelets 40 cm long, with funnel-form flowers rosy-purple. *Warm temperate.* p. 811

LINOSPADIX   *Palmae*
monostachya (New South Wales; Queensland), the "Walking stick palm"; clustering small palm (2-3 m) with cane-like smooth green stem less than 3 cm thick, partly covered by remnants of leaf-petioles; few pinnate fronds to 90 cm long, the leaflets dark green and variable in width. *Subtropic.* p. 784

LIPARIA   *Leguminosae*
spherica (So. Africa), "Mountain dahlia"; striking very leafy shrub of stiff habit, to 2½ m; leaves crowded and overlapping, 4-5 cm long; flowers in showy terminal heads to 10 cm wide, subtended by numerous bracts in a confused mass, bright yellow or orange. *Subtropic.* p. 558

LIQUIDAMBAR   *Hamamelidaceae*
styraciflua (Connecticut to Florida and Mexico), "American Sweet gum"; attractive tree, deciduous in colder climate, to 20 m or more high, with furrowed bark and corky wings on twigs, the leaves maple-like and deeply cut into 5-7 lobes, 8 to 15 cm wide, deep glossy green, but turning purple, yellow or deep red in autumn; flowers inconspicuous, but the dangling spiny fruits add ornament in winter. *Warm temperate.* p. 523

LIRIODENDRON   *Magnoliaceae*
tulipifera (Mass. to Florida and Mississippi), the noble "Tulip-tree"; becoming 60 m high, with curious lobed, almost square deciduous leaves to 12 cm each way, and bell-shaped flowers to 10 cm across, greenish yellow with orange at base, very fragrant. *Warm temperate.* p. 619

LIRIOPE   *Liliaceae*
platyphylla (China, Japan); robust, clustering plant with long linear leaves, and tall cylindrical inflorescence of blue-purple flowers. *Warm temperate.* p. 606

LITCHI   *Sapindaceae*
chinensis (Nephelium litchi) (So. China), the "Lychee nut"; evergreen fruiting tree, round-tipped 6 to 12 m high, with spreading branches; leaves compound with 2-4 pairs of shiny leathery, ovate leaflets 8-16 cm long; small flowers with greenish-white sepals, no petals, followed by the round, juicy red fruit, enclosed in brittle warty shell, 2 to 3 cm in dia., according to variety, in pendant clusters, looking like strawberries hanging from the end of the twigs. Opened with fingernails, the firm, whitish pulp is exposed, delicious to the taste and mildly acid. The dried fruit called "Litchi nuts" and with sweeter taste, is eaten like a raisin. Very popular in China. *Subtropic.* p. 457, 462, 477, 875

LITHOPS   *Aizoaceae*
bella (So. Africa), "Pretty stoneface"; small succulent plant with two thick leaves united, except for fissure across top and resembling pebbles which they mimic; brownish yellow with darker markings; flowers white. *Arid-subtropic.* p. 48

bromfieldii (Cape Prov.); cluster-forming split body almost round, red-brownish at sides, the flattened top strewn with gray translucent windows, and a network of brown and purple; flowers yellow. *Arid-subtropic.* p. 43

fulviceps (Southwest Africa); mostly single bodies 3 cm high, sides gray with reddish tint, top ochre with raised translucent dark dome-like dots with rusty red lines; flowers yellow, whitish beneath. *Arid-tropical.* p. 43

lesliei (Transvaal, Cape); split growths to 6 cm high, rust brown, top flat with network of deep gray, and islands of light brown; flowers bright yellow, pinkish on back. *Arid-subtropic.* p. 43

pseudotruncatella (S.W. Africa); tufted split-rocks, pale brownish-gray with network of brown lines; however, color of body changes according to soil color—brown over yellowish to chalky gray-white; flowers yellow. *Arid-tropical.* p. 43

turbiniformis (So. Africa: Cape Prov.), "Flat-top mimicry plant"; elegant round, split body to 4 cm high and broad, the sides gray, the top brownish and wrinkled, with dark brown branched grooves between the pebbly surface. The succeeding pair of leaves splits open the older body. Large 4 cm yellow flowers. *Arid-subtropic.* p. 49

villetii (So. Africa: Bushmanland), "Stoneface"; tiny succulent bodies to 2 cm high, with gray-green leafpairs united except for a cleft across the apex; the translucent window on the upper surface convex, with pale spots; flowers 3 cm across, with white petals. *Arid subtropic.* p. 42

LITTONIA   *Liliaceae*
modesta (So. Africa), "Climbing lily"; tuberous herb with climbing, flexuous leafy stem similar to Gloriosa, to 2 m; bright shining green leaves ending in a tendril, and axillary bell-shaped flowers rich orange, 3 cm long. *Subtropic.* p. 602

LIVISTONA   *Palmae*
australis (Queensland, New South Wales, Victoria), "Australian cabbage palm"; with slender, ringed trunk to 25 m or more, when younger covered with brown leaf bases and brown fiber; dense crown of soft-leathery palmate fronds rounded in outline, 1-1½ m across, divided to middle into narrow glossy green segments with yellow central nerve, and without threads between; the petiole with stout curved spines; the rib extends to 10 cm into base of leaf. *Subtropic.* p. 784, 785

chinensis (South China), the formerly widely grown "Latania borbonica" of horticulture, "Chinese fan palm"; spectacular, large fan palm with thick trunk that may grow to 10 m high, but extending in spread sideways to a diameter of 8 m, with gigantic, glossy, fresh green, plaited leaves more broad than long, to 2 m wide, cut halfway into many narrow, one-ribbed segments which are split again, the tips will hang like a fringe; petioles armed with small spines when the palms are young, usually disappear later; fruit metallic blue. Long popular in Europe and America in parlors, hotels and winter gardens before the advent of Howeias. Satisfied with medium-warm conditions but requiring lots of water and big space. *Subtropic.* p. 777, 780, 786

decipiens (Australia); handsome fan palm of medium size, with slender trunk, partially persistent leaf bases; loose crown of palmate leaves to more than 1 m across, glaucous beneath; center rib curves downward, the segments deeply cut to the base with tips freely pendulous; round 2 cm fruit. *Subtropic.* p. 784

mariae (Australia); a bronze fan palm of the semi-arid interior of Australia; trunks to 12 m high, with leaf bases persistent for several years; the crown of palmate fronds is of an open habit; leaf blades 1-2 m long and deeply cut; when young leaves and spiny petioles are bronzy red, later glaucous. *Tropical.* p. 786

saribus (cochinchinensis) (Cambodia, Vietnam, Malaya, Java, Philippines), a "Sugar palm"; noble fan palm with straight, slender trunk to 25 m tall, raising the dense globular rosette of palmate fronds well above canopy of the tropical rain forest; the ribbed, deeply segmented leaves to 1½ m long, on brown-red petioles with vicious thorns; small 1 cm blue fruit, in large clusters. *Tropical.* p. 786

LOBELIA *Lobeliaceae*
  cardinalis (E. and So. U.S.), "Cardinal flower"; perennial to 1 m, with purplish stems, oblong leaves 10 cm long, scarlet flowers in bracted racemes, 4 cm long. *Warm temperate.* p. 616
  erinus (So. Africa), "Bedding lobelia"; a popular bedding plant of dwarf habit; annual less than 20 cm high, with elliptic 2 cm leaves, and cheerful little blue or violet flowers with white throat. *Subtropic.* p. 616
  gibberoa (Central Africa, Uganda, Kenya), the "Giant lobelia"; interesting mainly because this species reaches almost tree-like dimensions, to 10 m high, in a genus that consists mostly of species of small habit; the Giant lobelia, really an out-grown herb, grows at higher altitudes, like tall candles, with erect solitary, hollow stalks, dense with oblanceolate, soft leaves to 60 cm long at base, becoming shorter toward top; long spike-like inflorescence of tiny greenish-white flowers. *Tropical.* p. 617

LOBIVIA *Cactaceae*
  aurea (Argentina: Cordoba), "Golden lily cactus"; lovely globular to cylindrical small plant similar to Echinopsis, dark green, to 10 cm high, with 14-15 acute ribs, spines yellowish brown; short funnel-form flowers freely borne, glossy lemon-yellow. *Arid-subtropic.* p. 261
  binghamiana (S.E. Perú), "Cob-cactus"; small depressed globe, to 10 cm dia., pale green with white dots, about 22 wavy ribs with orange radial spines; flowers purplish-red. *Arid-subtropic.* p. 260
  bruchii (Argentina: Tucuman); simple plant at first cylindric, becoming fairly large and globular with age, to 30 cm thick; up to 50 notched ribs or more and apex depressed, dark green, densely set with straw-colored spines and radials; small flowers deep red. *Arid-subtropic.* p. 260
  cylindrica (Argentina: Cordoba); small, deep green cylindrical plant 12 cm high with 11 ribs, bristling with whitish radials and brown needle spines; the large 5 cm flowers canary outside, yellow inside petals. *Arid-subtropic.* p. 261
  famatimensis (No. Argentina); small oval or elongated body to 15 cm high, with some 20 notched ribs somewhat spiral, and many yellowish short spines; flowering freely varying from yellowish to deep red. *Arid-subtropic.* p. 263
  huascha (Trichocereus) (W. Argentina); cylindrical erect stem 30-90 cm high and 5-8 cm thick, branching from the base, light green with 12-18 ribs, whitish areoles, and numerous needle spines yellow to brown; large yellow flowers with brownish hairs. *Arid-subtropic.* p. 261
  huascha rubriflora (W. Argentina); variety with tall clustering columns more slender, yellowish green, with 12 ribs and shorter spines; prolific bloomer with red flowers. *Arid-subtropic.* p. 261
  larabei (Perú: Urubamba, Cuzco); clustering small species similar to L. minuta; globes light green tinted with bronze, 4 cm dia., 9-15 knobbed ribs; the areoles with straw-colored spines; relatively large, beautiful crimson-red flowers. *Arid-tropical.* p. 260
  marsoneri (Northern Argentina); small clustering globes with short white radial spines; topped by large orange-yellow flowers. *Arid-subtropic.* p. 260
  pseudocachensis (No. Argentina); from Salta at 2500 m elev.; globular small green plant 3 cm high and 4 to 6 cm thick, with turnip-like root and appressed yellowish spines, forming clusters; funnel flowers 6 cm long, vivid dark red. *Arid-subtropic.* p. 262
  steinmannii (Rebutia) (Bolivia); clustering small globes 2 cm high, covered with tubercles; orange-red flowers. *Arid-subtropic.* p. 261

x LOBIVIOPSIS *Cactaceae*
  'Aurora' (Lobivia x Echinopsis); small globe with numerous ribs tipped by white glochids and sand-colored spiny hairs; very floriferous with long trumpet flowers, beautiful salmon-rose. *Arid-Subtropic.* p. 262

LOBULARIA *Cruciferae*
  maritima (Mediterranean), "Sweet alyssum"; a much branched little plant grown as an annual, with linear leaves, and many small pure white, very fragrant flowers blooming a long time. Very popular in summer garden beds. *Subtropic.* p. 379

LOCKHARTIA *Orchidaceae*
  acuta (Trinidad); small epiphyte with stem densely clothed with little shingled leaves folded flat, and axillary stalked, tiny flowers lemon yellow marked with red (summer). *Tropical.* p. 732
  lunifera (Brazil), "Braided orchid"; curious epiphyte having erect stems clothed with triangular leaves folded flat, out of which appear the tiny, golden yellow, long-lasting flowers with red-spotted lip. (Jan.-Aug.). *Tropical.* p. 732

LODOICEA *Palmae*
  maldivica (seychellarum) (Seychelles Isl. in the Indian Ocean), the "Double coconut"; slow-growing solitary fan palm to 30 m high, and trunk 30 cm dia., the large palmate fronds in dense crown on 4 m petioles, the deep glossy green blade 6 m long, cut 1/3 into segments whose ends are drooping; the female trees bear the famous "double coconuts", immense dark brown, woody fruits weighing to 20 kg., 45 cm long, taking 6 years to mature, the nut inside two-lobed or "double". *Tropical.* p. 476, 784, 789

LOLIUM *Gramineae*
  perenne (Europe, No. America), "Perennial ryegrass"; perennial grass of medium-coarse appearance, with reddish base, flat, glossy green leaf blades mostly 15-25 cm long if not mowed; 3 mm wide; usually grown as sod from seed, in improved varieties such as the pictured "Penn Blue" sold in California nurseries, and easy to grow, and inexpensive. *Temperate.* p. 510

LOMARIA: see Blechnum

LOMATOPHYLLUM *Liliaceae*
  purpureum (Mauritius); aloe-like succulent forming trunks to 2 m tall, topped by rosette of fleshy, channeled leaves to 80 cm long, deep green with red dentate margins; flowers red. *Tropical.* p. 582

LONCHOCARPUS *Leguminosae*
  sericeus (domingensis) (Sénégal, Trop. America); small tree with leaves of 7-13 dark green leaflets conspicuously veined; pinkish or purplish flowers in slender racemes. *Tropical.* p. 558
  violaceus (Trinidad, Tobago), "Lance-pod"; small tree with pinnate, fresh green foliage; racemes of rose-purplish flowers. *Tropical.* p. 558

LONICERA *Caprifoliaceae*
  heckrottii (americana x sempervirens?), "Gold-flame honeysuckle"; vine or small shrub with opposite, blue-green leaves 5 cm long; two-lipped flowers coral-pink outside, yellow inside, to 2 cm long. *Temperate.* p. 290
  hildebrandtiana (Burma, Thailand, So. China), "Giant honeysuckle"; evergreen climber, the woody stem reaching 18-24 m; ovate 15 cm leaves, and axillary, fragrant, creamy-white flowers deepening to orange-red, the corolla 10-15 cm long with 2-lipped limb. *Subtropic.* p. 290
  sempervirens (Connecticut to Florida and Texas), "Trumpet honeysuckle"; twining woody vine, with ovate leaves to 8 cm long, glaucous beneath; pretty, tubular flowers orange-scarlet, yellow inside, 4-6 cm long, in whorls at ends of branches in summer; scarlet fruit; evergreen in warmer climate. *Warm temperate.* p. 290

LOPEZIA *Onagraceae*
  lineata (coccinea) (Mexico), "Mosquito-flower"; shrubby plant 30-90 cm, with hairy stems and serrate ovate leaves, the small winged red flowers with 5 petals, the 2 upper ones bent away from center disclosing an apparent drop of honey, actually a glossy piece of hard tissue which deceives flies. *Tropical.* p. 700

LOPHOCEREUS *Cactaceae*
  schottii (Sonora, Arizona, Baja California), "Totem pole"; moss-green column 8 cm thick, with 5-9 acute ribs, the areoles set with small clusters of white wool, and short black spines, night-flowering red. The monstrose form L. mieckleyanus grows more slender, with irregular smooth ribs. *Arid-subtropic.* p. 251, 253
  schottii 'Monstrosus,' "Monstrose totem"; very peculiar yet attractive form consisting of a short column entirely composed of large, smooth knobs or remnants of ribs with spineless areoles, waxy moss-green. *Arid-subtropic.* p. 251

LOPHOPHORA *Cactaceae*
  williamsii (So. Texas, Mexico), the famous "Mescal" or "Peyote" of the ancient Mexicans, because of its powerful exhilarating and narcotic properties; small depressed globe to 8 cm dia., freely sprouting laterally, bluish-green, 5-13 low and wide ribs, tubercles white tufted; flowers pink to white. Venerated as a god by certain Indians in Mexico, at sacrifices or religious rites. *Arid-subtropic.* p. 267, 269

LORANTHUS *Loranthaceae*
  sansibarensis (Kenya, Tanzania, Zanzibar), the "Matchstick vine"; a parasitic straggler, which I have often encountered throwing a fiery splash over the colorless bush of the East African savannah; clambering shrub with oval 10 cm leaves, and axillary flowers 4 cm long with tubular corolla greenish at base, changing to orange-red and tipped black, and deep yellow stamen, the long slender tubes resembling a matchstick. The berry germinates on the host tree, and the roots appropriate its sap. *Tropical.* p. 618

LOTUS   Leguminosae
   berthelotii (peliorhynchus) (Cape Verde, Canary Isl.), "Winged pea"; silvery-haired shrub spreading along the ground with straggling branches, having pinnate, grayish foliage of thread-like leaflets, and butterfly-type scarlet flowers. *Subtropic.*   p. 557, 562

LOUREA: see Christia

LOURYA   Liliaceae
   campanulata (Vietnam); tropical evergreen herb with the habit of aspidistra, forming clusters; long petioled leathery leaves; bell-shaped flowers in tight racemes near the base, yellowish white with black disk. *Tropical.*   p. 595

LUCULIA   Rubiaceae
   grandifolia (tsetensis) (India); dense ornamental shrub to 6 m, with deep green ovate, corrugated leaves with reddish midrib 30-40 cm long, and showy clusters of tubular waxy flowers with spreading limb pure white. *Tropical.*   p. 863, 864

LUFFA   Cucurbitaceae
   cylindrica (Eastern hemisphere Tropics), the odd "Vegetable sponge", "Dishcloth gourd" or "Sauna sponge", is produced by a fast growing tropical climber of the cucumber family; 4-angled stems pull themselves up by coiling tendrils, the rough-hairy herbaceous angled or lobed green leaves cucumber-like; 10-15 cm wide; separate male and female 8 cm flowers golden yellow; when pollinized the pistillate flowers produce the swollen cylindric fruit to 60 cm long, which may be eaten when young, but the ripe yellow fruit contains the sponge-like fiber that is washed and dried, and marketed for scrubbing, stimulating massage and skin care. *Tropical.*   p. 379

LUISIA   Orchidaceae
   teretifolia (India to Monsoon Asia and New Caledonia), "Bee orchid"; remarkable epiphyte with slender, thin stems to 1 m long, set with pencil-like cylindric leaves; small heavy-textured 2 cm flowers nesting against the stem, yellowish green overlaid with bronze, the lip velvety brown-red (autumn). *Tropical.*   p. 741

LUMA   Myrtaceae
   apiculata (Eugenia) (Chile), "Temu"; densely branched evergreen shrub or tree 2 to 5 m high, with golden brown bark; small sharp-tipped ovate leaves 2-3 cm long; 2 cm flowers cream-white suffused with pink; small black berries. *Subtropic.*   p. 677

LUNARIA   Cruciferae
   annua (biennis) (Europe: Sweden, etc.), a biennial variously known as "Silver dollar", "Honesty", "Moonwort" or "Satin flower"; to 1 m high, with toothed ovate leaves, erect stems with violet to white flowers followed by moon-shaped seed pods opening to papery satiny, translucent disks which are cut and used for lasting winter bouquets. *Temperate.*   p. 379, 887

LUPINUS   Leguminosae
   polyphyllus 'Carmineus' (Pacific Northwest), "Garden lupine"; robust perennial to 1 m or more high; leaves palmately compound and deeply cut; inflorescence on stout stem, a column of pea-shaped flowers; carmine red in this cultivar, but other colors have been developed also, blue, rose, white. *Warm temperate.*   p. 556
   polyphyllus 'Roseus'; selected cultivar with erect spikes densely covered by rose-pink flowers. *Warm temperate.*   p. 556
   'Russell's hybrids'; strikingly handsome herbaceous perennials, the result of hybridizing L. polyphyllus (Wash. to Calif.) with the tree-lupine arboreus; this strain has massive spikes of pea-like flowers in many colors: pink, salmon, red, white, yellow, violet; leaves digitately compound. *Warm temperate.*   p. 556

LYCASTE   Orchidaceae
   aromatica (Mexico); dwarf epiphyte with oval pseudobulbs having 1-2 dark green, elliptic leaves; flower stalks clustered, each with a solitary, fragrant, waxy blossom, to 15 cm high, sepals greenish-orange, petals golden-yellow, the concave lip golden-yellow dotted red, (April-Oct.). *Tropical.*   p. 737
   candida (Costa Rica, Panama); floriferous epiphyte with 2-leaved pseudobulbs, the plicate foliage eventually deciduous; wiry stalks bearing fragrant 5 cm flowers, sepals pale green, petals white, lip white, tinted with yellow and dotted red-brown (spring). *Subtropic.*   p. 737
   cruenta (Guatemala); epiphyte with 2-3-leaved pseudobulbs and small 5 cm, long-lasting, individually stalked flowers, sepals greenish-yellow, petals deep golden-yellow, lip orange-yellow with blood-red in throat, (March-April, and longer). *Tropical.*   p. 737
   virginalis (skinneri) (Mexico, Guatemala, Honduras); handsome profusely flowering epiphyte with large pseudobulbs, each with 2-3 broad, plaited leaves to 60 cm long; large, waxy, solitary flowers to 15 cm across, rose-pink, shaded carmine-rose in the center, lip whitish, thickly spotted with rose and crimson, (Nov.-May). *Subtropic.*   p. 737

   virginalis alba (skinneri alba) (Guatemala), "White Nun orchid"; beautiful form with large waxy flowers pure white, crest of lip light yellow; distinguished as the national flower of Guatemala, (winter). *Subtropic.*   p. 737

LYCOPERSICON   Solanaceae
   lycopersicum cerasiforme, the "Cherry-tomato"; in cultivation since the 16th century, and before by the Incas of western South America, their native home; fleshy herb with strong-smelling, pinnately compound, hairy foliage, small yellow flowers, and clusters of globular red or yellow fruit 2-3 cm in dia., during the hot season, and used for ornament or in the kitchen; closely related to the potato, Solanum tuberosum, and I have proved for myself this relationship by successfully grafting tomatoes on potatoes, a combination known as 'Potomato'. *Tropical.*   p. 472
   lycopersicum 'Tiny Tim'; an edible miniature tomato, the little round scarlet fruit 1½-2 cm dia., in axillary clusters; the cute little decorative plant is of dwarf habit, about 30 cm high, and suitable for pot culture. *Tropical.*   p. 894

LYCOPODIUM   Lycopodiaceae
   phlegmaria (Queensland, and Tropics of Eastern hemisphere), "Queensland tassel-fern"; handsome tropical epiphyte, slender, wiry stems first erect then pendant, the pale rope-like, forking branches set closely with stiff-leathery, needle-like, glossy green ovate leaves 1 cm long in spirals or whorls, light green beneath, the fertile tasseled fronds thin, like catkins. *Humid-tropical.*   p. 618

LYCORIS   Amaryllidaceae
   radiata (China, Japan), "Red spider lily"; formerly sold as Nerine sarniensis or N. japonica; bulbous plant 45 cm high, good grower with linear basal leaves with silver band along middle, and disappearing before the flowers which are borne in an umbel, the recurved and crisped petals bright scarlet-red and edged with gold, the long stamens curving upward; late fall-blooming and hardy. *Warm temperate.*   p. 61
   squamigera (Japan), "Magic lily" or "Resurrection lily"; bulbous herb with leaves 2½ cm wide, the fragrant flowers in umbels, petals separated and a pretty lilac-pink, yellow in base, the long stamens turned up; hardy. *Warm temperate.*   p. 62

LYGODIUM   Schizaeaceae (Filices)
   circinatum (Malay Archipelago), "Malay climbing fern"; climbing fern with wiry stems, long and twining; leaves scattered and deeply 3 to 6 forked into linear segments, glossy green. *Humid-tropical.*   p. 448
   japonicum (Japan to Himalayas to No. Australia), "Climbing fern"; with twining, thread-like stems bearing pretty, pleasing green, pinnate, papery leaflets, the sterile pinnae with lobed segments (pinnules); the fertile pinnae narrow, 3-times divided. *Humid-subtropic.*   p. 448

LYSIMACHIA   Primulaceae
   clethroides (China, Japan), "Loose-strife"; erect, little branched herbaceous perennial to 1 m, with ovate, hairy leaves 8-15 cm; small 1 cm bell-shaped white flowers with spreading petals, dense in slender pyramidal spikes, in late summer. *Warm temperate.*   p. 830
   nummularia (C. Europe and Britain, nat. in E. No. America), "Creeping Jennie" or "Creeping Charlie", "Moneywort"; prostrate perennial creeper, with herbaceous dull light green, rounded leaves to 2½ cm long, in pairs along a thread-like, squarish, pink stem, forming rootlets opposite the axils; flowers bright yellow 2 cm across, winter-hardy. *Warm temperate.*   p. 830
   punctata (S.E. Europe, Asia Minor), "Garden loose-strife"; erect perennial hairy herb, branches in three's, 30-60 cm high, with ovate leaves, and starry yellow flowers in axillary whorls. *Subtropic.*   p. 831

LYTHRUM   Lythraceae
   salicaria (Old World), "Purple loosestrife"; downy perennial of wet places, to 2 m high; 10 cm lanceolate leaves; flowers purple in long leafy spikes, blooming all summer. *Warm temperate.*   p. 618

MACADAMIA   Proteaceae
   tetraphylla (Queensland, New South Wales), "Queensland nut" or "Macadamia nut"; evergreen tree to 15 m, with stiff, dark branches, sessile leaves in whorls mostly of 4, finely serrate in the juvenile stage; older leaves narrow, oblanceolate, stiff-leathery, 10-30 cm long; flowers pinkish; the hard-shelled fruit in pendant clusters, each 2-3 cm dia., and holding edible sweet nuts. *Subtropic.*   p. 467

MACARANGA   Euphorbiaceae
   grandifolia (Guatemala), "Coral tree"; growing into large tree, with rosettes of large peltate pointed, leathery leaves prominently veined, to 60 cm across; small petal-less flowers. *Tropical.*   p. 429

MACHAEROCEREUS *Cactaceae*
  eruca (Baja California), the "Creeping devil"; a ferocious creeper; prostrate stems to 10 cm thick and 3 m long, dark green with coppery tint, many-ribbed, ribs low and very spiny, with white central spine flattened, dagger-like and directed downward; dayblooming, creamy flowers. *Arid-tropical.* p. 251, 252, 254

MACKAYA *Acanthaceae*
  bella (Asystasia) (So. Africa); attractive erect shrub to 2 m, with ovate, slender-pointed, toothed leaves 12 cm long; 5 cm flowers bell-shaped with spreading segments, lilac, pencilled in throat. *Subtropic.* p. 38

MACROPIPER *Piperaceae*
  excelsum psittacorum (New Zealand: Poor Knight Isl.); "Lofty pepper"; small aromatic tree to 6 m high; cordate leathery, glossy green 12-15 cm leaves with pale veining, larger than the species; small flowers in spikes, and with edible yellow fruit. *Tropical.* p. 822

MACROZAMIA *Cycadaceae*
  denisonii (Lepidozamia peroffskyana) (New South Wales, Australia); trunk to 6 m or more, clothed with persistent leaf bases; topped by crown of leaves to 3 m long, pinnae 10-30 cm long, glossy green. *Subtropic.* p. 388
  moorei (Queensland, New South Wales); large plant with stem underground or to 1 m high; fronds 2 m long, pinnae spreading and directed forward, the lower ones progressively spine-like; cones cylindrical, the female salmon-pink inside. *Subtropic.* p. 386
  riedlei (Western Australia), "Zamia palm"; the only cycad in S.W. Australia; palm-like crown of fronds from thick, partially underground trunk 1 to 4 m high; the bluish green pinnate leaves 1 m or more long; rigid, well-spaced leaflets without distinct midvein and with needle-tips; nestling at base the large cone-like flowers, the male cones 25 cm wide. *Arid-subtropic.* p. 387, 388

MAGNOLIA *Magnoliaceae*
  grandiflora (Carolina to Florida and Texas), the "Southern magnolia"; noble evergreen pyramidal tree to 30 m high, with decorative thick-leathery, rich-green ovate oblong leaves to 20 cm long with shining surface and rusty tomentose beneath; large and beautiful 20 cm cup-shaped flowers creamy white and fragrant; rust-brown cone-like fruit. *Warm temperate.* p. 619, 621, 623
  liliflora (quinquepeta) (China); deciduous shrub to 4 m high, with obovate leaves to 20 cm long, pale beneath; flowers appearing before foliage, bell-shaped, purple outside, white inside to 10 cm long. *Warm temperate.* p. 619
  x soulangeana, (hybrid of M. denudata x liliflora) both from China, the "Saucer magnolia" is common in cultivation and one of the showiest spring-flowering small trees; obovate deciduous leaves and solitary 15 cm flowers white, tinged purplish rose outside, blooming ahead of foliage; (hardy). *Temperate.* p. 619, 621
  x soulangeana 'Verbanica'; cultivar with large cup-shaped flowers 15 cm across, purplish and pink, from long slender buds; blooming before leaves, later than M. soulangeana. *Warm temperate.* p. 617, 619
  stellata (halliana) (Japan), "Star magnolia"; deciduous, much branched shrub to 8 m high; dull green obovate 12 cm leaves; fragrant white flowers with narrow petals, spreading 8 cm across, later reflexed, and appearing before the foliage in early spring; red fruit; fairly hardy. *Temperate.* p. 619
  tripetala (tetrapetala) (Pennsylvania to Mississippi); small openheaded deciduous tree with large lanceolate leaves to 50 cm long, crowded at end of branches, giving an umbrella-like effect; large creamy-white flowers with obovate petals, to 25 cm across, followed by rosy fruit. *Warm temperate.* p. 619

MAHONIA *Berberidaceae*
  aquifolium (Brit. Columbia, Oregon to No. California), "Barberry", "Oregon grape" or "Holly mahonia"; handsome hardy evergreen thornless shrub to 1 m or more, with flexible stems and pinnate leaves with spiny-toothed, leathery leaflets 8 cm or more long, glossy dark green, the young foliage a pretty bronzyred; lemon-yellow flowers in dense clusters, followed by blue-black, edible berries with glaucous bloom. Very durable, hardy in So. New England. *Warm temperate.* p. 181
  repens (Berberis) (Brit. Columbia, No. California, East to Rocky Mountains), "Creeping mahonia"; small spreading evergreen creeping by underground stolons or stems, 30 cm or more high, dull bluish-green leaves have 3-7 spine-toothed leaflets, turning bronze in winter; yellow flowers in 8 cm clusters followed by dark blue, powdery berries. *Warm temperate.* p. 181

MALACOCARPUS: see Wigginsia

MALEPHORA: see Hymenocyclus

MALOPE *Malvaceae*
  trifida (Spain, No. Africa); annual to 1 m with broad ovate leaves, dentate at margins; flowers salver-shaped, 4 cm dia., rosepink with purple lines spreading out from center. *Subtropic.* p. 620

MALPIGHIA *Malpighiaceae*
  glabra (So. Texas to So. America), "Barbados cherry"; shrub or small tree to 4 m, with shining green ovate, leathery 8 cm leaves; pretty flowers carmine-rose with fimbriate petals, 1 cm across, followed by small 1 cm cherry-like edible fruit, of acid flavor and high in Vitamin C. *Subtropic.* p. 458, 620

MALUS *Rosaceae*
  baccata (Siberia, China), "Siberian crab apple"; tree with hard branchlets, and small ovate leaves, 2 cm white flowers; miniature 2 cm wax-like fruit yellow or red; trained into dwarfed bonsai trees in Japan. *Temperate.* p. 850
  floribunda 'Scheideckeri', beautiful "Flowering crab-apple"; of pyramidal habit, with ovate leaves and large, semi-double, 3 cm flowers of delicate rose-pink to pale tinged pink, borne with great profusion in large clusters during May, followed by small 2 cm yellow to reddish tart fruit, used for jellies. *Temperate.* p. 850
  pumila, (Europe and W. Asia), "Common apple"; roundheaded deciduous tree to 12 m high, with oval, leathery leaves 5-10 cm long, and flowers white or light pink, 3-5 cm across, appearing with first foliage, and followed by its large depressed globular edible fruit, with firm, tart-sweet flesh. *Temperate.* p. 851
  pumila 'Anna' (from Israel); an excellent apple photographed at Paul Thomson's, Bonsall, California; beautiful yellow fruit overlaid with red; of firm crisp textured flesh; producing well in subtropic climate. *Warm-temperate.* p. 468
  pumila 'Golden Delicious'; excellent multi-purpose apple with golden yellow skin and crisp, firm, aromatic flesh, ripening in late summer. Most important cultivar in U.S. *Temperate.* p. 465
  pumila paradisiaca (Asia to So. Russia), "Paradise apple"; charming apple tree of small stature, with bright pink flowers, and small fruit 2 cm in dia. *Temperate.* p. 846
  pumila 'Winter Banana'; a semi-tropic large, beautiful apple of pale color with red blush and waxy finish; distinctive, tangy aroma. Accepts mild winters; needs pollenizer. *Subtropic.* p. 468
  'Van Eseltine' (arnoldiana x spectabilis plena), "Doubleflowering apple"; deciduous tree with branches of very erect habit; large double flowers rich rose-pink; small 2 cm yellow fruit with red cheeks. Very striking when in bloom. *Temperate.* p. 847

MALVA *Malvaceae*
  sylvestris (So. Europe), "High mallow"; robust biennial to 1 m, with rounded lobed leaves; rosy-purple flowers with darker stripes, to 4 cm across. *Warm temperate.* p. 624

MALVAVISCUS *Malvaceae*
  arboreus (Mexico), "Wax mallow"; evergreen shrub 3 to 4½ m high, broadly ovate, crenate 3-lobed leaves to 12 cm long; rough above and soft-downy beneath; 3 cm flowers usually solitary in leaf axils, rich crimson red but petals do not open; long-protruding style. *Tropical.* p. 624
  arboreus mexicanus, "Turk's cap"; tall tropical shrub more or less hairy, with narrow ovate, toothed leaves, narrower than the species; flowers hibiscus-like 3 cm long, with scarlet corolla, but which do not open, and with protruding staminal column. *Tropical.* p. 623, 624, 625

MAMMEA *Guttiferae*
  americana (West Indies, So. America), the "Mammee apple"; handsome tropical tree 10 to 20 m high, with a broad crown of shiny leathery, oval leaves 10-20 cm long; fragrant white 2 cm male and female flowers; in spring the large globose, russet-brown fruit 8 to 20 cm dia., with rough, bitter skin and orange, apricot-flavored pulp surrounding 1 to 4 round seeds; the pleasantly sweet flesh is eaten raw or cooked. *Tropical.* p. 457, 475, 520, 522

MAMMILLARIA *Cactaceae*
  albilanata (Mexico: Guerrero); elongate plant 15 cm high dense with little grayish tubercles with tiny light brown tufts and grayishwhite radials, and light brown, short spines; flowers deep carmine. *Arid-tropical.* p. 276
  blossfeldiana (Mexico: Baja California); small globular plant somewhat elongate 4 to 5 cm thick, medium green; the short knobs with whitish radials and a central black hook-spine; flowers carmine-red. *Arid-subtropic.* p. 276
  bocasana (Mexico), "Powder puff"; like a bursting cotton boll, the little, mound-forming globes covered with snow-white silky hair as well as brown fish-hook spines, to 5 cm thick; 2 cm flowers cream-white. *Arid-tropical.* p. 276
  bocasana 'Ed Hummel', known as "Hummel's powder puff"; a California seedling, makes many offsets quickly forming a tall

pyramidal cluster; the individual little globes, dense with green tubercles, are clothed all over with silky white hair and half-hidden wiry brown fishhook spines. *Arid-tropical.* p. 277
camptotricha (Mexico), "Birdsnest"; small clustering globes to 5 cm, fresh green, with extended nipples and long yellow bristle-like spines often twisted; flowers white, greenish outside, hidden in axils. *Arid-tropical.* p. 276
candida (C. Mexico), "Snowball cactus"; an exquisite little globe closely tubercled and covered with a multitude of pure white radial spines; clustering flowers rose-colored. *Arid-tropical.* p. 276
celsiana (So. Mexico), "Showy pincushion"; attractive small deep green globe becoming cylindric and forming clusters with age, tubercles neatly arranged and white-woolly in axils, radials white, central spines pale yellow; flowers red. *Arid-tropical.* p. 276
chionocephala (Mexico: Coahuila); symmetrical, globular plant slowly becoming branched, 8 cm dia., dark green but the milky tubercles nearly hidden by bristly straw-white radials and short central spines tipped black; flowers rose. *Arid-subtropic.* p. 277
compressa (C. Mexico), "Mother of hundreds"; small globes, pale bluish green, in clumps; knobs short, axils woolly; white spines; flowers pinkish, 2 cm long. *Arid-tropical.* p. 279
"dolichocentra" in hort. "Ruby dumpling"; similar to tetracantha; bluish green; spines white; small 8 mm fl. rosy-red. May be var. of kewensis. *Arid-tropical.* p. 279
elongata (Mexico), "Golden stars"; small, clustering cylinders to 3 cm thick, light green, tubercles in spirals and covered with yellow, interlacing radial spines; flowers white. *Arid-tropical.* p. 279
geminispina (bicolor) (C. Mexico), "Whitey"; small clubshaped plant becoming cylindric and clustering, glaucous, with prominent knobs topped by white radials and needle-like white, black-tipped central spines; flowers red. *Arid-tropical.* p. 275, 279
geminispina nivea (Mexico); a beautiful form densely furnished with very long 5 cm needle spines glistening pure white tinted pink and tipped red, curving out and downward. *Arid-tropical.* p. 279
hahniana (Mexico), "Old lady cactus"; attractive little globe, rich green with long and curly snowy-white hair-like bristles and red-tipped spines; flowers violet-red. *Arid-tropical.* p. 279, 280
hemisphaerica (No. Mexico, Texas); flattened globe to 12 cm dia., tubercles elongated, dark green, woolly at top; yellowish radials needle-like, one straight brown central spine; flowers cream-white. *Arid-subtropic.* p. 279
hidalgensis (Mexico: Hidalgo); pretty, cylindrical plant to 30 cm high, dark green and rounded at top, woolly and with reddish spines; large tubercles conical, with floccose white wool in axils, few grayish spines; small carmine flowers. *Arid-tropical.* p. 279
lasiacantha (Chihuahua, Mexico); small globe to 4 cm high, gray-green, but covered with pale bristles; flowers white with rose center stripe. *Arid-subtropic.* p. 278
longicoma (Mexico: San Luis Potosi); small globe 5 cm wide, tubercles dense with hair-like and silky whitish radials, the central spines brown, some with hooks; flowers white suffused with rose. *Arid-tropical.* p. 278
macracantha (Mexico: San Luis Potosi); dark green globe with large oval, 4-angled knobs with few white radials but stiff yellowish brown central spines like needles, white-woolly in axils; flowers carmine red. *Arid-tropical.* p. 277
magnimamma (centricirrha) (C. Mexico), "Mexican pincushion"; clustering globe, to 10 cm, dark green and milky, with large conic tubercles topped with 3-5 recurved horn-colored spines; cream flowers. *Arid-tropical.* p. 278
nejapensis (Mexico: Oaxaca); beautiful elongate clustering globe to 15 cm high, with fresh green, prominent nipples tipped with white-woolly tufts, and sets of stout glistening white spines, red-brown at the extreme tip, the axils also filled with white wool especially near apex; yellow flowers shaded with red-purple. The curious crested form shown on pg. 277 is an old plant in California, with contorted ribs in shape of a smiling face and which the grower called "Smiling Jack". *Arid-tropical.* p. 277
parkinsonii (C. Mexico), "Owl's eyes"; branching and clustering little cylinder to 8 cm thick, glaucous green, tubercles neatly arranged and topped by white radials and prominent central spine; flowers white. *Arid-tropical.* p. 275, 279
parkinsonii cristata; a clustered old plant partially becoming deformed into fasciations and developing crests shaped like a down-turned mouth, accented by the straw-colored spines, giving rise to the name "Sadsack" in this particular plant 25 cm across; photographed at Crestview Nursery, Carlsbad, California. *Arid-tropical.* p. 277
pennispinosa (Coahuila, Mexico); small globe 3 cm high, white spines with one long hook spine; flowers white with rosy middle band. *Arid-subtropic.* p. 278
polythele (affinis) (Mexico: Hidalgo); globular to cylindric, to 10 cm thick and 50 cm high, dark bluish-green; prominent tubercles in symmetrical spirals, and with showy amethyst-brown spines; near the apex the reddish flowers 2 cm long. *Arid-tropical.* p. 276
pottsii (Coryphantha) (W. Texas, Chihuahua); densely spiny, cylindric plant to 12 cm high; numerous interlaced, slender white radials, and stouter central spines curving up, and tipped red-brown; small light purple flowers. *Arid-subtropic.* p. 279
seitziana (Mexico: Ixmiquilpan); attractive globe becoming cylindric to 25 cm high, prominent large pointed, conical knobs glaucous green, tipped by a few whitish spines, the centrals tipped black; flowers rose. *Arid-tropical.* p. 279
sempervivi (Cent. Mexico), a "Strawberry cactus"; the true species is an attractive globular plant dense with long angular nipples tipped by very short, hard spines, the axils toward apex filled with wool; flowers carmine rose. *Arid-tropical.* p. 278
tegelbergiana (Mexico: Chiapas); beautiful, perfect little globe dense with small dark green tubercles, completely clothed with interlacing, thin white radials and stouter central spines tipped black-brown; the axils toward apex filled with white wool; flowers rosy-pink. *Arid-tropical.* p. 279
theresae; small globe with high tubercles, tipped by pale radial spines; large flowers soft rose-pink. *Arid-subtropic.* p. 274
vaupelii cristata, "Silver brain"; fantastic fasciations of coiled and wavy crests growing into a large and tight symmetrical globe; fine brown spines silvery at crest over the fresh green body; small 2 cm pink flowers. *Arid-tropical.* p. 279
wildii (Mexico: Hidalgo), "Fish-hook pincushion"; cylindric plant to 8 cm thick; dark green tubercles thickly set with bristle-like white radials and honey-colored central spines, rosy hairs in axils; numerous white flowers. *Arid-tropical.* p. 278
zeilmanniana (Mexico: Guanajuato), "Strawberry cactus"; choice small globe 7 cm high, glossy green, with high cylindrical knobs covered by long white interlaced, soft radials, and reddish central spines, one hooked; flowers bright violet-purple. *Arid-tropical.* p. 274, 280

MANDEVILLA   Apocynaceae
x amabilis 'Alice du Pont' (Dipladenia), (formerly 'Splendens hybrid'); grown by Longwood Gardens (Baileya March 1962), a hybrid apparently derived from M. splendens, glabra, and superba. Woody climber growing to perhaps 10 m long; leaves opposite, dark green, rugose, lustrous, oblong-elliptic, 8-20 cm long; inflorescence a raceme from alternate leaf axils, each with to 20 blossoms; flowers funnel-shaped 6-10 cm dia., dawn pink with darker throat, turning dark rose. *Tropical.* p. 77
boliviensis (Bolivia), "White dipladenia"; free-blooming shrubby climber with slender branches; shining green oblong, slender pointed 5-8 cm leaves; and 5 cm funnel-form flowers white, with orange-yellow throat, in axillary racemes. *Subtropic.* p. 77
sanderi (Brazil), "Rose dipladenia"; woody vine with wiry stems having milky sap; small leathery 4-5 cm ovate leaves glossy green, bronzy beneath; the 6-8 cm flowers rose-pink with pure yellow throat. A beautiful climber, blooming throughout the year in good light, even as a smaller, shrublike plant; requires copious watering when growing, with good drainage. *Tropical.* p. 73, 77
sanderi 'Rosea' (Dipladenia), "Brazilian jasmine"; a cultivar of the Brazilian rose-flowered species; woody vine, wiry stems with milky sap; small leathery 3-5 cm ovate leaves glossy green, bronzy beneath; the 6-8 cm flowers salmon pink with pure yellow throat, blooming all year even as smaller plant. *Tropical.* p. 78
splendens (Dipladenia) (S.E. Brazil); woody twiner with stems finely hairy, and with milky sap; opposite, thin-textured leaves broadly elliptic, to 20 cm long; clusters of showy flowers 8-10 cm across, in a lovely rose-pink. *Tropical.* p. 78
suaveolens (Iaxa) (Argentina, Bolivia), the "Chilean jasmine"; woody vine with opposite, thin-leathery, ovate-cordate leaves 8-15 cm long, bright green and smooth, purplish to grayish beneath, on brownish wiry twining stems covered by rough warts; pure white funnel-shaped flowers 5 cm across and deliciously fragrant, in racemes of 6-8 or more, in summer. *Subtropic.* p. 77

MANDRAGORA   Solanaceae
officinarum (So. Europe), "Devil's apples", or "Mandrake"; herbaceous perennial steeped in superstition, with thick spindle-shaped tuberous roots often divided into two leg-like parts; the large wavy, ovate leaves to 30 cm long grow from the tips of the roots; bell-shaped, yellowish or purplish flowers cradled in the foliage, followed by the juicy berries. *Subtropic.* p. 898

MANETTIA   Rubiaceae
inflata (bicolor) (Paraguay, Uruguay), "Firecracker plant"; twining herb with threadlike stems, thin-fleshy, green, ovate leaves and attractive solitary, 2 cm, tubular, waxy flowers from the axils, flask-like and vivid yellow, the lower part of the tube densely covered with bright scarlet bristles, giving the appearance of a red corolla tipped yellow. *Subtropic.* p. 862, 868

MANFREDA *Amaryllidaceae*
   maculosa (Agave pubescens) (Mexico: Chiapas, Morelos); ornamental rosette of thin, soft-fleshy, sword-shaped concave leaves to 30 cm long, glaucous blue with large chocolate blotches; margins smooth; the inflorescence on long stalks to 1 m or more, with small pinkish-white flowers, opening at night. *Tropical.*   p. 56, 67

MANGIFERA *Anacardiaceae*
   indica (No. India, Burma, Malaya), the "Mango" tree; with large, spreading, evergreen crown, 18-30 m high, and grown for its delicious fruit all over the tropics; leathery, lanceolate leaves to 40 cm long; small pinkish flowers in terminal panicles, followed by large, variably yellow to reddish sweet-fleshy fruit averaging 12 cm long, containing the seed large adhering stone. *Tropical.*   p. 71, 454
   indica 'Haden', a "Mango"; good commercial variety with large 10 cm fragrant fruit, of good quality, greenish to red and covered with silvery glaucescence; low fiber content. *Tropical.* p. 455
   indica 'Joe Welch'; beautiful large mango fruit 10-15 cm long, yellow with red cheeks, and very sweet flesh. *Tropical.*   p. 455

MANIHOT *Euphorbiaceae*
   esculenta (Brazil), "Manioc"; evergreen bush with milky juice and long tuberous roots; leaves deeply parted into 3-7 lobes; flowers without petals. The sturdy roots yield tapioca, cassava and starch. The poison in the roots is destroyed by cooking. A most important root-crop in tropical cultivation. *Tropical.*   p. 428
   esculenta variegata (Brazil), "Cassava" or "Tapioca plant"; widely grown in tropical regions for the starchy tubers which yield tapioca; digitate, fresh green leaves in this form are beautifully variegated yellow along veins. *Tropical.*   p. 428
   peltata (Brazil); handsome foliage plant with milky juice, the peltate leaves glaucous bluish, deeply lobed and with pink veins. *Tropical.*   p. 419

MANILKARA *Sapotaceae*
   zapota (Achras) (C. America), "Sapodilla" or "Chicle tree"; evergreen tree to 20 m with milky sap; obovate leaves to 40 cm long; small white flowers; fruit ovoid to 15 cm long, russet-orange, with reddish, sweet, edible flesh. *Tropical.*   p. 457, 459, 880

MANILTOA *Leguminosae*
   gemmipara (New Guinea); handsome tree 5-20 m high, with short trunk, large pinnate leaves of leathery shining leaflets; when in new flush hanging limply on the young shoots, white or pink in color; white flowers in large clusters. *Tropical.*   p. 559

MANSOA *Bignoniaceae*
   difficilis (Brazil); woody vine, tendril-climbing, with leathery dark green, ovate leaves in pairs, 8-10 cm long; golden yellow crisped funnel flowers of crepy texture. *Tropical.*   p. 183

MARANTA *Marantaceae*
   arundinacea 'Variegata' ('Phrynium micholitzii'), (Mexico to S. America), "Arrow-root"; erect herb to 1 m high, with starchy roots and forking, zigzag branches; in this variegated form having the narrow lance-shaped, light green, papery leaves prettily variegated or margined with white; variegation passing through the leaf, showing underneath. *Tropical.*   p. 637, 638, 640
   bicolor (Brazil, Guayana); low plant with tuber-bearing roots; the oval leaves, to 15 cm long, usually horizontal, glossy, with broad, gray, feathered center, adjoining which is a zone of dark green, then grayish-green to margin, purple beneath; flowers white, lined violet. *Tropical.*   p. 637
   leuconeura (Brazil), "Rabbit's foot" or "Ten commandments"; ornamental small tropical plant with spreading branches; broad oval leaves to 12 cm long, the upper surface chartreuse with brownish patches turning moss-green with age; lower surface glaucous or purplish; the leafblades folding for the night; flowers white with purple spots. *Tropical.*   p. 637, 640
   leuconeura erythroneura (Brazil: Estado do Rio), the beautiful "Red-veined prayer plant" from the Organ Mts. near Petropolis; a low-growing herbaceous plant, the foliage more or less horizontal with the ground; 10 to 12 cm leaves obovate, on short winged petioles, patterned with a herringbone design of carmine-red veins over light yellow-green to dark velvety olive green, jagged silvery green along center; reddish beneath except green along center; flowers whitish with purple eye. *Tropical.*   p. 637, 638, 639, 640
   leuconeura kerchoveana (Brazil) "Prayer plant"; low-growing plant with 15 cm oval leaves mostly hugging the ground, and folding upward in the evening; the surface is vivid to pale-grayish green, more pronounced along the midrib and feathering along the veins, with a row of chocolate, later dark green, blotches on either side; blotched red beneath; small flowers white, striped purple, in a raceme. *Tropical.*   p. 637
   leuconeura 'Massangeana', (Brazil); low, strikingly beautiful plant with satiny bluish-green leaves, with feathery centerband of silver in thin stripes of pink, later silver, extending out toward the margin, after first passing a broad zone of reddish brown; red-purple beneath. *Tropical.*   p. 637

MARANTA: see also Calathea, Ctenanthe

MARATTIA   *(Filices) Marattiaceae*
   salicina (New Zealand, to Tahiti, Samoa, north to Hong Kong), the "Para" or "Horseshoe fern"; large tropical fern with starchy rootstock, and bipinnately divided dark green fronds, the leaflets leathery and smooth, on yellow axis and brown stalks; a good indoor plant in tubs in younger specimen. This species named according to Chas. Harrison of Palmerston North, and H.B. Dobbie, N.Z. Ferns; the plant known as M. fraxinea may be the same species. *Humid-subtropic.*   p. 437

MARCHANTIA   *Marchantiaceae*
   polymorpha (E. United States), a "Liverwort"; interesting plant looking like a prehistoric creation, botanically between the mosses and the algaes, useful for colonizing in rockgardens, also in conservatories and moist dishgardens; leaf-like flat scale-like plant body or thallus is 10-12 cm long and 2½ cm wide; the fruiting body on female plants looks like a tiny umbrella. *Warm temperate.*   p. 646

MARIANTHUS   *Pittosporaceae*
   pictus (Western Australia), "Bell climber"; lovely woody scrambler with narrow-linear leathery foliage; flowers with spreading petals lemon-yellow striped and spotted with purple. A genus dedicated to the Virgin Mary. *Subtropic.*   p. 823
   ringens (Western Australia), "Red bell climber"; twining subshrub with slender branches; lanceolate leaves scattered, 5-10 cm long; charming tubular flowers yellow at base, and spreading red limb 2 cm across. *Subtropic.*   p. 823

MARIPA   *Convolvulaceae*
   passifloroides (Trop. America); rambling clamberer with glossy ovate leaves; flowers open bell-shaped with purple flushed with pink, 5 cm across. *Tropical.*   p. 359

MARITIMOCEREUS   *Cactaceae*
   nanus (Loxanthocereus) (So. Peru); slender blue column to 10 cm long, with 12 ribs divided into nobs, these set with clusters of short stiff spines tipped red; flowers trumpet-like, orange scarlet. *Arid-subtropic.*   p. 254

MARLIEREA   *Myrtaceae*
   edulis (Brazil), "Cambuca"; evergreen fruiting tree with spreading branches; obovate corrugated, leathery glossy green leaves 15-20 cm long, with pale veins and lightly pubescent beneath; the globular, edible fruit orange-yellow 3-4 cm dia., of the shape of a pomegranate. *Tropical.*   p. 462, 686

MARNIERA   *Cactaceae*
   chrysocardium (Epiphyllum) (Costa Rica; So. Mexico: Chiapas); handsome epiphyte, with flattened branches deeply cut leaflike to strong midrib, to 30 cm wide; the large flowers 30 cm on long tube, outside pinkish, inner petals white, night blooming. *Tropical.*   p. 256

MARSILEA   *Marsileaceae (Filices)*
   drummondii (W. Australia), "Water-clover"; aquatic perennial tufted herb with creeping rhizome rooting at nodes, and floating 4-parted clover-like leaves 8 cm dia., the fan-like leaflets covered with whitish hairs, and with wavy margins, on long slender stalks; the bean-like spore cases or fruiting bodies at the base of leaf stalks. *Subtropic.*   p. 438

MARTYNIA   *Martyniaceae*
   annua (Timor, Moluccas), "Elephant's trunk"; according to Hortus widely distributed from Mexico, W. Indies, India, to Malaysia; hairy, herbaceous plant which I found in Timor, and identified by Dr. Burtt-Edinburgh; angled, heavily ribbed green leaves, with attractive tubular flowers white, the 5 flaring segments with lip blotched purple. *Tropical.*   p. 642, 646
   annua 'Purpurea' (Trop. America); cultivar with flowers red-purple, and orange in throat. *Tropical.*   p. 646

MASCARENA   *Palmae*
   lagenicaulis (Hyophorbe amaricaulis) (Mascarene Isl.), "Bottle palm"; grotesque, solitary palm to about 4 m high, the trunk very fat and bulging at base and quickly tapering upward, like a flask, topped by few heavy pinnate, arching fronds, 1-2 m long, the leaflets erect and rigid, yellowish green, the bases forming a prominent crown-shaft. *Tropical.*   p. 783
   revaughanii (Mauritius); another "Bottle palm", remarkable for its massive trunk thickly swollen from the 50 cm dia. base, becoming gradually smaller and forming an elongated neck; pinnate leaves few but beautifully arching on brown-red petiole, almost 2 m

long, the red-edged leathery leaflets 45 cm long; black fruit 2 cm dia. *Tropical.* p. 789

MASDEVALLIA   *Orchidaceae*
chimaera (Colombia); beautiful, very variable orchid of tufted habit with short-stalked leaves 12-25 cm long, and basal stalks with clusters of large flowers, brownish-yellow spotted with brownish-purple, the petals and lip small but the united 4 cm sepals large and showy and prolonged into reddish tails to 10 cm long; spring and summer. *Humid-subtropic.* p. 738

coccinea (Colombia, Perú); charming little epiphyte, with tufts of leathery oblong leaves; the solitary flowers on long stalks 30 cm long, with a short curved rose-pink tube, the sickle shaped lower petals usually in glowing scarlet; spring and summer blooming, in temperate to cool greenhouse. *Humid-subtropic.* p. 718

infracta (Brazil, Perú); small tufted epiphyte with spatulate leaves but without pseudobulbs; the flowers of fantastic shape, the prominent sepals extending into long tails, upper sepal whitish shaded yellow, lower sepals violet-purple inside and tails yellow, (May-July). *Humid-subtropic.* p. 737

x measuresiana (tovarensis x amabilis); small plant with tufts of obovate leaves and flowers with the lateral sepals partly united, the perianth white with nerves and margin lilac, and tails purplish-brown near the base, paler and greener toward the tips. *Humid-subtropic.* p. 737

militaris (Colombia); clustering species with rigid leaves to 10 cm long, stalked at base; handsome inflorescence, a long stalk 35 cm tall with solitary flowers 6 cm across, bright cinnabar-red, the dorsal sepal prolonged into linear tail (summer). *Subtropic.* p. 739

rosea (Colombia, Ecuador); small highland epiphyte with leaves 12 cm long; curious flowers, with upper sepals tail-like, yellow; the lateral sepals united at base, broad, rose pink; petals and lip reduced to tiny parts. *Humid-subtropic.* p. 736

veitchiana (Perú); a most beautiful species with densely tufted leathery, dark green leaves linear oblong and shining, and a wiry stalk bearing 1 or 2 showy, bright orange-scarlet flowers to 15 cm long, the dorsal sepal with long tail, the tailed lateral sepals partly grown together, closely studded with purple hairs (May-July). *Humid-subtropic.* p. 736, 737

MATTHIOLA   *Cruciferae*
incana, "Stocks" or "Gilliflower"; well-known for their spicy-sweet fragrance, by origin a semi-shrubby plant, usually grown as a biennial and lately as annual (10-14 weeks); at home around the Mediterranean from the Canary Is. to Asia Minor; cool-temperature plant highly developed, with brittle stems and grayish pubescent leaves, the popular branching stocks in the trade as 'Giants of California' or 'Bismarck', 45-60 cm high, with spike-like clusters of mostly double, quite fragrant flowers 3 cm wide, in many pastel or vivid colors from rose, apricot and red to blue, violet, lavender, even canary-yellow and white. Normally blooming from April to fall in cool climate. *Warm temperate.* p. 380, 381

incana 'Annua' (So. Europe), "Annual stocks"; grown as an annual, blooming in 10 weeks from seed. *Subtropic.* p. 380

incana 'Rosea', "Imperial stocks"; one of the many color forms, soft pink and a good commercial cultivar to 60 cm high. *Subtropic.* p. 380

sinuata (Greece), "Greek gilliflower"; short subshrub with white-felted gray leaves deeply lobed; small clusters of purple flowers. *Subtropic.* p. 380

MATUCANA   *Cactaceae*
haynei (Borzicactus) (Matucana, Perú); globose to oblong succulent body to 50 cm; 25-30 ribs, densely covered with woolly areoles and white spines; clusters of orange-red flowers. *Arid-tropical.* p. 270

MAURANDYA: see Asarina

MAXILLARIA   *Orchidaceae*
densa (Ornithidium) (Mexico to Honduras); epiphyte with small, compressed, single-leaved pseudobulbs; flowers borne in dense clusters on 3 cm stalks, sepals and petals creamy-white, often tinted with rose or green, lip white with red center, (Dec.-Sept.). *Tropical.* p. 748

picta (Brazil); epiphyte with ovoid pseudobulbs, each with 1 or 2 thick, strapshaped leaves; the individual, fragrant flowers yellow streaked and dotted purple and chocolate inside, petals incurved, lip white spotted purple, (Oct.-Aug.). *Tropical.* p. 739

punctata (Brazil); small epiphyte with ovoid pseudobulbs, solitary lanceolate leaf; flowers on short stalk, 6 cm across, light yellow, the lip with purple lines. *Tropical.* p. 739

rufescens (Cuba to Jamaica, Guatemala to Perú); clustered ovoid pseudobulbs with robust, solitary leaf to 20 cm long, short inflorescence of fragrant flowers 3 cm across, the sepals pinkish-brown, petals pale yellow, the yellow lip spotted dark red (winter). *Tropical.* p. 735

setigera (callichroma) (Colombia, Venezuela); miniature epiphyte with compressed pseudobulbs, solitary leathery leaves; fragrant flowers to more than 10 cm across, basically milk-white, changing to yellow and light brown toward tips of linear sepals and petals, lip yellow with purple throat. *Tropical.* p. 738

tenuifolia (Mexico); small clustering epiphyte, flattened pseudobulbs at intervals on ascending rhizomes, topped by solitary, linear leaves almost hiding the small, individually stalked, strongly scented flowers, sepals and petals dark rusty-red, lip spotted blood-red on yellow base, (Dec.-June). *Tropical.* p. 738

MAXIMILIANA   *Palmae*
maripa (regia) (Trinidad, N.E. So. America), "Cucurite palm"; solitary tropical feather palm with erect arching, pinnate fronds to 6 m or more long, dense with long pinnae borne in groups of 4 to 9 and in several layers. *Tropical.* p. 788

MAXIMILIANEA: see Cochlospermum

MAZUS   *Scrophulariaceae*
reptans (Himalayas), "Wart flower"; tiny mat-forming herb to 5 cm high, rooting at nodes, with toothed obovate 3 cm leaves, and purplish-blue bilabiate flowers, the lower lip spotted white, yellow and purple. *Subtropic.* p. 882

MEDINILLA   *Melastomataceae*
magnifica (Philippines, Java), "Rose grape"; gorgeous evergreen shrub to 2 m high, with angled woody branches and large, opposite, sessile, thick-leathery leaves to 30 cm long and with ivory midrib; striking inflorescence in a pendulous panicle to 30 cm long, of carmine-red flowers with purple anthers and great showy pink bracts. *Tropical.* p. 643, 645, 646, 647

sp. 'Rubra' (Finisterre Mts., New Guinea); shrubby epiphyte we collected on expedition, with woody branches, corrugated lanceolate leaves, the surface rugose and dark forest green, veins depressed; red inflorescence in elongate racemes with small purple flowers. *Tropical.* p. 645

MEDIOCALCAR   *Orchidaceae*
species (New Guinea); epiphyte with fleshy pseudobulbs formed at intervals from the apex of the older ones, each topped by succulent grayish leaves, and forming aerial roots; tiny buff-yellow flowers bell-like with welded sepals and sac-like lip; from the tropical cloud forest. *Humid-tropical.* p. 735

MEGACLINIUM   *Orchidaceae*
purpureorhachis (Bulbophyllum) (Zaire), the "Cobra orchid"; a sinister-looking, fantastic epiphyte from inner Africa, with clustered pseudobulbs and paired, rigid leaves to 15 cm long; the curious inflorescence 30 cm long, stalk a singular flattened green axis, densely spotted and overlaid with deep red, the tiny 1 cm deep brown flowers, borne in a single row on the flat sides; blooming August into winter. Snake-like in appearance, this is a striking curiosity plant in collections. *Tropical.* p. 745

MEGAKEPASMA   *Acanthaceae*
erythrochlamys (Venezuela), "Brazilian red-cloak"; showy tropical shrub to 3 m high, with appressed reddish hairs; stout stems with broad ovate leaves 12 to 30 cm long, dark green with pink midrib; inflorescence with conspicuous crimson bracts and two-lipped white flowers. *Tropical.* p. 39

MEIRACYLLIUM   *Orchidaceae*
trinasutum (Mexico); small, pretty epiphyte with creeping rhizome, little pseudobulbs, and a solitary, oval leaf; inflorescence short, flowers creamy-white changing to purple toward tips and on the concave lip, (summer). *Subtropic.* p. 738

MELALEUCA   *Myrtaceae*
genistifolia (Australia),"Bottle-brush"; tall shrub or tree to 12 m, with narrow almost linear, flat, stiff, sharp-pointed leaves to about 3 cm long, and showy spikes of white flowers with protruding bundles of stamens, resembling a bottle-brush. *Arid-subtropic.* p. 687

leucadendron (Australia, New Caledonia, Malaya), "Paper bark"; slender tree to 10 m, its undulate trunk with papery white bark, peeling off in broad strips; lanceolate 10 cm rigid bluish-green leaves, softly downy when young, with prominent veining, and bearing masses of honey-laden cream bottlebrush-like flowers, the stamens variable whitish, greenish-yellow, pink, or purple. *Arid-tropical.* p. 683, 685

quinquenervia (Australia), "Cajeput tree"; (usually sold as leucadendra); evergreen tree with thorny green trunk to 12 m; pendulous branches with narrow, stiff, pale green leaves 5-10 cm long, the young foliage covered with silky pubescence; fluffy flowers lemon-white, occasionally purple, with bundles of long stamens in short spikes. *Subtropic.* p. 680, 685

spathulata (W. Australia), "Honey-myrtle"; evergreen shrub to 1 m high; tiny flat 1 cm leaves, and covered by masses of pink stamen clusters. *Arid-subtropic.* p. 685

MELASTOMA *Melastomataceae*
decemfidum (sanguineum) (Sunda Islands); tropical shrub 1-2 m; with ovate-pointed, shining green leaves, and with 5 depressed parallel veins, ribs beneath and stalk red; large purplish-pink flowers. *Tropical.* p. 647

sanguineum (Malaya to Java); tropical shrub to 6 m high, with red-hairy branches; lanceolate leaves to 20 cm long, the five parallel nerves depressed; showy but fleeting purple flowers 5-8 cm across. *Tropical.* p. 642, 646

MELIA *Meliaceae*
azedarach (No. India, China) "Pride of India", "Bead-tree", "China berry", "Indian lilac"; spreading, partially deciduous tree to 20m, naturalized in Europe and America; handsome bipinnate foliage with feathery, toothed leaflets; clusters of fragrant, 2cm pale lavender flowers with purple stripes, followed by bright yellow but poisonous berries usually during a brief leafless stage. *Sub-tropic.* p. 641

azedarach umbraculifera, "Texas umbrella tree"; tree of weeping appearance, with foliage drooping from spreading branches giving an umbrella effect. *Warm temperate.* p. 642

MELIANTHUS *Melianthaceae*
major (South Africa), "Honey bush"; evergreen semi-woody shrub with widely creeping roots and herb-like stems erect to 3 m high or sprawling, with striking pinnate foliage 20-30 cm long, the deeply serrate, soft-fleshy leaflets in pairs, glaucous green above, grayish beneath; strong-scented when bruised. Reddish-brown 3 cm flowers in dense racemes secreting honey. *Subtropic.* p. 642

MELOCACTUS *Cactaceae*
intortus (Cactus inaguensis) (West Indies), known as "Turk's cap" in reference to the tall, cylindric cephalium on top of mature plants, its white woolly head setting off the red bristles making it look like a Turk's cap. Barrel-shaped plant to 1 m tall, with 14-20 ribs and yellow or brown stout spines; small pink flowers 2 cm long; 2-3 cm red fruit. *Arid-tropical.* p. 274, 275

matanzanus (Cuba), "Melon-cactus"; small domeshaped plant, deep green, 8-9 rounded ribs, yellowish clusters of recurved spines, "Turk's cap" of white wool and red bristles; flowers rose-pink. *Arid-tropical.* p. 274

neryi (Brazil: Amazonas); attractive small depressed globe 10-12 cm high, dark green with gray marbling, 10 acute ribs edged with clusters of mostly curving spines; at apex a low cushion-like cephalium on mature plants, white with red bristles; pale pink flowers 2 cm long, from top; fruit carmine-red. *Tropical.* p. 274

MENADENIUM *Orchidaceae*
labiosum (Venezuela, Guianas, No. Brazil); small epiphyte with flat pseudobulbs bearing two broad herbaceous leaves 25 cm long and with prominent veins; handsome large, 10 cm fragrant flowers heavy-textured, greenish-yellow tinged with copper, the large lip white with purple lines (spring). *Tropical.* p. 738

MENDONCELLA *Orchidaceae*
grandifloria (Mexico to Panama and Colombia); epiphyte with prominent pseudobulbs and paired folded leaves to 35 cm long; inflorescence from base with showy, fragrant, waxy flowers to 9 cm across, pointed sepals and petals yellowish green with purplish-brown stripes, lip white with red streaks. *Tropical.* p. 739

MERREMIA *Convolvulaceae*
holubii (Arabia); curious xerophyte with large beige bulb-like caudex, rough outside; from the apex sprouts a bouquet of ferney foliage, and small insignificant flowers. *Arid-tropical.* p. 360

MERYTA *Araliaceae*
denhamii (lanceolata), (New Caledonia); slender small tree to 5 m high; the leathery leaves variable in shape and size; on young plants linear, 15-30 cm long and 1 cm wide; on flowering trees oblanceolate to 1 m long, the margins with teeth. *Tropical.* p. 140

sinclairii (New Zealand), "Puka tree"; small evergreen tree with oblong, entire, leathery leaves crowded at ends of branches; blade glossy-green with irregular margin, pale, prominent veins, up to 50 cm long; petioles striped brown. *Subtropic.* p. 132, 135

MESEMBRYANTHEMUM: most species formerly listed under this genus are now referred to segregate genera in the Aizoaceae.

MESEMBRYANTHEMUM *Aizoaceae*
crystallinum (Cryophytum) (So. Africa, S.W. Africa; carried off to the shores of California), the California "Ice plant"; annual branching succulent with soft-fleshy stems creeping close to ground; spatulate soft-fleshy leaves 2-5 cm long, grayish green, thickly covered with crystal-clear bubbles filled with watery fluid and looking like ice-crystals; the starry flowers 1-3 cm across, translucent white tinted lavender. *Subtropic.* p. 42, 43

MESPILUS *Rosaceae*
germanica (W. Europe to Iran), "Medlar" or "Mispel"; small deciduous tree to 6 m, sometimes spiny, often of crooked, quaint habit; oblong leaves to 12 cm long, 3-5 cm white flowers; fruit apple-shaped, brown, 3-5 cm dia., open at top; edible after frost, or made into preserves. *Temperate.* p. 852

MESSERSCHMIDTIA *Boraginaceae*
argentea (So. Pacific Islands to Indian Ocean coral isles), "Tree heliotrope"; large shrub or small umbrella-shaped tree to 6 m high, at home on tropical shores; with deeply furrowed bark; thick ovate, fresh green leaves 15-20 cm long, covered with silky hairs, and densely clustered at ends of branches; small white flowers in silky clusters of coiled spikes; the fruit resembles small pointed brown peas. *Tropical.* p. 197

METASEQUOIA *Taxodiaceae (Coniferae)*
glyptostroboides (Central China: Hupeh) (Taxodiaceae), "Dawn redwood"; this "Fossil-age conifer" has been found growing to 30 m high near Chungking (1948) after having been thought extinct; handsome, deciduous, moisture loving tree with symmetric branches and soft textured, light green needles. *Temperate.* p. 352

METROSIDEROS *Myrtaceae*
collinus (Hawaii), "Ohia-Lehua"; the endemic bottlebrush of Hawaii, growing on the volcanic slopes to 3,000 m; a woody shrub or tree to 30 m; leathery oblong leaves to 8 cm long; conspicuous flowers with crimson stamens in bundles. *Subtropic.* p. 685

excelsus (tomentosus), (New Zealand), "Pohotukawa" or "New Zealand Christmas-tree"; handsome tree 10 to 20 m, with spreading branches; oval-pointed, leathery shining green leaves 5-10 cm long, the flowers in terminal clusters of showy brilliant scarlet stamens, exceeding the small petals; blooming at Christmas time during December-January (in N.Z.). *Subtropic.* p. 685, 687

tremuloides (prob. New Zealand); scandent shrub with twisting woody stems, thick-leathery elliptic leaves glossy green, and inflorescence with bundles of long, deep crimson stamens, at ends of branches. *Subtropic.* p. 687

villosus 'Variegatus' (hermadecensis) (New Zealand); attractive evergreen shrub or tree with oval leaves 3 cm long, grayish green with center area splashed with yellow; small red stamen flowers. *Subtropic.* p. 686

METROSIDEROS: see also Callistemon

METROXYLON *Palmae*
sagus (Indonesia, Philippines, Malaya), "Sago palm"; attractive tall growing feather palm to 12 m high, with smooth trunk becoming 40 cm thick; the pinnate fronds erect and gracefully arching, to 6 m long, the numerous leaflets glossy green; dull yellow 5 cm fruit; flowering only once, with tall branching inflorescence at apex, then leaves fall and tree dies. The trunks contain the edible flour-like sago, extracted from the pith. *Tropical.* p. 788

vitiense (Fiji); a handsome feather palm, with long, stiff-erect pinnate fronds on sturdy petioles with dense, glossy green pinnae; cultivated in Fiji for its sago, extracted from the pith of the trunk just before flowering; the 8 cm fruit yields hard ivory-nuts; leaves are used for roofing. *Tropical.* p. 788

warburgii (Tahiti), "Sago palm"; tall fan palm with trunk covered by old leaf bases; long arching fronds heavy with weight of the numerous leaflets, these glossy green and with drooping tips. *Tropical.* p. 788

MICHELIA *Magnoliaceae*
alba (longifolia) (Java), "Champac"; evergreen tropical tree resembling magnolia, but with axillary 5 cm white flowers, very fragrant; leaves ovate to 20 cm long. *Tropical.* p. 619

champaca (Himalayas), "Orange champak" or "Fragrant champaca"; tall evergreen tree with smooth gray trunk; narrow ovate, shiny, wavy leaves 20-25 cm long; orange-yellow flowers with 15-20 sepals and petals, 5 cm dia., and intensely fragrant, blooming much of the year and often mentioned in Indian poetry. Champaca oil is distilled from the blooms. *Tropical.* p. 619

MICONIA *Melastomataceae*
calvescens (magnifica) (Mexico), "Velvet-tree"; tropical foliage plant with woody stem and beautiful, long ovate, thin leaves to 75 cm long, velvety green with the sunken primary ribs ivory, and a network of pale green secondary veins; reddish purple beneath. Flowers insignificant. A show piece for the humid greenhouse. *Tropical.* p. 645, 646, 647

langsdorfii (Trop. America); shrubby bush with leathery elliptic leaves, glossy moss green with yellow parallel veins; flowers in terminal clusters of bundles of prominent white stamens; the fruit an edible berry. *Tropical.* p. 646

MICROCACHRYS *Podocarpaceae (Coniferae)*
tetragona (Mountains of Tasmania); low straggling bush with imbricated green scales; small brown cones at branch tips. *Subtropic.* p. 348

MICROCOELIA: see Gussonaea

MICROCOELUM *Palmae*
insigne (Brazil); small feather palm with short, thin trunk without crownshaft; the crown with long, gracefully recurving pinnate fronds, the glossy green leaflets 22 cm long and 2 cm wide, silvery underneath. *Tropical.* p. 790

MICROLEPIA *Polypodiaceae (Filices)*
strigosa (Japan, China, Trop. Asia, Polynesia); robust fern with stout creeping rhizome with wiry brown stalks and lacy bipinnate fronds 30-90 cm long, the oblique pinnae dentate, olive-green, thin but hard and somewhat glossy. *Humid-subtropic.* p. 444
hemispherica (Sao Paulo); twining tropical herb having opposite arrowshaped to hastate leaves with toothed margins, beautifully satiny green with silvery veining, purplish beneath; flowers flesh-colored. *Tropical.* p. 327

MIKANIA *Compositae*
ternata (Brazil), "Plush vine" or "Purple haze vine"; a most attractive and promising, rapid-growing trailer with brown stems, covered with lighter, felt-like hairs; opposite small herbaceous, palmately compound leaves 4 cm long, consisting of 5 lobed leaflets dark coppery green, purple beneath, densely covered all over with whitish hairs. Very graceful and eye-catching. *Tropical.* p. 324

MILLA *Liliaceae*
biflora (S.W. United States to Guatemala), "Mexican star flower"; cormous plant to 30 cm high, with thin grass-like leaves; flowers waxy white. *Subtropic.* p. 598

MILLINGTONIA *Bignoniaceae*
hortensis (Burma), "Indian cork tree"; graceful tropical evergreen tree to 25 cm high, with deeply cracked, spongy, cork-like bark; leaves opposite, unequally pinnate, with elliptic, crenate leaflets 3-5 cm long; long tubular, waxy flowers 8-10 cm long, pinkish-white, night-blooming and deliciously fragrant, in great profusion during the early hot months. The bark produces an inferior grade of cork. *Tropical.* p. 183, 564

x MILTASSIA *Orchidaceae*
'Cartagena' (Miltonia x Brassia); handsome bigeneric hybrid with large inflorescence of spidery, waxy white flowers spotted purple. (Moir-Honolulu). *Subtropic.* p. 736

MILTONIA *Orchidaceae*
candida (Brazil); beautiful epiphyte with flattened, pear-shaped, 2-leaved pseudobulbs; large, waxy, 8 cm flowers in loose, erect clusters, sepals and petals chestnut-brown tipped and barred with yellow, lip white tinged rose near base (Aug.-Nov.) *Tropical.* p. 739
candida 'Grandiflora'; differs from the typical species in having flowers twice as large, sepals and petals almost completely brown except for yellow tips, the lip white. *Tropical.* p. 739
candida 'Shelter Rock'; very floriferous form, with the waxy flowers larger, sepals and petals light brown with yellow markings, lip pinkish white. *Tropical.* p. 740
x hyeana (bleuana x vexillaria), "Pansy orchid"; nice colorful hybrid pink with deep blackish-crimson petals and white mask in center, (spring and fall). *Tropical.* p. 736
'Liberté' (Parnasse x Picadilly); lovely hybrid with purplish-red flower edged in white, center design and crest orange with yellow to white tracings; season in spring, but flowers usually again in autumn. *Tropical.* p. 738
spectabilis (Brazil); handsome epiphyte with flattened pseudobulbs bearing 2 thin, strapshaped leaves and showy large solitary flowers 8-10 cm across, the narrow petals and sepals pure white, large rosy lip with dark purple center and vein; blooming from spring to summer and into fall. This variable species has many color forms from white to deep purple. A spectacle when larger plants may bloom with dozens of flowers at the same time, lasting 6 weeks in beauty if kept cool and free from damp. The cv. 'Warneri' has petals and sepals richly blotched with purple. *Humid-subtropic.* p. 736
'Storm' (Mokadem x Piccadilly); small plant with light green, strapshaped leaves, and lovely large flowers spreading 9 cm, dark blood-red with velvety texture and a yellow brown face outlined in white, blooming in spring and again in autumn. *Tropical.* p. 738
vexillaria 'Volunteer'; very floriferous form, with flowers lavender pink veined purple, and a distinct pattern with a white eye stained in the center with deep yellow (May). *Tropical.* p. 738

x MILTONIDIUM *Orchidaceae*
'Aristocrat' (Miltonia schroederiana x Oncid. leucochilum); clustering plant with flattened pseudobulbs and strap-shaped leaves; inflorescence in loose arching racemes of small waxy flowers with narrow sepals and petals basically yellow but almost completely overlaid with deep maroon, the lip white with purple base (autumn). *Tropical.* p. 738

'Manhasset' (Miltonia 'Rio' x Oncidium leucochilum); large star-shaped waxy flowers with rosy petals and sepals, white at margins, and with purple blotches; lip pale pink at apex, purple striped in throat. *Tropical.* p. 738
'Yellow Monarch' (Miltonia x Oncidium); pendant raceme of charming, waxy flowers, petals and sepals orange-yellow and crisped, the lip overlaid with chocolate brown. *Tropical.* p. 736

MIMETES *Proteaceae*
cucullatus (S.W. Cape), "Scarlet bottlebrush"; in collection Kirstenbosch Botanic Garden; woody shrub with stiff, obovate leaves, grayish green, covered with velvety hairs and edged in red; pinkish flowers in terminal heads, subtended by showy, scarlet bracts. *Subtropic.* p. 837
hirta (S.W. Cape Mts., So. Africa), "Red and yellow bottlebrush"; rare and beautiful shrub to 1 m or more, with overlapping, short, ovate gray-green, ciliate leaves; from between the foliage appear the interesting flowers, set with white bristles at base, the petals copper-yellow and red. *Subtropical.* p. 838

MIMOSA *Leguminosae*
pudica (Brazil; naturalized in tropics), "Sensitive plant"; short-lived, spiny perennial; remarkable because of the ability of its pinnate leaves to go to sleep at the slightest touch, causing the leaflets to close and the petiole to fall; the flowers resemble little purplish puffs. An ubiquitous weed throughout the tropics, and it is thought that its mechanism is actuated by heat. I have seen it not only in Brazil, but in the West Indies where is is called "Mori-Vivi", through Africa, and in Viet Nam where it is known as the "Shame plant". *Tropical.* p. 546, 562

MIMULUS *Scrophulariaceae*
aurantiacus (Diplacus glutinosus) (Oregon, California), "Monkey flower"; branching shrub with narrow, leathery leaves toothed and turned down at the margins and sticky to the touch, pubescent beneath; the showy orange-salmon flowers 3 cm long, having notched, spreading lobes giving the effect of a monkey-face. *Subtropic.* p. 882
x burnetii (cupreus x luteus), "Monkey musk"; compact herbaceous perennial of Chilean parents forming tufts; ovate leaves palmately veined; flowers copper-yellow with yellow throat and brown spots inside. *Subtropic.* p. 886
cupreus (S. Chile), "Monkey flower"; compact summer-blooming annual to 20 cm high, with smooth, herbaceous leaves 4 cm long, toothed at margins; 4 cm open bell-flowers with spreading lobes, with a pretty face, yellow and boldly blotched with deep blood-red, the throat with crimson spots; other color variations are also seen. *Subtropic.* p. 882

MIMUSOPS *Sapotaceae*
commersonii (Madagascar), a "Spanish cherry"; dense evergreen tropical tree with milky sap, handsome obovate, leathery, glossy green leaves with yellow midribs, axillary small fragrant white flowers, followed by long-stalked pendant globular greenish to yellow fruit 3 cm dia., with yellow, edible pulp of mild flavor. *Tropical.* p. 880

MIRABILIS *Nyctaginaceae*
jalapa (Perú), "Four-o'clock"; deep-rooted, bushy, tuberous perennial herb but grown as an annual; with smooth, ovate leaves and large fragrant flowers opening in late afternoon, closing in morning, in shades of red, yellow and white, often striped, blooming profusely in late summer. *Subtropic.* p. 700
longiflora (Mexico), "Umbrellawort"; evergreen bush to 1 m high, with pubescent cordate ovate leaves; clusters of white flowers with long tubular 10 cm calyx, corolla like. *Subtropic.* p. 693

MISCANTHUS *Gramineae*
sinensis 'Variegatus', "Eulalia" or "Zebra grass"; robust perennial grass with long narrow, flat leaves, forming clumps, with age to 3 m high, green with yellowish stripes. *Warm temperate.* p. 512

MITCHELLA *Rubiaceae*
repens (E. No. America), "Partridge berry"; low evergreen creeper with thin, rooting stems to 30 cm long; tiny 2 cm oval, shining dark green leathery leaves often with white lines; twin white flowers with 4 spreading lobes, and fiery-red 1 cm berries characterized by having two navels. *Warm temperate.* p. 860

MITRARIA *Gesneriaceae*
coccinea (Chile); vining plant with woody stems, opposite toothed 2 cm leaves, smooth and green; flowers with tubular corolla to 3 cm long, scarlet red. *Subtropic.* p. 502

MOLUCCELLA *Labiatae*
laevis (Syria), "Bells of Ireland"; old-fashioned herbaceous annual interesting because its 60 cm stalks are closely set with whorls of the oversized, light-green, bell-shaped calyx in which nestle the little pinkish, bilabiate, fragrant flowers. *Subtropic.* p. 531

MONADENIUM *Euphorbiaceae*
  coccineum (Tanzania: Masai Dist.); semi-scandent or erect fleshy plant occasionally branched, with 5-angled green stem ⅓-1 m tall, 5-8 cm fleshy obovate leaves with keeled midrib and serrate margin; pretty, bright scarlet flowers, the lower petals green. *Tropical.* *p. 418*

MONODORA *Annonaceae*
  grandiflora (West Trop. Africa), "African orchid nutmeg"; tropical tree to 6 m high with oblong, glossy-green leaves purplish beneath, and pendulous orchid-like yellow flowers spotted red, the crisped outer petals 8-10 cm long, strikingly beautiful, blooming early in summer. *Tropical.* *p. 78*
  myristica (Sierra Leone, W. Africa, Angola), "Jamaica nutmeg", "Calabash nutmeg", "African false nutmeg"; ornamental tree becoming tall to 30 m; large leathery, oblong leaves with tendency to droop to 50 cm long; intriguing fragrant flowers with yellowish petals spotted with red, to 10 cm long, the inner ones shorter; aromatic seeds used like nutmeg. *Tropical.* *p. 72*
  tenuifolia (Trop. Africa), the "African nutmeg"; medium-sized shrub or tree to 12 m high, one of the prettiest of flowering trees; with narrow oblong leaves, and strangely beautiful flowers about 10 cm across, resembling orchids, with 3-pointed crinkly white sepals, and 6 united white and yellow petals with crimson spots, appearing from the underside of the twigs. *Tropical.* *p. 72*

MONOLENA *Melastomataceae*
  primulafolia (Bertolonia) (Colombia); exquisite plant of the rainforest, with thick rhizome; brownish pubescent petioles carry the ornate, fleshy leaves, cordate in outline, glossy deep green and with sunken veins, 10-15 cm long, purple beneath; the charming baby-pink flowers 2 cm across, in a bouquet. *Tropical.* *p. 647*

MONSTERA *Araceae*
  deliciosa (Philodendron pertusum) (So. Mexico, Guatemala), "Ceriman"; stout, woody-stemmed, close-jointed tree-climber forming long hanging, cord-like aerial roots, with large, thick, leathery leaves, glossy green, to 1 m, pinnately cut and perforated with oblong holes; bisexual spadix and boat-shaped, white spathe; cone-like, edible fruit with pineapple aroma; also known as "Mexican bread-fruit". The long-jointed, rapid-climbing juvenile stage with smaller, less perforated leaves is known as Philodendron pertusum. *Tropical.* *p. 110, 111, 112, 457*
  deliciosa 'Albo-variegata'; handsome mutant seen in South Africa, with large character leaves partly rich deep green and other sections a contrasting creamy white. *Tropical.* *p. 110*
  deliciosa borsigiana (Cordoba, Mexico); vining type with smaller, glossy leaves with pinnate lobes widely and evenly separated, few, if any holes; the leafstalk is wrinkled where it joins the leaf. *Tropical.* *p. 111*
  deliciosa 'Marmorata'; (Philodendron pertusum variegatum), "Variegated philodendron"; a mutant with irregular variegation where parts of the leaf may be entirely green, and other sections marbled cream to greenish-yellow, or entirely cream; new growth may be found to revert back to green. *Tropical.* *p. 110*
  deliciosa minima; photo shows tiny desk plant with 8 cm leaves, made of plastic in 1965, when plastic flowers were meant to replace lives plants in interior decoration. However, plastic fades in sunlight, and in time becomes unsightly and covered with dust. *Tropical.* *p. 112*
  epipremnoides (Costa Rica); interesting large climbing aroid with leaves somewhat resembling Epipremnum; thick, slowly elongating stem close-jointed, with long ovate leaves 60 cm long, on petioles winged two-thirds up from leave axils; the deep green blades deeply and widely perforated to a rounded sinus, the lobes with square tip; along the midrib a dual row of small or large oval holes; the spadix club-like, shielded by ivory, boat-shaped spathe. Photographed at Longwood Gardens, Pennsylvania 1976. *Tropical.* *p. 111*
  friedrichsthalii (Costa Rica); prolific climber with medium-size ovate, oblique, fresh green leaves, with many, mostly oval, perforations evenly distributed between veins; wavy margins; will not develop holes unless climbing up. *Tropical.* *p. 111*
  obliqua (Alto Amazonas); long climber with elliptic leaves with tapering apex, developing large; irregular, rather oval holes in older stage. *Tropical.* *p. 112*
  obliqua expilata (leichtlinii), "Window-leaf"; a form of obliqua; the perforated leaves of this interesting "laceleaf" are not much more than an ovate skeleton of veins, with very little green left of the blade. The leaves become 60 cm large, the holes in the blade are usually arranged in double rows. *Tropical.* *p. 113*
  pertusa (Panama, Guayana); lush climber with soft-textured, unequal-sided leaves perforated and pinnatisect; a poor keeper, from tropical lowland forests. *Tropical.* *p. 110*

MONTANOA *Compositae*
  grandiflora (Honduras); warm-climate flowering shrub with opposite foliage, the broad, rough-textured leaves deeply palmately lobed; charming double flowers with ray florets creamy white, and shaped like wide-open roses. *Tropical.* *p. 322*
  mexicana (Mexico), "Daisy tree"; sparry shrub with woody branches; large, opposite rugose leaves deeply indented or lobed; at the apex a cluster of white flowers with protruding yellow stamens in center. *Subtropic.* *p. 322*

MONTBRETIA: see Tritonia

MONTEZUMA *Malvaceae*
  speciosissima (Thespesia grandiflora) (Puerto Rico), "Hibiscus tree"; showy evergreen tree to 15 m; slender woody branches with heavy cordate, leathery leaves to 25 cm long, glossy green; large 15 cm waxy flowers, glossy crimson red, heavily veined, and with protruding style. *Tropical.* *p. 627, 629*

MONTRICHARDIA *Araceae*
  arborescens (Guayana); tree-like aroid, growing by the side of tropical rivers, with glossy-leathery, broad sagittate leaves 20-30 cm long, deep green with pale veins, on slender canes. Acording to J. Bogner of Munich Bot. Garden the photo shown on p. 109 is probably M. linifera, with leaves more cordate than arborescens. *Humid-tropical.* *p. 109*

MORAEA: see Dietes

MORINDA *Rubiaceae*
  citrifolia (S.E. Asia, Australia), "Indian mulberry" or "Noni"; small evergreen tree with angular branches and thick branchlets; thick, large conspicuous, broad elliptic leaves to 25 cm long, deeply veined; small white flowers in globose heads; strange ovoid, fleshy fruit 5 cm long, reputed to be poisonous. *Tropical.* *p. 862, 870*
  citrifolia 'Variegata'; cultivar with handsome foliage largely variegated with cream over shades of green. *Tropical.* *p. 868*

MORINGA *Moringaceae*
  oleifera (pterygosperma) (E. India), "Horseradish tree"; deciduous tree to 10 m high, with much divided 2 to 3 pinnate fern-like leaves to 50 cm long; panicles of 2 cm white flowers, followed by 3-6-angled pendulous pods to 45 cm long; seeds are roasted and eaten; the roots taste like horseradish. *Tropical.* *p. 662*

MORMODES *Orchidaceae*
  lineatum (Guatemala); showy epiphyte with plaited leaves on strong pseudobulbs; lightly arching raceme of very fragrant flowers, sepals and petals tongue-shaped, outside greenish, inside yellow with purple stripes, lip yellowish with red dots, (Jan.-March). *Tropical.* *p. 738*

MORUS *Moraceae*
  alba (China), "White mulberry"; deciduous tree to 25 m high; ovate, corrugated leaves 10 cm or more long, coarsely toothed, glossy green above, small flowers in drooping catkins; berry white to blackish purple, with insipid, sweet juice. *Temperate.* *p. 660*
  nigra (West Asia), "Black mulberry"; large shrub or tree to 10 m high, with rough, dull-green broad-serrate 20 cm leaves, crenate at margins; flowers in drooping catkins; luscious edible red to black berries somewhat hidden on underside of branches. Has the largest and juiciest fruit. *Warm temperate.* *p. 473, 660*

MUCUNA *Leguminosae*
  bennettii (New Guinea), "New Guinea creeper"; striking vine with glossy green, compound leaves on woody, twining stems, and long axillary pendant racemes of sickle-shaped, claw-like, waxy, fiery scarlet flowers; probably the most showy of all tropical climbers. *Tropical.* *p. 553, 562*

MUEHLENBECKIA *Polygonaceae*
  complexa (New Zealand), "Wire vine" or "Maidenhair-vine"; twining, threadlike, purplish-brown, wiry stems furnished with scattered, tiny round, fresh-green leaves; flowers greenish-white, in small spikes; a graceful basket plant. *Subtropic.* *p. 826*

MUNTINGIA *Tiliaceae*
  calabura (Trop. America), "Panama berry", "Capulin" or "Calabur"; tropical downy tree to 10 m high, with wide-spreading drooping branches bearing oblique oblong, rough leaves 12 cm long, serrate, and arranged in one plane; single or paired white 2 cm flowers, followed by globose red 2 cm berries of pleasant taste. *Tropical.* *p. 910*

MURRAYA *Rutaceae*
  paniculata (India), "Satinwood" and "Cosmetic bark tree"; related to M. exotica but more tree-like with strong and durable light yellow wood, and bark which is the source of a cosmetic; the large strongly scented 5-petaled white blooms in few-flowered clusters, and worn in the hair for their beauty and fragrance by the women of the East Indies. *Tropical.* *p. 872*

MUSA *Musaceae*
acuminata (cavendishii or nana) (So. China), "Dwarf banana", also known as "Dwarf Jamaica"; shapely stoloniferous plant of compact habit, to 1½-3 m high, with short stem and dense rosette of oblong, glaucous green, leathery leaves with satiny sheen, blotched with red when young, on short stout petioles; produces edible, smallish but deliciously fragrant yellow, 12 cm fruit; stands more cold than most bananas; good tub plant. *Subtropic.*   p. 465, 470, 478, 671, 672
basjoo (Southern Japan: Liu-Chiu archipelago), "Japanese fiber-banana"; cluster-forming to 4 m or more, slender reddish stem bearing shining thin leaves to 3 m long, bright green on both sides; arching inflorescence with yellowish flowers under reddish bracts; followed by bunches of 30 to 60 angled fruits, 8 cm long. Grown for its fiber in Japan. *Subtropic.*   p. 670
coccinea (Vietnam: Cochin), "Flowering banana"; showy flowering plant to 1½ m high and stoloniferous, with green pseudo-stems topped by spirally arranged, long-stalked, bright green leaves and erect flowering head with flaming fiery-red bracts yellow at tips and yellow flowers. Small fruit orange-yellow 5 cm long, and with white flesh. *Tropical.*   p. 666, 672, 673
Fehi (troglodytarum) (No. New Guinea), "Cooking banana"; tree-like, stoloniferous plant to 7 m high; stem cylindrical, formed of sheathing leaf-petioles, with violet sap; broad shining green leaves 2 m long, 60 cm wide, often windblown into shreds; inflorescence erect, the flowers hidden under colored bracts; small, angled fruit in large erect clusters, orange, 12-15 cm long, and eaten only after cooking. *Tropical.*   p. 470, 478
mannii (Assam); dwarf ornamental with pseudostems to 1 m high, tinged with black; the broad-oblong leaves glossy olive green with red midrib; short erect inflorescence 15 cm long, with male bracts light crimson. *Tropical.*   p. 668, 672
ornata (rosacea) (Bangladesh to Burma), "Flowering banana"; ornamental plant with pseudostems to 3 m high, pale green and waxy, with black blotches; leaves 2 m long, glaucous green; inflorescence erect with pink bracts tipped yellow. *Tropical.*   p. 668
x paradisiaca (x sapientum; acuminata x balbisiana) (India, Ceylon), "Common banana"; tree-like herb to 7 m high, with spirally arranged green leaves with reddish midrib, becoming frayed or broken by the wind, forming a slender trunk by their sheathing bases; flowers yellow, bracts violet, and the well known, yellow, edible fruit, about 20 cm long; after fruiting the stem dies but is replaced by new suckers. *Tropical.*   p. 470, 478, 671
x paradisiaca 'Champa', "Lady-fingers"; a slender banana bush seen in England, with pendant bunches of smaller size, finger-like fruit with thin skin, but with meat deliciously mild and tasty and of almost buttery consistency. *Tropical.*   p. 470
x paradisiaca 'French Plantain'; according to Hortus 3, the type of a good plantation banana; of medium height, with heavy pendant bunches of green to yellow aromatic fruit. *Tropical.*   p. 470
x paradisiaca 'Koae'; probably an Hawaiian bud mutation triploid of some Maoli banana of the Pacific; this most beautiful variegated-leaf banana is named after the Koae bird with "hair prematurely graying"; the leaves are striped or banded white and very light green on dark green, laterally; the midrib, petiole and trunk alternating white and green, even the immature fruit is variegated, but ripens to yellow, is short and roundish, with yellow flesh. *Tropical.*   p. 664
x paradisiaca 'Vittata', "Variegated banana"; beautiful plant with the light green leaves variegated milky green and white, the midrib white and edges red; loves shade and warmth. Short and roundish fruit ripens yellow, and is generally cooked to eat. *Tropical.*   p. 664
sumatrana (Sumatra), "Blood banana"; desirable plant to 2½ m high, not as tall as zebrina, leaves shorter and to 40 cm wide, on short petioles, deep grayish green blotched with dark wine-red, and red beneath as well, but losing this coloring with age; more tolerant to winter chills. *Tropical.*   p. 669
sumatrana 'Rubra'; form with leaves strongly overlaid with red, and reverse entirely blood-red. *Humid-tropical.*   p. 670
uranoscopus (Indonesia: Java); an ornamental banana which I found in the mountains of Tjibodas, Central Java; rhizomatous tree-like herb to 4 m tall, with vivid green elongate foliage, and beautiful inflorescence of red-bracted heads with purplish-blue flowers. *Humid-tropical.*   p. 671
velutina (Assam); dwarf species to 2 m or more high, slender, with pinkish stem, petioles and into midrib, leaves 1 m long and 30 cm wide; erect inflorescence with red bracts, pale yellow flowers and small, red velvety fruit. *Tropical.*   p. 671, 673
zebrina (Java), "Blood banana"; known in horticulture as sumatrana; slender plant to 4 m high with tall trunk bearing rather delicate long-stalked leaves, satiny bluish green richly variegated with blackish blood-red, the channeled midrib brown-red, underside reddish-wine; maintaining its coloring with age but sensitive to cold. *Tropical.*   p. 664, 670

MUSA: see also Ensete

MUSCARI *Liliaceae*
armeniacum (Turkey), "Grape hyacinth"; May-blooming bulbous plant suitable for forcing in pots, with narrow, channeled basal leaves, and delicate spires of nodding, urn-shaped flowers azure-blue, tipped white. *Subtropic.*   p. 592, 595
commutatum (Sicily to Palestine); a bulbous grape hyacinth with linear grooved flaccid leaves, and small cluster of blackish-blue flowers. *Subtropic.*   p. 598

MUSSAENDA *Rubiaceae*
erythrophylla (Zaire), "Ashanti blood"; beautiful spreading shrub or rambler with roundish ovate, bright green, silky-hairy, soft leaves, and tiny 1 cm creamy-white flowers with cushion of blood-red felt in center, each having appended one large, odd, ovate 10 cm showy sepal of rich vermilion-scarlet with parallel dark veining, which cover the bush almost entirely by their masses. *Tropical.*   p. 867, 870, 871
erythrophylla 'Rosea'; attractive color form with bracts a lovely carmine pink. *Tropical.*   p. 870
frondosa (E. Indies); semi-shrubby tropical evergreen, with dull green, large broad-ovate, corrugated leaves, orange-yellow star-flowers, subtended by ovate bracts of pure white. *Tropical.*   p. 871
philippica (Philippines, New Guinea); shrub with large ovate, soft-hairy or satiny leaves red-veined beneath, and terminal clusters of small star-like deep yellow flowers, subtended by one large ovate, pure white petaloid bract, actually a calyx-lobe expanded into a large sepal. *Tropical.*   p. 870, 871
philippica 'Aurorae' (Philippines: Luzon); an outstanding variety which I have seen offered at the flower market in Manila, as "Dona Aurora"; a bushy shrub with large corrugated, green, downy 8-15 cm leaves, each twig bearing a showy, pubescent inflorescence of small golden yellow flowers, surrounded by a circle of gleaming white, pendant, obovate enlarged sepals, almost the year round; this variety differs in having all the calyx lobes developed into 5 petal-like white bracts, not just a single one as in the species. *Tropical.*   p. 870, 871

MUTISIA *Compositae*
clematis (Colombia, Ecuador); pubescent climbing evergreen vine with pinnate leaves, the midrib extended into a tendril; nodding, woolly flower heads bright orange-scarlet with 9-10 bright red spreading rays.   p. 329

MYOPORUM *Myoporaceae*
laetum (New Zealand), the "Ngaio" or "Mouse-hole tree"; vigorous evergreen tree or shrub of exceptionally fast growth to 10 m high and 6 m spread, forming a dense, billowing mass of dark green; older growth stiff-woody, young shoots flexible, with narrow elliptic 8-10 cm soft-leathery, glossy leaves with translucent oil glands; clusters of 1 cm flowers white with purple markings; small purplish fruits. Favours the seashores in New Zealand and is very salt-resistant. *Subtropic.*   p. 674
parvifolium (Australia, Tasmania); low evergreen shrub spreading along the ground, and used as ground cover in California; leathery, linear leaves densely set 1-3 cm long, with crenate margins; small white honey-scented flowers. *Subtropic.*   p. 674

MYOSOTIS *Boraginaceae*
sylvatica (Europe, N. Asia), "Forget-me-not"; small tufted herb with oblong leaves, and racemes bearing the appealing little skyblue flowers with yellow eye. Often listed as alpestris. *Temperate.*

MYRICA *Myricaceae*
rubra (Japan, So. China, Korea, Philippine Islands), "Chinese strawberry tree"; small tree with obovate leathery leaves to 12 cm long; small flowers in short spikes; succulent edible fruit deep red-purple, to 3 cm dia. *Subtropic.*   p. 462, 673

MYRISTICA *Myristicaceae*
fragrans (Indonesia: Moluccas), the "Nutmeg tree"; evergreen, dioecious, 10-18 m high, with willowy green branches, and leathery, yellowish olive green ovate, aromatic leaves 5-12 cm long, bluish beneath; axillary pale yellow flowers; the yellow pear-like fleshy 5 cm fruit opens by two valves and showing red pulp—the spice—which surrounds a brown, hard-shelled seed; the kernel of the seed is nutmeg, borne on female trees. *Tropical.*   p. 466, 674

MYRSINE *Myrsinaceae*
africana (So. Africa, Arabia, to C. China), "African boxwood"; shrubby bush resembling boxwood but more graceful, to 1 m, with angled, downy, red shoots, dense with small rounded, shiny dark green, 1 cm leaves finely serrate; tiny pale brown axillary flowers and purplish-blue berries. *Subtropic.*   p. 674

salicina (New Zealand); evergreen tree to 8 m tall, with bark nearly black; obovate leathery leaves to 18 cm; inflorescence in dense clusters, small brownish flowers followed by red to purple berries. *Subtropic.* p. 674

MYRTILLOCACTUS *Cactaceae*
geometrizans (S. Mexico), "Blue myrtle"; branching tree type to 4 m high; smooth, slender columns six-ribbed, 8 to 10 cm dia., glaucous powder-blue; with practically no spines; small diurnal flowers greenish-white; edible blue fruit. *Arid-tropical.* p. 253, 254

MYRTUS *Myrtaceae*
communis (Mediterranean reg.), "Greek myrtle"; evergreen shrub to 4 m high, loosely leafy with leathery, rather broad-ovate, 5 cm leaves, dark lustrous green, spicy when bruised; fragrant white flowers with numerous stamens, and purple-black berries. *Subtropic.* p. 684

communis compacta, "Compact myrtle" or "Dwarf myrtle"; slow-growing, small, compact form with densely set small black-green leaves, 2-3 cm long, and which may be trimmed and sheared into various topiary or Bonsai forms; also for low edgings indoors or the patio. *Subtropic.* p. 684

communis 'Microphylla', "Dwarf myrtle"; the compact form grown by European plantsmen in pots and sheared into little globes and used for weddings; densely leafy shrub with brown twigs and small, needle-like, shining black-green leaves, and white flowers of aromatic fragrance. *Subtropic.* p. 684

communis 'Variegata', "Variegated myrtle"; small evergreen shrub with leathery, pointed, green leaves attractively variegated or margined with creamy-white. *Subtropic.* p. 688

pubescens (Argentina); dense evergreen shrub with small leathery, ovate leaves 3 cm long, and showy axillary flowers white with numerous stamens. *Subtropic.* p. 684

MYSTACIDIUM: see Angraecum

NAEGELIA: see Smithiantha

NANDINA *Berberidaceae*
domestica (China, Japan), "Heavenly bamboo"; attractive shrub usually low, but to 2½ m, slender cane-like stems with 2 to 3 pinnate leaves, the ultimate leaflets narrow; turning red in fall; small white flowers in large panicles, followed by bright red berries; fairly hardy. *Warm temperate.* p. 181

domestica 'Goshiki', "Five color rainbow"; curious cultivar photographed at Ono Nursery in Honolulu; evergreen plant with contorted branches and dense foliage; the linear to ovate leaflets pink to dark green, and margins turned under. *Subtropic.* p. 181

domestica longifolia 'Ito'; a curious form of "Heavenly bamboo", the woody stems with apparent leafless branches, the foliage reduced to ribs only, in red or green; photographed at Ono's Nursery, Honolulu. *Warm temperate.* p. 181

NAPOLEONA *Lecythidaceae*
heudelotii (W. Trop. Africa), "Napoleon's button"; tropical small tree with elliptic leaves 20 cm long; remarkable flowers with a double crown, in leaf axils or on trunk, the calyx forming a cup, the shorter corolla another cup inside and red. *Tropical.* p. 544, 762

NARCISSUS *Amaryllidaceae*
'Actaea' (Poeticus type); large flat, pure white petals, with yellow eye, broadly margined scarlet; derived form N. poeticus, the "Poet's narcissus", from France to Greece. *Warm temperate.* p. 68

cyclamineus 'February Gold', a hybrid of N. cyclamineus from Portugal with a "Trumpet" type; compact "Pot narcissus" about 25 cm high, with narrow linear leaves, and smallish but numerous flowers with lemon yellow perianth segments spreading 5 cm across, and orange yellow trumpet crenate at mouth. Normally early blooming, this pretty hybrid is ideal for pots from Valentine to Easter. *Warm temperate.* p. 66

'Gold Medal' (pseudo-narcissus); a true, short pot daffodil; outstanding large-flowered trumpet variety almost exactly like 'King Alfred' except that it is more stocky and shapely in pots for spring; the large 9 cm flowers clear yellow with richer colored trumpet, carried on solid stalks 30 cm high, 2 to 3 to a bulb; late blooming. *Warm temperate.* p. 68

'Gustav Mahler'; beautiful large-cupped daffodil with perianth or spreading petals vivid yellow, and corona cup contrasting scarlet-red. *Temperate.* p. 68

odorus 'Orange Queen' (Div. 7: Jonquils), a miniature daffodil to 30 cm tall, derived from odorus (Mediterranean, east to Yugoslavia); clustering, slender, narrow channeled leaves, and numerous stalks bearing 2-3 pretty, small 4 cm flowers with 6 segments and small cup, rich orange-yellow and very fragrant. Free flowering and very striking. *Temperate.* p. 66

NAUTILOCALYX *Gesneriaceae*
lynchii (Alloplectus) (Colombia), "Black alloplectus"; erect, stout, fibrous-rooted, succulent plant with bronze to blackish-red, shiny, elliptic crenate leaves; flowers in axillary clusters, creamy with purplish hair, not showy. *Tropical.* p. 498

NAVIA *Bromeliaceae*
arida (Venezuela); primitive mountain xerophyte of the "Lost world"; handsome small rosette with leathery, oblanceolate leaves covered with silvery scales and finely toothed at margins, the inner circle a glowing crimson; the sessile, nest-like inflorescence vivid yellow tipped red-purple. *Subtropic.*

NEANTHE: see Chamaedorea

NELUMBO *Nymphaeaceae*
nucifera (Nelumbium nelumbo) (Trop. East Asia to N.E. Australia), "East Indian lotus"; large, aquatic, stemless plant symbolic of perpetual life in Buddhism; long, milky, prickly petioles bear shield-like leaves above the water, and bold stalks with large, delicate pink flowers of a haunting fragrance; the petals soon fall leaving the prominent, flat-topped receptacle bearing edible seed. *Tropical.* p. 704, 707

nucifera 'Alba plena', "Shiroman"; a beautiful, large, double-flowered Japanese form, at first cream with greenish tinge, becoming pure white, and delightfully fragrant. *Tropical.* p. 707

NEMATANTHUS *Gesneriaceae*
fissus (Hypocyrta selloana), (Brazil); fibrous-rooted, robust plant with fleshy, oblong, deep green, hairy leaves with reddish midrib, purplish beneath; axillary, deep red, downy, long cylindric flowers with ventricose throat. *Humid-tropical.* p. 498, 502

NEMESIA *Scrophulariaceae*
strumosa (So. Africa), "Cape jewels"; floriferous, densely branching annual with opposite, sessile, linear, dentate leaves, and erect racemes of attractive, bilabiate flowers, with a pouch at base and bearded throat, borne in great profusion from June to Dec., in white, yellow, rose, orange, crimson, with spotted throat. *Subtropic.* p. 884, 885

NEOBENTHAMIA *Orchidaceae*
gracilis (E. Africa: Zanzibar); tall branching terrestrial to 1½ m high, with numerous 2-ranked linear 20 cm leaves and long-stalked, dense terminal clusters of white flowers, the lip spotted rose alongside a yellow center stripe (Jan.-April, Sept.). *Tropical.* p. 741

NEOBESSEYA *Cactaceae*
missouriensis (Dakota, Montana, Kansas, Oklahoma); globular cactus of the Great Plains, clustering with age, glaucous green with cylindrical knobs, the areoles and axils with white wool, gray needle spines; flowers outside reddish, inside yellow. *Temperate.* p. 280

NEOBINGHAMIA *Cactaceae*
mirabilis (Perú); succulent column to 2 m high and 8-10 cm dia, to 22 ribs and with woolly spines; tubular flowers 6 cm long, purple sepals and rose petals, blooming at night. *Arid-subtropic.* p. 254

NEOCOGNIAUXIA *Orchidaceae*
monophylla (Jamaica); charming epiphyte, with slender stems bearing solitary leaf; inflorescence arching, with spectacular flowers bright orange-scarlet, 2-5 cm across (autumn). *Tropical.* p. 740

NEODYPSIS *Palmae*
decaryi (So. Madagascar); peculiar feather palm 6-10 m high, with distinctive 3-sided appearance of its trunk caused by the bulging leaf bases attached in 3 ranks; the ascending pinnate fronds to 2½ m long, continue this 3-sided effect; leaflets gray-green and erect, the lower ones pendulous. *Tropical.* p. 790

lastelliana (Madagascar); very beautiful, elegant feather palm, the trunk with red crown shaft as of corduroy, the rigidly erect, long pinnate fronds with stiffish, rich green leaflets at right angles from the rachis; rare in cultivation. *Tropical.* p. 791

NEOFINETIA *Orchidaceae*
falcata (Angraecum falcatum) (Japan, Korea); lovely miniature epiphyte 8-15 cm high, with leathery linear light green, keeled leaves arranged in 2 ranks; freely blooming with clusters of pure white, waxy flowers 3 cm across, having a slender spur 5 cm long, and blooming in summer; intensely fragrant, mostly at night. *Subtropic.* p. 741

NEOLLOYDIA *Cactaceae: Echinocactaneae*
conoidea (Coryphantha, Echinocactus) (Texas to Cent. Mexico); small clustering globe to 10 cm high, with short straw-colored spines; pretty, silky flowers 6 cm across, carmine rose with deeper center and reflexed petals. *Arid subtropic.* p. 271

NEOMARICA *Iridaceae*
bicolor (Brazil), "Walking iris"; vigorous tropical iris-like plant with perennial rootstalk, the broad, fresh-green leaves arranged like

fans; from their sheathing base rise stout stalks bearing at an angle the attractive, if fleeting flowers lavender blue, the center segments yellow with brown cross-lines, their tips marked with blue. *Subtropic.* p. 526, 528

caerulea (So. Brazil), "Twelve apostles"; beautiful iris-like plant with large flowers having outer petals bright sky-blue, the center petals pale and marked with yellow and brown; unforgettable sight when I first saw such a field of blue on the Serra do Mar above Santos in Sao Paulo state. Flowers carried on clear stem not clasped by leaves, as on northiana. *Subtropic.* p. 525, 526, 529

gracilis (Mexico to Brazil), "Apostle plant"; swordlike leaves from short rhizome, stalks with iris-like but short-lived successive flowers, outer petals white, small inner petals recurved and blue, base marked brown. *Tropical.* p. 526

NEOPANAX *Araliaceae*

arboreus (Nothopanax) (New Zealand), known as "Five fngers"; evergreen tree with palmately compound leaves, the stalked segments leathery, dark green with coarsely serrate margins, pale center veins, and brownish stem. *Subtropic.* p. 133, 140

NEOPORTERIA *Cactaceae*

nidus (No. Chile), "Birdsnest cactus"; solitary stem at first globular, later almost cylindrical to 12 cm thick, entirely surrounded with long curved grayish spines, 16-18 ribs, the areoles set with wool; funnel-shaped reddish flowers on apex. *Arid-tropical.* p. 270

nidus senilis (Northern Chile), "Birdsnest"; globular, later elongate stem with to 18 ribs, densely covered with matted white spines; flowers half open, carmine-rose. The var. senilis has softer spines than the species, with larger flowers. *Arid-subtropic.* p. 269, 273

nigrihorrida (Chile); flattened globose, dark gray green, 16-18 ribs closely set with interwoven radials and thick and heavy, silvery central spines; carmine flowers 4 cm long. *Arid-tropical.* p. 270

NEOREGELIA *Bromeliaceae*

carolinae (Brazil); spreading rosette of strapshaped leaves 30 cm long and 4 cm wide, metallic copper over green, toothed margins; inflorescence formed by brilliant lacquer orange-red bract leaves surrounding the flowers with violet-purple petals edged white, deep in center. *Tropical.* p. 216, 219

carolinae 'Marechalii', "Blushing bromeliad"; dwarf variety with flattened metallic leaves; at flowering time the inner leaves are brilliant rosy-crimson, remaining so for six months or more, flowers lilac. *Tropical.* p. 219

carolinae 'Meyendorffii' (Karatas); broad rosette of flat olive green leaves with coppery tinting; at flowering time the inner leaves turn a dark maroon; flowers lilac deep in center. *Tropical.* p. 219

carolinae 'Meyendorffii variegata'; beautifully colored rosette of broad pale cream leaves 30 cm long, variegated with apple-green length-stripes, and tinted with rose, especially toward base. *Humid-tropical.* p. 216

carolinae 'Tricolor' (Brazil), "Striped blushing bromeliad"; very attractive variety with the glossy green leaves having ivory-white lengthwise bands becoming rose-tinted in good light; at flowering time they become shorter and carmine-red; fl. violet-purple edged white. *Tropical.* p. 216, 219

carolinae 'Volkaert's Favorit'; beautiful Belgian 'Marechalii' cultivar; a showy flattened rosette of broad leaves deep forest green, the margins a contrasting creamy yellow, center cup lacquer red; the nested inflorescence with violet flowers; photographed in Frankfurt Botanic Garden, Germany 1971. *Tropical.* p. 219

concentrica 'Plutonis'; an interesting cultivar found in Ghent, Belgium; low rosette of broad, light green leaves, contrasting vividly with the blood-red leaves around the center cup, the margins finely toothed; flowers pale lavender. *Tropical.* p. 216

cruenta (Brazil: Guanabara); stout rosette with broad leaves about 30 cm long and 8 cm wide, brownish-green with blood-red blotch at spine-tipped apex, the margins spiny also; flowers blue surrounded by bluish bracts, deep in center of plant, which turns rosy at blooming time. *Tropical.* p. 214, 218

farinosa (Espirito Santo, Rio), called "Crimson cup", because at flowering time the short inner leaves and bases of others turn a vivid crimson in advance of the small purple flowers; outer leaves deep olive to purplish. *Tropical.* p. 216

x 'Mar-Con' (marmorata x concentrica), "Marbled fingernail"; beautiful bold rosette 45 cm across, the broad leaves leathery, and 10 cm wide, glossy light fresh green overlaid with red-purple design leaving a pattern of apple green blotches; tips typically wine-red; margins with scattered red-brown teeth; inflorescence very slightly raised in center, flowers lavender. *Tropical.* p. 216

sarmentosa chlorosticta (Rio de Janeiro); small rosette with bright-green leaves painted maroon in such a way that the green shows as circular blotches; silver spotted or with a touch of silver beneath; sharp tips red; pale lavender flowers. *Tropical.* p. 220

spectabilis (Brazil), called "Fingernail plant" because of the red tips of the metallic olive green leaves; gray crossbands beneath; blue flowers in low cushion. *Tropical.* p. 218, 220

tristis x marmorata; tough rosette of good lasting quality, broad leaves green largely overlaid with deep purple marbling, tips blackish purple, and gray-scaly beneath; flowers deep in center cushion, azure-blue edged white. *Tropical.* p. 220

NEPENTHES *Nepenthaceae (Carnivorous Pl.)*

These tropical "Pitcher plants" are climbers of the open country as well as the jungle trees, in temperatures around 24°C (75°F) with high humidity. The prolonged midrib of the leathery leaf acts as a clinging tendril from which develops the hollow pitcher with thickened rim, and a lid to keep out the rain. Insects are attracted inside where they drown in pepsin liquid which digests them. The greenish flowers are not conspicuous, and male and female are on separate plants. A few species are non-climbing. *Humid-tropical.*

ampullaria (Malaya to New Guinea); contrary to others, this species grows mostly on the ground, and I have collected it in wet savannahs in Malaya where the little rounded green pitchers grow in matted clusters hidden by moss or clay. *Humid-tropical.* p. 294

bicalcarata (dyak) (Borneo); epiphyte with dark green, lanceolate leaves; pitchers bag-shaped, covered with rusty down; lip with 2 recurved spurs. *Humid-tropical.* p. 295

x coccinea (distillatoria x mirabilis); a very satisfactory old hybrid (1882), with gracefully pendulous flask-shaped pitchers pale yellowish green, richly splashed with red-brown, the ring around the top lined with maroon, and with ciliate wings in back; inside bluish with red spots. We have found this plant very willing to produce its colorful pitchers more so on young branches. *Humid-tropical.* p. 295

'Courtii' (psittacina x purpurea); a durable old hybrid, advertised by Roehrs Exotic Nurseries since 1911, and growing for years at our greenhouses in New Jersey; have always found this of easiest culture, tolerating lower than the most humid-warm conditions; firm leathery leaves and pitchers mostly green with maroon marbling toward apex, and fringed wings. *Humid-tropical.* p. 297

x intermedia (rafflesiana x gracilis); leathery leaves tapering to both ends; 15 cm pitchers green with purplish-red spots, and with broad wings fringed. p. 297

lowii (Malaysia: Borneo); very curious pitchers, rather graceful with inflated base yellow-green, opening into flaring mouth, richly marked with brown lid, erect with long spur inside. *Humid-tropical.* p. 295

x mastersiana (sanguinea x distillatoria); long pitchers cylindrical flaring, claret-red with green, and purple spots, red rim. *Humid-tropical.* p. 295

maxima (Celebes, Borneo, N. Guinea); slender, high-climbing species with colorful pitchers; the lower ones flaskshaped, the upper ones funnelshaped, pale green and heavily marbled wine-red. *Humid-tropical.* p. 293

mirabilis (Papua, Queensland); sturdy species which we found growing terrestrially in heavy pebbly red clay in light forest on the highlands of Papua, in company of Cycads and Dischidias; short plants with beet-like roots, the leafy stems bearing at their tips on prolonged wiry threads the pretty yellowish green constricted pitchers covered by rusty-red lids; the small flowers brown-red and white, in erect racemes. *Humid-subtropic.* p. 294

x mixta (maxima x northiana), a "Pitcher plant"; large funnelshaped pitchers to 30 cm long, yellow-green occasionally marked crimson, increasingly toward the ribbed rim. *Humid-tropical.* p. 293

rafflesiana (Sumatra to Borneo); straggling climber on low trees of the open savannah; the lower pitchers are urnshaped, the upper large funnelform, greenish-yellow marked purplish-brown; I have collected pitchers of giant size as long as 25 cm or more, in Malaya. *Humid-tropical.* p. 295

rajah (Borneo), "Giant pitcher plant"; epiphyte at 1,500 m elevation; largest pitchers of the genus, to 30 cm long; red-brown with green fringed wings and plaited collar. *Humid-tropical.* p. 295

sanguinea 'Superba'; (Malaya); high-climbing species with flask-shaped to cylindrical pitchers deep red to reddish green, the broad rim and lid glossy blood-red. *Humid-tropical.* p. 296

stenophylla (Borneo); handsome pitcher plant photographed in the collection of Osaka University, Kosobe, Japan; pitchers flask-like, 15 to 18 cm long, and 3-4 cm wide, bronzy-green with red markings, narrow rim and small lid, from long, strap-shaped leaves. *Humid-tropical.* p. 295

'Superba'; vigorous epiphyte with long soft-leathery, deep green leaves, and variable pitchers 15 cm or more long, some urn-shaped, later funnel-shaped, yellow-green blotched with wine-red, the glossy ribbed rim with wine-red and crimson, inside spotted red, the lid striped red, the fringe in back with red hairs. This may be a hybrid. *Humid-tropical.* p. 295

villosa (Kina Balu at 3,000 m); short pitcher, very interesting with collar red and with high cross-ridges all around and to the juncture with the lid. *Humid-subtropic.* p. 295

NEPETA: see Glechoma

NEPHELIUM *Sapindaceae*
lappaceum (Malaysia), "Rambutan" or "Hairy litchi"; large evergreen tropical fruit tree to 12 m high, similar to litchi, with leaves compound of 5-7 pairs of oblong leaflets each 10 cm long, shining dark green; small pubescent flowers with cleft calyx, but without petals, in axillary panicles or from branch tips; the fruit in clusters of ten or twelve, are oval, 5 cm long, crimson red and covered with soft fleshy spines; the outer covering is thin-leathery, easily torn off, exposing the white, juicy flesh; the flavor is somewhat acid like a grape; very popular in Malaya and Vietnam. *Tropical.* p. 457, 477

NEPHELIUM: see also Litchi

NEPHROLEPIS *Polypodiaceae (Filices)*
biserrata furcans (Cuba to Brazil, Africa, Hong Kong to Queensland), "Fishtail fern"; massive fern with long arching pinnate fronds, the segments widely spaced, broad, leathery yellow-green, and forked toward their tips. *Humid-tropical.* p. 445
duffii (New Zealand, Polynesia), "Pigmy sword fern"; densely crowded, compact fern with brown downy scales at base, the erect wiry stalks sometimes forked, and closely set with tiny rounded, toothed, leathery leaflets. *Humid-subtropic.* p. 445
exaltata 'Bostoniensis', the "Boston fern", a variety found in Boston in 1894, and an old time house plant; rich green fronds simple pinnate larger and wider than the basic species, to 1 m long; the leaflets not lobed and nearly flat, more graceful and pendant; entirely without fertile spores; therefore propagation by division or runners. Still a favorite for decoration, appreciating good light, 25 fc. min., better a 100 fc., but not burning sun; not too warm, and all the atmospheric moisture possible. Air too dry favors white scale, brown scale, white fly and mealy bugs. Active growth is fairly fast from May to October. Tolerates air conditioning but dislikes cold drafts of air. *Tropical.* p. 440
exaltata 'Bostoniensis compacta', "Dwarf Boston fern"; long used as a house plant, of more compact habit, the wide simply pinnate fronds fresh green and spreading; freely clustering and usually not over 30 cm high. *Tropical.* p. 442
exaltata 'Norwoodii' (1915), a compact "Lace fern"; a sport of 'Metropoli', a cultivar similar to 'Whitmanii', elegant commercial cultivar of short habit with broad, tripinnate leaves to 45 cm long, the fresh-green, dense pinnae set in even ranks behind each other, on wiry brown axis. *Humid-tropical.* p. 442
exaltata 'Verona', "Dwarf lace-fern"; a dwarf 3-pinnate variety of Boston fern with delicate, small, very finely lacy fronds of drooping habit, the tips unfolding from pearly buds; one of the best of the lace type for house conditions. *Humid-tropical.* p. 442
exaltata 'Whitmanii', a sport of 'Barrowsii', old fashioned "Lace-fern"; of open habit, the broad, light green fronds are relatively short and arching, or pendant when older, the segments deeply and evenly cut and not bunched; tripinnate, with small segments, leaves up to 45 cm long. *Humid-tropical.* p. 440, 442

NEPHTHYTIS *Araceae*
afzelii (Liberia, Sierra Leone); African herb with creeping rhizome with sagittate, papery leaf to 35 cm long, sinus acute; green spathe not dotted. *Tropical.* p. 102
gravenreuthii (Cameroon); broadly halberd-shaped, yellow-green leaf with dark veining and wide, open sinus on slender petiole; spathe dotted, spadix white with green dots; fruit orange. *Tropical.* p. 130
'Green Gold': see Syngonium podophyllum xanthophilum.
liberica: see Syngonium podophyllum.

NEPHTHYTIS: see also Syngonium, Rhektophyllum

NEPTUNIA *Leguminosae*
oleracea (Tropics of Southeast Asia and elsewhere), "Water-mimosa"; aquatic plant with long floating or anchored roots, branching stems with dark green bipinnate leaves, lightly sensitive to touch; yellow flower heads. *Tropical.* p. 560

NERINE *Amaryllidaceae*
bowdenii (So. Africa), a "Spider lily"; bulbous plant 45 cm high, with glossy green linear basal leaves rather thick; large umbels of beautiful soft pink flowers with a darker line on each segment, recurved at apex; blooming before foliage. *Subtropic.* p. 69
curvifolia fothergillii, "Curve-leaf Guernsey lily"; bulbous So. African plant with linear, glaucous leaves, and showy umbel of numerous 6 cm flowers, soft salmon-red to crimson, with straight stamens. *Subtropic.* p. 66
krigei (So. Africa); charming small "Spider lily"; blooming in late summer usually before foliage; strap-shaped basal leaves; the solid floral stalk topped by a cluster of spidery flowers crimson-red, with dark center line down each linear petal. *Subtropic.* p. 69
masonorum (So. Africa), "Dwarf pink Guernsey lily"; an exquisite small evergreen species, with grass-like leaves, and floral stalk hardly more than 15 cm high and bearing a small cluster of pale pink flowers with crisped, thread-like segments. Somewhat hardy in sheltered locations in warm-temperate zones, but very pretty in pots in the cool winter garden. *Subtropic.* p. 69
sarniensis (So. Africa); sometimes in hort. as Lycoris radiata, the "Guernsey lily"; bulbous herb with strap-shaped basal leaves and funnel-form, crimson flowers with green, crisped segments, and long nearly straight stamens, in umbels, appearing before the foliage; numerous hybrids have been developed, in various colors from white and pink to red. *Subtropic.* p. 69

NERINE: see also Lycoris

NERIUM *Apocynaceae*
oleander (Mediterranean), "Common oleander" or "Rose bay"; evergreen shrub from 2-6 m high, often used in tubs, with willowy branches set with pairs or whorls of linear-lanceolate, leathery leaves, and flowers in terminal cymes, rosy-red. All parts are poisonous if eaten. *Subtropic.* p. 77
oleander 'Album', "Sister Agnes oleander"; large flowered cultivar with white flowers; favored by So. California nurseries because the single varieties have a way of "cleaning" themselves, or "shed" their faded blooms. *Subtropic.* p. 77, 79
oleander 'Carneum florepleno', known in the trade as 'Mrs. Roeding', somewhat weaker in growth and with a slightly weeping habit, the long branches loaded with double salmon-pink blossoms; having a tendency to "hang on". *Subtropic.* p. 78, 79
oleander 'Roseum'; "Pink oleander"; charming color form profuse with single rose-pink flowers. *Subtropic.* p. 77
oleander 'Variegatum'; form with narrow, gray-green leaves edged creamy white; flowers carmine rose. *Subtropic.* p. 77, 78
oleander 'Variegatum plenum'; ornamental form with dark green leaves attractively edged with cream; the large double flowers 4-5 cm across, carmine-pink, with deep rose center. *Subtropic.* p. 77

NERTERA *Rubiaceae*
granadensis (depressa) (Andes, to Cape Horn, N.Z., Tasmania), "Coral-bead plant"; mat-forming, creeping ground cover with tiny, broad-oval, leathery, opposite leaves and inconspicuous, greenish flowers in June, followed by the attractive, pea-size, translucent, orange-red berries. I have collected this species along cold Milford Sound, in the Fiordland of New Zealand growing on dripping rocks, frozen stiff in winter—as well as in the mountains of New Guinea at 2,100 m under rippling water, with sphagnum. *Warm temperate.* p. 862, 865

NICOLAIA *Zingiberaceae*
elatior, long known in horticulture as Phaeomeria magnifica (Amomum) (Indonesia), the magnificent "Torch ginger"; gigantic herb forming clumps of robust, leafy, arching canes to 6 m high, with alternate, pointed leaves to 60 cm long in 2 ranks; the striking inflorescence of large, torch-like heads of brilliant red, formed of innumerable waxen bracts, on separate leafless stems 2 m high or more, subtended by red basal bracts, margined white, forming a nest for the red cone, brightened by yellow-margined lips of the small red flowers. *Tropical.* p. 927

NICOTIANA *Solanaceae*
alata (affinis) (Brazil, Uruguay, Paraguay), "Jasmine tobacco"; tender herbaceous perennial with tall, sticky-hairy stalks set with large, pubescent, ovate, soft leaves, terminated by loose racemes of long, trumpet-shaped flowers pale purple or white within, yellowish outside, closing in cloudy weather, and with a sugar-sweet perfume at night. *Tropical.* p. 895
tabacum (West Indies), "Common tobacco"; large herbaceous clammy-hairy plant growing to 1½ m high, closely furnished with huge, membranous, pale green leaves, and used in the manufacture of tobacco; loose terminal clusters of rosy, funnelshaped, fragrant flowers, open during daytime. *Tropical.* p. 895, 897

NIDULARIUM *Bromeliaceae*
"amazonicum": see N. innocentii var. innocentii
billbergioides (So. Brazil); rosette of dark metallic leaves with fine toothed edge; inflorescence a raised head on stalk, the stiff bracts are dark burnt-red and flowers white. *Subtropic.* p. 221
billbergioides 'Citrinum'; small rosette of glossy pea-green leaves; the star-like inflorescence of lemon-yellow bracts nesting in its center. *Tropical.* p. 220
billbergioides 'Flavum' (citrinum) (So. Brazil); rosette of thin-leathery, lance-shaped coppery-green leaves, finely toothed; in-

florescence a small raised center cup of yellow bracts with white flowers. *Subtropic.* p. 220
   fulgens (S.E. Brazil), "Blushing cup"; showy rosette with numerous flattened shiny leaves pea green with dark mottling and conspicuous spines; inflorescence cup in center bright crimson tipped nile-green, flowers blue. *Tropical.* p. 220
   innocentii (Karatas) (So. Brazil); bold open rosette of broad oblanceolate leaves 25 cm long, metallic-green above and tinted purple, especially toward base, deep wine-red underneath, the margins finely spiny; at blooming time a dull crimson center cup of shorter leaves forms a nest for white flowers. *Tropical.* p. 220
   innocentii var. innocentii (Brazil: Espirito Santo to Santa Catarina), "Black Amazonian birdsnest"; in the trade as Nid. "amazonicum"; showy rosette metallic purple to almost black, with finely toothed margins, glossy beneath; inflorescence a short cup of rusty-red bracts with white flowers. *Tropical.* p. 220
   innocentii 'Lineatum'; striking large rosette of broad leaves lined lengthwise with green and white bands and stripes, the inner cup of leaves glowing crimson as the blooming season approaches. *Tropical.* p. 218, 220
   innocentii 'Maureanum', a Belgian cultivar, possibly 'Morrenianum' (innocentii x ferdinandi-coburgii); bold rosette of broad olive-green leaves suffused with purple, glossy purple beneath; center cup burnt-rose toward tips, and white flowers. *Humid-tropical.* p. 218
   innocentii var. nana, "Miniature birdsnest"; small rosette of broad, thin-leathery leaves, matte olive-green, finely toothed, underside glossy purple; at flowering time, an inner nest of short leaves turns orange-red, flowers white. *Tropical.* p. 220
   innocentii striatum (Brazil), stocky rosette of broad light green recurved leaves striped lengthwise with yellow-ivory; finely toothed margins; white flowers deep in carmine-tipped center cup. Looking like Dracaena 'Massangeana' but sturdier. *Tropical.* p. 220
   innocentii viridis; flattened rosette with broad pea green to fresh green, somewhat mottled leaves and with marginal teeth; the shorter leaves surrounding the center are tipped carmine-red at flowering time. *Tropical.* p. 220
   regelioides (S.E. Brazil); compact rosette of rich-green leathery leaves mottled dark green; inflorescence a cup of rusty-red bracts and red flowers. *Subtropic.* p. 218
   'Souvenir de Mme. Morobé'; large hybrid rosette of good form, with numerous broad leaves, finely toothed at margins, glossy light green tinted bronze and with dark spots, the inner nest of floral leaves brilliant crimson. *Tropical.* p. 221

NIGELLA   *Ranunculaceae*
   damascena (So. Europe), "Love-in-a-mist" or "Devil-in-the-bush"; annual ornamental herb to 45 cm, with leaves pinnately cut into thread-like segments; showy flowers 3 cm dia., white or light blue, with prominent green pistils united at base. *Subtropic.* p. 844

NOLINA   *Liliaceae*
   bigelovii (Dasylirion) (California, Arizona, Sonora), "Bear grass"; leaves in large number arranged in a symmetrical way, their margins shredding away in brown fibers; with striking tall plumy panicles of numberless whitish-green minute flowers standing far above the leaves 2-3 m high. *Arid-subtropic.* p. 588

NOLINA: see also Beaucarnea

NOMOCHARIS   *Liliaceae*
   mairei (Southwest China); bulbous herb to 1 m; whorled lanceolate leaves; flowers flat, 10 cm across, white with purple spots, margins frilled. *Subtropic.* p. 602

NOPALEA   *Cactaceae*
   cochenillifera (Opuntia) (Puerto Rico, So. America), "Cochineal plant"; tree-like, to 5 m, with long fleshy, flattened joints usually spineless, glossy dark green; rosy flowers. Host to mealybug furnishing cochineal dye. *Arid-tropical.* p. 240

NOPALXOCHIA   *Cactaceae*
   ackermannii (Epiphyllum) (Mexico: Chiapas, Oaxaca 2000-2700 m); a species "Orchid cactus"; in habitat mostly epiphytic, with flattened green branches, sometimes 3-angled, the angles notched (crenate); large and showy funnelform flowers glowing red 12-16 cm across; good flowering plant. *Subtropic.* p. 282, 284
   phyllanthoides (Mexico); an old, free-flowering house plant widely grown under the name "Deutsche Kaiserin" (German Empress); an epiphyte from Puebla state at 1700 m; densely bushy with flattened, pendant, crenate branches bearing a profusion of day-flowering, carmine-rose flowers of medium size 5 cm across, and with pale tips and short tube; a lovely basket plant. Responds well to feeding. *Subtropic.* p. 281, 282, 283

NORANTEA   *Marcgraviaceae*
   braziliensis (Brazil); handsome tropical evergreen scandent shrub with leathery, obovate leaves; the singular reddish inflorescence at first with colored bracts covering the small flowers, and forming vessels holding nectar; later developing small berry-like blackish fruit. *Tropical.* p. 646
   guianensis (Guayana), "Red popcorn"; beautiful tropical clambering shrub, climbing by roots from the red branches, with oval, leathery foliage 12 cm long, and a striking inflorescence of long spikes of large, hollow or hooded, fleshy rosy-scarlet bracts, the small sessile flowers violet. The nectar-bearing bladders, open at top, attract small birds for cross-pollination. *Tropical.* p. 645

NORONHIA   *Oleaceae*
   emarginata (Malagasy Rep.), "Madagascar olive"; tropical evergreen tree, with paired, leathery 15 cm leaves dark green with cream midvein; clusters of fragrant yellow flowers from leaf axils, with thick, 4-parted corolla; purplish fruit 3 cm dia., with sweet-tasting edible pulp, enclosing a large seed. *Tropical.* p. 698

NOTHOFAGUS   *Fagaceae*
   menziesii (New Zealand), "Silver beech"; large evergreen tree with silvery bark, to 30 m high; small leathery leaves dark green and shining, 2 cm long, ovate or rounded, the margins crenate; produces downy nuts, 2-3 winged. *Subtropic.* p. 430

NOTHOPANAX   *Araliaceae*
   filicifolia (Aralia, Polyscias), (South Sea Islands), "Fernleaf aralia" or "Angelica"; evergreen shrub with flexuous stems and leathery but variable leaves, bright green and with purplish midrib, pinnate with leaflets cut into narrow lobes; fern-like in younger plants, broader and entire when older. *Tropical.* p. 140
   laetus (New Zealand); willowy shrub to 3 m high, with slender, smooth branches and fine leathery leaves palmately compound usually of 5 broad elliptic leaflets; very attractive with purplish-red petioles and midribs; robust, tough and vigorous. *Subtropic.* p. 140

NOTHOPANAX: see also Polyscias, Neopanax

NOTOCACTUS   *Cactaceae*
   apricus (Uruguay), "Ball cactus"; tiny, clustering globe to 5 cm thick, dark green densely covered with reddish, interlocking bristly spines; flowers yellow. *Arid-subtropic.* p. 270
   concinnus (So. Brazil, Uruguay); glossy green, depressed globe 6 cm high, with about 18 ribs notched into warts, recessed white woolly areoles and fine yellow spines; flowers red outside, satiny canary inside. *Arid-subtropic.* p. 270
   haselbergii (So. Brazil), "White-web ball"; lovely small globe to 8 cm, occasionally sprouting from the base; with about 30 low ribs, covered with soft glossy, silvery-white spines pale yellow at the top; flowers orange red to crimson. *Arid-tropical.* p. 273
   leninghausii (So. Brazil), "Golden ball"; attractive, small clustering, cylindrical column, to 1 m high and 10 cm thick, close-ribbed, covered with soft golden hair; flowers yellow at top. Beautiful, and of easy growth. *Arid-subtropic.* p. 273
   magnificus hort.; handsome deep green globe with 12 high ribs covered along their crest with woolly white areoles and brown needle spines. *Arid-subtropic.* p. 271
   mammulosus (Brazil to Argentina), "Lemon ball"; simple plant nearly globose, to 8 cm thick, shining green, with 18-25 high ribs, and yellowish to reddish spines from recesses on the knobs; fragrant yellow flowers to 4 cm long, appearing at an early stage. *Arid-subtropic.* p. 270, 271, 273
   ottonis (Brazil, Argentina), "Ball cactus"; small globe to 5 cm, glossy green, 10 broad ribs, short needle-like spines rising from notches; free flowering bright yellow. *Arid-subtropic.* p. 270
   rutilans (Argentina), "Pink ball cactus"; dark green globe with 25 rows of long knobs, the recessed areoles with straw-white radials and straight brownish central spines; flowers bright pink with yellow throat, shimmering like silk. *Arid-subtropic.* p. 272, 273
   scopa (Brazil, Paraguay), "Silver ball"; globular to cylindrical, 45 cm high, closely ribbed and nearly covered with short, soft-hairy white radials and long brown needle-spines; flowers silky canary-yellow deeper in center. *Arid-subtropic.* p. 271, 273
   scopa 'Cristata', "Spiralled silver ball"; a strangely beautiful fasciated form taking on the shape of a coiled snake, entirely covered with glistening, pure-white hair. *Arid-subtropic.* p. 273

NYCTOCEREUS   *Cactaceae*
   serpentinus (Mexico), a "Queen of the night" or "Snake cactus"; slender erect or clambering night-bloomer; cylindric many-ribbed stems, to 5 cm thick, deep green; woolly areoles and white to brownish spines; large white, sweet-scented funnel-form nocturnal flowers to 25 cm long; edible red fruit. *Tropical.* p. 251, 253

NYMPHAEA   *Nymphaeaceae*
   alba (Eurasia, No. Africa), "European white water-lily"; robust hardy water lily, with leaves red when young, crowded on rhizome; white flowers 8-12 cm across with yellow stigmas, open nearly all day, and floating on the water. *Warm temperate.* p. 706

caerulea (Egypt to C. Africa), "Blue lotus of Egypt"; tender, free blooming water lily, with large leathery, floating leaves glossy dark green, and light blue, faintly scented flowers with numerous narrow petals, borne well above the water. *Humid-tropical.* p. 704

capensis (So. Africa), "Cape blue waterlily"; subtropical day-blooming water-lily of robust, luxuriant habit, with sinuate leaves; stiff stems carry large flowers 15-20 cm across, sky blue with pale center, fading to nearly white. *Subtropic.* p. 706

capensis zanzibariensis (E. Africa), "Blue waterlily"; tender day-blooming, leaves 40 cm dia., with sinuate margins; light to deep blue flowers to 30 cm across. *Subtropic.* p. 706

colorata (Trop. Africa: Dar-es-Salaam); lovely pygmy tropical day-bloomer; from erect rhizome the vigorous, abundant foliage dark green, 25 cm dia.; light-blue broad-petalled 10 cm flowers with darker center; develops clusters of tiny tubers; good for tub culture. *Tropical.* p. 706

x daubeniana, a tropical viviparous "Pygmy water lily"; for confined spaces, very free blooming with small, fragrant, light blue flowers carried well above the water; young plants develop at the junction of petiole and leaf; possibly a hybrid of micrantha and caerulea. *Humid-tropical.* p. 706, 707

gigantea (New Guinea, Queensland), "Giant water-lily" or "Australian waterlily"; big leathery, glossy green leaves 50 cm dia.; flowers light blue with broad petals, tipped dark blue and with yellow center, 15-30 cm across; day-blooming for 7 days, and remaining open from the fourth day; very fragrant. *Tropical.* p. 704

lotus dentata (Egypt, Sierra Leone), "White lotus of Egypt"; robust, tender night-blooming species with smooth leaves 30-50 cm across; fragrant white flowers 12-25 cm dia., remaining open till noon. *Humid-tropical.* p. 705

x marliacea 'Chromatella' (mexicana x alba); free-blooming, hardy water lily with floating leaves, much blotched with brown, rising above the water when crowded; flowers bright yellow with concave petals and yellow stamens. *Humid-subtropic.* p. 706

micrantha (West Africa); a strange tropical daybloomer, with relatively small foliage; flowers the first season almost perfectly white, with bluish tinge the second, and definitely blue-white the third, 15 cm across; and with yellow anthers on blue filaments. Produces young plants viviparously. *Tropical.* p. 706

'Mme. Julien Chifflot'; extremely huge pink hardy water-lily with blossoms a rich shade of pink, floating on or near the surface of the water. *Warm temperate.* p. 704

'Pink Sensation'; perennial living through the winter even under ice, with round, smooth-edged leaves; fragrant flowers rich pink, opening to the sun, and remaining open for 2 or 3 hours after other blooms have closed, from May until frost. *Warm temperate.* p. 704

rubra (India), "India red waterlily"; beautiful tropical night-blooming, remaining open until nearly noon; flowers deep carmine-red 15-25 cm across; floating leaves red bronze and crisped, 30-50 cm dia. *Humid-tropical.* p. 704, 707

rubra 'Rosea'; flowers larger, rosy-carmine shading into pale pink inside, with longer, more pointed buds, leaves spotted brown and toothed at margins. *Humid-tropical.* p. 706

'Tashkent'; elegant day-blooming tropical water lily with bronzy leaves, and large rose-pink flowers with narrow petals, carried well above the water. *Humid-tropical.* p. 706

NYMPHOIDES Gentianaceae
indica (Pantropic), "Water snowflake"; perennial herb with kidney-shaped floating leaves to 20 cm across; fimbriate flowers white with yellow center. *Humid-tropical.* p. 483, 484

NYPA Palmae
fruticans (Philippines, Malaya, India, to Australia), "Nypa" or "Nipa palm"; low, shrubby palm usually growing more or less submerged in brackish water; trunk-like rootstock forming colonies with age, the pinnate fronds erect-recurving and rigid, 3-9 m tall, the leaflets 1-1½ m long, folded, and shiny bright green, grayish beneath; large compound fruit near base; the sap produces syrup and alcohol. *Humid-tropical.* p. 787, 790, 795

OBREGONIA Cactaceae
denegrii (Mexico), "Artichoke cactus"; low, interesting cactus with thick tap root, occasionally sprouting into groups, globular, flat on top, grayish to dark green, 8-12 cm across, with leaf-like tubercles spirally arranged, stiff and angled, without spines except at tips; white 4 cm flowers. *Arid-tropical.* p. 271

OCHAGAVIA Bromeliaceae
lindleyana (Chile); small succulent rosette, in habitat growing terrestrial or on rocks on dry mountain slopes; the linear recurving, spiny leaves to 50 cm long and 2½ cm wide, shiny green above and with grayish-mealy lines underneath; the hemispherical inflorescence nesting in the center with white-woolly bracts and rose-pink flowers 5 cm long, in dense cluster. This species may be the same as O. carnea. *Arid-subtropic.* p. 218

OCHNA Ochnaceae
kirkii (Trop. S.E. Africa), "Mickey-mouse plant"; evergreen smooth shrub with leathery oblong, finely toothed 5-8 cm leaves on woody branches; flowers bright yellow, the 5 petals soon falling, the glossy calyx lobes turning a glowing red and later the red to shining black fruit, this peculiar inflorescence looking like a fairytale Mickey Mouse. *Tropical.* p. 692, 698

madagascariensis (Malagasy Rep.); woody evergreen bush, with thick-leathery ovate leaves; at branch ends a cluster of yellow flowers subtended by glossy, waxy red calyx lobes; large fruit yellow-green to black. *Tropical.* p. 698

mossambicensis (Mozambique), "Birdseye bush"; bushy evergreen shrub to 3 m high, with obovate to oblanceolate stiff leaves to 22 cm long, lightly toothed; flowers in clusters on lateral branchlets; petals yellow and spreading red calyx, later forming black fruit. *Tropical.* p. 698

serrulata ('multiflora') (Natal), "Birdseye bush"; woody shrub to 1½m high with hard leathery, narrow-elliptic leaves to 12 cm long, serrate glossy green at the margins; the flower corolla yellow but quickly falling, the sepals at first greenish then turning bright red and persistent; interesting black berry-like fruit seated on a red receptacle. *Subtropic.* p. 694, 698

OCHROSIA Apocynaceae
elliptica (Kopsia arborea) (New Caledonia); evergreen tree to 6 m, with milky juice, smooth, green, obovate feathery leaves to 15 cm long; fragrant ivory-white, small flowers followed by scarlet red angled fruit 5 cm long; salt-resistant plant. *Tropical.* p. 78

marianensis (Pacific Islands); small tree with milky sap; whorled leathery leaves, long oblanceolate and glossy green; small white flowers; red flattened fruit with edible seeds. *Tropical.* p. 466

x ODONTIODA Orchidaceae
'Astargia' (Argia x 'Astoria', 1944); compact bigeneric with arching spray of roundish deep crimson-red long-lasting flowers. *Humid-subtropic.* p. 741

charlesworthii (Cochlioda roezliana x Odontoglossum harryanum, 1908); long-lasting waxy flowers, typical of this bigeneric hybrid group; long arching raceme of smallish flowers bronzy-red with yellow tips. *Humid-subtropic.* p. 740

x ODONTOCIDIUM Orchidaceae
'Surprise' (Odontoglossum pulchellum x Oncidium macrantha); exquisite small bigeneric, with waxy flowers white, shaded with a blush of pink. *Humid-tropical.* p. 741

ODONTOGLOSSUM Orchidaceae
'Alispum' (Alorcus x crispum); typical of many of the hybrids originated so successfully in England 50 years ago, featuring beautiful, large, rounded, 8-10 cm flowers with white base, blotched, barred or spotted brown, mauve, purple or crimson, and with crisped margins, (bl. various). *Humid-subtropic.* p. 741

bictoniense (Guatemala); attractive epiphyte with pseudobulbs surrounded by leaf-bearing sheaths, inflorescence in erect raceme to 75 cm tall, with 4 cm flowers, sepals and petals yellow-green blotched with brown, lip heart-shaped, shaded violet with yellow keels, (Aug.-Jan.). *Humid-subtropic.* p. 741

cervantesii roseum (So. Mexico, Guatemala); pretty, dwarf epiphyte with clustered pseudobulbs bearing a solitary petioled leaf; clusters of 4-6 cm roundish flowers pale rose with interrupted concentric circles of crimson (Oct.-May). *Humid-subtropic.* p. 741

citrosmum (pendulum) (Mexico); epiphyte with compressed pseudobulbs bearing 2 strap-shaped leaves, pendulous racemes of large 5 cm, fragrant, white flowers sometimes flesh-tinted, lip rose with yellow, red-spotted crest (May). *Humid-subtropic.* p. 743

crispum (Colombia); very handsome epiphyte with stout, 2-leaved pseudobulbs, and arching racemes of daintily crisped, waxy star-shaped 9 cm flowers of pure white, the lip yellow at base and blotched reddish toward front; variable, (Feb.-April). *Humid-subtropic.* p. 741

'Elise' (triumphans x 'Ascaria'); an exquisite, colorful flower yellow flushed orange toward the margins and irregularly splotched red-brown; the blooms are firm and on a close spray; flattened pseudobulbs with 2-3 leathery leaves. *Humid-subtropic.* p. 741

grande (Guatemala, Mexico), "Tiger orchid"; very beautiful epiphyte of compact habit and easy grower, with thick, 2-leaved pseudobulbs, and erect, 30 cm stalks with 12-15 cm flowers, sepals yellow barred with brown, petals half reddish-brown, tips yellow, lip cream spotted with brown, (Aug.-March). *Humid-tropical.* p. 745

insleayii (Mexico); noble epiphyte with 2-leaved compressed pseudobulbs, and an erect stalk with 5-10 cm flowers with oblong sepals and petals greenish-yellow transversely banded with chestnut red, the spoon-shaped lip bright yellow with a border of crimson spots (Aug.-Sept., Dec.-Jan.). *Humid-subtropic.* p. 741

krameri (Costa Rica); dwarf epiphyte with broad, pale green pseudobulbs topped by a solitary leaf; short, erect stalks with 2-5 fragrant, 3-5 cm, waxy flowers, freely produced, sepals and petals rose shading to white, lip purplish banded with white and red at base, (Aug.-Sept.). *Humid-tropical.* p. 743

laeve reichenheimii (Mexico); attractive variety; epiphyte with compressed 2-leaved pseudobulbs and slightly branched racemes of 5 cm fragrant flowers, with narrow sepals and petals yellowish green barred with purplish-brown, lip light purple without claw; willing bloomer (spring). *Humid-subtropic.* p. 742

'Pedrito' ('Aicard' x 'Mrs. Sanders'); charmingly colorful 1949 French hybrid; deep yellow flowers lavishly painted with crimson, and with frilled petals and lip. *Humid-subtropic.* p. 743

pulchellum (Guatemala), "Lily-of-the-valley orchid"; dainty epiphyte with dark green pseudobulbs topped by 2-3 grass-like leaves and clustering; erect racemes with small waxy, sweetly fragrant 2-3 cm flowers of crystalline white with yellow crest, (Dec.-May). *Humid-subtropic.* p. 743

rossii (Mexico, Guatemala); pretty epiphyte of dwarf habit, with 1-leaved pseudobulbs, delicate though wiry inflorescence with star-like, 5-8 cm flowers, linear sepals white, barred with brown, crisped petals, and lip white with yellow crest, (Feb.-April). *Humid-subtropic.* p. 742

ODONTONEMA *Acanthaceae*
strictum (C. America); handsome shrub 1 m high, with stiff erect stems and slender-pointed leaves shiny rich green; terminal inflorescence of waxy crimson-red flowers 1 cm long, and very attractive. *Tropical.* p. 36

ODONTOSPERMUM: see Asteriscus

OENOTHERA *Onagraceae*
fruticosa (Eastern U.S.), "Sundrops"; day-blooming perennial 30-50 cm high, with reddish stems, lance-shaped leaves, and yellow flowers nearly 5 cm wide. *Warm temperate.* p. 702

missouriensis (Central U.S.), "Evening primrose"; perennial evening-flowering garden plant to 30 cm high, with woody base; hairy lanceolate leaves to 10 cm long; showy bell-shaped yellow flowers nearly 10 cm wide. *Temperate.* p. 701

OLDENBURGIA *Compositae*
arbuscula (So. Africa: Cape Mountains); beautiful evergreen shrub to 2½ m, with spreading branches densely clothed with striking, big leathery, convex, oblanceolate leaves to 45 cm long, thickly covered with snow-white felt, with age shining dark green; fluffy clusters of purple and white flowers. *Subtropic.* p. 318, 322, 327

OLEA *Oleaceae*
europaea (E. Mediterranean reg.), "Olive tree"; small, sparry, evergreen tree with stiff-leathery, narrow lanceolate leaves gray-green above, silvery scurfy beneath; flowers yellowish-white and fragrant; oblong 3 to 4 cm fruit green turning shining black when ripe, used for its valuable oil. *Subtropic.* p. 458, 469, 695, 696

OLEA: see also Osmanthus

OLEANDER: see Nerium

OLIVERANTHUS *Crassulaceae*
elegans (Echeveria harmsii) (Mexico), "Red echeveria"; soft hairy succulent shrub, with fleshy leaves near ends of branches, obovate pointed 3 cm long, grass-green edged with red, in loose rosettes; beautiful flowers like lanterns, 1 to 3 to a spray, with large red ovary turning back to show its yellow lining. *Subtropic.* p. 374

OLYRA *Gramineae*
latifolia (Trop. Africa: Kenya, Uganda, Zaire, also E. Indies, Trop. America); a bamboo-like grass with underground running stolons sending up thin-wiry, cane-like yellowish stems from ½-2 m tall, set with alternate lanceolate thin-leathery, deep green leaves. When I found this species in the Ituri forest, in Pygmy country near the Mountains of the Moon in Central Africa, I had considered this to be the true Bamburanta which it closely resembles. *Tropical.* p. 516

OMPHALEA *Euphorbiaceae*
triandra (West Indies), "Pig-nut"; spreading tree to 4 m high, with obovate, leathery leaves; the flowers without petals, but 3 anthers; pendant, furrowed yellow fruit 4 cm dia., fleshy outside, 4-celled containing nuts, these edible after extracting the poisonous embryo. Juice extracted from this species turns black and is used for ink and glue. *Tropical.* p. 428

ONCIDIUM *Orchidaceae*
"alatum" in hort.; robust plant with large, flat pseudobulbs bearing pair of strap-like leaves 45 cm long; long arching spray 1½-2 m long, with 4 cm waxy flowers, brown petals and sepals, wavy-edged and tipped yellow; bright yellow lip. (Roehrs collection). *Tropical.* p. 743

ampliatum majus (Costa Rica, Guatemala); stiff epiphyte with 1-2 leaved pseudobulbs mottled brown, and a stout raceme to 1 m long, of 5 cm bright yellow flowers blotched with red-brown, and marked with white beneath, (Dec.-June). *Tropical.* p. 736

cheirophorum (Colombia), "Colombia buttercup"; compact, pretty epiphyte with small, compressed, 1-leaved pseudobulbs, and slender, arching panicles of numerous, small, bright yellow, sweetscented flowers. (Oct.-Dec.) *Subtropic.* p. 743

concolor (Brazil); lovely little epiphyte with 2-3 leaved flattened pseudobulb and cluster of large flowers 4-5 cm across, lemon-yellow and large yellow lip (Oct.-May). *Subtropic.* p. 743

crispum (Brazil); splendid epiphyte from the Organ mountains, with 2-3-leaved, compressed pseudobulbs; the panicled inflorescence ½-1 m high, with large, crisped flowers to 8 cm, chestnut-brown with center of golden yellow, (May-Nov.). *Subtropic.* p. 743

excavatum (Perú, Ecuador); attractive epiphyte with stout, compressed, 1-leaved pseudobulbs, and many-flowered inflorescence to 1 m long, the 4 cm flowers yellow, barred with red except the broad lip, (Oct.-March). *Tropical.* p. 742

flexuosum (Brazil, Paraguay), "Dancing doll orchid"; beautiful little epiphyte with 1-2-leaved pseudobulbs, and dainty, thin-wiry sprays of small "dancing doll"-like golden yellow flowers with center marked deep red, (Oct.-Aug.). *Tropical.* p. 740, 742

forbesii (Brazil); handsome epiphyte with compressed pseudobulbs and solitary leaf; many-flowered panicles of waxy, showy 5-6 cm blooms with crisped petals rich chestnut brown and broken golden borders (Mar.-April, Oct.-Nov.). *Subtropic.* p. 743

globuliferum (Costa Rica to Venezuela and Perú); climbing epiphyte with creeping, wiry, scandent rhizomes, at intervals forming small bulblet-like, compressed, leafy pseudobulbs with aerial roots; short-stalked flowers 3 cm long, bright yellow barred with red-brown, the large lip yellow. (Summer). *Tropical.* p. 740

grandiflorum (W. Ecuador); grand epiphyte which I collected in the jungle, down the Pacific slope along the road to Esmeraldas; big obovate leathery leaves; the inflorescence on long wiry stalk 2 to 3 m long, the beautiful large waxy flowers 6-8 cm across, brown over yellow with frilled margins, and white lip marked purple. *Humid-tropical.* p. 745

kramerianum (Ecuador, Colombia); beautiful epiphyte with small round pseudobulbs bearing a solitary leaf; flower stalks to 75 cm long and round, with several very curious, highly colored flowers in succession, the long narrow dorsal sepal and petals chocolate-brown, lateral sepals broad, orange-red mottled with yellow, lip lemon-yellow bordered red-brown, (March-May, Nov.-Dec.). *Tropical.* p. 743

lanceanum (Trinidad, Guayana), "Leopard orchid"; strikingly beautiful epiphyte with minute pseudobulbs, broad and thick, brown-spotted, solitary leaves and erect spikes with large, vanilla-scented flowers, the fleshy sepals and petals yellow shaded green, blotched with chocolate-brown, the large lip violet at base, rose in front, (May-Aug.). *Tropical.* p. 743

marshallianum (Brazil); two-leaved pseudobulbs; clustered inflorescence with numerous showy blooms; flowers 8 cm across, sepals greenish-yellow with pale reddish brown bars, petals undulate, canary-yellow with crimson base; lip 3-lobed with red dots. *Humid-tropical.* p. 742

ornithorhynchum (Mexico to Salvador); small epiphyte with twin leaves on small pseudobulbs, prolific with dainty, arching sprays of small, very fragrant flowers of soft rosy-lilac, with darker shading on lip and a yellow crest, (Aug.-Dec.). *Tropical.* p. 742

papilio (Trinidad, Venezuela, Brazil, Perú), "Butterfly orchid"; epiphyte with small pseudobulbs bearing a single leaf mottled purplish-brown; the large, unusual flowers developing successively on flat stalks to 1 m long, dorsal sepal and petals long linear, reddish-brown marked with yellow, lateral sepals oblong, brown barred with yellow, lip yellow with brown border. *Tropical.* p. 743

pardinum (Ecuador); similar to a large O. crispum but with many-flowered inflorescence to 1 m long, undulate flowers of medium size, bright yellow marked brown; compressed pseudobulbs with 2-3 leaves (Dec.-Jan.). *Tropical.* p. 742

reflexum (Mexico, Guatemala); large clustering epiphyte with compressed pseudobulbs, 1 or 2 leaves to 20 cm; long straggling inflorescence to 1 m long; 4 cm flowers with sepals and petals yellow green barred with red-brown, lip vivid yellow with red spots. *Humid-tropical.* p. 744

sarcodes (Brazil); vigorous epiphyte with dark green, tapering pseudobulbs with 2-3 shining green leaves; inflorescence to 2 m long, with short branches of large 5 cm flowers, sepals and petals chestnut-brown bordered yellow, lip bright yellow with red-brown spots, (April-July). *Tropical.* p. 742

sphacelatum (Mexico to Honduras), "Golden shower"; prolific epiphyte with elongate, flattened pseudobulbs with 2-3 leaves; the branched, loose inflorescence to 1½ m long, with many small, yellow flowers, marked with brown, 2 to 3 cm across. (Feb.-Sept.). *Tropical.* p. 736, 742

splendidum (Guatemala); handsome stout epiphyte with short pseudobulbs bearing a single stiff-fleshy, mahogany leaf, and erect, 1 m spike with large substantial flowers, 5 cm across, small sepals and petals yellow barred with brown, the large lip golden yellow, (Dec.-Feb.). *Tropical.* p. 742

uniflorum (Brazil); small epiphyte of bushy habit; the short inflorescence with 1-2 waxy flowers, sepals and petals greenish-yellow shaded brown, lip bright yellow (April-Dec.). *Tropical.* p. 740

varicosum (Brazil); very showy epiphyte with 2-3-leaved pseudobulbs, small by comparison with the long, branching sprays of numerous, large, 5 cm, delicate, bright yellow flowers, sepals and petals small, barred red-brown, and a large golden lip, (Sept.-July). *Tropical.* p. 742

wentworthianum (Guatemala); robust epiphyte with 1-2-leaved pseudobulbs; heavy arching flower-stalks to 2 m or more long, with lateral clusters of small, long-lasting, lemon-yellow flowers boldly blotched with red-brown, (April-July). *Tropical.* p. 744

ONCOBA *Flacourtiaceae*

kraussiana (So. Africa); evergreen shrub to 5 m, without spines; elliptic leaves 5 cm long; white 5 cm flowers; orange berry with oil-rich seed. *Subtropic.* p. 484

ONCOSPERMA *Palmae*

filamentosum (bot. tigillarum); O. filamentosum, so listed in Botanic Garden Singapore, is referred by Hortus III to tigillarum, which see. p. 792

horridum (Malaya to Philippines), the inland "Baya palm"; forming clusters, with very straight, slender trunks, armed with massive downward black spines; and crowned by the feathery fronds on spiny petioles, the leaflets held stiffly horizontal, drooping only at the tips. *Tropical.* p. 792

tigillarum (Malaya to Sumatra, Java and Philippines), "Nibong palm"; very ornamental forming extensive, many-stemmed clumps, the slender, spiny trunks to 20 m or more, topped by crowns of graceful feathery fronds 6 m long, on spiny petioles, the leaflets strongly pendant; inflorescence bright yellow; dark purple fruit; grows primarily along the coast. *Tropical.* p. 792

ONOCLEA *Polypodiaceae (Filices)*

sensibilis (Newfoundland to Louisiana, Siberia, Japan), called "Sensitive fern" because the herbaceous barren fronds are sensitive to cold or if cut, and fold their leaflets face to face; handsome sterile fronds of glaucous pale green pinnae with undulate, lobed margins; underground creeping rhizome; hardy. *Temperate.* p. 446

ONOPORDUM *Compositae*

nervosum (arabicum) (Spain), "Ornamental thistle"; coarse, thistle-like herb, 2-3 m high, with lobed, white-hairy leaves and spiny; flowers rose-purple within bracted heads, 5 cm across. *Subtropic.* p. 321

OPHIOGLOSSUM *Ophioglossaceae (Filices)*

"bulgatum" (E. Australia), "Ribbon fern"; interesting epiphyte with long narrow grass-like bluish dark green pendulous fronds 60-90 cm long, hanging like ribbons from baskets or trees. *Humid-subtropic.* p. 437

pendulum (Polynesia, Queensland, Philippines, Ceylon, Madagascar), a curious epiphytic "Ribbon-fern"; with fleshy roots and ribbon-like pendulous, fleshy grass-green sterile fronds from 1 to 4 m long, sometimes forking; the small, slender fertile frond, to 15 cm long, rises from the middle of the barren blade. *Humid-tropical.* p. 438

OPHIOPOGON *Liliaceae*

jaburan 'Variegatus', "Variegated mondo grass"; attractive form which I have collected in Java, with long linear, symmetrically arranged leaves friendly milky-green, striped and edged in white; the drooping little, waxy-white flower bells on erect raceme. The species O. jaburan, or "White lily-turf" is from Japan, an evergreen clump-forming perennial with cord-like roots, and dark green, thick-leathery recurving leaves 30-60 cm long and 1 cm wide. *Subtropic.* p. 607

japonicus 'Kyoto Dwarf', (Japan); "Snake's-beard"; lawn-forming stemless perennial with long underground stolons with tufted, recurving leaves only 4 cm high, dark-green; small pale lilac flowers. Seen planted at old Imperial Palace in Kyoto, dating from 780 A.D. *Warm temperate.* p. 606

planiscapus nigrescens, in hort. as "arabicus" (So. Japan), the "Black dragon"; small grass-like clustering plant 10-15 cm or more high, with narrow-linear, leathery, curving leaves 4-6 mm wide, arranged in opposite ranks, at first bright green and glossy, later almost black; lavender flowers followed by black berries. *Subtropic.* p. 606

OPHIOPOGON: see also Liriope

OPHRYS *Orchidaceae*

fuciflora (Spain to Albania, No. Africa), "Late spider orchid"; terrestrial with leafy stems, to 40 cm high; stout, broad leaves; the stalk with several flowers, the sepals and petals velvety white or pink; large lip dark brown with white center, and a shining eye-like knob each side of column. *Subtropic.* p. 744

fusca (South Europe), "Bee orchid"; handsome terrestrial orchid to 30 cm tall, from rhizomatous tuber; with leaves at base and along the spike of flowers 4 cm long, yellow-green and hairy, the lip covered with velvety brown hairs and two small, yellow-edged, mirror-like areas, in all resembling bees or flies. *Subtropic.* p. 744

lutea, (Mediterranean Reg.), "Insect-orchid"; terrestrial rosette of dagger-shaped fleshy foliage, miniature cupped, waxy flowers 15 mm dia., yellow with black eye *Subtropic.* p. 744

lutea minor, (Greece: Delos), "Mirror orchid"; variety with smaller flowers and lip; found growing amongst the ruins of the ancient temple of Apollo on Delos. *Subtropic.* p. 744

speculum (So. Europe, to Greece), "Mirror of Venus"; miniature terrestrial 10-25 cm high, with tuberous rhizome producing a leafy stalk with remarkable 3 cm flowers, sepals pale green outside and light brown within, the shorter petals violet-brown, brown fringed lip with disk a steel-blue glassy mirror edged in gold and margined maroon (spring). *Subtropic.* p. 744

OPLISMENUS *Gramineae*

hirtellus variegatus (Panicum variegatum) (W. Indies), "Basket grass"; weak, creeping grass, rooting at nodes, with flowering culms generally erect, the rather broad, lanceolate, thin leaves daintily striped white and pink. *Tropical.* p. 510

OPUNTIA *Cactaceae: Opuntieae*

arbuscula (Arizona, Sonora); dense succulent, much branched shrub becoming 1 m high; the pencil-size olive-green joints 1 cm thick and 10 cm long, with prominent warts and long needle-spines; small flowers bronze to yellowish; club-shaped fruit. *Arid subtropic.* p. 244

basilaris (S.W. U.S., Mexico), "Beaver tail"; growing in clumps, broadly obovate fleshy pads a bluish coppery color, almost spineless; large purple flowers 5 to 8 cm across, variable to rose or yellow. *Arid-subtropic.* p. 240, 242, 244

bigelovii, a typical "Cholla cactus"; characteristic of the stony deserts from New Mexico to Southern California, erect trunks to 1 m high with short light green branches to 5 cm thick, densely set with glistening cream spines; purple flowers 4 cm across, and yellow fruit. *Arid-subtropic.* p. 242

brasiliensis (Brazil, Argentina, Bolivia), "Tropical tree-opuntia"; tree-like, to 4 m high; trunk and branches cylindrical, the terminal joints flat and leaf-like, glossy fresh green, with few spines; pale yellow flowers. A good and attractive house plant because of its tropical origin, resembling a miniature tree even as a young plant. *Subtropic.* p. 238

clavarioides (Austrocylindropuntia) (Chile), "Sea coral"; a very interesting and curious plant known because of its shape and color as "Black fingers", "Nigger hand", or "Fairy castles"; low straggling plant with cylindrical grayish-brown joints but usually growing fan-shaped or in other fasciated forms, covered with short white, hair-like spines; rarely flowering, pale greenish-brown. *Arid-subtropic.* p. 242, 243, 244

compressa (humifusa) (Ontario, Mass. to Alabama, Missouri); low and spreading succulent with grass-green joints, almost smooth, oblong to 12 cm long, with few spines; yellow flowers 5-8 cm across. *Temperate.* p. 240

decumbens (Mexico, Guatemala); low-growing tuna-cactus with spreading, often creeping branches; flat joints oval-oblong 12-15 cm long, dark green, minutely pubescent, reddish around the areoles; yellowish glochids but few spines; flowers yellow to reddish, 5 cm across; 3 cm red-purple fruit. *Arid-tropical.* p. 244

dillenii (S.E. U.S., W. Indies, Spanish Main), "Tuna"; either low spreading, or tall branched, the long fleshy, flat joints bright green when young, glaucous bluish later, and heavy orange-yellow spines; flowers yellow and edible red fruit. *Subtropic.* p. 242

elata (Brazil, Paraguay), "Orange tuna"; erect bush to 1 m with fat obovate, waxy-smooth pads rich green with brown purple blotches around areole; very occasional straw-colored to gray needle spines; 5 cm flowers orange-yellow. *Subtropic.* p. 242

elata elongata (Paraguay), "Green wax cactus"; very attractive with more elongate, smooth, fat but flattened joints, 20 cm or more long, nearly spineless, rich nile green to dark olive, and purplish brown below each areole, and with satiny, waxy surface; pretty flowers with 2 layers of petals rich orange. *Subtropic.* p. 243

erinacea ursina (California), "Grizzly bear"; a form with tawny white hairs or flexuous spines 12-25 cm long, more numerous on smaller joints; very beautiful, found in the Mojave Desert. *Arid-subtropic.* p. 241, 242, 243

exaltata (Ecuador, Perú, Bolivia); branching, 2 to 4 m high, with stems or joints cylindrical, grayish green and with brownish needle-spines; large reddish flowers; similar to subulata. *Arid-subtropic.* p. 240

ficus-indica (Trop. America: prob. Mexico), the "Indian fig"; a flat-jointed cactus that may grow bushy, or with woody stems to 3 m or more high; widely spread into warm-climate countries and cultivated for its pear-shaped juicy, orange-red fruit 6-8 cm long, which is peeled and the pulp eaten raw or cooked for its flavor and food value, from Mexico to Spain and Southern Italy and Eastward. The oblongish flat joints are green or glaucous bluish from 30 cm to almost 60 cm long, in some forms spineless but with irritating yellow bristles; the flowers are yellow to 10 cm across. These are monstrous plants, but durable with sculptured and exotic decorative effect. *Arid-tropical.* p. 240, 241, 459, 464

ficus-indica 'Burbank's Spineless', "Spineless Indian fig"; a lightly glaucous bluish-green, tree-like form to 4 m, with long flattened joints almost 60 cm long, with yellow flowers and edible orange fruits. *Arid-subtropic.* p. 241

fragilis (Wisconsin to Brit. Columbia and south to Texas and Arizona), "Pigmy tuna"; low spreading plant not over 5 cm high, with fresh green, roundish or cylindrical joints very fragile, dropping off easily; small white areoles and brownish spines; pale yellow flowers. Also found as far north as Cache Creek, Alberta. *Temperate.* p. 243, 244

fulgida (Arizona), "Jumping cholla"; cylindrical joints 10-20 cm long, with age becoming tree-like 1 to 4 m tall, the branches of sparry habit; the raised knobs extremely spiny; pink flowers 2 cm across. *Arid-subtropical.* p. 239

fulgida mamillata 'Monstrosa' (mamillata in hort.) (Arizona, Mexico), "Boxing glove"; succulent knobby joints to 5 cm thick growing irregular and forming monstrose crests toward tips; sharp yellowish spines; flowers pink. *Arid-subtropic.* p. 243

leptocaulis (frutescens) (S.W. U.S., Mexico), "Tesajo"; brushlike, to 2 m, pencil-thin branched joints, dull green and woody, slender spines and small deciduous leaves; flowers yellowish, 1 to 2 cm across. *Arid-subtropic.* p. 244

linguiformis (So. Texas); bushy freely-branching cactus, to 1 m, with flat joints elongated, to 50 cm long, broad at base, tapering upward, rich green; brown areoles with yellow glochids and 2-4 slender yellow spines; yellow flowers, and purplish fruit. *Arid-subtropic.* p. 242

linguiformis 'Maverick', "Maverick cactus"; very attractive California mutation, a flat joint forming monstrose branches, the fleshy pad rich green, with multitudes of bulb-like and finger-like growths sticking out from the areoles, and tipped by beige bristles and glochids; resembling a miniature pyramidal tree. *Arid-subtropic.* p. 243

littoralis (So. California coast), a "Prickly pear"; with rounded or oblong fresh-green pads to 15 cm long, with numerous long white to yellow spines; large 12 cm golden yellow flowers, and red, juicy fruit. *Arid-subtropic.* p. 242

mamillata in hort. see fulgida mamillata 'Monstrosa'.

microdasys 'Albata', "Angelwings" or "Angora bunny-ears"; freely clustering baby opuntia with miniature pads or "wings", and covered with snow-white hair-like glochids. *Arid-subtropic.* p. 243

microdasys 'Albispina' (albescens), "Polka dots"; vigorous normal large type with areoles having soft white glochids (tufted barbed hairs) prominently arranged in neat rows; flowers pale yellow, 3-5 cm across. *Arid-subtropic.* p. 243

monacantha: see vulgaris

pachypus (Tephrocactus) (Peru); candelabralike plant to 1 m, cylindric branches marked with spiralled tubercles and tiny deciduous leaves, awlshaped yellow spines; inconspicuous scarlet flowers. *Arid-subtropic.* p. 238

polyacantha (Brit. Columbia and Alberta, Dakota to Arizona); prostrate, spreading prickly-pear with rounded, thin joints 5-10 cm wide, close-set areoles and deflexed spines; flowers mostly yellow 5 cm across. *Temperate.* p. 239

polyacantha 'Rubra'; Heavy, roundish glaucous pads with long spines; large red flowers. *Arid-subtropic.* p. 239

ramosissima (Cylindropuntia) (Southwestern U.S.; Mexico, Sonora, B. Calif.), a "Pencil cactus"; photographed in the Mohave Desert, California, growing into a densely branched bush of slender, cylindric, rigid, angled stems 2 to 4 cm thick, bluish green to gray, and with long reddish spines; flowers greenish-yellow 4 cm across. *Arid-subtropic.* p. 244

robusta (Central Mexico); robust "Tuna" cactus becoming tree-like to 6 m high; the pad-like joints elongate or disk-shaped, to 30 cm long, grayish-green or blue-glaucous and smooth, and few yellowish needle-spines; yellow flowers 5 cm across; dark red fruit 8 cm long and edible. *Arid-tropical.* p. 238

rufida (microdasys rufida) (Texas, No. Mexico), "Cinnamon-cactus"; bushy plant to 2 m high, eventually forming trunk; fleshy pads velvety grayish-green covered with tufts of short brown bristles (glochids), which rub off easily and cause itching under the skin; 5 cm flowers yellow. Will rot if too wet. *Arid-subtropic.*

schickendantzii (N. Argentina), "Lion's tongue"; shrub-like and much branched, elongate, warted, flattened, rather thin and narrow joints to 20 cm long, fresh green with reddish spines; yellow flowers 4 to 5 cm across. *Arid-subtropic.* p. 238

spinosior (Mohave Desert); tree-like cactus with spiny, glaucous, cylindric joints 3 cm thick, with raised knobs; large crimson flowers 5 cm across. *Arid-subtropic.* p. 239

stricta (inermis) (Cuba, Forida to S. Texas); naturalized in Chile, So. France and W. Australia; spreading semi-prostrate bush with obovate, flattened joints to 15 cm long, green or grayish green, with few spines; flowers yellow, and red pear-shaped fruit. *Arid-subtropic.* p. 239

verschaffeltii (Northern Bolivia); clustering plant with cylindrical stems branching with little fingers of tubercled joints, light dull-green, the knobs supporting small fleshy, deciduous leaves, slightly woolly areoles and occasional straw-colored needle-spines; deep red or orange flowers very beautiful. *Arid-subtropic.* p. 244

violacea santa-rita (glosseliniana var.), (Texas to Arizona), "Blue blade" or "Dollar cactus"; handsome branching bush to 2 m, with flat joints rounded in outline, to 20 cm across, bluish to coppery purple, with glochids but few spines; yellow flowers 10 cm across, red at base inside. *Arid-subtropic.* p. 243, 244

vulgaris (monacantha) (So. Brazil to Argentina), "Irish mittens"; treelike to 2½ m, flattened fleshy, glossy green joints almost spineless; flowers yellow; the unripened fruit will root and grow forming little ears, and offered as "Eared buds". *Subtropic.* p. 238

vulgaris 'Variegata', "Joseph's coat"; variegated form with the long smooth joints beautifully patterned green and white; large yellow flowers 8 cm across. *Subtropic.* p. 243

ORBIGNYA *Palmae*

cohune (Attalea), (Mexico to Costa Rica), "Cohune palm"; large feather palm with trunk 6 m long and to 60 cm thick, the characteristically erect pinnate fronds to 9 m long, with glossy-green stalk and folded, leathery leaflets; abundant egg-shaped brown fruit to 8 cm long, in large clusters with nuts furnishing cohune oil. *Tropical.* p. 768, 791

ORCHIDANTHA *Musaceae*

maxillarioides (Malaya), "Orchid flower"; perennial herb with tufted foliage of the feel of aspidistra and with flowers resembling orchids; lanceolate leathery leaves on wiry petiole, light moss-green with some mottling, depressed midrib; flowers with violet calyx and green lip variegated purple. *Tropical.* p. 672

ORCHIS *Orchidaceae*

latifolia (Europe), "Marsh orchid"; terrestrial 30 cm high, with lanceolate leaves; dense spike of many flowers, purple or red and with conical spur. *Warm temperate.* p. 744

papilionacea (Portugal, Algeria, Greece, Turkey); "Pink butterfly orchid"; tuberous terrestrial forming a rosette of broad, dark green fleshy leaves 8-10 cm long, from the center a spike with pale purple flowers 3 cm long, the lip veined with crimson, the petals forming a hood; photographed amongst ancient ruins on Rhodes, Greece. *Subtropic.* p. 744

OREOCEREUS *Cactaceae*

celsianus (Andes of Bolivia, Perú, Chile), "Old man of the Andes"; growing in clumps, creeping when young upright to 1 m, areoles with long white hairs and long thin red spines; dark red flowers 10 cm long. *Subtropic.* p. 251

hendrickseniana (So. Perú to No. Chile); clustering columns to 1 m, branching from base, and with 5 to 10 ribs and matted hair between; large flowers to 7 cm long, rosy-red with white center. *Arid-tropical.* p. 254

OREODOXA: see Roystonea

OREOPANAX *Araliaceae*

capitatus, also known as nymphaeifolius (Mexico, C. America, Indies); small evergreen tree having glossy-green, broad ovate, leathery leaves very variable, juvenile stage having peltate base, adult stage obtuse or cordate, on slender stalks. *Subtropic.* p. 140

salvinii (peltatus) (Mexico); evergreen tree with thin-leathery, palmately lobed leaves, the lobes fresh green with pale veins and toothed or lobed again; rough beneath; stalks with loose hairs. *Subtropic.* *p. 140*

xalapensis (Mexico, C. America); evergreen shrub with palmately compound, thin-leathery leaves, the segments obovate, light green and corrugated, serrate, on brown petioles. *Tropical.* *p. 140*

ORNITHOCEPHALUS *Orchidaceae*
bicornis Panama), "Mealybug orchid"; miniature epiphyte 5-6 cm high with bright green leaves spreading and overlapping fan-like; small greenish flowers with white petals and waxy greenish lip (winter and various). *Tropical.* *p. 742*

ORNITHOGALUM *Liliaceae*
caudatum (So. Africa), "False sea-onion"; an old-fashioned window-sill plant, with ovate, green bulb to over 10 cm thick, usually showing above the soil, 5-6 basal strap leaves and a stalked raceme 45-90 cm long with 50-100 small white flowers with petals having a green median stripe. The filament is wide at base (narrow in Urginea); the narrow channeled leaves 4 cm wide, and form a tube at base. Also known as "Healing onion", or "Meerzwiebel", crushed leaves are tied over cuts and bruises, and used as cooked syrup with rock-candy, against colds. *Subtropic.* *p. 592, 598*
montanum (Sicily to Palestine); small bulbous plant with lanceolate leaves and short, flattened head with a few long stalked white flowers, the outside of each petal with a median green band. *Subtropic.* *p. 598*
saundersiae (So. Africa), the "Giant chincherinchee"; tall-growing 1-2 m high; broad, sword-shaped basal leaves to 60 cm long; 2½ cm flowers grouped in a flat-topped cluster on long erect stalk, each bloom with 6 spreading, creamy petals and prominent black center. *Subtropic.* *p. 598*
thyrsoides (So. Africa), "Chincherinchee"; winterblooming, tender bulb with fleshy lanceolate leaves and strong 60 cm racemes with numerous long-lasting flowers with spreading segments 3 cm across, white with buff eye. *Subtropic.* *p. 595, 598, 599*
umbellatum (Mediterranean), "Star-of-Bethlehem"; small spring-blooming bulb with grass-like leaves and 15 cm stems with clusters of numerous star-like flowers satiny-white inside, green striped white outside. *Subtropic.* *p. 598*

ORONTIUM *Araceae*
aquaticum (Atlantic No. America), "Golden club"; aquatic perennial growing in ponds or along streams, with fleshy rootstocks and leaf-stalks 25-50 cm long, the parallel-veined, oblong, dark green leaves floating or ascending, 15-30 cm long; long bright yellow, club-like spadix, and small inconspicuous spathe, on white, 60 cm stalks. *Warm temperate.* *p. 103*

OROSTACHYS *Crassulaceae*
erubescens japonicus (W. Japan); starry succulent rosette with fleshy narrow pointed leaves to 3 cm long, yellowish green, extending upward into a spike with inflorescence to 10 cm long; flowers whitish pink. *Warm temperate.* *p. 374*
iwarenge fa. 'Fuji'; variegated cultivar from Japan; beautiful suculent rosette 12-15 cm across; persistent through winter; obovate leaves milky green with broad cream bands along sides; later pushing up a tall spike to 35 cm high with small white flowers. *Warm temperate.* *p. 374*

ORPHIUM *Gentianaceae*
frutescens (So. Africa), "Sticky flower"; shrubby perennial to 60 cm, with narrow-linear, leathery, light green leaves, and showy, shining rosy, star-like flowers 4 cm dia. *Subtropic.* *p. 484*

ORTHOPHYTUM *Bromeliaceae*
maracasense (Brazil: Bahia); star-like rosette with thick-fleshy tapering leaves 30 cm long, coppery to amethyst red, edged with prominent reversed spines, green tipped brown, the reverse pencil-grooved and thickly covered by mealy scales; erect inflorescence with spreading bracts and white flowers. *Tropical.* *p. 224*
vagans (Brazil: Espirito Santo); semi-succulent trailing plant rambling over rocks forming large mats; rosettes of narrow metallic-green, thorny leaves, toward apex these becoming floral bracts brilliantly colored from orange to red, and forming a colorful red head with green and white flowers. *Tropical.* *p. 222*

x ORTHOTANTHUS *Bromeliaceae*
'What' (Cryptanthys 'It' x Orthophytum saxicola rubra); bigeneric hybrid; star of spiny leaves olive-green with cream stripes and margins. Photographed at Seaborn-Del Dios, California. *Tropical.* *p. 221*

ORYZA *Gramineae*
clandestina, "Wild rice". I photographed this plant in the aquatic greenhouse of Nymphenburg Botanic Gardens in Munich where it is listed as from the North Temperate zone, growing in shallow water. However, Oryza are primarily tropical from Asia or Africa, as is the true rice, Oryza sativa; the temperate climate wild rice is Zizania aquatica which provided grain for American Indians. The handsome grass shown is perennial and clustering with slender canes about 1 m high, with narrow, matte rich green leaves, very ornamental at the edge of ponds. Its valid name may be Leersia oryzoides, the "Cut grass". *Tropical.* *p. 513, 573*
sativa (S.E. Asia), "Rice"; annual tropical swamp grass, 1-2 m high, with flat elongate leaves; fruiting panicle drooping, producing the yellow rice kernels; of great economic importance in tropic regions, yielding 2 or 3 crops a year. *Tropical.* *p. 512, 513*

OSMANTHUS *Oleaceae*
armatus (China), "Chinese osmanthus"; attractive and shapely evergreen shrub with holly-like narrow-elliptic, leathery, glossy green leaves with ivory midrib, 16 cm long, the margins with spines, slow-growing and with stiff reddish branches; tiny white, fragrant flowers. *Warm temperate.* *p. 696*
fragrans (Olea) (Himalayas, China, So. Japan), "Sweet olive"; small tree to 10 m high, with wiry twigs and holly-shaped, stiff-leathery, olive-green leaves to 10 cm long, finely toothed at margins; the small white flowers in clusters, strongly and deliciously fragrant. *Subtropic.* *p. 696, 700*
heterophyllus 'Variegatus' (ilicifolius var.) "False holly"; extremely attractive, slow-growing, dense evergreen shrub resembling variegated holly but a better keeper; the spiny, glossy-leathery leaves somewhat smaller, 4 to 6 cm long, fresh green to bluish-gray-green, edged and variegated creamy-white, tinted pink when young; should be grafted on privet for best growth. The green-leaved species is from Japan and Taiwan. *Subtropic.* *p. 696, 700*

OSMUNDA *Osmundaceae (Filices)*
cinnamomea (E. No. America, Mexico to Brazil, E. Asia), "Cinnamon-fern"; coarse but attractive deep fibrous rooted fern with large crowns of 1½ m fronds, the fertile fronds 2-pinnate and becoming brown as spores mature; its roots are used as a growing medium for orchids. *Humid-subtropic.* *p. 437*

OSTEOSPERMUM *Compositae*
fruticosum (So. Africa: Cape, Natal), "Trailing African daisy" or "Burgundy mound"; semi-shrubby plant spreading rapidly by trailing, rooting branches; small fleshy leaves with several points; large ray-flowers to 5 cm across, lilac above, fading nearly white by second day, deeper purple beneath, and dark purple center cushion; excellent ground cover for sunny, mild climate. *Tropical.* *p. 328*
fruticosum 'Album', "Trailing African daisy"; variety with flowers white, 5 cm across; very floriferous. *Subtropic.* *p. 328*

OTHONNA *Compositae*
herrei (So. Africa: Namaqualand); curious succulent with short, thickened stem, and knotty from the persistent leaf bases; soft-fleshy leaves 5 cm long, irregularly obovate and undulate, bluish glaucous, at the ends of branches; deciduous when resting; small yellow flowers. *Arid-subtropic.* *p. 329*

OURATEA *Ochnaceae*
groussordyi (Venezuela); attractive small evergreen shrub with thin-leathery elliptic leaves 8-10 cm long; the inflorescence in tight clusters usually at the end of short lateral branchlets; the buds enclosed by brown calyx, flowers when open rich yellow, 3 cm across. *Subtropic.* *p. 689*

OUVIRANDRA: see Aponogeton

OXALIS *Oxalidaceae*
adenophylla (Chile); hardy stemless perennial 10-15 cm high, from a roundish, bulb-like base; long-stalked leaves in basal rosette, with 12-22 obcordate leaflets 1 cm long, glaucous grayish-green; 2½ cm flowers lilac pink with deeper veins and orange throat and with blackish-red spots at base of each of 5 petals; solitary or in umbels. *Warm temperate.* *p. 762*
braziliensis (Brazil); bulbous plant without stem, 8-15 cm high, with stalked bright green basal leaves having 3 bluntly obcordate leaflets 1 cm long and wide; fowers 2½ cm dia., bright rosy-red with darker veins, and yellow in throat; in winter and spring. One of the "Shamrocks" sold by florists. *Tropical.* *p. 762*
deppei (esculenta; Ionoxalis) (So. Mexico), "Lucky clover"; bulbous plant with edible tuber, large leaves having 4 truncate segments (cut off straight at the apex) 4 cm long, and crossed by a purplish-brown zone; flowers rosy-red with yellow base. Attractive pot plant for winter bloom. *Tropical.* *p. 762, 763*
gigantea (Chile); erect shrub becoming tree-like, 2 to 4 m high, with curious growth resembling columnar cactus; tiny green trifoliate, clover-like leaves with obcordate 1 cm leaflets; small 2 cm yellow flowers. *Subtropic.* *p. 763*

hedysaroides rubra (Colombia, Venez., Ecuador), "Firefern"; beautiful plant with erect, shrubby, wiry stem and thin, fern-like foliage of glowing, satiny wine-red, each petiole with 3 stalked ovate leaflets which are sensitive to the touch; many, little, bright yellow flowers in attractive contrast to the showy leaves. Subshrub to 1 m tall. *Tropical.* p. 762, 763

hirta (So. Africa); winter-blooming plant with branching leafy pubescent stem from a large brown bulb, at first erect but becoming procumbent; with feathery foliage of 3 small spatulate leaflets nearly sessile, alternate and scattered; axillary flowers deep rose with yellow tube, and silky sepals. *Subtropic.* p. 762, 763

lasiandra (Ionoxalis) (Mexico); large, stemless plant with scaly bulbs and long-stalked, fleshy leaves pendant by their weight, having 5 to 11 wedge-shaped leaflets, going to "sleep" in darkness; slender stalks with umbels of rosy-carmine flowers opening over a period of 6 weeks. *Subtropic.* p. 762

martiana 'Aureo-reticulata' (corymbosa var.) (Trop. America); attractive ornamental "Sour-clover", with scaly bulb producing numerous prostrate and ascending stalks, bearing large herbaceous leaves to 8 cm across, the 3 leaflets obcordate, fresh to deep green, beautifully veined and reticulated with yellow; long-stemmed flowers carmine-rose with red lines from white throat; goes "to sleep" at night. *Tropical.* p. 762

pes-caprae (cernua) (So. Africa), "Bermuda buttercup"; perennial with thickened roots and deep scaly bulbs, with many basal, long-stalked leaves of 3 obcordate leaflets hairy beneath; nodding bell-shaped bright yellow flowers 2½-4 cm across, in spring; a weed in Bermuda and Florida and other mild districts. *Subtropic.* p. 763

purpurea (variabilis) (Cape Prov.), known in the cultivated form 'Grand Duchess'; a low, spreading bulbous perennial with succulent, fresh green foliage of 6-8 cm across, of three ciliate leaflets not notched, on rosy-red stalks pubescent with soft white hair, barely topped by large and showy, pretty flowers to 5 cm, bright rose with yellow base, winter-blooming in the sun. *Subtropic.* p. 762, 763

succulentum (Chile); creeping succulent with small light green, fleshy leaves in threes on flexuous stalks; the stems reddish. Seen at Royal Botanic Gardens, Sydney. *Subtropic.* p. 763

vulcanicola (siliquosa) (Costa Rica to Panama); mat-forming creeper with red stalks, the foliage with 3 reddish obovate or obcordate leaflets, and golden yellow flowers. *Tropical.* p. 762

OXERA *Verbenaceae*
pulchella (New Caledonia); climbing shrub with opposite oblong leathery leaves to 12 cm long, and trumpet-shaped 4-lobed white flowers 5 cm long, with conspicuous calyx and long protruding stamens. *Tropical.* p. 914

OYEDAEA *Compositae*
verbesinioides (Trop. So. America); tall branching shrub with grayish, ovate leaves, having cream primary veins; spreading clusters of flowers with brownish cushion, and scattered yellow ray flowers. *Tropical.* p. 327

PACHIRA *Bombacaceae*
aquatica (So. America), the "Guinea chestnut" or "Oje"; large tropical tree with palmately compound leaves of 5-7 leathery-oblong leaflets 8-20 cm long; large flowers with narrow pinkish petals 10-15 cm long, which drop off and expose numerous long white stamens; the large green or brown woody ovoid five-valved fruit is 10-30 cm long, containing seeds edible raw or roasted. Known in Hawaii as the "Malabar chestnut". *Tropical.* p. 197, 457, 468

PACHYCEREUS *Cactaceae*
pecten-aboriginum (Chihuahua, Sonora, Baja California), the "Hairbrush cactus"; waxy-green column growing into a branched tree 8 m high and 25-45 cm thick; 10-12 acute ribs, spines gray to fawn, 1-2 large central spines tipped black toward apex, the upper areoles with tufts of some white wool; day-flowers red outside, white inside, 10 cm long; woolly 8 cm fruit. *Arid-subtropic.* p. 245

pringlei (Mexico: Sonora, Baja California), "Mexican giant"; one of the most massive cacti, stout branching tree to 11 m high and 1 m thick at base, the olive green columns to 40 cm dia., with 10-16 prominent but rounded ribs, closely studded with large oval areoles with short white or grayish wool especially toward apex, numerous ash-gray radials and 1-3 long central spines at first reddish, afterwards gray; bell-shaped flowers 10 cm long, greenish-red outside, white inside. *Arid-tropical.* p. 245, 254

PACHYCEREUS: see also Lemaireocereus

PACHYPHYTUM *Crassulaceae*
bracteosum (Mexico), "Moonstones"; branching succulent with stout stem to 30 cm high; rosette of obovate swollen club-shaped leaves 5-10 cm long, gray-glaucous with rosy tint; 20-30 cm stalk with red flowers. *Arid-tropical.* p. 378

longifolium (Mexico: Hidalgo); rosette of fleshy, thick oblanceolate leaves 6-10 cm long, glaucous bluish tinted with purple; flowers dark red inside. *Arid-subtropic.* p. 375

oviferum (San Luis Potosi), "Pearly moonstones"; fine colored succulent reminiscent of "Moon stones", the swollen, long egg-shaped leaves are beautifully glaucous in delicate silver-white, and shading to pink and amethyst, 3 to 5 cm long; bellshaped red flowers. *Arid-tropical.* p. 376

PACHYPODIUM *Apocynaceae*
baronii var. **windsori** (So. Madagascar); dwarf succulent shrub with thick, globose trunk to 8 cm dia., with a few thick, rough-skinned, cylindrical branches, with tiny spines; pairs of leathery obovate leaves to 6 cm long, shining surface and felted reverse; 5 cm red flowers between the foliage. *Arid-tropical.* p. 82

lamerei (Madagascar: Fort Dauphin), "Club-foot"; weird succulent a thick, spiny gray column 1-3 m high and scarcely branched; at the base spindle-shaped, the pinkish spines 3 cm long; toward the top the spirally arranged strap-like leathery 20 cm leaves, dark shining green with white midrib; small funnel-shaped white flowers. A member of the queer vegetation prevailing in Southern Madagascar. *Arid-tropical.* p. 75, 79, 82, 83

namaquanum (So. Africa: Cape Prov., Namibia), called "Ghostman" by the Hottentots; stout succulent column to 1½ m or more high, armed with 5 cm spines as protection against animals; in winter when growing, a tuft of leaves forms at apex, 10-12 cm long and 5-6 cm wide; flowers from the leaf axils velvety and reddish, striped yellow inside. The stems are leaning toward the sun. *Arid-tropical.* p. 83

PACHYSANDRA *Buxaceae*
terminalis (Japan), "Japanese spurge"; low evergreen perennial herb spreading by means of creeping rootstocks, with fleshy obovate leaves grouped whorl-like, coarsely toothed toward apex and 3-8 cm long; terminal spikes of greenish-white flowers. *Warm-temperate.* p. 237

terminalis 'Variegata', has its soft-leathery evergreen leaves bordered and variegated ivory white. *Warm temperate.* p. 237

PACHYSTACHYS *Acanthaceae*
coccinea (Jacobinia) (Trinidad, So. America), "Cardinal's guard"; herbaceous shrub 1-2 m high, with elliptic, rugose leaves to 20 cm; scarlet flowers in terminal heads, the calyx with 5 linear segments, 5 cm corolla two-lipped and reflexed. *Tropical.* p. 39

lutea (Perú, "Lollypops"; introduced in Europe as Beloperone "Super Goldy"; semi-woody plant 25 cm high with lanceolate, depressed-veined herbaceous, matte dark green leaves; contrasting with a striking, erect inflorescence of hops-like, shingled, orange-yellow bracts, bursting with creamy-white flowers in late summer. *Tropical.* p. 39

PACHYSTACHYS: see also Jacobinia, Justicia

x PACHYVERIA *Crassulaceae*
haagei, "Jewel plant"; compact star-like hard rosette of short fleshy, boatshaped leaves flat on top, bluish green dipped in purplish red toward apex and bluish glaucous; flowers yellow and orange red. Possibly P. glauca. *Subtropic.* p. 378

nigra (Pachyphytum sp. x Echeveria multicaulis or metallica); handsome Hummel-California cultivar known as "Black Magic"; low, flat rosette 12 cm across, of obovate-pointed, concave leaves 4 cm long, waxy green but turning black-purple in strong sun. *Arid-subtropic.* p. 378

PAEONIA *Ranunculaceae*
lactiflora (albiflora sinensis) (China), the "Chinese garden peony"; hardy perennial with roots a collection of narrow tubers; herbaceous leaves, twice compound; stems to 1 m high; large fragrant flowers, 10-12 cm across, typically white; the variety sinensis shown in photo on page 843, has double, very large crimson flowers 15 cm dia. *Temperate.* p. 843

lactiflora 'Festiva' (Siberia, China, Japan), "Double-flowered peony" or "Chinese peony"; outstandingly beautiful hardy perennial herb to 1 m high, with spindle-shaped dahlia-like tuberous roots, large ornamental, herbaceous, compound leaves, and magnificent rose-type fragrant flowers to 15 m across, with double incurved white petals with pink sheen; blooming normally in our gardens in late May or early June. *Temperate.* p. 844, 847

lactiflora 'Requiem'; free-blooming cultivar with medium-size single white flowers, showing a center cushion of yellow stamens. *Temperate.* p. 844

officinalis 'Rubra plena' (France to Albania origin), "Piney"; compact bush with twice compound, herbaceous leaves, the leaflets cut into narrow elliptic segments 10 cm long; flowers crimson red to 12 cm across, and with red filaments. *Warm temperate.* p. 844

suffruticosa 'Angelet', "Tree peony"; beautiful tree-peony of Chinese heritage; deciduous shrub with picturesque woody stems, to 2 m high; large divided bluish green leaves; very showy flowers 12-15 cm across, with ruffled petals orange-yellow and crimson red at base. *Temperate.* p. 844

suffruticosa 'Vesuvian', "Red tree-peony"; deciduous shrub with woody stems; large semi-double to double flowers glistening crimson-red, the petals undulate and crisped. Developed from species at home in China to Tibet. *Temperate.* p. 844

PAGIANTHA   *Apocynaceae*
dichotoma (So. India, Sri-Lanka, So. China), "Forbidden fruit of India"; large evergreen shrub dense with shiny green, leathery, ovate leaves to 20 cm long; very exotic with waxy white flowers 4 cm across, their 5 long linear petals curved sickle-shaped windmill-like, twisted and with margins turned under, and very sweetly fragrant. *Tropical.* p. 76

PALICOUREA   *Rubiaceae*
rigida (Trop. So. America); handsome evergreen shrub to 1 m high, the branches 4-angled, with broadly oblong, leathery leaves having undulate margins, and showing yellow ribs; small, downy yellow flowers in dense, erect panicles. *Tropical.* p. 868

PALISOTA   *Commelinaceae*
barteri (Fernando Po); dark green rosette with spreading, stalked elliptic glossy leaves with parallel veins and edged with hairs; flowers purplish, near base, followed by a dense cluster of showy orange-red, pointed berries. *Tropical.* p. 306

PALIURUS   *Rhamnaceae*
spina-christi (So. Europe to No. China), "Christ-thorn" or "Jerusalem-thorn"; large shrub or tree to 6 m, branches armed with twin spines, and ovate, 3-veined leaves 3 cm long, finely toothed, alternate in 2 ranks; small greenish yellow flowers; 2 cm fruit brownish yellow; cultivated more for religious interest than as ornamental. *Warm temperate.* p. 832

PANAX: see Pseudopanax, Nothopanax, Polyscias

PANCRATIUM   *Amaryllidaceae*
zeylanicum (Trop. Asia); summer-blooming bulbous plant; from the large pear-shaped bulb emerge the linear basal leaves to 30 cm long, developing with the inflorescence; the solitary flower on solid stalk shorter than the foliage, the waxy white petals recoiling; long stamens united at base into a corona with teeth. *Needs dormancy. Tropical.* p. 306

PANDANUS   *Pandanaceae*
baptistii (New Britain Isl. near New Guinea), "Blue screwpine"; symmetrical plant with stiff channeled leaves spirally arranged, gracefully arching, and tapering to a long point, blue-green with several yellow center stripes, without thorns at the margins. *Tropical.* p. 810

brosimus (New Guinea), "Pandanus palm"; tall erect tree with stout trunk to 30 m high, and gigantic stilt-roots; broad-leathery, saw-edged leaves 5 m long with drooping tips; important economically to the mountain people of New Guinea at elevations of 2,000-3,000 m as a source of food; to highland Papuans, the pineapple-like fruit is their coconut. *Subtropic.* p. 809

leram (Sri Lanka), "Nicobar bread-fruit"; large tree dividing into numerous twisting branches and supported by stilt-roots; topped by long pendant bluish-green, prickly leaves arranged in spirals; notable for its large ovoid fruits made up of many wedge-shaped 12 cm drupes (ripened ovary) containing sugar and starch, and nut-like seeds. *Tropical.* p. 474, 809

odoratissimus (Hawaii, Polynesia to New Hebrides, Queensland, Monsoon Asia), the "Hala screw-pine" or "Walking tree"; forming groves, but favoring the seashores; picturesque tree branching into twisted woody and flexuous, ringed stems to 6 m high, and with straight, supporting aerial roots as if standing on stilts; the thin-leathery, pliable sword-shaped leaves arranged spirally are grayish green 1 m or more long, and prickly at margins; flowers on male trees fragrant; brownish 20 cm fruit looks like a pineapple, consisting of nut-like, edible seeds. *Tropical.* p. 477, 809

pristis (Madagascar); clustering rosette 2 to 4 m high, dark green glossy leathery foliage with prominent pale spines, the leaves spreading out horizontally and recurving beyond the middle; beginning to flower when 1 m high. *Tropical.* p. 810

rockii (Hawaii); tall, freely branching screw pine, forming stilt-roots, with smooth branches topped by rosettes of leathery, lanceolate leaves, the margins finely serrate. *Tropical.* p. 809

sanderi (Moluccas, Polynesia); a handsome, very ornamental screwpine with short green stem, but suckering freely with great spiralling rosettes of magnificent long leaves, glossy green largely variegated with ivory-white to yellow, green band toward finely spiny margins, with age breaking above the middle and laxly pendant. *Tropical.* p. 808

sanderi 'Roehrsianus' (Timor, Solomon Isl.); robust, beautiful rosette with long leathery 1 m sword-like leaves often pendant beyond the middle, friendly green and shining, transversed lengthwise with stripes and bands of light golden yellow, the young growth golden orange, margins finely spiny; tufting with suckers. *Tropical.* p. 808

tectorius (So. trop. Asia to Polynesia), a "Pandanus palm"; widely cultivated for its economic values, especially the pendulous clusters of seed; slender, branching, 6 m tree with flexuous trunk supported by brace-roots, light green fibrous leaves pendant above the middle, with white spines. *Tropical.* p. 808

utilis (Madagascar), a useful as well as ornamental "Screw pine"; spiral rosette of long curving, strap-like, thick-leathery leaves to 2 m long and 10 cm wide, keeled beneath, deep olive-green with showy red spines; with age becoming a branching tree to 18 m high, with stilt-like brace roots; the leaves are used for making hats and baskets. *Tropical.* p. 810

veitchii (Polynesia), "Variegated screwpine"; shapely and attractive house plant; rosette of thin-leathery, recurving leaves to 8 cm wide, narrowing to a long point, shining light to deep green lined and broadly margined with creamy-white, the edges and keel beneath with small spines; with age developing stilt-like, thick aerial roots. *Tropical.* p. 810

PANDOREA   *Bignoniaceae*
jasminoides (Bignonia) (New So. Wales, Queensland), "Bower plant" or "Bower of beauty"; tall flowering climber with compound leaves feather-fashion, with 5-9 glossy green elliptic leaflets 3-5 cm long; the Tecoma-like trumpet flowers 5 cm long, opening to 5 crepy lobes; pinkish-white streaked with pink or red in throat. *Subtropic.* p. 183

jasminoides 'Alba'; charming cultivar popular in California, very floriferous with showy white flowers. *Subtropic.* p. 183

jasminoides 'Rosea'; a beautiful cultivar in California nurseries; twiner with glossy, dark green leaves, and salverform flowers rose-pink with crimson-red in throat. *Tropical.* p. 183

PANICUM: see Oplismenus

PAPAVER   *Papaveraceae*
aurantiacum (Switzerland), "Alpine poppy"; miniature alpine annual photographed at Schynige Platte Alpine Garden near Interlaken; deeply divided glaucous leaves, and large yellow flowers. *Temperate.* p. 812

nudicaule (Arctic to Colorado), "Iceland poppy"; perennial grown as annual in warmer regions; divided leaves with coarse pubescence; hairy stems 30-50 cm high; flowers cup-shaped 3-8 cm across, fragrant, sometimes double; white or cream with yellow at base, salmon, orange, pink, rose to scarlet red. *Temperate.* p. 812, 813

orientale (Mediterranean to Iran), "Oriental poppy"; a showy perennial with milky-colored juice and white-hairy sparsely leafy stem to 1 m high, the thick, green leaves pinnately dissected and toothed, and large 10-15 cm flowers scarlet with black spot in the base, followed by an ornamental glaucous capsule; summer-blooming. *Subtropic.* p. 812, 813

PAPHINIA   *Orchidaceae*
cristata (Colombia to Guianas); small epiphyte; clustered pseudobulbs with 2-3 fleshy, pleated leaves 15 cm long; pendant inflorescence of strange flowers to 10 cm dia., spreading petals and sepals pale yellow tigered and banded brown-maroon, the clawed lip chocolate-purple with tuft of white hair at apex (autumn). *Tropical.* p. 747

PAPHIOPEDILUM   *Orchidaceae*
'Albion' (Astarte x niveum); chaste hybrid with leaves lightly mottled gray, and demure, waxy-white flower thinly dotted with red and green disk in center (spring and fall). *Tropical.* p. 748

x aureum 'Oedipe' (spicerianum x nitens); an old free-blooming ladyslipper, with faintly mottled leaves, and sturdy, waxy flowers glossy coppery brown, the dorsal white with purple stripes and shading. *Tropical.* p. 749

bellatulum (Burma, Thailand); dwarf terrestrial with fleshy dark green leaves mottled with pale green, purple beneath; solitary 5-8 cm waxy flowers on very short stalk, shell-shaped, creamy-white, covered with raised spots and blotches of purple-maroon, (April-Sept.). *Tropical.* p. 749

callosum (Thailand, Vietnam); colorful species with leaves tesselated bright green on deep green, and large 9-10 cm flowers with greenish petals deflexed and strap-like, purple at tip and with several black warts, the dorsal white lined with purple and green veins, and a brown-purple pouch. *Tropical.* p. 749

'Camarin' ('Harefield Hall' x spicerianum); and old English hybrid with silver gray leaves spotted with dark; the waxy flowers with frilled petals and dorsal yellow and spotted maroon, the dorsal white on margins; slipper yellow, flushed beige. *Subtropic.* p. 749

dayanum (Cypripedium) (Borneo); attractive leaves pale green mottled with dark green; waxy flower 10-12 cm dia., dorsal white with green and black-purple lines, petals flushed purplish brown and with blackish spots, fringed with black hairs, pouch a lacquered blackish-brown (spring). *Tropical.* p. 747

exul (Thailand); distinct species with stout green leaves to 25 cm long, and 8 cm flowers with dorsal green edged with white, and brown-spotted petals yellow-green with dark spots at base, lip brownish-yellow (Feb.-July). *Tropical.* p. 748

fairrieanum (Himalayas, Bhutan, Assam); dwarf terrestrial with pale green leaves and uniquely pretty flowers, dorsal white, greenish at base and with purple lines, petals similar, sickle-shaped, pouch green flushed red and veined with purple, (July-Jan.). *Tropical.* p. 749

x harrisianum 'C.S. Ball'; distinct form with large 6-8 cm, sinister, very dark vibrant deep wine red, waxy flower having dorsal greenish yellow and white striped with red; leaves mottled light and dark green (summer). *Tropical.* p. 746, 749

insigne (Cypripedium) (Nepal, Assam), terrestrial "Lady slipper"; growing on rocks in the Himalayas at 2000 m; small plant with green leaves, soon forming tufts; the flowers waxy, dorsal yellow-green with purple spots at base, white at apex, petals yellowish-green veined with brown, slipper reddish-brown, (Oct.-March). *Subtropic.* p. 749

insigne 'Harefield Hall' (Assam); perhaps the best of many named forms, exceptionally large, waxy flowers, with dorsal yellowish, margined white, and heavily blotched with chocolate, (winter). *Subtropic.* p. 749

insigne 'Sanderae' (Assam); desirable form with flowers of a beautiful primrose-yellow, except the upper part of the dorsal which is pure white, (winter). *Subtropic.* p. 749

lawrenceanum (No. Borneo); robust terrestrial with long leaves brightly mottled pale green, and bold flowers, dorsal white with shining purple stripes, greenish at base, the horizontal petals greenish, shaded purple and black-warted, lip brown-purple, (April-Aug.). *Tropical.* p. 749

lowii (No. Borneo); beautiful epiphyte on trees or limestone rocks, with straplike, light green leaves and stalks to 1 m, with 2 to 6 hairy flowers, dorsal greenish with purple stripes, the narrow, obovate petals greenish yellow blotched brown and purple-tipped, pouch coppery green with purple lines, (Feb.-July). *Tropical.* p. 749

x maudiae (Cypripedium) (callosum x lawrenceanum); beautiful and highly desirable hybrid with marbled leaves yellow-green and bluish-gray, the usually single flowers with a white dorsal striped with green, the slender petals greenish closely lined with green, pouch yellowish-green, (April-Aug.). *Tropical.* p. 746, 748, 749

niveum (Malaya); lovely dwarf species, with close-set oblong 15 cm leaves purple beneath and dark green above, tesselated with lighter oblong spots; the 6-9 cm flowers are satiny white inside, minutely speckled with dots of red-purple. *Tropical.* p. 748

'Olivia' (niveum x tonsum); dainty hybrid with leaves mottled yellowish green on dark green, and purple underside; hairy stalk with waxy flower white flushed pink and regularly lined and dotted with purple (winter). *Tropical.* p. 746

praestans (New Guinea); beautiful species; with green leaves to 30 cm; a purple stalk carries several interesting flowers; the dorsal sepal whitish streaked with purple; the petals linear and twisted, 12 cm long, yellow green with brown veins and black warts; lip yellow. *Tropical.* p. 746

rothschildianum (Papua, Borneo); remarkable terrestrial, strong growing, with leathery green leaves and racemes with 2-5 flowers having long pointed petals yellowish green, spotted red-purple, dorsal yellowish, striped with dark purple, pouch red-brown, with yellow at the opening, (Jan.-Aug.). *Tropical.* p. 748

sukhakulii (Thailand); checkered leaves gray-green and pale green; flowers with dorsal sepal pale green and dark lines, petals blotched blackish purple, the glossy pouch with purple veins. *Tropical.* p. 747

villosum (So. Burma); large terrestrial with light green strap-leaves to 45 cm long; large flowers, nearly 12-15 cm across and very glossy, dorsal green with white border and lined purplish-brown, petals yellowish-brown and purple midline, pouch brownish-yellow (Oct.-Mar.). *Subtropic.* p. 748

PARADRYMONIA *Gesneriaceae*
hypocertifolia (W. Ecuador); herbaceous epiphyte with scandent stems forming roots; long reddish petioles with elliptic leaves 10-15 cm long, toothed at margins and grayish beneath; clusters of small waxy but showy, brilliant crimson-red flowers from the leaf axils. *Tropical.*

PARINARI *Rosaceae*
mobola (Sénégal), "Nef-fruit"; large xerophytic tree with sparry branches, broad, thick-leathery ovate leaves covered with white fuzz, 15 cm long; rough ovoid fruit light brown with white tubercles, 6 cm long, and eaten by the Wolofs of West Africa. *Arid-tropical.* p. 467, 852

PARKIA *Leguminosae*
biglobosa (Trop. Africa), "African locust"; tropical tree to 15 m, wide-spreading with fern-like bipinnate foliage, tiny 1 cm leaflets; pear-shaped orange or red flower heads 4 cm dia.; flat beans to 30 cm long. *Tropical.* p. 561

javanica (India to Java); grand tree to 50 m high, with remarkable finely divided bipinnate leaves to nearly 1 m long; countless yellow flowers are borne in club-shaped long-stemmed heads; the thick, flat, black hanging pods 30 cm long. Seeds are roasted and eaten. *Tropical.* p. 561

PARKINSONIA *Leguminosae*
aculeata (Trop. America to Cape Verde Is.), "Jerusalem thorn" or "Mexican palo verde"; small spiny ornamental tree, with narrow bipinnate, fern-like foliage; small yellow flowers sweet-scented, in loose racemes; 12 cm pods constricted between the seeds. *Subtropic.* p. 561

PARMENTIERA *Bignoniaceae*
cereifera (Panama), "Candle tree"; interesting small tree with rough bark; the leaves with winged petiole and 3 obovate leaflets; bell-shaped, curved white flowers to 8 cm long; curious with cylindrical fleshy, greenish yellow fruit pendant direct from old woody branches 30 cm to 1 m long, with apple-like odor, and resembling wax candles. *Tropical.* p. 185, 187

edulis (Mexico, Guatemala), "Guajilote tree"; tropical tree of dense, thorny growth to 10 m high, with ovate leaves; large, funnelform flowers greenish-yellow; oblong, grooved greenish-yellow fruit 10-15 cm long, edible but of poor quality. *Tropical.* p. 186, 456

PARMENTIERA: see also Crescentia

PARODIA *Cactaceae*
aureispina (N. Argentina), "Tom Thumb"; tiny 3 cm bluish-green globe with spiralled ribs, and covered with bright yellow spines; golden flowers at an early stage. *Arid-subtropic.* p. 267, 273

camargensis (camblayana) (Bolivia); small globe to 13 cm high, with spiralled ribs, thickly covered with sand-brown spines; flowers carmine-red, edged yellow. *Arid-subtropic.* p. 271

chrysacanthion (Argentina); small cylindric plant to 8 cm high, pale green, with 24-30 tubercled, spiral ribs, set with numerous bristle-like, glossy lemon-yellow spines; flowers golden yellow. *Arid-subtropic.* p. 273

mutabilis (Argentina: Salta); small globe to 9 cm, glaucous to olive green, with tubercles densely set with bristly white radials, and a hooked orange spine; white-woolly at apex; flowers golden yellow. *Arid-subtropic.* p. 273

sanguiniflora (Argentina: Salta), "Red Tom Thumb"; solitary little soft green globes woolly on top, with spiralled tubercles set with bristly white radials and brownish central spines, one hooked; numerous flowers a silky blood red. *Arid-subtropic.* p. 271

PARTHENOCISSUS *Vitaceae*
inserta (quinquefolia vitacea) (Québec to Arizona and Mexico), known in California greenhouses as "Cissus sicyoides", a miniature Virginia creeper, and grown there on treefern poles, an attractive decorative plant, with lustrous deep green, palmately compound leaves only 6 cm across; the 5 leaflets lanceolate and deeply toothed; vines climbing by tendrils without disks; fruit bluish-black. Covering walls of buildings, aided by wires, in the Zona Rosada in Mexico City. *Warm-temperate.* p. 922

quinquefolia (Ampelopsis, Vitis) (U.S.; New England to Florida and Texas), "Virginia creeper", or "American ivy"; vigorous climber with smooth stems having tendrils with 5-12 branches; sticky-tipped; large palmate foliage with usually 5 coarsely toothed dull green leaflets 4-15 cm long, glaucous beneath, turning crimson-red in autumn; small blue-black berries; hardy. *Temperate.* p. 920

tricuspidata (Vitis veitchii) (Japan, China), "Boston-ivy" or "Japanese ivy"; popular, quick-growing deciduous climber attaching itself to walls and houses by their sticky-tipped tendrils; densely shingled, variable smallish foliage, mostly broadly ovate irregularly lobed and toothed, or trifoliolate, 5-12 cm wide; when adult occasional large leaves are produced 15-25 cm across; the shining green color changes to autumn shades of crimson; small blue berries. *Warm temperate.* p. 920

tricuspidata 'Lowii'; attractive variety dense with 5 to 7-lobed leaves, often forked and resembling a butterfly, each wing with about 3 main lobes; the bronzy green changing to red in autumn. *Warm temperate.* p. 921

tricuspidata 'Veitchii', "Japanese ivy"; cultivar of compact, dense habit in climbing; young leaves 3-5 cm across, purplish. *Warm temperate.* — p. 920

PASPALUM Gramineae
notatum (W. Indies, Mexico, So. America), "Bahia grass"; perennial lawn grass spreading by rhizomes; although coarser than Zoysia or Bermuda, is widely planted in Florida for its toughness and ability to thrive in sandy soil and disease resistance; leaf blades flat or folded, 5 mm wide, green or grayish green. Usually established by seed; photo shows the trade-named "Argentine Bahia". *Subtropic.* p. 510

PASSIFLORA Passifloraceae
x alato-caerulea (pfordtii) (alata x caerulea), "Showy passion flower"; free-blooming hybrid well-known because of its large and showy, fragrant, axillary, 10 cm flowers with sepals white, petals pink, and a fringed crown purple, white and blue; with trilobed leaves. *Tropical.* p. 817

amethystina (Brazil); luxuriant climber with very slender stem, smooth leaves deeply 3-lobed, 10 cm wide; flowers 7 cm across, lapis-lazuli blue, corona rays deep purple. *Tropical.* p. 815

antioquiensis (Colombia), "Banana passion fruit"; tendril-climber with slender stems, leaves either lanceolate and unlobed, or deeply 3-lobed with long slender-pointed segments; beautiful flowers rich rosy-red to 12 cm across, with long tube, small violet corona; fruit edible. *Tropical.* p. 817

aurantia (mixta) (New Guinea to New Hebrides); vigorous climber with glossy green, deeply 3-lobed leaves; lovely starry flowers with rosy, convex petals, and blood-red corona; photographed in the rainforest in the mountains of S.E. Papua. *Tropical.* p. 815

bomareifolia (Venezuela); rapidly climbing tendril vine with waxy green, leathery ovate leaves, 6-7 cm long; the flowers 7 cm dia., with linear sepals and petals scarlet red. *Tropical.* p. 814

caerulea (Brazil), a showy "Blue passion flower"; religious symbol to early missionaries who saw in its greenish-white petals the 10 apostles at the crucifixion, in the blue, white and purple rays of the corona the crown of thorns, in the 5 anthers the wounds, the 3 stigmas the nails; cords and whips in the coiling tendrils of the vine, and in 5-lobed leaves the cruel hands of the persecutors. The 6-10 cm flowers keep best floating on water. *Tropical.* p. 816, 817

capsularis (W. Trop. America); climber with 3 to 5-angled stem, two-lobed leaves, downy beneath; 2½-5 cm flowers greenish-white or yellow-green, pale yellow spreading corona. *Tropical.* p. 814

cinnabarina (Australia); slender climber with 12 cm 3-lobed foliage; scarlet 6 cm flowers star-like with 5 long sepals, the petals shorter, the corona yellow; green, aromatic fruit. *Tropical.* p. 816

coccinea (Trop. So. America), a "Red passion flower"; climbing with grooved purplish downy stems, leaves 8-15 cm long, ovate and coarsely crenate; free-blooming with flowers of medium size, with glowing scarlet petals, red sepals yellowish outside, the crown filaments deep purple, pink to white base; ovoid yellow or orange fruit 5 cm dia., and edible. *Tropical.* p. 816, 817

coriacea (So. Mexico to Perú), "Bat-leaf vine"; vigorous climber with red stems, interesting mainly because of its hard to being brittle, transversely oblong-peltate leaves, more broad than long like a butterfly, blue-green blotched with silver-gray; small 2½ cm flowers in clusters, with pale green petals, a yellow ray-crown, and purplish-chocolate base. *Tropical.* p. 814, 816

edulis (Brazil), the "Purple granadilla"; sturdy climber with angular stems, large 15 cm leaves deeply lobed, with wavy edges; 6 cm white flowers with corona white banded with purple, mostly summer-blooming; the 8 cm aromatic, edible fruit thickly purple-dotted and quite ornamental; widely grown in warm climates for its delicious flavor in beverages, fruit salads, sherbets, also jam and marmelade; fruit in spring and fall. *Tropical.* p. 473, 815, 816

edulis fa. flavicarpa; form with fruit maturing yellow. p. 473

holosericea (Central America); tall-climbing, densely downy, with smooth leaves lightly lobed to 8 cm wide; fragrant flowers 4 cm across, in axillary double clusters, the sepals white or lemon-yellow, petals white spotted with red, corona rays deep yellow with purple base; several flowers at each node. *Tropical.* p. 814

'Imperatrice Eugenie' (alata x caerulea); famous hybrid; tendril-climbing, with deeply 3-lobed rich green leaves; flowers 50% larger than caerulea, measuring 10-16 cm across; white sepals, rosy-purple petals, ray-crown purple and white. *Tropical.* p. 815

maculifolia (Venezuela), "Blotched-leaf passion vine"; attractive climber with lightly hairy, wiry stem and curious, nearly triangular leaves, with three brief lobes at the cut-off apex, the quilted surface green, irregularly spotted and marbled with yellow; the underside purple and glandular-hairy; small creamy-white flowers. Also known as maculosa and organensis. *Tropical.* p. 816

maliformis (Trop. America), (naturalized on Tonga, South Pacific); strong tendril-climber to 6 m; ovate green leaves 15 cm long; fragrant flowers to 10 cm across, sepals green, petals greenish-white spotted purple, rays of corona white and violet; edible 5 cm yellow fruit within hard green capsule. Juice with grape-like flavor. *Tropical.* p. 815

novo-guineensis (New Guinea); collected on Chimbu Pass at 2,500 m, in the Goroka highlands of New Guinea; robust climber with beautiful, large clear pink flowers 10 cm across, followed by oblong yellow fruit. *Subtropic.* p. 816

quadrangularis (Trop. America), "Giant granadilla"; robust climber with winged stems and oval leaves, much grown in the tropics for its edible fruit, which is oblong, 12-25 cm long, yellowish green and pulpy; 8-12 cm fragrant flowers with oval, white sepals, reddish petals, and crown with 5 rows of white and purple rays. *Tropical.* p. 817

racemosa (princeps) (Brazil), a "Red passion-vine"; climbing by tendrils, with deeply 3-lobed or occasionally ovate leaves; 10 cm flowers in pendulous racemes, the narrow petals rosy-crimson and spreading, the fringed crown with outer rays purple tipped white, the short inner rays red. *Tropical.* p. 815, 816

rubra (West Indies); climber 3 to 5 m, stems softly downy; 2-lobed or sometimes 3-lobed 5-12 cm quilted leaves purplish beneath; flowers 5 cm across, sepals and smaller petals light carmine rose, outer rays of corona reddish-purple; 2 cm red fruit. *Tropical.* p. 817

seemannii (Trop. America); robust tendril climber, with smooth, broad ovate leaves 5-12 cm long, lighter below; large flowers blue and white to 10 cm across; oblong fruit 5 cm dia. *Tropical.* p. 815

trifasciata (Venezuela, Brazil, Perú), "Three-banded passion vine"; ornamental climber with wiry stems and broad, beautifully colored, lightly 3-lobed leaves, satiny olive to deep bronze-green with 3 broad, pink to silvery-green zones along the purple veins, and purple beneath; small, fragrant, yellowish flowers 3 to 4 cm across; globose 2½ cm fruit. *Tropical.* p. 815

violacea (Brazil, Paraguay, Bolivia), "Purple passion flower"; tall climber with trifoliate leaves, the segments occasionaly lobed again, gray-green beneath; fragrant 10 cm, pretty flowers pendant from long stalks, sepals pinkish-lilac, petals violet, crown with violet and white rays. *Tropical.* p. 815

vitifolia (Tacsonia) (Nicaragua to Perú), "Crimson passion flower"; climber with rusty-downy stems and grape-like leaves, coarsely toothed; showy, 10-15 cm flowers orange-scarlet to blood-red, with bristle-tipped sepals, outer rays of corona bright red, the shorter, inner ones pale red; shy bloomer. *Tropical.* p. 814, 816

PAULOWNIA Scrophulariaceae
tomentosa (China), "Empress tree"; deciduous, hairy tree 12-20 m tall, resembling Catalpa, with horizontal branches; light green, cordate leaves 12-30 cm long; flowers before leaves, forming upright clusters of trumpet-shaped, fragrant flowers of lilac blue with dark spots and yellow stripes inside, 5 cm long. *Warm temperate.* p. 884

PAUROTIS Palmae
wrightii (Acoelorrhaphe in Hortus 3) (So. Florida, West Indies, C. America), "Everglades palm"; cluster palm at home in the Everglades swamps, with slender trunks to 12 m, covered by red-brown matting and bases of leaf stalks; palmate nearly round leaves 60-90 cm dia., divided more than halfway, with stiff segments light green above, silvery beneath, and split at apex; young plants have entire leaves not divided. *Tropical.* p. 792

PAVETTA Rubiaceae
angustifolia (So. Africa); evergreen bush with corrugated leaves; attractive rounded, 8-10 cm heads dense with slender tubular, sweetly scented greenish-white flowers with long protruding styles. *Subtropic.* p. 866

lanceolata (So. Africa), "Christmas bush"; evergreen to 3 m, with leathery, narrow long-lanceolate leaves; clusters of showy white flowers. *Subtropic.* p. 866

opaca (So. Africa), "Christmas bush"; photographed in Kirstenbosch Botanic Garden, Feb., 1977; evergreen shrub to 3 m; glossy dark green lanceolate foliage; large clusters of sweetly fragrant white flowers with long ivory styles, blooming in summer. *Subtropic.* p. 865

revoluta (obovata) (So. Africa); neat evergreen bush to 1 m, with glossy green, rounded leaves; sweetly scented white flowers with long protruding white styles. *Subtropic.* p. 866

PAVONIA *Malvaceae*

intermedia (Brazil); possibly a form of multiflora, very variable; shrubby plant with spear-shaped glossy, rich green leaves, rough beneath, and with smooth or barely toothed margins, 15-20 cm long; inflorescence with small purple flowers and showy crimson red bracts, unfolding into flaring narrow filaments and revealing pink style and purple stamens; *Tropical.* p. 624

intermedia 'Kermesina'; a commercial form of compact habit, the inflorescence with scarlet petals, and feathery carmine-red bracts. Charming winter bloomer. *Tropical.* p. 624, 625

multiflora (wiotii) (Brazil); robust tropical shrub with lanceolate, toothed, slender-pointed leaves and unusual flowers in terminal clusters, with many hairy, narrow red bracts in whorls below calyx, the calyx purplish, and purple corolla rolled together; September. *Tropical.* p. 624

PEDDIEA *Thymelaceae*

africana (Cape Prov. to E. Transvaal); evergreen branching shrub to 4 m high; elliptic, thin-leathery 8 cm leaves shining dark green; inflorescence in umbels of 12-16 pale green, slightly scented, tubular flowers. *Subtropic.* p. 908

PEDILANTHUS *Euphorbiaceae*

bracteatus (Mexico); succulent bush with smooth stems, fleshy oblong leaves 8-10 cm long, occasionally notched at tips and rounded at base; the inflorescence with large crimson lacquered bracts subtending the greenish flowers. *Subtropic.* p. 419

tithymaloides retusus (Amazonas); succulent with curious fleshy leaves oval, but notched at apex; small red flowers. *Tropical.* p. 418

tithymaloides 'Variegatus' (W. Indies), "Zigzag plant"; branching succulent bush with milky juice, gray green, fleshy stems bent with each waxy, pale green, ovate leaf, highly variegated white, and tinged carmine-red; flowers red. *Tropical.* p. 418

PEDIOCACTUS *Cactaceae*

knowltonii (Colorado); tiny globe 4 cm high, with conical tubercles tipped by white radial spines; small pink flowers 2 cm across. *Temperate.* p. 272

PELAGODOXA *Palmae*

henryana (Marquesas, French Polynesia), "Vahana palm"; unusual feather palm with brown trunk to 8 m tall and 20 cm thick, without crown-shaft; spectacular leaves 3 m long, pinnately veined but undivided except for apex and notched margins, vivid green above, silvery underneath; 10 cm fruit. *Tropical.* p. 791

PELARGONIUM *Geraniaceae*

crassicaule (So. Africa), "Succulent geranium"; a succulent species with thick, fleshy stems and small, pubescent, lobed leaves; flowers white, the upper petals spotted purple. *Subtropic.* p. 490

x domesticum; "Martha Washington geranium" or "Regal geranium"; complex hybrids involving P. grandiflorum and other species; soft-hairy brittle stems, to 50 cm or more high; toothed leaves 5-8 cm across; large showy flowers 3-6 cm, in clusters in colors white, pink, red, purple, the two upper petals with dark blotches and veins. *Subtropic.* p. 485

x domesticum 'Circus Day'; a premium cultivar with big heads of very large flowers rosy pink, the upper petals deep pink with velvety brown-black blotch flushed with salmon. *Subtropic.* p. 486

x domesticum 'Easter Greeting'; popular early and long-blooming variety; rosy-carmine flowers 6 cm across, each petal with a long black blotch. *Subtropic.* p. 486

x domesticum 'Fire Dancer'; brilliant cultivar with plain green leaves setting off the vivid crimson of the well-rounded flowers; dark maroon blotch in center, and a narrow white edge all around. *Subtropic.* p. 486

x domesticum 'Gay Nineties'; good commercial cultivar with large 6 cm frilled flowers white with crimson eye. *Subtropic.* p. 486

x domesticum 'Grand Slam'; excellent, free-blooming premium variety of bushy habit, and with handsome umbels of large, good-holding flowers flame-red, with wavy margins, the top petals with velvety red-brown eye. *Subtropic.* p. 486

x domesticum 'Grossmama Fischer'; favorite salmon pot plant of compact habit, early blooming with large wavy flowers of clear rich salmon, marked brown-black. *Subtropic.* p. 486

x domesticum 'Jessie Jarrett'; flowers of an exceptional shade of dark violet rose to purple with black maroon blotch on all petals. *Subtropic.* p. 486

x domesticum 'Lavender Queen'; popular, long-blooming variety with large flower clusters lavender blotched black-purple on all petals. *Subtropic.* p. 486

x domesticum 'MacKay'; well-shaped commercial plant with flowers dark salmon-pink, and crimson blotches in center. *Subtropic.*

x domesticum 'Marie Vogel' (grandiflorum hyb.), "Lady Washington pelargonium"; popular, free-blooming variety with large, showy flowers of salmon-rose, with crimson blotches on upper petals; toothed, glabrous leaves on hairy, brown-woody stems. *Subtropic.* p. 486

x domesticum 'Mrs. Mary Bard'; compact grower and early bloomer, with medium-size flowers pure white, striped purple at base of petals. *Subtropic.* p. 490

x domesticum 'Orchid Edith North'; color variation of the free-blooming 'Edith North' with petals pale purple, heavily overlaid and striped with dark purple in upper petals. *Subtropic.* p. 486

x domesticum 'Pansy'; "Pansy geranium"; pretty hybrid with 'Mrs. Layal' but with flowers 3 times as large; the lower petals white with purple lines, the two upper petals almost entirely violet-purple edged white. *Subtropic.* p. 487

x domesticum 'Springtime'; lovely, long-blooming variety with white flowers prettily edged bright rose, and dainty, ruffled white margin. *Subtropic.* p. 486

x domesticum 'Swabian Maid'; old commercial variety, with medium-sized 5cm flowers carmine-rose, with darker upper petals and blackish-crimson blotches. *Subtropic.* p. 486

echinatum (Cape Prov.), "Sweetheart geranium", a "Cactus geranium"; succulent stem with hooked, soft, thorn-like stubbles, and gray-green, lobed leaves, white hairy beneath; single white flowers marked maroon in the shape of a heart. Deciduous when resting. *Arid-subtropic.* p. 487

grandiflorum (So. Africa); a species responsible for the form and size of the modern 'domesticum' group; shrubby plant with small toothed, lobed leaves and relatively large pale purplish flowers with purple stripes on upper petals. *Subtropic.* p. 490

graveolens (Cape Prov.), old fashioned "Rose geranium"; the rose-scented leaves of which are used in cookery, sachet, and for making perfume; bushy plant with deeply lobed leaves appearing gray-green because of their covering of soft white hair; blooms lavender pink marked purple. *Subtropic.* p. 490

x hortorum (So. African origin), "Zonal geranium"; as grown by commercial nurseries today, these are of complex hybrid origin, largely derived from P. zonale and inquinans, familiar as potplants, in window boxes or in garden beds. Succulent stems 30-50 cm high with rounded or kidney-shaped leaves 6-10 cm across, often colorfully zoned or variegated; the flowers in clusters, with petals nearly equal, in many vivid colors, primarily scarlet red, purple, soft pink or salmon, pure white, or bicolor, also double-flowered and miniatures; in many fancy-named cultivars. *Subtropic.* p. 483

x hortorum 'Antares'; miniature geranium with dark foliage and large, single flowers in dark burning scarlet. *Subtropic.* p. 488

x hortorum 'Better Times'; commercial cultivar with large semi-double flowers salmon-scarlet fading lighter, and with crimson lines; low growing, compact plant with medium green foliage. *Subtropic.* p. 488

x hortorum 'California Appleblossom', or 'Appleblossom single'; eye-catching geranium with large single flowers fiery cherry-red changing to white in center. *Subtropic.* p. 488

x hortorum 'Carefree'; all-American selection as an outstanding color-form of a new F_1 hybrid or heterosis seed strain of single and double garden geraniums by Pan-American Seed Co.; grown directly from seed, and available also in shades of pink and red, also bicolors and white. Geraniums from seed are not as early as those propagated from cuttings, but begin to bloom in June-July and are at their best from August to October. Seed has to be started warm at a constant 21-24°C; for early bloom in Mid-May sow beginning of January; grow with all the light or sun possible, keep warm to initiate buds, then intermediate or cool, although they tolerate high summer temperatures. *Subtropic.* p. 489

x hortorum 'Daybreak Salmon'; compact geranium with zoned leaves, and large clusters of semi-double flowers of salmon pink. *Subtropic.* p. 488

x hortorum 'Enchantress Fiat'; bushy plant with zoned leaves and beautiful double flowers blush pink with deeper rose in center. *Subtropic.* p. 488

x hortorum 'Genie Irene'; attractive, vigorous cultivar very free-blooming, large semi-double flowers on unusual clear pink. *Subtropic.* p. 488

x hortorum 'Improved Ricard'; excellent commercial cultivar, freely branching, of shapely habit, foliage only lightly zoned; and with good heads of large semi-double flowers light brick-red or orange-scarlet. *Subtropic.* p. 488

x hortorum 'Irene'; good commercial, bushy plant with fine, zoned foliage, and free-blooming with semi-double red flowers close to American beauty cerise. Introduced 1942 and progeny of many excellent named varieties. *Subtropic.* p. 489

x hortorum 'Madame Salleron', "Carpet-bed geranium"; a unique, old French hybrid which I remember my father using in fancy carpet bedding; a small plant sprouting multiple short branches from the base, with little papery, crenate leaves glistening gray-green with white border on long, thin petioles; rarely blooming, with small salmon-red flowers. *Subtropic.* p. 489

x hortorum 'Miss Burdett Coutts'; a compact old tricolor with beautiful, firm, rounded leaves differing from others by its silvery hue, the leaf center is grayish-green, margins ivory-white, and the circular zoning deep blood-red splashed with rose-pink; single scarlet flowers. English hybrid 1860. *Tropical.* p. 487

x hortorum 'Mr. Wren'; a very striking variety with single, bright red flowers edged in white, but not very freely produced, on strong plants. *Subtropic.* p. 487

x hortorum 'Mrs. Henry Cox'; medium-size plant with tricolored foliage, grayish green center, followed by a rosy-red zone bordered in creamy yellow; salmon flowers. *Subtropic.* p. 487

x hortorum 'Mrs. Parker'; an excellent old free-blooming English variety of compact habit with shimmering green, rather thin, quilted leaves beautifully bordered in white; small semi-double, soft pink flowers. *Subtropic.* p. 487

x hortorum 'Mrs. Pollock'; bushy plant having tricolor leaves with yellow margin, red and crimson break from the broad brown zone; single scarlet flowers. Famous fancy-leaved England 1858. *Subtropic.* p. 487

x hortorum 'Mrs. Strang'; beautiful multicolored foliage silvery-green zoned with black-maroon and red, edged with gold and canary-yellow. *Subtropic.* p. 487

x hortorum (zonale) 'Olympic Red' (Ricard x Radio Red); popular commercial geranium of early-blooming, free-flowering and branching habit; beautifully zoned leaves and large heads of semi-double, fiery red flowers. *Subtropic.* p. 485, 488

x hortorum 'Penny Irene'; excellent bloomer, with large heads of semi-double, dark pink flowers. *Subtropic.* p. 488

x hortorum 'Radio Red'; tall grower with slender stems, light green foliage, and with a profusion of medium-sized double flowers intense clear crimson. *Subtropic.* p. 488

x hortorum 'Red Fiat'; rich red sport of the salmon-pink 'Fiat'; constant bloomer with semi-double flowers, on self-branching plant with pubescent foliage. *Subtropic.* p. 488

x hortorum 'Red Rambler'; vigorous plant with green velvety, lobed leaves, and large clusters of fully double flowers with white and red petals. *Subtropic.* p. 485

x hortorum 'Salmon Supreme'; distinctive commercial variety of the French type, more vigorous than 'Poitevine' and a freer bloomer but of the same fine, large semi-double flowers of clear light salmon pink. *Subtropic.* p. 488

x hortorum 'Skies of Italy', "Mapleleaf tricolor geranium"; small plant of tricolored, maple-like leaves with pointed lobes, carried on slender petioles, green with creamy marginal variegation, and a brown zone tinted with orange-red and crimson; single scarlet-red flowers. *Subtropic.* p. 487

x hortorum 'Velma', "Tricolor geranium"; gaily-colored, very compact geranium with small tricolored foliage, the rounded, near waxy leaves grayish green in center, surrounded by a rosy-red zone and bordered in creamy-yellow; single salmon 3 cm flowers; of stronger growth, larger leaves than Mrs. H. Cox. *Subtropic.* p. 487

inquinans (South Africa); spreading species with hairy, brown branches and plain green velvety, fleshy 7-lobed leaves, toothed at margins; profuse with stalked clusters of vivid brownish-red single flowers, with petals almost 2½ cm long. *Subtropic.* p. 485

x limoneum 'Variegatum', "English finger-bowl geranium"; attractive crispum hybrid with lemon-scented small leaves lobed and crisped, grayish pubescent and with white margins; flowers pink with purple stripes. *Subtropic.* p. 490

peltatum (Eastern South Africa), called the "Ivy geranium", because of the ivy-like, pointed 5-lobed leaves which are fresh green, waxy and rather succulent, 5-8 cm across, on trailing, zigzag branches; flowers normally single and rose-carmine; but now highly hybridized into many types and colors with single or double flowers, in rounded clusters, in white, pink, rose, red and lavender; the 2 upper petals usually blotched or striped. Ivy geraniums are ideally suited for hanging baskets, window boxes, patio containers or as ground cover. *Subtropic.* p. 489

peltatum 'Galilee'; old cultivar with clusters of double flowers a soft pink. *Subtropical.* p. 489

peltatum 'L'Elegante', an old French variety known as "Sunset ivy"; with beautiful waxy gray-green leaves edged in white, later adding a pink tinting or red border; small single white flowers with lavender tints, and darker markings in throat. *Subtropic.* p. 483

peltatum 'Lumiere du Matin' (peltato-zonale hyb.); floriferous French hybrid with long creeping or pendant stems and waxy green, ivy-leaf type foliage; similar to 'Ville de Paris' but with single flowers crimson-red; very successful in sun-starved, cool regions; ideal for window boxes and containers outdoors. *Subtropic.* p. 489

peltatum 'Mexico', bicolor "Ivy geranium"; striking cultivar seen in Germany 1976; free growing with long branches; fresh green waxy, lobed leaves, and large double flowers with petals white in center, red around the margins. *Subtropic.* p. 489

peltatum 'Variegatum', "Variegated ivy geranium"; charming variety with silvery-green waxy foliage, having margins beautifully bordered with creamy-white. *Subtropic.* p. 489

'Prince Rupert variegated', also known as "French lace"; lemon-scented; habit of P. crispum but bushier and more prolific, the small lobed and crisped leaves are light green and prettily edged in white. *Subtropic.* p. 490

tomentosum (Cape of Good Hope), "Peppermint-geranium"; strongly scented species of sprawling habit, with large, soft velvety, emerald green leaves, triangular-heartshaped, shallowly lobed, with a felt-like covering of white hairs; small, fluffy white flowers. *Subtropic.* p. 490

'Ville de Paris' (peltato-zonale hyb.), "Strassbourg geranium"; color form of this new French hybrid race, introduced about 1965; profusely blooming ivy-leaf type geranium with thin-leathery leaves, and 3-4 cm single flowers with narrow-linear or obovate petals glowing carmine-rose with blood red stripes, in small clusters but forming a blanket of color with cascades to 1 m long, during summertime, even in sun-poor regions, where it has become immensely popular in window boxes and outdoor urns. *Subtropic.* p. 485, 489

violarium (So. Africa), "Viola geranium"; odd species with low spreading branches; long-stalked powdery-gray ovate, toothed leaves; and pansy-faced flowers, with lower petals white, the two upper petals ruby red with dark blotch. Nice in baskets and hanging pots, blooming in spring and summer. *Subtropic.* p. 485

zonale (So. Africa), "Zonal geranium"; principal parent of P. x hortorum; fleshy stem with rounded, lobed leaves fresh green with brown-red zone; flowers pale pink. *Subtropic.* p. 490

PELECYPHORA   *Cactaceae*

aselliformis (Mexico: San Luis Potosi), "Hatchet cactus"; globular to cylindric plant to 5 cm thick, with grayish green tubercles strongly flattened sidewise and in spirals, the minute spines arranged comblike and not prickly; flowers purple. *Arid-tropical.* p. 280

pseudopectinata (Thelocactus) (Northern Mexico); globular plant to 6 cm, at base nearly square, flattened on top, the areoles elongate with spines arranged comb-like; flowers pale pink with red center band. *Arid-subtropic.* p. 280

PELLAEA   *Polypodiaceae (Filices)*

rotundifolia (New Zealand), "Button fern"; small rockloving fern with creeping rhizome and pubescent stems, fronds nearly uniform, and staying near ground; simply pinnate, evenly spaced leaflets, round when young, later oblong, dark green and waxy leathery. *Humid-subtropic.* p. 442

viridis macrophylla, better known in horticulture as Pteris adiantoides; variety with bipinnate fronds to 45 cm high, having thin-leathery leaflets much larger and broader than in the species, although less in number on each pinna, and resembling Holly-fern, but leaves are not as leathery. Used in small fern-dishes as a "Table-fern". From S.E. Africa. *Humid-subtropic.* p. 444

PELLIONIA   *Urticaceae*

daveauana (So. Vietnam, Malaya, Burma), "Trailing watermelon begonia"; depressed herbaceous creeper with succulent, pinkish stems and 2-ranked, flattened, alternate, thin-fleshy leaves oval when small, lanceolate-pointed, and 2-5 cm long in older plants, brown-purple to blackish with pale green to gray center area, base oblique. *Tropical.* p. 912

pulchra (Vietnam: Cochin), "Satin pellionia"; attractive fleshy creeper hugging its support or pendant from a basket, with pinkish stems and stipules, obliquely oval leaves light green to grayish, entirely covered with a network of blackish or brownish veins, pale purple beneath on gray. *Tropical.* p. 912

PELTANDRA   *Araceae*

virginica (Maine to Florida and Missouri), "Arrow arum"; hardy aquatic perennial herb with thick fibrous roots and large bright green, firm arrow-shaped leaves 15 to 75 cm long, on long sheathing petioles; undulate, fleshy, greenish-white spathe to 20 cm long; the spadix with female flower in lower third; male, staminate flowers on upper two-thirds. *Temperate.* p. 110

PELTOPHORUM   *Leguminosae*

pterocarpum (Malaysia to Australia), "Yellow flame tree"; stately tree to 15m, with heavy luxuriant, bipinnate foliage to 60cm long; large erect panicles of round, rust-colored buds opening into orange yellow fragrant flowers 4cm across. *Tropical* p. 561, 563

PENNISETUM Gramineae
  setaceum 'Cupreum' (Africa: Ethiopia), "Red fountain grass"; perennial grass with simple stems 1m high, forming clumps; narrow arching leaves 60cm long, reddish brown, and reddish spikes. *Subtropic* p. 512

PENSTEMON Scrophulariaceae
  barbatus (Utah to Mexico), "Scarlet beard-tongue"; smooth perennial to 2m high, with glaucous stems and narrow lanceolate leaves; tall stems with racemes of elongate two-lipped scarlet flowers. *Warm temperate* p. 883
  x gloxinioides, "Beard-tongue"; garden hybrid of Mexican P. hartwegii; strong herbaceous perennial with lance-shaped leaves and dense spikes of large, gloxinia-like, bell flowers to 5cm across, in many brilliant colors, blooming in autumn. *Temperate* p. 887
  hartwegii (Mexico), "Beard-tongue"; smooth perennial to 1 m high, narrow lanceolate leaves to 10cm; stiff stalk with tubular flowers 5cm long, more or less two-lipped, scarlet-red with white inside tube. *Subtropic* p. 887

PENTAPTERYGIUM Ericaceae
  serpens (E. Himalayas, Khasias); epiphytic shrub 60-90cm high, from a tuberous rootstock, with slender pendant stems, small lanceolate 2cm leathery leaves, and tubular nodding flowers bright red with darker markings. *Subtropic* p. 397

PENTAS Rubiaceae
  lanceolata (Trop. Africa, Arabia), "Egyptian star-cluster"; herbaceous flowering plant with woody base, downy branches and soft or limp, ovate, bright green, hairy leaves, with sunken veins; the tubular flowers in showy clusters, hairy in the throat, purplish rose. *Tropical* p. 862, 866
  lanceolata 'Pallida'; cultivar with clusters of pale pink flowers. *Tropical* p. 866

PEPEROMIA Piperaceae
  arifolia litoralis (Sao Paulo); compact rosette with hard, flat, cordate leaves dark green with lustrous silver bands between the primary veins, looking like a shining shield. *Tropical* p. 819
  bicolor (Ecuador), "Silver velvet peperomia"; beautiful plant with red-brown stem covered with white hair, and broadly oval, velvety leaves olive to gray metal-green, with broad silver center band as well as parallel silver stripes and edge, rosy-red underneath. *Tropical* p. 819
  campylotropa (Mexico); small tuberous, branching plant with tiny 1½-3½cm orbicular-peltate leaves glossy green, cupped; erect spikes of greenish, insignificant flowers; in the collection Botanic Garden, Munich, Germany. *Tropical* p. 818
  caperata (Brazil), "Emerald Ripple"; sturdy, very useful little species with short, branching stem developing dense clusters of roundish, heartshaped or peltate leaves deeply corrugated and quilted like a washboard, waxy forest-green, the valleys tinted chocolate, the ridges often grayish, reverse pale green, the pink petioles striped red; the slender flowering catkins greenish-white. *Tropical* p. 818
  caperata 'Red Ripple'; attractive new (1977) horticultural cultivar with succulent brownish stem, red petioles, and corrugated broad cordate leaves 6-8cm long, and with depressed parallel veins, the surface glossy forest green with traces of silver on the high ridges, reverse red between veins. *Tropical* p. 818
  caperata 'Variegata', "Variegated ripple"; attractive variety with the little succulent, quilted leaves waxy deep green and broadly margined with white, on pink stalks.*Tropical* p. 819
  cerea (in hort. as 'Dr. Goodspeed') (Perú), introduced in California; succulent little bush to 45 cm high; stem red of numerous stripes, set dense with fat little ovate, 2½cm leaves in whorls of three, matte olive green; the underside keeled and brown-purple. *Tropical* p. 819
  clusiaefolia (West Indies), "Red-edged peperomia"; stocky, slow-growing plant with thick-fleshy, rather narrow-obovate, concave leaves, metallic olive-green with broad, red-purple margin, light green beneath except for the purple midrib. *Tropical* p. 820
  cubensis (rotundifolia) (West Indies); known as 'rotundifolia' or "Yerba Linda" from Puerto Rico, this little succulent has soft, waxy, friendly fresh-green broad ovate leaves with depressed parallel veins, on reddish stems. *Tropical* p. 822
  cuspidilimba (Perú); attractive trailer with flexible wiry stems, narrow elliptic, pointed leaves moss-green with pale parallel veining, and red underside, set in whorls at intervals on branches. *Tropical* p. 818
  galioides (Colombia); small branching plant with succulent light green stems with a dot of red at joints and glands beneath and waxy fresh green needle-like leaves 2½cm long in whorls of 4-5, depressed along center. *Tropical* p. 819
  glabella 'Variegata', "Variegated wax privet"; dainty, freely branching plant with slender rosy-red stems and small elliptic leaves light green or milky green broadly bordered or variegated creamy-white. to 4 cm long. *Tropical* p. 819
  griseo-argentea (Brazil), known in Brazil and introduced in Europe as hederaefolia. "Ivy peperomia"; very attractive, bushy rosette with long pink petioles bearing round cordate, shield-like, thin, quilted leaves painted with glossy silver, while the sunken veins are purplish olive; long, erect, greenish-white catkins. Somewhat difficult in cultivation. *Tropical* p. 822
  hoffmanii (C. America); low succulent creeper hugging the ground, with reddish stems and tiny 1cm obovate waxy leaves dull olive green set in whorls of 4 along branches; collection Brooklyn Botanic Garden. *Tropical* p. 822
  incana (Brazil), "Felted pepperface"; amazing little plant that should be rated as a succulent; I have found it growing on granite rocks between cacti and bromeliads on the Restinga in South-east Brazil; stiff green stem with broadly heart-shaped stiff-fleshy, gray leaves entirely covered with white felt. *Arid-Tropical* p. 820
  maculosa (Santo Domingo), "Radiator plant"; ornamental, fleshy species with long pendant, narrow-lanceolate leaves to 16cm long, waxy bluish gray-green, with silvery-green to ivory ribs; petioles prettily spotted red-purple; inflorescence of tiny flowers in spikes to 30cm long. *Tropical* p. 820
  magnoliaefolia (West Indies), "Desert privet"; robust species with large, fleshy obovate-elliptic leaves 10-12cm long, glossy fresh-green with depressed veins, on brownish stem; stalk of flower spike not hairy. *Tropical* p. 820
  margaretifera (Juan Fernandez, Chile), "Pepper face"; spreading bush with glossy green elliptic leaves on erect stems; collection Brooklyn Botanic Garden. *Subtropic* p. 822
  obtusifolia (Venezuela), "Pepper face"; long cultivated as a good dishgarden plant; succulent stem with reclining base and short petioles both striped maroon-brown, with waxy-green, fleshy obovate or spatulate, 5-8cm, concave leaves, obtuse or notched at apex, pale green beneath, growing to 30cm high; stalk of flower spike minutely hairy. *Tropical* p. 820
  obtusifolia 'Sensation'; (in the collection of Brooklyn Botanical Garden); robust, succulent plant with thick-fleshy obovate leaves glossy green, and irregularly variegated and marbled with cream, some leaves almost entirely cream, and with fine red edging. *Tropical* p. 820
  obtusifolia 'Variegata', "Variegated peperomia"; beautiful, small, succulent plant with pale stems blotched bright red; alternate, rounded or obovate-elliptic, waxy leaves, light green variegated with milky-green, and from the margin inward, a broad area of creamy-white. *Tropical* p. 819
  polybotrya (pericatii) (Colombia), "Coin-leaf peperomia"; succulent plant with erect, fleshy, green stem with stiff reddish petioles, the thick, shining, waxy, smooth, peltate-pointed, shield-like leaves of vivid green with fine purple edge, gray-green beneath; branched white catkins. *Tropical* p. 820
  puteolata (Perú), "Parallel peperomia"; gorgeous hanging plant with angled stems and slender, lanceolate, leathery leaves 10cm long, waxy dark green with 5 contrasting, yellowish, sub-translucent, parallel veins depressed on the surface and raised on the light green reverse. *Tropical* p. 818
  rauhii (prob. Trop. America); succulent with stiff-erect, fleshy stems, dense with thick ovate 4cm leaves keeled beneath, all around the stem, smooth velvety olive green. *Tropical* p. 818
  sandersii (argyreia) (Brazil), "Watermelon peperomia"; attractive rosette, almost stemless, with deep-red petioles bearing fleshy, broad peltate-pointed, concave leaves 8 to 10 cm across, glossy fresh-green to bluish, painted with showy bands of silver radiating from their upper center, pale beneath; minute flowers in long, whitish catkins. *Tropical* p. 818, 819
  sarcophylla (pseudovariegata) (Ecuador, Colombia); heavy plant with large, broadly lanceolate, thick-fleshy, pendant, 20cm leaves of shining forest-green, ivory-green ribs and a band of gray alongside them, on thick petioles heavily blotched red; long brown catkins. *Tropical* p. 818
  scandens (Perú), "Philodendron peperomia"; scandent creeper with fleshy, reddish stem and petioles, the waxy fresh-green, small heartshaped leaves resembling a philodendron, and well spaced, with long internodes. *Tropical* p. 820
  scandens 'Variegata'; colorful, scandent, semi-erect, fleshy creeper with small cordate leaves gracefully slender-pointed, light green to milky-green, irregularly bordered with creamy-white, on red petioles and reddish strems. Somewhat rank in growth, but useful for baskets. *Tropical* p. 820

verschaffeltii (Alto Amazonas), "Sweetheart peperomia"; a beautiful shapely plant which I rediscovered on the upper Amazon; a short-stemmed rosette of fleshy, oval-heartshaped leaves 10 cm long, similar to marmorata but basal lobes not overlapping, and alternate on short branching stem, the waxy surface is bluish-green with broad silver bands between the recessed yellowish veins, on petioles red with dots, (I.H. 1869). *Tropical* p. 819, 821

PERESKIA Cactaceae
bleo (Panama, Colombia), "Wax rose"; a tree-like primitive cactus with leaves, to 6 m high; the young branches red; thin oblanceolate leaves to 20 cm long, black needle spines; rosy flowers. *Tropical* p. 238

corrugata (Rhodocactus corrugatus) (Trop. America); tall shrub or small tree to 3 m high, with stiff-erect woody stems covered by bundles of slender spines; grass-green elliptic, fleshy leaves 8-32 cm long; flowers orange-red, 2½ cm across. *Tropical* p. 238

grandifolia (Brazil), a "Rose cactus"; a shrub or tree to 4 m with very spiny trunk, the elliptic fleshy waxy, rich green leaves to 15 cm long; flowers like wild roses rose-pink, in terminal clusters. *Tropical* p. 238, 241, 243

PERISTERIA Orchidaceae
pendula (Panama to Guayana); remarkable epiphyte with robust pseudobulbs, with elliptic, plaited leaves to 35 cm long, deciduous before flowering; the inflorescence a short, pendulous raceme of hooded flowers 5 cm dia., lemon-white and heavily spotted with blackish red (winter). *Tropical* p. 747

PERISTROPHE Acanthaceae
hyssopifolia 'Aureo-variegata' (angustifolia var.) (Java), "Marble-leaf"; colorful, spreading herb, with small lanceolate fresh-green leaves to 8 cm long, variegated yellow in center; small rosy flowers. *Tropical* p. 36

PERNETTYA Ericaceae
mucronata (So. America: Magellan region to Chile), "Chilean pernettya"; bushy little evergreen shrub becoming ½-1 m high, with woody branches densely set with small stiff glossy green 2 cm ovate leaves lightly toothed, and tipped by a sharp translucent spine; numerous tiny urnshaped nodding white or pink flowers in late spring, followed on female plants by little 1 cm stalked, depressed globose berries, persistent through the winter; depending on variety these may be white, pink, or brilliant red. Very pretty for tubs or window boxes, but note that berries are poisonous. More fruit is set if more plants are grouped together. *Warm temperate* p. 394

mucronata 'Alba'; rigid, much branched small evergreen shrub, dense with glossy green leaves; in this cultivar the attractive little berries are a glistening ivory white. *Warm temperate* p. 394

PERSEA Lauraceae
americana (W. Indies, Guatemala and Mexico), the "Avocado" tree; also called "Alligator pear"; a round-headed tropical and subtropical tree 6 to 10 m high or more, spreading wide with large leathery, elliptic or oval 10-20 cm leaves, glaucous beneath; and small 6 mm greenish flowers, forming in winter. The Avocado tree grows fast in well-drained soil containing humus, beginning to bear when 4 to 8 years old, the large fleshy apple or pear-shaped edible fruit, for which it is usually cultivated, is about 10 cm across with green or purplish skin, its flesh is buttery and of high nutritional value rich in vitamins and containing 7 to 23% fat; it is served in salads. The cultivar 'Fuerte' belongs to the "Mexican race" which includes the hardiest types (Persea drymifolia x americana); fruit 10 cm long, dull green, with flesh cream yellow; season Jan.-August in California. *Subtropic* p. 463, 536

PESCATORIA Orchidaceae
cerina (Costa Rica); pretty epiphyte from Volcano Chiriqui at 2,300 m, with tufted leaves tapering to the base, the short-stalked, solitary 8 cm flowers fragrant and showy, the waxy sepals and petals lemon-yellow, the convex lip bright yellow with red-brown ridges in throat, (June-Dec.). *Tropical* p. 747

PETREA Verbenaceae
arborea (Trinidad, Guayana, Venezuela), "Queen's wreath tree"; shrub or low tree to 8 m high, cordate-elliptic leaves to 15 cm long; axillary raceme of purplish-blue flowers, calyx with 5 linear lobes, the inner corolla with oval petals. *Tropical* p. 918

volubilis (racemosa) (Mexico to Panama), "Purple Wreath"; one of the most beautiful of tender twiners, climbing perhaps 10 m high, with woody or wiry stems, long brittle-hard, rough leaves to 20 cm long, and showy racemes of lovely, star-like flowers of long, lilac-blue sepals and small violet corolla; in March-April. *Tropical* p. 913, 915, 918

PETRONYMPHE Amaryllidaceae
decora (Mexico); tuberous plant with shoe-string-like narrow, succulent, channeled leaves 60 cm long and 6 mm wide; long-stalked inflorescence of tubular flowers in pendulous umbels, the 6 cm corolla pale creamy-yellow with green length-stripes to tip of corolla lobes. *Tropical* p. 66

PETUNIA Solanaceae
x hybrida 'California Giant'; herbaceous plant with small oval leaves, dwarfed by giant single flowers ruffled toward the margin, 10-15 cm across, in shades from white, rose, orchid, purple, combined with showy centers in contrasting colors or designs; desirable potplant for Mother's Day. *Tropical* p. 896

x hybrida 'Celestial Rose'; dwarf herbaceous, floriferous plant, whose ancestery goes back to the So. American species violacea and nyctaginiflora; of bushy habit, with small, light green, downy leaves, and single fl. carmine-rose with pale eye. *Tropical* p. 896

x hybrida fl. pl. 'Caprice'; compact plant with clear carmine-rose flowers densely double with frilled petals pale beneath, early flowering; a named variety of the Panamerican strain, which also comes in white, lavender, purple, or variegated forms, all of which make superb potplants. *Tropical* p. 896

x hybrida fl. pl. 'Sonata'; favorite commercial pot variety for large fully-double, pure white flowers, beautifully fringed and crisped at margins; should be pinched to encourage bushy growth; Panamerican strain. *Tropical* p. 896

x hybrida grandiflora 'Bingo'; huge bicolor flowers up to 12 cm across, richly variegated wine-red and white in shape of a star; part of the blooms are only 8 cm in dia., but are equally colorful; of compact habit. *Tropical* p. 896

x hybrida grandiflora 'Crusader'; large single flowers gaily striped or banded starlike with bright carmine-rose on white base with nicely frilled white edge. *Tropical* p. 896

x hybrida grandiflora 'Elk's Pride', "Blue balcony petunia"; so reminiscent of the deep blue single balcony petunias in the Swiss and Bavarian Alpine villages, but of more erect and shorter habit; still the favorite rich velvety deepest violet shade, though some strains become tall and lanky. *Tropical.* p. 896

x hybrida grandiflora 'Pink Magic'; F-1 or first generation hybrid, a favorite bedding and pot plant of uniform, dwarf, bushy habit, loaded with numerous medium size single, smooth flowers of bright carmine-rose with darker veining and small white eye; larger than 'Celestial Rose'. *Tropical.* p. 896

x hybrida grandiflora 'Popcorn'; good pot variety because of its shorter, bushy habit, yet with large white, single blooms with quilted surface and ruffled margins. *Tropical.* p. 896

x hybrida multiflora 'Glitters'; an All-America winner for 1957; floriferous F-1 hybrid of the smaller single flowered, bushy multiflora strain ideal for bedding and pots, a striking combination of a salmon-red star alternating with white. *Tropical.* p. 896

x hybrida multiflora 'Satellite'; lovely free-blooming F1 hybrid with brilliant bright rose flowers with a perfect white star; excellent bedding or pot plant. *Tropical.* p. 896

x hybrida multiflora 'Summer Fun'; F1 hybrid, a smaller-flowered, but free-blooming compact bedding type, in unusual shade of bright yellow, 5 cm across. (Geo. Ball, W. Chicago, 1976). *Tropical.* p. 896

PHAEDRANTHUS: see Distictis

PHAEOMERIA: see Nicolaia

PHACELIA Hydrophyllaceae
campanularia (Colorado to So. California: Mohave), "California bluebell"; bristly-hairy annual to 50 cm high, with elliptic, coarsely toothed leaves on long petioles; bell-shaped, bright blue flowers for spring and summer bloom. *Temperate.* p. 523

PHAIUS Orchidaceae
tankervilleae (grandifolius) (No. India, So. China, Malaysia to No. Australia), "Nun's orchid"; charming, robust terrestrial from grassy savannahs, with plaited leaves on stout pseudobulbs, and erect spikes to 1 m high with spreading, fleshy flowers, sepals and petals light brown, silvery white behind, lip rose with darker throat. (Feb.-April). *Tropical.* p. 748

PHALAENOPSIS Orchidaceae
'Alice Bowen'; handsome hybrid with clusters of waxy flowers in an unusual shade of deep rosy red. *Humid-tropical.* p. 750

amabilis (Malaya, Sunda Isl.), "Moth orchid"; exquisite epiphyte without pseudobulbs, fleshy, light green, deflexed leaves, and a pendant spray of flowers, glistening snowy-white, except for its yellow crest spotted with red, (Oct.-Jan.). *Tropical.* p. 750, 751

amboinensis (Moluccas); small epiphyte with rosette of small leaves, and waxy starry flowers white, each sepal and petal cross-banded with light purple markings. *Humid-tropical.* p. 750

'Doris' ('Elizabethae' x 'Katherine Siegwart'); a tetraploid of excellent keeping quality; magnificent sprays of heavy-textured flowers 10-12 cm across, glistening white; derived from P. amabilis and rimestadiana parentage (Duke Gardens). *Tropical.*   p. 752

'Hellé' (Marmouset x Adonis); excellent French hybrid with schilleriana and sanderiana blood; flowers of perfect shape, and very fine deep rose flushed pink toward the edges, and maroon spots in the center (early spring). *Tropical.*   p. 750

lueddemanniana (Philippines); compact epiphyte with pale green, fleshy leaves and short racemes with thick, waxy flowers whitish and beautifully marked with cinnamon-brown and bars of amethyst; stalks often bearing offshoots, (May-June and various). *Tropical.*   p. 750

x rothschildiana (schilleriana x amabilis); old James Veitch hybrid with leaves dark green mottled with silvery gray; well-rounded, dainty flowers in large sprays, with white petals, the sepals pale sulphur, tinted with rosy-pink, the lobed lip spotted with purple. *Tropical.*   p. 750

schilleriana (Philippines), "Rosy moth orchid"; beautiful epiphyte with flat roots and long, flat, tongue-like leaves transversely blotched with silvery gray, purplish-red beneath; arching, branched inflorescence with delicate, 5-8 cm flowers of dainty rose in varying tints, (Feb.-May). *Tropical.*   p. 750, 752

PHALERIA   *Thymelaeaceae*

capitata (Java); a small tree with beautiful waxy-white flowers, sweet-scented like daphne, blooming in clusters directly from trunk or heavy branches. This is an example of tropical cauliflory where flower buds form deep within the tissues of the tree and then burst out through the bark. *Tropical.*   p. 909

PHASEOLUS   *Leguminosae*

caracalla (Vigna in Hortus 3) (Trop. So. America), "Snailflower"; perennial twiner with leaves usually of 3 leaflets, and light purple, fragrant flowers tinted yellow, to 5 cm long and with contorted standard, the keel and wings spirally coiled like a snail shell. Also known as "Corkscrew flower". *Tropical.*   p. 557, 560

coccineus (Trop. So. America), "Scarlet runner bean"; an ornamental bean; deep green leaves and bright scarlet flowers held well above the foliage, blooming over a long period and followed by the 20 cm beanpods; when ripe the 2½ cm seed is mauve-pink spotted and striped with black. *Tropical.*   p. 557

PHELLODENDRON   *Rutaceae*

amurense (China, Japan), "Amur cork"; deciduous tree to 12 m, interesting with corky bark, branchlets with pinnate leaves, 5-13 ovate leaflets 10 cm long, glossy above, glaucous beneath; greenish, pubescent flowers; black, berry-like fruit. *Temperate.*   p. 872

PHILADELPHUS   *Saxifragaceae*

lewisii (N.W. No. America); deciduous shrub to 3 m high; ovate leaves 6 cm long, stiff-hairy and with occasional teeth; large white, satiny flowers 4 to 5 cm across and fragrant. State flower of Idaho. *Temperate.*   p. 880

x virginalis (lemoinei x nivalis), "Mock orange"; showy floriferous, deciduous shrub with curved branches, the brown bark peeling, leaves ovate to 8 cm long, pubescent beneath, double white blooms in 3-7 flowered clusters; late spring. *Temperate.*   p. 880

PHILODENDRON   *Araceae*

andreanum (Colombia), "Velour philodendron"; beautiful climber from the moist coastal forest, with iridescent, velvety, oblong sagittate leaves dark olive, suffused with copper, and ivory-white veins; translucent edge. *Tropical.*   p. 115

bipinnatifidum (Rio to Mato Grosso); stout tree with a formal head of upright, waxy green, stiff leaves to 1 m long, bipinnate with 10-12 segments each side of prominent midrib, the lobes are narrow, and lobed again, with long lobe at apex; spathe chestnut-brown; pale yellow berries; tender. *Tropical.*   p. 112, 123

cf borsighiana (Trop. America); slow climber with handsome stalked, oblanceolate leaves to 50 cm long, of leathery texture and glossy deep green; the inflorescence with broad waxy white spathe and white spadix. Name of uncertain standing. *Tropical.*   p. 116

cannifolium (Guayana), "Flask philodendron"; epiphyte with slowly creeping stem bearing leathery, lanceolate leaves with tapering leaf base, swollen leafstalks are channeled. *Tropical.*   p. 115

"colombianum" (Colombia); attractive, compact climber from Villavicencio with large and heavy-textured, heartshaped leaves a deep, glossy green. *Tropical.*   p. 112

cordatum in hort.: see scandens

x corsinianum (lucidum x coriaceum, Florence 1888), "Bronze shield"; Italian historical hybrid; slow creeper with large, broadly cordate, coppery green, quilted leaves having sinuate edge, the veins light green, and purplish-red beneath. *Tropical.*   p. 112

cruentum (Ecuador, Perú), upright growing creeper with waxy green, oblong-pointed leaves with cordate base and depressed veins; back of leaf a beautiful wine-red, hence the name "Red-leaf"; petioles winged. *Tropical.*   p. 112

distantilobum (Amazonas); an unusual climber which I found growing in the light forest up the Rio Negro from Manaus. The glossy leaves are pinnately parted with the pointed segments narrowed toward base; spathe whitish-green. *Tropical.*   p. 112

domesticum (hastatum hort.) (Brazil), "Elephant's ear"; lush fresh green arrow-shaped leaves, later hastate and undulate; pale veins raised and ascending; gorgeous inflorescence with tubular, pale green spathe, red inside. *Tropical.*   p. 112

domesticum 'Variegatum'; a striking mutant with the fleshy, light to dark green leaves irregularly variegated and splashed nile-green, yellow and cream white. *Tropical.*   p. 114

eichleri (Minas Geraes, Alto da Serra), "King of tree philodendrons," with magnificent, pendant leaves to 2½ m long, sagittate, metallic glossy and with scalloped edges; spathe rosy-red, hooding a white spadix. *Tropical.*   p. 113

elegans (Trop. So. America); high climber sending out aerial roots freely from internodes; large leaves thin-leathery, deep green, deeply pinnatifid with the finger-like segments barely more than ribs; very distinctive and ornamental. *Tropical.*   p. 113

'Emerald Duke' (Bamboo Nurseries, Florida 1975); huge self supporting hybrid, with large, broad cordate leaves of rich green about 30 cm long. *Tropical.*   p. 116

'Emerald King'; new hybrid developed and patented by R. McColley of Bamboo Nurseries of Florida; huge, spade-shaped medium green leaves similar to 'Emerald Duke' but with leaves more pointed, 30 cm or more long; trained against a pole; more resistant to disease than P. domesticum (hastatum); recommended for home or office (photographed Feb. 1976). *Tropical.*   p. 116

'Emerald Queen'; an F1 hybrid of two unidentified species, of good keeping quality, and disease resistant to "shot-gun" fungus and bacterial rot; deep green, vigorous plant with short petioles and close internodes, hastate, medium sized, shiny leaves; an excellent totem pole subject because foliage stays about same size; cold resistant. *Tropical.*   p. 114

erubescens (Colombia), "Blushing philodendron"; clamberre rooting at every joint, with arrow shaped, 25 cm waxy leaves bronzy green edged red, wine red beneath; petioles green with red, occasionally winged. *Tropical.*   p. 114, 116

'Espirito Santo' (Brazil), the juvenile form of williamsii, which see.

x evansii (selloum x speciosum); semi-selfheading, showy plant with 1m, glossy leaves like elephant's ears, lobed and wavy; young growth pinkish beneath. *Tropical.*   p. 112

'Florida' (laciniatum x squamiferum); attractive hybrid with slender climbing stem with very little hairy fuzz, the petioles slender, round and rough-warty, the soft-leathery, deep green leaves usually cut into 5 pointed main-lobes; pale midrib, ribs depressed and brown-red on reverse. *Tropical.*   p. 114

'Florida compacta' (quercifolium x squamiferum); an excellent hybrid needing no support as it is non-vining, or only very slowly creeping; the tough petioles are round, and marked purplish-red, the interestingly lobed leaves are thick-leathery, deep waxy-green with veins barely recessed. *Tropical.*   p. 114

fragrans (So. Brazil); slow creeper with heavy, broad, cordate leaves having full round basal lobes, glossy green with primary veins sunken; stocky, triangular petioles from fibrous base. *Tropical.*   p. 113

frits-wentii (C. America); sturdy, slow vining species with thick, heart-shaped leaves, waxy-glossy grass-green, vaginate petioles halfway up the stalk; small cylindric spathe encloses the spadix. *Tropical.*   p. 123

giganteum (Puerto Rico to Trinidad), "Giant philodendron"; a giant with climbing trunk and beautifully lacquered leaves to 1 m long, ovate in outline, with cordate base and pale veins, on closely bunched, fleshy petioles. *Tropical.*   p. 113

giganteum x imbe; climber with lush spear-shaped leaves glossy green and with pale veins, erect or pendant, occasionally tinted with gold along lateral ribs, and more than 30 cm long; the reverse reddish; long slender petioles. *Tropical.*   p. 112

glaziovii (So. Brazil: Guanabara); tall climber from the forest at the foot of the Corcovado, and on Monte Gavea; deep green, lanceolate leathery leaves 30-45 cm long with pale midrib; the axillary inflorescence with spathe crimson inside the pale yellowish-green tube. *Tropical.*   p. 115

'Golden erubescens'; rampant climber with slender round stems; variant with solid golden yellow arrow-shaped leaves, the new growth and underside of leaves pinkish. *Tropical.*   p. 117

'Goldiana'; a beautiful Florida clone selected by Bamboo Nurseries from seedlings of imbe with wendlandii and x mandaianum; compact grower with deep green long-ovate leaves flecked throughout with deep yellow; the underside red; short petioles; the developing leaves are bright gold. *Tropical.* p. 114

hastatum: see domesticum

ilsemannii (Brazil); may be a variegated form of sagittifolium; spectacular climber with leathery, sagittate leaves almost entirely white or cream with gray and dark green marbling; petioles vaginate at base. *Tropical.* p. 114

imbe (Pernambuco to Sao Paulo); "Imbe" is the Brazilian Indian's name for most climbing philodendron. The species has leaves oblong sagittate, parchment-like and with veins nearly at right angles, some reddish beneath; petioles marked red. *Tropical.* p. 119, 120

imbe 'Variegatum' (Guanabara); waxy sagittate leaves with pointed basal lobes, and irregularly variegated; one side of leaf may be dark green with cream marbling, the other side cream with nile green and dark green spots. *Tropical.* p. 119

"imperialis" (speciosum?) (Brazil); spectacular species with scandent trunk bearing sagittate leaves to 2 m long, silvery in juvenile stage, later dark green, very much corrugated and margins wavy; bold fresh-green veins. *Tropical.* p. 119

karstenianum (Oaxaca), "Mexican philodendron"; climbing species known as 'Mex', with brittle-glazed stem, oblong cordate, thin-leathery leaves, fresh green, on reddish petioles, some round, some vaginate; the 20 cm leaves always stay about the same size. *Tropical.* p. 114

lacerum (Cuba, Haiti, Jamaica); stem climber with juvenile leaves ovate, entire and with undulate margin; later leaves crenately lobed and mature leaves deeply incised, glossy-green with light green veins. *Tropical.* p. 115, 119, 123

linnaei (Surinam); slowly creeping and scandent stems with close joints, forming a crown, the thick-leathery leaves long oblanceolate 50-60 cm long, waxy green above, tinged bronze, reddish underside, and with thick midrib; spathe inflated below, constricted in the middle, and with white hood. *Tropical.* p. 114

longistilum (Brazil); resembling a self-heading wendlandii but sends out creeping runners; glossy-green, leathery, oblanceolate leaves with bold midrib and distinctive wine-red back, on abbreviated stalks. *Tropical.* p. 117

'Lynette' (wendlandii x elaphoglossoides), "Quilted birdsnest"; very attractive rosette in form of a birdsnest, the leathery, fresh-green leaves with showy, ribbed depressions. *Tropical.* p. 117

x magnificum (selloum x eichleri); a selfheader with the long, waxy leaf of eichleri but more deeply cut, with the lobes widely spaced. *Tropical.* p. 115

'Majesty'; new Florida hybrid of more compact habit, with large spear-shaped glossy leaves 20-25 cm long, deep coppery-green with wine-red underside and red petioles and stocky stems; good table plant with several cuttings per pot. Bred and patented by R. Mc Colley of Orlando. *Tropical.* p. 116

mamei (Ecuador), "Quilted silver-leaf"; slow creeper with large arrow-shaped, waxy, cordate-ovate, quilted leaves grass-green to grayish green marbled with silvery areas; flattened petioles green suffused with pink, with horny edges, smooth and with whitish length-stripes, rounded at bottom of petiole. *Tropical.* p. 114, 118

x mandaianum (hastatum x erubescens), "Red leaf philodendron"; the best clone of this hybrid, selected by Manda for darkest red coloring, both the glossy, arrowshaped leaves and the stems being deep wine to metallic, purplish-red; the first philodendron hybrid in the United States, 1936. *Tropical.* p. 114, 118

martianum (Brazil); epiphytic clusters of waxy green, ovate leaves with pale, inflated stalks growing on procumbent stem; differing from cannifolium, according to Burle-Marx and Blossfeld, by the boatshaped, merely flattened stalks, and a leaf with obtuse or cordate base, and red edge. *Tropical.* p. 117

melanochrysum (Colombia, Costa Rica), "Black gold"; so-called because of the beautiful, almost black-olive, velvety leaves, shimmering with pink, veins pinkish, smaller than andreanum. Exquisite but delicate. *Tropical.* p. 114

melinonii (Guayana), "Red birdsnest"; shapely rosette of ovate, fresh green leaves with pale veins, and marked by channeled, swollen, red leafstalks. One can see these spots of red a far distance, high on 45 m forest trees, when traveling in a dugout canoe deep in the interior of Surinam. *Tropical.* p. 109

mello-barretoanum (Acre, N.W. Brazil); majestic tree forming a stout trunk, with a crown of large, broadly hastate leaves deeply cut, and the pointed segments lobed again and overlapping; wide basal sinus; has robust thorns on the trunk between leaf scars. *Tropical.* p. 121

nobile (Venezuela, Guayana); birdsnest-like rosette of thick-leathery leaves broadly oblanceolate, with bold midrib, on thick, slowly creeping stem. *Tropical.* p. 118

oxycardium: see scandens

panduraeforme (bipennifolium) (So. Brazil), "Fiddle-leaf"; climber with an unusual "fiddleleaf", the basal lobes extended, central lobes narrowed toward middle, of leathery texture, dull olive green. Very decorative trained to poles. *Tropical.* p. 118

"pertusum" (So. Mexico), "Split-leaf philodendron"; the fast-climbing juvenile stage of Monstera deliciosa; the smallest leaves are roundish entire, later growth will be more pinnatisect and forming occasional perforations. *Tropical.* p. 125

pinnatifidum (Venezuela, Amazonas); selfheader with leathery leaves pinnately parted, the lobes well apart with wide sinus metallic green, veins sunken; the channeled petioles spotted red. *Tropical.* p. 115, 121

pittieri (microstictum) (Costa Rica); slow climber with broad heartshaped, thick, glossy, apple-green leaves, attached to round petioles at edge of shallow sinus, giving a pleasing appearance. *Tropical.* p. 118

poeppigii (Alto Amazonas); semi-selfheading tree-dweller with cupped, heartshaped leaves on long, stiff petioles with edges rolled up; base fibrous. *Tropical.* p. 115

pseudoradiatum (Mexico: Chiapas); slow climber with large 60 cm attractive, sagittate leaves glossy green with prominently contrasting pale ribs and undulate margins; on cylindrical petioles; tubular spathe green outside, brilliant purple inside. *Tropical.* p. 121

'Red Duchess'; patented new hybrid bred by Bob Mc Colley of Orlando, Florida, to withstand the rigors of office environment; dark glossy green, heart-shaped leaves with reddish underside, 20-25 cm long, and with red petioles and slowly climbing stem, for training along a supporting pole. (Photographed Feb. 1976). *Tropical.* p. 116

'Red Emerald'; robust Florida erubescens hybrid with long ruby-red, round petioles, and long-cordate leaves 30-40 cm long, dark glossy green with ribs red on reverse; smooth stems also wine-red. *Tropical.* p. 118

'Red Princess' (Bamboo Nurseries, Florida); climbing hybrid of smaller habit, with broad spade-shaped angular leaves green suffused with red, on red petioles and red stem; *Tropical.* p. 116

'Royal Queen' (Rob. McColley - Florida); graceful hybrid more compact than 'Majesty', with blood-red stem and petioles; smallish ovate leaves bronze when young, later glossy rich green. *Humid-tropical.* p. 116

sagittifolium (So. Mexico); climber with arrow to halberd-shape leaves, the basal lobes in advanced leaves widely spread, the blade wavy, veins depressed. *Tropical.* p. 121

'Santa Leopoldina' (Espirito Santo); an important species having the most beautiful leaf I have seen in Brazil; it is slowly climbing, the stems and petioles are red, with elegant, long sagittate, leathery leaves 1 m long, glossy dark green with ivory ribs and red margins; spadix enveloped in a reddish spathe. *Tropical.* p. 116

scandens oxycardium (Puerto Rico to Jamaica and Central America), known in horticulture as "cordatum"; pre-Linnaean as hederaceum. The most popular and widely sold vining Philodendron, known as "Heartleaf philodendron" or "Parlor ivy", or simply "Cordatum vine". A tall tropical, rapid climber by aerial roots, with glossy deep green, broadly heart-shaped, soft-leathery leaves in juvenile stage 10-15 cm long; in maturity or flowering stage to 30 cm long. In habitat when very young the foliage is apparently velvety, from observations I made in Costa Rica. This species may be used in many ways, as a cascading vine in pots, baskets, window boxes, or room dividers; or it may be trained against support, preferably on mossed poles, bark slabs, or milled treefern pillars. *Tropical.* p. 115, 121

scandens oxycardium 'Variegatum', heartshaped leaves marbled ivory-white and gray-green on dark, glossy green. *Tropical.* p. 118

selloum (S.W. Brazil), "Lacy tree-philodendron"; self-header, tree-like or scandent on trees. In the moist forests of western Parana I have seen it growing epiphytic, sending down aerial roots to strike the ground. The lush, dark green, pendant 60 cm leaves are bipinnate with short lobe at tip; juvenile leaves are merely lobed; spathe greenish-white. *Tropical.* p. 113, 120, 121, 123

sellowianum (Brazil); fast growing selfheader with deeply cut leaves which stand considerable cold. According to Blossfeld similar to bipinnatifidum, but has green seed pod sheaths instead of black ones. *Tropical.* p. 122

sodiroi (laucheanum) (Brazil), "Silver-leaf philodendron"; in juvenile stage vining with small cordate, pointed, bluish-green glossy leaves largely covered with silver, ribs are red underneath; petioles wine-red and winged, in later stage petiole becomes flat on top and rugose with green puckers, and without wings; leaves are larger and rounded, internodes close. *Tropical.* p. 118, 123

speciosissimum (Brazil: Guanabara); handsome climber of the rainforest, with sagittate leaves 75 cm long, similar but smaller than speciosum, with sinus more flaring, wavy margins, glossy deep green. *Tropical.* p. 122

speciosum (Minás Geraes, S. Paulo, Mato Grosso), "Imperial philodendron"; majestic arborescent species becoming tree-like with age, with huge sagittate leaves 1½ m or more long, rich green, thin-leathery, the veins sunken and margins wavy and almost frilled; flowers beautiful with fleshy spathe green with purple margins, carmine-red inside. *Tropical.* p. 122

tweedianum (Argentina, Paraguay); probably the southernmost philodendron which I found growing in the Paraná delta near Buenos Aires; large, hastate deep green leaves irregularly lobed, and wavy margined; forming trunk. *Subtropic.* p. 117

undulatum (Paraguay); a smaller selfheader, and I have seen these small trees in the dry savannahs of Paraguay, the sagittate wavy and lobed leaves somewhat cupped, carried on erect stalks. *Tropical.* p. 120, 123

variifolium (Perú); slow vine with small, leathery, heartshaped leaves, greenish-brown when young, bluish-green when older, with silver feather-bands between lateral veins; winged petioles red, also veins beneath. *Tropical.* p. 118

verrucosum (Costa Rica, Colombia), "Velvet leaf"; long-vining, with delicate, undulate, heart-shaped leaves, shimmering velvety, dark, bronzy green; pale green vein areas and margins emerald green, salmon violet beneath; petioles red and covered with green hairs; showy inflorescence with spathe an ovoid rose-purple tube 8 cm long, densely white-hairy and with white margin, and enclosing the slender spadix. Gorgeous. *Tropical.* p. 92, 123

warmingii (Sao Paulo); dense clusters of swollen stalks with ovate, yellow-green leaves having their base obtuse; similar to martianum, but the petioles are channeled. *Tropical.* p. 117

warscewiczii (Mexico, Guatemala, Honduras, Panama); scandent species forming snaky trunks, with lush, very soft leaves, fresh green, triangular sagittate and pinnately parted, the pointed segments well separated in mature leaves, and with overlapping lobes; goes deciduous in dry season. *Tropical.* p. 120

x wend-imbe (wendlandii x imbe); semi-selfheading hybrid with many waxy green, oblong-pointed leaves carried stiffly on flattened stalks; new growth pink underneath. *Tropical.* p. 118

wendlandii (Costa Rica, Panama), "Birdsnest philodendron"; selfheading rosette of thick, waxy-green, long obovate leaves with thick midrib, arranged like a birds-nest; the short petioles are spongy; spathe cream-white. *Tropical.* p. 118

williamsii (Bahia, Espirito Santo); a magnificent epiphyte, also known as 'Espirito Santo', with arborescent stem, and fresh green, deeply hastate leaves almost 1 m long and with undulate margins and reddish veins; spathe pale green outside, yellowish inside. *Tropical.* p. 122, 123

PHILODENDRON: see also Monstera

PHLOX   *Polemoniaceae*

diffusa (Oregon to So. Dakota); perennial with prostrate branches, to 15 cm high; linear 2 cm leaves; 2 cm flowers pink or white. *Temperate.* p. 826

drummondii (Texas), "Dwarf annual phlox"; dwarf branching, pretty annual with fresh-green, small lanceolate and flat, terminal clusters of brightly colored, salverform, 2½ cm flowers in shades of rose-red; cultivars in white, buff, pink, red, purple, and blue, with colorful eyes; good plant for pots. *Warm temperate.* p. 826

drummondii 'Sternenzauber'; charming German cultivar with star-like flowers having lacy margins, rosy red in center, white along pointed lobes. *Warm temperate.* p. 826

paniculata (U.S.: New York to Arkansas), "Summer phlox"; herbaceous perennial phlox, to 1 m, with lanceolate leaves, the leafy stems topped by panicles of large 2½ cm purple flowers; varying in colors white, salmon, scarlet, lilac, in summer and autumn; hardy. *Temperate.* p. 826

PHOENIX   *Palmae*

canariensis (Canary Islands), "Canary Islands date palm"; stately feather palm widely planted in subtropical regions as an ornamental; compact, robust and stiff when young, with age forming thick, straight trunks, becoming 15 m high, with arching pinnate leaves 6 m long, the short stalk armed with yellow spines, the leaflets glossy-green, in various directions; on female trees small yellow fruit in large clusters. *Subtropic.* p. 787, 793, 794, 797

dactylifera (Arabia, No. Africa), the fruiting "Date palm" of Egypt and North Africa, and its descendants in the Coachella Valley in the California desert; a massive tree becoming 30 m high, dense with stiff pinnate fronds spiny at the base, with narrow rigid folded pinnae in double rows when older, bluish-glaucous, to 45 cm long and sharp-pointed; female trees will set delicious oblong edible fruit in great, heavy clusters. *Tropical.* p. 471, 474, 787, 797

dactylifera 'Deglet Noor'; a favorite variety of date palm cultivated in California for its excellent large and meaty, "semi-soft" fruit, which packs and keeps well; each female tree producing, after artificial pollination, 90-120 kg of dates a year, on irrigated land, where the trees "have their feet in water and their heads in fire", according to an Arab saying; to produce good fruit, just hot days are not good enough, and the desert climate of the Coachella valley, in interior California near the Mexican border, with temperatures of 40°C at night and to 52°C in daytime suits them best. I remember the custom of spraying my bed with a water hose, before going to sleep at night, while living in Palm Springs, before the advent of air-conditioning. *Tropical.* p. 471, 787, 793

loureirii (India to China); nearly stemless, clustering feather palm to 2 m or more; pinnate leaves rather stiff and recurving, the leaflets glaucous and in groups along the rachis; small red fruit. Somewhat resembling P. roebelenii, but more coarse and less graceful. *Tropical.* p. 794

reclinata (Trop. Africa from Sénégal to Natal), "Senegal date palm"; a leaning date palm somewhat resembling Cocos, in habit, and which will live in the subtropics; solitary trunks 12 m high, or shorter if allowed to cluster, the pinnate lustrous green leaves rather stiff and curving downward; small red fruit. *Tropical.* p. 793, 797

roebelenii (humilis loureiri) (Assam to Viet Nam), "Pigmy date palm"; very graceful both as a miniature potplant or when with slender, rough 4 m trunk, topped by a dense round crown of feathery leaves, the pinnae narrow and folded and dark green, glossy when rubbed; berry-like black fruit in large clusters; female trees often clustering. *Tropical.* p. 793, 794, 804

rupicola (Himalayas of Nepal, Bhutan); attractive feather palm growing amongst rocks, with slender trunk to 6 m high, and a crown of feathery leaves appearing gracefully soft, the bright green pinnae decurved and limp; shining yellow fruit. *Subtropic.* p. 794

sylvestris (India), "East Indian wine palm"; stout erect palm to 12 m, with rough trunk, and arching pinnate, grayish green or glaucous fronds 3-4 m long, with rigid, somewhat clustered leaflets, on spiny stalk; olive-like 2 cm reddish fruit; produces sugar from the sap. *Tropical.* p. 794, 797

taiwaniana (Formosa); a compact, ornamental palm seen on the streets of Taipei, with short stout trunk swollen toward top, crown of rather short, arching, coarse pinnate leaves, with rigid leaflets dark green. *Tropical.* p. 794

PHOENOCOMA   *Compositae*

prolifera (Cape Prov.); small rigid shrub to 60 cm; hard scale or needle-like leaves; flower heads with leathery ray petals white, pink or rose. *Subtropic.* p. 322

PHORADENDRON   *Loranthaceae*

flavescens (New Jersey to New Mexico), an American "Mistletoe"; a green parasite on many deciduous trees from New Jersey to Florida and westward, forming dense bunches ½ to 1 m across, with brittle-woody cylindrical, forking twigs, thick oval and opposite yellowish evergreen leaves to 5 cm long. Male and female flowers are on separate plants, borne in short spikes or catkins, the females developing amber-white, small round berries. Cut branches are used for Christmas decoration, a custom in Europe where the similar Old World mistletoe, Viscum album, is used and considered an invitation to a kiss, going back in history to the Druids. *Warm temperate.* p. 618

PHORMIUM   *Liliaceae*

colensoi 'Tricolor', "Mountain flax"; large clusters of arching, leathery, green strap-leaves colorfully variegated lengthwise with irregular banding of cream to golden yellow, the margins reddish. *Subtropic.* p. 585, 588

tenax (New Zealand), "New Zealand flax"; large tufting plant with 2-ranked, tough-leathery leaves which may grow to 3 m long, dark or brownish green with reddish margin, clasping at base, and splitting at apex; flowers dull red in tall panicle. Seen in N.Z. growing even in the cold water of glacial lakes. Very dramatic and tolerant as container plant. *Subtropic.* p. 588, 611

tenax 'Tricolor'; stiffly erect foliage beautifully striped and banded green with gray, creamy yellow and salmon red. p. 588

tenax 'Variegatum', "Variegated flax"; attractive variant with the usually brownish-green leaves striped and margined with creamy-yellow and white. *Subtropic.* p. 588, 607

PHOTINIA   *Rosaceae*

x fraseri (glabra x serrulata), "Redleaf photinia"; colorful evergreen to 3 m, with glossy leaves to 12 cm long, new growth a showy bronzy red; small white flowers. *Warm temperate.* p. 852

PHOTINIA: see also Heteromeles

PHRAGMIPEDIUM   *Orchidaceae*

caudatum (Selenipedium) (Perú, Ecuador), "Mandarin orchid"; remarkable terrestrial of robust habit, with long straplike leaves, racemes loosely 1-4 flowered, to almost 1 m high, the dorsal yellowish with green veins, petals ribbon-like, twisted, to 60 cm long, brownish-crimson shaded yellow, pouch bronzy green, (April-Aug.). *Humid-subtropic.* p. 746

x grande (Selenipedium) (longifolium x caudatum), "Spiralled lady-slipper"; plant similar to caudatum but flowers with tail-like, spiralled petals shorter, at the base yellowish-white, changing to carmine-red, dorsal pale yellow veined green, waxy pouch greenish-yellow spotted red inside, (summer). *Humid-subtropic.* p. 748

longifolium (Colombia); large plant with leaves to 60 cm long and racemes to 1 m high with numerous elegant greenish flowers flushed coppery red, the pouch mahogany red, lateral sepals lacquer red, the dorsal pale green with rose veins and edged white, the petals pale yellow green with rose margins (Mar.-May). *Humid-subtropic.* p. 748

philippinense (Philippines); terrestrial species with leathery, glossy green leaves to 30 cm long; floral stalks to 50 cm high, bearing 3-5 striking flowers, with dorsal sepal white, striped with brown purple, the long thread-like red-purple petals pendulous and twisted, 15 cm or more long; lip tawny marked with brown. *Humid-tropical.* p. 747

vittatum (Selenipedium) (Brazil), a "Mandarin orchid"; of terrestrial habit, with flowers greenish and overlaid or striped with brown, the long narrow petals twisted and undulate at margins; slipper-like lip light brown outside. *Humid-tropical.* p. 748

PHYGELIUS *Scrophulariaceae*
aequalis (So. Africa), "River bells"; herbaceous pubescent perennial to 1 m high, with dark green, oval soft leaves 12 cm long, on winged, square stems; flowers scarlet, with flaring limbs, yellow inside. *Subtropic.* p. 886

capensis (So. Africa),"Cape fuchsia"; shrub with herbaceous, angled, purple branches, opposite, smooth, crenate leaves, and pyramidal panicles of tubular, coral-scarlet flowers, pendulous from horizontal stalks, blooming in summer. *Subtropic.* p. 882, 887

PHYLLAGATHIS *Melastomataceae*
rotundifolia (Sumatra); ornamental herbaceous shrub 30-60 cm, with stout, square stems, roundish, corrugated leaves 15-30 cm long, glossy dark green tinted metallic-blue and purple, coppery-red beneath, and with prominent veins; small flowers with magenta-pink petals and filaments, in tight axillary clusters. *Tropical.* p. 643

PHYLLANTHUS *Euphorbiaceae*
acidus (So. Asia), "Otaheite" or "Gooseberry tree"; handsome tree to 10 m high and grown for its edible fruit; leaves ovate to 8 cm long, two-ranked on twigs; small reddish flowers; yellow 2 cm fruit 6-8 ridged, clustered on the trunk and older branches, and made into preserves. *Tropical.* p. 429

angustifolius (Xylophylla) (Jamaica), "Foliage flower"; interesting shrub with long flattened waxy green branches looking like leaves, the tiny reddish flowers borne on the edges of the branches. *Tropical.* p. 421

arbuscula (speciosus) (Jamaica); shrub or tree to 6 m high, with lanceolate leaflike branchlets (phyllodia) 5-8 cm long and 2½ cm wide, arranged in two ranks on a branch, appearing like pinnate leaves; tiny flowers with whitish calyx. *Tropical.* p. 428

PHYLLANTHUS: see also Breynia

PHYLLITIS *Polypodiaceae (Filices)*
scolopendrium (Scolopendrium vulgare) (Europe, Madeira, E. No. America), "Hart's-tongue fern"; rhizomatous, hardy fern with stout rhizome and long straight or curved strap-shaped leathery fronds 15-45 cm long, pale yellow green to bright green, quilted and undulate at margins; spore masses in thick strips at right angles to the bold midrib, running into a short black stalk. *Humid-subtropic.* p. 445, 446

scolopendrium 'Crispum', "Crisped heart's-tongue"; attractive fern with long narrow, strap-like light green fronds nicely and regularly undulated and wavy, 30 cm long by 2-3 cm wide, crested at the base, on stalks covered with hair-like brown scales. *Humid-subtropic.* p. 445

scolopendrium 'Undulatum'; shapely cultivar of compact habit, the stalked fronds linear oblong, fresh glossy green 20 cm long, the blades folded and undulate. *Humid-tropical.* p. 442

PHYLLOCACTUS: see Epiphyllum

PHYLLORRHACHIS *Gramineae*
sagittata (Trop. Africa); ornamental bamboo-like clustering perennial which I found growing in the tropical pool at Munich Botanic Garden; thin reed-stems 50-80 cm high, set with leathery rich green, lanceolate leaves corrugated lengthwise, and with stem-clasping base. *Tropical.* p. 516

PHYLLOSTACHYS *Gramineae*
aurea (China), the "Golden bamboo" or "Fish pole bamboo"; tall woody grass with wide-ranging rhizomes; hollow canes flattened on one side, to 4 m or more high and 3 cm thick, brilliant yellow, the internodes at the base very short; very straight and stiffly erect, very hard and bonelike when matured and used for fishing poles; leaves usually 5-10 cm long and long-pointed, light green, and glaucous beneath. Usually hardy around New York City. Young shoots appearing in mid-spring are edible. *Subtropic.* p. 517, 518

bambusoides (China), "Giant timber-bamboo"; one of the largest and most valuable "hardy" timber bamboos, with green culms to about 20 m high, thick-walled and more than 12 cm dia., striped or yellow in some forms; two unequal branches of each branch-bearing node, oblong pointed leaves 6 to 15 cm long; culm-sheaths greenish to reddish. Very versatile in its uses, especially for construction, and its strength is exceeded only by the Tonkin cane (Arundinaria amabilis). Hardy north to about Norfolk, Virginia. *Subtropic.* p. 514, 515

nigra (So. China), "Black bamboo"; graceful black-culmed bamboo to 8 m high and 3 cm thick, green at first later speckled then all black; slim branchlets with small leaves commonly 8 cm long, culm-sheaths greenish to buff; thin-walled; hardy to about Norfolk, Virginia, and where temperatures do not go below -20°C. *Subtropic.* p. 518

sulphurea (viridis var.) (China, Japan), "Yellow running bamboo" or "Moso bamboo"; a very hardy running bamboo 4-9 m high, stems yellow or green and 5-8 cm thick; leaves to 12 cm long and 2 cm wide, glaucous beneath, on purple petioles. Very handsome with arching yellow canes, and effective in containers in entryways or the patio. Hardy in the Pacific Northwest, or Washington D.C., tested to -23°C. *Warm temperate.* p. 515

PHYMATIDIUM *Orchidaceae*
tillandsioides (Brazil); miniature epiphyte forming a tufted green rosette of grass-like leaves resembling Tillandsia, the wiry stalks bearing tiny white flowers 3 cm across (summer). *Tropical.* p. 752

PHYSALIS *Solanaceae*
alkekengi (franchetii) (Japan), "Chinese lantern plant" or "Winter cherry", known since the 12th century; herbaceous perennial with creeping underground stems, often grown as an annual, 1 m high, with large ovate leaves, inconspicuous axillary yellowish flowers; the calyx, after fertilization, becomes inflated like a lantern, bright orange-red, enclosing the scarlet berry-like fruit; when cut in fall the lanterns will keep for a long time. *Warm temperate.* p. 891, 895, 897

PHYSOSTEGIA *Labiatae*
virginiana (Eastern U.S.), "Obediance"; stoloniferous perennial to 1 m high, with lanceolate leaves, and spikes of bell-shaped flowers rose-purple. *Warm temperate.* p. 531

PICEA *Pinaceae (Coniferae)*
abies (excelsa) (North and Central Europe), "Norway spruce"; evergreen coniferous forest tree to 45 m high, of pyramidal habit, reddish-brown scaly bark, and whorled branches, the usually pendulous branchlets with linear 4-angled, shiny dark green needles 2 cm long, spirally arranged; pistillate flowers bright purple; drooping light brown cones 10-17 cm long. *Temperate.* p. 344, 346

glauca albertiana (C. No. America, especially Alberta), "Alberta white spruce"; evergreen tree of dense narrow pyramidal form, slow growing; thin 4-angled glaucous-green needles 1½ cm long, with sharp point, set radially around thin, much branched twigs; roundish cones to 4 cm long. *Temperate.* p. 346

glauca 'Conica' (Canadian Rockies), "Dwarf Alberta spruce"; evergreen of compact conical or pyramidal habit, very dense, to 2 m high, branches erect and stiff; the branchlets thin, with needles set radially, nearly round, 1 cm long, light glaucous green. *Temperate.* p. 346

glauca densata (South Dakota), "Black Hills spruce"; slow-growing, dense symmetrical tree of great hardiness; the linear needles glaucous green on upper side, to 1 cm long; cylindrical cones to 6 cm long, green to light brown. *Temperate.* p. 345

orientalis 'Aurea' (Caucasus and Asia Minor), "Oriental spruce"; branches spreading horizontally, with younger needles golden yellow. *Warm temperate.* p. 346

pungens (Wyoming, Colorado, Utah, New Mexico), "Colorado spruce" or "Blue spruce"; noble pyramidal tree broad at base, to 30 m tall; horizontal branches with stout branchlets, needles 4-angled, stiff, sickle-shaped, bluish-green, to 3 cm long, with spiny point, arranged radially around twig; oblong light brown cones 5-10 cm long. *Temperate.* p. 341, 343, 344, 346

pungens 'Kosteriana', "Koster's blue spruce"; world-famous cultivar first distributed by Koster (Boskoop, Holland 1885); forming magnificent pyramids with branches in whorls; stiff, curving needles distinct silvery-glaucous, 2-2½ cm long, dense on orange-brown branchlets. *Temperate.* p. 346

PIERIS Ericaceae
japonica (Andromeda) (Japan), "Lily-of-the-valley bush"; evergreen shrub to 3 m high; with glossy green, leathery leaves 5-8 cm long, in dense whorls; small waxy-white 1 cm urn-shaped flowers in pendulous clusters, the calyx lobes tinted red. *Temperate.* p. 394

PIGAFETTA Palmae
filaris (Indonesia: Celebes, Moluccas); very handsome, solitary-growing spiny feather palm, with ringed trunk; the base of crown and petioles densely covered with erect hairs; the beautiful pinnate fronds spreading with long leaflets rich green. David Fairchild considered Pigafetta his favorite palm. Photographed at Dr. Darian's palmarium, Vista, California. *Tropical.* p. 784, 787, 791

PILEA Urticaceae
cadierei (Vietnam: Annam), "Aluminum plant"; also known as "Watermelon pilea"; a rapid growing, succulent plant, with thin fleshy, opposite, rather large 8 cm obovate, quilted foliage remotely toothed or crenate, attractively painted shining silvery aluminum over the vivid green to bluish-green blade; tiny flowers in stalked heads. *Tropical.* p. 912

depressa (Puerto Rico), "Miniature peperomia"; freely branching, low, succulent creeper with tiny 6 mm roundish obovate, fleshy leaves, light pea green and glossy, with the apex crenate, opposite and dense on thin green stems, rooting at nodes where touching the ground. *Tropical.* p. 912

grandis (Jamaica); herbaceous plant with large green, ovate, corrugated leaves with depressed parallel veins and dentate margins, 10 cm or more long, resembling Coleus; tiny white flowers in flat clusters. *Tropical.* p. 911

microphylla (muscosa) (West Indies), "Artillery plant"; small plant densely branched, with suberect, fleshy stems thick with tiny, watery-succulent, oblong, green leaves to 6 mm long, having a tapering, cuneate base; flower clusters with staminate flowers discharging a cloud of pollen when dry or shaken. *Tropical.* p. 912

mollis (Costa Rica, Colombia), "Moon Valley Green"; charming tropical plant similar to 'Moon Valley' which is probably a mutant heavily overlaid with brown; in the species the foliage is deeply quilted like a carpet and a soft light emerald green, and covered with white hairs; veins olive green. *Tropical.* p. 911

'Moon Valley Green': see mollis

serpyllacea (Mexico), "Creeping Charley"; dwarf but stronger plant than microphylla, with heavy succulent branches having orbicular leaves rounded at the base; little greenish flower heads on long stalks. *Tropical.* p. 912

'Silver Tree' (Caribbean), "Silver and bronze"; a copyrighted name 1957 by Mulford for this species from the Caribbean area; herbaceous branching plant with white-hairy stalks, quilted ovate leaves with depressed veins and crenate margins, bronzy-green, with broad silverband along center, silver dots on sides; reddish beneath. *Tropical.* p. 912

spruceana 'Norfolk', "Angel wings" or "Panamiga"; lovely English (Mason) cultivar of the species from Perú; free-growing, small ornamental herb, dense with somewhat fleshy, deeply quilted, broad-oval leaves 6-8 cm long, metallic bronze to blackish green with raised silver bands, wine-red beneath; the tiny rosy flowers clustered closely in the axils of foliage. *Tropical.* p. 912

superba hort., in Florida nursery trade; robust tropical herbaceous plant with brownish stems, opposite long-stalked elliptic, deep metallic green, thin leaves 6-8 cm long, corrugated and with depressed coppery parallel ribs, and sharply toothed margins. *Humid-tropical.* p. 912

PIMENTA Myrtaceae
dioica (officinalis) (Jamaica to C. America), "All-spice"; small tree to 12 m high, with oblong feathery, aromatic leaves to 16 cm long, small white flowers, followed by clusters of small green to brown pleasantly spicy 1 cm pea-sized berries; these are picked green and dried and called "All-spice" because they seem to have the combined flavors of cinnamon, nutmeg and cloves, and are used to flavor food. *Tropical.* p. 684

racemosa (acris) (West Indies to Guayana), "Bay-rum tree"; aromatic tree to 15 m high, with leathery, obovate leaves 15 cm long, showing reticulated veins; small white 1 cm flowers, and dark brown berry-like fruit. Oil-of-bay is distilled from leaves and twigs. *Tropical.* p. 682, 684

PINANGA Palmae
patula (Sumatra, Borneo); small clustering palm 2 m high, with cane-like stems 3 cm dia., the pinnate fronds with few curved leaflets. *Tropical.* p. 796

PINGUICULA Lentibulariaceae (Carnivorous Plants)
caudata (bakeriana) (Mexico), "Tailed butterwort"; carnivorous flattened rosette from moist bogs, with fleshy obovate pale green leaves which are sticky with a digestive fluid and capture insects, gradually absorbing them; flowers carmine, with long spur. *Humid-subtropic.* p. 292

lutea (No. Carolina to Florida, Louisiana); a southern "Butterwort"; terrestrial rosette from boggy places, with ovate, fleshy, yellow green leaves 6 to 8 cm long, and sticky to the touch; flowers ochre-yellow with curved spur, on long stalks. The foliage is covered with a sticky secretion on which insects become stuck. *Warm temperate.* p. 292, 296

vulgaris (N. No. America, Europe, Asia), "Butterwort"; small rosette to 15 cm across, with oblong obtuse, succulent leaves greasy to the touch from a sticky glandular fluid which holds and digests insects; flowers purple. *Warm temperate.* p. 292

PINUS Pinaceae (Coniferae)
canariensis (Canary Islands), "Canary Island pine"; tall tree to 30 m, with drooping branchlets; very long, string-like, pendant needles, glossy green, in bundles of 3, to 30 cm long; ovoid 20 cm cones. *Tropical.* p. 349

cembroides (Arizona, Mexico), "Pinyon pine" or "Mexican stone pine"; shrub or small tree with gnarled, scarred trunk, and twisted branches; the dark green needles stiff and generally curved, in bundles of 3, but occasionally 2, 4, or 5; small cones 3 cm dia. *Subtropic.* p. 347, 349

halepensis (Portugal to Afghanistan), "Aleppo pine" or "Jerusalem pine"; characteristic pine of rugged character, older trees irregular and round-topped, 15-25 m high; soft needles in two's, to 10 cm long, light green; oblong reddish cones to 8 cm. Yields turpentine. *Subtropic.* p. 345, 349

halepensis stankewiczii (Russia: Southern Crimea); round or flat-topped tree with green needles 14 to 18 cm long, moderately stiff; pointed solitary cones. *Subtropic.* p. 347

montezumae (Mexico, Guatemala), "Montezuma pine"; tall tree to 40 m, with spreading crown and rough bark; long stiffish needles 16 to 24 cm long, in bundles of five, spreading or drooping, green or bluish green, the margins finely toothed; variable cones 6-24 cm long, usually conical. *Subtropic.* p. 347

palustris (U.S.: Virginia to Florida and Mississippi), "Longleaf-pine"; large coniferous tree 25-30 m high, with thread-like, graceful dark green needles to 45 cm long, arranged in clusters of three; cylindric 25 cm cones. Cut branches are used frequently for decoration. *Warm temperate.* p. 349

parviflora (pentaphylla) (Japan) (Pinaceae), "Japanese white pine"; widely adopted in Japan for the culture of Bonsai, or dwarfed trees, and I have seen them 3 centuries old less than 60 cm high; the slender bluish-green needles are arranged in clusters of five; often grafted on P. thunbergii, the black pine. *Warm temperate.* p. 348, 349

pinea (So. Europe and Turkey), "Italian stone pine"; the characteristic broad and flat-topped pine of the South Italian landscape, 15 to 26 m high; with stiff, bright to gray-green needles in two's, 10 to 20 cm long. In juvenile stage, needles are short, 3 cm long and silvery glaucous. Handsome decorator, when container grown; takes heat and drought. *Warm temperate.* p. 345

ponderosa (Brit. Columbia to Baja California, So. Dakota to Texas), "Western yellow pine" or "Ponderosa pine"; majestic tree to 50 or 70 m high, with red bark to nearly black, fissured into large plates, branches spreading; needles in bundles of three, densely crowded on branchlets, 12-26 cm long, dark green; ovoid cones 8-15 cm long, shining brown. *Temperate.* p. 349

radiata (California), "Monterey pine"; legendary tree of the California coast, 24-40 m high, with irregularly open crown, becoming wind-blown and characteristically one-sided where exposed to ocean winds, thick dark brown bark on old trees; needles in bundles of three, in dense clusters 10-14 cm long; fresh green nut-brown cones 7-14 cm long, 5-6 cm broad, in sessile clusters. *Warm temperate.* p. 349

roxburghii (Himalayan foothills), "Emodi pine" or "Indian longleaf pine"; decorative graceful tree when young, with long, drooping foliage; later broad, with round top, to 50 m high in habitat; arching needles in bundles of three, shining green, to 30 cm long; ovoid cones to 18 cm, the scales reflexed. Similar to Canary Island Pine. *Subtropic.* p. 348

strobus (Newfoundland and Manitoba south to Georgia), "White pine" or "Weymouth pine"; well-known, widely distributed graceful conifer 30 m or more high, fast-growing and ornamental, with a smooth, gray bark when younger, and a feathery effect because of its thin, soft needles in bundles of five, and bluish green; the pendant cones long cylindrical, often curved and very resinous, 8-15 cm long. Used in Japan for training as dwarfed Bonsai in containers. *Temperate.* p. 349

thunbergiana (thunbergii) (Japan), "Japanese black pine"; handsome spreading tree 8 to 30 m high, resembling, and perhaps a form of the Austrian pine but quicker growing; with somewhat shorter, darker green leaves rarely over 10 cm long, the sharp-pointed stiff needles in pairs. May be pruned; excellent in planters and as Bonsai. *Temperate.* p. 345, 348, 349

PIPER *Piperaceae*
auritum (Mexico); a woody pepper with beige-gray scandent branches, broadly ovate-cordate leaves 15 to 50 cm long, dull grass-green, grayish beneath on short, usually pubescent petioles; flower spikes to 25 cm long. *Tropical.* p. 822

betle (Bali, E. Indies), "Betel-leaf"; commonly used in Indonesia and India for chewing with betel-nut; stems trailing; and in Bali I have seen it climbing high on trees, with foliage in neat ranks, the leaves are fleshy, broadly heartshaped 8-15 cm long, dark green with depressed veins; flowers in stalked catkins, opposite of the leaves. *Tropical.* p. 821

crocatum (Perú), "Ornamental pepper"; rich-looking, beautiful, ornamental climber, sometimes confused with P. ornatum; with thin-wiry stem, and peltate highly glossy, slender-pointed leaves blackish olive-green, with silver-pink marbling and spotting along veins and veinlets, and corrugated; deep purple underneath. *Tropical.* p. 818, 821, 822

futokadsura (Japan), "Japanese pepper"; shrubby plant climbing by aerial roots, much like nigrum but nearly hardy and said to be deciduous, stands some frost; thick-leathery, ovate, slender-pointed leaves with 5 depressed veins, waxy, blackish green, and light green beneath; flowers greenish and berries red; slow grower. *Subtropic.* p. 819

magnificum (bicolor) (Perú), "Lacquered pepper-tree"; beautiful, erect branching foliage plant with corky stem and winged, clasping leafstalks bearing large, fleshy, quilted, oval leaves lacquered forest-green of a metallic sheen, with ivory veins and edge, wine-red beneath. *Tropical.* p. 822

nigrum (Malabar coast, Malaya, Java), "Black pepper"; tropical climber with flexuous stems dense with leathery, glossy blackish-green, ovate or elliptic leaves, and bearing long clusters of green berries turning first red, then black; furnishing, when dried, black pepper. *Tropical.* p. 466, 821

ornatum (Celebes, New Britain), "Celebes pepper"; very attractive, tall climber with slender reddish stems and petioles, and broad, peltate-pointed, shield-like, waxy leaves 8-10 cm long, deep green, beautifully etched with markings of silvery pink, becoming white in older leaves, pale green beneath. *Tropical.* p. 819

porphyrophyllum (Cissus) (Indonesia), "Velvet cissus"; strikingly beautiful climber without tendrils, stems red with lines of white bristles; with roundish-cordate, recurved, 8-10 cm quilted leaves, velvety moss-green with yellow veins, and pink markings mainly along the veins, wine-red underneath. *Tropical.* p. 822

sylvaticum (Burma), "Silver cissus"; attractive, ornamental vine, with thin-wiry stems, and heavy, leathery, ovate leaves more or less with cordate base, corrugated between the sunken dark veins, dark steel-green, the raised areas covered with stippled silver, and with metallic pink sheen. *Tropical.* p. 819

PISONIA: see Heimerliodendron

PISTACIA *Anacardiaceae*
atlantica (Canary Is., Medit. reg.) "Mt. Atlas mastic tree" or "Pistache"; semi-evergreen ornamental tree to 18 m tall; odd-pinnate leaves of 7 to 11 leaflets; small flowers without petals, followed by berry-like fruit first red, later purple when ripe, 1 cm dia. *Arid subtropic.* p. 71

lentiscus (Mediterranean Reg.), "Mastic tree"; shrubby evergreen bush or small tree to 4 m high; pinnate leaves with leathery leaflets in 3 to 5 pairs; small red flower clusters without petals, followed by ornamental reddish fruit turning black. Cultivated for mastic, a high grade resin. *Arid subtropic.* p. 71

vera (Syria), "Pistachio nut"; widely planted throughout the Mediterranean region and Northern India; spreading tree to 10 m high, partially deciduous and suited only for dry regions where the olive grows; the leaves of 1 to 5 pairs of thick, oval leaflets, and an extra one at the tip; tiny brownish-green flowers unisexual, without petals on female trees; clusters of oblong red fruit, 2-3 cm long, becoming wrinkled; inside the husk a stone contains the pistachio kernel, of rich taste, and used for flavoring cake and candy. The nuts are prepared in brine while still in the shell and are also favored for eating and nibbling. *Subtropic.* p. 73, 467

PISTIA *Araceae*
stratiotes (Trop. America), "Water lettuce"; water-floating leaf rosettes bright green and velvety, hairy, with hanging roots; small green flowers hidden between leaves. Apparently naturalized in other parts of the tropical world, I have seen them colonizing lagoons and rivers in West Africa, in Kenya and Uganda, and along the upper Nile. *Tropical.* p. 129

PITCAIRNIA *Bromeliaceae*
andreana (Colombia: Choco); small terrestrial, with leafy stems to 20 cm long; recurving, narrow-lanceolate leaves, green and speckled with silver dots above, heavily frosted beneath with white scales; inflorescence with orange petals tipped with yellow. *Tropical.* p. 224

corallina (Colombia), "Palm bromeliad"; terrestrial plant to 1 m, with stalked leaves dark green and corrugated, petioles with brown spines; inflorescence in prostrate raceme with red stem and coral-red flowers. *Tropical.* p. 224

flammea (Brazil: Espirito Santo to Sao Paulo); terrestrial rosette with petioled, narrow linear leaves to 1 m long and 3 cm wide, scaly beneath and without spines; unbranched slender inflorescence with red flowers 3 cm long. *Tropical.* p. 224

heterophylla (Mexico, C. America, to Ecuador); small epiphytic rosette, with bulbous base formed by outer leaves, the grass-like linear, green foliage to 20 cm long x 1 cm wide; flowers in a near sessile spike with reddish bracts and sepals, and bright red petals occasionally white. *Tropical.* p. 221

tabuliformis (Mexico: Chiapas); terrestrial rosette with oblong spatulate, papery 15 cm leaves light green showing darker, parallel veining, and lying flat on the ground; showy bright red flowers in a sessile head. *Tropical.* p. 222

PITTOSPORUM *Pittosporaceae*
eugenioides 'Variegatum'; New Zealand cultivar with attractive long-elliptic, leathery leaves grayish-green, with wavy ivory-white margins. The species P. eugenioides is a decorative evergreen growing into a tree to 12 m high, with narrow, glossy-green foliage 10 cm long, and small yellow flowers scented like honey. *Subtropic.* p. 823

moluccanum (ferrugineum) (Indonesia); small evergreen tree with spreading crown, grayish bark, on slender brownish twigs, the long obovate glossy green leaves with undulate margins; small 1 cm yellowish flowers with honey fragrance; clusters of 1 cm orange fruits gaping in halves and showing the scarlet mass of pulpy seeds. *Subtropic.* p. 823

tenuifolium 'Variegatum' (New Zealand), "Kohutu"; evergreen shrub or small tree, with gray bark; elliptic thin-leathery leaves 5-8 cm long, grayish green with undulate, cream-white margins; small purple flowers. *Subtropic.* p. 824

tobira (China, Japan) "Mock orange"; tough, evergreen shrub branching into a rather flat-topped, shapely bush, with thick-leathery obovate, dark lustrous green leaves to 10 cm long, arranged in dense pseudowhorls, and with terminal clusters of small, creamy-white flowers, very fragrant. *Subtropic.* p. 823

tobira 'Variegatum', "Variegated mock-orange"; attractive variegated form with leathery leaves slightly thinner, milky or grayish-green raggedly margined creamy-white; the little, fragrant flowers resembling orange-blossoms. *Subtropic.* p. 823

tobira 'Wheeler's Dwarf'; miniature form of tobira; very compact low mound of glossy, dark green foliage. *Subtropic.* p. 824

undulatum (Australia), "Sweet pittosporum"; loosely branching tree growing to 12 m high, with soft-leathery, long elliptic or oblanceolate leaves to 15 cm long, lightly pendant, and with wavy margins, dark green. *Subtropic.* p. 823

wrightii (Seychelles); evergreen shrub with slender woody branches; thin-leathery obovate leaves with margins convex; clusters of numerous small creamy-white fragrant flowers. *Tropical.* p. 823

PITYROGRAMMA *Polypodiaceae (Filices)*
chrysophylla (calomelanos aureo-flava) (So. America, West Indies), a "Gold fern"; of graceful habit, with wide-spreading tripinnate, somewhat fleshy fronds to 60 cm long, dull green above, thickly covered with golden yellow, waxy powder on the underside, carried on purplish-brown channeled stalks; the young shoots also waxy golden-yellow. *Humid-tropical.* p. 444

PLATYCERIUM *Filices: Polypodiaceae*
"alcicorne" hort. (Madagascar, Comores, Mauritius). "Elkhorn fern"; distinct from bifurcatum in having the fertile fronds shorter, more rigidly erect, a bright green, and only thinly hairy, widening to short forks, 2 to 3-lobed, and lightly recurving, the brown soral patches are located at or beside the last bifurcation, sometimes extending to the lower portions of the lobes instead of on the tip of the distal segments as in bifurcatum; the young basal fronds are rounded becoming somewhat crenulate with age, prolonged above into a few finger-like lobes. *Tropical.* p. 450, 451, 452

andinum (E. Perú, E. Bolivia), "American staghorn"; this sole So. American, subandine species is a large epiphyte with mighty erect, lobed barren fronds and much forked long pendant fertile leaves to 3 m long, with long ribbon-like lobes, and sporangia placed at third fork back. *Tropical.* p. 449, 452

angolense (Trop. Africa), "Elephant's ear fern"; large epiphytic fern, close to stemaria, with ascending sterile fronds having purplish veining and wavy crest, broad wedge-shaped fertile fronds not divided into lobes, and with a felt-like covering of rust-colored wool underneath; the sporangia along the apex. I have encountered this species all across Africa from the Usambaras in Tanzania, on Lake Victoria, Uganda, the Ituri forest of Zaire to the Niger Delta in Nigeria. *Tropical.* p. 449

bifurcatum (E. Australia, New Guinea, New Caledonia, Sunda Isl.), the common "Staghorn fern"; easy growing epiphyte freely producing young plants on its roots; the basal fronds are kidney-shaped, in old specimen lobed; the usually laxly pendant, leathery, grayish dark green fertile fronds to 1 m long are thinly covered with white, stellate hairs, and usually twice long forked; soral patches only on distal segments, being the tips of the ultimate forks. Sterile fronds round, but feathered on back. Reverse silvery or green. Known as "Elkshorn" in Australia. On Mt. Boss, New South Wales, at 900 m this species tolerates -9°C cold, rainfall is 320 cm yr.; near Sydney they are found growing between 240-450 m altitude. *Subtropic.* p. 450, 451

bifurcatum 'Majus' (Polynesia); a variety of more robust habit with larger foliage; the barren fronds are roundish, convex, and overlap each other; the broad, rich green fertile fronds tend to be erect, with the broad forking lobes pendulous. *Subtropic.* p.450

bifurcatum cv. 'Netherlands' (alcic. 'Regina Wilhelmina'); from Holland, with soft-leathery fertile fronds broader than the type, well divided into numerous lobes, and the habit of growth is with bright green fronds in all directions star-like. *Subtropic.* p. 450

coronarium (biforme) (Burma, Thailand, Malaya, Java, Philippines), "Crown staghorn"; a glorious epiphyte of which I have seen immense clusters growing high in trees of the rainforest in Malaya, the long, fresh green pendulous fronds are to 4 m long, several times widely forked, and the lobes gracefully twisted; the thick barren fronds are tall and lobed; spore is curiously borne on a separate fertile reniform disk. *Tropical.* p. 451

"diversifolium" hort. "Erect elkhorn"; dwarf plant attractive because of the erect habit of its fronds which are broadly spreading into twice-divided, pendant lobes, covered with whitish stellate hairs; basal fronds reniform, neatly covering the fibrous roots. *Tropical.* p. 450

grande (superbum) (E. Australia, Singapore, Philippines), "Regal elkhorn"; magnificent epiphyte with a regal crown of upright spreading sterile fronds of glossy vivid green, the upper lobes doubly forked and staghorn-like with dark venation; pendulous, forked pairs of fertile fronds appear with age, holding between them the wedge-shaped disk bearing the sporangia. I have found P. grande as far south as New South Wales at 600 m, where the temperature ranges from 3 to 49°C hosted on Stinging trees; also near Sydney at 200-270 m. Mature specimen with fan-like sterile fronds spread 1 m; fertile fronds 1½ m long. Spores are produced yearly after some 20 years of age. *Tropical.* p. 451, 453

hillii (Queensland), "Stiff staghorn"; handsome, fresh green species with basal leaves always round, covering the rootstock; the several fertile fronds are rigidly erect, gradually broadening fanlike before dividing into numerous pointed lobes, green beneath, with sori carried at base of ultimate tips. *Tropical.* p. 449

madagascariense (Madagascar); small, very attractive species mainly because of its deeply quilted, rounded basal frond, with the network of bluish veins raised high and forming tough elevated ridges; small leathery, bright green, wedge-shaped fertile frond is lightly lobed and rudimentary, and set with sori on the upper edge. *Humid-tropical.* p. 449, 453

quadridichotomum (W. Madagascar); epiphytic plant with basal frond irregularly oblong and nest-like, appressed and round in the lower part, and erect and free, undulate and laciniate; the normal fertile fronds pendant, regularly in pairs 3 or 4 times, the divisions strap shaped, the underside covered with yellowish stellate hairs. Sporangia area between first and third forks. *Tropical.* p. 452

"ridleyii" hort. (Malaya, Borneo, Sumatra); epiphyte at home in high trees, allied to coronarium, but more compact; rounded basal fronds mostly appressed to the support; normal frond rather erect, fresh green 30-60 cm long, about 5 times irregularly forked in pairs and sterile, but one branch sometimes carrying at its base a concave fertile lobe bearing the sporangia; the divisions are characteristically short and wide-spreading. The existence of this species is discussed in Baileya Sept. 64, but the photo shown is not yet proven correct. *Tropical.* p. 452

stemaria (aethiopicum), (W. Africa to Madagascar), "Triangle staghorn"; curious species with basal fronds convex and elongated into lobes; the triangular grayish green fertile fronds to 45 cm long, thick-leathery, with prominent ribs, and divided twice, the main fork spreading wide, with a sinus, around which follow the spore masses; the underside densely covered with silvery-white felt. *Tropical.* p. 449, 452

superbum; large and showy epiphytic species photographed in the collection of Munich Botanic Garden, Germany 1971; a spreading crown of long bright green, deeply lobed sterile fronds flaring gracefully outward; the heavy, thick-fleshy fertile fronds forked into two sections of long, laxly pendant lobes; the sporangia patch in the rounded sinus of the main fork. *Humid-tropical.* p. 452

wallichii (Burma, Indochina); attractive epiphyte, with sterile basal fronds first circular and cupping, then with lobes upwards; the fertile fronds broad and several times slender-forked into concave segments, thick in all parts, very green and slightly yellowish tomentose; looks like grande but more diminutive. *Tropical.* p. 453

wilhelminae-reginae (newly valid: *wandae*) (New Guinea), "Queen elkhorn"; with large crown of feathered sterile fronds spreading 1½ m; not as deeply lobed as grande but fuller; the long, gracefully pendant fronds to 2 m long, in pairs each with 3-4 long lobes, flanked on both outsides by one or two sets of separate 1 m obliquely broad-triangular spore-blades; glossy dark green above, silvery beneath, with prominent dark veins; lettuce serration at base; young growth covered with silvery scales. I have found this species widespread from southern Papua near sea level to the forbidding mountain ranges of northern New Guinea, at 1,100 m. *Tropical.* p. 450, 452, 453

willinckii (Java), "Silver staghorn"; a distinct epiphyte with uneven, forked basal leaves and densely silvery-pubescent fertile fronds, erect at first, later completely pendant, very narrow and several times forked into long slender lobes, sporangia-bearing at tips. *Tropical.* p. 452

PLATYCLADUS: see Thuja orientalis

PLATYCLINIS: see Dendrochilum

PLECTRANTHUS   *Labiatae*

coleoides 'Marginatus' (tomentosus), "Candle plant"; low, bushy plant dense with opposite, ovate, hairy herbaceous, 5-8 cm leaves dark green and grayish, the crenate and scalloped margins creamy-white; 4-angled stem; flowers white with purple; the type is from the Nilghiris (So. India). *Subtropic.* p. 532

fruticosus (So. Africa), "African spurflower"; erect shrub to 1 m high, with ovate, crenate leaves to 10 cm long, sparsely hairy; blue flowers in terminal clusters. *Subtropic.* p. 532

madagascariensis (S. E. Africa, Madagascar), "Mintleaf"; small creeping shrub with brownish, angled stem, with opposite, firm green, wrinkled leaves 3-4 cm long, of fleshy texture, crenate margins and covered with whitish bristle, strongly scented of mint when bruised; flowers in erect small spikes, very pale lavender, almost white. I found and photographed this in 1973 at Kirstenbosch near Cape Town. *Subtropic.* p. 532

nummularius (australis) (Australia, Pacific Islands), the "Swedish ivy"; vigorous creeping perennial herb with small leathery, thickish, metallic-green, waxy leaves almost round, 3-6 cm across and deeply crenate, glaucous gray-green beneath and with purplish veins; small white 2-lipped flowers in spikes. A tough trailer tolerating abuse; good for hanging pots or wall containers. *Tropical.* p. 530, 532

nummularius 'Variegatus', "Variegated Swedish ivy"; variety with the glossy leaves in greens and white. *Subtropic.* p. 532

oertendahlii (Natal), "Prostrate coleus"; fleshy creeper with 4-angled stem and small, broad leaves, friendly green to bronzy, patterned with an attractive network of silvery veins, the lightly crenate margins purple, the surface short-hairy, older leaves beneath, petioles purple; flowers pale pink. *Subtropic.* p. 530

saccatus (So. Africa: Natal, Zululand), the "Edging spurflower"; lovely flowering species suitable for hanging baskets; small herbaceous plant to 30 cm high, with dark matte green, rather thin, crenate leaves, on branches spreading horizontally and bearing ornamental sprays of bilabiate flowers 2-3 cm long, the buds are deep purple and open to a lighter shade of lavender. I found this charming plant at Canterbury University Botanic Gardens in Christchurch, New Zealand, covered with blooms in February. *Subtropic.* p. 532

'Variegated Mintleaf' (prob. madagascariensis cv, from S. E. Africa), in hort. also as "minima" and "Iboza"; pretty creeper with brownish, hirsute stems; small 3-4 cm fleshy, crenate leaves milky-green bordered white, strongly mint-scented; white flowers flushed lilac. Ideal for baskets. *Subtropic.* p. 532

PLEIONE *Orchidaceae*
forrestii (China: W. Yunnan); dwarf plant, rock-dwelling or epiphytic, cluster-forming, with flask-shaped pseudobulbs, deciduous folded foliage, and solitary, proportionately large and handsome flowers 8 cm across, which arise with the new growth; blooms yellow to orange with frilled lip marked purple. *Subtropic.* *p. 748*

PLEIOSPILOS *Aizoaceae*
bolusii (Cape Prov.: Karroo), "Living rock cactus"; small stemless succulent with pairs of thick, stone-like keeled leaves 5 cm long and broad, flattened inside, light gray-green with numerous dark-green dots; flowers deep yellow. *Arid-subtropic.* *p. 49*
nelii (Cape Prov.); succulent "Mimicry plant" in form of a split globe with thick leaves in pairs, gray and with raised, dark dots; showy yellow flowers. *Arid-subtropic.* *p. 48, 49*

PLEOMELE *Liliaceae*
angustifolia honoriae (Solomons, Torres Straits); beautiful ornamental with willowy, scandent stems densely clothed with clasping flexible-leathery, lanceolate leaves 15-25 cm long, shining grass-green with distinct ivory-yellow borders, and turning metallic red in the sun. *Tropical.* *p. 581, 582, 588*
reflexa (Dracaena) (Madagascar, Mauritius, India), "Malaysian dracaena"; ornamental rosette of densely clustering short and narrow, leathery leaves, deep glossy green, without midrib, wavy and reflexed, persistently clasping the willowy, selfbranching stem, to 3½ m high if given support; I have noticed them widely used in India and Thailand as a most satisfactory pot plant; flowers whitish. *Tropical.* *p. 580, 582*
reflexa angustifolia (Malaya, Java); self-branching but more tree like, with long oblanceolate, very narrow, thin leathery, strap-like leaves medium green, crowded around the slender stem and spreading. *Tropical.* *p. 582*
reflexa 'Variegata' (South India, Ceylon), "Song of India"; a beauty I fell in love with when I first saw it in Ceylon; a tropical evergreen; self-branching with slender, flexuous stems eventually becoming scandent to 3 m long, densely furnished with clasping, narrow lanceolate, leathery leaves 12 cm long, beautifully margined by two wide bands of golden yellow or cream and framing the green center; very slow-growing. *Tropical.* *p. 577, 581, 611*

PLEUROTHALLIS *Orchidaceae*
cardiothallis (Mexico, Nicaragua, Costa Rica); curious epiphyte of the humid cloud forest, with slender stems bearing a solitary leathery leaf 10-18 cm long and heart-shaped in outline; the tiny fleshy flowers 2 cm long, appearing out of the base axil of the leaf, deep brownish-red. *Humid-tropical.* *p. 751*
quadrifida (ghiesbreghtiana) (Mexico, C. America, Panama, W. Indies); epiphyte with solitary leaf, and elongate raceme of small lemon-yellow flowers with pointed sepals and petals in ranks along the wiry stem, slightly fragrant, (winter). *Tropical.* *p. 752*
sonderiana (Trop. America); small species forming clusters; short and narrow channeled, fleshy leaves; pretty flowers with linear sepals and petals yellow into apricot salmon, lip lavender pink. *Humid-tropical.* *p. 750*

PLUMBAGO *Plumbaginaceae*
auriculata (capensis) (So. Africa), "Cape leadwort"; straggling, shrubby perennial with small oblong, scattered leaves and wiry raceme of salvershaped, azure-blue flowers having a very slender tube and phlox-like lobes. *Subtropic* *p. 811*
indica (coccinea, rosea) (E. Indies), "Scarlet leadwort"; showy perennial, with wiry, zigzag stems, elliptic leaves and long terminal spike of scarlet-red salvershaped flowers. *Tropical* *p. 811*

PLUMERIA *Apocynaceae*
obtusa (Hispaniola, Cuba, Yucatán Penin.), "Temple tree" or "Singapore plumeria"; small evergreen tree to 8 m high, the heavy oblanceolate, dark green leaves with milky sap, to 18 cm long; large white, waxy flowers with spreading petals, 5 to 7 cm across and with yellow center, intensely fragrant. *Tropical.* *p. 81*
pudica (Colombia, Venezuela, Curacao); "Bashful frangipani"; small tree to 3 m, photographed in Curacao, and remarkable because of its narrow-spoon-shaped, thin-leathery drooping foliage to 30 cm long, and large 8 cm snow-white flowers with pale yellow center. *Tropical.* *p. 83*
rubra (acuminata) (Mexico to Ecuador), "Frangipani tree"; large, waxy, single blossoms carmine-rose with yellow eye, very fragrant; thick, soft branches with latex-like, sticky juice, dark-green leaves, shedding in dry season. *Tropical.* *p. 78, 79, 81, 83*
rubra acutifolia (Mexico), the "West Indian jasmine", and the "Temple tree" of India; leaves wedge-shaped; flowers waxy-white with yellow throat, sweetly fragrant, funnelform. *Tropical.* *p. 81, 83*
'Rubra hybrid', "Hawaiian flag"; robust plant photographed near Diamond Head, Honolulu, with dense, long-elliptic stiff leaves, and clusters of soft rosy-red flowers variegated with white. *Tropical.* *p. 81*
rubra 'Tricolor'; a tri-colored frangipani seen in Venezuela; robust densely branching tree with large clusters of lovely waxy flowers in colors pink, shading to salmon red and crimson, orange-yellow and white. *Tropical.* *p. 83*
'Singapore hybrid', "Pink plumeria"; nowhere have I seen larger-flowered "frangipani" trees than this type in Singapore, in colors from ivory to rosy-purple, with glossy leaves that are said to be non-deciduous but this may be aided by an equable moist-tropical climate. *Tropical.* *p. 81*

PODALYRIA *Leguminosae*
calyptrata (So. Africa); much branched shrub to 1½ m; shoots and 5 cm obovate leaves downy; clusters of flowers pale rose. *Subtropic.* *p. 556*

PODOCARPUS *Podocarpaceae*
elongatus (spinulosus) (Western Africa), "African yellow-wood", coniferous evergreen tree to 20 m high, with gracefully pendant branches, densely pinnate, with long, leathery, narrow-linear, tapering leaves, bright green; short male catkins, and globose crimson fruit. *Tropical.* *p. 350*
falcatus (So. Africa), "Oteniqua yellow-wood"; evergreen tree rather loose in habit, with long, narrow, grayish-green needles. *Subtropic.* *p. 350*
gracilior (Kenya, Uganda, Ethiopia), "African fern pine"; subtropical coniferous tree to 18 m high, common on the slopes of Mt. Kenya at 2,100 to 2,800 m; a valuable timber tree; graceful willowy branches with long, narrow-lanceolate, needle-like leathery leaves glossy deep green, to 10 cm long on young trees, and loosely arranged; 6 cm long and dense on older specimen; glaucous, purple berries. This may be the same as the cultivated P. "elongata". *Tropical.* *p. 350*
henkelii (Transvaal), "Long-leaved yellow-wood"; luxuriant rounded tree dense with glossy, deep green, pendant or recurving leaves 12 to 15 cm long; male catkins cylindrical and erect from leaf axils. *Subtropic.* *p. 350*
macrophyllus (chinensis) (China, Japan), "Buddhist pine"; dioecious coniferous tree to about 12 m high, with horizontal branches, and numerous crowded, leafy twigs, the leathery, deep green, narrow linear-lanceolate leaves needle-like 5-6 cm long as seen in China, in cultivation 8-10 cm long; with a single midrib prominent on both sides; male, axillary flowers resembling catkins; berries bluish purple. *Subtropic.* *p. 350, 351*
macrophyllus 'Maki' (China); a compact "Southern yew"; popular evergreen shrub; for hedges in the South, and grown as a superb decorative container plant in the North; lends itself well to shearing and shaping; as a tree attaining 15 m in this variety with rather erect branches, dense with waxy, blackish green, linear lanceolate leaves spirally arranged, 4-8 cm long and about 1 cm wide, with distinct midrib; pale green beneath. Male flowers in 4 cm catkins; the fleshy oval fruit on female trees glaucous-purple. *Subtropic.* *p. 348, 350, 351, 353*
milanjianus (Uganda); bold evergreen tree with thick leathery, glossy green leaves 10 cm long, having a furrowed midrib. *Subtropic.* *p. 350*
Nagi (China, Japan, Formosa), "Broadleaf podocarpus"; tall conifer to 30 m high, with smooth, purplish bark, elegant spreading branches and slender, semi-pendant branchlets having shiny green, rigid-leathery, elliptic leaves to 2½ cm wide; very durable decorator but not as dense as macrophyllus. *Subtropic.* *p. 350*
neriifolius (China to New Guinea); evergreen tree to 20 m or more tall, of straggly growth; lance-shaped, leathery leaves to 15 cm long, glossy green above, somewhat glaucous beneath, with raised midrib both sides. *Tropical.* *p. 350, 351*
totara (New Zealand); the "Totara" is a lofty, coniferous timber tree, to more than 30 m high, and 5-7 m girth; with small needle-like leaves 1-2 cm long, spiny-pointed, and rigid-leathery, bronzy-green; fruit is a small nut set on a fleshy, crimson base. *Subtropic.* *p. 351*

PODOPHYLLUM *Berberidaceae*
emodi (Himalayas), "May-apple"; perennial woodland herb with creeping rootstock and thick roots, with peltate, palmately lobed leaves on fleshy stalks 15-30 cm high, and white, waxy flowers to 4 cm across, followed by 5 cm red, edible fruit. *Warm temperate* *p. 181*

PODRANEA *Bignoniaceae*
ricasoliana (So. Africa), "Port St. Johns creeper"; showy evergreen climber with pinnate, deep glossy green leaves divided in-

to 7-11 ovate leaflets 5 cm long, with toothed margins; flowers with inflated calyx and trumpet-like corolla pinkish-lavender with red veining, 5 cm long and opening into 5 rounded lobes, on blackish petioles. *Subtropic.* p. 186

POINCIANA: see Caesalpinia, Delonix

POINSETTIA: see Euphorbia pulcherrima

POLIANTHES *Amaryllidaceae*
tuberosa (Mexico), the famed "Tuberose"; widely cultivated in the tropics where it is esteemed for the purity and powerful fragrance of its blooms as a cut flower and in gardens; a beautiful summer or fall blooming herb having a bulb-like tuberous root stock covered with the broadened bases of the grass-like, channeled leaves; the leafy floral spikes are wiry and ½ to 1 m high, bearing numerous funnel-shaped waxy-white flowers 4-6 cm long, in pairs. The tuberose may be had in flower throughout most of the year by potting bulbs in succession, taking 4 to 5 months to bloom; grow warm but keep dry until the leaves appear, then water freely; dry off after blooming when foliage turns yellow. *Tropical* p. 64, 68

POLYALTHIA *Annonaceae*
longifolia pendula (India), the "Ashoka tree" or "Mast tree" in India; a lofty evergreen, graceful column of symmetrical pyramidal growth 15 m or more tall, with willowy, weeping pendulous branches, long narrow lanceolate leaves to 20 cm long, shiny green with undulate margins; in spring the tree is covered with delicate, star-like pale green flowers with wavy petals, followed by the ovoid black 2 cm fruit, loved by bats and flying foxes. It is held in great esteem by Hindus, and planted near their temples. *Tropical.* p. 72

POLYGALA *Polygalaceae*
x dalmaisiana (oppositifolia x myrtifolia), "Sweet-pea shrub"; evergreen shrub to 2 m high or more, with small ovate to linear leaves to 3 cm long; flowers purplish or rosy-red, blooming almost continuously. *Subtropic.* p. 811
myrtifolia (So. Africa), "Milk-wort"; woody shrub to 2½ m high, with small obovate 3 cm leaves; flowers greenish-white veined with purple, the lower petal with fringed crest, borne in erect terminal racemes. *Subtropic.* p. 811
oppositifolia (So. Africa), "Purple broom"; densely leafy evergreen shrub 1 m or more high, small ovate, grayish green leaves; unusual flowers with sepals wing-like light purple, the petals united and with yellowish green keel. *Subtropic.* p. 811

POLYGONUM *Polygonaceae*
capitatum (Himalayas: No. India), "Knot-weed"; pretty perennial with trailing branches, the wiry, rooting brownish stems set with small alternate elliptic leaves 4 cm long, indistinctly crenate, matte-green with pale or light brown midrib and prominent, acute-angled design in brown-purple; the small pink flowers in globular heads. *Subtropic.* p. 825

POLYPODIUM *Polypodiaceae (Filices)*
aureum (Phlebodium) (W. Indies to Brazil, Australia); named "Hare's foot fern" because of the stout creeping rhizomes clothed with bright rusty brown hair-like scales, the wiry stalks bearing bold, metallic light green, thin-leathery fronds, lobed with broad linear pinnae, separated by a rounded sinus and not cut to center; epiphytic. *Humid-tropical.* p. 436
aureum 'Glaucophyllum' (W. Indies to Colombia and Ecuador); pretty little species with slender, wide-creeping rhizomes; oblong spear-shaped, leathery fronds 10 to 25 cm long, glossy bright green above, bluish beneath, contrasting with the golden-yellow spore-masses. *Humid-tropical.* p. 442
aureum 'Mandaianum', "Crisped blue fern"; a beautiful crested form having graceful bluish glaucous fronds of broad, pendulous, wavy pinnae with margins irregularly lobed, crisped and lacerated. (Manda 1912) *Humid-tropical.* p. 446
bifrons, (Ecuador); a very unusual creeping fern with wiry slender rhizomes carrying small oblanceolate deeply lobed, both sterile and fertile fronds but besides, forming curious chestnut-like vessels hollow inside pocket-like, similar to Dischidia, separate from its leaves. Rare and difficult in cultivation, best trained to a piece of tree fern trunk and hung from a greenhouse rafter; needs warmth and humidity and should be dunked in water daily. *Humid-tropical.* p. 447
crassifolium (West Indies, Mexico to Perú and Brazil); fern with very leathery, simple, narrow oblanceolate fronds 30-90 cm long and 2½-12 cm broad, the upper surface with a few white dots, the margins wavy; short-stalked on stout, creeping, woody rhizome covered with brown scales. *Humid-tropical.* p. 448
irioides: see punctatum
musifolium (Drynariopsis, Phymatodes) (Malaya, Philippines, New Guinea); handsome epiphyte with woody rhizome with stalkless, oblanceolate, thin-leathery leaves of pea green, prettily marked with a network of dark veining. *Humid-tropical.* p. 441
persicarifolium (Trinidad); small fern with wiry creeping rhizomes, the leaves long lanceolate 12-18 cm long, leathery and glossy green. *Humid-tropical.* p. 448
punctatum (polycarpon) (New South Wales, Natal, Angola, Guinea), "Climbing birdsnest"; singular-looking, succulent fern with stout rhizome and stalkless, thick-fleshy, yellow-green, simple fronds to 1 m long, gradually narrowed on both ends and irregularly indented or undulate at margins. *Tropical.* p. 441, 447
punctatum 'Grandiceps' (irioides grandiceps), "Fish-tail"; clustering fern with odd-shaped, thick leathery, almost succulent, waxy yellow-green fronds 30-60 cm high with prominent midrib, and tips forking to points or broad crests. Very curious, yet attractive and durable house plant. *Humid-tropical.* p. 446, 447
subauriculatum 'Knightiae' (Australia), "Lacy pine fern"; an excellent, slow growing and durable basket fern, with glossy yellow-green, pinnate fronds at first upright, later pendulous and long, the linear pinnae deeply serrate and sliced into narrow, pointed lobes. *Humid-tropical.* p. 435
vulgare (Newfoundland to Alaska, Alabama; Eurasia), "Common polypody", or "Adder's fern"; ornamental, hardy evergreen fern, often epiphytic, growing on walls, roofs or trees; mat-forming on stout, rusty-scaly rhizomes, straw-colored stalks bearing 15-30 cm pinnate fronds, the papery leaflets toothed and wavy. *Temperate.* p. 448

POLYPODIUM: see also Drynaria

POLYSCIAS *Araliaceae*
balfouriana (Aralia) (New Caledonia), "Dinner plate aralia"; leafy, bushy tropical shrub in habitat to 8 m high, branching with willowy stems; the large leathery leaves variable but at first entire, later usually of 3 rounded, coarsely toothed glossy green leaflets to 10 cm across, on bronzy stems speckled gray. *Tropical.* p. 141
balfouriana 'Blackie'; as grown in Florida nurseries; compact bush of sinister appearance, the pinnate leaves with coarsely puckered leaflets, glossy blackish or bronzy green, the margins with soft teeth. This may properly be a cultivar of P. guilfoylei, also known as 'Crispa' or 'Amazonica'. *Tropical.* p. 142
balfouriana 'Marginata', (New Caledonia), "Variegated Balfour aralia"; variegated form with the grayish-green, leathery leaflets having an irregular, white border. Much planted for hedges in tropical regions. *Tropical.* p. 142
balfouriana 'Pennockii', "White aralia"; attractive cultivar by Pennock's, Puerto Rico; the waxy leaves olive-green with pronounced vein areas of creamy-white. *Tropical.* p. 142
balfouriana 'Variegata' (Aralia) (New Caledonia), "Dinner plate aralia"; leafy, bushy tropical shrub in habitat to 8 m high, branching with willowy stems; the large leathery leaves variable but at first entire, later usually of 3 rounded, coarsely toothed leaflets to 10 cm across, variegated with cream, on bronzy stems speckled gray. Needs moisture with nourishing soil and good drainage. *Tropical.* p. 142
filicifolia (Aralia) (South Sea Is.); evergreen shrub with leathery, fern-like leaves bipinnately cut into narrow lobes, bright green, midrib purplish. *Tropical.* p. 141
fruticosa (Nothopanax) (Polynesia, Malaysia, India), "Ming aralia"; evergreen shrub 1½-2½ m high, with spotted, willowy branches, and very feathery leaves irregularly 3-pinnately cut, to 30 cm long, the segments spiny-toothed, often edged with white. *Tropical.* p. 141
guilfoylei 'Laciniata'; small evergreen shrub somewhat resembling 'Victoriae' but with thin-leathery, grayish green leaflets larger and with jagged margins. *Tropical.* p. 142
guilfoylei 'Quinquefolia'; "Celery-leaved panax"; a variety grown in Florida, with its hard-leathery leaves more or less irregularly cut, with five divisions or lobes, deep coppery olive green. *Tropical.* p. 142
guilfoylei 'Victoriae' (Polynesia), "Lace aralia"; charming tropical evergreen dense with slender, willowy branches and grayish-green, thin-leathery, lacy, bipinnate leaves, the small, pendant, feathery segments toothed and bordered white. Lovely small foliage plant enjoying warmth and moisture, with its tasseled variegated foliage the most exquisite of the variegated leaf forms. *Tropical.* p. 142
paniculata 'Variegata'; attractive variegated form of the species from Mauritius; willowy shrub with pinnate leaves, the leaflets leathery and deeply serrate, deep green and richly splashed with cream and greenish-white, glossy on both sides. *Tropical.* p. 142
scutellaria (Nothopanax) (Java), "Panax"; large decorative shrub with heart-shaped to oval, saucer-like leaves simple or with 2-5 leaflets 5-12 cm dia. margins wavy. In Malaysia young leaves are cooked as a vegetable. *Tropical.* p. 142

POLYSCIAS: see also Nothopanax

POLYSTACHYA *Orchidaceae*
stricta (Kenya, Uganda, Zaire); cylindrical pseudobulbs thickened at base, 3-4 lanceolate leaves; branched inflorescence with small green flowers marked mauve inside, the yellow lip marked brown. *Tropical.* p. 746

POLYSTICHUM *Polypodiaceae (Filices)*
aculeatum (Old World and So. America), "Prickly shield fern"; rosette of rigid, dark green, bipinnate leaves 30-60 cm long, on brown-scaly rachis. *Humid-subtropic.* p. 447
adiantiformis: see Rumohra
setiferum, sometimes spelled setigerum (Trop. and Temp. Zones of both hemispheres), the variable, hardy "Hedge-fern"; feathery pinnate fronds 30-60 cm long, covered with brown hair-like scales, borne on shaggy, stout stalks; the fresh-green pinnae close together, and deeply lobed; hardy. *Warm temperate to subtropical.* p. 447
setiferum 'Proliferum' (viviparum) (Australia), "Filigree fern"; tufted fern, scaly at base, the fleshy, brown-woolly stalk bearing the pinnate, light green fronds, the pinnae deeply cut or lobed, and bud-bearing, giving rise to young plantlets. *Subtropic.* p. 447
strigosum (Dryopteris rigida) (Mediterranean to Afghanistan, Alaska to California), "Wood-fern"; spreading fern with long stalks, bearing bipinnate herbaceous fronds, the pinnae well separated on thin wiry rachis. *Humid-subtropic.* p. 447
tsus-simense (Aspidium); from the island of Tsus-sima in the Straits of Korea; dwarf and shapely tufted fern suitable for terrariums, with small leathery, lanceolate, dark green fronds, bipinnate in the lower part, the segments becoming gradually smaller toward the slender point and sharply toothed. *Subtropic.* p. 447

POMETIA *Sapindaceae*
pinnata (Malaysia to Polynesia), "Langsir"; luxuriant evergreen tree with glossy green pinnate leaves 20 cm or more long, the large corrugated leaflets in a dozen or more pairs; tiny greenish flowers hanging in long panicles, followed by nearly globular fruit with brownish rind 3 to 5 cm dia.; the seeds are edible when roasted. Photographed on Nuku Hiva, Marquesas. *Tropical.* p. 474, 875

PONCIRUS *Rutaceae*
trifoliata (North China), "Hardy orange"; small, stiff-growing spiny deciduous tree to 4 m, dark green flattened branches and long stout spines 6 cm long; blooms in spring on bare branches in axils of large spines; white flowers opening flat, to 5 cm across; trifoliate leaves, with thin-leathery shining green leaflets, on winged petiole; small orange-like, aromatic fruit to 5 cm dia, with acid pulp; fairly hardy north to Philadelphia and N.J. *Temperate.* p. 872

PONTEDERIA *Pontederiaceae*
cordata (Nova Scotia to Florida and Texas south), "Pickerelweed"; aquatic perennial herb to 1 m, with thick parallel-veined long-stalked leaves from a rootstock, heart- or arrow-shaped to 25 cm long; blue flowers in spikes; used in ponds and water-gardens. *Warm temperate.* p. 828
lanceolata (Southeast U.S.), "Pickerel-weed"; aquatic similar to P. cordata with shorter spikes and fruits, and hairy blue flowers. *Warm temperate.* p. 828

POPULUS *Salicaceae*
tremuloides (No. America), "Quaking aspen"; deciduous tree 6-20 m tall, with grayish-white trunk; dainty, light green round or ovate leaves to 8 cm long, and which tremble in slightest air movements; turning golden yellow in autumn; flowers in drooping catkins, appearing before the leaves. *Temperate.* p. 875, 876

PORANA *Convolvulaceae*
paniculata (No. India to Upper Burma), "Snow vine"; twiner with ovate 15 cm leaves, clusters of small white flowers, with enlarged membranous sepals, in great masses. *Tropical.* p. 356

PORPHYROCOMA *Acanthaceae*
pohliana (lanceolata) (Brazil); beautiful herbaceous plant with opposite, lanceolate leaves 12-15 cm long, rich green with contrasting cream-yellow veining; the inflorescence in cone-like spike with large crimson-red overlapping bracts, 2-lipped tubular flowers purple. *Tropical.* p. 38

PORPHYROSTACHYS *Orchidaceae*
pilifera (Ecuador, Perú); terrestrial of Andean highlands, to 60 cm high, bearing a flaring cluster of rose pink or brilliant scarlet red flowers. *Subtropic.* p. 750

PORTEA *Bromeliaceae*
petropolitana (Coastal S.E. Brazil); shapely rosette of shiny yellow-green leaves broadened at base and with blackish spines; erect stem with dense panicle of orange-pink ovaries and sepals and white-lavender petals. *Tropical.* p. 221
petropolitana extensa (Espirito Santo, Rio); a form with the inflorescence on a striking coral-red arching stalk, the brilliant coloring extending to the slender green ovaries, tipped purple; flowers lilac. *Tropical.* p. 221

PORTLANDIA *Rubiaceae*
grandiflora (West Indies), "Glorias floridas de Cuba"; shiny evergreen shrub 3-4 m high, with elliptic, leathery leaves, and large white, solitary bell-shaped flowers 12 cm long, reddish inside, very fragrant at night. *Tropical.* p. 870

PORTULACA *Portulacaceae*
grandiflora (Brazil), "Rose-moss"; succulent herb with low spreading branches, scattered, cylindrical leaves and colorful, sun-blooming flowers in rose, red, purple, yellow, or white, surrounded by whorls of leaves and tufts of hairs; grown as an annual. *Arid-tropical.* p. 826, 831
grandiflora flore pleno 'Jewel', "Double-flowered moss-rose"; fancy double cultivar with cluster-forming succulent branches, the large flowers 5 cm across, with several layers of carmine-red petals. *Subtropic.* p. 831

PORTULACARIA *Portulacaceae*
afra (So. Africa), "Elephant bush"; succulent shrub to 3 m high, with thick-fleshy, brown stems, sparry branches, and opposite, small obovate, glossy green leaves to 2 cm long, thick and juicy, round beneath; dainty pink, inconspicuous flowers. *Arid-subtropic.* p. 830
afra 'Macrophylla', "Giant elephant bush"; handsome succulent neatly branching from thick brown-red stem, the fleshy, glossy-green obovate leaves to 3 cm long, much bigger though fewer than the species. Grown by Los Angeles Plant Co., Vista, California. *Subtropic.* p. 830
afra 'Variegata', "Rainbow bush"; lovely little succulent with red-brown stems and sparry, opposite branches dense with pretty leaves milky-green broadly margined ceamy-white, and with a thin carmine-red edge. *Arid-subtropic.* p. 826

POTENTILLA *Rosaceae*
verna (W. Europe), "Spring cinquefoil"; dainty, bright green, tufted creeper to 15 cm high; leaves divided into 5-7 toothed leaflets; rich yellow flowers 1 cm wide, in spring and summer. Good ground cover in California. *Warm temperate.* p. 854

POTHOS *Araceae*
scandens (Ceylon to Vietnam); branching creeper, broadly winged petiole with small, ovate blade attached, appearing as one leaf, arranged in 2 ranks along wiry stem. *Tropical.* p. 120, 125, 127

POTHOS: see also Scindapsus

PRIMULA *Primulaceae*
malacoides (Yunnan, China), "Fairy primrose" or "Baby primrose"; small, bushy herbaceous plant with numerous light green, smallish, papery leaves, white-hairy beneath, and toothed at margins, several straight stalks, bearing small flowers in successive umbels above each other, lavender or rose-pink to crimson-red or white, flowering late winter-spring. *Subtropic.* p. 831
malacoides 'Glory of Riverside'; excellent commercial "Fairy primrose" representative of the progress made in breeding large-flowering strains of malacoides; the freely produced flower clusters are dense and bushy, the individual blooms measure 3 cm or more across, in lighter and darker shades of salmon-rose with deeper eye and orange-yellow center, the buds a pretty white-farinose; its foliage does not irritate skin. *Subtropic.* p. 833
obconica (Hupeh, China); winter-blooming pot-primrose with fresh-green, brittle, broad-cordate leaves sparsely covered with irritating hairs, and showy umbels of large flowers in pastel shades of rose-pink, lavender, lilac to carmine-red, or even white; with greenish eye. *Subtropic.* p. 833
obconica 'Friesdorf Salmon', a "German primrose"; new German color strain with large flowers a soft rosy salmon, 3-5 cm across. *Subtropic.* p. 831
obconica 'Grandiflora', "Florist's primrose"; highly developed cultivar for florists' sales, with rosy-carmine flowers 5 cm or more across; the foliage may cause skin irritation. *Subtropic.* p. 832
× polyantha (a hybrid group of P. vulgaris probably with veris and elatior), the hardy "Polyanthus primrose", "Lady's fingers" or "English primrose"; popular in American gardens and blooming in early spring; clump-forming perennial 20-30 cm high with long obovate leaves narrowed into winged petioles; the 2½-5 cm flowers sweetly fragrant like roses, in many colors, yellow, red and yellow, orange, bronze, maroon, or with white, borne in clusters well above the foliage; (those of vulgaris are solitary). *Warm temperate.* p. 833

x polyantha 'Pacific Giant', "Polyanthus primrose"; the Pacific strain has large early flowers in brilliant shades of blue, yellow, red, pink, or white; not hardy in cold climate. *Warm temperate.* p. 831

PRITCHARDIA Palmae
arecina (Hawaii); endemic palm of Maui; handsome fan-palm, with big, pleated leaves, cut about 1/3 into linear segments, covered with light yellow wool beneath; slender erect trunk to 11 m high. *Tropical.* p. 794

beccariana (Hawaii); beautiful fan palm, 10 m tall, largest of the genus in Hawaii; palmate rich green leaves very broad, 1¼ m across, of stiff texture, densely pleated, and cut toward apex only; tolerates cooler locations. *Tropical.* p. 794

hillebrandii (Hawaii: Molokai), "Loulou-lelo palm"; slender straight trunk to 7 m tall, crowned by firm palmate leaves 1 m or more long, smooth above, powdery glaucous beneath, on downy leaf-stem; bluish black 2 cm fruit. *Tropical.* p. 794

macrocarpa (Eupritchardia) (Polynesia), "Loulou palm"; handsome fan palm with robust trunk, large palmate leaves, deeply folded, glossy green on stocky petioles, scale-like surface underneath, typical of the genus. *Tropical.* p. 798

pacifica (Fiji, Tonga, Samoa), "Fiji fan palm"; impressive with slender, clean trunk 9 m tall, to 30 cm in dia., the numerous short-stalked palmate fronds forming a large round crown to 2½ m across, the leaves 1 m wide, bright glittering olive-green and deeply folded, very leathery, covered with brownish-white fuzz when young, only lightly cut at apex; the 1 m stalk unarmed, with brown fiber at base; 1 m spadix amongst foliage, with brownish fragrant flowers, and 1 cm lustrous blue-black fruit. *Tropical.* p. 797

remota (Hawaii: Bird Is.), "Loulu palm"; tall fan palm with straight, slender trunk, the grayish green fronds on stout petioles, the tips pendant. *Tropical.* p. 796

thurstonii (Fiji); smaller fan palm to 5 m tall, with folded leaves incised into many pointed segments, underside with thin waxy covering, a single axis exceeding the leaves. *Tropical.* p. 798

PROBOSCIDEA Martyniaceae
fragrans (So. Mexico), "Unicorn flower"; photographed in habitat at the Toltec temple of Tlaxcala; small sticky-pubescent herb with stout stem, and rough, deeply lobed leaves to 15 cm wide; pretty, fragrant flowers 5 cm long, rosy-purple with dark blotch, the throat white with yellow lines; the curious woody fruit to 15 cm long, terminating in a long curved, splitting beak. *Tropical.* p. 642

PROSOPIS Leguminosae
glandulosa (Kansas to N.E. Mexico), "Mesquite"; large shrub to 10 m, with deep roots, branches crooked, spines to 3 cm long; leaves bipinnate, with small leaflets bright green; small yellowish flower heads; long, somewhat curved fruit, with seeds made into flour by Indians of the desert. *Warm temperate.* p. 558, 562

PROTEA Proteaceae
barbigera (So. Africa: S.W. Cape Prov.), "Giant woollybeard"; most handsome, sparry evergreen shrub 1½ m high; white-hairy, light gray green, oval, leathery leaves 15 cm long, undulate and hairy at margins; the beautiful, large inflorescence to 20 cm across, balls resembling pine cones; the outer bracts soft pink or rose, tipped with fine silvery-white hairs, and surrounding a soft mass of white woolly flowers which become black-violet in the raised center of the flower heads. *Subtropic.* p. 839

compacta (Cape Prov.), "Bot river protea"; slender evergreen shrub to 3 m high, with smooth, narrow, light green shingled leaves 10 cm long; terminal, silvery, soft pink 10 cm flower heads, the colored bracts are a lovely clear pink, and covered with fine, velvety hair. *Subtropic.* p. 839

cynaroides (So. Africa), "King protea"; characteristic, showy shrub ½-1½ m high, with varying thick-leathery stalked leaves 5-12 cm long edged red, on woody red stem, the flowers packed in large 20 cm heads surrounded by numerous shingled series of stiff leathery bracts as in a cup, white to delicate pink and silvery silky-downy (summer). *Subtropic.* p. 837, 839

eximia (latifolia) (So. Africa), "Ray protea"; large-flowered protea from the S.W. Cape mountains; a dense shrub to 2½ m high, with broad oval, red-edged silvery leaves to 9 cm long, densely clasping and clothed around the upright branches, topped by the flower heads 12 cm long, with deep rose spoon-shaped bracts fringed with white silky hairs, the cup is filled with a mass of flowers tipped with old rose bristles. *Subtropic.* p. 838

grandiceps (So. Africa), "Peach protea"; most beautiful neat, dense shrub from the high Cape mountains; its elliptic, leathery, 12 cm highly decorative grayish leaves edged in red, and arranged closely and regularly on woody branches; the 10 cm flower head with 7 or 8 whorls of incurving rosy-salmon bracts arranged like in an artichoke, and with soft white hairs. *Subtropic.* p. 840

mundii (So. Africa: Cape Prov.); big rounded, quick-growing shrub to 3 m or more; leathery, elliptic, deep green leaves on reddish stems; flower heads with erect styles, straw-white or pink, to 1 cm long, with silky greenish-pink bracts; attractive, but not of long duration. *Subtropic.* p. 838

neriifolia (So. Africa: Cape), the "Oleander-leaved protea"; large floriferous bush to 3 m high, with long narrow linear, leathery leaves, and a showy terminal inflorescence 12 cm long, of cupping, shingled bracts pale salmon to deep rose with silvery sheen, the tips white and incurving and bearded with purplish black, set off against the mass of tawny-colored little flowers within the head. *Subtropic.* p. 839

repens (mellifera) (So. Africa: Cape); "Sugarbush" or "Honey-protea"; a shapely evergreen one cannot help admiring when travelling down the Cape Peninsula; large rounded bush to 3 m, its woody branches with dense whorls of silvery-glaucous, leathery, elliptic leaves 8-12 cm long; terminal inflorescence, 12 cm long, of shingled, smooth, shiny and silky, sticky bracts, varying in color from white to nearly red. In the early morning these cupped heads are half filled with nectar which the early colonists boiled into syrup in the absence of sugar. *Subtropic.* p. 838, 840

susannae (So. Africa: Cape); profusely blooming bush to 2 m, with narrow gray-green leaves to 12 cm long; terminal inflorescence shaped like a pincushion 10 cm across, with its central mass of hairy pinkish-white filiform flowers nestling in a basket of brownish pink bracts tipped rusty red. *Subtropic.* p. 840

PRUNUS Rosaceae
armeniaca (China), "Apricot"; small tree with reddish bark and smooth twigs; ovate leaves 5-8 cm long; flowers before the leaves, pinkish white; yellow sweet edible fruit flushed with red, with ridge on one side, 3-4 cm dia. *Warm temperate.* p. 850

armeniaca 'Moorpark' (China), "Moorpark apricot"; small round-crowned tree wth reddish bark and smooth twigs; leaves subcordate 5-8 cm long; 2 cm flowers pinkish, before the leaves; large, smooth pubescent fruit yellow flushed with red, 5 cm dia., excellent rich flavor, ideal for home garden. *Warm temperate.* p. 469

armeniaca 'Royal', "California apricot"; medium to large 3-4 cm fruit yellow to orange-yellow, aromatic orange flesh; California standard commercial variety. *Subtropic.* p. 469

cerasifera 'Atropurpurea' (Balkans to C. Asia), "Purple-leaf plum"; ornamental deciduous shrub or small tree to 8 m; ovate leaves 4-8 cm long, dark purple becoming greenish bronze in late summer, margins crenate; white flowers, and small red plums of sweet taste, and known as "Cherry plums". Planted in California for its colorful foliage. *Warm temperate.* p. 850

cerasus (S.E. Europe to W. Asia), "Sour cherry"; shrub or small tree suckering from the roots; leaves 5-8 cm long, doubly serrate; 2 cm flowers white to pink; bright red 2 cm glossy fruit with sour flavor; self-fruitful. *Temperate.* p. 469, 851

cerasus 'Montmorency', "Sweet-tart cherry"; preferred variety with small 1½ cm bright red, soft, juicy, sweet-tart fruit, and red juice. *Temperate.* p. 469

domestica (Eurasia), "Dwarf Italian plum"; small unarmed deciduous tree with dull green ovate leaves 10 cm long, serrate at margins; whitish or pinkish flowers 2 cm dia., before the leaves; ovoid fruit. *Warm temperate.* p. 851

dulcis (amygdalus) (W. Asia), the "Almond"; deciduous peach-like tree 6-10 m high, with gray bark; lanceolate firm, shining leaves, finely serrate; large, showy 3 cm pink flowers appearing before the leaves; fruit a compressed stone-fruit with hard flesh, 4 cm long, splitting open at maturity and freeing the pitted stone (almond). Limited to areas with less frost than tolerated by peach; confined mainly to California. *Warm temperate.* p. 470, 851, 853

laurocerasus (S.E. Europe to Iran), "English laurel"; decorative, quick-growing, evergreen bush with smooth, pale green shoots and broad, handsome, leathery, dense foliage, dark glossy green, oblong-pointed, to 15 cm long; small, white flowers, and dark purple fruit. *Warm temperate.* p. 852

laurocerasus 'Otto Luyken'; excellent cultivar by Hesse, and used in German landscaping as a very compact form of English laurel; short and bushy 50 cm high, the foliage dense and glossy dark green, to 10 cm long and 2-3 cm wide, slender pointed; free-blooming with spikes of white flowers in May; winter-hardy in C. Europe. *Warm temperate.* p. 852

mume (China), "Japanese apricot"; deciduous tree to 10 m, with thin green twigs; ovate 4-10 cm leaves; flowers before foliage, 3 cm dia., white or pinkish with red eye; small bitter fruit. *Warm temperate.* p. 851

persica (Amygdalus) (China), "Peach tree"; normally a bushy tree, but trained by pruning and starving into dwarf forms in Japan and China, with masses of solitary, delicate pink flowers in advance of the lanceolate foliage, early in spring. *Warm temperate.* p. 846, 850

persica 'Elberta', "Elberta peach"; medium to large yellow freestone peach, skin blushed red; high quality; midseason; needs winter chill. *Temperate.* p. 851

persica flore albo-plena, "Double white-flowering peach"; bushy tree loaded with double white flowers in spring. *Warm temperate.* p. 846

persica 'Golden Jubilee'; a standard variety of freestone peach originating in New Jersey but also grown in California; fruit of medium size with firm yellow flesh of fair flavor; skin yellow and mottled bright red; early, ripening 3 weeks before 'Elberta'. *Temperate.* p. 468, 856

persica 'Rubra plena', "Double flowering peach"; small deciduous tree, blooming before the foliage appears in spring, then literally covered with masses of brilliant carmine-red blossoms with double petals, 3 cm across. *Warm temperate.* p. 853

persica 'Santa Rosa', "Freestone peach"; freestone peach for Southern California; medium size with yellow flesh, fruiting in August. *Warm temperate.* p. 469

persica 'Sim's Cling'; large clear yellow cling peach of good canning quality; flesh yellow to the pit; late August. *Temperate.* p. 468

persica 'Ventura', "Freestone peach"; medium sized attractive yellow freestone peach with smooth skin; mid-season. *Warm temperate.* p. 468

salicifolia (C. America), "Capulin cherry"; small tree to 10 m, with obovate leaves toothed at margins, glaucous beneath; white flowers; sweet-tasting globular fruit 2 cm dia., maroon-purple with green flesh. *Subtropic.* p. 469

salicina 'Great Yellow' (China), "Great yellow plum"; small smooth tree to 7 m high; obovate, pointed leaves, serrate at margins; 2 cm white flowers; sweet fruit yellow or light red, 5-6 cm dia. *Warm temperate.* p. 469

salicina 'Satsuma' (Japan), "Japanese plum"; small to medium fruit deep dull red; dark red, solid, meaty flesh, mild, sweet. *Warm temperate.* p. 469

serrulata (E. Asia), "Japanese flowering cherry"; deciduous tree to 20 m, with shining ovate leaves, serrate at margins; double white flowers with or before foliage; small black fruit. *Warm temperate.* p. 846, 850

serrulata 'Kwansan' (Japan), "Japanese double flowering cherry"; deciduous tree to 12 m high, with stiffly ascending branches, ovate leaves with bristly teeth, turning reddish in late autumn; large double, deep rose-pink flowers in clusters of 2 to 5. *Temperate.* p. 850, 851

triloba plena (China), a graceful shrub known as "Double flowering almond"; often grown as a small standard and forced, with serrate, deciduous leaves, the dainty, double pink flowers flushed with rose, unfolding ahead of the foliage from tight round buds set closely along woody branches of last year's growth, in spring. *Temperate.* p. 851

PSAMMOPHORA   Aizoaceae

modesta (Namibia); tiny tufted, very succulent plant to 5 cm high; with fat, triangular gray-green leaves 12 mm long, rough and usually sand-encrusted because of sticky surface; smallish starry, pale purple flowers. *Arid-tropical.* p. 49

PSEUDERANTHEMUM   Acanthaceae

alatum (Mexico), "Chocolate plant"; low growing herb with copper-brown papery leaves on flat, winged petioles, silver blotching near midrib; gray beneath; small salverform, purple flowers in racemes. *Tropical.* p. 34

atropurpureum 'Tonga' (South Pacific); handsome shrub with willowy branches, leathery oval leaves 10-15 cm long, shaded blackish-purple varying to deep metallic green; small flowers in erect terminal racemes, rose pink and covered with purple spots; crimson center. *Tropical.* p. 37

atropurpureum 'Variegatum' (Polynesia); colorful woody evergreen shrub with dense erect branches; attractive waxy, elliptic leaves 10-15 cm long, variegated coppery purple with rose, gray-green, yellow and white; small rose-purplish flowers with red markings at branch tips. *Tropical.* p. 37

reticulatum (New Hebrides); tropical shrub with attractive lanceolate smooth foliage, slightly fleshy, green with reticulation of golden-yellow veins; 3 cm flowers with wine-purple in throat, and dots of same color on lower lip. *Tropical.* p. 37

sinuatum (New Caledonia); shrubby plant with linear leaves 8 cm long, olive-green mottled with gray, the margins deeply scalloped, purplish beneath; white flowers in terminal racemes, 3 cm across, freely spotted rosy-lavender. *Tropical.* p. 37

PSEUDOBOMBAX: see Bombax

PSEUDOCALYMMA   Bignoniaceae

alliaceum (Guyana, Brazil), "Garlic vine" or the true "Garlic-scented vine"; ornamental woody climber with leaves of two ovate, leathery leaflets 10 cm long, with or without tendril between them, the leaf-tips curling downward; 5-lobed flowers 6 cm across, light purple with creamy throat, in clusters. Crushed leaves and blooms have a faint garlic odor. *Tropical.* p. 193

PSEUDOESPOSTOA   Cactaceae

melanostele (Perú); shapely dark green column with rows of brown-tipped needle spines and dense covering of white woolly hair especially with age. *Arid-tropical.* p. 248

PSEUDOHYDROSME   Araceae

gabunensis (Equatorial Africa: Gabon); curious tropical aroid with fleshy tuber, the large and showy, short-stalked inflorescence chalice-like 25 cm long bright yellow, inside red-purple; the dissected leaf appears after blooming. *Tropical.* p. 110

PSEUDOLARIX   Pinaceae (Coniferae)

kaempferi (Chrysolarix amabilis), (E. China), "Golden larch"; deciduous conifer 15 to 20 m high, often almost as wide at base, with spreading branches, pendulous at tips; linear needles 4-6 cm long, mainly clustered in tufts, bluish green, turning golden yellow in autumn; pendant 8 cm cones. *Warm temperate.* p. 348

PSEUDOLITHOS   Asclepiadaceae

cubiforme (Lithocaulon) (Somalia); small, dull green, globular succulent with surface in cubical checkers, the fleshy caudex 5 cm dia., the inflorescence in downy, grayish rays sessile to the body, holding small red, cupshaped flowers. *Arid-tropical.* p. 154

PSEUDOPANAX   Araliaceae

'Adiantifolius' (New Zealand); robust shrub of flexible stems with spatulate lobed, leathery green leaves; small whitish flowers in heavy cluster. *Subtropic.* p. 140

crassifolius (New Zealand), "Lance-wood"; evergreen growing into large, variable tree; when young the leaves are long linear, rigid and finely toothed, with thick midribs, dark green, blotched yellow above, purple beneath, 30-90 cm long; at a later stage, 3-5 foliate. *Subtropic.* p. 141

ferox (New Zealand); slender evergreen tree with variable, thick and stiff leaves, linear, later obovate, chocolate-brown with pale brown midrib; margins jaggedly toothed. *Subtropic.* p. 140

PSEUDOSASA   Gramineae

japonica (Bambusa Metake), (Japan), the "Female arrow bamboo" or "Hardy Metake bamboo"; a running bamboo of moderate size; with hollow round stems 2 to 4 m high and to 1½ cm dia., from creeping rootstocks, with broad deep green foliage 10-30 cm long, glaucous beneath. Fairly hardy in New Jersey, has withstood freezing to -18° C. One of the best for decoration in tubs, holding leaves better than other species; for cool bright rooms; keep moist. *Warm temperate.* p. 514

PSEUDOTSUGA   Pinaceae (Coniferae)

menziesii (Brit. Columbia to California), "Douglas fir"; graceful tree to 60 m high, pyramidal when young, the crown very broad in older trees, with trunk to 4 m in dia., bark brown, fissured into broad ridges; the spreading branches slightly pendulous; needles shining dark green 2-3 cm long, and smelling of camphor; cones pendulous, elongate to 10 cm long, 3-4 cm dia. *Warm temperate.* p. 347, 348

PSIDIUM   Myrtaceae

cattleianum (littorale longipes in Hortus 3) (Brazil), "Strawberry guava"; dense shrub to 7½ m high, smooth branches with obovate, leathery leaves 5-8 cm long, white flowers with many stamens, and 4 cm berry-like purplish-red fruit with strawberry flavour. *Subtropic.* p. 460, 686

cattleianum lucidum (littorale) (Brazil), "Yellow strawberry guava"; variety of the "Purple guava", with juicy fruit, sulphur-yellow. *Subtropic.* p. 460, 686

guajava (W. Indies, Mexico to Perú), "Guava"; small branching tree, reaching 9 m; 4-angled branchlets with light green, elliptic leaves, hairy beneath; large white flowers, and producing edible, globose yellow, sweet flavored fruit, used for making jam. *Tropical.* p. 460, 686

guineense (Trop. America), "Guyana guava"; evergreen shrub or small tree to 6 m high; oblong leathery leaves to 12 cm long, rusty hairy beneath; fragrant white flowers; large luscious-looking red to yellow fruit 4 cm dia., slightly acid to the taste. *Tropical.* p. 686

littorale (as listed in Hortus 3) see cattleianum lucidum (as known in horticulture). p. 460

PSILOTUM   Psilotaceae

nudum (triquetrum), (New Zealand and subtrop. and trop. So. and No. hemisphere), "Whisk-fern"; curious club-moss botanically interesting as a very primitive vascular plant, supposedly the first step of evolution from liverworts toward higher plants; they resemble more closely than any other living forms the extinct Psylophytales, the most primitive vascular plants yet discovered and regarded as the parent stock from which all other vascular plants may have evolved; the plant is a herbaceous perennial devoid of roots and true leaves, growing both as epiphyte and terrestrial, with the underground and ribbed aerial stems with numerous scale leaves; erect or pendant branched green shoots to 50 cm long, with yellow, globose sporangia looking like flowers in leaf axils. We see them in Florida and Hawaii, on tree ferns etc. *Tropical.* p. 829

PSITTACANTHUS *Loranthaceae*
  calyculatus (So. Mexico to Oaxaca, Chiapas and Veracruz), "Parrot flower" or "Mexican mistletoe"; striking woody parasite on branches of tropical trees, sometimes growing so thick that the host appears to be a flowering tree. The thick-leathery leaves are ovalish or sickle-shaped 6 cm long; showy flowers in large clusters, slender-tubular, divided into linear segments, vivid salmon-red, 5 cm long, and with protruding red stamens. *Tropical.* p. 617

PTERIS *Polypodiaceae (Filices)*
  adiantoides: see Pellaea
  cretica 'Albo-lineata'; a very pretty, useful, variegated form of low habit, with small, clean-cut, leathery fronds differing from the species only in the broad band of creamy white down the center of each linear lanceolate leaflet which are toothed and wavy-margined; the fertile fronds are taller and more slender. *Humid-tropical.* p. 442
  cretica 'Childsii'; handsome cultivar with its pinnate fronds consisting of broad, fresh green leaflets prettily frilled, lobed, or wavy along margins. *Humid-tropical.* p. 445
  cretica 'Wimsettii', a popular "Table fern"; robust, variable form of medium height, making the first break into cresting away from the plain species, and desirable because of its almost leathery toughness, the slender fresh green leaf segments irregularly toothed or pointedly lobed, and some of the tips terminate in small forks or crests. *Humid-tropical.* p. 442
  ensiformis 'Victoriae', "Victoria fern"; an elegant, graceful little fern with both the short, broad sterile fronds and the abundant, erect, slender fertile fronds having leaflets beautifully banded white, bordered by a wavy margin of rich green. *Humid-tropical.* p. 442
  longipinna; fibrous-rooted fern with large fronds divided into pairs of broad pinnae, each cut deeply to the rachis; the lowest pair of pinnae forked. *Humid-tropical.* p. 445

PTEROSPERMUM *Sterculiaceae*
  acerifolium (Burma, Java), "Bayur tree"; tall evergreen tree, covered with rusty-brown wool; peltate leathery yellow-green leaves with dark veins, 15 cm dia., white hairy below, scented fleshy flowers with silky sepals and yellowish petals; timber similar to teak and oak. *Tropical.* p. 902
  lancifolium (E. Indies); evergreen hairy tree with leathery, long-ovate leaves; white flowers with long protruding stamens and recoiled petals, and sweetly fragrant. *Tropical.* p. 902
  semi-sagittatum (India; Assam); large evergreen tree, photographed in Durban Botanic Garden, So. Africa, with dark green leathery leaves half-arrow-shaped, alternately 2-ranked, 15-20 cm long; large showy flowers creamy-white, with reflexed linear greenish sepals, 10 cm or more across. *Tropical.* p. 902

PTEROSTYLIS *Orchidaceae*
  banksii (New Zealand), "Hooded orchid"; terrestrial to 45 cm high, with small underground tuber, straplike narrow-linear leaves, sheathing the stalk, bearing solitary green flowers striped with white, 5-8 cm long; the concave dorsal sepal arched forward, with the petals forming a hood or helmet; the lateral sepals running into long slender tails; linear lip. *Humid-subtropic.* p. 746
  baptistii (Australia: Queensland to Victoria); handsome terrestrial of slender habit, flat basal rosette of leaves with prominent network of veins; long wiry stalks 20 to 50 cm tall, with solitary hooded flowers to 8 cm long, light green with light brown bands, the dorsal sepal curving forward to a point (autumn); deciduous after bloom. *Subtropic.* p. 746

PTERYGODIUM *Orchidaceae*
  catholicum (S. Africa); small tuberous-rooted terrestrial with leafy stems to 20 cm high; fleshy succulent lanceolate leaves clasping the stem; axillary and toward apex with small greenish-yellow flowers. *Subtropic.* p. 752

PTYCHOSPERMA *Palmae*
  elegans (Seaforthia elegans) (Queensland), "Solitaire palm"; handsome solitary feather palm to 6 m tall, with gracefully slender trunk, topped by 6 to 8 rather short pinnate fronds 1 to 2 m long; about 20 pairs of bright green pinnae, cut off and jagged at apex; bushy, white, fragrant flowers, and small bright red fruits. *Tropical.* p. 796
  macarthurii (Actinophloeus, Kentia) (New Guinea), "Hurricane palm"; suckering feather palm with several slender grayish trunks to 8 m high; pinnate leaves in a sparse crown, the pinnae glossy-green and rather soft, with the apex jagged and toothed as if bitten off; fruit bright red. *Tropical.* p. 796

PTYCHOSPERMA: see also Archontophoenix

PULMONARIA *Boraginaceae*
  saccharata (So. Europe), "Lungwort", or "Bethlehem sage"; clustering perennial herb with creeping rootstock, 30-40 cm high, with stalked ovate basal leaves, bristly hairy, dark green, freely spotted and blotched with silvery-white; clusters of funnel-shaped flowers first rose then blue; fairly hardy. *Subtropic.* p. 196

PUNICA *Punicaceae*
  granatum (S.E. Europe to Himalayas), "Pomegranate"; shrub or small tree becoming 6 m high, with shining oblong leaves 2 to 8 cm long; flowers orange-red, with crumpled petals, the calyx purple, and with its lobes persistant on the developing edible fruit, growing as large as an orange, outside deep yellow to red, inside the juicy flesh crimson, of delicious somewhat acid flavor. *Subtropic.* p. 458, 464, 829
  granatum florepleno; variety with large double flowers crimson red. *Subtropic.* p. 824
  granatum 'Legrellei'; popular, free-blooming, ornamental form of the pomegranate tree, distinguished by its fully-double, showy flowers of coral-red, striped yellowish-white outside, and pendant at the end of wiry branches, dense with lanceolate, wavy-margined, glossy leaves. *Subtropic.* p. 824
  granatum 'Nana' (Iran to Himalayas), "Dwarf pomegranate"; a miniature version of the pomegranate tree, in form of a shrub to 2 m high, with shining vivid-green, narrow leaves and scarlet flowers with salmon calyx at end of thin branchlets, producing orange-red fruit with hard rind and juicy, edible pulp; attractive as a small potplant. *Subtropic.* p. 458, 824

PUYA *Bromeliaceae*
  alpestris (Chile); dense rosette of narrow spiny-margined leaves shiny light gray-green and recurving; 2-3 cm wide, silvery-gray beneath; branched inflorescence, with metallic blue flowers and orange anthers, in up to 20 branches to 1 m high. *Subtropic.* p. 222
  raimondii (Cordilleras of Bolivia, Perú); a giant plant, and found on the Puna to 3,800 m altitude; dense rosette with bayonet-shaped, leathery leaves grass-green but covered with white scales especially beneath, and edged with brown spines; forms trunk and, when about 150 years old, develops 5 m inflorescence, with thousands of whitish flowers within green bracts. Growing to 12 m tall, this is probably the largest monocarpic herb, dying after producing flowers and seed. *Subtropic.* p. 214, 221, 225

PYCNOSPATHA *Araceae*
  soerensenii (arietina) (Thailand); a tropical "Dragon-lily"; with tuberous roots and long stalk marbled with purple and pink, carrying umbrella-like a large, several times dissected leaf; from the base a rather small inflorescence with hooded, curving spathe, later deciduous. *Tropical.* p. 110

PYRACANTHA *Rosaceae*
  coccinea 'Lalandei'; a robust cultivar of the "Firethorn" from So. Europe and Asia Minor; evergreen, woody, thorny shrub with oval oblong, shiny, leathery, dark green leaves, finely toothed; the numerous small, white flowers followed by dense clusters of waxy, orange-red berries which stand out from the stem. *Temperate.* p. 846, 858
  fortuneana 'Graberi' (crenulata yunnanensis), "Chinese firethorn"; this evergreen shrub from China has been developed into outstanding forms such as 'Graberi', a vigorous grower with narrow, oblanceolate, thick-leathery leaves rounded at the apex, and great clusters of large orange-red berries somewhat appressed to stem, from September to winter. *Warm temperate.* p. 852
  koidzumii 'Victory' (crenato-serrata cv.), an excellent "Red firethorn"; robust evergreen shrub of Chinese (Taiwan) origin, becoming large and spreading to 2 m with thorny, rambling branches and deep green leathery foliage, very ornamental as a smaller container plant bearing large clusters of glistening brilliant, scarlet, long-lasting 1 cm berries into winter and for Christmas. Small white, fragrant flowers in May, on spurs along wood of last year's growth. Timely decorator for the Christmas season, the branches often trained into pyramids or against trellis. Not as hardy as P. coccinea. *Warm temperate.* p. 852

PYROSTEGIA *Bignoniaceae*
  venusta (Bignonia ignea) (Brazil, Paraguay), "Flame vine" or "Flaming trumpet"; gorgeous flowering woody vine high-climbing by tendrils over fence and roof-tops, with leaves of 2 or 3 ovate

leaflets; red-orange slender-tubed flowers 8 cm long, in heavy clusters hanging brilliantly from eaves or arbors. *Tropical.* p. 188

PYRROSIA *Polypodiaceae (Filices)*
 lingua (Cyclophorus) (Japan, China, Vietnam, Taiwan), "Tongue fern"; creeping fern close to Asplenium, with thin scaly rhizome; stalked lanceolate, wavy, tomentose leaves, 22 cm long, covered with gray felt beneath, becoming pendant; in horticulture as Diplazium lanceum. *Humid-subtropic.* p. 448
 nummularifolia (Cyclophorus) (Malaysia), "Felt fern"; scandent epiphytic fern, the rhizomes covered with stalked scales, simple leathery leaves, covered with appressed hairs, brown beneath; leaves oval and 2-3 cm long. *Humid-tropical.* p. 444

PYRUS *Rosaceae*
 communis (Europe and W. Asia), "European pear"; grown in America since the earliest settlement; large long-lived deciduous, hardy tree with oval or ovate, dark green leathery leaves; flowers 3 cm across, white sometimes tinged pink, appearing with the first foliage; fruit pyriform and edible, with gritty cells, in many sweet-tasting varieties. Dwarf pears or espaliers are commonly grafted on quince. *Temperate.* p. 468
 communis 'Doyenne du Comice'; fruit large, yellow with crimson, fine aromatic flesh; successfully fruited in Southern California by Paul Thomson of Bonsall. *Temperate.* p. 468
 communis 'Kieffer' (Europe, W. Asia), "Kieffer pear tree"; long-lived deciduous tree, with rough bark, leathery, ovate leaves 3-6 cm long; 3 cm flowers with first leaves, white or tinged pink; fruit medium to large greenish yellow, gritty in texture; requires winter chill. *Temperate.* p. 850
 communis 'Lincoln'; a pear grown successfully in Bonsall, So. California by Paul Thomson; medium sized light brown fruit in large bunches. *Temperate.* p. 468
 kawakami (Taiwan), "Evergreen pear"; shrub or tree to 10 m high, evergreen in California, partially deciduous in colder zones; branchlets drooping, at end the large glossy, leathery, obovate leaves to 10 cm long or more; few white flowers; small inedible fruit, seldom seen. Willowy young branches are often fastened to trellis for ornamental espalier. *Warm temperate.* p. 851, 853

QUAMOCLIT *Convolvulaceae*
 lobata (Mina) (Mexico), "Star-glory"; annual twining herb vigorously climbing to 6 m with leaves deeply 3-lobed 8 cm across; brilliant boat-shaped baggy 5-angled flowers fiery scarlet-red when in bud, when opening changing to creamy-yellow and orange; stamens and style protruding; on the upper side of long-stalked axillary racemes. *Tropical.* p. 359

QUASSIA *Simaroubaceae*
 amara (Surinam), "Bitterwood"; tropical shrub or small tree to 8 m, with odd-pinnate leaves to 25 cm long, rachis conspicuously winged, 3-7 elliptic leaflets 15 cm long; inflorescence a raceme of slender crimson flowers 5 cm long. *Tropical.* p. 895

QUERCUS *Fagaceae*
 agrifolia (California), "California live oak"; great, picturesque wide-spreading tree to 30 m high, more or less evergreen; with rough black bark, and long, persistent, dark green, hard leathery, oval convex leaves scalloped or lobed, and with spiny teeth, 8 cm long, light green and glossy beneath; the conical 2 cm acorns partly enclosed in the silky cup. *Subtropic.* p. 430
 dentata (Japan), "Daimyo oak"; deciduous tree to 25 m high, large ovate leaves to 30 cm long, the margins lobed, pubescent beneath; cup partially enclosing a nut. Produces tan bark. *Warm temperate.* p. 430
 robur (Europe, Africa, W. Asia), "English oak"; deciduous tree to 30 m high, obovate leaves 12 cm long, deeply lobed along sides; cup enclosing half of nut. *Temperate.* p. 430
 suber (So. Spain, Portugal, No. Africa), the "Cork oak"; an evergreen related to our California "Live oak", to 15 m or more high, with broad round-topped head and thick, deeply furrowed bark which is spongy and possessing elastic properties; shining dark green, ovate 8 cm leathery leaves with toothed margins, grayish-tomentose beneath. Their light-weight bark is removed in sections around the Western Mediterranean for use in insulation and other economic purposes, and trees are selected in rotation or when ready for the stripping of their grayish bark down to the cambium layers about every 8 to 10 years, following which it grows back again without seriously harming the tree. A curiosity plant which may be grown in containers. *Subtropic.* p. 430

QUESNELIA *Bromeliaceae*
 arvensis (Sao Paulo); formal rosette of leathery green leaves with gray cross-bands and black spines; inflorescence a thick stalk with dense head of rose-pink bracts and blue and white flowers. *Subtropic.* p. 221
 quesneliana (French Guiana); large regular rosette of fresh-green leaves banded gray beneath; inflorescence a gray stalk with sheathing white bract leaves topped by a cylindrical head of shingled, papery, rosy-red bracts dusted white, red calyx leaves, and petals white with blue edge. *Tropical.* p. 222
 testudo (So. Brazil); spreading rosette with leathery glossy light green, channeled leaves to 60 cm long by 3 cm wide, grayish scaly beneath; the margins dense with fine brown spines; small erect brush-like inflorescence with surrounding bracts rosy-pink, flowers electric blue. *Subtropic.* p. 236

QUISQUALIS *Combretaceae*
 indica (Burma to Philippines), "Rangoon creeper"; tropical clambering shrub with liana-like, vining, woody stems; soft, light green, pubescent leaves, and beautiful, drooping flowers having slender green tubes with petals red when in bud, opening white but later changing to pink and crimson-red; very fragrant. *Tropical.* p. 303, 304

RAFFLESIA *Rafflesiaceae*
 arnoldii (Sumatra), the world's largest flower; enormous fleshy parasite, with a solitary giant flesh-colored flower 1 m across, rising from a superficial rhizome, and without leaves; the foul-smelling inflorescence weighs 7 kg, the central cup is intense purple and holds 6 litres of water. Its seeds lodge in the surface roots of Cissus angustifolia where they penetrate the bark and germinate. *Humid-tropical.* p. 830

RAMONDA *Gesneriaceae*
 nathaliae (Yugoslavia, Bulgaria); basal rosette of small obovate, crinkled hairy, deep green leaves, with wavy-toothed margins, about 5 cm long, in an overlapping, flat pattern; densely hairy stalks to 12 cm high, bear 1-3 lavender-blue, 4-6 lobed flowers, with orange-yellow eye; fairly hardy. *Warm temperate.* p. 501

RANDIA *Rubiaceae*
 maculata (W. Africa: Sierra Leone); small unarmed tropical tree with shining green, quilted and undulate leaves; solitary tubular purple flowers with white spreading lobes, the corolla sometimes downy. *Tropical.* p. 862

RANUNCULUS *Ranunculaceae*
 asiaticus (S.E. Europe, Syria, Iran), "Persian buttercup"; slender perennial with tuberous roots, alternate leaves, divided into narrow segments, on erect stalks, each bearing 1-4 flowers, usually double; many various colors, the wild type yellow; cultivated varieties white, yellow, orange, pink, scarlet, crimson, (spring). *Subtropic* p. 842, 843
 asiaticus 'Tecolote', "Turban ranunculus" of florists; very fine, vigorous, California tuberous strain developed about 1930; neat and compact of habit, with giant double or semi-double flowers 6-10 cm across, in prevailing colors white, cream, yellow, rose to crimson, also bicolor, yellow with red margins. *Subtropic.* p. 842, 846
 lyallii (South Island, New Zealand), "Mount Cook lily"; a most beautiful plant of the New Zealand Alpine flora; to 1½ m tall, with fleshy roots; peltate, fleshy leaves 12-30 cm dia., and saucer-shaped; pure white flowers 5 cm across. *Warm temperate.* p. 844

RAPHIA *Palmae*
 farinifera (monbuttorum) (Uganda), a "Raffia palm"; beautiful, gregarious feather palm of the swampy tropical rainforest, to 12 m high, with short, thick trunk sometimes to 6 m; very showy, plumy and gracefully arching pinnate fronds to 8 m long, the large midrib with long flexible, glossy dark green leaflets 3 cm wide, fluttering in the wind; the female trees with pendulous terminal inflorescence, bearing cone-like fruit 10 cm long covered with shiny brown scales. After flowering the tree will die. *Tropical.* p. 799
 pedunculata (Madagascar); giant feather palm without trunk; tall erect pinnate fronds 5 m or more long, the glossy dark green, narrow leaflets set almost at right angles to midrib. Photographed in the collection of Heidelberg University Botanic Garden, Germany. *Tropical.* p. 799

RAPHIDOPHORA *Araceae*
 decursiva (Ceylon to Indochina); tree-climber with stem stiffly scandent; large, glossy, dark green leaves pinnately divided to midrib to 60 cm long; decorative, but slow and stubborn, and not easy to train. *Tropical.* p. 110, 112

RAPHIOLEPIS *Rosaceae*
 indica (So. China), "Indian hawthorne"; attractive, evergreen shrub to 1½ m high, with alternate, shining leathery, lanceolate, 8 cm leaves bluntly toothed, and loose clusters of small pink flowers blooming intermittently from February to August, more profusely through winter. *Subtropic.* p. 854, 855
 indica 'Enchantress'; "India-hawthorne"; charming evergreen shrub to 1½ m high; dark green, leathery elliptic leaves to 8 cm

long, bluntly toothed at margins; large pretty flowers appleblossom-like, rosy-pink with white eye, 2½ cm across, in loose clusters, carried in profusion from late winter to late spring and into summer; dark blue berries follow. A compact-growing form of the species indica which has smaller white flowers tinged with pink and comes from South China. *Subtropic.* p. 854, 856

RAUWOLFIA *Apocynaceae*
  verticillata (chinensis), (China); smooth shrub with long-elliptic, glossy green, thin-leathery leaves in whorls; the flowers with flaring whitish septals; berry-like red fruit. *Warm temperate.* p. 80

RAVENALA *Musaceae*
  guyanensis (Pará, Guayana, Llanos de Colombia), the western hemisphere counterpart to R. madagascariensis, of more dwarf habit, and which I found in swampy jungles in the interior of Surinam as well as along the Amazon; usually only 1½-2 m high, with long-stalked, thick-leathery leaves; large boatshaped bracts with white flowers, the seed-covering scarlet. *Tropical.* p. 670
  madagascariensis (Madagascar), "Travelers tree"; striking tree with palm-like trunk to 30 m high, topped by leathery, banana-like leaves with pale midrib, arranged like a fan on long petioles, and sheltering the great flower bracts with white blooms and sky-blue seed; the cup-shaped leaf bases hold healthy drinking water for thirsty travelers. *Tropical.* p. 662, 669, 670, 672

REBUTIA *Cactaceae*
  calliantha (Northern Argentina); small globe with scattered spines; showy flowers 4½ cm across, with narrow petals a vivid crimson red. *Arid-subtropic.* p. 261
  crispata (South America); small globe entirely covered with long white spines, mostly appressed; flower crested and glowing carmine-rose. *Arid-subtropic.* p. 262
  deminuta (Aylostera) (No. Argentina); clustering diminutive globes with spiralled tubercles and numerous white or brown rigid spines; free flowering orange red; keep dry and cold to induce blooming; water when buds show. *Arid-subtropic.* p. 261
  digitiformis (So. America); small globes, dark green with tubercles tipped by small spine bundles; large rosy-crimson flowers. *Arid-subtropic.* p. 261
  fiebrigii (Aylostera), (Bolivia), "Crown cactus"; small globose to cylindrical, clustering plant, spiralled tubercles, with white hairlike spines; vermilion-red flowers. *Arid-subtropic.* p. 257
  grandiflora (Argentina), "Scarlet crown cactus"; flattened globe to 5 cm high, covered with knobs and short spines; large crimson-red flowers. *Arid-subtropic.* p. 260
  hybrida, "Red crown cactus"; low globe entirely covered with soft white bristles; large scarlet red flowers. *Arid-subtropic.* p. 261
  kupperiana (Aylostera) (Bolivia); small globe with depressed top, glaucous gray, acute tubercles in about 20 rows covered with thin spines; grows to 10 cm high but blooms freely even as a tiny 2½ cm plant, with showy scarlet flowers. *Arid-subtropic.* p. 261
  minuscula (N.W. Argentina); tiny bright green flattened globe to 5 cm dia., and becoming tufted; tubercles in many spirals, very small whitish spines; bears its scarlet-red flowers freely, 6 cm long, lasting several days; *Arid-subtropic.* p. 262
  pseudodeminuta (Argentina: Salta); little globes, freely clustering around the base, 5 cm thick, grass-green; the prominent tubercles set with 11 radials and 2-3 central spines all glassy-white tipped brown; flowers golden yellow shading to red. *Arid-subtropic.* p. 262
  senilis (Argentina, Chile), "Fire crown"; depressed clustering, bluish globe, to 25 cm high, but usually very small; spiralled tubercles, covered with interlocking white to yellow hairlike spines; red flowers 3½ cm across. *Arid-subtropic.* p. 262
  violacaeflora (Northern Argentina), "Rosy crown cactus"; low globe with sharp needle spines, very tiny and barely 2 cm dia.; the flowers relatively large 3 cm across, a beautiful rose-pink. *Arid-subtropic.* p. 260

RECHSTEINERIA: see Sinningia

REGELIA *Myrtaceae*
  megacephala (Western Australia); low shrub with short, broad, appressed deep green leaves in ranks; the prominent stamens and flowers carmine-red, in showy clusters. *Subtropic.* p. 687

REHMANNIA *Gesneriaceae (formerly Scrophul.)*
  angulata (C. China), "Foxglove gloxinia"; perennial herb for the cool greenhouse, with a rosette of soft, irregularly lobed, obovate leaves; a sticky-hairy, leafy stalk with showy, large, bilabiate flowers, rosy-red with yellow throat and spotted purple. Previously included with the foxgloves (Digitalis), family Scrophulariaceae. *Warm temperate.* p. 882

RENANTHERA *Orchidaceae*
  imschootiana (Assam, Burma, Vietnam); brilliant epiphyte with short leaves along the erect stem, the branching inflorescence arching, to 60 cm long, with numerous showy 6 cm flowers, the small dorsal and the petals yellow, spotted scarlet, the large lower sepals scarlet-red, lip yellow marked red, (May-July).*Subtropic.* p. 751
  monachica (Burma), "Fire orchid"; colorful epiphyte with stiff leafy stems, short mottled leaves and an arching raceme of orange-yellow, spreading flowers marked with blotches and bars of fiery-red, (April). *Tropical.* p. 759, 760

RESEDA *Resedaceae*
  odorata (No. Africa, Egypt), "Mignonette"; branching herb, at first upright but becoming spreading, with oblanceolate leaves and yellowish-white, inconspicuous flowers with contrasting saffron-red anthers, in pyramidal, terminal racemes; much loved for their sweet fragrance. *Subtropic.* p. 830

RESTREPIA *Orchidaceae*
  antennifera (Andes of Venezuela and Colombia); small epiphyte to 30 cm high; stems in tufts, with fleshy leaves 12 cm long; showy flowers 3 cm long, sepals and upper petals thread-like, lateral sepals united and boat-shaped, yellow brown with maroon stripes. *Subtropic.* p. 760
  guttulata (Andes of Venezuela to Ecuador); miniature epiphyte with stems 5 cm high, and in tufts, with solitary leaves; the conspicuous flowers, having upper sepals and the petals linear, greenish white with crimson line, lateral sepals united into a boat shape, greenish yellow with purple spots. *Subtropic.* p. 758

RHABDOTHAMNUS *Gesneriaceae*
  solandri (New Zealand: North Island); sole gesneriad in N.Z.; twiggy shrub ½-1½ m high, with grayish hairy branches; small roundish, membranous, rough-hairy, 3 cm leaves coarsely toothed; showy 2½ cm tubular bell-shaped flowers, with oblique limb, yellowish outside and lined with red, the spreading lobes salmon red, yellow in throat and lined with dark crimson.*Subtropic.* p. 504

RHAMNUS *Rhamnaceae*
  alaternus 'Variegata' (So. Europe), a variegated "Buckthorn"; bushy shrub with twiggy, beige-brown stem, obovate 6 cm leaves, attractively colored green in center, margins ivory-white, slightly toothed; small greenish flowers, followed by bluish-black fruit. *Subtropic.* p. 832

RHAPIS *Palmae*
  excelsa (flabelliformis) (So. China), "Large lady palm"; miniature fan-palm with bamboo-like canes 3½ m or more, the thin stems densely matted with coarse fiber, forming clumps from underground suckers, the leathery leaves glossy-green divided into 3-10 broad segments; widely used in China and Japan as a durable potted palm. *Subtopic.* p. 795, 798, 799, 802
  excelsa (flabelliformis) 'Variegata'; very attractive Japanese cultivar having palmate leathery leaves with segments banded and striped lengthwise, with ivory-white alternating with green. *Subtropic.* p. 798
  humilis (So. China), "Slender lady palm"; clustering stems thinner and more graceful than excelsa, less vigorous, likewise covered with dark brown fibers; the deep green palmate leaves more slender and divided into 9 to 20, narrower segments. *Subtropic.* p. 798, 799

RHEKTOPHYLLUM *Araceae*
  mirabile (Nephthytis picturata) (Nigeria, Cameroon, Zaïre); creeping and climbing aroid sending out long rooting internodes; the large thin-leathery, arrow-shaped, hastate leaves 20-30 cm long, dark green between the veins, variegated silvery-cream in form of a fern-leaf, becoming green in older leaves; the maturity-stage leaves broad-heartshaped in outline, 30-50 cm long and deeply sliced into broad, obtuse, glossy green segments; the 10 cm spathe green outside, red-purple inside. *Tropical.* p. 108, 125

RHEUM *Polygonaceae*
  palmatum (China), "Chinese rhubarb"; stout perennial herb with woody rhizome, 1½ m high; large, palmately lobed leaves on cylindrical stalks; flowers deep red in large cluster. *Warm temperate.* p. 825

RHIGOZUM *Bignoniaceae*
  trichotomum (Namibia); spiny, sparry shrub with woody, tiny hairy leaves; large cream-white frilled flowers with yellow center. *Arid-subtropic.* p. 186

RHIPSALIDOPSIS *Cactaceae*
  gaertneri (Schlumbergera, Epiphyllopsis) (So. Brazil), the "Easter cactus"; bushy epiphyte with stiffish spreading branches of long flattened joints, dull green with purplish crenate margins, a few bristles at apex; starlike regular flowers, deep scarlet in March-April; the ovaries angled. *Tropical.* p. 285, 286

x graeseri (Epiphyllopsis) (gaertneri x rosea), a free-blooming hybrid which I first saw in Brazil where it flowered in their spring (Sept.), while in the Northern hemisphere it blooms in March, indicating its being influenced by day-length; wide open, starshaped regular flowers rosy-red with double row of broad petals; freely branching compact plant. *Tropical.* p. 285

x graeseri 'Rosea' (Epiphyllopsis), a lovely cultivar with flowers of a lighter shade: clear pink with the center flushed deep rose. *Tropical.* p. 285

rosea (Brazil: Paraná), "Dwarf Easter cactus"; densely bushy, shrub-like dwarf cactus rather erect, with small 2 cm flattened joints 3-5 angled or keeled, almost pencil-like, waxy green tinted purple, with minute hair-like bristles; small wide open regular flowers of delicate rosy pink with orchid eye, 2½ cm dia. *Subtropic.* p. 285

RHIPSALIS   Cactaceae

burchellii (Brazil: Sao Paulo); a "mistletoe" cactus of vigorous, free-branching habit, with pendant smooth green pencil-stems; rose-pink berry-like fruit. 'Burchellii hybrid' has fruits varying from white to red. *Tropical.* p. 286

capilliformis (E. Brazil), "Old man's head"; epiphytic cactus with long, branching cylindrical, stringlike, hanging stems; many cream-colored flowers along sides. *Tropical.* p. 285

cassutha (Florida to Perú to Brazil, Ceylon and trop. Africa) "Mistletoe cactus"; growing on trees or rocks hanging in many strands to 10 m; branches thin-cylindrical, somewhat bristly when young; flowers cream; with mistletoe-like white fruit, as have many Rhipsalis. In Africa, this is found on Lake Kivu, and I have seen them in eastern Kenya, and the Usambara Mts. of Tanzania. *Tropical.* p. 284, 286

clavata (Rio de Janeiro); erect when young but soon hanging;much branched, joints all similar narrowly club-shaped, deep green; flowers white. *Tropical.* p. 285

mesembryanthemoides (Brazil: Rio), "Clumpy mistletoe cactus"; pretty epiphyte erect with two kinds of branches, long slender 10-20 cm woody joints, dense with numerous lateral short fleshy, light green, needle-like fruiting joints set with sparse white bristles near tips; small white flowers. *Tropical.* p. 285

neves-armandii (Brazil); stiff, pendant stems, angular cylindric, 1 cm dia., small areoles without bristles; large 2 cm flowers, white with pink sheen; fruit greenish. *Tropical.* p. 286

quellebambensis (Perú), "Red mistletoe"; epiphytic cactus with pendant thin cylindric branches, dull green with occasional purple markings and lightly grooved; glossy carmine-red berry-like fruit at the tips. *Tropical.* p. 285

zanzibarica (Kenya, Tanzania, Mozambique); miniature epiphyte with thin, thread-like dark green, circular branches, first erect but quickly pendant; areoles close and with fine white hairs; the depressed globular fruit, as in all Rhipsalis, sticky, adhering to birds or driftwood, and probably in this manner brought to Africa from America thousands of years ago. *Tropical.* p. 286

RHIZOPHORA   Rhizophoraceae

mangle (Florida, W. Indies to So. America), the "American mangrove" or "Red mangrove", so common along the shores and bayous of the Florida Keys; growing into small trees; producing many trunks or rooting shoots forming dense thickets by the many arching aerial roots; thick-leathery opposite, dark green leaves 5 to 15 cm long; yellow, long-stemmed flowers with 4 calyx lobes and 4 narrow hairy, pale yellow petals; fruit 3 cm long; before dropping into the wet soil, the fruit usually germinates and develops a root 30 cm long. *Humid-tropical.* p. 841

RHODELEIA   Hamamelidaceae

championii (Hong Kong); a small evergreen tree with reddish stem and petioles, glossy dark green, ovate, thick-leathery leaves, glaucous beneath; nodding 6 cm flowers with rosy petals in bracted heads. *Subtropic.* p. 522

RHODOCHITON   Scrophulariaceae

volubile (Mexico), the "Purple bell-vine"; graceful vine climbing to 3 m assisted by coiling petioles, similar to Asarina but more vigorous, and usually treated as a tender annual; alternate heart-shaped downy leaves to 8 cm long, and pendulous flowers 5 cm long with tubular corolla dark blood-red and spreading bell-like calyx pale reddish, on red stems; June. *Tropical.* p. 886

RHODOCODON   Liliaceae

urginioides (C. Madagascar); bulbous plant with long and narrow flexible leaves spreading from top of swollen base; a wiry stalk bearing a raceme of small bell-like, white flowers. *Tropical.* p. 603

RHODODENDRON   Ericaceae

arboreum (Himalayas), "Tree rhododendron"; evergreen tree to 15 m high, with leathery, lanceolate leaves 20 cm long, green and glossy above, silvery beneath; flowers scarlet with darker spots, 5 cm across. *Warm temperate.* p. 395

auriganum (N.E. New Guinea); strikingly colorful epiphytic evergreen which we collected in the Finisterre Mts. at 2,100 m alt. near the frost-line; clusters of large 12 cm trumpet-shaped flowers with fleshy, yellow tube and salmon-rose petal-lobes; small elliptic, leathery leaves. *Subtropic.* p. 398

'Bow Bells' ('Corona' x williamsianum); miniature evergreen about 30 cm high, with hard-leathery, 4 cm ovate, smooth leaves, pale beneath; and clusters of 4 cm open funnel single flowers of good substance, dainty salmon-pink with red spots in throat. *Temperate.* p. 405

canadense (Rhodora) (No. America: Newfoundland to Pennsylvania); attractive deciduous shrub which can be seen in profuse bloom in the Pocono Mountains of N.E. Pennsylvania into May; the clustered flowers 4 cm wide, with 2-lipped corolla rosy-lavender with darker purple tips; the foliage follows after blossom time, the oval leaves 3-5 cm long, downy beneath. *Temperate.* p. 406

catawbiense (Eastern U.S.), "Mountain rose-bay"; evergreen to 3 m; shining, leathery elliptic leaves 15 cm long; flowers in clusters, lilac-purple, 6 cm across. *Temperate.* p. 406

'Christmas Cheer'; compact bush to 1 m; tight clusters of medium-size white flowers. *Warm temperate.* p. 405

'Dexter hybrid' (fortunei cultivar); beautiful hybrids incorporating the attractive fragrant flowers and clear colors of R. fortunei from China; fairly winter hardy. *Warm temperate.* p. 404

'Eureka Maid' (Countess of Derby), (Pink Pearl x Cynthia); very good Easter forcing variety of more regular compact growth and with large flowers darker pink than 'Pink Pearl', pale in throat and with crimson spots on upper petal, the buds deep crimson rose, in large trusses. *Warm temperate.* p. 405

ferrugineum (Mountains of C. Europe), "Alpine rose"; shrubby evergreen, growing to 1 m high, with 4 cm elliptic, leathery leaves with rusty scales; salmon-rose flowers 2 cm long. *Temperate.* p. 406

'Ghent hybrid' (Azalea); a hybrid class of winter-hardy, deciduous azaleas of the European Rhododendron luteum and the American occidentalis; beloved in gardens for their bright colored, fragrant flowers on the leafless shrubs; late-blooming, substantial 4-6 cm flowers with narrow, somewhat crisped petals delicate salmon-pink with rosy stripes the lip richly tinged with orange; other cultivars in shades from bright pink to soft yellow; the leaves herbaceous and wrinkled. *Temperate.* p. 405

indicum (lateritium) (Japan). "Satsuki azalea", "Macranthum azalea"; low evergreen or semi-evergreen densely branched shrub, seldom to 2 m high, and lending itself to dwarfing as bonsai; small 2½-3 cm lanceolate, slightly hairy leaves with finely toothed margins; large funnel-form flowers 8 cm across, bright red or rosy with crimson markings in throat, June blooming unless forced; hardy. This is not the Azalea (Rhod.) "indica" of florists, which is a group name for many large-flowered tender hybrids derived from Azalea (Rhod.) simsii, the so-called "Belgian indica", forced in greenhouses during winter and spring in many named cultivars in a wide assortment of lovely colors. *Warm temperate.* p. 399, 406

indicum 'Shinnyo-no-tsuki' ('Moon of the Real Moon') (Azalea); a spectacular hybrid with large single flowers of 7 cm dia., white, with broad crimson border, May flowering, and developed as one of the 'Satsuki' (Fifth Moon) race of azaleas for beauty of leaves and flowers both, from the Japanese species indicum (lateritium). *Warm temperate.* p. 398

javanicum (Java); tropical evergreen shrub with showy flowers a rich orange, borne in clusters, and the most beautiful species on the island; I photographed this plant growing epiphytic on a tree in the mountains of Java at an altitude above 1,250 m. *Tropical.* p. 404

'Jean Marie de Montague'; excellent griffithianum cultivar suitable for Easter forcing indoors; compact, shapely plant budding well with firm, solid heads of bell-shaped flowers to more than 8 cm across, glowing bright crimson and very charming with wavy petals; smallish, dull-green elliptic foliage. Beautiful color for the outdoor garden on the Pacific Coast, but not hardy enough in the northern Atlantic States. *Warm temperate.* p. 405

'Kaempferi hybrid', "Torch azalea"; cultivar of Japanese species; tall deciduous bush to 2½ m high, elliptic, rough textured leaves 6 cm long, pubescent on both sides; flowers 4 cm across, brilliant orange-red, blooming before leaves. *Temperate.* p. 406

lochae (Queensland); the only known Australian rhododendron, found on Bellenden Ker mountain above 1,500 m, not far from tropical Cairns; evergreen shrub often epiphytic, usually grown as a dwarf plant; small leathery broad elliptic leaves 5-8 cm long, dark glossy green; and hanging bells of waxy 4-5 cm flowers deep rosy crimson with salmon sheen. *Subtropic.* p. 405

luteum (Azalea pontica) (E. Europe to Caucasus), "Pontic azalea" winter hardy shrub to 4 m high spring flowering before foliage; leaves rugose and partly hairy; showy flowers 5 cm across, orange yellow and very fragrant. *Temperate.* p. 404, 405, 406

macgregoriae (New Guinea); straggly evergreen shrub with clustered elliptic leaves; charming little waxy flowers 2 cm across, creamy white with orange center. *Subtropic.* p. 399

maximum; the photograph on pg. 404 is a true-to-life imitation of the living plant, made of glass by the famous father and son Blaschka glass artist• in Dresden, Germany in the early 1900's, for the Botanical Museum of Harvard University, Cambridge, Mass. The species is North American, from Nova Scotia to Alabama, an evergreen shrub to 12 m high, with glossy green leaves 25 cm long, tomentose beneath; the bell-shaped 5 cm flowers are pale rose and spotted with green, in large terminal clusters, blooming June into summer. *Temperate.* p. 404

mucronatum (ledifolia alba) (Japan), "Snow azalea"; spreading shrub to 2 m; elliptic leaves, corrugated, 6 cm long; funnel-shaped fragrant white flowers 5 cm across. *Warm temperate.* p. 406

mucronulatum (Azalea) (No. China); shrub to 2 m, with erect branches, deciduous or half evergreen, the narrow leaves 3-8 cm long; bell-like flowers appearing before leaves, rose-purple, to 4-5 cm dia., in profusion. *Temperate.* p. 404

obtusum (Azalea) (Japan), "Kirishima azalea"; spring-blooming evergreen becoming dense and bushy, 1 m or more high; leathery elliptic leaves 2-3 cm long, glossy dark green; small funnelform flowers carmine-rose with darker eye, 2½ cm across. *Warm temperate.* p. 394, 399

obtusum 'Kaempferi hybrid', "Torch azalea"; cultivar of Japanese species; tall deciduous bush to 2½ m high; elliptic rough-textured leaves 6 cm long, pubescent on both sides; flowers brilliant orange-red before leaves. *Temperate.* p. 406

occidentalis (Oregon), "Western azalea"; deciduous shrub to 3 m and more, with rough, obovate leaves 3 cm long; funnelform flowers 3 cm across, pale pink to purple. *Warm temperate.* p. 406

ponticum 'Superbum' (Spain, Portugal); late blooming evergreen, to 5 m high; leaves 15 cm long; flowers funnel-shaped 5 cm across in clusters; purple in the species; white with purple center in 'Superbum'. *Warm temperate.* p. 405

pulchrum phoeniceum (China); tender evergreen to 2 m high, with elliptic leaves 6 cm long; flowers funnelform 6 cm across, rosy purple in the species, magenta red in phoeniceum. *Subtropic.* p. 406

'Roseum elegans'; popular old hybrid known for its hardiness, bred with R. catawba, a good evergreen species from mountainous regions of Virginia to Georgia; compact globular bush with leathery decurved, olive green leaves, a heavy budder with numerous clusters of rosy lilac flowers of good substance, deeper at margins, and pale in the center, marked with purple spots on upper petals; mid-season. *Temperate.* p. 404

'Scintillation' (Dexter hybrid); shapely evergreen with deep green beautiful foliage; flowers of good substance, 8 cm in dia., rich pink when opening, later blush-pink in center and with wavy edges, in large trusses 18 cm across. *Warm temperate.* p. 405

'Simsii hybrids' or "Belgian indica" azaleas, see listed under Azalea 'Simsii hybrids'.

simsii 'Vittatum' (China); native of the moist, warm, subtropical valleys of Yunnan and the Yangtse; parent of the "Belgian indica" azalea hybrids; semi-evergreen shrub to 1½ m or more; oblong elliptic, hairy leaves to 5 cm long; showy funnelform flowers 5-6 cm across, variable from rose to dark red, spotted darker; the cv. 'Vittatum' has white flowers striped with purple. *Subtropic.* p. 399

thomsonii (Sikkim, Tibet, Bhutan); evergreen becoming treelike to 6 m high, with leathery, elliptic leaves 20 cm long; bell-shaped flowers deep blood-red, often spotted, to 8 cm across. *Warm temperate.* p. 398

'Trilby'; excellent hybrid of dense, compact habit, with numerous medium-large flowers per head, rich deep crimson spotted with black in throat; thick-leathery leaves dark green and pointed; good for for Easter forcing. (Queen Wilhelmina x Stanley Davies). *Warm temperate.* p. 405

'Unknown Warrior' (Stanley Davies x Queen Wilhelmina); exceptionally fine arboreum hybrid of bushy, floriferous habit, small elliptic or obovate leaves and perfectly round heads of flowers bright crimson red. *Warm temperate.* p. 405

westlandii (China); small evergreen with narrow elliptic, leathery leaves; semi-double lavender flowers with orange center. *Subtropic.* p. 405

RHODODENDRON: see also Azalea

RHODOMYRTUS *Myrtaceae*
tomentosa (India, Malaya), "Rose myrtle"; evergreen shrub, sometimes a small tree to 3 m, nearly all parts densely downy; obovate, leathery leaves 3-8 cm long, with 3 prominent veins; small axillary rosy flowers 3 cm wide, with pink stamens and downy outside; tiny 1 cm ovoid, berry-like purple fruit, pleasantly flavored. *Subtropic.* p. 684

RHODOPHIALA: see Habranthus
RHODORA: see Rhododendron

RHOEO *Commelinaceae*
spathacea (discolor) (Mexico),"Moses-in-the-cradle"; fleshy rhizomatous rosette of stiff waxy lanceshaped, metallic dark green leaves, vivid glossy purple beneath; in the leaf-bases, little white flowers are peeking from boatshaped bracts. *Subtropic.* p. 307

spathacea 'Vittata', "Variegated boat-lily"; variegated form with leaves striped lengthwise on the surface pale yellow, tinted red. *Tropical.* p. 307

RHOICISSUS *Vitaceae*
capensis (Vitis) (So. Africa), "Evergreen grapevine"; strong clambering vine with globular ground tubers; brown-hairy, somewhat woody stems, and long-stalked, thickish leathery, metallic green, glossy leaves nearly round or kidney-shaped, to 20 cm across, deeply lobed and wavy-toothed, rusty-tomentose beneath and at the margin; red-black glossy fruit. *Subtropic.* p. 920

RHOMBOPHYLLUM *Aizoaceae*
nelii (Hereroa) (So. Africa), "Elkhorns"; clustering succulent with spreading, gray-green leaves, two-lobed at apex; flowers yellow. *Arid-subtropic.* p. 49

RHOPALOSTYLIS *Palmae*
sapida (New Zealand, Norfolk Is.), the "Nikau palm" also known as the "Feather-duster"; representing the southern limit of palms; attractive palm to 10 m or less high usually with straight trunk strongly ringed, 10-20 cm thick, and topped by a prominent bulbous crownshaft; the pinnate fronds 1-4 m long, stand stiffly erect in a crown like a brush-like tuft, the erect, channeled leaflets glossy green ,with split apex; purplish flowers at base of crown-shaft, and small vivid red fruit. *Subtropic.* p. 801

RHUS see Cotinus

RHYNCHOSPERMUM: see Trachelospermum

RHYNCHOSTYLIS *Orchidaceae*
coelestis (Thailand), a "Foxtail orchid"; delightful dwarf epiphyte of vandaceous habit; stout stem to 20 cm, clothed with rigidly fleshy leaves closely spaced in ranks; cord-like roots developing from the base; the inflorescence dense with 2 cm fragrant waxy flowers, greenish-white and marked with purple (through summer). *Tropical.* p. 750

RIBES *Saxifragaceae*
sativum (W. Europe), "Red currants"; deciduous shrub to 1½ m, unarmed, with erect canes; palmate, lobed leaves; flowers green or purplish; glossy little, juicy red fruit in pendant clusters 5-8 mm dia., with tart flavor. *Warm temperate.* p. 472

speciosum (California), "Fuchsia-flowered gooseberry"; woody evergreen shrub to 4 m, bristly and spiny, fuchsia-like foliage; drooping inflated flowers bright red, the long red stamens protruding; bristly red fruits. *Subtropic.* p. 880

RICINUS *Euphorbiaceae*
communis (Trop. Africa), "Castor-oil plant" or "Palma Christi"; striking gigantic tree-like annual herb to 5 m high, or in the tropics where it is widely naturalized, a tree to 12 m; in gardens often planted for foliage effects; from a stout hollow stem, the large handsome peltate leaves, palmately divided into 5 to 11 crenate lobes, 15 to 90 cm across, and with metallic luster; small greenish-white, unimpressive flowers in clusters on long stalks, followed by attractive 3 cm prickly husks covered by brown spines and containing the poisonous seeds, source of a valuable oil. *Tropical.* p. 428

communis 'Coccineus', "Red-leaf Palma Christi"; ornamental form with dark red stems and coppery foliage. *Tropical.* p. 428

RIVINA *Phytolaccaceae*
humilis (laevis) (West Indies, Mexico, etc.), "Rouge plant"; soft-leaved herb with thin stem and branches; membranous, green, lightly pubescent, ovate foliage and pendant little sprays of tiny pinkish-white flowers, forming lustrous little berries of bright crimson, soon dropping. p. 812

ROBINIA *Leguminosae*
hispida (Southeastern U.S.), "Moss locust" or "Rose acacia"; semi-deciduous shrub 2 m or more high, of spreading habit, branches with bright red bristles; leaves divided into 3-6 pairs of leaflets; inflorescence in pendant racemes of deep rose flowers each 3 cm long. *Temperate.* p. 556, 564

ROCHEA *Crassulaceae*
coccinea (Crassula rubicunda) (S. Africa); branched succulent with small pointed, closely set leaves on fleshy stem, green above,

red beneath; tubular flowers bright scarlet and fragrant, in terminal clusters. *Subtropic.* p. 367
    coccinea 'Rosea' (*Crassula rubicunda* of hort.); a cultivar seen in Germany, with striking clusters of fragrant flowers brilliant crimson, the centers a star of rose-pink to white. *Subtropic.* p. 377

RODRIGUEZIA   *Orchidaceae*
    decora (Brazil); dwarf epiphyte with wire-like rhizomes to 1 m long, the scattered short, flat pseudobulbs 3 cm high, bearing a solitary hard 10 cm leaf; inflorescence gracefully arching; the flowers 4 cm long, sepals and petals white with purple spots, large white lip with purple red keels; the lateral sepals forming a spur. *Tropical.* p. 760
    secunda (Colombia, Panama, Trinidad, Guayana), "Coral orchid"; epiphyte with compressed pseudobulbs and narrow leaves; arching stalks to 30 cm high with rosy flowers usually on one side, (Feb.-Oct.). *Tropical.* p. 750, 752
    speciosa (Perú); attractive epiphyte with clustered pseudobulbs bearing solitary leaves; pendant inflorescence of silky pink flowers with pale yellow throat and broad lip. *Tropical.* p. 746
    strobelii (Ecuador); small epiphyte with prominent pseudobulbs and lanceolate leathery leaves; arching raceme with long, pure white flowers having large rounded lip. *Tropical.* p. 758
    venusta (*Burlingtonia fragrans*) (Brazil); dwarf epiphyte with small pseudobulbs, and pendulous racemes of very fragrant, 4 cm flowers pure white or flushed with rose, with yellow blotch on lip, (Jan.-May). *Tropical.* p. 713

ROHDEA   *Liliaceae*
    japonica (Japan, China), "Sacred lily of China"; extremely durable, modest plant with thick rhizome; basal rosette of oblanceolate, arching, channeled or plaited, thick-leathery leaves, densely arranged somewhat in two ranks, matte green; white flowers aroid-like; fruit a red berry. *Subtropic.* p. 606
    japonica 'Marginata' (Japan), "Sacred Manchu lily"; sheathing, channeled, leathery foliage black-green with white border; a favorite in Japan where hundreds of named varieties are cultivated by fanciers. *Subtropic.* p. 607

RONDELETIA   *Rubiaceae*
    odorata (West Indies, Panama); evergreen shrub 1-1½ m high, with downy, straggling branches; ovate dark green leaves 5 cm long, in opposite, distant pairs, the margins wavy or turned down, dentate toward apex; clusters of fragrant, tubular flowers with 5 expanding lobes, 1½ cm across, cinnabar-red with conspicuous yellow eye. *Tropical.* p. 862, 868
    strigosa (Guatemala); luxuriant bush with small green ovate leaves in pairs along slender woody branches; the flowers deep bronzy crimson with orange cup. *Tropical.* p. 868

ROSA   *Rosaceae*
    banksiae (China), "Lady Banks rose"; charming climber with vigorous canes rambling to 6 m, very spiny, evergreen foliage; deciduous in cold winters; pretty 3 cm double flowers pale buff-yellow, in large clusters, slightly fragrant; good for arbors, in mild climates. *Warm temperate.* p. 855, 857
    'Blaze' (Climber) ('Paul's Scarlet' x 'Gruss an Teplitz'); rambling canes 3-5 m long, dark green leathery foliage; medium large 5 cm double cupped flowers crimson-red, in large clusters, faintly fragrant; recurrent bloomer (Kalley/Jackson-Perkins 1932). *Temperate.* p. 859
    'Bonfire' (1928) (*multiflora* x *wichuraiana*); a good, vigorous climbing rose for Easter forcing in pots as a "trained rambler", more upright and slightly later than 'E. Jacquet', but its densely double flowers are more crimson-red and not as bluish carmine; the heavy flower clusters appearing as short axillary branches from the upper end of frost-ripened canes. *Temperate.* p. 857
    borboniana 'Magna Charta' (1876), "Hybrid-perpetual"rose; long favored for forcing in pots because of its prolific, bushy habit, producing numerous, erect canes with light green foliage, topped by clusters of great, very double, fragrant globular flowers with numerous, carmine-rose petals usually opening more or less at the same time and making a grand show. *Temperate.* p. 857
    canina (Europe, W. Asia), "Dog rose" or "Brier"; ancient rose with arching stems to 3 m and strong hooked spines; pinnate leaves with toothed 2-3 leaflets; single flowers pink to white; scarlet fruit, used medicinally. *Temperate.* p. 854
    chinensis 'Minima' (*rouletii*) (China), "Pygmy rose"; well-loved miniature, averaging about 20-25 cm high, vigorous, hardy and long-lived, with appealing 4 cm double flowers of lively rose-pink with pale eye, in continuous bloom; once thought lost to cultivation, it turned up again on the window-sill of a Swiss cottage in 1918. Also known as "Fairy rose." *Warm temperate.* p. 857
    'Double Paul's Scarlet' (1943); 'Fern Roehrs' rose, beautiful, very double sport of 'Paul's Scarlet Climber', a pillar rose with vigorous canes with deep green, leathery leaves, bearing axillary clusters abundant with large, fragrant, very double flowers 6-8 cm across, with 80-100 rolled, persisting petals, four times the number of the parent, and of a glowing deep scarlet-red with velvety sheen. *Temperate.* p. 856, 859
    'Eugene Jacquet' (1916) (*wichuriana* x *multiflora*); a free-blooming, symmetrical "Rambler" rose much used for forcing in pots for Easter, the long, rambling canes trained into globes, baskets or on trellis, with fresh-green, pinnate leaves and big clusters of small, fragrant, carmine-rose, double flowers on short, axillary branches appearing from the upper parts of winter-hardened canes of the previous season. *Temperate.* p. 858
    Floribunda class; a group of hybrids which was created by inbreeding the "Polyanthas" (*rehderiana*) with "Hybrid Tea" roses; more free-flowering over a longer season, with larger blooms than the "Polyanthas" and more vigorous. *Temperate.* p. 855, 858
    (Floribunda) 'Crimson Rosette'; dwarf hybrid (1948) with dark leathery leaves; open flowers deep crimson, slightly fragrant, 3-4 cm across; profuse bloomer. *Temperate.* p. 857
    (Floribunda) 'Europeana' ('Ruth Leuwerik' x 'Rosemary Rose'), a "Baby rose" from Holland; shapely bush to 70 cm with bronze-green foliage, and masses of satiny crimson double flowers 8 cm dia., faintly fragrant; repeated growth producing continuous bloom. (de Ruiter 1964: All-American Award 1968). p. 859
    (Floribunda) 'Fashionette'; a spreading, rather sparry bush with bright green foliage and long-stemmed replicas of fine tea roses in smaller size about 8 cm dia., in soft shell pink veined with red and flushed with yellow, quickly opening; delightfully fragrant, but we have found it a shy bloomer in pots. *Temperate.* p. 857
    (Floribunda) 'Korham'; remarkable cultivar seen in Germany with relatively large flowers salmon-yellow, shading to orange-red toward reflexed margins. *Temperate.* p. 858
    (Floribunda) 'Rumba' (Poulsen 1958) ('Masquerade' x 'Poulson's Bedder' x 'Floradora'); vigorous, bushy plant with dark glossy leaves; abundant bloomer with double flowers (35 petals), cupped, slightly fragrant, center yellow, margins poppy red and of 6 cm dia. *Temperate.* p. 859
    (Floribunda) 'Scarlet Marvel'; sturdy bush with large 6 cm double, rather flat flowers in brilliant scarlet; subject to mildew. *Temperate.* p. 856
    (Floribunda) 'Spice'; free grower with spreading branches bearing heavy clusters of fully double salmon-red flowers. p. 856
    Grandiflora class; this new group of hybrids approach in size of flowers the Hybrid Teas; developed by breeding of Floribundas with Hybrid Teas; characterized by more flowers per stem, in larger clusters than the HT, and individual stems longer than in the Floribunda, making them suitable for cutting; more nearly everblooming, tallish, with erect branches, reaching to 1½ m high. *Temperate.*
    (Grandiflora) 'Queen Elizabeth'; vigorous hybrid of Hybrid Tea and Floribunda, to 1½ m or more high, with Hybrid Tea type 9 cm flowers borne singly or in long-stemmed clusters, soft rose-pink and very fragrant; very close to Hybrid Tea roses, this type is valuable for its large number of flowers appearing at one time. *Warm temperate.* p. 856
    noisettiana 'Maréchal Niel' (1864) (*chinensis* x *moschata*); a famous old "Noisette" rose of climbing habit, with long rambling canes which I remember my father training on wires following the ridge of his greenhouse, and from which developed weak, axillary, pendant branches with large, most beautiful, hauntingly fragrant blooms of a delicate pale yellow, resembling a fine Tea rose; not hardy north. *Subtropic.* p. 859
    odorata 'Better Times' (1932) (hybrid tea); sport of 'Briarcliff', the first red rose for greenhouse cut flower growers that grew relatively easily, was free-blooming, and gave a satisfactory cut flower, proving exceedingly popular; dark green, leathery foliage and long crimson buds opening into large durable brilliant cerise red flowers with 36-50 petals, on long stems; with enchanting fragrance. *Warm temperate.* p. 857
    odorata 'Golden Rapture' ('Geheimrat Duisberg') (1933) (hybrid tea); seedling of 'Rapture' (Sport of 'Butterfly', which in turn was a sport of 'Ophelia') x Julien Potin; the first good growing greenhouse cut-flower yellow, with fresh-green glossy foliage, and gracefully slender, clear golden yellow flowers of medium size with deeper shading in the center, not fading, and with an old-rose fragrance, 40 petals. *Warm temperate.* p. 857
    odorata 'Happiness' ('Rouge Meilland') (1951) (hybrid tea); seedling involving 'Rome Glory', 'Tassin', etc.; an exhibition bloom and at the present time the leading "fancy" red rose as a greenhouse cut flower; large double flowers with 35-40 heavy, substantial petals glowing deep red unfolding to velvety crimson with blackish sheen, the outer petals curl to points in layer after layer

about the solid center, 12-18 cm across; very long lasting and slightly fragrant; not too free but very rewarding. *Warm temperate.* p. 857

odorata 'Joseph's Coat' (Armstrong 1964) ('Buccaneer' x 'Circus'); large double flowers 8 cm across, slightly fragrant, yellow and red multicolor; dark glossy foliage; vigorous pillar rose; recurrent bloom. *Warm temperate.* p. 859

odorata 'Mrs. Pierre S. Dupont' (Ophelia hybrid, 1929); elegant-hybrid tea of moderate growth and rich green foliage; long pointed buds reddish-gold, open to fully double (40 petals) golden yellow flowers and with fruity fragrance. *Warm temperate.* p. 857

odorata 'Mrs. W.C. Miller' (1909); a dependable old "Hybrid tea" rose of robust habit, and with leathery leaves, used for forcing in pots as it remains stocky, with a thick neck producing elegant, tightly double, dainty pink flowers flushed with rose, the petals are of good substance and unfold gracefully like a true Tea rose, with the same intense fragrance, blooming "monthly" into autumn. *Temperate.* p. 857

odorata 'Peer Gynt' ('Koenigin der Rosen' x 'Golden Giant'); a hybrid tea rose of impressive qualities, large double yellow flowers (50 reflexing petals), slightly fragrant; of vigorous, bushy habit, and free bloomer. *Warm temperate.* p. 856

odorata 'Red American Beauty' (hybrid tea) ('Happiness' x 'San Fernando'); vigorous, upright growth, with deep green leaves; ovoid bud, double, fragrant flowers, large to 12 cm dia., deep crimson red, prolific bloom. (Morey/Jackson x Perkins 1959). *Warm temperate.* p. 859

polyantha 'Margo Koster' (Sunbeam) (1935); excellent commercial "Baby-rose" (rehderiana), sport of 'Dick Koster', of short compact habit about 30 cm high, and large rather globe-shaped double flowers with incurved petals like 'Dick Koster' but a delicate soft light orange shading to salmon-red in the heart; good pot-forcer for Easter. *Temperate.* p. 856

polyantha 'Mothersday' (1949); excellent "Baby rose" (rehderiana) for pots, sport of 'Dick Koster', well-shaped, of small compact habit, never rambling, with glossy, pinnate leaves and short wiry branches profusely bearing clusters of relatively large, globular flowers of firm texture, deep crimson-red. *Temperate.* p. 856, 859

polyantha 'Triomphe Orleanais' (1912); a dependable vigorous old "Baby-rose" with glossy rich green, healthy foliage, freely branching with strong shoots 50-90 cm high bearing large clusters of medium size, long-lasting, semi-double flowers bright carmine or cherry-red with white eye, slightly fragrant; a good commercial pot plant for early Easter forcing. *Temperate.* p. 856, 858

rubiginosa 'Magnifica' (Hortus III: eglantaria); much used for hybridizing; flowers single or semi-double, bright rose 3 cm dia.; scarlet fruit; thorny canes 1-2 m long; foliage hairy beneath. *Temperate.* p. 859

villosa (Europe to Iran), "Apple rose" or "Hedge rose"; vigorous wild rose with scandent stems covered with slender prickles; leaves with 2 or more pairs of leaflets 2-4 cm long, hairy on both sides; single flowers pink to carmine-rose, 3-5 cm across. Fruit is eaten and used for beverages. *Warm temperate.* p. 856

ROSMARINUS *Labiatae*
officinalis (Mediterranean), "Rosemary"; evergreen shrub with downy shoots well known as a sweet herb, and grown for its aromatic leaves which are needle-like and grayish, shiny above, white downy beneath; flowers light blue. *Subtropic.* p. 533

ROSULARIA *Crassulaceae*
pallida (Armenia, Asia Minor); succulent rosette 3-4 cm across, with flat cylindric 2 cm leaves, glossy coppery to vivid crimson in the stem; erect inflorescence with cream-white flowers. *Warm temperate.* p. 376

ROTALA *Lythraceae*
macrandra (E. Indies); robust shrub growing in wet places; red stems with opposite oval, perfoliate, leathery green leaves; inflorescence in small spikes with tiny rosy flowers. *Tropical.* p.618

ROYSTONEA *Palmae*
elata (Oreodoxa) (South Florida), "Florida royal palm"; majestic feather palm at home in the Everglades, on wet ground; similar to R. regia but taller; smooth gray trunk to 32 m tall, swollen above the middle, graced by a heavy crown of arching plume-like pinnate leaves to 6 m long, the leaflets glossy deep green arranged in several rows, on bright green petiole; long inflorescence with fragrant flowers followed by round, circular 2 cm dark red fruit. *Tropical.* p. 801

oleracea (Trinidad, N. So. America), the "South American royal palm"; very tall feather palm 30 m or more high, with slender erect, smooth trunk not bulging, bearing the large glossy green crownshaft, and crown of gracefully arching pinnate fronds 3-6 m long, the numerous glossy leaflets attached to rachis in opposite, horizontal rows; fragrant flowers, and small purplish-black fruit. *Tropical.* p. 801

regia (Oreodoxa) (Cuba), "Cuban royal palm"; smooth erect gray trunks somewhat swollen above the middle, to 20 m high or more, with a terminal crown of gracefully arching feathery fronds regularly pinnate, 2-3 m long, the pinnae to 75 cm long and 2 cm wide, bright green and prominently ribbed and arranged in double rows, in 2 planes on either side of the axis. *Tropical.* p. 801

venezuelana (No. South America), "Venezuelan Royal palm"; stately feather palm with smooth, straight gray trunk; the fronds when young on purplish petioles; leaflets broader and more shiny than R. oleracea; smaller in habit with shorter trunk, but relatively large crown. *Tropical.* p. 800

RUBUS *Rosaceae*
idaeus (Eurasia), "Red raspberry"; scandent prickly canes propagated by suckers; leaves with 3 or 5 toothed leaflets; small whitish flowers; deliciously sweet; aromatic red berry fruit. *Warm temperate.* p. 472

laciniatus 'Prof. Rudloff' (Europe), "Thornless blackberry"; trailing shrub with perennial canes, without thorns in this European cultivar; leaves cut into toothed segments, flowers white or pink; juicy black berries. *Warm temperate.* p. 472, 852

reflexus (So. China, Hong Kong), "Trailing velvet plant"; attractive robust, somewhat sparry woody clamberer with rambling stems having occasional thorns, the young growth, petioles and underside of the foliage covered with cinnamon-colored wood; sturdy, pubescent, toothed and lobed leaves 10-20 cm long, vivid emerald-green painted with chocolate-brown along primary ribs, followed by a zone of splashed silver. Handsome foliage vine especially striking in younger shoots and when kept warm; best planted out but if in pots, then trained on trellis or wire frame. *Subtropic.* p. 856, 858

rosifolius 'Coronarius' (E. Asia), "Mauritius raspberry"; evergreen shrub with prickly, scandent canes 2 m or more long; pinnate leaves, with corrugated leaflets double-serrate, 3-9 cm long; flowers white, 4 cm across, very double in form coronarius; edible red berries. *Warm temperate.* p. 854

ursinus (Oregon to Baja California), "Pacific dewberry"; scandent or vining canes with stout prickles, dull green foliage tomentose beneath and with 3 leaflets, flowers white; berries deep red to black. *Warm temperate.* p. 472

RUDBECKIA *Compositae*
fulgida (Pennsylvania to Florida and Texas), "Orange cone flower"; perennial ½-1 m high, with 3-nerved, lanceolate leaves, and typical 4 to 5 cm cone-flowers with 12-14 orange-yellow rays, surrounding the high, black-purple disk. *Temperate.* p. 328, 329

hirta (Ontario to Florida and Texas), "Black-eyed Susan"; annual or biennial, bristly-hairy, 30-60 cm high; sessile leaves spatulate to lanceolate; solitary flower heads with about 14 golden-yellow rays to 5 cm long, and high purple-brown disk of tubular florets. *Temperate.* p. 328

hirta 'Double Gloriosa', "Gloriosa daisy"; the Gloriosa Double daisy strain has flowers 10-12 cm across, orange yellow with dark central cone. *Warm temperate.* p. 329

RUELLIA *Acanthaceae*
affinis (speciosa) (Brazil); tropical shrub with pendant flexuous branches, in habitat up to 6 m long; small oval leaves; showy funnel-shaped scarlet flowers 9 cm long, solitary in leaf axils. *Tropical.* p. 37

amoena (graecizans) (So. America), "Red-spray ruellia" or "Red Christmas pride"; herbaceous sub-shrub 30-60 cm high, with lanceolate, glossy green leaves 8-12 cm long and becoming smaller toward the top; brilliant red, swollen tubular 3 cm flowers, streaked yellow inside, blooming in summer or winter. Sun-loving flowering house plant best grown from fresh cuttings annually, or from seed. *Tropical.* p. 34, 37

colorata (Brazil: Amazon); strikingly beautiful shrubby plant with smooth gray-brown stem; opposite elliptic, quilted leaves 12-15 cm long, green beneath; showy inflorescence with carmine-red bracts, 6 cm trumpet flowers with 5 spreading lobes fiery orange-scarlet with yellow in throat and with long protruding stamens. *Tropical.* p. 37

macrantha (Brazil), "Christmas pride"; bushy plant with erect stems; opposite lanceolate leaves matte dark green, to 15 cm long, veins depressed; large trumpet-shaped flowers carmine-rose, with pale throat lined red. *Tropical.* p.37

makoyana (Brazil), "Monkey plant"; low-spreading herb with small, ovate leaves, satiny olive-green shaded violet and silvery veins; rosy-carmine flowers. *Tropical.* p. 37

squarrosa; (Dipteracanthus) (Okinawa); attractive spreading, low herbaceous shrub with 3-5 cm ovate, pubescent leaves; bright blue flowers with pale eye, 3 cm across. Grown in South Florida as ground cover. *Subtropic.* p. 38

RUMOHRA *Polypodiaceae (Filices)*
adiantiformis (Polystichum coriaceum) (So. America, So. Africa, New Zealand, Polynesia), "Leather fern", a spreading fern in dense clusters with creeping brown rhizome, similar to Davallia, with fresh green fronds to 1 m, thick-leathery, 1-3 pinnate, with oblong segments coarsely toothed. also known as Aspidium capense, the "Leather-leaf-fern". *Subtropic.*

RUSCHIA *Aizoaceae*
granitica (So. Africa: Namaqualand); low succulent creeper forming dense mats; the small bluish green leaves sharply angled and keeled, ½ cm long, the angles red; little purplish pink flowers 1 cm across. *Arid subtropic.* p. 44

RUSCUS *Liliaceae*
hypoglossum (Hungary, Italy, Asia Minor), "Mouse-thorn"; evergreen, tufted shrub with creeping rootstock and rigid stems to 45 cm high, the flexible, leaflike branches narrow and tapering to both ends, and bearing small yellow flowers on their centers. *Subtropic.* p. 588, 606, 607

RUSSELIA *Scrophulariaceae*
equisetiformis (Mexico), "Coral plant" or "Fountain plant"; shrubby plant with whip and rush-like, 4-angled stems to 1 m long, arching or pendulous; the normally lanceolate dentate leaves mostly reduced to small scaly bracts on the branches; tubular two-lipped flowers with fiery scarlet-red corolla 3 cm long, in nearly continuous bloom. *Subtropic.* p. 886
sarmentosa (Mexico, C. America), "Coral blow"; tropical shrub with slender, arching 4-angled stems with whorls of 2 to 4 wedge-shaped nettle-like leaves with crenate margins, 5 cm long; at leaf axils the tubular flowers velvety scarlet-red. *Tropical.* p. 886

RUTTYA *Acanthaceae*
fruticosa (So. Africa); shrubby plant with wrinkled ovate leaves, and flowers 4½ cm long, tubular two-lipped corolla orange-red with a black blotch, and with small linear bracts. *Subtropic.* p. 38, 40
speciosa scholesei (Ethiopia, Upper Nile, Rift Valley of Africa); a yellow variation of speciosa; shrub with glossy ovate leaves and showy 2-lipped flowers orange yellow, blotched and spotted with glossy black. *Tropical.* p. 38, 40

SABAL *Palmae*
causiarum (Puerto Rico, Virgin Islands), "Puerto Rican hat-palm"; stocky fan-palm 9-18 m tall, with gray, smooth trunk to 75 cm thick, bearing a heavy crown of palmate leaves 2 m or more long, many threads at sinuses, divided ⅔ to base; the segments stiff and bright green or bluish; fragrant white flowers and blackish fruit. *Tropical.* p. 800
minor (No. Carolina, Florida to Georgia and Texas), the "Dwarf palmetto", "Blue palmetto", or "Scrub palmetto"; wide-ranging dwarf fan palm usually without trunk, the rigid palmate leaves 1-1½ m wide, stiff and flat, glaucous or grayish green, the segments cut halfway or more; fronds generally upright but older blades kink at the junction of petioles, hanging downward and folding; erect inflorescence, black-glossy fruit. *Subtropic.* p. 798
palmetto (Carolina coast to Florida), "Cabbage palm", or "Palmetto"; variable fan palm 6 to 20 m high or more, with stout trunk either almost smooth and brown, or covered with a criss-cross pattern of leaf-bases, slightly curving; the palmate leaves 1-1½ m long, divided into many slender, hanging segments, green or bluish above and gray on the underside, with numerous thread-like fibers; the midrib extending through the blade; white flowers and blackish fruit. *Subtropic.* p. 795

SABINEA *Leguminosae*
carinalis (Dominica), "Carib wood"; small tree with pinnate leaves, leaflets in 6-8 pairs, 2 cm long; showy flowers crimson red, to 4 cm long. *Tropical.* p. 561

SACCHARUM *Gramineae*
officinarum (China, East Indies), "Sugarcane"; very tall stout perennial grass to 4 m high, with solid yellowish-green canes 2-5 cm thick; rich green, arching, clasping leaves to 1 m long, with broad midrib and rough edge; inflorescence in spikelets in large terminal fluffy silky plumes; its sap is a major source of sugar; the fermented and distilled juice becomes a well-known intoxicating drink, otherwise known as rum. *Subtropic to tropical.* p. 511, 512

SADLERIA *Polypodiaceae (Filices)*
cyatheoides (Hawaii), "Pigmy cyathea"; attractive, vigorous small tree fern forming a trunk 1-1½ m high, with a crown of fleshy, soft-leathery, light green, bipinnate fronds 1 m long, with neatly regular, linear segments crenate and turned under at edges; leaf stalks stout and fleshy. *Humid-tropical.* p. 438

SAGINA *Caryophyllaceae*
subulata (W. Europe to Italy and Greece), "Scotch moss" or "Pearlwort"; small mat-forming perennial herb with low, mossy foliage; leaves yellowish-green, linear, awl-shaped to 1 cm long, and bristle-tipped; small white flowers. Planted in California as ground cover in drier places. *Warm-temperate.* p. 510

SAGITTARIA *Alismataceae*
latifolia (No. America), "Arrowhead"; aquatic herb with variable leaves from linear lanceolate to broad, arrow-shaped; flowers white; freely adapting itself to all sorts of growing conditions. *Warm temperate.* p. 70

SAINTPAULIA *Gesneriaceae*
'**Blue Boy-in-the-Snow**'; meritorious leaf sport having light green foliage splashed with creamy-white throughout the leaf and maintaining its variegation; flowers violet-blue on pink petioles. The cultivar 'Blue Boy' was the first primary hybrid raised in America, between S. ionantha x confusa, by Armacost & Royston, Los Angeles 1930. *Humid-tropical.* p. 500
'**Blue Caprice**', typical "double light blue"; charming stocky, compact plant and willing early bloomer, with "girl" type scalloped leaves dark green, except for the characteristic pale center near the base, and with 3-4 cm light lavender, double flowers in profusion. One of the best commercial cultivars, with flowers that stand up and last. *Humid-tropical.* p. 499
'**Blue Peak**'; of flat habit; free blooming though short, the double flowers are violet-blue with each fringed little petal daintily edged in white; ovate leaves bronzy-green. *Humid-tropical.* p. 499
'**Blushing Bride**', a superb, typical "Double pink African violet", charming Roehrs (1966) hybrid; an open rosette with fleshy, crinkled leaves, and beautiful cupped, double and semidouble 3-4 cm flowers, good clear pink with deep rose center and prettily frilled petals; a welcome addition with its dainty charm. *Humid-tropical.* p. 500
'**Corinne**', shapely plant with light green foliage and medium-size flowers fully double; buds greenish opening to pure white; keeps well. *Humid-tropical.* p. 499
'**Danube Waves**'; free-blooming with heavy double flowers light blue, the inner petals curled like a Camellia. *Humid-tropical.* p. 501
'**Diana**', typical "European single violet" of the Englert 'Harmonie' strain; bushy, evenly formed, and compact cultivar much grown in Europe, with rich green leaves, and a full bouquet of large single flowers, velvety purple with darker violet center carried very prolifically on stiff erect stalks. This 'Harmonie' strain is distinguished by its vigorous growth and floriferous blooming habit, combining with it the non-dropping feature of its long-lasting flowers, even the single ones; the visible pollen creates a charming contrast. Roots are comparatively not salt-sensitive and help them to grow under adverse conditions. *Humid-tropical.* p. 500
'**Double Delight**'; an excellent, floriferous plant of flat habit with double, medium-blue flowers which will open well even during the heat of summer; bronzy leaves, wine-red beneath. *Humid-tropical.* p. 499
'**Elfriede**' (Rhapsodie strain); one of the successful German (Holtkamp) triploid hybrids characterized by their "Biedermeier" habit of growth forming closed, rather flat rosettes of dark green, firm leaves, and a formal mound of long-lasting, thick-textured, non-dropping blooms; 'Elfriede' has single round, dark blue 4 cm flowers; a prolific bloomer. *Humid-tropical.* p. 500, 504
'**Flash**'; low plant with small waxy leaves; the flowers daintily double carmine-red with pale margins. *Humid-tropical.* p. 499
'**Fuchsia Red**'; shapely cultivar with bouquet of semi-double flowers purplish red. *Humid-tropical.* p. 499
grotei (Tanzania); from deep shade in the E. Usambara Mts., near Amani at 900 m on rocks close to a waterfall; robust trailing or straggling species growing into a large plant full of fresh-green, short-hairy, rounded, fleshy leaves with notched edges, on long, flexible, brown petioles; small flowers 2½ cm across, pale violet-blue with darker center and edges, in axillary clusters. Excellent for hanging baskets. *Humid-subtropic.* p. 500
'**Icefloe**', typical "large double white"; an excellent Granger introduction of flat habit, with large dark green leaves, and a prolific bloomer with double 3-3½ cm flowers pure white but changing to bluish pink with aging. *Humid-tropical.* p. 501
ionantha (Tanzania), "African violet", found near Tanga at 30 m and higher, warm-humid; parent with S. confusa of most hybrids; upright species with large flowers a pretty violet-blue, to 2½ cm across, and dark, coppery-green, pubescent leaves reddish beneath. On the steep and breezy rocks towering up from the Amboni Caves, north of Tanga, I have seen plants hanging from nar-

row clefts or footholds surviving the seasonal dry periods by means of rhizomes almost bare of foliage. *Humid-tropical.* *p. 499*

ionantha alba; the blue-flowered species ionantha from time to time gives rise to another color, such as white or rose-pink. *Humid-tropical.* *p. 500*

'Kenya violet-blue' (Roehrs 1964), typical long-lasting double violet-blue, the "Kenya double violet"; a new break-through in Saintpaulia hybrids, incorporating parentage of the blue S. rupicola which I found in Kenya, with the wine-red cv. 'Flash', resulting in a vigorous, shapely plant with large 9 cm shiny green, spoon-type leaves, and 3-4 cm flowers double like a rose and a clear violet-blue; the non-dropping blossoms drying up on the plant instead of rotting when past blooming. *Humid-tropical.* *p. 499*

'Pink Miracle'; large sturdy plant, rather flat, with round green leaves with red backs; single pink flowers of fair size, with dark eye and edged with dark rose, and lightly ruffled; free blooming. *Humid-tropical.* *p. 499*

'Pocono'; low platter-like rosette with waxy cordate leaves; the giant 5 cm single to semi-double flowers of firm substance, purple with fringed pinkish margins. *Humid-tropical.* *p. 500*

'Ruffled Queen'; exquisite variety with large, firm flowers dark amethyst, the edges frilled as if with lace; large round, crenate, long-hairy leaves. *Humid-tropical.* *p. 499*

rupicola (Kenya), an "African violet"; pretty species which I collected north of Mombasa, growing on perpendicular limestone rocks in crevices, when older often hanging by a long, thick rhizome which sustains the plant through the dry seasons; grass green, thinnish, lightly wrinkled, faintly crenate foliage 7-8 cm long, glossy pale silvery-green beneath; both short and long hairs erect or slightly bent near top; pretty flowers 3 cm across, wisteria-blue with darker center. *Humid-tropical.* *p. 495, 499*

'Savannah Sweetheart', typical "holly-leaf, double bright rose"; low rosette with fleshy rich green leaves holly-like curly and ruffled, with margins turned under forming irregular points; free-blooming with heavy, large double flowers 4 cm across, with frilled petals rosy pink; hanging because of their weight. *Humid-tropical.* *p. 501*

'Show Queen'; sturdy compact variety with stiff leaves variegated yellow toward margins, and large firm violet flowers charmingly ruffled at edges. *Humid-tropical.* *p. 500*

'Snow Prince'; excellent, free-flowering, snow-white, single variety of neat habit, with large, full round, glistening flowers held high on strong stems; light green foliage. *Humid-tropical.* *p. 499*

SALIX *Salicaceae*

caprea 'Pendula' (Europe), "Pussy willow" or "Kilmarnock willow"; deciduous small tree with oblong leaves to 10 cm long, gray-pubescent beneath, in cv. 'Pendula' on crooked, drooping, willowy branches; usually grafted on regular species; catkins appear before foliage. *Temperate.* *p. 876*

matsudana 'Tortuosa', "Dragon-claw willow; a form of the species from No. China and Korea; small tree with twisted, contorted branches, and wiry pendulous branchlets with curiously twisted linear olive-green leaves 4-8 cm long, grayish glaucescent beneath. *Temperate.* *p. 876*

SALPIGLOSSIS *Solanaceae*

sinuata (Chile, Perú), "Painted tongue"; showy, branching clammy-hairy annual with sinuately toothed leaves and large solitary, 6 cm, funnel-shaped flowers brilliantly colored yellow through scarlet and primrose, and nearly to blue, and with a variation in veining in the wide throat. *Subtropic.* *p. 898*

SALVIA *Labiatae*

farinacea (New Mexico, Texas), "Mealy-cup sage"; perennial or annual to 1 m high, forming mounds; gray-green 10 cm leaves with serrate margins; erect, slender spikes of small, violet-blue or lavender flowers. *Subtropic.* *p. 533*

icterina; low shrubby perennial with long oval leaves 4-6 cm long, very rugose and pebbly, grayish green with irregular cream borders. In collection Botanic Gardens Kew. *Subtropic.* *p. 533*

involucrata (Mexico, Central America) "Rose-leaf-sage"; subshrub to 1 m or more high, with ovate, toothed leaves 12 cm long, the bract-like floral leaves colored; calyx purplish, and 2½ cm rose corolla, in whorls. *Subtropic.* *p. 531, 533*

leucantha (Mexico), "Mexican bush-sage"; shrub about 60 cm high with woolly branches and narrow wrinkled leaves 15 cm long, white-downy beneath; flowers white and woolly, and the calyx covered with dense purple wool, in long racemes. *Subtropic.* *p. 533*

mexicana (Mexico), "Ramona"; perennial to 1 m high, with rugose ovate leaves to 8 cm long, tomentose beneath; deep blue flowers in erect raceme. *Subtropic.* *p. 533*

officinalis (Mediterranean reg.), "Garden sage"; hardy perennial subshrub, white woolly, with slender gray-green pebbly leaves; flowers in terminal racemes light purple or white. Use: for seasoning poultry, cheese, veal, sausage, tomatoes; also medicinal. *Subtropic.* *p. 533*

splendens (Brazil), "Scarlet sage"; a well-known shrub, cultivated as an annual, with ovate, rich green, glabrous leaves on erect spikes bearing showy, scarlet-red flowers with scarlet calyx, in late summer. *Tropical.* *p. 533*

SALVINIA *Salviniaceae (Filices)*

auriculata (Tropical America), "Floating fern"; small, flowerless, aquatic fern-ally (Cryptogam), floating on water, with 1 cm, oval leaves pale yellowish-green and warty-haired on the surface, set along thread-like floating rhizomes, soon forming clusters. *Humid-tropical.* *p. 446, 448*

SAMANEA *Leguminosae*

saman (Pithecellobium) (West Indies, Central America), the "Rain-tree", "Saman", or "Monkey-pod"; great tropical shade tree to 24 m high, with spreading head like an open umbrella, to 30 m wide; bipinnate or 4-pinnate leaves with oblique ovate or roundish leaflets to 4 cm long, shiny above, downy beneath; silky-yellow flowers with long pink stamens, and flat pods 15-20 cm long. *Tropical.* *p. 560, 563, 564*

SAMBUCUS *Caprifoliaceae*

nigra (Europe, No. Africa, West Asia), "European elderberry"; deciduous shrub to 10 m long; foliage with usually 5 elliptic leaflets 12 cm long; flowers yellowish-white in umbels; edible berries shining black. *Warm temperate.* *p. 291*

SAMUELA *Liliaceae*

carnerosana (Coahuila, Mexico), "Palma de San Pedro"; bold tree with thick trunk 1½-6 m high, crown of thick gray-green, dagger-shaped leaves 1 m long, 6 cm wide, with marginal filaments and corky edge and sharp brown point. *Arid-subtropic.* *p. 586*

SANCHEZIA *Acanthaceae*

nobilis glaucophylla (Ecuador); handsome tropical shrub cultivated for its large lanceolate, soft-leathery leaves to 22 cm long, glossy-green with bold, contrasting yellow veins; large yellow flowers with bright red bracts in showy terminal panicle. Foliage tends to become green in older plants. *Tropical.* *p. 35, 38*

SANDERSONIA *Liliaceae*

aurantiaca (So. Africa: Natal), "Christmas-bells", or "Chinese lantern-lily"; tuberous climbing plant 45-60 cm high, with alternate, 10 cm ribbed, lanceolate leaves, often tipped with a tendril, along slender wiry stems; axillary bright orange-yellow, inflated urnshaped, nodding flowers 2½ cm long, of shiny, papery texture. *Subtropic.* *p. 602*

SANSEVIERIA *Liliaceae*

aethiopica (Zimbabwe, Transvaal); rosette of open habit; broad leaves to 40 cm long, dark green mottled lighter green, and with red corky margins; flowers white. *Subtropic.* *p. 610*

arborescens (E. Trop. Africa); tree-like, erect growing succulent, forming stem to 1 m high, from which curve the horizontal, slender linear, channeled leaves, plain green, with horny white edge. *Arid-tropical.* *p. 612*

caespitosa (Namibia); small rosette of rather flat, lanceolate recurving leaves tapering to a slender point, dark green, mottled grayish; freely suckering. *Arid-tropical.* *p. 608*

cylindrica (sulcata) (So. Trop. Africa, Natal), "Spear sansevieria"; round, arching, but rigid leaves to 1 m long and 3 cm thick, circular in dia., and usually furrowed or grooved, grouped several to a shoot, and tapering to a point, dark green with gray-green crossbanding which disappears with age; flowers pinkish. *Arid-tropical.* *p. 610*

dawei (Trop. Africa); broad, oblanceolate leaf, deep green, faintly mottled light green; corky edge; broad silver crossbands beneath. *Tropical.* *p. 610*

desertii (Southern Africa: Botswana, Zimbabwe, North Transvaal), "Rhinograss"; dweller from the Kalahari desert to the dry Veld, one of the stoutest plants in Africa; hard-succulent, erect cylindric leaves sunburnt green, with 9-12 shallow grooves, and narrow channel on top, 2½-3 cm thick, ½-1 m long, fan-like arranged; of stoloniferous habit, forming dense masses; insignificant orchid-like white flowers. Shunned by wild animals because of sharp tipped apex. *Arid-subtropic.* *p. 610*

dooneri (Rift Valley, Kenya); shapely rosette of lax, lanceolate leathery leaves 10-50 cm long, on narrow petiole, very smooth, dark green with pale cross markings and dark pebbling; flowers purplish outside, white within. *Arid-tropical.* *p. 608*

ehrenbergii (Ethiopia, Kenya, Tanzania), "Blue sansevieria"; plant with leaves alternately arranged as in a large fan, concealing the stem; the blue-green foliage above with triangular channel with white papery edge, flat on sides and rounded below. *Arid-tropical.* *p. 610, 612*

fasciata (Zaire); erect to recurving cylindrical, stiff leaves to 80 cm long, frequently in pairs, dark green with light green zigzag cross marbling. *Tropical.* p. 610

gracilis (Mazeras, Kenya; Tanzania); short-stemmed with spreading branches; tufts of slender leaves 20 to 90 cm long, cylindrical except for concave channel down the face, grayish green with faint gray cross-bands; flowers white, in pairs. *Tropical.* p. 610

grandis (Somaliland), "Grand Somali hemp"; epiphytic succulent sending out runner-like rhizomes producing 2-4 broad-obovate leaves to 15 cm wide, spreading near the ground, dull green with broad bands of deeper green, margins red; white flowers in dense raceme. *Arid-tropical.* p. 610

grandis zuluensis (South Africa: Zululand); succulent with creeping rootstock, producing long fleshy, oblanceolate leaves 6-10 cm wide, narrower than in grandis; concave and irregularly twisted, dark green with rows of gray blotches across the leafblade, the margins corky; erect cylindrical raceme of white flowers, opening in late afternoon. *Arid-subtropic.* p. 609

guineensis (West Africa: Guinea coast of Nigeria, Cameroun), a "Bowstring hemp"; robust, attractive rosette with 8-10 extremely broad, sword-shaped leaves 8-10 cm wide, and 45-90 cm high, arranged in birdsnest fashion, glossy deep green, irregularly cross-banded light grayish green. I photographed this striking species along the Guinea coast, in Southern Nigeria. *Tropical.* p. 608

guineensis 'Marginata', "Variegated bowstring hemp"; striking cultivar with broad leaves having contrasting, wide bands of yellow along margins. *Arid-tropical.* p. 608

kirkii pulchra (Zanzibar); handsome flattened rosette of wavy leaves dark green, dusted gray beneath, prettily marbled with whitish-green to reddish; corky red-brown margins; older plants more erect and less colorful; flowers white tinged with green. *Tropical.* p. 609

raffillii (Tsavo desert, Kenya); spreading by rhizomes; broad, stiff lanceolate leaves becoming 60-150 cm long, dark green with yellow blotches. *Arid-tropical.* p. 609

scabrifolia (S.E. Africa); rosette of stiff, recurving, narrow leaves tapering to long point, channelled on surface, keeled beneath, grayish green with gray crossbands; corky margins. *Subtropic.* p. 608

senegambica (cornui) (Sénégal, Gambia); sub-erect, oblanceolate leaves, with a concave channel at the base, flattening out higher up and ending in a slender point, matte medium green, only little striped on the outside. *Tropical.* p. 608

singularis (Voi, Tsavo National Park, Kenya); rosette of rigid, thick leaves ½ to 2½ m high, 4-8 cm wide; when young concave on surface, later cylindric 4 cm dia., dull grayish or bluish-green, with irregular silver crossbands. *Arid-tropical.* p. 609

splendens (Trop. Africa); long, thick leaves concave on surface, 15 cm wide, blackish green, faint marbling on back; inflorescence a tall raceme. *Arid-tropical.* p. 608

stuckyi (Zimbabwe); creeping rosette with cylindrical leaves to 90 cm high, which I saw near Victoria Falls growing between rocks, forming spreading colonies from underground rhizomes, the round columns are dark green with light crossbands, and lightly grooved all around except for a deep channel on the lower part of the inside. *Arid-tropical.* p. 610

thyrsiflora (S. E. Africa), "Bowstring hemp"; thick, creeping rhizome sending up full rosettes about 45-60 cm high, on some 20 stiff leaves fairly narrow, angularly channeled, keeled beneath, deep grayish-green with few markings, the margins white-translucent; fragrant flowers greenish-white. *Subtropic.* p. 608

trifasciata (Transvaal, Natal, E. Cape), "Snake plant"; erroneously called "zeylanica" in horticulture; erect plant with leathery linear-lanceolate, concave leaves to 1 m long, deep grass to almost blackish-green with light green to gray-white cross bands, on fleshy rhizomes bearing an average of 6-8 leaves; flowers greenish white, in loose raceme. Fragrant at night. *Subtropic.* p. 607, 609

trifasciata 'Bantel's Sensation' (sport of laurentii 1927); "White sansevieria"; unusual variety with slender, oblanceolate leaves, some of which have one side with the typical crossbars of trifasciata, the other half a golden yellow band like laurentii, and adjoining this are stripes of white, edged with green, in other leaves either one or the other character may predominate. *Tropical.* p. 609

trifasciata 'Craigii;' pleasing variety with erect, long obovate, flexible leaves having broad creamy yellow bands alongside the margin, the center milky green marked gray, and narrow dark green edging. *Tropical.* p. 607, 609

trifasciata 'Golden Hahnii' (pat. 1953); very showy plant when fully variegated; a sport of hahnii with firm, broad-elliptic leathery leaves in a low rosette, grayish-green with broad cream to golden-yellow bands alongside the margin, and more or less cross-banded in gray. *Tropical.* p. 607, 609

trifasciata 'Hahnii' (U.S. Pat. 1941), "Birdsnest"; sport of laurentii found at New Orleans in 1939, entirely different in habit, forming a low, vase-like rosette of broad, elliptic leaves spirally arranged, spreading and reflexed, dark green with pale green crossbanding, robust and freely suckering. *Tropical.* p. 596

trifasciata 'Hahnii variegata'; low birdsnest rosette variegated lengthwise with cream and yellow bands between green. Very attractive and ornamental as house plant *Tropical.* p. 596

trifasciata 'Laurentii' (Zaïre: Eastern Province), "Variegated snake plant"; leading commercial variety because of its elegant, stiff, swordshaped leaves having yellow bands on either side of the deep green, light banded center, its good keeping qualities, and nicely turned rosettes, soon clustering from the fleshy rhizomes. *Tropical.* p. 607, 609, 611

trifasciata 'Laurentii compacta' ('goldiana'); an excellent commercial mutant forming compact rosette with extremely stiff, numerous, blackish-green leaves of even height, not much over 40 cm and with a golden band along the margins. *Tropical.* p. 607

SANTALUM *Santalaceae*

ellipticum (Hawaii), the "Iliahi" or Hawaiian "Sandalwood"; small evergreen tree with opposite thick, glossy pale green oval leaves; clusters of pink to red flowers with red or green calyx at branch ends or leaf axils, and black berries. Santalum is partially parasitic, sending out roots with sucking organs into the root system of neighboring trees such as Casuarina or Acacia koa, stealing their food. Of economic importance during the height of the sandalwood trade with China mostly from 1810 to 1820, where the wood was used for incense and small furniture. *Tropical.* p. 876

SANTOLINA *Compositae*

chamaecyparissus (incana) (Mediterranean), "Lavender cotton"; low shrubby plant with silvery-gray tomentose lacy leaves; yellow globular flower heads. Used in carpet-bedding and for edging. *Subtropic.* p. 318

SANVITALIA *Compositae*

procumbens (Mexico), "Hussars heads"; hairy perennial grown as annual, with wide-spreading trailing stems, ovate leaves to 2½ cm long, and 1 cm heads of flowers having light yellow ray flowers and dark purple disk. *Tropical.* p. 326

SARACA *Leguminosae*

declinata (Malaya, Indonesia), "Red saraca"; tropical tree with pinnate leaves to 50 cm long, the leaflets to 30 cm; small 2 cm flowers without petals, but brightly colored calyx and bracts, yellow turning blood-red, and bracts scarlet, growing from old branches and twigs. *Tropical.* p. 563

indica (Mysore, India), the "Ashoka tree" or "Sorrow-less tree"; dense pyramidal tree, to 10 m tall, pinnate with long narrow, shining, undulate leaflets to 20 cm long; beautiful flower heads rich orange, fragrant, with long crimson stamens; sacred to Hindus and Buddhists, as Buddha is believed to have been born under it. *Tropical.* p. 563, 565

thaipingensis (Malay Peninsula), "Yellow saraca"; tropical tree to 10 m, with large pinnate leaves to 45 cm long, the lanceolate leaflets thick and leathery; showy floral clusters to 30 cm across, from old wood, with yellow bracts and petal-less flowers, becoming red at mouth of calyx tube. *Tropical.* p. 563

SARCOCAULON *Geraniaceae*

rigidum (Namibia), "Bushman's candles"; low succulent shrub from the coastal desert, with a water-storing stem 3-4 cm thick and horizontal, and with spiny branches; leaves spoon-shaped and grayish green, soon deciduous; large bright rose flowers. The branches are covered by a resinous coat and are used as fuel by the Africans. *Arid-tropical.* p. 487

SARCOCOCCA *Buxaceae*

saligna (Himalayas) "Willow-sweet-box"; handsome evergreen shrub 1 m or more high, photographed in Opatija, Yugoslavia; gracefully arching branches with glossy dark green, leathery oblanceolate leaves 8-15 cm long; small greenish-yellow flowers in clusters, followed by 2 cm black fruit. *Warm temperate.* p. 237

SARCODES *Monotropaceae*

sanguinea (Oregon, California), "Snow plant"; curious plant lacking chlorophyll, arising on the forest floor from a thick fleshy mat of roots and living on dead organic matter, 20-60 cm high; bright red pubescent herbage with fleshy stem and scale-like leaves, the upper strap-shaped and 5-10 cm long; small red flowers in a stout terminal cluster. *Warm temperate.* p. 643

SARCOPODIUM: see Epigeneium

SARITAEA *Bignoniaceae*

magnifica (Arrabidaea, Bignonia) (Colombia); climbing tropical vine with opposite, bifoliate leaves, obovate leaflets 10 cm long, smooth and leathery; the flowers tubular bell-shaped 8 cm long, mauve to purple, throat yellow. *Tropical.* p. 184, 187, 188

SARMIENTA *Gesneriaceae*
repens (Chile); shrubby plant with creeping, wiry stems, dense with opposite, small fleshy leaves; the numerous small flowers with inflated crimson tubes and spreading lobes; in habitat climbing over trees and rocks, requiring abundant water. Needs 3 months winter rest at 5 to 8° C. to bloom. (Dr. Messick, American Gloxinia Society 1967). *Subtropic.* p. 491

SARRACENIA *(Carnivorous pl.) Sarraceniaceae*
x chelsonii (purpurea x rubra); erect, fluted trumpets yellow green, marbled with white toward apex, the frilled lids whitish with red network. *Humid-subtropic.* p. 297
flava (Virginia to Florida), "Yellow pitcher plant"; tall and slender tubes to 1 m long, light green with mouth and lid edged yellow green and crimson throat; large nodding yellow flowers. *Warm temperate.* p. 296
leucophylla (S.E. No. America); handsome pitcher plant of the passive type, with slender, long green tubes 40 cm high, toward the inflated apex beautifully marbled with white, the lid white with green veining; photographed at Munich Botanic Gardens, Germany. *Warm temperate.* p. 293
'Melanorhoda' (purpurea x stevensii); showy pitcher plant taller than purpurea, with inflated tubes light green at base, heavily netted with red on the upper part and the flaring mouths. *Humid-subtropic.* p. 297
minor (N. Carolina to Florida), "Hooded pitcher plant"; pitchers short and also growing to 60 cm tall, fresh-green to purple, with white translucent spots near the yellowish top, the lid arching over mouth, purple netted inside; flowers pale yellow. *Warm temperate.* p. 297
'Moorei' (drummondii x flava); tall, slender trumpets green with red stripes lengthwise, the lid erect and wavy; flowers greenish-yellow. *Humid-subtropic.* p. 296
psittacina (in sandy swamps of Georgia, Florida, Alabama and Louisiana), the "Parrot pitcher plant"; low rosette with leaves 5-15 cm long, the tube club-shaped, with an erect, broad obovate, flat wing, green with red and white veins, the inflated top with incurved beak, like a parrot; flowers green and red. *Humid-subtropic.* p. 294, 296
purpurea (Labrador to Maryland and Rocky Mts.), "Sweet pitcher plant"; low rosettes usually found in sphagnum bogs of prostrate pitchers broadly winged, throat and lid hairy, and beautifully veined crimson; nodding purple flowers. *Temperate.* p. 294, 296, 297
rubra (N. Carolina to Florida), "Red pitcher plant"; slender tubular leaves green to reddish-brown, 3 to 10 cm long, the lid forming a hood and veined purple; flowers crimson and fragrant, on elongate stalks. *Warm temperate.* p. 296, 297

SASA *Gramineae*
fortunei (Arundinaria variegata, Pleiobastus) (Japan), "Miniature bamboo"; I photographed this plant in Brazil, and consider it the most attractive small bamboo in cultivation; its growth habit is low with gracefully arching branches and leaves much broader than pygmaea, strikingly banded white; the underside, unlike pygmaea, is finely pubescent. *Subtropic.* p. 516
tessellata (China); a dwarf "Running" bamboo with creeping rootstocks, and slender culms ¾-1 m high, short-jointed, with brownish-green culm-sheath, and ovate to lanceolate, leathery leaves 8 to 12 cm long and 3-4 cm broad, rich-green above, glaucous beneath. *Subtropic.* p. 516

SATYRIUM *Orchidaceae*
odorum (So. Africa: Cape); small tuberous-rooted terrestrial with candle-like fleshy stalk bearing clasping, succulent lanceolate leaves and a conical raceme of small fragrant lemon-yellow flowers. *Subtropic.* p. 752

SAUROMATUM *Araceae*
venosum (East Indies); tropical tuberous herb with solitary dissected leaf of 7 to 11 segments, carried on one 45 cm long stalk, produced after blooming, the segments 25 cm long; the inflorescence with an inflated tube 5-10 cm long, continuing into a purple blade 30-60 cm long, the spadix very long and slender. Probably a form of Sauromatum guttatum. *Tropical.* p. 129

SAXIFRAGA *Saxifragaceae*
longifolia (Pyrenees); attractive gray-hairy dense formal rosette to 60 cm across; linear spoon-shaped leaves; white flowers sometimes spotted purple. *Warm temperate.* p. 881
rosacea 'Arendsii'; "Kumomaso" in Japan; of European hybrid origin, with caespitosa as one parent; evergreen low herbaceous perennial forming moss-like mats 4-6 cm high; rosettes of finely divided fresh-green leaves, and literally covered with charming star-like flowers 15 mm across, carmine to crimson-red with white center. *Warm temperate.* p. 881
sarmentosa 'Tricolor', (stolonifera in Hortus 3), "Magic carpet"; beautiful variety smaller and more tender than the type, with leaves dark green and milky green, variegated inward from the margin with ivory-white, tinted pink or even rosy-crimson in younger leaves, with red edging, and purplish-rose beneath. S. sarmentosa, popularly known as "Strawberry geranium", is from China and Japan; a loosely tufted perennial spreading near the ground, strawberry-like, by thread-like runners, bearing young plantlets with rounded, bristly leaves 3-5 cm across; small white flowers on erect panicles. *Subtropic.* p. 877, 881

SAXIFRAGA: see also Bergenia

SCAEVOLA *Goodeniaceae*
frutescens (South Pacific), "Beach Naupaka" or "Half-flower"; spreading shrub to 3 m high, with fleshy, obovate leaves to 15 cm long, flowers white with purple streaks, 2 cm long; corolla 5-lobed all to one side. *Tropical.* p. 508
sericea (frutescens) (Malaya to South Pacific), the "Half-flower", "Naupaka" or "Sea-lettuce tree"; common on tropical shores; succulent shrub or small tree from 1 to 6 m high, with spreading stout branches, and long fleshy, bright green, obovate leaves 8-12 cm long; the small white fragrant flowers, streaked with purple are curious because the 5 petals are arranged to one side looking like only half a flower; small white berries resembling hail stones. *Tropical.* p. 523

SCHEELIA *Palmae*
liebmanii (Mexico); handsome impressive feather palm, some 15 m tall, with massive trunk about 1 m thick; the giant pinnate fronds erect to 10 m long, the numerous dense, long leaflets dark green, rough and rigid, and arranged to appear standing on edge; inflorescence with tremendous clusters of fragrant yellowish flowers, followed by roundish 6 cm brownish fruit. *Tropical.* p. 803
osmantha (Brazil); tall feather palm with ringed gray trunk, crowned by long plumy, very erect pinnate fronds 5 m long, the leaflets glossy green, very dense and drooping; similar to Attalea. *Tropical.* p. 803

SCHEFFLERA *Araliaceae*
actinophylla: see Brassaia
arboricola (Heptapleurum) (Taiwan), "Hawaiian Elf"; freely branching plant of dwarf habit, resembling when young a miniature Brassaia; wiry stems flexible and becoming scandent with age; the palmate foliage glossy green, to 15 cm across, arranged in a circle of 7 to 8 leaflets; inflorescence in erect, terminal cluster of orange-red to blackish berries; very charming in appearance and a good decorative plant. *Subtropic.* p. 139, 144
delavayi (China: Yunnan, Honan); spreading shrub 4 m high, with large palmately compound leaves on long petiole, the thin-leathery ovate leaflets matte-brownish green, brown pubescent underneath, faintly toothed, or lobed at margins, and with prominent midrib; long inflorescence with white flowers; resistant to some cold. *Subtropic.* p. 133
digitata (New Zealand), "Seven fingers"; bush or small tree 3-6 m high, densely branching, sometimes growing epiphytic, with thin-leathery leaves palmately compound, 5-10 foliolate obovate leaflets to 17 cm long, dull satiny green above, shiny light green beneath, densely ciliate and undulate at margins, and with yellowish, depressed veins; greenish-yellow flowers in panicles, purplish-black fruit berry-like. S. digitata prefers cool locations and I saw large colonies of this evergreen bush in the chilly southern Fiordland, especially near Milford Sound. *Warm temperate.* p. 139, 144
farinosa (Malaya, Sumatra, Java); very ornamental tree with shiny stems covered by scales; the leaves palmate-compound 35 cm across, with green leathery leaflets covered with mealy wax, and beautifully arranged in a formal circle; small whitish flowers in terminal inflorescence consisting of numerous incurving spikes. *Tropical.* p. 139
octophylla (Hong Kong), an "Umbrella tree"; branching small evergreen tree which one can see growing abundantly in the forest near the Peak station of the tram above Victoria; palmate leaves with 8-9 sturdy leathery, shining green leaflets having smooth, entire margins. *Subtropic.* p. 144
venulosa (Heptapleurum venulosum) (Queensland, China, Indochina, India), a "Starleaf"; branching tree with palmately compound leaves; the 7-8 stalked leaflets lanceolate when young, obovate or elliptic in maturity, soft-leathery, semi-glossy on both sides, to 15 cm long; mature leaves entire, dark green, but lightly toothed in juvenile stage; inflorescence in panicles with whitish flowers, followed by small red fruit. *Subtropic.* p. 144
venulosa erystrostachys (Heptapleurum stelzerianum hort.) (Trop. Asia to Australia), a "Starleaf"; scandent evergreen to 6 m

high, inclined to become semi-climbing, forming adventitious roots; leaves alternate, palmate with 6-7 leaflets, oval or ovate, fleshy, to 15 cm long, yellowish-green; globular flowers dark red, in dense clusters. *Tropical.* p. 144

volkensii (polyscidia) (Tanzania, Uganda); great evergreen tree to 25 m high from mountains at 2,200-2,700 m, such as growing in the mist-forest belt on Mt. Kilimanjaro at 2,700 m. Palmately compound leaves with 5-7 glossy-leathery leaflets to 15 cm long. *Subtropic.* p. 133, 144

SCHEFFLERA: see also Brassaia

SCHINUS *Anacardiaceae*

molle (Ecuador, Perú), "Peppertree" or "California pepper tree"; dioecious evergreen tree with gracefully weeping branches, 6-10 m high, with age developing a rough, gnarled trunk 60-90 cm thick; pinnate, feathery leaves 12-20 cm long, of numerous leathery, linear, deep green leaflets; yellowish-white flowers; on female trees the rosy 6 mm fruit resembling pepper-corns, in pendulous terminal clusters. *Subtropic.* p. 72

terebinthifolius (Brazil, Paraguay), "Brazilian pepper-tree" or "Christmas-berry tree"; ornamental evergreen tree 6-9 m high, of more rigid habit and less pendulous than S. molle, the California pepper tree; broadly spreading with willowy to woody branches densely clothed with pinnate leaves 10-17 cm long, of 5 to 9 broad leathery leaflets, dark glossy green and long-persistant. Small white flowers, followed by bright red berries on female trees, very showy in winter. *Subtropic.* p. 72

SCHISMATOGLOTTIS *Araceae*

novo-guineensis (New Guinea); stoloniferous tropical plant forming clusters; large ovate leaves with cordate base, deeply corrugated, to 25 cm long, bright green; spathe with pale green tube and white limb. *Tropical.* p. 108

SCHISMATOGLOTTIS: see also Aglaonema

SCHIZANTHUS *Solanaceae*

retusus (Chile), "Butterfly flower" or "Poor-man's orchid"; herbaceous annual to 1 m, with sticky stems and foliage; leaves finely divided, to 12 cm long; attractive irregular orchid-like flowers along the stalks, flaring wide 4-5 cm across, peach-colored with deep yellow center and lilac on lip. *Subtropic.* p. 898

x wisetonensis, "Butterfly-flower," or "Poor man's orchid"; beautiful hybrid of pinnatus x grahamii (Chile), bushy herbaceous plant with slender, sticky branches, pale green, divided leaves, and a profusion of showy, irregular, pansy-like flowers in shades of lilac, purple, pink, carmine, reddish-brown, or white, the upper lip often marked with purple and yellow, blooming spring or summer. *Subtropic.* p. 900

SCHIZOCAPSA: see Tacca

SCHIZOCASIA *Araceae*

lauterbachiana (bot. Xenophya) (New Guinea); fleshy rosette of stiff lanceolate leaves with lobed margins and hastate base, metallic, bronzy green and paler veins; spathe blade persistant, opening briefly in the middle only, opposite the staminate flowers. *Tropical.* p. 129

portei (Alocasia) (New Guinea, Philippines); large, triangular, waxy leaf deeply lobed, fresh green, on strong stem mottled brown. Bunting (Baileya Sept. 62) refers this species to Alocasia, having the same reflexed spathe blade characteristic of that genus. *Tropical.* p. 124

SCHIZOCENTRON *Melastomataceae*

elegans (Heterocentron in Hortus 3) (Mexico), "Spanish shawl"; creeping herb forming a dense mat, reddish stems rooting at nodes; small 1 cm ovate leaves, deep green and lightly hairy; covered with purple flowers during summer; good basket plant. *Subtropic.* p. 646

SCHIZOLOBIUM *Leguminosae*

parahybum (excelsum) (Panama, Brazil), the "Bacurubu"; or "Tower tree"; slender, fast-growing deciduous tree more than 30 m tall, with a crown of fronds resembling treefern; the feathery leaves 1 m or more long are finely bipinnate, the leaflets lobed, whitish beneath; masses of yellow flowers form erect 30 cm clusters. *Tropical.* p. 559

SCHIZOSTACHYUM *Gramineae*

brachycladum (Moluccas); tropical bamboo with erect culms forming open clumps; canes green and yellow, 3-4 cm thick; long lanceolate foliage. *Tropical.* p. 515

sp. 'Finisterre' (Northeastern New Guinea); a giant bamboo with hollow culms 8 to 10 cm dia., turning a polished bright yellow when mature. Long sections of the stems are cut and the dividing sectional cross-walls are rodded out except for the lowest one, and so used by the Melanesian tribes to carry water from the river valleys to their mountain villages. *Tropical.* p. 513

SCHLUMBERGERA *Cactaceae*

bridgesii (Epiphyllum truncatum) (Bolivia?), "Christmas cactus"; branching epiphyte with small glossy green leaflike joints, crenate and with blunt apex; pendant flowers in December, with flaring petals carmine-red tinged purple in center; angled ovaries. *Tropical.* p. 282

russelliana (Organ Mts., Brazil), "Shrimp cactus"; dwarf growing epiphyte and one of the oldest known species; will respond to grafting on such as Selenicereus; small leaves with crenate edge; regular, starshaped small flowers with several rows of orange-red petals, March blooming. *Tropical.* p. 281

SCHOMBURGKIA *Orchidaceae*

undulata (Laelia) (Trinidad, Venezuela, Colombia); bold epiphyte with tall, spindle-shaped, 2-3 leaved pseudobulbs, bearing the 1 m reedy stalk topped by a many-flowered dense raceme, the waxy sepals and petals longer and narrower than crispa, much twisted and crisped, wine-purple with a rosy lip, (Dec.-July). *Tropical.* p. 759

SCHOTIA *Leguminosae*

brachypetala (Transvaal, Natal, Zululand), "Tree fuchsia"; notable tree on the dry, ochreous bushveld of the Transvaal, to 12 m high, loaded with ponderous, crowded clusters of glowing deep crimson-red flowers, with 4-lobed leathery calyx and minute bristlelike petals, and protruding red stamens; pinnate green leaves partially deciduous, dropping most of them before flowering. *Subtropic.* p. 558, 561, 565

SCIADOPYTIS *Taxodiaceae (Coniferae)*

verticillata (Central Japan), the "Umbrella pine"; slow growing, very ornamental evergreen to 20 or 30 m high, of pyramidal habit; dense with slender linear, dark green needles 8-15 cm long, with 2 white bands below and deflected in graceful whorls; oblong woody 8 to 12 cm cones. *Warm temperate.* p. 352

SCILLA *Liliaceae*

peruviana (Mediterranean reg.), erroneously called the "Cuban lily"; bulbous plant very showy with dense cluster of small star-like lilac-blue flowers with petals edged in rose, blue stamen and yellow pollen, on thick-fleshy stalk, above broad, soft succulent, fresh-green leaves; Spring. *Subtropic.* p. 592

violacea (So. Africa), "Silver squill"; small suckering bulbous plant with swollen base, attractive because of its variegated foliage; strap-like fleshy leaves olive-green with silver blotching and banding, glossy wine-red beneath; small green and blue flowers on slender racemes; in winter. *Subtropic.* p. 592, 606

SCINDAPSUS *Araceae*

aureus (Epipremnum in Hortus 3) (Solomon Islands), "Devil's ivy", known commercially as "Pothos"; fleshy vine climbing tall by rootlets; juvenile leaves broad ovate, waxy, dark green with yellow variegation; mature leaves to 60 cm long, the blades becoming lobed or slashed; bisexual flowers on short spadix within the boat-shaped spathe. (Birdsey 1962 to Raphidophora; Bunting 1964 to Epipremnum). *Tropical.* p. 124, 125, 131

aureus 'Marble Queen', "Taro vine"; mutant with the green leaves richly variegated and streaked nearly pure white when grown in good light; resents chills and wetness. *Tropical.* p. 128

aureus 'Wilcoxii', "Golden pothos"; a California form found and selected by Wilcox, with sturdy green leaf variegated with golden-yellow, the variegations not blending into the green portion of the leaf but terminating abruptly; petioles and portions of stems often ivory-white in color. More durable than the more highly variegated cultivars. *Humid-tropical.* p. 128

sp. "Exotica" (New Guinea); graceful, very pretty, tall tree climber with channeled, partially winged and sheathing petioles bearing lance-shaped narrow, oblique-lanceolate, slender-pointed, thin-leathery leaves 10-20 cm long, grayish deep green, lavishly splashed and painted in the direction of the lateral veins with silvery gray; glossy pale gray beneath. *Tropical.* p. 131

pictus argyraeus (Pothos argyraeus) (Borneo), "Satin pothos"; beautiful creeper with the smaller, cordate leaves satiny, bluish-green with markings and edge of silver; probably the juvenile stage of pictus. *Tropical.* p. 128

SCIRPUS *Cyperaceae*

cernuus (Isolepis gracilis) (E. Indies, naturalized in Europe); grasslike, graceful tufted plant with numerous round, threadlike, fresh glossy-green stems becoming pendant, tipped with little white flower heads as in bullrush. *Subtropic.* p. 390

SCOLOPENDRIUM: see Phyllitis
SCUTELLARIA *Labiatae*
mociniana (Mexico), the "Scarlet skullcap"; herbaceous shrub of robust habit, to 50 cm high, with purplish-brown square stem, opposite, ovate-cordate, quilted leaves 10 cm long, dark metallic green, thin-leathery, and with crenate margins, grayish green beneath; striking erect, terminal spikes with a dense burst of brilliant scarlet-red long-tubular flowers 4-5 cm long, with orange-yellow lip and showing white stamens; the individual flowers will last only 6-10 days but succeeding clusters of flowers appear from upper leaf axils, extending blooming period from January to July. *Tropical.* p. 530

SCUTICARIA *Orchidaceae*
mooreana (Amazonian Perú); handsome epiphyte, with long, whip-like, pendulous foliage; waxy, fragrant flowers with small petals greenish and striped with purple; the larger sepals greenish-yellow; wide-flaring lip white with purple spots, and deep crimson in lip. *Humid-tropical.* p. 760

steelei (hadwenii) (Brazil to Ven.); peculiar but charming epiphyte with branching rhizome bearing on each branch a solitary channeled, whiplike, pendant cylindric leaf to 1 m long, and short stalked clusters of handsome fragrant, waxy flowers 5-8 cm or more across, pale yellow spotted with brown-purple, the lip with an orange crest; the long-lasting flowers appearing in summer or fall. Because of their pendulous habit, plants must be grown on rafts, treefern slabs or in baskets hung sideways; best in osmunda or treefern fiber. *Tropical.* p. 760

SEAFORTHIA: see Archontophoenix, Ptychosperma
SEDUM *Crassulaceae*
adolphii (Mexico), "Golden sedum"; small branching rosette of plump, fleshy keeled leaves waxy yellowish green with reddish margins; flowers white. *Subtropic.* p. 378

cauticolum (Japan); hardy, spreading succulent perennial with prostrate purple stems; the rounded, cupped, fleshy leaves 3 cm long, glaucous bluish-gray and dotted with red; the inflorescence with carmine-rose flowers in loose clusters. *Warm temperate* p. 376

dendroideum praealtum (Mexico); large handsome branching succulent to 1½ m or more high, oblanceolate spoon-shaped, incurving leaves arranged in loose rosette at apex of stems, 5-7 cm long, pale green flushed red at tips; flowers light yellow, in loose inflorescence. *Arid-subtropic.* p. 375, 376

morganianum (Mexico), "Burro-tail", a lovely hanging succulent plant, with tassels of short spindle-shaped leaves yellowish-green covered with silvery-blue bloom; terminal flowers pale pink. The heavy branches are quite pendulous and a beautiful sight when grown like long queues. *Arid-subtropic.* p. 377, 378

pachyphyllum (Oaxaca), "Jelly beans"; small shrubby succulent with cylindric clubshaped fleshy leaves curved upward, light green and glaucous blue, with red tips; flowers yellow. *Subtropic.* p. 378

populifolium (Siberia), "Hollyleaf sedum"; interesting succulent subshrub with branches to 25 cm long, later creeping; the deeply lobed, waxy, dull-green leaves 2 cm long on brown-striped stems; deciduous in autumn; loose inflorescence with rosy or white flowers delicately scented. *Temperate.* p. 375

sieboldii (Japan), "October plant"; graceful perennial creeper suitable for hanging baskets; flexible stems set with whorls of 3 roundish, notched leaves, glaucous-blue changing to copper, and edged with red; flowers pink, blooming in October. *Warm temperate.* p. 377

sieboldii 'Medio-variegatum'; an attractive variegated form with the grayish green leaves having a cream-yellow center, and stems are pink. *Warm temperate.* p. 375

sieboldii 'Variegatum', an "October plant"; leaves glaucous, blue and variegated with cream. *Warm temperate.* p. 375

spathulifolium 'Capa Blanca' (Brit. Columbia to No. California); branching succulent with stems topped by small rosettes, leaves spatulate 2-3 cm long, glaucous bluish with red margins; flowers yellow. *Warm temperate.* p. 375

spectabile (Japan, Central China), hardy "Live forever" or "Showy sedum"; strong growing, tough winter-hardy perennial 30-60 cm high, freely suckering at base with erect thick fleshy stems and soft leathery, glaucous light green or grayish green obovate leaves 6-8 cm long, toothed toward apex, and generally set in twos and threes along the stems, which terminate in large flat clusters of rosy-lavender or red flowers in late summer. Propagation by division of the root-stock or cuttings. Probably the showiest of Sedums, good for tubs which may be left outdoors during the cold season with plants going dormant in winter. *Warm temperate.* p. 375

spurium 'Splendens'; clustering perennial with branching stems; opposite fleshy green obovate leaves 2-3 cm long, ciliate at margins; starry pink flowers in the species from the Caucasus; crimson in this cultivar. *Warm temperate.* p. 376

telephium (W. Europe), "European ice plant" or "European live-forever"; robust perennial with thickened carrot-like roots; strong stems erect to 45 cm high, set with oblong-ovate bluish-green soft-leathery, toothed leaves 5-10 cm long and rarely opposite, not in twos and threes as spectabile; in early autumn each leafy stem is topped by a large cluster of bronzy, rosy or red-purple flowers. Winter-hardy and going dormant. *Warm temperate.* p. 377

weinbergii: see Graptopetalum paraguayense
SEEMANNIA *Gesneriaceae*
latifolia (Bolivia at 600-1000 m); a pretty, very floriferous branching plant from scaly rhizomes, closely related to Kohleria; by appearance smooth, but covered with a coating of very fine hairs; the stems first erect, then reclining with tips erect; elliptic to lanceolate bright green leaves 12-15 cm long, on clasping petiole; from the axils in whorls around the main stem the bell-shaped swollen 2 cm flowers brick-colored or cinnabar-red outside, the short lobes deep red inside, with the throat yellow speckled red-brown; in constant bloom for months, beginning in early spring. *Tropical.* p. 501

SELAGINELLA *Selaginellaceae*
apoda (Eastern No. America), "Basket selaginella"; stems delicate and weak, prostrate, forming mats to 40 cm across; minute leaves in 4 ranks, pale green. *Warm temperate.* p. 888

caulescens japonica (Mountains of Japan); dwarfed plant forming a fountain-like rosette of firm fronds with appressed leaves, tipped by yellowish inflorescence spikes. *Warm temperate.* p. 889

emmeliana (cuspidata emiliana) (So. America), "Sweat plant"; lacy, small rosette of fern-like, erect, bright green fronds, revelling in high humidity; if allowed to dry the tips will curl and turn brown and won't recover. *Humid-tropical.* p. 889

galleottii (Mexico, Costa Rica); stems 30-50 cm long, sub-erect, with root fibers from lower half, sometimes whip-like at tip, finely pinnate; leaves on lower plane close on branchlets, leaves on upper plane more shingled. *Humid-tropical.* p. 888

kraussiana (Lycopodium denticulatum) (So. Africa, Cameroons) "Spreading clubmoss"; a charming, moss-like herb with matforming, creeping stems rooting as they grow, with tiny, crowded, bright green, pinnate, scale-like leaves; very useful as a quickly spreading ground cover in terrarium or conservatory. *Humid-tropical.* p. 889

kraussiana 'Aurea', "Golden trailing club-moss"; attractive spreading form with creeping branches having leaves of yellow-green, perhaps a little heavier than the type, and the young growth especially is golden yellow. *Humid-tropical.* p. 889

kraussiana 'Brownii' (apus elegans) (Azores), "Dwarf club-moss"; shapely, moss-like cushions of densely clustering, short branches of vivid emerald green, supported by translucent aerial roots; very attractive in terrariums. *Humid-tropical.* p. 888

lepidophylla (Texas, Mexico, Perú), "Resurrection plant" or "Rose of Jericho"; flat rosette of densely tufted, branched stems with fairly hard, scale-like leaves, red-brown with age; when the plants dry out the branches curl up into a tight ball, but will unfold to fresh emerald-green when placed into water. *Subtropic.* p. 889

martensii 'Variegata'; a beautiful form of coarse, rather stiff, branching habit, largely supported by long stilt roots, with broad, glossy-green leaves irregularly but attractively tipped or variegated white. S. martensii is a bold species from Mexico, with stems to 30 cm long. *Tropical.* p. 888

plumosa (E. Himalayas, Ceylon, Burma, So. China); crawling ground cover, like kraussiana, but more dense, with short flat fronds, bright green, fairly firm in texture, and overlapping the pale stems; the fruiting spikes are square. *Subtropic.* p. 888

serpens (variabilis mutabilis) (West Indies); densely matted, spidery, trailing stems, reaching to 22 cm long, pinnate with slightly compound, slim and firm branches; leaves bright green with distinct midrib, and with spore-bearing spikes. *Tropical.* p. 888

tamariscina (Honshu, Kyushu); evergreen tufted rosette of ribbon-like linear firm fertile fronds green with yellow tip; the tiny scale-like leaves in shingled ranks closely appressed to the flattened axis; forms miniature tree with age. Photographed at an exhibition of Japanese fanciers in Ikeda Park, Tokyo. *Subtropic.* p. 889

umbrosa (Yucatán to Colombia); erect red stems to 30 cm high or more, unbranched in lower part, branched above; leaves ascending, dark green and firm. *Tropical.* p. 888

uncinata (caesia) (So. China), "Rainbow fern"; exquisite low creeper with straw-colored, slender, rambling stems rooting along the ground, alternately set with tiny branched lateral leaves which are metallic blue and iridescent in the shade. *Humid-tropical.* p. 888

willdenovii (caesia arborea) (Vietnam, Malaysia, Himalayas), "Peacock fern"; robust growing and most beautiful, shade-loving rambler at first erect, but soon climbing between shrubs to 6 m high, the light brown stems supported by stiff stilt-roots, and bearing spreading fronds of magnificent, shimmering peacock-blue. *Humid-tropical.* p. 888

SELENICEREUS *Cactaceae*
  grandiflorus (Jamaica, Cuba), "Queen of the night"; climbing epiphyte with large flowers 15 to 25 cm long, salmon outside, white inside, and blooming by moonlight, with a powerful vanilla perfume, earning it the name "Queen of the night"; the flowers expand at sunset or by moonlight and fade off in the morning; stems 3 cm thick, light grayish green to purplish. *Tropical.* p. 255, 256
  hamatus (So. and E. Mexico, Lesser Antilles), "Moon cactus"; epiphytic climber with vigorous stems glossy grass-green 2 cm thick and ribbed, with few aerial roots; flowers white, pale green outside, to 25 cm long. *Tropical.* p. 257
  pteranthus (Mexico), "King of the night", or "Princess of the night"; trailing epiphytic climber to 3 m long; the jointed, stiff, rooting stems 1½-4½ cm thick, 4 to 6-angled, dull glaucous green, usually flushed purple, the areoles with white wool and short thick spines; night-blooming, funnel form flowers 30 cm across, with narrow purplish-yellow sepals and broad creamy-white petals; reddish fruit. *Tropical.* p. 256
  urbanianus (Cuba, Haiti), "Moon-cereus"; clambering nightbloomer, epiphytic with aerial roots, and 4 to 5-angled joints of 5 cm dia; giant 30 cm fragrant flowers white inside, tan outside. *Tropical.* p. 256
  vagans (Mexico); epiphytic climber, with creeping stems 1½ cm thick, 5-6 ribbed, and raised brown areoles; flowers very freely borne at the top of stems, sepals brownish or greenish, petals, white; night-flowering. *Tropical.* p. 256

SELENIPEDIUM: see Phragmipedium

SEMELE *Liliaceae*
  androgyna (Ruscus) (Canary Isl.), "Climbing butcher's broom"; shrubby vine having leaves represented by scales, the apparent leaves being leaf-like branches, to 10 cm long, ovate, leathery; the tiny greenish-yellow flowers with purple center borne in clusters on their margins. *Subtropic.* p. 606

SEMPERVIVUM *Crassulaceae*
  arachnoideum (Mts. of So. Europe), "Cobweb hen-and-chicks"; tiny rosettes to 2 cm dia., clustering by stolons and forming mounds, and covered by a cobweb of white hairs from tip to tip of leaves; flowers reddish. *Temperate.* p. 367, 375, 378
  tectorum calcareum (French Alps), "House leek"; attractive small leathery rosettes forming clusters, glaucous light gray-green leaves broadly painted red-brown at apex; flowers pale red on spikes. *Warm temperate.* p. 378

SENECIO *Compositae*
  cineraria (Cineraria maritima) (Mediterranean); beautiful, white-woolly perennial to 1 m tall, with thick leaves at first oak-leaved, later pinnately cut, the pinnae well separated and broad, crenate at their broadening apex; inflorescence on beautiful white-felted stalks, in cymes of small 2 cm thistle-like flower heads with short bright yellow rays. Listed in seed catalogs as Cineraria maritima 'Diamond'; much used as a border plant. *Subtropic.* p. 310
  cineraria 'Silver Dust'; very dwarf "Dusty Miller" to 20 cm high, with finely cut silvery-white foliage. *Subtropic.* p. 310
  citriformis, (So. Africa: Cape Prov.); charming hanging basket plant; long thread-like stems furnished like a string of beads with globular grass-green shiny balls, coming to a short point at one end; with numerous pale gray or translucent longitudinal lines; flowers creamy-yellow in stalked heads. *Subtropic.* p. 325
  confusus (Mexico), "Mexican Flame vine"; colorful tender vine or scandent shrub with fresh green, fleshy, ovate leaves 4 to 10 cm long, coarsely toothed at margins; striking daisy-like double flame-scarlet flowers in clusters at ends of branches. *Subtropic.* p. 325
  fulgens (So. Africa: Natal), "Scarlet kleinia"; succulent plant with tuberous root and thick fleshy stem to 45 cm long, wholly covered with glaucous bloom; the fleshy obovate leaves 8-12 cm long and toothed toward apex, light green and waxy-powdered violet-gray; flower heads with orange red florets. Promising house plant. *Subtropic.* p. 324
  glastifolius (So. Africa: S. W. Cape), "Maple aster"; tall perennial to 1 m high, with small light green, prickly, serrate leaves 8 cm long; and clusters of daisy-like flowers soft pink to mauve, with yellow center. *Subtropic.* p. 328
  grandifolius (Mexico); semi-woody shrub 2-4 m high, with erect, stout, purplish stem; large ovate leaves to 45 cm long, coarsely toothed, dark green above, downy beneath; winter flowering with large cluster of yellow heads. *Subtropic.* p. 325
  haworthii, in horticulture as Kleinia tomentosa (So. Africa), "Coccoon plant"; striking small succulent semi-shrub about 30 cm high, entirely clothed with appressed soft pure white wool; the fleshy cylindrical-pointed leaves 2½-5 cm long; flower heads orange-yellow. Beautiful in their snowy dress, the foliage unfortunately drops off easily when handled. *Arid-subtropic.* p. 324
  herreianus (Kleinia gomphophylla) (Namibia), "Green marble vine"; clustering succulent with creeping, rooting branches, the distant, fleshy green leaves berry-like, to 2 cm long and pointed, with translucent stripes serving as windows. *Arid-subtropic.* p. 325
  hybridus (hybrid of cruentus), from the Canary Islands; after much modification and possibly hybridization with S. heritieri and populifolia, have become the widely cultivated "Cineraria" of florists; a showy, herbaceous cool house plant with handsome, triangular-ovate, large turgid leaves rich grass-green above, purplish beneath and grouped around a large rounded, and dense truss of starry flowers with variously colored rays from white to shades of pink to red, purple and blue, surrounding the usually purple center cushion; often with white eye. *Humid-subtropic.* p. 328
  hybridus 'Grandiflora'; large flowered class 5-6 cm flowers of medium-tall habit, blue with white eye. *Subtropic.* p. 326
  hybridus 'Maxima', "Exhibition cineraria"; showy heads of beautifully formed giant flowers measuring 10 cm or more across. *Subtropic.* p. 326
  hybridus 'Multiflora grandiflora'; showy heads of numerous large flowers 5-6 cm dia., of dwarf habit. *Subtropic.* p. 326
  hybridus 'Multiflora nana'; medium small-flowered class of low, compact habit; the many smallish, often white-eyed flowers, primarily in bright reds, purples and blues, make up in big showy heads what they lack in individual size; a favorite commercial and widely cultivated group for early spring bloom. *Subtropic.* p. 326
  johnstonii, (Tanzania), the "Giant groundsel"; mainly of interest becauase of the large, tree-like habit of this species; I have found these above tree-line on Mt. Kilimanjaro where they occur between 2,700 and 5,000 m altitude; tree-like to 4 m high, with several woody stems topped by dense rosettes of large 30-45 cm cordate-ovate, turgid, recurved leaves, fresh-green, somewhat gray felted underneath, and on channeled leafstalks; small flowers in terminal 1 m branched spikes. *Subtropic.* p. 309, 327
  macroglossus 'Variegatum', "Variegated wax-vine"; very attractive creeper, of which the green type comes from E. Cape Prov., but I found this lovely variegated form in Kenya; densely branching and mat-forming, with small 3-4 cm ivy-like lobed, waxy, thin-succulent leaves with cordate base, green to milky-green and bordered or variegated with cream; pretty daisy-like ray flowers with 12-14 white florets and yellow center. *Subtropic.* p. 323, 324, 325
  mikanioides (So. Africa), "Climbing senecio"; glabrous twiner with fresh-green, ivy-shaped, lobed leaves rather soft-fleshy; fragrant, yellow disk-flowers. *Subtropic.* p. 323, 326
  pendulus (Kleinia) (Arabia), "Inchworm"; succulent shrub with cylindrical stems, decumbent, snake-like, 2 cm thick, grayish over bronze; flowers red to orange. *Arid-tropical.* p. 324
  petasitis (So. Mexico), the "California-geranium"; shrub-like perennial to 2 m high with roundish, wavy-lobed grayish downy leaves 15-30 cm long; 3 cm flower heads in large terminal panicles, the 5-6 ray florets bright yellow, disk brownish. *Tropical.* p. 322
  petraeus (Notonia) (Kenya, Tanzania); succulent, erect when small, later creeping upwards, with alternate obovate fleshy, sessile nerveless leaves 8 cm long, glossy green, the stems rooting at the joints; flower heads orange. *Arid-tropical.* p. 324
  picticaulis (Tanzania and Sudan); succulent plant with tuberous root stock, erect fleshy, cylindric stems to 20 cm long, and 2 cm thick, pale green with dark lines; flower heads pink with red lobes. *Arid-tropical.* p. 324
  radicans (Cape Prov.: Karroo; Namibia); creeping succulent with flexuous stems rooting at the nodes, somewhat angular; leaves thick and round, tapering to both ends, 1½-2½ cm long, directed upwards, dark green, often reddish and wrinkled; flowers white with yellow anthers; related to S. radicans but with leaves more spindle-shaped rather than round as marbles. *Arid-subtropic.* p. 325
  rowleyanus (So. Namibia), "String of pearls"; creeping thin, flexible stems with adventitious roots forming dense mats; furnished as in a string of beads the globular pointed, succulent leaves shiny pale to dark green, with a translucent stripe, about 1 cm dia.; the flowers cinnamon-scented, without ray-florets, the corolla white and with brownish-violet anthers. *Subtropic.* p. 324, 325
  serpens (better known as Kleinia repens) (So. Africa: Cape Prov.), "Blue chalk sticks"; low succulent branching shrub 20-30 cm high, with fleshy, nearly cylindrical leaves to 3 cm long, grooved above, bluish-gray with blue waxy coating; flower heads pale yellow. Attractive but has a tendency to drop leaves. *Arid-subtropic.* p. 324

SEQUOIA *Taxodiaceae (Coniferae)*
  sempervirens (Oregon and Calif. Coast ranges), "Redwood"; the tallest tree in the world, to 110 m, with red bark, horizontal

branches spreading in flat sprays, needles deep green, bluish beneath, and persistent. Their knotty burls, cut from trunks, will sprout young growth in a shallow dish of water. *Subtropic.* p. 355

SEQUOIADENDRON *Taxodiaceae (Coniferae)*
   giganteum (California), the famous "Giant sequoia", "Giant redwood" or "Big tree"; at home on the high western slopes of the Sierra Nevadas at an elevation of 1,400 to 1,800 m, where venerable old trees have lived for 5000 years; diameter of trunk to 11 m, and reaching a height of 81 m; the pendulous green shoots are smooth and cord-like; closely covered by spine-like leaf bases, leaves widely awl-shaped, spirally densely arranged closely overlapping, deep green, glaucous when young; grayish hanging cones 5 to 9 cm long. The wood of the Giant sequoia is a beautiful red brown, and extremely durable, never invaded by destructive insects; the fissured bark is divided into cinnamon-brown ridges and to 50 cm thick. Young trees may be grown in pots from small seedlings; it is fairly winter-hardy. *Warm temperate.* p. 352, 355

SERENOA *Palmae*
   repens (So. Carolina to Florida Keys), "Saw palmetto"; scrubby, variable palm with creeping stems often underground, and forming wide-spread colonies; heads of palmate leaves 1 m across, deeply cut almost to base into 18-24 widely separated rigid, pendant segments, powdery blue green or bright yellow-green, on thorny petioles; fragrant white flowers, and edible blackish fruit. *Subtropic.* p. 795

SERRURIA *Proteaceae*
   florida (So. Africa; Cape Mts.), "Blushing bride"; slender, evergreen shrub to 1½ m high; distinctive feathery foliage finely divided into smooth, needle-shaped segments; nodding inflorescence 5 cm across, consisting of a showy nest of papery petal-like pointed, creamy-white bracts flushed with pink, surrounding the true flowers which appear in a mass of silky pinkish hair. *Subtropic.* p. 838

SESBANIA *Leguminosae*
   grandiflora (Agati) (Trop. Asia to W. Africa), "Vegetable hummingbird" or "Red wisteria"; soft-wooded tree to 12 m; pinnate leaves to 30 cm long, leaflets in 10 to 30 pairs, to 5 cm long; remarkable sickle-shaped red to white flowers 10 cm long, hanging in two's to four's from leaf axils; followed by flat beans to 50 cm long. *Tropical.* p. 563

SETCREASEA *Commelinaceae*
   purpurea (Mexico), "Purple heart"; so named because of the striking purple color of this plant in strong sun, aided by a pubescence of pale hair covering the lanceshaped leaves, on erect fleshy stems; large 3-petaled orchid flowers. *Subtropic.* p. 306, 307

SETCREASEA: see also Callisia, Tradescantia

SEVERINIA *Rutaceae*
   buxifolia (Taiwan, So. China), "Boxthorn" or "Chinese Box orange"; small tree with dense foliage; small leathery, obovate, parallel-veined leaves 3-4 cm long; small white flowers in terminal clusters, followed by shiny black 1 cm berries; pulp of insipid flavor. *Subtropic.* p. 873

SEYRIGIA *Cucurbitaceae*
   gracilis (Madagascar); twining plant with angular, reddish, fleshy stem; gray-felted, tiny tri-lobed leaves with simple tendrils; the usually missing flowers cream colored; obconical salmon-red fruit. *Tropical.* p. 382
   humbertii (Madagascar); clustering succulent with slender, furrowed reddish stems conspicuously covered with a felt of gray hairs; later clambering or climbing, and with simple tendrils; tiny 3 mm, 3-lobed leaves usually absent; flowers greenish-white; ovoid fruit with red flesh. *Arid-tropical.* p. 380

SIDA *Malvaceae*
   fallax (Pacific Islands), the "Ilima"; a large shrub, sometimes prostrate and creeping; ovate, gray-felted leaves with crenate margins; hibiscus-like 3 cm flowers bright yellow and semi-double; variations to rich orange and dull-red; the yellow "Ilima" is the flower of Oahu and widely used for leis. *Tropical.* p. 623, 624, 629

SIDERASIS *Commelinaceae*
   fuscata (Tradescantia, Pyrrheima) (Brazil), "Brown spiderwort"; clustering rosette of broad and oblong olive green leaves with silvery center band, and covered with brown hair as is the purple reverse; large lavender blue flowers at base. *Subtropic.* p. 306

SIGMATOSTALIX *Orchidaceae*
   hymenantha (Costa Rica, Panama); small epiphyte with clustered compressed pseudobulbs and solitary, thin-leathery leaves; slender, erect inflorescence to 12 cm long, with tiny 5 mm flowers translucent greenish to light brown, lemon-yellow lip and curving column (winter). *Tropical.* p. 760

SILENE *Caryophyllaceae*
   colorata (N.W. Africa, So. Europe, S.W. Asia), "Catchfly"; annual with mealy-white stems, obovate to linear leaves; flowers with bi-lobed pink petals and white crown. *Arid-subtropic.* p. 298

SILYBUM *Compositae*
   marianum (Mediterranean region), "Holy thistle"; thistle like biennial to 1 m high, with shining dark green leaves netted with white, deeply cut, and with spiny lobes; flower heads purple 4-8 cm across. Seen in Delos, Greece as Nicus benedictus. *Subtropic.* p. 325

SINNINGIA *Gesneriaceae*
   cardinalis (Rechsteineria) (C. America), "Cardinal flower"; brilliantly flowered, tuberous plant with round cordate, emerald green, velvety leaves, topped by large curved, tubular, bilabiate flowers, white downy over brightest scarlet, throat marked purple. *Tropical.* p. 501, 502
   cardinalis 'Innocence' (Rechsteineria); an exquisite cultivar photographed in Germany in 1974; compact plant with bright green, pubescent foliage, and a veritable bouquet of flowers glistening white as snow. *Humid-tropical.* p. 502
   concinna (Brazil); miniature tuberous plant, with stems, stalks and veins red, the small hairy leaves roundish-ovate with crenate margins; 2½ cm flowers purple above, yellowish beneath, and spotted inside the inflated tube. *Tropical.* p. 503
   'Doll Baby' (S. pusilla x eumorpha); a pretty miniature rosette to 8 cm across, with thin, dark olive-green crenate leaves with bronze veins; from the heart rise charming small 3 cm slipper flowers lavender-blue with purple eye and pale lemon throat; blooming for a long time, best in summer. *Humid-tropical.* p. 503
   hirsuta (Brazil); tuberous rosette of broad ovate leaves 6-10 cm long, with a network of sunken veins, covered with white hairs, the margins crenate and purplish beneath; from the center a large bouquet of smallish flowers 3 cm across, lilac white, violet inside tube and with violet spots on expanded limb. *Humid-tropical.* p. 507
   leucotricha (Rechsteineria) (Brazil: W. Paraná), a breathtaking new species which I saw when first on exhibition in Sao Paulo in 1954; huge 30 cm tubers sprouting happily without soil in glazed bowls, the glistening silvery foliage suggesting to me the name "Brazilian edelweiss". Found on cliffs near the waterfall 'Salto Apucarazinho' at 1100 m, it is called locally "Rainha do Abismo". The stout, densely matted, white, later brown hair stems to 25 cm high, carry one or two whorls of 3-4 large obovate leaves to 15 cm long, densely covered with shimmering, long silvery-white hair, with margins entire or obscurely crenate; slender tubular inflated 3 cm flowers soft rosy coral, entirely covered outside with silky white hair, the lobes sometimes marked with crimson; blooming spring and summer. *Tropical.* p. 495, 501
   'Longiflora' (Rechsteineria) (cardinalis x eumorpha); attractive hybrid with large fleshy, pubescent leaves, and a pyramid of tubular flowers deep rose at base, the flaring lobes and inside pale pink to white. *Humid-tropical.* p. 501
   macropoda (Rechsteineria) (So. Brazil), "Vermillion helmet flower"; charming tuberous herb with unbranched hairy stem 15-22 cm high, bearing opposite, rather thin, rugose, velvety bright green leaves almost round, 8-12 cm broad; small nodding flowers in clusters, the slender 2½-3 cm tubes vermillion-red with the lower lobes marked brown-red; in March-April. *Humid-tropical.* p. 501
   pusilla (Brazil); miniature rosette only 5 cm high, of little, oval, puckered leaves olive-green with brown veins, hugging the ground, slender stem bearing a 6 mm, attractive tubular flower with five spreading lobes, orchid-colored with darker veins and lemon-yellow throat. *Humid-tropical.* p. 503
   pusilla 'White Sprite'; a darling miniature, seen at the Munich Botanic Garden in Germany; only 3-5 cm high, the tiny tubular flowers with spreading lobes only 12 mm across, and glistening pure white. At high humidity this tropical species is willing to bloom constantly; in the home it will grow successfully in small glass brandy snifter covered with clear plastic. If grown open, the little plant will go into a short dormancy in autumn until January. *Humid-tropical.* p. 504
   regina (Brazil), "Cinderella slippers"; tuberous species related to speciosa, with ovate pointed, bronzy green, red-backed, velvety leaves beautifully patterned with ivory veins; and a profusion of nodding "slipper" type, violet flowers, shorter and more slender than S. speciosa 'Macrophylla'. *Humid-tropical.* p. 503
   'Rex' (x Gloxinera); free-blooming hybrid between Sinningia eumorpha and S. (Rechsteineria) macropoda, resembling a slipper type gloxinia; inflated trumpet flowers scarlet red. *Humid-tropical.* p. 502

speciosa fyfiana, the "Gloxinia" of florists; commercial hybrid strain involving S. crassifolia hort., with erect bell-shaped, regular flowers in white, violet, rose or red, variously variegated or spotted, and with 5-12 lobes. *Humid-tropical.* p. 504

speciosa fyfiana 'Defiance'; large leaf "crassifolia" type, medium-large flowers completely dark cerise-red, with crimson sheen, the margins lightly wavy. *Humid-tropical.* p. 503

speciosa fyfiana 'Double Chicago'; a "Double gloxinia" of the habit of the 7 to 8-petalled frilled 'Switzerland'; produced following meticulous breeding in U.S.; very exciting in having large 9 cm flowers featuring 2 or more rows of frilled petals glowing crimson with crisped white borders, blooming best from spring to summer. *Humid-tropical.* p. 503

speciosa fyfiana 'Emperor Frederick'; a leading commercial hybrid "gloxinia" bred with large upright bell-shaped flowers velvety dark ruby red bordered white, and the typical, over-sized, robust, somewhat brittle, and horizontal leaves of the "crassifolia" class; forming large tubers. *Humid-tropical.* p. 503

speciosa fyfiana 'Emperor William'; popular leading commercial variety of the large leaf "crassifolia" type gloxinia; flowers deep velvety violet-blue with regular white border, early flowering. *Humid-tropical.* p. 503

speciosa fyfiana 'Switzerland', a superb "grandiflora" gloxinia, inheriting its prolific nature from the soft-leaved regina hybrid 'Gierth's Red', and its intensive scarlet with white border from 'Emperor Frederick'; in addition, the large erect flowers have 7-8 lobes which are daintily ruffled. *Humid-tropical.* p. 504

speciosa fyfiana 'Tigrina'; large-flowered "grandiflora" hybrid with regina blood, flexible leaves with light veins, and erect flowers having white throat speckled in the same color as the limb, usually red or blue. *Humid-tropical.* p. 503

'Tom Thumb' (speciosa x pusilla); the most charming little Baby gloxinia that ever appeared for the pleasure of the gesneriad fancier; in every detail resembling its big brother the florists (fyfiana) gloxinia, except for its diminutive size no larger than an African violet. First of its kind to flower in 6-8 cm pots, and a perfect window sill plant with little bell-shaped, velvety red blossom edged in white, 4 cm across. The plant is only 10 cm high, and its small ovate, crenate leaves 5 cm long, deep green above, silvery-haired beneath; the bell flowers are erect and not oblique and slipper-like as in speciosa. This hybrid was developed by Fischer's in New Jersey. *Humid-tropical.* p. 503

SINOARUNDINARIA *Gramineae*
nitida (Arundinaria) (Korea, China); running bamboo with slender canes 3 to 6 m high, forming robust clumps, the stems purplish-black and very hollow, only 1 cm thick, and leafless the first year; branchlets dense with leaves bright green above, glaucous beneath, to 18 cm long, the margins bristly on one side. Photographed in Bot. Garden Copenhagen. *Warm temperate.* p. 518

SIPHOCAMPYLUS *Campanulaceae*
manettiaeflorus (Cuba); small shrubby plant to 30 cm high with netted elliptic leaves and slender tubular flowers 2½ cm long, scarlet with pointed yellow tips. *Tropical.* p. 288

SKIMMIA *Rutaceae*
japonica 'Dwarf Female'; evergreen, broad compact shrub, dioecious, with thick-leathery, elliptic, glossy green leaves, clustered at end of branchlets, and tipped by erect clusters of small-creamy-white, fragrant flowers, followed by coral-red berries, which last for months, during winter and spring. Originally from Japan, this dwarf type is a selected seedling resembling reevesiana but not as tall and leggy; developed by Teufel-Oregon. *Warm temperate.* p. 873

SMILAX: see Asparagus

SMITHIANTHA *Gesneriaceae*
cinnabarina (Naegelia) (Mexico, Guatemala); called "Temple bells" because of their bright display of nodding bells, scarlet red outside, with creamy belly and red spotted throat, from a red central stalk, above beautiful red plush, cordate leaves; tall growing, from scaly rhizomes. *Humid-tropical.* p. 506

'Exoniensis'; a prolific growing and free blooming hybrid with leaves plush red on green, and masses of orange-yellow nodding "slipper" flowers. *Humid-tropical.* p. 506

'Golden King'; a cultivar with deep green velvety leaves mottled and painted purple along the veins; the saccate yellow flowers deep golden in the throat and dotted red inside. *Humid-tropical.* p. 506

x hybrida 'Compacta', "Temple bells"; low bushy zebrina hybrid, with large deep green leaves covered with purplish velvety hair; freeblooming with nodding, foxglove-like bells lemon-yellow, red on top, and veined with red. *Humid-tropical.* p. 506

'Orange King', "Orange temple bells"; hybrid with beautiful large velvety, emerald-green leaves with crenate edge, overlaid with a pattern of red along veins; the flower bells orange-red, yellow beneath, inside yellow with red spots. *Humid-tropical.* p. 506

zebrina (Naegelia) (Vera Cruz, Mexico); tall plant to 1 m and willing bloomer, with dark green, flexible leaves with red-brown markings and covered with silky hairs; numerous flowers on slender stalks with bells brilliant scarlet above, yellow below, throat pale yellow spotted with red. *Humid-tropical.* p. 506

'Zebrina discolor', "Temple bells"; known in horticulture as S. x hybrida; a cultivar with more contrastingly colored purplish-red and green foliage; flowers scarlet with yellow, brown-spotted inside. These commercial hybrids combine the beauty and colors of S. zebrina and multiflora, and are easy to grow and willing to bloom. *Humid-tropical.* p. 505

SOBRALIA *Orchidaceae*
decora (galleottiana) (Mexico, Guatemala), "Reed orchid"; vigorous terrestrial with slender, reedlike stems to 60 cm high, clothed with scattered, plaited leaves, and bearing large, cattleya-like flowers, of short duration, sepals and petals creamy-white with light rose blush, lip purplish rose, (April-July). *Tropical.* p. 760

leucoxantha (Costa Rica); reedy terrestrial with leafy stems 1 m high, and large solitary flowers with white sepals and petals, and broad golden yellow crisped lip flushed orange in throat, and bordered white (April-Sept.) *Tropical.* p. 759

macrantha (Mexico to Costa Rica); reed-like terrestrial with stems 1½-2 m high, furnished with plaited, slender pointed leaves and large fragrant 15 cm flowers rich purple and crimson, the broad tip with cream or yellow throat (summer). *Tropical.* p. 759

xantholeuca (Central America); "Reed orchid"; leafy stems to 60 cm high, lanceolate sessile leaves and solitary flowers lemon yellow with golden throat streaked with darker yellow; sepals and petals 10 cm long; lip with crisped margins. *Tropical.* p. 760

SOEHRENSIA *Cactaceae*
formosa (Lobivia) (Argentina: Mendoza); elongate globe to 50 cm or more high, densely covered with bristly hairs; small orange-yellow flowers. to 8 cm long. *Arid-subtropic.* p. 261

ingens (No. Argentina); handsome globe with age 50 cm or more dia., resembling Echinocactus grusonii but with many more ribs; dark green with numerous ridges and covered with long, fine straw-colored spines on white woolly areoles; trumpet flowers 4-5 cm across, yellow or orange-red. *Arid-subtropic.* p. 262

oreopepon (No. Argentina: Mendoza); broadly globular, to 30 cm dia, grayish olive green, with 18 to 30 ribs lightly notched; soft, thin, flexible spines yellowish to reddish; gray-haired funnel flowers 8-10 cm long, golden yellow. *Arid-subtropic.* p. 262

SOLANDRA *Solanaceae*
longiflora (Jamaica), "Trumpet plant"; evergreen shrub to 1½ m high, woody branches with small hard, oval or obovate leaves on purple petioles, and large, showy, stiff upright, trumpet-like flowers 22-30 cm long, greenish-white, showing purplish-brown venation, contracted at the throat, the limb turned back and frilled. *Tropical.* p. 894

nitida (guttata) (Mexico), "Chalice vine" or "Cup-of-gold"; clambering to 6 m or more, with leathery, elliptic, glossy leaves; the large 25 cm long, chalice-shaped flowers with corolla-lobes reflexed and frilled, yellow with purplish stripes. *Tropical.* p. 894, 900

SOLANUM *Solanaceae*
aviculare (New Zealand, Australia), "Kangaroo-apple"; shrub 1½ to 3 m high, with variable dark green membranous leaves lanceolate or pinnately lobed, 15-30 cm or more long; few violet flowers 2-3 cm across; the rather large 2½ cm ovoid edible berry green or yellow. *Subtropic.* p. 899

capsicastrum 'Variegatum', "Variegated false Jerusalem-cherry"; colorful form with foliage variegated or edged with creamy-white. The type species is from So. Brazil and Uruguay, a small evergreen subshrub 25-60 cm high, densely branched, branches, young shoots and leaves grayish and soft-hairy, the elliptic undulate leaves to 8 cm long, in unequal pairs; white flowers, and orange-scarlet ovoid fruit 1 cm or less in dia., smaller than in pseudo-capsicum. *Subtropic.* p. 894

giganteum (Colombia, Ecuador), "African holly"; shrub 1-3 m high, with stout prickles; large ovate leaves 10-20 cm long, white-hairy, mainly beneath; small clusters of flowers 2 cm across, pale violet to blue; showy 1 cm fruit bright red. *Subtropic.* p. 898

hispidum (warzcewiczii) (Mexico), "Devil's fig"; evergreen rusty-hairy shrub to 2 m, with short stout spines; leaves to 30 cm long or more and deeply lobed; clusters of 3-4 cm white flowers; small glossy yellow fruit. *Subtropic.* p. 898

jasminoides (Brazil), "Potato-vine"; shrubby deciduous twiner with twiggy stems, ovate leaves to 8 cm long, sometimes 2 to 5 parted; star-shaped flowers white tinged blue, 2 cm across, in branching clusters. *Tropical.* p. 899

SOLANUM

jasminoides 'Grandiflorum'; bluish-white flowers to 3 cm across, in larger clusters. *Subtropic.* p. 899

macranthum (Brazil), "Brazilian potato-tree"; spiny pubescent shrub or tree to 10 m high, with ovate leaves 30 cm long, sinuately lobed, prickly on veins; flowers corn flower-blue, 3-6 cm across, in racemes paling with age. *Tropical.* p. 899

mammosum (C. America), "Nipple-fruit"; a thorny, sparry shrub with few pubescent leaves and weirdly formed, orange-colored, waxy fruit, large like a tomato, but shaped somewhat like an inverted pear with nipples near base. *Tropical.* p. 891

marginatum (North Africa: Ethiopia); spiny white-hairy shrub to 1 m high, with broad ovate, sinuately lobed leaves prickly on both sides, with white margins, and densely white tomentose beneath, to 20 cm long; flowers white lined with blue, 2½ cm across; spiny, yellow fruit. *Subtropic.* p. 900

martii (Brazil: Panaceix); handsome small tree with woody trunk, and large, leathery leaves 30 cm long, glossy green with beige colored ribs; between the foliage the white star-flowers hidden in woolly cushions from the axil of the leaf. Photographed at Jardim Botanico, Rio de Janeiro. *Tropical.* p. 898

melongena esculentum (India), "Egg plant" or "Jew's apple"; pubescent prickly herb or woody shrub, to 1 m high; ovate, angled or lobed leaves 8-22 cm long, with starry wool beneath; the purplish flowers 3 cm across; the edible fruit blackish-purple, 15 cm long. The common egg-plant usually grown as an annual is the var. esculentum, with shining fruit to 30 cm long, purple, white, yellowish or striped, and has fewer prickles. *Tropical.* p. 894

melongena 'Moorea', "Ornamental eggplant"; woody bush with large lobed, silvery-hairy leaves; flowers pale lavender; and showy, large oblong fruit orange-yellow 15 cm long. Photographed in a native garden on Moorea in French Polynesia. *Tropical.* p. 897

pseudo-capsicum (Madeira), "Jerusalem cherry"; robust, shrubby plant with flexible branches dense with lanceolate, turgid-firm, deep green leaves wavy at the margins and smooth beneath, 5-6 cm long, glabrous-smooth, though with velvety feel above; branches smooth or with slight fuzz; small, white, star-like flowers followed by large globular, lustrous orange-scarlet, cherry-like fruit of 1½-2 cm dia.; much grown as a potplant for Christmas; there is also a yellow-fruited strain. *Subtropic.* p. 900

quitoense (Perú), "Naranjilla"; ornamental woody bush with brittle branches, large pubescent, ovate leaves prettily lobed and with ivory ribs; purple flowers followed by large tomentose fruit, 4 cm dia. *Subtropic.* p. 899

rantonnetii (Paraguay, Argentina), "Blue potato tree"; rambling evergreen or deciduous shrub to 2 m high, unarmed and nearly smooth; ovate or oval, bright green undulate leaves to 10 cm long; charming 3 cm flowers dark blue or violet with yellow eye, in clusters blooming throughout the warm summer and fall, sometimes nearly all year; the red fruit like small apples. Blooms as small plant but may be grown into treeform by staking, or trained on support as a vine. *Subtropic.* p. 899, 900

seaforthianum (So. America), "Star potato vine"; shrubby climber to 3 m with mostly pinnate leaves to 20 cm long, the leaflets unequal; large star-shaped lavender or purple 2½ cm flowers with yellow anthers; scarlet fruit. *Tropical.* p. 899

uporo (So. China); an ornamental type of tomato, seen in Kwangtung Prov., China; woody shrub with ovate, hairy leaves and bearing showy globular, soft-fleshy fruit 6 cm dia., ridged at base and colored brilliant scarlet. *Subtropic.* p. 897

wendlandii (Costa Rica), "Potato-vine"; shrubby climber, to 6 m, with a few scattered, hooked prickles; variable leaves bright green, the upper ovate or tri-lobed, the lower pinnate or lobed; large lilac-blue flowers 6 cm across, in branched clusters. *Tropical.* p. 899

SOLIDAGO   *Compositae*
nemoralis (Canada to Arizona), "Golden-rod"; clump-forming perennial to 1 m high, densely pubescent, narrow leaves; small yellow flowers in one-sided branches. *Temperate.* p. 329

SOLISIA   *Cactaceae*
pectinata (Mexico: Tehuacán), "Lace bugs"; very small globe 3 cm thick, with dense warts almost entirely covered with white spines arranged comb-like; flowers around the sides yellow; juice milky. *Arid-tropical.* p. 274

SONERILA   *Melastomataceae*
margaritacea (Java), "Pearly sonerila"; small tropical herbaceous plant with pubescent red stem and ovate, deep copper-green leaves with bristly surface, puckered with pearly spots and bristles that glisten in the sunlight like frosted silver; glowing red-purple beneath; flowers rosy lavender. *Humid-tropical.* p. 646

margaritacea 'Argentea'; attractive variety having the pointed leaves strongly overlaid with silver while the sunken veins are olive-green and often barely showing. *Humid-tropical.* p. 644

SOPHORA   *Leguminosae*
secundiflora (Texas to Mexico), "Mescal-bean"; evergreen tree of chapparal deserts, to 10 m high; pinnate leaves 15 cm long; leaflets notched at apex; flowers 2½ cm long, violet-blue and violet-scented. *Subtropic.* p. 560

SOPHRONITIS   *Orchidaceae*
coccinea (grandiflora) (So. Brazil); dwarf epiphyte with small pseudobulb and a stiff, 8 cm leaf; large solitary flowers to 8 cm across, showy scarlet with salmon sheen, the throat yellow with red stripes. I will never forget these brilliant spots of red in the chilly rain-forest of the Serra do Mar, (Sept.-Feb.) *Tropical.* p. 753, 759

SORBARIA   *Rosaceae*
aitchisonii (Afghanistan, Kashmir), "False spiraea"; deciduous flowering shrub to 3 m; tall stems with few branches, large pinnate leaves of bright green, toothed leaflets to 8 cm long; small white flowers in feathery clusters 25 cm long. *Warm temperate.* p. 854

SORBUS   *Rosaceae*
americana (No. America), "Missey-moosey"; smooth deciduous shrub or tree to 10 m high, with pinnate leaves to 25 cm long, 13-15 narrow leaflets, with sharp marginal teeth, grayish beneath; small white flowers in terminal clusters, followed by bright red 1 cm berries, loved by birds. *Temperate.* p. 855

aucuparia (Europe, Asia Minor), "Mountain ash"; deciduous tree to 20 m, with narrow crown; gray-brown twigs, pinnate leaves 10-25 cm long, glaucous beneath; small white flowers in dense flat clusters, followed by orange-red 1 cm berries liked by birds; also used to make brandy. *Temperate.* p. 854

intermedia (No. Europe), "Swedish white-beam"; deciduous tree to 10 m; deeply lobed and toothed leaves 10 cm long, gray-hairy underneath; spring-flowering, small white flowers in large clusters 10 cm across, followed by bright red berries. *Temperate.* p. 854

SPARAXIS   *Iridaceae*
tricolor 'Firebrand' (So. Africa), "Scarlet wandflower" or "Velvet-flower"; dainty cormous plant with linear leaves to 30 cm long; spectacular, large brilliant scarlet flowers 5 cm across, with yellow ring surrounding purple center. *Subtropic.* p. 528

SPARMANNIA   *Tiliaceae*
africana (So. Africa), "African hemp"; much cultivated in German homes, and known there as "Zimmer-Linde", as it resembles a miniature linden tree, with large lobed, light green, soft leaves, white hairy on both sides and on the sparry stems; flowers with white petals and yellow filaments; rapidly growing into a tree-like shrub with many trunks, to 6 m high. *Subtropic.* p. 907, 909

SPARTIUM   *Leguminosae*
junceum (Genista) (Mediterranean, Canary Isl.), "Spanish broom"; ornamental shrub, to 3 m, with almost leafless, reedlike branches, the small leaves linear; fragrant, showy yellow, butterfly-like flowers in loose, terminal racemes. *Subtropic.* p. 556

SPATHICARPA   *Araceae*
sagittifolia (Bahia), "Fruit-sheath plant"; interesting herb with tuberous rhizome having small, arrowshaped, membranous leaves waxy green; inflorescence on stiff stalks with recurved green spathe, and spadix attached along its center. *Tropical.* p. 128

SPATHIPHYLLUM   *Araceae*
blandum (gardneri) (W. Indies, Jamaica, Surinam); robust plant with large, lanceolate, leathery, deep green leaves and sunken veins; spathe spoonshaped and pointed, pale green; spadix white with elevated knobs. *Tropical.* p. 126

cannaefolium (dechardii) (Venezuela, Guayana); leathery plant with thick dull, black-green, corrugated leaves tapering toward base of ribbed petioles; thick, fleshy spathe green outside, white inside; long, free, cream spadix. *Tropical.* p. 126, 131

'Clevelandii' (kochii), "White flag", also known as "Peacelily"; freely branching and free flowering commercial plant close to wallisii but larger in all parts; thin-leathery, glossy-green, lanceolate leaves with undulate margin; the inflorescence on reed-like stems with ovate-pointed, white, papery spathe 10-15 cm long, turning apple green with age, and having a green line on back; maze-like spadix white. *Tropical.* p. 126, 128

commutatum (Indonesia, Philippines); elegant, strong growing plant with broad, corrugated, fresh green, elliptic leaves on smooth, vaginate petioles; the short, thick spadix is white, becoming green, subtended by a broad, leathery, cream-white spathe with green edge, and turning green beneath with age. *Tropical.* p. 126

floribundum (multiflorum) (Colombia), "Spathe flower"; dwarf, compact plant, freely suckering and forming clusters; with matte, satiny green, leathery leaves obovate or elliptic, with pale center band, on broadly winged petioles; small spathe white; short spadix green and white. *Tropical.* p. 126

'Marion Wagner' (cochlearispathum x wallisii); large, decorative hybrid by Wagner-Hollywood 1946, with satiny-glossy quilted leaves blooming intermittently April-July with large spathe to 20 cm long lasting 6-8 weeks, greenish-white, shading to pale chartreuse; spadix attached. Sweetly fragrant. *Tropical.* p. 126

'Mauna Loa'; a diploid hybrid developed by Griffith-Los Angeles, and so far as I could determine in discussing its origin with him, a seedling of S. floribunda x a Hawaiian hybrid, probably S. 'McCoy'; robust plant of compact habit, tending to divide from the base; leaves dark glossy green; very floriferous over an extended period, with pure white spathes 10-12 cm and even 20 cm long in older plants in large pots, somewhat cupping and of soft-leathery texture, slightly scented. We have found the strain variable, leaning toward 'Clevelandii'. *Tropical.* p. 126

'McCoy' (Takahashi); hybrid of probably cochlearispathum x 'Clevelandii'; Mrs. Lester McCoy related to me at her gardens near Honolulu how her late gardener, Mr. Takahashi in 1942, crossed a local 'Clevelandii' plant with a Guatemalan species probably cochlearispathum, which her husband brought back, resulting in this vigorous, large-growing very showy plant to 1½ m high, with large and long, glossy green leaves; the spathes white or creamy changing to light green with age, 20-30 cm long, of good texture; the slender club-like white spadix attached 2 cm up from base of spathe, and having pointed knobs. *Tropical.* p. 126, 128

ortgiesii (Mexico: Vera Cruz); evergreen perennial herbaceous, clustering plant with fibrous roots and thin-leathery, bright green oblanceolate leaves 35-60 cm long, 13-26 cm wide, on long petioles; the blade closely furrowed in feather fashion, and sharply bent at base; inflorescence with broad-elliptic pure white spathe 12-30 cm long, and 5-10 cm spadix with perfect flowers. *Tropical.* p. 128

phryniifolium (Costa Rica, Panama), "Peace plant"; found in the trade as S. friedrichsthalii, a robust plant with large fresh-green, corrugated, fleshy leaves having long-cuneate base; the inflorescence with broad, pale yellow-green, papery spathe, and enclosed, white, knobby spadix. *Tropical.* p. 128

wallisii (Colombia, Venezuela), vigorous plant with glossy green, thin-leathery, oblong-lanceolate leaves; inflorescence on reedlike stems, the ovate spathe white turning green with age, maze-like white spadix; very close to 'Clevelandii' (kochii) but smaller in all parts. *Tropical.* p. 126

SPATHODEA *Bignoniaceae*
campanulata (nilotica) (Trop. Africa: W. Kenya, Uganda), "African tuliptree", or "Flame of the forest"; spectacularly showy evergreen tree to 20 m high, with odd-pinnate foliage, the leaflets 6-12 cm long; large bell-shaped flowers 8-12 cm long, swollen on one side and 5-lobed, crimson red with yellow frilled edge, carried facing upwards in terminal clusters. *Tropical.* p. 187, 188

SPATHOGLOTTIS *Orchidaceae*
papuana (Papua, New Guinea); terrestrial orchid with long, ribbed, plicate leaves; inflorescence tall and slender, bearing a succession of small, handsome carmine-rose flowers over many months. *Tropical.* p. 754

plicata (Malaysia); terrestrial with small corm-like pseudobulbs and grass-green plaited leaves, to 1 m tall; erect racemes with 2½-5 cm flowers of rosy-purple, (April-June). *Tropical.* p. 753, 754

plicata var. pallida; lovely variety with flowers light pink, with red mark in center. *Tropical.* p. 754, 758

vieillardii (New Caledonia); terrestrial similar to plicata, but larger in all parts, the flowers to nearly 5 cm across, light pinkish purple, the lip callus yellow. *Tropical.* p. 754

SPIGELIA *Loganiaceae*
splendens (Mexico, Guatemala), "Mexican pinkroot"; fleshy-rooted, showy perennial 45 cm high, with slightly hairy obovate leaves 10-12 cm long; bright scarlet tubular flowers 2½ cm long, tipped with white, on one-sided recurved spikes. *Tropical.* p. 616

SPIRAEA *Rosaceae*
x vanhouttei (cantoniensis x trilobata), "Bridal wreath"; beautiful deciduous, hardy shrub to 2 m high, with numerous, gracefully arching stems, coarsely toothed leaves bluish beneath, and tiny, white flowers in masses of dense, lacy clusters forming a veritable blanket of snow in May-June. *Temperate.* p. 854

SPIRAEA: see also Astilbe

SPIRANTHES *Orchidaceae*
nudicaulis, "Ladies-tresses"; terrestrial rosette of oblanceolate leaves, from the center several erect stalks carry a slender inflorescence of numerous small pinkish flowers. *Subtropic.* p. 754

SPIRONEMA: see Callisia

SPONDIAS *Anacardiaceae*
dulcis (cytherea) (Society Islands: Tahiti), "Otaheite apple"; smooth gray-barked tropical tree to 20 m high, with bright green pinnate foliage 20-80 cm long, clustered at branch ends; tiny greenish-white flowers in large clusters; the large ovoid-fleshy fruit with tough orange skin, 5 to 8 cm long, and though unpleasant smelling, has apple-flavored yellow pulp surrounding the fibrous core with a spiny seed; fruiting October to January. Eaten fresh, or in preserves. *Tropical.* p. 71, 455

venulosa (mombin), (Trop. America), the "Mombin", "Hog plum", "Spanish plum", or "Caja Mirim", a most important and valued fruit tree in tropical America; a tree to 8 m high, with furrowed bark, and long pinnate leaves 20 cm or more long; small 2 cm purplish-maroon flower clusters, followed by globular or ovoid light brown or purple fruit with soft, yellow flesh, 3 to 4 cm diameter, and when eaten fresh has an acid spicy flavor resembling cashew but less aromatic. Fruit is also boiled or dried. Photographed in Rio de Janeiro. *Tropical.* p. 71, 455

SPREKELIA *Amaryllidaceae*
formosissima (Mexico), "Jacobean lily"; beautiful bulbous herb with linear leaves; solitary, showy crimson flower, with a spathe-like bract, appearing in June, before the foliage. *Subtropic.* p. 64

STACHYS *Labiatae*
grandiflora (Caucasus); rosette-forming perennial with ovate, hairy leaves 6 cm long, crenate; spike-like inflorescence with orchid-purple flowers. *Warm temperate.* p. 531

STACHYTARPHETA *Verbenaceae*
speciosa (Trop. America); in Botanic Garden Rio de Janeiro; shrubby herbaceous plant with ovate, wrinkled, rugose leaves crenate at margins; flowers carmine-rose along a spike, 1-2 cm dia., opening from below a few at a time. *Tropical.* p. 916

STACHYURUS *Stachyuraceae*
praecox (Japan), "Spike-tail"; deciduous shrub 2-4 m high, with reddish shoots and willow-like branches; lanceolate, toothed leaves 8 to 15 cm long; small 1 cm bell-shaped flowers pale yellow, borne charmingly in stiffly pendant racemes 10 cm long, like a chain of pearls, in spring before the leaves. *Warm temperate.* p. 906

STANGERIA *Cycadaceae*
eriopus (So. Africa); fern-like perennial cycad with underground stem, each bearing 1-4 pinnate leaves to 2 m long; cones silvery-pubescent. *Subtropic.* p. 386

STANHOPEA *Orchidaceae*
graveolens straminea (Brazil), "Horned orchid"; curious epiphyte with ovoid, clustered pseudobulbs bearing a solitary leathery, plaited leaf; the pendulous inflorescence oddly appears through the bottom of the orchid basket; exquisite, waxy flowers are very fragrant, pure white; the fleshy lip with two lateral horns. *Humid-tropical.* p. 759

hasseloviana (Amazonian Perú); "Horned orchid"; fascinating epiphyte with clustered conical pseudobulbs topped by solitary plaited leaf; the pendulous waxy, heavy-textured flowers very fragrant, to more than 12 cm long, the reflexed sepals and petals whitish or pale rose spotted with purple (summer). *Tropical.* p. 755

tigrina (Mexico), "El toro"; from apprentice days in my father's greenhouses have I remembered this fantastic epiphyte as my most impressive orchid; the memory of its large, 15 cm waxy flowers with sepals and petals deep blood-red marked with yellow, an orange-yellow lip blotched with maroon and its ivory horns, together with an overpowering fragrance of vanilla, haunts me still, (May-July). Valid name may be S. hernandezii. *Tropical.* p. 759

STAPELIA *Asclepiadaceae*
ambigua (So. Africa: Cape Prov.); clustering succulent with 4-angled soft pubescent olive green stems, toothed along angles; large fleshy star-flowers with deep cup, 12-15 cm across; ribbed pinkish or flesh-colored base and with red cross-ridges; soft purplish hair in center and along margins. *Arid-subtropic.* p. 155

bella, possibly hyb. of revoluta or nobilis; a succulent with erect, branching stems to 20 cm high, light green tinted red, 4-angled with short teeth; 5 cm flowers deep purplish-red, yellow center, and violet marginal hairs. *Arid-subtropic.* p. 155

gigantea (red form) (Zululand to Zimbabwe); ribbed, fat stems pale green and velvety, with gigantic flowers to 20 cm or more across, pale yellow with transverse crimson lines, variable in color, with forms more reddish, and covered with purplish or crimson hairs; with offensive odor. *Arid-tropical.* p. 154

gigantea pallida (Trop. Africa: Malawi), "Giant toad plant"; clustering succulent with fat, angled stems 10-20 cm high, silky olive green and toothed along angles; big star-flowers somewhat cupped and with long slender petal tips, lemon-yellow covered with greenish yellow ridges and purple pubescent. *Arid-tropical.* p. 155

hirsuta (Cape Province), "Hairy star-fish flower"; clustering fingers sooty-green, with 10 cm flowers purple-brown and transverse lines of cream or purple, ciliate margins. *Arid-subtropic.* p. 155

nobilis (Transvaal); branching, tufted, light green stems; flowers star-like, reflexed, reddish purple on back, yellow on face with crimson lines, covered with purple hairs; flowers 15 cm or more across, and with a deep depression in center. *Arid-subtropic.* p. 155

semota lutea (Tanzania); small clustering succulent to 7 cm tall, 4-angled stems grayish green, with long tubercles tapering to sharp tip; charming little star-flowers 4 cm dia, bright yellow with blood-red center star, and with ciliate margins. *Arid-tropical.* p. 155

variegata (Cape Province), the "Star flower"; very showy, fleshy flowers 5-8 cm across, greenish-yellow with purple-brown spots on petals; branching green stems clustering and fingerlike. Unfortunately the flowers, like those of other Stapelias, exude a carrion smell to attract flies which, in the hope of finding ripe meat, succeed only in transferring pollen from one flower to another. *Arid-subtropic.* p. 155

STATICE: see Limonium

STEMMADENIA *Apocynaceae*
galeottiana (Cent. America), "Lecheso"; tropical shrub to 3 m high, with opposite soft-leathery, glossy green long-elliptic leaves to 12 or 15 cm long, and showy salver form cream-white flowers with 5 broad petals, yellow inside tube, and sweetly fragrant; small 1 cm orange fruit in pairs. *Tropical.* p. 82, 83

STENANDRIUM *Acanthaceae*
lindenii (Perú); low tropical herb with leaves papery-smooth, broad elliptic, metallic coppery green with beautifully contrasting yellow-green vein area, purplish beneath; flowers yellow, in 8 cm spikes. *Tropical.* p. 38

STENOCACTUS *Cactaceae*
coptonogonus (Mexico), "Permanent-wave cactus"; globose, to 10 cm thick, with 10-14 knife-edged high ribs, dark green, set with occasional curved spines; flower petals purplish, bordered white. *Arid-tropical.* p. 272

STENOCARPUS *Proteaceae*
sinuatus (Queensland, New South Wales), "Wheel of fire"; evergreen tree 10-30 m high with leaves either oblong-lanceolate and unlobed or pinnately cut into 1 to 4 pairs of oblong lobes, 30-45 cm long, leathery, glossy light green with pale midrib and lighter beneath; blooming in summer or fall; 8 to 10 cm flowers brilliant scarlet with yellow stamens, resembling a pinwheel; orange-red when young; the inflorescence explodes like fireworks into fiery-red when mature. *Subtropic.* p. 837, 840

STENOCHLAENA *Polypodiaceae (Filices)*
tenuifolia (palustris of hort.) (India, China, Australia), "Liane fern"; tropical epiphytic fern climbing by its slender woody rhizome, which is covered with occasional brown scales, and bearing leathery, pinnate fronds to 1 m long, shining green, finely serrate at margins, coppery when young; as a pot plant likes to climb on tree-fern slabs. *Humid-subtropic.*

STENOGLOTTIS *Orchidaceae*
longifolia (Natal); African terrestrial with tuberous roots producing a tuft of soft, brown-purple, spotted leaves from the center of which springs an erect spike to 60 cm long, with small, pretty, star-shaped flowers, lavender-pink and spotted purple, (Aug.-Oct.). *Subtropic.* p. 754

STENOLOBIUM: see Tecoma

STENOMESSON *Amaryllidaceae*
variegatum (Ecuador, Perú, Chile); tender bulbous plant, with inflorescence arising before the leaves; floral stalk 40-50 cm tall, with tubular-funnel-shaped flowers red, the outer segments with green markings. *Tropical.* p. 64

STENOSPERMATION *Araceae*
sessile (Costa Rica); tropical ornamental with erect or scandent stem rooting at internodes; the oblong leathery, grass green leaves in ranks; white, ovate spathe deciduous, and white, cylindric spadix. *Tropical.* p. 120

STENOTAPHRUM *Gramineae*
secundatum (So. Carolina to Florida and Texas), "St. Augustine grass"; creeping perennial subtropical grass with branching, compressed stems coarse-textured, with wide blades of dark green, less than 15 cm long; makes a serviceable lawn, fairly pest-free and salt-tolerant; cut 2 to 4 cm high. *Subtropic.* p. 510

secundatum 'Variegatum' (So. U.S., Trop. America); creeping stoloniferous grass with flattened stems and firm linear leaves prettily banded creamy white, the tips round. *Subtropic.*

STEPHANOTIS *Asclepiadaceae*
floribunda (Madagascar), "Madagascar jasmine" or "Wax flower", evergreen wiry climber with milky juice, twining to 5 m high, with opposite elliptic, thick-leathery glossy dark green leaves to 10 cm long, producing axillary clusters of very beautiful, exquisite waxy, white tubular flowers 5 cm wide, and intensely fragrant. A favorite for pots on wire or trellis, for the winter garden, or light, warm window; keep cooler and drier in winter. Propagate by cuttings. *Tropical.* p. 145, 153

STERCULIA *Sterculiaceae*
ceramica (Celebes), "Fairchild's sterculia"; interesting large tropical tree with leathery, cordate leaves showing net of palmate yellow veins; inflorescence without petals, but with colored calyx, developing into the peculiar rusty-red fruit, splitting open to display shiny black seed inside. *Tropical.* p. 902

foetida (E. Africa to No. Australia), "Indian almond"; tall, noble tree to 20 m high, with gray bark faintly ridged; reddish branches usually horizontal, crowded at end with large, digitate leaves 10-30 cm long and divided into 5-7 leaflets; appearing usually on bare branches, before the new leaves, are the small calyx flowers of crimson and yellow 2 cm long, with offensive odor. *Tropical.* p. 903

STEREOSPERMUM *Bignoniaceae*
chelonoides (India, So. China, Malaya), "Yellow snake tree"; semi-evergreen tree to 18 m, with pinkish-gray bark; the leaves 40 cm long, divided feather-fashion into 7-13 elliptic leaflets which are paired, with one at tip; fragrant flowers ochre-buff, tinged and lined with red, 3 cm long; snake-like, curved thin fruiting capsules extend to 75 cm. *Tropical.* p. 188

STEWARTIA *Theaceae*
sinensis (China); evergreen or deciduous small tree to 10 m, with obovate, toothed, leathery leaves 10 cm long; showy white, solitary flowers 5 cm across, with many stamens, and fragrant; young growth fine-hairy. *Warm-temperate.* p. 906

STICTOCARDIA *Convolvulaceae*
beraviensis (Tropical Africa); woody twiner with ovate leaves 15-20 cm long, having prominent depressed veins; flowers bright crimson, yellow in throat. *Tropical.* p. 359

tiliifolia (Trop. E. Africa: Kenya); beautiful twining perennial herb with broad-cordate, quilted leaves bright waxy green; from the axils the striking large bell-shaped flowers 8 cm across, brilliant crimson red with yellow throat, and a design of flame-like scarlet red lines from the center, the limb recurved; apple-like fruit. Photographed in Bombay, India. *Tropical.* p. 359

STIGMAPHYLLON *Malpighiaceae*
ciliatum (Brazil), "Golden vine"; slender twiner with thread-like stems, and glabrous, heartshaped, thin-leathery leaves, oblique at base and with ciliate margins; large 4 cm golden yellow flowers in clusters of 3-7. *Tropical.* *

STRELITZIA *Musaceae*
alba (augusta) (Natal), "Great white strelitzia"; palm-like trees of which I have seen extensive groves inland west of Durban, to 10 m high, with woody trunk bearing shining green, leathery leaves 1-1½ m long and arranged fan-like, frequently cut into ribbons by the wind; the curious, large inflorescence on short stalks between the foliage; from a rigid, boat-shaped pointed purplish bract or spathe, rises a row of white sepals and petals. *Subtropic.* p. 665, 669

caudata (No. Transvaal, Swaziland); growing tree-like, from 2-5 m high, with great, banana-like leathery wind-torn leaves set fan-like in opposite ranks; large axillary inflorescence with pink boat-shaped spathe nesting the all-white flowers, similar to S. alba. *Subtropic.* p. 670

nicolai (So. Africa); trunk forming, clustering tree to 4 m high, with banana-like, shining green, leathery leaves having an obtuse base, arranged in 2 ranks; inflorescence with boatshaped, reddish bracts cradling white sepals and light blue petals united tongue-like. *Subtropic.* p. 664

parvifolia (South Africa: E. Cape), the "Small-leaved bird-of-paradise"; about 1 m high, very similar to S. reginae, trunkless, but this species may be recognized by its leaves which are reduced to very small, spoon-shaped, thin blades at the tips of tall stiff, reed-like stalks; the flowers with yellow sepals, and the united petals or tongue vivid electric blue, borne in clasping, horizontal green bracts edged with red. *Subtropic.* p. 672

parvifolia juncea (South Africa: E. Cape), the "Rush-like strelitzia", from the Port Elizabeth area; very curious form which I was amazed to find because it has no leaves at all; just a dense cluster 1-1½ m high, of spiky tufts of cylindrical, fleshy but rigid, reed-like grayish stems tapering to a needle-point; inflorescence with bright orange sepals and blue tongue. *Subtropic.* p. 670

reginae (Transkei, South Africa), "Bird-of-paradise"; trunkless, compact, clustering but slow-growing plant to 1½ m high, with stiff-leathery, concave, oblong, bluish-gray leaves with pale or red midrib; strikingly exotic, long-stemmed flowers emerging from the green boatshaped bracts bordered in red, the numerous pointed petals brilliant orange contrasting with an arrow-shaped tongue of vivid blue. *Subtropic.* p. 664, 665, 669, 670

STREPTOCALYX *Bromeliaceae*
fuerstenbergii (No. Brazil); epiphytic dense rosette with recurving grayish green leaves to 75 cm long and 8 cm wide, spiny at margins; bold cylindric inflorescence 45 cm long with boat-shaped shingled bracts sheltering the blue flowers. *Tropical.* p. 221, 222
holmesii (Amazonian Perú); wonderful discovery, and most spectacular with its inflorescence measuring to 60 cm, lush with tiers of cherry-red 15 cm bracts, and crowned by panicles of flowers with fleshy orange calyx and white petals, over rich green leaves. *Tropical.* p. 222
poeppigii (Amazonas to Bolivia); large rosette with stiff, strongly armed matte-green leaves with gray pencil-lines beneath; large flower spike with scarlet bracts and violet flowers; growing as epiphyte. *Tropical.* p. 223
poitaei (Amazonian Perú); dense rosette of linear leaves coming to a short point at apex, the underside somewhat scaly; the bold inflorescence with showy crimson bracts. *Tropical.* p. 223

STREPTOCARPUS *Gesneriaceae*
caulescens (Trop. East Africa), "Violet nodding bells"; succulent branching plant with light green, fleshy stem covered with white hair, small blunt-cordate leaves, with slender axillary stalks bearing panicles of tiny 1 cm slipper flowers with spreading petals, beautiful violet with white throat. *Humid-tropical.* p. 501, 506
grandis (Zululand, Natal); curious stemless plant with a single large sessile, long ovate, hairy leaf 60-90 cm long, with sunken veins; small light blue flowers dark in center, and white on lower petals, on clustered stalks emerging from the base of the leaf. *Humid-subtropic.* p. 507
x hybridus, "Hybrid cape primrose"; complex group of hybrids with a long line of parents including S. dunnii for color, rexii for bushy habit, and wendlandii for stem-length; light green fleshy, quilted leaves; free-blooming with large trumpet-like flowers, in a wide range of color, from white with purple veining, through rose, orchid, mauve, blue to purple. *Humid-subtropic.* p. 506, 507
x hybridus 'Farbenwunder'; an excellent German (Meisert) strain forming shapely plants of compact habit, early-blooming, and with large wide-open flowers 8-12 cm across, in colors carmine-red to purple with dark eye blooming 6 to 8 months after sowing the seed; their main flowering season is from spring through autumn. *Humid-subtropic.* p. 506
x hybridus 'Veitch'; English hybrid characterized by its compact habit, 25 cm high, with tufting foliage, and large flowers purple with light throat. *Tropical.* p. 506
x kewensis (dunnii x rexii); English hybrid seen at Longwood Gardens, Pennsylvania, with several corrugated basal leaves; long-stalked oblique trumpet flowers with slender tube, 4 cm across, pale lavender with purple stripes. *Tropical.* p. 507
phyllanthus (So. Africa); showy tropical plant with a large and fleshy, solitary leaf 20-25 cm long and almost as wide, the surface grayish green, furrowed by a network of depressed veins, and covered with soft white hair; the inflorescence from the leaf base; branched slender stalks carry a pendant shower of flowers pale lavender with yellow center. Photographed at Longwood Gardens, Kennett Square, Pennsylvania. *Humid-subtropic.* p. 505
rexii (So. Africa), "Cape primrose"; small fibrous-rooted, stemless plant with long narrow, quilted and pubescent leaves in rosette hugging the ground, with several flower stalks bearing trumpets of pale lavender lined with purple in the throat. *Humid-subtropic.* p. 506
saundersii (Natal); single large corrugated, recurving leaf to 35 cm long, 20-22 cm wide, heart-shaped, rich green above, rose-purple and hairy beneath; stalks from base of leaf with flowers pale lilac with two purple patches, and a median yellow band in the curved tube with spreading limb, 2½ cm across, 2 lobes of the corolla shorter than the rest. *Humid-subtropic.* p. 505
saxorum (Tanzania), "False African violet"; small bushy plant from the Usambara mts., with fleshy, pubescent, elliptic, 3 cm leaves in crowded whorls; flowers with white tube and oblique limb of large spreading pale lilac lobes, on long thin stems, blooming over many months. Have seen them clinging to perpendicular, exposed cliffs, at 1,100 m in the Usambara Mountains of Tanzania. *Humid-subtropic* p. 505
synandrus (Zimbabwe 2,400 m); perennial with lanceolate leaves 10-15 cm long, crenate at margins; flowers trumpet-like with long lobes, pale lavender with magenta stripes. *Subtropic.* p. 507

x STREPTOGLOXINIA *Gesneriaceae*
hybrida (x Stroxinia) (Streptocarpus x Sinningia); attractive bigeneric hybrid as grown by Roehrs in New Jersey 1962; shapely plant of the habit of Sinningia speciosa, the quilted, soft olive green leaves with pale to pinkish veins, 15 to 18 cm long, reddish beneath; substantial curved trumpet-shaped flowers 6 cm long, with 5 spreading lobes, 5 cm across, strong carmine-red, darker crimson inside and spotted red outside; pale yellow throat. *Humid-tropical.* p. 503, 507

STREPTOSOLEN *Solanaceae*
jamesonii (Colombia, Ecuador), "Marmalade bush"; rough-pubescent, floriferous, rambling shrub with small oval, wrinkled leaves on flexuous branches, and terminal clusters of tubular, bell-shaped, orange flowers 3 cm long, in spring. *Tropical.* p. 898

STROBILANTHES *Acanthaceae*
dyerianus (Perilepta) (Burma), "Persian shield"; beautiful tropical herbaceous shrub with magnificent iridescent 15 cm leaves, long-ovate and toothed, purple with silver above and curiously shimmering; glowing purple beneath; pale blue flowers. Best in a moist-warm greenhouse. *Humid-tropical.* p. 35
lactatus (Brazil); attractive tropical herbaceous plant with shrubby base; leaves dark green handsomely variegated with white; at Longwood Gardens. *Tropical.* p. 38
maculatus (Himalayas); charming herbaceous foliage plant with ovate, quilted grass-green leaves decorated with two rows of silver blotches; margins finely serrated. Photographed at Royal Botanic Garden, Edinburgh. *Tropical.* p. 35

STROMANTHE *Marantaceae*
porteana (Brazil); large perennial herb to 1½ m high, with thick rootstocks, long paddle-like, oblique lanceolate leaves 45 cm long, glossy-leathery with lightly corrugated surface with feather-design of rich green lateral bands on nerve ridges alternating with silver-gray bands in the furrows between; reverse deep purple with gray cast; flowers blood red. *Humid-tropical.* p. 637
sanguinea (Calathea discolor) (Brazil); stiff plant growing to 1½ m tall; thick-fleshy, long-lanceolate leaves glossy dark olive-green, beautiful blood-red beneath, the juvenile basal leaves have a whitish midrib; stems branching, topped by a showy head of waxy salmon-red bracts and white flowers. *Humid-tropical.* p. 637, 638

STROMBOCACTUS *Cactaceae*
disciformis (Central Mexico); depressed globe only 6 cm wide, cut into overlapping tubercles, spirally arranged; flowers yellow. *Arid-tropical.* p. 272

STRONGYLODON *Leguminosae*
macrobotrys (Philippines), "Philippine jade vine"; a most beautiful and striking woody twiner, with large pinnate leaves, best trained over pergola or frame support to allow the large racemes, 30 to 90 cm long, a free display of the curious 8 cm flowers with their slender-pointed, upturned beak, and rolled, recurved standard, colored entirely in an unusual shade of bluish-green jade over yellow-green, the beak a pale blue-green. *Tropical.* p. 557, 559, 560

STROPHANTHUS *Apocynaceae*
gratus (West Tropical Africa), "Climbing oleander"; robust clambering evergreen shrub, with opposite leathery, ovate to obovate olive green leaves somewhat puckered, 10 cm or more long, on brown-purple woody stems; the waxy, trumpet-shaped flowers 4 cm diameter, with crinkled lobes, flushed purplish-red outside and pinkish-white inside, and with a prominent pale rose-purple inner crown, blooming late spring. *Tropical.* p. 83
speciosus (So. Africa: Cape, Natal, Zululand), "Corkscrew-flower"; rambling evergreen shrub, in habitat climbing into the trees; narrowly oval, leathery leaves to 9 cm long; curious flowers at tip of branches, a wide-mouthed corolla opens into 5 long, 3-5 cm narrow, spirally twisted lobes radiating in all directions, deep yellow with large red spot at base. *Subtropic.* p. 80

SUCCULENTS: see Aizoaceae, Amaryllidaceae, Asclepiadaceae, Bromeliaceae, Compositae, Crassulaceae, Euphorbiaceae, Geraniaceae, Liliaceae, Piperaceae, Portulacaceae, Vitaceae, in 'Pictorial section'.

SUTERA *Scrophulariaceae*
grandiflora (Transvaal, Swaziland), "Purple glory plant"; woody subshrub with long, waving stalks 1 m high, covered with wrinkled soft, gray-green aromatic leaves 4 cm long; topped by clusters of 3 cm light purple flowers. *Subtropic.* p. 883

SUTHERLANDIA *Leguminosae*
frutescens (floribunda) (So. Africa), "Cancer bush" or "Balloon pea"; downy shrub with pinnate leaves and rich scarlet flowers in axillary racemes, developing into numerous decorative puffed, papery seed pods. *Subtropic.* p. 556

SWAINSONA *Leguminosae*
galegifolia (Queensland, New So. Wales), "Swan orchid" or "Winter sweet-pea"; ornamental, freely blooming greenhouse

plant; semi-climber with scandent branches, bearing unequally pinnate leaves, and small pea-like flowers in long-stalked showy racemes, deep red, or other shades. *Subtropic.* p. 560

SWIETENIA Meliaceae
macrophylla (Trop. America), "Bigleaf mahogany"; lowland tree to 50 m high, with pinnate leaves to 30 cm long, thin rich green leaflets; woody fruit 12-15 cm long. Major source of true mahogany timber. *Tropical.* p. 642
mahagoni (Florida Keys and West Indies), "Mahogany"; tropical evergreen tree to 25 m, with dark red wood which furnishes the mahogany of commerce for furniture, etc.; dark glossy green, leathery, pinnate leaves to 30 cm long; small white flowers; showy green woody fruit capsule 10 cm dia. *Tropical.* p. 641

SYAGRUS Palmae
coronata (Cocos coronata hort.) (Brazil), "Licuri palm"; handsome palm to 10 m tall, the trunk partially covered by persistent leaf bases arranged in an attractive spiral pattern; pinnate fronds to 3 m long, erect and arching, arranged in 5-ranked spiral; leathery, dark lustrous green leaflets folded at base, set in clusters and at various angles; yellow flowers in typical boat-shaped spathe, and fleshy 2 cm orange, edible fruit. *Tropical.* p. 798
oleracea (Brazil); small feather palm with bushy crown of densely pinnate fronds, the leaflets irregularly arranged in groups of 2 or more, in several planes along the rachis. *Tropical.* p. 803
vagans (Brazil); clustering shrubby feather palm with yellowish petioles, the pinnate fronds erect and rather crowded; with long leaflets presenting an unkempt appearance. *Tropical.* p. 803
weddelliana (Brazil: Guanabara), "Baby cocos palm"; according to Blossfeld, this should be Microcoelum martianum, but in the trade as Cocos weddelliana; attractive little feather palm in 5 cm pots, which grows in the humid Organ Mountains to 2 m, making slender, solitary trunks with graceful pinnate fronds, the narrow, stiff segments glossy yellow-green and neatly spaced; small orange fruit. *Humid-tropical.* p. 798, 803

SYMBEGONIA Begoniaceae
sanguinea (strigosa) (New Guinea); beautiful member of the begonia family, with branches spreading and scandent from fibrous roots; the handsome foliage is oblique-ovate, deeply quilted and rugose with bristles and serrate at margins, and of red-bronze coloring; the inflorescence is hairy, with flowers carmine-red; petals free in male flowers, but in female flowers forming a tube, one of the characteristics of this genus; in begonias the petals are all free. Found at 300 m in the Milne Bay District, Papua. *Tropical.* p. 171

SYNADENIUM Euphorbiaceae
grantii rubra (Euphorbia) (Tanzania), "Red milk bush"; ornamental shrub with thick, succulent branches, milky juice; fleshy obovate leaves beautiful wine red, margins finely toothed, reverse vivid red purple; flowers red. *Tropical.* p. 408, 428

SYNANDROSPADIX Araceae
vermitoxicus (No. Argentina); handsome aroid from the Andes near Tucumán and Salta; thick tuber with fleshy petioles striped with dark green, carrying sagittate-cordate leaves 20-45 cm long, 15-20 cm wide; showy spathe pale greenish to brownish inside, lined green, 10 cm long; spadix brown-red. *Subtropic.* p. 129

SYNGONIUM Araceae
erythrophyllum (Panama), "Copper syngonium"; dainty creeping plant with small, arrow-shaped, waxy leaves having 2 earlike basal lobes; blade metallic coppery-green and covered with tiny pink dots; reddish beneath; mature leaf trifoliate. *Tropical.* p. 130
macrophyllum (Mexico to Panama), "Big-leaf syngonium"; climber with large, heartshaped, emerald-green, fleshy leaves with velvet sheen, becoming divided in the maturity stage, much larger than podophyllum. *Tropical.* p. 127
podophyllum 'Albolineatum' (earliest name: angustatum) (Mexico, Nicaragua), "Arrow-head vine"; known commercially as Nephthytis triphylla; the juvenile leaves heartshaped or 3-lobed, very ornamental with silver white center and veining; the mature palmate leaves are green. *Tropical.* p. 130
podophyllum 'Albo-virens'; a mutant with slender juvenile hastate leaves shaded ivory to greenish-white, blade edged green. *Tropical.* p. 130
podophyllum 'Emerald Gem variegated'; of compact habit and creeping very slowly; dark green arrow-shaped leaves with pure white and milky-gray, irregular variegation on the glossy, thin-leathery blades to varying degree. *Tropical.* p. 130
podophyllum 'Imperial White'; an "Arrowhead-vine" with arrow-shaped leaves deep bluish green, and a contrasting network of ivory-white veining. Widely grown in California greenhouses. *Tropical.* p. 125
podophyllum 'Trileaf Wonder'; cultigen with a varying amount of ashgreen on the leaf, principally on the midrib and lateral veins; also the mature, segmented leaves are produced more quickly. *Tropical.* p. 130
podophyllum 'Variegatum'; an "African evergreen", slowly trailing or climbing, with firm leaves medium green, irregularly splashed and variegated with milky white. *Tropical.* p. 125
wendlandii (Costa Rica); dainty creeper with tri-lobed, deep green, velvety leaves and sharply contrasting white veins in the juvenile foliage, the divided maturity-stage leaves plain green. *Tropical.* p. 130
xanthophilum (Mexico), the cultivated plant 'Green Gold' probably belongs here; the juvenile arrowshaped leaves are suffused and marbled with yellow-green. *Tropical.* p. 130

SYRINGA Oleaceae
vulgaris (S.E. Europe), "Lilac" or "Common lilac"; bushy deciduous shrub with crooked woody stems to 3 m or more high, the flexible twigs with opposite, thin ovate leaves to 12 cm long; blooming outdoors in late spring with terminal or lateral pyramidal clusters of sweetly fragrant small flowers, normally pale purple tinged with violet. *Temperate.* p. 699

SYZYGIUM Myrtaceae
aromaticum (Eugenia) (Moluccas), the "Clove tree"; evergreen to 10 m high, with elliptic, glandular dotted, aromatic leaves; yellow tubular flowers 1 cm across, with red base. The sun-dried flower buds are the commercial "cloves". *Tropical.* p. 678
'Jambeiro' (Eugenia) (Brazil); so labeled in Botanic Garden Rio; tree with spreading, brittle branches, large oval, thick-leathery leaves 20 cm long, with cordate base; showy carmine red stamen bundles. *Tropical.* p. 677
jambos (S.E. Asia), "Malay rose-apple" or "Malabar plum"; evergreen tree to 12 m high, with dark green, lanceolate leathery leaves to 20 cm; greenish-white flowers 5-6 cm across; fruit cream-yellow to 4 cm long, fragrant and edible; prized for jellies. *Tropical.* p. 460, 677
malaccense (Malaya), "Rose apple"; beautiful tropical tree 5-12 m high; glossy ovate oblong, leathery leaves 15-30 cm long; showy flowers of many stamens purplish-red, from old wood; pear-shaped red fruit, edible raw or cooked. *Tropical.* p. 460, 675, 677
megacarpa (India, Malaya); sparry tree with long ovate, leathery leaves, and colorful inflorescence of crimson stamen flowers growing directly from the branches and bare wood. *Tropical.* p. 677
paniculatum (Eugenia myrtifolia) (Australia), "Brush cherry"; small vigorous shrub which lends itself to shearing into pyramids; slender branches dense with small elliptic foliage vivid red when young, later shining green; fluffy white flowers followed by red berries. *Subtropic.* p. 677
samarangense (Eugenia javanica) (Malay Archipelago), "Java apple"; evergreen tree to 10 m; leathery elliptic leaves 15 cm long; white flowers 4 cm across; pear-shaped fruit 4 cm long, white to glossy red; edible, but insipid. *Tropical.* p. 460, 677, 688

SYZYGIUM: see also Eugenia

TABEBUIA Bignoniaceae
argentea (Tecoma argentea) (Paraguay); the "Silver trumpet-tree", or "Golden bell"; showy tropical flowering tree with crooked trunk and corky bark, to 8 m high, covering itself in the leafless stage with a profusion of rich-yellow trumpet-flowers 5-8 cm long; after bloom appears the foliage, leaves palmately divided into 5-7 narrow leaflets to 15 cm long, and covered with silvery scales; oblong woody dark brown fruit 15 cm long. *Subtropic.* p. 189, 190
avellanedae (ipe) (Paraguay, Argentina); erect trumpet tree to 18 m tall, usually evergreen with palmate leaves of smooth, dark green leaflets 15 cm long; flowers clustered profusely at branch tips, lavender-rose with white throat banded yellow and purple. *Subtropic.* p. 191
chrysantha (C. America, So. America); tropical flowering tree 4-5 m high, with leaves digitately compound of 5 hairy, ovate leaflets; trumpet-shaped yellow flowers 5 cm long in showy clusters at the end of leafless branches. *Tropical.* p. 186
chrysotricha (Colombia, Brazil), "Golden trumpet-tree"; semi-evergreen tree to 8 m or more; palmate leaves with 5 leaflets to 10 cm long, the underside with yellowish fuzz; bright yellow trumpet flowers 6 cm long, blooming heaviest in spring, when leaves drop for brief period. *Subtropic.* p. 190
donnell-smithii (Cybistax) (Mexico, Guatemala), "Primavera" or "Gold tree"; magnificent deciduous tree to 20 m tall, with smooth gray trunk; leaves palmately compound with 5-7 leaflets to 25 cm long; when the foliage has fallen, masses of golden yellow blooms crowd branch ends, a most beautiful sight; the flowers are bell-shaped and crepy, 3 cm long. *Tropical.* p. 189, 190
heterophylla (West Indies); shrub or small tree to 8 m high; palmately compound leaves of 3 to 6 elliptic leaflets to 4 cm long,

dark green and thin-leathery; great clusters of inflated tubular flowers 3-4 cm long, carmine pink outside and on slightly flaring lobes, creamy inside. *Tropical.* p. 190

impetiginosa (Brazil); flamboyant deciduous small tree to 6 m; palmate leaves of 5 oblong to ovate leaflets to 12 cm long; large purplish-pink trumpet flowers with salmon sheen, 5 cm long; white and purple with yellow markings in throat. Blossoms cover bare branch tips in spring. *Subtropic.* p. 191, 192

pallida (W. Indies, C. and No. America), the "Cuban pink trumpet-tree"; showy flowering tree to 15 m or more high; leaves with 3-5 elliptic leaflets 10-15 cm long, often renewed after blooming; handsome flowers in dense terminal clusters, slender tube with flaring pale lilac pink limb, yellow in throat, the corolla 6-8 cm long; winterblooming; the fruit a long pod. *Tropical.* p. 190

rosea (Mexico, C. America, Colombia), the "Rosy trumpet-tree"; small to medium winter-flowering tree with digitately compound leaves of 3-5 leathery leaflets, simple in young plants; large clusters of showy flaring tubular flowers a pretty lilac-rose. *Tropical.* p. 189, 190, 191

roseo-alba (Brazil, Paraguay, Bolivia); small evergreen tree to 8 m high; leaves with 3 or more oval leaflets glossy rich green and leathery; large rose-pink frilled trumpet flowers of crepy texture. *Tropical.* p. 190

serratifolia (Puerto Rico, Trinidad), "Yellow pui"; spectacularly showy flowering trumpet-tree about 6-9 m high, of densely branched habit, with papery leaves of 4-5 ovate leaflets to 12 cm long; the narrowly funnel-shaped 6 cm flowers in brilliant yellow, an unforgettable sight for the visitor to lovely Puerto Rico. *Tropical.* p. 189, 190

umbellata (fosteri) (Brazil); small deciduous tree to 5 m high; compound leaves of 3-5 oblong or elliptic leaflets to 8 cm long; clusters of golden-yellow trumpet flowers with crepy limbs 6 cm long, blooming during leafless stage. *Tropical.* p. 191, 192

TABERNAEMONTANA   *Apocynaceae*

coronaria (Ervatamia) (India), the "Crape jasmine", "Clavel de la India" or "East Indian rose-bay"; a handsome shrub 2 to 3 m high with glossy green, elliptic, thinnish leaves 8-15 cm long, and waxy white, undulate flowers 4 to 5 cm wide, the lobes prettily crisped, very fragrant at night; blooming in summer; widely cultivated in tropical gardens. *Tropical.* p. 76

coronaria plena (Ervatamia), "Fleur d'amour" or "Butterfly gardenia" with waxy, lanceolate leaves and pure white, lacy, double flowers, intensely fragrant. *Tropical.* p. 76

corymbosa (Ervatamia) (No. India to Malaya), "Great Rosebay", "Red bay" or "Flower of Love"; evergreen shrub with milky sap to 10 m; obovate leaves 8 to 20 cm long, glossy green and depressed at the veins; sweet-scented creamy-white, waxy gardenia-like 3 cm flowers with flaring, curving petals. The latex is poisonous. *Tropical.* p. 73, 76

crassa (Ervatamia), "Adams apple"; large evergreen shrub containing latex, with opposite broad-ovate, deep green leaves; large terminal clusters of small white, waxy flowers with sickle-shaped, curving petals; large fleshy twin-fruit 15 cm dia., splitting along the lower edge, opening out and disclosing the rows of seeds in the scarlet pulp. *Tropical.* p. 76

divaricata (Ervatamia coronaria), (No. India), the "Paper gardenia"; dense evergreen, smooth, gardenia-like bush to 2 m or more high; shiny green, thin-leathery, paired leaves 8-16 cm long; waxy-white, nearly scentless tubular flowers with 5 or 6 crinkled lobes spreading to 3-5 cm across. *Tropical.* p. 76

TACCA   *Taccaceae*

chantrieri (Malaya), "Bat-flower"; curious, bat-like inflorescence both in shape and color, with wide-spreading, wing-like bracts of rich maroon-black, accompanied by long trailing filaments or "whiskers"; the small black flowers are succeeded by heavy berries; corrugated olive-green leaves with oblique base. *Tropical.* p. 902

plantaginea (Schizocapsa) (South China), "Cat's whiskers"; perennial herb with large lanceolate leaves 30 cm long; the whitish green flowers, with green bracts, subtended by long thread-like filaments; fruit not berry-like, but a dry-skinned capsule. *Subtropic.* p. 902

TACCARUM   *Araceae*

weddellianum (Brazil, Paraguay, Bolivia); tropical tuberous plant with finely dissected leaf 40 cm diameter, triangular in outline, on the stout petiole 40-80 cm long; short stalked, very curious inflorescence with ovate spathe yellowish-green, cupped at base of raceme-like cylindric spadix 20 cm high when fully grown, the upper part containing the male flowers. *Tropical.* p. 129

TAENIOPHYLLUM   *Orchidaceae*

fasciola (syn. Epidendrum, Limodorum, Vanilla); one of the tiniest of orchids, a leafless epiphyte which I found on the trunk of a large tree in the rain forest of Naduruloulou, in Fiji; the highly developed, brittle strands of roots only 2 mm wide but 8-15 cm long, containing chlorophyll and delegated to carry out the function of the absent leaves; these roots are flattened against the bark, and radiate from a common center from which rise stems reduced to a minimum, bearing little pale cream flowers under 4 mm long; capsule bright yellow. *Tropical.* p. 753

TAGETES   *Compositae*

patula (Mexico, Guatemala), "French marigold"; bushy small annual, strongly scented, originally from Mexico, with finely dissected leaves and solitary heads of 3-4 cm double flowers, golden yellow, marked with red-brown. *Tropical.* p. 321

patula 'Diamond Jubilee'; All America selection for excellence; compact bush 50 cm high, dark green foliage and huge double flowers, 10 cm dia., primrose yellow. *Tropical.* p. 330

patula 'Gay Ladies'; large bush with huge ball-shaped flowers, orange yellow, fully double to 10 cm across. *Tropical.* p. 330

patula 'Gold Rush'; dwarf marigold strain to 25 cm high, the fully double flowers with frilled orange and red center crest, and crimson red outer guard petals. *Tropical.* p. 330

patula 'Naughty Marietta', "Single French marigold"; very pretty single marigold of semi-dwarf habit, to 40 cm high, bushy, with finely netted leaves; 5 cm flowers golden-yellow, painted in the center with a deep red, velvety eye. *Tropical.* p. 330

patula 'Orange Lady'; 8-10 cm flowers fully double, bright deep orange carnation-like blooms. *Tropical.* p. 330

patula 'Petite Yellow'; extra dwarf, fully double strain with yellow flowers, 3-4 cm across, for edging. *Tropical.* p. 330

petula 'Rusty Red'; strong scented perennial originally from Mexico, now highly developed and grown as an annual "French marigold", with finely dissected leaves and solitary heads of 5 cm double flowers, in this medium dwarf cultivar a velvety brown. *Tropical.* p. 330

patula 'Spry'; popular free-blooming, dwarf "Bedding marigold" to 25 cm high, the double flowers with a brightly contrasting golden yellow anemone crest over deep rich mahogany guard petals. *Tropical.* p. 330

TAIWANIA   *Taxodiaceae (Coniferae)*

cryptomerioides (Mountains of Taiwan); evergreen tree, to 50 m in habitat; allied to Cryptomeria, but differing in having foliage of two kinds: juvenile leaves curved awl-shaped 2 cm long, adult leaves scale-like and imbricate; small 2 cm cones. *Subtropic.* p. 354

TALINUM   *Portulacaceae*

guadalupense (Guadalupe Island, Mexico), "Fame flower"; succulent rosette with elongate, fleshy stem, and obovate, spoon-shaped leaves 5 cm long, blue-glaucous, edged with red; pink flowers. *Tropical.* p. 829

TAMARINDUS   *Leguminosae*

indica (Trop. Africa), the "Tamarind tree"; immense, picturesque evergreen tree to 25 m high, important both as an ornamental shade tree and economically for its brown 20 cm pods containing fleshy pulp of pleasing acid flavor; feathery pinnate, pendant leaves with leathery leaflets 2 cm long; and pale yellow 2 cm flowers, striped red. Bark, wood, leaves, flowers, and seeds are also useful. *Tropical.* p. 556, 565

TAMARIX   *Tamaricaceae*

africana (W. Europe, Canary Is., No. Africa), "Salt-cedar"; shrub or bushy tree with black bark; small scale-like leaves; inflorescence in plume-like sprays of small white or pink flowers, mostly on branches of previous year. *Subtropic.* p. 911

parviflora (S.E. Europe),"Tamarisk"; spreading shrub 2 to 5 m high, with graceful airy, arching branches with reddish bark; scale-like leaves, and small pink flowers in lateral sprays in spring, on last year's twigs; hardy in So. New England; twigs and leaves are shed in autumn. *Warm temperate.* p. 903

ramosissima (pentandra) (S.E. Europe, Asia), "Salt-cedar" or "Summer tamarisk"; spreading feathery shrub 2-10 m high, with tiny scale-like blue-green leaves; small rose-pink flowers in great plumes on current year's growth. *Warm temperate.* p. 903

TAPEINOCHILUS   *Zingiberaceae*

ananassae (Malaysia, to Ceram and Queensland), "Giant spiral ginger"; perennial herb related to Costus, to 2½ m high, with bamboo-like canes, bearing leathery 15 cm green leaves arranged spirally; terminal inflorescence of cone-like, hard, recurved crimson bracts 10 cm across, nesting the yellow flowers; an ovoid cone resembling a pineapple, covered with brown bracts on leafless stalk rises directly from the rootstock. *Tropical.* p. 923

TAVARESIA   *Asclepiadaceae*

angolensis (Angola: Luanda); tufting succulent 10-15 cm high, with swollen stems sharply angled, the ridges set with spreading teeth each bearing 3 short spines; large tubular flowers from the base, to 8 cm long, outside yellowish with brown-red spots and mottling. *Arid tropical.* p. 155

TAXODIUM *Coniferae: Taxodiaceae*
distichum (New Jersey, Delaware to Florida, west to Texas), the "Bald cypress" or "Swamp cypress", also known as the "Tidewater red cypress"; tall deciduous coniferous tree, becoming 50 m high, with buttressed trunk usually 1-1½ m but sometimes 4 m or more in diameter, usually hollow in old age; the spreading branches with delicate and feather-like light yellow-green 1 cm needles in graceful sprays. At home in cypress swamps and tidewater bayous along the Gulf of Mexico and the Florida Everglades; this is not only a very decorative tree but valuable as the "Red" or "Yellow cypress" for its durable red or yellow lumber used in greenhouse construction and benches, for boats, piling, shingles, and wherever wood must withstand warm and wet conditions. The base of the tree is flared to help absorb oxygen from the water; it is also known for its curious "Cypress knees" which are modified roots of a very light, soft spongy wood, above water. *Subtropic.* p. 355

mucronatum (So. Mexico, Guatemala), "Montezuma cypress"; evergreen tree reaching great dimensions and age, a tree in Chapultepec, Mexico City is 51 m high and 700 years old; the famous tree of Tule near Oaxaca is estimated 4000 years old, with trunk dia. of 12 m and spreading 42 metres. Trunk covered with reddish, shredded bark, with roots often growing in water; needles 1 cm long, partially deciduous, on weeping branches; small 2 cm cones. *Subtropic.* p. 353

TAXUS *Coniferae: Taxaceae*
baccata (Europe, No. Africa, W. Asia), "English yew"; densely branched evergreen shrub or tree growing slowly to 15 m, with wide-spreading branches forming a broad crown; flat needles dark green and glossy above, paler beneath, 2-3 cm long; female plant bearing fleshy scarlet, cup-shaped berries containing a single, poisonous seed. *Warm temperate.* p. 352, 353

cuspidata (Japan, Korea, Manchuria), "Japanese yew"; somber evergreen tree to 15 m high, cultivated in a number of horticultural forms; T. cuspidata is the most important of ornamental yews, best for dense bush or formal hedges, as it lends itself to shearing; far hardier than the English yew (T. baccata) and faster-growing; needle-like flat leaves about 2 cm long, suddenly tapering to a sharp point, leathery in texture, blackish-green, 2 yellowish bands below, and with prominent midrib, arranged on one plane along either side of the flexuous branchlets; berry-like scarlet fruit. *Temperate.* p. 352

cuspidata 'Capitata' hort., "Upright Japanese yew"; botanically synonymous with T. cuspidata, but because of its pyramidal habit, American nurserymen have attached the term "Capitata" to the species name in order to distinguish it from the spreading type. There is no mystery about the background of "Capitata" however: all true seed collected in its Japan habitat and imported since about 1910 has produced seedlings of typical pyramidal tree shape, and this character can also be perpetuated if using cuttings from only leader tips of older plants. *Temperate.* p. 352, 353

TECOMA *Bignoniaceae*
stans (Stenolobium) (West Indies, Mexico to Perú), "Yellow-bells", or "Yellow elder"; ornamental shrub or tree-like to 6 m high; unequally pinnate leaves with serrate oval leaflets to 10 cm long; a profusion of showy 5 cm flowers bright yellow, in pendulous clusters, with funnel-shaped corolla hairy inside, the stamens curved in two pairs; long narrow pod-like fruits. *Tropical.*
p. 187, 193

TECOMA: see also Bignonia, Distictis, Tabebuia

TECOMANTHE *Bignoniaceae*
dendrophylla (New Guinea); large woody vine, with dark green leaves 16 cm long, pinnately compound, with 4 paired oblong leaflets and one at apex; attactive flowers with inflated bell-shaped corolla 10 cm long, deep rose to maroon outside and tipped with 5 short lobes, lighter inside, the lobed calyx purple. *Tropical.*
p. 191, 192

speciosa (Three Kings Isl. off N.W. New Zealand); curious woody twiner, discovered 1945, capable of ascending to 10 m, and in habitat growing over tall specimen of Leptospermum trees. Having seen this novel plant in N.Z., I feel it may become a desirable windbreak over fences, and ornamental even without support. The glossy leathery compound leaves, 15-30 cm long, of obovate leaflets; 6 cm trumpet flowers creamy tinged green. *Tropical.* p. 191

venusta (New Guinea); beautiful, rare tropical woody climber to 3 m with compound leaves, the ovate leaflets light green and lightly quilted; the striking tubular, waxy flowers burnt-red or crimson outside and cream inside, in clusters from the woody stems, in early spring. *Tropical.* p. 191

TECOMARIA *Bignoniaceae*
capensis (So. Africa to Transvaal and north), "Cape honeysuckle"; rambling, evergreen shrub 2-3 m high, with leaves of 7-9 ovate, shining green, toothed leaflets to 5 cm long, and bearing bunched masses of curved funnel-form flowers fiery orange-scarlet, 5 cm long, and with protruding stamens. *Subtropic.* p. 191

capensis 'Aurea' (So. Africa); yellow-flowered form of the "Cape honeysuckle", a rambling evergreen shrub, sending out runners over the ground and up supports; dark glossy green leaves 10-15 cm long divided feather-fashion into 7 to 9 toothed leaflets; the showy slender funnel-shaped curved flowers 5 cm long, with 4 stamens protruding in close clusters. *Subtropic.* p. 187

TECTONA *Verbenaceae*
grandis (India to Malaysia and Java); the "Teakwood"; a great tree of Monsoon Asia, to 50 m tall, with quadrangular shoots more or less woolly; leaves arranged in two's or three's, ovate, of leathery texture, 30 to 80 cm long, rough above, gray or brownish woolly beneath; small white flowers in terminal clusters, the calyx longer than the petals; dry, 3 cm fruit pale green to brown. The strong, durable wood is valued as highly as mahogany, and used in furniture. *Tropical.* p. 915, 916

TELIPOGON *Orchidaceae*
species (Colombia); a delicate miniature from the cloud forests, without pseudobulbs, clustered little leaves; from their axils a raceme of fragile but lovely blossoms, to 5 cm across, the broad petals creamy white to greenish, the center heavily painted and traced with purple. *Humid-tropical.* p. 758

TELOPEA *Proteaceae*
speciosissima (New South Wales, Queensland), "Waratah"; striking flowering shrub related to Protea, of erect habit, 3-4 m high, with narrow obovate, leathery leaves toothed in upper part; magnificent terminal inflorescence of coral-red tubular flowers with protruding styles, packed in a dense globose head 8-10 cm across, and subtended by numerous narrow bracts of brilliant crimson. *Subtropic.* p. 838, 840

TERMINALIA *Combretaceae*
catappa, (East Indies), "Tropical almond" or "Olive-bark tree"; small or large tree to 25 m high, much planted in tropical countries near sea-shores for ornament and shade; wide-spreading branches arranged in tiers with large obovate, leathery leaves to 30 cm long, becoming red before they fall twice a year; flowers greenish-white in spikes, followed on female flowers by greenish or reddish 2-angled fruit 3-5 cm long, with oil-bearing seed, which can be eaten raw or roasted. *Tropical.* p. 304, 467

TERNSTROEMIA *Theaceae*
gymnanthera (India, Malaya, Japan); smooth evergreen shrub or small tree to 10 m; the foliage is spirally arranged and crowded near tips of new growth on red twigs; elliptic glossy-leathery leaves 4-8 cm long, handsome with new growth bronzy-red, later green and purplish; small 2 cm fragrant flowers creamy-yellow; small red-orange fruit splitting to reveal shiny black seeds. *Subtropic.* p. 906

TESTUDINARIA: see Dioscorea

TETRAPANAX *Araliaceae*
papyriferus (Aralia) (China, Taiwan), "Rice paper plant"; a small tree with large, ornamental foliage; the woody trunk and branches filled with white pith used to make paper; the lobed leaves are covered with white felt while young. *Subtropic.* p. 143

papyriferus 'Variegatus' (N.S. Wales); form with the large, palmately lobed, and toothed leaves recurved umbrella-like and beautifully variegated bright green to dark, emerald-green with ivory to clear white. *Subtropic.* p. 137

TETRAPLASANDRA *Araliaceae*
meiandra (Hawaii), "Hawaiian Ohe tree"; evergreen tree with stout willowy branches and thick, soft-leathery glossy green leaves with pale veining, pinnately divided into 7 oblique ovate leaflets, pale beneath, on coppery petioles. On Hilo, trees grow to 24 m tall. *Tropical.* p. 143

TETRASTIGMA *Vitaceae*
obovatum (So. China); tendril-climbing herbaceous shrub with large palmately divided, corrugated leaves of obovate leaflets, and covered with light brown felt. *Subtropic.* p. 920

voinierianum (Vitis) (Vietnam), "Chestnut vine"; robust climber with woody stems and clambering, fleshy, brown-hairy branches having coiled wiry tendrils and gigantic, digitate, thick-fleshy leaves with 3-5 shining green, stalked, broad-obovate or oblique leaflets to 25 cm long, wavy toothed at margins and pale green, pubescent underneath. *Tropical.* p. 920

THEA: see Camellia

THECOPHYLLUM *Bromeliaceae*
insigne in hort. (Guzmania 'Insignis'); showy rosette with soft, olive green leaves laxly pendant, carmine-rose toward base; beautiful inflorescence of ovate, rosy-red bracts, the flower-heads yellow. Photographed at Palmengarten Frankfurt, Germany. According to Victoria Padilla, a hybrid made by Dutrie of Guzmania lingulata splendens with G. zahnii, more beautiful and larger than the species Tillandsia (Guzmania) insignis. *Tropical.* p. 223

THECOPHYLLUM of hort.: see Vriesea sintenisii

THELOCACTUS *Cactaceae*
bicolor (So. Texas, Mexico), "Glory of Texas"; globular to cylindrical plant to 10 cm thick, bluish green, with 8 oblique notched ribs furnished with arched spines variously colored white, yellow and red; flowers purplish. *Arid-subtropic.* p. 280
heterochromus (Mexico: Chihuahua to Coahuila); depressed globe to 15 cm dia., bluish-green, with bold tubercles, tipped by stiff red-tipped spines; 10 cm flowers purplish. *Subtropic.* p. 280
pottsii (heterochromus) (Mexico: Chihuahua to Coahuila); small globe to 15 cm high, with 8-9 high ribs, needle spines and central spine; flowers pale rose with purple center. *Arid-subtropic.* p. 280
saussieri (Mexico); globular-depressed cactus to 20 cm in dia., spiral ribs, cut into light green conical knobs 4-sided at the base; silvery-white, spreading radial spines, and needle-like brown central spines; 4 cm flowers purple-red. *Arid-subtropic.* p. 266

THELYMITRA *Orchidaceae*
ixioides (Australia, New Zealand), "Women's cap orchid"; terrestrial with long-linear leaves, flat or channeled; inflorescence on 30 cm stalk with masses of corn-flower-blue star-like flowers with light centers, 3-5 cm across. *Subtropic.* p. 756

THENARDIA *Apocynaceae*
floribunda (Mexico: Michoacán to Oaxaca), "Petatillo"; woody climber with milky sap; thin ovate leaves 5 to 12 cm long; dense clusters of attractive, small 3 cm star-shaped fragrant flowers purplish pink with greenish-white center. *Sub-tropical.* p. 78

THEOBROMA *Sterculiaceae*
cacao (C. America, Trinidad, Guayana), "Cacao-tree"; wide-branching evergreen tree to 8 m high, with attractive, satiny, hard-papery, pendant leaves to 30 cm long; the small, yellowish flowers in axillary clusters, or curiously even from the trunk, succeeded there by the large, ribbed fruit containing bean-like seed which is the source of chocolate. *Tropical.* p. 474, 477, 902

THEOPHRASTA *Theophrastaceae*
americana (fusca) (Santo Domingo); evergreen shrub to 1 m high; with stout, simple stem; glossy green, leathery, linear, oblanceolate leaves in whorls, to 45 cm long, the margins with spiny teeth; bell-shaped flowers white to rusty-brown in hairy raceme; fleshy fruit apple-shaped. *Tropical.* p. 906

THESPESIA *Malvaceae*
lampas (E. Africa to Philippines); tropical shrub to 2½ m, with 3-5 lobed leaves, to 20 cm long or more, the upper leaves smaller, often unlobed; flowers with yellow petals and dark purple center, falling early; globular seed capsules split open. *Tropical.* p. 628
populnea (Pantropic), "Portia tree" or "Milo"; tropical tree to 20 m high, with cordate-ovate unlobed leaves 15 cm long or more; solitary bell-shaped flowers 5-8 cm dia., yellow with purple center, fading to orange-yellow and withering into pink during the day. *Tropical.* p. 628
populnea acutiloba; seaside tree with triangular foliage having wide open sinus and pronounced lobes; flowers yellow changing to orange and pink-purple. *Tropical.* p. 628

THEVETIA *Apocynaceae*
peruviana (W. Indies, Mexico), "Yellow oleander" or "Be-still tree"; tropical evergreen shrub 2-3 m high with linear, shining green, 10-15 cm leaves with edges rolled under; large, funnel-shaped lemon-yellow flowers 5-8 cm long, shading to pinkish or orange-apricot, and sweetly fragrant like a tea-rose; blooming anytime, mostly June to November. Takes heat and sun with ample water; may be trained into small tree. Poisonous like their relatives, the oleanders. *Tropical.* p. 78, 80, 82
thevetioides (Mexico: Michoacán to Veracruz), "Narciso amarillo"; bushy shrub or small tree to 5 m high, with milky sap; linear dark green leaves 10 cm or more long, with the margins turned under and hairy beneath; large trumpet-like orange-yellow or pinkish yellow flowers to 9 cm long. *Subtropical.* p. 80

THRINAX *Palmae*
ekmanii (morrisii) (West Indies: Navassa Is.); attractive fan palm with fresh green leaves almost a complete circle, cut into broad segments, but not quite to base. *Tropical.* p. 804
microcarpa (So. Florida and nearby Keys, Greater Antilles, Yucatan, B. Honduras, Panama), "Key palm"; a thatch palm of robust habit, with trunk usually 3-4 m but reaching 10 m and 20-30 cm dia.; the petiole with fibrous webbing, supporting palmate leaves about 1 m across, divided for about ½ their length, the straight segments forming a semi-circle, shiny grayish-green above, light gray beneath; small white sessile fruit. *Tropical.* p. 804
radiata (Florida, W. Indies, to Honduras), "West Indian thatch palm"; moderately sized unarmed fan palm to 12 m tall, the palmate leaves glossy green and folded, deeply sliced into drooping segments; flowers white; ivory-colored fruit pea-like. *Tropical.* p. 804

THRYALLIS: see Galphimia

THUJA *Cupressaceae (Coniferae)*
occidentalis (E. No. America; Quebec to Hudson Bay, New Jersey to N. Carolina, west to Minnesota), "American arbor-vitae" or "White cedar"; ornamental evergreen tree of pyramidal habit, about 20 m high, with reddish-brown buttressed trunk divided near ground into several secondary trunks, the branches densely arranged, flat fan-like, branchlets like fern fronds but with hard scale-like, shingled leaves, needle-shaped when young, dark green above, yellowish green beneath, with a strong resinous odor; oblong cones erect when young, brown and pendulous when mature, 2 cm long. *Temperate.* p. 340
occidentalis 'Rheingold'; dwarf form of German origin, of globular shape, to 1.5 m high; branchlets crowded, thin and flexible; leaves linear and scale-like, golden yellow with tips tinted rose. Common in Europe. *Temperate.* p. 341
orientalis (Biota; Platycladus in Hortus 3), (China: Manchuria; Korea), "Oriental arborvitae"; dense conical evergreen usually consisting of several stems; differing from Thuja occidentalis in having branchlets 2-ranked fan-like in a flat, vertical plane; the tiny scaly, juvenile leaves needle-like, glossy yellowish green; female cones 2 cm long, fleshy, and bluish before ripening. Warm temperate, but widely planted in the tropics, as far south as Oaxaca, Mexico. *Warm-temperate.* p. 339, 340
orientalis 'Aurea nana' (Biota), "Dwarf Golden arborvitae"; handsome compact form pointed globe-shaped, usually 1 m high, with the flat, 2-ranked branchlets having lacquered fresh green foliage tipped with golden-yellow. *Warm-temperate.* p. 340
plicata (No. California to Alaska), "Western arborvitae" or "Giant cedar"; noble pyramidal tree 30 to 60 m high; buttressed trunk with cinnamon-red bark divided into wide ridges; branches horizontal, pendant at the ends, and set with shiny green scale-like leaves in 2 ranks, forming flat, graceful, lacy sprays; erect brown cones 2 cm long. *Temperate.* p. 340

THUNBERGIA *Acanthaceae*
alata (So. E. Africa, but naturalized in Tropics), the "Black-eyed Susan"; twining perennial herb, with herbaceous, toothed, triangular ovate leaves to 8 cm long, on winged petioles; funnel-shaped showy flowers 4 cm long, creamy-yellow or orange with or without black-purple throat, blooming late summer to autumn. Attractive vine for the cool, light window. Can be grown as an annual from seed. *Subtropic.* p. 38, 40
erecta (Trop. West Africa), "King's-mantle"; erect evergreen shrub to 2 m; thin branches with almost glossy ovate leaves 3-6 cm long; axillary trumpet-shaped flowers 6 cm long, with large violet lobes, and yellow inside tube; July. *Tropical.* p. 40
fragrans (India, Sri Lanka); woody twiner with leathery, triangular ovate leaves to 8 cm long; fragrant, pure white flowers with spreading lobes, 5 cm across. *Tropical.* p. 40
grandiflora (India), "Clock vine"; woody twiner with rough, toothed, ovate leaves; bell-like flowers somewhat two-lipped, lavender-blue; throat white, solitary in the leaf axils. *Tropical.* p. 40
grandiflora 'Alba', "Trumpet vine" or "Sky flower"; vigorous climber forming woody stems; lush green ovate leaves to 20 cm long, usually 3-5 nerved from base; showy tubular-flaring flowers 6-8 cm across, pure white. *Tropical.* p. 40
laurifolia (India); choice woody twiner with lanceolate leathery leaves to 12 cm long; funnel-form flowers, light blue with white throat, to 8 cm across, several to a raceme. *Tropical.* p. 40
mysorensis (So. India); tall-climbing, vigorous tropical vine from the Nilghiri Mountains; lance-shaped 3-nerved leaves 10-15 cm long; the attractive funnel flowers golden yellow 5-6 cm across, with yellow and red-brown spreading limb, in long pendant racemes. *Tropical.* p. 40

THUNIA *Orchidaceae*
alba (India, Burma, China); robust epiphyte, occasionally, becoming terrestrial, with tufted, tapering leafy stems 60 cm high, topped by a nodding raceme of large, pure white flowers to 8 cm across, with white lip inside yellow, striped with orange and shortly spurred and crested, (June-July). *Tropical.* p. 754

TIBOUCHINA   *Melastomataceae*

grandifolia (Brazil); tropical tree with large, parallel-veined velvety foliage; flowers purple with inside white, turning orange-red with age, 4 cm dia., and smaller than semidecandra. *Tropical.* p. 647

granulosa (Brazil, Bolivia), "Purple glory-tree"; Andean tree 5 to 12 m tall, in cultivation as a smaller shrub; square, winged branches with long ovate, dark green pubescent leaves with 5 depressed veins, bristly on margins; large flowers purple or violet-blue, 5-7 cm across. *Subtropic.* p. 643, 647

heteromalla (Brazil); small evergreen shrub to 1½ m high, with round stems, ovate leaves soft-hairy above; flowers purplish-violet 2½-4 cm across. *Tropical.* p. 647

semidecandra (S. Brazil), "Glory bush"; free-branching, tree-like shrub growing to 8 m high, with 4-angled stems and fresh-green, 10 cm ovate leaves densely covered with soft white hairs; large violet-purple flowers blooming over a long period of time. *Tropical.* p. 642

TIGRIDIA   *Iridaceae*

pavonia (Mexico, Guatemala), "Tiger flower"; gay summer-blooming bulbs; erect stems with leaflike spathes bearing large brilliantly colored cup-like flowers spotted with yellow and purple in the center, 8 to 15 cm across, the spreading segments red or yellow, lasting only a day but succeeded by others. *Tropical.* p. 525

TILIA   *Tiliaceae*

x euchlora (cordata x dasystyla), "Crimean linden"; deciduous tree to 20 m, rounded ovate leaves 10 cm long, serrate at margins; inflorescence in pendant clusters of yellowish flowers, principally stamens, and partially attached linear bract. *Temperate.* p. 909

x europaea (intermedia x vulgaris), "Common lime" or "Basswood"; large deciduous tree to 40 m high; soft leaves broadly ovate to 10 cm long, and sharply serrate; fragrant yellowish flowers with numerous stamens, subtended by a strap-like pale green bract. *Temperate.* p. 909

TILLANDSIA   *Bromeliaceae*

albertiana (Argentina: Salta Prov.); small, pretty species to 20 cm high, in habitat growing on rocks; elongate branched stems forming clusters; the leaves to 15 cm long, narrow concave, and covered with gray scales; solitary large showy flowers 4 cm long, vivid cinnabar red. *Arid-subtropic.* p. 223

andreana (Colombia, Venezuela); small, stiff, clustering epiphytic rosette, stemless but eventually forming leaning stems; tapering linear leaves to 5 cm long, densely brown scurfy-scaly, keeled beneath, and recurving; compound, shingled inflorescence without stalk, with short membranous floral bracts, papery sepals and erect red petals 3 cm long. *Subtropic.* p. 223

bergeri (Uruguay, Argentina); small succulent rosette of stiff, gray-green leaves to 10 cm long, spirally arranged and forming stems and densely matting with continuous growth, the foliage channeled and covered with silvery scales; pretty inflorescence with pink bracts and spreading 3 cm purplish petals with white base. *Subtropic.* p. 226

caput-medusae (Mexico and south); attractive small rosette with bulb-like base; thick channeled, tapering and twisting leaves glistening with silky-gray hairs; short panicles with pale blue flowers. *Tropical.* p. 224, 226

crispa (Panama, Colombia, Ecuador, Perú); miniature epiphyte 10-25 cm high, with ovoid bulbous base, bearing brown-scurfy narrow leaves with undulate margins, green with maroon spots; slender stalk bearing spiked inflorescence with scurfy, red floral bracts and purplish flowers. *Tropical.* p. 224, 226

cyanea (morreniana) (Ecuador: Manabi to Loja Prov.), "Pink quill"; excellent, suckering rosette of linear, channeled leaves with red-brown lines; short spike with broad flattened, clear pink bracts and large violet-blue flowers. *Tropical.* p. 224, 230

duratii (So. Brazil, Bolivia, Uruguay, Argentina); curious epiphyte or xerophytic rock-dweller, forming curved and twisted stems to 30 cm long; the white-scurfy, channeled leaves to 20 cm long, spreading or coiled downward or around its host; the stout floral stalk with very fragrant lilac flowers. *Subtropic.* p. 223

fasciculata (Florida, W. Indies, C. America), "Wild pine"; epiphytic dense rosette with hard linear-lanceolate leaves, gray and recurved; branched inflorescence with greenish bracts tinged red, flowers blue. *Subtropic.* p. 226, 227

fendleri (No. Perú); beautiful species growing on rocks of the precipitous mountains in the Western Cordilleras and found on the ruins of the ancient Inca fortress of Machu Picchu; with inflorescence 1 to 2 m high, broad green leaves laxly pendant, forming rosette 60-100 cm in dia., the floral stalk branched with numerous flattened spikes of red to greenish bracts and blue petals. *Tropical.* p. 225

flabellata (Guatemala), "Red fan"; beautiful rosette of narrow, recurved leaves fresh green turning red in good light; giving rise to a 60 cm inflorescence branching into flattened spikes arranged fan-like, the bracts vivid red, flowers blue. *Tropical.* p. 226

flexuosa (aloifolia) (So. Florida to So. America), "Spiralled air plant"; hard rosette with leaves starting off at the base with a twist, broad but tapering, thick leathery, concave, silvery gray over green, with indistinct silver bands outside; 2-ranked inflorescence with rose bracts and white flowers. *Subtropic.* p. 224

geminiflora (So. Brazil, Paraguay, Argentina); small stiff rosette of purplish gray pointed leaves with a short pendant raceme with coppery-red bracts, yellow or lavender flowers. *Subtropic.* p. 224

imperialis (Mexico: Oaxaca, Puebla, Veracruz), "Christmas candle"; showy epiphyte at home at 1,500-2,600 m altitude, and largely used by Mexicans at Christmas time to decorate for their "Natividad", because of the festive spirit radiated by the flaming red central inflorescence, looking like a candle or slender cone; remaining in brilliant color through summer into winter; flowers purple; the dense, formal rosette of broad, smooth leathery leaves a pleasing light green, about 45 cm long. *Tropical.* p. 227

ionantha (erubescens) (So. Mexico to Nicaragua), "Sky plant"; tufting, miniature rosette only 5-10 cm high with numerous closely overlapping leaves recurving, thick-fleshy, channeled, fresh green but covered on outside with silvery bristles; sessile inflorescence with violet flowers. *Subtropic.* p. 224, 227

jalisco-monticola (Jalisco, Mexico); striking epiphyte with mealy-gray recurving, concave leaves; inflorescence a flattened spike of boat-shaped, imbricated bracts, red at base, to green near apex; flowers yellow. *Tropical.* p. 227

lindenii (N.W. Perú), "Blue-flowered torch"; attractive, formal rosette of recurved linear channeled leaves green with red-brown pencil lines becoming more prominent toward base; inflorescence a long spike of flattened carmine-rose bracts with large royal-blue flowers with white eye in center. *Tropical.* p. 224

mooreana (Amazonian Perú); open olive-green rosette with giant branched inflorescence similar to T. wagneriana but larger, the inflated bracts a deeper shade of pink, and with large lavender-blue flowers. *Tropical.* p. 226

multicaulis (Mexico, Costa Rica, Panama), the "Gold fish bromeliad"; beautiful small epiphyte from humid mountain forests to 2,500 m; resembling Vriesea with soft, shiny green leaves to 30 cm long forming a birdsnest-like rosette; from between the foliage rise several flattened spikes of gold and scarlet red looking like shimmering gold fish; the petals violet-blue. *Tropical.* p. 230, 232

oerstedii (Costa Rica); magnificent, large species; rosette of stiff-erect, grayish-green leaves, from the center a tall, stout stalk with multitudes of lateral, flattened spear-shaped spikes of shingled canary-yellow bracts and light purple flower petals. *Tropical.* p. 223

plumosa (Mexico, Guatemala); small epiphytic rosette 15 cm high, forming dense masses; thread-like, recurving, soft leaves to 18 cm long, silvery gray-green; inflorescence barely stalked, with scurfy, spreading floral bracts, and narrow violet petals. *Tropical.* p. 223

ponderosa (Guatemala); large rosette of sword-shaped leaves; central raceme of clustered spikes of yellow bracts, purple petals. *Tropical.* p. 223

prodigiosa (Mexico: Chiapas, etc.); striking bromeliad, in habitat epiphytic in pine and oak forests at 1,800-2,400 m; somewhat tubular, flaccid rosette of broad grayish-green thin-leathery leaves spotted purple; so spectacular because of the massive, pendant inflorescence 1-1½ m, and even 2 m long, composed of little vriesea-like flower heads, subtended by large elliptical floral bracts in brilliant red, corolla yellowish. *Tropical.* p. 227

punctulata (Mexico); birdnest-like dense rosette 30 cm high, with broad ovate base suddenly narrowing to tapering concave leaves, fresh green and like flexible leather; dense spike with bright red bracts with lilac flowers tipped white. *Tropical.* p. 228

recurvata (Subtropical and Tropical America: So. U.S., Florida to Key Largo, to No. Argentina and Chile), the "Ball-moss"; epiphytic rosette forming dense masses, on occasion adapting itself to a strange life perched high on telephone wires, a familiar sight in Florida; thread-like, stiff-leathery, curving leaves 3-18 cm long, covered with silvery-gray or brownish scales, on short, tufting stems, with some roots present; few-flowered inflorescence with scurfy bracts, and pale violet petals. *Subtropic.* p. 225, 236

streptophylla (Jamaica), "Twist plant"; leaves in dense basal rosette with sharply recurved leathery leaves gradually tapering to a coiling tip; gray-green thickly covered with silvery scurf, turning red-purple in strong light; branched inflorescence with rosy bracts and lilac flowers. *Tropical.* p. 227

stricta (Brazil, Paraguay, Argentina, Guayana, Venez.), "Hanging torch"; small rosette becoming caulescent, with narrow, thin-leathery, tapering leaves recurving, green with silvery scurf, and short-stalked inflorescence with red-tinged bracts; flowers violet turning red. *Tropical.* p. 227

TILLANDSIA

tenuifolia (pulchella) (No. Argentina); small rosette of narrow leaves with arching inflorescence of rosy bracts and purplish flowers. *Subtropic.* p. 230

tricolor (So. Mexico, Guatemala, Nicaragua, Costa Rica); grass-like densely tufting, epiphytic rosettes spreading by branching rhizomes, the stiff leaves dark glaucous green, purplish at base, and finely scaly, from a linear 2 cm width tapering into an awl-shaped point; inflorescence on slender stalks with single two-ranked flattened spike, boat-shaped green floral bracts, red toward base, and yellow apex; flowers violet tipped with white. *Tropical.* p. 229

usneoides (S.E. United States to Argentina and Chile), "Spanish moss"; growing from trees as silvery-gray threadlike masses to 6 m long, densely covered by the gray scales which are a means of receiving and holding atmospheric moisture, and which helps to enable the plant to dispense with roots; small axillary flowers with petals 1 cm long, in changing colors yellowish-green to blue. *Subtropic.* p. 228, 230

utriculata (Florida, W. Indies), "Big wild pine"; rosette of spreading linear leaves 60 cm long, gradually tapering from an ovate base, and recurved top; compound spike with two-ranked green bracts edged red, and erect flowers with greenish white petals; plant dies after flowering without off-setting. *Subtropic.* p. 227, 229

x 'Victoriae' (ionantha x brachycaulon); by M. Foster; miniature rosette 10 cm dia., recurving narrow leaves, reddish with gray scurf; blue flowers. *Subtropic.* p. 230

wagneriana (Amazonian Perú), "Flying bird"; possibly the most outstanding and beautiful of all Tillandsias with its branched inflorescence of long-lasting flattened bracts in silver pink, and blue flowers, rising ½ m high from a rosette formed by the soft, thin-leathery, shining green leaves. *Tropical.* p. 227

xerographica (So. Mexico: Oaxaca; El Salvador); stiff epiphytic rosette with narrow concave, recurving silver-gray leaves; branched inflorescence with rose bract-leaves, flattened lateral greenish spikes on red stems; petals purple. *Tropical.* p. 225

TIPUANA *Leguminosae*
tipu (So. Brazil, Bolivia, Argentina), "Rosewood" or "Pride of Bolivia"; semi-deciduous, wide-crowned tree to 30 m high; the light green, odd-pinnate fern-like leaves with 6-11 pairs of oval leaflets 3-4 cm long; profuse sprays of golden-yellow or apricot-colored peaflowers, in profusion during summer like tiny butterflies; winged pods 6 cm long; a source of rosewood timber. *Subtropical.* p. 563

TITANOTRICHUM *Gesneriaceae*
oldhamii (Taiwan, So. China); erect perennial herb to 1 m high, with fleshy rhizome; rough green, white-hairy toothed, ovate leaves decreasing in size up the elongate stem to the nodding, swollen tubular hairy golden flowers, bold brown-red inside on spreading lobes; scale-like reproductive bodies often replace flowers toward the apex of the inflorescence. *Humid-subtropic.* p. 506

TITHONIA *Compositae*
rotundifolia (speciosa) (Mexico, Cent. America), "Mexican sunflower"; shrubby annual forming bushes 1 to 3 m high, with ovate crenate leaves 15 cm long, sometimes 3-lobed; on hairy, stout blackish stalks, the inflorescence sunflower-like 8 cm across, a disk ringed by oval florets vivid orange-scarlet, orange beneath; in autumn. *Tropical.* p. 332

TODEA *Osmundaceae (Filices)*
barbara (New Zealand, Australia, So. Africa); thick, erect rhizome with crown of leathery bipinnate leaves to 1 m long, the pinnules toothed. *Humid-subtropic.* p. 438

TOLMIEA *Saxifragaceae*
menziesii (Alaska to California Coast), "Piggy-back plant"; pubescent perennial herb with soft, fresh-green, lobed and toothed leaves covered with scattered white bristles, carried in a basal rosette, grown in pots as a curiosity, as it produces young plantlets out of the base of mature leaves, which can be cut off and rooted; small greenish nodding flowers, lined with maroon, in long erect, slender raceme. *Warm temperate.* p. 877, 881

TORENIA *Scrophulariaceae*
fournieri (Vietnam), "Wishbone plant"; delicate, small, herbaceous annual with ovate, fresh-green, serrate leaves and scattered, bilabiate, attractive, pale violet flowers with lower lip having 3 lobes of velvety deep violet and a yellow blotch in the middle of the lower lobe, blooming almost continuously. *Tropical.* p. 886

TRACHELOSPERMUM *Apocynaceae*
jasminoides (Rhynchospermum) (Himalayas), the "Star jasmine" or "Confederate jasmine" of the South; small woody evergreen with wiry stems and milky sap, slowly climbing and twining, with 5-8 cm leathery leaves; small white, star-like 2 cm flowers with wavy lobes. A pretty, free-blooming jasmine for pots and intensely fragrant; can be kept bushy by trimming. *Subtropic.* p. 80

jasminoides 'Variegatum'; an attractive sport of the "Confederate jasmine", with variegated leaves dark green, milky-green and white, and often tinted carmine red. *Subtropic.* p. 78

TRACHYCARPUS *Palmae*
fortunei (Chamaerops excelsa hort.) (China, Japan), "Windmill palm"; somewhat hardy, with solitary, shaggy trunk 3-12 m high, covered with a mat of long, dark brown fibers, topped by dense crown of tough fan-shaped dark green leaves 50 to 75 cm across, divided into stiffish, folded segments nearly to base, glaucous beneath; fruit lustrous blue. *Subtropic.* p. 795, 802, 804, 805

fortunei 'Nana' "Dwarf windmill palm"; dwarf form cultivated in Hawaii and Japan, with stiff fronds 30 cm across. *Subtropic.* p. 804

wagnerianus (takil in hort.) (Western Himalayas), "Takil palm"; handsome fan palm of compact habit and slow growth, with short robust trunk clothed with a mat of furry brownish fiber; the stiffly rigid, deeply plaited palmate leaves average 40 cm across, rough dull green on horizontal unarmed petioles; very decorative in small plantings or containers. Trees at Los Angeles Arboretum in Arcadia are about 6 m high, shorter than T. fortunei. (see Myron Kimnach, Principes Oct. 1977). *Subtropic.* p. 805

TRADESCANTIA *Commelinaceae*
albiflora 'Albo-vittata' (C. America), "Giant white inch plant"; vigorous plant with succulent creeping branches; fairly large, fleshy lanceshaped leaves delicate bluish-green, striped and borderd white; 3-petaled white flowers. *Tropical.* p. 306

fluminensis, the "Rio tradescantia"; wide-spreading succulent creeper with fleshy, small bluish leaves, purplish beneath; white flowers. *Tropical.* p. 306

fluminensis 'Variegata' (Argentina, Brazil), "Wandering Jew"; lively little creeper rooting at nodes, generally smaller and weaker than albiflora, with shining ovate leaves fresh green and striped and banded yellow and cream; flowers white. *Tropical.* p. 305, 306

hirta (Setcreasea hirsuta) (Namibia); clustering plant with grass-like rosettes of narrow, succulent leaves concave above, rounded beneath, to 20 cm long, glaucous green and thickly covered with long hair; flowers purplish pink. *Subtropic.* p. 307

multiflora: see Gibasis geniculata

sillamontana (villosa) (N.E. Mexico), "White velvet creeper"; introduced by the trade as Tradescantia 'White Velvet' and 'White Gossamer'; fleshy trailer with clasping ovate leaves in ranks, deep green with parallel veins but entirely covered with fluffy white wool, underside and stems purplish; tripetaled flowers rich orchid; erroneously listed as Cyanotis veldthoutiana. *Subtropic.* p. 305, 306

TRADESCANTIA: see also Callisia, Dichorisandra, Rhoeo, Siderasis, Gibasis, Hadrodemas, Zebrina.

TREVESIA *Araliaceae*
palmata 'Micholitzii' (Yunnan, China), "Snowflake plant"; small evergreen tree with puckered leaves, 40 cm across or more, palmately lobed, the segments irregularly pinnate, thin-leathery, glossy green and covered with silvery dots, which earned it the name "Snowflake plant"; the petioles are spiny. *Subtropic.* p. 143

TRICHANTHA: see Columnea

TRICHOCAULON *Asclepiadaceae*
meloforme (Namibia, Cape Prov.); thick solitary spherical succulent 8-10 cm high gray-green, with flattened tubercles; small 1 cm cup-shaped flowers near the apex, dark outside, yellow inside and spotted maroon. *Arid-subtropic.* p. 154

TRICHOCENTRUM *Orchidaceae*
albo-purpureum (No. Brazil); pretty, dwarf epiphyte having minute pseudobulbs with solitary leaves 8-15 cm long; stalks with relatively large, waxy flowers 5 cm across; narrow, pointed sepals and petals buff-brown tipped with yellowish-green, and spotted red-brown; the wide fan-like lip white marked purple on either side of the crest; summer. *Tropical.* p. 759, 760

tigrinum (Costa Rica, Ecuador); desirable dwarf epiphyte with shining green 10-15 cm leaves on very small pseudobulbs; large flowers 5 cm across, sweet-scented and waxy, with sepals and petals tawny yellow and blotched with brown, the large lip white with two large purple spots near base (May-June). *Humid-subtropic.* p. 759

TRICHOCEREUS *Cactaceae*
johnsonii (So. America); deeply ribbed green tall column, nearly spineless; large bell flowers, inside white, outside greenish, 12-14 cm across. *Arid-tropical.* p. 251

pachanoi (Ecuador), "Night-blooming San Pedro"; slender column to 5 m high, forming many vertical branches 8 cm or more thick, light to dark green, glaucous while young, 6-8 rounded ribs, wanting spines; large white nocturnal flowers very fragrant, 20 cm long. *Arid-tropical.* p. 251, 252, 253

pasacana (Bolivia, Argentina), "Torch cactus"; giant stout tree to 10 m high, sparingly branched; columns to 30 cm thick, dull green, closely ribbed, freely yellow-spined; nocturnal flowers white, 10 cm long. Edible greenish fruits called "Pasacana". Barrel-shaped in seedling stage. *Subtropic.* p. 246
peruvianus (Perú), "Peruvian torch-cactus"; branching with stout columns, to 4 m tall, and 8-10 cm thick, dull green, glaucous when young, 6-8 broad and rounded ribs, brown spines; large white flowers to 25 cm long. *Tropical.* p. 254
racaquiensis (Andes of So. America); slender, tall column with 10-12 deep ribs, long red spines; toward apex densely covered with white wool. *Arid-subtropic.* p. 254
spachianus (W. Argentina), the beautiful "Torch cactus"; short slender columns to 2 m high, close-ribbed, 6 cm thick, clustering, short brown spines; flowers white, 20 cm long. *Subtropic.* p. 252
weberbaueri (Weberbauerocereus) (Perú); distinctive cluster of slender spiny columns to 4 m high, branching from the base; the limbs to 10 cm dia., gray green, with 15-22 ribs; 5 cm flowers light brown; fruit orange yellow 4 cm dia. *Arid-tropical.* p. 245

TRICHODIADEMA   *Aizoaceae*
densum (Cape Province: Karroo), "Miniature desert rose"; spreading shrub-like small succulent with fleshy roots and short stems, the crowded green leaves 2 cm long topped by long radiating white hairs; flowers violet-red 5 cm across. *Arid-subtropic.* p. 49
stellatum (Cape Prov.); turf-forming shrubby succulent to 10 cm high, with long fleshy roots, leaves 1 cm long and nearly cylindric, gray-green and rough with stiff white bristles at apex; flowers violet-red. *Arid-subtropic.* p. 44

TRICHOMANES   *Hymenophylleae (Filices)*
reniforme (New Zealand), "the Kidney-fern"; a little fern of such peculiar habit that it does hardly look like a fern; usually growing on the floor of beech-forest but often epiphytic; with naked wiry rhizomes, bearing on thin stalks 5-20 cm long, chalice-like or kidney-shaped fronds 5-10 cm broad and notched at base, very transparent and light green when young, thickish and glossy dark green, when older; the margins ciliate and with spore masses all around, on thin stalks 5-20 cm long. *Humid-subtropic.* p. 437

TRICHOPILIA   *Orchidaceae*
suavis (Costa Rica); beautiful small epiphyte with thin pseudobulb and solitary broad leaves, and clusters of relatively large flowers delicately hawthorn-scented, creamy-white, with large frilled lip, yellow in throat and spotted purplish rose (Dec.-May, Oct.) *Tropical.* p. 759

TRICHOSANTHES   *Cucurbitaceae*
cucumeroides (Japan), "Snake-gourd"; herbaceous vine with thick, tuberous root, climbing by tendrils 4 to 5 m; leaves palmately lobed and lightly crenate, finely pubescent beneath; white flowers; the fleshy long-ovoid fruit light gray, with deep green markings turning red. Dried fruit is used as soap substitute. *Subtropic.* p 383

TRICHOSPORUM: see Aeschynanthus

TRICHOSTACHYS   *Rubiaceae*
aurea (Trop. Africa); low creeping plant with small cordate green leaves 2-3 cm long, the surface puckered and shaded with bronze; small scarlet red berries are carried upright on wiry stalks. *Tropical.* p. 862

TRICYRTIS   *Liliaceae*
hirta (Japan), "Toad lily"; rhizomatous herb to 1 m, hairy all over; leafy stems with ovate foliage, 15 cm long; flowers in clusters, white and spotted purple inside, to 3 cm long. *Warm temperate.* p. 602

TRIFOLIUM   *Leguminosae*
dubium (minus) (native in Ireland and elsewhere in Europe); naturalized in No. America), the "Yellow clover", and widely grown as "Irish Shamrock"; annual trefoil with branching creeping stems 15-45 cm long and hued brown; the 3 small leaflets matte satiny green, obovate and obcordate 1 cm long, the terminal one attached by individual stalklets to the petiole, the petioles and stipules being hairy, unlike in repens which are not hairy and has leaflets rounded at summit; T. dubium has canary-yellow or greenish-yellow small flowers 1 cm long, in loose heads, while repens has corolla white or tinged with pink. *Temperate.* p. 546

TRIPOGANDRA: see Hadrodemas

TRISTANIA   *Myrtaceae*
conferta 'Aurea-variegata' (E. Australia), "Brisbane box"; evergreen tree with beautiful foliage; the green species 10-20 m high, and widely planted as a street tree in Sydney and Melbourne; reddish bark peeling to show smooth pale bark beneath; handsome leathery, ovate leaves 15 cm long, clustered in whorls at end of branchlets, glossy green with golden yellow variegation along center; creamy-white 2 cm flowers. *Subtropic.* p. 680, 684, 688

TRISTELLATEIA   *Malpighiaceae*
australasiae (Malaysia, Australasia), "Galphimia vine"; climbing shrub with dull green ovate leaves 5-10 cm long; bright yellow flowers with red stamens in striking racemes. *Tropical.* p. 620, 621

TRITOMA: see Kniphofia

TRITONIA   *Iridaceae*
crocata (So. Africa), "Kalkoentje"; cormous herb 60 cm high with linear sword-shaped leaves, and wiry stems with expanded bell-shaped flowers tawny-yellow or orange 5 cm wide, May-June; somewhat winter-hardy with protection. *Subtropic.* p. 526, 527
x crocosmaeflora (Crocosmia aurea x Tritonia pottsii), known in horticulture as "Montbretia"; a popular hybrid much cultivated as cut flower and widely planted in mild climate gardens; charming cormous plant with 4 soft light green, sword-shaped leaves 2 cm wide on either side of the floral stalk, with many parallel ribs; branched sprays 50 cm or more high, with curved funnel flowers in long one-sided spikes, the waxy corolla 3-4 cm long, with segments flaring wide and a flaming orange-crimson, lasting a long time on the plant or 2 weeks and more as cut flowers; somewhat winter-hardy with protection. *Subtropic.* p. 527, 528
pottsii (So. Africa: Natal, Zululand), the "Slender tritonia"; this species resembles the garden montbretia, T. crocosmaeflora, but the flowers are more narrow and are a rich crimson; each flower has a curved slender tube about 4 cm long with 6 short petals which do not open very wide; the flower spike has numerous flowers open at the same time and buds continue to open in midsummer; lateral spikes spring from the main stalk, growing to nearly 1 m tall, subtended by the bluish-green leaves. *Subtropic.* p. 529

TRITONIA: see also Crocosmia

TROCHODENDRON   *Trochodendraceae*
aralioides (Mountains of Japan, So. Korea, Taiwan), "Wheel tree"; evergreen tree to 20 m; obovate leathery leaves 10 cm or more long, with serrate margins, in clusters at end of branches; small 1 cm green flowers without petals. *Warm temperate.* p. 908

TROPAEOLUM   *Tropaeolaceae*
majus (Perú), Brazil), known as "Nasturtium" or "Indian cress"; quick-growing, pretty, somewhat succulent glabrous annual herb, climbing by means of coiling petioles, to a height of 2-4 m, with waxy peltate leaves, and long-stemmed, fragrant 5-6 cm irregular flowers usually bright orange, sometimes in shades of red. *Subtropic.* p. 907, 908, 911
speciosum (Chilean Andes); climbing perennial to 3 m with fleshy roots; leaves divided into 5-6 leaflets; beautiful flowers to 4 cm long, vivid scarlet red, yellow at base. *Subtropic.* p. 908
tricolor (Chile), "Tricolored Indian cress"; slender climbing perennial herb with fleshy tubers; leaves circular, of 6 leaflets; curious flowers 2-3 cm long, not wide-mouthed nor spreading limb, but with fiery scarlet spur, purplish calyx lobes or sepals, and bright yellow petals. Very pretty summer-flowering twiner on wire frames, for the sunny cool window. *Subtropic.* p. 908

TSUGA   *Pinaceae (Coniferae)*
canadensis (E. No. America: Nova Scotia to Alabama), "Hemlock-spruce" or "Canada hemlock"; a very prolific coniferous often branching evergreen tree with slender horizontal branches, gracefully drooping in age; the flat needles lustrous dark green above, bluish beneath, about 1 cm long mostly arranged in opposite rows on branchlets; small brown pendulous cones. Prefers moisture, sun and wind protection but very tolerant to situations wet or dry, sunny or shade, and may be clipped into dense columns, or hedges. Best from seed. *Temperate.* p. 345, 348

TULBAGHIA   *Liliaceae*
fragrans (Transvaal), known as "Pink agapanthus"; with wide glaucous leaves less than 30 cm high, and slender stem with umbels of numerous small lavender very fragrant flowers, but the foliage does not have the garlic odor of T. violacea. *Subtropic.* p. 595
violacea (So. Africa), "Society garlic"; cormous plant with a garlic-like odor; linear channeled, soft-fleshy leaves, and umbels of bright lilac, star-like flowers, the spreading segments with deeper purplish median stripe, on stems to 60 cm high. *Subtropic.* p. 598
violacea 'Variegata', "Variegated society garlic"; attractive variety with linear leaves milky green and white margins, pink at base; starry, pretty 2 cm flowers pinkish lavender. *Subtropic.* p. 592

TULIPA   *Liliaceae*
gesnerana hybrids or "Common garden tulips", widely used for garden planting; hybrids of this group are best for early spring and Easter-flowering in pots. Derived from T. gesnerana (Armenia, Iran), and inbred with several other species in Asiatic Turkey, So.

Russia and Iran, they were brought to Western Europe from 1554 on, reaching Holland sometime after 1573. *Warm temperate. p. 604*
  '**Apeldoorn**' (Darwin hybrid); large and beautiful flowers of noble form, orange-scarlet, their black base outlined in yellow. *p. 604*
  '**Bartigon**' (Darwin); popular pot variety with medium large flowers bright cochineal to crimson red, normally mid-May blooming; but can be forced from Jan. 15 on; 60 cm tall. Darwin flowers have squarish base in profile. *p. 604*
  '**Blenda**' (Triumph); stiff-erect Triumph tulip with cupped flowers, pointed petals, deep rose with white base, early. *p. 605*
  '**Blizzard**' (Triumph); heavy-bodied Triumph tulip 40-45 cm high, with stout and heavy foliage; substantial large, full flowers creamy-white 7 cm long; good for Easter pots. *p. 605*
  '**Karel Doorman**' (Parrot); also known as "Dragon tulip", large cherry red flowers mingled with green, and edged with golden yellow, the perianth segments variously frilled; 40 cm. *p. 605*
  '**Kees Nelis**' (Triumph); mid-season Triumph tulip about 34-46 cm tall, of somewhat flexuous habit but responding fast to forcing; very striking in dark flame red, edged with orange-yellow. *p. 605*
  '**Makassar**' (Triumph); dark canary yellow, excellent late-blooming Triumph tulip for Easter pots, about 50 cm tall, with stiff stem and foliage, the cupped flowers 7 cm high, clear yellow; often producing more than one bloom per stalk. *p. 605*
  '**Paris**' (Triumph); a very colorful, recommended pot variety, deep orange-red, each segment edged with contrasting yellow, on stiff stalks, 45 cm high. *p. 605*
  '**Peony**'; "Double Early" tulip with robust stems carrying the large very double flowers of deep rose pink. *p. 604*
  '**Princess Irene**' (Single Early); sport of 'Couleur Cardinal'; coral or soft orange with bronze-red flaming from base. *p. 605*
  '**Red Giant**'; an excellent "Triumph" tulip for Easter pots; a group resulting from crosses of "Darwin" and "Early" tulips; large, substantial, long-lasting, deep scarlet flowers, often 2-3 on the stiff robust stem. *p. 605*
  '**Robinea**' (Triumph), a superior "Triumph tulip" for Easter flowering; a healthy companion to 'Red Giant', even stockier, 30-36 cm above pot, with firm-fleshy, large flowers deep crimson, often producing more than one bloom per stalk; resistant to "rotting off" or "fire". *p. 604, 605*
  '**Rose Beauty**' (Triumph); lovely bicolor, deep pink maturing to cherry red against a white base; the flowers becoming quite large as they mature; 38 cm. *p. 605*
  '**United Europe**', a "Triumph tulip" of sturdy habit, strong erect stalks with large, solid flowers scarlet red with salmon sheen, the petals with narrow yellow margin. Excellent keeper. *p. 605*
  '**Ursa Minor**' (Single Early); shapely oval medium size flower clear deep yellow, on slender stem, a good commercial pot variety once included with the "Triumphs". *p. 605*
  greigii '**Royal Orange**'; hybrid with wide cup-shaped flowers with recurving, pointed outer segments 9 cm long, vivid orange-scarlet edged with yellow, base black; the glaucous leaves with purple-brown mottling. The species is bright scarlet (Turkestan). *Warm temperate. p. 604*
  saxatilis (Crete); stoloniferous species 30-45 cm high, with long, narrow, shining green leaves, and stalks with 1-3 cup-shaped fragrant flowers opening flat, the pointed segments 5 cm long, pale lilac with yellow bottom; midseason. *Subtropic. p. 599*

TUPIDANTHUS *Araliaceae*
  calyptratus (Assam: Khasia Hills; Burma), "Mallet flower"; small evergreen tree which later becomes a tall climber to 6 m high; leaves palmately divided into stalked, somewhat pendant, obovate segments to 18 cm long, glossy green and fleshy; reddish petioles and stem. A trim patio plant or for the solarium, resembling Brassaia but branches from base and grows into broad, denser shrub, and more tough-leathery. *Subtropic. p. 132, 143, 144*

TUPISTRA *Liliaceae*
  macrostigma (tupistroides) (India); perennial herb with thick creeping rhizome, long-stalked, lanceolate bright green 30 cm leaves; inflorescence a pendant spike with dark purple 1½ cm flowers. *Tropical. p. 590*

TURBINA *Convolvulaceae*
  holubii (So. Africa); curious xerophyte with large woody, brown caudex, from the apex springs the finely divided foliage. *Arid-subtropic. p. 360*

TURBINICARPUS *Cactaceae*
  polaskii (Mexico: San Luis Potosi); a midget cactus of globular shape only 1 cm high, to 2.7 cm across, resembling a small Lophophora; olive green with wrinkles, and with a single hook spine; starry flowers relatively large; 1½ cm across, white inside flushed with pink. *Arid-tropical. p. 272*

TURNERA *Turneraceae*
  ulmifolia (W. Indies, Mexico to Argentina), "West Indian holly" or "Sage rose"; pretty herbaceous shrub ½-1¼ m high, with scandent stems, and alternate narrow-elliptic leaves 8-10 cm long, nettle-like, deep glossy green, white-hairy underneath; axillary golden-yellow, 5 cm flowers with 5 petals, blooming from March to September. *Tropical. p. 910*

TURRAEA *Meliaceae*
  obtusifolia (So. Africa: Natal, E. Cape), "South African honeysuckle", or "Bluntleaf star-bush"; attractive, broad, slow-growing, more or less evergreen shrub 1-1½ m high, with small obovate, recurved dark green leaves 3-5 cm long; numerous axillary solitary, star-like white flowers with 5 narrow petals, 2½ cm across, and sweetly scented; red berries. *Subtropic. p. 642*

TYDAEA: see Kohleria

TYPHA *Typhaceae*
  latifolia, (N. America, Europe, Asia), "Cat-tail"; decorative acquatic perennial, with creeping root-stock, tall erect unbranched reed-like stems to 2½ m high, the leaves linear, almost flat and glaucous; cylindric flower spike dark brown, 30 cm long, and 3 cm thick; dried for indoor decoration and known in England as "Reed-mace". *Warm temperate. p. 916*

TYPHONIUM *Araceae*
  trilobatum (India, Malaya); tropical aroid with near-globular tuber 4 cm thick; slender 25-30 cm petioles with hastate leaves deeply cut into 3 lobes, midlobe to 16 cm long; short-stalked, showy spathe with ovate blade 15 cm long, green outside, rose-purple inside. *Tropical. p. 129*

THYPHONODORUM *Araceae*
  lindleyanum (Madagascar, Zanzibar); handsome tree-like, smooth aroid, in habitat growing in warm lowland rivers, and sometimes becoming gigantic, 3-4 m high; forming thick stems 1-3 m long; the large, thick-fleshy, bright green sagittate leaves ½ to 1 m long, with triangular sinus; suberect creamy-white spathe 45-60 cm long, enclosing golden-yellow spadix to 40 cm long, female in lower part, male toward apex. *Humid-tropical. p. 127, 129*

UEBELMANNIA *Cactaceae*
  pectinifera (Brazil); small globular cactus to 10 cm or more dia., later becoming elongate, with about 15 ribs; at first blackish brown tinted purple, later with tiny white scales; small ½-1½ cm spines, blackish near apex, arranged in rows comb-like, the areoles dense along ridges; flowers yellow. *Arid-tropical. p. 267, 271, 272*

ULEX *Leguminosae*
  europaeus, (Western Europe, including Britain), "Gorse", "Whin" or "Furze"; densely spiny shrub to 1 m as seen wild, twice as high in gardens; shoots hairy and flowers golden-yellow 1½ cm long. *Warm temperate. p. 537*

ULMUS *Ulmaceae*
  parvifolia, (China, Japan), "Evergreen elm" or "Chinese elm"; graceful, open-headed tree to 20 m high, evergreen in mild climates; dense with ovate leaves small and firm, 2 to 8 cm long, glossy grass green and finely crenate, on pendant, weeping willowy branches with foliage arranged rather flat; blooming in late summer or autumn, with inconspicuous greenish flowers. Beautiful tree in California plantings. *Warm temperate. p. 908*

UMBELLULARIA *Lauraceae*
  californica, (California to Oregon), "Myrtlewood" or "California laurel"; strongly aromatic evergreen tree, 15-25 m; leaves alternate, glossy green above, leathery, 5-12 cm long, narrowly oval or oblong, tapered to both ends; flowers yellowish-green in stalked umbels 1 cm wide; fruit roundish 1-2½ cm long, green becoming purplish. *Warm temperate. p. 534*

UNCARINA *Pedaliaceae*
  grandidieri (Madagascar); curious succulent shrub with thick, knobby branches, small trilobed, deciduous leaves, and clusters of yellow flowers with spreading limb. *Arid-tropical. p. 812*

URBINIA: see Echeveria

URCEOLINA *Amaryllidaceae*
  peruviana (Perú), "Urn flower"; interesting low species with small 3 cm bulb; 1 or 2 oblanceolate leaves 15-20 cm long, striated and with reflexed margins; the slender floral stalk rising at base, not from crown of bulb, 20 to 40 cm high, bearing several nodding, tubular-inflated, scented flowers 4 cm long, bright scarlet, blooming in early summer. *Subtropic. p. 64, 67*

URGINEA *Liliaceae*
  maritima (Canary Islands to Syria, Brittany, Normandy), the "Sea-onion", or "Squills"; an old house plant, forming a very

large ovoid red-brown bulb 10-15 cm thick, partially above ground; 10-20 fleshy, glaucous green strap-shaped leaves 30-45 cm long, wide above the middle, in the spring, the old leafbases remaining for a time; during the leafless time in summer, a 30-90 cm long slender stalk bears a short, pyramidal raceme of whitish flowers 1½ cm wide, each segment with an indistinct green median stripe, the filament is thread-like, anthers green. The bulb furnishes syrup of squills. *Warm temperate.* p. 603

UROSKINNERA *Scrophulariaceae*
spectabilis (Mexico); softly gray hairy herb 30-50 cm high; ovate, toothed leaves 5-10 cm long; dense cluster of funnel-shaped flowers rosy-violet, 4 cm long. *Subtropic.* p. 886

URSINIA *Compositae*
versicolor (Transvaal); showy small tree in South Africa, covered in September with masses of white flowers in small heads. *Subtropic.* p. 323

UTAHIA *Cactaceae*
sileri (Pediocactus) (Utah); small globular cactus 10 cm dia., slightly flattened at the white-woolly top; 13-16 prominent, warty ribs set with 13-15 awl-shaped whitish radial spines and 3-4 dark brown central spines 2 cm long; 3 cm yellow flowers. *Warm temperate.* p. 272

VACCINIUM *Ericaceae*
corymbosum (Maine to Florida), "Highbush blueberry"; deciduous dense shrub to 5 m high, with leathery, usually pubescent elliptic leaves to 8 cm long; white to pinkish urn-shaped small 1 cm flowers, followed by the glaucous blue-black edible berries full of sweet, juicy, black flesh. *Temperate.* p. 394, 472

VALLOTA *Amaryllidaceae*
speciosa (purpurea) (So. Africa: Cape Prov.), the "Scarborough lily"; a charming evergreen plant with large brown bulb; strap-shaped bright green leaves 45-60 cm long, the fleshy, hollow ½-1 m stalk carrying a cluster of funnel-shaped bright scarlet, long-lasting flowers 8 cm across, with stamens attached to each petal; blooming from June on. Strong undisturbed bulbs produce several flower stalks in succession. An old, good house plant, but which must be kept moderately moist even during its cool rest period in winter. *Subtropic.* p. 64, 67

VANDA *Orchidaceae*
coerulea (Himalayas, Assam, Burma), the beautiful "Blue orchid" with stems 30-90 cm high, two-ranked with 20 cm, strap-type, channeled leaves, the axillary racemes with large round, membranous flowers 5-8 cm across, sepals and petals light blue with a network of deep azure, the small lip blue, (July-Jan.). *Humid-subtropic.* p. 755, 756

'Josephine' (teres aurorea x hookeriana alba); hybrid strain originated in Singapore Botanic Garden 1938, tall growing with long fleshy, channeled leaves densely arranged in two ranks; racemes of large waxy flowers pink and peach, the lip yellow overlaid with purple. *Tropical.* p. 757

lamellata (Philippines, Marianas); robust stems very densely leafy, to 50 cm tall, narrow folded leaves and recurved, 20 cm long, and cut at tip; inflorescence with waxy flowers very fragrant, to 3 cm across and long lasting; sepals and petals yellowish or greenish blotched and striped with red-brown; small lip yellow with brown. *Tropical.* p. 761

'Miss Agnes Joaquim' (hookeriana x teres) (Singapore 1893), "Corsage orchid"; terete hybrid, the famous orchid grown in Hawaii for leis and corsages, flowering in succession throughout the year; large 8 cm blooms lasting a long time, sepals white tinged with rose, the larger petals mauve purple, the broad lip purple, with yellow throat spotted red. *Tropical.* p. 755, 756, 757

'Onomea' (rothschildiana x sanderiana); large-flowered Hawaiian hybrid, 1948, typical of the large, round flowers of modern hybridizing; pink flowers flushed with peach and with a network of purple lines. *Tropical.* p. 756

parishii marriottiana (So. Burma); distinct, dwarf compact epiphyte with the short stems closely-set with fleshy leaves in 2 ranks, and axillary raceme of 5 cm flowers with sepals and petals bronzy brown suffused with rose and magenta purple lip (summer). *Tropical.* p. 756

x rothschildiana (coerulea x sanderiana); gorgeous strapleaf hybrid with clusters of large, flat and round flowers 8 to 15 cm across, of a beautiful amethyst-blue with darker netting, and of long-lasting qualities, (fall and winter on). *Tropical.* p. 756

sanderiana (Euanthe, Esmeralda) (Philppines: Mindanao); a wonderful epiphyte with stout, 2-ranked leafy stem 60 cm or more high, and axillary racemes of large round, 8-10 cm blossoms with upper segments white or pink, the lower greenish or canary-yellow tesselated with brown-crimson and small lip canary-yellow streaked with red; (Sept.-Oct.). *Tropical.* p. 756

'Sunset'; of compact habit, the stems with 2-ranked fleshy leaves; basally a wiry stalk bears a showy cluster of rose-pink flowers. *Tropical.* p. 754

teres (N.E. India, Upper Burma); showy epiphyte with terete (round) stems and leaves, to 2 m high, climbing trees in hot plains by aerial roots; flowers few but large, to 8 cm across, with sepals white, tinged with rose, the larger petals rose-magenta, lip 3-lobed carmine red with orange throat, (May-Sept). *Tropical.* p. 757, 761

teres 'Alba'; variety with slender leafy stems, the flowers 6-8 cm across, pure white. *Tropical.* p. 761

tricolor (Java), spectacular epiphyte with stems to 1 m long, dense with two ranks of recurving strap-shaped leaves; the inflorescence in lateral racemes of fragrant waxy flowers 5-8 cm across, sepals and petals lemon-yellow spotted with reddish brown, lip white with purple, (Oct.-July). *Tropical.* p. 755, 756, 757

tricolor suavis (Java, Bali); a beautiful variation, with the fragrant waxy flowers having a white base, spotted with blood-purple; boys and girls in the mountains of Bali, at 1000 m, not far from the temple of Besakih, offered me plants in flower in July, one naked boy asking 2 Rupiahs, then 20¢. *Tropical.* p. 755, 756

x VANDACHNIS *Orchidaceae*
'Premier' (Arachnis flos-aeris x Vandopsis lissochiloides); an exotic Hawaiian bigeneric "Spider orchid", with waxy flowers having curving sepals and petals creamy-yellow boldly cross-banded brown-red. *Tropical.* p. 756

x VANDACOSTYLIS *Orchidaceae*
'Dawn' (Rhynchostylis gigantea x Vanda 'Colorful'); narrow, fleshy channeled leaves; flowers with waxy petals and sepals rosy-pink; a bigeneric Hawaiian hybrid. *Tropical.* p. 761

x VANDAENOPSIS *Orchidaceae*
'Frank C. Atherton' (bigeneric Vandaenopsis x 'Jawaii' x Vanda sanderiana); lovely, robust hybrid with stout stem and fleshy leaves in ranks; the waxy flowers tawny-yellow and copper, the small lip with red. *Tropical.* p. 755

VANDOPSIS *Orchidaceae*
gigantea (Burma, Thailand); handsome epiphyte with short, stout stems to 30 cm high, with few tongue-shaped, heavy leaves 2-lobed at apex, to 60 cm long; inflorescence arching, with numerous waxy flowers to 8 cm dia., thick and heavy-textured, creamy yellow with rings of brown-red, lip yellow; lasting for several months (spring). *Tropical.* p. 758

lissochiloides (Vanda batemanii) (Philippines, Bali, Malacca); beautiful epiphyte with stout stems to 1½ m high, 2-ranked strap-shaped leaves; inflorescence waxy-white, axillary stalks with numerous large 5-8 cm, waxy flowers, sepals and petals inside golden yellow spotted with crimson, outside violet-rose, lip with yellow side lobes and orchid midlobe striped purple, (April-Oct.). *Tropical.* p. 758

VANILLA *Orchidaceae*
fragrans (planifolia or aromatica) (Eastern Mexico), "Common vanilla"; tall climbing orchid said to attain a length of 100 m, the light green, cylindrical stem bears 2 ranks of succulent, green, fleshy, elliptic leaves as well as aerial roots, the 5 cm flowers in axillary clusters, sepals and petals greenish-yellow, and wavy-edged lip almost white, deep yellow in throat; its dried seed pod provides vanilla for flavoring (Dec.-June). *Tropical.* p. 473, 753

fragrans 'Marginata'; attractive variety if grown on bark, for ornamental purposes, having nicely draped fleshy leaves of milky-green, bordered on each margin by a broad band of creamy-white. *Tropical.* p. 751, 759

fragrans 'Variegata'; a variegated leaf-form cultivated by Lecoufle in France, with the shiny green leaves banded and striped longitudinally with creamy-yellow. *Tropical.* p. 761

pompona (lutescens) (Mexico to Panama to Venezuela); a large-leaved climber with heavy cylindrical stem and broad ovate, thick-fleshy, dark green leaves 8-18 cm, with marked veining on the surface; large 15 cm flowers greenish yellow with bright yellow lip; freer blooming than fragrans. *Tropical.* p. 759, 761

ramosa (C. Africa, Nigeria, Guinea, Ivory Coast); rambling epiphyte with long climbing cylindric zigzag stems forming clinging or aerial roots; the fleshy green, elliptic leaves to 20 cm long; axillary clusters of waxy flowers, sepals and petals 2-3 cm long, yellowish-white, the small pointed lip lined red inside; bean-like fruit to 11 cm long. *Tropical.* p. 761

VEITCHIA *Palmae*
arecina (New Caledonia); elegant, medium-size feather palm with slender, smooth trunk to 10 m high and 15 cm dia.; above the long glossy green crown shaft an umbrella of 8-10 pinnate fronds on stiff petioles spreading horizontally, 2½ m long; vivid green broad leaflets 8 cm wide and 60 cm long, cut off obliquely at apex; 3 cm red fruit in large clusters on female trees. *Tropical.* p. 806

VEITCHIA

joannis (Fiji); magnificent solitary feather palm 15 m tall, and in habitat to 30 m; the smooth straight trunk prominently ringed, 25 cm dia., with green crownshaft topped by handsome pinnate fronds gracefully arching, to 3 m long, on short petioles, the dense leaflets lustrous green and prettily pendant, cut off obliquely at their apex; inflorescence from below crownshaft, with red fruit 3 cm long. *Tropical.* p. 806

merrillii (formerly Adonidia) (Philippines), the "Christmas palm", or "Manila palm"; attractive, erect palm to 6 m high, with rather slender, prominently ringed single trunk; the 1½ m fronds above a glossy green crown-shaft in handsome rigidly arching crown; bright green sword-shaped, leathery, broad leaflets many and closely placed, feathered almost to base of petiole; lustrous, attractive red fruit in pendulous clusters below the crown, a striking sight during our winter season. *Tropical.* p. 806

winin (New Hebrides); graceful feather-palm to 20 m high; the erect, ringed trunk bears a long glossy green crownshaft topped by spreading pinnate fronds, not as pendant as in V. joannis, nor are the lustrous dark green leaflets as gracefully pendulous but more horizontal; large inflorescence bearing clusters of ornamental bright red, small 1 cm fruit. *Tropical.* p. 806

VELTHEIMIA *Liliaceae*
viridifolia (capensis) (So. Africa), "Forest lily"; bulbous plant having broad lanceshaped bright green leaves with undulate margins, and arching; long tubular, nodding flowers to 4 cm long, yellowish-green shading to dusty-red, and spotted, tipped green, on long red-spotted stalk; winter blooming. *Subtropic.* p. 592

x VENIDIO-ARCTOTIS *Compositae*
'Champagne' (Arctotis x Venidium); half-hardy hybrid genus of Arctotis grandis and brevispatha with Venidium fastuosum; erect branched specimen bearing 8-10 cm daisy-like flowers; the cv. 'Champagne', white with purple zone. *Subtropic.* p. 332

VENIDIUM *Compositae*
fastuosum (So. Africa: Namaqualand), "Cape-daisy", or "Namaqualand-daisy"; showy annual herb 60 cm or more high, cobwebby when young, with small grayish-green leaves irregularly lobed; solitary large, daisy-like flowers 8-10 cm across, with orange-yellow florets chocolate-brown at base, appearing as if in two rows, arranged around the flat, shining blackish-purple disk, and opening to the sun. *Subtropic.* p. 332

VERBASCUM *Scrophulariaceae*
thapsus (Europe to Asia), "Common mullein"; tall biennial, to 2 m high, with large tomentose leaves arranged as in rosette, to 30 cm long; branched inflorescence of numerous spikes of yellow flowers 2 cm across. *Temperate.* p. 885

VERBENA *Verbenaceae*
x hortensis, "Rainbow vervain"; hybrids of teucrioides and others from Brazil to Chile; herbaceous bedding plants with spreading stems more or less rooting near the base; hairy, narrow-lobed leaves, soft to the touch; broad, showy clusters of salver-form flowers in pink, red, yellow, white, blue, salmon, purple, lilac, some with white eye, blooming profusely from spring to October. *Tropical.* p. 918

peruviana 'Flame', "Scarlet vervain"; low carpet-forming prostrate perennial, in its original form with crimson flowers, at home in Perú, Uruguay, and So. Brazil; the first creeping and rooting, then ascending branches with crenate, rough leaves 2½-5 cm long, and showy clusters of salver-form flowers 1 cm wide; brilliant scarlet in this color-form, and nearly everblooming, especially in summer. *Tropical.* p. 918

tenera var. maonettii (So. Brazil to Argentina); semi-shrubby creeping perennial with spreading stems rooting at nodes; hairy leaves finely divided; flat clusters of small 1-2 cm carmine-rose flowers with distinct white margins. *Subtropic.* p. 918

VERNONIA *Compositae*
colorata (Uganda), a tropical "Ironweed"; sparry shrub with woody branches, broad hairy leaves, and showy umbels of massed small white flower heads. *Tropical.* p. 323

natalensis (So. Africa), "Ironweed"; shrubby bush with felty elliptic leaves to 10 cm long; the branches terminated by clusters of tubular crimson red flowers. *Subtropic.* p. 329

VERONICA: see Hebe

VERSCHAFFELTIA *Palmae*
splendida (Seychelles), "Stilt-root palm"; unique fanleaf palm, beautiful as a slender mature tree, to 25 m tall; trunk to 15 cm dia., spiny when young, and supported by aerial roots at the base; the deep green leaves pinnately veined, quilted and more or less entire, especially when young; green fruit 3 cm dia., carried between leaves. *Tropical.* p. 804, 805

VERTICORDIA *Myrtaceae*
nitens (Western Australia), "Feather flower"; erect evergreen, heath-like shrub to 1 m high; with slender stems and long-linear, fleshy, needle-like leaves 3 cm long; the inflorescence in striking terminal clusters of deep yellow fringed cups, and masses of prominent orange-yellow anthers. *Subtropic.* p. 687

VIBURNUM *Caprifoliaceae*
x hillieri (erubescens x henryi); semi-evergreen shrub to 2 m; young foliage coppery, bronze in winter; cream-white flowers; berries red, later blackish. *Warm temperate.* p. 291

lantana rugosum (Europe, West Asia), "Wayfaring tree"; deciduous shrub to 5 m, sometimes tree-like; finely toothed ovate leaves; flowers white; red berry-like fruit, becoming black. *Warm temperate.* p. 291

odoratissimum 'Irvinii' (Himalayas to Japan), "Sweet viburnum"; evergreen shrub to 6 m high with lacquered elliptic leaves 15-28 cm long, bright green turning red; clusters of white, slightly fragrant flowers. *Warm temperate.* p. 291

opulus (Europe, No. Africa, No. Asia), "Cranberry bush"; deciduous shrub 3-4 m high, with opposite 3 to 5-lobed maple-like green leaves, rugose above and pubescent beneath, to 10 cm long, turning red in autumn; flowers white in stalked clusters, the white marginal flowers sterile, the center filled with fertile flowers producing the scarlet fruit. *Temperate.* p. 291

plicatum (China, Japan), "Japanese snowball"; deciduous shrub to 3 m high, with 10 cm ovate toothed leaves, pubescent beneath; clusters of white flowers in 2 rows. *Temperate.* p. 291

tinus (S.E. Europe, Mediterranean reg.), "Laurustinus"; evergreen thickly branched and luxuriantly leafy shrub, with ovate deep green, stiff-leathery 8 cm foliage with rough underside, on reddish petioles; dense 5 to 8 cm clusters of tiny pinkish-white, very fragrant flowers; blooming May to August. Known in Europe as "Laurustinus", it is one of the most popular durable decorator tub plants, and with its handsome foliage and compact shape ideal for cooler areas, the patio, or roof garden. *Subtropic.* p. 290, 291

VICTORIA *Nymphaeaceae*
cruziana (trickeri) (Paraná, Paraguay, No. Argentina), "Santa Cruz water lily"; large perennial aquatic with a thick rhizome, thorny petioles and round, floating leaves not as large as regia but with higher upturned margins to 20 cm high and green; flowers white turning deep pink on second day; requires only moderate heat. *Humid-subtropic.* p. 705

regia (amazonica) (Guayana, Amazon, Bolivia), "Royal water lily"; gigantic, floating, 1½ m fresh green leaves with upturned red edges; projecting air-filled ribs beneath give the leaves great buoyancy, sufficient to support great weight; the fragrant floating flowers are white turning deep rose the following day; at home in quiet "Igarapes" of warm 30°C water. *Humid-tropical.* p. 705

VIGNA: see Phaseolus

VINCA *Apocynaceae*
major 'Variegata' (So. Europe, No. Africa), "Band plant"; trailing evergreen basket plant or for window boxes, with long, thin, wiry vines with opposite, oval, green leaves beautifully edged in cream, to 5 cm long; flowers blue. *Subtropic.* p. 74

minor (So. Europe to Asia Minor), "Periwinkle"; trailing evergreen subshrub with glossy dark green, ovate leaves; flowers bluish-purple with white throat, 2 cm dia. *Warm-temperate.* p. 75

rosea: see Catharanthus

VIOLA *Violaceae*
hederacea (Erpetion reniforme) (New South Wales, Victoria, Tasmania), "Australian violet", "Trailing violet", or "Ivy-leaved violet"; attractive trailing species; the vertical rhizome putting out long, thread-like stolons with well separated tufts of leaves; these kidney-shaped or rounded, 2-4 cm across, fresh green and herbaceous; small 2 cm flowers with petal-tips white, center area violet except for white eye; scarcely spurred. *Subtropic.* p. 910

tricolor (No. Europe to C. Asia), "Wild pansy", "Johnny-jump-up" or "Kiss-me-love"; pretty annual of tufted habit, to 15 cm high, with ovate, crenate, fresh waxy-green leaves; small and cute miniature pansy flowers 1-2 cm across, with faces yellow and purple. *Temperate.* p. 910

tricolor hortensis 'Maxima' (origin Europe), "Pansy"; charming shortlived, small herbaceous perennial, with long branching stems bearing glaucous leaves, and giant flowers to 10 cm across, usually in 3 colors, violet-blue, yellow or white, and a blackish "face", as well as many other shades and colors including brown; blooming best in the cool of spring. *Warm temperate.* p. 910

VISCUM *Loranthaceae*
album (Europe, No. Asia), the Old world "Mistletoe" of legend; parasitic evergreen shrub attached to host trees such as ap-

ples, poplars, maples or pines stealing their food; forming pendulous clusters of twiggy little branches of woody texture, dichotomously forked; sickle-shaped grayish-leathery 4-9 cm leaves; berry-like fruit yellowish to translucent white, on female plants. In Europe, cut branches are an invitation to a kiss. *Temperate.* p. 618

VITEX *Verbenaceae*
lucens (New Zealand: northern No. Island), the "Puriri"; shrub or tree becoming 20 m high, with strong, oak-like timber; compound leaves of broad-elliptic, folded, leathery, bright glossy green leaflets 8-10 cm long; axillary clusters of pink, trumpet-like flowers with spreading lobes, yellow in throat, and with protruding stamens. *Subtropic.* p. 916

VITIS *Vitaceae*
vinifera 'Black Alicante' (Alicante or Black Tokay), European "Vinous" grape or "Dessert grape"; vigorously vining canes with large, deep-green leaves, covered with down underneath looking silvery; the oval fruit thick-skinned, black and covered with dense blue bloom, semi-sweet, with squashy flesh of a strong wine-flavor and earthy; for late season, and an excellent keeper, free-fruiting with large bunches of splendid appearance and weighing 1 to 2½ kg. Foremost variety long used for forcing under glass in northern Europe, and even in U.S.A. where during the time of the great conservatory ranges, it was listed in Roehrs catalogs 50 years ago. *Warm temperate.* p. 465, 473
vinifera 'Riesling Sylvaner', the "Wine grape", originally believed from the Caucasus region, known in ancient Egypt, and cultivated for centuries; woody deciduous vine moderately climbing by tendrils, with rather thin, scarcely toothed, 3-5 lobed leaves and with intermittent tendrils; small, greenish unisexual flowers in long clusters followed by delicious, fleshy, glaucous green berries, tender and sweet. The tawny yellow, high acid Riesling grape is most widely grown in temperate Europe for high production of the best white wines. *Temperate.* p. 465, 473
vinifera 'Thompson Seedless'; sport of the English white grape, discovered in Sutter County, California, from an old country budsport (from cutting); world's leading raisin, table and wine variety; large clusters of medium-sized, elongated greenish-white to light golden seedless berries, ripening in August in California, and of delicious sweet taste as a table grape. *Subtropic.* p. 921

VITIS: see also Cissus, Parthenocissus, Rhoicissus, Tetrastigma

VRIESEA *Bromeliaceae*
ampullacea ('Espirito Santo') (Brazil); an interesting new plant found by Burle-Marx in Espirito Santo; star-shaped xerophytic rosette growing on rocks, stiff daggershaped leaves silvery-pearled and tinted amethyst; edges as if cut off squarely; spiked inflorescence yellowish. *Tropical.* p.228
bituminosa (Brazil); large stocky rosette with broad blue-green leaves purple-tipped; inflorescence on erect spike with scattered bracts on either side of stalk; flowers yellow. *Tropical.* p. 231
carinata (S.E. Brazil), "Lobster claws"; dainty plant with pale green foliage; colorful and striking, flattened spike with spreading bracts deep yellow with crimson base, the yellow dotted green; flowers yellow. *Tropical.* p. 228
carinata x poelmanii; dark green glossy leaves; inflorescence branched with large flattened spikes red in center and yellow toward apex; flowers yellow. *Tropical.* p. 235
erythrodactylon (S.E. Brazil); dense rosette with a singularly beautiful inflorescence which caught my eye as growing in low rainforest in the moist-cool Serra do Mar above Santos; flattened spike green in center, curving out to spreading, coxcomblike, rose-pink tips. *Tropical.* p. 228
'Favorite' (ensiformis hybrid); vigorously growing rosette of shiny rich-green leaves; inflorescence a slender stem, usually branched, with maroon keel-shaped bracts darker than poelmanii, and arranged separated along stalk, flowers yellow. *Tropical.* p. 228
fenestralis (Brazil), "Netted vriesea"; compact rosette of broad recurved leaves arranged spirally, yellow-green leaves ornamented by numerous dark green lines and network of cross lines, purplish circles underside; sulphur yellow flowers scattered on pale spike. *Humid-tropical.* p. 228, 231
fenestralis 'Variegata'; elegant formal epiphytic rosette with broad recurving leaves 40 cm long and 6-8 cm wide, the foliage glossy rich green with a network of yellowish markings with some of the leaves having a contrasting cream border; purplish circles underside; simple inflorescence with night-blooming sulphur-yellow, fragrant flowers. *Tropical.* p. 228
fosteriana (Espirito Santo); showy rosette of stiff, nile-green to bluish leaves attractively marked with irregular dark green pencil lines across the surface, lines maroon beneath and maroon border; showy inflorescence to 2 m high, with scattered greenish bracts and pale yellow flowers. *Tropical.* p. 231

fosteriana 'Red Chestnut'; beautiful cultivar photographed at the German National Bromeliad Show in Frankfurt 1973; large rosette with broad leaves deep green with irregular cream crossbands, coppery underneath. *Tropical.* p. 228
fosteriana 'Seideliana'; handsome formal rosette of olive green leaves crossed by irregular, narrow yellow bands, center and reverse bronze. Photographed at Botanic Garden Frankfurt, Germany. *Tropical.* p. 228
gigantea (tessellata) (S.E. Brazil); strong rosette of broad leaves to 75 cm long, glabrous bluish-green when older, the younger leaves yellow-green and marked with many dark length and crosslines like a checker-board, purplish edge and tip; tall branched spike with green bracts and yellow flowers. *Tropical.* p. 231, 234
grande in hort.; showy rosette with dark foliage; branched raceme with wiry stalk; boat-shaped bracts with red base and yellow flowers. Exhibited at the German National Horticultural Show, Cologne. *Tropical.* p. 234
guttata (So. Brazil), "Dusted feather"; small compact rosette of glaucous bluish-green leaves liberally marked with maroon spots; inflorescence a pendant, lightly flattened spike of greenish beige bracts, covered with silver pink bloom, flowers lemon-yellow. *Subtropic.* p. 236
heliconoides (Guatemala, Costa Rica, Colombia, Brazil, Guayana, Bolivia); striking flowering plant, with rosette of plain, glossy green leaves 20 cm long, suffused with red underneath; the erect flattened inflorescence heliconia-like with lateral triangular boat-shaped floral bracts bright red above the middle, greenish-yellow at the apex; the flowers peeking out with creamy white petals. *Tropical.* p. 234
hieroglyphica (Brazil: Espirito Santo to Paraná), "King of bromeliads"; large epiphytic rosette with broad yellow-green leaves beautifully cross-banded with hieroglyphic marks dark green above and purplish-brown beneath; inflorescence a tall branched spike with sulphur-yellow flowers. *Humid-tropical.* p. 230, 231, 234
imperialis (Estado do Rio), "Giant vriesea"; gigantic terrestrial rosette which I found growing on the dry west slopes of the Organ Mountains; leathery green leaves in good light are deep wine-red, and even young plants produce seedling-like suckers at the base; the inflorescence a tall branched spike 2 m or more, the large bract leaves glossy maroon-red, and from which extend the arching bracted spikes with yellow flowers. *Tropical.* p. 228, 233, 234, 236
incurvata (S.E. Brazil), "Sidewinder vriesea"; light green rosette of soft leaves; the inflorescence on a leaning stem with the bract head recurving upward, fleshy bracts red and edged yellow; flowers yellow. *Subtropic.* p. 228
x intermedia (hieroglyphica x fenestralis); beautiful rosette with thin-leathery waxy leaves pale yellow-green attractively cross-barred and marbled with dark green hieroglyphics on both sides; raised inflorescence with pale yellow flowers. *Tropical.* p. 234
macrostachia (Cordillera Central, Puerto Rico); bold epiphyte with broad, hard, gray leaves; inflorescence on a central stalk, with spikes of grayish bracts in opposite ranks. *Tropical.* p. 234
patula (Tillandsia) (Perú: Tarma, to 2400 m); small tillandsia-like gray rosette with pendant inflorescence of scarlet bracts with yellow flowers. *Subtropic.* p. 234
'Perfecta' (carinata x poelmanii); magnificent flowering plant, photographed at the World's Fair Expo 1970 in Osaka, Japan; rosette with spreading glossy green leaves tinted with copper; the stout erect, long lasting, branching inflorescence of several flattened spikes of imbricated scarlet bracts with yellow flowers; a stunning display of elegance and color. *Tropical.* p. 233, 235
philippo-coburgii (S.E. Brazil), "Vagabond plant"; small rosette which sends out its offshoots from travelling stems, leaves light green with black bases; inflorescence a flattened spike with yellow and red bracts and yellow flowers. *Subtropic.* p. 232
platynema (West Indies to Brazil); rosette of flaccid strap leaves, deep glossy green; inflorescence with stalk bearing opposite spikes of yellow boat-shaped bracts, and red at base. *Tropical.* p. 232
x poelmanii (gloriosa x vangeertii); vigorous, shapely rosette of light green leaves; the inflorescence stalk usually branching into flattened spikes; bracts crimson-red with greenish-yellow apex, and yellow flowers. *Tropical.* p. 232, 236
'Poelmanii hybrid'; a highly developed, beautiful cultivar of German origin; compact rosette of broad, glossy fresh green leaves; from the center a bold and stocky, branched inflorescence of several flattened spikes of imbricated bracts, brilliant crimson red with yellow base, and yellow flowers; much more compact than V. poelmanii, and long-lasting as a decorative plant. *Tropical.* p. 235
'Polonia' (kitteliana x vigeri); small shapely rosette of little green leaves, with attractive inflorescence a branched stem with several miniature flattened heads of glowing rosy-red bracts and deep yellow flowers. *Tropical.* p. 232, 236

x retroflexa (Brazil) (natural hyb. of psittacina x simplex); small rosette with thin-leathery, waxy, pale green leaves; inflorescence pendulous, with open or loose, yellow floral bracts laterally depressed, and yellow flowers. *Tropical.* p. 236

ringens (W. Indies, Costa Rica, Colombia); small rosette of light green leaves with speckled maroon crossbands, and a slender spike with scattered rusty red bracts and yellow flowers; I found this on the Rio Mameyes in Puerto Rico, and interesting because so far north. Flowers fragrant, open at night. *Tropical.* p. 232

rodigasiana (So. Brazil), "Wax shells"; dwarf rosette of soft, dull green leaves with base tinged purple; inflorescence on branched stem with waxy lemon-yellow bracts and flowers. *Tropical.* p. 236

'Rubin'; small, open rosette of green leaves; large inflorescence of flattened spikes, glossy bracts crimson-red; petals yellow; very striking in its compact shapeliness. *Tropical.* p. 235

rubyae (Brazil); small rosette of gray green leaves; pendant inflorescence a flattened spike of shingled scarlet bracts and yellow petals. *Tropical.* p. 236

scalaris (Brazil: Espirito Santo to Santa Catarina); small rosette of thin-leathery, obovate light green leaves 15 cm long; the charming inflorescence pendulous with scattered flowers attached as if to a snaky wire; floral bracts red-brown, flowers yellow. Ideal for wall bracket or hanging basket. *Tropical.* p. 236

sintenisii (Thecophyllum) (Puerto Rico); upright rosette of glossy wine-red soft-leathery leaves; inflorescence on spike with showy red bracts and yellow flowers. *Tropical.* p. 224, 229, 233, 235

splendens (Guayana), "Flaming sword"; leathery rosette of slender bluish-green leaves marked with broad deep purple crossbands; underneath grayish with the purple bands very bold; flower spike long and swordshaped with flattened fiery-red bracts and yellow flowers. *Tropical.* p. 232, 235, 236

splendens 'Major'; clone more robust; wider leaves with bold purplish crossbands and broader, swordlike spike a flaming-red suffused with copper. The similar cultivar **'Flaming Sword'** is a cross between V. splendens 'Major' and V. splendens longibracteata. *Tropical.* p. 233

splendens 'Variegata'; a remarkable cultivar on exhibit at the German National Flower Show in Cologne 1971, with leaves blackish-green cross-banded in sharply contrasting yellow-green to creamy-white. *Tropical.* p. 232

viridiflora (Costa Rica); rosette of green leaves to 1 m long; brilliantly colored erect inflorescence with yellow to glossy red bracts; petals greenish white. *Tropical.* p 236

WALLICHIA *Palmae*
densiflora (E. Himalaya: Assam); tropical, clustering small palm without noticeable trunk; very fibrous base, and erect-arching pinnate fronds to 4 m long, the numerous lanceolate leaflets dark glossy-green above and whitish underneath, the midrib brown, and margins lobed and toothed; the male spathes dark purple streaked yellow; flowers violet. *Tropical.* p. 804

disticha (Himalayas, Burma); handsome, straight-erect fan palm 3-6 m high, forming solitary trunk 12-15 cm thick, the pinnate leaves 2-3 m long, peculiarly arranged on opposite sides in two ranks almost like the Traveller's tree; the stiff, dark green leaflets 30-60 cm long, notched at the middle, glaucous beneath; flowers green, and dull-red fruit. *Tropical.* p. 805

WARREA *Orchidaceae*
costaricensis (Costa Rica, Panama); handsome terrestrial orchid with pseudo bulbs topped by strong-veined leaves; erect floral spike 50 cm tall, with unusual flowers cup-shaped, burnished pink to flesh-colored and with purple lines, the dorsal sepal forming a hood. *Tropical.* p. 754

warreana (Colombia, Venezuela, Guianas, Brazil); robust terrestrial with clustered spindle-shaped pseudobulbs; plicate, heavy-textured leaves 50 cm long and 12 cm wide; inflorescence erect on purplish stalk to 70 cm high; flowers not opening fully, faintly fragrant, 4 cm across, sepals and petals white tinged with yellow, the lip inside yellow and red (late summer). *Tropical.* p. 758

WARSZEWICZIA *Rubiaceae*
coccinea (Trinidad, C. America to Brazil), "Wild poinsettia"; large tropical shrub to 6 m high, with sparry, woody branches bearing big obovate, corrugated leaves 20-50 cm long; at end of arching shoots the striking elongate inflorescence 30 cm or more long, serving as a rachis (axis) for sessile clusters of small 1 cm orange flowers set at intervals and subtended on both sides by the bract-like crimson-red calyx lobe 8-12 cm long. Photographed at Papeari Botanic Garden, Tahiti. *Tropical.* p. 863, 865, 868

coccinea cv 'David Auyong' (Trinidad cultivar); spreading branches, with large corrugated leaves extending into a long rachis carrying a dense row of small yellow flowers sitting along the top, on both sides a dual row of oblong glowing red bracts or calyx leaves. *Tropical.* p. 867

WASHINGTONIA *Palmae*
filifera (So. California desert, S.W. Arizona, Baja California), "Desert fan palm" or "Petticoat palm"; in habitat in Agua Caliente canyon, Palm Springs, California; bold solitary, erect fan palm with massive grayish trunk to 1 m thick and 18 to 25 m tall, usually clothed by the densely shingled older leaves and looking like a skirt, unless burnt off; the top with a crown of palmate, gray-green, leathery fronds 2 m or more across, divided more than halfway to base, and with many long threads attached to segments, on thorny, long, green petioles. For dry-hot climate. *Subtropic.* p. 795, 807

robusta (N.W. Mexico: Sonora, Baja California), the "Mexican fan palm"; more slender than filifera and faster growing, to nearly 34 m tall, the upper part dense with brown dead, and living glossy bright green foliage, the plaited fan-leaves are stiff and lightly cut, to 1¼ m long, and with some fibrous threads in juvenile stage; fruit black-brown. Rarely used as container plant, but much planted in warm-arid climate along avenues and homes for tropical effect. *Subtropic.* p. 807

WATSONIA *Iridaceae*
beatricis (So. Africa); evergreen Watsonia with rich orange-red tubular flowers in great profusion on erect spikes. *Subtropic.* p. 527

longifolia (Cape Province); charming cormous plant with sword-like leaves to 1 m; spike many-flowered with white to soft pink slender trumpets with spreading petals. *Subtropic.* p. 526

meriana (So. Africa); herbaceous plant from a corm, the stems with sword-shaped flat leaves glaucous green, and stalks to 1 m tall with spikes bearing successive curved tubular flowers with spreading lobes, rosy pink with carmine in center; summer-blooming. *Subtropic.* p. 526

pyramidata (rosea) (South Africa), "Pink watsonia"; breathtaking to see a slope in the Hottentot-Holland Mountains, or down the Cape Peninsula, covered in spring with these rosy-flowered Watsonias; cormous plant 1-1½ m high, with nearly linear basal leaves, and the long, branched spike-like inflorescence carrying showy rose-pink funnel-shaped flowers with curved tube, the spreading lobes to 5 cm across. *Subtropic.* p. 526, 529

WEDELIA *Compositae*
trilobata (West Indies, N. So. America), "Creeping daisy"; prolific herb with slender, flexible trailing stems; elliptic, fresh green, notched and lightly lobed, somewhat fleshy leaves, 5 to 10 cm long; and attractive marigold-like flowers with golden yellow florets. A cheerful basket plant or ground cover. *Tropical.* p. 325, 326

WEIGELA *Caprifoliaceae*
florida 'Bristol Ruby'; deciduous shrub 2 m tall, the species from China and Korea; small ovate 10 cm leaves; trumpet flowers ruby-red (rose-colored in the species), blooming late spring. *Temperate.* p. 291

WEINGARTIA *Cactaceae*
neocumingii (Gymnocalycium) (Bolivia); small globe to 10 cm dia., with chin-like tubercles, tipped by yellowish radial spines; flowers orange. *Arid-subtropic.* p. 274

WELWITSCHIA *Gnetaceae*
mirabilis (bainesii) (Namibia, So. Angola), "Tree tumbo"; curious succulent of the foggy Namib desert; from a low caudex to 1 m thick, with big taproot, spread 2 thick-fleshy, monstrous, strap-like leaves to 6 m long and 20 cm wide, glaucous or waxy green, corrugated lengthwise, with parallel veining, and continuing to grow from the base, not from the apex, over the surface of the ground or curling; the ends usually become split and shredded into leathery thongs if exposed to desert storms. Inflorescence cone-like, the females green, 5 cm long, on branched stalk, the male cones red-brown, only 3 cm long. Said to become 1000-2000 years old, these desert dwellers survive on the surface dew which daily settles along the southwestern coast of Africa, with cold nights and hot days. *Arid-tropical.* p. 509

WIDDRINGTONIA *Cupressaceae (Coniferae)*
whytei (Tanzania, Malawi); noble, symmetrical coniferous evergreen, first shapely pyramidal, later becoming a large tree to nearly 45 m high and wide-spreading; the juvenile plant with rising, plumy branches densely set with bluish-green fine needles to 2½ cm long, later scale-like, pressed close to shoot; small cones to 3 cm long. *Tropical.* p. 340

WIGANDIA *Hydrophyllaceae*
caracasana (Venezuela); robust subshrub to 3 m high, silky pubescent; the large ovate leaves 40 cm long, covered with glistening, irritating hairs; small violet flowers. *Subtropic.* p. 508

WIGGINSIA *Cactaceae*
arechavaletai (Uruguay, Argentina); small moss-green globe with 13-21 high ribs, tipped by short radial spines; flowers golden yellow, to 5 cm across. *Arid-subtropic.* p. 271

tephracantha (Malacocarpus) (So. Brazil, Uruguay, Argentina); globe-shaped, later elongate cactus to 15 cm dia., deeply furrowed with 16-18 acute ribs, dark green or grayish, the edges with areoles and small yellow spine clusters, and woolly at the top; flowers 4 cm across, with inner petals canary yellow, style reddish; pink berries. *Arid-subtropic.* p. 272

WILCOXIA *Cactaceae*
   poselgeri (Texas, Coahuila); low spiny cactus with tuberous, dahlia-like roots; slender cylindrical dark green stems 60 cm high and 2 cm thick, with 8-10 ribs almost hidden by appressed whitish spines; pink flowers lasting several days. *Arid-subtropic.* p. 252

WISTERIA *Leguminosae*
   floribunda 'Alba' (Glycine) (Japan), "Japanese wisteria"; woody liane with twining stems to 10 m or more long; fresh-green pinnate leaves, and beautiful pendulous racemes to 45 cm long, with fragrant 2 cm flowers, violet-blue in the species; white in cv. 'Alba'. Magnificent when trained over pergolas. *Temprate.* p. 564
   floribunda 'Macrobotrys' (multijuga) (Japan); tall-climbing woody vine with glossy green pinnate leaves 25-30 cm long; the showy terminal racemes of 2 cm pea-like, purple flowers, in extremely long, hanging clusters to 1 m long, followed by velvety pods; the most popular of cultivated Japanese wisterias; the flowers appearing before, or with the leaves. *Warm temperate.* p. 564
   sinensis (China), "Chinese wisteria"; deciduous twining woody vine, of great size, with pinnate leaves divided into 7 to 13 leaflets; blooming before leaves, in great pendant clusters to 30 cm long; flowers 3 cm long, bluish violet, not fragrant. *Temperate.* p. 560

WITTROCKIA: see Nidularium

WORMIA: see Dillenia

WORSLEYA: see Hippeastrum

WRIGHTIA: see Holarrhena

XANTHORRHOEA *Liliaceae*
   arborea (No. New So. Wales, Queensland), the ancient and curious "Grass-tree"; long-lived perennial plant with a strong woody trunk 15-22 cm thick, rough with old leaf-bases, topped by a dense multitude of narrow whip-like leathery, dark green leaves, flat or triangular, 1-1½ m long and less than 5 mm wide. Have noticed colonies of grass-trees 2-3 m high, on grassy slopes in eucalypt forests from south of Brisbane to the table lands above Cairns, taking ages to grow tall. *Subtropic.* p. 615
   preissii (Western Australia), the famous "Blackboy"; conspicuously dominating the landscape of the interior of the dry West, like the occasional aboriginal one meets; the massive black trunks becoming 60 cm or more thick, the old leaf bases cemented together by black resinous gum; on top a tuft of hard-leathery rigid, reed-like leaves ½-1 m long; the inflorescence a long, candle-like spike, with small flowers, in habitat always opening on the north (sun) side first. *Arid-subtropic.* p. 589, 615
   undulata; seen in Zuerich Botanic Garden and labeled as from New Zealand; appearing like a bursting fountain of narrow, stiff-leathery, glossy grass-green leaves. *Subtropic.* p. 588

XANTHOSOMA *Araceae*
   jacquini lineatum (Brazil), "Malanga"; large, sagittate leaves with open sinus and basal lobes, friendly green with white lines between veins. *Tropical.* p. 127
   lindenii (Phyllotaenium) (Colombia), "Yautia"; ornamental, evergreen herb growing from rhizome; showy, arrowshaped, thin-leathery leaves with hastate base, 20 cm long, matte green with grayish sheen, and silver-white veins. *Tropical.* p. 130
   lindenii 'Magnificum' (Phyllotaenium), "Indian kale"; horticultural form of X. lindenii (Colombia) differing by having its yellowish to deep-green leaves beautifully and broadly veined cream to white, and with pale line just inside of margin. This ornamental, evergreen herb, growing from rhizome, with its showy, arrow-shaped, thin-leathery leaves, is one of the most beautiful of warm greenhouse exotics. *Tropical.* p. 125, 130
   sagittaefolium (W. Indies, S. America), "Yautia"; forming trunk to 1 m high; mossy green, broad, arrowshaped, soft leaves with pale veins and naked sinus, grayish beneath; fleshy stalks winged at base, not channeled. *Tropical.* p. 124

XEROSICYOS *Cucurbitaceae*
   danguyi (S.W. Madagascar); tendril-climbing liane to 5 m tall, branching from the base, with cylindrical stems; thick succulent, orbicular grayish leaves 4 cm long, opposite to a long tendril; flowers greenish-yellow, and small fruit on female plants. *Arid-tropical.* p. 380

XIMENIA *Olacaceae*
   americana (El Salvador), "Sour plum"; small shrub of the subtropical zone, with leathery ovate leaves; the small yellow fruits are edible, with acid flavor. *Subtropic.* p. 482

XIPHIDIUM *Hamamelidaceae*
   caeruleum (West Indies); low rosette of oblanceolate leathery leaves 30 cm or more long, glossy dark green; inflorescence a tall erect raceme of small white flowers. *Temperate.* p. 523

YUCCA *Liliaceae*
   aloifolia (S.E. United States, West Indies, Mexico), "Spanish bayonet"; tree-forming stiff rosette, becoming 8 m tall, with usually solitary woody trunk; thick-fleshy, sharp-pointed, dagger-like concave leaves to 75 cm long and 6 cm wide, glaucous green, margins faintly toothed but not thread-bearing; cup-shaped flowers creamy-white tinged with purple, in 60 cm clusters. In Puerto Rico, one can see this yucca decorated with egg-shells, a custom carried over from the Carib Indians, and believed to ward off ill fortune; called locally "Mata de Rosa Blanca". *Subtropic and tropical.* p. 615
   aloifolia 'Marginata' (S.E. United States, W. Indies, Mexico); attractive, tree-forming, stiff rosette of thick fleshy, sharp-pointed, dagger-like leaves 6 cm wide, glaucous green with creamy-yellow margins, on usually single trunks to 6 m high; cup-shaped flowers cream-white tinged with purple. *Subtropic.* p. 607
   aloifolia 'Quadricolor'; beautiful plant seen at Royal Botanic Gardens, Edinburgh; broad, dark green leaves variegated red and yellow down the middle. *Subtropic.* p. 596
   aloifolia 'Tricolor', "Red dagger"; colorful variety with the pungent dagger-leaves margined ivory, partly yellow in the center and with a tinge of red when young. *Subtropic.* p. 614
   baccata (Colorado, Arizona, California), "Datil yucca"; heavy succulent rosette of thick, stiff-erect, rough, concave leaves to 90 cm, spine-tipped and with peeling fiber at margins; large white flowers; dried fruit is eaten by Indians. *Warm temperate.* p. 614
   brevifolia, native to high deserts in Southern California, Nevada, Utah and Arizona, the extraordinary "Joshua tree"; hard to believe but this is a member of the lily family, growing into a succulent tree to 22 m high, with palm-like trunk to 1⅓ m thick branching into tortuous arms dense with rosettes of short and rigid dagger-shaped leaves; in bloom from February to April with greenish white 5 cm cup-shaped flowers in dense clusters 30 cm long. Its grotesque silhouette is very characteristic in dry-sunny desert landscapes. *Arid-subtropic.* p. 611, 613
   elata (Arizona), the "Soaptree yucca"; tree-forming to 6 m; dense rosettes of narrow-linear spine-tipped leaves 60 cm long and 1 cm wide, yellow-green and smooth, margins white; flowers white in tall inflorescence. *Arid-subtropic.* p. 613
   elephantipes (guatemalensis) (Mexico, Guatemala), "Spineless yucca"; round-headed "Palm-lily" with trunks springing from a swollen base, branching above with age, and reaching 15 m high, topped by rosettes of leaves to 10 cm wide, glossy grass-green with rough margins and soft tip; flowers ivory-white, in large clusters. *Subtropic.* p. 612, 614
   filamentosa, or better, smalliana (Carolina to Florida and Mississippi), "Adam's needle"; bold rosette, nearly stemless, with stiff sword-shaped leaves 75 cm long, bluish-glaucous, and with marginal, curly white threads; pendulous flowers nearly white in tall panicle to 4 m high. *Warm temperate.* p. 614
   filamentosa 'Variegata'; beautiful rosette of broad bayonet-shaped leaves with cream or white bands along margins. *Subtropic.* p. 596, 614
   gigantea in hort. see elephantipes
   gloriosa (shores of Carolina to Florida), "Spanish dagger"; to 2½ m high, with short, thick trunk topped by dense rosette of sword-shaped, flat glaucous gray-green, rough leaves 5 cm wide, with reddish margins and spiny point; white bell-like flowers striped purple outside. *Subtropic.* p. 614
   recurvifolia (pendula) (Georgia, Alabama, Mississippi), "Lord's candle"; stem-forming lax rosette of broad, flexible, glaucous dark green leaves 6 cm wide and recurving; large panicle of white flowers. *Warm temperate.* p. 614
   schidigera (Mohave Desert, California); shrubby rosette or with trunks to 2½ m; leaves sword-shaped to 75 cm long, yellow green; creamy flowers, tinged with purple. *Arid-subtropic.* p. 614
   whipplei (California, Baja California), "Our Lord's candle"; stemless rosette of stiff glaucous leaves to 50 cm long, with terminal spine and finely serrate margin; fragrant flowers creamy-white. *Arid-subtropic.* p. 614

YUSHANIA *Gramineae*
   aztecorum (Mexico), "Evergreen Mexican bamboo"; beautiful clump-forming bamboo, forming tall green culms 2-3 cm dia., the shorter and younger ones with narrow linear fresh green leaves 30 cm long, in dense masses and gracefully bending downward like giant plumes. *Subtropic.* p. 511, 518

ZAMIA *Cycadaceae*
   fischeri (Mexico); small spindle-shaped trunk, bearing several fern-like pinnate fronds 30 to 45 cm long, with thin-leathery,

lanceolate leaflets shiny grass-green, serrate along margins; male cone 4 to 8 cm long, female shorter. *Tropical.* p. 387

furfuracea (Mexico and Colombia), "Jamaica sagotree"; stem more or less tuberous, sometimes branched, bearing a tangled profusion of pinnate leaves 1-1⅓ m long, on prickly stalks, the thick-leathery leaflets oblanceolate 20 cm long, more or less toothed and overlapping, densely brown scurfy beneath, or on both sides when young; male cones 10 cm long, the female ones shorter. An excellent, "different" decorator plant, hard and durable as iron, and of relatively small size. *Subtropic.* p. 386

ZAMIOCULCAS *Araceae*
zamiifolia (Zanzibar; E. Africa); tropical herb with thick, horizontal rhizome, sending up swollen stalks with pinnately arranged, small, dark, waxy leaves with yellow-green veins; short inflorescence near base with boatshaped, green spathe. *Tropical.* p. 127

ZANTEDESCHIA *Araceae*
aethiopica (Calla, Richardia) (South Africa and north), "White calla", "White arum-lily", or as known in South Africa: "Pig-lily"; robust marsh-loving herb with thick rhizome; forming a tuft of fleshy-stalked, glossy-green leaves 60-90 cm high, succulent; a stout basal stalk 1-1½ m high bearing the large, funnel-shaped rolled-flaring waxy-white spathe 20-25 cm long, surrounding a bright yellow spadix. To the visitor of the moister parts of the South African Cape it is a common sight to see whole fields of these beautiful callas in bloom during summertime; but I have also seen them at 2,300 m altitude in Kenya. *Subtropic.* p. 128, 129, 131

aethiopica 'Childsiana', a "Dwarf calla lily"; a very dwarf, bushy form 30 cm high, with short leaves only 10 cm long; very floriferous, with the white spathes ony 6 cm wide; occasionally the leaves are partly spathe-like with a large white basal area. *Subtropic.* p. 128

'Green Lily' ('Green Goddess'); very charming flowering plant, probably a cultivar of Z. aethiopica, of luxuriant growth with lush green arrow-shaped leaves, and strong-stalked inflorescence with flaring, fleshy spathe 15-20 cm long, white at base, but moss-green at margins and across apex. *Subtropic.* p. 129

elliottiana (So. Africa: Transkei); known as "Yellow calla" and growing from a flattened rhizome; the succulent, cordate, bright green leaves have translucent white spots; the obliquely flaring, tubular spathe is rich yellow, to 15 cm long. *Subtropic.* p. 128, 131

rehmannii (Natal); the shapely "Pink calla" with lanceolate leaves bright green with linear, translucent white spots; pale rosy-purple spathe an obliquely flaring tube lined with cream. *Subtropic.* p. 128

'Striped hybrida'; striking descendants of South African species including Z. aethiopica; flaring spathes in pink, orange-yellow and red, some with contrasting white bands. Photographed at the mercado in Oaxaca, Southern Mexico. *Subtropical.* p. 124

ZAPOTA: see Manilkara

ZEA *Gramineae*
mays (America, prob. Mexico or C. America), the "Indian corn" or "Maize"; its exact origin has been lost in pre-Mayan, Aztec or Incan antiquity; cultivated since and to them the most important food, with the exception of the potato to the Incas in South America; robust, tall annual grass 3 to 4 m high, suckering at base, with jointed, solid stem bearing broad, glossy green, sword-shaped leaves; male flowers in terminal spike with pollen, the female blooms in small clusters borne below the males, their long styles are the showy corn-silk, and producing the ears of edible corn kernels containing starch, heavily sheathed in husks. For farm production, highly refined cultivars have been developed, containing less starch and more sugar. *Tropical.* p. 512, 513

mays gracillima (Trop. America), "Decorative maize"; robust annual grass, suckering at base, long sword-shaped fresh green leaves, and stiff canes to 1 m high, bearing a tassel of tiny flowers; an attractive decorative plant in outdoor bed. *Subtropic.* p. 512

mays rugosa, "Sweet corn"; stems short or of medium size; the sweet grains much wrinkled when mature and dried. The common table corn used in the tender stage. *Subtropic.* p. 512

ZEBRINA *Commelinaceae*
pendula (Mexico), "Silvery wandering Jew"; fleshy trailing plant rooting at joints, small ovate leaves fairly succulent, deep green to purple with two broad, glistening silver bands, vivid purple beneath; flowers rosy-purple. *Tropical.* p. 305

pendula 'Quadricolor', "Happy wandering Jew"; an exquisitely colorful form with the small leaves purplish green broadly banded glistening white, alternating with pink and red; edged purple as well as purple beneath; pretty but delicate. *Tropical.* p. 306

ZEPHYRANTHES *Amaryllydaceae*
citrina (Guayana, Trinidad), a "Rain lily"; small bulbous herb with channeled linear, grass-like leaves 20-30 cm long; on hollow stalks the solitary, funnel-shaped lemon-yellow flowers to 4 cm long, during summer; resting in winter; somewhat winter-hardy. *Tropical.* p. 70

grandiflora (Mexico, Guatemala), a pretty "Zephyr-lily"; with flat, linear basal leaves and large, deep rose-pink, funnel-form flower, to 10 cm across, at end of the hollow stalk; flowering through spring and summer. *Tropical.* p. 575

ZEPHYRANTHES: see also Habranthus

ZINGIBER *Zingiberaceae*
capitatum (S.E. Asia), "Roundhead ginger"; robust rhizomatous plant forming clumps, the canes furnished with glossy green, lanceolate leathery leaves; the inflorescence a striking roundish head of boat-shaped, flaring orange bracts with yellow tips, and creamy flowers. *Tropical.* p. 924

officinale (India to Pacific Isl.), "Common ginger"; slender, reed-like stems ½-1 m high, from tuberous rhizomes which are used as a pleasantly flavored rootspice; the scattered, sessile leaves are glossy deep green and narrow, almost grass-like; flowers in dense spike with pale green bracts, yellowish corolla, and purple lip marked yellow. *Tropical.* p. 928

zerumbet (India, Malaya to Polynesia), the "Wild ginger", or "Bitter ginger"; well-known and widely cultivated tropical ginger with knobbed rootstock at first taste aromatic, then becoming bitter; leafy shoots 45-60 cm high, the lanceolate, thin leaves 10-20 cm long, more or less hairy beneath; in the late summer an oblong flowering head, 5-8 cm long, appears on a stalk about 30 cm long, separate from the leaves, consisting of large green to red overlapping bracts, and small yellowish flowers. *Tropical.* p. 927

zerumbet 'Variegata' (India); attractive tropical ginger of low habit, 30-50 cm high, with leafy stems from underground knobby aromatic rootstock; leathery leaves rich grass green and beautifully variegated with cream; the inflorescence in cone-like heads of overlapping fleshy bracts a vivid scarlet red, with small yellow flowers. Photo at Botanic Gardens Trinidad. *Tropical.* p. 928

ZINNIA *Compositae*
elegans (Mexico), "Youth-and-old-age"; gaily colored annual garden favorite, 50 cm to 1 m high, stiff growing and covered with short hairs; bright green leaves ovalish, clasping the stem; flowers in solitary heads 6-8 cm across in white, pink, rose, red, yellow, orange, purple; distinctly hot-weather plants. *Tropical.* p. 331

elegans 'Cherry Ruffles'; both cut flower and bedding plant, 50-75 cm high, with 6-7 cm flowers, in a new shade of cherry red. *Tropical.* p. 332

elegans 'Dwarf Salmon Rose'; pumila cut-and-come-again type; flowers 5-6 cm across on compact, bushy plant, a strain between giants and lilliputs. *Tropical.* p. 332

elegans 'Red Sun'; large double flowers brilliant bronzy red, 10 cm across. *Tropical.* p. 332

elegans 'Thumbelina' (miniature); extra dwarf zinnia only 15 cm high, very compact and ideal for bedding or window-boxes; the bright and colorful semi-double flowers 4 cm across in shades including yellow, salmon, rose or red; blooming all summer, a rare breeding achievement by Bodgar Seeds, El Monte, California. *Tropical.* p. 332

elegans 'Yellow Ruffles'; free-blooming bushy plant 50-75 cm tall, with stiff stems carrying 6 cm double blooms. *Tropical.* p. 332

ZIZIPHUS *Rhamnaceae*
mauritiana (India), "Indian jujube"; evergreen shrub or small tree, with the pendant zigzag twigs and undersides of foliage rusty tomentose; the leathery 3-nerved leaves broad-oval, 6 cm long, lacquered rich green above; small greenish-white flowers; fleshy fruit 2-3 cm dia. first green, later brown; edible, with acid taste but usually sweet-pickled, candied or stewed. *Tropical.* p. 462, 832

ZOMICARPA *Araceae*
riedeliana (Brazil); tropical herb with tuberous rhizome, producing leaves and flowers at same time; the ornamental leaves divided into 3 to 7 segments, the middle one to 10 cm long; glaucous green and splashed with silver gray, the margins serrate; spathe 8-10 cm long, glaucous outside, yellowish green inside, the long ovate blade arching forward. *Tropical.* p. 130

ZOYSIA *Gramineae*
tenuifolia (Mascarene Islands), the "Mascarene grass" or in Hawaii as "Temple grass"; perennial creeping grass with rich green leaves thread-like and finer than japonica, and forming a turf; widely planted in California and Hawaii, but as ground cover grass somewhat lumpy; with its fine texture, it makes a beautiful tapestry, especially on slopes. Winter-hardy to -12°C., but turns brown after first frost. *Tropical.* p. 510

ZYGOCACTUS *Cactaceae*

'Llewellyn'; a hybrid "Crab-cactus" of upright habit, branching with flattened, toothed joints; the large zygomorphic flowers an iridescent orange-red and contrasting pale lavender base; will bloom for Christmas. *Tropical.* p. 281

truncatus (Organ Mts., Rio), "Thanksgiving cactus", also called "Crab-cactus"; branching epiphyte with flattened joints dark glossy green and having two prominent teeth or claws at apex; Oct.-Nov. blooming with irregular (zygomorphic) scarlet flowers with round ovaries. *Tropical.* p. 281, 282, 284

truncatus delicatus; variety of upright habit, reluctant to branch, long dark green joints sharply toothed, and with irregular white flowers delicately tinged pink; more rose-pink in good light; the ovaries round; Nov.-Dec. (winter) bloom. *Tropical.* p. 282

ZYGOPETALUM *Orchidaceae*

mackayi (So. Brazil: Sao Paulo); although said to be an epiphyte, I have seen it grown in quantity in red clay on high savannah of the Serra do Mar above Santos; robust species with clustered pseudobulbs and 45 cm leaves having raised veins; erect racemes of beautiful 5-8 cm fragrant flowers, sepals and petals yellow-green blotched with brown, the large lip white streaked with blue, (Nov.-June). *Tropical.* p. 756

109. Montrichardia arborescens — *is probably* M. linifera *(J. Bogner, Munich)*
124. etc. Scindapsus aureus — *has been transferred to* Epipremnum aureum *(Hortus 3)*
132. etc. all Brassaias *may be reclassified as* Schefflera *(David Frodin, Papua - New Guinea)*
142. Polyscias balfouriana 'Blackie' — *may be* P. guilfoylei 'Crispa'
183. Pandorea jasminoides — *not jasminioides*
215. Hechtia stenopetala in hort. — *may be* H. ghiesbreghtii, *or* glomerata highly colored by sun; *possibly* tehuacanensis
217. Guzmania sang. brevipedunculata — *not brevipenduculata*
232. Vriesea philippo-coburgii — *not phillipo*

291. heading CARICACEAE — *not CARIACACEAE*
329. Asteriscus — *not Astericus*
360. Ipomoea tuberosa — has been transferred to Merremia in Hortus 3
361. Adromischus maculatus of horticulture; is bot. A. rupicola fa. maculatus *(P. Hutchison, Escondido)*
361. Adromischus rupicola — *not rupicolus*
364. Crassula portulacea 'Hobbit' — *changed from* 'Convoluta'
390. Fuirena — *not Fuirema*
430. Cercis — *belongs to* LEGUMINOSAE *family*
445. Phyllitis scolopendrium 'Crispum' — *may be* Asplenium, possibly var. of nidus
530. Lamium galeobdolon variegatum — *spell properly cv* 'Variegatum'
548. Calliandra 'Minima' — *is* C. emarginata
558. Laburnum anagyroides — *not anagryoides*
565. Tamarindus indica (Trinidad) — *could be* Enterolobium
586. Dasylirion serratifolium — *not serratifolia*
617. Lagerstroemia is "Crape-myrtle" — *not* "Grape myrtle"
618. Lagerstroemia is "Crape-myrtle" — *not* "Grape myrtle"
686. Psidium cattleianum lucidum — *same as p. 460* Psidium littorale *(as per Hortus 3)*
829. Talinum guadalup. is "Fame flower" — *not* "Flame flower"

Other changes in nomenclature as listed in HORTUS Third (1976) but not substituted in TROPICA captions will generally be found referred to in the descriptive text.

PRONUNCIATION of BOTANICAL NAMES

Botanical nomenclature is basically Latin, or words adopted from other languages, with Latin endings, and conceived to be understandable internationally. Correctly pronounced and clearly enunciated, the recital of botanical names has a stately and noble sound. However, in the words of L. H. Bailey, there is no standard agreement on rules for the pronunciation of botanical binomials. Many English-speaking people pronounce generic and descriptive specific names simply as if the words were English, in what is known as the Traditional English system.

Alternately there is the Restored Academic, or phonetic pronunciation of classical scholars, which comes close to the manner of speech of the ancient Romans. Their idiom has been conserved through the centuries. The Florentine vernacular Latin of Dante in the 14th Century is the touchstone of modern Italian. Castilian Spanish also is an enhanced but faithful perpetuation of spoken Latin practically unchanged since the 5th Century. Anyone conversant with these languages will have no difficulty to articulate Botanical Latin in the classical tradition.

In spoken Latin, much the same as in Italian and Spanish, also in German or even Japanese, the vowels are pronounced precisely and uniformly: a as in apart; e as in pet; i as in pin; o as in note; u as in full; y in phyllus as in the French rue, the German or Chinese ü. Typical of the clear sound of spoken Latin is the Spanish expression "Te amo!".

Combinations of two vowels, or imperfect diphthongs found in Latin or Greek are enunciated separately as two syllables: aë (ah-eh = Gr. aer, Aërides); ai (ah-ee = eye); au (house); ei (eight); eo (areole); eu (eh-oo = aureus); ie (ee-eh = variegata); oi (oh-ee = deltoides); iu (ee-uu = folius); ue (uu-eh = cruentus); ui (ruin).

Exceptions are the perfect diphthongs or inseparable ligatures æ or ae (in caeruleus, Linnaeus, Caesarea), sounded as one vowel halfway between ah and eh, as in hat or fair, in French père, or the German "Umlaut" ä; œ or oe (in Coelogyne, coelestis) as in the French heureux, also the German or Swedish "Umlaut" ö.

Consonants: In the classical Latin used by Cicero in the first century B.C., the Romans never pronounced C like an English s, or G as j, but always like k and g (in get). By 180 A.D. however, the classical standard was gradually lost, and while C was still being pronounced as k before a, o, u— it changed to sound as z or s before ae, e, i, oe, y. G remained hard before a, o, u—as in Gardenia, but became a soft j as in joy before e and i.

Many botanical names or epithets are derived from foreign-root personal or geographical names with Latin endings. To be recognizable, these are best pronounced in the idioms of their source, with accent on the preferred syllable.

BOTANICAL TERMS CHART

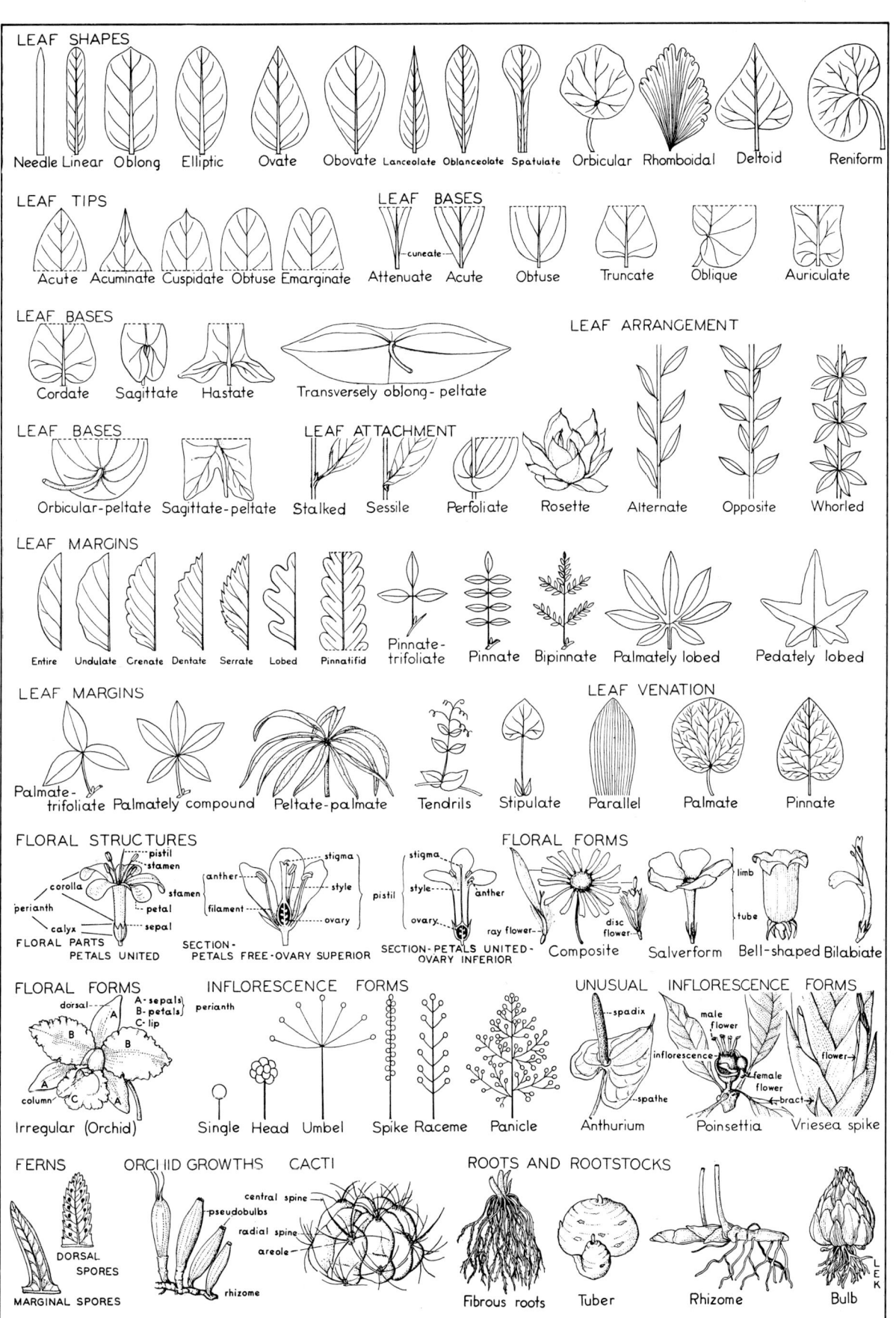

GLOSSARY of SCIENTIFIC TERMS

acicular — needle-like
acuminate — tapering to a point
acute — sharply pointed, but not drawn out
adventitious — other than usual place
alternate — arranged along a stem at different levels
anther — pollen bearing top of stamen
apex — the tip of an organ (as a leaf)
apiculate — with short, not stiff point
areole — cushion-like structure out of which can arise spines, branches, and flowers, a characteristic confined to cacti
articulated — jointed, separating freely by a clean scar
asexual — propagates without benefit of sex
attenuate — becoming narrow, tapered
auriculate — with ears at base
axil — the point just above the leaf where it rises from the stem
basal — at the base of an organ
bifid — divided halfway into two
bifurcate — forked
bilabiate — divided into two or equal lips
bipinnate — both primary and secondary divisions with separate leaflets
bipinnatisect — 3 times divided leafblade whose 3 parts are again several times divided.
bipinnatifid — twice pinnately cut
bisexual — possessing perfect (hermaphrodite) flowers having both stamens and pistils
blade — the expanded portion of a leaf
bract — modified leaves intermediate between flower and the normal leaves, frequently colored
bristly — bearing stiff strong hairs or bristles
bulb — a growth bud with fleshy scales, usually underground
bullate — blistered or puckered
calyx — outer circle or cup of floral parts (usually green)
campanulate — bell-shaped flower with broad base
carpel — division in a compound fruit (section)
caudex — upright root stock or trunk
caulescent — becoming stalked
cauliflorous — production of flowers or fruit directly out of old wood (Ficus)
cephalium — woolly cap at the apex of cacti
channeled — hollowed out like a gutter
chloroplast — the granules of protoplasm which are of a green color
chromosomes — microscopic rodlike bodies in the plant cell, bearing the hereditary material
ciliate — fringed with eyelash hairs
cladodes — branchlet simulating a leaf, leaf-like (Asparagus)
clasping — leaf surrounding stem
cleft — cut halfway down
column — combined stamens and style into one body (as in orchids)
compound — similar parts aggregated into a common whole
compound leaf — a leaf of two or more leaflets
concave — hollowed out
connate — united
convex — umbrella-like
cordate — heart-shaped
corm — bulb-like but solid; enlarged fleshy base of a stem
corolla — complete circle of petals
corymb — a flat-topped open flower-cluster blooming from the outside in
creeper — a trailing shoot rooting at intervals
crenate — with teeth rounded, scalloped
crested — with elevated and irregular ridge
culm — the peculiar hollow stem or stalk of grasses and bamboo
cultivar — special form originating in cultivation
cuneate — wedge-shaped, triangular
cuspidate — tipped with a sharp and stiff point
cyme — a broad, usually flat-topped flower cluster with center flowers opening first
decumbent — reclining, but summit ascending
deltoid — triangular
dentate — with coarse teeth, usually directed outward
dichotomous — forked, parted by pairs
digitately lobed — fingered and main veining radiating from more than one point
dioecious — unisexual; the male and female reproductive organs in different plants
diploid — having the basic chromosome number twice the number in normal germ-cells, characteristic of a species
disc — (in orchids) rounded structure on lip
dissected — several times cleft into small segments
distichous — two-ranked; in two vertical rows

diurnal — daytime
divided — separated at the base
dorsal — back; in orchids usually a top sepal
downy — clothed with soft short hairs
drupe — ripened ovary containing nuts
elliptical — oblong, with widest point at center
elongate — drawn out in length
emarginate — notched at the end
endemic — native to a restricted region
entire — margin without toothing or division
epiphyte — air-plant; a plant growing on another, but not taking food from its host
F 1 hybrid — first generation hybrid obtained by artificial cross-pollination between two dissimilar parents (F meaning filial), each from a pure line or race, and each bearing hereditary factors (genes), which characteristics will be transmitted, according to their dominant genes, to an F 1 hybrid; as a rule imparting also greater vigor (heterosis).
F 2 generation — second generation from a given cross, usually obtained through self pollination within the F 1 hybrids, and which can then segregate into the various types present in the family lines, according to Mendel's Law.
farinose — covered with a mealiness, or starchy matter
ferns — plants without flowers
fertile — spore bearing or seed bearing
fibrous — with fibers, or thread-like parts
filament — thread-like stalk of an anther or stamen
filiform — thread shaped; very slender
floccose — with locks of soft hair or wool
frond — leaf of fern
funnelform — a tubular flower gradually widening upward and spreading into disk
geniculum — thickened joint or node of a stem
glabrous — smooth, not hairy nor rough
glaucous — covered with a white powder that rubs off
glochid — barbed hair, or bristle, as in cacti
glutinous — sticky
hairy — having longer hairs
haploid — having the basic chromosome number, or half the diploid number characteristic of a species
hastate — halberd-shaped, with basal lobes turned outward, or flared
head — a short dense flower spike
herb — a plant with no persistent stem above ground, usually contrasted with woody plants
herbaceous — non-woody
hermaphrodite — stamens and pistils in same flower
hirsute — hairy, with long rather stiff hairs
hybrid — a plant resulting from a cross between parents that are unlike
hypochil — lower or basal part of the lip in some orchids
imbricated — overlapping (as tiles on a roof)
inferior ovary — one that is below the calyx leaves
inflorescence — the flowering portion of a plant, or more precisely the mode of its arrangement
insectivorous — plants which capture insects and absorb nutriment from them
intergeneric — hybrid between genera
internode — space between two joints
involucre — bracts surrounding flowers or their support
irregular flower — a flower which cannot be halved in any plane, or in one plane only
laciniate — slashed into narrow irregular pointed lobes
lanceolate — lance-shaped; tapering toward the tip
lateral — from the side
lenticel — lens-like corky elevations on young bark giving vent to breathing pores
lepidote — beset with small scurfy scales
limb — the border or expanded part of corolla (or spathe) above the throat
line — 1/12 of an inch or 1.2 mm
linear — narrow and flat, margins parallel
lip — the principal lobes of a bilabiate corolla; in orchids a much modified petal
lobe — any projection of a leaf, rounded or pointed
lobed — leaf cut less than halfway to the base
marcottage — airlayering (from Latin mergus)
marginal — at the edge
membranous — thin, semi-transparent
monocarpic — a plant that flowers but once
monoecious — the stamens and pistils in separate flowers but borne on the same plant
mutant — form derived by sudden change from a species

GLOSSARY of SCIENTIFIC TERMS

needle-shaped — long, slender and rigid
node — a joint in a stalk where leaves or their vestiges are born
obcordate — inversely heartshaped, the notch being at the apex
oblanceolate — broad end near tip, long tapering toward base
oblique — slanting of unequal sides
oblong — much longer than broad, with parallel sides
obovate — inverted ovate, the broad end upward
obtuse — blunt or rounded at the end
opposite — opposite each other
orbicular — leaf with circular outline
ovary — that part of the pistil which contains the future seed
ovate — a leaf broadest near base, tapering upward
palmate — veins or leaflets radiating from tip of petiole
palmately compound — more than 3 leaflets borne at tip of petiole
palmately lobed — palmately divided leaf not cut to base
panicle — an open and branched flower cluster
parallel — equally distant at every part
parasite — organism subsisting on another living organism
parted — leaf cut ¾ or more
pectinate — comb-like, merely fringed, with spines
pedate — footed; palmately divided or parted
pedicel — stalk of each flower and cluster
peduncle — primary flower stalk
peltate — leaf-blade attached to stalk inside its margin
peltate-palmate — palmate leaf completely circular in outline
pendant — hanging down from its support
perianth — the calyx, or corolla, or both
perfoliate — petiole in appearance passing through the leaf
petal — a flower-leaf
petiole — the supporting stalk of a leaf; leaf stem
petiolate — furnished with a petiole
petiolule — a small petiole
phenotype — of similar physical make-up as the type species, influenced by environment
phylloclade — a flattened branch assuming the function of foliage
phyllode — petiole taking on the form and functions of a leaf
phyllodia — leaf-like stems and no blades (as in Acacia or Epiphyllum)
pilose — shaggy with soft hairs
pinnae — primary division of a pinnate leaf, its leaflets
pinnate — feather formed; separate leaflets arranged along side of leaf stalk; separation complete
pinnatifid — feathered; cut halfway to midrib
pinnatisect — pinnately divided down to the rachis; a feathered leaf cut down to the midrib
pinnule — secondary pinna or segment
pistil — the female organ of a flower, consisting of ovary, style and stigma
pistillate — flower having pistils only; female
plicate — pleated; folded like a fan or ribbed; plaited
pollen — the fertilizing powder contained in the anther
polymorphic — variable as to habit
procumbent — lying along the ground; leaning
prostrate — lying flat on the ground
prothallus — first stage of germination of fern spore into flat shield, bearing the sexual organs
pseudobulb — thickened and bulb-like portion of stem in epiphytic orchids
pubescent — covered with short, soft hairs, downy
punctate — having tiny translucent glands, appearing like dots
raceme — elongated simple inflorescence with stalked flowers
rachis — axis bearing flowers or leaflets
ray — marginal portion or floret of a Compositae flower when distinct from the disk
recurved — bent backward or downward
regular flower — with the parts in each set alike
reniform — kidney-shaped
rhizome — creeping rootstock, on or under the ground
rhombic — irregularly slanting rectangle
rosette — a cluster of leaves radiating in a circle from a center usually near the ground
rosulate — bearing a rosette, or basal cluster of spreading leaves covered with wrinkles
rugose — covered with wrinkles
runner — a slender prostrate shoot, rooting at the end or at joints
saccate — bag-shaped
sagittate — arrow-shaped, with basal ears turned straight downward or inward
salverform — slender tube abruptly expanded into disk-like limb
saxicolous — living on rock
scabrous — rough or harsh to the touch
scale — usually small, dry leaves or bracts

scaly rhizome — a rhizome with closely appressed, much modified leaves, scale-like in appearance
scandent — climbing, in whatever manner
scape — leafless flower stalk arising from the ground (root)
scorpioid — curved or coiled at the end
segment — one of the divisions into which a plant organ may be cleft
sepal — each segment of a calyx, or outer floral envelopes
serrate — notched like saw; finely toothed
sessile — sitting close, without stalk
setose — covered with bristles
simple leaf — one blade; opposite of compound
single flower — flower with one set of petals
sinuate — with a deep wavy margin, curved
sinus — the curve between two lobes of a leaf
slipper-shaped — tubular ventricose
sori — spore masses (in ferns)
spadix — a fleshy spike bearing tiny flowers as in aroids
spathe — a flower-like bract partly surrounding the inflorescence, often colored or showy
spatulate — oblong, broadly rounded at tip but tapering to narrow base
spike — elongated flower stem, with flowers not stalked
spine — a sharp woody outgrowth from stem
sporangium — a sac producing spores — a spore-case in ferns
spore — in ferns a reproductive cell, somewhat corresponding to seed in flowering plants
spur — a tubular projection from the base of a petal or sepal
stamen — the pollen-bearing or "male" organ
staminate — flower wholly male
stellate — star-form; stellate hairs have radiating branches
stigma — that part of the pistil or style which receives the pollen
stipe — "leafstalk" of a fern
stipule — a leaf-like appendage at base of a petiole
stoloniferous — sending out, or propagating itself by stolons
style — the connecting stalk between the ovary and stigma
sub-cordate — indented a trifle
subtend — to extend under, or be opposite to
subulate — awl shaped, tapering from broad or thick base to a sharp point
succulent — juicy, or storing water in stems or leaves
sulcate — grooved or furrowed
superior ovary — when all petals and sepals are inserted below it
synonym — a name rejected in favor of another
tendril — a thread-shaped shoot used for climbing
terete — circular, rounded in cross section; cylindric and usually tapering
terrestrial — plants growing in the ground
tetraploid — having four sets of chromosomes
thallus — plant body showing no differentiation into distinct members, as stem, leaves and roots
throat — the opening of the flower
tomentose — densely covered with matted wool
transverse — directed across (as on a leaf); crosswise
transversely oblong-peltate — long target-like leaf lying crosswise
trapeziform — no two lines parallel
trifoliate — three-leaved
trifoliolate — with three leaflets, as in clover—commonly, but incorrectly, termed "trifoliate"
triploid — having 3 times the haploid chromosome number
truncate — as if cut off at the end
tube — the united portion of calyx or corolla
tuber — modified underground stem; the thickened portion of subterranean stem, provided with "eyes"
tubercle — a wart-like or knobby projection
tubular — having form of a hollow cylinder
turgid — inflated; swollen
umbel — inflorescence in which flower stalks or cluster arise from same point
undulate — wavy, or wavy-margined
unisexual — of one sex; staminate (male), or pistillate (female) only (see dioecious)
vaginate — sheathed; surrounded by a sheath, usually of leaf stems
ventral edge — belly side
ventricose — swollen on one side
viable — capable of germinating or living
viviparous — producing young, while attached to parent
whorled — leaves in circle around stem (above)
woolly — clothed with long and entangled soft hairs
xerophytic — growing in dry situation, subsisting with little moisture
zygomorphic — can be divided into two symmetrical halves only a single longitudinal plane passing through the axis

CHARACTERISTICS of PLANT FAMILIES

The plant listed at the end of each description is shown by a line-drawing as typical of this family.

ACANTHACEAE (Acanthus family): chiefly tropical herbs with opposite simple leaves, many strikingly colored or with showy bilabiate flowers. (Beloperone)

AIZOACEAE (Figmarigold family): mostly subtropical succulent herbs; leaves fleshy; single flowers look like daisies; seed pod opens when wet. (Faucaria)

ALISMATACEAE (Water plantain family): marsh or aquatic perennial herbs of warm and temperate zones. Stalked basal leaves, very variable, and frequently completely submerged. (Sagittaria)

AMARANTHACEAE: usually herbs, leaves often highly colored; flowers showy in mass, each flower forms only one seed. (Amaranthus)

AMARYLLIDACEAE: herbs mostly bulbous; many with lily-like flowers in umbels but differ from Liliaceae in having ovary below flowers (inferior ovary). (Amaryllis)

ANNONACEAE (Custard Apple family): tropical trees and shrubs with simple alternate leaves, arranged in such a way as to appear pinnate; producing odd flowers often in tones of brown, or edible fruit. (Monodora)

APOCYNACEAE (Dogbane family): mostly warm-climate plants with opposite leaves; juice prevailingly milky. (Allamanda)

APONOGETONACEAE: warm climate aquatic perennial herbs with tuberous rhizomes and floating or submerged leaves. (Aponogeton)

AQUIFOLIACEAE (Holly family): trees or shrubs with alternate simple leaves, often evergreen; leaf margins usually with sharp teeth. Fruits berry-like, borne only on female trees (dioecious) (Ilex)

ARACEAE (Aroid family): largely tropical herbs, with fleshy or woody stems, others rootclimbers; varied and variable leaf-forms; characterized by inflorescence composed of densely flowered spadix, and subtended by a spathe or bract often colored and showy. (Anthurium)

ARALIACEAE: temperate and tropical trees, shrubs or vines often having aromatic foliage, mostly palmately divided or lobed. (Hedera)

ARAUCARIACEAE: resinous evergreen coniferous trees with whorled branches; leaves alternate awl-shaped to broadly ovate, usually leathery. (Araucaria)

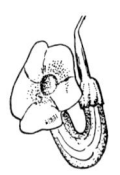
ARISTOLOCHIACEAE: mostly vines from warm regions with alternate usually open base leaves; peculiarly-shaped flowers with tubes usually bent. (Aristolochia)

ASCLEPIADACEAE (Milkweed family): herbs, vines or succulent shrubs frequently from the tropics, mostly with milky juice; flowers waxy appearing and commonly bearing a curious internal crown. (Stapelia)

BALSAMINACEAE: mostly warm climate succulent herbs with watery stems and simple leaves; flowers showy and always spurred. (Impatiens)

BEGONIACEAE: chiefly tropical herbs more or less succulent, usually with lopsided leaves; male flowers with 2 petals, female flowers 3-5 petals having usually 3-angled ovary. (Begonia)

BIGNONIACEAE: mostly tropical trees, shrubs and vines with thin mostly opposite simple or compound leaves; flowers large and showy. (Campsis)

BORAGINACEAE: herbs and shrubs with alternate leaves; inflorescence coiling in the bud. (Heliotropium)

CHARACTERISTICS of PLANT FAMILIES

BROMELIACEAE (Pineapple family): American herbs usually of warm countries, mostly epiphytic, with generally stiff or fleshy leaves, channeled above, forming rosettes or funnels holding water; inflorescence often with showy colored bracts. (Billbergia)

BUXACEAE (Boxwood family): cool climate herbs and evergreen shrubs with leathery simple leaves. (Buxus)

CACTACEAE (Cactus family): Succulents almost entirely American, mostly tropical, with modified fleshy stems having characteristic areoles from which all growth takes place; tree-like, creeping, or epiphytic; mostly spiny; flowers nearly always showy. (Opuntia)

CAMPANULACEAE (Bellflower family): usually herbs of tropical and temperate regions, mostly with milky juice. Leaves mostly alternate; regular flowers bell-shaped or saucer-shaped (campanulate) and prevailingly blue, sometimes white. (Campanula)

CANNACEAE: tropical herbs growing from rhizomes, with banana-like leaves; mostly showy flowers. (Canna)

CAPRIFOLIACEAE (Honeysuckle family): shrubby plants, some climbing, mainly from north temperate zone, a few from tropical mountains. Leaves opposite, mostly oval. Ornamental in flower and fruit; flowers usually fragrant. (Lonicera)

CARYOPHYLLACEAE (Carnation family): herbs grown for their flowers, mostly of temperate and cold regions, with stems usually swollen at the joints, opposite entire narrow leaves with parallel veins; regular mostly bright colored flowers often fringed or toothed. (Dianthus)

CELASTRACEAE: trees and shrubs with simple, often leathery leaves, and small regular flowers. (Celastrus)

CERATOPTERIDACEAE (Floating Fern family): true aquatic succulent ferns, floating or rooting in mud, of warm regions. Leaves borne in rosettes; sterile fronds floating, fertile fronds erect. (Ceratopteris)

COMBRETACEAE: tropical vines, shrubs and trees with simple leaves; small flowers usually showy in mass. (Quisqualis)

COMMELINACEAE (Wandering Jew family): warm climate watery-stemmed herbs and creepers, alternate, usually showy leaves, simple, and parallel-veined. (Rhoeo)

COMPOSITAE (Aster family): chiefly herbs with aster or thistle-like flowers — individually small, but combined into composite head. (Gerbera)

CONVOLVULACEAE (Morning Glory family): mostly twisting herbs with alternate leaves; flowers funnel-shaped, often large and bright colored. (Ipomoea)

CORNACEAE (Dogwood family): temperate climate trees and shrubs with opposite simple leaves; small 4-petaled flowers; fruits usually showy. (Cornus)

CRASSULACEAE (Stonecrop family): succulent herbs and pliable shrubs of temperate and tropical regions; usually fleshy foliage and stems; leaves often in rosettes. (Sempervivum)

CRUCIFERAE (Mustard family): nonwoody (herbaceous) plants; mostly of temperate regions. Flowers with 4 petals arranged as in a cross, and a characteristically splitting dry pod. (Aubrieta)

CYATHEACEAE (Tree-fern family): ferns with distinct trunks, from moist-warm regions; apex of trunk, young fronds and bases of leaf-stalks bearing flattened scales as well as hairs; fronds pinnate, spore-bearing. (Cyathea)

CYCADACEAE (Sago-palm family): palm-like trees or shrubs of slow growth, mostly moist-tropical; stiff pinnate leaves in rosettes. (Cycas)

CYCLANTHACEAE (Panama-hat family): moist tropical palm-like plants with very short trunk, long-stalked, wedge-shaped or cleft leaves. (Carludovica)

CYPERACEAE (Sedge family): herbs often growing in swamps; leaves grass-like or assembled at top of long slender stem. (Cyperus)

CHARACTERISTICS of PLANT FAMILIES

DICKSONIACEAE: tropical and subtropical treeferns; apex of trunk, young fronds and bases of leaf-stalks bearing hairs, but not scales; fronds numerous, large, pinnately divided, usually leathery, in a crown. (Cibotium)

DILLENIACEAE: tropical trees and shrubs, often climbing. Leaves alternate entire or dentate, strongly pinnately veined; yellow, white or red regular flowers with 5 petals, usually showy. (Hibbertia)

DIOSCOREACEAE (Yam family): twining non-woody vines with woody or tuberous roots from tropical and temperate regions. Leaves with a middle vein and strong side veins from the base; netted-veined flowers small and green. (Dioscorea)

DROSERACEAE (Sundew family): low herbs of swampy places, with leaves in basal rosettes; insect-eating by means of their sticky hairs. (Dionaea)

ELAEAGNACEAE: trees or shrubs of temperate and subtropic regions of the northern hemisphere, simple and entire leaves; covered with silvery or golden scales usually on the underside. Flowers without petals, but sepals imitating them. (Elaeagnus)

EPACRIDACEAE: shrubs and small trees mostly in Australia and New Zealand; sister of the Heath family and looking like Ericas, with showy flowers and small, stiff, heath-like leaves. (Epacris)

ERICACEAE (Heath family): chiefly cool-climate hardwood shrubs and small trees with simple leaves usually alternate; flowers mostly showy. (Rhododendron)

EUPHORBIACEAE (Spurge family): primarily tropical herbs, shrubs and trees, frequently cactus-like; sap often milky, and often poisonous; some have showy colored bracts. (Euphorbia)

FUMARIACEAE (Bleeding heart family): annual and perennial herbs of the north temperate zone and So. Africa, with heartshaped or one-spurred flowers having four petals in very different pairs. (Dicentra)

GENTIANACEAE: mostly cool-climate herbs, usually small, with opposite simple and entire leaves and showy flowers, especially in blue. (Exacum)

GERANIACEAE: widely distributed herbs, sometimes semi-woody; leaves alternate and frequently scented; flowers mostly in umbels usually showy. (Pelargonium)

GESNERIACEAE: chiefly moist tropical herbs and creepers; opposite simple leaves frequently colored above or underneath and hairy; stems watery; flowers usually showy. (Sinningia)

Gnetum

GRAMINEAE (Grass family): herbs and shrubs, usually sprouting from underground rhizomes, sometimes reedy-stemmed. (Bambusa)

GUTTIFERAE: tropical trees or shrubs with resinous juice; leaves simple, mostly opposite or whorled often thick and usually evergreen. Flowers regular; fruit berry-like, usually on female specimen. (Clusia)

HALORAGIDACEAE (Water Milfoil family): mostly aquatic herbs with ornamental foliage, which is very diverse in character, from small-leaved aquatics to giant marsh-herbs. (Myriophyllum)

IRIDACEAE (Iris family): non-woody plants with corms, bulbs or rootstocks, from moist locations; flattened leaves, flowers often distinctly iris-shaped. (Iris)

LABIATAE (Mint family): herbs and shrubs with leaves or sap usually aromatic; foliage of horticultural varieties mostly colored; square stems; flowers bilabiate. (Monarda)

LAURACEAE (Laurel family): aromatic trees and shrubs primarily in warm climates, mostly evergreen; leathery simple leaves; small flowers. (Laurus)

LEGUMINOSAE (Pea family): trees, herbs, shrubs, vines, usually with pinnate compound leaves. Characteristic of the family is the bean-like fruit or true pod. (Acacia)

LENTIBULARIACEAE (Bladderwort family): aquatic or marsh herbs, many floating; with insect traps on their leaves. (Pinguicula)

CHARACTERISTICS of PLANT FAMILIES

LILIACEAE: herbs, many of them bulbous, sometimes tree-like. Parallel veined leaved; regular flowers 6-parted with ovary inside of flower. (Lilium)

LOGANIACEAE: Chiefly shrubs from warm regions; commonly opposite simple leaves; regular flowers usually slender tubular. (Buddleja)

LYTHRACEAE: herbs and shrubs with alternate or whorled leaves; flowers usually showy and with tubular calyx (petals widely separated). (Cuphea)

MALPIGHIACEAE: mostly tropical shrubs and woody herbs with opposite leaves often holly-like; flower petals widely separated, and having long claws. (Malpighia)

MALVACEAE (Mallow family): herbs, shrubs or trees from temperate and tropical regions, with alternate leaves palmately veined; the showy flowers are widely bell-shaped; stamens united in long tube. (Hibiscus)

MARANTACEAE (Arrow root family): tropical American herbs, many having tubers; leaves typically patterned feather design, showy, mostly large and sheathing; flowers irregular and surrounded by spathe-like bracts. (Maranta)

MELASTOMATACEAE: mostly tropical herbs, shrubs and trees, with simple opposite or whorled leaves, often showy, and strong parallel nerves. (Tibouchina)

MORACEAE (Mulberry-Fig family): trees, shrubs and vines, often with milky juice; alternate simple leaves; small flowers united into spikes — in Ficus inside a fruit-like body. (Ficus)

MUSACEAE (Banana family): tropical herbs, often very large and tree-like, with huge leaves on long stalks; showy inflorescence with flowers mostly in boat-shaped bracts. (Strelitzia)

MYRSINACEAE: tropical and subtropical trees and shrubs with alternate and simple, leathery leaves; regular flowers; fruit usually showy and with single seed. (Ardisia)

MYRTACEAE: tropical and subtropical trees and shrubs; simple mostly opposite and evergreen leaves with translucent dots; regular flowers with multiple stamens. (Callistemon)

NEPENTHACEAE: tropical insectivorous plants more or less climbing; alternate leaves with midrib prolonged into tendril bearing at end a hollow pitcher with lid. (Nepenthes)

NYCTAGINACEAE (Four-o'clock family): known for herbs and woody vines from warm regions; simple leaves, flowers without petals but often with showy bracts imitating the flower. (Bougainvillea)

NYMPHAEACEAE (Water-lily family): aquatic plants, with usually large floating leaves rising from a submerged rootstock; regular, often showy flowers. (Nymphaea)

OLEACEAE (Olive family): trees and shrubs from tropical and temperate regions with leaves mainly opposite; the regular flowers frequently fragrant and with 2 stamens only. (Jasminum)

ONAGRACEAE (Evening-primrose family): herbs and some shrubs, largely American; usually 4-parted, showy flowers having ovary below flower. (Fuchsia)

ORCHIDACEAE (Orchid family): Epiphytic and terrestrial plants mostly with bulbous or thickened stems or pseudo-bulbs; the showy or fragrant flowers having 3 sepals and 3 petals of which one is notably different in form of a larger or smaller lip or pouch; the stamens and pistils united to form a column. (Cattleya)

OXALIDACEAE (Wood-sorrel family): known for herbs, some bulbous or tuberous, chiefly with palmately divided leaves which close at night. (Oxalis)

PALMAE: The palms are chiefly tropical evergrowing woody plants, tree-like, some climbing and others bushy; large leathery ornamental leaves either fan-shaped or feather-leaved and with parallel veins. (Cocos)

PANDANACEAE (Screw-pine family): tropical, rather woody plants, related to palms, stiff sword-like leaves usually saw-edged and arranged in screw-like spirals. (Pandanus)

CHARACTERISTICS of PLANT FAMILIES

PAPAVERACEAE (Poppy family): herbs widely distributed, often showy in bloom, having colored juice and regular flowers with sepals soon falling, and mostly four free, shortlived petals, numerous stamens, and superior one-celled ovary; fruit in capsules having valves or pores. (Eschscholtzia)

PASSIFLORACEAE (Passion-flower family): tropical woody vines with simple tendrils, alternate leaves; plate-shaped flowers with fringed center crown. (Passiflora)

PHYTOLACCACEAE (Pokeweed family): mostly tropical shrubs and herbs with alternate entire leaves; flowers have no petals. (Rivina)

PINACEAE (Pine family): coniferous trees and shrubs having linear needle-like leaves, usually evergreen and cone-bearing; sap resinous. (Pinus)

PIPERACEAE (Pepper family): mostly tropical herbs and vines, with alternate entire leaves often ornamental; very small flowers without petals or sepals; if climbing, not by tendrils. (Piper)

PITTOSPORACEAE: an old world family of evergreen trees and shrubs with alternate simple leaves usually leathery; flowers small but sometimes showy. (Pittosporum)

PLUMBAGINACEAE (Leadwort family): herbs and small shrubs with alternate leaves; usually small, yet showy flowers, on spikes. (Plumbago)

POLEMONIACEAE (Phlox family): annual and perennial herbs and shrubs having showy and bright-colored flowers with 5-lobed tubular or salverform corolla. (Phlox)

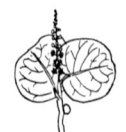
POLYGONACEAE (Knotweed family): trees, shrubs, herbs and vines with peculiar jointed stems; simple leaves with tiny sheaths encircling stems; flowers without petals. (Coccoloba)

POLYPODIACEAE (Common-fern family): most of the common ferns without distinct trunk. Non-flowering plants propagating from spores. Fronds often pinnate with peculiar net veining. (Platycerium)

PONTEDERIACEAE (Pickerel-weed family): swamp plants and aquatics, mostly tropical, with watery leaves; irregular flowers often showy but short-lived. (Eichhornia)

PORTULACACEAE: succulent shrubs and herbs, mostly with fleshy stems and leaves; short-lived flowers with 4-5 petals and only 2 sepals. (Portulaca)

PRIMULACEAE (Primrose family): cool-climate herbs; flowers regular and attractive, their 5 petals more or less united, and mostly salver-form. (Primula)

PROTEACEAE: trees and shrubs of the southern hemisphere, alternate, hard leaves. Flowers in racemes or heads, and very peculiarly constructed. (Grevillea)

PUNICACEAE (Pomegranate family): subtropical shrubs with mostly opposite, simple, narrow leaves; large flowers and edible fruit. (Punica)

RANUNCULACEAE (Buttercup family): herbs of the temperate and cold zones furnishing many favorite flower-garden subjects such as Delphinium, Columbine, Clematis. (Helleborus)

RESEDACEAE (Mignonette family): annual or perennial herbs of the Mediterranean; irregular very fragrant flowers in spikes or terminal racemes. (Reseda)

ROSACEAE (Rose family): herbs, shrubs and tres with mostly alternate leaves and regular 4-5 parted flowers, a wide-spread family containing many fruit trees and ornamental plants. (Rosa)

RUBIACEAE (Madder family): herbs, shrubs and trees, sometimes climbing; a family mostly tropical, of great economic value. Leaves opposite and never lobed or toothed; petioles uniting at base. Flowers usually 4-parted. (Ixora)

RUTACEAE: tropical and substropical trees and shrubs, many with scented juice and edible fruit, mostly alternate leaves with translucent dots. (Citrus)

CHARACTERISTICS of PLANT FAMILIES

SALVINIACEAE: tiny free-floating warm climate aquatic ferns with soft spore-bearing bodies borne on an under-water stalk beneath the leaves. (Salvinia)

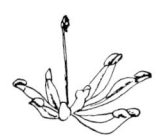
SAPINDACEAE (Soapberry family): warm-climate trees and shrubs, some vines; leaves usually alternate and pinnately or palmately compound; flowers mostly small; seed pods balloon-like. (Koelreuteria)

SARRACENIACEAE (Pitcher plant family): temperate and tropical insectivorous herbs; mostly swamp plants with tubular pitcher-like or trumpet-form leaves. (Sarracenia)

SAXIFRAGACEAE: primarily cool-climate shrubs and herbs with opposite leaves often lobed or toothed; flowers showy in mass. (Saxifraga)

SCHIZAEACEAE: a fern family mainly tropical, often climbing, with simple or pinnate leaves, some thread-like, others palmate: spore bearing. (Lygodium)

SCROPHULARIACEAE (Figwort family): herbs and shrubs from temperate and tropical regions, mainly with opposite, soft leaves on square stems; flowers usually irregular. (Calceolaria)

SELAGINELLACEAE (Spike moss family): Fern allies; herbs, moss-like in appearance and related to ferns, propagating from spores. (Selaginella)

SOLANACEAE (Night shade family): vines, shrubs and trees abounding in poisonous plants; leaves usually alternate; flowers commonly wheel-shaped. (Petunia)

STERCULIACEAE: mostly tropical trees and shrubs furnishing such as cocoa and cola-nut; alternate leaves mostly large; regular, flowers usually showy and which sometimes appear directly on the trunk, odd looking fruit. (Theobroma)

TACCACEAE: tropical herbs with fleshy roots and large basal leaves; green or brownish-purple flowers with long threads between the flowers. (Tacca)

TAXACEAE (incl. PODOCARPACEAE) (Yew family): temperate and subtropical, coniferous trees and shrubs, with evergreen leaves usually needle-like. Fruit mostly red or blue fleshed with one seed. (Podocarpus)

THEACEAE (Tea family): mostly evergreen trees and shrubs from warm regions; leaves commonly leathery, simple and alternate; the regular flowers usually large and showy. (Camellia)

THYMELACEAE (Daphne family): trees or shrubs, rarely herbs, temperate and tropical, with simple leaves; the regular flowers having four to five lobed calyx, colored like a corolla but without petals; some genera have medicinal and economic uses, others are very ornamental. (Daphne)

TILIACEAE (Linden family): mostly trees and shrubs, the usually alternate simple leaves frequently large and hairy. (Sparmannia)

TROPAEOLACEAE: quick-growing soft herbs, often climbing by coiling petioles; succulent stems, leaves alternate; irregular showy flower, one sepal forming a slender spur. (Tropaeolum)

URTICACEAE (Nettle family): creeping herbs, shrubs and trees, many tropical; simple leaves, some showy; flowers very small and usually green. (Pilea)

VERBENACEAE: mostly tropical shrubs and trees, herbs and woody vines, with opposite leaves; the irregular flowers are usually showy and ornamental. (Clerodendrum)

VIOLACEAE (Violet family): herbs, shrubs and trees of temperate and tropical regions; irregular showy flowers one petal of which is spurred. (Viola)

VITACEAE (Grape family): mostly woody vines climbing by branched tendrils, with alternate simple or divided leaves; the small flowers are usually greenish. (Cissus)

ZINGIBERACEAE (Ginger family): tropical herbs, rhizomatous; mostly with cane-like stems and alternate elongated leaves; juice often scented; the irregular flowers usually showy. (Alpinia)

BIBLIOGRAPHY and LITERATURE REFERENCES

Abrams-Ferris: Flora of the Pacific States *(Stanford 1960)*
Ackerson: Book of Chrysanthemums *(New York 1957)*
Adams: Flowering Plants of Jamaica *(Jamaica 1972)*
Allan: Flora of New Zealand *(Wellington 1961)*
American Hort. Society: Cultivated Palms *(Washington 1961)*
American Orchid Society: Bulletin *(Cambridge 1931-1978)*
American Rhododendron Society: Rhododendrons for your Garden *(Portland, Oregon 1961)*
Arbelaez: Plantas Utiles de Colombia *(Bogotá 1947)*
Archbold, Rand, Brass: New Guinea Expedition *(New York 1948)*
Aristeguieta: El Genero Heliconia *(Caracas, 1961)*
Backeberg: Das Kakteen Lexikon *(Stuttgart 1966)*
Backeberg: Die Cactaceae *(Jena 1958-1962)*
Backer-Posthumus: Varenflora voor Java *(Bogor 1939)*
Bailey: Hortus Third *(New York 1976)*
Bailey: Manual of Cultivated Plants *(New York 1949)*
Bailey: Standard Cyclopedia of Horticulture *(New York 1928)*
Baileya, Journal of Horticultural Taxonomy *(Ithaca 1953-1976)*
Baker: Handbook of Fern Allies *(London 1887)*
Barrett: Exotic Trees of South Florida *(Gainesville 1956)*
Barron: Vines and Vine-Culture *(London 1883)*
Beard: West Australian Plants *(Perth 1970)*
Beddome: Ferns of British India *(London 1883)*
Begonia Society Bulletin (Los Angeles 1932-1978)
Bellair & Saint-Léger: Les Plantes de Serre *(Paris 1939)*
Berger: Die Agaven *(Jena 1915)*
Bernstiel: Die Farnpflanzen *(Stuttgart 1936)*
Birdsey: The Cultivated Aroids *(Berkeley 1951)*
Boehmig: Die Gattung Begonia *(Berlin 1955)*
Bohnstedt: Kalt und Warmhauspflanzen *(Berlin 1934)*
Booth: Encyclopedia of Annuals & Biennials *(London 1957)*
Borg: Cacti *(London 1956)*
Botanical Register (London 1815-)
Bravo: Las Cactaceas de Mexico *(Mexico 1937)*
Breitung: The Agaves *(Reseda, Calif. 1968)*
Brilmeyer: All About Begonias *(Garden City, N.Y. 1960)*
Britton & Brown: Flora of the N.E. United States and Canada *(New York 1958)*
Britton & Rose: The Cactaceae *(Washington 1919)*
Britton & Wilson: Botany of Puerto Rico and the Virgin Islands *(New York 1923-1930)*
Bromeliad Society Bulletin (Orlando 1951-1978)
Brown: Florida's Beautiful Crotons *(Indialantic, Fla. 1960)*
Brown: Flora of Southeastern Polynesia *(Honolulu 1931)*
Browne: Forest Trees of Sarawak and Brunei *(Kuching 1955)*
Bruenner: Wasserpflanzen *(Braunschweig 1953)*
Bruggeman: Indisch Tuinboek *(Amsterdam 1948)*
Burgeff: Marchantia *(Jena 1943)*
Buxbaum: Cactus Culture *(London 1958)*
Buxton: Begonias and How to Grow Them *(New York 1946)*
Buxton: Check List of Begonias *(Los Angeles 1957)*
Cabrera: Flora de Buenos Aires *(Buenos Aires 1953)*
Cactus & Succulent Journal (Pasadena 1879-1978)
Chabouis: Flore de Tahiti *(Paris 1972)*
Chidamian: Book of Cacti and Other Succulents *(New York 1958)*
Chittenden: Royal Hort. Society Dictionary of Gardening *(Oxford 1951)*
Christ: Die Geographie der Farne *(Jena 1910)*
Christensen: Index Filicum *(Copenhagen 1906-1933)*
Clay: Tropical Shrubs, Hawaii *(Honolulu 1977)*
Codd: Trees and Shrubs of Kruger National Park *(Pretoria 1951)*
Condit: Ficus *(Arcadia, Calif. 1969)*
Constantin: Atlas en Couleurs des Orchidées *(Paris 1913)*
Corner: Wayside Trees of Malaya *(Singapore 1952)*
Cowen: Flowering Trees and Shrubs in India *(Bombay 1950)*
Craig: The Mammillaria Handbook *(Pasadena 1945)*
Curtis Botanical Magazine (London 1787-1974)
Cutak: Cactus Guide *(Princeton 1956)*
Cutak: All About Sansevierias *(St. Louis 1966)*
Dale, Greenway: Kenya Trees and Shrubs *(Nairobi 1961)*
Davies: New Zealand Native Plant Studies *(Wellington 1961)*
Davis and Steiner: Philippine Orchids *(New York 1952)*
Decker: Cultura das Orquideas no Brasil *(Sao Paulo 1946)*
Degener: Flora Hawaiiensis *(New York 1946)*

Den Ouden, Boom: Manual of Cultivated Conifers *(The Hague 1965)*
DeWit: Aquarienpflanzen *(Stuttgart 1971)*
Dobbie: New Zealand Ferns *(Auckland 1952)*
Eggeling: The Indigenous Trees of Uganda *(Entebbe 1951)*
Eliovson: South African Flowers *(Cape Town 1955)*
Eliovson: Flowering Shrubs and Trees *(Cape Town 1953)*
Eliovson: Shrubs, Trees and Climbers *(Johannesburg 1975)*
Encke: Parey's Blumengaertnerei, 2nd Ed. *(Berlin 1961)*
Encyclopedia Britannica (Chicago 1955)
Engler: Das Pflanzenreich *(Leipzig 1900-)*
Engler: Botanische Jahrbuecher *(Leipzig 1881-)*
Engler-Prantl: Die Natuerlichen Pflanzenfamilien *(Leipzig 1899-)*
Erikson: Flowers and Plants of Western Australia *(Sydney 1973)*
Everett: Begonias *(New York 1939)*
Everett: The American Gardeners Book of Bulbs *(New York 1954)*
Everett: Encyclopedia of Gardening *(New York 1960)*
Fassett: Manual of Aquatic Plants *(Madison 1957)*
Flowering Plants from Cuban Gardens (New York 1958)
Foster: Ferns to Know and Grow *(New York 1972)*
Foster: Bromeliads — A Cultural Handbook *(Orlando 1953)*
Foster: Brazil — Orchid of the Tropics *(Lancaster 1945)*
Fotsch: Die Begonien *(Stuttgart 1933)*
Free: All About House Plants *(New York 1948)*
Gardeners Chronicle (London 1841-)
Gartenwelt (Berlin-Hamburg 1896-)
Genders: Bulbs *(London 1973)*
Gerbing: Camellias *(Fernandina 1945)*
Giddy: Cycads of South Africa *(Cape Town 1974)*
Gooding: Flora of Barbados *(London 1965)*
Graf: Exotica Series 3, 9th Ed. Rev. *(E. Rutherford, NJ 1978)*
Graf: Exotic Plant Manual, 5th Ed. *(E. Rutherford, NJ 1978)*
Graf: Exotic House Plants, 10th Ed. *(E. Rutherford, NJ 1976)*
Graf: Tropica *(E. Rutherford, NJ 1978)*
Gray Herbarium Card Index (Cambridge 1873-1959)
Grootendorst: Rhododendrons en Azaleas *(Boskoop 1954)*
Guenther: A Naturalist in Brazil *(London 1931)*
Haage: Cacti and Succulents *(New York 1963)*
Harris: Australian Plants for the Garden *(Sydney 1953)*
Harrison: Bulbs and Perennials *(New Zealand 1967)*
Harrison: Climbers and Trailers *(New Zealand 1975)*
Harrison: Ornamental Conifers *(New Zealand 1975)*
Harrison: Trees and Shrubs for the Southern Hemisphere *(Wellington 1967)*
Haselton: Cacti for the Amateur *(Pasadena 1938)*
Haselton: Succulents for the Amateur *(Pasadena 1939)*
Hawkes: Encyclopedia of Cultivated Orchids *(Miami-London 1965)*
Hawkes: The Major Kinds of Palms *(Coconut Grove 1950)*
Hawkes: Orchids, Their Botany and Culture *(New York 1961)*
Hay, Synge, Kalmbacher: Color Dictionary of Flowers and Plants *(New York 1969)*
Hay, Synge, Herklotz: Das Grosse Blumenbuch *(Stuttgart 1973)*
Hellyer: Sanders Encyclopedia of Gardening *(London 1952)*
Herklotz: The Hong Kong Countryside *(Hong Kong 1959)*
Hertrich: Palms and Cycads *(San Marino 1951)*
Hibberd: New and Rare Beautiful-leaved Plants *(London 1891)*
Higgins: Crassulas in Cultivation *(London 1964)*
Hilliard and Burtt: Streptocarpus *(Pietermaritzburg 1971)*
Hoehne: Iconografia de Orchidaceas do Brasil *(Sao Paulo 1949)*
Honig-Verdoorn: Science in Netherlands Indies *(New York 1945)*
Hong Kong Shrubs, Trees (Hong Kong 1976)
Hooker: Flora of British India *(London 1875-1897)*
Hoshizaki: Fern Growers Manual *(New York 1975)*
Hoyt: Check Lists, Ornamental Plants for Subtropical Regions *(Calif. 1958)*
Hulme: Wild Flowers of Natal *(Pietermaritzburg 1954)*
Hume: Azaleas, Kinds and Culture *(New york 1954)*
Hutchinson: British Flowering Plants *(London 1948)*
Hutchinson: Flora of West Tropical Africa *(London 1958)*
Hutchinson: The Families of Flowering Plants *(London 1926, 1934)*
International Code of Botanical Nomenclature (Utrecht 1958)
Index Kewensis and Supplements (Oxford 1895-1959)

BIBLIOGRAPHY and LITERATURE REFERENCES

Index Londinensis to Illustrations *(Oxford 1929-1941)*
Jackson: Glossary of Botanic Terms *(New York 1950)*
Jacobsen: Handbook of Succulent Plants *(London 1960)*
Jacobsen: Handbuch der Sukkulenten Pflanzen *(Jena 1954)*
Jacobsen: Das Sukkulenten Lexikon *(Stuttgart 1970)*
Jex-Blake: Gardening in East Africa *(Nairobi 1948)*
Johnson's Gardener's Dictionary *(London 1846, revised 1917)*
Julius Roehrs Company: Exotic Catalogs *(Rutherford 1911, 1913)*
Kanehira: Formosan Trees *(Fukuoka 1936)*
Kelly: Eucalypts *(Melbourne 1969)*
Kelsey: Standardized Plant Names *(Harrisburg 1942)*
Kerchove: Les Palmiers *(Paris 1878)*
Kew Bulletin *(London 1887-1969)*
Kidd: Wild Flowers of the Cape Peninsula *(Cape Town 1950)*
Koeppen: Grundriss der Klimakunde *(Berlin 1931)*
Krauss: Begonias for American Homes *(New York 1947)*
Krauss: Geraniums for the Home *(New York 1955)*
Kruessmann: Die Laubgehoelze *(Berlin 1957)*
Kruessmann: Die Nadelgehoelze *(Berlin 1955)*
Kruessmann: Rosen *(Berlin 1974)*
Kuck and Tong: Hawaiian Flowers and Flowering Trees *(Honolulu 1958)*
Laing and Blackwell: Plants of New Zealand *(Auckland 1940)*
Lamb: Cacti and Other Succulents *(New York 1955)*
Lamb: Stapeliads in Cultivation *(London 1957)*
Latif: Bunga Anggerik (Orchids) *(Bandung 1953)*
Lawrence: The Cultivated Hederas *(Ithaca 1942)*
Lawrence: Taxonomy of Vascular Plants *(New York 1951)*
Le Bon Jardinier, Encyclopedie Horticole *(Paris 1964)*
Lecomte: Flore Générale de l'Indochine *(Paris 1942)*
Lecoufle & Rose: Orchids *(Paris 1957)*
Lee: The Azalea Book *(Princeton 1958)*
Letty: Wild Flowers of the Transvaal *(Pretoria 1962)*
L'Illustration Horticole *(Ghent 1854-1896)*
Linnaeus: Species Plantarum *(Uppsala 1753)*
Liu: Illustrations of Ligneous Plants of Taiwan *(Taipei 1960)*
Little, Woodbury, Wadsworth: Trees of Puerto Rico and the Virgin Islands *(Washington 1964, 1974)*
Lloyd: The Carnivorous Plants *(Waltham 1942)*
Lord: Shrubs and Trees for Australian Gardens *(Melbourne 1960)*
Lowe: Ferns, British and Exotic *(London 1864)*
Lowe: Beautiful Leaved Plants *(London 1872)*
MacMillan: Trop. Planting and Gardening, Ceylon *(London 1956)*
Macself: Ferns for Garden and Greenhouse *(London 1952)*
Maerz and Paul: Dictionary of Color *(New York 1930)*
Makino: Illustrated Flora of Japan *(Tokyo 1967)*
Marnier-Lapostolle: Le Genre Kalanchoe *(Paris 1964)*
Marshall & Bock: Cactaceae *(Pasadena 1941)*
Mathias, McClintock: Woody Ornamental Plants of California *(Univ. of Calif. 1963)*
Matuda: Las Araceas Mexicanas *(Mexico 1954)*
McClure: The Bamboos *(Cambridge, Mass. 1966)*
McCurrach: Palms of the World *(Palm Beach 1959)*
McFarland: American Rose Annual *(Harrisburg 1916-1959)*
McFarland: Modern Roses *(Harrisburg 1969)*
Menninger: Flowering Trees of the World *(Stuart, Florida 1961)*
Menninger: Flowering Vines of the World *(New York 1970)*
Moering: Die Hortensien *(Aachen 1956)*
Moldenke: Plants of the Bible *(New York 1952)*
Moore: African Violets, Gloxinias, and Relatives *(New York 1957)*
Moore: Flora of New Zealand, Vol. II *(Wellington 1970)*
Morton: 500 Plants of South Florida *(Miami 1974)*
Neal: In Gardens of Hawaii *(Honolulu 1948)*
Nel: Lithops *(Stellenbosch 1946)*
Nicholson: Dictionnaire Pratique d'Horticulture *(Paris 1893)*
Nicholson: Illustrated Dictionary of Gardening *(London 1887)*
Noble and Merkel: Plants Indoors *(New York 1954)*
North American Flora *(New York 1949-)*
Northen: Home Orchid Growing *(New York 1970)*
O'Gorman: Mexican Flowering Trees & Plants *(Mexico City 1961)*
Ohwi: Flora of Japan *(Washington 1965)*
Ospina Hernandez: Orquideas Colombianas *(Bogotá 1958)*
Padilla: Bromeliads *(New York 1973)*
Palgrave: Trees of Central Africa *(Salisbury 1957)*
Parodi: Enciclopedia Argentina de Agricultura *(Buenos Aires 1959)*
Pesman: Meet Flora Mexicana *(Arizona 1962)*

Pertchick: Flowering Trees of the Caribbean *(New York 1951)*
Piers: Orchids of East Africa *(Nairobi 1968)*
Phytologia *(Yonkers, N.Y. 1956-)*
Pittier: Plantas Usuales de Venezuela *(Caracas 1926)*
Polunin: Flowers of the Mediterranean *(Boston 1966)*
Polunin: Guide des Plantes et Fleurs de l'Europe *(Paris 1974)*
Popenoe: Manual of Tropical and Subtropical Fruits *(New York 1934)*
Principes, Journal of the Palm Society *(Miami, Florida 1956-1978)*
Rauh-Heidelberg: Bromelien *(Stuttgart 1970)*
Rehder: Manual of Cultivated Trees and Shrubs *(New York 1954)*
Reiter-Boehmig: Schnittblumen und Topfpflanzen *(Berlin 1958)*
Reynolds: Aloes of South Africa *(Johannesburg 1950)*
Reynolds: Aloes of Trop. Africa and Madagascar *(Swaziland 1966)*
Rice and Compton: Wild Flowers of the Cape *(Kirstenbosch 1950)*
Richards: New Zealand Trees and Flowers *(Christchurch 1956)*
Richter: Bromeliaceen *(Radebeul 1962)*
Rickett, New York Botanical Garden: Wild Flowers of the United States, Vol. 1 to 6 *(New York 1966-1973)*
Rockwell: The Complete Book of Bulbs *(New York 1953)*
Russell: Mosses and Liverworts *(London 1908)*
Sanders Orchid Guide and List of Hybrids *(St. Albans, 1927-1963)*
Schlechter: Die Orchideen *(Berlin 1927)*
Schimper: Pflanzen-Geographie *(Jena 1898)*
Scheirlinck: Tuinbouw: De Azalea Indica *(Antwerpen 1938)*
Schneider: Book of Choice Ferns *(London 1892-1894)*
Schultz: Gesneriads *(Grandview, Mo. 1967)*
Schwantes: Flowering Stones *(London 1957)*
Scott: The Florists Manual *(Chicago 1899)*
Sibree: A Naturalist in Madagascar *(London 1915)*
Silva-Tarouca: Unsere Freiland-Stauden *(Wien 1934)*
Small: Flora of the Southeastern U.S. *(New York 1913)*
Smith: Ferns, British and Foreign *(London 1896)*
Smith: The Bromeliaceae of Brazil *(Washington 1955)*
Smith: The Bromeliaceae of Colombia *(Washington 1957)*
Smith: Notes on Bromeliaceae (Phytologia) *(Baltimore 1971)*
Smith and Downs: Pitcairnioideae *(New York 1974)*
Spalding: Pelagonium Checklist *(California 1972)*
Sprechman, Dugdale, Cole, DeBoer: Lithops *(Rutherford 1970)*
Spuy: South African Shrubs and Trees *(Johannesburg 1971)*
Standley: Trees and Shrubs of Mexico *(Washington 1926)*
Step: Favorite Flowers of Garden and Greenhouse *(London 1896)*
Stevenson: Palms of So. Florida *(Miami 1964)*
Stocken: Andalusian Flowers *(Devon 1969)*
Stodola: Encyclopedia of Water Plants *(Jersey City 1967)*
Sventenius: Floram Canariensem *(Madrid 1960)*
Taeckholm, Drer, Fadeel: Flora of Egypt *(Cairo 1956)*
Tagawa: The Japanese Pteridophyta *(Osaka 1959)*
Taylor: Encyclopedia of Gardening *(Boston 1961)*
Thompson: Begonia Guide *(New York 1977)*
Thrower: Plants of Hong Kong *(Hong Kong 1971)*
Tobler: Die Gattung Hedera *(Jena 1912)*
Van Pelt Wilson: African Violet Book *(New York 1970)*
Van Tubergen: Catalog of Bulbs *(Haarlem 1961-1976)*
Veitch: Manual of Orchidaceous Plants *(London 1894)*
Verdoorn: Plant Science in Latin America *(Waltham 1945)*
Vogts: Proteas *(Johannesburg 1958)*
Walther: Echeveria *(San Francisco 1972)*
Wayside Gardens: Catalog *(Mentor, Ohio 1959-1977)*
Webber and Batchelor: The Citrus Industry *(Berkeley 1948)*
Weberbauer: El Mundo Vegetal de los Andes Peruanos *(Lima 1945)*
Western Garden Book (Lane) *(Menlo Park, Calif. 1969)*
Wettstein: Handbuch der Botanik *(Leipzig-Wien 1923)*
White: American Orchid Culture *(New York 1939)*
White and Sloane: The Stapelieae *(Pasadena 1937)*
Williams: Orchid Growers Manual, 7th Ed. *(London 1894)*
Williams: Orchids of Mexico *(Honduras 1965)*
Williams: Useful and Ornamental Plants of Zanzibar *(Zanzibar 1949)*
Willis: Dictionary of the Flowering Plants and Ferns, 8th Ed. *(Cambridge 1973)*
Wilson: Geraniums—Pelargoniums *(New York 1946-1956)*
Withner: The Orchids *(New York 1959)*
Wood: A Fuchsia Survey *(London 1956)*
Wright: Orquideas de Mexico *(Mexico 1958)*
Zohary: Plant Life of Palestine *(Jerusalem 1962)*

COMMON NAMES of EXOTIC PLANTS

There are some 3000 Common names listed in TROPICA, of which 1600 of the most important are used in this index, generally referring them to their generic name only. Page numbers to photos are not exclusive, as the same plant is often shown elsewhere and can be located by checking the Text-Index.

Adam's needle—*Yucca* 614
African boxwood—*Myrsine* 674
African daisy—*Arctotis* 332
African daisy—*Gerbera* 318
African evergreen—*Syngonium* 125
African iris—*Dietes* 529
African locust—*Parkia* 561
African nutmeg—*Monodora* 72
African tulip tree—*Spathodea* 187
African violet—*Saintpaulia* 495
Agave cactus—*Leuchtenbergia* 269
Airplant—*Kalanchoe pinnata* 373
Air-potato—*Dioscorea bulbifera* 393
Aki tree—*Blighia sapida* 455
Algerian ivy—*Hedera canariensis* 136
Alligator pear—*Persea* 536
All-spice—*Pimenta dioica* 684
Almond—*Prunus amygdalus* 470
Alpine violet—*Cyclamen* 829
Amaryllis—*Hippeastrum* 66
Amazon lily—*Eucharis* 64
Anchor plant—*Colletia* 857
Angel's-trumpet—*Datura* 892
Angelwing begonia—*Begonia coccinea etc.* 179
Apostle-plant—*Neomarica* 525
Apricot—*Prunus armeniaca* 850
Arabian jasmine—*Jasminum sambac* 699
Arbor-vitae—*Thuja* 340
Archangel—*Lamium galeobdolon* 530
Areca palm—*Chrysalidocarpus* 774
Arrow bamboo—*Pseudosasa jap.* 514
Arrowhead—*Sagittaria* 70
Arrowroot—*Zamia furfuracea* 386
Artichoke—*Cynara* 331
Artillery plant—*Pilea* 912
Ashoka tree—*Polyalthia* 72
Asiatic poison bulb—*Crinum* 60
Asoka tree—*Saraca* 563
Assai palm—*Euterpe edulis* 774
Atlas cedar—*Cedrus atlantica* 342
Australian fuchsia—*Epacris* 398
Australian pine—*Casuarina* 300
Autumn crocus—*Colchicum* 575
Avalanche lily—*Erythronium* 594
Avacado pear—*Persea* 536

Baboon flower—*Babiana* 526
Baby breath—*Gypsophila* 298
Baby primrose—*Primula malacoides* 831
Baby rose—*Rosa polyantha* 856
Baby rubber plant—*Peperomia* 820
Baby tears—*Helxine* 907
Baby toes—*Fenestraria* 44
Bahia grass—*Paspalum* 510
Bald cypress—*Taxodium* 355
Ball cactus—*Notocactus* 270
Balloon vine—*Cardiospermum* 876
Balsam apple—*Clusia rosea* 523
Bamboo palm—*Chamaedorea* 773
Banana—*Musa* 470
Banyan tree—*Ficus benghalensis* 651
Baobab—*Adansonia* 55
Barbados cherry—*Malpighia* 458
Basket grass—*Oplismenus* 510
Bat flower—*Tacca* 902
Bay-rum tree—*Pimenta* 684
Baytree—*Laurus nobilis* 535
Beach morning-glory—*Ipomoea* 360
Beach strawberry—*Fragaria* 849
Bear-berry—*Arctostaphylos* 466
Beard-tongue—*Penstemon* 883
Bear grass—*Nolina* 588
Beaver-tail—*Opuntia basilaris* 240
Beefsteak begonia—*Begonia erythrophylla* 160
Beefsteak plant—*Iresine* 50
Bee orchid—*Ophrys* 744
Belladonna lily—*Amaryllis* 65
Bell-climber—*Marianthus* 823
Bent grass—*Agrostis* 510
Bermuda buttercup—*Oxalis* 763
Bermuda grass—*Cynodon* 510
Betel-leaf pepper—*Piper* 821
Betelnut palm—*Areca* 765
Birch—*Betula* 182
Bird-catcher-tree—*Heimerliod.* 692
Bird-of-paradise—*Strelitzia* 665
Birdseye bush—*Ochna* 698

Birdsnest fern—*Asplenium nidus* 443
Bitter Aloe—*Aloe vera* 571
Bitter ginger—*Zingiber* 927
Bishop's cap—*Astrophytum* 264
Bishop's weed—*Aegopodium* 910
Black bamboo—*Phyllostachys* 518
Black banana—*Ensete maurelii* 662
Black-boy—*Xanthorrhoea* 615
Black-eyed Susan—*Thunbergia* 38
Black mulberry—*Morus nigra* 473
Black pepper—*Piper nigrum* 821
Black sapote—*Diospyros* 463
Bleeding heart—*Dicentra* 484
Bloodleaf—*Iresine* 50
Blood lily—*Haemanthus multiflorus* 66
Blue amaryllis—*Hippeastrum procerum* 65
Blueberry—*Vaccinium corymbosum* 472
Blue cedar—*Cedrus atlantica* 343
Blue daisy—*Felicia* 315
Blue ginger—*Dichorisandra* 305
Blue gum—*Eucalyptus globulus* 679
Blue hibiscus—*Hibiscus huegelii* 628
Blue lotus—*Nymphaea caerulea* 704
Blue marguerite—*Felicia* 315
Blue myrtle—*Myrtillocactus* 254
Blue orchid—*Vanda coerulea* 755
Blue potato bush—*Solanum* 899
Blue spruce—*Picea pungens* 346
Blue star creeper—*Isotoma* 616
Blue stars—*Aristea ecklonii* 528
Blue trumpet vine—*Thunbergia grandiflora* 40
Boat-lily—*Rhoeo spathacea* 307
Boston ivy—*Parthenocissus* 920
Bo-tree—*Ficus religiosa* 650
Bottle brush—*Callistemon* 675
Bottle palm—*Beaucarnea* 586
Bower plant—*Pandorea* 183
Bowing hemp—*Sansevieria* 608
Box-elder—*Acer negundo* 41
Boxing glove—*Opuntia* 243
Boxwood—*Buxus* 237
Brazilian edelweiss—*Sinningia leucotricha* 501
Brazilian pepper tree—*Schinus* 72
Brazilian plume—*Jacobinia* 39
Brazilian rose—*Cochlospermum* 303
Brazilwood—*Caesalpinia* 548
Breadfruit—*Artocarpus* 649
Bread palm—*Cycas* 388
Bridal wreath—*Spiraea* 854
Brisbane box—*Tristania* 680
Prism cactus—*Leuchtenbergia* 269
Broom—*Spartium* 556
Brush cherry—*Syzygium* 677
Bucket orchid—*Coryanthes* 720
Buddha'a belly bamboo—*Bambusa ventricosa* 514
Buddhist pine—*Podocarpus macrophyllus* 350
Bugleweed—*Ajuga* 533
Bunny-ears—*Opuntia* 243
Bunya-bunya—*Araucaria bidwillii* 333
Burgundy mound—*Osteospermum* 328
Burning bush—*Combretum* 302
Burnt plume—*Celosia* 52
Burro-tails—*Sedum morganianum* 377
Bush cherry—*Eugenia myrtifolia* 688
Busy-Lizzie—*Impatiens* 156
Butter-cup tree—*Cochlospermum* 304
Butterfly flower—*Schizanthus* 898
Butterfly gardenia—*Tabernaemontana* 76
Butterfly lily—*Hedychium* 927
Butterfly orchid—*Oncidium* 743
Butterfly palm—*Chrysalidocarpus* 773
Butterfly pea—*Clitorea* 544
Butterfly tree—*Bauhinia* 541
Butterwort—*Pinguicula caudata* 292
Button cactus—*Epithelantha* 263

Cabbage palm—*Sabal palmetto* 795
Cabbage tree—*Cussonia* 132
Cajeput—*Melaleuca* 680
Calabash gourd—*Lagenaria* 383
Calabash tree—*Crescentia* 456
Calamondin—*Citrus mitis* 481
Calico flower—*Aristolochia* 145

Calico hearts—*Adromischus* 361
Calico plant—*Alternanthera* 50
California boxwood—*Buxus japonica* 237
California Christmas tree—*Cedrus* 346
California holly—*Heteromeles* 852
Calif. laurel—*Umbellularia* 534
California pitcher—*Darlingtonia* 297
California poppy—*Eschscholtzia* 812
California privet—*Ligustrum ovalifolium* 696
Calla begonia—*Begonia semp. albo-foliis* 176
Calliopsis—*Coreopsis tinctoria* 326
Camboge tree—*Garcinia* 520
Camphor tree—*Cinnamomum* 534
Canary date palm—*Phoenix* 787
Candelabra cactus—*Lemaireocereus* 249
Candle bush—*Cassia alata* 550
Candle tree—*Parmentiera* 187
Candle plant—*Plectranthus* 532
Candytuft—*Iberis* 379
Cannonball tree—*Couroupita* 535
Cantaloupe—*Cucumis melo* 382
Canterbury bells—*Campanula* 287
Cape belladonna—*Amaryllis* 65
Cape chestnut—*Calodendrum* 873
Cape cowslip—*Lachenalia* 599
Cape daisy—*Venidium* 332
Cape fuchsia—*Phygelius* 887
Cape honeysuckle—*Tecomaria* 191
Cape jewels—*Nemesia* 884
Cape primrose—*Streptocarpus* 507
Cape weed—*Arctotheca* 308
Capulin cherry—*Prunus salicifolia* 469
Carambola—*Averrhoa* 463
Cardamon ginger—*Amomum* 928
Caribbean agave—*Agave angustifolia* 58
Caricature plant—*Graptophyllum* 36
Carob—*Ceratonia* 474
Carpet begonia—*Begonia imperialis* 170
Carpet plant—*Episcia* 496
Carrion flower—*Stapelia* 155
Carrot wood—*Cupaniopsis* 876
Casaguate—*Ipomoea arborescens* 358
Cashew-nut—*Anacardium* 71
Cassia-bark tree—*Cinnamomum* 534
Cast iron plant—*Aspidistra* 590
Castor-bean begonia—*Beg. ricinif.* 162
Catherine wheel—*Leucospermum* 836
Cats-claw—*Doxantha* 183
Cat's whiskers—*Tacca* 902
Cat-tail—*Typha* 916
Cayenne pepper—*Capsicum* 472
Cedar of Lebanon—*Cedrus* 341
Century plant—*Agave americana* 55
Ceriman—*Monstera deliciosa* 112
Ceylon gooseberry—*Dovyalis* 458
Chalice vine—*Solandra* 894
Champac—*Michelia alba* 619
Chaulmoogra tree—*Hydnocarpus* 484
Checkered lily—*Fritillaria* 602
Cherimoya—*Annona cherimola* 455
Cherry tomato—*Lycopersicon* 472
Chestnut vine—*Tetrastigma* 920
Chicken gizzard—*Iresine* 50
Chicle—*Manilkara zapota* 880
Chilean myrtle—*Pernettya* 394
Chilean wine palm—*Jubaea* 780
Chile bells—*Lapageria* 591
Chin cactus—*Gymnocalycium* 268
China aster—*Callistephus* 314
China fir—*Cunninghamia* 353
China flower—*Adenandra* 872
Chinaman's hat—*Holmskioldia* 917
Chincherinchee—*Ornithogalum* 599
Chinese evergreen—*Aglaonema* 86
Chinese fan palm—*Livistona* 777
Chinese lantern—*Abutilon* 622
Chinese lantern—*Physalis* 897
Chinese holly—*Ilex cornuta* 84
Chinese quince—*Cydonia* 470
Chinese rain tree—*Koelreuteria* 875
Chinese rhubarb—*Rheum* 825
Chinese taro—*Alocasia cucullata* 88
Chinese trumpet-creeper—*Campsis* 184
Chives—*Allium schoenoprasum* 575
Chocolate plant—*Pseuderanthemum* 34
Cholla—*Opuntia fulgida* 239
Christmas bells—*Blandfordia* 602

COMMON NAMES INDEX

Christmas berry—*Heteromeles* 849
Christmas bush—*Pavetta* 866
Christmas cactus—*Schlumbergera* 282
Christmas candle—*Till. imperialis* 227
Christmas heather—*Erica melanth.* 396
Christmas jewels—*Aechmea racinae* 202
Christmas palm—*Veitchia* 806
Christmas pepper—*Capsicum* 894
Christmas pride—*Ruellia amoena* 34
Christmas-rose—*Helleborus* 843
Cigar box cedar—*Cedrela* 641
Cigar flower—*Cuphea* 620
Cineraria—*Senecio hybrida* 326
Cinnamon tree—*Cinnamomum* 534
Cleft stone—*Pleiospilos nelii* 48
Cliff rose—*Cowania* 546
Climbing aloe—*Aloe ciliaris* 573
Climbing butcher's broom—*Semele* 606
Climbing lily—*Littonia* 602
Climbing onion—*Bowiea* 603
Climbing pandanus—*Freycinetia* 810
Clove—*Syzygium aromaticum* 678
Clove pinks—*Dianthus* 299
Clown fig—*Ficus aspera* 648
Club foot—*Pachypodium* 75
Coast redwood—*Sequoia* 355
Cob-cactus—*Lobivia* 260
Cobra orchid—*Megaclinium* 745
Cobra plants—*Darlingtonia* 293
Cochineal plant—*Nopalea* 240
Cocoa tree—*Theobroma* 902
Coco de mer—*Lodoicea* 784
Coconut palm—*Cocos nucifera* 769
Cohune palm—*Attalea* 768
Cohune palm—*Orbygnia* 791
Cola nut—*Cola nitida* 902
Colorado spruce—*Picea pungens* 341
Columbine—*Aquilegia* 843
Common fig—*Ficus carica* 456
Confederate rose—*Hibiscus* 629
Cook pine—*Araucaria columnaris* 335
Copper pinwheel—*Aeonium* 361
Coral-bead plant—*Nertera* 865
Coral bells—*Heuchera* 879
Coral-berry—*Aechmea fulgens* 198
Coral berry—*Ardisia* 673
Coral gems—*Lotus berthelotii* 562
Coral gum—*Eucalyptus torquata* 681
Coral orchid—*Rodriguezia* 752
Coral pea—*Kennedia* 557
Coral plant—*Russelia* 886
Coral tree—*Erythrina* 555
Coral vine—*Antigonon* 825
Corkscrew flower—*Phaseolus* 560
Cornflower—*Centaurea cyanus* 308
Cornish heath—*Erica vagans* 396
Corn lily—*Ixia* 527
Cornstalk plant—*Dracaena mass.* 580
Corsage orchid—*Vanda* 755
Cotton-rose—*Hibiscus mutabilis* 628
Coxcombs—*Celosia cristata* 51
Coyote bush—*Baccharis* 322
Crab-apple—*Malus floribunda* 850
Crane lily—*Strelitzia* 670
Crape jasmine—*Tabernaemontana* 76
Crape-myrtle—*Lagerstroemia* 618
Crawflower—*Calothamnus* 678
Crazyleaf begonia—*Beg. templinii* 178
Creeping Charlie—*Plectranthus* 532
Creeping daisy—*Wedelia* 325
Creeping devil—*Machaerocereus* 251
Creeping euonymus—*Euonymus fortunei* 301
Creeping fig—*Ficus pumila* 655
Creeping gloxinia—*Asarina* 886
Crossvine—*Bignonia capreolata* 187
Crown beauty—*Hymenocallis* 70
Crown cactus—*Rebutia* 261
Crown imperial—*Fritillaria* 602
Crown plant—*Calotropis* 146
Cuban lily—*Scilla peruviana* 592
Cub's paws—*Cotyledon ladismith.* 363
Cudjoewood—*Jacquinia* 674
Cup of gold—*Solandra* 894
Cushion bush—*Calocephalus* 310
Cushion moss—*Selaginella* 888
Custard apple—*Annona muricata* 454

Daffodil—*Narcissus* 68
Daisy tree—*Montanoa* 322
Dancing doll orchid—*Oncidium* 740
Dasheen—*Colocasia esculenta* 103
Date palm—*Phoenix dactylifera* 471
Daughter-of-the-West—*Brahea* 781
Dawn redwood—*Metasequoia* 352

Day-jessamine—*Cestrum diurnum* 895
Day-lily—*Hemerocallis* 575
Deodar cedar—*Cedrus deodara* 344
Desert candle—*Eremurus* 603
Desert spoon—*Dasylirion* 586
Desert rose—*Adenium* 73
Devil's apples—*Mandragora* 898
Devil's backbone—*Kalanchoe daigr.* 372
Devil's ivy—*Scindapsus aureus* 124
Devil's tongue—*Amorphophallus* 91
Devil tree—*Alstonia* 74
Dhoum palm—*Hyphaene* 782
Dipladenia—*Mandevilla* 77
Divi-divi tree—*Caesalpinia coriaria* 547
Dogwood—*Cornus florida* 356
Double coconut—*Lodoicea* 784
Douglas fir—*Pseudotsuga* 347
Dove tree—*Davidia* 673
Dracaena fig—*Ficus pseudopalma* 661
Dragon lily—*Amorphophallus* 90
Dragon tree—*Dracaena draco* 586
Drunkard's dream—*Hatiora* 285
Dumbcane—*Dieffenbachia* 106
Durian—*Durio zibethinus* 461
Dusty miller—*Centaurea* 310
Dutchman's pipe—*Aristolochia* 151
Dwarf banana—*Musa acuminata* 671
Dwarf ginger lily—*Kaempferia* 925
Dwarf palmetto—*Sabal minor* 798
Dwarf pineapple—*Ananas nanus* 207
Dwarf poinciana—*Caesalpinia* 561

Easter cactus—*Rhipsalidopsis* 285
Easter lily—*Lilium longiflorum* 599
Easter lily cactus—*Echinopsis* 261
Easter lily vine—*Beaumontia* 80
Easter tree—*Holarrhena* 76
Easter orchid—*Cattleya mossiae* 716
Eastern redbud—*Cercis* 552
Edelweiss—*Leontopodium* 331
Eggplant—*Solanum melongena* 894
Egyptian paper plant—*Cyperus* 389
Egyptian star-cluster—*Pentas* 866
Elderberry—*Sambucus nigra* 291
Elephant-apple—*Dillenia* 463
Elephant bush—*Portulacaria* 830
Elephant-ear—*Kalanchoe beharensis* 373
Elephant's foot—*Dioscorea elephantipes* 393
Elephant's-foot—*Pachypodium* 83
Elephant's trunk—*Martynia* 646
Elkhorns—*Rhombophyllum nelii* 49
Elmleaf begonia—*Begonia ulmifolia* 180
Empress tree—*Paulownia* 884
English daisy—*Bellis* 308
English holly—*Ilex aquifolium* 84
English ivy—*Hedera helix* 135
English laurel—*Prunus laurocerasus* 852
English lavender—*Lavandula* 530
English wallflower—*Cheiranthus* 380
English walnut—*Juglans* 467
English yew—*Taxus baccata* 352
Eulalia—*Miscanthus* 512
European fan-palm—*Chamaerops* 772
Evening primrose—*Oenothera* 701
Everglades palm—*Paurotis* 792
Evergreen ash—*Fraxinus* 696
Evergreen elm—*Ulmus parvifolia* 908
Evergreen grapevine—*Rhoicissus* 920
Evergreen pear—*Pyrus kawakamii* 851
Everlastings—*Helipterum roseum* 320
Eyelash begonia—*Begonia bowerae* 160

Fairy-carpet beg.—*Beg. versicolor* 169
Fairy-lily—*Chlidanthus* 60
Fairy primrose—*Primula malacoides* 831
Fairy rose—*Rosa chinensis* 857
Falling stars—*Campanula isophylla* 287
False aralia—*Dizygotheca* 132
False olive—*Cassine orientalis* 302
False sea-onion—*Ornithogalum* 592
False spiraea—*Sorbaria* 854
False spiraea—*Astilbe* 877
Farewell-to-spring—*Clarkia* 702
Fat pork tree—*Clusia* 521
Feather flowers—*Verticordia* 687
Felt-bush—*Kalanchoe beharensis* 373
Felt plant—*Calotropis* 146
Fern-asparagus—*Aspar. setaceus* 574
Fernleaf aralia—*Polyscias filicifolia* 141
Fernleaf begonia—*Begonia foliosa* 180
Fern palm—*Cycas circinalis* 387
Fiddle leaf fig—*Ficus lyrata* 652
Fig tree—*Ficus carica* 459
Fingernail plant—*Neoreg. spect.* 218

Fire barrel—*Ferocactus acanthodes* 264
Firebird—*Heliconia bihai* 663
Firecracker—*Cleistocactus* 246
Firecracker flower—*Brodiaea* 60
Firecracker vine—*Manettia* 868
Fire crown—*Rebutia senilis* 262
Fire-fern—*Oxalis hedysaroides* 763
Fire lily—*Cyrtanthus herrei* 67
Fire orchid—*Renanthera monachica* 759
Firethorn—*Pyracantha* 852
Fireweed—*Epilobium* 702
Firewheel tree—*Stenocarpus* 840
Fishhook cactus—*Ferocactus* 263
Fish poison tree—*Barringtonia* 536
Fishpole bamboo—*Phyllostachys* 517
Fishtail palm—*Caryota urens* 773
Five fingers—*Neopanax arboreus* 133
Flamboyant—*Delonix regia* 537
Flame gold—*Koelreuteria* 875
Flame of the forest—*Spathodea* 188
Flame of the forest—*Butea frondosa* 542
Flame-of-the-woods—*Ixora* 869
Flame tree—*Brachychiton* 901
Flame violet—*Episcia cupreata* 497
Flaming Katy—*Kalanchoe bloss.* 373
Flaming sword—*Vriesea splendens* 232
Flaming torch—*Guzmania bert.* 216
Flaming trumpet—*Pyrostegia* 188
Flamingo flower—*Anthurium scherzerianum* 92
Flamingo plant—*Jacobinia* 39
Flannel-bush—*Fremontodendron* 903
Flannel flower—*Actinotus* 910
Fleabane—*Erigeron* 315
Fleur d' amour—*Tabernaemontana* 76
Floating heart—*Nymphoides* 484
Floradora—*Stephanotis floribunda* 153
Floripondio tree—*Datura* 892
Floss silk tree—*Chorisia* 193
Flowering almond—*Prunus triloba* 851
Flowering banana—*Musa coccinea* 672
Flowering cabbage—*Brassica* 379
Flowering gum—*Eucalyptus ficifolia* 678
Flowering kale—*Brassica* 379
Flowering peach—*Prunus persica* 846
Flowering quince—*Chaenomeles* 848
Fly-trap—*Dionaea muscipula* 292
Forbidden fruit of India—*Pagiantha* 76
Forest lily—*Veltheimia* 592
Fountain grass—*Pennisetum* 512
Four o'clock—*Mirabilis* 700
Foxglove—*Digitalis* 887
Foxglove gloxinia—*Rehmannia* 882
Fox-tail orchid—*Rhynchostylis* 750
Fragrant olive—*Osmanthus fragrans* 700
Frangipani tree—*Plumeria* 78
Freckleface—*Hypoestes* 34
French heather—*Erica hyemalis* 397
French marigold—*Tagetes* 321
Frilled panties—*Protea barbigera* 839
Fringed orchid—*Habenaria* 734
Fruit-sheath plant—*Spathicarpa* 128

Galphimia vine—*Tristellateia* 620
Garlic vine—*Pseudocalymna* 193
Garlic vine—*Cydista* 187
Geiger tree—*Cordia* 196
Genista of florists—*Cytisus* 538
Geranium of hort.—*Pelargonium* 489
German myrtle—*Myrtus communis* 684
German primrose—*Primula obconica* 831
Ghost plant—*Graptopetalum* 374
Giant bamboo—*Dendrocalamus* 517
Giant orchid—*Grammatophyllum* 730
Giant redwood—*Sequoiadendron* 352
Giant reed—*Arundo donax* 511
Ginger—*Zingiber officinale* 928
Gingerbread palm—*Hyphaene* 783
Ginger-lily—*Hedychium* 924
Globe amaranth—*Gomphrena globosa* 52
Globe spear lily—*Doryanthes* 61
Glory-bower—*Clerodendrum* 915
Glory bush—*Tibouchina* 643
Glory lily—*Gloriosa* 593
Glory-of-the-snow—*Chionodoxa* 602
Glory-lily—*Gloriosa* 592
Glory pea—*Clianthus* 544
Glossy privet—*Ligustrum lucidum* 698
Goat's beard—*Aruncus* 847
Gold-dust dracaena—*Drac. gods.* 583
Gold-dust tree—*Aucuba japonica* 357
Golden ball—*Notoc. leninghausii* 273
Golden bamboo—*Phyllost. aurea* 518
Golden barrel—*Echinoc. grusonii* 267
Golden chain tree—*Laburnum* 553

COMMON NAMES INDEX

Golden club—*Orontium* 103
Golden dewdrop—*Duranta* 916
Golden larch—*Pseudolarix* 348
Golden mimosa—*Acacia baileyana* 539
Golden rain tree—*Cassia fistula* 551
Golden-rod—*Solidago* 329
Golden sedum—*Sedum adolphii* 378
Golden trumpet—*Allamanda* 75
Golden wattle—*Acacia longifolia* 539
Goldfish vine—*Columnea* 495
Goldflower—*Hypericum* 524
Gold-lace orchid—*Haemaria* 745
Good luck plant—*Cordyline term.* 578
Good-luck plant—*Oxalis deppei* 763
Gorse—*Ulex europaeus* 537
Gourd—*Cucurbita pepo* 383
Governor's plum—*Flacourtia* 458
Grapefruit—*Citrus x paradisi* 481
Grape hyacinths—*Muscari* 595
Grape ivy—*Cissus rhombifolia* 922
Grass tree—*Xanthorrhoea* 615
Green carpet—*Herniaria* 510
Green ebony—*Jacaranda* 185
Green lily—*Chlorophytum* 584
Green platters—*Aeonium* 361
Grizzly bear—*Opuntia erinacea* 241
Ground ivy—*Glechoma hederacea* 532
Guadalupe palm—*Brahea* 780
Guava—*Psidium* 460
Guernsey lily—*Nerine sarniensis* 69
Guinea chestnut—*Pachira* 468
Guinea goldvine—*Hibbertia* 391

Half-flower—*Scaevola* 523
Hardy begonia—*Begonia grandis* 167
Hardy fuchsia—*Fuchsia magellanica* 703
Hardy orange—*Poncirus trifoliata* 872
Hatchet cactus—*Pelecyphora* 280
Hawaiian elf—*Schefflera arboricola* 144
Heart of fire—*Bromelia balansae* 207
Heart of Jesus—*Caladium bicolor* 96
Heavenly bamboo—*Nandina* 181
Hedgehog cactus—*Echinocereus* 257
Helmet flower—*Sinningia macrop.* 501
Helmet orchid—*Coryanthes* 720
Hemlock—*Tsuga canadensis* 348
Hen-and-chickens—*Echeveria* 368
Hen-and-chicks—*Sempervivum* 367
Herald's trumpet—*Beaumontia* 73
Hibiscus tree—*Montezuma* 629
Hidden lily—*Curcuma* 923
Himalayan cedar—*Cedrus deodara* 343
Hindustan gentian—*Chirita* 491
Holly—*Ilex* 84
Hollyhock—*Alcea rosea* 625
Hollywood juniper—*Jun. 'Torulosa'* 337
Honesty—*Lunaria annua* 379
Honey bush—*Melianthus* 642
Honey locust—*Gleditsia* 558
Honey mesquite—*Prosopis* 562
Honey-myrtle—*Melaleuca* 685
Honeysuckle—*Lonicera* 290
Honeysuckle fuchsia—*F. triphylla* 700
Hongkong hawthorn—*Raphiolepis* 855
Honolulu Queen—*Hylocereus und.* 256
Hooded orchid—*Pterostylis* 746
Hopseed bush—*Dodonaea* 876
Horned orchid—*Stanhopea* 759
Horse chestnut—*Aesculus* 524
Horseradish tree—*Moringa* 662
Horsetail trees—*Casuarina* 300
Hortensia—*Hydrangea macrophylla* 846
Hottentot's fig—*Carpobrotus* 45
Houseleek—*Sempervivum* 378
Hummingbird tree—*Sesbania* 563
Hurricane palm—*Ptychosperma* 796
Hyacinth bean—*Dolichos* 557
Hyacinth orchid—*Bletilla* 711
Hybrid tea rose—*Rosa odorata* 857

Iceland poppy—*Papaver nudicaule* 812
Iceplant—*Lampranthus etc.* 47
Ice plant—*Mesembryanthemum* 42
Ifafa lily—*Cyrtanthus* 53
Ilima—*Sida fallax* 623
Immortelle—*Helichrysum* 310
Impala lily—*Adenium* 75
Imperial morning-glory—*Ipomoea nil* 358
Incense cedar—*Libocedrus* 339
Inch plant—*Tradescantia* 306
Inchworm—*Senecio pendulus* 324
Indian almond—*Sterculia* 903
Indian bean—*Catalpa* 184
Indian cork tree—*Millingtonia* 564
Indian corn—*Zea mays* 513

Indian cress—*Tropaeolum* 908
Indian fig—*Opuntia ficus-indica* 240
Indian ginger—*Alpinia* 924
India hawthorne—*Raphiolepis* 854
Indian jujube—*Ziziphus* 832
Indian laurel—*Ficus nitida* 654
Indian mulberry—*Morinda* 862
Indian paintbrush—*Castilleia* 883
Indian privet—*Clerodendrum* 913
India rubber plant—*Ficus elastica* 650
Indian rubber vine—*Cryptostegia* 145
Indian shot—*Canna indica* 288
Indigo—*Indigofera* 563
Indoor linden—*Sparmannia* 907
Insect orchid—*Ophrys lutea* 744
Irish bells—*Moluccella* 531
Irish mittens—*Opuntia vulgaris* 238
Irish moss—*Arenaria* 510
Irish moss—*Helxine* 907
Irish shamrock—*Trifolium* 546
Ironweed—*Vernonia* 323
Italian aster—*Aster amellus* 314
Italian bellflower—*Camp. isophylla* 287
Italian cypress—*Cupressus sempervirens* 338
Ivy geranium—*Pelarg. peltatum* 489

Jacobean lily—*Sprekelia* 64
Jack-bean—*Canavallia* 544
Jackfruit—*Artocarpus heterophyllus* 478
Jack-in-the-pulpit—*Arisaema* 97
Jackwood—*Cordia alba* 196
Jade plant—*Crassula argentea* 364
Jade vine—*Strongylodon* 559
Jamaica sago—*Zamia furfuracea* 386
Jamaican nutmeg—*Monodora* 72
Japan plum—*Eriobotrya* 849
Japanese apricot—*Prunus mume* 851
Japanese aralia—*Fatsia japonica* 133
Japanese boxwood—*Buxus micr.* 237
Japanese cedar—*Cryptomeria* 352
Jap. flowering cherry—*Prunus serr.* 850
Japanese holly—*Ilex crenata* 84
Japanese ivy—*Parthenocissus* 920
Japanese lantern—*Physalis* 895
Japanese laurel—*Aucuba* 357
Japanese maple—*Acer palmatum* 41
Japanese plum—*Prunus salicina* 469
Japanese privet—*Ligustrum jap.* 695
Japanese quince—*Chaenomeles* 470
Japanese rose—*Kerria* 850
Japanese spurge—*Pachysandra* 237
Japanese yew—*Taxus cuspidata* 352
Jasmine tobacco—*Nicotiana alata* 895
Java almond—*Canarium* 474
Java apple—*Syzygium samarang.* 688
Java fig—*Ficus benjamina* 651
Jelly beans—*Sedum pachyphyllum* 378
Jelly palm—*Butia capitata* 766
Jerusalem cherry—*Solanum pseudo-capsicum* 900
Jerusalem thorn—*Parkinsonia* 561
Jesuit's bark—*Cinchona* 861
Jewel orchid—*Haemaria* 741
Jewel weed—*Impatiens hawkeri* 157
Joseph's coat—*Amaranthus tricolor* 52
Joshua tree—*Yucca brevifolia* 611
Judas tree—*Cercis* 552
Jungle flame—*Ixora* 863

Kafir lily—*Clivia miniata* 64
Kahili ginger—*Hedychium gardner.* 928
Kangaroo paw—*Anigozanthos* 59
Kangaroo thorn—*Acacia armata* 538
Kangaroo vine—*Cissus antarctica* 922
Kapok—*Ceiba pentandra* 195
Kapok tree—*Bombax ceiba* 194
Kashmir bouquet—*Clerodendrum* 913
Kauri pine—*Agathis* 334
Kei-apple—*Dovyalis caffra* 458
Kentia palm—*Howeia forsteriana* 780
Key palm—*Thrinax microcarpa* 804
King anthurium—*Anthurium veitchii* 89
King ixora—*Ixora macrothrysa* 861
King jasmine—*Jasminum rex* 700
King palm—*Archontophoenix alexandrae* 764
King protea—*Protea cynaroides* 837
Kiss-me-quick—*Brunfelsia* 900
Kiwi vine—*Actinidia chinensis* 473
Knot-weed—*Polygonum capitatum* 825
Kolomikta vine—*Actinidia* 41
Korean grass—*Zoysia tenuifolia* 510
Kumquat—*Fortunella japonica* 874

Lace-bark—*Hoheria populnea* 622
Lace bugs—*Solisia* 276
Laceleaf—*Aponogeton madagascariensis* 82
Ladies-tresses—*Spiranthes* 754
Lady of the night—*Brassavola* 713
Lady of the night—*Brunfelsia* 900
Lady palm—*Rhapis* 802
Lady's eardrop—*Fuchsia splendens* 702
Lady's pocketbook—*Calceolaria* 883
Lady slipper—*Paphiopedilum* 749
Lance-pod—*Lonchocarpus* 558
Langsir—*Pometia pinnata* 474
Larkspur—*Delphinium* 844
Latania borbonica—*Livistona* 780
Laurustinus—*Viburnum tinus* 290
Lavender cotton—*Santolina* 318
Lavender starbush—*Grewia* 909
Lawn-leaf—*Dichondra micrantha* 510
Lemon—*Citrus limon* 479
Lemon gum—*Eucalyptus citriodora* 681
Leopard orchid—*Ansellia* 718
Levant cotton—*Gossypium* 620
Lemon ball—*Notocactus* 270
Lenten-rose—*Helleborus orientalis* 843
Leopard plant—*Ligularia* 326
Liane fern—*Stenochlaena*
Life-plant—*Kalanchoe* 372
Lignum vitae—*Guaiacum sanctum* 928
Lilac—*Syringa vulgaris* 699
Lilly-pilly—*Acmena smithii* 460
Lily cactus—*Echinopsis* 258
Lily of the fields—*Anemone* 842
Lily of the Incas—*Alstroemeria* 61
Lily-of-the-Nile—*Agapanthus* 575
Lily of the valley—*Convallaria* 592
Lily-of-the-valley bush—*Pieris* 394
Lily-of-the-valley orchid—
  Odont. pulchellum 743
Lily of the valley tree—*Clethra* 303
Lily-thorn—*Catesbaea* 864
Lilyturf—*Liriope* 606
Lion's ear—*Leonotis* 531
Lion's tongue—*Opuntia schick.* 238
Lipstick plant—*Aeschynanthus* 492
Lipstick tree—*Bixa* 181
Litchi nut—*Litchi chinensis* 477
Live-forever—*Sedum spectabile* 375
Liverwort—*Marchantia polymorpha* 646
Living rock-cactus—*Pleiospilos* 49
Living star—*Ariocarpus* 264
Livingstone-daisy—*Dorotheanthus* 46
Lobster claw—*Heliconia* 664
Locust tree—*Robinia* 564
Lollypops—*Pachystachys* 39
Looking-glass tree—*Heritiera* 901
Loose-strife—*Lysimachia* 830
Loquat—*Eriobotrya japonica* 456,849
Lords and ladies—*Arum* 95
Lotus—*Nelumbo nucifera* 704
Loulu palm—*Pritchardia* 796
Love-in-a-mist—*Nigella* 844
Love-lies-bleeding—*Amaranthus* 52
Love plant—*Anacampseros* 830
Love tree—*Cercis siliquastrum* 552
Love-vine—*Antigonon* 827
Lucian palm—*Kentiopsis* 782
Lucky clover—*Oxalis deppei* 762

Madagascar dragon tree—*Dracaena marginata* 582
Madagascar jasmine—*Stephanotis* 145
Madagascar olive—*Noronhia* 698
Madagascar periwinkle—*Catharanthus* 74
Madre de cacao—*Gliricidia* 553
Magic flower—*Achimenes* 493
Magic flower—*Cantua* 826
Magic lily—*Lycoris* 62
Maguey—*Agave americana* 57
Mahoe—*Hibiscus tiliaceus* 628
Mahogany—*Swietenia* 641
Maidenhair tree—*Ginkgo* 508
Maidenhair vine—*Muehlenbeckia* 826
Maize—*Zea mays* 512
Malabar chestnut—*Pachira* 457
Malabar ebony—*Diospyros* 393
Malabar plum—*Syzygium jambos* 677
Malacca rattan—*Calamus* 772
Malay apple—*Syzygium malaccense* 677
Malaysian dracaena—*Pleomele* 580
Mallet flower—*Tupidanthus* 132
Mallow—*Hibiscus* 627
Mammee apple—*Mammea* 475
Mandarin hat—*Holmskioldia* 917

COMMON NAMES INDEX

Mandarin orchid—*Phragmipedium* 746
Mandarin orange—*Citrus reticulata* 481
Manga del Nino—*Pandorea* 183
Mango—*Mangifera indica* 71, 454
Mangosteen—*Garcinia mangostana* 471
Mangrove—*Bruguiera conjugata* 841
Mangrove—*Rhizophora mangle* 841
Manuka—*Leptospermum scoparium* 685
Manzanita—*Arctostaphylos* 395
Mapleleaf begonia—*Beg. 'Cleopatra'* 160
Mapleleaf begonia—*Beg. welt.* 178
Marble-leaf—*Peristrophe* 36
Marble vine—*Senecio herreianus* 325
Marguerite—*Chrys. frutescens* 314
Marigold—*Tagetes* 330
Marmalade bush—*Streptosolen* 898
Martha Washington Geranium—*Pelarg. domesticum* 487
Mask flower—*Alonsoa* 887
Mastic tree—*Pistacia* 71
Matchstick vine—*Loranthus* 618
Mauritius hemp—*Furcraea* 58
Medeola—*Aspar. asparagoides* 574
Medicine plant—*Aloe vera* 573
Melon cactus—*Melocactus* 274
Mescal—*Agave parryi* 56
Mescal—*Lophophora williamsii* 269
Mescal bean—*Sophora* 560
Mesquite—*Prosopis glandulosa* 558
Mexican apple—*Casimiroa edulis* 463
Mexican aster—*Cosmos* 326
Mexican bamboo—*Yushania aztecorum* 518
Mexican blue palm—*Brahea armata* 780
Mexican breadfruit—*Monstera delic.* 110
Mexican fan palm—*Washingtonia robusta* 807
Mexican flame vine—*Sen. confusus* 325
Mexican foxglove—*Allophyton* 882
Mexican giant—*Pachycereus pringlei* 254
Mexican grass tree—*Dasylirion* 586
Mexican handflower—*Chiranthod.* 193
Mexican horncone—*Ceratozamia mexicana* 384
Mexican line—*Citrus aurantifolia* 479
Mexican mistletoe—*Psittacanthus* 617
Mexican orange-blossom—*Choysia* 872
Mexican pinkroot—*Spigelia* 616
Mexican rose—*Dombeya* 903
Mexican rose—*Echeveria simulans* 372
Mexican snowball—*Echev. elegans* 368
Mexican star flower—*Milla* 598
Mexican sunflower—*Tithonia* 332
Mexican violet—*Exacum affine* 484
Meyer lemon—*Citrus limon 'Meyer'* 874
Mickey mouse plant—*Ochna* 692
Mignonette—*Reseda odorata* 830
Ming aralia—*Polyscias fruticosa* 141
Miniature bamboo—*Sasa fortunei* 510
Miniature caladium—*Caladium humboldtii* 98
Miniature calla—*Callopsis* 128
Miniature fishtail—*Chamaedorea metallica* 770
Miracle-leaf—*Kalanchoe pinnata* 373
Mirror of Venus—*Ophrys* 744
Mirror plant—*Coprosma* 864
Mistletoe—*Viscum album* 618
Mistletoe cactus—*Rhipsalis cass.* 284
Mistletoe fig—*Ficus diversifolia* 653
Mocassin flower—*Cypripedium* 714
Mock orange—*Philadelphus* 880
Mock orange—*Pittosporum* 823
Mock strawberry—*Duchesnea* 849
Mombin—*Spondias venulosa* 455
Monastery-bells—*Cobaea* 826
Mondo grass—*Ophiopogon jaburan* 607
Moneywort—*Lysimachia* 830
Monkey-apple—*Clusia rosea* 523
Monkey-bread tree—*Adansonia* 454
Monkey flower—*Mimulus* 882
Monkey plant—*Ruellia* 37
Monkey-pot tree—*Lecythis* 536
Monkey-puzzle—*Araucaria araucana* 333
Monk's-hood—*Astrophytum* 264
Montbretia—*Tritonia crocosmaeflora* 527
Montezuma cypress—*Taxodium* 353
Moon cactus—*Harrisia* 251
Moon cactus—*Selenicereus hamatus* 257
Moon flower—*Calonyction* 360
Moonstones—*Pachyphytum* 376
May apple—*Podophyllum* 181
Meadow sweet—*Astilbe* 877
Mealybug orchid—*Ornithocephalus* 742
Moon Valley plant—*Pilea mollis* 911

Moonwort—*Lunaria annua* 887
Morning-glory tree—*Ipomoea murucoides* 358
Morning-noon-and-night—*Brunfelsia* 900
Mosaic plant—*Fittonia verschaffeltii* 34
Moses-in-the-cradle—*Rhoeo* 307
Mosquito flower—*Lopezia* 700
Moss-rose—*Portulaca* 831
Mother-in-law tree—*Albizia* 540
Moth orchid—*Phalaenopsis* 752
Mountain ash—*Sorbus* 855
Mountain bottle brush—*Greyia* 643
Mountain dahlia—*Liparia* 558
Mountain daisy—*Celmisia* 308
Mountain flax—*Phormium colensoi* 588
Mountain grape—*Mahonia* 181
Mountain laurel—*Kalmia* 396
Mountain palm—*Euterpe globosa* 781
Mountain thistle—*Acanthus* 36
Mouse-thorn—*Ruscus* 606
Mud-plantain—*Heteranthera* 828
Mulberry—*Morus alba* 660
Mulesfoot fern—*Angiopteris* 733
Myrtle—*Myrtus communis* 684
Myrtlewood—*Umbellularia calif.* 534

Napoleon's hat—*Napoleona* 762
Nasturtium—*Tropaeolum majus* 908
Natal paintbrush—*Haemanthus* 69
Natal plum—*Carissa grandiflora* 74, 462
Nerve plant—*Fittonia* 34
New Guinea creeper—*Mucuna* 553
New Zealand Christmas tree—*Metrosideros* 685
New Zealand flax—*Phormium* 588
New Zeal. grasstree—*Dracophyllum* 395
New Zeal. tea tree—*Leptospermum* 683
Ngaio—*Myoporum laetum* 674
Nibong palm—*Oncosperma* 792
Nicobar breadfruit—*Pandanus leram* 809
Night-blooming cereus—*Selenicer.* 256
Night-jessamine—*Cestrum noct.* 895
Nikau palm—*Rhopalostylis* 801
Nipa palm—*Nypa fruticans* 795
Nipple-fruit—*Solanum mammosum* 891
Nodding pincushion—*Leucospermum nutans* 836
Noni—*Morinda citrifolia* 870
Norfolk Island pine—*Araucaria heterophylla* 333
Norway maple—*Acer platanoides* 41
Norway spruce—*Picea abies* 344
Nun's orchid—*Phaius* 748
Nutmeg tree—*Myristica* 674

Oatgrass—*Arrhenatherum* 512
Obedience plant—*Maranta arund.* 638
October plant—*Sedum sieboldii* 375
Octopus plant—*Aloe arborescens* 573
Oilcloth-flower—*Anthurium* 93
Oil palm—*Elaeis guineensis* 778
Oiti—*Licania tomentosa* 463
Old lady cactus—*Mamm. hahniana* 278
Old man cactus—*Cephaloc. senilis* 247
Old man of the Andes—*Oreocereus celsianus* 251
Oleander—*Nerium oleander* 78
Olive tree—*Olea europaea* 469, 695
Orange bell-climber—*Marianthus* 823
Orange champac—*Michelia champac* 619
Orange jessamine—*Murraya* 872
Orchid cactus—*Epiphyllum* 281
Orchid plant—*Orchidantha* 672
Orchid tree—*Bauhinia* 541
Oregon grape—*Mahonia* 181
Organ pipe—*Lemaireoc. marginatus* 246
Oriental arborvitae—*Thuja orientalis* 339
Oriental poppy—*Papaver orientale* 812
Ornamental banana—*Ensete* 662
Ornamental hops—*Humulus* 660
Ornamental yam—*Dioscorea* 392
Otaheite apple—*Spondias dulcis* 455
Otaheite orange—*Citrus taitensis* 873
Owl eyes—*Huernia* 147
Owl's eyes—*Mammillaria parkinsonii* 275
Ox-eyes—*Heliopsis* 329
Ox-tongue—*Gasteria* 597

Pagoda tree—*Fatshedera* 138
Paint brush—*Haemanthus* 66
Painted lady—*Gladiolus* 528
Painted nettle—*Coleus blumei* 531
Painted tongue—*Salpiglossis* 898
Palm lily—*Yucca gloriosa* 614
Palmira palm—*Borassus* 474, 766

Palm leaf begonia—*Beg. luxurians* 170
Palo verde—*Cercidium* 562
Pampas grass—*Cortaderia* 511
Pamplemousse—*Citrus maxima* 479
Panama berry—*Muntingia* 910
Panama hat plant—*Carludovica* 391
Panamiga—*Pilea spruceana* 912
Panda plant—*Kalanchoe tomentosa* 372
Pansy—*Viola tricolor* 910
Pansy orchid—*Miltonia* 736
Papaya—*Carica papaya* 291
Paperbark tree—*Melaleuca* 680
Paper flower—*Bougainvillea* 694
Parachute plant—*Ceropegia* 148
Paradise flower—*Solan. wendlandii* 899
Paradise palm—*Howeia forsteriana* 783
Parakeet flower—*Heliconia* 666
Parlor ivy—*Philodendron scandens* 121
Parlor maple—*Abutilon* 623
Parlor palm—*Chamaedorea* 777
Parrot flowers—*Heliconia* 664
Parrot-leaf—*Alternanthera* 50
Parrot's-beak—*Clianthus* 553
Parrot's beak—*Lotus berthelotii* 557
Partridge berry—*Mitchella* 860
Partridge plant—*Aloe variegata* 573
Passion flower—*Passiflora* 815
Patient Lucy—*Impatiens* 156
Peace plant—*Spathiphyllum* 128
Peach—*Prunus persica* 856
Peach palm—*Bactris gasipaes* 766
Peach protea—*Protea grandiceps* 840
Peacock fern—*Selaginella* 888
Peacock flower—*Caesalpinia* 548
Peacock flower—*Delonix regia* 543
Peacock hyacinth—*Eichhornia* 825
Peacock plant—*Calathea makoyana* 639
Peanut—*Arachis hypogaea* 472
Peanut cactus—*Chamaec. sylvestri* 257
Pearly dots—*Haworthia* 597
Pearl flower—*Heterocentron* 647
Pecan nut—*Carya pecan* 467
Peepul tree—*Ficus religiosa* 651
Pencil cactus—*Opuntia ramosissima* 244
Pen-wiper—*Kalanchoe marmorata* 372
Pepper face—*Peperomia* 822
Pepper tree—*Schinus molle* 72
Perennial ryegrass—*Lolium* 510
Periwinkle—*Vinca minor* 75
Persian buttercup—*Ranunculus* 842
Persian ivy—*Hedera colchica* 138
Persian shield—*Strobilanthes* 35
Persian violet—*Exacum* 484
Persimmon—*Diospyros* 463
Peruvian apple—*Cereus peruvianus* 459
Peruvian daffodil—*Hymenocallis* 70
Peruvian lily—*Alstroemeria* 60
Peruvian old man—*Espostoa lanata* 248
Petticoat palm—*Copernicia* 776
Petticoat palm—*Washingtonia* 795
Peyote—*Astrophytum* 265
Pheasant's eye—*Adonis* 842
Philippine jade vine—*Strongylodon* 557
Phillipine violet—*Barleria* 34
Pickeral rush—*Pontederia* 828
Pigeon berry—*Duranta* 916
Piggyback—*Tolmiea menziesii* 881
Pigmy date—*Phoenix roebelenii* 793
Pigtail plant—*Anthurium scherz.* 92
Pimento—*Pimenta dioica* 841
Pincushion—*Leucospermum nutans* 836
Pineapple—*Ananas comosus* 459
Pineapple flower—*Eucomis* 603
Pineapple guava—*Feijoa sellowiana* 686
Pineapple lily—*Eucomis* 603
Pineapple quince—*Cydonia oblonga* 470
Pinguin—*Bromelia pinguin* 212
Pink calla—*Zantedeschia rehmannii* 128
Pink paintbrush—*Haemanthus* 69
Pink Pui—*Tabebuia rosea* 189
Pink quill—*Tillandsia cyanea* 230
Pink shower tree—*Cassia grandis* 551
Pink trumpet—*Tabebuia rosea* 190
Pinwheel—*Aeonium haworthii* 362
Pinwheel flower—*Tabernaemontana* 76
Pinyon pine—*Pinus cembroides* 347
Pistachio nut—*Pistacea* 467
Pitcher plant—*Nepenthes* 293
Pitcher plant—*Sarracenia* 297
Plantain lily—*Hosta* 590
Plover eggs—*Adromischus* 361
Plume poppy—*Bocconia* 648
Plush vine—*Mikania ternata* 324
Poet's jasmine—*Jasm. grandiflorum* 695
Poison bulb—*Crinum asiaticum* 63

COMMON NAMES INDEX

Poker plant—*Kniphofia uvaria* 591
Pomegranate—*Punica granatum* 458
Pompon tree—*Dais* 909
Pony-tail—*Beaucarnea recurvata* 582
Poor-man's orchid—*Schizanthus* 900
Popcorn bush—*Cassia didymob.* 549
Porcelain flower—*Hoya* 153
Porcelain lily—*Alpinia zerumbet* 923
Portia tree—*Thespesia populnea* 628
Potato tree—*Solanum macranthum* 899
Potato vine—*Solanum jasminoides* 899
Pot-aster—*Callistephus* 314
Pot-marigold—*Calendula* 308
Powder puff—*Calliandra* 547
Powder puff—*Mammillaria bocasana* 276
Prayer plant—*Maranta* 640
Prickly pear—*Opuntia* 242
Prickly poppy—*Argemone* 812
Prickly water lily—*Euryale* 705
Pride of Barbados—*Caesalpinia* 561
Pride of India—*Melia azedarach* 641
Pride of Madeira—*Echium* 196
Pride of Table Mountain—*Disa* 738
Primavera—*Tabebuia donnell-smithii* 189
Primrose jasmine—*Jasmin. mesnyi* 697
Princes feather—*Amaranthus* 51
Princess flower—*Tibouchina* 642
Princess vine—*Ipomoea horsfalliae* 359
Puas tree—*Harpullia* 875
Puka tree—*Meryta* 132
Pulque agave—*Agave atrovirens* 57
Pumpkin—*Curcubita pepo* 382
Punting pole bamboo—*Bambusa tuldoides* 514
Purple allamanda—*Allam. violacea* 75
Purple allamanda—*Cryptostegia* 80
Purple-bell vine—*Rhodochiton* 886
Purple broom—*Polygala* 811
Purple glory plant—*Sutera* 883
Purple granadilla—*Passiflora edulis* 473
Purple heart—*Setcreasea purpurea* 307
Purple-leaf plum—*Prunus cerasifera* 850
Purple passion vine—*Gynura* 324
Purple sand-olive—*Dodonaea* 876
Purple waffle plant—*Hemigraphis* 38
Purple wreath—*Petrea volubilis* 913
Pussy-ears—*Kalanchoe tomentosa* 373
Pussy willow—*Salix caprea* 876
Pyramid tree—*Lagunaria* 624

Quaking aspen—*Populus tremul.* 875
Queen agave—*Agave victoria-reginae* 56
Queen anthurium—*Anthurium warocqueanum* 92
Queen Emma lily—*Crinum augustum* 62
Queen palm—*Arecastrum* 765
Queen protea—*Protea barbigera* 839
Queen sago—*Cycas circinalis* 385
Queen of dracaenas—*Dracaena goldieana* 577
Queen of flowering trees—*Amherstia* 542
Queen of the night—*Selenicereus grandiflorus* 255
Queens tears—*Billbergia nutans* 208
Queensland bottle—*Brachychiton* 901
Queensland kauri—*Agathis robusta* 334
Queensland nut—*Macadamia* 467
Queensland tassel.—*Lycopodium* 618
Queensland umbrella tree—*Brassaia* 132
Quilted taffeta plant—*Hoffmannia* 862
Quince—*Cydonia oblonga* 849
Quinine—*Cinchona officinalis* 861

Rabbit's-tracks—*Maranta leuconeura* 640
Radiator plant—*Peperomia* 820
Rainbow-bush—*Portulacaria* 826
Rainbow cactus—*Echinocereus dasyacanthus* 259
Rainbow fern—*Selaginella uncinata* 888
Rainbow orchid—*Epidendrum prismatocarpum* 729
Rainbow pinks—*Dianthus chinensis* 299
Rainbow plant—*Dracaena marginata* 'Tricolor' 583
Rain lily—*Zephyranthes* 70
Rain tree—*Samanea saman* 560
Rambutan—*Nephelium lappaceum* 457
Rangoon creeper—*Quisqualis* 303
Raspberry—*Rubus idaeus* 472
Rat-tail cactus—*Aporocactus* 255
Rattan palm—*Calamus rotang* 772
Rattle-box—*Crotalaria retusa* 546
Rattlesnake plant—*Calathea insignis* 632
Red cap—*Melocactus* 274
Red cap—*Gymnocalycium mihan.* 268

Red cap gum—*Eucalyptus erythrocorys* 679
Red cassia—*Cassia marginata* 551
Red cedar—*Juniperus virginiana* 341
Red dracaena—*Cordyline terminalis* 576
Red echeveria—*Oliveranthus elegans* 374
Red ginger—*Alpinia purpurata* 923
Red hot poker—*Kniphofia* 591
Red ivy—*Hemigraphis colorata* 36
Red kapok—*Bombax ceiba* 194,195
Red popcorn—*Norantea* 645
Red valerian—*Centranthus* 916
Reed orchid—*Epidendrum ibaguense* 726
Reed orchid—*Sobralia decora* 760
Reed palm—*Chamaedorea seifrizii* 770
Resin bush—*Euryops* 315
Resurrection lily—*Kaempferia* 924
Resurrection plant—*Selaginella lepidophylla* 889
Rex-begonia vine—*Cissus discolor* 919
Ribbon bush—*Hypoestes* 34
Ribbon plant—*Dracaena sanderiana* 583
Ribbon plant—*Chlorophytum com.* 584
Rice—*Oryza sativa* 512
Rice paper plant—*Tetrapanax* 143
River bells—*Phygelius* 886
Rock palm—*Brahea dulcis* 780
Rock rose—*Cistus* 302
Roman laurel—*Laurus nobilis* 534
Rosary vine—*Crassula repestris* 366
Rosary vine—*Ceropegia woodii* 148
Rose acacia—*Robinia* 556
Rose apple—*Syzygium jambos* 460
Rose balsam—*Impatiens* 156
Rose begonia—*Beg. Lady Frances* 176
Rose cactus—*Pereskia* 243
Rose geranium—*Pelargonium graveolens* 490
Rose heath—*Erica gracilis* 396
Rose plaid cactus—*Gymnocalycium* 268
Rose of Sharon—*Hibiscus syriacus* 630
Rose of Venezuela—*Brownea* 542
Rosemary—*Rosmarinus officinalis* 533
Rosewood—*Tipuana tipu* 563
Rouge plant—*Rivina humilis* 812
Royal heath—*Erica regia* 397
Royal palm—*Roystonea* 800
Royal poinciana—*Delonix regia* 540
Royal water lily—*Victoria regia* 705
Rubber tree—*Ficus elastica* 652
Rubber vine—*Cryptostegia* 80,145
Ruby dumpling—*Mammillaria dolichocentra* 279
Running bamboo—*Phyllostachys* 515
Rusty fig—*Ficus rubiginosa* 652

Sacred bamboo—*Nandina* 181
Sacred Bo-tree—*Ficus religiosa* 660
Sacred lily of China—*Rohdea* 606
Sacred lotus—*Nelumbo nucifera* 707
Saffron spike—*Aphelandra squarrosa* 35
Sage—*Salvia officinalis* 533
Sago palm—*Metroxylon* 788
Saguaro—*Carnegiea gigantea* 245
Salt-cedar—*Tamarix parviflora* 903
Sandal wood—*Santalum* 876
Sand dollar—*Astrophytum asterias* 265
Sand olive—*Dodonaea* 876
Sapodilla—*Manilkara zapota* 880
Sapote—*Casimiroa* 463
Sapphire flower—*Browallia* 890
Satin flower—*Clarkia* 703
Saucer plant—*Aeonium tabul.* 361
Sausage tree—*Kigelia pinnata* 454
Scarborough lily—*Vallota* 64
Scarlet bush—*Hamelia patens* 867
Scarlet leadwort—*Plumbago indica* 811
Scarlet paint brush—*Crassula falcata* 367
Scarlet runner—*Phaseolus* 557
Scarlet sage—*Salvia splendens* 533
Scarlet skullcap—*Scutellaria* 530
Scarlet trumpet vine—*Distictis* 192
Scented boronia—*Boronia meg.* 873
Scorpion orchid—*Arachnis* 710
Scotch moss—*Sagina* 510
Screw pine—*Pandanus* 808
Scrub palmetto—*Serenoa* 795
Sea-buckthorn—*Hippophae* 393
Sea cups—*Hernandia* 523
Seagrapes—*Coccoloba uvifera* 467
Sea lavender—*Limonium* 811
Sealing wax palm—*Cyrtostachys* 779
Sea onion—*Urginea maritima* 603
Sea onion—*Ornithogalum caudatum* 598
Seersucker plant—*Geogenanthus* 306

Senegal date—*Phoenix reclinata* 793
Sensitive plant—*Mimosa pudica* 546
Sentry palm—*Howeia belmoreana* 777
Seven fingers—*Schefflera digitata* 139
Seven stars—*Ariocarpus* 264
Shaddock—*Citrus maxima* 479
Shamrock—*Oxalis braziliensis* 762
Shasta-daisy—*Chrysanthemum maximum* 308
Shaving brush—*Bombax* 193
Sheep laurel—*Kalmia* 396
Shell ginger—*Alpinia zerumbet* 923
Shooting star—*Dodecatheon* 830
Shower orchid—*Congea* 918
Shrimp begonia—*Beg. limmingh.* 164
Shrimp plant—*Beloperone guttata* 35
Shrimp tree—*Koelreuteria elegans* 875
Shrub vinca—*Kopsia* 80
Silk oak—*Grevillea robusta* 834
Silkweed—*Asclepias physocarpa* 151
Silver-berry—*Elaeagnus* 393
Silver crown—*Cotyledon undulata* 361
Silver dollar—*Lunaria annua* 379
Silver dollar—*Eucalyptus cinerea* 679
Silver dollar tree—*Eucal. polyanth.* 680
Silver squill—*Scilla violacea* 606
Silver sword—*Argyroxiphium* 309
Silver torch—*Cleistocactus strausii* 248
Silver tree—*Leucadendron argent.* 837
Silver vase—*Aechmea fasciata* 198
Sisal hemp—*Agave sisalana* 55
Sky-flower—*Duranta repens* 916
Slipper flower—*Calceolaria* 883
Smilax—*Asparagus asparagoides* 574
Smoke tree—*Cotinus* 71
Snail flower—*Phaseolus* 557
Snake cactus—*Nyctocereus* 253
Snake gourd—*Trichosanthes* 383
Snake plant—*Sansevieria trifasciata* 607
Snakewood tree—*Cecropia* 648
Snapdragon—*Antirrhinum* 882
Snowball—*Viburnum plicatum* 291
Snowball cactus—*Mamm. candida* 276
Snowballs—*Hydrangea macrophylla* 888
Snowdrop—*Galanthus nivalis* 70
Snowflake plant—*Trevesia palmata* 143
Snow plant—*Sarcodes sanguinea* 643
Snow vine—*Porana paniculata* 356
Society garlic—*Tulbaghia* 598
Solitaire palm—*Ptychosperma* 796
Sorrowless tree—*Saraca indica* 565
Sour cherry—*Prunus cerasus* 469
Sour plum—*Ximenia* 482
Soursop—*Annona muricata* 454
Spanish bayonet—*Yucca baccata* 614
Spanish cherry—*Mimusops* 880
Spanish jasmine—*Jasminum grandiflorum* 695
Spanish moss—*Tillandsia usneoides* 228
Spanish plum—*Spondias venulosa* 455
Spanish shawl—*Schizocentron* 646
Spathe flower—*Spathiphyllum* 126
Speedy Henry—*Tradescantia* 306
Spider aloe—*Aloe humilis* 512
Spider cactus—*Gymnocalycium* 268
Spider flower—*Cleome* 290
Spider lily—*Crinum* 63
Spider lily—*Hymenocallis* 70
Spider lily—*Lycoris radiata* 61
Spider lily—*Nerine bowdenii* 69
Spider orchid—*Arachnis* 708
Spider plant—*Chlorophytum* 584
Spiketail—*Stachyurus praecox* 906
Spineless yucca—*Yucca elephant.* 612
Spine palm—*Aiphanes caryotaefolia* 764
Spiny club—*Bactris* 766
Spiny stars—*Coryphantha* 274
Spiraea—*Astilbe japonica* 877
Spiral flag—*Costus spiralis* 926
Spiral ginger—*Costus speciosus* 926
Split-leaf philodendron—*Monstera deliciosa* 110
Spreading clubmoss—*Selaginella kraussiana* 889
St. Augustine grass—*Stenotaphrum* 510
St. John's bread—*Ceratonia siliqua* 544
St. John's-wort—*Hypericum* 524
Star-begonia—*Begonia heracleifolia* 166
Star cactus—*Haworthia retusa* 597
Star fish flower—*Stapelia* 155
Star-glory—*Quamoclit lobata* 359
Star-jasmine—*Jasm. gracillimum* 695
Star jasmine—*Trachelospermum* 80
Star of Bethlehem—*Camp. isophylla* 287
Star of Bethlehem—*Ornithogalum* 598

COMMON NAMES INDEX

Star-pine—*Araucaria heterophylla* 335
Statice—*Limonium sinuatum* 811
Stepladder plant—*Costus* 926
Stocks—*Matthiola incana* 380
Stoneface—*Lithops* 48
Strangler fig—*Ficus gibbosa* 657
Strawberry guava—*Psidium cattl.* 686
Strawberry tree—*Arbutus unedo* 462
Strawflower—*Helichrysum bract.* 320
String of hearts—*Ceropegia woodii* 149
Striped dracaena—*Drac. 'Warneckei'* 577
Sugar apple—*Annona squamosa* 72
Sugar bush—*Protea repens* 838
Sugar cane—*Saccharum* 512
Sugar maple—*Acer saccharum* 41
Sugar palm—*Arenga pinnata* 768
Summer cypress—*Kochia* 302
Summer snowflake—*Ornithogalum* 598
Sun-cactus—*Heliocereus speciosus* 281
Sundew—*Drosera* 292
Sunflower—*Helianthus annuus* 320
Sun plant—*Portulaca grandiflora* 831
Surinam cherry—*Eugenia uniflora* 677
Swamp lily—*Crinum* 62
Swanflower—*Swainsona galegifolia* 560
Swan orchid—*Cycnoches* 724
Sweat plant—*Selaginella emmeliana* 889
Swedish ivy—*Plectr. nummularius* 530
Sweet alyssum—*Lobularia* 379
Sweet bay—*Laurus nobilis* 534
Sweet flag—*Acorus calamus* 90
Sweet gum—*Liquidambar* 523
Sweet olive—*Osmanthus fragrans* 696
Sweet pea—*Lathyrus odoratus* 557
Sweet pea bush—*Podalyria* 556
Sweet pea shrub—*Polygala* 811
Sweet potato—*Ipomoea batatas* 465
Sweet William—*Dianthus barbatus* 298
Sweetheart geranium—*Pelarg. echinatum* 487
Sweetheart hoya—*Hoya kerrii* 153
Sweetheart peperomia—*Pep. verschaffeltii* 819
Sweetshade—*Hymenosporum* 823
Sword lily—*Gladiolus* 528
Sycamore fig—*Ficus sycomorus* 659
Symbol flower—*Gardenia taitensis* 867
Syrup palm—*Jubaea chilensis* 782

Taffeta plant—*Hoffmannia* 862
Tahitian bridal veil—*Gibasis* 307
Tailflower—*Anthurium andraeanum* 93
Takil palm—*Trachycarpus wagner.* 805
Talipot palm—*Corypha* 778
Tamarind—*Tamarindus indica* 556
Tangerine—*Citrus reticulata* 874
Tapeworm plant—*Homalocladium* 824
Taro—*Colocasia esculenta* 105
Taro vine—*Scindapsus* 128
Tarragon—*Artemisia dracunculus* 310
Tassel flower—*Amaranthus caudatus* 52
Tea plant—*Camellia sinensis* 904
Tea tree—*Melaleuca leucadendron* 685
Tea tree—*Leptospermum scoparium* 687
Teak tree—*Tectona grandis* 915
Teddy bear—*Opuntia bigelovii* 242
Tembusu—*Fagraea fragrans* 616
Temple bells—*Smithiantha* 506
Temple tree—*Plumeria* 81
Thanksgiving cactus—*Zygocactus* 281
Thatch palm—*Thrinax* 804
Thread agave—*Agave filifera* 58
Tiare Tahiti—*Gardenia taitensis* 866
Tiger flower—*Tigridia* 525
Tiger jaws—*Faucaria* 48
Tiger lily—*Lilium tigrinum* 600
Tiger orchid—*Odontogloss. grande* 745
Toad lily—*Tricyrtis* 602
Tobacco—*Nicotiana tabacum* 897
Toddy palm—*Borassus* 771
Tomato tree—*Cyphomandra* 473
Tom Thumb—*Parodia aureispina* 267
Touch-me-not—*Mimosa pudica* 546
Torch cactus—*Trichocer.spach.* 252
Torch ginger—*Nicolaia elatior* 927
Torch lily—*Kniphofia* 615
Tortoise cactus—*Deamia* 256
Tortoise plant—*Dioscorea macrost.* 392
Totem pole—*Lophocereus schottii* 251
Toyon—*Heteromeles arbutifolia* 852
Trailing violet—*Viola hederacea* 910
Transvaal daisy—*Gerbera* 318
Traveler's joy—*Clematis vitalba* 842
Travelers tree—*Ravenala* 662
Tree aloe—*Aloe bainesii* 568

Tree anemone—*Carpenteria* 880
Tree dahlia—*Dahlia imperialis* 317
Tree datura—*Datura arborea* 893
Treasure flower—*Gazania* 319
Tree fuchsia—*Schotia* 558
Tree gloxinia—*Kohleria* 502
Tree heliotrope—*Messerschmidtia* 197
Tree hibiscus—*Bombycidendron* 620
Tree mallow—*Lavatera* 624
Tree of heaven—*Ailanthus* 891
Tree of kings—*Cordyline* 587
Tree peony—*Paeonia suffruticosa* 844
Tree tumbo—*Welwitschia* 509
Tree wisteria—*Bolusanthus* 542
Trembling aspen—*Populus* 876
Tropical almond—*Terminalia* 467
Tropical apricot—*Dovyalis* 458
Troutleaf begonia—*Beg. argenteo-guttata* 160
Trumpet-tree—*Tabebuia* 190
Trumpet vine—*Campsis radicans* 184
Tuberose—*Polianthes tuberosa* 64
Tulip tree—*Liriodendron* 619
Tulip wood—*Harpullia* 875
Tuna—*Opuntia* 242
Turk's cap—*Melocactus* 274
Turk's cap—*Malvaviscus* 623
Turmeric—*Curcuma domestica* 926
Twelve apostles—*Neomarica* 526
Twin flower—*Bravoa* 60

Umbrella pine—*Pinus pinea* 345
Umbrella pine—*Sciadopitys* 352
Umbrella plant—*Cyperus* 390
Umbrella tree—*Brassaia* 139
Umbrella tree—*Melia azedarach* 642
Unicorn flower—*Proboscidea* 642
Urn flower—*Urceolina* 67
Urn vine—*Dischidia* 149

Vahana palm—*Pelagodoxa* 791
Variegated ginger—*Alpinia sanderae* 924
Varieg. ivy—*Hedera can. varieg.* 136
Varieg. screwpine—*Pand. veitchii* 810
Variegated wax-vine—*Senecio* 323
Vegetable sponge—*Luffa* 379
Velour philodendron—*P. andreanum* 115
Velvet plant—*Gynura aurantiaca* 327
Velvet rose—*Aeonium canariense* 362
Velvet tree—*Miconia* 645
Venus fly-trap—*Dionaea* 292
Vertical leaf—*Kalanchoe thyrsiflora* 378
Violet bush—*Iochroma* 900
Virginia creeper—*Parthenocissus* 920
Virgin's bower—*Clematis* 842
Virgin's palm—*Dioon edule* 384

Wait-awhile palm—*Desmoncus* 774
Wait-awhile vine—*Calamus* 772
Walking iris—*Neomarica* 526
Walking-stick palm—*Linospadix* 784
Wallflower—*Cheiranthus* 380
Wandering Jew—*Tradescantia* 305
Wand-flower—*Sparaxis* 527
Waratah—*Telopia speciosissima* 840
Wart flower—*Mazus* 882
Wart plant—*Haworthia* 607
Warty aloe—*Gasteria* 596
Water arum—*Calla palustris* 102
Water hyacinth—*Eichhornia* 825
Water lettuce—*Pistia* 129
Water lily—*Nymphaea* 706
Watermelon—*Citrullus* 382
Watermelon peper.—*Pep. sandersii* 818
Water mimosa—*Neptunia* 560
Water platter—*Victoria* 705
Water poppy—*Hydrocleis* 237
Water snowflake—*Nymphoides* 483
Water star-grass—*Heteranthera* 828
Water trumpet—*Cryptocoryne* 103
Wattle—*Acacia* 537
Wax begonia—*Beg. semperflor.* 159
Wax flower—*Anthurium andraeanum* 92
Waxgourd—*Benincasa* 383
Wax-leaf privet—*Ligustrum lucidum* 698
Wax mallow—*Malvaviscus* 624
Wax palm—*Ceroxylon* 774
Wax plant—*Hoya* 153
Wax rose—*Pereskia* 238
Wax rosette—*Echeveria gilva* 371
Wayfaring tree—*Viburnum* 291
Wedding flower—*Dombeya* 903
Weeping fig—*Ficus benjamina* 654
West Indian creeper—*Wedelia* 326
West Indian gherkin—*Cucumis* 382

West Indian holly—*Turnera* 910
West Indian jasmine—*Plumeria* 81
Weymouth pine—*Pinus strobus* 349
Wheel-tree—*Trochodendron* 908
Whisk-fern—*Psilotum* 829
White calla—*Zantedeschia aeth.* 129
White fir—*Abies concolor* 342
White flag—*Spathiphyllum* 126
White fluff-post—*Eulychnia* 251
White gossamer—*Trad. sillamontana* 306
White lotus—*Nymphaea lotus* 705
White nun orchid—*Lycaste* 737
White-plush plant—*Echev. leucot.* 370
White shrimp plant—*Justicia* 39
White sugar bush—*Protea repens* 840
White sugar bush—*Leucospermum* 838
White velvet—*Trad. sillamontana* 305
White-web ball—*Notocactus has.* 273
Whitey—*Mammillaria geminispina* 279
Wild almond—*Prunus dulcis* 851
Wild garlic—*Tulbaghia* 595
Wild iris—*Dietes* 526
Wild parsnip—*Angelica* 908
Wild pear—*Dombeya* 901
Wild phlox—*Sutera* 883
Wild pine—*Tillandsia fasciculata* 226
Wild poinsettia—*Warscewiczia* 863
Wild pomegranate—*Burchellia* 866
Wild rice—*Oryza clandestina* 513
Wild tamarind—*Leucaena* 556
Willow—*Salix* 876
Willow jasmine—*Cestrum parqui* 895
Willow sweet box—*Sarcococca* 237
Windmill palm—*Trachycarpus* 795
Window orchid—*Cryptophoranthus* 722
Window plant—*Fenestraria* 42
Window-leaf—*Monstera* 113
Wine cups—*Geissorhiza* 525
Wine-glass vine—*Ceropegia* 148
Winged pea—*Lotus berthelotii* 557
Winter begonia—*Bergenia crassifolia* 881
Winterberry—*Ilex verticillata* 84
Winter spice—*Hymenocallis* 70
Wintersweet—*Acokanthera* 74
Winter sweet—*Chimonanthus* 287
Wish-bone plant—*Torenia* 886
Woman's cap orchid—*Thelymitra* 756
Women's tongue—*Albizia lebbeck* 539
Wonder flower—*Ornithogalum* 598
Woodland p. petals.—*Lapeirousia* 525
Wood rose—*Ipomoea tuberosa* 360
Woolly bear—*Begonia leptotricha* 169
Woolly rose—*Echev. 'Doris Taylor'* 370
Woolly torch cactus—*Cephalocereus* 247
Wormwood—*Artemisia arborescens* 310

Yautia—*Xanthosoma* 124
Yellow bell—*Fritillaria* 598
Yellow bells—*Tecoma stans* 187
Yellow calla—*Zanted. elliottiana* 128
Yellow-elder—*Tecoma stans* 193
Yellow flag—*Iris pseudacorus* 527
Yellow flame tree—*Peltophorum* 561
Yellow gentian—*Gentiana lutea* 483
Yellow ginger—*Hedychium flavum* 928
Yellow jasmine—*Gelsemium* 616
Yellow oleander—*Thevetia* 78
Yellow pagoda—*Aphelandra* 33
Yellow pine—*Pinus ponderosa* 349
Yellow poinciana—*Peltophorum* 563
Yellow Pui—*Tabebuia* 189
Yellow sage—*Lantana camara* 918
Yellow-wood—*Podocarpus henkelii* 350
Yesterday-tomorrow—*Brunfelsia* 890
Ylang-Ylang—*Cananga* 72
Youth-and-old-age—*Aeonium domesticum* 361
Youth-and-old-age—*Zinnia* 331
Youth-on-age—*Tolmiea* 881

Zanzibar balsam—*Impatiens* 156
Zebra plant—*Calathea zebrina*—639
Zebra plant—*Cryptanthus zonatus* 212
Zebra plant—*Aphelandra squarrosa* 33
Zephyr lily—*Zephyranthes* 575
Zig-zag begonia—*Begonia parilis* 168
Zigzag shrub—*Asparagus retrofr.* 574
Zulu fig—*Ficus nekbudu* 653
Zulu giants—*Stapelia gigantea* 154

GENERIC BOTANICAL INDEX

Page numbers of photos of individual species, varieties and cultivars will be found alphabetically at the end of each listing in the text, serving as a quick-reference to plants illustrated.

Abelia 290
Abies 342
Abutilon 621-623
Acacia 537-540
Acalypha 408,409,423
Acanthocalycium 264,265
Acanthostachys 200
Acanthus 33,36
Acca 676
Acer 41
Achillea 308
Achimenes 492,493
Acmena 460,766
Acokanthera 74
Acorus 88,90
Acrocomia 764
Acrostichum 438
Actinidia 41
Actinotus 910
Ada 708
Adansonia 195,454
Adenandra 872
Adenia 814
Adenium 73,75
Adhatoda 35
Adiantum 439
Adonis 842
Adromischus 361
Aechmea 198-206,208,211
Aegle 455
Aegopodium 910
Aeonium 361-363
Aerangis 708-710
Aerides 709
Aeschynanthus 491,492
Aesculus 520,522,524
Aethionema 379
Agapanthus 575
Agastache 531
Agathis 334
Agave 55-58
Ageratum 314
Aglaonema 85-87
Agrostemma 298
Agrostis 510
Aichryson 363
Ailanthus 891
Aiphanes 764, 790
Ajuga 533
Alberta 861,865
Albizia 539,540,552
Alcea 625
Aleurites 407
Allamanda 73,75
Allium 575
Allophyton 882
Alloplectus 492
Alluaudia 389,391
Alocasia 85,86,88,89
Aloe 566-573,576
Alonsoa 887
Alpinia 923-926,928
Alsophila 431-433,436
Alstonia 74
Alstroemeria 60,61
Alternanthera 50
Amaranthus 51,52,54
Amaryllis 65,69
Amherstia 542,543
Amomum 928
Amorphophallus 90,91
Anacampseros 830
Anacardium 71,467
Ananas 205-207,209,459,464
Anchomanes 90
Andryala 308
Anemone 842,843
Anemopaegma 184
Angelica 908
Angelonia 883
Angiopteris 433,438
Angraecum 708-710
Anigozanthos 59
Annona 72,454,455
Ansellia 710,711,718
Anthurium 89,92-95,118
Antigonon 825,827
Antirrhinum 882,884,885
Apeiba 910
Aphelandra 33,35

Apaleia 306
Aponogeton 82
Aporocactus 255
x Aporophyllum 255
Aptenia 42,49
Aquilegia 843
Arabis 379
Arachis 472
Arachnis 708,710,718,719
Aralia 132
x Aranda 710
Araucaria 333,335,336
Arbutus 394,395,458,462
Archontophoenix 764-766,774

Arctostaphylos 394,395,466
Arctotheca 308
Arctotis 332
Ardisia 673
Areca 474,765,766,802
Arecastrum 765-768
Arenaria 510
Arenga 768
Argemone 812
Argyreia 357
Argyroderma 46
Argyroxiphium 309,323
Arisaema 97
Arisarum 96,97
Aristea 527
Aristolochia 145,151
Arophyton 97
Arpophyllum 709,710
Arrhenatherum 512
Artemisia 310
Arthrocereus 248
Arthropodium 575
Artocarpus 475,478,648,649
Arum 95,96
Aruncus 847,848
Arundina 709
Arundinaria 509,516,517
Arundo 511,512
Asarina 886
Asclepias 146,151
Ascoglossum 711
Asparagus 574,584
Asphodelus 585,603
Aspidistra 577,590
Asplenium 438-440,443,445,447
Astelia 584,586
Aster 314
Asteriscus 329
Asterostigma 102
Astilbe 877,879,881
Astrocaryum 769
Astrophytum 263-265
Attalea 768
Aucuba 357
Averrhoa 463,763
Azalea 400-403
Babiana 526,527
Baccharis 322
Bactris 471,766
Baikiaea 543
Bambusa 509,514-519
Banksia 834-836
Barbosa 767
Barleria 34
Barlia 711
Barringtonia 536
Bauhinia 541
Beaucarnea 582,585,586,612
Beaufortia 676
Beaumontia 73,80,82
Begonia 159-180
Bellis 308
Beloperone 35
Benincasa 383
Bergenia 881
Bertolonia 644
Beschorneria 59
Betula 182
Bifrenaria 709,722
Bignonia 187
Billbergia 208-214
Biophytum 762
Bischofia 407
Bismarckia 767
Bixa 181
Blakea 870
Blandfordia 602
Blechnum 441
Bletilla 711
Blighia 455,876
Bocconia 648
Boea 493
Boeriagiodendron 132,133
Boliviocereus 248,250
Bolusanthus 542
Bomarea 53,60,61
Bombax 193-195
Bombycidendron 620
Borassus 474,766-771
Boronia 872,873
Bougainvillea 689-694
Bourreria 196
Bowiea 603
Bowkeria 884
Bouvardia 860
Bowenia 384
Brachychilum 924
Brachychiton 901
Brachycorythis 712,714
Brahea 780,781
Brassaia 132-144
Brassavola 708-713

Brassia 709,711
Brassica 378-381
x Brassocattleya 713
Bravoa 60
Breynia 407,419
Brodiaea 60
Bromelia 206-212
Brosimum 648
Browallia 890
Brownea 538-545
Bruguiera 841
Brunfelsia 890,900
Brunsvigia 54
Buckinghamia 834
Buddleja 543,616
Bulbine 584
Bulbinella 603
Bulbophyllum 711-716
Burchellia 866
Butea 542,543
Butia 766,771
Buxus 237
Byblis 292
Caesalpinia 547,548,561
Caladium 96-101
Calamus 770,772
Calanthe 712,714
Calathea 631-640
Calceolaria 882-884
Calendula 308
Calicotome 548
Calla 102
Calliandra 547,548
Callisia 307
Callistemon 675,676,688
Callistephus 314
Callitris 342
Callopsis 128
Calocephalus 310
Calodendrum 873
Calonyction 360
Calophyllum 520,521
Calostemma 60
Calothamnus 676,678
Calotropis 146,543
Calvoa 644
Calycophyllum 863,870
Calymanthium 248
Camassia 602
Camellia 904-906
Camoensis 544
Campanula 287
Campelia 306
Campsis 184,191
Camptotheca 673
Cananga 72
Canarina 287
Canarium 237,304,474
Canavallia 544
Canistrum 210,211
Canna 288,289
Cantua 825,826
Capsicum 472,894,900
Carallia 147,154
Cardiospermum 876
Carex 390
Carica 291,459,475
Carissa 74,462
Carlephyton 102
Carlina 322
Carludovica 391
Carnegiea 245,247,253
Carpenteria 881
Carpobrotus 45,46
Carya 467
Caryota 770,773
Casasia 864
Casimiroa 461,463,872
Cassia 538-552
Cassine 302
Castilleja 883,886
Casuarina 300
Catalpa 184
Catasetum 712
Catesbaea 462,864
Catharanthus 74
Catrophractes 184
Cattleya 712-720
Cavendishia 397
Cecropia 648
Cedrela 641
Cedrus 341-346
Ceiba 195
Celmisia 308,309
Celosia 51,52,54
Centaurea 308,310
Centranthus 916
Centropogon 616
Cephaelis 867
Cephalocereus 246-254,571
Cephalophyllum 44
Cephalotaxus 352
Ceratonia 474,544,561
Ceratopteris 438
Ceratostigma 811

Ceratostylis 715
Ceratozamia 384
Cerbera 74
Cercidium 562
Cercis 430,552
Cereus 247-253,459,464
Ceropegia 148,149
Ceroxylon 774
Cestrum 891,895,900
Ceterach 443
Chaenomeles 470,848
Chamaecereus 257
Chamaecyparis 337
Chamaedorea 770,773,777
x Chamaelopsis 258
Chamaeranthemum 34,35,38
Chamaerops 769-777
Chambeyronia 774
Cheiranthus 380
Chimonanthus 287
Chionanthes 696
Chionodoxa 602
Chiranthodendron 193
Chirita 491,498
Chlidanthus 60
Chlorophytum 576,584,590
Chordospartium 549
Chorisia 193,197
Chorizema 544
Choysia 872
Christia 546
Chrysalidocarpus 770,773,774
Chrysanthemum 308-316
Chrysophyllum 463,880
Chrysothemis 491,493,497
Chysis 712,715
Cibotium 435,436
Cinchona 861
Cinnamomum 534
Cissus 919-922
Cistus 302
Citrullus 382,472
Citrus 479-482,873,874
Clarkia 702,703
Clavija 907
Cleistocactus 246-252
Clematis 842-847
Cleome 290
Clerodendrum 913-917
Clethra 303
Clianthus 544,545,553
Clidemia 647
Clitorea 544
Clivia 64
Clusia 520-523
Clytostoma 184
Cnidoscolus 407
Cobaea 826
Coccoloba 467,824,827
Coccothrinax 774
Cocculus 642,646
Cochemiea 274
Cochleanthes 758
Cochliostema 307
Cochlioda 711,715
Cochlospermum 303
Cocos 465,474,476,775,802
Codiaeum 410-416
Codonanthe 491,494
Coelogyne 715,716,721
Coffea 466,471,860-864
Cola 466,902
Colchicum 526,575
Coleus 530,531
Colletia 832,857
Colletogyne 102
Colocasia 85,103,105
Columnea 491-498,502,507
Colvillea 540,545
Combretum 302,303
Congea 909
Conicosia 46
Conocarpus 303
Conophyllum 42
Conophytum 48
Convallaria 592,595
Copernicia 776,778
Coprosma 864
Cordia 196,459
Cordyline 586-587,611
Coreopsis 326
Cornus 356,357,456
Coronilla 546
Correa 872
Cortaderia 511
Coryanthes 720,722
Corypha 769,778
Coryphantha 265,274,276
Cosmos 315,326
Costus 924,926
Cotinus 71
Cotoneaster 848
Cotyledon 361-366

Couroupita 535
Cowania 546
Crassocephalum 316
Crassula 362-375
Crescentia 185-187,456,457
Crinum 54-68
Crocosmia 525,527
Crocus 526
Crossandra 34,35
Crotalaria 546
Croton 411
Cryptanthus 212,213
Cryptocereus 255,256
Cryptocoryne 103,104
Cryptomeria 352,354
Cryptophoranthus 722
Cryptostegia 80,145,146
Ctenanthe 637-640
Cucumis 382,383
Cucurbita 382,383
Culcasia 97
Cunninghamia 352,353
Cunonia 391
Cupaniopsis 876
Cuphea 620,621
Cupressocyparis 338
Cupressus 337,338,341
Curcuma 923-926
Cussonia 132-134
Cyathea 431-436
Cycas 384-388
Cyclamen 829-833
Cycnoches 724
Cydista 187
Cydonia 470,482,849
Cymbidiella 718,720,722
Cymbidium 714,723,724
Cynara 331
Cynodon 510
Cypella 527
Cyperus 389,390
Cyphomandra 473,894
Cyphostemma 919
Cypripedium 714
Cyrtandra 495
Cyrtanthus 53,67
Cyrtomium 445
Cyrtopodium 715,731
Cyrtorchis 720
Cyrtosperma 102,108
Cvrtostachvs 779
Cytisus 538,546,562
Dacrydium 348
Dahlia 316,317,321
Dahlstedtia 543
Dais 909
Dalechampia 407
Dammaropsis 648,651
Daphne 909
Darlingtonia 293,297
Dasylirion 586,588
Datura 891-894
Davallia 440,444
Davidia 673
Deamia 256
Delonix 537,540,543
Delosperma 44,46
Delphinium 844,845
Dendrobium 718,721-731
Dendrocalamus 517,519
Dendrochilum 724
Dendropanax 134
Dendrosicyos 380
Derris 560
Desmoncus 774
Deuterocohnia 214
Deutzia 880
Dianthus 298,299
Diastema 497
Dicentra 484
Dichondra 357,510
Dichorisandra 275,305-307
Dicksonia 432,434,436
Dictyosperma 779
Didierea 389
Dieffenbachia 105-108
Dierama 527
Dierea 389
Dietes 526,529
Digitalis 887
Dillenia 391,392,463
Dimorphanthera 398
Dionaea 292
Dioon 384
Dioscorea 392,393
Diospyros 392,393,461,463
Diplothemium 804
Disa 720,721,738
Dischidia 146,149
Disocactus 283
Distictis 183,192
Dizygotheca 132
Dodecatheon 830
Dodonaea 876
Dolichandrone 183
Dolichos 557

GENERIC BOTANICAL INDEX

Dolichothele 276
Dombeya 901,903
Dorotheanthus 42,46
Dorstenia 650
Doryanthes 61
Dovyalis 456,458
Doxantha 183
Dracaena 577-589,613
Dracontium 103
Dracophyllum 394,395
Dracunculus 102
Drimiopsis 606
Drosanthemum 44,46
Drosera 292,296
Drymoglossum 442,444
Drymonia 492
Drynaria 440,441,443
Duchesnea 849
Dudleya 362,365,366
Duranta 916
Durio 194,461
Dyckia 212

Eccremocarpus 184
Echeveria 361,368-372
Echidnopsis 147
Echinocactus 249,263-267,275
Echinocereus 257-260
Echinomastus 266
Echinopsis 258-263
Echites 78
Echium 196
Edithcolea 147
Eichhornia 825,827,828
Elaeagnus 393,462
Elaeis 778,802
Elaphoglossum 446
Elleanthus 720
Embothrium 838
Emilia 315
Emmenosperma 832
Encephalartos 384-388
Ensete 662,666,668
Epacris 394,398
Eperua 544
Epidendrum 724-732
Epilobium 702
Epigenium 732,750
Epiphronitis 732
Epiphyllanthus 281
Epiphyllum 281-284
Epipremnopsis 104
Epipremnum 105,109
Episcia 496,497
Epithelantha 263,265
Epithema 504
Eranthemum 38
Eremurus 603
Eria 723,732
Erica 396-398
Erigeron 315
Eriobotrya 456,462,849
Eriopsis 734
Erodium 490
Erycina 732
Eryngium 908,911
Erythrina 545,554,555,564
Erythronium 594
Erythroxylum 407
Escallonia 880
Eschscholtzia 812,813
Escobaria 277,278
Escontria 246,250
Espostoa 248
Eucalyptus 675,678-682,688
Eucharis 64,66,67
Eucomis 603
Eugenia 460,677,688
Eulophia 720,731
Eulophiella 735
Eulychnia 251
Euonymus 301,302
Eupatorium 315
Euphorbia 408,417-427
Eurya 904,906
Euryale 705
Euryops 315
Euterpe 774,781
Exacum 484
Excoecaria 421

Fagraea 616
Fatshedera 138
Fatsia 132,133,137
Faucaria 48
Feijoa 463,686
Felicia 315
Fenestraria 42,44,48,49
Ferocactus 263-268
Festuca 510
Ficus 456,459,648-661
Filipendula 849
Firmiana 902
Fittonia 34
Flacourtia 458,484
Fockea 149
Fortunella 482,873,874
Fouquieria 430
Fragaria 849
Fraxinus 696
Freesia 526
Fremontodendron 903

Freycinetia 808,810
Frithia 48
Fritillaria 598,602
Fuchsia 700-705
Fuirena 390
Furcraea 55,56,58,61

Gaillardia 319,321
Galanthus 70
Galtonia 603
Garcinia 457,471,520,521
Gardenia 860,861,866,867
Gasteria 591,596,597
Gastorchis 730,732
x Gastrolea 597
Gazania 318,319
Geissorhiza 525
Gelsemium 616
Gentiana 483,484
Geoffroea 545
Geogenanthus 305,306
Geranium 490
Gerbera 318,331
Gerrardanthus 381
Gesneria 498,502
Gibasis 307
Gibbaeum 44
Ginkgo 508
Gladiolus 528
Glechoma 532
Gleditsia 558
Gleichenia 437
Gliricidia 553,558
Globba 925
Gloriosa 591-594
Glottiphyllum 44
Gloxinia 501
Glyptostrobus 354
Gmelina 917
Goethea 620
Gomphrena 52
Gonatanthus 102,104,109
Gonatopus 102
Gongora 734,735
Gordonia 905
Gossypium 620
Govenia 734
Grammatophyllum 730-734
Graphorkis 732
Graptopetalum 369,374
Graptophyllum 36
Greenovia 374
Greigia 215
Grevillea 834,840
Grewia 909
Greyia 643,646
Griselinia 357
Guadua 518,519
Guaiacum 928
Gunnera 508
Gurania 382
Gussonaea 736
Guzmania 215-217
Gymnocalycium 266-269
Gynura 324,327
Gypsophila 298

Habenaria 734
Habranthus 67
Hadrodemas 305
Haemanthus 63,66,69
Haemaria 732,741,745
Haematoxylum 558
Hakea 834,836
Hamatocactus 266,268,269
Hamelia 867
Hapaline 130
Hardenbergia 560
Harpullia 875
Harrisia 251
Hatiora 285
Haworthia 596,597,607
Hebe 882,887
Hechtia 215
Hedera 135-138
Hedychium 924-928
Heimerliodendron 692,694,698
Helenium 320
Heliamphora 297
Helianthus 320
Helichrysum 310,320,321
Heliconia 663-668,673
Heliocereus 281
Heliopsis 320,329
Heliotropum 196
Helipterum 320
Helleborus 843
Helxine 907,911
Hemerocallis 575
Hemigraphis 36,38
Hemionitis 448
Hemitelia 436
Heracleum 908
Hereroa 49
Heritiera 901
Hernandia 523
Herniaria 510
Heteranthera 828
Heterocentron 647
Heteromeles 849,852
Heteropteris 620
Heuchera 879,881

Hevea 429
Hibbertia 391
Hibiscus 623-630
Hippeastrum 65-67
Hippobroma 616
Hippomane 407
Hippophae 393
Hoffmannia 862
Hohenbergia 219
Hoheria 627
Holarrhena 76,78,80
Holcoglossum 761
Holmskioldia 914,917
Homalocladium 824
Homalomena 96,106,108
Hoodia 147,154
Houlletia 734
Hosta 590
Howeia 774,777,780,783
Hoya 145,150-153
Huernia 147
Humata 442
Humulus 650,660
Hunnemannia 812
Huntleya 734
Hyacinthus 599
Hydnocarpus 484
Hydrangea 877-880
Hydrocleis 237
Hydrosme 91
Hydrostachys 522
Hylocereus 256
Hymenocallis 61,66,70
Hymenocyclus 45,49
Hymenosporum 823
Hyoscyamus 898
Hypericum 524
Hyphaene 777,782,783
Hypocyrta 498
Hypoestes 34
Hypoxis 522

Iberis 379
Ibervillea 382
Iboza 532
Idria 430
Ilex 84
Impatiens 156-158
Indigofera 563
Inga 563
Iochroma 890,900
Ipomoea 358-360,465
Iresine 50,52
Iris 525,526,528
Isochilus 735
Isotoma 616
Ixia 528
Ixora 861-869

Jacaranda 185,186
Jacobinea 39
Jacquinia 674,906
Jasminum 694-700
Jatropha 418,419,421,428
Juanulloa 895
Jubaea 780,782
Juglans 467,520
Juniperus 337-341
Jurinea 322
Justicia 39

Kaempferia 924-926
Kalanchoe 365,372-378
Kalmia 396
Kennedia 557
Kentiopsis 780,782
Kerria 850
Kielmeyera 520,521
Kigelia 185,187,454,459
Kniphofia 591,595,615
Kochia 302
Koelreuteria 875
Kohleria 498,502
Kopsia 76,80

Laburnum 553,558
Lachenalia 591,599
Laelia 732-734,737
x Laeliocattleya 715,737
Lagenandra 104
Lagenaria 382,383
Lagerstroemia 617,618,622
Lagunaria 624
Laguncularia 303
Lamium 530,531
Lamourouxia 885
Lampranthus 43,45,47
Lantana 918
Lapageria 591,593
Lapeirousia 525
Lasia 106
Latania 780,785
Lathyrus 557,559
Laurus 534,535
Lavandula 530
Lavatera 624
Lawsonia 620
Lecythis 536
Leea 920
Lemaireocereus 246,249,251-253
Leochilus 734
Leonotis 531
Leontopodium 323,331
Lepanthes 735

Lepismium 285
Leptopteris 437
Leptospermum 683,685,687
Leschenaultia 508
Leucadendron 836,837,839
Leucaena 556
Leuchtengergia 269
Leucojum 70
Leucospermum 836,838,839
Liatris 322
Libocedrus 339
Licania 303,463
Licuala 777,784,785,789
Ligularia 326,327,329
Ligustrum 695,696,698-700
Lilium 594,599-601
Limnobium 522
Limonium 811
Linospadix 784
Liparia 558
Liquidambar 523
Liriodendron 619
Liriope 606
Litchi 457,462,477,875
Lithops 42,43,48,49
Littonia 602
Livistona 777,780,784-786
Lobelia 616,617
Lobivia 260-263
x Lobiviopsis 262
Lobularia 379
Lockhartia 732
Lodoicea 476,784,789
Lolium 510
Lomatophyllum 582
Lonchocarpus 558
Lonicera 290
Lopezia 700
Lophocereus 251,253
Lophophora 267,269
Loranthus 618
Lotus 557,562
Lourya 595
Luculia 863,864
Luffa 379
Luisia 741
Luma 677
Lunaria 379,887
Lupinus 556
Lycaste 737
Lycopersicon 472,894
Lycopodium 618
Lycoris 61,62
Lygodium 448
Lysimachia 830,831
Lythrum 618

Macadamia 467
Macaranga 429
Machaerocereus 251,252,254
Mackaya 38
Macropiper 822
Macrozamia 386-388
Magnolia 617,619,621,623
Mahonia 181
Malope 620
Malpighia 458,620
Malus 465,468,846-851
Malva 624
Malvaviscus 623-625
Mammea 457,475,520,522
Mammillaria 274-280
Mandevilla 73,77,78
Mandragora 898
Manettia 862,868
Manfreda 56,67
Mangifera 71,454,455
Manihot 419,428
Manilkara 457,459,880
Maniltoa 559
Mansoa 183
Maranta 637-640
Marchantia 646
Marianthus 823
Maripa 359
Maritimocereus 254
Marlierea 462,686
Marniera 256
Marsilea 438
Martynia 642,646
Mascarena 783,789
Masdevallia 718,736-739
Matthiola 380,381
Matucana 270
Maxillaria 735,739,748,788
Mazus 882
Medinilla 643,645-647
Mediocalcar 735
Megaclinium 745
Megakepasma 39
Meiracyllium 738
Melaleuca 680,683,685,687
Melastoma 642,646,647
Melia 641,642
Melianthus 642
Melocactus 274,275
Menadenium 738
Mendoncella 739
Merremia 360
Meryta 132,135,140
Mesembryanthemum 42,43

Mespilus 852
Messerschmidtia 197
Metasequoia 352
Metrosideros 685-687
Metroxylon 788
Michelia 619
Miconia 645-647
Microcachrys 348
Microcoelum 790
Microlepia 444
Mikania 324,327
Milla 598
Millingtonia 183,564
x Miltassia 736
Miltonia 736,738-740
Miltonidium 736,738
Mimetes 837,838
Mimosa 546,562
Mimulus 882,886
Mimusops 880
Mirabilis 693,700
Miscanthus 512
Mitchella 860
Mitraria 502
Moluccella 531
Monadenium 418
Monodora 72,78
Monolena 647
Monstera 110-113,457
Montanoa 322
Montezuma 627,629
Montrichardia 109
Morinda 862,868,870
Moringa 662
Mormodes 738
Morus 473,660
Mucuna 553,562
Muehlenbeckia 826
Muntingia 910
Murraya 872
Musa 465,470,478,664-673
Muscari 592,595,598
Mussaenda 867,870,871
Mutisia 329
Myoporum 674
Myosotis —
Myrica 462,673
Myristica 466,674
Myrsine 674
Myrtillocactus 253,254
Myrtus 684,688
Nandina 181
Napoleona 544,762
Narcissus 66,68
Nautilocalyx 498
Navia —
Nelumbo 704,707
Nematanthus 498,502
Nemesia 884,885
Neobenthamia 741
Neobesseya 280
Neobinghamia 254
Neocogniauxia 740
Neodypsis 790,791
Neofinetia 741
Neolloydia 271
Neomarica 525,526,528,529
Neopanax 133,140
Neoporteria 269,270,273
Neoregelia 214,216,218-220
Nepenthes 293-297
Nephelium 457,477
Nephrolepis 440,442,445
Nephthytis 102,130
Neptunia 560
Nerine 66,69
Nerium 77-79
Nertera 862,865
Nicolaia 927
Nicotiana 895,897
Nidularium 218,220,221
Nigella 844
Nolina 588
Nomocharis 602
Nopalea 240
Nopalxochia 281-284
Norantea 645,646
Noronhia 698
Nothofagus 430
Nothopanax 140
Notocactus 270-273
Nyctocereus 251,253
Nymphaea 704-707
Nymphoides 483,484
Nypa 787,790,795
Obregonia 271
Ochagavia 218
Ochna 692,694,698
Ochrosia 78,466
x Odontioda 740,741
Odontocidium 741
Odontoglossum 741-743,745
Odontonema 36
Oenothera 701,702
Oldenburgia 318,322,327
Olea 458,469,695,696
Oliveranthus 374
Olyra 516
Omphalea 428
Oncidium 736,740,742-745
Oncoba 484

GENERIC BOTANICAL INDEX

Oncosperma 792
Onoclea 446
Onopordum 321
Ophioglossum 437,438
Ophiopogon 606,607
Ophrys 744
Oplismenus 510
Opuntia 238-244,459,464
Orbygnia 768,791
Orchidantha 672
Orchis 744
Oreocereus 251,254
Oreopanax 140
Ornithocephalus 742
Ornithogalum 592,595, 598,599
Orontium 103
Orostachys 374
Orphium 484
Orthophytum 222,224
x Orthotanthus 221
Oryza 512,513
Osmanthus 696,700
Osmunda 437
Osteospermum 328
Othonna 329
Ouratea 689
Oxalis 762,763
Oxera 914
Oyedaea 327
Pachira 197,457,468
Pachycereus 245,254
Pachyphytum 375,376,378
Pachypodium 75,79,82,83
Pachysandra 237
Pachystachys 39
Pachyveria 378
Paeonia 843,844,847
Pagiantha 76
Palicourea 868
Palisota 306
Paliurus 832
Pancratium–
Pandanus 474,477,808-810
Pandorea 183
Papaver 812,813
Paphinia 747
Paphiopedilum 746-749
Paradrymonia –
Parinari 467,852
Parkia 561
Parkinsonia 561
Parmentiera 185-187,456
Parodia 267,271,273
Parthenocissus 920-922
Paspalum 510
Passiflora 473,814-817
Paulownia 884
Paurotis 792
Pavetta 865,866
Pavonia 624,625
Peddiea 908
Pedilanthus 418,419
Pediocactus 272
Pelagodoxa 791
Pelargonium 483,485-490
Pelecyphora 280
Pellaea 442,444
Pellionia 912
Peltandra 110
Peltophorum 561,563
Pennisetum 512
Penstemon 883,887
Pentapterigium 397
Pentas 862,866
Peperomia 818-822
Pereskia 238,241,243
Peristeria 747
Peristrophe 36
Pernettya 394
Persea 463,536
Pescatorea 747
Petrea 913,915,918
Petronymphe 66
Petunia 896
Phacelia 523
Phaius 748
Phalaenopsis 750-752
Phaleria 909
Phaseolus 557,560
Phellodendron 872
Philadelphus 880
Philodendron 92,112-125
Phlox 826
Phoenix 471,474,787,793- 797,804
Phoenocoma 322
Phoradendron 618
Phormium 585,588,607,611
Photinia 852
Phragmipedium 746-748
Phygelius 882,886,887
Phyllagathis 643
Phyllanthus 421,428,429
Phyllitis 442,445,446
Phyllorhachis 516
Phyllostachys 514,515, 517,518
Phymatidium 752
Physalis 891,895,897
Physostegia 531

Picea 343-346
Pieris 394
Pigafetta 784,787,791
Pilea 911,912
Pimenta 682,684
Pinanga 796
Pinguicula 292,296
Pinus 345,347-349
Piper 466,818,819,821,822
Pistacia 71,73,467
Pistia 129
Pitcairnia 221,222,224
Pittosporum 823,824
Pityrogramma 444
Platycerium 449-453
Plectranthus 530,532
Pleione 748
Pleiospilos 48,49
Pleomele 577,580-582,588, 611
Pleurothallis 750-752
Plumbago 811
Plumeria 78,79,81,83
Podalyria 556
Podocarpus 348,350,351,353
Podophyllum 181
Podranea 186
Polianthes 64,68
Polyalthia 72
Polygala 811
Polygonum 825
Polypodium 435,441-448
Polyscias 141,142
Polystachya 746
Polystichum 447
Pometia 474,875
Poncirus 872
Pontederia 828
Populus 875,876
Porana 356
Porphyrocoma 38
Porphyrostachys 750
Portea 221
Portlandia 870
Portulaca 826,830,831
Portulacaria 826,830
Potentilla 854
Pothos 120,125,127
Pritchardia 794,796,798
Proboscidea 642
Prosopis 558,562
Protea 837-840
Prunus 468-470,846,850- 853,856
Psammophora 49
Pseuderanthemum 34,37
Pseudocalymma 193
Pseudoespostoa 248
Pseudohydrosme 110
Pseudolarix 348
Pseudolithos 154
Pseudopanax 140,141
Pseudosasa 514
Pseudotsuga 347,348
Psidium 460,686
Psilotum 829
Psittacanthus 617
Pteris 442,445
Pterogodium 752
Pterospermum 902
Pterostylis 746
Ptychosperma 796
Pulmonaria 196
Punica 458,464,824,829
Puya 214,221,222,225
Pycnospatha 110
Pyracantha 846,852,858
Pyrostegia 188
Pyrrosia 444,448
Pyrus 468,850,851,853
Quamoclit 359
Quassia 895
Quercus 430
Quesnelia 221,222,236
Quisqualis 303,304
Rafflesia 830
Ramonda 501
Randia 862
Ranunculus 842-844,846
Raphia 799
Raphidophora 110,112
Raphiolepis 854-856
Rauwolfia 80
Ravenala 662,669,670,672
Rebutia 257,260-262
Regelia 687
Rehmannia 882
Renanthera 751,759,760
Reseda 830
Restrepia 758,760
Rhabdothamnus 504
Rhamnus 832
Rhapis 795,798,799,802
Rhektophyllum 108,125
Rheum 825
Rhigozum 186
Rhipsalidopsis 285,286
Rhipsalis 284-286
Rhizophora 841
Rhodeleia 522

Rhodochiton 886
Rhodocodon 603
Rhododendron 394,395, 398-406
Rhodomyrtus 684
Rhoeo 307
Rhoicissus 920
Rhombophyllum 49
Rhopalostylis 801
Rhynchostylis 750
Ribes 472,880
Ricinus 428
Rivina 812
Robinia 556,564
Rochea 367,377
Rodriguezia 713,746,752, 758,760
Rohdea 606,607
Rondeletia 862,868
Rosa 854-859
Rosmarinus 533
Rosularia 376
Rotala 618
Roystonea 800,801
Rubus 472,852,854,856,858
Rudbeckia 328,329
Ruellia 34,37,38
Rumohra —
Ruschia 44
Ruscus 588,606,607
Russelia 886
Ruttya 38,40
Sabal 795,798,800
Sabinea 561
Saccharum 511,512
Sadleria 438
Sagina 510
Sagittaria 70
Saintpaulia 495,499-501,504
Salix 844,876
Salpiglossis 898
Salvia 531,533
Salvinia 446,448
Samanea 560,563,564
Sambucus 291
Samuela 586
Sanchezia 35,38
Sandersonia 602
Sansevieria 596,607-612
Santalum 876
Santolina 318
Sanvitalia 326
Saraca 563,565
Sarcocaulon 487
Sarcococca 237
Sarcodes 643
Saritaea 184,187,188
Sarmienta 491
Sarracenia 293,294,296,297
Sasa 516
Satyrium 752
Sauromatum 129
Saxifraga 877,881
Scaevola 508,523
Scheelia 803
Schefflera 133,139,144
Schinus 72
Schismatoglottis 108
Schizanthus 898,900
Schizocasia 124,129
Schizocentron 646
Schizolobium 559
Schizostachyum 513,515
Schlumbergera 281,282
Schomburgkia 759
Schotia 558,561,565
Sciadopitys 352
Scilla 592,606
Scindapsus 124,125,128,131
Scirpus 390
Scutellaria 530
Scuticaria 760
Sedum 375-378
Seemannia 501
Selaginella 888,889
Selinicereus 255-257
Semele 606
Sempervivum 367,375,378
Senecio 309,310,322-328
Sequoia 355
Sequoiadendron 352,355
Serenoa 795
Serruria 838
Sesbania 563
Setcreasea 306,307
Severinia 873
Seyrigia 380,382
Sida 623,634,629
Siderasis 306
Sigmatostalix 760
Silene 298
Silybum 325
Sinningia 495,501-504,507
Sinoarundinaria 518
Siphocampylus 288
Skimmia 873
Smithiantha 505,506
Sobralia 759,760
Soehrensia 261,262
Solandra 894,900
Solanum 891,894,897-900

Solidago 329
Solisia 274
Soneria 644,646
Sophora 560
Sophronitis 753,759
Sorbaria 854
Sorbus 854,855
Sparaxis 528
Sparmannia 907,909
Spartium 556
Spathicarpa 128
Spathiphyllum 126,128,131
Spathodea 187,188
Spathoglottis 753,754,758
Spigelia 616
Spiraea 854
Spiranthes 754
Spondias 71,455
Sprekelia 64
Stachys 531
Stachytarpheta 916
Stachyurus 906
Stangeria 386
Stanhopea 755,759
Stapelia 154,155
Stemmadenia 82,83
Stenandrium 38
Stenocactus 272
Stenocarpus 837,840
Stenochlaena-
Stenoglottis 754
Stenomesson 64
Stenospermation120
Stenotaphrum 510
Stephanotis 145,153
Sterculia 902,903
Stereospermum 188
Stewartia 906
Stictocardia 359
Stigmaphyllon-
Strelitzia 664,665,669-672
Streptocalyx 221-223
Streptocarpus 501,505-507
x Streptogloxinia 503,507
Streptosolen 898
Strobilanthes 35,38
Stromanthe 637,638
Strombocactus 272
Strongylodon 557,559,560
Strophanthus 80,83
Sutera 883
Sutherlandia 556
Swainsona 560
Swietenia 641,642
Syagrus 798,803
Symbegonia 171
Synadenium 408,428
Synandriospadix 129
Syngonium 125,127,130
Syringa 699
Syzygium 460,675,677, 678,688
Tabebuia 186,189-192
Tabernaemontana 73,76
Tacca 902
Taccarum 129
Taeniophyllum 753
Tagetes 321,330
Taiwania 354
Talinum 829
Tamarindus 556,565
Tamarix 903,911
Tapeinochilus 923
Tavaresia 155
Taxodium 353,355
Taxus 352,353
Tecoma 187,193
Tecomanthe 191,192
Tecomaria 187,191
Tectona 915,916
Telipogon 758
Telopia 838,840
Terminalia 304,467
Ternstroemia 906
Tetrapanax 137,143
Tetraplasandra 143
Tetrastigma 920
Thecophyllum 223
Thelocactus 266,280
Thelymitra 756
Thenardia 78
Theobroma 474,477,902
Theophrasta 906
Thespesia 628
Thevetia 78,80,82
Thrinax 804
Thuja 339-341
Thunbergia 38,40
Thunia 754
Tibouchina 642,643,647
Tigridia 525
Tilia 909
Tillandsia 223-230,232,236
Tipuana 563
Titanotrichum 506
Tithonia 332
Todea 438
Tolmiea 877,881
Torenia 886
Trachelospermum 78,80
Trachycarpus 795,802-805

Tradescantia 305-307
Trevesia 143
Trichocaulon 154
Trichocentrum 759,760
Trichocereus 245,246, 251-254
Trichodiadema 44,49
Trichomanes 437
Trichopilea 759
Trichosanthes 383
Trichostachys 862
Tricyrtis 602
Trifolium 546
Tristania 680,684,688
Tristellateia 620,621
Tritonia 526-529
Trochodendron 908
Tropaeolum 907,908,911
Tsuga 345,348
Tulbaghia 592,595,598
Tulipa 599,604,605
Tupidanthus 132,143,144
Tupistra 590
Turbina 360
Turbinicarpus 272
Turnera 910
Turraea 642
Typha 916
Typhonium 129
Typhonodorum 127,129
Uebelmannia 267,271,272
Ulex 537
Ulmus 908
Umbellularia 534
Uncarina 812
Urceolina 64,67
Urginea 603
Uroskinnera 886
Ursinia 323
Utahia 272
Vaccinium 394,472
Vallota 64,67
Vanda 754-757,761
Vandachnis 756
x Vandacostylis 761
x Vandaenopsis 755
Vandopsis 758
Vanilla 473,751,753,759,761
Veitchia 806
Veltheimia 592
x Venidio-arctotis 332
Venidem 332
Verbascum 885
Verbena 918
Vernonia 323,329
Verschaffeltia 804,805
Verticordia 687
Viburnum 290,291
Victoria 705
Vinca 74,75
Viola 910
Viscum 618
Vitex 916
Vitis 465,473,921
Vriesea 224,228-236
Wallichia 804,805
Warrea 754,758
Warszewiczia 863,865, 867,868
Washingtonia 795,807
Watsonia 526,527,529
Wedelia 325,326
Weigela 291
Weingartia 274
Welwitschia 509
Widdringtonia 340
Wigandia 508
Wigginsia 271,272
Wilcoxia 252
Wisteria 560,564
Xanthorrhoea 588,589,615
Xanthosoma 124,125, 127,130
Xerosicyos 380
Ximenia 482
Xiphidium 523
Yucca 596,607,611-615
Yushania 511,518
Zamia 386,387
Zamioculcas 127
Zantedeschia 124,128, 129,131
Zea 512,513
Zebrina 305,306
Zephyranthes 70,575
Zingiber 924,927,928
Zinnia 331,332
Ziziphus 462,832
Zomicarpa 130
Zoysia 510
Zygocactus 281,282,284
Zygopetalum 756

MORE INFORMATION

on Decorative Plants and for research,
the world's most complete reference work
—for the serious plant lover

EXOTICA

Pictorial Cyclopedia of Exotic Plants

By A. B. GRAF

**MORE THAN 12,000 ILLUSTRATIONS, ABOUT 300 IN COLOR,
TO AID VISUAL PLANT IDENTIFICATION**

listing Scientific and Common names
Descriptions and Origin of plants
Key to care in five languages
Culture and Use of plants indoors
Propagation; Pest control
Botanical Terms chart and Glossary

Plant geography and Climatic backgrounds comprehensively and interestingly presented, and illustrated with habitat photos of the Author's travels.

Horticultural color chart, matching names in everyday use. Nearly 2,000 pages, $8\frac{1}{2} \times 11$ in., $(22 \times 28 cm)$ printed on enameled paper. Beautifully bound in buckram.

---

by the same author—in handy, more compact desk format

Exotic Plant Manual

*Fascinating Plants to live with
—their Requirements, Propagation and Use—*

a readily portable reference book of College dictionary size for homegardeners, library, students, decorators travelers. $6\frac{1}{2} \times 9\frac{1}{2}$ in. $(16 \times 24$ cm$)$, 840 pages.

4200 illustrations of plants
in practical groupings: Foliage plants for warm or cool interiors, Blooming plants, Orchids, Bromeliads, Succulents, Palms, Ferns, Vines; including every plant likely used indoors.

Comprehensive descriptive and cultural text
about requirements, progagation and tolerance of plants for decoration or as a hobby, all well illustrated.

Also modern chapters on—
Biology, Physiology and Genetics, Light—Temperature—Water—Soils, Hydroculture of plants in water, Gardening in the Tropics, Control of Pests and Diseases, the Pronunciation of Botanical Names.

ROEHRS COMPANY - Publishers
East Rutherford, New Jersey 07073, U.S.A.

PRONUNCIATION of BOTANICAL NAMES

Botanical nomenclature is basically Latin, or words adopted from other languages, with Latin endings, and conceived to be understandable internationally. Correctly pronounced and clearly enunciated, the recital of botanical names has a stately and noble sound. However, in the words of L. H. Bailey, there is no standard agreement on rules for the pronunciation of botanical binomials. Many English-speaking people pronounce generic and descriptive specific names simply as if the words were English, in what is known as the Traditional English system.

Alternately there is the Restored Academic, or phonetic pronunciation of classical scholars, which comes close to the manner of speech of the ancient Romans. Their idiom has been conserved through the centuries. The Florentine vernacular Latin of Dante in the 14th Century is the touchstone of modern Italian. Castilian Spanish also is an enhanced but faithful perpetuation of spoken Latin practically unchanged since the 5th Century. Anyone conversant with these languages will have no difficulty to articulate Botanical Latin in the classical tradition.

In spoken Latin, much the same as in Italian and Spanish, also in German or even Japanese, the vowels are pronounced precisely and uniformly: a as in apart; e as in pet; i as in pin; o as in note; u as in full; y in phyllus as in the French rue, the German or Chinese ü. Typical of the clear sound of spoken Latin is the Spanish expression "Te amo!".

Combinations of two vowels, or imperfect diphthongs found in Latin or Greek are enunciated separately as two syllables: aë (ah-eh = Gr. aer, Aërides); ai (ah-ee = eye); au (house); ei (eight); eo (areole); eu (eh-oo = aureus); ie (ee-eh = variegata; oi (oh-ee = deltoides); iu (ee-uu = folius); ue (uu-eh = cruentus); ui (ruin).

Exceptions are the perfect diphthongs or inseparable ligatures æ or ae (in caeruleus, Linnaeus, Caesarea), sounded as one vowel halfway between ah and eh, as in hat or fair, in French père, or the German "Umlaut" ä; œ or oe (in Coelogyne, coelestis) as in the French heureux, also the German or Swedish "Umlaut" ö.

Consonants: In the classical Latin used by Cicero in the first century B.C., the Romans never pronounced C like an English s, or G as j, but always like k and g (in get). By 180 A.D. however, the classical standard was gradually lost, and while C was still being pronounced as k before a, o, u—it changed to sound as z or s before ae, e, i, oe, y. G remained hard before a, o, u—as in Gardenia, but became a soft j as in joy before e and i (Geranium).

Many botanical names or epithets are derived from foreign-root personal or geographical names with Latin endings. To be recognizable, these are best pronounced in the idioms of their source, with accent on the preferred syllable.

The accent for nomenclatural Latin names with two syllables is on the first syllable; in words with several syllables the stress is usually on the next-to-the-last. If in doubt, pronounce all syllables with equal emphasis.

With a background of European schooling, I find it appropriate to use the pleasing phonetics of Continental Latin; those employing the English inflections may have difficulty being understood in non-English speaking areas. However, since English is increasingly a universally understood world language, it may well be employed, wherever found to be more convenient, in the pronunciation of Botanical names. I feel that a language should be our servant, not our master.